Residential Construction Academy

House Wiring

Affiliate

HBI
BUILDING CAREERS

NAHB
National Association
of Home Builders

RESIDENTIAL CONSTRUCTION ACADEMY

House Wiring

Third Edition

GREG FLETCHER

DELMAR
CENGAGE Learning™

Australia • Brazil • Japan • Korea • Mexico • Singapore • Spain • United Kingdom • United States

**Residential Construction Academy:
House Wiring, Third Edition**
Greg Fletcher

Vice President, Career and Professional Editorial: Dave Garza

Director of Learning Solutions: Sandy Clark

Senior Acquisitions Editor: Jim DeVoe

Managing Editor: Larry Main

Product Manager: Brooke Wilson, Ohlinger Publishing Services

Editorial Assistant: Cris Savino

Vice President, Career and Professional Marketing: Jennifer Baker

Marketing Director: Deborah Yarnell

Marketing Manager: Katie Hall

Marketing Coordinator: Mark Pierro

Production Director: Wendy Troeger

Production Manager: Mark Bernard

Content Project Manager: Michael Tubbert

Art Director: Casey Kirchmayer

Technology Project Manager: Joe Pliss

For product information and technology assistance, contact us at
Cengage Learning Customer & Sales Support, 1-800-354-9706

For permission to use material from this text or product,
submit all requests online at www.cengage.com/permissions
Further permissions questions can be e-mailed to
permissionrequest@cengage.com

Library of Congress Control Number: 2011923006

ISBN-13: 978-1-111-30621-2

ISBN-10: 1-111-30621-4

Delmar
5 Maxwell Drive
Clifton Park, NY 12065-2919
USA

Cengage Learning is a leading provider of customized learning solutions with office locations around the globe, including Singapore, the United Kingdom, Australia, Mexico, Brazil, and Japan. Locate your local office at:
international.cengage.com/region

Cengage Learning products are represented in Canada by Nelson Education, Ltd.

To learn more about Delmar, visit **www.cengage.com/delmar**

Purchase any of our products at your local college store or at our preferred online store **www.cengagebrain.com**

Printed in China
4 5 6 7 17 16 15 14

Table of Contents

Preface xiii
Introduction xxiv

SECTION 1

Preparing and Planning a Residential Wiring Job 1

CHAPTER 1

Residential Workplace Safety 3

Objectives 3
Glossary of Terms 4
Understanding Electrical Hazards 5
National Electrical Code® 8
NEC® Applications 13
NFPA 70E Standard for Electrical Safety in the Workplace 15
Occupational Safety and Health Administration (OSHA) 16
Personal Protective Equipment 18
Ground Rules for General and Electrical Safety 20
Classes of Fires and Types of Extinguishers 28
Summary 30
Procedure 1-1: Suggested Procedure to Find Information in the NEC® 31
Procedure 1-2: Suggested Lockout/Tagout Procedure 32

Procedure 1-3: Procedure to Follow When a Coworker is Being Shocked 32
Review Questions 34

CHAPTER 2

Hardware and Materials Used in Residential Wiring 36

Objectives 36
Glossary of Terms 37
Nationally Recognized Testing Laboratories ... 39
Electrical Boxes 40
Conductors and Cable Types 44
Raceways 51
Devices 54
Overcurrent Protection Devices 63
Panelboards, Loadcenters, and Safety Switches 65
Fasteners 66
Summary 71
Procedure 2-1: Ganging Metal Device Boxes 72
Procedure 2-2: Installing Aluminum Conductors 73
Procedure 2-3: Connecting Wires Together with a Wirenut 74
Procedure 2-4: Installing Toggle Bolts in a Hollow Wall or Ceiling 75
Procedure 2-5: Installing a Lead (Caulking) Anchor 77
Procedure 2-6: Installing Plastic Anchors 78
Procedure 2-7: Installing Tapcon Masonry Screws.. 79
Review Questions 82

CHAPTER 3

Tools Used in Residential Wiring.................83

Objectives.. 83
Glossary of Terms.................................... 84
Common Hand Tools 85
Specialty Tools 92
Power Tools.. 99
Summary.. 105
Procedure 3-1: Using a Screwdriver................106
Procedure 3-2: Using Lineman Pliers
to Cut Cable 108
Procedure 3-3: Using a Wire Stripper 109
Procedure 3-4: Using a Knife to Strip the
Sheathing from a Type NM Cable 110
Procedure 3-5: Using a Knife to Strip
Insulation from Large Conductors...............112
Procedure 3-6: Using an Adjustable
Wrench...113
Procedure 3-7: Using a Manual Knockout
Punch to Cut a Hole in a Metal Box 114
Procedure 3-8: Setting Up and Using a
Hacksaw 116
Procedure 3-9: Using a Torque
Screwdriver 117
Procedure 3-10: Using a Rotary Armored
Cable Cutter 118
Procedure 3-11: Using a Cordless Pistol
Grip Power Drill 119
Procedure 3-12: Drilling a Hole in a Wooden
Framing Member with an Auger Bit and
a Corded Right-Angle Drill 121
Procedure 3-13: Cutting a Hole in a Wooden
Framing Member with a Hole-Saw and
a Corded Right-Angle Drill 122
Procedure 3-14: Drilling a Hole in Masonry
with a Corded Hammer Drill......................124
Procedure 3-15: Using a Corded
Circular Saw 126
Procedure 3-16: Using a Corded
Reciprocating Saw 128
Review Questions................................... 130
What's Wrong With This Picture?................ 132

CHAPTER 4

Test and Measurement Instruments Used in Residential Wiring..............133

Objectives.. 133
Glossary of Terms.................................... 134
Continuity Testers 135
Voltage Testers and Voltmeters.................. 135
Ammeters... 139
Ohmmeters, Megohmmeters, and Ground
Resistance Meters 141
Multimeters.. 144
Watt-Hour Meters 146
Safety and Meters 148
Meter Care and Maintenance 149
Summary.. 149
Green Checklist....................................... 150
Procedure 4-1: Using a Continuity Tester........151
Procedure 4-2: Using a Voltage Tester............152
Procedure 4-3: Using a Noncontact
Voltage Tester 153
Procedure 4-4: Using a Voltmeter 154
Procedure 4-5: Using a Clamp-On Ammeter 156
Procedure 4-6: Using a Digital Multimeter 157
Procedure 4-7: Reading a Kilowatt-Hour
Meter 161
Review Questions 162

CHAPTER 5

Understanding Residential Building Plans.....................164

Objectives.. 164
Glossary of Terms.................................... 165
Overview of Residential Building Plans........ 167
Common Architectural Symbols 176
Electrical Symbols 176
Residential Framing Basics........................ 180
Summary.. 184
Review Questions 185

CHAPTER 6

Determining Branch Circuit, Feeder Circuit, and Service Entrance Requirements........186

Objectives................................. 186
Glossary of Terms........................ 187
Determining the Number and Types of
   Branch Circuits 188
Determining the Ampacity of a Conductor.... 198
Branch-Circuit Ratings and Sizing the
   Overcurrent Protection Device 205
Sizing the Service Entrance Conductors........ 207
Sizing the Loadcenter 213
Sizing Feeders and Subpanels 215
Summary.................................. 215
Procedure 6-1: Calculation Steps: The
   Standard Method for a Single-Family
   Dwelling 216
Procedure 6-2: Calculation Steps: The
   Optional Method for a Single-Family
   Dwelling 219
Review Questions........................ 222

SECTION 2

Residential Service Entrances and Equipment 225

CHAPTER 7

Introduction to Residential Service Entrances.................227

Objectives................................. 227
Glossary of Terms........................ 228
Service Entrance Types................... 230
Service Entrance Terms and Definitions 231
Residential Service Requirements
   (Article 230)......................... 233
Grounding and Bonding Requirements for
   Residential Services (Article 250) 243

Working with the Local Utility Company 254
Summary.................................. 257
Review Questions........................ 258
Know Your Codes........................ 259

CHAPTER 8

Service Entrance Equipment and Installation..........................260

Objectives................................. 260
Glossary of Terms........................ 261
Overhead Service Equipment and Materials... 262
Overhead Service Installation 270
Underground Service Equipment, Materials,
   and Installation 278
Intersystem Bonding Termination 280
Voltage Drop 280
Service Equipment Installation 281
Subpanel Installation 288
Service Entrance Upgrading 289
Summary.................................. 291
Review Questions........................ 292
Know Your Codes........................ 293

SECTION 3

Residential Electrical System Rough-In ...295

CHAPTER 9

General Requirements for Rough-In Wiring 297

Objectives................................. 297
Glossary of Terms........................ 298
General Wiring Requirements 299
General Requirements for Conductors 308
General Requirements for Electrical Box
   Installation 311
Communications Outlet.................... 321
Summary.................................. 322
Review Questions........................ 323

CHAPTER 10

Electrical Box Installation ...325

Objectives.................................... 325
Glossary of Terms........................... 326
Selecting the Appropriate Electrical
  Box Type.................................. 327
Sizing Electrical Boxes...................... 329
Installing Nonmetallic Device Boxes........... 337
Installing Metal Device Boxes................. 338
Installing Lighting Outlet and Junction
  Boxes 341
Installing Boxes in Existing Walls and
  Ceilings 343
Summary..................................... 345
Green Checklist.............................. 346
Procedure 10-1: Installing Device Boxes
  in New Construction 347
Procedure 10-2: Installing a Handy Box
  on a Wood or Masonry Surface 350
Procedure 10-3: Installing Outlet Boxes
  with a Side-Mounting Bracket............. 352
Procedure 10-4: Installing Outlet Boxes
  with an Adjustable Bar Hanger 353
Procedure 10-5: Installing Old-Work
  Electrical Boxes in a Wood Lath and
  Plaster Wall or Ceiling 355
Procedure 10-6: Installing Old-Work
  Metal Electrical Boxes in a Sheetrock
  Wall or Ceiling 357
Procedure 10-7: Installing Old-Work
  Nonmetallic Electrical Boxes in a
  Sheetrock Wall or Ceiling 358
Review Questions............................. 360
Know Your Codes.............................. 361

CHAPTER 11

Cable Installation 362

Objectives.................................... 362
Glossary of Terms............................ 363
Selecting the Appropriate Cable Type 364
Requirements for Cable Installation 364
Preparing the Cable for Installation 374
Installing the Cable Runs 376
Starting the Cable Run....................... 380

Securing and Supporting the Cable Run 381
Installing Cable in Existing Walls and Ceilings... 384
Summary..................................... 385
Green Checklist.............................. 386
Procedure 11-1: Starting a Cable Run
  from a Loadcenter 387
Review Questions............................. 391
Know Your Codes.............................. 392

CHAPTER 12

Raceway Installation 393

Objectives.................................... 393
Glossary of Terms............................ 394
Selecting the Appropriate Raceway Type
  and Size................................. 395
Introduction to Cutting, Threading, and
  Bending Conduit 412
Installation of Raceway in a Residential
  Wiring System 417
Raceway Conductor Installation 419
Summary..................................... 421
Procedure 12-1: Cutting Conduit with a
  Hacksaw 422
Procedure 12-2: Cutting and Threading RMC
  Conduit 424
Procedure 12-3: Bending a 90-Degree
  Stub-Up................................. 427
Procedure 12-4: Bending a Back-to-Back
  Bend 429
Procedure 12-5: Bending an Offset Bend 431
Procedure 12-6: Bending a Three-Point
  Saddle.................................. 434
Procedure 12-7: Box Offsets.................. 437
Review Questions............................. 440
Know Your Codes.............................. 441

CHAPTER 13

Switching Circuit
Installation..........................442

Objectives.................................... 442
Glossary of Terms............................ 443
Selecting the Appropriate Switch Type........ 444

Installing Single-Pole Switches.................. 448

Installing Three-Way Switches................... 453

Installing Four-Way Switches.................... 457

Installing Switched Duplex Receptacles 461

Installing Double-Pole Switches.................. 466

Installing Dimmer Switches...................... 468

Installing Ceiling Fan Switches.................. 470

Summary.. 472

Green Checklist.................................... 473

Review Questions................................. 474

CHAPTER 14

Branch-Circuit Installation ..476

Objectives.. 476

Glossary of Terms................................. 477

Installing General Lighting
Branch Circuits 478

Installing Small-Appliance
Branch Circuits 480

Installing Electric Range Branch Circuits...... 481

Installing the Branch Circuit for Counter-
Mounted Cooktops and Wall-Mounted
Ovens 485

Installing the Garbage Disposal Branch
Circuit 486

Installing the Dishwasher Branch Circuit...... 489

Installing the Laundry Area Branch
Circuits.. 490

Installing the Electric Clothes Dryer
Branch Circuit................................. 492

Installing the Bathroom Branch Circuit........ 495

Installing a Water Pump Branch Circuit........ 495

Installing an Electric Water Heater Branch
Circuit 499

Installing Branch Circuits for Electric
Heating 500

Installing Branch Circuits for
Air-Conditioning 504

Installing the Branch Circuit for Gas and
Oil Central Heating Systems 509

Installing the Smoke Detector Branch
Circuit 510

Installing the Carbon Monoxide Detector
Branch Circuit................................. 512

Installing the Low-Voltage Chime Circuit..... 514

Summary.. 518

Review Questions................................. 519

Know Your Codes................................. 520

CHAPTER 15

Special Residential Wiring Situations 521

Objectives.. 521

Glossary of Terms................................. 522

Installing Garage Feeders and Branch
Circuits 523

Installing Branch-Circuit Wiring for a
Swimming Pool 528

Installing Outdoor Branch-Circuit Wiring 540

Installing the Wiring for a Standby Power
System 546

Summary.. 551

Green Checklist.................................... 552

Procedure 15-1: Stripping the Outside
Sheathing of Type UF Cable.................... 553

Procedure 15-2: Connecting a Generator's
Electrical Power to the Critical Load
Branch Circuits 555

Review Questions................................. 556

Know Your Codes................................. 557

CHAPTER 16

Video, Voice, and Data Wiring Installation 558

Objectives.. 558

Glossary of Terms................................. 559

Introduction to EIA/TIA 570-B Standards 560

Installing Residential Video, Voice,
and Data Circuits 567

Testing a Residential Structured Cabling
System 577

Summary.. 578

Procedure 16-1: Installing a Crimp-On
Style F-Type Connector on an RG-6
Coaxial Cable 579

Procedure 16-2: Installing a Compression
Style F-Type Connector on an RG-6
Coaxial Cable 580

Procedure 16-3: Installing an RJ-45 Jack on a Four-Pair UTP Category 5e Cable...........581

Procedure 16-4: Installing an RJ-45 Jack on a Four-Pair UTP Category 6 Cable 584

Procedure 16-5: Assembling a Patch Cord with RJ-45 Plugs and Category 5e UTP Cable 586

Review Questions.........................589

SECTION 4

Residential Electrical System Trim-Out...591

CHAPTER 17

Lighting Fixture Installation...........................593

Objectives.................................593
Glossary of Terms........................594
Lighting Basics 595
Overview of Lamp Types Found in Residential Lighting.................596
Selecting the Appropriate Lighting Fixture 605
Installing Common Residential Lighting Fixtures 614
Summary..................................622
Green Checklist...........................624
Procedure 17-1: Installing a Light Fixture Directly to an Outlet Box 625
Procedure 17-2: Installing a Cable-Connected Fluorescent Lighting Fixture Directly to the Ceiling 626
Procedure 17-3: Installing a Lighting Fixture to an Outlet Box with a Strap..........627
Procedure 17-4: Installing a Lighting Fixture to a Lighting Outlet Box Using a Stud and Strap 628
Procedure 17-5: Installing a Fluorescent Fixture (Troffer) in a Dropped Ceiling..........629
Procedure 17-6: Installing a Ceiling-Suspended Paddle Fan with an Extension Rod and a Light Kit 631
Review Questions.........................634

CHAPTER 18

Device Installation................636

Objectives.................................636
Glossary of Terms........................637
Selecting the Appropriate Receptacle Type 638
Ground Fault Circuit Interrupter Receptacles 642
Arc Fault Circuit Interrupter Devices 644
Transient Voltage Surge Supressors 645
Installing Receptacles 647
Selecting the Appropriate Switch 651
Installing Switches........................652
Summary..................................655
Procedure 18-1: Using Terminal Loops to Connect Circuit Conductors to Terminal Screws on a Receptacle or Switch 656
Procedure 18-2: Installing a Duplex Receptacle in a Nonmetallic Electrical Device Box with One Cable 658
Procedure 18-3: Installing a Duplex Receptacle in a Metal Electrical Device Box with One Cable..................................661
Procedure 18-4: Installing a Duplex Receptacle in a Nonmetallic Electrical Device Box with Two Cables..............................663
Procedure 18-5: Installing Feed-Through GFCI Duplex Receptacles in Nonmetallic Electrical Device Boxes............................666
Procedure 18-6: Installing Single-pole Switches in a Single-Gang Nonmetallic Box 667
Procedure 18-7: Installing Single-pole Switches in a Multi-Ganged Nonmetallic Box......669
Review Questions 670

CHAPTER 19

Service Panel Trim-Out.......671

Objectives.................................671
Glossary of Terms........................672
Understanding Residential Overcurrent Protection Devices 673
GFCI and AFCI Circuit Breakers 678
Installing Circuit Breakers in a Panel 680
Summary..................................684

Procedure 19-1: Installing a Single-Pole Circuit Breaker.................685

Procedure 19-2: Installing a Single-Pole GFCI Circuit Breaker.................687

Procedure 19-3: Installing a Single-Pole AFCI Circuit Breaker.................689

Procedure 19-4: Installing a Two-Pole Circuit Breaker for a 240-Volt Branch Circuit..........691

Procedure 19-5: Installing a Two-Pole Circuit Breaker for a 120/240-Volt Branch Circuit 693

Review Questions........................ 695

SECTION 5

Maintaining and Troubleshooting a Residential Electrical Wiring System 697

CHAPTER 20

Checking Out and Troubleshooting Electrical Wiring Systems 699

Objectives............................... 699

Glossary of Terms....................... 700

Determining if All Applicable *NEC*® Installation Requirements Are Met........... 701

Determining if the Electrical System Is Working Properly 701

Troubleshooting Common Residential Electrical Circuit Problems 710

Service Calls............................ 714

Summary............................... 714

Procedure 20-1: Testing 120-Volt Receptacles with a Voltage Tester to Determine Proper Voltage, Polarity, and Grounding.........................716

Procedure 20-2: Testing 120/240-Volt Range and Dryer Receptacles with a Voltage Tester to Determine Proper Voltage, Polarity, and Grounding 718

Procedure 20-3: Testing a Standard Three-Way Switching Arrangement...............722

Procedure 20-4: Using a Continuity Tester to Troubleshoot a Receptacle Circuit That has a Ground Fault 724

Procedure 20-5: Using a Circuit Tracer to Find a Circuit Breaker That Protects a Specific Circuit...........................726

Review Questions........................ 728

SECTION 6

Green House Wiring Techniques 731

CHAPTER 21

Green Wiring Practices........733

Objectives................................. 733

Glossary of Terms......................... 734

Energy Efficiency 735

Selecting and Installing Energy Efficient Electrical Equipment 735

Selecting and Installing Electric Controls 740

Maintaining the Integrity of the Air Barrier and Insulation 741

On-Site Electric Power Generation............... 742

Durability and Water Management.............. 745

Green Product Selection.............................. 745

Reduce Material Use and Recycle Waste 746

Home owner Education and Reference Manual 747

Summary.. 748

Review Questions....................................... 749

CHAPTER 22

Alternative Energy System Installation............................750

Objectives...................................... 750

Glossary of Terms..................................... 751

Introduction to Alternative Energy
   Systems 752
Introduction to Photovoltaic Systems 752
Understanding Photovoltaic System
   Electricity Basics 757
Solar Fundamentals 758
PV System Components 761
PV System Wiring and the *NEC*® 776
PV System Installation Guidelines 791
Safety and Photovoltaic System
   Installation 792

Small Wind Turbine Systems 794
Summary 805
Review Questions 807

Glossary 809

Index 825

NEC Index 845

Preface

HOME BUILDERS INSTITUTE RESIDENTIAL CONSTRUCTION ACADEMY: HOUSE WIRING, THIRD EDITION

About the *Residential Construction Academy* Series

One of the most pressing problems confronting the building industry today is the shortage of skilled labor. The construction industry must recruit hundreds of thousands of new craft workers each year to meet future needs. This shortage is expected to continue because of projected job growth and a decline in the number of available workers. At the same time, the training of available labor is an increasing concern throughout the country and is affecting construction trades, threatening the ability of builders to construct quality homes.

These challenges led to the creation of the innovative *Residential Construction Academy Series*. The *Residential Construction Academy Series* is the perfect way to introduce people of all ages to the building trades while guiding them in the development of essential workplace skills, including carpentry, electrical wiring, HVAC, plumbing, masonry, and facilities maintenance. The products and services offered through the Residential Construction Academy are the result of cooperative planning and rigorous joint efforts between industry and education. The program was originally conceived by the National Association of Home Builders (NAHB)—the premier association in the residential construction industry—and HBI (Home Builders Institute)—a national leader for career training in the building industry and the NAHB Federation's workforce development partner.

For the first time, construction professionals and educators created national skills standards for the construction trades. In the summer of 2001, NAHB, through the HBI, began the process of developing residential craft standards in six trades: carpentry, electrical wiring, HVAC, plumbing, masonry, and facilities maintenance. Groups of employers from across the country met with an independent research and measurement organization to begin the development of new craft training standards. Builders and remodelers, residential and light commercial, custom single family and high production or volume builders are represented. The guidelines from the National Skills Standards Board were followed in developing the new standards. In addition, the process met or exceeded American Psychological Association standards for occupational credentialing.

Next, through a partnership between HBI and Delmar / Cengage Learning, learning materials—textbooks, videos, and instructor's curriculum and teaching tools—were created to teach these standards effectively. A foundational tenet of this series is that students *learn by doing*. Integrated into this colorful, highly illustrated text are Procedure sections designed to help students apply information through hands-on, active application. A constant focus of the *Residential Construction Academy* is teaching the skills needed to be successful in the construction industry and constantly applying the learning to real-world applications.

An enhancement to the Residential Construction Academy Series is industry Program Credentialing and Certification for both instructors and students by HBI. National Instructor Certification ensures consistency in instructor teaching/ training methodologies and knowledge competency when teaching to the industry's national skills standards. Student Certification is offered for each trade area of the Residential Construction Academy Series in the form of rigorous testing. Student Certification is tied to a national database that will provide an opportunity for easy access for potential employers to verify skills and competencies. Instructor and Student Certification serve the basis for Program Credentialing offered by HBI. For more information on HBI Program Credentialing and Instructor and Student Certification, please visit HBI.org.

About this Book

The third edition of *House Wiring* covers the basic electrical wiring principles and practices used in the installation of residential electrical wiring systems. It is based on the *2011 National Electrical Code®*. Wiring practices that are commonly used in today's residential electrical market are discussed in detail and presented in a way that not only tells what needs to be done, but also shows how to do it. Both general safety and electrical safety are stressed throughout the textbook.

This textbook provides a valuable resource for the knowledge and skills used in house wiring that are required of an entry-level residential electrician. This includes the basic "hands-on" skills as well as the more advanced theoretical knowledge needed to achieve job proficiency. In addition to important topics such as using the National Electrical Code®, sizing electrical boxes, sizing circuit conductors, sizing fuses or circuit breakers, and sizing service entrance conductors, this text also focuses on "hands-on" wiring skills. These "hands-on" wiring skills include things like the proper use of hand and power tools, splicing wires together properly, attaching electrical boxes to building framing members, fishing a cable in an existing wall, and installing an overhead service entrance. This edition also includes coverage of green wiring practices, reflecting the *National Green Building Standard ICC 700-2008,* and the installation of photovoltaic solar electric systems and small wind turbine systems. The format is intended to be easy to learn and easy to teach.

ORGANIZATION

This textbook is organized in the same way that a typical residential wiring project unfolds. The first five sections cover the installation of a residential wiring system from start to finish. The sixth section covers green house wiring practices and the installation of photovoltaic and small wind turbine systems. It is recommended that the sections be covered in the order that they appear in the textbook; however, all sections and section chapters are designed so that they can be covered in any order an instructor chooses.

- *Section 1: Preparing and Planning a Residential Wiring Job* is designed to show students how to apply common safety practices; how to use materials, tools and testing instruments; and how to read and understand residential building plans. Determining the requirements for branch circuits, feeder circuits, and service entrances are also covered.

- *Section 2: Service Entrances and Equipment* includes material on how to install the necessary equipment to get electrical power from the electric utility to the dwelling unit.

- *Section 3: Residential Electrical System Rough-In* demonstrates how to install electrical boxes and run cable or raceway according to the electrical circuit requirements.

- *Section 4: Residential Electrical System Trim-Out* involves installing all of the switches, receptacles, and luminaires (lighting fixtures) throughout a house. Installing circuit breakers and fuses are also covered.
- *Section 5: Maintaining and Troubleshooting a Residential Electrical Wiring System* explains how to test each circuit to make sure they are installed according to the *NEC®* and are in proper working order. It also shows how to troubleshoot and correct problems to ensure a satisfied customer.
- *Section 6: Green House Wiring Techniques* covers the material that every electrician should know so that the home they are wiring has the most up-to-date and efficient green electrical system possible. It also covers the information needed for a good understanding of how a solar or wind electricity producing system is installed.

NEW FEATURES

GREEN TIP is boxes, added throughout the chapters, that highlight green building techniques and practices.

GREEN CHECKLISTS have been added at the end of chapters to highlight the chapter's green coverage and to provide a quick reference for students.

Know Your Codes is a new end of chapter feature that prompts the readers to research their local and regional codes for selected chapter topics.

What's Wrong with This Picture? is a new end of chapter feature in each section that highlights common mistakes in a photo of a situation in which one or more things are wrong. The companion photo shows the situation corrected along with text explaining both the problem and the solution.

KEY FEATURES

This innovative series was designed with input from educators and industry and informed by the curriculum and training objectives established by the Standards Committee. The following features aid in the learning process:

Learning Features such as the **Introduction, Objectives,** and **Glossary** set the stage for the coming body of knowledge and help the learner identify key concepts and information. These learning features serve as a road map for the chapter. The learner may also use them as a reference later.

Active Learning is a core concept of the *Residential Construction Academy Series.* Information is heavily illustrated to provide a visual of new tools and tasks encountered by the learner. In the **Procedures,** various tasks used in the plumbing installation and service are grouped in a step-by-step approach. The overall effect is a clear view of the task, making learning easier.

SAFETY is featured throughout the text to instill safety as an "attitude" among learners. Safe jobsite practices by all workers is essential; if one person acts in an unsafe manner, then all workers on the job are at risk of being injured, too. Learners will come to appreciate that safety is a blend of ability, skill, and knowledge that should be continuously applied to all they do in the Construction industry.

From Experience provides tricks of the trade and mentoring wisdom that make a particular task a little easier for the novice to accomplish.

Review Questions complete each chapter. They are designed to reinforce the information learned in the chapter as well as give the learner the opportunity to think about what has been learned and what they have accomplished.

CAUTION features highlight safety issues and urgent safety reminders for the trade.

NEW **Green Tips** are included throughout the book to highlight existing or new information pertaining to Green Building.

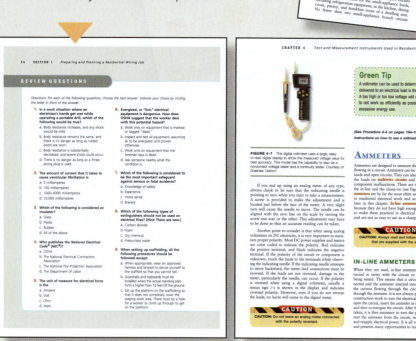

NEW **Know Your Codes** feature is included in many chapters allowing students to research their local codes. This teaches code book use and strengthens the knowledge of local codes and research capabilities required on a job site.

NEW **Green Checklists** have been added to the end of chapters to highlight the chapter's green coverage and provide a quick reference tool for students.

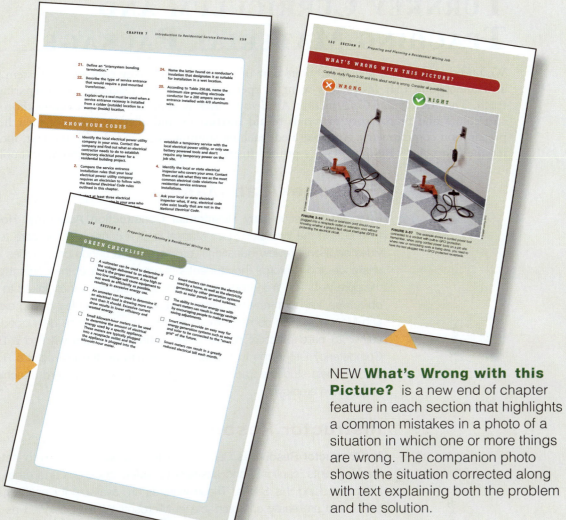

NEW **What's Wrong with this Picture?** is a new end of chapter feature in each section that highlights a common mistakes in a photo of a situation in which one or more things are wrong. The companion photo shows the situation corrected along with text explaining both the problem and the solution.

TURNKEY CURRICULUM AND TEACHING MATERIAL PACKAGE

We understand that a text is only one part of a complete, turnkey educational system. We also understand that Instructors want to spend their time on teaching, not preparing to teach. The *Residential Construction Academy Series* is committed to providing thorough curriculum and preparatory materials to aid Instructors and alleviate some of their heavy preparation commitments. An integrated teaching solution is provided with the text, including the Instructor's Resource CD, a printed Instructor's Resource Guide, and Workbook.

Workbook/Lab Manual

Designed to accompany **Residential Construction Academy: House Wiring** third edition the Workbook/Lab Manual is an extension of the core text and provides additional review questions, problems, and hands-on activities designed to challenge and reinforce the student's comprehension of the content presented in the core text. The Workbook section includes study outlines linked to the chapter objectives, key term reviews, chapter quizzes, and troubleshooting exercises. The Lab Manual section contains more than 50 hands-on labs developed by the text author to build practical house wiring skills. Instructors will find a Lab Grading Competency List in the Instructor's Guide and on the Instructor's Resource CD that can help students track their progress and build a record of achievement for prospective employers.

Instructor Resources

The **Instructor Resources** CD contains lecture outlines, notes to instructors with teaching hints, cautions, and answers to review questions and other aids for the Instructor using this *Series*. These features are available for each chapter of the book, and are easily customizable in Microsoft Word. Designed as a complete and integrated package, the Instructor is also provided with suggestions for when and how to use the accompanying **PowerPoint®, Computerized Test Bank, Video,** and **CD Courseware** package components. There are also print and pdf versions of the **Instructor's Resource Guide** available, as well as other aids for the Instructor using this *Series*.

The **Computerized Testbank** in ExamView makes generating tests and quizzes a snap. With hundreds of questions and different styles to choose from, you can create customized assessments for your students with the click of a button. Add your own unique questions and print rationales for easy class preparation.

Customizable **PowerPoint® Presentations** focus on key points for each chapter through lecture outlines that can be used to teach the course. Instructors may teach from this outline or make changes to suit individual classroom needs.

Use the hundreds of images from the **Image Library** to enhance your PowerPoint® Presentations, create test questions, or add visuals wherever you need them. These valuable images are pulled from the accompanying textbook, are organized by chapter, and are easily searchable.

DVDs

The **House Wiring DVD Series** is an integrated part of the **Residential Construction Academy House Wiring**, third edition, package. The series contains a set of eight, 20-minute lessons that provide step-by-step instruction for wiring a house. All the essential information is covered in this series, beginning with the important process of reviewing the plans and following through to the final phase of testing and troubleshooting. Need to know *NEC*® articles are highlighted, and Electrician's Tips and Safety Tips offer practical advice from the experts.

The complete set includes the following: DVD #1-Safety and Safe Practices, DVD #2: Hardware, DVD #3: Tools, DVD #4: Initial Review of Plans, DVD #5: Rough-In, DVD #6: Service Entrance, DVD #7: Trim-Out, and DVD #8: Testing & Troubleshooting.

About the Author

The author of this textbook, Greg Fletcher, has over 30 years of experience in the electrical field as both a practicing electrician and as an electrical instructor. His practical experience has been primarily in the residential and commercial electrical construction field. He has been a licensed electrician since 1976; first as a Journeyman Electrician and then as a Master Electrician. He has taught electrical wiring practices at both the secondary level and the post-secondary level. He has taught apprenticeship electrical courses and has facilitated many workshops on topics such as Using the *National Electrical Code*, Fiber Optics for Electricians, Understanding Electrical Calculations, and Introduction to Photovoltaics. The knowledge gained over those years, specifically on what works and what does not work to effectively teach electrical wiring practices, was used as a guide to help determine the focus of this textbook.

Since 1988 he has been Department Chairman of the Trades and Technology Department and an Instructor in the Electrical Technology program at Kennebec Valley Community College in Fairfield, Maine. He holds an Associate of Applied Science Degree in Electrical Construction and Maintenance, a Bachelor of Science Degree in Applied Technical Education, and a Master of Science Degree in Industrial Education.

Mr. Fletcher is a member of the International Association of Electrical Inspectors and The National Fire Protection Association. He lives in Waterville, Maine with his wife and daughter. When not teaching or writing textbooks he enjoys reading, golfing, motorcycling, and spending time with his family.

Acknowledgments

HOUSE WIRING NATIONAL SKILL STANDARDS

The NAHB and the HBI would like to thank the many individual members and companies that participated in the creation of the House Wiring National Skills Standards. Special thanks are extended to the following individuals and companies:

Stephen L. Herman, Lee College

Roy Hogue, TruRoy Electrical

Fred Humphreys, Home Builders Institute

Mark Huth, Delmar Learning

John Gaddis, Home Builders Institute Electrical Instructor

Ray Mullin, Wisconsin Schools of Vocational, Technical and Adult Education

Clarence Tibbs, STE Electrical Systems, Inc.

Jack Sanders, Home Builders Institute

Ron Rodgers, Wasdyke Associates/Employment Research

Ray Wasdyke, Wasdyke Associates

In addition to the standards committee, many other people contributed their time and expertise to the project. They have spent hours attending focus groups, reviewing and contributing to the work. Delmar Learning and the author extend their sincere gratitude to:

Jerry W. Caviness, Talking Leaves Job Corps

David Gehlauf, Tri-County Career Center

Jason Ghent, York Technical College

Ben Henry, Cassadaga Job Corps Academy

Gary Reiman, Dunwoody College of Technology

Martin A. Stronko, Keystone Job Corps

The author would like to express a special thanks to David Gehlauf of Tri-County Career Center in Nelsonville, Ohio, for lending his time and expertise to this writing project. Many of David's ideas and observations found their way into this textbook. David's contributions helped the author meet his goal of writing a residential wiring textbook that was up-to-date, easy-to-use, and technically accurate.

Finally, the author would like to thank his wife, Mary, and daughter, Hannah, for their continued support and help during the writing of this textbook.

Introduction

ORGANIZATION OF THE INDUSTRY

The residential construction industry is one of the biggest sectors of the American economy. According to the U.S. Department of Labor, construction is one of the Nation's largest industries, with 7.2 million wage and salary jobs and 1.8 million self-employed and unpaid family workers in 2008. About 64 percent of wage and salary jobs in construction were in the specialty trade contractors sector, primarily plumbing, heating and air-conditioning, electrical, and masonry. The National Association of Home Builders (NAHB) reports that home building traditionally accounts for 50-55 percent of the construction industry. Opportunities are available for people to work at all levels in the construction industry, from those who handle the tools and materials on the job site to the senior engineers and architects who spend most of their time in offices. Few people spend their entire lives in a single occupation, and even fewer spend their lives working for one employer. You should be aware of all the opportunities in the construction industry so that you can make career decisions in the future, even if you are sure of what you want to do at this time.

Construction Personnel

The occupations in the construction industry can be divided into four categories:

- Unskilled or semiskilled labor
- Skilled trades or crafts
- Technicians
- Design and management

Unskilled or Semiskilled Labor

Construction is labor-intensive. That means it requires a lot of labor to produce the same dollar value of end products by comparison with other industries, where labor may be a smaller part of the picture. Construction workers with limited skills are called *laborers*. Laborers are sometimes assigned the tasks of moving materials, running errands, and working under the close supervision of a skilled worker. Their work is strenuous, and so construction laborers must be in excellent physical condition.

Construction laborers are construction workers who have not reached a high level of skill in a particular trade and are not registered in an apprenticeship program. These laborers often specialize in working with a particular trade, such as mason's tenders or carpenter's helpers (Fig. I-1). Although the mason's tender may not have the skill of a bricklayer, the mason's tender knows how to mix mortar for particular conditions, can erect scaffolding, and is familiar with the bricklayer's tools. Many laborers go on to acquire skills and become skilled workers. Laborers who specialize in a particular trade are often paid slightly more than completely unskilled laborers.

Skilled Trades

A *craft* or *skilled trade* is an occupation working with tools and materials and building structures. The building trades are the crafts that deal most directly with building construction (see Fig. I-2).

The building trades are among the highest paying of all skilled occupations. However, work in the building trades can involve working in cold conditions in winter

FIGURE I-1 Mason.

Carpenter
    Framing carpenter
    Finish carpenter
    Cabinetmaker
Plumber
    New construction
    Maintenance and repair
Roofer
Electrician
    Construction electrician
    Maintenance electrician
Mason
    Bricklayer (also lays concrete blocks)
    Cement finisher
HVAC technician
Plasterer
    Finish plaster
    Stucco plaster
Tile setter
Equipment operator
Drywall installer
    Installer
    Taper
Painter

FIGURE I-2 Building trades.

© Cengage Learning 2012.

or blistering sun in the summer. Also, job opportunities will be best in an area where a lot of construction is being done. The construction industry is growing at a high rate nationwide. Generally, plenty of work is available to provide a comfortable living for a good worker.

Apprenticeship

The skill needed to be employed in the building trades is often learned in an apprentice program. Apprenticeships are usually offered by trade unions, trade associations, technical colleges, and large employers. *Apprentices* attend class a few hours a week to learn the necessary theory. The rest of the week they work on a job site under the supervision of a *journeyman* (a skilled worker who has completed the apprenticeship and has experience on the job). The term "journeyman" is a gender neutral term that has been used for decades. It is worth noting that many highly skilled building trades' workers are women. Apprentices receive a much lower salary than journeymen, often about 50 percent of what

a journeyman receives. The apprentice wage usually increases as stages of the apprenticeship are successfully completed. By the time the apprenticeship is completed, the apprentice can be earning as much as 95 percent of what a journeyman earns. Many apprentices receive college credit for their training. Some journeymen receive their training through school or community college and on-the-job training. In one way or another, some classroom training and some on-the-job supervised experience are usually necessary to reach journeyman status. Not all apprentice programs are the same, but a typical apprenticeship lasts four or five years and requires between 100 and 200 hours per year of classroom training along with 1200 to 1500 hours per year of supervised work experience.

Technicians

Technicians provide a link between the skilled trades and the professions. Technicians often work in offices, but their work also takes them to construction sites. Technicians use mathematics, computer skills, specialized equipment, and knowledge of construction to perform a variety of jobs. Figure I-3 lists several technical occupations.

Most technicians have some type of college education, often combined with on-the-job experience, to prepare them for their technical jobs. Community colleges

Technical Career	Some Common Jobs
Surveyor	Measures land, draws maps, lays out building lines, and lays out Roadways
Estimator	Calculates time and materials necessary for project
Drafter	Draws plans and construction details in conjunction with architects and engineers
Expeditor	Ensures that labor and materials are scheduled properly
Superintendent	Supervises all activities at one or more job sites
Inspector	Inspects project for compliance with local building codes at various stages completion
Planner	Plans for best land and community development

FIGURE I-3 Technicians.

© Cengage Learning 2012.

often have programs aimed at preparing people to work at the technician level in construction. Some community college programs are intended especially for preparing workers for the building trades, while others have more of a construction management focus. Construction management courses give the graduate a good overview of the business of construction. The starting salary for a construction technician is about the same as for a skilled trade, but the technician can be more certain of regular work and will have better opportunities for advancement.

Design and Management

Architecture, engineering, and contracting are the design and management professions. The *professions* are those occupations that require more than four years of college and a license to practice. Many contractors have less than four years of college, but they often operate at a very high level of business, influencing millions of dollars, and so they are included with the professions here. These construction professionals spend most of their time in offices and are not frequently seen on the job site.

Architects usually have a strong background in art, so they are well-prepared to design attractive, functional buildings. A typical architect's education includes a four-year degree in fine art, followed by a master's degree in architecture. Most of their construction education comes during the final years of work on the architecture degree.

Engineers generally have more background in math and science, so they are prepared to analyze conditions and calculate structural characteristics. There are many specialties within engineering, but civil engineers are the ones most commonly found in construction. Some civil engineers work mostly in road layout and building. Other civil engineers work mostly with structures in buildings. They are sometimes referred to as structural engineers.

Contractors are the owners of the businesses that do most of the building. In larger construction firms, the principal (the owner) may be more concerned with running the business than with supervising construction. Some contractors are referred to as general contractors and others as *subcontractors* (Fig. I-4). The general contractor is the principal construction company hired by the owner to construct the building. A general contractor might have only a skeleton crew, relying on subcontractors for most of the actual construction. The general contractor's superintendent coordinates the work of all the subcontractors.

It is quite common for a successful journeyman to start his or her own business as a contractor, specializing in the field in which he or she was a journeyman. These are the subcontractors that sign on to do a specific part of the construction, such as framing or plumbing. As the contractor's company grows and the company works on several projects at one time, the skilled workers with the best ability to lead others may become foremen. A foreman is a working supervisor of a small crew of workers in a specific trade. All contractors have to be concerned with business management. For this reason, many successful contractors attend college and get a degree in construction management. Most states require contractors to have a license to do contracting in their state. Requirements vary from state to state, but a contractor's license usually requires several years of experience in the trade and a test on both trade information and the contracting business.

An Overall View of Design and Construction

To understand the relationships between some of the design and construction occupations, we shall look at a scenario for a typical housing development. The first people to be involved are the community planners and the real estate *developer*. The real estate developer has identified a 300-acre tract on which he would like to build nearly 1000 homes, which he will later sell at a good profit. The developer must work with the city planners to ensure that the use he has planned is acceptable to the city. The city planner is responsible for ensuring that all building in the city fits the city's development plan and zoning ordinances. On a project this big, the developer

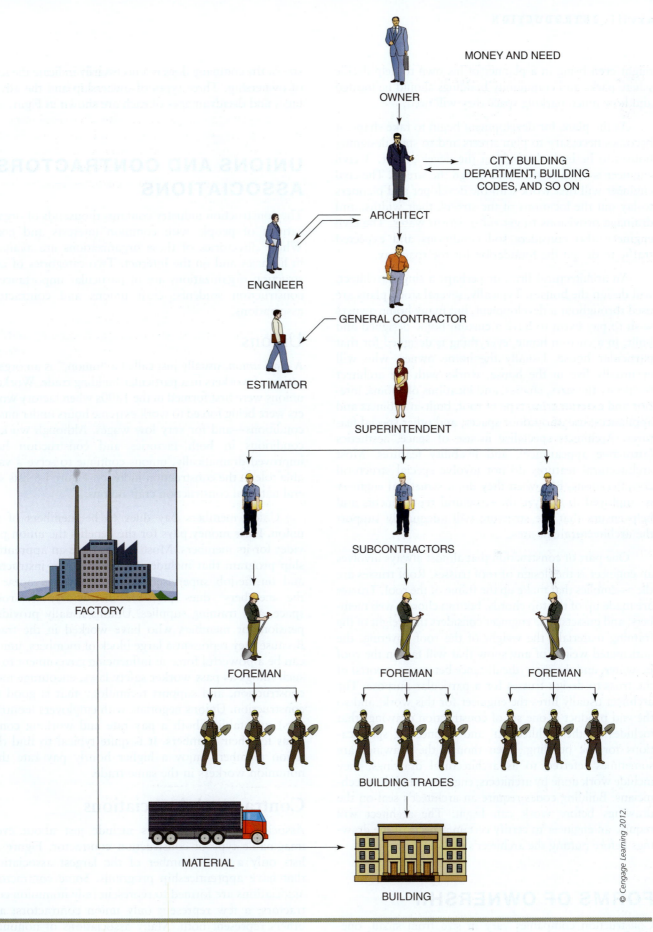

MONEY AND NEED

OWNER

CITY BUILDING
DEPARTMENT, BUILDING
CODES, AND SO ON

ARCHITECT

ENGINEER

GENERAL CONTRACTOR

ESTIMATOR

SUPERINTENDENT

SUBCONTRACTORS

FOREMAN   FOREMAN   FOREMAN

BUILDING TRADES

FACTORY

MATERIAL

BUILDING

© Cengage Learning 2012.

FIGURE I-4 Organization of the construction industry.

might even bring in a planner of his own to help decide where parks and community buildings should be located and how much parking space they will need.

As the plans for development begin to take shape, it becomes necessary to plan streets and to start designing houses to be built throughout the development. A civil engineer is hired to plan and design the streets. The civil engineer will first work with the developer and planners to lay out the locations of the streets, their widths, and drainage provisions to get rid of storm water. The civil engineer also considers soil conditions and expected traffic to design the foundation for the roadway.

An architectural firm, or perhaps a single architect, will design the houses. Typically, several stock plans are used throughout a development, but many home owners wish to pay extra to have a custom home designed and built. In a custom home, everything is designed for that particular house. Usually the home owner, who will eventually live in the house, works with the architect to specify the sizes, shapes, and locations of rooms, interior and exterior trim, type of roof, built-in cabinets and appliances, use of outdoor spaces, and other special features. Architects specialize in use of space, aesthetics (attractive appearance), and livability features. Most architectural features do not involve special structural considerations, but when they do, a structural engineer is employed to analyze the structural requirements and help ensure that the structure will adequately support the architectural features.

One part of construction that almost always involves an engineer is the design of roof trusses. Roof trusses are the assemblies that make up the frame of the roof. Trusses are made up of the top chords, bottom chords, web members, and gussets. The engineer considers the weight of the framing materials, the weight of the roof covering, the anticipated weight of any snow that will fall on the roof in winter, and the span (the distance between supports) of the truss to design trusses for a particular purpose. The architect usually hires the engineer for this work, and so the end product is one set of construction drawings that includes all the architectural and engineering specifications for the building. Even though the drawings are sometimes referred to as architectural drawings, they include work done by architects, engineers, and their technicians. Building codes require an architect's seal on the drawings before work can begin. The architect will require an engineer to certify certain aspects of the drawings before putting the architect's seal on them.

FORMS OF OWNERSHIP

Construction companies vary in size from small, one-person companies to very large international organizations that do many kinds of construction. However, the size of the company does not necessarily indicate the form of ownership. Three types of ownership and the advantages and disadvantages of each are shown in Figure I-5.

UNIONS AND CONTRACTORS' ASSOCIATIONS

The construction industry contains thousands of organizations of people with common interests and goals. Whole directories of these organizations are available in libraries and on the Internet. Two categories of construction organizations are of particular importance to construction students: craft unions and contractors' associations.

Unions

A *craft union*, usually just called a "union," is an organization of workers in a particular building trade. Workers' unions were first formed in the 1800s when factory workers were being forced to work extreme hours under unsafe conditions—and for very low wages. Although working conditions in both factories and construction have improved dramatically, unions continue to serve a valuable role in the construction industry. Figure I-6 lists several national construction craft unions.

Union members pay dues to be members of the union. Dues money pays for the benefits the union provides for its members. Most unions have an apprenticeship program that includes both classroom instruction and on-the-job supervised work experience. Some of the members' dues pay for instructors, classroom space, and training supplies. Unions usually provide a pension for members who have worked in the trade. Because they represent a large block of members, unions can be a powerful force in influencing government to do such things as pass worker safety laws, encourage more construction, and support technology that is good for construction. Unions negotiate with employers (contractors) to establish both a pay rate and working conditions for their members. It is quite typical to find that union members enjoy a higher hourly pay rate than nonunion workers in the same trade.

Contractors' Associations

Associations of contractors include just about every imaginable type of construction contractor. Figure I-7 lists only a small number of the largest associations that have apprenticeship programs. Some contractors' associations are formed to represent only nonunion contractors; a few represent only union contractors; and others represent both. Many associations of nonunion contractors were originally formed because the contractor members felt a need to work together to provide

Forms of Ownership	What it Means	Advantages	Disadvantages
Sole Proprietorship	A sole proprietorship is a business whose owner and operator are the same person.	The owner has complete control over the business and there is a minimum of government regulation. If the company is successful, the owner receives high profits.	If the business goes into debt the owner is responsible for that debt. The owner can be sued for the company, and the owner suffers all the losses of the company.
Partnership (**General** and **Limited Liability Partnership (LLP)**)	A partnership is similar to a sole proprietorship, but there are two or more owners. *General*: In a general partnership, each partner shares the profits and losses of the company in proportion to the partner's share of investment in the company. *LLP*: A limited liability partner is one who invests in the business, receives a proportional share of the profit or loss, but has limited liability.	*General Partnership*: The advantage is that the partners share the expense of starting the business and partnerships are not controlled by extensive government regulations. *LLP*: A limited liability partner can only lose his or her investment	*General Partnership:* Each partner can be held responsible for all the debts of the company. *LLP*: Every LLP must have one or more general partners who run the business. The general partners in an LLP have unlimited liability and they can be personally sued for any debts of the company
Corporation	In a corporation a group of people own the company. Another, usually smaller, group of people manage the business. The owners buy shares of stock. A share of stock is a share or a part of the business. The value of each share increases or decreases according to the success of the company.	In a corporation, no person has unlimited liability. The owners can only lose the amount of money they invested in stock. The owners of a corporation are not responsible for the debts of the corporation. The corporation itself is the legal body and is responsible for its own debts.	The government has stricter regulations for corporations than for the other forms of ownership. Also, corporations are more expensive to form and to operate than are proprietorships and partnerships.

FIGURE I-5   Three types of ownership.

some of the benefits that union contractors receive—such as apprentice training and a lobbying voice in Washington, D.C.

BUILDING CODES

Most towns, cities, and counties have building codes. A *building code* is a set of regulations (usually in the form of a book) that ensure that all buildings in that jurisdiction (area covered by a certain government agency) are of safe construction. Building codes specify such things as minimum size and spacing of lumber for wall framing, steepness of stairs, and fire rating of critical components. The local building department enforces the local building codes. States usually have their own building codes, and state codes often require local building codes to be at least as strict as the state code. Most small cities and counties adopt the state code as their own, meaning that the state building code is the one enforced by the local building department.

Until recently, three major model codes were published by independent organizations. (A model code is a suggested building code that is intended to be adopted as is or with revisions to become a government's official code.) Each model code was widely used in a different region of the United States. By themselves model codes have no authority. They are simply a model that a

International Association of Bridge, Structural, Ornamental and Reinforcing Iron Workers (www.ironworkers.org/)

International Association of Heat and Frost Insulators and Asbestos Workers (www.insulators.org/)

International Brotherhood of Boilermakers, Iron Ship Builders, Blacksmiths, Forgers and Helpers (www.boilermakers.org/)

International Brotherhood of Electrical Workers (www.ibew.org/)

International Brotherhood of Teamsters (www.teamster.org/)

International Union of Bricklayers and Allied Craftworkers (www.bacweb.org/)

International Union of Elevator Constructors (www.iuec.org/)

International Union of Operating Engineers (www.iuoe.org/)

International Union of Painters and Allied Trades (www.iupat.org/)

Laborers' International Union of North America (www.liuna.org/)

Operative Plasterers' and Cement Masons' International Association of the United States and Canada (www.opcmia.org/)

Sheet Metal Workers' International Association (www.smwia.org/)

United Association of Journeymen and Apprentices of the Plumbing and Pipefitting Industry of the United States and Canada (www.ua.org/)

United Brotherhood of Carpenters and Joiners of America (www.carpenters.org/)

United Union of Roofers, Waterproofers and Allied Workers (www.unionroofers.com/)

Utility Workers Union of America (www.uwua.org/)

FIGURE I-6 Construction craft unions.

Air Conditioning Contractors of America (http://www.acca.org)

Air Conditioning Heating and Refrigeration Institute (http://www.ahrinet.org/))

Associated Builders and Contractors (http://www.abc.org)

National Association of Home Builders (http://www.nahb.org)

Home Builder's Institute (http://www.hbi.org)

Independent Electrical Contractors Association (http://www.ieci.org)

National Electrical Contractors Association (http://www.necanet.org)

National Utility Contractors Association (http://www.nuca.com)

Plumbing-Heating-Cooling Contractors Association (http://www.phccweb.org)

The Associated General Contractors (AGC) of America (http://www.agc.org)

FIGURE I-7 These are only a few of the largest construction associations.

government agency can choose to adopt as their own or modify as they see fit. In 2009, the International Code Council published a new model code called the *International Building Code.* They also published the *International Residential Code* to cover home construction (Fig. I-8). Since publication of the first *International Building Code,* states have increasingly adopted it as their building code.

Other than the building code, many codes govern the safe construction of buildings: plumbing codes, fire protection codes, and electrical codes. Most workers on the job site do not need to refer to the codes much during construction. It is the architects and engineers who design the buildings that usually see that the code requirements are covered by their designs. Plumbers and electricians do, however, need to refer to their respective codes frequently. Especially in residential construction, it is common for the plans to indicate where fixtures and outlets are to be located, but the plumbers and electricians must calculate loads and plan their work so it meets the requirements of their codes. The electrical and plumbing codes are updated

© Cengage Learning 2012.

FIGURE I-8   The *International Residential Code* and the *International Building Code*.

frequently, so the workers in those trades spend a certain amount of their time learning what is new in their codes.

WORKING IN THE INDUSTRY

Often success in a career depends more on how people act or how they present themselves to the world than it does on how skilled they are at their job. Most employers would prefer to have a person with modest skills but a great work ethic than a person with great skills but a weak ethic.

Ethics

Ethics are principles of conduct that determine which behaviors are right and wrong. The two aspects of ethics are values and actions. *Values* have to do with what we believe to be right or wrong. We can have a very strong sense of values, knowing the difference between right and wrong, but not act on those values. If we know what is right but we act otherwise, we lack ethics. To be ethical, we must have good values and act accordingly.

We often hear that someone has a great work ethic. That simply means that the person has good ethics in matters pertaining to work. Work ethic is the quality of putting your full effort into your job and striving to do the best job you can. A person with a strong work ethic has the qualities listed in Figure I-9. Good work ethics become habits, and the easiest way to develop good work ethics is to consciously practice them.

Common Rationalizations

We judge ourselves by our best intentions and our best actions. Others judge us by our last worst act. Conscientious people who want to do their jobs well often fail to consider their behavior at work. They tend to compartmentalize ethics into two parts: private and occupational. As a result, sometimes good people think it is okay to do things at work that they know would be wrong outside of work. They forget that everyone's first job is to be a good person. People can easily fall prey to rationalizations when they are trying to support a good cause. "It is all for a good cause" is an attractive rationale that changes how we see deception, concealment, conflicts of interest, favoritism, and violations of established rules and procedures. In making tough decisions, do not be distracted by rationalizations.

A person with a strong work ethic:

- Shows up to work a few minutes early instead of a few minutes late.
- Looks for a job to do as soon as the previous one is done. (This person is sometimes described as a self-starter.)
- Does every job as well as possible.
- Stays with a task until it is completely finished.
- Looks for opportunities to learn more about the job.
- Cooperates with others on the job.
- Is honest with the employer's materials, time, and resources.

© Cengage Learning 2012.

FIGURE I-9 Characteristics of a good work ethic.

Good work ethics yield great benefits. As little children, most of us learned the difference between right and wrong. As adults, when we do what we know is right, we feel good about ourselves and what we are doing. On the other hand, doing what we know is wrong is depressing. We lose respect for ourselves, knowing that what we have done is not something we would want others to do to us. Employers recognize people with a good work ethic. They are the people who are always doing something productive, their work turns out better, and they seem cheerful most of the time. Which person do you think an employer will give the most opportunities to: a person who is always busy and whose work is usually well done or a person who seems glum and must always be told what to do next?

Working on a Team

Constructing a building is not a job for one person acting alone (Fig. I-10). The work at the site requires cooperative effort by carpenters, masons, plumbers, painters, electricians, and others. Usually several workers from each of these trades collaborate. A construction project without teamwork would have lots of problems. For example, one carpenter's work might not match up with another carpenter's work. There could be too much of some materials and not enough of others. Walls may be enclosed before the electrician runs the wiring in them.

Teamwork is very important on a construction site, but what does being a team player on a construction team mean? Effective team members have the best interests of the whole team at heart. Each team member has to carry his or her own load, but it goes beyond that. Sometimes a team member might have to carry more

than his or her own load, just because that is what is best for the team. If you are installing electrical boxes and the plumber says one of your boxes is in the way of a pipe, it might be in the best interests of the project to move the electrical box. That would mean you would have to undo work you had just completed and then redo it. It is, after all, a lot easier to relocate an outlet box than to reroute a sink drain.

The following are six traits of an effective team:

- *Listening.* Team members listen to one another's ideas. They build on teammates' ideas.
- *Questioning.* Team members ask one another sincere questions.
- *Respect.* Team members respect one another's opinions. They encourage and support the ideas of others.
- *Helping.* Team members help one another.
- *Sharing.* Team members offer ideas to one another and tell one another what they have learned.
- *Participation.* Team members contribute ideas, discuss them, and play an active role together in projects.

Communication

How could members function as a team without communication? Good communication is one of the most important skills for success in any career. Employers want workers who can communicate effectively; but more importantly, you must be able to communicate with others to do your job well and to be a good team member. How many of the six traits of an effective team require communication?

Many forms of communication exist, but the most basic ones are speaking, listening, writing, reading, and body language. If you master those five forms of communication, you will probably succeed in your career.

Speaking

To communicate well through speech, you need a reasonably good vocabulary. It is not necessary, or even desirable, to fill your speech with a lot of flowery words that do not say much or that you do not really understand. What is necessary is to know the words that convey what you want the listener to hear, and it is equally necessary to use good enough grammar so those words can be communicated properly. Using the wrong word or using it improperly can cause two serious problems: For one thing, if you use the wrong word, you will not be saying what you intended to say. This is also often true if you use a great word wrong since you still might not be saying what you thought you were saying. For another thing (the second serious problem),

Courtesy of Louisiana-Pacific Corporation.

FIGURE I-10 Work on the job requires cooperative efforts by different individuals from different trade areas.

using a poor choice of words or using bad grammar gives the listener the impression that you are poorly educated or that maybe you just do not care about good communication skills. As a businessperson, you will find that communicating is critical to earning respect as a professional as well as to gaining people's business. Three important steps of effective communication are:

- Looking your listeners in the eye.
- Asking yourself if you think they understand what you are saying. If it is important, ask them if they understand.
- Trying a different approach if they do not understand.

Listening

Good listening is an important skill. Have you ever had people say something to you, and after they were finished and gone, you wondered what they said or you missed some of the details? Perhaps they were giving you directions or telling you about a school assignment. If only you could listen to them again! If possible, try paraphrasing. Paraphrasing means to repeat what they said but in different words. If someone gives you directions, wait until the person is finished. Then, repeat the directions to the person, so he or she can tell you if you are correct. Look at the speaker and form a mental picture of what the speaker is saying. Make what the speaker is saying important to you. Good listening can mean hearing and acting on a detail of a job that will result in giving a competitive edge in bidding.

Writing

Writing is a lot like speaking, except you do not have the advantage of seeing if the person seems to understand or of asking if the person understands. That means you really have to consider your reader. If you are giving instructions, keep them as simple as possible. If you are reporting something to a supervisor, make your report complete, but do not take up his or her time with unrelated trivia. Penmanship, spelling, and grammar count. Always use good grammar to ensure that you are saying what you intend and that your reader will take you seriously. Use standard penmanship, and make it as neat as possible. Do not invent new ways of forming letters, and do not try to make your penmanship ornate. You will only make it harder to read. If you are unsure of how to spell a word, look it up in a dictionary. Next time, you will know the word and will not have to look it up. After you write something—read it, thinking about how your intended reader will take it. Make changes if necessary. Your writing is important! Sole proprietors have to demonstrate good writing skills in proposals and contracts.

If either of these is poorly written, it can cost the business a lot of money.

Reading

You will have to read at work. That is a fact no matter what your occupation. You will have to read building specifications, instructions for use of materials and tools, safety notices, and notes from the boss (Fig. I-11). To develop reading skills, find something you are interested in and spend at least 10 or 15 minutes every day reading it. You might read the sports section of the newspaper, books about your hobby, hunting and fishing magazines, or anything else that is interesting to you. What is important is that you read. Practicing reading will make you a better reader. It will also make you a better writer and a better speaker. When you come across a word you do not know how to pronounce or you do not know the meaning of, look it up or ask someone for help. You will find that you learn pronunciation and meaning very quickly, and your communication skills will improve faster than you expect. In practically no time, you will not need help very often.

Body Language

Body language is an important form of communication. How you position your body and what you do with your hands, face, and eyes all convey a lot of information to the person you are communicating with. Whole books are written about how body language is used to communicate and how to read body language. We will only discuss a couple of key points here.

When you look happy and confident, the message you convey is that you are honest (you have nothing to hide or to worry about) and you probably know what

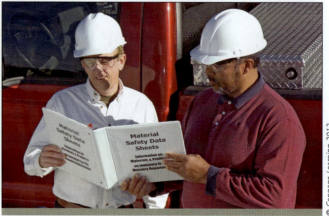

FIGURE I-11 Copies of material data safety sheets.

© Cengage Learning 2012.

you are talking about. If you look unhappy, unsure of yourself, or uninterested, your body language tells the other person to be wary of what you are saying—something is wrong. The following are a few rules for body language that will help you convey a favorable message:

- Look the other person in the eye. Looking toward the floor makes you look untrustworthy. Looking off in space makes you seem uninterested in the other person.

- Keep your hands out of your pockets, and do not wring your hands. Just let your hands rest at your sides or in your lap if you are sitting. An occasional hand gesture is okay, but do not overdo it.

- Dress neatly. Even if you are wearing work clothes, you can be neat. Faddish clothes, extra baggy or extra tight fitting clothes, and T-shirts with offensive messages on them all distract from the real you.

- Speak up. How loudly you speak might not seem like body language, but it has a lot to do with how people react to you. If they have to strain to hear what you are saying, they will think that either you are not confident in what you are saying or you are angry and not to be trusted. If you see your listeners straining to hear you or if they frequently ask you to repeat what you are saying, speak a little louder.

Customer Service

In any industry, you will only be as successful as you are good at building your reputation for doing quality work and to the degree that your customers are happy with you and your job. On the job site, your customer might be a crew chief, a foreman, a subcontractor, or a contractor. If you are the contractor or subcontractor, the customer will be whoever hired you. It does not actually matter who hired you, though—your role will always be to do the very best job you can for whomever it is that you are working.

Good customer service also includes providing a good value for your fees, being honest, communicating clearly, being cooperative, and looking to provide the best possible experience your customer can have in working with you. Just as when you practice good ethics, when you provide great customer service you will enjoy your job much more. You will be proud of your work, others will want to hire you more often, and your career will be much easier to build. Think about how you like being treated when you are a customer—and always try to treat your customers at least as well.

Lifelong Learning

Lifelong learning refers to the idea that we all need to continue to learn throughout our entire lives. We have greater opportunities to learn and greater opportunities

to move up a career ladder today. Our lives are filled with technology, innovative new materials, and new opportunities. People change not only jobs, but entire careers several times during their working life. Those workers who do not understand the new technology in the workplace, along with those who do not keep up with the changes in how their company is managed, are destined to fall behind economically. There is little room in a fast-paced company of this century for a person whose knowledge and skills are not growing as fast as the company. To keep up with new information and to develop new skills for the changing workplace, everyone must continue to learn throughout life.

CONSTRUCTION TRENDS

Every industry has innovations, and construction is no exception. As a construction professional, it is important to be aware of new technologies, new methods, and new ways of thinking about your work. This is as important for a worker's future employment as being aware of safety and ethical business practices. Some of the key technological trends include disaster mitigation, maintenance, building modeling, and green building.

Disaster Mitigation

Both new and existing buildings need to be strengthened and improved to deal with earthquakes, floods, hurricanes, and tornados. Actions like improving wall bracing or preparing moisture management reduces damage and improves safety when these events occur. These actions are increasingly required by building regulations (especially in disaster-prone areas) and requested by property owners and insurers.

Maintenance

Unlike single natural disasters, preventing long-term wear and tear is also an important industry trend. Property owners are more concerned about the costs, effort, and time required to repair and to maintain their homes and buildings. So, there has been significant research into materials that are more durable, construction assemblies that manage moisture, air and elements better, and overall higher quality construction work.

Building Modeling

One of the biggest new trends in construction technology does not include construction materials at all: it includes being able to design, simulate, and manage buildings with the use of computer and information technology. Some of these tools, like Computer-Aided Drafting (CAD) and Computer-Aided Manufacturing

(CAM), have been around for decades. Others, like energy modeling and simulation software or project management tools, are being used more and more. Still others, like Building Information Modeling (BIM) are gathering many of these previous tools into single computing platforms. In all cases, the ability to use computers and professional software is becoming mandatory among workers.

Green Building

Probably the biggest trend in the construction industry over the last decade has been *green building*—that is, planning, design, construction, and maintenance practices that try to minimize a building's impact on the environment throughout its use. Although a set definition of green building is still evolving, everyone agrees on a few key concepts that are important and that in themselves are also major construction trends.

Occupant Health and Safety

The quality of indoor air is influenced by the kinds of surface paints and sealants that are used as well as the management of moisture in plumbing lines, HVAC equipment, and fixtures. Long-term maintenance and care by home owners and remodelers also can shape the prevalence of pests, damage, and mold. Builders and remodelers are becoming more aware of the products and assemblies they use that could have an effect on indoor environments.

Water Conservation and Efficiency

Many builders and property owners are attempting to collect, efficiently use, and reuse water in ways that save the overall amount being used. From using collectors of rainwater to irrigate lawns, to installing low-flow toilets and water-conserving appliances, to feeding used "greywater" from sinks and showers into secondary non-occupant water needs, water efficiency is a trend in all green building but especially where water shortages or droughts are prevalent.

Low-Impact Development

Builders concerned with the effect of the construction site on the land, soil, and water underneath are incorporating storm water techniques, foundation and pavement treatments, and landscaping preservation methods to minimize disturbances to the land and surrounding natural environments.

Material Efficiency

Builders are becoming more aware of the amount of waste coming from construction sites, and inefficiency in the amount of materials (like structural members) that they install in buildings. Many of the materials that are used in construction also do not come from naturally renewable sources or from recycled content materials. Using materials from preferred sources, using them wisely, and then appropriately recycling what is left is a big industry trend (Fig. I-12).

Energy Efficiency and Renewable Energy Sources

The most widely known of all green building trends involves the kind and amount of energy that buildings use. Oftentimes, builders can incorporate the use of renewable energy sources (like solar photovoltaics) or passive solar orientation into their designs. Then, the combination of good building envelope construction and efficient equipment and appliances can all reduce utility costs for property owners, much like the maintenance trend reduces repair costs (Fig. I-13).

There are many ways to keep track of the latest trends in the construction industry. Trade or company journals, on-line resources and blogs, and the latest research coming out of government and university laboratories are several ways to keep informed and up-to-date on the latest industry trends.

FIGURE I-12 Construction site waste recycling. *Courtesy of Carl Seville*

© Cengage Learning 2012.

FIGURE I-13 Duct insulation increases system energy efficiency.

JOB OPPORTUNITIES

Because of the widespread need for electrical services, jobs for residential electricians are found in all parts of the country. Most states and localities require electricians to be licensed. Although licensing requirements differ from area to area, residential electricians generally have to pass an examination that tests their knowledge of electrical theory, the National Electrical Code, local electrical codes, and the various types of wiring methods and materials. Experienced electricians can advance to jobs as estimators, supervisors, project managers, or even electrical inspectors. They may also decide to start their own electrical contracting business.

Employment

As key members of the residential construction industry, residential electricians install, maintain, and troubleshoot electrical wiring systems in a house. A residential electrician's job requires skills and knowledge in the following areas:

- Understanding and following both general and electrical safety procedures.
- Knowing the common hardware and materials used in a residential electrical system.
- Knowing how to read and understand residential building plans.
- Knowing how to use hand tools, power tools, and testing equipment.

- Understanding and applying the National Electrical Code as well as State and local codes when installing electrical wiring.
- Installing service entrances so that electrical power can be provided from the electric utility to a building's electrical system.
- Installing branch and feeder circuits with various types of cable and conduit.
- Installing metal or nonmetallic electrical boxes to the framing members of a house.
- Connecting conductors to circuit breakers, lighting fixtures, receptacles, switches, paddle fans, electric motors, and other electrical equipment.
- Installing video, voice, and data wiring and equipment.
- Rewiring a home or replacing an old fuse box with a new circuit breaker loadcenter.
- Going on "service calls" to identify and then fix electrical problems.
- Installing, maintaining, and troubleshooting photovoltaic and small wind turbine renewable energy systems.

Working as a residential electrician can be very rewarding and is usually a lot of fun. There are many positive aspects about the job, such as:

- Residential electricians have a great sense of accomplishment once a house has been wired and everything works correctly.
- Residential electricians work on different sizes and types of houses that make the job very interesting and never boring.
- Residential electricians often work outdoors.
- Being a residential electrician requires a higher level of technical skill than many other construction trades.
- Residential electricians receive higher pay than many other construction trade jobs.

Like any job or career, there are challenges as well. Some of the challenges of being a residential electrician are:

- The work is sometimes strenuous and may require standing or kneeling for long periods of time.
- Residential electricians may be subject to bad weather conditions when working outdoors.
- Residential electricians may have to travel long distances to job sites.

- The work may involve frequently working on ladders and scaffolding.
- There is a risk of injury from electrical shock, burns, falls, and cuts.

Residential electricians should be in good health and have decent physical strength. It also helps to have good agility and manual dexterity. Good color vision is a requirement because electricians often must identify electrical wires by color. It is also important that residential electricians have good people and communication skills.

Job Outlook

Employment for residential electricians is expected to increase approximately 12 percent from now through the year 2018 according to the Bureau of Labor Statistics, U.S. Department of Labor. As the population and the economy grow, more residential electricians will be needed to install and maintain electrical equipment in homes. Also, new technologies are expected to continue to spur demand for residential electricians. Efforts to increase conservation of energy in existing homes and in new home construction will boost demand for residential electricians because electricians are key to installing some of the latest energy savers, such as solar panels and motion sensors for turning on lights.

Because of their training and relatively high earnings, a smaller proportion of electricians than other trade workers leave their occupation each year. In addition to jobs created by the increased demand for electrical work, many openings are expected to occur over the next two decades as a large number of older electricians retire. The average hourly wage for all electricians as of 2008 (the newest data available) was $22.32. The lowest 10 percent earned less than $13.54, and the highest 10 percent earned more than $38.18. Residential electricians may earn more or less per hour, depending on a number of factors such as experience, education, licenses, and worker demand. Beginning electricians usually start at between 30% and 50% of the rate paid to fully licensed and trained electricians.

Charles Martz

TITLE

Master Electrician, Electrical Inspector, SME (Subject Matter Expert) and Theatre Master Electrician for Iraq and Chief of Electrical Training

EDUCATION

Associate of Applied Science degree in Industrial Electrical/Electronics Technology from Kennebec Valley Community College in Fairfield, Maine.

HISTORY

Charles had always been fascinated by electricity and he needed a way to provide for his family. Becoming an electrician was an ideal career path for Charles to pursue.

ON THE JOB

Since becoming an electrician his responsibilities have varied greatly. Charles spent 2009 in Iraq as an electrical inspector and eventually the SME/Theatre Master Electrician. It was there that he had the opportunity to brief Generals and Colonels on electrical issues in Iraq. Charles also had the opportunity to train third country nationals, electricians and soldiers on proper grounding and bonding. Since coming home, he has worked for a company where he travels around the country working on various electrical installations, such as Federal Prisons and most recently a clean coal production plant.

CHALLENGES

According to Charles, the most difficult challenge for an electrician is staying up to date with the newest technologies, latest equipment and the most recent National Electrical Code changes.

IMPORTANCE OF EDUCATION

Charles believes that a proper education is essential to becoming an electrician and working in the electrical industry. His knowledge of electricity and how it works has not only aided him in the proper installation of electrical equipment, but it has also kept him safe from electric shock.

FUTURE OPPORTUNITIES

Charles currently has two job offers in Afghanistan as an Electrical Inspector and one working here in the United States doing electrical inspections for the FBI.

WORDS OF ADVICE

"Learn all that you possibly can and ask lots of questions. Do not just take what you hear from another electrician as being gospel, but take the initiative to find out if it is the truth...do not be afraid to challenge and question! Go above and beyond. Do not be satisfied with 'good enough.' Soak up all the knowledge you can when you have the opportunity."

CAREER PROFILE

Franqee Higgins

TITLE

Electrical Instructor Home Builder Institute, Clearfield Job Corps, Clearfield, Utah

EDUCATION

Franqee completed Home Builders Institute Electrical Program in Guthrie, Oklahoma. He has an ICC certified combination inspector license and is currently attending Western Governors University in Salt Lake City, Utah, where he is working on his Bachelor of Science degree in business management.

HISTORY

Franqee always wanted to work in the construction industry. After high school he decided to attend a Job Corps program for Plumbing. While receiving his training he realized he had an interest in the electrical industry. Franqee switched gears and worked toward completing an electrical course at Guthrie Job Corps through HBI. He went on to work for the second largest electrical contractor in Texas, eventually becoming a job foreman. That opportunity led him into the inspection side of construction, where he then worked as an electrical inspector in the Dallas metro area for four years. In 2007, Franqee was awarded the National Association of Home Builders President Award and is now an electrical instructor with the Home Builders Institute.

ON THE JOB

Every day for Franqee is different, but what is consistent is that each day he has the opportunity to inspire youth and help them to learn a craft that can potentially change their lives.

CHALLENGES

Transitioning from student to teacher was challenging for Franqee. Now he not only has to teach a trade, but also instill a sense of pride and honor in his students. Many of his students come from troubled backgrounds and have no concept of success. Once they grasp that concept, they begin to excel at an extraordinary pace.

IMPORTANCE OF EDUCATION

As an instructor Franqee constantly teaches the importance of education. Education is the key to opening doors that most of his students did not know existed. Franqee was motivated to go back to college in order to show students that regardless of age, education never ends.

FUTURE OPPORTUNITIES

Franqee was awarded the HBI rookie instructor of the year for 2009. Although not sure what exactly the future holds for him, he does see himself playing an integral part in the learning process of inner-city youth for years to come. Be it as a teacher, motivational speaker, or even as a director at a job corps center, Franqee is passionate about the value and importance of education.

WORDS OF ADVICE

"Don't Quit. Your future is already laid out for you; it is up to you to obtain the skill required to claim it."

CAREER PROFILE

Todd Noel

TITLE

Maine Licensed Master Electrician and business owner of Wire Guys Electric

EDUCATION

Todd received his Certificate in Electrical Construction Technology and earned an Associate of Applied Science Degree in Industrial Electrical/ Electronic Technology from Kennebec Valley Community College in Fairfield, Maine.

HISTORY

Having an electrician for a father meant that Todd had an early introduction into the electrical trade at the age of 13. After high school Todd enrolled at Kennebec Valley Community College where he received his certificate and Associates Degree. Following college he spent the next 10 years working in various aspects of the electrical industry while still working part-time for his father doing primarily residential electric construction. In 2004, after he was laid off by his current employer, Todd decided to start his own business: Wire Guys Electric. He has been self-employed now for six years and established a solid and ever-expanding client base.

ON THE JOB

A typical day for Todd starts in his office checking and replying to emails, returning phone calls, ordering parts, etc. By 8:00 a.m., he is typically at the job site and, depending upon the job, may have a helper/journeyman with him. He meets with the home owner or general contractor about

© Cengage Learning 2012.

job specifics for that day, assigns tasks and supervises while working on his own wiring job. By mid-afternoon he is usually moving on to look at new jobs and returning phone calls. At the end of his work day and in the evening Todd works on estimates and invoicing.

CHALLENGES

Todd finds there are many challenges as a self-employed electrical contractor. It can be difficult to juggle efficiency with the different job requirements, customer schedules, and employee's schedules. In addition, finding time for continuing education, meeting with advertisers, meeting with inspectors, keeping and doing all of the paperwork can also be challenging.

IMPORTANCE OF EDUCATION

Todd's education proved to be very important for many reasons. He needed an education in order to take the tests to get his licenses, but it has also prepared him for other opportunities and challenges he had no idea he would face. Todd's education has given him the knowledge and foundation to act and react in difficult and unfamiliar circumstances, while still completing jobs on schedule. Todd truly believes that he would not have gotten any of the jobs he has had since college if he did not have his education.

FUTURE OPPORTUNITIES

Since he graduated Todd has taken his continuing education seriously. He has continued his education by taking more CAD courses, a supervisory management course, PLC programming courses, servo programming courses, a pneumatic course, an alternative energy course, and a lead smart course. Todd even attended and graduated from a Radon Mitigation program. Taking these courses and keeping up to date on developments in the field is the best way, he believes, to be prepared for future opportunities.

WORDS OF ADVICE

"It is important to treat everyone you deal with respect. You never know who is going to get the opportunity to say something good or bad about you. Make sure you leave a positive impression."

LOOKING FORWARD

Before you can start a career in the residential construction industry as a residential electrician you must acquire the knowledge, skills, and work attitudes required for the job. The chapters that follow in this textbook cover all of the areas needed for you to become proficient as a residential electrician. Your job now is to be a student and study the material presented in this textbook to the best of your ability. Your instructor will guide you through the process. To a large degree, the better you do your job now as a student will determine how well you eventually do your job as an electrician in the residential construction industry.

Preparing and Planning a Residential Wiring Job

CHAPTER 1
Residential Workplace Safety

CHAPTER 2
Hardware and Materials Used in Residential Wiring

CHAPTER 3
Tools Used in Residential Wiring

CHAPTER 4
Test and Measurement Instruments Used in Residential Wiring

CHAPTER 5
Understanding Residential Building Plans

CHAPTER 6
Determining Branch Circuit, Feeder Circuit, and Service Entrance Requirements

Residential Workplace Safety

OBJECTIVES

Upon completion of this chapter, the student should be able to:

- Demonstrate an understanding of the electrical hazards associated with electrical work.

- Demonstrate an understanding of the purpose of the *National Electrical Code®*.

- Demonstrate an understanding of the arrangement of the *National Electrical Code®*.

- Cite examples of rules from the *National Electrical Code®* pertaining to common residential electrical safety hazards.

- Demonstrate an understanding of the purpose of NFPA 70E Standard for Electrical Safety in the Workplace.

- Identify common electrical hazards and how to avoid them on the job.

- Demonstrate an understanding of the purpose of OSHA.

- Cite specific OSHA provisions pertaining to various general and electrical safety hazards associated with residential wiring.

- Demonstrate an understanding of the personal protective equipment used by residential electricians.

- List several safety practices pertaining to general and electrical safety.

- Demonstrate an understanding of material safety data sheets.

- Demonstrate an understanding of various classes of fires and the types of extinguishers used on them.

GLOSSARY OF TERMS

ampere the unit of measure for electrical current flow

arc the flow of a high amount of current across an insulating medium, like air

arc blast a violent electrical condition that causes molten metal to be thrown through the air

arc-flash a dangerous condition associated with the possible release of energy caused by an electric arc

circuit (electrical) an arrangement consisting of a power source, conductors, and a load

conductor a material that allows electrical current to flow through it; examples are copper, aluminum, and silver

current the intensity of electron flow in a conductor

double insulated an electrical power tool type constructed so the case is isolated from electrical energy and is made of a nonconductive material

electrical shock the sudden stimulation of nerves and muscles caused by electricity flowing through the body

grounding an electrical connection to an object that conducts electrical current to the earth

ground fault circuit interrupter (GFCI) a device that protects people from dangerous levels of electrical current by measuring the current difference between two conductors of an electrical circuit and tripping to an open position if the measured value exceeds approximately 5 milliamperes

hazard a potential source of danger

insulator a material that does not allow electrical current to flow through it; examples are rubber, plastic, and glass

load (electrical) a part of an electrical circuit that uses electrical current to perform some function; an example would be a lightbulb (produces light) or electric motor (produces mechanical energy)

Material Safety Data Sheet (MSDS) a form that lists and explains each of the hazardous materials that electricians may work with so they can safely use the material and respond to an emergency situation

National Electrical Code® (NEC®) a document that establishes minimum safety rules for an electrician to follow when performing electrical installations; it is published by the National Fire Protection Association (NFPA)

Occupational Safety and Health Administration (OSHA) since 1971, OSHA's job has been to establish and enforce workplace safety rules

ohm the unit of measure for electrical resistance

Ohm's law the mathematical relationship between current, voltage, and resistance in an electrical circuit

personal protective equipment (PPE) equipment for the eyes, face, head, and extremities, protective clothing, respiratory devices, and protective shields and barriers

polarized plug a two-prong plug that distinguishes between the grounded conductor and the "hot" conductor by having the grounded conductor prong wider than the "hot" conductor prong; this plug will fit into a receptacle only one way

power source a part of an electrical circuit that produces the voltage required by the circuit

resistance the opposition to current flow

scaffolding also referred to as staging; a piece of equipment that provides a platform for working in high places; the parts are put together at the job site and then taken apart and reconstructed when needed at another location

shall a term used in the *National Electrical Code®* that means that the rule *must* be followed

ventricular fibrillation very rapid irregular contractions of the heart that result in the heartbeat and pulse going out of rhythm with each other

volt the unit of measure for voltage

voltage the force that causes electrons to move from atom to atom in a conductor

afety should be the main concern of every worker. Each person on the job should work in a safe manner, no matter what the occupation. Too often, failure on the part of workers to follow recommended safe practices results not only in serious injury to themselves and fellow workers but also in costly damage to equipment and property. The electrical trades, perhaps more than most other occupations, require constant awareness of the hazards associated with the occupation. The difference between life and death is a very fine line. There is no room for mistakes or mental lapses. Trial-and-error practices are not acceptable! Electricity plays a big part in our lives and serves us well. Being able to control electricity allows us to make it do the things we want it to do. Control comes with a good understanding of how electricity works and an appreciation of the hazards and consequences involved when this control is not present. A good residential electrician will, in addition to being proficient in the technology of the trade, possess and display respect for the hazards associated with the occupation. Residential electrical workers must realize from the beginning of their training that if they do not observe safe practices when installing, maintaining, and troubleshooting an electrical system, there will be a good chance that they could be injured on the job. Both general and electrical safety is serious business. It is very important that you make workplace safety a part of your everyday life.

UNDERSTANDING ELECTRICAL HAZARDS

There are three general categories of electrical hazards that a residential electrician could be exposed to. They are electrical shock, arc-flash, and arc-blast. In the following paragraphs all three hazards will be discussed. It is very important that an electrician understands these hazards and learns how to avoid them.

ELECTRICAL SHOCK

Electrical shock is considered the biggest safety hazard associated with doing electrical work. Approximately 30,000 nonfatal electrical shock accidents happen each year. Around 1000 people are killed each year from electrical shock. Roughly half of those happen while working on systems of less than 600 volts. Many residential electricians think that the voltages encountered in residential work will not really hurt them. Others think that residential wiring just does not present the same opportunities for an electrical shock that commercial or industrial electrical work does. They are wrong. The shock hazard exists in residential wiring to the same degree that it exists in other wiring areas. To understand and appreciate the shock hazard in residential wiring, a review of basic electrical theory is provided in the following paragraphs.

Electricity refers to the flow of electrons through a material. The force that drives the electrons and makes this electron flow possible is known as the voltage. Any material or substance through which electricity flows is called a conductor. Examples of conductors used in electrical work include copper and aluminum. These substances offer very little resistance to electron flow. Some materials offer very high resistance to electron flow and are classified as insulators. Examples are plastic, rubber, and porcelain. Electricity flows along a path or circuit. Typically, this path begins with a power source and follows through a conductor to a load. The path then flows back along another conductor to the power source (Figure 1-1).

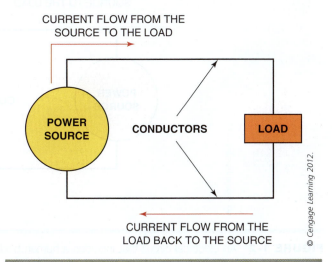

CURRENT FLOW FROM THE SOURCE TO THE LOAD

POWER SOURCE   CONDUCTORS   LOAD

CURRENT FLOW FROM THE LOAD BACK TO THE SOURCE

© Cengage Learning 2012.

FIGURE 1-1 A basic electrical circuit showing the location of the power source, the conductors, and the load.

A very important point to consider at this time is that the human body can, under certain conditions, readily become a conductor and a part of the electrical circuit (Figure 1-2). When this happens, the result is often fatal. Electrons flowing in the circuit have no way of detecting the difference between human beings and electrical equipment.

There is a certain relationship between the **current**, voltage, and **resistance** in an electrical circuit. Georg Simon Ohm discovered this relationship many years ago. His discovery resulted in a mathematical formula that became known as **Ohm's law:**

Current = Electrical Force/Resistance

Current flow is measured in **amperes**, electrical force is measured in **volts**, and electrical resistance is measured in **ohms**. The lower the circuit resistance, the greater the current that a voltage can push through a circuit.

Table 1-1 shows some common resistance values for the human body. Dry skin offers much more opposition to current flow through the human body than does a perspiring or wet body. Remember that the higher the ohm value, the more opposition to current flow there is.

The human body reacts differently to the level of current flowing through it. Table 1-2 shows some typical reactions when a body is subjected to various

TABLE 1-1 Human Body Resistance under Wet and Dry Conditions

SKIN CONDITION	RESISTANCE
Dry skin	100,000 to 500,000 ohms
Perspiring (sweaty hands)	1000 ohms
In water (completely wet)	150 ohms

amounts of current. Remember that a milliampere (mA) is 1/1000 (0.001) of an ampere. It is a very small amount of current.

Consider a situation where a residential electrician is operating an electric-powered drill to bore holes through 2 × 6-inch wall studs. It is a hot and humid day in the middle of the summer. Like many of us, the electrician will perspire heavily and end up using the drill with wet, sweaty hands. According to Table 1-1, the electrician's body resistance would be reduced to about 1000 ohms. If the electrician becomes part of the electrical circuit because of a faulty drill, the amount of current flowing through the electrician's body could be dangerously high. Assuming that the voltage is 120 volts and applying Ohm's law:

Current in Amperes = 120 volts/1000 ohms
= 0.12 ampere, or 120 mA

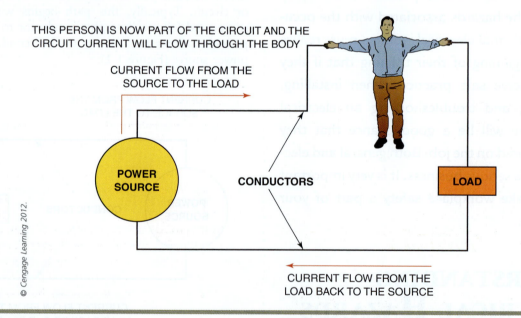

© Cengage Learning 2012.

THIS PERSON IS NOW PART OF THE CIRCUIT AND THE CIRCUIT CURRENT WILL FLOW THROUGH THE BODY

CURRENT FLOW FROM THE SOURCE TO THE LOAD

POWER SOURCE

CONDUCTORS

LOAD

CURRENT FLOW FROM THE LOAD BACK TO THE SOURCE

FIGURE 1-2 An electrical circuit that includes a human body as part of the circuit. Because of the way the body is in the circuit, current will flow into the right hand and travel through the right arm, through the heart and lungs, through the left arm, and back into the circuit through the left hand. The amount of current that flows through the body depends on the body's resistance and the voltage of the power source.

TABLE 1-2 Effect of Current on the Human Body*

CURRENT FLOW	EFFECT ON THE HUMAN BODY
Less than 1 milliampere	No sensation
1 milliampere	Possibly a tingling sensation
5 milliamperes	Slight shock felt; not painful, but disturbing; most people can let go; strong involuntary reactions may lead to injuries
6 to 30 milliamperes	Can definitely feel the shock; it may be painful and you could experience muscular contraction (which could cause you to hold on)
50 to 150 milliamperes	Painful shock, breathing could stop, severe muscle contractions; death is possible
1000 to 4300 milliamperes	Heart convulsions (**ventricular fibrillation**), paralysis of breathing; usually means death
10,000 milliamperes	Cardiac arrest and severe burns; death is probable

*The effects are for currents lasting one second at the voltage levels found in residential wiring systems. Higher voltages will cause severe burns.

With this amount of current, the electrician could receive a shock that is sufficient to cause his or her breathing to stop. The electrician may not survive without proper medical treatment.

CAUTION

CAUTION: High voltage (over 600 volts) is more likely to cause death than low voltage. However, more electrocution deaths occur from low voltage because electricians are exposed more often to low voltages and do not usually use the same caution around low voltage that they do around high voltage.

CAUTION

CAUTION: The longer a body is in an electrical circuit, the greater the severity of the injury.

Let's look at another situation, this time in the home. After a hard day working in the heat, an electrician goes home and decides to cool off by taking a dip in his new aboveground pool. The house the electrician lives in is an older home and does not have the electrical safety features that newer homes have, such as ground fault circuit interrupter receptacles. As the electrician cools off in the pool, he reaches to adjust the volume on a small television setting by the pool with a ballgame on. The television is accidentally knocked into the pool. Because the television is plugged into an outdoor 120-volt receptacle located close to the pool, the outcome is probably death or, at the very least, a serious shock. Body resistance in this case is about 150 ohms, and the current amount is well in excess of the shock required to cause paralysis of breathing (see Tables 1-1 and 1-2):

$$\text{Current in Amperes} = 120 \text{ volts}/150 \text{ ohms}$$
$$= 0.800 \text{ ampere or } 800 \text{ mA}$$

Do not become a shock victim like the electricians described in the preceding paragraphs. By understanding the shock hazard and following electrical safety procedures at all times, your chance of becoming another victim is greatly reduced.

ARC-FLASH

Burns caused by electricity are another hazard encountered by electricians. An **arc-flash** is defined as a dangerous condition associated with the possible release of energy caused by an electric **arc**. When an electric current arcs through the air, the temperature can reach as high as 35,000 degrees Fahrenheit. If you are exposed to these temperatures, direct burns to the skin will occur. Also, unless you are wearing the proper clothing, your clothes could catch on fire.

Usually the severity of the burn depends on the voltage of the circuit. While the chance of an arc-flash burn in residential electrical work is certainly present, the chances for burn injuries tend to be greater on commercial and industrial electrical job sites, where the voltages are typically higher.

An electrical burn is sometimes a result of getting an electrical shock. Burns occur whenever electrical current flows through bone or tissue. This is a very serious type of injury since it happens inside the body. It may not look that severe from the outside, but you need to be aware that severe tissue damage could have taken place under the skin where you cannot see it. Seek medical help whenever you get an electrical shock, especially if it could result in internal burning.

ARC-BLAST

Another hazard associated with an arc is the **arc-blast**. When an arc occurs, the extremely high temperature causes an explosive expansion of both the surrounding air and the metal in the arc path. This explosive expansion causes molten metal to be thrown through the air and onto the skin or into the eyes of an electrical worker. The speed of the molten metal flying through the air is estimated to be around 700 mph which is fast enough for the metal pieces to completely penetrate the human body. Also, the high pressure that results from the arc-blast can rupture eardrums and knock people off ladders and scaffolding. Poor electrical connections or insulation that has failed are the usual causes of arcing that can result in personal injury. Proper personal protective equipment must be used where the possibility of arcing exists. This type of protection is covered later in this chapter.

CAUTION

CAUTION: Always wear the proper personal protective equipment and test the equipment for voltage before working on electrical equipment. Never work on energized equipment unless it is absolutely necessary and you have permission from your supervisor!

NATIONAL ELECTRICAL CODE®

The *National Electrical Code®* (*NEC®*) is the guide for safe wiring practices in the electrical field. Every three years the *NEC®* is updated to reflect the latest changes and trends in the electrical industry. It contains specific rules to help safeguard people and property from the hazards arising from the use of electricity. Its content should become very familiar to all individuals involved with the design, installation, or maintenance of electrical wiring in residential buildings. Many residential electricians refer to the *National Electrical Code®* as "the electrician's bible," which demonstrates their idea of the importance of knowing and applying the *NEC®*. It is used in all fifty of the United States, all U.S. territories, and in many other countries around the world. It should be noted that only those *NEC®* areas that pertain to residential wiring practices are presented in this textbook, although many of the residential *NEC®* rules can also apply to commercial and industrial wiring applications.

HISTORY AND DEVELOPMENT OF THE *NEC®*

Knowing a little bit about the history of the *NEC®* will help you understand why it is structured the way it is and ultimately will help you in understanding how to use the Code in an effective manner. The first edition of the *National Electrical Code®* was published in 1897 and came about primarily because of the many electrical fires that were being caused by electric lighting systems installed during the latter part of the nineteenth century. By 1881, sixty-five textile mills in the New England area had been destroyed or badly damaged from fires that were started by faulty electric lighting systems. Because the electrical construction and operation field was relatively new, updates and revisions occurred on a regular basis during the early years of the *NEC®*. Starting after the 1975 *NEC®*, new editions containing additions, revisions, and deletions began to come out in a three-year cycle that continues to this day. The *National Electrical Code®* is considered to be among the finest building code standards in use today.

The *National Electrical Code®* is developed and published by the National Fire Protection Association (NFPA). The NFPA develops and publishes many different codes and standards, not just the *NEC®*. As a matter of fact, the NFPA develops and publishes some 300 safety codes and standards that deal with a range of subjects related to fire, building, life safety, and of course, electrical. The *NEC®* is document #70 out of all of the NFPA safety codes and standards and is actually titled *NFPA 70: National Electrical Code*. Other NFPA codes of interest to residential electrical system installers include *NFPA 70A Electrical Code for One- and Two-Family Dwellings and Mobile Homes* and *NFPA 70E Standard for Electrical Safety in the Workplace*. To learn more about the National Fire Protection Association and the various codes and standards it develops and publishes, visit http://www.nfpa.org.

THE ORGANIZATION OF THE NEC®

When new electricians first take a look at the *National Electrical Code®*, they are usually intimidated. It is not the easiest book to read, and it even tells us in the introduction that it is not intended as a "how-to" book or a book to use for someone who has no electrical training. The best way for a new electrician to learn how to use the *NEC®* is to use it in conjunction with a textbook. However, the first step in learning how to use the *NEC®* is to understand how it is organized (Figure 1-3). The following sections cover how the *NEC®* is organized and written.

Chapters

The *National Electrical Code®* is divided into nine chapters (Figure 1-4). Chapters 1 through 4 consist of rules that apply generally to all electrical installations. **Chapter 1: General** contains definitions and rules covering the basic requirements for electrical wiring installations. **Chapter 2: Wiring and Protection** contains rules that apply to the installation of things such as branch circuits, feeders, services, fuses, circuit breakers, and grounding. **Chapter 3: Wiring Methods and Materials**

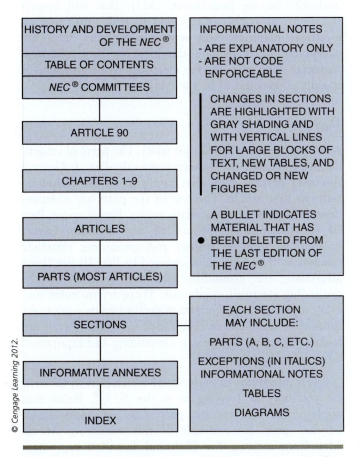

FIGURE 1-3   The layout of the *National Electrical Code®*.

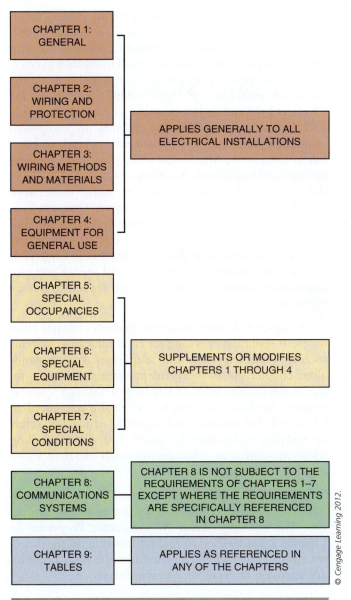

FIGURE 1-4   The arrangement of the chapters in the *National Electrical Code®*.

contains rules that apply to various wiring methods, such as nonmetallic sheathed cable, metal-clad cable, rigid metal conduit, and electrical metallic tubing. **Chapter 4: Equipment for General Use** contains rules that apply to the installation of equipment and materials such as receptacles, panelboards, lighting fixtures, electric motors, and transformers.

Chapters 5, 6, and 7 apply to special occupancies and equipment and other special wiring conditions. **Chapter 5: Special Occupancies** contains rules that apply to wiring in special areas such as gas stations, aircraft hangers, hospitals, movie theaters, agricultural buildings, and mobile homes. **Chapter 6: Special Equipment** contains rules that apply to the installation of special equipment such as electric signs, cranes, elevators, electric welders, swimming pools, and solar photovoltaic systems. **Chapter 7: Special Conditions** contains rules that apply to special wiring installations such as emergency power systems, fire alarm systems, and optical fiber cables. Chapters 5, 6, and 7 often amend the rules found in Chapters 1 through 4. For example, in Chapter 2, Article 250 Grounding and Bonding permits an equipment grounding conductor to be insulated, covered, or bare, but in Chapter 6, Article 680 Swimming Pools requires an *insulated* equipment grounding conductor. Whenever the requirements of Chapters 5, 6, and 7 differ from Chapters 1 through 4, the requirements in Chapters 5, 6, and 7 apply. For example, in Chapter 2, Articles 210 and 220 contain requirements for branch circuits and service-entrance calculations for dwelling units, but in Chapter 5, Article 550 contains branch circuit and service-entrance calculation requirements that must be applied to mobile home dwellings.

Chapter 8: Communications Systems contains rules for the installation of wiring for communication systems such as cable television and telephones. It is a stand-alone chapter. As an independent chapter, it is not subject to the requirements of Chapters 1 through 7 unless material in those chapters is specifically referenced in Chapter 8. For example, in Chapter 8, Article 800 refers an installer to Chapter 3, Article 300 when installing telephone wiring in a building through metal or wood framing members.

Chapter 9: Tables has only tables. The tables contain information such as conduit fill and properties of conductors. For example, **Table 8: Conductor Properties** in Chapter 9 can be used to find information on the different sizes of conductors recognized by the *NEC*®. Tables are discussed in detail later in this chapter.

Articles

Chapter 1 through 8 are broken down into a series of articles. Each chapter heading describes the general subject area covered in the chapter. Articles cover specific subjects that fall under the general subject area of a chapter. For example, "Chapter 2, Wiring and Protection" is a general subject area. The articles within this chapter cover topics such as branch circuits, load calculations, grounding and bonding, services, and overcurrent protection. Articles are very specific in their coverage. The article number will always start with the number of the chapter in which it is found. For example, "Article 250, Grounding and Bonding" is found in Chapter 2; and "Article 422, Appliances" is found in Chapter 4 (Figure 1-5).

Article 90, Introduction is the only article that is not actually located in a chapter. It stands alone at the beginning of the *NEC*® and contains information on

Examples of Article Locations in Specific Chapters

Article <u>1</u>10 Requirements for Electrical Installations is located in Chapter <u>1</u> General.

Article <u>2</u>30 Services is located in Chapter <u>2</u> Wiring and Protection

Article <u>3</u>30 Metal Clad Cable: Type MC is located in Chapter <u>3</u> Wiring Methods and Materials.

Article <u>4</u>22 Appliances is located in Chapter <u>4</u> Equipment for General Use.

Article <u>5</u>17 Health Care Facilities is located in Chapter <u>5</u> Special Occupancies.

Article <u>6</u>80 Swimming Pools, Fountains, and Similar Installations is located in Chapter <u>6</u> Special Equipment.

Article <u>7</u>60 Fire Alarm Systems is located in Chapter <u>7</u> Special Conditions.

Article <u>8</u>10 Radio and Television Equipment is located in Chapter <u>8</u> Communications Systems.

© Cengage Learning 2012.

FIGURE 1-5   The first digit of the article number indicates the chapter in which it is located.

such things as the purpose of the *NEC*®, what kind of electrical installation the *NEC*® covers and doesn't cover, and how the rules listed in the *NEC*® are to be enforced. Article 90 clearly points out that it is the authority having jurisdiction (AHJ) that will inspect the electrical system installation to make sure that all electrical materials and wiring techniques used are acceptable. It is usually the local or state electrical inspector who inspects the electrical work and who is usually considered to be the AHJ.

Parts

Most articles are divided into parts. The parts of an article are indicated by Roman numerals. Part I: General discusses general rules that apply to all the following parts in that article. The rest of the parts are independent and apply to very specific topics. For example, Part III of Article 250, Grounding and Bonding covers grounding electrode system and grounding electrode conductor. The *only* rules covered in Part III are those addressing the grounding electrode system and the grounding electrode conductor. There are no rules covering topics such as equipment grounding or equipment grounding conductors. The rules for these topics are found in Part VI: Equipment Grounding and Equipment Grounding Conductors.

It is important to understand which part of the article you may be looking at in the *NEC*®. Rules that apply under one specific part may not apply under another part. An example of this occurs in Article 110, Requirements for Electrical Installations, where Part II covers applications of 600 volts or less and Part III covers applications of over 600 volts. Applying a rule found in Part II "600 volts or less" to a Part III "over 600 volts" application could cause a serious safety hazard.

Sections

Each rule found in the *National Electrical Code®* is called a section. Articles contain the *NEC®* rules covering a specific electrical area. As a result, articles (except Article 100, Definitions) consist of many sections. Sections are identified by a number that consists of the article number, a dot, and then the section number. For example, Section 110.3 is located in Chapter 1, Article 110, and is Section 3. Note that older editions of the *NEC®* used a dash instead of a dot, as in Section 110-3. Some sections are further divided into parts identified by letters in parentheses: (A), (B), (C), and so on. For example, Section 110.3(B) is located in Part (B) of Section 110.3. Sections may further be divided into numbers in parentheses: (1), (2), (3), and so on. There is even a third level of division, which is identified with lower-case letters: (a), (b), (c), and so on.

Many residential electricians who use the *NEC®* make the mistake of using the term "article" when they are actually referring to a "section." For example, it should be said that standard sizes of fuses and nonadjustable circuit breakers are found in *Section 240.6(A)*, not in *Article 240.6(A)*.

Tables

There are numerous tables found in the *NEC®*, and a lot of important information is contained in them. Many tables include numerical data that can be applied to a particular wiring installation. Make sure you read the title of each table very carefully so that you can understand completely what the table actually contains, where its information can be applied, and what its limitations are. When a table is used, make sure to read all footnotes and notes to the tables, because the material there must also be applied.

Figures

You have most likely heard the old adage that "a picture is worth a thousand words." Unfortunately, the *NEC®* does not include many pictures to help us understand how to apply the code rules. However, there are a few basic illustrations used, and they are intended to help explain a section that may be somewhat confusing. A good example is Section 410.2, which defines storage space in a clothes closet, and Figure 410.2, which is used to illustrate the storage space in a clothes closet. Knowing what is considered storage space in a clothes closet is important for an electrician know because Section 410.16 lists certain installation requirements that apply to luminaires (light fixtures) located in clothes closets, and they include minimum mounting distances from the storage space.

Exceptions

When an *Exception* to a code rule is used, it is printed in *italics*. *Exceptions* always follow the rule that they amend, and they only apply to the code rule that they follow. They are in the *NEC®* to provide an alternate method to a specific requirement. Two types of *Exceptions* are found in the *NEC®*. A mandatory *Exception* is identified by the use of the terms "shall" or "shall not" and means that you *must* apply the rule in a specific way. A permissive *Exception* is identified by a phrase like "is permitted." This means that you could apply the rule as written but you may decide not to. Sometimes there is more than one *Exception* to a rule. When this is the case, they are shown as *Exception No. 1*, *Exception No. 2*, and so on.

There has been an effort in the last few code cycles to rewrite the code in a more positive language. As a result, there are fewer *Exceptions* found in the *NEC®* than there used to be. It is felt that fewer *Exceptions* in the *NEC®* will make the *NEC®* easier to understand and use.

Informational Notes

Informational notes provide explanatory material and are found following the section or table to which they apply. They often refer the reader to other *NEC®* sections that give additional information on the same subject. Informational notes may also refer the reader to other NFPA documents for more information. Informational notes are informational only and, as such, are not enforceable as requirements of the *NEC®*. Some sections have several informational notes after them and are simply displayed as Informational Note No. 1, Informational Note No. 2, and so on.

Footnotes to tables, mentioned in an earlier section of this chapter, are not considered to be explanatory material unless they are identified by the wording "Informational Note." Table 1 in Chapter 9 is a good example of a table having Informational Notes under it. The information contained in the Informational Notes is informational only.

In 2008 and earlier editions of the *National Electrical Code®* explanatory material was identified as Fine Print Notes (FPN).

Extractions

Many of the requirements contained in the *National Electrical Code®* have been taken or extracted from other existing NFPA codes and standards. Brackets that contain section references to other NFPA documents are used to indicate where the extracted material comes from. The bracketed references immediately follow the extracted text in the *NEC*®. In some earlier editions of the *National Electrical Code*®, a superscript^X was used to indicate extracted material and was typically placed after the section number.

Table of Contents

The table of contents (TOC) lists each article in numerical order starting with Article 90. Except for Article 90, each article is listed in the TOC under the chapter in which it resides. The title of each article is given and a corresponding page number is listed. Under each article in the TOC, the various parts of the article are listed with corresponding page numbers. The parts are numbered with a Roman numeral and have specific titles. Just a reminder: The *NEC*® is publication number 70 of the National Fire Protection Association's published works. Therefore, all page numbers start with the number 70, a dash, and then the actual page number. The only other item that should be mentioned at this time about the TOC is that for the section titled *Chapter 9 Tables,* no articles are listed—just tables in numerical order. Each table number has a corresponding title and page number.

When locating information in the *NEC*®, the table of contents is a good place to begin if you are unable to think of a particular *NEC*® subject by name or if you need to look up material on a subject such as grounding and bonding but can't remember the article number. Another good use for the TOC is that when a section states something like, "Any of the wiring methods in Chapter 3 may be used," it won't take long to see which wiring methods are included in Chapter 3 when you look at the Chapter 3 area in the table of contents.

Index

The index is located in the back of the *National Electrical Code*®, immediately after the Informative Annexes. The index contains the major topic areas that are covered in the *NEC*® and conveniently lists them in alphabetical order. Several sub-topics are listed under most of the major topic areas. The sub-topics are also listed in alphabetical order. The major topics are printed in bold print to help locate them easily, while the sub-topics are printed in regular print. Many, but not all, of the major topic areas have *NEC*® articles, parts, or sections listed after them to let you know where specific information on the major topic area can be found. All of the sub-topics have article, part, section, or table listings that contain information about the sub-topic. No page numbers are given in the index, so knowing how the *NEC*® is organized becomes very important when looking up information on a specific topic area. Starting with the

2008 *NEC*®, and continuing in the 2011 *NEC*®, key words are included at the top of each page. Be aware that some major topic areas and sub-topics simply have references to other topic areas where you will find article, part, section, or table listings that contain information about your specific subject.

The index has been greatly expanded in recent editions of the *NEC*®, but unfortunately, it still does not cover all of the terminology electricians often use. Sometimes you have to try to find another word that could mean the same thing or try different terminology when you are looking for a main topic area in the index.

Informative Annexes

The informative annexes are located in the back of the *NEC*® after Chapter 9. They contain additional explanatory material that is designed to help users of the *NEC*®.

Terms and Definitions

Article 100 contains definitions of terms essential to the proper use of the *NEC*®. The definitions contained in Article 100 are generally those terms that are used in two or more *NEC*® articles. The article does not include commonly defined general terms or common technical terms from related codes or standards. The terms listed in Article 100 are in alphabetical order and are displayed in bold print so that they can be located easily.

Terms that are unique to an article are located and defined within that article. Generally, these definitions are at the beginning of the article. Article 517, Health Care Facilities is a good example of terms and definitions being in the article and not in Article 100. Section 517.2 lists several terms and definitions that apply only to health care facilities like hospitals and nursing homes. Because these terms and definitions are not used throughout the *NEC*®, they are not included in Article 100, Definitions.

Scope

The scope of an article, outlined in the first section, describes what the article covers and is very important to anyone using the *NEC*®. Reading the scope of an article can help you determine whether the article will likely contain the information you are looking for. Otherwise, you may find yourself looking for specific information in an article which does not even include the information which you are looking.

Bold Print

The writing style used in the *NEC*® is designed to allow for information to be found quickly and easily. One of the writing styles used is to include bold print lettering in certain locations. Bold print lettering is used for the chapter titles, article titles, part titles, each section number and title, and each sub-part. Table titles and row/column titles in the tables also use bold print lettering. This makes locating the information you are looking for much quicker and easier.

Vertical Lines and Grey Highlighting

In 2005 and earlier editions, changes to the current *NEC®* from the previous edition are indicated by a vertical solid line beside the paragraph, table, or figure in which the change occurred. Sometimes these changes are quite extensive and the vertical line is quite long. Other times the changes are nothing more than adding or changing a word or two and the vertical line is quite short. This writing style is designed to allow a user of the *NEC®* to recognize quickly where a change has been made from the last code. Starting in the 2008 *NEC®*, changes are shown in grey highlighting within sections and with vertical lining for large blocks of changed or new text and for new tables and changed or new figures. This is designed to make the code more user-friendly.

Bullets

A "bullet" or black dot (•) is used to denote where one or more complete paragraphs have been deleted from the previous edition of the *NEC®*. The dot (•) is found on the page where the material from the last *NEC®* was located.

(See Procedure 1-1 on page 31 for a suggested procedure to find specific information in the *NEC®*.)

NEC® APPLICATIONS

The following paragraphs cover some examples of common residential wiring applications and the *NEC®* rules that should be followed. The purpose of this section is simply to show you, the beginning electrician, how the *NEC®* can be applied to residential wiring situations.

GROUND-FAULT CIRCUIT-INTERRUPTER PROTECTION OF PERSONNEL

The *NEC®* requires **ground fault circuit interrupter (GFCI)** protection at various locations. GFCI protection is usually provided with a GFCI receptacle

FIGURE 1-6   A ground fault circuit interrupter (GFCI) receptacle.

© Cengage Learning 2012.

(Figure 1-6) but could also be provided with a GFCI circuit breaker. This protection is for personnel and is not designed specifically to protect equipment. Look up Section 210.8(A) in your *NEC®*, and let's look at a few examples.

1 All 125V, 15A, and 20A receptacles installed in bathrooms shall have GFCI protection.

2 All 125V, 15A, and 20A receptacles installed outside of a dwelling unit shall have GFCI protection.

3 All 125V, 15A, and 20A receptacles located to serve the kitchen countertop shall have GFCI protection.

There are many other receptacle locations in house wiring that require GFCI protection and are covered in detail later in this book.

In addition, ground-fault protection for personnel on construction site wiring installations must be provided that complies with Section 590.6(A) and (B) of the *NEC®*. This section applies only to temporary wiring installations used to supply temporary power to equipment used by personnel during construction or remodeling of buildings. All 125-volt, single-phase, 15-, 20-, and 30-ampere receptacle outlets that are used by electricians to plug in portable power tools or lighting equipment on a construction site must have GFCI protection for personnel. One way to meet the intent of this *NEC®* requirement is for a residential electrician to use cord sets with built-in GFCI protection and to plug their portable power equipment and lighting into the GFCI-protected cord sets (Figure 1-7).

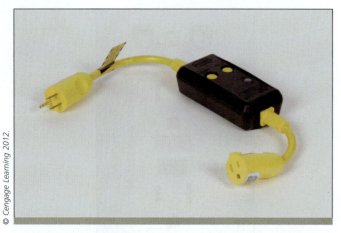

FIGURE 1-7   A cordset with built-in GFCI protection. It can be used to meet the *National Electrical Code®* requirement that all 125-volt, single-phase, 15-, 20-, and 30-amp receptacle outlets used at a construction site to provide electrical power to portable power tools or lighting equipment be GFCI protected.

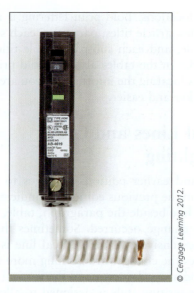

FIGURE 1-8   An arc-fault circuit-interrupter (AFCI) circuit breaker is used to provide arc fault protection for 120-volt, 15- and 20-amp branch circuits.

ARC-FAULT CIRCUIT-INTERRUPTER (AFCI) PROTECTION

An arc-fault circuit-interrupter is defined as a device intended to provide protection from the effects of arc faults and to de-energize a circuit when an arc fault is detected. While a GFCI is designed to protect you from electrical shock, an AFCI is designed to protect you from fires started by faulty electrical wiring. Section 210.12(A) of the *NEC®* requires arc-fault protection for 120-volt, 15- and 20-amp branch circuits supplying outlets in almost all locations of a house. The outlets that must be protected are receptacle outlets and lighting outlets. The typical way that electricians meet this requirement is to protect branch circuits with an AFCI circuit breaker (Figure 1-8). AFCI protection is covered in greater detail later in this book.

GROUNDING

Grounding is covered in Article 250 of the *NEC®*. Grounding of electrical tools and equipment is one of the most important methods of controlling the hazards of electricity. If the insulation in electrical equipment deteriorates or if a wire works loose and comes into contact with the frame or some other part that does not normally carry current, these parts can become energized. The electricity is no longer controlled. It is ready to follow a path to ground—any path! The unprotected human body will work just fine! As we discussed earlier in this chapter, the consequences could be fatal.

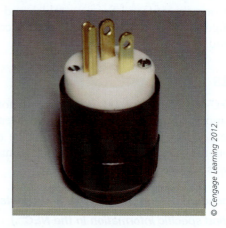

FIGURE 1-9   A three-prong electrical plug. This device is considered to be a grounding type.

It is the green or bare grounding wire of an electrical circuit that is connected to the noncurrent-carrying metal parts of electrical equipment. This wire provides a low-resistance path to ground for more precise control of the electrical current flow. This results in fuses and circuit breakers working faster in the event of an electrical problem, causing the circuit to open and stopping the flow of electricity until the problem can be found and corrected. If the metal parts of electrical equipment were not grounded and a loose energized wire came in contact with the metal, the metal would now be energized. If a human touches the equipment, a fatal shock could be delivered. Grounding the equipment properly can all but eliminate the shock hazard. This discussion should alert you to the importance of the third prong on an approved cord with a three-prong plug (Figure 1-9).

CAUTION

CAUTION: Never cut off the grounding prong on an electrical plug!

These are just a few of the many *NEC*® articles and sections that have a direct bearing on the safe installation of electrical wiring. Ongoing study of the *NEC*® is a must for every electrician.

NFPA 70E
Standard for Electrical Safety in the Workplace

The purpose of NFPA 70E (Figure 1-10) is to provide a practical safe working area for employees relative to the hazards arising from the use of electricity. It is concerned with "workplace" safety for electricians, while the *National Electrical Code*® is concerned with a safe electrical "installation." NFPA 70E addresses electrical safety concerns in all of the places that the *National Electrical Code*® addresses safe installation practices, including residential construction.

An employer must provide an electrical safety program and safety training. The employees must follow the program's requirements. This will apply to all employees who could be exposed to electrical shock, electrocution, or the thermal hazard associated with arc flash. This certainly applies to residential electricians. The type of training can be either classroom training or on-the-job (OJT). Usually, the training involves both types to make sure an employee really understands the topics presented. The safety training should include the following items:

- The type of electrical hazards present in the workplace
- How each electrical hazard affects body tissue
- How to determine the degree of each hazard
- How to avoid exposure to each hazard
- How to minimize risk by body position
- What personal protective equipment (PPE) is needed
- How to select the proper PPE
- What employer-provided procedures the employee must follow
- How increased length of exposure time to an electrical hazard results in more injuries
- How to determine approach boundaries and recognize that these boundaries are related to protection from exposure to electrical shock and electrocution

NFPA 70E covers the PPE required when an electrical worker is exposed to an electrical hazard. In residential work the maximum voltage you would typically be exposed to is 240 volts. When working on energized equipment rated 240 volts and below, NFPA 70E states that the Hazard/Risk category is a 0 or 1 depending on whether the energized parts are exposed or not. The type of PPE you need to wear depends on the Hazard/Risk Category. For example, if the Hazard/Risk Category is 0, you need to wear protective clothing that consists of a long sleeved shirt, long pants, safety glasses, hearing protection, and leather gloves. If the Hazard/Risk Category is 1, you need to wear FR (Flame Resistant) clothing and protective equipment consisting of:

- Arc-rated long sleeve shirt,
- Arc-rated pants,
- Arc-rated face shield,
- Hard hat,
- Safety glasses,
- Hearing protection,
- Leather gloves,
- And leather work shoes.

Remember, your employer will determine what the Hazard/Risk Category is and will provide you with the proper personal protective clothing and equipment.

There is a lot more material covered in NFPA 70E than what was presented here. The intent of this section was to introduce you to NFPA 70E and tell you a little bit about

© Cengage Learning 2012.

FIGURE 1-10   NFPA 70E Standard for Electrical Safety in the Workplace.

what it covers. It is very important that you study and learn more about NFPA 70E as you continue your career as an electrician.

OCCUPATIONAL SAFETY AND HEALTH ADMINISTRATION (OSHA)

Another set of safety rules and regulations that apply to both general and electrical safety are those specified by the Occupational Safety and Health Act of 1970. The Code of Federal Regulations (CFR) is published and administered by the Occupational Safety and Health Administration (OSHA). The purpose of OSHA is to ensure safe and healthy working conditions for working men and women on the job site by authorizing enforcement of the standards developed under the act; by assisting and encouraging the states in their efforts to ensure safe and healthy working conditions; and by providing for research, information, education, and training in the field of occupational safety and health and for other purposes. CFR Title 29, Part 1910, covers the OSHA regulations for general industry, and CFR Title 29, Part 1926, covers the regulations for the construction industry. The full text of these standards is available on OSHA's Web site: http://www.osha.gov. Additionally, regulations on general electrical safety are covered in CFR Title 29, 1910, Subpart S—Electrical, and electrical safety on the construction site is covered in CFR Title 29, 1926, Subpart K—Electrical (see Figure 1-11).

The National Fire Protection Association's Standard 70E, Standard for Electrical Safety in the Workplace, also contains safety regulations that an electrician should be aware of. The NFPA 70E standards are incorporated into the OSHA regulations. NFPA 70E was introduced to you in the previous section.

It is not the intent of this book to cover in detail all the regulations on safety outlined in OSHA CFR Title 29, 1910 and CFR Title 29, 1926 or NFPA 70-E, but a residential electrician should be familiar with some of the more important regulations. The following paragraphs cover some specific OSHA provisions that have a bearing on proper safety practices in residential electrical work.

SAFETY TRAINING

OSHA 29 CFR Part 1910.332 requires employees to be trained in safety-related work practices and procedures that are necessary for safety from electrical hazards.

Code of Federal Regulations (CFR), Title 29, Part 1910, Subpart S – Electrical
(abbreviated as 29 CFR 1910)

General
1910.301   Introduction

Design Safety Standards for Electrical Systems
1910.302   Electrical utilization systems
1910.303   General requirements
1910.304   Wiring design and protection
1910.305   Wiring methods, components, and equipment for general use
1910.306   Specific purpose equipment and installations
1910.307   Hazardous (classified) locations
1910.308   Special systems

Safety Related Work Practices
1910.331   Scope
1910.332   Training
1910.333   Selection and use of work practices
1910.334   Use of equipment
1910.335   Safeguards for personnel protection

Code of Federal Regulations (CFR), Title 29, Part 1926, Subpart K – Electrical
(abbreviated as 29 CFR 1926)

General
1926.400   Introduction
Installation Safety Requirements
1926.402   Applicability
1926.403   General requirements
1926.404   Wiring design and protection
1926.405   Wiring methods, components, and equipment for general use
1926.406   Specific purpose equipment and installation
1926.407   Hazardous (classified) locations
1926.408   Special systems

Safety-Related Work Practices
1926.416   General requirements
1926.417   Lock-out and tagging circuits

Safety-Related Maintenance and Environmental Considerations
1926.431   Maintenance of equipment
1926.432   Environmental deterioration of equipment

Safety Requirements for Special Equipment
1926.441   Batteries and battery charging

Definitions
1926.449   Definitions applicable to this subpart

FIGURE 1-11   A list of OSHA standards related to electrical safety. Subpart S relates to General Industry and Subpart K relates to the Construction Industry. You can read the full text of these standards on the OSHA Web site: http://www.osha.gov.

Employers are required to provide safety training to those employees who face the greatest risk of injury from electrical hazards. This certainly includes electricians. OSHA is fairly flexible and allows this training to consist of classroom and on-the-job training.

The OSHA Outreach Training Program is the primary way to train workers in the basics of occupational safety and health. Through the program, individuals who complete a one-week OSHA trainer course are authorized to teach 10-hour and 30-hour courses in construction or general industry safety and health hazard recognition and prevention. Authorized trainers can receive OSHA course completion cards for their students. The cards verify that you have completed an OSHA 10 or OSHA 30 course. Many employers around the country require people to have the OSHA 10 or OSHA 30 card before they are hired.

LOCKOUT/TAGOUT

This rule is one of the more important safety regulations that you need to follow. It is covered in OSHA 29 CFR Part 1910.333 and 1926.417 and outlines the requirements for locking out and then tagging out the electrical source for the circuit or equipment you may be working on. OSHA provides only the minimum requirements for lockout/tagout. Each company is required to establish its own lockout/tagout procedure, so make sure to follow the procedure of the company you are working for.

Only devices that are designed specifically for locking out and tagging out are to be used (Figure 1-12). When using padlocks, they must be numbered and then assigned to only one person. No duplicate or master keys to the lockout devices are to be available, except possibly the site supervisor. The tags are supposed to be of white, red, and black and need to include the employee's name, the date, and company for which you work. The information you put on the tag should be in permanent marker.

FIGURE 1-12 *Examples of lockout/tagout devices. An electrician must use devices of this type when a lockout/tagout procedure is being followed.*

© Cengage Learning 2012.

CAUTION

CAUTION: Always follow the lockout/tagout procedure that is used by the company for which you work.

(See Procedure 1-2 on page 32 for a suggested procedure for lockout/tagout.)

INITIAL INSPECTIONS, TESTS, OR DETERMINATIONS

OSHA 29 CFR Part 1926.416 requires that existing conditions be determined by inspection or testing with a test instrument before work is started (test and measurement instruments are covered in Chapter 4). All electrical equipment and conductors are to be considered energized, until determined de-energized by testing methods. Work is not to proceed on or near energized parts until the operating voltage is determined.

CAUTION

CAUTION: Always assume a circuit to be energized until you verify otherwise.

POWER TOOLS

OSHA 29 CFR Part 1926.404 contains some important regulations for power tools. The requirements for electrical power tools stipulate that they be equipped with a three-wire cord. The ground wire (green wire) must be permanently connected to the tool frame, and a three-prong plug used as a means for grounding must be at the other end. If a tool is constructed and labeled as **double insulated**, a two-prong **polarized plug** can be used (Figure 1-13).

TRENCHING AND EXCAVATING

OSHA CFR Title 29, Part 1926, Subpart P 1926.651 covers excavation and sets out specific requirements for trenching and excavating that are of interest to the residential electrical worker who may be involved in the installation of underground wiring. Of special interest is the requirement for sloping, or shoring up, the sides of trenches. Without shoring, the threat of an excavated

© Cengage Learning 2012.

FIGURE 1-13 A two-prong polarized electrical plug. This device is considered to be a nongrounding type of plug. Notice that one prong is wider than the other. This will enable the plug to be inserted into a receptacle only in the proper way.

hole caving in while you are in the hole is very real and also very sobering. There have been many workers who have been buried alive. Do not put yourself in a position where this could happen to you.

As you can see from the preceding paragraphs, OSHA provides many regulations for safe work practices that all electricians need to follow. A more in-depth study of OSHA regulations is a good idea for anyone doing electrical work. Understanding and following these regulations can mean the difference between being hurt seriously on the job and enjoying a long, injury-free career as a residential electrician. Don't forget to get an OSHA 10 or OSHA 30 card as soon as possible!

PERSONAL PROTECTIVE EQUIPMENT

Selecting and using the proper **personal protective equipment (PPE)** is very important for anyone installing a residential electrical system. OSHA defines PPE as "equipment for the eyes, face, head, and extremities, protective clothing, respiratory devices, protective shields and barriers." It is considered to be the last line of defense between you and an injury (Figure 1-14). PPE requirements for the construction industry are found in 29 CFR Part 1926. OSHA 29 CFR Part 1910.335 also contains some PPE requirements that apply to any worker performing electrical installations. OSHA requires electricians working where there is a potential electrical or

© Cengage Learning 2012.

FIGURE 1-14 Examples of common personal protective equipment (PPE) used by residential electricians.

other safety hazard wear to personal protective equipment. It is not a choice for a residential electrician—it is a requirement! OSHA requires that all PPE must be maintained and periodically inspected to ensure that it is always in good working condition. The following rules should be followed when determining if PPE is required and what type of PPE should be used:

- Safety glasses, goggles, or face shields must be worn anytime a hazard exists that can cause foreign objects to get in your eyes from the front or the sides. This would include using hand or power tools to drill, cut, or secure items during a residential electrical system installation. It is extremely important to wear eye protection when there is an electrical hazard that could produce an electrical arc or an arc blast. If you wear prescription glasses and must wear the glasses to effectively see what you are doing, goggles or some other protective device that fit over the prescription glasses will have to be worn.

- Head protection must be worn whenever there is a potential for objects to fall from above, bumps to your head from objects fastened in place, or from accidental contact with electrical hazards. Do not use a metal hard hat as it could conduct electricity. Look for certain markings on the inside of a hard hat to make sure it meets the minimum requirements for the type of work you are doing. Hard hats that are approved for electrical work are marked as "Class E." Prior to 1997, hard hats were marked "Class A" if they protect you from falling objects and electrical shock by voltages up to 2000 volts and marked with "Class B" to indicate protection from falling objects and voltages up to 20,000 volts. Newer hard hats may still use the older markings but may also be marked "Type 1" or "Type 2." A Type 1 hard hat protects you from objects falling on the top of your head. A Type 2 hard hat also protects you from objects striking the sides of your head.

- Residential electricians should wear work shoes or boots with slip-resistant and puncture-proof soles. Safety-toed footwear should be worn to prevent crushed toes when working around heavy equipment or falling objects. If possible, wear safety-toed footwear that uses a nonconductive material like fiberglass or Kevlar instead of steel. Many electricians wear off the material from the toes of steel-toed work boots. This exposes the steel and presents a safety hazard when working on electrical circuits. Footwear should be American National Standards Institute (ANSI) approved and meet the ANSI Z41 protection standard. A marking of "EH" means that the footwear has been approved for electrical work. The markings are typically found inside the tongue of the shoe or boot.

- Hand protection should be worn anytime your hands are exposed to a potential hazard. Use insulated gloves with a proper voltage rating when working on or near energized electrical circuits. When cutting and stripping conductors, or working with equipment that has sharp edges, many electricians wear gloves made with a material like Kevlar, which are designed to protect their hands from cuts and scrapes.

- Hearing protection in the form of earplugs or earmuffs should be used in high-noise areas. When power tools are used during a residential electrical system installation, they can produce noise levels that can result in hearing loss if proper hearing protection is not used.

In addition to using the proper PPE, residential electrical workers should:

- Wear long-sleeved cotton shirts and pants made with a sturdy, comfortable material like denim. In many parts of the country, residential electricians typically wear short-sleeved T-shirts, and even shorts, in an effort to keep cool when working in higher temperatures. While this practice is not recommended, as long as you are using the proper PPE, it is allowed.

- Wear clothing that fits closely to your body and is not loose or flapping. Loose or flapping clothing could get caught in machinery or on other stationary objects as you are moving around on the job site.

- Not wear clothing with exposed metal zippers, buttons, or other metal fasteners.

- Remove your rings, wristwatch, and any other metal jewelry before beginning work. This is especially true when working on or near exposed current-carrying parts.

By dressing appropriately and following the personal protective equipment information outlined here, residential electrical workers will minimize the possibility of an injury happening to them on the job.

FALL PROTECTION

Fall protection is an area that many residential electricians do not think should concern them. The fact is that OSHA 29 CFR Part 1926, Subpart M requires fall protection to be used whenever workers are on a walking or work surface that is 6 feet or more high and has unprotected sides. These areas include finished or unfinished floors and roof areas in a residential application. If you are working above unguarded dangerous equipment, fall protection must be used at all times, no matter what the height. The type of fall protection used is typically chosen by your supervisor and can include items such as guardrails, personal fall arrest systems, or safety nets (Figure 1-15).

When working at a residential construction site, you should always follow these fall protection tips:

- Identify all potential tripping and fall hazards before you begin work.

- Practice good housekeeping. Pick up all construction debris that could be a tripping hazard. Keep electrical cords out of walkways and other travel areas.

- Look for unprotected floor openings, floor edges, skylight openings, stairwell openings, roof openings, and roof edges.

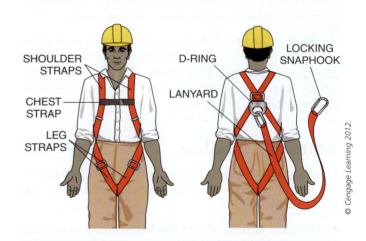

SHOULDER STRAPS
CHEST STRAP
LEG STRAPS
D-RING
LANYARD
LOCKING SNAPHOOK

© Cengage Learning 2012.

FIGURE 1-15 A personal fall arrest system.

- Always inspect any fall protection equipment you are planning on using for defects.

- Select, wear, and use fall protection equipment that is appropriate for the job you are doing.

- All warnings and instructions must be read and understood before using fall protection equipment.

GROUND RULES FOR GENERAL AND ELECTRICAL SAFETY

Residential electricians must follow industry-suggested practices for both general safety and electrical safety on the job site. The rules for each area of safety are similar; however, general safety rules should be followed at all times, while electrical safety rules are followed when the hazard of electricity is present.

GENERAL SAFETY RULES

General safety in residential electrical work takes into account the proper planning of your work and includes choosing the proper materials and the right tools to do the wiring job. It also includes how an electrician should behave on the job. Be serious about your work and never fool around on the job. This type of behavior can result in serious injury to you or a coworker. Intoxicating drugs or alcohol must never be used on the job. They impair judgment, and the consequences could be fatal. Remember, you are working with electricity, and it takes a very small quantity to kill you. Do not talk unnecessarily when working. A lapse in concentration could be fatal when engaged in electrical work.

During the installation, an electrician must practice safe work methods and exercise good judgment in both giving and receiving instructions. The following paragraphs cover some of the more important rules for general safety.

Material Handling

Handling the materials that you use to install a residential wiring system requires practicing the following safety rules:

- Learn the correct way to lift: Get solid footing, stand close to the load, bend your knees, and lift with your legs, not your back (see Figure 1-16).

- Never get on or off moving equipment like delivery trucks or company vans.

- When two or more persons are carrying long objects like a ladder, the object should be carried on the same shoulder (right or left) of each person. One person should give the signal for raising or lowering (Figure 1-17).

- Use hand lines to raise or lower bulky materials and heavy tools from a scaffold or staging. Never drop down or throw up an item to another worker (Figure 1-18).

Ladders

Improper use of step ladders and straight or extension ladders cause many workplace injuries. The following rules should be followed when using ladders:

- OSHA stipulates that portable metal ladders should not be used near energized lines or equipment. Use a nonconductive ladder made of a material such as fiberglass.

- Inspect the ladders before use. Look for any missing, loose, or cracked parts. If the ladder is not in good shape, do not use it!

- Always place a straight or extension ladder at the proper angle. It is strongly suggested that you place the ladder so that the bottom of the ladder is about one-fourth the vertical height from the structure it is up against (Figure 1-19).

- The ladder should extend at least 3 feet above the top support when placed against a structure that is not as tall as the ladder (Figure 1-20).

- Set ladders on firm footing and tie them off where possible. If there is a danger of a ladder moving, station someone to hold on to it as you climb.

- Use both hands and face the ladder when going up or down.

- Keep hands free of tools; materials should be hoisted or lowered with a hand line, never carried up or down.

- Make sure to open a step ladder all the way and lock the spreaders in place.

- Never use a step ladder as a straight ladder. If an OSHA inspector catches you doing this, your company is in violation and subject to a fine (Figure 1-21).

- The top cap and top rung are not for standing on a step ladder. Standing on them may cause the ladder to fall, resulting in serious injury to you (Figure 1-22).

- Do not leave materials or tools on the top of a step ladder. They can fall off and injure someone.

from experience...

A local electrical contracting company was assessed a fine by an OSHA inspector because a new electrical worker placed a step ladder up against a wall like a straight ladder to climb up and hang a lighting fixture. The inspector observed this behavior and quickly cited the employer. How happy do you think the new electrician's boss was about this?

Scaffolding

Scaffolding, or staging as it is sometimes called, is used regularly in residential wiring (Figure 1-23). For example, it may be used to stand on when hanging a ceiling paddle fan in a residence with a high cathedral ceiling. The following rules should be followed when using staging (note that some of the following information is derived from OSHA 29 CFR Part 1926, Subpart L regulations):

• All workers must have scaffold training arranged by their employers. The training must include:

  • The proper methods for constructing the scaffold

  • The hazards for the type of scaffolding being used

1	2	3	4
APPROACH THE LOAD AND SIZE IT UP AS TO WEIGHT, SIZE, AND SHAPE. CONSIDER YOUR PHYSICAL ABILITY TO HANDLE THE LOAD.	PLACE FEET CLOSE TO THE OBJECT TO BE LIFTED AND 8 TO 12 INCHES APART FOR GOOD BALANCE.	BEND THE KNEES TO THE DEGREE THAT IS COMFORTABLE AND GET A HANDHOLD. THEN USING BOTH LEG AND BACK MUSCLES . . .	LIFT THE LOAD STRAIGHT UP, SMOOTHLY AND EVENLY. PUSH WITH YOUR LEGS AND KEEP THE LOAD CLOSE TO YOUR BODY.

5	6	7
LIFT THE OBJECT INTO CARRYING POSITION, MAKING NO TURNING OR TWISTING MOVEMENTS UNTIL THE LIFT IS COMPLETED.	TURN YOUR BODY WITH CHANGES OF FOOT POSITION AFTER LOOKING OVER YOUR PATH OF TRAVEL, MAKING SURE IT IS CLEAR.	SETTING THE LOAD DOWN IS JUST AS IMPORTANT AS PICKING IT UP. USING LEG AND BACK MUSCLES, COMFORTABLY LOWER LOAD BY BENDING YOUR KNEES. WHEN LOAD IS SECURELY POSITIONED, RELEASE YOUR GRIP.

© Cengage Learning 2012.

FIGURE 1-16 How to lift safely. As part of their jobs, electricians are required to lift heavy objects that are sometimes of a shape that make them hard to handle. Following proper lifting techniques will help keep your body injury-free.

FIGURE 1-17   The proper technique for two people carrying a long object, like a ladder, is to carry it on the same side and to designate one person to signal when the object is to be lifted or lowered.

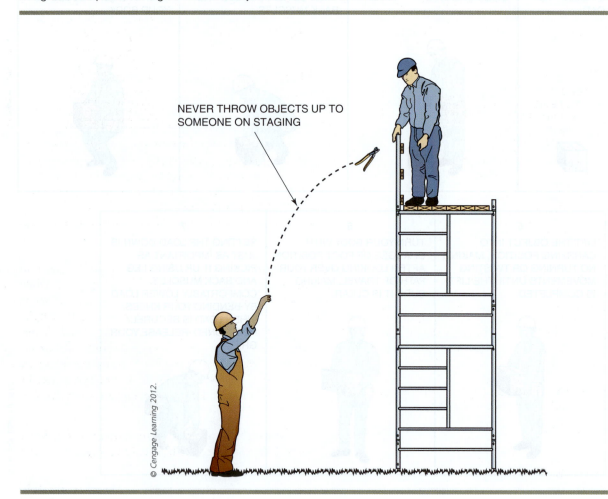

NEVER THROW OBJECTS UP TO
SOMEONE ON STAGING

FIGURE 1-18   Never throw objects up to a person standing on staging. Hand lines should be used to haul up the tools and materials that are needed. Electricians can get injured from being hit by the thrown object or, even worse, fall off the staging while reaching for the thrown object.

USE BLOCKING TO LEVEL THE LADDER WHEN NECESSARY.

4 FT.

16 FT.

CORRECT BASE POSITION IS ¹⁄₄ THE VERTICAL HEIGHT.

FIGURE 1-19   The safe ladder angle results from having the base position no more or less than one-fourth of the vertical height. In this example, the vertical height is 16 feet, and one-fourth of 16 feet is 4 feet. Therefore, the location of the base position should be 4 feet out from the building.

- Maximum intended load and capacity for the scaffolding being used
- Recognizing and reporting defects
- Fall hazards
- Electrical hazards
- Falling object hazards
- Any other hazards that may be encountered
- Guardrails are required when workers are on platforms 10 feet or higher.

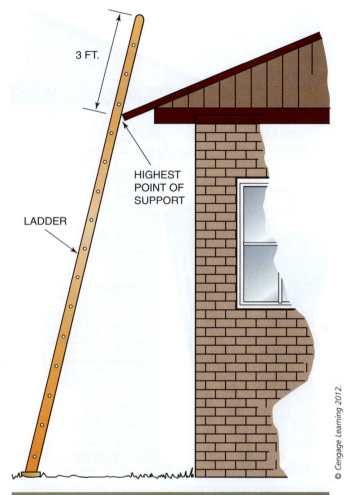

3 FT.

HIGHEST POINT OF SUPPORT

LADDER

FIGURE 1-20   When using a straight or extension ladder, place it so that the top of the ladder extends 3 feet above the highest point of support. This will allow you to mount or dismount the ladder in a safe manner.

- Personal fall arrest systems could also be used at heights of 10 feet or more and *must* be used when the scaffolding platform is less than 18 inches wide, guardrails are less than 36 inches high, or if a single- or two-point suspension scaffold is being used. A free fall from a scaffold while using a personal fall arrest system is limited to 6 feet. The anchor point where the personal fall arrest system is attached must be able to support at least 5000 lbs for each worker attached to it.

- A personal fall arrest system includes a full body harness, a lanyard with a locking snaphook, and an anchor point.

- The platform must completely cover the scaffold work area. Working platforms must be planked close to the guardrails. The basic rule is that planks must be placed so there is no more than a 1-inch gap between planks. There must be no holes large

STEP LADDER LEANING
AGAINST A WALL

© Cengage Learning 2012.

FIGURE 1-21   Step ladders should not be used as a straight ladder and placed up against a wall. OSHA can cite electrical contractors for using step ladders this way and assess a fine.

enough for people to possibly fall through. If planks are to be overlapped, they must overlap on a scaffold support by at least 6 inches, but not more than 12 inches.

- Legs, posts, frames, poles, and uprights must be on base plates or on a firm foundation. When using caster wheels, they must have a mechanism that can lock the wheels in place for stationary work and unlock the wheels when moving the scaffolding.

- Toeboards must be installed on all open sides of scaffold platforms when they are located more than 10 feet high so that tools or other material cannot fall on people below.

- Safe work practices for scaffolding include:
  - Never extend your body over the edge of the platform or guardrail.
  - Keep both feet on the platform surface at all times.

- Always keep at least three points of contact when climbing scaffolding ladders.

- Keep all walkways and working surfaces clear of tools and other debris.

- If a scaffold is covered with snow, ice, or other slippery material, clean it off before you start to work while standing on it.

- Always be sure to stay at least 10 feet away from energized power lines.

- Use a ladder or stairs to access a scaffold when there is more than 2 feet above or below the access point. Never use the crossbraces to climb onto the scaffold or to go to another level of the scaffolding.

- Scaffolds must be able to bear their own weight as well as four times the maximum intended load. Make sure the scaffolding you are using meets this criteria.

- Don't work on scaffolding located outside when there are high winds or storms in the area.

Tools

The safe and proper use of the tools used to install residential electrical systems is very important. If a tool comes with an owner's manual, read it! The place where you bought the tool is also an excellent source to find out how to safely and properly use the tool. A residential electrician should consider the following items:

- Use the right tools for the job. Be sure that they are in good working order.

- Use only those power tools for which you have received proper instruction.

- Use tools for their intended purpose (for example, pliers are not to be used for hammering) (Figure 1-24).

- Portable power tools should be provided with a three-prong grounding plug for attachment to a ground unless it is a double-insulated power tool.

- Sharp-edged or pointed tools should be carried in tool pouches and not in clothing pockets.

CAUTION

CAUTION: Don't carry tools like screwdrivers or other sharp tools in your pants pockets. One slip or stumble could result in the tool causing serious injury to you.

- Tools should not be left on any overhead workspace or structure unless held in suitable containers that will prevent them from falling.

SAFETY PRACTICES
FOR STEP LADDERS

CORRECT

INCORRECT

© Cengage Learning 2012.

FIGURE 1-22 Safe practices for using step ladders include not standing on the top cap. Two steps down from the top cap is the highest point on the step ladder that is suitable for standing.

- Any tools used on or around energized electrical equipment must be of the nonconductive type. The tools must be designed for use on "live" circuits and must be insulated at or higher than the circuit voltage.

from experience...

A residential electrician was mounting electrical boxes on the side of studs and using nails to attach them. He was using lineman pliers to hammer the nails. Since the side of the pliers is not designed to take the tremendous force that hammering nails produces, a piece of metal broke off from the pliers and embedded itself in the electrician's arm. He was fortunate that the metal did not go into his eye. As it was, a visit to the doctor was required to have the metal piece removed from his arm. The electrician learned the hard way that you must always use a tool for what it was designed to do. If you do not, serious injury could result.

> **CAUTION**
>
> **CAUTION:** Electrical tools such as screwdrivers and pliers with rubber or vinyl grips supplied by the manufacturer are *not* considered nonconductive. The grips are for comfort only. You must use tools that are specifically made to be nonconductive when working on energized circuits.

Material Safety Data Sheets

An electrician may have to use or may come in contact with hazardous materials on the job site. Solvents used to clean tools and other equipment are examples of materials that are considered hazardous. Electricians must follow specific procedures and methods for using, storing, and disposing of most solvents and other chemicals. **Material safety data sheets (MSDS)** must be made available to a worker using any hazardous material.

Material safety data sheets (Figure 1-25) are designed to inform the electrician of a certain material's physical properties as well as the effects on health that make the material dangerous to handle. They instruct you on the proper type and style of protective equipment needed when using the material and on proper first aid treatment if you are exposed to the material

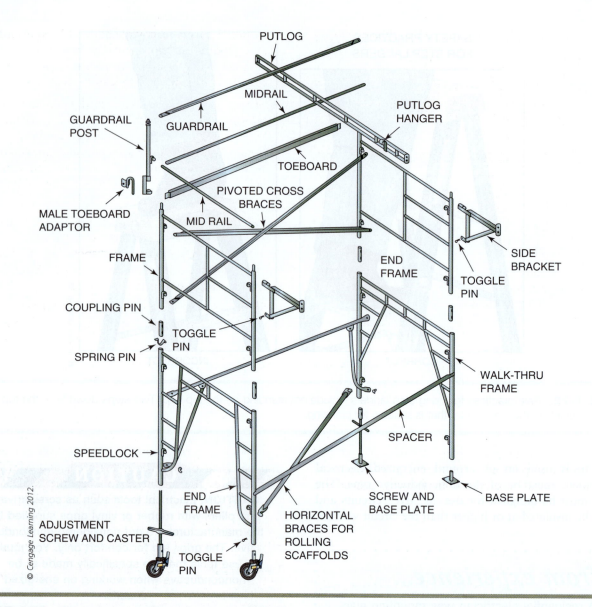

FIGURE 1-23 The typical parts that make up a scaffold. Electricians should be given training on how to construct a scaffolding or staging system.

or its hazards. Other information given includes proper storage methods, how to safely handle spills of material, and how to properly dispose of the material.

It is important for a residential electrician to be able to know how to read an MSDS. There are two basic formats used: the OSHA format and the ANSI (American National Standards Institute) format. OSHA's recommended format is used most widely. It is broken down into eight different sections. The first thing the OSHA format covers is the name of the chemical. It includes the product's chemical name as well as its more common name, if there is one. The sections contain the following information:

- Section 1 covers the supplier's information and gives the name, address, and contact information of the company that manufactures the chemical. The date the MSDS was prepared is also included in this section.

- Section 2 covers hazardous ingredients/identity information and lists the hazardous components of the chemical by both scientific and common name. Sometimes a chemical component is not listed because it might be a "trade secret" and the company doesn't want any other competing company to know what it is.

- Section 3 covers physical/chemical characteristics and includes things such as the boiling or melting points,

FIGURE 1-24 Use tools properly. Do not hammer with a pair of pliers like this electrician is doing. Personal injury can result from not using a tool properly.

© Cengage Learning 2012.

vapor pressure, vapor density, evaporation rate, specific gravity, and the chemical's solubility in water. This information tells you what conditions change the chemical's form. This affects the type and severity of the chemical's hazard.

• Section 4 covers fire and explosion data and includes the chemical's flash point, as well as the chemical's flammable or explosion limits. It will also tell you what to use to put out a fire started by the chemical and any special hazards or firefighting procedures you need to know.

• Section 5 covers reactivity data and contains information about what happens when the chemical is mixed with air, water, or other chemicals. It also covers what conditions and other chemicals to stay away from.

• Section 6 covers health hazard data and provides information on how the chemical can get into your body and what health hazards that could result from being exposed to it. It also provides information on symptoms of exposure and gives emergency and first aid procedures to follow if you are accidentally exposed to the chemical.

• Section 7 covers precautions for safe handling and use. Information is provided that covers the correct way to handle, store, and dispose of the chemical. Information is also included on what to do if there is a chemical spill or leak.

• Section 8 covers control measures and lists the personal protective equipment (PPE) to use when working with the chemical. It also provides ventilation requirements and proper hygiene practices to prevent accidental exposure.

Each company is required by law to develop a hazard communications (HazCom) program that must contain, at a minimum, warning labels on containers of hazardous material, employee training on the safe use and handling of hazardous material, and MSDS. The law requires the employer to keep the MSDS up-to-date and to keep them readily accessible so that

FIGURE 1-25 Material safety data sheets (MSDS) contain information about hazardous materials that you may need to know. A typical MSDS will have eight parts. The different parts are shown in this illustration.

employees can access them quickly when they are needed. Check with your supervisor on the job site to find out where your company's MSDS are located.

ELECTRICAL SAFETY RULES

As we discussed earlier in this chapter, when electricity is present, a whole new set of hazards are present for an electrical worker. To minimize the chance of an accident occurring, a residential electrician should practice the following electrical safety rules:

- Install all electrical wiring according to the *NEC*®.

- Whenever possible, work with a buddy. Avoid working alone.

(See Procedure 1-3 on page 32 for the procedure to follow when a coworker is being shocked.)

- Always turn off the power and lock it out before working on any electrical circuits or equipment. If possible, when working on electrical equipment, stand on a rubber mat or a wooden floor or wear rubber-soled shoes.

- Never cut off the grounding prong from a three-prong plug on any power extension cord or from a power cord on any piece of equipment.

- Assume all electrical equipment to be "live" and treat them as such.

- Do not defeat the purpose of any safety devices, such as fuses or circuit breakers. Shorting across these devices could cause serious damage to equipment and property in addition to serious personal injury.

- Do not open and close switches under load unless absolutely necessary. This practice could result in severe arcing.

- Clean up all wiring debris at the end of each workday (or more often if it presents a safety hazard). Keeping the job site free of debris will help eliminate tripping hazards.

CLASSES OF FIRES AND TYPES OF EXTINGUISHERS

Fire is an ever present danger when working with electricity. The following three components must be present for a fire to start and sustain itself:

- Fuel: Any material that can burn

- Heat: Raises the fuel to its ignition temperature

- Oxygen: Is required to sustain combustion

These three components are often referred to as the sides of the "fire triangle" (Figure 1-26). If any one of

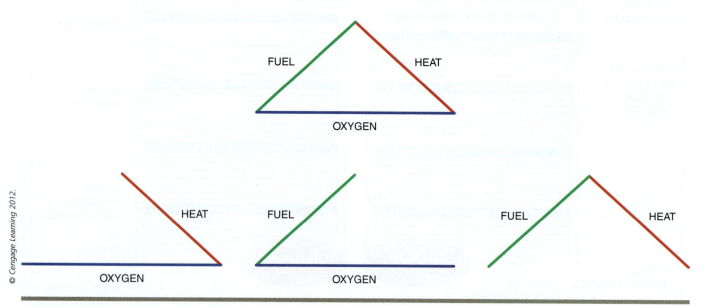

FIGURE 1-26   The three components necessary to start and sustain a fire are fuel, heat, and oxygen. Together they make up the "fire triangle," shown here. By removing any one of the components, a fire can be extinguished.

© Cengage Learning 2012.

the three is missing, a fire cannot be started. With the removal of any one of them, the fire will be extinguished.

There are different types of fires that an electrician could encounter. Recognizing the different fire classifications and knowing the proper type of extinguisher used to combat the fire is very important.

Fires are classified as follows (see Figure 1-27):

- Class A: Fires that occur in ordinary combustible materials, such as wood, rags, and paper
- Class B: Fires that occur with flammable liquids, such as gasoline, oil, grease, paints, and thinners
- Class C: Fires that occur in or near electrical equipment, such as motors, switchboards, and electrical wiring
- Class D: Fires that occur with combustible metals, such as powdered aluminum and magnesium

There are many types of fire extinguishers. Each one is used to put out a specific class of fire (see Figure 1-28).

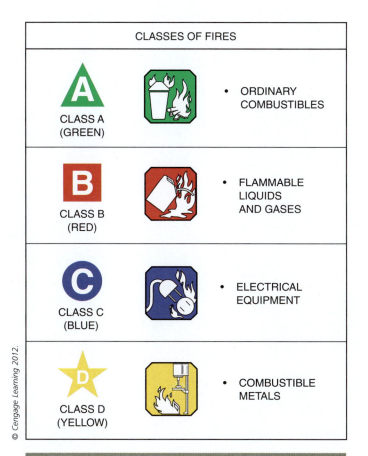

© Cengage Learning 2012.

FIGURE 1-27   There are four classes of fires. This illustration shows what they are and the symbols that are used to represent them.

© Cengage Learning 2012.

FIGURE 1-28   Fire extinguishers that you will encounter should have symbols on them like those illustrated here. These markings will let you know what type of fire you can use the extinguisher on.

You should become familiar with the fire extinguisher type that is available at the job site and learn how to properly use it. The extinguishers can be broken down as follows:

- Pressurized water: Operates usually by squeezing a handle or a trigger and spraying a stream of water onto the fire; used on Class A fires.
- Carbon dioxide (CO_2): Operates usually by squeezing a handle or trigger and spraying first at the base of the flames and then moving upward; used on Class B and C fires.
- Dry chemical: Operates usually by squeezing a handle, trigger, or lever and spraying at the base of the fire and then on top of any remaining materials that are burning; used on Class A, B, and C fires. This is the type of extinguisher found on most job sites and in existing buildings.

- Foam: Typically operates by turning the extinguisher upside down and spraying out the foam so that it lands on top of the fire; used on Class A and B fires.

SUMMARY

There is a lot to know when it comes to workplace safety. You now have some knowledge about the concerns of the *NEC*® and OSHA for safety. But these rules and regulations are of little effect by themselves. You must give meaning to them by performing competently, by being knowledgeable, and most of all by being a safe, mentally alert, and responsible electrical worker. You must be able to do your work so that no injury is inflicted on yourself or your fellow workers. Following safe wiring practices will help keep you safe

and will prevent equipment and property from damage. Knowledge of both general safety and electrical safety is the most important safeguard against serious or fatal accidents on the job site.

The *NEC*®, OSHA, and common safe electrical trade practices provide the guidelines for residential electrical construction safety. The stakes for unsafe practice are high. Death or injury to you or your colleagues and damage to or destruction of equipment are the unfortunate consequences of unsafe electrical installation practice.

The safety rules outlined in this chapter should serve as a guide to you as you seek to become a knowledgeable and skilled residential electrician. Learn them and the values that they imply. Practice them as you continue your career as a residential electrician. Your life may depend on it.

PROCEDURE 1-1

Suggested Procedure to Find Information in the NEC®

The following procedure can be used to find information in a quick and accurate manner in the National Electrical Code®. The author's comments follow each step. Remember, as you become better at using the NEC®, there may be steps in the procedure that you could skip, but for now it is best if you follow each of the seven steps. The steps are:

1 Determine the main topic area for the information you want.

*This is a very important first step. If you are not able to identify the proper **main topic area,** the information you eventually locate in the code probably will not be the information you want. Identifying the proper **main topic area** gets easier with practice.*

2 Locate the main topic area in the index.

*The index lists **main topic areas** in bold print and in alphabetical order to help you find them easier. If you are unable to find the **main topic area** in the index, try alternative wording. For example, the topic area "electric motors" is not listed in the index but the topic area "motors" is.*

3 Determine the appropriate sub-topic.

*In most cases, the index lists several **sub-topics** under the main topic area heading. The **sub-topics** are listed in alphabetical order. Identifying the proper **sub-topic** gets easier with practice. Note: If the **main topic area** you have chosen does not have any **sub-topics** listed under it, skip this step and proceed to Step 4.*

4 Determine which article, part, section, or table is referenced for the topic area.

If more than one reference is given, choose the one that will most likely have information pertaining to your application. If you find that the reference you chose doesn't result in the information you are looking for, simply try the next reference. Continue to try the references until you find the proper information.

5 In the table of contents, find the page number of the article that contains the reference.

This step is optional if you have a good feel for where an article is located in the code. If this is the case, you can simply open up to the article and skip to Step 6. Also, you can look at the top left and right corners in the NEC® and see the section number, which begins the left page and ends the right page. By using this information, you can quickly scan to a specific area in the code that was referenced in the index. Another method that can be used to quickly locate an article is to use "Code Tabs." Tabs are commercially available from several companies and when installed on your NEC®, make finding the major articles and tables in the code much easier.

6 Turn to the article and scan through until you find the referenced part, section, or table.

As you become more experienced with using the code, you may be able to simply jump to the part, section, or table in the article instead of scanning through the article from the beginning.

7 Read all of the material in the referenced area until you find the information that applies to your specific application.

Now that you have located the area in the code that was referenced in the index, you must be able to determine what material in this area applies to your particular application. Once again, the more experience you have finding information in the code will make this step much easier.

PROCEDURE 1-2

Suggested Lockout/Tagout Procedure

- Let all supervisors and affected workers know that you are performing a lockout/tagout on a particular electrical circuit.

- De-energize the equipment or circuit in the normal manner.

- Lock out the electrical source to the equipment or circuit in the OFF position with an approved lockout device.

- Check that after the lockout device is attached, the electrical switch cannot be placed in the ON position.

- Place a tag with your name, date, and for whom you work on the switch.

- Use a proper test instrument to verify that all parts of the equipment you are working on are de-energized.

- Once you have completed the necessary work and made sure that all tools and meters are removed from the equipment you were working on, get ready to re-energize the circuit.

- Let everyone affected by the lockout know that you are re-energizing the circuit and equipment.

- Remove the lockout device and tags from the energy source.

- Turn on the electrical source to restore power.

PROCEDURE 1-3

Procedure to Follow When a Coworker is Being Shocked

1 Shut off the electrical current supply to the victim.

- If possible, have someone call for help while you do this.

- If you cannot quickly find or get to the switch controlling the current flow, pry the victim from the circuit using something that is not a conductor of electricity.

Warning: Do not touch the victim yourself without using an insulating material or you will become a shock victim as well!

2 Stay with the victim unless there is no other option but go for help yourself.

3 Once you are sure there is no more current flowing through the victim, use a loud but calm voice; call out to the victim to see if there is a response.

- If the victim is conscious, tell them not to move.

4 Examine the victim for signs of major bleeding.

- If there is major bleeding, use something like a shirt or handkerchief to put pressure on the wound and try to stop the bleeding.

- If the wound is on an arm or leg, elevate those parts while keeping pressure on the wound.

5 If the victim is unconscious, check for signs of breathing.

- If she is not breathing, someone trained in CPR should begin artificial breathing. CPR is most effective if done within 4 minutes of the shock.

6 Keep the victim warm and continue to talk to her until help arrives.

REVIEW QUESTIONS

Directions: For each of the following questions, choose the best answer. Indicate your choice by circling the letter in front of the answer.

1. **In a work situation where an electrician's hands get wet while operating a portable drill, which of the following would be true?**
   a. Body resistance increases, and any shock would be mild.
   b. Body resistance remains the same, and there is no danger as long as rubber boots are worn.
   c. Body resistance is substantially decreased, and severe shock could occur.
   d. There is no danger as long as a three-prong plug is used.

2. **The amount of current that it takes to cause ventricular fibrillation is**
   a. 5 milliamperes
   b. 100 milliamperes
   c. 1000–4300 milliamperes
   d. 10,000 milliamperes

3. **Which of the following is considered an insulator?**
   a. Glass
   b. Plastic
   c. Rubber
   d. All of the above

4. **Who publishes the *National Electrical Code® (NEC®)*?**
   a. OSHA
   b. The National Electrical Contractors Association
   c. The National Fire Protection Association
   d. The Department of Labor

5. **The unit of measure for electrical force is the**
   a. Ampere
   b. Volt
   c. Ohm
   d. Watt

6. **Energized, or "live," electrical equipment is dangerous. How does OSHA suggest that the worker deal with this potential hazard?**
   a. Work only on equipment that is marked or tagged "dead."
   b. Inspect and test all equipment, assuming all to be energized until proven otherwise.
   c. Work only on equipment that the foreman says is dead.
   d. Ask someone nearby what the condition is.

7. **Which of the following is considered to be the most important safeguard against serious or fatal accidents?**
   a. Knowledge of safety
   b. Experience
   c. Horse sense
   d. Bravery

8. **Which of the following types of extinguishers should not be used on electrical fires? (Hint: There are two.)**
   a. Carbon dioxide
   b. Foam
   c. Dry chemical
   d. Pressurized water

9. **When setting up scaffolding, all the following procedures should be followed except:**
   a. When appropriate, wear an approved harness and lanyard to secure yourself to the scaffold so that you cannot fall.
   b. Guardrails and toeboards must be installed when the actual standing platform is higher than 10 feet off the ground.
   c. Set up the platform on the scaffolding so that it does not completely cover the staging work area. There must be a hole for a worker to climb up through to get on the platform.

d. Keep scaffold platforms clear of unnecessary materials, tools, and scrap; they may become a tripping hazard or be knocked off, endangering people.

10. Fires that occur in or near electrical equipment, such as motors, switchboards, and electrical wiring, are classified as

a. Class A fires

b. Class B fires

c. Class C fires

d. Class D fires

Directions: Circle T to indicate true statements or F to indicate false statements.

T F 11. The higher the resistance in a circuit, the lower the amount of current that can flow in the circuit.

T F 12. The resistance of the human body is fixed regardless of the conditions.

T F 13. OSHA requires voluntary compliance, and employers may comply at their pleasure.

T F 14. An energized line is the same as a "live" or "hot" line.

T F 15. Lower voltages (120–240 V) cannot kill.

T F 16. Copper is not a conductor of electricity since only materials that are magnetic can conduct electricity.

T F 17. The third prong (grounding prong) on a three-prong plug is optional and may be removed.

T F 18. Safety eyeglasses and goggles should only be worn when working on "live" circuits or equipment.

T F 19. The *NEC*® gives minimum safety standards for electrical work and is not a how-to manual.

T F 20. Electricians should wear metal helmets on job sites to avoid injury from falling objects.

Directions: Answer the following items with clear and complete responses.

21. How should a hard hat manufactured for electrical work in 2010 be marked?

22. How should footwear approved for electrical work be marked?

23. Name five types of personal protective equipment (PPE).

24. Explain the proper procedure for manually lifting an object.

25. Name the three components of the fire triangle.

26. What do the letters OSHA stand for?

27. List three rules to follow when handling material.

28. List three rules to follow when using ladders.

29. List three rules to follow when using scaffolding.

30. Explain the purpose of a material safety data sheet (MSDS).

31. What does it mean when you receive an OSHA 10 card?

32. What is the purpose of NFPA 70E?

Hardware and Materials Used in Residential Wiring

OBJECTIVES

Upon completion of this chapter, the student should be able to:

- List several nationally recognized testing laboratories and demonstrate an understanding of the purpose of these labs.

- Identify common box and enclosure types used in residential wiring.

- Identify common box covers and raised rings used in residential wiring.

- Identify common conductor and cable types used in residential wiring.

- Identify types of cable connectors, conductor terminals, and lugs.

- Identify common raceway types used in residential wiring.

- Identify common devices used in residential wiring.

- Identify common types of fuses and circuit breakers used in residential wiring.

- Describe the operation of a fuse and a circuit breaker.

- Identify common panelboards, loadcenters, and safety switches used in residential wiring.

- Identify common types of fasteners, fittings, and supports used in residential wiring.

GLOSSARY OF TERMS

American Wire Gauge (AWG) a scale of specified diameters and cross sections for wire sizing that is the standard wire-sizing scale in the United States

ampacity the current in amperes that a conductor can carry continuously under the conditions of use without exceeding its temperature rating

antioxidant a special compound that is applied to exposed aluminum conductors; its purpose is to inhibit oxidation

approved when a piece of electrical equipment is approved, it means that it is acceptable to the authority having jurisdiction (AHJ)

bimetallic strip a part of a circuit breaker that is made from two different metals with unequal thermal expansion rates; as the strip heats up, it will tend to bend

cabinet an enclosure for a panelboard that is designed for either flush or surface mounting; a swinging door is provided

cable a factory assembly of two or more insulated conductors that have an outer sheathing that holds everything together; the outside sheathing can be metallic or nonmetallic

circuit breaker a device designed to open and close a circuit manually and to open the circuit automatically on a predetermined overcurrent without damage to itself when properly applied within its rating

circular mils the diameter of a conductor in mils (thousandths of inches) times itself; the number of circular mils is the cross-sectional area of a conductor

connector a fitting that is designed to secure a cable or length of conduit to an electrical box

copper-clad aluminum an aluminum conductor with an outer coating of copper that is bonded to the aluminum core

device a piece of electrical equipment that is intended to carry but not use electrical energy; examples include switches, lamp holders, and receptacles

device box an electrical device that is designed to hold devices such as switches and receptacles

disconnecting means a switch that is able to de-energize an electrical circuit or piece of electrical equipment; sometimes referred to as the "disconnect"

fitting an electrical accessory, like a locknut, that is used to perform a mechanical rather than an electrical function

fuse an overcurrent protection device that opens a circuit when the fusible link is melted away by the extreme heat caused by an overcurrent

ganging joining two or more device boxes together for the purpose of holding more than one device

ground fault an accidental connection of a "hot" electrical conductor and a grounded piece of equipment or the grounded circuit conductor

handy box a type of metal, surface-mounted device box used to hold only one device

insulated a conductor that is covered by a material that is recognized by the National Electrical Code® as electrical insulation

junction box a box whose purpose is to provide a protected place for splicing electrical conductors

knockout (KO) a part of an electrical box that is designed to be removed, or "knocked out," so that a cable or raceway can be connected to the box

loadcenter a type of panelboard normally located at the service entrance in a residential installation and usually containing the main service disconnect switch

mil 1 mil is equal to .001 inch; this is the unit of measure for the diameter of a conductor

multiwire circuit a circuit that consists of two or more ungrounded "hot" conductors that have a voltage between them, and a grounded neutral conductor that has the same voltage between it and each of the ungrounded conductors

National Electrical Manufacturers Association (NEMA) an organization that establishes certain construction standards for the manufacturers of electrical equipment; for example, a NEMA Type 1 box purchased from Company X will meet the same construction standards as a NEMA Type 1 box from Company Y

new work box an electrical box without mounting ears; this style of electrical box is used to install electrical wiring in a new installation

old work box an electrical box with mounting ears; this style of electrical box is used to install electrical wiring in existing installations

outlet a point on the wiring system at which current is taken to supply a piece of electrical equipment such as a light fixture or appliance

outlet box a box that is designed for the mounting of a receptacle or a lighting fixture

overcurrent any current in excess of the rated current of equipment or the ampacity of a conductor; it may result from an overload, a short circuit, or a ground fault

overload a larger-than-normal current amount flowing in the normal current path

panelboard a panel designed to accept fuses or circuit breakers used for the protection and control of lighting, heating, and power circuits; it is designed to be placed in a cabinet and placed in or on a wall; it is accessible only from the front

plug the device that is inserted into a receptacle to establish a connection between the conductors of the attached flexible cord and the conductors connected to the receptacle

pryout (PO) small parts of electrical boxes that can be "pried" open with a screwdriver and twisted off so that a cable can be secured to the box

raceway an enclosed metal or nonmetallic channel designed to hold wires or cables; examples are rigid metal conduit, electrical metallic conduit, electrical nonmetallic conduit, and flexible metal conduit

Romex™ a trade name for nonmetallic sheathed cable (NMSC); this is the term most electricians use to refer to NMSC

safety switch a term used sometimes to refer to a disconnect switch; a safety switch may use fuses or a circuit breaker to provide overcurrent protection

service entrance the part of the wiring system where electrical power is supplied to the residential wiring system from the electric utility; it includes the main panelboard, the electric meter, overcurrent protection devices, and service conductors

sheath the outer covering of a cable that is used to provide protection and to hold everything together as a single unit

short circuit a low-resistance path that results in a high value of current that flows around rather than through a circuit; usually, short circuits are unintended and happen when two "hot" conductors touch each other by mistake

spliced connecting two or more conductors with a piece of approved equipment like a wirenut; splices must be done in approved electrical boxes

switch box a name used to refer to a box that just contains switches

torque the turning force applied to a fastener

utility box a name used to refer to a metal single-gang, surface-mounted device box; also called a handy box

wirenut a piece of electrical equipment used to mechanically connect two or more conductors together

One of the things that a new residential electrician finds difficult about the job is recognizing and knowing when to use the large and varied amount of hardware and materials used to install a residential electrical system. There is a lot to learn, and it is no wonder that some electricians find it overwhelming. However, when it comes to common hardware and materials, recognizing what they are and where to use them is a very important skill for a new electrician. This chapter introduces you to many of the common hardware and materials used to install a residential wiring system. Hardware and materials used in special applications are covered as those situations come up in future chapters.

FIGURE 2-1 The Underwriters Laboratories (UL) label. If a piece of electrical equipment has this label on it, it means that the item has met the specific standards it was tested for and has been listed and labeled by UL.

NATIONALLY RECOGNIZED TESTING LABORATORIES

Residential electricians must use hardware and materials that are **approved**. The *National Electrical Code®* (*NEC®*) tells us in Article 90 that it is the authority having jurisdiction (AHJ) who has the job of approving the electrical materials used in an electrical system. The AHJ is usually the local or state electrical inspector in your area. The *NEC®* in Section 110.3(B) also requires electricians to use and install equipment and materials according to the instructions that are included in the listing or labeling. Many AHJs base their approval of electrical equipment on whether it is listed and labeled by a nationally recognized testing laboratory (NRTL).

So what does it mean when a product is listed or labeled? When an electrical product is listed, it means that the product has been put on a list published by an NRTL that is acceptable to the authority having jurisdiction. The independent testing laboratory evaluates the product, and if the product meets a series of designated standards, it is found suitable for its intended purpose and is placed on the list. A label is put on the equipment (or on the box that it comes in if the item is too small to put a label on) to serve as the identifying mark of a specific testing laboratory. The label allows an electrician or, more important, the AHJ to see that the product complies with appropriate standards. If the product is installed and used according to the installation instructions, it will work in a safe manner.

Underwriters Laboratories (UL) is the most recognizable of the NRTLs. Most manufacturers of electrical products submit their products to UL, where the equipment will be subjected to several types of tests. The tests will determine if the product can be used safely under a variety of conditions. Once UL determines that the product complies with the specific standards it was tested for, the manufacturer is allowed to put the UL label on the product (Figure 2-1). The product is then listed in a UL directory.

By listing a product in its directory, UL is not saying that it approves the product, only that when they tested the product to nationally recognized safety standards, the product performed adequately in terms of fire, electric shock, and other related safety hazards.

For residential wiring, there are three UL directories that the electrician should consult: the *Electrical Construction Equipment Directory* (Green Book), the *Electrical Appliance and Utilization Equipment Directory* (Orange Book), and the *General Information for Electrical Equipment Directory* (White Book) (Figure 2-2). The White Book is the directory that will be used the most. It gives information on specific requirements, permitted uses, and limitations of the product.

Two more NRTLs that you should be aware of are CSA International and Intertek Testing Services. CSA International used to be known as the Canadian Standards Association (CSA). This lab tests, evaluates, and lists electrical equipment in Canada. Many electrical products that you may use will have the CSA International label (Figure 2-3) or the older CSA label on them. This label and listing, like the UL label, is a basis for the AHJ approval of equipment required by the

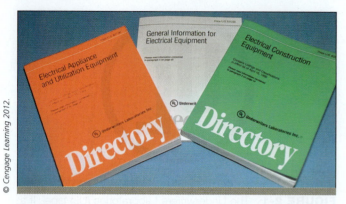

FIGURE 2-2 The UL Green, White, and Orange Books. Electricians should consult these books to determine specific information, such as permitted uses and installation instructions, for a listed electrical product.

FIGURE 2-3 This CSA International label is found on many electrical products and indicates that the product has been tested and meets the specific testing standards of CSA International.

NEC®. Intertek Testing Services (ITS), formerly known as Electrical Testing Laboratories (ETL), also tests and evaluates electrical products according to nationally recognized safety standards. Again, the AHJ will often use the ITS labeling and listing as a basis for approval.

While the **National Electrical Manufacturers Association (NEMA)** is not actually a testing laboratory, it is an organization of which you should be aware. It develops electrical equipment standards that all member manufacturers follow when designing and producing electrical equipment. More than 500 manufacturers of electrical equipment belong to NEMA. As an example of how NEMA affects an electrician, consider that a residential electrician in Maine may use a NEMA 5-20R receptacle made by Company X that will look like, install like, and work like a NEMA 5-20R receptacle that an electrician in California is using that was manufactured by Company Y.

ELECTRICAL BOXES

The residential electrical system wiring connections must be contained in approved electrical boxes. The design of these boxes helps prevent electrical sparks and arcs from causing fires within the walls, floors, and ceilings of a house. Boxes can be made of metal or a nonmetallic material, like plastic. A **device box** is designed to contain switches and receptacles and is the *NEC's®* term used for this type of box. Many electricians refer to this type of box as an outlet box, or a "wall case" box. The term **outlet box** is used to indicate a box that may contain switches, receptacles, or a lighting fixture. Boxes that contain just switches are sometimes referred to as simply **switch boxes.**

Some boxes are designed to contain only **spliced** conductors. This type of box is referred to as a **junction box.** Junction boxes, or J-boxes as they are sometimes called, are either octagonal or square in shape. They are also larger in total volume than a device box and can usually contain more conductors. Sizing electrical boxes is covered in detail in Chapter 10. Outlet boxes can also serve as a junction box. This occurs when splices are made in the same box that also is used to mount a lighting fixture or hold a receptacle or a switch.

METAL DEVICE BOXES

Metal device boxes are still used widely in residential wiring, even though nonmetallic device boxes have become the box of choice across the country. The standard metal device box has a 3 × 2-inch (75 × 50 mm) opening and can vary in depth from 1½ to 3½ inches (38 to 90 mm). The most common metal device box used is the 3 × 2 × 3½-inch (75 × 50 × 90 mm) box because it can contain several wires and a device (Figure 2-4).

A metal device box has many features for the electrician to use. One common feature involves taking the sides off the box and **ganging** the boxes together to make a box that can accommodate more than one device (Figure 2-5). For example, two switches may be required beside the front door of a house. One switch is required to control an outside light, and the other switch is necessary to control the lighting in the room. For this situation, the electrician would need to gang together two single-gang metal device boxes to make a two-gang device box.

(See Procedure 2-1 on page 72 for a procedure to gang metal device boxes together.)

Figure 2-6 shows the parts of a typical metal device box. Boxes with mounting ears are typically called **old work boxes** because the ears allow them to be installed

FIGURE 2-4   A 3 × 2 × 3½-inch (75 × 50 × 90 mm) metal device box with a side-mounting bracket and internal cable clamps.

FIGURE 2-5   Two 3 × 2 × 2¾-inch (75 × 50 × 70 mm) metal device boxes ganged together. A side-mounting bracket is left on one box for mounting the ganged boxes to a building framing member.

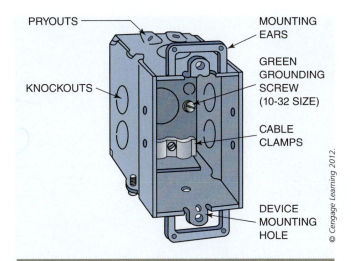

FIGURE 2-6   Parts of a typical metal device box. Mounting ears on a device box allow the box to be used in remodel or "old work." The mounting ears are not used on metal boxes used in "new work."

in an existing wall. On the other hand, boxes without mounting ears are referred to as **new work boxes** and are the kind used when installing an electrical system in a new home. The device mounting holes are designed to accommodate the screws of a switch or receptacle and are used to secure the device to the box. The holes are tapped for a 6-32 screw size. In the back of the box, you will find a tapped hole that accommodates a 10-32 grounding screw. This screw will be hex headed and green in color, and its purpose is to attach the circuit electrical grounding conductor to the metal box. **Knockouts (KO)** are included on the sides, the top, and the bottom of the box. When a KO is removed, an opening exists for a **connector** to be used to secure a wiring method to the box. The **pryouts (PO)** located at the rear top and bottom of the box are "pried" off and provide an opening through which a cable can be secured to the box. Most of these boxes come with internal cable clamps that are tightened onto the cable after it has been inserted through the pryout opening. The cable clamps secure the cable to the box.

CAUTION

CAUTION: The *NEC®* requires wiring methods, like cables and raceways, to always be secured to an electrical box with a connector designed for the purpose. The design may include an internal or an external clamping mechanism.

Another type of metallic device box recognized by the *NEC®* is the **handy box,** sometimes also referred to as a **utility box.** This style of box is used primarily for

© Cengage Learning 2012.

FIGURE 2-7   A handy box is used normally for surface mount applications. Another name for this box style is "utility box."

© Cengage Learning 2012.

FIGURE 2-8   (A) Single-gang, (B) two-gang, and (C) three-gang nonmetallic device boxes. All three box styles are secured to wood framing members with nails.

surface mounting and can accommodate one device, such as a receptacle or a switch. The size of the opening for this type of box is 4 × 2⅛ inches (100 × 54 mm) and the depth of the box can range from 1½ to 2⅛ inches (38 to 54 mm) (Figure 2-7).

NONMETALLIC DEVICE BOXES

The most popular style of device box used in residential wiring is the nonmetallic device box. Typically, these boxes are made of polyvinyl chloride (PVC), phenolic, or polycarbonate. They are lightweight, strong, and very easy to install. Another feature that makes them so popular is that, when compared to metal device boxes, they are very inexpensive. In residential wiring, this box type is used only with a nonmetallic sheathed cable wiring method.

Nonmetallic device boxes are available in a variety of shapes and sizes (Figure 2-8). Common sizes include single, two-, and three-gang boxes. The boxes come from the manufacturer with nails already attached. This makes it easy to mount nonmetallic device boxes on wooden framing members, like wall studs.

Some newer nonmetallic device boxes can be attached to a building framing member without the use of nails or screws (Figure 2-9). These boxes have special mounting brackets that, when pushed onto the edge of a wooden framing member like a 2″ × 4″ stud, grip the stud and hold the box securely in place. Although this type of box is more expensive than regular nonmetallic device boxes, they install more quickly and may even save you money in the long run because of reduced labor costs.

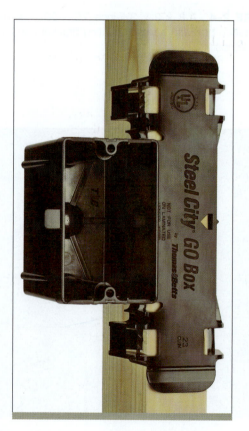

FIGURE 2-9   This nonmetallic device box uses a special attachment bracket to hold the box securely in place once it has been pushed onto a framing stud at the desired location. No nails or screws are needed. *Courtesy of Thomas and Betts.*

The single-gang nonmetallic box is made so that no cable clamp is needed. This feature contributes to the popularity of this box type. The electrician simply uses a tool like a screwdriver to open a knockout on the top or bottom of the box, and the cable is put into the box through the resulting hole. The *NEC®* requires the electrician to secure (staple) the cable within 8 inches (200 mm) of the single-gang nonmetallic box. Two or more ganged nonmetallic device boxes have built-in cable clamps and, when used, require securing the cable within 12 inches (300 mm) of the box.

OUTLET AND JUNCTION BOXES

When installing lighting fixtures in a ceiling or on a wall, an outlet box is typically used. This box style can also be used when connecting appliances that require a large receptacle for cord and plug connection. They are larger than device boxes and provide more room for different wiring situations than do device boxes. In residential wiring, you will find these boxes in round, octagon, or square shapes. The round boxes are typically nonmetallic (Figure 2-10), while the octagon and square boxes are typically metallic (Figure 2-11). Devices can be installed in this type of box but only when special mounting covers called "raised plaster rings" or "surface covers" are used (Figure 2-12).

When several conductors are spliced together at a point on the wiring system, the *NEC®* requires a junction box to be used. The *NEC®* also requires junction boxes to be accessible after installation without having to alter the finish of a building. Attics, garages, crawl spaces, and basements are typical areas where electricians locate junction boxes. Like outlet boxes, junction

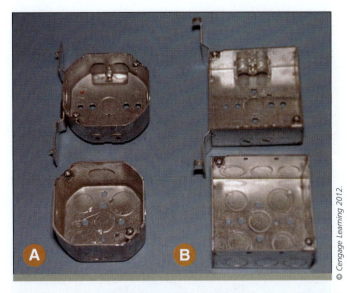

FIGURE 2-11 (A) Metal octagon and (B) square boxes are used in two common sizes for residential work: 4 × 1½ inches (100 × 38 mm) and 4 × 2⅛ inches (100 × 54 mm). They may have side-mounting brackets or may be attached directly to framing members with screws.

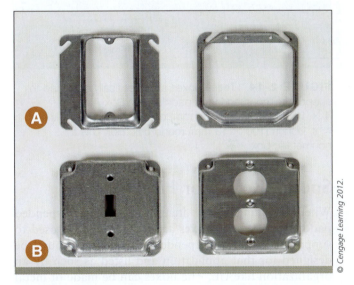

FIGURE 2-12 Raised covers must be used with square outlet boxes when you want to install a switch or receptacle in them. (A) Raised plaster rings are used when the square box is used in a wall or ceiling location. (B) Surface covers are used in a surface-mount situation.

boxes are found in round, octagon, and square shapes. Junction boxes must always be covered and blank covers are available for this (Figure 2-13).

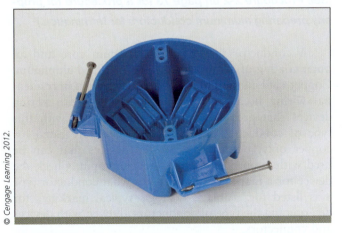

FIGURE 2-10 A nonmetallic round outlet box. This box is used in walls and ceilings and is designed to support luminaires. Nails are used to secure this box to a wood framing member.

CAUTION

CAUTION: Never bury a junction box in a wall or ceiling. Junction boxes must be accessible after they are installed.

FIGURE 2-13 When conductors are spliced in a junction box, a cover must be used. Flat blank covers are available in an (A) octagon or a (B) square shape.

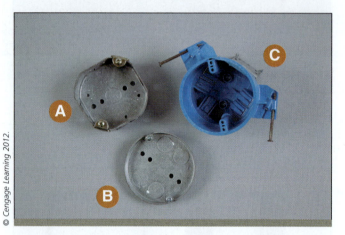

FIGURE 2-14 These boxes are specifically designed to support ceiling-suspended paddle fans. Box A is a 4 × 1½ inches (100 × 38 mm) octagon box. Box B is a 4 × ½ inch (100 × 13 mm) round box. Box C is a nonmetallic box.

Special Boxes for Heavy Loads

When installing heavy items like ceiling suspended paddle fans, special boxes are required. Boxes used to support a ceiling suspended paddle fan must be designed for that application and listed as such by an organization like UL. The boxes can be made of metal or nonmetal and are specifically designed and tested to support heavy loads (Figure 2-14).

CONDUCTORS AND CABLE TYPES

When residential electricians install a wiring system in a dwelling, the electrical conductors required for the circuits are most often installed as part of a cable assembly. A **cable** is described as a factory assembly of two or more conductors that has an overall outer covering. The covering is referred to as a **sheath**, sheathing, or

jacket and can be nonmetallic or metallic. The sheathing holds the cable assembly together and provides protection to the conductors. This section of the chapter introduces you to the types of cables and conductors used in residential wiring.

CONDUCTORS

Conductors in residential wiring are usually installed in a cable assembly. A few situations in residential wiring call for conductors to be installed in **raceways.** (Conductors installed in raceways are discussed later in this chapter.) As you already know, conductors conduct the current that is delivered to various loads in a residential wiring system. The installed conductors are usually made of copper but may also be made of aluminum or copper-clad aluminum. Copper is the preferred material because of its great ability to conduct electricity, its strength, and its proven record of having very few, if any, problems over the long-term when correctly installed.

Aluminum conductors are typically used in larger conductor sizes because of their lower cost and lighter weight, which makes them easier to handle. Because aluminum has a higher resistance value and does not conduct electricity as well as copper, a larger-size aluminum conductor must be used to carry the same amount of current as a copper conductor. Aluminum also tends to oxidize much quicker than copper, and electricians must use an **antioxidant** compound on exposed aluminum at all terminations whenever the manufacturer's installation instructions require it. Without the antioxidant, the exposed aluminum will oxidize, and a white powder (aluminum oxide) will develop at the conductor termination. This will cause high resistance at the termination and result in excessive heating and possible damage to the conductor and electrical equipment.

(See Procedure 2-2 on page 73 for a procedure for properly preparing aluminum conductors for termination.)

A conductor that has an aluminum core and an outer coating of copper is called **copper-clad aluminum.** Copper-clad aluminum is not used very often in residential wiring because it can be used only with wiring devices that are listed and labeled for use with this type of conductor. Most of the devices are rated for copper only, but there are listed devices that can be used with copper or copper-clad aluminum. Check the device for a marking that indicates what conductor type it is listed for. Table 2-1 shows typical markings for determining the type of conductors allowed with devices and connectors.

The **American Wire Gauge (AWG)** is the system used to size the conductors used in residential wiring. In this system: the smaller the number, the larger the conductor;

TABLE 2-1 Terminal Identification Markings

TYPE OF DEVICE	MARKING ON TERMINAL OR CONDUCTOR	CONDUCTOR PERMITTED
15- or 20-ampere receptacles and switches	CO/ALR	Aluminum, copper, copper-clad aluminum
15- and 20-ampere receptacles and switches	NONE	Copper, copper-clad aluminum
30-ampere and greater receptacles and switches	AL/CU	Aluminum, copper, copper-clad aluminum
30-ampere and greater receptacles and switches	NONE	Copper only
Screwless pressure terminal connectors of the push-in type	NONE	Copper or copper-clad aluminum
Wire connectors	AL	Aluminum
Wire connectors	AL/CU or CU/AL	Aluminum, copper, copper-clad aluminum
Wire connectors	CC	Copper-clad aluminum only
Wire connectors	CC/CU or CU/CC	Copper or copper-clad aluminum
Wire connectors	CU or CC/CU	Copper only
Any of the above devices	COPPER or CU ONLY	Copper only

and the larger the number, the smaller the conductor. For example, an 8 AWG conductor is larger than a 10 AWG conductor, and a 12 AWG conductor is smaller than a 10 AWG conductor. The *NEC*® recognizes building wire from 18 AWG up to 4/0. 1/0, 2/0, 3/0, and 4/0 are sometimes shown as 0, 00, 000, and 0000, respectively. Electricians pronounce these wire sizes as "1 aught," "2 aught," "3 aught," and "4 aught."

Conductor sizes larger than 4/0 are measured in **circular mils.** A **mil** is equal to .001 inch. If the diameter of a conductor is measured in mils (thousandths of inches) and then squared, you get circular mils. The next size up from a 4/0 is a 250,000-circular-mil conductor. The *NEC*® refers to conductor sizes over 4/0 as having a certain number of kcmils. A kcmil is 1000 circular mils; therefore, a 250,000-circular-mil conductor is referred to as 250 kcmil. It is the size of the cross-sectional area of the conductor that is actually being described when its size in kcmils is given. The largest wire size recognized by the *NEC*® is 2000 kcmil. Table 2-2 shows some common conductor sizes and applications in residential wiring. Figure 2-15 shows a comparison of the common conductor sizes used in residential wiring.

In residential wiring, 14, 12, and 10 AWG conductors are solid when used in a cable assembly and are usually stranded when installed in a raceway. Once in a while, 8 AWG will be used as a solid conductor but normally it will be stranded. Conductor sizes that are 6 AWG and larger will always be stranded. Stranding makes the conductors easier to work with, especially in the larger sizes.

The **ampacity** of a conductor refers to the ability of a conductor to carry current. The current in amperes that a conductor can carry continuously under the conditions of use without exceeding its temperature rating is how the term "ampacity" is defined in the *NEC*®. The larger the conductor size, the higher the ampacity of the conductor. A residential electrician must be able to choose the correct conductor size depending on the ampacity needed for each circuit that is being installed. The ampacity of a conductor depends not only on the size of the conductor but also on what insulation type the conductor has. An **insulated** conductor has an outer covering that is approved and recognized as electrical insulation by the *NEC*®. Letters in certain combinations identify the type of conductor insulation and where the conductor with that particular insulation can be used.

TABLE 2-2 Conductor Applications in Residential Wiring

COPPER CONDUCTOR SIZE	OVERCURRENT PROTECTION (FUSE OR CIRCUIT BREAKER)	TYPICAL APPLICATIONS
18 AWG	Circuit transformers provide overcurrent protection.	Low-voltage wiring for thermostats, chimes, security, remote control, home automation systems, and so on.
16 AWG	Circuit transformers provide overcurrent protection.	Same applications as above. Good for long runs to minimize voltage drop.
14 AWG	15 amperes	Typical lighting branch circuits.
12 AWG	20 amperes	Small appliance branch circuits for the receptacles in kitchens and dining rooms. Also, laundry receptacles and workshop receptacles. Often used for lighting branch circuits, as well as some smaller electric water heaters.
10 AWG	30 amperes	Most electric clothes dryers, built-in ovens, cooktops, central air conditioners, some electric water heaters, heat pumps.
8 AWG	40 amperes	Electric ranges, ovens, heat pumps, some large electric clothes dryers, large central air conditioners, heat pumps.
6 AWG	50 amperes	Electric furnaces, heat pumps, feeders to subpanels.
4 AWG	70 amperes	Electric furnaces, feeders to subpanels.
3 AWG and larger	100 amperes	Main service entrance conductors, feeders to subpanels, electric furnaces.

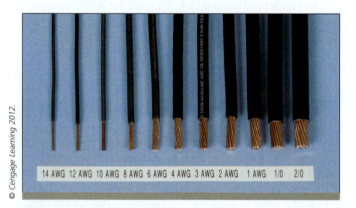

14 AWG 12 AWG 10 AWG 8 AWG 6 AWG 4 AWG 3 AWG 2 AWG 1 AWG 1/0   2/0

FIGURE 2-15 A comparison of the common conductor sizes used in residential wiring.

© Cengage Learning 2012.

The letters are written on the insulation of the conductor or on the cable itself. Table 310.104(A) of the *NEC®* lists the recognized insulation types. Table 2-3 shows some typical insulation types used in residential wiring.

Table 310.15(B)(16) of the *NEC®* is where residential electricians go to determine the ampacity of the conductors they will be using. This table shows the ampacities of conductors from size 18 AWG to 2000 kcmil. It is broken down into copper wire sizes and aluminum/copper-clad aluminum wire sizes. The ampacity columns are headed by the insulation temperature rating. The ratings are 60, 75, and 90 degrees Celsius (140, 167, and 194 degrees Fahrenheit). Choosing the correct conductor size for a specific application is covered later in this book.

Conductor functions are identified with a certain color coding of their insulation. The following color coding applies to residential wiring applications:

- Black: Used as an ungrounded, or "hot," conductor and carries the current to the load in 120-volt circuits.

- Red: Also used as an ungrounded, or "hot," conductor and carries current to the load in 120/240-volt circuits, like an electric clothes dryer circuit.

- White: Used as the grounded circuit conductor and returns current from the load back to the source. It is sometimes referred to as the "neutral" conductor but is truly neutral only when used with a black and a red wire in a **multiwire** circuit, like a 120/240-volt electric clothes dryer circuit.

TABLE 2-3 Typical Conductor Insulations Used in Residential Wiring

TRADE NAME	TYPE LETTER	OPERATING TEMPERATURE	MAXIMUM APPLICATION PROVISIONS	INSULATION	AWG	OUTER COVERING
Heat-resistant thermoplastic	THHN	194°F (90°C)	Dry and damp locations	Flame-retardant, heat-resistant thermoplastic	14–1000 kcmil	Nylon jacket or equivalent
Moisture- and heat-resistant thermoplastic	THHW	167°F (75°C) 194°F (90°C)	Wet location Dry location	Flame-retardant, moisture- and heat-resistant thermoplastic	14–1000 kcmil	None
Moisture- and heat-resistant thermoplastic	THWN Note: If marked THWN-2, okay for 194°F (90°C) in dry or wet locations.	167°F (75°C)	Dry and wet locations	Flame-retardant, moisture- and heat-resistant thermoplastic	14–1000 kcmil	Nylon jacket or equivalent
Moisture- and heat-resistant thermoplastic	THW Note: If marked THW-2, okay for 194°F (90°C) in dry or wet locations.	167°F (75°C)	Dry and wet locations	Flame-retardant, moisture- and heat-resistant thermoplastic	14–2000 kcmil	None

- Bare: Used as an equipment-grounding conductor that bonds all non-current-carrying metal parts of a circuit together; it never carries current.

- Green: Used as an insulated equipment-grounding conductor; could be green with yellow stripes; never carries current.

Conductors are terminated to electrical equipment and each other with a variety of connector types, terminals, and lugs. A wirenut is the most common wire connector used by electricians to splice together two or more conductors. Figure 2-16 shows several types of wire connectors used in residential wiring.

(See Procedure 2-3 on page 74 for the procedure to correctly install a wirenut.)

CABLE TYPES

The conductors installed for circuits in residential wiring are usually installed as part of a cable. Nonmetallic sheathed cable, underground feeder cable, armored-clad cable, metal-clad cable, and service entrance cable are the types of cable commonly used. Each has certain advantages and uses in residential wiring. Residential electricians need to recognize these cable types and know when and where to use them. This section introduces you to these cable types, and later chapters discuss their installation procedures.

Nonmetallic Sheathed Cable: Types NM, NMC, and NMS

Residential electricians refer to nonmetallic sheathed cable (Type NM) as Romex™. This is the name first given to Type NM cable by the Rome Wire and Cable Company. Even though other companies manufacture Type NM cable, electricians call it "Romex™." It is the least expensive wiring method to purchase and install, which is the main reason why it is used more than other wiring methods in residential applications.

Article 334 of the *NEC*® covers Type NM cable, which is defined as a factory assembly of two or more insulated conductors having an outer sheathing of moisture-resistant, flame-retardant, nonmetallic material. It is available with two or three conductors in sizes from 14 through 2 AWG copper and 12 through

CRIMP CONNECTORS USED TO SPLICE AND TERMINATE 20 AWG TO 500 KCMILS ALUMINUM-TO-ALUMINUM, ALUMINUM-TO-COPPER, OR COPPER-TO-COPPER CONDUCTORS.

PROPERLY CRIMP
THEN TAPE

CONNECTORS USED TO CONNECT WIRES TOGETHER ON COMBINATIONS OF 18 AWG THROUGH 6 AWG CONDUCTORS. THEY ARE TWIST-ON, SOLDERLESS, AND TAPELESS.

*WIRE-NUT® and WING-NUT® are registered trademarks of IDEAL INDUSTRIES, INC. Scotchlok® is a registered trademark of 3M.

WIRE CONNECTORS VARIOUSLY KNOWN AS
WIRE-NUT,® WING-NUT,® AND SCOTCHLOK.®

INSULATED CONNECTOR USED TO SPLICE OR TAP WIRES TOGETHER IN COMBINATIONS OF 14 AWG-750 KCMIL. KNOWN IN THE ELECTRICAL TRADE AS A 'UNITAP' OR 'POLARIS' CONNECTOR

UNITAP INLINE SPLICER

SOLDERLESS CONNECTORS ARE AVAILABLE IN SIZES 14 AWG THROUGH 500 KCMIL CONDUCTORS. THEY ARE USED FOR ONE SOLID OR ONE STRANDED CONDUCTOR ONLY, UNLESS OTHERWISE NOTED ON THE CONNECTOR OR ON ITS SHIPPING CARTON. THE SCREW MAY BE OF THE STANDARD SCREWDRIVER SLOT TYPE, OR IT MAY BE FOR USE WITH AN ALLEN OR SOCKET WRENCH.

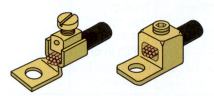

SOLDERLESS CONNECTORS
(LUGS)

COMPRESSION CONNECTORS ARE USED FOR 8 AWG THROUGH 1000 KCMIL CONDUCTORS. THE WIRE IS INSERTED INTO THE END OF THE CONNECTOR, THEN CRIMPED ON WITH A SPECIAL COMPRESSION TOOL.

COMPRESSION CONNECTOR

SPLIT-BOLT CONNECTORS ARE USED FOR CONNECTING TWO CONDUCTORS TOGETHER, OR FOR TAPPING ONE CONDUCTOR TO ANOTHER. THEY ARE AVAILABLE IN SIZES 10 AWG THROUGH 1000 KCMIL. THEY ARE USED FOR TWO SOLID AND/OR TWO STRANDED CONDUCTORS ONLY, UNLESS OTHERWISE NOTED ON THE CONNECTOR OR ON ITS SHIPPING CARTON.

SPLIT-BOLT CONNECTOR

FIGURE 2-16 Examples of wire connectors.

2 AWG aluminum. Two-wire cable has a black and a white insulated conductor along with a bare grounding conductor. The three-wire cable will have a black, red, and a white conductor along with a bare grounding conductor. It is used in residential applications for general-purpose branch circuits, small-appliance branch circuits, and individual branch circuits. The ampacity of Type NM cable is found in the "60 degrees Celsius" column of Table 310.15(B)(16) in the *NEC®*.

Manufacturers of nonmetallic sheathed cable are now color-coding their cables. The cable colors are white for 14 AWG Type NM cable, yellow for

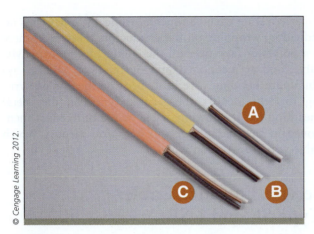

FIGURE 2-17 An example of nonmetallic sheathed cable (Type NM). The outside sheathing is color coded so that an electrician (or inspector) can tell at a glance what size conductor is in the cable. (A) Shows a 14 AWG cable, (B) shows a 12 AWG cable, and (C) shows a 10 AWG cable.

© Cengage Learning 2012.

12 AWG Type NM cable, orange for 10 AWG Type NM cable, and black for larger sizes (Figure 2-17). This allows ready identification of the cable size. An electrician (or an electrical inspector) can look around a house that has been roughed-in with Romex™ and easily determine where the 14 AWG, 12 AWG, 10 AWG, or larger cables have been installed.

> **CAUTION**
>
> **CAUTION:** Even though the ampacity for 14, 12, and 10 AWG Type NM cable is found in the "60 degrees Celsius" column of Table 310.15(B)(16), Section 240.4(D) requires a maximum fuse or circuit breaker size of 15 amps for 14 AWG, 20 amps for 12 AWG, and 30 amps for 10 AWG conductors used in residential circuits.

Underwriters Laboratories lists three types of nonmetallic sheathed cable. The type will be written on the outer jacket of the cable:

- Type NM-B is by far the most common type used and has a flame-retardant, moisture-resistant, nonmetallic outer jacket. It can be used in dry locations only. The conductor insulation is rated at 194 degrees Fahrenheit (90 degrees Celsius). However, the ampacity of the Type NM-B is based on 140 degrees Fahrenheit (60 degrees Celsius).

- Type NMC-B is not used often in residential work. It has a flame-retardant and moisture-, fungus-, and corrosion-resistant nonmetallic outer jacket and can be used in dry or damp locations. The conductor insulation and ampacity are rated the same as the Type NM-B cable.

- Type NMS-B is used in new homes that have home automation systems using the latest technology. The cable contains the power conductors, telephone wires, coaxial cable for video, and other data conductors all in the same cable. It has a moisture-resistant, flame-retardant, nonmetallic outer jacket.

Underground Feeder Cable: Type UF

Type UF cable is covered in Article 340 of the *NEC*® (Figure 2-18). It is used for underground installations of branch circuits and feeder circuits. It can also be used in interior installations but must be installed following the installation requirements for Type NM cable. It is available in wire sizes from 14 through 4/0 AWG copper and from 12 through 4/0 AWG aluminum. The ampacity for this cable type is found in the "60 degrees Celsius" column of Table 310.15(B)(16). The burial depth for Type UF cable is found in Table 300.5 of the *NEC*®. Additionally, Type UF can be used for outside residential installations but only if the cable is listed and marked as sunlight resistant.

Armored-Clad (Type AC) and Metal-Clad (Type MC) Cable

There are certain locations in the United States where the AHJ may not allow nonmetallic sheathed cable in residential construction. The alternative wiring method most often used is armored-clad or metal-clad cable (Figure 2-19). Both have a metal outer sheathing and provide very high levels of physical protection for the conductors in the cable.

FIGURE 2-18 An example of underground feeder cable (Type UF). *Courtesy of Southwire Company.*

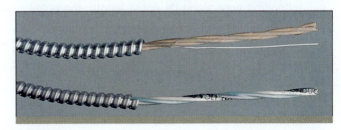

FIGURE 2-19 Top: Armored-clad cable (Type AC); bottom: metal-clad cable (Type MC). *Courtesy of AFC Cable Systems, Inc.*

Armored-clad cable has been around for a long time and has a proven track record. Electricians usually refer to it as "BX" cable—a trademark owned by the General Electric Company that has become a generic term used as a trade name for any company's armored-clad cable.

Article 320 of the *NEC®* covers Type AC cable. It is defined as a fabricated assembly of insulated conductors in a flexible metallic enclosure. It is available with two, three, or four conductors. A small aluminum bonding wire is included and is used to ensure electrical continuity of the outside flexible metal sheathing. Because of the bonding wire, the outside metal sheathing can be used as the grounding conductor. Type AC is also available with a green insulated grounding conductor. This wiring method is available in wire sizes 14 through 1 AWG copper and 12 through 1 AWG aluminum. Like Romex™, Type AC ampacity is found in the "60 degrees Celsius" column of Table 310.15(B)(16).

Metal-clad cable looks very much like armored-clad cable. However, there are a few differences. Article 330 covers Type MC and defines it as a factory assembly of one or more insulated circuit conductors enclosed in an armor of interlocking metal tape or a smooth or corrugated metallic sheath. There is no limit to the number of conductors found in Type MC, and it is available in wire sizes from 18 AWG through 2000-kcmil copper and 12 AWG through 2000-kcmil aluminum. Unlike BX cable, the outer metal sheathing cannot be used for a grounding conductor unless it is specifically listed as such. A green insulated grounding conductor will always be included in the cable assembly. The ampacity is found using the "60 degrees Celsius" column of Table 310.15(B)(16), but for 1/0 AWG and larger conductors, you can use the "75 degrees Celsius" column.

> ## *from experience...*
>
> At least one company makes a style of Type MC cable that is manufactured with an internal aluminum bonding wire just like Type AC cable. Because it contains the aluminum bonding wire along its entire length, the outer metal sheathing is listed as an acceptable grounding means.

Electricians sometimes find that it is hard to distinguish Type AC from Type MC cable. One of the easiest ways to do this is to look at how the conductors in the cables are wrapped. Type AC cable has a light brown paper covering each conductor, while Type MC has no individual wrapping on the conductors. However, there is a polyester tape over all the conductors in Type MC cable, and as mentioned previously, Type MC has no aluminum bonding wire in it. Because there is no writing to help identify these cable types on the metal sheathing itself, knowing a few of the differences mentioned in this paragraph will help you identify which cable you are working with.

Service Entrance Cable: Type SE and USE

Service entrance (SE) cables are installed to supply the electrical power from the utility company to the building's electrical system. They are also sometimes used to supply power to a large appliance, like an electric range. This cable type is covered in Article 338 of the *NEC®* and is defined as a single conductor or multiconductor assembly provided with or without an overall covering. Because of the larger size requirements and expense, most residential installations use aluminum conductors instead of copper in service entrance cables.

Electricians should be familiar with three types of service entrance cable. The most commonly used type is SEU. Type SEU cable is used for service entrance installations (Figure 2-20). It contains three conductors enclosed in a flame-retardant, moisture-resistant outer covering. The service neutral conductor is made up of several bare conductors wrapped around the insulated ungrounded conductors. The electrician needs to twist these bare conductors together for termination.

Type SER cable is similar to Type SEU but has four conductors wrapped in a round configuration (Figure 2-21). The neutral conductor is insulated, and the cable also contains a bare grounding conductor. SER cable is what you would use to wire a new electric range or electric dryer if you chose service entrance cable as the wiring method. Type SER is also used as a feeder from the main service panel to a subpanel.

© Cengage Learning 2012.

FIGURE 2-20 Service entrance cable (Type SEU) with two black insulated ungrounded conductors and one wraparound bare neutral grounded conductor.

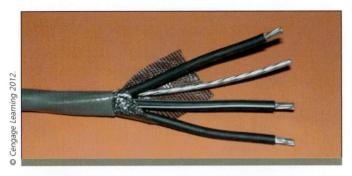

FIGURE 2-21 Service entrance cable (Type SER) with two black insulated ungrounded conductors, one white identified insulated grounded conductor, and one bare grounding conductor.

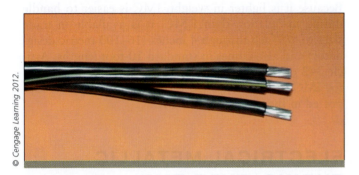

FIGURE 2-22 Service entrance cable (Type USE) is used to install underground service entrances. Unlike Type SEU and Type SER service entrance cables, the insulation on Type USE cable is suitable for direct burial. The conductors are simply wrapped together and do not have an overall outer sheathing.

FIGURE 2-23 Examples of common cable connectors used in residential wiring. (A) Type NM nonmetallic cable connectors; (B) Type NM metal cable connectors; (C) Type AC and Type MC cable connectors.

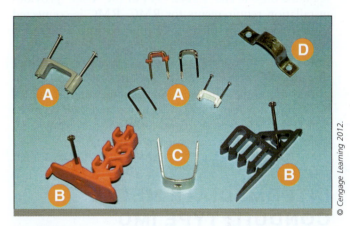

FIGURE 2-24 Examples of common items used to support and secure cables in residential wiring. (A) Staples are the most common type of supporting and securing method; (B) Cable stackers used to keep multiple cables in the center of a stud; (C) Fold over clip for Type SEU cable; (D) Two-hole strap for Type SEU cable.

Type USE cable is used for underground service entrance installations (Figure 2-22). It has a moisture-resistant covering but, because it is buried, is not flame retardant. It can also be used for underground feeder or underground branch circuits. It is usually installed as an assembly of three or four conductors wrapped around each other with no overall outer sheathing. This configuration is called a URD (underground residential distribution) cable.

Cable Fittings and Supports

There are a large variety of fittings used with the cables described in the preceding paragraphs. **Fittings** allow cables to be installed in compliance with the *NEC®*. When a cable is terminated at an electrical box, a connector must be used to secure it to the box. Figure 2-23 shows some examples of cable connectors commonly used in residential wiring.

The *NEC®* also requires cables to be properly secured and supported during their installation. Figure 2-24 shows some common items used to secure and support cables.

RACEWAYS

Article 100 of the *NEC®* defines a raceway as an enclosed channel of metal or nonmetallic materials designed expressly for holding wires or cables. Raceways that may be used in residential wiring include rigid metal conduit, rigid PVC conduit, intermediate metal conduit, liquid tight flexible conduit, flexible metal conduit, electrical nonmetallic tubing, and electrical metallic tubing.

Many residential wiring installations do not have a need for the installation of any raceway. However, some installations may use a raceway for parts of a service entrance or to protect conductors going to an outside air-conditioning unit. Because raceways are used in some installations, residential electricians need to recognize common raceway types.

RIGID METAL CONDUIT: TYPE RMC

Rigid metal conduit (RMC) is typically made of steel and is galvanized to enable it to resist rusting. It is sometimes referred to as "heavywall" conduit. RMC is often used as a mast for a service entrance. A mast will allow the service conductors from the electric utility to be located high enough from the ground that the danger to pedestrians is minimized. This conduit is threaded on each end and is available in trade sizes of 1/2 inch through 6 inches (16–155 metric designator). It is available in a standard length of 10 feet, and a coupling is included on one end. Article 344 of the *NEC*® provides installation requirements for RMC. RMC is connected to an enclosure by threading it directly into a threaded hole in the enclosure or by using two locknuts, one inside and one outside the box, to secure it to a nonthreaded KO hole. The *NEC*® requires RMC to be supported at regular intervals. The general support rule is to support it within 3 feet of each box and then no more than every 10 feet thereafter. Figure 2-25 shows an example of RMC with some associated fittings.

INTERMEDIATE METAL CONDUIT: TYPE IMC

Intermediate metal conduit (IMC) (Figure 2-26) is a lighter version of RMC but can be used in all the locations that the heavier RMC may be used. It is threaded

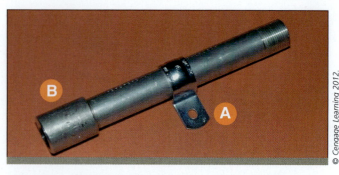

FIGURE 2-26 *Intermediate metal conduit (IMC) and associated fittings. (A) One-hole strap; (B) Coupling.*

on each end and also comes with a coupling on one end. Because it is lighter in weight, IMC is easier to handle during installation than RMC. It is available in trade sizes of 1/2 inch through 4 inches (16–103 metric designator) and also comes in standard 10-foot lengths. Article 342 of the *NEC*® provides installation requirements for IMC. IMC is attached to electrical boxes and supported in the same way as RMC.

ELECTRICAL METALLIC TUBING: TYPE EMT

Electrical metallic tubing (EMT) is often referred to as "thinwall" conduit because of its very thin walls. Article 358 of the *NEC*® covers the installation requirements for EMT. It is much lighter and easier to install than both RMC and IMC. EMT cannot be threaded. Connectors and couplings used with EMT utilize setscrew or compression tightening systems. EMT is available in trade sizes of 1/2 inch through 4 inches (16–103 metric designator) and comes in standard 10-foot lengths. Figure 2-27 shows an example of EMT and various fittings used with it.

RIGID POLYVINYL CHLORIDE CONDUIT: TYPE PVC

Rigid polyvinyl chloride conduit (PVC) is inexpensive and can be used in a variety of residential wiring applications. Article 352 of the *NEC*® covers the installation requirements of PVC. It is available in trade sizes of 1/2 inch through 6 inches (16–155 metric designator) and, like the other raceways mentioned in this section, comes in standard 10-foot lengths. There are two types of PVC conduit that electricians typically will use. Schedule 40 has a heavy wall thickness, and Schedule 80 has an extra-heavy-duty wall thickness. The outside diameters of these two types are the same, but the inside diameter of Schedule 40 is greater than Schedule 80. Schedule 40

FIGURE 2-25 *Rigid metal conduit (RMC) and associated fittings. (A) One-hole strap; (B) Conduit hanger; (C) Erickson coupling; (D) Insulated bonding bushing; (E) Coupling.*

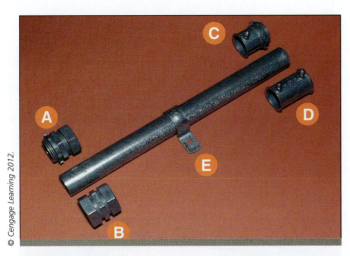

FIGURE 2-27 Electrical metallic tubing (EMT) and associated fittings. (A) Compression connector; (B) Compression coupling; (C) Set-screw connector; (D) Set-screw coupling; (E) One-hole strap.

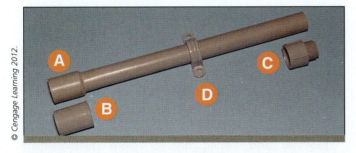

FIGURE 2-28 Rigid polyvinyl chloride conduit (PVC) and associated fittings. (A) Integral coupling; (B) Separate coupling; (C) Connector; (D) Two-hole PVC strap.

is used where it is not subject to physical damage, while Schedule 80 is used where it is subject to physical damage. Connectors and couplings are attached to the conduit with a PVC cement in the same manner that plumbers use to connect fittings to their PVC pipes. However, there is a difference between plumbing PVC pipe and electrical PVC pipe. The white plumbing pipe is designed to withstand water pressure from the inside, while the gray electrical PVC is designed to withstand forces from the outside. Figure 2-28 shows an example of PVC and some associated fittings.

FLEXIBLE METAL CONDUIT: TYPE FMC

Flexible metal conduit (FMC) is a raceway of circular cross section made of a helically wound, formed, interlocked metal strip. Article 348 of the *NEC®* covers the use and installation requirements of this raceway type. The trade name for this raceway type is "Greenfield."

It was invented in 1902 by Harry Greenfield and Gus Johnson and when it was listed by the Sprague Electric Co. it was called "Greenfield flexible steel conduit." It looks very similar to BX cable but does not have the conductors already installed. It is up to the electrician to install the conductors in FMC. It is designed to be very flexible and is often used to connect appliances and other equipment in residential applications. For example, a built-in oven will come from the factory with a short length of FMC enclosing the oven conductors. The electrician secures the FMC to a junction box and makes the necessary connections. FMC is available in trade sizes of 1/2 inch through 4 inches (16-103 metric designator). A trade size of 3/8 inch (12 metric designator) can be used for lengths of not more than 6 feet (1.8 m). Also, if listed for grounding, FMC contains conductors protected by a fuse or circuit breaker rated 20 amps or less, is used in lengths not longer than 6 feet, and is used with connectors that are listed for grounding (the outside metal sheathing can be used as an equipment-grounding conductor). Otherwise, a green equipment-grounding conductor must be installed. Figure 2-29 shows an example of FMC and some associated fittings.

ELECTRICAL NONMETALLIC TUBING: TYPE ENT

Electrical nonmetallic tubing (ENT) is a flexible nonmetallic raceway that is being used more and more in residential wiring. Electricians often refer to it as "Smurf Tube" because its blue color reminds some electricians of the blue Smurf cartoon characters. Article 362 of the *NEC®* provides the installation requirements for ENT and defines it as a nonmetallic pliable corrugated raceway of circular cross section with integral or associated couplings, connectors, and fittings for the installation of electric conductors. ENT is composed of a

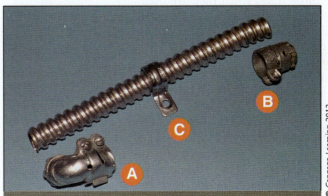

FIGURE 2-29 Flexible metal conduit (FMC) and associated fittings. (A) 90 degree connector; (B) Straight connector; (C) One-hole strap.

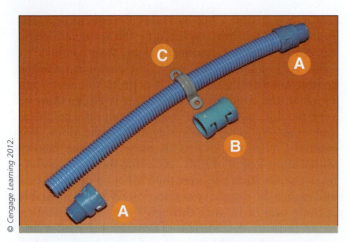

© Cengage Learning 2012.

FIGURE 2-30 Electrical nonmetallic tubing (ENT) and associated fittings. (A) Snap-on connector; (B) Snap-on coupling; (C) Two-hole strap.

© Cengage Learning 2012.

FIGURE 2-31 Liquidtight flexible metal conduit (LFMC) and associated fittings. (A) 90 degree connector; (B) Straight connector.

FIGURE 2-32 Liquidtight flexible nonmetallic conduit (LFNC) and associated fittings. (A) Straight connector; (B) 90 degree connector.

© Cengage Learning 2012.

material that is resistant to moisture and chemical atmospheres and is flame retardant. It is available in trade sizes of 1/2 inch through 2 inches (16–53 metric designator). Figure 2-30 shows an example of ENT and some associated fittings.

LIQUIDTIGHT FLEXIBLE METAL CONDUIT: TYPE LFMC AND LIQUIDTIGHT FLEXIBLE NONMETALLIC CONDUIT: TYPE LFNC

Liquidtight flexible metal conduit (LFMC) and liquidtight flexible nonmetallic conduit (LFNC) are raceway types that are used where flexibility is desired in wet locations, such as outdoors. While they are not used often in residential wiring, it is important for an electrician to be able to recognize these raceway types for those times when they are used.

LFMC is defined as a raceway of circular cross section having an outer liquidtight, nonmetallic, sunlight-resistant jacket over an inner flexible metal core with associated couplings, connectors, and fittings for the installation of electric conductors (Figure 2-31). Article 350 of the *NEC*® covers the installation requirements for LFMC. When combined with proper connectors, an installation that does not allow liquid into the raceway and around the conductors is accomplished. It is available in trade sizes of 1/2 inch through 4 inches (16–103 metric designator). A 3/8-inch (12 metric designator) trade size is available for special situations outlined in Article 348. A common application for this wiring method is the connection to a central air-conditioning unit located outside a dwelling.

Article 356 of the *NEC*® covers the installation requirements for liquidtight flexible nonmetallic conduit (LFNC). It is defined as a raceway of circular cross section of various types as follows:

1 A smooth seamless inner core and cover bonded together and having one or more reinforcement layers between the core and covers, designated as Type LFNC-A

2 A smooth inner surface with integral reinforcement within the conduit wall, designated as Type LFNC-B

3 A corrugated internal and external surface without integral reinforcement within the conduit wall, designated as LFNC-C

LFNC is flame resistant and with proper fittings is approved for the installation of electrical conductors (Figure 2-32). LFNC-B is the type most often used. LFNC is used in the same applications as LFMC. It is also available in trade sizes of 1/2 inch through 4 inches (16–103 metric designator), and a 3/8-inch (12 metric designator) trade size also is available.

DEVICES

The *NEC*® defines a **device** as a unit of an electrical system that is intended to carry but not utilize electric energy. Components such as switches, receptacles, attachment plugs, and lamp holders are considered devices because they distribute or control, but do not consume, electricity.

RECEPTACLES

Receptacles are probably the most recognizable parts of a residential electrical system. They provide ready access to the electrical system and are defined as a contact device installed at the outlet for the connection of an attachment plug. Even though many people, including electricians, refer to a receptacle as an "outlet," it is the incorrect term to use. An **outlet** is the point on the wiring system at which current is taken to supply equipment. Many people also refer to a receptacle as a "plug." This is also an incorrect term to use. A **plug** is defined as the device that is inserted into a receptacle to establish a connection between the conductors of the attached flexible cord and the conductors connected to the receptacle. A receptacle is the device that allows the electrician to access current from the wiring system and deliver it through a cord and attachment plug to a piece of equipment (Figure 2-33).

A single receptacle is a single contact device with no other contact device on the same yoke. This type of receptacle is sometimes used in residential wiring. An example would be the receptacle installed for a washing machine, which is usually a single receptacle with a 20-ampere, 125-volt rating. However, the most common type of receptacle used in residential wiring is a duplex receptacle rated for 15 amperes at 125 volts. It consists of two single receptacles on the same mounting strap. The short contact slot on the receptacle receives the "hot" conductor from the attached cord. The long contact slot receives the grounded conductor from the attached cord. There is also a U-shaped grounding contact that receives the grounding conductor. Silver screws are located on the side with the long contact slot and are used to terminate the white, grounded circuit conductor. Brass- or bronze-colored screws are located on the same side as the short contact slot and are used to terminate the "hot," or ungrounded, circuit conductor. A green screw is located on the duplex receptacle for terminating the circuit bare or green grounding conductor (Figure 2-34).

There are several other features on a duplex receptacle that you should be familiar with. The following features are usually found on the front of a receptacle:

- Connecting tabs: These are used to connect the top half and the bottom half of the duplex receptacle. They can be taken off to provide different wiring configurations with "split" receptacles. Wiring split receptacles is covered in a later chapter.

- Mounting straps: These are used to attach the receptacle to a device box. New receptacles will have 6-32 screws held in place by small pieces of cardboard or plastic in the mounting straps.

- Ratings: Both the amperage and the voltage rating of the receptacle are written on the receptacle.

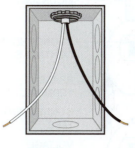

A RECEPTACLE OUTLET IN WHICH ONE OR MORE RECEPTACLES WILL BE INSTALLED.

A RECEPTACLE OUTLET WITH A SINGLE RECEPTACLE. (ONE CONTACT DEVICE)

A RECEPTACLE OUTLET WITH A MULTIPLE (DUPLEX) RECEPTACLE. THIS IS TWO RECEPTACLES. (TWO CONTACT DEVICES)

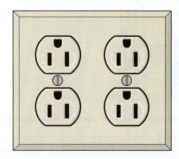

A RECEPTACLE OUTLET WITH TWO MULTIPLE (DUPLEX) RECEPTACLES. THIS IS FOUR RECEPTACLES. (FOUR CONTACT DEVICES)

© Cengage Learning 2012.

FIGURE 2-33 Electricians sometimes confuse the term "outlet" with the term "receptacle." Article 100 of the *NEC®* defines a receptacle "outlet" as the branch-circuit wiring and the box where one or more receptacle devices are to be installed. A strap (yoke) with one, two, or three contact devices is defined as a "receptacle."

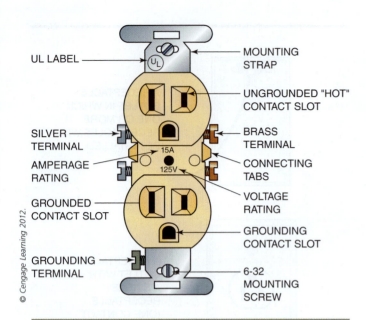

FIGURE 2-34   The parts of a duplex receptacle.

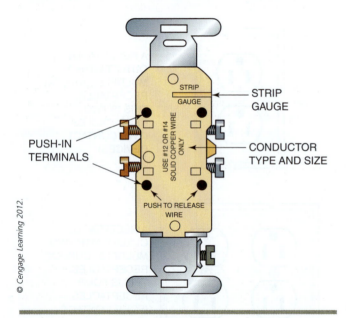

FIGURE 2-35   The back of a receptacle device contains information that should be checked before installation. There are also some receptacle features located on the back of the device.

- NRTL label: A label from an NRTL like UL will be on the receptacle. This is used as a basis for product approval by the AHJ.

The following features are usually found on the back of a duplex receptacle (Figure 2-35):

- Push-in terminals: Sometimes called "back-stabs," these are used when electricians strip a conductor and push it into the hole rather than terminating on the screws.

- Strip gauge: This gauge is used to let the electrician know how much insulation needs to be stripped off the conductor when using the push-in terminals.

- Conductor size: This will tell the electrician what the maximum conductor size is for this device. Most duplex receptacles are rated for 14 or 12 AWG conductors.

- Conductor material markings: These markings will indicate what conductor material is okay to use with the device. "CU" indicates that only copper conductors can be used, "Cu-Clad Only" indicates that only copper-clad aluminum can be used, and "CO/ALR" indicates that copper, aluminum, or copper-clad aluminum can be used.

from experience...

Starting with the 2008 National Electrical Code almost all receptacles in a house must be tamper-resistant. The style of tamper-resistant receptacle (Figure 2-36) most used has shutters across the slots of the receptacle that will not allow individual items like a paper clip to be stuck into the slot. The only way the shutters will open is when a two-prong or three-prong attachment plug is pushed into the slots. The intent of this requirement is to protect children from shocks and burns that could occur if they stick something conductive into the slots of a receptacle.

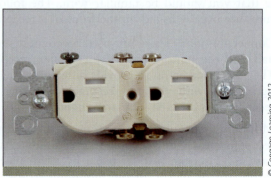

FIGURE 2-36   A tamper-resistant receptacle. Notice that the slots have shutters behind them that keep objects from being inserted into a slot.

Ground Fault Circuit Interrupter Receptacles

Duplex receptacles installed in residential wiring are also available in a ground fault circuit interrupter (GFCI) type (Figure 2-37). In addition to their use as duplex receptacles, GFCI receptacles also protect people from electrical shock. Several locations in residential wiring are required by the *NEC®* to have GFCI protection. Section 210.8(A) outlines those areas that require GFCI protection for all 15- and 20-ampere-, 125-volt-rated receptacles. Some of these locations include kitchens, bathrooms, and basements. Installation of GFCI receptacles in these locations is covered in later chapters.

The GFCI receptacle can be used to provide protection only at its location or may provide GFCI protection to other "regular" duplex receptacles connected "downstream" from the GFCI receptacle. GFCI receptacles have the same features as a regular duplex receptacle with a few exceptions. There is a "reset" and a "test" button on the front of the GFCI receptacle. They are used for testing the GFCI on a regular basis and then resetting it. An important feature to know for new electricians on the GFCI receptacle is the location of the words "LOAD" and "LINE." It is very important to make sure that the incoming electrical power, or the "line," is connected to the proper LINE terminals and that the outgoing electrical power, or the "load," wires are connected to the proper LOAD terminals. The color coding is the same as for a regular receptacle: silver for the white, grounded conductor and bronze or brass for the "hot," ungrounded conductor.

Special Receptacle Types

Some circuits are intended to feed only one piece of equipment. Usually, this equipment requires a special receptacle that is larger and has a different configuration than the single or duplex receptacles previously covered. Appliances can be classified into three groups:

1 Portable appliance: This is a small appliance, like a toaster or coffeemaker, and is plugged into 15- and 20-ampere receptacles like those previously discussed. They do not require a special kind of receptacle.

2 Stationary appliance: This is an appliance like an electric range, an electric clothes dryer, or a room air conditioner. They usually require large amounts of current and are connected to receptacles that are designed specifically for the amperage and voltage that these appliances need to operate.

3 Fixed appliance: This type of appliance is fastened in place and is not easily moved. Examples are built-in ovens, built-in cooktops, electric water heaters, and furnaces. This type of appliance usually is "hardwired" and is not cord and plug connected.

Special receptacles are available in a flush-mount style and a surface-mount style. Flush mount is used in new construction and requires first mounting an electrical box and then installing the flush-mount receptacle in the box. It gets its name because it is "flush," or even, with the finished wall when the installation is complete. A surface-mount receptacle is not attached to an electrical box. It is a self-contained piece of equipment and is installed by attaching a back plate to the floor or wall surface, connecting the wiring to the proper terminals, and securing a plastic cover over the installation. It can be used in new construction but is most often used in remodel work when an electrical box is not easily installed in a wall.

In residential wiring, there are two appliances that typically require a special receptacle installation: the electric range and the electric clothes dryer (Figure 2-38). The range requires a heavy-duty, 50-ampere, 250-volt-rated receptacle and attachment

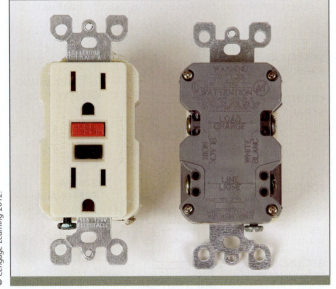

© Cengage Learning 2012.

FIGURE 2-37 A GFCI receptacle is equipped with test and reset buttons. The back of the device contains information that is useful when installing the device in a circuit, such as which terminals are used to connect the incoming (LINE) wires and the outgoing (LOAD) wires.

© Cengage Learning 2012.

FIGURE 2-38   (A) A four-wire 50-amp-rated range receptacle and (B) a four-wire 30-amp-rated dryer receptacle. Both are available in a surface- or flush-mount design. The attachment plug used with each receptacle is also shown. Prior to the 1996 *NEC®*, three-wire receptacles and cords were permitted for the connection of electric ranges and clothes dryers. Three-wire circuits are no longer permitted for electric ranges and clothes dryers.

plug for its installation. The dryer requires a heavy-duty, 30-ampere-, 250-volt-rated receptacle and attachment plug.

The range and dryer receptacles and plugs have special letter designations that you should know. The letter "G" indicates the location of the equipment-grounding conductor. The letter "W" indicates the location of the white, or grounded, conductor. The letters "X" and "Y" indicate the location of the "hot" ungrounded conductors.

The NEMA has compiled a chart that shows all of the general-purpose receptacle and plug configurations (see Table 2-4).

SWITCHES

Devices called switches are used to control the various lighting outlets installed in residential wiring. This device type can be called many different names, such as "toggle switch," "snap switch," or "light switch," but in this section we refer to these devices simply as "switches." Single-pole, double-pole, three-way, and four-way switches are the switch types commonly used in residential wiring. This section shows you how to recognize these different switch types.

The most common type of switch used in residential wiring is called a single-pole switch. This switch type is used in 120-volt circuits to control a lighting outlet or outlets from only one location. An installation example would be in a bedroom where the single-pole switch is

located next to the door and allows a person to turn on a lighting fixture when the person enters the room. Figure 2-39 shows the parts of a basic single-pole switch. The main parts include the following:

- Switch toggle: Used to place the switch in the ON or OFF position when moved up or down.

- Screw terminals: Used to attach the lighting circuit wiring to the switch. On a single-pole switch, the two terminal screws are the same color, usually bronze.

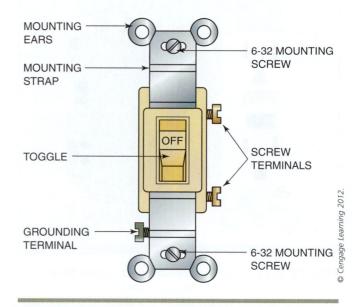

FIGURE 2-39   The parts of a single-pole switch.

TABLE 2-4 NEMA General-Purpose Nonlocking Plugs and Receptacles

NEMA RECEPTACLE AND PLUG CHART CONFIGURATIONS

	NEMA No.		15 AMPERE		20 AMPERE		30 AMPERE		50 AMPERE		60 AMPERE	
			RECEPTACLE	PLUG	RECEPTACLE	PLUG	RECEPTACLE	PLUG	RECEPTACLE	PLUG	RECEPTACLE	PLUG
TWO-POLE TWO-WIRE	1	125V	1-15R	1-15P								
	2	250V		2-15P	2-20R	2-20P	2-30R	2-30P				
TWO-POLE THREE-WIRE GROUNDING	5	125V	5-15R	5-15P	5-20R	5-20P	5-30R	5-30P	5-50R	5-50P		
	6	250V	6-15R	6-15P	6-20R	6-20P	6-30R	6-30P	6-50R	6-50P		
THREE-POLE TWO-WIRE	7	277V AC	7-15R	7-15P	7-20R	7-20P	7-30R	7-30P	7-50R	7-50P		
	10	125/250V			10-20R	10-20P	10-30R	10-30P	10-50R	10-50P		
THREE-POLE FOUR-WIRE GROUNDING	11	3Ø 250V	11-15R	11-15P	11-20R	11-20P	11-30R	11-30P	11-50R	11-50P		
	14	125/250V	14-15R	14-15P	14-20R	14-20P	14-30R	14-30P	14-50R	14-50P	14-60R	14-60P
FOUR-POLE FOUR-WIRE	15	3Ø 250V	15-15R	15-15P	15-20R	15-20P	15-30R	15-30P	15-50R	15-50P	15-60R	15-60P
	18	3ØY 120/208V	18-15R	18-15P	18-20R	18-20P	18-30R	18-30P	18-50R	18-50P	18-60R	18-60P

- Grounding screw terminal: Used to attach the circuit-grounding conductor to the switch. It is green in color.

- Mounting ears: Used to secure the switch in a device box with two 6-32 size screws.

> **CAUTION**
>
> **CAUTION:** When installing single-pole switches, always mount them so that when the toggle is in the up position, the writing on the toggle says "ON." You can tell when you have installed a single-pole switch upside down because when the switch is in the ON position, it will read as "NO."

Single-pole switches, as well as all other switch types, will have their amperage and voltage ratings written on them. They, like receptacles, may also have push-in terminals on the back.

A switch that is used on 240-volt circuits to control a load from one location is called a double-pole switch (Figure 2-40). It is similar in construction to a single-pole switch but has four terminal screws instead of two. The top two screws are usually labeled as the "line" terminals and have the same color. The bottom two screws are labeled as the "load" terminals and are colored the same but in a color that is different from the top two screw terminations. An installation example of this switch is as a **disconnecting means** for a 240-volt electric water heater. The double-pole switch works like a single-pole switch except that there are two sets of internal contacts that are connected in such a way that when the toggle is in the ON position, both sets of contacts are closed, and when the toggle is in the OFF position, both sets of contacts are open.

Three-way switches are used to control a lighting outlet or outlets from two locations. An installation example would be in a living room that has two doorways. One three-way switch would be located next to one doorway and another three-way switch next to the other doorway. A person entering or leaving the living room can turn the lighting outlet(s) in that room ON or OFF from either doorway. Three-way switches get their name from the fact that they have three screw terminals on them. Figure 2-41 shows the common parts of a three-way switch. There are some characteristics about three-way switches that make them different from single-pole switches. One characteristic, mentioned earlier, is that three-way switches have three terminals. Two of the terminals, called "traveler terminals," typically have the same brass color and are located directly across from each other on opposite sides of the switch. The other screw terminal is usually black in color and is called the "common terminal." Some electricians may refer to this as the "point" or "hinge" terminal. Identifying the common and the traveler terminals will enable you to correctly connect three-way switches in a lighting circuit.

> **CAUTION**
>
> **CAUTION:** The terminals may be located differently on three-way switches made by different manufacturers, but the color coding will be the same. Always look for the two brass terminals, which will typically be the traveler terminals, and the black terminal, which will usually be the common terminal. It does not matter where they are located on the switch.

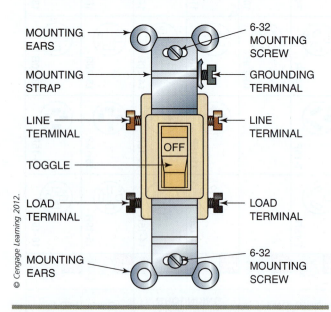

FIGURE 2-40   The parts of a double-pole switch.

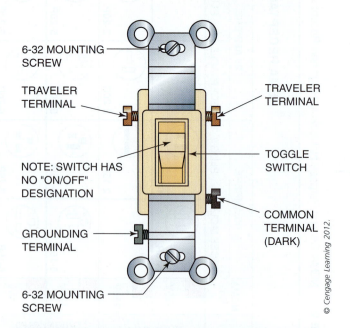

FIGURE 2-41   The parts of a three-way switch.

Another distinguishing characteristic of a three-way switch is that the toggle, unlike a single-pole switch, does not have any ON or OFF position written on it. In other words, there is no up or down on a three-way switch, and it does not make any difference which way the switch is mounted in a device box.

Three-way switches must always be installed in pairs, and a three-wire cable or three wires in a raceway must always be run between the two three-way switches for proper connection.

Four-way switches are used in conjunction with three-way switches to allow control of a lighting outlet or outlets from more than two locations. An installation example would be in a room with three doorways. A three-way switch would be located at two of the doorways and a four-way switch at the third doorway. When such switches are wired correctly, a person could turn ON or OFF the room lighting outlet(s) from any of the three locations. The four-way gets its name from the fact that there are two sets or four total screw terminals on the switch (Figure 2-42). Most manufacturers distinguish the two sets of terminals with different colors. One set will typically be black in color, and the other set will be a lighter color, like brass. The two screws in a set are located opposite each other on most four-way switches. Four-way switch terminals are also called "traveler terminals." There is no common terminal as found on a three-way switch. However, like three-way switches, there is no ON or OFF designation written on the toggle of a four-way switch. Therefore, there is no up or down mounting orientation when a four-way is installed in a device box.

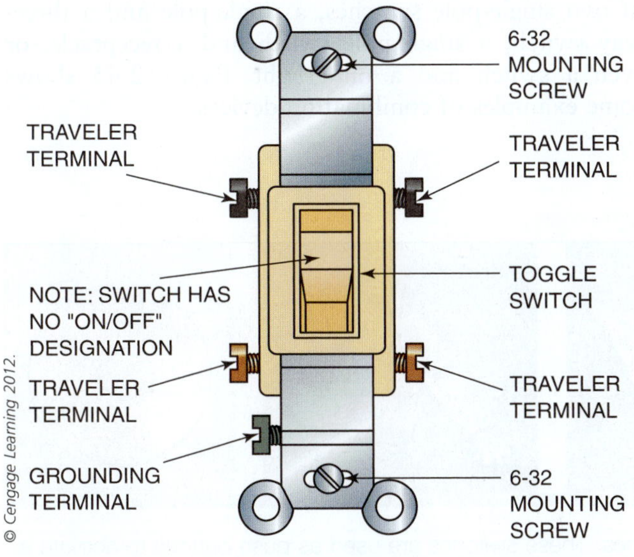

from experience...

Four-way switches are always installed in combination with three-way switches and are wired between them. There is no limit to the number of four-way switches that could be wired between two three-way switches. If a large room had five doorways and switch control of the room lighting outlets was required from each of the doorways, the residential electrician would use two three-way switches and three four-way switches.

Dimmer switches are used to dim or brighten the light output of a lighting fixture. They come in both single-pole and three-way configurations. A single-pole dimmer might be used, for example, to control a large lighting fixture located over a dining room table. Figure 2-43 shows some common styles of dimmer switches. Today's dimmers used in residential applications use electronic circuitry to provide dimming capabilities. They are usually rated at 125 volts and 600 watts, although larger wattage ratings are available.

© Cengage Learning 2012.

6-32 MOUNTING SCREW

TRAVELER TERMINAL

TRAVELER TERMINAL

TOGGLE SWITCH

NOTE: SWITCH HAS NO "ON/OFF" DESIGNATION

TRAVELER TERMINAL

TRAVELER TERMINAL

GROUNDING TERMINAL

6-32 MOUNTING SCREW

FIGURE 2-42 The parts of a four-way switch.

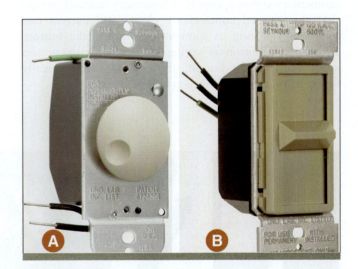

FIGURE 2-43 Examples of dimmer switches. (A) A rotating knob-style dimmer. (B) A sliding knob-style dimmer. Dimmer switches are available as single-pole or three-way models. Both dim or brighten a lamp by varying the applied voltage to the lamp. *Courtesy of Pass & Seymour/Legrand.*

from experience...

The four-way switches made by most of today's manufacturers have vertical configurations as described in this section. However, some manufacturers make four-ways with a horizontal configuration. The residential electrician should always check the color coding and read the instructions that come with the switch to determine the proper switch configuration.

FIGURE 2-45 Examples of combination switches. This switch type has two devices on one strap. Typical combinations include two single-pole switches, a single-pole and a three-way switch, or two three-way switches.

Examples of low-voltage switches used in residential signaling systems are shown in Figure 2-44. A residential electrician would install this switch type next to a front and a rear doorway so that a signal can be sent to a chime that lets people inside the house know that there is someone at the door. This switch type has a push button rather than a toggle and is referred to as a momentary-contact switch. Momentary contact means that the internal contacts are closed and current can flow only as long as someone is pushing the button. Because the button is spring loaded, when a person stops pushing the button, the contacts automatically revert to their normal open position.

COMBINATION DEVICES

Combination devices are used when more than one device is needed at one location. This type of device has a combination of two devices, both of which are mounted on the same strap. It may be a combination of two single-pole switches, a single-pole and a three-way switch, a single-pole switch and a receptacle, or even a switch and a pilot light. Figure 2-45 shows some examples of combination devices.

FIGURE 2-44 Examples of low-voltage momentary-contact switches. These switches are used as push buttons to activate a door chime. *Courtesy of Nutone, Inc.*

from experience...

Sometimes there are locations in residential wiring where two devices are needed but there is not enough room between the studs to mount a two-gang device box to hold the two devices. An example might be in a bathroom where one switch is needed to control a lighting outlet and another switch is needed to control the ventilation fan. There is not enough room at the desired location for a two-gang electrical box to be installed. A common practice to accomplish this installation is to mount a single-gang device box in the narrow space between the two studs and install a combination device with two single-pole switches. This installation will meet the two-device requirement and can be installed in the space available.

FIGURE 2-46 Typical molded-case circuit breakers. The circuit breaker on the left is a single-pole breaker and is used to provide overcurrent protection for 120-volt residential circuits. The circuit breaker on the right is a two-pole breaker and is used to provide overcurrent protection for 240-volt residential circuits.

OVERCURRENT PROTECTION DEVICES

In residential wiring, **overcurrent** protection devices consist of **fuses** or **circuit breakers.** The *NEC*® states that overcurrent protection for conductors and equipment is provided to open the circuit if the current reaches a value that will cause an excessive or dangerous temperature in the conductors or conductor insulation. Recognizing common fuse and circuit breaker types is important for anyone doing residential wiring.

Circuit breakers are the most often used type of overcurrent protection device in residential wiring. A circuit breaker is an automatic overcurrent device that trips into an open position and stops the current flow in an electrical circuit. An **overload, short circuit,** or **ground fault** can cause the circuit breaker to trip. Circuit breakers used in residential wiring are of the thermal/magnetic type. Thermal tripping is caused by an overload. When a larger current flow than the breaker's rating occurs, a **bimetallic strip** in the breaker gets hot and tends to bend. If the larger-than-normal current flow

continues, the temperature of the bimetallic strip increases, and the strip continues to bend. If it bends enough, a latching mechanism is tripped, and the breaker contacts open, causing current flow to stop. When a short circuit or ground fault occurs, circuit resistance is drastically reduced, and large amounts of current flow through the breaker. This high current causes the magnetic part of the breaker to react. A strong magnetic field is created from the high current, which causes a metal bar attached to the latching mechanism to trip, opening the circuit breaker contacts and stopping current flow.

Circuit breakers used in residential wiring are available in single-pole for use on 120-volt circuits and two-pole for use on 240-volt circuits (Figure 2-46). They are rated in both amperes and voltage. Common residential amperage ratings are 15, 20, 30, 40, and 50 amperes. Larger sizes, such as a 100 or 200 amp used as the main service entrance disconnecting means, are available. The voltage rating on residential circuit breakers is usually 120/240 volts. The slash (/) between the lower and higher voltage rating in the marking indicates that the circuit breaker has been tested for use on a circuit with the higher voltage between "hot" conductors (240 V) and with the lower voltage from a "hot" conductor to a grounded conductor (120 V).

Circuit breakers can also be found in ground fault circuit interrupter (GFCI) and arc fault circuit interrupter (AFCI) models. These breakers will provide regular overcurrent protection to circuits and equipment and will also provide GFCI and AFCI protection to the

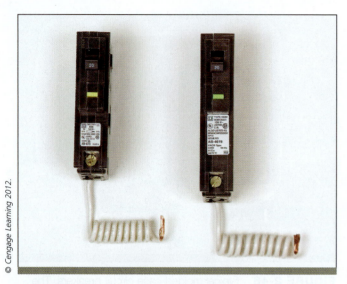

FIGURE 2-47 A GFCI and an AFCI circuit breaker. They are easily recognizable from a regular circuit breaker by the push-to-test button on the front of the breakers and the length of white insulated conductor attached to them.

FIGURE 2-48 A 15-amp Type W Edison-base plug fuse is shown on the left. Notice the hexagonal shape of the area around the window. The hexagonal shape indicates a fuse rated at 15 amps or smaller. The fuse on the right is a 20-amp-rated Type T time delay Edison-base plug fuse. *Courtesy of Cooper Bussman, Inc.*

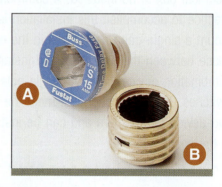

FIGURE 2-49 (A) A 15-amp-rated (blue in color) Type S plug fuse and (B) adapter. A 20-amp (orange in color) or a 30-amp (green in color) Type S fuse will not work in the 15-amp-rated adapter. This type of fuse will help prevent overfusing. *Courtesy of Cooper Bussman, Inc.*

whole circuit. They are easily distinguishable from regular circuit breakers by the test button on the front of the breakers and the length of white wire that is attached to them (Figure 2-47).

Fuses are not often used in residential wiring, but electricians will certainly encounter them from time to time on new installations or when doing remodel work. A fuse is an overcurrent protection device that opens a circuit when the fusible link is melted away by the extreme heat caused by an overcurrent. There are two styles of fuses that you will see on a residential installation: plug fuses and cartridge fuses. Both of these fuse types are available in a time-delay or non-time-delay configuration. Electrical circuits that have electric motors in them require a time-delay fuse. The reason for this is that motors draw a lot of current to get started, but once started and turning their load, they draw their normal current amount. Time-delay fuses allow the motors to start and get up to running speed. If non-time-delay fuses were used on electric motor circuits, the fuse would blow every time the motor tried to start. Circuit breakers used in residential wiring have time-delay capabilities built in.

Plug fuses are broken down into two types: the Edison base and the Type S. The Edison-base plug fuse (Figure 2-48) can be used to replace only existing Edison-base fuses. New installations cannot use Edison-base fuses. This is because all sizes of Edison-base plug fuses have the same base size and will all fit in the same fuse holder. Overfusing can occur with this fuse type when a person puts in a 30-ampere

Edison-base fuse in the fuse holder that should have a 15-ampere fuse. Type S, or safety, fuses (Figure 2-49) are used in all new installations that require plug fuses. They are considered nontamperable and can fit and work only in a fuse holder with the same amperage rating as the fuse. In other words, a 30-ampere Type S fuse cannot be put in a 15-ampere Type S adapter and work. This feature is designed to prevent overfusing of circuits when using plug fuses.

Cartridge fuses are used sometimes in residential wiring. They come in two styles: the ferrule style and the blade-type style. The ferrule style (Figure 2-50) is basically a cylindrical tube of insulating material, like cardboard, with metal caps on each end. Common ampacities for ferrule-type cartridge fuses are 15, 20, 30, 40, 50, and 60 amps. Cartridge fuses with amperage ratings over 60 amps are found in the blade-type style (Figure 2-51). This fuse has a cylindrical body of insulating material with protruding metal blades on each end.

FIGURE 2-50 Examples of ferrule-type cartridge fuses. This fuse style is found in fuse sizes that rated 60 amps and less. *Courtesy of Cooper Bussman, Inc.*

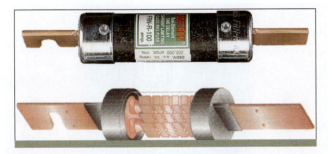

FIGURE 2-51 A blade-type cartridge fuse. This cartridge fuse style is found in fuses that are rated larger than 60 amps. *Courtesy of Cooper Bussman, Inc.*

PANELBOARDS, LOADCENTERS, AND SAFETY SWITCHES

A **panelboard** is defined as a single panel that includes automatic overcurrent devices used for the protection of light, heat, or power circuits. It is designed to be placed in a **cabinet** located in or on a wall, partition, or other support. It is accessible only from the front. A **loadcenter** is a type of electrical panel that contains the main disconnecting means for the residential service entrance as well

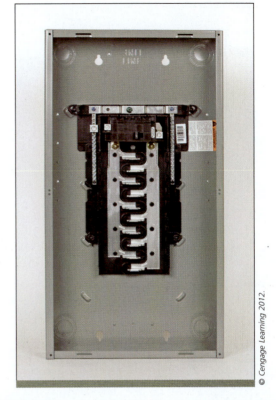

FIGURE 2-52 A typical 120/240-volt, single-phase, three-wire main breaker loadcenter. A cover is installed on this loadcenter after all the connections have been made.

as the fuses or circuit breakers used to protect circuits and equipment like electric water heaters, ranges, dryers, and lighting (Figure 2-52). Most loadcenters are placed inside a dwelling unit and are typically a NEMA Type 1 enclosure, which is the most commonly used type of enclosure in residential wiring, being designed for use indoors under usual service conditions. Sometimes an electrical enclosure, such as a meter enclosure or a loadcenter, must be installed outside. A NEMA Type 3, which is a weatherproof enclosure designed to give protection from falling dirt, rain, snow, sleet, and windblown dust, will need to be used outside. In residential work, a NEMA Type 3R is often used. The "R" stands for "rain-tight" and means that the enclosure will provide the same protection as a NEMA Type 3, except for windblown dust.

Sometimes a residential wiring system has a large concentration of electrical circuits in one area of the house, like a kitchen area. If the kitchen is located some distance from the main loadcenter, the circuits that are run from the main loadcenter to the kitchen can be quite long. This makes them more expensive to install and may result in some low-voltage problems for the kitchen equipment. It is common practice to run a set of larger-current-rated conductors from the main

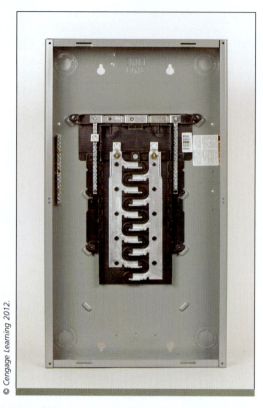

© Cengage Learning 2012.

FIGURE 2-53   A typical 120/240-volt main-lug-only (MLO) loadcenter. This panel does not have a main circuit breaker.

© Cengage Learning 2012.

FIGURE 2-54   Typical 120/240-volt-rated safety switches. The fusible safety switch on the left requires cartridge fuses for 60-amp-rated and larger switches. Lower-rated safety switches will require plug fuses. The nonfusible safety switch on the right is used as a disconnecting means only and does not provide any overcurrent protection.

loadcenter to a smaller loadcenter located closer to the kitchen area. This type of loadcenter is called a "subpanel" (Figure 2-53) and allows shorter runs for the circuits going to the kitchen area. Since a subpanel does not have a main circuit breaker or main set of fuses, it is often referred to as a "main-lug-only" or simply "MLO" panel.

A **safety switch** is used as a disconnecting means for larger electrical equipment found in residential wiring. It is typically mounted on the surface and is operated with an external handle or with a pullout fusible device (Figure 2-54). It can be classified as a fusible type or a nonfusible type. The fusible type usually contains plug fuses in the smaller sizes and cartridge fuses in the larger sizes. The nonfusible safety switch is strictly a disconnecting means and contains no fuses for overcurrent protection. Disconnect switch enclosures can also contain a circuit breaker (Figure 2-55). Safety switch enclosures are available in a variety of NEMA classifications, including NEMA Type 1 and NEMA Types 3 and 3R.

© Cengage Learning 2012.

FIGURE 2-55   A disconnect switch that uses a circuit breaker to provide overcurrent protection.

FASTENERS

There are several different types of fasteners used in residential wiring. They are used to assemble and install electrical equipment. Fasteners used in residential wiring include items such as nails, anchors, screws, bolts, nuts, and tie wraps. This section covers the most common fastener types.

NAILS

The nail is a very common fastener type used in residential wiring. Nails are used for such things as mounting electrical boxes to studs and fastening running boards to the underside of joists, as well as fastening other items used in the installation of a residential electrical system. Nails are made from a variety of materials including aluminum, brass, copper, and steel; however, nails used by electricians are usually made of steel. Sometimes steel nails are coated with zinc to prevent rusting. This type of nail would be suitable for use to fasten items outdoors or in areas where moisture is present. This process is called galvanizing and the nails are called galvanized nails. A special hardened nail is sometimes used to fasten wood materials to masonry walls and is called a masonry nail.

Nails are available in many different styles including roofing, finish, spiral shank, box, common, and masonry (Figure 2-56). The style used most often by electricians is the common nail. Nails are typically sized based on their length and diameter. The penny system is used to designate the different sizes. The symbol for penny is "d." For example, a reference of 16d means that it is a 16 penny nail with a length of 3½ inches. Common penny (d) sizes for nails are shown in Figure 2-57.

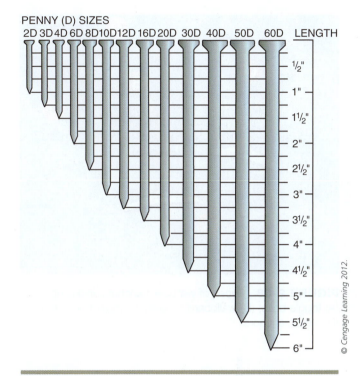

FIGURE 2-57 Nails are usually sized according to the penny system.

CAUTION

CAUTION: Always wear eye protection when driving nails. Pieces of the nail can break off while the nail is being hit with a hammer and fly through the air into your eyes. Also, watch out for nails that fly out when you first start to hammer them in.

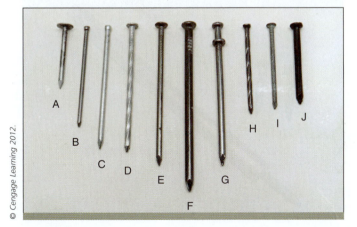

FIGURE 2-56 Kinds of commonly used nails: (A) roofing, (B) finish, (C) galvanized finish, (D) galvanized spiral shank, (E) box, (F) common, (G) duplex, (H) spiral shank, (I) coated box, and (J) masonry.

ANCHORS

Anchors are used to mount electrical boxes and enclosures on solid surfaces like concrete as well as hollow wall surfaces like wallboard. Some common anchors used in residential wiring to mount items on wallboard are toggle bolts, wallboard anchors, and molly bolts (also called a sleeve-type anchor) (Figure 2-58). Plastic anchors are sometimes used, but care must be taken not to mount very heavy items with them since they pull out of a sheetrock wall quite easily. Make sure to follow the manufacturer's recommendations when installing anchors in hollow walls and ceilings.

(See Procedure 2-4 on pages 75–76 for the procedure for installing toggle bolts in a hollow wall or ceiling.)

Plastic anchors, lead (caulking) anchors, and drive studs (Figure 2-59) are used to mount electrical enclosures and equipment to masonry surfaces in residential applications. Plastic and lead anchor installation requires drilling a hole with a masonry bit and inserting the anchor. Drive studs are installed using a powder-actuated fastening system that uses a tool specifically designed to use the force of a gunpowder load to drive

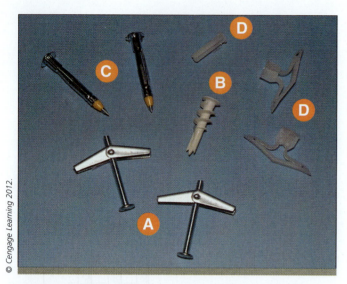

© Cengage Learning 2012.

FIGURE 2-58 Types of wallboard anchors, including (A) toggle bolts, (B) wallboard screws, (C) molly bolts, and (D) plastic anchors.

© Cengage Learning 2012.

FIGURE 2-59 Types of masonry anchors, including (A) plastic anchors, (B) caulking anchors, (C) drive studs used with (D) a powder-actuated fastening system, (E) lead anchors, and (F) TapCons.

a fastener into concrete or steel. The depth to which the stud or pin is driven depends on the strength of the gunpowder load and the density of the material you are installing on. The different powder loads are color coded, and it depends on the specific manufacturer as to what color code is being used. Consult the manufacturer's instructions.

(See Procedure 2-5 on page 77 for the procedure for installing a lead [caulking] anchor in a concrete wall.)

(See Procedure 2-6 on page 78 for the procedure for installing plastic anchors.)

CAUTION

CAUTION: OSHA 29 CFR 1926.302(e) governs the use of powder-actuated fastening systems and requires special training and certification before an electrician is allowed to use this type of fastening system. If an OSHA inspector catches you on a jobsite using a powder-actuated tool without a license to operate it, you will likely be made to stop working and be escorted off the jobsite. Usually, a manufacturer of this type of system will provide training and a test you can take for a license to operate their powder-actuated fastener.

Many electricians use masonry screw anchors to mount items on concrete, block, and brick walls. This is a system that uses a steel fastener that screws directly into the masonry and eliminates the need for caulking anchors (Figure 2-60). The masonry screws are often installed with a special tool. The tool is used with an electric drill motor and has a masonry bit that drills a proper size hole in the concrete. The masonry bit is then retracted back into the tool, and the masonry screw is inserted in the tool's hexagonal socket head and driven into the previously drilled hole. It is an easy two-step installation, and this quick-and-easy system makes this a very popular fastening system for mounting items on masonry. Electricians commonly refer to these masonry screws as "TapCons." TapCon® is actually a registered trademark for the masonry screws made by Greenlee Textron.

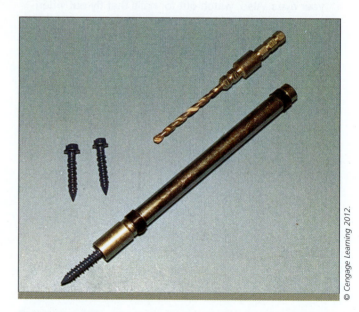

© Cengage Learning 2012.

FIGURE 2-60 A popular anchor type used for securing electrical equipment to a masonry wall is a masonry screw. Electricians refer to these screws as "TapCons."

(See Procedure 2-7 on pages 79–80 for the procedure for installing TapCons in a masonry wall.)

SCREWS

Screws used to mount enclosures to wood surfaces include wood screws, sheet metal screws, and sheetrock screws (Figure 2-61). Sheet metal screws work very well because the thread extends for the full length of the screw. Sheetrock screws work well for the same reason, and they are also very easy to use with a screw gun, which saves a lot of time and energy versus having to screw them in by hand. Mounting items on metal studs or on other thin metal surfaces is accomplished using sheet metal screws or self-drilling screws (sometimes called "Tek-Screws") (Figure 2-62).

Bolts and machine screws are types of threaded fasteners used in residential wiring. Bolts are rarely used, but residential electricians use machine screws all the time. For this reason, we keep our discussion of threaded fasteners focused on machine screws. Machine screws are used to attach devices to electrical boxes, to secure covers to devices, and to secure covers to junction boxes. The most common sizes are 6-32, 8-32, and 10-32. The 6-32 screw size is used to attach switches

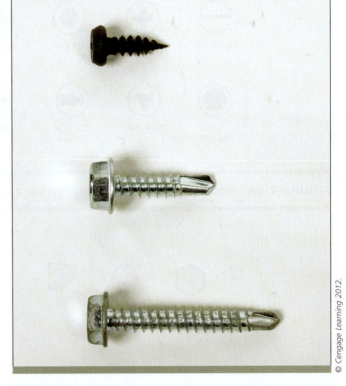

FIGURE 2-62 Self-drilling screws, or "Tek-Screws" as they are called by some electricians, are used to secure electrical equipment to metal studs or other sheet metal surfaces.

and receptacles to electrical device boxes, and 8-32 screws are used to attach covers to octagonal and square electrical boxes; a 10-32 machine screw is the size of the green grounding screw used to secure an equipment-grounding conductor to an electrical box. Machine screws are available in either a slotted or a Phillips head and come in different lengths. Other common screwhead shapes used in residential wiring applications are shown in Figure 2-63.

NUTS AND WASHERS

Many applications require the use of nuts and washers with threaded fasteners. Nuts are found in either a square or hexagonal shape and are threaded onto a bolt or machine screw to help secure an electrical item in place. Washers fit over a bolt or screw and distribute the pressure of the bolt or screw over a larger area. They also provide a larger area for the bolt head or nut to tighten against when the fastener is being used in a hole that is larger than the diameter of the fastener. Lock washers are used to help keep bolts or nuts from working loose. Figure 2-64 shows several of the most common nuts and washers used in residential work.

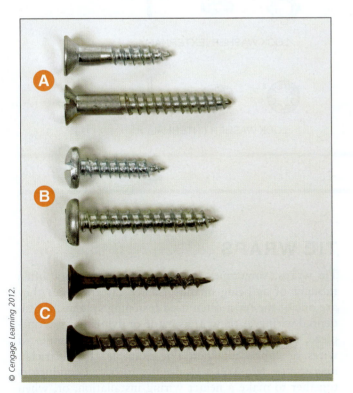

FIGURE 2-61 Examples of various fasteners for securing electrical equipment to a wood surface include: (A) wood screws, (B) sheet metal screws, and (C) sheetrock screws.

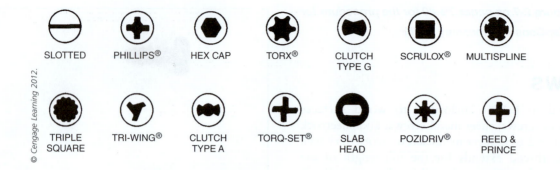

© Cengage Learning 2012.

FIGURE 2-63   Examples of the different heads found on machine screws.

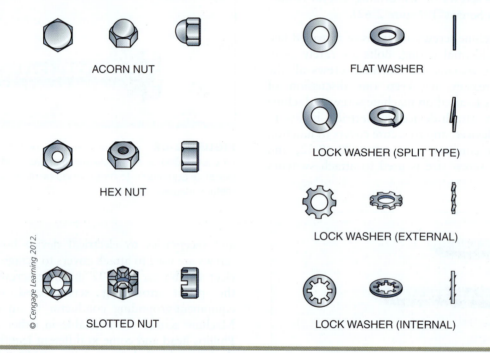

© Cengage Learning 2012.

FIGURE 2-64   Examples of common nuts and washers.

TIE WRAPS

Tie wraps, sometimes called cable ties, are not often thought of as being in the fastener category, but they are one of the most often used fastening systems in residential wiring. A tie wrap is made of nylon and is a one-piece, self-locking device used to fasten a bundle of wires or cables together. For example, an electrician may use tie wraps in a loadcenter to fasten conductors together to make a neater wiring installation, or, when several Romex™ cables are bundled together, tie wraps can be used to hold them together. Tie wraps designed for outdoor use are often black in color and are made to resist the harmful effects of sunlight and outdoor

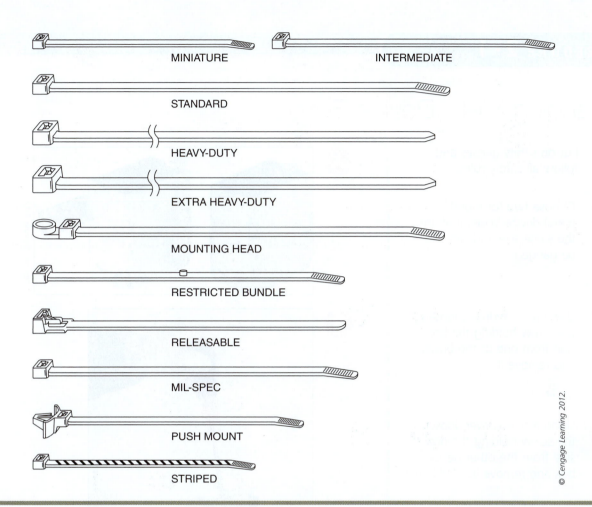

MINIATURE

INTERMEDIATE

STANDARD

HEAVY-DUTY

EXTRA HEAVY-DUTY

MOUNTING HEAD

RESTRICTED BUNDLE

RELEASABLE

MIL-SPEC

PUSH MOUNT

STRIPED

© Cengage Learning 2012.

FIGURE 2-65   Cable ties, or tie wraps as they are sometimes called, are used often to secure wires and cables. They are available in many styles and sizes.

atmospheric conditions. Indoor-use tie wraps come in a variety of colors and styles. Figure 2-65 shows some common cable tie examples.

SUMMARY

It is very important to be familiar with the correct names and applications for electrical hardware and materials. Using the proper name and selecting the right part for the job helps avoid confusion and will help ensure that the electrical system is properly installed.

There is a wide variety of hardware and materials used in residential electricity. This chapter covered only the most common items. You will run into other hardware and material types as you continue your residential electrical career. However, they will be similar to the items discussed in this chapter, and by following the manufacturer's instructions and using your newly acquired hardware and material knowledge, you should have no problem when using them.

PROCEDURE 2-1

Ganging Metal Device Boxes

- Put on safety glasses and follow all safety rules.

A Choose two (or more) metal device boxes that are the same size and can be ganged.

B Using a screwdriver, loosen the screw holding the box side from one of the boxes and remove it.

C Using a screwdriver, loosen the screw holding the box side from the other device box and remove it.

D Align the two boxes together and tighten all screws to hold the boxes securely together.

- Put all tools and materials away.

A

© Cengage Learning 2012.

B

© Cengage Learning 2012.

C

© Cengage Learning 2012.

D

© Cengage Learning 2012.

PROCEDURE 2-2

Installing Aluminum Conductors

- Observe proper safety procedures and always follow the manufacturer's installation instructions.

A Strip off the required amount of conductor insulation on the end you wish to terminate.

© Cengage Learning 2012.

B Apply a liberal amount of antioxidant to the newly exposed aluminum.

© Cengage Learning 2012.

C Scrape the conductor end with a wire brush. This breaks down the oxidation layer, and the antioxidant keeps the air from contacting the conductor, preventing further oxidation.

- Attach the conductor to the termination and tighten to the proper torque requirements.

- Clean up any excess antioxidant.

© Cengage Learning 2012.

PROCEDURE 2-3

Connecting Wires Together with a Wirenut

- Put on safety glasses and observe all applicable safety rules.

- Using a wire stripper, remove approximately 3/4 inch of insulation from the ends of the wires.

A Place the ends of the wires together and then place the wirenut over the ends of the conductors. When splicing two wires together, there is no need to twist the wires together before putting on the wirenut. When splicing three or more wires together, it is a good practice to twist the wires together before putting on the wirenut.

B While exerting a slight inward pressure on the wirenut, twist the wirenut in a clockwise direction. Continue to twist the wirenut until it is tight on the wires.

C Test the splice by pulling on the conductors. If any of the wires come loose, re-do the splice. The completed splice should be secure and have no bare wire showing below the bottom of the wirenut.

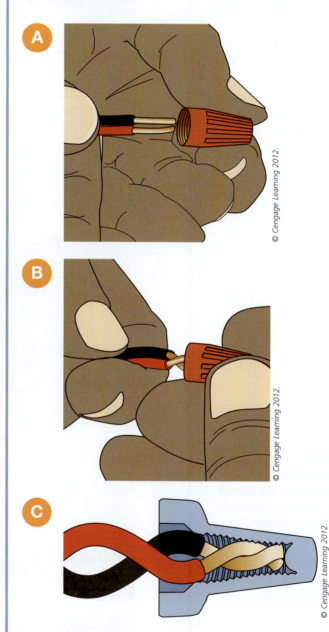

A

© Cengage Learning 2012.

B

© Cengage Learning 2012.

C

© Cengage Learning 2012.

PROCEDURE 2-4

Installing Toggle Bolts in a Hollow Wall or Ceiling

- Put on safety glasses and observe all applicable safety rules.

- Mark the location on the wall or ceiling where the toggle bolt will be installed.

A Using the proper size drill bit or punch tool, drill or punch a hole completely through the wall or ceiling in the desired location.

- Insert the toggle bolt (without the wings) through the hole of the item you are mounting.

- Thread the toggle wing onto the end of the toggle bolt, making sure that the wing can fold back toward the bolt head.

B While folding the wings back, insert the toggle bolt into the hole until the spring-loaded wing opens up.

A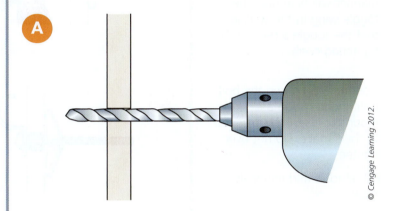

© Cengage Learning 2012.

B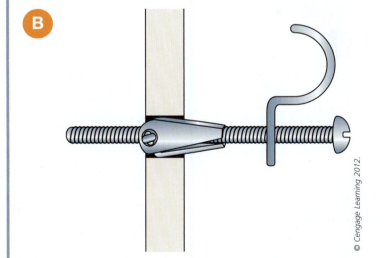

© Cengage Learning 2012.

PROCEDURE 2-4

Installing Toggle Bolts in a Hollow Wall or Ceiling (Continued)

C Tighten the toggle bolt with a screwdriver until the mounted item is secure. Pulling back on the item to be mounted in a careful but firm manner will help hold the toggle wing in the wall so that the toggle screw can be tightened easily.

D Continue to tighten the bolt until the item you are mounting is securely fastened to the surface. Do not over tighten.

- Put all tools and materials away.

C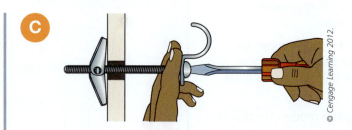

© Cengage Learning 2012.

D

© Cengage Learning 2012.

Installing a Lead (Caulking) Anchor

- Put on safety glasses and observe all applicable safety rules.

- Mark the location on the wall where the caulking anchor will be installed.

A Using the proper size masonry drill bit as found in the manufacturer's instructions, drill a hole to the required depth (usually the depth of the anchor itself for a flush installation) in the desired location.

B Insert the anchor, with the conical threaded end first, and tap it in with a hammer until it is flush with the surface.

C Using a setting tool, hit the lead outer sleeve with sharp blows.

D Position the item to be mounted and insert a screw or bolt into the caulking sleeve and tighten.

- Put all tools and materials away.

E **Note:** For installing caulking anchors in hollow masonry walls like cinder block or for setting all anchors at the same depth no matter how deep the initial hole was drilled, a screw anchor expander tool is used.

A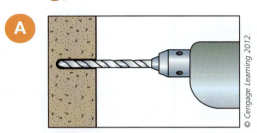

© Cengage Learning 2012.

B

© Cengage Learning 2012.

C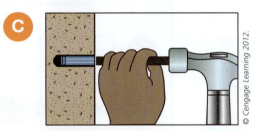

© Cengage Learning 2012.

D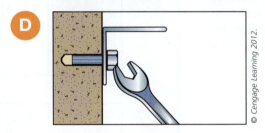

© Cengage Learning 2012.

E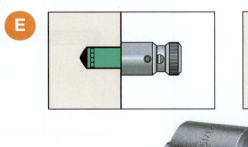

© Cengage Learning 2012.

PROCEDURE 2-6

Installing Plastic Anchors

- Put on safety glasses and observe all applicable safety rules.

A Using the proper masonry bit size, drill a hole to the depth of the anchor or up to 1/8-inch deeper.

A
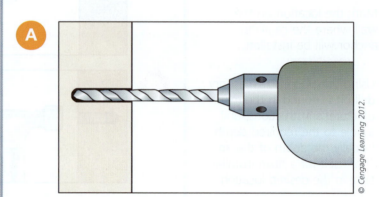

B With a hammer or other appropriate setting tool, tap the plastic anchor into the hole until it is flush with the surface.

- Insert a screw of the size designed to be used with the anchor through the hole in the item to be mounted and into the plastic anchor.

B
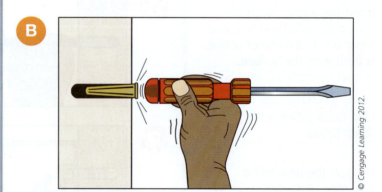

C Tighten the screw until the item being mounted is secure. Be careful not to overtighten.

- Put all tools and materials away.

C
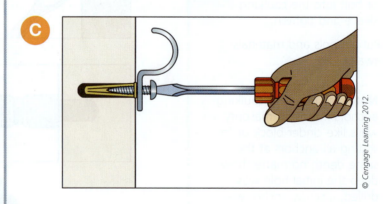

PROCEDURE 2-7

Installing Tapcon Masonry Screws

- Put on safety glasses and observe all applicable safety rules.

- Determine the length of screw needed. To do this, measure the thickness of the material you are securing and add 1 to 1.75 inches.

 Using a hammer drill and properly sized masonry drill bit, drill a hole in the masonry at the spot where you want to install the masonry screw. The hole should be ½ inch deeper than the length of screw that will be screwed into the hole.

- Clean the masonry dust and debris from the hole using a brush, compressed air, or vacuum.

© Cengage Learning 2012.

PROCEDURE 2-7

Installing Tapcon Masonry Screws (Continued)

 Insert the pointed end of the masonry screw through a hole in the item you are fastening and then into the hole you previously drilled in the masonry.

© Cengage Learning 2012.

C Using a drill, drive masonry screw slowly into the hole allowing the threads to tap the masonry.

- Continue to tighten until the item is secure.

- Put all tools and materials away.

C

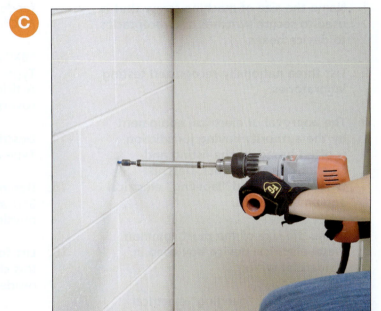

© Cengage Learning 2012.

REVIEW QUESTIONS

Directions: Answer the following items with clear and complete responses.

1. Name the size of machine screw that is used to secure switches and receptacles to device boxes.

2. List three nationally recognized testing laboratories.

3. The approval of electrical equipment by the authority having jurisdiction is often based on the listing and labeling of the product. Describe what "listing" and "labeling" electrical equipment means.

4. Name the size of the most common metal electrical device box used in residential wiring.

5. Describe what "ganging" electrical device boxes is and what purpose ganging electrical device boxes serves.

6. Discuss when an electrician would use a "new work" box and an "old work" box.

7. An electrician installs a 4-inch square box in a location that requires a duplex receptacle. Describe what must be done so that the duplex receptacle can be installed in the square box.

8. Name and describe three cable types that can be used in residential electrical installations.

9. Conductors used in residential wiring are selected for a specific circuit according to their ampacities. Define the term "ampacity."

10. Name the system that is used to size the conductors used in electrical work.

11. Name the insulation colors of the wires found in a three-wire, nonmetallic sheathed cable.

12. Type UF cable is used in _____ installations for branch circuits and

feeders. Article_____ of the *NEC®* covers Type UF cable.

13. "BX" cable is the trade name used for Type _____ cable. Article _____ of the *NEC®* covers this cable type.

14. Describe how an electrician can tell Type AC and Type MC cable apart.

15. Describe the difference between Type SEU and Type SER cable in terms of physical characteristics.

16. List four raceway types described in this chapter that could be used in residential wiring.

17. Name the trade name used for flexible metal conduit.

18. Define the term "receptacle."

19. Describe when three-way switches would be used in a residential lighting circuit.

20. Describe a combination device.

21. List the two types of plug fuses.

22. A single-pole circuit breaker is used on a circuit voltage of _____ volts, and a two-pole circuit breaker is used on a circuit voltage of _____ volts in residential wiring.

23. Describe a loadcenter.

24. An electrician needs to mount a chime on a sheetrock wall. Name three anchor types described in this chapter that could be used.

25. Describe a powder-actuated fastening system.

Tools Used in Residential Wiring

OBJECTIVES   *Upon completion of this chapter, the student should be able to:*

- Identify common electrical hand tools and their uses in the residential electrical trade.

- Identify common specialty tools and their uses in the residential electrical trade.

- Identify common power tools and their uses in the residential electrical trade.

- List several guidelines for the care and safe use of electrical hand tools, specialty tools, and power tools.

- Demonstrate an understanding of the procedures for using several common hand tools, specialty tools, and power tools.

GLOSSARY OF TERMS

auger a drill bit type with a spiral cutting edge used to bore holes in wood

bender a tool used to make various bends in electrical conduit raceway

chuck a part of a power drill used for holding a drill bit in a rigid position

chuck key a small wrench, usually in an L or T shape, used to open or close a chuck on a power drill

crimp a process used to squeeze a solderless connector with a tool so that it will stay on a conductor

cutter a hardened steel device used to cut holes in metal electrical boxes

die the component of a knockout punch that works in conjunction with the cutter and is placed on the opposite side of the metal box or enclosure

hydraulic a term used to describe tools that use a pressurized fluid, like oil, to accomplish work

knockout punch a tool used to cut holes in electrical boxes for the attachment of cables and conduits

level perfectly horizontal; completely flat; a tool used to determine if an object is level

plumb perfectly vertical; the surface of the item you are leveling is at a right angle to the floor or platform you are working from

reciprocating to move back and forth

strip to damage the threads of the head of a bolt or screw

tempered treated with heat to maximize a metal's hardness

torque the turning or twisting force applied to an object when using a torque tool; it is measured in inch-pounds or foot-pounds

wallboard a thin board formed from gypsum and layers of paper that is used often as the interior wall sheathing in residential applications; commonly called "sheetrock"

Residential electricians must become skilled in selecting and using the right tool for the job. There are many common hand tools and specialty tools that are used to install electrical systems. Power tools come in various styles and are used often by residential electricians. In this chapter, we take a look at the common hand tools, specialty tools, and power tools that residential electricians use on a regular basis.

COMMON HAND TOOLS

Using hand tools to install a residential electrical system can be hazardous. Injuries to yourself and others can result if you are not careful. It is very important for electricians to be aware of some guidelines for the care and safe use of electrical hand tools. The following points should be considered whenever you are using hand tools:

- Use the correct tool for the job. Using a tool that is not designed for the job you are trying to do may result in serious damage to you as well as to the equipment you are working on.

- Keep all cutting tools sharp. It is a fact that more injuries result from dull cutting tools than from sharp ones. The reason is simple: Dull cutting tools require much more force to do their job. With the extra force being applied, a tool can more easily slip and cause serious damage to you or the equipment you are working on.

- Keep tools clean and dry. Hand tools that are kept clean simply work better. When they are free of rust and dirt, they open and close better and are easier to grip.

- Lubricate tools when necessary. This goes along with keeping the tools clean and dry. The lubrication will allow the tools with a hinged joint to work easily.

- Inspect tools frequently to make sure they are in good condition. Do not use tools that are not in good working condition. A tool could malfunction and cause serious damage. An example would be using a hammer with a loose head. The head could fly off, striking someone and causing serious injury.

- Repair damaged tools promptly and dispose of broken or damaged tools that cannot be repaired. Some hand tools are relatively easy to repair. For example, a new wooden handle can easily replace a broken wooden handle on a hammer. However, it has been this author's experience that most electrical hand tools cannot be easily repaired and need to be disposed of. Only after buying a new tool will you be assured of the tool working in a safe and reliable manner.

- Store tools properly when not in use. Cutting tools especially need to be protected so that their cutting edges will not be dulled and, more important, so that their cutting edges cannot injure someone.

- Pay attention when using tools. Using a hand tool is serious business. Always pay strict attention to what you are doing.

- Wear eye protection when necessary. This is a no-brainer. Using hand tools may cause pieces of wire to fly through the air. If you are not wearing proper eye protection, the pieces can easily fly into your eye.

There are many different types of hand tools used in the residential electrical trade. Hand tools are commonly referred to as "pouch" tools because of their ability to fit easily into an electrical tool pouch. The most common hand tools used by a residential electrician to install electrical systems are outlined in the following paragraphs.

SCREWDRIVERS

Screwdrivers are available in a variety of styles, depending on the type of screw that it is designed to fit (Figure 3-1). The styles used most often in residential electrical work include the Keystone tip, cabinet tip, and Phillips-tip screwdrivers. The Keystone- and cabinet-tip screwdrivers (Figure 3-2) are used to remove and install slot-head screws and to tighten and loosen slot-head lugs.

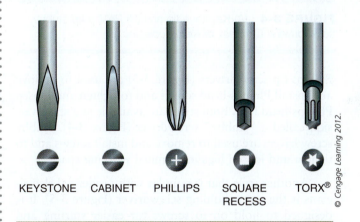

KEYSTONE   CABINET   PHILLIPS   SQUARE RECESS   TORX®

© Cengage Learning 2012.

FIGURE 3-1 Screwdrivers are available with many different tip styles. Keystone, cabinet, and Phillips tip screwdrivers are used the most in residential electrical work. Other screwdriver tip styles that are used sometimes in residential electrical work include the square recess tip (also called a Robertson tip) and TORX® tip.

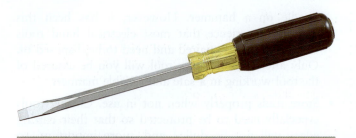

FIGURE 3-2   A Keystone tip screwdriver with a comfort grip. *Courtesy of IDEAL INDUSTRIES, INC.*

FIGURE 3-3   A Phillips tip screwdriver with a number 2 head. *Courtesy of IDEAL INDUSTRIES, INC.*

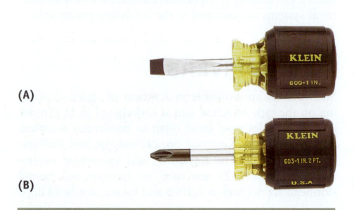

(A)

(B)

FIGURE 3-4   (A) Keystone tip and (B) Phillips tip stubby screwdrivers. *Courtesy of Klein Tools, Inc.*

Phillips-tip screwdrivers (Figure 3-3) are used to remove and install Phillips-head screws and to tighten and loosen Phillips-head lugs. Both styles are available in a short version called a "stubby" screwdriver (Figure 3-4). Stubby screwdrivers are used to remove and install screws and to tighten and loosen lugs in a limited working space.

Another style that is popular with residential electricians is the screw-holding screwdriver (Figure 3-5). It is designed to hold on to screws for easier starting and setting of screws in hard to reach places. This style works well when you have to install a grounding screw in the bottom of a 3 × 2 × 3½-inch (75 × 50 × 90 mm) device box. The depth of the box makes it very difficult to reach to the bottom and start the screw with your fingers. Using a holding screwdriver lets you start the screw in the bottom of the box very easily.

FIGURE 3-5   A screw-holding screwdriver. *Courtesy of IDEAL INDUSTRIES, INC.*

Remember that this screwdriver is designed just to start screws and that you need to use a regular screwdriver to finish tightening the screw.

from experience...

Many residential electricians use ratcheting screwdrivers that utilize multiple tips (Figure 3-6). This allows an electrician to carry just one screwdriver because changing the tips allows them to work on different headed screws. Ratcheting screwdrivers also result in less finger and hand fatigue than traditional screwdrivers. Some residential electricians also like to use a quick-rotating screwdriver (Figure 3-7). This type of screwdriver is great for quickly tightening or loosening screws in switches, receptacles, and lighting fixtures. Quick-rotating screwdrivers are available in both a Keystone tip and Phillips tip style.

FIGURE 3-6   A ratcheting screwdriver that uses multiple tips. *Courtesy of IDEAL INDUSTRIES, INC.*

FIGURE 3-7 A quick-rotating screwdriver. *Courtesy of IDEAL INDUSTRIES, INC.*

Each screwdriver type has three parts: the handle, the shank, and the tip (Figure 3-8). The handle is designed to give you a firm grip. The shank is made of hardened steel and is designed to withstand a high twisting force. The tip is the end that fits into a specific screw head. A good screwdriver tip is made of **tempered** steel to resist wear and prevent bending and breaking.

> **CAUTION**
>
> **CAUTION:** The comfort-grip handles on screwdrivers used in electrical work are designed for comfort and to provide a secure grip. They are not designed to act as an insulator from electrical shock and are not to be used on "live" electrical equipment.

It is important to have the screwdriver tip fit snugly in the screw head, as using the wrong size screwdriver can cause the head of the screw to **strip**. Screwdrivers can be dangerous to use. They are very sharp, so do not point the tip toward yourself or anyone else and do not carry them in your back pocket, especially with the tip sticking up toward your back. To make sure that a screwdriver does not become damaged from misuse, do not use it as a chisel, punch, or pry bar.

(See Procedure 3-1 on pages 106–107 for the correct procedure for using a screwdriver.)

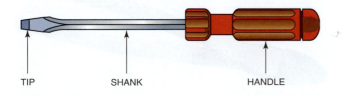

TIP   SHANK   HANDLE

FIGURE 3-8 The three parts of a screwdriver are the tip, the shank, and the handle.

© *Cengage Learning 2012.*

PLIERS

A residential electrician uses many different pliers styles. These are considered pouch tools, and the most common styles include lineman pliers, long-nose pliers, diagonal cutting pliers, and pump pliers.

Lineman pliers, sometimes referred to as "sidecutter" pliers, are used to cut cables, conductors, and small screws; to form large conductors; and to pull and hold conductors. It is important to use a pliers that is large enough for the job. Typically, the handles should be around 9 inches in length so that a minimum of hand pressure is required to cut the conductor or cable. In most cases, one hand is all that is needed to make the cut, although at times, when the conductor or cable is fairly large, two hands may be required to provide enough pressure to make the cut. Lineman pliers are available in a variety of styles and sizes, but the 9¼-inch handle with a New England nose is probably the best style for residential electrical work (Figure 3-9). Like all of the pliers used in electrical work, the handles are covered with vinyl or some other material designed for comfort and to provide a secure grip.

(See Procedure 3-2 on page 108 for the correct procedure for using lineman pliers to cut cable.)

> **CAUTION**
>
> **CAUTION:** The covering on the handles of electrical pliers is designed for comfort and to provide a secure grip. It is not designed to function as an insulator from electrical shock, so do not use pliers with regular grips on "live" circuits.

Long-nose pliers, sometimes called "needle-nose" pliers, are used to form small conductors, to cut conductors, and to hold and pull conductors (Figure 3-10). The

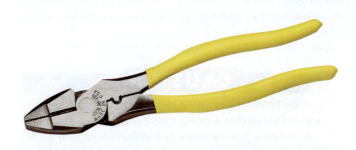

FIGURE 3-9 Lineman pliers with 9¼-inch handles and a New England nose with a crimping die is a good choice for a residential electrician. *Courtesy of IDEAL INDUSTRIES, INC.*

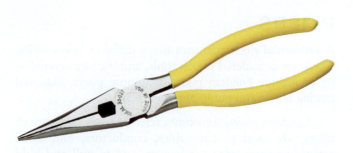

FIGURE 3-10　Long-nose pliers with 8½-inch handles. *Courtesy of IDEAL INDUSTRIES, INC.*

FIGURE 3-11　Diagonal cutting pliers with 8-inch high-leverage handles and an angled head. *Courtesy of IDEAL INDUSTRIES, INC.*

FIGURE 3-12　Pump pliers with 10-inch handles. This tool is sometimes called "tongue-and-groove pliers." *Courtesy of Klein Tools, Inc.*

FIGURE 3-13　Cable cutter pliers. These pliers can cut cables without the distortion of the cable ends that you can get when cutting a cable with lineman pliers. *Courtesy of IDEAL INDUSTRIES, INC.*

FIGURE 3-14　Crimping pliers. These pliers are used to crimp bare or insulated terminals and splices. *Courtesy of IDEAL INDUSTRIES, INC.*

narrow head allows working in tight areas. Residential electricians should use a long-nose pliers with at least an 8-inch handle size.

Diagonal cutting pliers, sometimes called "dikes," are used to cut cables and conductors in limited spaces (Figure 3-11). Electricians can cut conductors off much closer to the work with this pliers type than with other cutting pliers types. Diagonal cutting pliers are available in a straight-head or an angled-head design.

Pump pliers, often called "channel-lock pliers" or "tongue-and-groove" pliers, are used to hold and tighten raceway couplings and connectors and to hold and turn conduit and tubing (Figure 3-12). They can also be used to tighten locknuts and cable connectors. Because of their unique design, the gripping heads can be adjusted to several different sizes.

There are two other types of pliers that some residential electricians use. The first one is cable cutter pliers (Figure 3-13). This type of pliers can cut cables without the distortion of the cable ends that you can get when cutting a cable with lineman pliers. They also have the ability to cut larger sizes of conductors and cables than a lineman plier. The second type is crimping pliers (Figure 3-14). This type of pliers is used to **crimp** bare or insulated terminals and splices. Crimping pliers work great when an electrician is using a crimp sleeve to connect several grounding conductors together in an electrical box.

CAUTION

CAUTION: Never use a regular pair of pliers to squeeze together a crimp type connector or splice. A crimping pair of pliers or lineman pliers with a crimping feature must be used. If the crimp type connection or splice is not made with a proper crimping tool, an electrical inspector is likely to not pass your installation.

WIRE AND CABLE STRIPPERS

Wire strippers are used to strip insulation from conductors and to cut and form conductors. They are available in several different styles. Probably the most popular style of wire stripper used in residential electrical work is the nonadjustable type, commonly called a "T-stripper" (Figure 3-15). The T-stripper is designed to strip the insulation from several different wire sizes without having to be adjusted for each size. A good choice for residential electricians is a wire stripper designed to strip the insulation from 10 through 18 AWG solid and 12 through 20 AWG stranded, which are the wire sizes used most often in residential electricity.

(See Procedure 3-3 on page 109 for the correct procedure for stripping the insulation from a conductor using a wire stripper.)

Cable rippers are used to more easily remove the outside sheathing from a nonmetallic sheathed cable (Figure 3-16). They are designed to slit the cable sheathing. The electrician then uses a knife or cutting pliers to cut off the sheathing from the cable.

FIGURE 3-15 A T-stripper-style wire stripper. This model strips 10- to 18-AWG solid wire and 12- to 20-AWG stranded wire. *Courtesy of IDEAL INDUSTRIES, INC.*

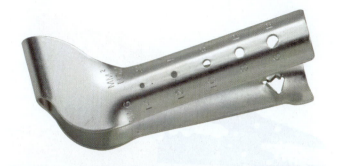

FIGURE 3-16 Cable rippers are used to slit the insulation for stripping on Romex™-type cables. The standard model requires inserting the cable through the back, while the side-entry model allows the electrician to place the ripper onto the cable at the point to be stripped. *Courtesy of IDEAL INDUSTRIES, INC.*

from experience...

Electrical conductors are available in either a solid or a stranded configuration. Stranded wire has a slightly larger diameter than does solid wire. For this reason, wire strippers that are designed to strip solid wire will "catch" on stranded wire and not strip the insulation off smoothly. Manufacturers make wire strippers for both solid and stranded wire. In the past, manufacturers made yellow-handled strippers for solid wire and red-handled strippers for stranded wire. Now manufactures make their wire strippers so that the stripping holes on one side are labeled with solid wire sizes and the holes on the other side are labeled with stranded wire sizes. This allows an electrician to use only one tool to strip insulation from both solid and stranded conductors.

A multipurpose tool, sometimes called a "six-in-one tool," is extremely versatile. (Figure 3-17). As its name implies, it is designed to do several different things such as strip insulation, cut conductors, crimp smaller solderless connectors, cut small bolts and screws, thread small bolts and screws, and can be used as a wire gauge.

An electrician's knife is a valuable tool and is used often by residential electricians. It is used to open cardboard boxes containing electrical equipment and to strip

FIGURE 3-17 This multipurpose tool is able to do seven things: strip 6- to 16-AWG stranded wire, crimp solderless connectors, cut wire, cut small bolts and screws, reform threads on small bolts and screws, ream and deburr conduit, and gauge wire. *Courtesy of IDEAL INDUSTRIES, INC.*

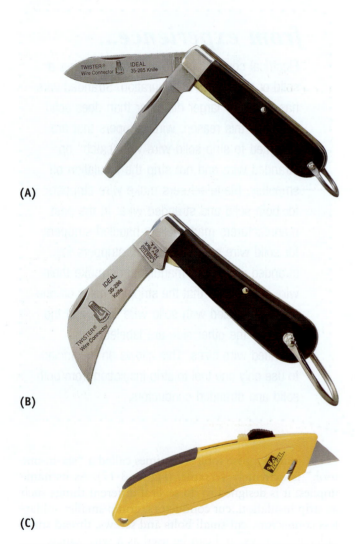

(A)

(B)

(C)

FIGURE 3-18 (A) An electrician's knife, (B) a hawkbill knife, and (C) a utility knife. *Courtesy of IDEAL INDUSTRIES, INC.*

large conductors and cables. Some models of an electrician's knife include both a cutting blade and a screwdriver blade that can be used to tighten and loosen small screws. Many electricians prefer a knife with a curved blade, called a "hawkbill" knife. Others prefer using a utility knife with a retractable blade. Figure 3-18 shows some examples of knives used in the residential electrical trade.

(See Procedure 3-4 on pages 110–111 for the procedure for stripping insulation from Type NM cable using a knife.)

(See Procedure 3-5 on page 112 for the procedure for stripping insulation from large conductors using a knife.)

OTHER COMMON HAND TOOLS

In addition to screwdrivers, pliers, and wire strippers, there are a few other important hand tools that a residential electrician should be familiar with.

A tap tool is used to tap holes for securing equipment to metal, enlarging existing holes by tapping for larger screws, re-tapping damaged threads, and determining screw sizes. The most common tap tool is called a "triple tap" tool (Figure 3-19) and typically has the capability to tap 10-32, 8-32, and 6-32 hole sizes. This tool comes in handy when the 6-32 screw holes on a device box are stripped and you cannot secure a switch or receptacle to the box. By using the tap tool to redo the stripped 6-32 threads, you are now able to secure the device to the box.

An adjustable wrench (Figure 3-20) is used to tighten couplings and connectors, tighten pressure-type wire connectors, and remove and hold nuts and bolts. The modern adjustable wrench was first developed by the Crescent Tool and Horseshoe Company, and is often referred to as a "crescent wrench", regardless of the actual manufacturer. This wrench has one fixed jaw and one movable jaw. A thumbscrew is turned to adjust

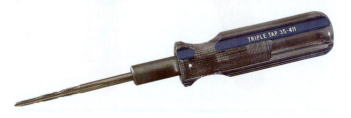

FIGURE 3-19 This triple-tap tool can tap or retap 6-32, 8-32, and 10-32 sizes. *Courtesy of IDEAL INDUSTRIES, INC.*

FIGURE 3-20   A 10-inch adjustable wrench. *Courtesy of Klein Tools, Inc.*

FIGURE 3-21   An awl with a cushioned rubber grip. *Courtesy of IDEAL INDUSTRIES, INC.*

FIGURE 3-22   An electrician's hammer. *Courtesy of IDEAL INDUSTRIES, INC.*

the jaws to any number of head sizes. Common sizes in residential electrical work include an 8- and a 10-inch size.

(See Procedure 3-6 on page 113 for the correct procedure to follow when using an adjustable wrench.)

An awl, sometimes called a "scratch awl," is used to start screw holes, make pilot holes for drilling, and mark on metal (Figure 3-21). Starting wood screws or sheet metal screws in wood for the purpose of mounting electrical equipment can be difficult. Using an awl to make a small pilot hole will allow the screws to be started easily.

> ## CAUTION
>
> **CAUTION:** The end of an awl is very sharp. Care must be used when using and transporting an awl so the sharp end does not cause personal injury.

A very important hand tool for the residential electrician is an electrician's hammer (Figure 3-22). It is used to drive and pull nails, pry boxes loose, break **wallboard**, and strike awls and chisels. A hammer used by an electrician should have long, straight claws to simplify the removal of electrical equipment. A strong, shock-absorbent fiberglass handle is recommended for electrical work. An 18- or 20-ounce hammer size is very common and works well in residential work.

A residential electrician has to take measurements for the correct location of electrical equipment. A folding rule or tape measure (Figure 3-23) can be used to check measurements on blueprints, determine electrical box location, measure lengths of wire and cable, and determine the depth and setout of electrical boxes. A folding rule usually opens up to a maximum length of 6 feet and works well when measuring shorter distances. Electricians use tape measures more often than folding rules. The tape measures come in standard 12-, 16-, 20-, and 25-foot lengths. The tape should be at least ¾-inch wide. If the tape is too narrow, it has very little strength and will be more likely to "break down" when it is extended.

(A)

(B)

FIGURE 3-23   (A) A 25-foot tape measure. (B) A 6-foot folding rule. *Courtesy of IDEAL INDUSTRIES, INC.*

Some measuring tapes have special features like a magnetic tip that stays "stuck" to a metal object when you are measuring some dimension of that object. Other tape measures have a tape lock feature that automatically keeps the tape from retracting back into the case once it is extended. To get it to retract all you need to do is push a button.

A tool pouch is used to hold, organize, and carry an electrician's hand tools (Figure 3-24). It is made from leather or a strong fabric like nylon polyester or heavyweight canvas. A good-quality belt should be used with the pouch. Because the tool pouch is quite heavy when all of your tools are in it, a wide belt works better than a narrow one to distribute the weight around your waist.

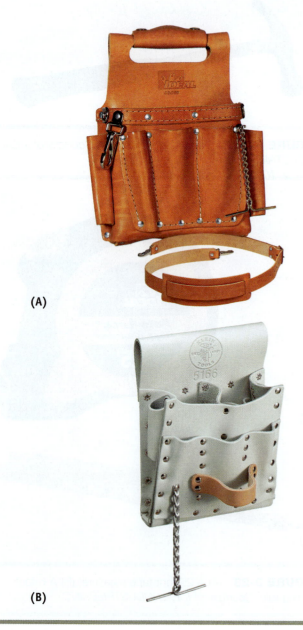

(A)

(B)

FIGURE 3-24   Two common tool pouch styles. *(A) Courtesy of IDEAL INDUSTRIES, INC. and (B) Klein Tools, Inc.*

from experience...

Many electricians carry their tools in a canvas or polyester tote bag. Some also use an empty 5-gallon drywall mud bucket (Figure 3-25). It's a good idea to carry only those tools you are using for a specific job in your tool pouch and to leave the tools you do not need in your tote or bucket. This makes the weight of the tool pouch much lighter and will allow you to save a lot of energy over the course of a workday by not having to carry all that unnecessary weight of a full tool pouch.

FIGURE 3-25   A bucket bag tool carrier. *Courtesy of IDEAL INDUSTRIES, INC.*

SPECIALTY TOOLS

Specialty tools are tools that are not typically used on a regular basis and are not usually found in a tool pouch. They are used to install specific parts of an electrical system. Using specialty tools when installing a residential electrical system can be just as hazardous as using common hand tools. The guidelines for the care and safe use of specialty electrical tools are the same as for the guidelines for common hand tools that were discussed earlier in this chapter. The following paragraphs describe the different types of specialty tools and their common trade uses.

FIGURE 3-26 A seven-piece nut driver set with ³⁄₁₆-, ¼-, ⁵⁄₁₆-, ¹¹⁄₃₂-, ³⁄₈-, ⁷⁄₁₆-, and ½-inch sizes. *Courtesy of IDEAL INDUSTRIES, INC.*

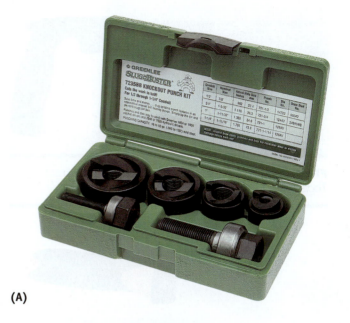

(A)

(B)

FIGURE 3-27 (A) ½- through 1¼-inch manual knockout set and (B) a hydraulic knockout set. *Courtesy of Greenlee Textron.*

A set of nut drivers (Figure 3-26) is used to tighten and loosen various sizes of nuts and bolts. They look like screwdrivers but have heads designed to fit on various sizes of nuts and bolts with hexagonal heads. One special feature of a nut driver is that the shank is hollow so that the tool can fit down over long bolts to loosen or tighten the nut on the bolt. A typical set of nut drivers comes with seven sizes ranging from ³⁄₁₆- through ½-inch trade size.

A **knockout punch** (Figure 3-27) is used to cut holes for installing cable or conduit in metal boxes, equipment, and appliances. This job usually requires the electrician to drill a hole through the enclosure at the spot where the hole needs to be made. The drilled hole should be just large enough for the threaded stud of the knockout set to fit easily through. The knockout set is then used to make a hole that will match the trade size of the conduit or cable connector that is going to be used. Typically, a manual knockout set is used to make trade-size holes from ½ inch through 1¼ inches. For these sizes and for sizes up to 6 inches, there are also hydraulic knockout sets. An electrician should be familiar with both the manual and the **hydraulic** types of knockout tools.

(See Procedure 3-7 on pages 114–115 for the procedure for cutting a hole in a metal box with a manual knockout punch.)

A keyhole saw, sometimes called a "compass saw," is used in remodel work to cut holes in wallboard for installing electrical boxes (Figure 3-28). This saw can also be used to cut exposed lath for "old work" box installation in an older lath-and-plaster installation. The standard blade is 6 to 12 inches long and has seven to eight teeth per inch. A newer style of saw that can do the same jobs as a keyhole saw is called a jab saw (Figure 3-29). The jab saw has replaceable blades for a variety of jobs and comes with an ergonomic grip for easy handling.

A hacksaw (Figure 3-30) is used to cut some conduit types and to cut larger conductors and cables. Today's hacksaws use rugged frames that are

FIGURE 3-28 A keyhole saw. *Courtesy of Klein Tools, Inc.*

FIGURE 3-29 A jab saw. *Courtesy of IDEAL INDUSTRIES, INC.*

FIGURE 3-30 A heavy-duty hacksaw that uses 12-inch blades. *Courtesy of IDEAL INDUSTRIES, INC.*

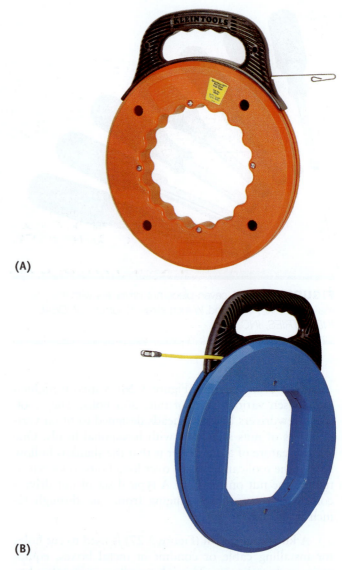

(A)

(B)

FIGURE 3-31 (A) Steel fish tape. *Courtesy of Klein Tools, Inc.* (B) Fiberglass fish tape. The cases are made from impact-resistant plastic. *Courtesy of IDEAL INDUSTRIES, INC.*

lightweight but provide ample rigidity for exceptional control when cutting. Hacksaw blades for electrical work are available in configurations of 18, 24, and 32 teeth per inch. The best all-around blade has 24 teeth per inch.

(See Procedure 3-8 on page 116 for the correct procedure for setting up and using a hacksaw.)

A fish tape and reel (Figure 3-31) is used to pull wires or cables through electrical conduit and to pull or push cables in wall or ceiling cavities. The tape itself can be made of stainless steel, standard blued steel, nonconductive fiberglass, nonconductive nylon, or multistranded steel for greater flexibility. Modern fish tapes are housed in a metal or nonmetallic case called the reel, which provides a convenient way to play out the fish tape to the length you need and to then "reel" it back in.

A conduit **bender** (Figure 3-32) is used to bend electrical metallic tubing (up to 1.25 inches) or rigid metal conduit (up to 1 inch). The bender consists of a bending head and a handle. The heads can be made of iron for durability or of aluminum for less weight. Special markings on the bender head help the electrician make the

FIGURE 3-32 This conduit hand bender has many markings to assist the electrician in making accurate bends in conduit. It is used with a 38-inch-long handle. *Courtesy of IDEAL INDUSTRIES, INC.*

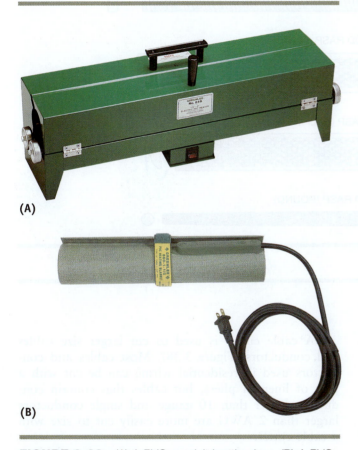

(A)

(B)

FIGURE 3-33 (A) A PVC conduit heating box. (B) A PVC conduit heating blanket. *Courtesy of Greenlee Textron.*

desired bend accurately. An introduction to basic conduit bending with a hand bender is given in Chapter 12 of this text.

When bending PVC conduit, a heater box or heating blanket is used (Figure 3-33). The box uses electricity to produce a high temperature that causes the inserted length of pipe to soften. After the pipe has "cooked" for the

FIGURE 3-34 A 9-inch torpedo level with an aluminum frame and a magnetic edge on one side. *Courtesy of IDEAL INDUSTRIES, INC.*

© Cengage Learning 2012.

FIGURE 3-35 A plumb bob.

manufacturer's recommended length of time, the electrician pulls it out and bends the pipe by hand into the desired shape. A bending blanket is wrapped around the pipe and softens it in the same manner. Again, after the blanket has heated the pipe for a certain length of time, it is taken off, and the pipe is bent into the desired shape.

Levels are used to **level** or **plumb** conduit, equipment, and appliances. Electricians usually use a small level called a "torpedo level" (Figure 3-34). A torpedo level is approximately 9 inches long and has a plastic or aluminum frame. One side of the level is magnetized so that it will stay on metal electrical equipment while the equipment is being installed.

A plumb bob (Figure 3-35) is a rather unusual tool for an electrician to use, but it can help make an electrical installation go much easier. It can transfer location points from ceiling to floor or floor to ceiling. For example, say you are installing a light fixture in the middle of a bedroom ceiling. You can easily find the center of the room by taking measurements on the floor and making a mark at that spot. Then, while on a step ladder, use the plumb bob to transfer the location from the floor to the ceiling and mark the location.

Metal files are used to deburr conduit, sharpen tools, and cut and form metal. Wood files or rasps can be used to enlarge drilled holes in wooden framing members (Figure 3-36).

A chisel is a metal tool with a sharpened and angled edge. It can be used to notch wood for boxes and cables and to cut and shape masonry or metal. Wood chisels often have to be used to "fine-tune" an opening for an electrical box in a remodel job (Figure 3-37).

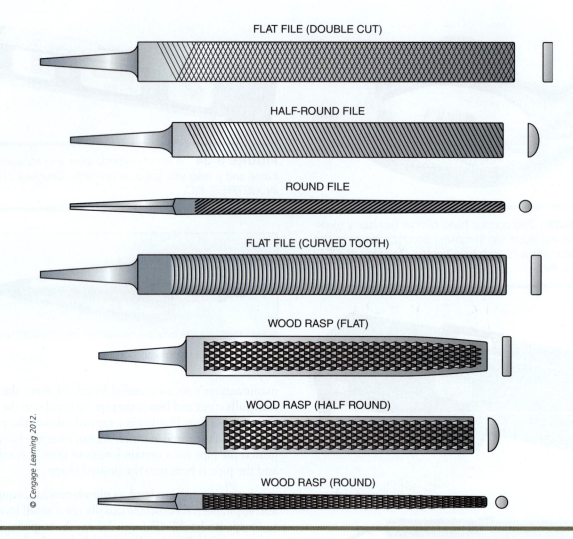

FLAT FILE (DOUBLE CUT)

HALF-ROUND FILE

ROUND FILE

FLAT FILE (CURVED TOOTH)

WOOD RASP (FLAT)

WOOD RASP (HALF ROUND)

WOOD RASP (ROUND)

© Cengage Learning 2012.

FIGURE 3-36   Examples of common file types.

(A)

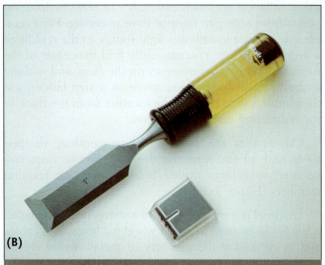

(B)

FIGURE 3-37   (A) A metal chisel (sometimes called a "cold chisel"). (B) A wood chisel. *Courtesy of Klein Tools, Inc.*

A cable cutter is used to cut larger size cables and conductors (Figure 3-38). Most cables and conductors used in residential wiring can be cut with a pair of lineman pliers, but cables that contain conductors larger than 10 gauge and single conductors larger than 2 AWG are more easily cut to size with a cable cutter.

Residential electricians use a sledgehammer to install ground rods. The *NEC*® requires that ground rods be 8 feet in the ground. Depending on the soil conditions, driving a ground rod can be a tough job. Sledgehammers may be found in sizes from 2 through 20 pounds. A 3- to 5-pound sledgehammer is recommended for driving a ground rod (Figure 3-39).

A hex key set, often called Allen wrenches, is used to tighten and loosen countersunk hexagonal setscrews. Larger electrical panelboards will require an Allen wrench to tighten the main conductor lugs (Figure 3-40).

FIGURE 3-40   An Allen wrench set provides several different hex key sizes in one tool. *Courtesy of IDEAL INDUSTRIES, INC.*

FIGURE 3-38   This cable cutter can cut conductors up to 350 kcmil in size. It is only 18 inches long and is very easy to store. *Courtesy of Greenlee Textron.*

FIGURE 3-41   Fuse pullers with notched handles for a safe grip. These pullers can handle cartridge fuse sizes from ½ to 1 inch in diameter, 1 to 100 amperes. *Courtesy of IDEAL INDUSTRIES, INC.*

FIGURE 3-39   A 3-pound sledgehammer. *Courtesy of Klein Tools, Inc.*

A fuse puller is used to remove cartridge-type fuses from electrical enclosures. They are constructed of a strong nonconductive material, such as glass-filled polypropylene (Figure 3-41). Some models are hinged so that one end accommodates a certain range of fuse sizes and the other end accommodates another range of fuse sizes.

Torque screwdrivers are designed to tighten smaller screws and lugs to the manufacturer's recommended torque requirements. Torque wrenches are used to tighten Allen-head and bolt-type lugs to the manufacturer's torque recommendations. Section 110.3(B) of the *NEC®* requires that all equipment be installed according to the instructions that come with the equipment. Most electrical equipment will have instructions that list the torque

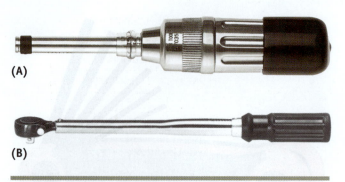

FIGURE 3-42 (A) This torque screwdriver has a capacity to torque from 2 to 36 inch-pounds and can be used on both slotted and Phillips-type heads. (B) The torque wrench can torque from 40 to 200 inch-pounds. *Courtesy of Klein Tools, Inc.*

FIGURE 3-43 This rotary armored cable cutter can be used on common sizes of Type AC and Type MC cable as well as flexible metal conduit. *Courtesy of IDEAL INDUSTRIES, INC.*

specifications in inch-pounds or foot-pounds. It is mandatory that an electrician uses the proper torque tools to tighten the equipment to the stated requirements (Figure 3-42).

(See Procedure 3-9 on page 125 for the procedure to follow when using a torque screwdriver.)

A rotary armored cable cutter is a tool that is used to strip the outside armor sheathing from Type AC cable and Type MC cable. It can also be used to cut flexible metal conduit (Figure 3-43). It uses a rotating handle to turn a small cutting wheel that is designed to cut through the cable without damaging the inner conductors. This tool is a real time-saver when installing a residential electrical system using a metal-sheathed cable.

(See Procedure 3-10 on page 126 for the procedure to follow when using a rotary armored cable cutter.)

Whenever possible, an electrician must work on circuits and equipment that have been de-energized. Chapter 1 covered the correct way to lock-out and tag-out electrical equipment and circuits so that they are safe to work on. However, it may be necessary to work on energized circuits or equipment in special situations. OSHA 1910.335 requires insulated tools to be used whenever an electrician is working on exposed energized circuits or equipment. Pliers, screwdrivers, and knives are available that have been UL-tested and certified for work on 1000-volt or less circuits. The cost of these tools is expensive compared to "regular" tools, so quite often an electrical contractor will have a set of these tools (Figure 3-44) available for the company electricians to use when they are working on "live" circuits or equipment.

FIGURE 3-44 A basic tool kit that includes a variety of insulated tools to meet job-site requirements. These tools are double insulated and are 1000-volt certified. They have special grips that provide exceptional safety, comfort, and control. All of the tools in this kit comply with OSHA 1910.335 regulations and NFPA 70E standards. *Courtesy of IDEAL INDUSTRIES, INC.*

CAUTION

CAUTION: Work on energized circuits and equipment should only be done when absolutely necessary and only by qualified electricians. Always follow OSHA 1910 S and NFPA 70E safety requirements when performing electrical work on energized parts. This includes wearing the proper personal protective equipment (PPE). Use insulated tools that are specifically made and certified for use on energized circuits. Remember, the cushion grips on "regular" tools are for comfort and are not made to protect the user from electrical shock.

POWER TOOLS

Residential electricians use various types of electric power tools when they install electrical systems. Electric power tools include those powered by 120-volt AC and those powered by low-voltage DC electricity. The power tools that use 120-volt AC electricity have a cord and an attachment plug that is plugged into a wall receptacle or extension cord. A "double-insulated" power tool will have a two-prong attachment plug. All others will have a three-prong grounding attachment plug. A battery supplies the DC voltage necessary to power cordless power tools.

Whether you are using a corded power tool or a cordless power tool, it is important to review the information on the tool's nameplate. Typically, the information found on the nameplate includes the catalog number of the tool, the voltage, the amperage, the revolutions per minute (rpm), and the weight of the tool. If the power tool is corded, then you must use the amperage from the nameplate to determine the correct extension cord size. Figure 3-45 shows the minimum wire gauge needed for an extension cord based upon the nameplate amperes and the length of the extension cord.

This section introduces you to the common electrical power tools used by residential electricians. Guidelines for the safe use of power tools are also discussed.

> **CAUTION**
>
> **CAUTION:** This section provides an overview of common power tools used by residential electricians and should not be considered as the only instruction and training you need before you use a power tool. Never use a power tool before you have received the proper training on how to use the power tool and have read the owner's manual associated with the power tool.

POWER TOOL SAFETY

The most important thing to remember when using any power tool is to read the operator's manual that comes with every power tool. The manual will contain information about the tool's applications and limitations as well as hazards associated with that tool. The following safety guidelines should be followed when using power tools:

- Always use proper personal protective equipment when using power tools. Safety glasses or goggles need to be worn at all times. A dust mask, nonskid safety shoes, a hard hat, or hearing protection must be worn when appropriate.

Recommended Minimum Wire Gauge for Extension Cords*

Nameplate Amperes	Extension Cord Length				
	25'	50'	75'	100'	150'
0–2.0	18	18	18	18	16
2.1–3.4	18	18	18	16	14
3.5–5.0	18	18	16	14	12
5.1–7.0	18	16	14	12	12
7.1–12.0	16	14	12	10	
12.1–16.0	14	12	10		
16.1–20.0	12	10			

*Based on limiting the line voltage drop to five volts at 150% of the rated amperes.

© Cengage Learning 2012.

FIGURE 3-45 The minimum wire gauge needed for an extension cord based upon the nameplate amperes and the length of the extension cord.

- Keep the work area clean and well illuminated.

- Do not operate electric power tools in explosive atmospheres such as in the presence of flammable liquids like gasoline. Power tools create sparks that could cause the gasoline fumes to ignite.

- Make sure that people not working on the job are kept away from the work area when you are using a power tool. Flying debris could hurt them.

- Make sure that grounded tools are plugged into a properly installed grounded receptacle outlet. Never remove the grounding prong from a grounding-type attachment plug.

- Double-insulated power tools use a polarized attachment plug. Make sure that it is plugged into a correctly installed polarized receptacle.

- Do not use electric power tools in wet conditions. Water entering a power tool will increase the risk of electric shock.

- All 120-volt power tools with cords must be plugged into ground fault circuit interrupter (GFCI)-protected receptacles on a residential construction site.

- Do not abuse the cord on a power tool. Never carry the tool by hanging on to the cord.

- If using a power tool outside, be sure to use an extension cord that is designed for outdoor use. The cord will be marked with "W-A" or "W" if it is rated for outdoor use.

- Stay alert when using a power tool. Never operate a power tool when you are tired or under the influence of drugs or alcohol.

- Do not wear loose clothing or wear jewelry. Longer hair needs to be put up inside a hat. Loose clothes, jewelry, or long hair can get caught in a power tool, and serious injury can result.

- Be sure that the power tool switch is turned OFF before plugging in the tool.

- Make sure that the **chuck key** or other tightening wrench is removed before turning a power tool on. A wrench or key can be thrown from the tool at high speed and cause personal injury.

- Make sure you have firm footing when using a power tool. Do not overreach. Slipping or tripping when using a power tool can result in serious injury.

- Always try to keep your hands and other body parts as far away as possible from all cutting edges and moving parts of power tools.

- Always secure the material that you are using a power tool on. If the part is unstable, you may lose control of the power tool.

- Always use the correct power tool for the job.

- Do not force a power tool. If it is working too hard to do the job, it either is malfunctioning or is the wrong tool for the job.

- Always unplug a power tool or take out the batteries before you make any adjustments, change accessories, or store the tool.

- Always store power tools in a dry and clean location away from children and other untrained people.

- Maintain tools with care. Properly maintained tools are less likely to malfunction and are easier to control.

- Do not use a damaged power tool. You should tag a damaged tool with a "Do Not Use" message until it is repaired.

POWER DRILLS

A power drill is used with the appropriate bit to bore holes for the installation of cables, conduits, and other electrical equipment in wood, metal, plastic, or other material. The power drill is the power tool used most often by residential electricians. Like other power tools, models are available with a cord and plug, or are cordless. Power drills can be broken down into pistol grip drills, right-angle drills, hammer drills, and cordless drills.

PISTOL GRIP DRILLS

Pistol grip electric drills (Figure 3-46) are very popular because they are small, relatively lightweight, and easy to use. This style gets its name from the fact that it looks

FIGURE 3-46   This pistol grip drill has a ½-inch chuck. It reverses easily and has a trigger switch for speed control of 0 to 850 rpm. *Courtesy of Milwaukee Electric Tool Corporation.*

like and is held like a pistol. They are available in three common **chuck** sizes: ¼, ⅜, and ½ inch. A chuck is the part of the drill that holds the drill bit securely in place. Most drills have chucks that are tightened and loosened with a chuck key or wrench. Some newer models are using a keyless chuck.

Electricians usually use the ⅜- or ½-inch size. The ⅜-inch size is used for small hole boring at higher speed. The ½-inch size is used for larger hole boring at lower speeds. The speed in revolutions per minute (rpm) of this drill type is inversely proportional to the chuck size. In other words, the larger the chuck size, the slower the speed of the drill. The turning force, or the **torque**, of the drill is proportional to the chuck size. That is, the larger the chuck, the more torque is available. A trigger switch controls the speed of this drill type. The harder you squeeze the trigger, the faster the drill turns. The drill can also be reversed.

The pistol grip drill can be used with a wide variety of drill bits. Drill bits are the tools that are attached to the drill and actually do the hole boring. Bits used with a pistol grip drill should be designed for use at higher speeds. A twist bit is designed to drill wood or plastic at high speed and metal at a lower speed. A flat-bladed spade bit, sometimes called a "speed-bore bit," is used to drill holes in wood at high speed. A masonry bit is used to drill holes in concrete, brick, and other masonry surfaces. An **auger** bit is used for drilling wood at a relatively slow speed. Figure 3-47 shows examples of drill bit types used with a pistol grip drill. A newer type of drill bit that is very popular is called a step bit

FIGURE 3-49 This right-angle drill is designed for use in electrical system installation. The drill has a ½-inch chuck that is attached to a 360-degree swivel head that allows drilling in virtually any direction. The drill is reversible and has speed control from 0 to 600 rpm. *Courtesy of Milwaukee Electric Tool Corporation.*

FIGURE 3-47 Examples of common bit types and accessories used with a pistol grip drill: (A) twist bit, (B) flat-blade spade bit, (C) masonry bit, and (D) bit extension. *Courtesy of Milwaukee Electric Tool Corporation.*

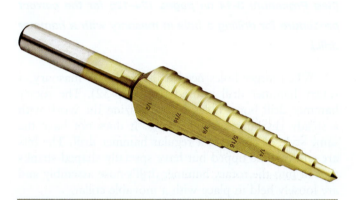

FIGURE 3-48 Step bits can be used for precise, repetitive drilling though a variety of tough materials. *Courtesy of IDEAL INDUSTRIES, INC.*

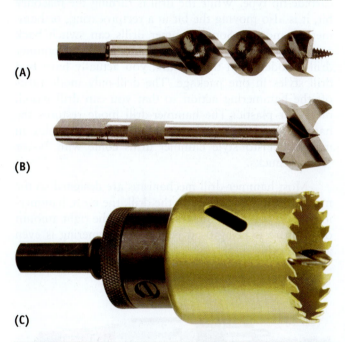

FIGURE 3-50 Examples of common bit types and accessories used with a right-angle drill: (A) auger bit, (B) forstner bit, and (C) hole saw. *Courtesy of Milwaukee Electric Tool Corporation.*

(Figure 3-48). Step bits can be used for precise, repetitive drilling through a variety of tough materials. Step bits are "stepped' so that one bit can drill several different size holes.

(See Procedure 3-11 on pages 119–120 for the correct procedure for using a cordless pistol grip power drill.)

RIGHT-ANGLE DRILLS

In residential wiring, electricians need to drill holes for the installation of cables through wooden framing members. The tight space between wall studs can make drilling the required holes with a pistol grip drill a slow and awkward process. To allow easier drilling of wood framing members in tight spaces, the right-angle drill (Figure 3-49) was introduced by the Milwaukee Electric Tool Company in 1949. The head of the drill is at a right angle (90 degrees) in relation to the rest of the drill, which allows the drill body to be located away from the material being drilled. Right-angle drills are usually used with a ½-inch chuck and work well with auger bits, forstner bits, or hole saws to drill holes in wood framing members. Figure 3-50 shows examples of drilling bits used with a right-angle drill.

(See Procedure 3-12 on page 121 for the correct procedure for drilling a hole in a wood framing member with an auger bit and a right-angle drill.)

(See Procedure 3-13 on pages 122–123 for the correct procedure for cutting a hole in a wood framing member with a hole saw and a right-angle drill.)

HAMMER DRILLS

Hammer drills (Figure 3-51) are used to drill holes in masonry or concrete walls and floors. When installing anchors to hold electrical equipment on a masonry wall, hammer drills are used to drill the anchor holes. Hammer drills are used with special masonry bits to bore holes in masonry. The bits used are of the percussion carbide-tip type. While the drill is turning the masonry bit, it is also moving the bit in a reciprocating, or hammering, motion. Some hammer drills can switch back and forth from a "drill only" mode to a "hammer drill" mode. With this feature, you actually have two drill styles in one package. The drill-only mode turns off the hammering action so that you can drill wood, metal, or plastic. The hammer-drill mode restores the hammering action and allows you to drill holes in masonry. Pistol-style hammer drills come with ⅜- or ½-inch chucks.

Most hammer-drill mechanisms are designed so the more pressure you exert on the drill, the more hammering action you get. You will know that the right amount of pressure is being used when the hammering is even and smooth.

FIGURE 3-51 A hammer drill used for percussion carbide-bit drilling in concrete and masonry and drilling without hammering in wood or metal. *Courtesy of Milwaukee Electric Tool Corporation.*

FIGURE 3-52 A rotary hammer drill. *Courtesy of Milwaukee Electric Tool Corporation.*

(See Procedure 3-14 on pages 124–125 for the correct procedure for drilling a hole in masonry with a hammer drill.)

When larger holes need to be drilled in masonry, a rotary hammer drill is used (Figure 3-52). The rotary hammer drill bores holes by pulverizing the work with a steady rhythm of heavy blows. It does not have the same style of chuck as a regular hammer drill. The bits are still carbide tipped but have specially shaped shanks that fit into the rotary hammer drill's nose assembly and are loosely held in place with a movable collar.

Residential electricians need to decide whether to buy and use a regular hammer drill or a rotary hammer drill. The deciding factor is usually frequency of use. Electricians who drill only an occasional hole in masonry can get by very nicely with a regular hammer drill. Electricians who plan on drilling small or large holes in concrete on a daily basis should lean toward a rotary hammer drill.

CORDLESS DRILLS

Pistol grip drills, right-angle drills, and hammer drills are all available in cordless models (Figure 3-53). The source of electrical power for these drills usually is the Ni-Cad battery. This battery type is rechargeable, long lasting, virtually maintenance free, and relatively cost effective. It can provide the power needed to operate any of the power tools that an electrician may use. Ni-Cad is short for nickel-cadmium, the materials responsible for the electrochemical reaction in the battery that produces a voltage. Manufacturers are also using Lithium batteries, especially in the larger sizes. The voltage of

FIGURE 3-53   A cordless drill. This drill is rated for 18 volts and comes with a case, a battery charger, and an extra battery. *Courtesy of Milwaukee Electric Tool Corporation.*

today's cordless power tools usually will be 12, 14.4, 18, or 28 volts. Some manufacturers are now offering 36-volt power tools. The larger voltage will result in more power available for the power tool. A cordless drill kit usually comes with the drill, a battery charger, and at least one battery.

Many companies are making their cordless drills with a "keyless" chuck. No wrench is needed to tighten the bit into the chuck. To secure a drill bit in a cordless drill with a keyless chuck, make sure the battery is removed and then open the chuck with your hand until it is open enough to receive the drill bit. Insert the bit and tighten the chuck by hand until the shank of the bit is firmly gripped. The last thing you need to do is a little tricky. Reinstall the battery and, with the drill set to turn in the clockwise direction, firmly grip the chuck with one hand and lightly squeeze the drill trigger with the other hand. The drill will start to turn, but the resistance that you are applying by holding onto the chuck will allow the bit shank to be locked tightly into the chuck.

Most cordless drills also have an adjustable clutch feature that allows the amount of turning force to be increased or decreased. This allows the cordless drills to act like a power screwdriver. Electricians often use this feature for installing receptacles and switches in device boxes and for other tasks that normally involve using a manual screwdriver. Using the cordless drill as a power screwdriver is much faster than using a manual screwdriver. It also is physically easier on the installing electrician.

POWER SAWS

Residential electricians are required to use power saws on the job for cutting to size such things as plywood backboards for mounting electrical equipment and for cutting building framing members to facilitate the installation of the electrical wiring. There are two styles of power saws you should be familiar with: the circular saw and the reciprocating saw.

Circular Saws

Many electricians call this saw type a "skilsaw." This name comes from the first portable electric power saw, which was made by a company named Skil. A circular saw (Figure 3-54) is designed to cut wood products, such as studs and plywood, to a certain size. Most circular saws still use a cord and plug connection, but several manufacturers have recently introduced powerful cordless models. Electricians do not use this saw type very often but should be familiar with it for those times when it is used.

The size of this saw is measured by the diameter of the saw blade. The most common size is 7¼ inches. These saws are fairly heavy and can weigh as much as 12 or 15 pounds. The handle of the saw has a trigger switch that starts the blade turning. The harder you squeeze the trigger, the faster the blade turns. It is important to make sure that the blade's teeth are pointing in the direction of rotation. Blades typically have an

FIGURE 3-54   A circular saw with a 7¼-inch blade. *Courtesy of Milwaukee Electric Tool Corporation.*

arrow on them that indicates their required direction of rotation. A two-part guard protects the blade. The top half is stationary, and the bottom half is hinged and spring-loaded. As you push the saw forward when cutting, the lower half of the guard is hinged up and under the top guard, allowing the saw to continue to cut.

Using a circular saw is very dangerous. You should always receive proper instruction on its use before you attempt to cut anything with it. Try to follow these guidelines:

- Always wear the proper personal protective equipment.
- Always check to see that the blade guard is working properly.
- Make sure you know what is behind or below the piece of wood you are cutting.
- Try to always keep two hands on the saw at all times.
- Do not force the saw through the work. Apply a steady but firm pressure and let the saw do its job.
- Always secure the work you are cutting.
- If the saw has a cord, be aware of where it is when you are sawing. Do not cut off the cord.
- Do not reach under the work while the saw is operating. You may lose your fingers.
- Try to position yourself so you are standing to one side of the work as you cut, not directly behind it.

(See Procedure 3-15 on pages 126–127 for the correct procedure for using a corded circular saw.)

Reciprocating Saws

A **reciprocating** saw has a blade that moves back and forth to make a cut rather than using a rotating blade (Figure 3-55). It can make a straight or curved cut in many materials, including wood and metal. When cutting wood, a wood blade must be used; when cutting metal, a metal blade must be used. Electricians usually refer to this saw type as a "Sawzall®." The name Sawzall® is a trademark of the Milwaukee Electric Tool Corporation. Reciprocating saws are usually found with a cord and plug, but, like circular saws, manufacturers are now producing well-engineered cordless models.

Residential electricians use the reciprocating saw much more often than a circular saw. It can be used to cut heavy lumber and to cut off conduit, wood, and fasteners flush to the surface. It works well when doing new or remodel work for such things as cutting box holes in plywood or fine-tuning framing members so that an electrical box can be properly mounted. It is great for sawing in tight locations where other saw types cannot reach.

When using a reciprocating saw, the following cutting tips should be followed:

- Always have at least three teeth in a cut. Less than three teeth will result in the blade getting hung up on the material being cut. More than three teeth slow down the cutting process. The solution is to use the coarsest blade possible while still having three teeth in the cut at all times.
- Turning the blade 180 degrees makes it easier to make a flush cut. This way, the head of the saw does not get in the way as much.
- When cutting wood, splintering can be limited by placing the finished side down when cutting. You can also cover the cutting line with masking tape and then make the cut.
- If a blade breaks when cutting a thin material, switch to a blade with more and finer teeth.
- Drilling a starter hole larger than the widest part of the blade is the best way to get a cut started that is away from the edge of a material. Plunge cutting can

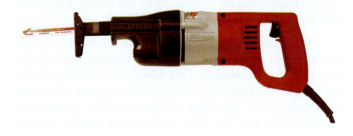

FIGURE 3-55   A Sawzall® reciprocating saw. *Courtesy of Milwaukee Electric Tool Corporation.*

be done in wood only by starting the saw at a shallow angle and then increasing the angle of the cut until the blade goes through the material.

- When cutting metal, cut at a lower speed and use a good-quality cutting oil.

(See Procedure 3-16 on pages 128–129 for the correct procedure for using a corded reciprocating saw.)

SUMMARY

This chapter has covered the common electrical hand tools, specialty tools, and power tools used in residential wiring. It has looked at how to identify the many tool types used in installing a residential wiring system as well as how to safely use them. You may run across hand tools, specialty tools, and power tools that are not covered in this chapter. Don't worry. Simply look over the owner's manual that comes with the tool and ask an experienced electrician to show you the proper way to use the tool. Attend trade shows and vendor days at your local electrical distributor to stay up to date on the newest tools for electricians. Remember, a good residential electrician knows which tool is right for the job and how to use that tool in a safe and proper manner.

PROCEDURE 3-1

Using a Screwdriver

- Follow proper safety procedures and wear appropriate personal protective equipment.

- Choose the proper blade type for the screw head and make sure that the screwdriver tip fits the screw head correctly.

A If you are installing a screw in a new location, make a pilot hole at the spot where you are installing the screw. An awl can be used in softwood. Drill a pilot hole for the screw in sheet metal or hardwood.

A

© Cengage Learning 2012.

B Insert the tip of the screw in the pilot hole and then insert the screwdriver tip in the head of the screw. Hold the screwdriver tip steady and position the shank perpendicular to the screw head.

B

© Cengage Learning 2012.

While using one hand to keep the screwdriver steady, use the other hand to turn the handle of the screwdriver. Apply firm, steady pressure to the screw head and turn the screw clockwise to tighten or counterclockwise to loosen. (A good way to remember which way a screw is turned to tighten or loosen it is to use either of the following sayings: "Right Is Tight, Left Is Loose" or "Righty Tighty, Lefty Loosey.")

- Continue turning the screwdriver until the screw is completely tightened or loosened.

- Clean up the work area and return all tools and materials to their proper locations.

© Cengage Learning 2012.

PROCEDURE 3-2

Using Lineman Pliers to Cut Cable

- Follow proper safety procedures and wear appropriate personal protective equipment.

- Determine how much cable you want to cut off.

A Open the lineman pliers and place the cable into the cutting jaws at the spot where you want to cut it. Place the cable as close to the pliers joint as possible.

© Cengage Learning 2012.

B Squeeze the handles together and cut off the length of cable. One hand should be used to make the cut.

- Clean up the work area and return all tools and materials to their proper locations.

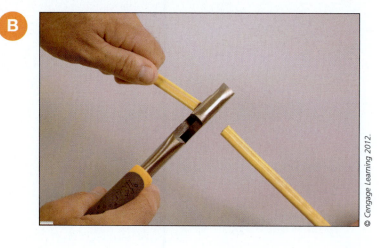

© Cengage Learning 2012.

PROCEDURE 3-3

Using a Wire Stripper

- Follow proper safety procedures and wear appropriate personal protective equipment.

A Insert the conductor to be stripped in the proper stripping slot.

- Close the jaws until you feel that you have reached the conductor and then open the jaws slightly.

B With an even pressure, pull back the strippers and remove the conductor insulation.

C Check the conductor for a ring or nick. If a nick occurs, cut off the nicked piece of conductor and restrip until the insulation is removed without any conductor damage.

- Clean up the work area and return all tools and materials to their proper locations.

A

© Cengage Learning 2012.

B

© Cengage Learning 2012.

C

© Cengage Learning 2012.

PROCEDURE 3-4

Using a Knife to Strip the Sheathing from a Type NM Cable

- Follow proper safety procedures and wear appropriate personal protective equipment.

- Determine how much cable sheathing you want to strip off.

 A Place the cable end on a sturdy surface and using your knife make about a 3-inch slit in the sheathing. Be extra careful doing this on a 3-wire cable because it is easier to cut the insulation on the conductors.

B Using the knife, carefully cut around the sheathing, making sure not to cut deep enough to cut the conductor insulation.

© Cengage Learning 2012.

With one hand holding the cable securely just before the cut, use your other hand to pull off the sheathing. With a 3-wire cable you may need to twist the sheathing as you pull it off.

© Cengage Learning 2012.

 Trim off any paper filler, if necessary.

- Clean up the work area and return all tools and materials to their proper locations.

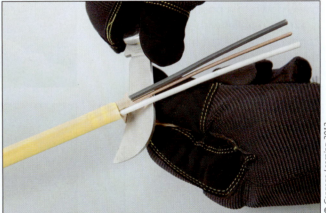

© Cengage Learning 2012.

PROCEDURE 3-5

Using a Knife to Strip Insulation from Large Conductors

- Follow proper safety procedures and wear appropriate personal protective equipment.

- Mark the conductor to indicate the amount of insulation to be stripped.

A Place the knife blade on the insulation at the point marked and carefully cut around the conductor to a depth that is just short of touching the actual conductor.

B Place the blade across the conductor and lay the knife blade down to a position that is almost parallel with the conductor at the cut made in the previous step.

C Using a pushing motion away from you, cut the insulation to a depth that is as close as possible to the conductor, all the way to the end. Be careful not to nick the conductor in this step.

D Using your fingers (or possibly your long-nose pliers), peel off the remaining insulation.

- Clean up the work area and return all tools and materials to their proper locations.

A

© Cengage Learning 2012.

B

© Cengage Learning 2012.

C

© Cengage Learning 2012.

D

© Cengage Learning 2012.

PROCEDURE 3-6

Using an Adjustable Wrench

- Follow proper safety procedures and wear appropriate personal protective equipment.

- Determine the size of the nut or bolt and turn the thumbscrew to set the jaws to the correct size.

- Whether you're tightening or loosening a nut or bolt, always position the movable jaw toward you.

- Before you turn the wrench, make sure its jaws are snug against the nut or bolt and touching it on at least three sides. Retighten the jaws each time you position the wrench on a new nut or bolt.

- Using the proper turning force, turn the wrench in the direction required for tightening or loosening. Remember that pressure must be put on the stationary jaw, not the movable jaw.

A Tightening

© Cengage Learning 2012.

B Loosening

- Continue turning the wrench until the nut or bolt is completely tightened or loosened.

- Clean up the work area and return all tools and materials to their proper locations.

© Cengage Learning 2012.

PROCEDURE 3-7

Using a Manual Knockout Punch to Cut a Hole in a Metal Box

- Follow proper safety procedures and wear appropriate personal protective equipment.

- Set up a portable power drill for use. Remember to do a complete check of the power tool before using it. Choose a metal drill bit that is slightly larger than the threaded stud of the knockout tool and install the bit into the drill. Make sure it is tightened securely in the chuck using the chuck key. You may want to drill a pilot hole with a small drill bit and then use the larger bit to enlarge the hole to the size required for easy insertion of the knockout set's threaded stud.

A Drill a hole slightly larger than the knockout set's threaded stud in the center of the space you are going to punch. A center punch can be used to make an indentation for your drill to start in. This will reduce the tendency of the drill bit to move off line when you start to drill. Hold the drill firmly while drilling. A loose grip could cause an accident. Remember that the drill bit will be hot, so use caution around it until it cools. Note: A step bit is shown being used in this procedure.

A

© Cengage Learning 2012.

B Insert the threaded stud through the drilled hole and put the cutting **die** back on the stud. Make sure that the cutting die is aligned so that the cutting edge is toward the metal enclosure.

B

© Cengage Learning 2012.

C Tighten the drive nut with a properly sized wrench. An adjustable wrench works well.

- Continue to tighten the drive nut until the **cutter** is pulled all the way through the metal. Remove the knockout punch when the cutter is finally pulled through.

C

© Cengage Learning 2012.

D Remove the cutter from the threaded stud and shake out the punched metal.

- Replace the cutter onto the threaded stud and place the knockout punch in its proper storage area.

- Clean up the work area and return all tools and materials to their proper locations.

D

© Cengage Learning 2012.

PROCEDURE 3-8

Setting Up and Using a Hacksaw

- Follow proper safety procedures and wear appropriate personal protective equipment.

A Insert the blade in the frame of the hacksaw and tighten it securely in place. Be sure the teeth angles are pointed toward the front of the saw. There is usually an arrow on the blade that indicates the proper direction for the blade to be installed.

- Mark the point on the material where you want to cut and secure the item being sawed in a vise.

- Set the blade of the hacksaw on the point to be cut.

B Push gently forward until the cut is started. Do not exert too much pressure on the saw.

- Make reciprocal strokes until the cut is finished. Remember that the cutting stroke is the forward stroke. Your cut should be straight and relatively smooth. Be aware that excessive speed while cutting can ruin blades.

- Clean up the work area and return all tools and materials to their proper locations.

A

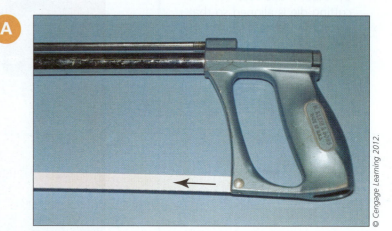

© Cengage Learning 2012.

B

© Cengage Learning 2012.

PROCEDURE 3-9

Using a Torque Screwdriver

- Follow proper safety procedures and wear appropriate personal protective equipment.

 Look at the equipment instructions and determine the required torque specifications.

 Adjust the torque screwdriver to the desired number of inch-pounds

- Place the torque screwdriver tip on the object to be tightened.

 Keeping everything properly aligned, turn the torque screwdriver in the direction for tightening.

- Continue to tighten until you hear a click, which means that you have tightened the fastener to the required torque specification.

- Clean up the work area and return all tools and materials to their proper locations.

Panelboard Lug Torque Data

Line Lugs and Main Breaker

	Wire Range (awg)	Torque (in. lbs.)
Line Neutral Lug	4 – 2/0 CU or AL	50
Main Lugs	6 – 2/0 CU or AL	50
Main Breaker	4 – 2/0 CU or AL	(see circuit breaker)

Branch Circuit Neutral and Equipment Ground Bar

Wire Range (awg)	Torque (in. lbs.)
3 – 1/0 CU or AL	50
4 CU or AL	45
6 CU or AL	40
8 CU or AL	35
14 – 10 CU; 12 – 10 AL	30

Branch Circuit Breakers

See the individual circuit breakers for wire-binding screw torque.

© Cengage Learning 2012.

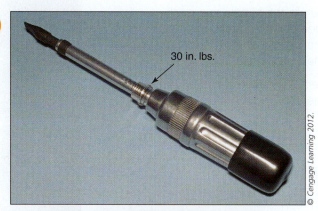

30 in. lbs.

© Cengage Learning 2012.

© Cengage Learning 2012.

PROCEDURE 3-10

Using a Rotary Armored Cable Cutter

- Follow proper safety procedures and wear appropriate personal protective equipment.

- Determine how much armored cable sheathing you want to strip off.

A Place the cable end in the rotary cutter and position it so the cutting wheel is lined up with the spot you want to cut.

- Tighten the mechanism on the rotary cutter to hold the cable in place. Depending on the rotary cutter, this could be a thumbscrew to tighten or a lever to squeeze.

B Rotate the handle in a clockwise direction until you feel the cutting wheel go through the armor sheathing.

- Remove the rotary cutter from the end of the cable.

C Pull the end of the armor sheathing off the cable.

- Trim off any paper from Type AC cable or plastic wrapping from Type MC cable as necessary.

- Clean up the work area and return all tools and materials to their proper locations.

A

© Cengage Learning 2012.

B

© Cengage Learning 2012.

C

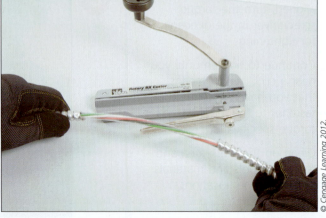

© Cengage Learning 2012.

PROCEDURE 3-11

Using a Cordless Pistol Grip Power Drill

- Follow proper safety procedures and wear appropriate personal protective equipment.

(A) Choose the proper bit type for the job and place it in the drill chuck in the following manner:

- Open the chuck by turning it with your hand in a counterclockwise direction until it can accommodate the shank of the drill bit.

- Insert the bit shank into the chuck and tighten the chuck by hand as much as you can.

- Using the chuck key, tighten the bit securely in the chuck. If a chuck has more than one tightening hole, use the key to tighten at each hole. If the drill has a keyless chuck, tighten the bit securely in the chuck by hand.

  - Remember to remove the chuck key if one was used.

- Make a mark at the location you want to drill and, using an awl for wood or a center punch for metal, make a

  small indent exactly where you want to drill. Note: For soft wood there is no need to make an indent before drilling.

(A)

© Cengage Learning 2012.

PROCEDURE 3-11

Using a Cordless Pistol Grip Power Drill (Continued)

- If necessary, securely clamp the material that is being drilled.

- Firmly grip the drill and place the tip of the bit at the indent.

B Hold the drill perpendicular to the work so that the bit will not make a hole at an angle. Start the drill slowly at first so that the bit will not stray from the spot where you want the hole. Squeeze the trigger harder and speed the drill up while applying moderate pressure to the drill.

- When you sense that the bit is about to go through the material you are drilling, reduce the amount of pressure you are exerting on the drill and let the drill bit complete boring out the hole.

- While the drill is still turning the bit in a clockwise direction, slowly pull the drill and bit out and then release the trigger.

- Remove the bit from the chuck and place everything in its proper storage area.

B

© Cengage Learning 2012.

PROCEDURE 3-12

Drilling a Hole in a Wooden Framing Member with an Auger Bit and a Corded Right-Angle Drill

- Follow proper safety procedures and wear appropriate personal protective equipment.

- Make a mark at the location where you want to bore the hole.

 Set up the right-angle drill with the auger bit. Make sure that the bit is secure in the drill chuck by tightening the chuck evenly with the chuck key. Remember that the cord must be unplugged while working on the drill.

- Plug the drill in and place the tip of the bit at the spot that was previously marked.

 Keeping an even force on the drill, start the drill and allow the auger bit to feed itself completely through the wooden framing member. Make sure the drill is turning in a clockwise direction.

- Once the auger bit has gone through the wood, stop the drill; reverse it, and, while pulling back with an even pressure, back out the bit.

- Unplug the cord and take the auger bit out of the drill chuck. Place everything in its proper storage area.

© Cengage Learning 2012.

© Cengage Learning 2012.

PROCEDURE 3-13

Cutting a Hole in a Wooden Framing Member with a Hole-Saw and a Corded Right-Angle Drill

- Follow proper safety procedures and wear appropriate personal protective equipment.

 Choose the proper hole saw and arbor for the size hole you wish to cut and assemble it. Set up the hole saw so that the pilot bit on the hole saw extends past the end of the hole saw by approximately 1 inch.

© Cengage Learning 2012.

B Make sure the drill is unplugged and secure the hole saw in the drill chuck by tightening the chuck evenly with the chuck key.

- Mark the location on the material where you want the hole to be made.

- Plug the drill in and place the tip of the hole saw pilot bit at the marked spot.

© Cengage Learning 2012.

Keeping an even force on the drill, start the drill and allow the hole saw to feed itself completely through the wooden framing member. Make sure the drill is turning in a clockwise direction.

- Once the hole saw has gone through the wood, stop the drill; reverse it, and pulling back with an even pressure, back out the hole saw.

© Cengage Learning 2012.

 Unplug the drill and remove the wood from the hole saw by using a screwdriver or some other narrow tool to dislodge the wood.

- Disassemble the hole saw and put everything back in the proper storage area.

© Cengage Learning 2012.

PROCEDURE 3-14

Drilling a Hole in Masonry with a Corded Hammer Drill

- Follow proper safety procedures and wear appropriate personal protective equipment.

A Choose the proper size masonry bit for the job and place it in the drill chuck in the following manner:

- Open the chuck by turning it with your hand in a counterclockwise direction until it can accommodate the shank of the drill bit.

- Insert the bit shank into the chuck and tighten the chuck by hand as much as you can.

- Using the chuck key, tighten the bit securely in the chuck. If a chuck has more than one tightening hole, use the key to tighten at each hole.

- Remember to remove the chuck key if one was used.

- Check to be sure that the hammer drill is in the hammer-drill mode.

- Make a mark at the location where you want to drill the hole.

- With the drill plugged in, grip the drill firmly and place the drill bit at the mark.

A

© Cengage Learning 2012.

Apply some pressure to the drill and slowly squeeze the trigger to increase the speed.

- Continue to apply pressure until the hammering is smooth and even and drill-out the hole to the desired depth.

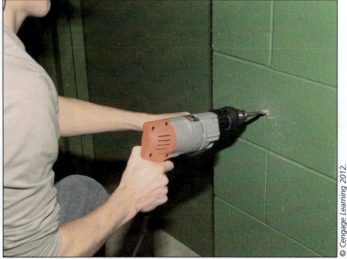

© Cengage Learning 2012.

 Discontinue putting pressure on the drill, and the hammering will stop. While the drill is still turning, slowly pull the bit from the hole. This will help clean the hole of masonry debris.

- Unplug the drill, remove the bit, and put everything away.

© Cengage Learning 2012.

PROCEDURE 3-15

Using a Corded Circular Saw

- Follow proper safety procedures and wear appropriate personal protective equipment.

A Secure the material to be cut. Use clamps whenever possible.

B Mark along the track you want to cut with a pencil.

C With the saw unplugged, adjust the saw's blade depth to a value that is slightly greater than the thickness of the material being cut.

- Plug the saw in and with the front edge of the saw base plate resting on the work, line up the guide slot with your cutting mark.

A

© Cengage Learning 2012.

B

© Cengage Learning 2012.

C

© Cengage Learning 2012.

D Grip the saw firmly with both hands and slowly exert pressure on the trigger switch. Bring the blade speed up to full speed and slowly push the saw forward to begin the cut.

- Continue to use firm and steady force to cut along the marked line, still using two hands on the saw.

E While maintaining the saw speed, push the saw completely through the end of the work. Make sure the blade continues to saw along the cutting line since at the end of the work, the guide slot is off the work and cannot be used as a guide.

F Release the trigger switch and pull the saw up and away from the work. The blade should stop, and the lower guard should now be covering the bottom of the blade.

- Unplug the saw and store it in an appropriate location.

© Cengage Learning 2012.

PROCEDURE 3-16

Using a Corded Reciprocating Saw

- Follow proper safety procedures and wear appropriate personal protective equipment.

A Set up the saw with a cutting blade suitable for cutting metal or wood. Make sure that the blade is secure in the saw head. The cutting blade in this saw is tightened with a setscrew and an Allen wrench. Some reciprocating saws have hand operated mechanisms for securing a cutting blade. Note: Always unplug the cord while working on the saw.

- Mark the location where you want to cut on the work material.

- If possible, secure the work with a vise or clamp to reduce vibration while cutting.

B Plug in the saw and set the blade on the work at the point to be cut as indicated by your mark. Always try to position the tool so that you are cutting in a downward direction.

© Cengage Learning 2012.

Pull gently back on the trigger switch to start the saw and begin the cut. Do not exert too much pressure on the saw. It will cut through very easily by allowing the weight of the tool itself to provide the necessary pressure.

- Let the saw make reciprocal strokes until the cut is finished. Your cut should be straight and relatively smooth.

- Unplug the tool, remove the blade, and store the tool in an appropriate location.

© Cengage Learning 2012.

REVIEW QUESTIONS

1. **Match the following hand tools to their common residential electrical trade use. Write the correct number for the tool in the blanks.**

1 Lineman pliers	7 Triple-tap tool
2 Adjustable wrench	8 Long-nose pliers
3 Wire strippers	9 Multipurpose tool
4 Electrician's knife	10 Flat-blade screwdriver
5 Phillips screwdriver	11 Awl
6 Hammer	12 Diagonal pliers

___ a.   Used to remove and install slot-head screws and to tighten and loosen slot-head pressure connectors

___ b.   Used to cut cables and conductors, form large conductors, and pull and hold conductors

___ c.   Used to form small conductors, cut conductors, and to hold and pull conductors

___ d.   Used to cut cables and conductors in a limited space

___ e.   Used to strip insulation from conductors

___ f.   Used to strip insulation, crimp solderless connectors, cut and size conductors, and cut and thread small bolts

___ g.   Used to strip large conductors and cables

___ h.   Used to tap 6-32, 8-32, and 10-32 holes for securing equipment to metal, retap damaged threads, and determine screw sizes

___ i.   Used to start screw holes, make pilot holes for drilling, and mark metal

___ j.   Used to drive and pull nails, pry boxes loose, and strike chisels

2. **Match the following specialty tools and power tools to their common residential electrical trade use. Write the correct number for the tool in the blanks.**

1 Portable drill	8 Knockout punch
2 Keyhole saw	9 Reciprocating saw
3 Plumb bob	10 File
4 Hacksaw	11 Chisel
5 Cable cutter	12 Auger bit
6 Torpedo level	
7 Fish tape	

___ a.   Used to cut conduit, large conductors, and cables

___ b.   Used with appropriate bits to bore holes for cables and conduits

___ c.   Used in a drill to bore holes through larger wooden framing members

___ d.   Used to cut holes for cable or conduit in metal boxes, equipment, and appliances

___ e.   Used to cut holes in wallboard for boxes and to cut lath and plaster for box installation in remodel work

___ f.   Used to cut lumber and to cut off pipe, wood, and fasteners flush to the surface

___ g.   Used to pull wires or cables through conduit and to pull cables in insulated walls

___ h.   Used to level or plumb conduit, equipment, and appliances

___ i.   Used to transfer location points from ceiling to floor or floor to ceiling and to plumb conduit and equipment

___ j.   Used to cut large cables and conductors

Directions: Answer the following questions with clear and complete responses.

3. List five guidelines for the care and safe use of electrical hand tools.

4. Name the three styles of screwdrivers used most often in residential electrical work.

5. Name the tool used to strip the outside armor of Type AC or Type MC cable.

6. Name the tool used to remove cartridge fuses from an electrical enclosure.

7. Name the tool used to tighten smaller screws and lugs to the manufacturer's recommended torque requirements.

8. Name the drill bit type that can drill several different sized holes in metal.

9. Name the power tool used to drill holes in masonry or concrete walls and floors.

10. Name the power tool that electricians use to drill holes through building framing members in tight spaces.

WHAT'S WRONG WITH THIS PICTURE?

Carefully study Figure 3-56 and think about what is wrong. Consider all possibilities.

✗ WRONG

✓ RIGHT

FIGURE 3-56   A tool or extension cord should never be plugged into a receptacle outlet or extension cord without knowing whether a ground fault circuit interrupter (GFCI) is protecting the electrical circuit.

FIGURE 3-57   This example shows a corded power tool connected to a cordset with built-in GFCI protection. Remember, when using corded power tools on a job site where new or remodeling work is being done, you need to have the tool plugged into a GFCI protected receptacle.

© Cengage Learning 2012.

CHAPTER 4

Test and Measurement Instruments Used in Residential Wiring

OBJECTIVES

Upon completion of this chapter, the student should be able to:

- Demonstrate an understanding of continuity testers and how to properly use them.

- Demonstrate an understanding of the differences between a voltage tester and a voltmeter.

- Connect and properly use a voltage tester and a voltmeter.

- Demonstrate an understanding of the differences between an in-line ammeter and a clamp-on ammeter.

- Connect and properly use a clamp-on ammeter.

- Demonstrate an understanding of ohmmeters, megohmmeters, and ground resistance meters.

- Demonstrate an understanding of multimeters.

- Connect and properly use a multimeter to test for voltage, current, resistance, and continuity.

- Demonstrate an understanding of the uses for a true RMS meter.

- Demonstrate an understanding of how to read a kilowatt-hour meter.

- Demonstrate an understanding of safe practices to follow when using test and measurement instruments.

- Demonstrate an understanding of the proper care and maintenance of test and measurement instruments.

GLOSSARY OF TERMS

ammeter (clamp-on) a measuring instrument that has a movable jaw that is opened and then clamped around a current-carrying conductor to measure current flow

ammeter (in-line) a measuring instrument that is connected in series with a load and measures the amount of current flow in the circuit

analog meter a meter that uses a moving pointer (needle) to indicate a value on a scale

auto-ranging a meter feature that automatically selects the range with the best resolution and accuracy

continuity tester a device used to indicate whether there is a continuous path for current flow through an electrical circuit or circuit component

digital meter a meter where the indication of the measured value is given as an actual number on a liquid crystal display (LCD)

DMM a series of letters that stands for "digital multimeter"

harmonics a frequency that is a multiple of the 60 Hz fundamental; harmonics cause distortion of the voltage and current AC waveforms

icon a symbol used on some electrical meters to indicate where a selector switch must be set so the meter can measure the correct type of electrical quantity

kilowatt-hour meter an instrument that measures the amount of electrical energy supplied by the electric utility company to a dwelling unit

manual ranging a meter feature that requires the user to manually select the proper range

megohmmeter a measuring instrument that measures large amounts of resistance and is used to test electrical conductor insulation

multimeter a measuring instrument that is capable of measuring many different electrical values, such as voltage, current, resistance, and frequency, all in one meter

multiwire circuit a circuit in residential wiring that consists of two ungrounded conductors that have 240 volts between them and a grounded conductor that has 120 volts between it and each ungrounded conductor

noncontact voltage tester a tester that indicates if a voltage is present by lighting up, making a noise, or vibrating; the tester is not actually connected into the electrical circuit but is simply brought into close proximity of the energized conductors or other system parts

nonlinear loads a load where the load impedance is not constant, resulting in harmonics being present on the electrical circuit

ohmmeter an instrument that measures values of resistance

open circuit a break in an electrical conductor or cable

polarity the positive or negative direction of DC voltage or current

short circuit a connection in an electrical circuit or device that results in a very low resistance; this is normally not a desired condition

smart meter a meter that measures electrical consumption in more detail than a conventional kWH meter; it also has the ability to communicate that information back to the local electric utility for monitoring and billing purposes

true RMS meter a type of meter that allows accurate measurement of AC values in harmonic environments

voltage tester a device designed to indicate approximate values of voltage or to simply indicate if a voltage is present

voltmeter an instrument that measures a precise amount of voltage

VOM a name sometimes used in reference to a "multimeter"; the letters stand for "volt-ohm milliammeter"

Wiggy a trade name for a solenoid type of voltage tester

Residential electricians must be familiar with a variety of test and measurement instruments. These instruments, usually referred to as "testers" or "meters," can provide a lot of great information to a residential electrician. For example, an electrician can use a meter to measure the current draw of an air-conditioning unit or measure the voltage value at a certain point on the wiring system. A residential electrician needs to know how to use test and measurement instruments in a safe and proper manner for a variety of house wiring applications. This chapter looks at how to properly use several of the most common types of meters that are encountered in residential wiring. Safe meter practices, as well as common meter care and maintenance techniques, are discussed in detail.

CONTINUITY TESTERS

A **continuity tester** is a device used to indicate whether there is a continuous path for current flow in an electrical circuit or electrical device. Continuity testers are the least complicated of all the instruments that a residential electrician may use, but their use in residential wiring is extensive. For example, they can be used to test continuity in electrical conductors, to test for faulty fuses, to test for malfunctioning switches, to identify individual wires in a cable, and many other applications. Residential electricians should have some type of continuity tester as part of their complement of tools. Figure 4-1 shows a commercially available continuity tester that is inexpensive and very easy for an electrician to use. This particular model has a light that comes on to indicate continuity.

CAUTION: Never attach a continuity tester to a circuit that is energized.

(See Procedure 4-1 on page 151 for step-by-step instructions on how to use a continuity tester.)

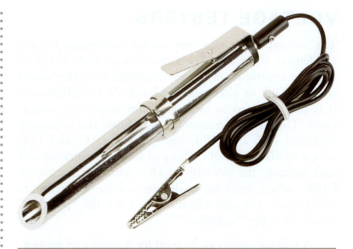

FIGURE 4-1 A basic continuity tester. This tester has a lamp at the end of the tester that lights up to indicate continuity. *Courtesy of IDEAL INDUSTRIES, INC.*

VOLTAGE TESTERS AND VOLTMETERS

Measuring for a certain amount of voltage or simply testing to see if a voltage is present are two of the most often done test and measurement tasks performed by a residential electrician. The instruments used to test or measure for voltage are similar in how they are connected into an electrical circuit but are quite different in other areas, such as the accuracy of the meter reading. This section explains the differences between a voltage tester and a voltmeter. It also explains how to use each of them in a variety of residential electrical applications.

from experience...

After the electrical circuit conductors have been "roughed in," the ceiling and wall coverings go on and hide the wiring. If the conductors terminating in each electrical box have not been identified, the electrician installing the switches and receptacles will not know which wire goes where. An easy way to identify them is to use a continuity tester to track down where each wire originates and where it ends.

VOLTAGE TESTERS

The **voltage tester** is a very useful instrument for the residential electrician. This instrument is designed to indicate approximate values of voltage for either direct current (DC) or alternating current (AC) applications. It is not designed to be a precision measuring instrument. The more common types indicate the following voltages: 120, 240, 480, and 600 volts AC, and 125, 250, and 600 volts DC. Many of these instruments will also indicate the **polarity** of DC, which can come in handy when working with low-voltage DC circuits and you are not sure which conductor is the "positive" or which is the "negative."

A voltage tester can be used for a variety of applications, such as identifying the grounded conductor of a circuit, checking for blown fuses, and distinguishing between AC and DC. However, the voltage tester's main job is to test for whether a voltage is present and to provide an approximate value of what that voltage is. The voltage tester is small and rugged, making it easy to carry and store. It fits easily in a tool pouch or may have its own carrying case that can clip onto your belt for easy access.

CAUTION

CAUTION: Remember that a voltage tester indicates an approximate voltage amount.

The most common style of voltage tester (Figure 4-2), or **Wiggy** as it is often called in the electrical trades, operates by having the circuit voltage that is present applied to a solenoid coil. A magnetic field is developed that is proportional to the amount of voltage present, and a movable core in the solenoid moves a certain distance, depending upon the strength of the magnetic field. There is an indicator, usually a colored band on the movable core, which will indicate the approximate value of voltage on a scale printed on the face of the tester. This solenoid action also causes the tester to vibrate in your hand. The more voltage present, the harder the tester vibrates. Small lamps may also light to indicate a voltage is present on some models.

Figure 4-3 shows a common solenoid type of voltage tester that is an excellent choice for residential wiring applications. It can indicate AC voltage up to 600 volts and DC voltage up to 600 volts, and, as an added feature, it can also be used as a continuity tester. This combination voltage tester and continuity tester will allow an electrician to carry one meter for these two jobs instead of two separate meters.

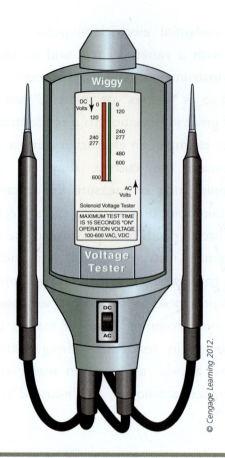

FIGURE 4-2 A very common style of voltage tester. It is most often referred to as a "Wiggy."

FIGURE 4-3 An example of a voltage tester that also can be used as a continuity tester. This tester is an excellent choice for residential electricians. *Courtesy of IDEAL INDUSTRIES, INC.*

CAUTION

CAUTION: Make sure to never use a voltage tester on circuits that could exceed the tester's voltage rating. You could be seriously injured or even killed.

from experience...

Many electricians use just the feeling of the vibration to verify that there is a voltage present. They are able to do this because they know that in residential wiring you will encounter either 120-volt or 240-volt AC. The electrician only has to recognize three different vibration levels to know what is going on. No vibration means that no voltage is present, a medium amount of vibration means that 120 volts is present, and a high level of vibration means that 240 volts is present. It is still always best for the electrician to look at the tester to see what approximate value of voltage is indicated.

FIGURE 4-4 An example of a digital voltage tester. This tester uses digital circuitry for more accurate voltage measurements. This model can also test for continuity. *Courtesy of IDEAL INDUSTRIES, INC.*

Some newer styles of voltage testers are classified as digital voltage testers (see Figure 4-4). They tend to be more accurate since they use digital circuitry rather than the electromechanical mechanism found in solenoid testers. Typically, they use light-emitting diodes (LEDs) that light up to indicate a particular measured voltage level. Some manufacturers are even equipping their voltage testers with a liquid crystal display (LCD) digital readout. Of course, the fancier the voltage tester gets, the more it typically costs. A good-quality solenoid voltage tester is an instrument that a residential electrician can use over and over again for many years.

(See Procedure 4-2 on page 152 for step-by-step instructions on how to use a voltage tester.)

CAUTION

CAUTION: Always read and follow the instructions that are supplied with the voltage tester.

Another type of voltage tester that has become very popular with residential electricians is the **noncontact voltage tester** (see Figure 4-5). As the name implies, this tester does not need to actually be connected to the electrical circuit to get a reading. All you have to do is bring the tester in close proximity of a conductor, and it will indicate whether it is energized. The indication can occur with the tester lighting up, vibrating in your hand, making an audible tone, or some combination of all three. This type of tester is inexpensive, easy to use, and fits easily into your pocket or tool pouch. It is really no wonder that it has become so popular with residential electricians.

(See Procedure 4-3 on page 153 for step-by-step instructions on how to use a noncontact voltage tester.)

VOLTMETERS

Voltmeters can be used for the same applications as voltage testers. However, voltmeters are much more accurate than voltage testers, so more precise voltage information can be obtained. For example, if the supply voltage to a building is slightly below normal, the voltmeter can indicate this problem where the voltage tester could not. The voltmeter can also be used to determine the exact amount of voltage drop on feeder- and branch-circuit conductors.

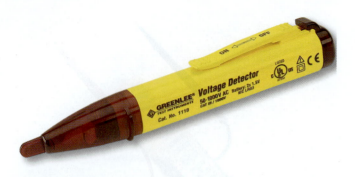

FIGURE 4-5   A noncontact voltage tester like this does not need to be connected directly to the circuit. It is brought close to an electrical conductor and will light up or make a sound if the conductor is energized. It can be used to detect the presence of voltage at receptacle outlets, lighting fixtures, wires, and cables. *Courtesy of Greenlee Textron.*

FIGURE 4-6   An example of an analog meter. The needle points to the value of the quantity being measured. This meter has a multicolored analog display with a mirrored scale. *Courtesy of IDEAL INDUSTRIES, INC.*

Because voltmeters are connected in parallel with the circuit or the component being tested, it is necessary that they have relatively high resistance. The internal resistance of the meter keeps the current flowing through the meter to a minimum. The lower the value of current through the meter, the less effect it has on the electrical characteristics of the circuit.

CAUTION

CAUTION: Remember to always connect a voltmeter across (in parallel with) the load or power source.

A meter that uses a moving pointer or needle to indicate a value on a scale is called an **analog meter** (see Figure 4-6). Analog meters used to be the only type of meter available. Now they are essentially a thing of the past. It is virtually impossible to purchase an analog meter from any of the major meter manufacturers. **Digital meters**, a meter where the indication of the measured value will be given as an actual number in a liquid crystal display (LCD), are the style of meter most often made by meter manufacturers, and are the style of choice for today's electrician. However, there are still analog meters being used in the electrical field, and if you ever have to use an analog meter, there are a few things you need to remember. For example, an accurate meter reading is obtained only by standing directly in front of the meter face and looking directly at it. An error (called "parallax error") can occur in your meter reading if you view the meter face from the side. Some higher-quality analog meters have a mirror behind the scale to help compensate for parallax error. Simply adjust the angle of your sight until there is no

reflection of the indicating needle in the mirror. This means that you are looking at the meter face "dead on" and that the reading you see is accurate.

The average analog voltmeter is generally between 95 and 98 percent accurate. This range of accuracy is satisfactory for most applications. It is very important, however, that the electrician strives to obtain the most accurate reading possible. Today's digital voltmeters (Figure 4-7) are extremely accurate, and an **auto-ranging** feature always gives the electrician the best accuracy possible in response to the electrical application being tested.

CAUTION

CAUTION: Always read and follow the instructions that are supplied with the voltmeter.

Voltmeters often have more than one scale. Auto-ranging voltmeters will pick the proper scale for you, but many older voltmeters are **manual ranging**, meaning that they have a range selector switch that requires you to select the scale manually. It is very important to select the scale that will provide the most accurate measurement. The range selector switch is provided for this purpose. It is advisable to begin with a high scale and work down to the lowest scale so that you do not exceed the range limit of any scale. Setting the range selector switch on the lowest usable scale possible will provide the most accurate reading. It is rare for a digital voltmeter to not be auto-ranging. The auto-ranging feature is a big reason why digital meters are so popular in electrical work.

(See Procedure 4-4 on pages 154–155 for step-by-step instructions on how to use a voltmeter.)

FIGURE 4-7 This digital voltmeter uses a large, easy-to-read digital display to show the measured voltage value for best accuracy. This model has the capability to also be a noncontact voltage tester and a continuity tester. *Courtesy of Greenlee Textron.*

If you end up using an analog meter of any type, always check to be sure that the indicating needle is pointing to zero when you start to take a measurement. A screw is provided to make the adjustment and is located just below the face of the meter. A very slight turn will cause the needle to move. The needle can be aligned with the zero line on the scale by turning the screw one way or the other. This adjustment may have to be done so that an accurate reading can be taken.

Another point to consider is that when using analog voltmeters on DC electricity, it is very important to maintain proper polarity. Most DC power supplies and meters are color coded to indicate the polarity. Red indicates the positive terminal, and black indicates the negative terminal. If the polarity of the circuit or component is unknown, touch the leads to the terminals while observing the indicating needle. If the indicating needle attempts to move backward, the meter lead connections must be reversed. If the leads are not reversed, damage to the meter, particularly the needle, can occur. If the polarity is reversed when using a digital voltmeter, usually a minus sign (–) is shown in the display and indicates reversed polarity. However, even if you do not reverse the leads, no harm will come to the digital meter.

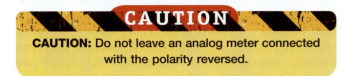

CAUTION

CAUTION: Do not leave an analog meter connected with the polarity reversed.

AMMETERS

Ammeters are designed to measure the amount of current flowing in a circuit. Ammeters can be used to locate overloads and open circuits. They can also be used to balance the loads on **multiwire circuits** and to locate electrical component malfunctions. There are two ammeter types: the in-line and the clamp-on (see Figure 4-8). **Clamp-on ammeters** are by far the most often used type of ammeter in residential electrical work and are covered in detail later in this chapter. **In-line ammeters** are seldom used because they do not have high enough amperage ratings to make them practical in electrical construction work and are not as easy to use as a clamp-on ammeter.

CAUTION

CAUTION: Always read and follow the instructions that are supplied with the ammeter.

IN-LINE AMMETERS

When they are used, in-line ammeters are always connected in series with the circuit or circuit component being tested. This means that a circuit must be disconnected and the ammeter inserted into the circuit so that the current flowing through the circuit is also flowing through the ammeter. It is not always practical in electrical construction work to turn the electrical power OFF, break open the circuit, insert the ammeter in series with the load, and then re-energize the circuit. After the reading has been taken, it is then necessary to turn the power OFF, disconnect the ammeter from the circuit, reconnect the circuit, and reapply electrical power. It is all very time consuming and presents many opportunities to make a mistake.

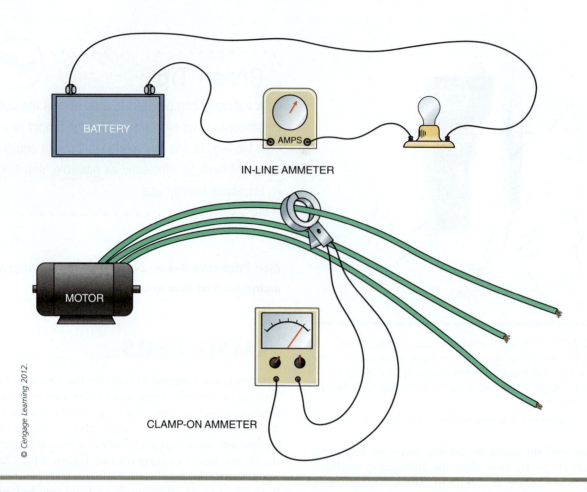

© Cengage Learning 2012.

FIGURE 4-8   Two types of ammeters are available to electricians: the in-line and the clamp-on. This figure shows the proper connection techniques for both.

In-line ammeters require that the resistance of the meter be extremely low so that it does not restrict the flow of current through the circuit. When measuring the current flowing through very sensitive equipment, even a slight change in current caused by the ammeter may cause the equipment to malfunction.

Analog in-line ammeters, like analog voltmeters, have an adjustment screw to set the indicating needle to zero. Each time a reading is taken, make sure that the needle starts at zero. Adjust if necessary. Many analog in-line ammeters have mirrors located behind the needle to assist the user in obtaining an accurate reading by overcoming parallax error.

CAUTION

CAUTION: In-line ammeters should always be connected in series with the circuit or component being tested. If DC is being measured, always check the polarity.

CLAMP-ON AMMETERS

Clamp-on ammeters are much easier to use than in-line ammeters. They are designed with a movable jaw that can be opened, which allows the meter to be clamped around a current-carrying conductor to measure current flow. When current is flowing through a conductor, a magnetic field is set up around that conductor. The more current flowing through the conductor, the stronger the magnetic field will be. The clamp-on ammeter picks up the strength of the magnetic field and converts the magnetic field's strength into a proportional value of current. The display can be analog or digital (see Figure 4-9).

A clamp-on ammeter is a valuable tool for any residential electrician. Meters are now available that provide clamp-on ammeter capabilities as well as being able to measure voltage and resistance. Many of these meters have other capabilities, such as capacitor testing and frequency measurement. These meters are most often referred to as "multimeters" because they can measure several different quantities. Multimeters are covered in greater detail later in this chapter.

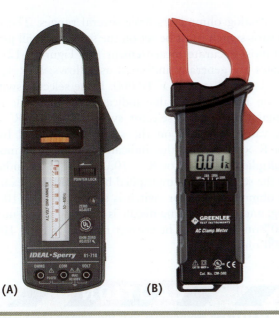

FIGURE 4-9 Examples of: (A) an analog clamp-on amme-ter *Courtesy of IDEAL INDUSTRIES, INC.* and (B) a digital clamp-ammeter. *Courtesy of Greenlee Textron.*

(See Procedure 4-5 on page 156 for step-by-step instructions on how to use a clamp-on ammeter.)

from experience...

A common application for a clamp-on amme-ter is for a residential electrician to measure each "hot" conductor of the service entrance when all loads are connected to see how much current is flowing on each conductor. It is good wiring practice to balance the current flowing on each service conductor. Knowing what the current is on each conductor will allow the electrician to adjust the loads accordingly for proper balancing.

Green Tip

An ammeter can be used to determine if an elec-trical load is drawing more current than it should. Excessive current draw results in lower efficiency and wasted energy.

OHMMETERS, MEGOHMMETERS, AND GROUND RESISTANCE METERS

Ohmmeters, Megohmmeters, and Ground Resistance Meters are the types of instruments used to take a resis-tance measurement in house wiring. It isn't often that a residential electrician will need to take a resistance measurement, but for those times when it is necessary, they should be familiar with the information presented in this section.

OHMMETERS

An **ohmmeter** is used to measure the resistance of a circuit or circuit component (see Figure 4-10). Batteries located in the meter case furnish the power for the operation of an ohmmeter, and just like the other meters discussed in this chapter, there is an ana-log and a digital model. In the analog model, the ohm-meter scale is designed to be read in the direction opposite other meters—in other words, from right to left, not left to right (see Figure 4-11). When the elec-trical circuit or component being measured is "open," the indicating needle should point to infinity (indicated with a symbol that looks like the number "8" lying on its side [∞]). The infinity symbol is located on the far left side of the scale. The needle can be aligned with the infinity mark by turning the adjustment screw located in the face of the meter in the same manner as adjusting the needle to point to zero on analog voltmeters and analog ammeters.

> **CAUTION**
>
> **CAUTION:** Always read and follow the instructions that are supplied with the ohmmeter.

> **CAUTION**
>
> **CAUTION:** It is very important to be sure that the circuit or component is disconnected from its regu-lar power source before connecting an ohmmeter. Connecting an ohmmeter to a circuit that has not been de-energized can result in damage to the meter and possible injury to the user.

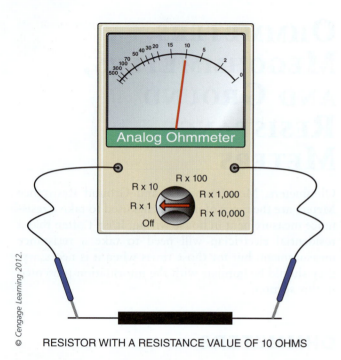

RESISTOR WITH A RESISTANCE VALUE OF 10 OHMS

FIGURE 4-10 An analog ohmmeter being used to measure the resistance of a 10-ohm resister. The range selector switch is on R × 1, and the needle is pointing to 10 ohms. Multiply the indicated amount (10 ohms) by the range switch value (1) to get the actual value being shown by the meter. 10 ohms × 1 = 10 ohms.

Analog ohmmeters have several ranges. The range selector switch must be set on the scale that will provide the most accurate measurement (see Figure 4-12). The ranges are generally indicated as follows: R × 1, R × 10, R × 100, and R × 10,000. If the selector switch is set on R × 1, the value indicated on the scale is the actual value. If the selector switch is set on R × 10, the value indicated on the scale is multiplied by 10. For R × 100, the value indicated on the scale must be multiplied by 100. For R × 10,000, the value indicated on the scale must be multiplied by 10,000. For example, an ohmmeter selector switch is set on the R × 100 setting, and the indicating needle is pointing to the number "10" on the scale. The correct reading is 10 × 100, or 1000 ohms.

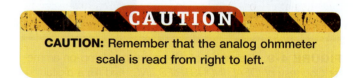

CAUTION: Remember that the analog ohmmeter scale is read from right to left.

The analog ohmmeter is also an excellent continuity tester. Connect the ohmmeter across the circuit or component to be tested. If there is continuity through the circuit or component, the needle will move. A reading of zero generally indicates a **short circuit.** A reading of infinity indicates an **open circuit**.

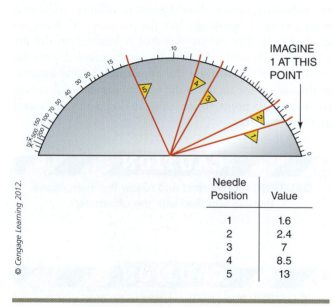

Needle Position	Value
1	1.6
2	2.4
3	7
4	8.5
5	13

FIGURE 4-11 An example of an analog ohmmeter scale. Notice that the scale reads from the right to the left. Take a look at the five needle positions shown and their values as indicated in the chart. Make sure you understand why each needle position has the value it does.

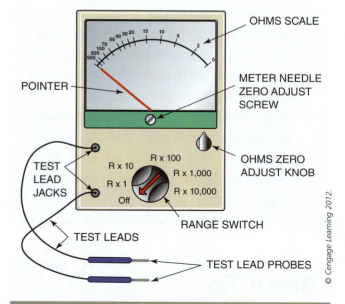

FIGURE 4-12 The common parts of an analog ohmmeter. Notice that the range selector switch has an OFF position. Make sure that when the meter is not in use, the switch is set to the OFF position. This will make sure that the batteries do not lose their charge.

MEGOHMMETER

A **megohmmeter**, commonly known by the trade name "Megger," is an instrument used to measure very high values of resistance (see Figure 4-13). Residential electricians will seldom be asked to use a megohmmeter, but they should still be familiar with them. For this reason, the following information is offered.

> ### CAUTION
> **CAUTION:** Always read and follow the instructions that are supplied with the megohmmeter.

Megohmmeters may be used to test the resistance of the insulation on circuit conductors, transformer windings, and motor windings. A megohmmeter is designed to measure the resistance in megohms. One megohm ($M\Omega$) is equal to 1 million ohms.

Visual inspection of insulation and leakage tests with voltmeters are not always reliable. A megohmmeter test is one of the most reliable tests available to an electrician. Insulation tests should be made at the time of the installation and periodically thereafter. For residential circuits and equipment rated at 600 volts or less, the 1000-volt setting can be used.

A small generator called a "magneto" is contained within the megohmmeter housing. The magneto furnishes the power for the instrument just as batteries do for an ohmmeter. The magneto can be hand powered or driven by batteries. Some megohmmeters use batteries with electronic circuitry to produce the high voltage levels required to measure high values of resistance. Megohmmeters have many different voltage ratings. Some of the most common are designed to operate on one of the following values: 250, 500, and 1000 volts.

> ### CAUTION
> **CAUTION:** Never touch the test leads of a megohmmeter while a test is being conducted. Also, isolate whatever it is on which you are conducting the test. High voltage is present and could injure you or the item you are testing.

> ### CAUTION
> **CAUTION:** Before a megohmmeter is connected to a conductor or a circuit, the circuit must be de-energized. When testing the circuit insulation, the testing is generally done between each conductor and ground. A good ground is a vital part of the testing procedure. The ground connection should be checked with the megohmmeter and with a low-range ohmmeter to ensure good continuity.

(A)

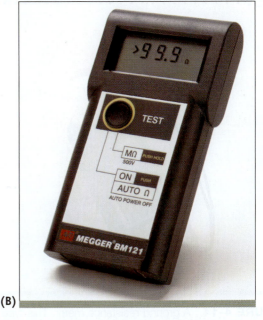
(B)

FIGURE 4-13 (A) A crank-type and (B) a battery-type Megohmmeter. Many electricians refer to any megohmmeter as a "Megger," but "Megger" is a registered trademark of Megger (formerly AVO International), and only their megohmmeters actually bear the Megger® name. *Courtesy of Megger.*

GROUND RESISTANCE METERS

Ground resistance meters are used to determine the amount of resistance to the earth when the resistance to earth through a grounding electrode is unknown. The various types of grounding electrodes are introduced and discussed in Section II of this textbook. A common type of grounding electrode used in house wiring is an 8-foot-long copper-clad or steel ground rod driven into the earth. The resistance through the ground rod to the earth must be 25 ohms or less. If the resistance is more than 25 ohms, another ground rod installed at least 6 feet from the first one is required. The best way to determine if the ground rod you installed has a resistance of 25 ohms or less to the earth is by using a ground resistance meter. The easiest model to use has a clamp that opens and goes around the driven ground rod. A digital display will show the resistance value in ohms (Figure 4-14).

from experience...

Meggers and ground resistance meters are quite expensive and are not used that often. As a result, most residential electricians do not own them. In many cases, an electrical contracting company will buy a Megger and a ground resistance meter. The electricians who work there will use them as needed.

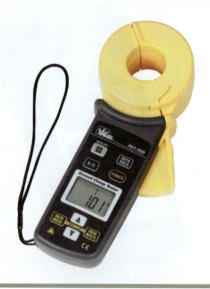

FIGURE 4-14   A ground resistance meter. This type of meter can be clamped around a ground rod driven into the earth and will indicate the amount of resistance to the earth. *Courtesy of IDEAL INDUSTRIES, INC.*

MULTIMETERS

Multimeters are designed to measure more than one electrical value. For example, the basic volt-ohm milliammeter (**VOM**) can measure DC and AC voltages, DC and AC current, and resistance. The major advantage of this type of meter is that several different types of tests and measurements can be taken with only one meter. This means that electricians can get by with one meter rather than having separate meters for each value that they want to measure. Various accessories such as clamp-on ammeter adapters and temperature probes are available for most multimeters. Analog (Figure 4-15) and digital (Figure 4-16) models are available. However, most major meter manufacturers offer only the digital version. These manufacturers refer to their meters as **DMMs**, or digital multimeters.

CAUTION

CAUTION: Always read and follow the instructions that are supplied with the multimeter.

Measurements taken with an analog multimeter are basically done the same way as with an analog voltmeter, analog ammeter, or analog ohmmeter. These operations were all discussed earlier in this chapter. Remember that with the analog multimeter, several electrical quantities

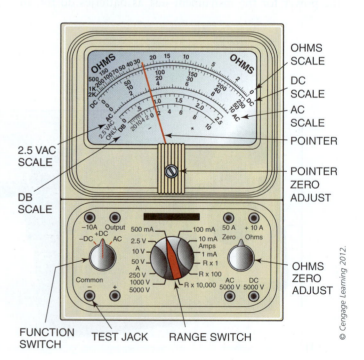

© Cengage Learning 2012.

FIGURE 4-15   An analog multimeter showing the major parts and their locations on the meter.

FIGURE 4-16 A digital multimeter. *Courtesy of IDEAL INDUSTRIES, INC.*

ICON	MEASUREMENT
V $=$	DC VOLTAGE
A $=$	DC AMPERAGE
V $\sim$	AC VOLTAGE
A $\sim$	AC AMPERAGE
Ω	OHMS
→|)))	CONTINUITY

© Cengage Learning 2012.

FIGURE 4-17 Common digital multimeter icons.

FIGURE 4-18 An example of a true RMS multimeter. This type of meter must be used to take accurate measurements on electrical circuits and equipment that have harmonics present. *Courtesy of IDEAL INDUSTRIES, INC.*

can be measured with the meter. You must specify which quantity you are measuring by setting a selector switch on the meter to the desired electrical quantity.

Measurements taken with a digital multimeter are quite easy to do. A selector switch must be used to indicate which electrical quantity you wish to measure. The selector switch is set to point at an **icon** (Figure 4-17) on the meter face that represents the particular electrical quantity you are measuring.

(See Procedure 4-6 on pages 157–160 for step-by-step instructions on how to use a digital multimeter.)

Digital multimeters are used extensively in electrical construction work. They are very rugged and take up less room to store than analog multimeters. The DMMs have advanced features, such as auto-zeroing, auto-polarity, auto-ranging, and an automatic shut off. These features help make DMMs easier to use than analog multimeters. They are also much easier to read since the electrical value is displayed directly in digits. An analog multimeter is considered a good "bench meter" and is often found on a repair bench in the shop, but residential electricians will find that a DMM will be the multimeter of choice in the field.

Nonlinear loads, harmonics, and true RMS meters are discussed next. **Nonlinear loads** are load types where the load impedance is not constant. This type of load can cause harmonics to occur on electrical circuits. Examples of these loads include computers and fluorescent lighting fixtures with electronic ballasts. **Harmonics**

are frequencies that are multiples of the fundamental frequency (60 hertz). For example, a third harmonic would be at 180 hertz (3 × 60 Hz). Harmonics cause distortion of the basic alternating current waveforms. Regular measuring instruments, called "average" reading instruments, do not respond fast enough to accurately read values caused by harmonic distortion. **True RMS meters** (Figure 4-18) provide accurate measurement of AC values in environments with harmonics, but true RMS meters can still be used when harmonics are not present. True RMS meters tend to be more expensive than "average" reading meters. Residential electrical systems rarely have the types of loads that produce harmonics.

However, as computers and other loads that produce harmonics are used more often in residential settings, electricians should be aware of what harmonics are and the type of meter that must be used to accurately measure electrical values when harmonics are present.

CLAMP METERS

A relatively new type of test tool is available that combines a clamp-on ammeter with a multimeter. It is often called a clamp meter and typically can measure alternating current amperage, both direct current and alternating current voltage, and resistance. Many of them can also measure values of capacitance and frequency. The meter shown in Figure 4-19 can also be used as a noncontact voltage tester. The clamp meter is reasonably priced and

is a great choice for a residential electrician because it eliminates the need to carry around several separate test and measurement instruments.

WATT-HOUR METERS

Residential electricians install the meter enclosure as part of the service entrance that connects the dwelling unit electrical system to the electric utility. (Service entrances and equipment are covered in Section II of this book.) The meter enclosure contains a **kilowatt-hour meter** (Figure 4-20). The local electric utility meter department usually installs the meter into the meter enclosure once the service entrance is done and the dwelling is ready to receive electrical power. This meter measures the amount of electrical energy used by the dwelling's electrical system. Residential electricians should have a basic understanding of how these meters work and how to read them.

Electrical energy is the product of power and time. Electrical power is measured in watts. Larger amounts of electrical power are measured in kilowatts (1000 watts is equal to 1 kilowatt). The watt-hour meter

FIGURE 4-19 A good example of a clamp meter. It combines a clamp-on ammeter with a multimeter and can be used for several different electrical measurements. *Courtesy of IDEAL INDUSTRIES, INC.*

FIGURE 4-20 A typical single-phase kilowatt-hour meter with dials. This is the most common style of watt-hour meter found in residential applications. *Courtesy of Landis+Gyr Energy Management Inc.*

measures the amount of power consumed over a specific amount of time. It is a meter that registers the amount of watt-hours delivered by the electric utility to the customer. Because residential customers require a large amount of electrical energy, the standard meter is designed to indicate kilowatt-hours (1 kilowatt-hour is equal to 1000 watt-hours.)

The kilowatt-hour meter works on the principle of magnetic induction. Moving magnetic fields cause currents to flow in an aluminum disk. These currents, called "eddy currents," produce magnetic fields that interact with the moving magnetic fields, causing the disk to rotate like a small electric motor. The rotating disk drives a gear, which in turn drives a series of smaller gears, which ultimately position an indicating needle on a dial. Kilowatt-hour meters have either four or five dials. Each dial has an indicating needle and a scale from 0 to 9. Many electric utility companies are now using kilowatt-hour meters that have a digital readout (Figure 4-21). This makes the meter easier to read.

FIGURE 4-21 A digital kilowatt-hour meter. This meter is programmable and uses a digital display to show the total number of kilowatt-hours used by a residential customer. *Courtesy of Landis+Gyr Energy Management Inc.*

Green Tip

Small kilowatt-hour meters can be used to determine the amount of electrical energy used by a specific appliance. These meters are typically plugged into a receptacle outlet and then the appliance is plugged into the kilowatt-hour meter. Some appliances like refrigerators can cost a small fortune to run, especially if they are an older model, so it's very important to check such items every so often to ensure that they're not wasting electrical energy and costing too much money to use.

(See Procedure 4-7 on page 161 for step-by-step instructions on how to read a kilowatt-hour meter.)

SMART METERS

A new type of kilowatt-hour meter, called a **smart meter** (Figure 4-22), is now being installed by electric utility companies in many parts of the country. These meters fit into the same meter enclosures as the "regular" kilowatt-hour meters discussed previously. Smart meters have two-way communications and are able to send and receive information to and from home owners and the electric utility company. These meters can measure not

FIGURE 4-22 A smart meter. This meter has two-way communication between the customer and the electric utility.

only the electricity used by a home, but also the electricity generated by other generation systems such as solar panels or wind turbines. The utility company can read these meters remotely, which means that meter readers are not needed. The data is transmitted from the smart meters to cell relays, which are wireless devices installed on utility poles. Cell relays pick up signals from smart meters and transmit the data to a "collection point." The data is then typically sent via a power line carrier, microwave, or fiber optic cable to the local electric utility's data center.

Smart meters can automate not only meter reading but also electric service connection and disconnection. This will help reduce the time to have new electric service connected. These meters also automatically notify the electric utility when the power goes out, resulting in less time for the utility workers to respond to the outage. As an added benefit, smart meters will also be able to communicate with future smart appliances in homes.

Green Tip

Smart meters open new possibilities for energy conservation. The ability to monitor energy use can result in energy savings by encouraging people to make energy-saving adjustments like turning off unneeded appliances, changing to more efficient lighting, or adjusting their thermostats. Additionally, smart meters will create an easy way for energy generation systems, such as wind and solar to be connected to the "smart grid" of the future. Smart meters measure electricity generated, as well as received, eliminating the need for installation of expensive specialized metering. This "net metering" of electricity can result in a greatly reduced electrical bill each month. Alternative energy systems connected to a home's electrical system can help reduce the need for new fossil-fuel generated capacity, now and in the future, and will benefit the environment in many ways. Alternative energy systems and net metering are covered in detail later on in this textbook.

SAFETY AND METERS

Safety is very important for a residential electrician when using test and measurement instruments. Many times, a meter is used to make a test or measurement while the electrical circuit is energized. Be sure to follow regular safety procedures when working on energized

circuits or equipment. A review of Chapter 1 may be appropriate at this time. Any meter that is not in good working order should not even be taken to the job site, where it could be used and somebody could get seriously injured or killed. Equipment could also become badly damaged or destroyed if an electrician does not follow proper safety procedures when using test and measurement instruments.

Meters that are used on electrical construction sites tend to lose their accuracy over time. This is usually because of the rough handling that sometimes occurs with these meters. Meters that are exposed to hot or cold temperature extremes are also likely to become inaccurate over time. Recalibration is necessary from time to time to bring a meter back to its intended level of accuracy. Stickers are placed on a meter so that an electrician can see when the meter was last calibrated (Figure 4-23). It is up to the electrician to make sure that the meters being used are recalibrated on a regular basis.

Although the electrical dangers encountered in residential wiring may not be as severe as those found in commercial and industrial wiring, they still present a serious safety hazard. Residential electricians should always follow safe meter practices when using meters of any kind. Here are a few safety reminders:

- Always wear safety glasses when using test and measurement instruments.

- Wear rubber gloves when testing or measuring "live" electrical circuits or equipment.

- Never work on energized circuits unless absolutely necessary.

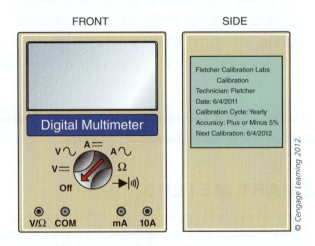

FIGURE 4-23 An example of a meter showing a calibration sticker. The calibration sticker will tell an electrician when the meter was last calibrated and when the next calibration should take place.

- If you must take measurements on energized circuits, make sure you have been properly trained to work with "live" circuits.

- Do not work alone, especially on "live" circuits. Have a buddy work with you.

- Keep your clothing, hands, and feet as dry as possible when taking measurements.

- Make sure that the meter you are using has a rating that is equal to or exceeds the highest value of electrical quantity you are measuring.

- Look for two or more labels from independent testing labs such as UL (USA), CSA (Canada), or TUV (Europe) to verify that your meters have been tested and certified.

- NFPA 70E requires that you visually inspect test and measurement tools frequently to help detect damage and ensure proper operation.

- NFPA 70E requires the use of IEC (International Electrotechnical Commission) rated meters. Look for 600 volt or 1000 volt CAT III or 600 volt CAT IV ratings on the front of meters and testers. Also, look for the "double insulated" symbol or wording on the back of the meter.

from experience...

A residential electrician who got a job in a local paper mill was seriously injured when he used a voltage tester with a 600-volt maximum voltage rating to do a test on a 4160-volt motor circuit. The meter exploded in his hands, and he suffered serious burns on his face, neck, hands, and arms that took many months to heal. He was lucky that he was not killed.

METER CARE AND MAINTENANCE

The final subject covered in this chapter has to do with the care and maintenance of test and measurement instruments. The meters previously discussed in this chapter run in price from around $20 for a basic non-contact voltage tester to over $400 for a good-quality true RMS DMM. To make sure that a meter lasts in the harsh environment of residential electrical construction, a few rules should be followed:

- Keep the meters clean and dry.

- Do not store analog meters next to strong magnets; the magnets can cause the meters to become inaccurate.

- All meters are very fragile and should be handled with care. Do not just throw them in your toolbox or on the dashboard of the company truck.

- Do not expose meters to wide temperature changes. Too much heat or too much cold can damage a meter.

- Make sure you know the type of circuit you are testing (AC or DC).

- Never let the value being measured exceed the range of the meter.

- Multimeters and ohmmeters will need to have their batteries changed from time to time.

- Many meters have fuses to protect them from exposure to excessive voltage or current values. Replacement of these fuses may have to be done. Check the meter's instruction manual for replacement fuse sizes and the location of the fuses.

- Have measuring instruments recalibrated once a year by a qualified person.

SUMMARY

This chapter presented information on the many measuring instrument types used in residential wiring. The proper use of these meters was also presented. Safe meter practices, as well as common meter care and maintenance techniques, were discussed in detail. Residential electricians must be familiar with a variety of test and measurement instruments. These instruments can provide a wealth of information to a residential electrician. All residential electricians need to know how to use test and measurement instruments in a safe and proper manner for a variety of residential wiring applications.

GREEN CHECKLIST

☐ A voltmeter can be used to determine if the voltage delivered to an electrical load is the proper amount. A too high or too low voltage will cause equipment to not work as efficiently as possible, resulting in excessive energy use.

☐ An ammeter can be used to determine if an electrical load is drawing more current than it should. Excessive current draw results in lower efficiency and wasted energy.

☐ Small kilowatt-hour meters can be used to determine the amount of electrical energy used by a specific appliance. These meters are typically plugged into a receptacle outlet and then the appliance is plugged into the kilowatt-hour meter.

☐ Smart meters can measure the electricity used by a home, as well as the electricity generated by other generation systems such as solar panels or wind turbines.

☐ The ability to monitor energy use with smart meters can result in energy savings by encouraging people to make energy-saving adjustments.

☐ Smart meters provide an easy way for energy generation systems such as wind and solar to be connected to the "smart grid" of the future.

☐ Smart meters can result in a greatly reduced electrical bill each month.

PROCEDURE 4-1

Using a Continuity Tester

- In this example, an electrician will determine if there is a break in a roll of nonmetallic sheathed cable using a continuity tester.

- Put on safety glasses and observe regular safety procedures.

- Strip about 1 inch of insulation from the conductors at each cable end.

(A) On one end, wirenut the conductors together.

(B) Turn the continuity tester on and connect it across the conductors at the other end. Note: Some continuity testers do not need to be turned on or off.

- If the continuity tester indicates continuity, there is no break in the cable. Note: Continuity will be indicated by a lamp coming on or an audible tone.

- If there is no indication of continuity, there is a break in the cable.

- Remove the tester and turn it off.

- Properly store the tester.

A

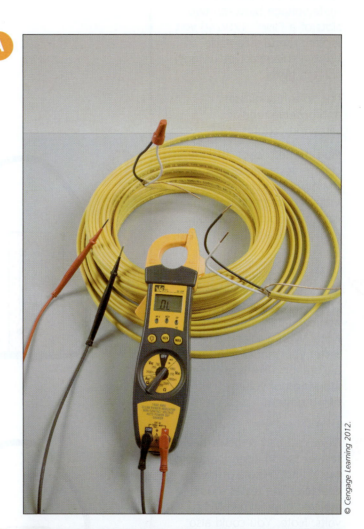

© Cengage Learning 2012.

B

© Cengage Learning 2012.

PROCEDURE 4-2

Using a Voltage Tester

- In this example, an electrician will determine the approximate voltage between two slots of a "live" 120/240 volt electric range receptacle using a solenoid type (Wiggy) voltage tester.

- Put on safety glasses and observe regular safety procedures.

- Plug the test leads into the voltage tester according to the owner's manual. Note: Voltage testers do not typically have to be turned on and off like a voltage meter.

A Insert the test leads into the X and Y slots of the receptacle. The tester should indicate approximately 240 volts. Note: With a solenoid-type tester you should also feel a vibration. This is another indication of voltage being present.

B Next, carefully remove the test leads and insert them into the X and W slots of the receptacle. The tester should indicate approximately 120 volts. Note: You could also use the Y and W slots for this test.

- Remove the tester leads.

- Properly store the tester.

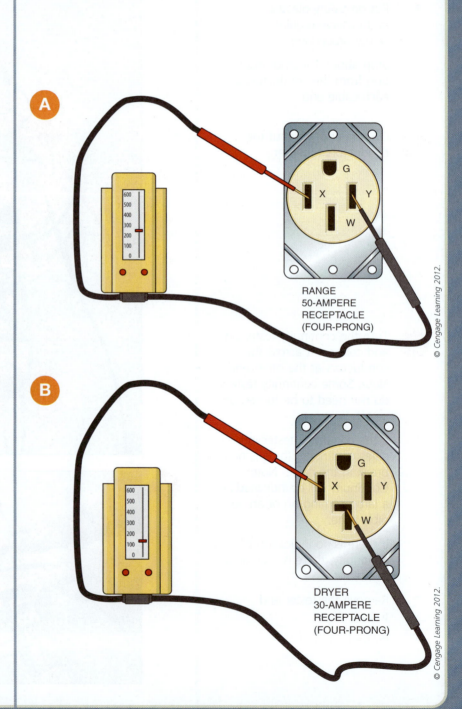

A

RANGE
50-AMPERE
RECEPTACLE
(FOUR-PRONG)

© Cengage Learning 2012.

B

DRYER
30-AMPERE
RECEPTACLE
(FOUR-PRONG)

© Cengage Learning 2012.

PROCEDURE 4-3

Using a Noncontact Voltage Tester

- In this example, an electrician will determine if an electrical conductor is energized using a noncontact voltage tester.

- Put on safety glasses and observe regular safety procedures.

- Identify the conductor to be tested.

(A) Bring the noncontact voltage tester close to the conductor. Note: Some noncontact voltage testers may have to be turned on before using.

- Listen for the audible alarm, observe a light coming on, or feel a vibration to indicate that the conductor is energized.

- Move the noncontact voltage tester away from the conductor and turn it off if necessary.

- Properly store the tester.

(A)

© Cengage Learning 2012.

PROCEDURE 4-4

Using a Voltmeter

- In this example, an electrician will measure the actual voltage between two slots of a "live" 120 volt duplex receptacle.

- Put on safety glasses and observe regular safety procedures.

- Plug the test leads into the voltmeter according to the owner's manual.

- Turn the voltmeter on if necessary.

A Set the selector switch on the voltmeter to AC volts. Make sure the voltage range selected will be at least 120 volts.

A

© Cengage Learning 2012.

B Insert the meter test leads into the slots of the receptacle.

B

C Read the value measured on the digital display.

- Turn the meter off.
- Properly store the meter.

C

PROCEDURE 4-5

Using a Clamp-On Ammeter

- In this example, an electrician will be measuring current flow through a conductor with a clamp-on ammeter. Note: You can take a current reading with a clamp-on ammeter clamped around only one conductor. For example, a clamp-on meter will not give a reading when clamped around a two-wire Romex™ cable.

- Put on safety glasses and observe regular safety procedures.

- Plug the test leads into the clamp-on ammeter according to the owner's manual.

- Turn the clamp-on ammeter on if necessary.

- If the meter has a scale selector switch, set it to the highest scale. Skip this step if the meter has an auto-ranging feature.

(A) Open the clamping mechanism and clamp it around the conductor.

(B) Read the displayed value.

- Turn the meter off.

- Properly store the meter.

A

© Cengage Learning 2012.

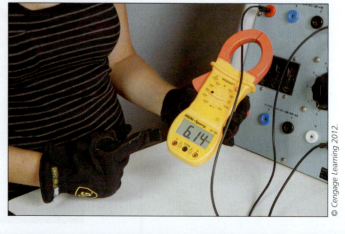

B

© Cengage Learning 2012.

PROCEDURE 4-6

Using a Digital Multimeter

Measuring alternating current voltage between two conductors.

- Put on safety glasses and observe regular safety procedures.

(A) Plug in the test leads—the black lead to the common jack and the red lead to the volt/ohm jack.

- Set the selector switch to the AC voltage icon.

- Connect the test leads in parallel with the conductors.

- Read the displayed voltage.

- Turn the meter off.

Measuring the current draw of a small (200-milliamp or less) AC load using an in-line connection.

- Put on safety glasses and observe regular safety procedures.

(B) Plug in the test leads—the black lead to the common jack and the red lead to the mA jack.

- Set the selector switch to the AC current icon.

- Turn the electrical power off and connect the meter in series with the load.

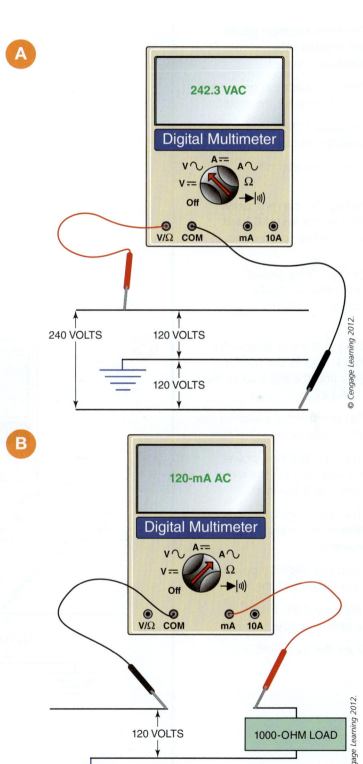

PROCEDURE 4-6

Using a Digital Multimeter (Continued)

- Turn the electrical power on and read the displayed current.

- Turn the electrical power off and disconnect the meter.

- Reconnect the circuit and re-energize.

- Turn the meter off.

Measuring the current draw of a 10-amp or less AC load using an in-line connection.

- Put on safety glasses and observe regular safety procedures.

C Plug in the test leads—the black lead to the common jack and the red lead to the 10A jack.

- Set the selector switch to the AC current icon.

- Turn the electrical power off and connect the meter in series with the load.

- Turn the electrical power on and read the displayed current.

- Turn the electrical power off and disconnect the meter.

- Reconnect the circuit and re-energize.

- Turn the meter off.

C

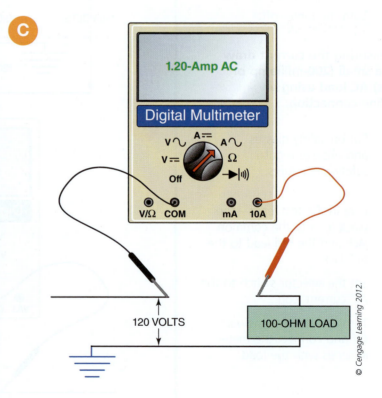

© Cengage Learning 2012.

Measuring the current draw of an AC load using a clamp-on ammeter accessory.

- Put on safety glasses and observe regular safety procedures.

D Plug in the clamp-on ammeter accessory according to the manufacturer's instructions.

- Set the selector switch to the AC current icon.

- Open the clamping mechanism and clamp around the conductor.

- Read the displayed value.

- Turn the meter off.

Measuring resistance.

- Put on safety glasses and observe regular safety procedures.

E Plug in the test leads—the black lead to the common jack and the red lead to the volt/ohm jack.

- Set the selector switch to the "ohms" icon.

- Connect the leads across the item to be tested.

- Read the displayed value.

- Turn the meter off.

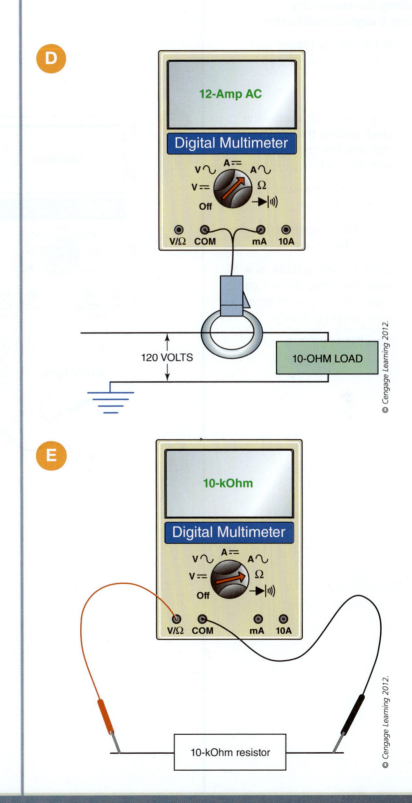

© Cengage Learning 2012.

PROCEDURE 4-6

Using a Digital Multimeter (Continued)

**Testing for continuity
using a digital multimeter.**

- Put on safety glasses and observe regular safety procedures.

- Plug in the test leads—the black lead to the common jack and the red lead to the volt/ohm jack.

- Set the selector switch to the "continuity" icon.

- Connect the leads across the item to be tested.

- Listen for the audible tone. If there is a tone, there is continuity. If there is no tone, there is no continuity.

- Turn the meter off.

A TONE WILL SOUND IF THERE IS CONTINUITY AND AN OHM READING WILL BE DISPLAYED.

.125 Ohm

Digital Multimeter

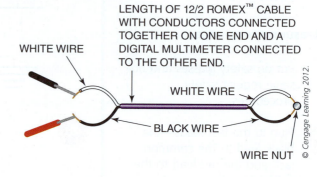

LENGTH OF 12/2 ROMEX™ CABLE WITH CONDUCTORS CONNECTED TOGETHER ON ONE END AND A DIGITAL MULTIMETER CONNECTED TO THE OTHER END.

WHITE WIRE

WHITE WIRE

BLACK WIRE

WIRE NUT

© Cengage Learning 2012.

PROCEDURE 4-7

Reading a Kilowatt-Hour Meter

- Note: To calculate the kilowatt-hours used since the last meter reading, simply subtract the previous reading from the new reading. This is how local electric utility companies determine your kilowatt-hour usage per month. Your electric bill is based on this value.

- Put on safety glasses and observe regular safety procedures.

 Begin with the right-hand dial and, working to the left, identify the number indicated. The dials indicate units, tens, hundreds, thousands, and ten thousands. Remember, if the indicating needle is between two numbers, the smaller of the two numbers is always read.

- Once all of the numbers have been identified, read the number from left to right. This is the number of kilowatt-hours that the meter has registered.

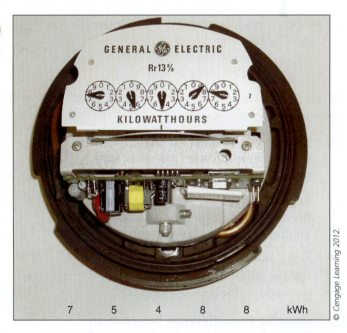

© Cengage Learning 2012.

REVIEW QUESTIONS

Directions: Answer the following items by circling the "T" for a true statement or the "F" for a false statement.

T F 1. An analog ohmmeter scale is read from left to right.

T F 2. A voltmeter is used to measure the amount of current flowing in a circuit.

T F 3. Multimeters can measure a variety of DC and AC electrical values.

T F 4. A voltage tester is a very accurate meter when compared to a voltmeter.

T F 5. A kilowatt-hour meter is used to measure the value of voltage delivered to a dwelling by an electric utility.

Directions: Answer the following questions with clear and complete responses.

6. Explain the purpose of the adjustment screw found on the face of analog measuring instruments.

7. Explain the reason for having a mirror behind the indicating needle on some analog measuring instruments.

8. State the most important rule to follow when using an ohmmeter or megohmmeter.

9. Describe the difference between an "in-line" ammeter and a "clamp-on" ammeter.

10. List three important rules to follow in the care and maintenance of meters.

1.

2.

3.

Directions: Circle the letter of the word or phrase that best completes the statement.

11. **An ammeter is used to measure _____.**
   a. resistance
   b. current
   c. voltage
   d. ohms

12. **Voltage is measured using a (n) _____.**
   a. ammeter
   b. megohmmeter
   c. continuity tester
   d. voltmeter

13. **A megohmmeter measures high values of resistance in _____.**
   a. megohms
   b. milliohms
   c. megawatts
   d. kilovolts

14. **A Wiggy is an electrical trade name for a(n) _____.**
   a. voltmeter
   b. voltage tester
   c. ammeter
   d. ohmmeter

15. **Kilowatt-hour meters are used to measure _____.**
   a. electrical power
   b. electrical energy
   c. large amounts of voltage
   d. large amounts of amperage

16. A ground resistance meter is used to determine the amount of resistance _____.
  a. through a grounding electrode to the earth
  b. though a grounding electrode to the metal frame of an electrical cabinet
  c. from one end of a grounding conductor to the other end
  d. from a grounded conductor to the earth

17. An ohmmeter is used to measure _____.
  a. voltage
  b. current
  c. continuity
  d. resistance

18. A noncontact voltage tester will indicate whether a _____ is present on a conductor.
  a. current
  b. voltage
  c. resistance
  d. short circuit

19. A continuity tester will indicate whether there is _____ through a conductor.
  a. voltage
  b. current
  c. continuity
  d. resistance

20. Recalibration of a meter is necessary from time to time in order to _____.
  a. make sure the warranty is still in place
  b. bring the meter back to its intended level of accuracy
  c. make sure it is the right caliber
  d. determine whether you have the right meter for the job

Understanding Residential Building Plans

OBJECTIVES

Upon completion of this chapter, the student should be able to:

- Demonstrate an understanding of residential building plans.

- Identify common architectural symbols found on residential building plans.

- Determine specific dimensions on a building plan using an architect's scale.

- Demonstrate an understanding of residential building plan specifications.

- Demonstrate an understanding of basic residential framing methods and components.

GLOSSARY OF TERMS

architectural firm a company that creates and designs drawings for a residential construction project

architect's scale a device used to determine dimensions on a set of building plans; it is usually a three-sided device that has each side marked with specific calibrated scales

balloon frame a type of frame in which studs are continuous from the foundation sill to the roof; this type of framing is found mostly in older homes

band joist the framing member used to stiffen the ends of floor joists where they rest on the sill

blueprint architectural drawings used to represent a residential building; it is a copy of the original drawings of the building

bottom plate the lowest horizontal part of a wall frame, which rests on the subfloor

break lines lines used to show that part of the actual object is longer than what the drawing is depicting

bridging diagonal braces or solid wood blocks installed between floor joists, used to distribute the weight put on the floor

ceiling joists the horizontal framing members that rest on top of the wall framework and form the ceiling structure; in a two-story house, the first-floor ceiling joists are the second floor's floor joists

centerline a series of short and long dashes used to designate the center of items, such as windows and doors

detail drawing a part of the building plan that shows an enlarged view of a specific area

dimension a measurement of length, width, or height shown on a building plan

dimension line a line on a building plan with a measurement that indicates the dimension of a particular object

draft-stops also called "fire-stops"; the material used to reduce the size of framing cavities in order to slow the spread of fire; in wood frame construction, it consists of full-width dimension lumber placed between studs or joists

electrical drawings a part of the building plan that shows the electrical supply and distribution for the building electrical system

elevation drawing a drawing that shows the side of the house that faces in a particular direction; for example, the north elevation drawing shows the side of the house that is facing north

extension lines lines used to extend but not actually touch object lines and that have the dimension lines drawn between them

floor joists horizontal framing members that attach to the sill plate and form the structural support for the floor and walls

floor plan a part of the building plan that shows a bird's-eye view of the layout of each room

footing the concrete base on which a dwelling foundation is constructed; it is located below grade

foundation the base of the structure, usually poured concrete or concrete block, on which the framework of the house is built; it sits on the footing

framing the building "skeleton" that provides the structural framework of the house

girders heavy beams that support the inner ends of floor joists

hidden line a line on a building plan that shows an object hidden by another object on the plan; hidden lines are drawn using a dashed line

leader a solid line that may or may not be drawn at an angle and has an arrow on the end of it; it is used to connect a note or dimension to a part of the building

legend a part of a building plan that describes the various symbols and abbreviations used on the plan

object line a solid dark line that is used to show the main outline of the building

platform frame a method of wood frame construction in which the walls are erected on a previously constructed floor deck or platform

plot plan a part of a building plan that shows information such as the location of the house, walkway, or driveway on the building lot; the plan is drawn with a view as if you were looking down on the building lot from a considerable height

rafters part of the roof structure that is supported by the top plate of the wall sections; the roof sheathing is secured to the rafters and then covered with shingles or other roofing material to form the roof

ribbon a narrow board placed flush in wooden studs of a balloon frame to support floor joists

scale the ratio of the size of a drawn object and the object's actual size

schedule a table used on building plans to provide information about specific equipment or materials used in the construction of the house

sectional drawing a part of the building plan that shows a cross-sectional view of a specific part of the dwelling

section line a broad line consisting of long dashes followed by two short dashes; at each end of the line are arrows that show the direction in which the cross section is being viewed

sheathing boards sheet material like plywood that is fastened to studs and rafters; the wall or roofing finish material will be attached to the sheathing

sill a length of wood that sets on top of the foundation and provides a place to attach the floor joists

specifications a part of the building plan that provides more specific details about the construction of the building

subfloor the first layer of floor material that covers the floor joists; usually 4-by-8-plywood or particleboard

symbol a standardized drawing on the building plan that shows the location and type of a particular material or component

top plate the top horizontal part of a wall framework

wall studs the parts that form the vertical framework of a wall section

Every residential electrician must be familiar with residential building plans. Most new construction and remodel electrical wiring jobs require an electrician to follow a building plan when installing the electrical system. This chapter introduces you to residential building plans and the common architectural symbols found on them. Building plan specifications and basic residential framing are also covered.

OVERVIEW OF RESIDENTIAL BUILDING PLANS

A residential building plan consists of a set of drawings that craft people use as a guide to build the house. The building plan is often called various names, such as prints, blueprints, drawings, construction drawings, or working drawings. It is not that important what name is used to refer to the building plans, but it is very important for a residential electrician to know what the various parts of a typical building plan are and how to read and interpret the information found on them. The main parts of a building plan that a residential electrician should be familiar with are the plot plan, floor plan, elevation drawings, sectional drawings, detail drawings, electrical drawings, schedules, and specifications. Understanding the types of drawing lines used on building plans and understanding the scale of a drawing is also very important.

PLOT PLANS

A plot plan is usually the first sheet in a set of building plans. It shows information such as the location of the house, walkway, or driveway on the building lot. The plot plan is drawn with a view as if you were looking down on the building lot from a considerable height. Electricians can get some good information from looking at a plot plan. For example, if the house is going to have an underground service entrance, the plot plan can be used to help determine where the trench for the underground wiring will need to be dug. Figure 5-1 shows a good example of a plot plan.

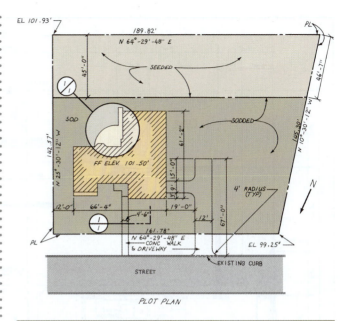

FIGURE 5-1   A plot plan shows the location of the house on the building lot. *Courtesy of PTEC-Clearwater-Architectural Drafting Department.*

FLOOR PLANS

The floor plan is a drawing that shows building details from a view directly above the house. A "bird's-eye" view is how it is often described. It is drawn to show the house as if a horizontal cut was made through the building at window height and then the top taken off. You are left with a view of the bottom half. Usually, each floor and the basement (if included) will have a floor plan drawing. The floor plans for each floor are called "first-floor plan," "second-floor plan," and so on. The floor plan for the basement is called the "basement plan." Figure 5-2 shows a typical floor plan.

Floor plans show the length and the width of the floor it is depicting. Dimensions are drawn on the floor plans and can be used by electricians to determine the exact size and location of various parts of the building structure, like doors, windows, and walls. This information is necessary when determining where to install wiring and electrical equipment.

ELEVATION DRAWINGS

An elevation drawing shows the side of the house that is facing a certain direction. It may show the height, length, and width of the house. Electricians can use elevation drawings to determine the heights of windows, doors, porches, and other parts of the structure. This type of information is not available

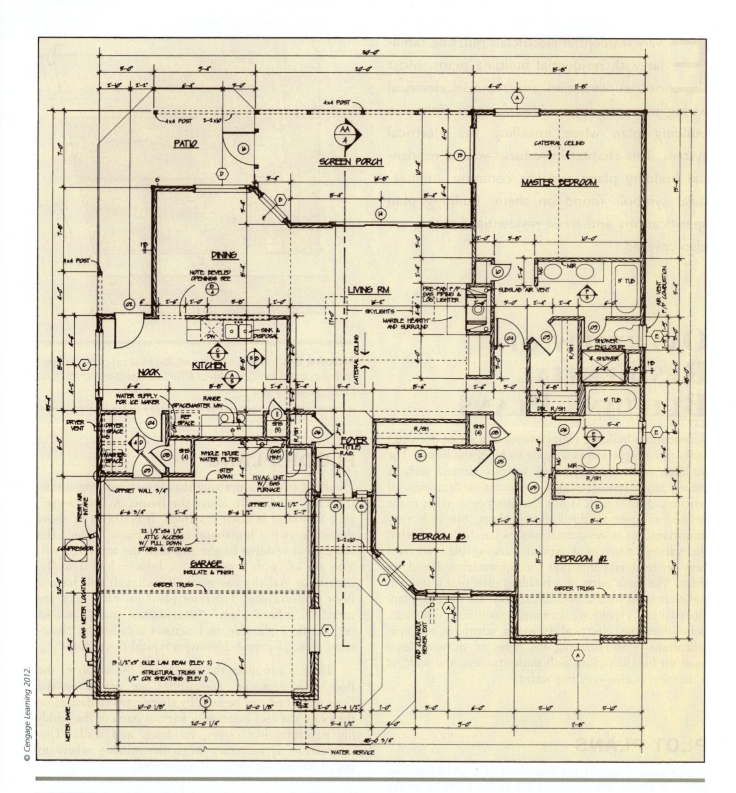

FIGURE 5-2   A floor plan shows the location of walls, partitions, windows, doors, and other building features. Building dimensions are found on the floor plan.

on floor plans but is needed when installing electrical items such as outside lighting fixtures and outside receptacles. Figure 5-3 shows an example of an elevation drawing.

Interior elevation drawings are often included in a set of building plans. The most common are kitchen and bathroom wall elevations. These drawings will show the design and size of cabinets and appliances located on or

FIGURE 5-3 An elevation drawing shows a side view of the structure.

against the walls. An interior elevation drawing of the kitchen or bathroom can help an electrician locate where the receptacles, switches, and lighting outlets can be located. Remember, much of the wiring and boxes will have to be installed before the cabinets and appliances are in, so it is very important to locate the electrical equipment properly. This will ensure that when everything is installed switches or receptacles are not covered up and that lighting outlets are in the right place. Figure 5-4 shows an example of an interior elevation drawing.

SECTIONAL DRAWINGS

A sectional drawing is a view that allows you to see the inside of a building. The view shown by a sectional drawing can be described as follows. Imagine that a Sawzall® has been used to cut off the side of a house. When the side of the house is moved away, you are left with a cutaway view of the rooms and the structural members of that part of the house. Figure 5-5 shows an example of a sectional drawing.

The point on the floor plan or the elevation drawing that is depicted by the sectional drawing is shown with a dashed line with arrows on the ends and is called a section line. Because a building plan may have several sectional drawings, the section lines are distinguished with letters or numbers located at the end of the arrows on the section lines. A typical sectional drawing may be labeled as "Section A–A" or "Section B–B."

The sectional drawings contain information that is important to an electrician. For example, a wall section drawing can allow an electrician to determine how he or she may run cable, or a sectional drawing of a floor may show an electrician how thick the wood will be for him or her to drill.

DETAIL DRAWINGS

A detail drawing shows very specific details of a particular part of the building structure. It is an enlarged view that makes details much easier to see than in a sectional drawing. They are usually located on the same plan sheet where the building feature appears. If they are shown on a separate sheet, they are numbered to refer back to a particular location on the building plan. Figure 5-6 shows an example of a detail drawing.

ELECTRICAL DRAWINGS

The most important part of the building plan for electricians is the electrical drawings. They show exactly what is required of an electrician for the complete installation of the electrical system. Electrical symbols are used on the plans to depict electrical equipment and devices. They are used as a type of shorthand to show an electrician which electrical items are required and where they are located. Using the symbols makes the plan less cluttered and easier to read. Specific electrical symbols are covered later in this chapter. Figure 5-7 shows an example of an electrical plan.

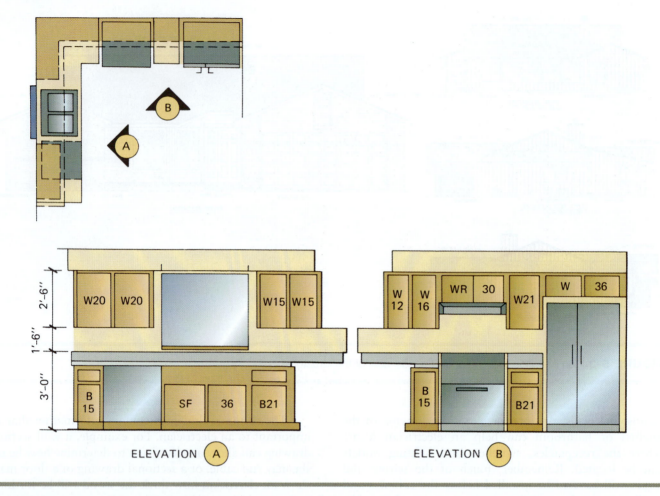

ELEVATION (A)

ELEVATION (B)

FIGURE 5-4   A typical interior wall elevation showing the location of cabinets and appliances in a kitchen. This type of drawing will allow an electrician to accurately locate where switch, receptacle, and lighting boxes can be installed. *Courtesy of PTEC-Clearwater-Architectural Drafting Department.*

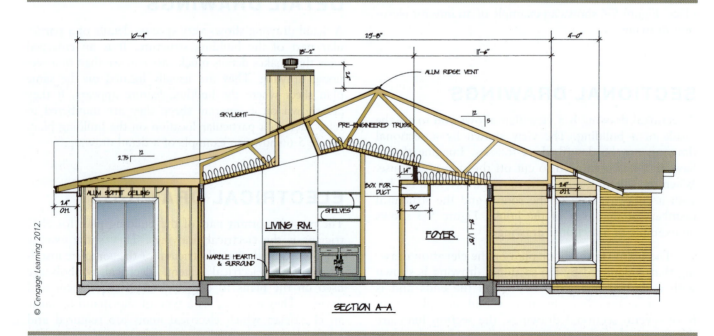

SECTION A-A

FIGURE 5-5   A sectional drawing shows a cutaway view of a certain part of the structure. You can see what part of the house this drawing is showing by looking at the section line in Figure 5-2.

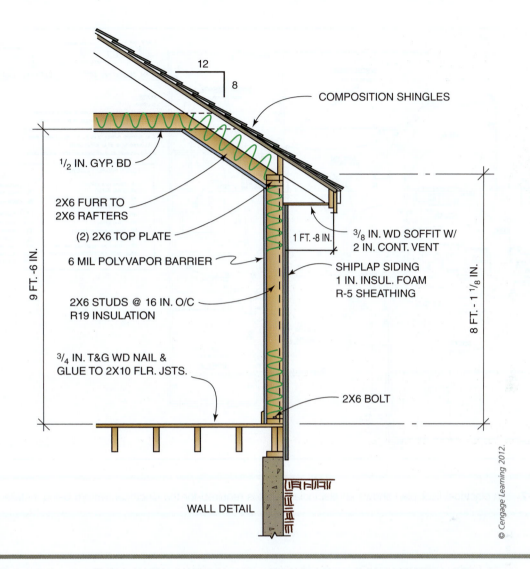

12
8

COMPOSITION SHINGLES

1/2 IN. GYP. BD

2X6 FURR TO
2X6 RAFTERS

(2) 2X6 TOP PLATE

6 MIL POLYVAPOR BARRIER

2X6 STUDS @ 16 IN. O/C
R19 INSULATION

3/4 IN. T&G WD NAIL &
GLUE TO 2X10 FLR. JSTS.

9 FT. -6 IN.

1 FT. -8 IN.

3/8 IN. WD SOFFIT W/
2 IN. CONT. VENT

SHIPLAP SIDING
1 IN. INSUL. FOAM
R-5 SHEATHING

8 FT. - 1 1/8 IN.

2X6 BOLT

WALL DETAIL

© Cengage Learning 2012.

FIGURE 5-6   A detail drawing shows an enlarged and more detailed view of a specific part of the structure.

Electrical contractors use the electrical plans to estimate the amount of material and labor needed to install the electrical system. This amount is used to project a total cost for the electrical system installation and is used in the bidding process. The electrical plans also provide a good map of the electrical system and can be consulted in the future, if and when problems arise.

Electricians rely on schedules to provide specific information about electrical equipment that needs to be installed in the building. For example, a lighting fixture schedule lists and describes the various types of lighting fixtures used in the house. This schedule also tells the electrician what type and how many lamps are used with each lighting fixture. Figure 5-8 shows an example of a lighting fixture schedule.

SCHEDULES

Schedules are used to list and describe various items used in the construction of the building. The schedules are usually set up in table form. For example, door and window schedules list the sizes and other pertinent information about the various types of doors and windows used in the building.

SPECIFICATIONS

The building plan **specifications** help provide clarity to the building plans. Only so much information can be included in a floor plan or an electrical plan. Specifications provide the extra details about equipment and construction methods that are not in the regular building plans.

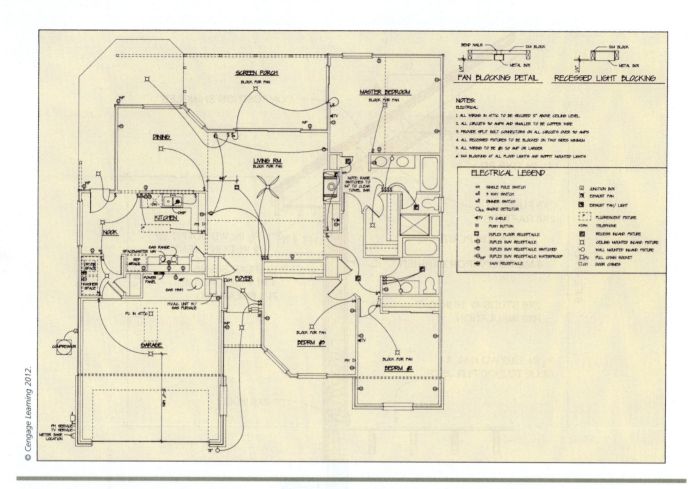

FIGURE 5-7   An electrical floor plan shows an electrician what is required for the electrical system being installed in the house.

Symbol	Number	Manufacturer and Catalog Number	Mounting	Lamps Per Fixture
A	2	Lightolier 10234	Surface	2 40-watt T-8 CWX Fluorescent
B	4	Lightolier 1234	Surface	4 40-watt T-8 WWX Fluorescent
C	1	Progress 32-486	Surface	1 60-watt medium base LED
D	1	Progress 63-8992	Pendant	5 60-watt medium base incandescent
E	3	Lithonia 12002-10	Recessed	1 75-watt medium base reflector
F	3	Hunter Paddle Fan 1-3625-77	Surface	3 25-watt medium base CFL
G	2	Nutone Fan/Light Model 162	Recessed	1 60-watt medium base incandescent

FIGURE 5-8   A lighting fixture schedule. Schedules provide more detailed information on certain pieces of equipment and materials being installed in a house. In this example, detailed information is given on the lighting fixtures to be installed as part of the electrical system.

Specifications provide detailed information to all the construction trades involved with the building. Of course, it is the electrical specifications that we electricians are most interested in. The electrical "specs" often include the specific manufacturer's catalog numbers and other information so that the electrical items will be the right size and type as well as having the proper electrical rating. The specifications are also used by the electrical contractor to help with the estimate of what it will cost to install the proposed electrical system. For example, the electrical specifications may state that all wiring in the house be no smaller than 12-AWG copper. If an electrical contractor based his estimate on using 14-AWG wire where possible, the bid would end up being much smaller than what it should be because 12-AWG costs more than 14-AWG wire to purchase and install. It pays to read the specifications through from beginning to end. Figure 5-9 shows an example of typical electrical specifications for a residential wiring job.

TYPES OF DRAWING LINES

There are many different line types used on a set of building plans. Recognizing what the lines represent will make it easier for a residential electrician to understand the building plan. Figure 5-10 shows several of the common drawing lines used on building plans. These lines are used as follows:

- An **object line** is a solid dark line that is used to show the main outline of the building. This includes exterior walls, interior partitions, porches, patios, and interior walls.

---

1. **GENERAL:** The "General Clause and Conditions" shall be and are hereby made a part of this division.

2. **SCOPE:** The electrical contractor shall furnish and install a complete electrical system as shown on the drawings and/or inthe specifications. Where there is no mention of the responsible party to furnish, install, or wire for a specific item on the electrical drawings, the electrical contractor will be responsible completely for all purchases and labor for a complete operating system for this item.

3. **WORKMANSHIP:** All work shall be executed in a neat and workmanlike manner. All exposed conduits shall be routed parallel or perpendicular to walls and structural members. Junction boxes shall be securely fastened, set true and plumb, and flush with finished surface when wiring method is concealed.

4. **LOCATION OF OUTLETS:** The electrical contractor shall verify location, heights, outlet and switch arrangements, and equipment prior to rough-in. No additions to the contract sum will be permitted for outlets in wrong locations, in conflict with other work, and so on. The owner reserves the right to relocate any device up to 10 feet (3.0 m) prior to rough-in, without any charge by the electrical contractor.

5. **CODES:** The electrical installation is to be made in accordance with the latest edition of the National Electrical Code (NEC®), all local electrical codes, and the utility company's requirements.

6. **MATERIALS:** All materials shall be new and shall be listed and bear the appropriate label of Underwriters Laboratories, Inc., or another nationally recognized testing laboratory for the specific purpose. The material shall be of the size and type specified on the drawings and/or in the specifications.

7. **WIRING METHOD:** Wiring, unless otherwise specified, shall be nonmetallic-sheathed cable, metal clad cable, or electrical metallic tubing (EMT), adequately sized and installed according to the latest edition of the NEC® and local ordinances.

8. **PERMITS AND INSPECTION FEES:** The electrical contractor shall pay for all permit fees, plan review fees, license fees, inspection fees, and taxes applicable to the electrical installation and shall be included in the base bid as part of this contract.

9. **TEMPORARY WIRING:** The electrical contractor shall furnish and install all temporary wiring for handheld tools and construction lighting per latest OSHA standards and Article 590, NEC®, and include all costs in the base bid.

10. **NUMBER OF OUTLETS PER CIRCUIT:** In general, not more than 10 lighting and/or receptacle outlets shall be connected to any one lighting branch circuit. Exceptions may be made in the case of low-current-consuming outlets.

11. **CONDUCTOR SIZE:** General lighting branch circuits shall be 14-AWG copper-protected by 15-ampere overcurrent devices. Small appliance circuits shall be 12-AWG copper-protected by 20-ampere overcurrent devices. All other circuits: conductors and overcurrent devices as required by the NEC®.

© Cengage Learning 2012.

FIGURE 5-9 An example of a plan's electrical specifications. Specifications provide additional written information that help explain the building drawings to the craft people constructing the building. Electricians are interested mainly in the electrical specifications, but other areas of the plan specifications may be helpful.

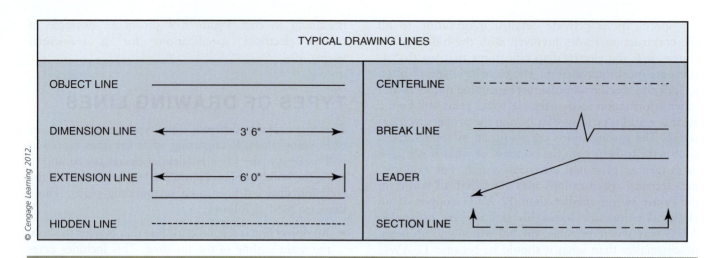

FIGURE 5-10   These lines represent some of the common drawing lines used on a set of building plans.

- **Dimension lines** are thin unbroken lines that are used to indicate the length or width of an object. Arrows are usually placed at each end of the line, and the dimension value is placed in a break of the line or just close to the line.

- **Extension lines** are used to extend but not actually touch object lines and have the dimension lines drawn between them.

- **Hidden lines** are straight dashed lines that are used to show lines of an object that are not visible from the view shown in the plan.

- A **centerline** is a series of short and long dashes used to designate the center of items, such as windows and doors. Sometimes you will see the dashed line going though the letter "C" to specify the center of an object. They provide a reference point for dimensioning.

- **Break lines** are used to show that part of the actual object is longer than what the drawing is depicting. The full length of the object may not be able to be drawn in some building plans.

- A **leader** is a solid line that is usually drawn at an angle and has an arrow at the end of it. It is used to connect a note or dimension to a part of the building shown in the drawing.

- A **section line** is often a broad line consisting of long dashes followed by two short dashes. At each end of the line are arrows. The arrows show the direction in which the cross section is being viewed. Letters are used to identify the cross sectional view of a specific part of a building.

SCALE

If a building were drawn to its true size, it obviously would not fit on a piece of paper. If a building were drawn just so it would fit on a piece of paper, the drawing would be very much out of proportion. To enable a drawing of a building to be put on a piece of paper and still keep everything in proportion, the building is drawn to some reduced **scale**. All dimensions of the building will be drawn smaller than the actual size and will be reduced in the same proportion. Most residential plans are drawn to a scale of 1/4 inch = 1 feet, 0 inches. This means that each 1/4 inch on the drawing would equal 1 foot on the actual building. If a part of the building plan used a 1/8 inch = 1 foot, 0 inch scale, 1/8 inch on the drawing would equal 1 foot of the actual building.

Electricians need to know what the scale of the drawing is so that they can get an accurate measurement for where electrical equipment will be located. The scale to which the drawing has been done can be usually found in the title block (Figure 5-11). The title block is often located in the lower right-hand corner of the drawing but may be found in other locations on the building plans. In addition to the drawing scale, a title block contains other information such as the name of the building project, the address of the project, the name of the **architectural firm**, the date of completion, the drawing sheet number, and a general description of the drawing.

Architect's Scale

An **architect's scale** is used to determine dimensions on a set of building plans (Figure 5-12). Architect's scales can be made of wood but are now more often made of aluminum or plastic. It is usually a three-sided device that has each side marked with specific calibrated scales. Single sided architect's scales are also available. The scales are generally grouped in pairs and arranged so that one scale is read from the right, and the other scale is read

⚠	TYPE OF REVISION OR ENGINEERING STATUS		INT.
⚠			
⚠			
REV	DATE	DESCRIPTION	APPD

YOUR COMPANY NAME YOUR COMPANY ADDRESS & PHONE	SEAL

SHEET TITLE **DRAWING TITLE**			ISSUED **FINAL DATE**		
DESIGNED	DRAWN	CxD	JOB NO	DRAWING NO	REV	SHEET
SCALE		DATE	FILE NO		OF

© *Cengage Learning 2012.*

FIGURE 5-11 A title block is typically located in the lower right-hand corner of a building drawing. It contains some very important information for an electrician, such as the scale of the drawing.

© *Cengage Learning 2012.*

FIGURE 5-12 An architect's scale can be used to determine actual measurements from the building plans.

from the left. Typical architect's scales used in the United States have the following scales:

- Full scale, with inches divided into sixteenths of an inch.
- $3'' = 1'\text{-}0'' / 1\text{-}1/2'' = 1'\text{-}0''$
- $1'' = 1'\text{-}0'' / 1/2'' = 1'\text{-}0''$
- $3/4'' = 1'\text{-}0'' / 3/8'' = 1'\text{-}0''$
- $1/4'' = 1'\text{-}0'' / 1/8'' = 1'\text{-}0''$
- $3/16'' = 1'\text{-}0'' / 3/32'' = 1'0''$

To use an architect's scale you must first determine the scale of the drawing. Then, determine which side of the architect's scale has that same scale marked on it and line up the zero mark of the scale with the beginning of the length you wish to measure. Then, determine at what point on the scale the end of the length you wish

to measure is located. If the distance you are measuring falls exactly on a line of the architect's scale, then that is the measurement in feet. If the distance you are measuring does not fall exactly on a line, it means that the measurement will include a certain number of inches. Determining the inches will require a second measurement. The part of the scale located before the zero mark is subdivided into twelfths, which corresponds to inches. Before removing the architect's scale, use a pencil to lightly mark the location on the drawing of the last foot mark before the end of the length you are measuring. Then, use the inches section of the scale to determine the number of inches from your pencil mark to the end of the length you are measuring. The full measurement will be the number of feet plus this number of inches. This method may seem a little confusing at first but the more you use the architect's scale the easier it will become.

from experience...

Because most residential construction plans are drawn to a $1/4$ inch = 1 foot, 0 inch scale, electricians are able to use a folding rule or their tape measure to get accurate dimensions from the building plans. Each $1/4$ inch on the tape or rule represents 1 foot, each $1/8$ inch (half of $1/4$ inch) equals 6 inches and each $1/16$ inch (half of $1/8$ inch) equals 3 inches. Going the other way, 1 inch (4 × $1/4$ inch) on the tape or rule will equal 4 feet; 2 inches (8 × $1/4$ inch) on the tape or rule will equal 8 feet. See Figure 5-13.

FIGURE 5-13   An electrician can use a tape measure to determine actual measurements on a building plan with a $1/4$ inch = 1 foot scale.

COMMON ARCHITECTURAL SYMBOLS

There are many architectural **symbols** used on building plans to depict everything from a kitchen sink to a window. Notes are used to provide additional information about a particular symbol. The symbols and notes help keep the plan from becoming so cluttered with information that it would be impossible to read. Each trade has its own set of symbols that are used to identify items associated with that trade. You should learn to recognize the common symbols used by other trades on residential building plans so that you can make sure that these items do not interfere with your installation of the electrical system. Figure 5-14 shows some common architectural symbols that residential electricians should be familiar with.

ELECTRICAL SYMBOLS

As we discussed earlier in this chapter, the electrical drawing is the most important part of the building plan for an electrician. The electrical drawing contains many electrical symbols that show the location and type of electrical equipment required to be installed as part of the electrical system. Recognizing what each symbol represents is a very important skill for an electrician to have.

The American National Standards Institute (ANSI) has approved a standard titled *Symbols for Electrical Construction Drawings,* which is published by the National Electrical Contractors Association (NECA). This document shows standard electrical symbols for use on electrical drawings. Figure 5-15 shows the common electrical symbols for lighting outlets, Figure 5-16 shows the common electrical symbols for receptacle outlets, and Figure 5-17 shows common symbols used to show switch types. A number of other symbols used to represent common pieces of electrical equipment are shown in Figure 5-18. Most people who draw the electrical plans will use the ANSI electrical symbols. However, plans may have symbols that are not standard. If this is the case, a symbols **legend** is usually included in the plans, listing the symbols used on the building plans and what they all mean.

There are a few symbols used on electrical plans to indicate electrical wiring. A curved dashed line on a plan that goes from a switch symbol to a lighting outlet symbol indicates that the outlet is controlled by that switch (Figure 5-19). This line is usually curved to eliminate confusion about whether it is a hidden line. You may remember that a hidden line is also dashed, but it is always drawn straight. A curved solid line is used in a cabling diagram to represent the wiring method used. If the curved solid line has an arrow on the end of it, it is a "home run" and indicates that wiring from that point goes all the way to the loadcenter. A number next to the "home run" symbol will indicate its circuit number. Slashes on the curved solid line are sometimes used in a cable diagram to indicate how many conductors are in the cable. A good cabling diagram can make the installation of the wiring very easy. It also is helpful in troubleshooting a circuit in the event of a problem anytime in the future. Figure 5-20 shows the difference between a cabling diagram and a wiring diagram.

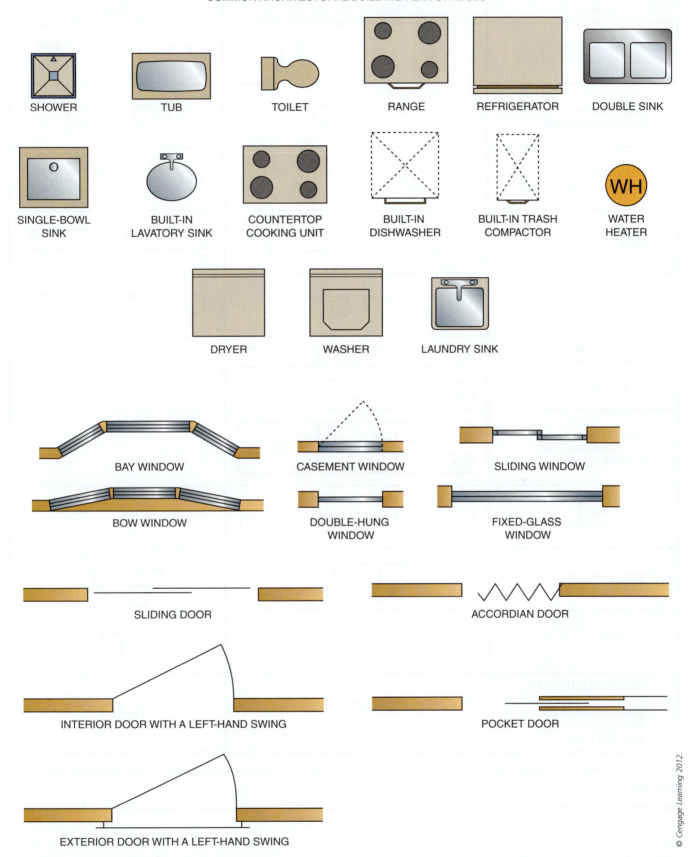

FIGURE 5-14   Common residential architectural symbols.

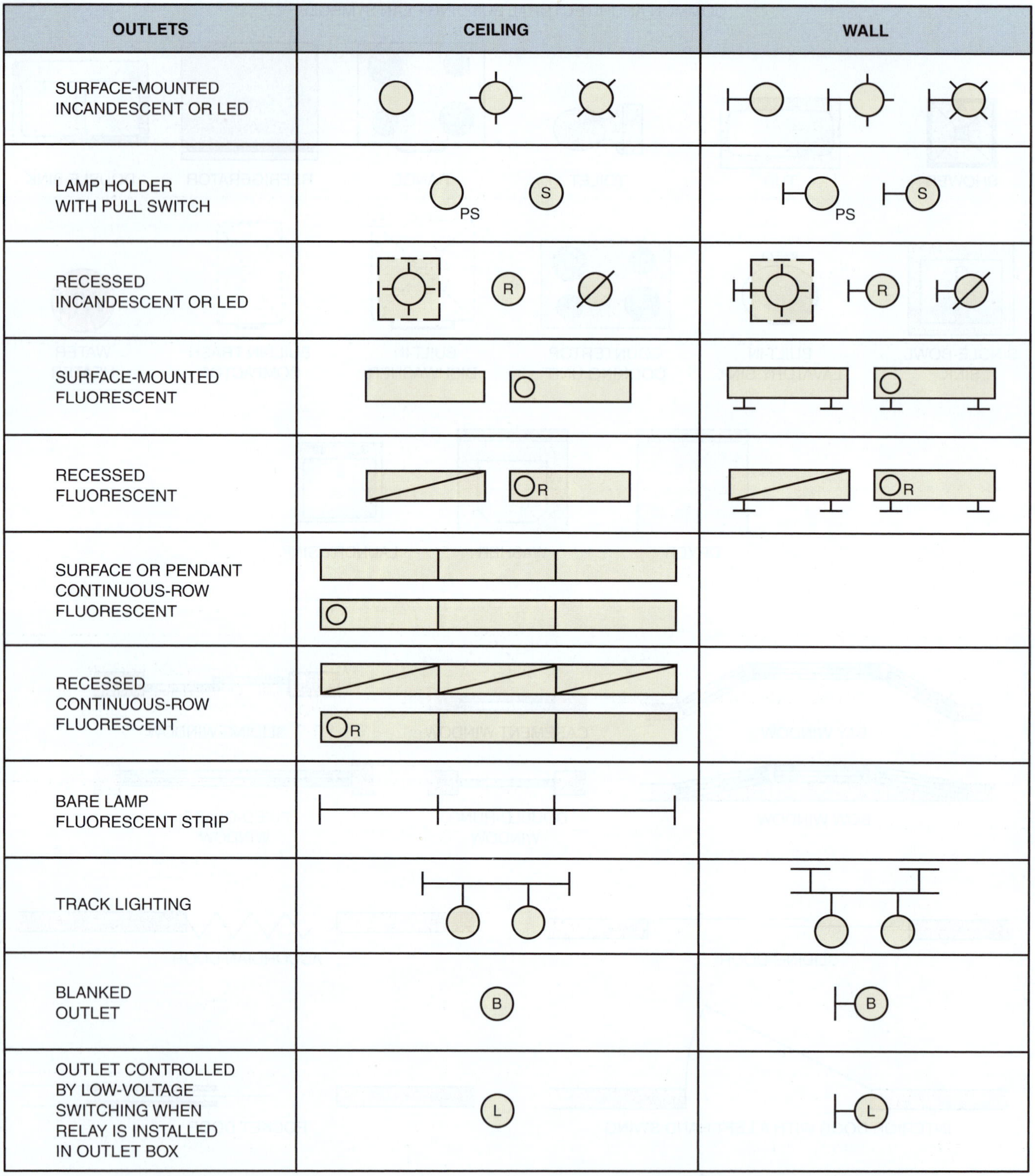
OUTLETS
CEILING
WALL
SURFACE-MOUNTED INCANDESCENT OR LED
LAMP HOLDER WITH PULL SWITCH
PS
S
PS
S
RECESSED INCANDESCENT OR LED
R
R
SURFACE-MOUNTED FLUORESCENT
RECESSED FLUORESCENT
R
R
SURFACE OR PENDANT CONTINUOUS-ROW FLUORESCENT
RECESSED CONTINUOUS-ROW FLUORESCENT
R
BARE LAMP FLUORESCENT STRIP
TRACK LIGHTING
BLANKED OUTLET
B
B
OUTLET CONTROLLED BY LOW-VOLTAGE SWITCHING WHEN RELAY IS INSTALLED IN OUTLET BOX
L
L

FIGURE 5-15 Common lighting outlet symbols.

RECEPTACLE OUTLETS		
SINGLE-RECEPTACLE OUTLET		ELECTRIC CLOTHES DRYER OUTLET
DUPLEX-RECEPTACLE OUTLET		FAN OUTLET
TRIPLEX-RECEPTACLE OUTLET		CLOCK OUTLET
DUPLEX-RECEPTACLE OUTLET, SPLIT CIRCUIT		FLOOR OUTLET
DOUBLE-DUPLEX RECEPTACLE (QUADPLEX)		MULTIOUTLET ASSEMBLY; ARROW SHOWS LIMIT OF INSTALLATION. APPROPRIATE SYMBOL INDICATES TYPE OF OUTLET, SPACING OF OUTLETS INDICATED BY "X" INCHES.
WEATHERPROOF RECEPTACLE OUTLET		FLOOR SINGLE-RECEPTACLE OUTLET
GROUND FAULT CIRCUIT INTERRUPTER RECEPTACLE OUTLET		FLOOR DUPLEX-RECEPTACLE OUTLET
RANGE OUTLET		FLOOR SPECIAL-PURPOSE OUTLET
SPECIAL-PURPOSE OUTLET (SUBSCRIPT LETTERS INDICATE SPECIAL VARIATIONS: DW = DISHWASHER. A, B, C, D, ETC., ARE LETTERS KEYED TO EXPLANATION ON DRAWINGS OR IN SPECIFICATIONS).		

FIGURE 5-16 Common receptacle outlet symbols.

SWITCH SYMBOLS	
S OR S_1	SINGLE-POLE SWITCH
S_2	DOUBLE-POLE SWITCH
S_3	THREE-WAY SWITCH
S_4	FOUR-WAY SWITCH
S_D	DOOR SWITCH
S_{DS}	DIMMER SWITCH
S_G	GLOW SWITCH TOGGLE— GLOWS IN OFF POSITION
S_K	KEY-OPERATED SWITCH
S_{KP}	KEY SWITCH WITH PILOT LIGHT
S_{LV}	LOW-VOLTAGE SWITCH
S_{LM}	LOW-VOLTAGE MASTER SWITCH
S_{MC}	MOMENTARY-CONTACT SWITCH
 Ⓜ	OCCUPANCY SENSOR—WALL MOUNTED WITH OFF-AUTO OVERRIDE SWITCH
 Ⓜ P	OCCUPANCY SENSOR—CEILING MOUNTED "P" INDICATES MULTIPLE SWITCHES WIRE-IN PARALLEL
S_P	SWITCH WITH PILOT LIGHT ON WHEN SWITCH IS ON
S_T	TIMER SWITCH
S_R	VARIABLE-SPEED SWITCH
S_{WP}	WEATHERPROOF SWITCH

FIGURE 5-17 Common electrical switch symbols.

RESIDENTIAL FRAMING BASICS

Residential electricians need to become familiar with the basic structural framework of a house. During the rough-in stage of the residential wiring system installation, electrical boxes will need to be mounted on building **framing** members, and wiring methods will have to be installed on or through building framing members.

Knowing how a house is constructed will allow an electrician to install the wiring system in a safe and efficient manner.

Wood construction is still the most often used framing type, but metal framing is being used more and more in residential construction. Whether the framing type is wood or metal, the structural parts that electricians should know are pretty much the same. There are two construction framing methods that electricians will encounter most often. The **platform frame**, sometimes called the "western frame," is the most common method used in today's new home construction (Figure 5-21). In this type of construction, the floor is built first, and then the walls are erected on top of it. If there is a second floor, it is simply built the same way as the first floor and placed on top of the completed first floor. The other common framing method is called the **balloon frame** (Figure 5-22). This construction method is not used today but was used extensively in the construction of many existing homes. In balloon framing, the wall studs and the first-floor joists both rest on the sill. If there is a second floor, the second-floor joists rest on a 1-by-4-inch **ribbon** that is cut flush with the inside edges of the studs.

Knowing the names of the structural parts and the role they play will help you better understand where and how to install the electrical system. The following major structural parts are used in houses that are built with either the platform framing or the balloon framing method:

- **Rafters** are used to form the roof structure of the building and are supported by the top plate.

- **Ceiling joists** are horizontal framing members that sit on top of the wall framing. They form the structural framework for the ceiling. If there is a floor above, these ceiling joists become **floor joists** and also form the structural framework for the floor above.

- **Draft-stops**, commonly called "fire-stops," are used to curtail the spread of fire in a house. They are usually pieces of lumber that are placed between studs or joists to block the path of fire through the cavities formed between studs or joists.

- The **top plate** is located at the top of the wall framework. It is usually a 2-by-4-inch or a 2-by-6-inch piece of lumber. Some construction methods call for two pieces of lumber to form the top plate. When two pieces of lumber are used, the common name for this structural member is the "double plate."

- **Wall studs** are used to form the vertical section of the wall framework. Interior walls are usually 2-by-4-inch lumber but may be 2-by-6-inch.

- The **bottom plate** is the bottom of the wall framework and rests on the top of the subfloor. It is usually a 2-by-4-inch or a 2-by-6-inch piece of lumber.

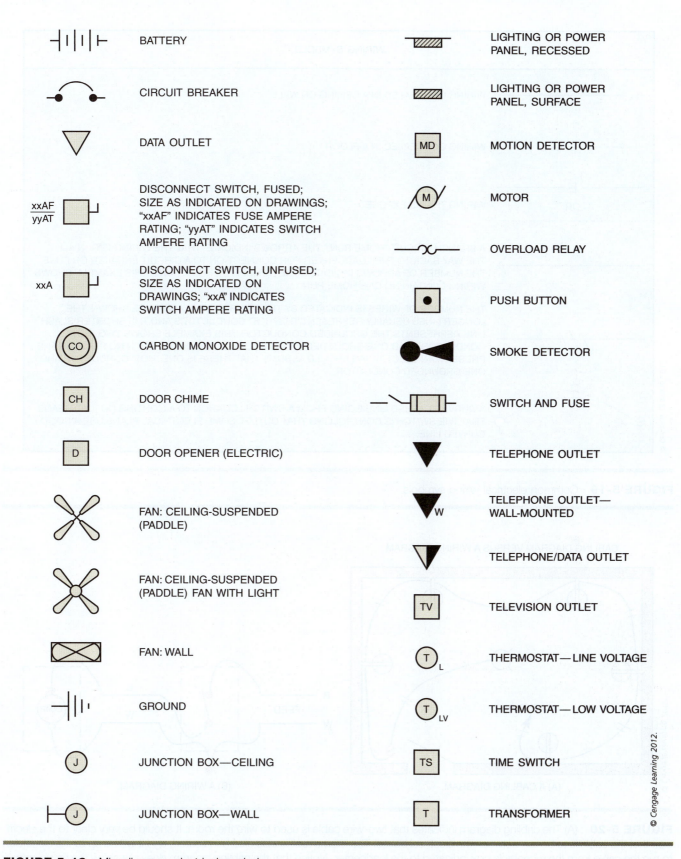

FIGURE 5-18   Miscellaneous electrical symbols.

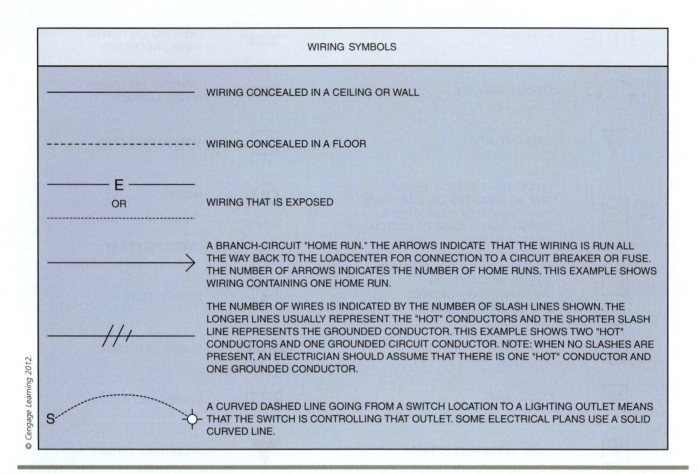

FIGURE 5-19 Common electrical wiring symbols.

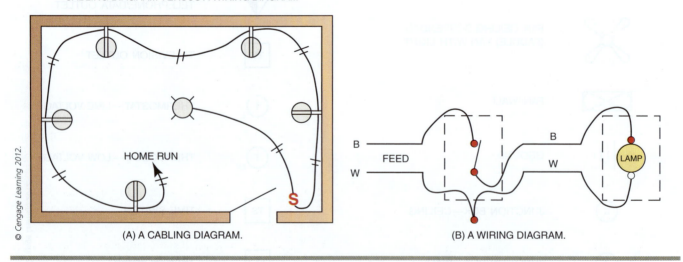

FIGURE 5-20 (A) The cabling diagram indicates that two-wire cable is used to wire the room. It should be very clear to the electrician that the cable is to be run between the electrical boxes. The "home run" is also clearly indicated, and it will be up to the electrician to run the cable from the receptacle box indicated to the loadcenter. Notice that the cabling diagram does not show the actual connections to be made at the electrical boxes. (B) The wiring diagram will show an electrician exactly where the conductors terminate. New electricians find wiring diagrams helpful when wiring circuits and equipment. As you become more experienced, you will not need to rely on wiring diagrams as much.

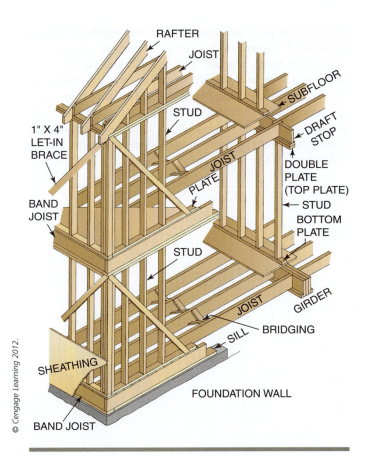

FIGURE 5-21   Platform frame construction showing the location and names of the various framing members.

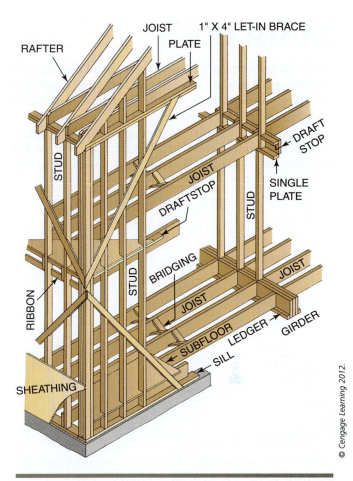

FIGURE 5-22   Balloon frame construction showing the location and names of the various framing members.

- The **subfloor** is the first layer of flooring that covers the floor joists and is usually made of 4-by-8-foot sheets of plywood or particleboard. Some areas of the country may still be using 1-by-6-inch boards for the subfloor.

- **Floor joists** are horizontal framing members that attach to the sill and form the structural support for the floor and walls. Floor joists are usually 2-by-8-, 2-by-10-, or 2-by-12-inch pieces of lumber.

- The **girders** in a house are the heavy beams that support the inner ends of the floor joists. A girder can be metal or wood. A wooden girder is usually made from several 2-by-10-inch or 2-by-12-inch lengths of lumber fastened together. There are two ways that a girder can support the floor joists: The joists can rest on top of the girder, or joist hangers can be nailed to the sides of the girder, and the hangers support the joists.

- The **band joist** (sometimes called a rim joist) is the framing member in platform framing that is used to stiffen the ends of the floor joists. It is normally the same size as the floor joists.

- The **sill** is a piece of wood that lies on the top of the foundation and provides a place to attach the floor joists.

- **Sheathing boards** are sheet material (like plywood) that are attached to the outside of studs or rafters and add rigidity to the framed structure. The roof and outside wall finish material is attached to the sheathing.

- A **foundation**, usually made of poured concrete or concrete block, is the part of the house that supports the framework of the building.

- The **footing** is located at the bottom of the foundation and provides a good base for the foundation to be constructed.

- The diagonal braces or solid wood blocks installed between floor joists are called **bridging**, which are used to distribute the weight put on the floor.

SUMMARY

This chapter introduced you to the parts of a residential building plan that are most important to the electrician installing the electrical system. Common residential architectural symbols were covered with special emphasis on electrical symbols. An overview of building specifications was given, and basic residential construction framing was covered. Reading and interpreting the information on a building plan are important skills for a residential electrician to have. As with most acquired skills, the more practice you have doing it, the more proficient you become. Your skill in reading and interpreting building plans will increase the more you practice. Remember the basics as covered in this chapter, and you should be able to figure out what most residential plans are showing. When in doubt, always ask your job-site supervisor.

REVIEW QUESTIONS

Directions: Answer the following items with clear and complete responses.

1. Describe an elevation drawing.

2. Describe a detail drawing.

3. Describe a floor plan.

4. Describe an electrical floor plan.

5. What is the purpose of a legend on a set of building plans?

6. The scale on a building plan is ¼ inch = 1 foot, 0 inch. An electrician uses a tape measure to measure the length of a wall on the plan. If the length on the tape measure is 4½ inches, determine the actual length of the wall.

7. Describe the purpose of a dimension line.

8. A hidden line shows part of the building that is not visible on the drawing. How are hidden lines drawn?

9. On an electrical floor plan, dashed lines are sometimes drawn from switches to various outlets. What do the dashed lines represent?

10. Describe the purpose of the plan specifications.

Directions: In the spaces provided, draw the electrical symbol for each of the items listed.

11. _____ Duplex-receptacle outlet

12. _____ Single-pole switch

13. _____ Clock outlet

14. _____ Three-way switch

15. _____ Four-way switch

16. _____ Ceiling lighting outlet

17. _____ Ceiling junction box

18. _____ Floor duplex receptacle outlet

19. _____ Single-receptacle outlet

20. _____ Special-purpose outlet (dishwasher)

21. _____ Smoke detector

22. _____ Split duplex receptacle

23. _____ A homerun

24. _____ Television outlet

25. _____ Dimmer switch

26. _____ Key operated switch

27. _____ Fan outlet

28. _____ Range outlet

29. _____ Track lighting outlet (ceiling)

30. _____ Recessed LED outlet (ceiling)

Determining Branch Circuit, Feeder Circuit, and Service Entrance Requirements

OBJECTIVES

Upon completion of this chapter, the student should be able to:

- Determine the minimum number and type of branch circuits required for a residential wiring system.

- Demonstrate an understanding of the basic *NEC®* requirements for calculating branch-circuit sizing and loading.

- Calculate the minimum conductor size for a residential service entrance.

- Determine the proper size of the service entrance main disconnecting means.

- Determine the proper size for a loadcenter used to distribute the power in a residential wiring system.

- Calculate the minimum-size feeder conductors delivering power to a subpanel.

- Demonstrate an understanding of the steps required to calculate a residential service entrance using the standard or optional method as outlined in Article 220 of the *NEC®*.

GLOSSARY OF TERMS

ambient temperature the temperature of the air that surrounds an object on all sides

ampacity the current, in amperes, that a conductor can carry continuously under the conditions of use without exceeding its temperature rating

bathroom branch circuit a branch circuit that supplies electrical power to receptacle outlets in a bathroom; lighting outlets may also be served by the circuit as long as other receptacle or lighting outlets outside the bathroom are not connected to the circuit; it is rated at 20 amperes

branch circuit the circuit conductors between the final overcurrent device (fuse or circuit breaker) and the outlets

bundled cables or conductors that are physically tied, wrapped, taped, or otherwise periodically bound together

dwelling unit one or more rooms arranged for complete, independent housekeeping purposes, with space for eating, living, and sleeping, facilities for cooking, and provisions for sanitation

feeder the circuit conductors between the service equipment and the final branch-circuit overcurrent protection device

general lighting circuit a branch-circuit type used in residential wiring that has both lighting and receptacle loads connected to it; a good example of this circuit type is a bedroom branch circuit that has both receptacles and lighting outlets connected to it

general purpose branch circuit a branch circuit that supplies two or more receptacles or lighting outlets, or a combination of both; household appliances such as vacuum cleaners, small room air conditioners, televisions, or stereo equipment can also receive electrical power from this type of branch circuit

individual branch circuit a circuit that supplies only one piece of electrical equipment; examples are one range, one space heater, or one motor

laundry branch circuit a type of branch circuit found in residential wiring that supplies electrical power to laundry areas; no lighting outlets or other receptacles may be connected to this circuit

lug a device commonly used in electrical equipment to terminate a conductor

nipple an electrical conduit of 2 feet or less in length used to connect two electrical enclosures

outlet a point on the wiring system at which current is taken to supply electrical equipment; an example is a lighting outlet or a receptacle outlet

service disconnect a piece of electrical equipment installed as part of the service entrance that is used to disconnect the house electrical system from the electric utility's system

small-appliance branch circuit a type of branch circuit found in residential wiring that supplies electrical power to receptacles located in kitchens and dining rooms; no lighting outlets are allowed to be connected to this circuit type

volt-ampere a unit of measure for alternating current electrical power; for branch-circuit, feeder, and service calculation purposes, a watt and a volt-ampere are considered the same

Before the actual installation of a residential wiring system can begin you need to determine the requirements for the number and size of the branch circuits that will distribute electrical power throughout the house. There is also a need to determine the number and size of any feeder circuits required. Last, but certainly not least, you need to calculate the minimum size service entrance that will provide the correct amount of electrical energy from the electric utility to the house wiring system. The *National Electrical Code®* (*NEC®*) outlines the procedures for calculating the minimum size of branch circuits, feeders, and service entrance conductors required for a residential electrical system. It can initially be a confusing task for new electricians to determine exactly which steps are required for the calculations. However, once the steps are identified, the calculations are not that difficult to perform. This chapter covers the calculation of branch-circuit and feeder loads as well as how to determine the minimum-size conductor and maximum-size overcurrent device for these circuits. It presents in detail the steps required to determine what size service entrance needs to be installed as part of a residential electrical system. Sizing service entrance main electrical panels and subpanels is also covered.

DETERMINING THE NUMBER AND TYPES OF BRANCH CIRCUITS

Before an electrician can determine the size of the service entrance required to supply the house electrical system with power, the number and types of circuits supplying power to the various electrical loads in the house must be determined. An understanding of the circuit types used in residential wiring is necessary for this. The *NEC®* defines a **branch circuit** as the circuit conductors between the final overcurrent device (fuse or circuit breaker) and the power or lighting outlets (Figure 6-1). Types of branch circuits used in residential wiring include the following:

- Lighting branch circuits
- Receptacle branch circuits
- General purpose branch circuits
- General lighting branch circuits
- Small-appliance branch circuits
- Laundry branch circuits
- Bathroom branch circuits
- Individual branch circuits

GENERAL LIGHTING CIRCUITS

A lighting branch circuit will have only lighting outlets on it. A receptacle branch circuit will only have receptacle outlets on it. In residential electrical systems, these two branch circuit types are usually combined into what the *NEC®* calls a **general purpose branch circuit**. A general purpose branch circuit is a branch circuit that supplies two or more receptacles or lighting outlets, or a combination of both. Household appliances such as vacuum cleaners, small room air conditioners, televisions, or stereo equipment can also be cord-and-plug connected to this type of branch circuit. A good example of this circuit type is a bedroom branch circuit that has several receptacles and a wall switch-controlled, ceiling-mounted lighting fixture. The main goal of general purpose branch circuits installed in a house is to provide general lighting in rooms throughout the house. This is done by having lamps plugged into the circuit receptacles or by having switch-controlled wall and/or ceiling lighting fixtures connected to the circuit. As a result, many people refer to this type of circuit as a **general lighting circuit** (Figure 6-2). This textbook will refer to a branch circuit that has both receptacles and lighting outlets connected to it as a general lighting circuit. General lighting circuits usually make up the majority of branch circuits installed in a residential electrical system.

To determine the minimum number of general lighting circuits required in a house, a calculation of the habitable floor area is required. The calculated floor area is then multiplied by the unit load per square foot for general lighting to get the total general lighting load in volt-amperes. Section 220.12 of the *NEC®* states that a unit load of not less than that specified in Table 220.12 (Figure 6-3) for occupancies listed in the table will be the minimum lighting load. The unit load for a **dwelling unit**, according to Table 220.12, is 3 **volt-amperes** per square foot (33 volt-amperes per square meter). The

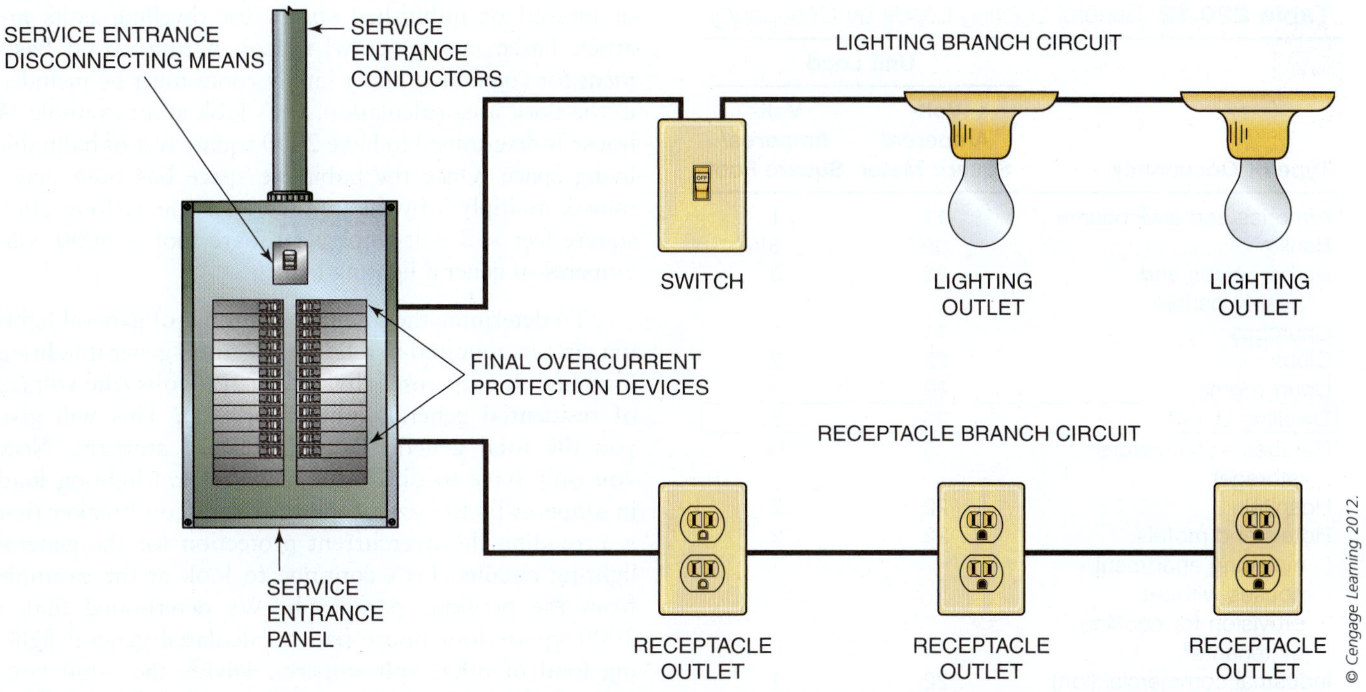

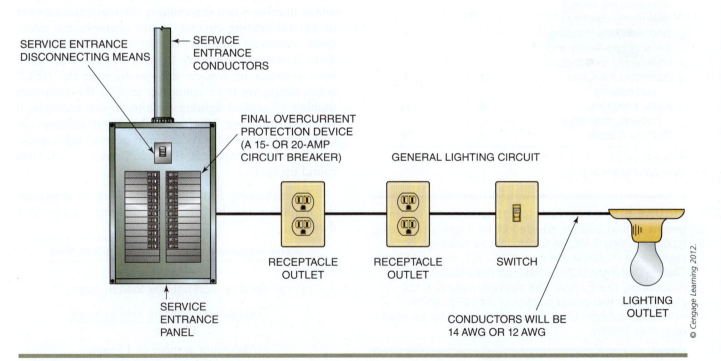

FIGURE 6-1 A branch circuit is defined as the circuit conductors between the final fuse or circuit breaker and the power or lighting outlets.

FIGURE 6-2 A general lighting circuit is a circuit that supplies electrical power to both receptacle outlets and lighting outlets in a residential wiring application. They will be installed with 14 AWG wire and a fuse or circuit breaker rated at 15 amperes or 12 AWG wire and a fuse or circuit breaker rated at 20 amperes.

Table 220.12 General Lighting Loads by Occupancy

Type of Occupancy	Unit Load	
	Volt-Amperes/ Square Meter	Volt-Amperes/ Square Foot
Armories and auditoriums	11	1
Banks	39[b]	3½[b]
Barber shops and beauty parlors	33	3
Churches	11	1
Clubs	22	2
Court rooms	22	2
Dwelling Units[a]	33	3
Garages—commercial (storage)	6	½
Hospitals	22	2
Hotels and motels, including apartment houses without provision for cooking by tenants[a]	22	2
Industrial commercial (loft)	22	2
Lodge rooms	17	1½
Office buildings	39[b]	3½[b]
Restaurants	22	2
Schools	33	3
Stores	33	3
Warehouses (storage)	3	¼
In any of the preceding occupancies except one-family dwellings and individual dwelling units of two-family and multifamily dwellings:		
Assembly halls and auditoriums	11	1
Halls, corridors, closets, stairways	6	½
Storage spaces	3	¼

[a]See 220.14(J).
[b]See 220.14(K).

FIGURE 6-3 *NEC®* Table 220.12 provides the minimum general lighting load per square foot for a variety of building types, including 3 VA per square foot for dwelling units. *Reprinted with permission from NFPA 70®, National Electric Code®, Copyright © 2010, National Fire Protection Association, Quincy, MA. This reprinted material is not the complete and official position of the NFPA on the referenced subject, which is represented only by the standard in its entirety.*

of unused or unfinished spaces for dwelling units are attics, basements, or crawl spaces. A finished-off basement for something like a family room must be included in the floor area calculation. Let's look at an example. A house is determined to have 2000 square feet of habitable living space. Once the habitable space has been determined, multiply it by the unit load per square foot: 2000 square feet × 3 volt-amps per square foot = 6000 volt-amperes of general lighting load.

To determine the minimum number of general lighting circuits required in a house, the total general lighting load in volt-amperes is divided by 120 volts (the voltage of residential general lighting circuits). This will give you the total general lighting load in amperes. Now you only have to divide the total general lighting load in amperes by the size of the fuse or circuit breaker that is providing the overcurrent protection for the general lighting circuits. Let's continue to look at the example from the previous paragraph. We determined that a 2000-square-foot house had a calculated general lighting load of 6000 volt-amperes. Divide the 6000 volt-amperes by 120 volts to get the general lighting load in amperes: 6000 volt-amperes/120 volts = 50 amps. Now, if 15-ampere-rated general lighting circuits are being installed, divide the 50-ampere general lighting load by 15 amps: 50 amp/15 amp = 3.333. Therefore, a minimum of four 15-amp general lighting circuits must be installed. Always round up to the next higher whole number when determining the minimum number of general lighting circuits. In our example, the calculated minimum number is 3⅓ circuits. Obviously, there is no such thing as ⅓ of a circuit—you either have a circuit or you do not—so to meet the *NEC®* requirement, we must round up to 4 as the minimum number of general lighting circuits. In our example, if we were installing 20-amp-rated general lighting circuits, a minimum of three 20-amp general lighting circuits would be required: 50 amp/20 amp = 2.5, then round up to 3.

The following formulas can be used to determine the minimum number of general lighting circuits in a dwelling:

$$\frac{3 \text{ volt-amperes} \times \text{calculated square-foot area}}{120 \text{ volts}}$$
$$= \text{Total general lighting load in amps}$$

$$\frac{\text{Total general lighting load in amps}}{15 \text{ amp}}$$
$$= \text{Minimum number of 15-amp general lighting circuits}$$

$$\frac{\text{Total general lighting load in amps}}{20 \text{ amp}}$$
$$= \text{Minimum number of 20-amp general lighting circuits}$$

floor area for each floor is to be computed from the *out-side* dimensions of the dwelling unit. The computed floor area does not include open porches, garages, or unused or unfinished spaces not adaptable for future use. Examples

SMALL-APPLIANCE BRANCH CIRCUITS

A small-appliance branch circuit is a type of branch circuit found in residential wiring that supplies electrical power to receptacles located in kitchens, dining rooms, pantries, and other similar areas (Figure 6-4). Section 210.11(C)(1) states that two or more 20-ampere-rated small-appliance branch circuits must be provided for all receptacle outlets specified by Section 210.52(B). Section 210.52(B) requires two or more 20-ampere circuits for all receptacle outlets for the small-appliance loads, including refrigeration equipment, in the kitchen, dining room, pantry, and breakfast room of a dwelling unit. No fewer than two small-appliance branch circuits

must supply the countertop receptacle outlets in kitchens. These circuits may also supply receptacle outlets in the pantry, dining room, and breakfast room as well as an electric clock receptacle and receptacles for gas-fired appliances, but these circuits are to have no other receptacle outlets. In addition, no lighting outlets are allowed to be connected to small-appliance branch circuits.

Section 220.52(A) states that in each dwelling unit, the load must be computed at 1500 volt-amperes for each two-wire small-appliance branch circuit. This means that a minimum load for small-appliance branch circuits in a house would be 3000 volt-amperes (minimum of 2 small-appliance branch circuits × 1500 volt-amperes = 3000 volt-amperes). If a residential wiring system is to have more than two small-appliance branch circuits installed, each circuit over two must also be calculated at 1500 volt-amperes. For example, a dwelling unit is to have four small-appliance branch circuits installed. The total load for the small-appliance branch circuits would be 6000 volt-amperes (4 small-appliance branch circuits × 1500 VA = 6000 VA).

LAUNDRY BRANCH CIRCUITS

A laundry branch circuit is a type of branch circuit found in residential wiring that supplies electrical power to laundry areas (Figure 6-5). The area may be a separate laundry room, or it may be just an area in a basement or a garage. The purpose of this circuit type is to provide electrical power to a washing machine and other laundry-related items, like a clothes iron. If the house will have a gas clothes dryer, the laundry circuit is used to also provide power (120 volt) for that appliance. An electric clothes dryer will require a separate electrical circuit (120/240 volt), and its load will be calculated separately. (We will look at electric clothes dryer loads later in this chapter.) No lighting outlets or receptacles in other rooms may be connected to this circuit. Section 220.52(B) tells us that a load of not less than 1500 volt-amperes is to be included for each two-wire laundry branch circuit installed. Usually, there is one two-wire laundry circuit installed in each dwelling unit, so the laundry circuit load is normally just 1500 volt-amperes. Of course, if more two-wire laundry circuits are installed, 1500 volt-amperes must be calculated for each circuit. Section 210.11(C)(2) states that the laundry circuit must have a 20-ampere rating.

BATHROOM BRANCH CIRCUITS

A bathroom branch circuit is a circuit that supplies electrical power to a bathroom in a residential application (Figure 6-6). A bathroom is defined as an area including

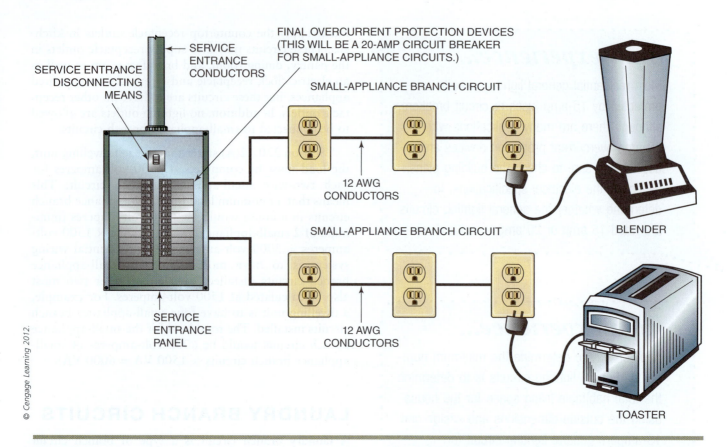

FIGURE 6-4   A small-appliance branch circuit provides electrical power to the receptacles in kitchens, dining rooms, and other similar areas. Section 210.11(C)(1) requires the installation of at least two 20-ampere-rated small-appliance branch circuits in each dwelling unit.

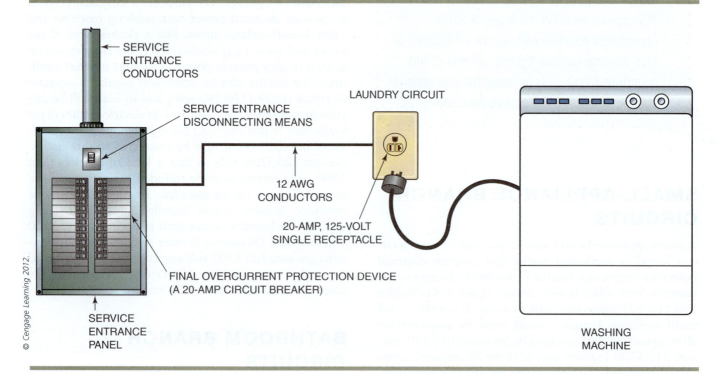

FIGURE 6-5   A laundry branch circuit is a circuit that supplies electrical power to a laundry room or area in a house. Section 210.11(C)(2) requires the installation of at least one 20-ampere-rated laundry circuit in each dwelling unit.

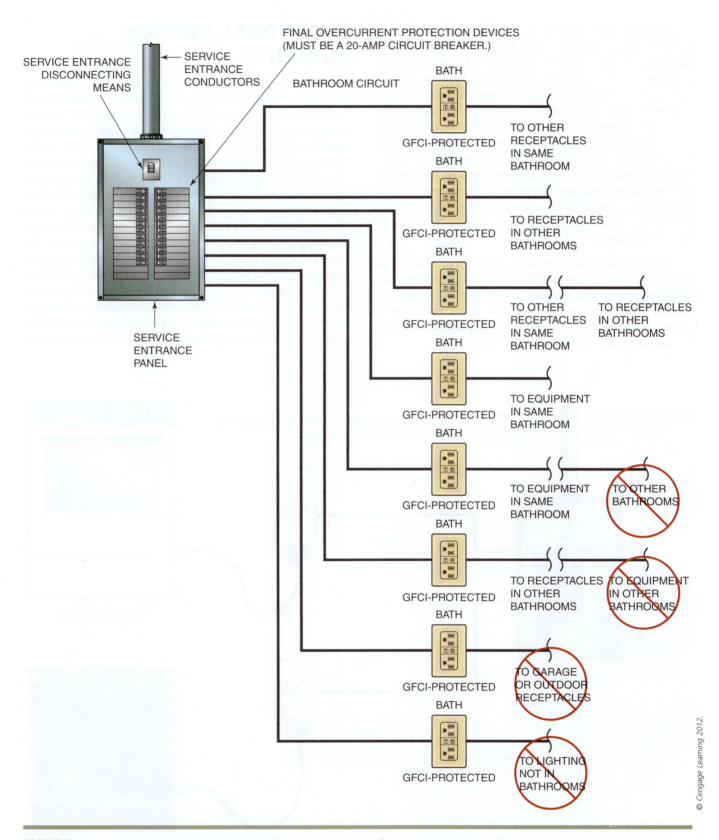

FIGURE 6-6   A bathroom branch circuit supplies electrical power for receptacles in a bathroom. Luminaires in the bathroom may also be connected to the bathroom circuit as long as the circuit does not feed other bathrooms. Section 210.11(C)(3) requires the installation of at least one 20-ampere-rated bathroom circuit in each dwelling unit.

a basin with one or more of the following: a toilet, a urinal, a tub, a shower, a foot bath, a bidet, or similar plumbing fixtures. Section 210.11(C)(3) states that in addition to the number of general lighting, small-appliance, and laundry branch circuits, at least one 20-ampere branch circuit must be provided to supply the bathroom receptacle outlet(s). This circuit cannot have any other lighting or receptacle outlets connected to it. An *Exception* allows lighting outlets and other equipment in a bathroom to be connected to it as long as the 20-amp circuit feeds only items in that one bathroom. If the circuit is installed so that it feeds more than one bathroom, only receptacle outlets can be fed by the bathroom circuit.

There is no load allowance given by the *NEC®* to the required bathroom circuit. Be aware that at least one 20-ampere bathroom circuit must be provided for a residential electrical system, but there is no volt-ampere value for this circuit type included in your calculation for the total electrical load of a house.

INDIVIDUAL BRANCH CIRCUITS

An **individual branch circuit** is a circuit that supplies only one piece of electrical equipment (Figure 6-7). An example is the circuit that supplies electrical power to one electric range or one electric clothes dryer. A circuit rating for a specific appliance is to be computed on the basis of the ampere rating of the appliance. This information is found on the appliance nameplate. For example, a garbage disposal being installed in a house may have a nameplate rating of 6 amps at 120 volts. By multiplying the amperage of the appliance by the voltage you will get the total load for that appliance in volt-amperes: 6 amps × 120 volts = 720 volt-amperes. The individual branch circuit for the garbage disposal in this example would need a capacity of at least 720 volt-amperes. This value will also be included in the service entrance calculation. Some appliances will have a volt-amp rating or wattage rating listed on their nameplate instead of an amperage rating, and this is the rating you would use.

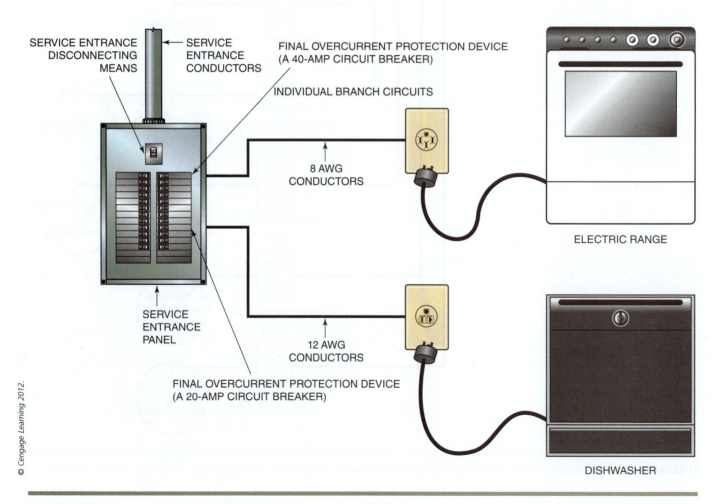

FIGURE 6-7   An individual branch circuit supplies only one piece of electrical equipment. Examples of individual branch circuits include an electric range circuit or a dishwasher circuit.

EXAMPLE 1

A house will have an electric range that has a nameplate rating of 11.5 kW. The first sentence in Note 4 to Table 220.55 says that it is permissible to use Table 220.55 to calculate the branch circuit load for one range. Use column C of Table 220.55 to get the maximum demand in kilowatts of a range not over 12 kW in rating.

Table 220.55 Note 4 Branch-Circuit Load. It shall be permissible to compute the branch-circuit load for one range in accordance with Table 220.55. The branch-circuit load for one wall-mounted oven or one counter-mounted cooking unit shall be the nameplate rating of the appliance. The branch-circuit load for a counter-mounted cooking unit and not more than two wall-mounted ovens, all supplied from a single branch circuit and located in the same room, shall be computed by adding the nameplate rating of the individual appliances and treating this total as equivalent to one range.

MAXIMUM kW FOR ONE RANGE RATED 12 kW OR LESS ACCORDING TO COLUMN C IS 8 kW (8,000 W)

THE MINIMUM CONDUCTOR SIZE IS FOUND BY CALCULATING THE CURRENT AND MATCHING IT TO A WIRE SIZE THAT WILL HANDLE IT.
- 8,000 W ÷ 240 V = 33.33 AMPS
- 8 AWG COPPER IS THE MINIMUM WIRE SIZE ACCORDING TO TABLE 310.15(B)(16)

THE MAXIMUM OVERCURRENT PROTECTION DEVICE WILL BE RATED AT 40 AMPS ACCORDING TO SECTIONS 210.19(A)(3), 240.4, AND 240.6

The first row is used when there is one range in a house. Because the range being installed is rated at 11.5 kW, use row 1, column C, which says that 8 kW is the value to be used for one range that has a rated value of 12 kW or less (Figure 6-8).

120/240-VOLT 11.5-kW RATING

© Cengage Learning 2012.

FIGURE 6-8   A range calculation for a single electric range rated not more than 12 kW.

EXAMPLE 2

A house will have an electric range that has a nameplate rating of 14 kW. Because this value is over 12 kW, column C of Table 220.55 cannot be used without a slight modification. The modification comes from Note 1 to Table 220.55, which says that for every kilowatt over 12 kW, the maximum demand for one range (8 kW) must be increased by 5%. Here is how this is done. The nameplate rating of 14 kW is 2 kW over 12 kW. At an increase of 5% of each kilowatt over 12 kW, the total percentage increase will be 10% (2 × 5% = 10%). Increase the maximum demand for one range (8 kW) by 10% (8 kW × 1.10 = 8800 W). So the actual value used for loading calculations for a 14-kW-rated range is 8.8 kW (Figure 6-9).

EXAMPLE 2 CONTINUED

Table 220.55 Note 1 Over 12kW through 27 kW ranges all of same rating. For ranges individually rated more than 12 kW but not more than 27 kW, the maximum demand in Column C shall be increased 5 percent for each additional kilowatt of rating or major fraction thereof by which the rating of individual ranges exceeds 12 kW.

MAXIMUM kW FOR A RANGE RATED MORE THAN 12 kW IS FOUND BY:

- INCREASING THE MAXIMUM kW FOR ONE RANGE NOT RATED MORE THAN 12 kW BY 5% FOR EACH kW OVER 12 kW:
  - ⋆ 14 kW − 12 kW = 2 kW
  - ⋆ 2 kW × 5% = 10%
  - ⋆ 8 kW × 1.10 = 8.8 kW

THE MINIMUM CONDUCTOR SIZE IS FOUND BY CALCULATING THE CURRENT AND MATCHING IT TO A WIRE SIZE THAT WILL HANDLE IT.

- 8.8 kW ÷ 240 V = 36.67 AMPS
- 8 AWG CU IS THE MINIMUM WIRE SIZE ACCORDING TO TABLE 310.15(B)(16)

THE MAXIMUM OVERCURRENT PROTECTION DEVICE WILL BE RATED AT 40 AMPS ACCORDING TO SECTIONS 240.4 AND 240.6.

120/240-VOLT 14-kW RATING

© Cengage Learning 2012.

FIGURE 6-9 A range calculation for a single electric range rated more than 12 kW.

EXAMPLE 3

A house will have a countertop cook unit rated at 5 kW and a built-in oven rated at 7 kW. Both will be connected to the same branch circuit. In this case, Note 4 to Table 220.55 will be used and states that if a single branch circuit supplies a counter-mounted cooking unit and not more than two wall-mounted ovens, all of which are located in the same room, the nameplate ratings of these appliances can be added together and the total treated as the equivalent of one range. In this example, the 5 kW countertop unit and the 7 kW oven are added together (5 kW + 7 kW = 12 kW). Now use column C of Table 220.55 to get the maximum demand in kilowatts of a range not over 12 kW in rating. Because the combined rating of the cooktop and oven being installed is 12 kW, use row 1, column C, which says that 8 kW is the value to be used for this combination (Figure 6-10).

EXAMPLE 3 CONTINUED

Table 220.55 Note 4 Branch-Circuit Load. It shall be permissible to compute the branch-circuit load for one range in accordance with Table 220.55. The branch-circuit load for one wall-mounted oven or one counter-mounted cooking unit shall be the nameplate rating of the appliance. *The branch-circuit load for a counter-mounted cooking unit and not more than two wall-mounted ovens, all supplied from a single branch circuit and located in the same room, shall be computed by adding the nameplate rating of the individual appliances and treating this total as equivalent to one range.*

MAXIMUM kW FOR A COOKTOP AND NOT MORE THAN 2 OVENS IS FOUND BY:

   ADDING TOGETHER THE NAMEPLATE RATINGS AND
   TREATING THEM LIKE ONE RANGE.
   * 7 kW + 5 kW = 12 kW
   THE MAXIMUM kW FOR ONE RANGE RATED 12 kW OR
   LESS ACCORDING TO COLUMN C IN TABLE 220.55 IS 8 kW.

THE MINIMUM CONDUCTOR SIZE IS FOUND BY CALCULATING THE CURRENT AND MATCHING IT TO A WIRE SIZE THAT WILL HANDLE IT.

- 8,000 W ÷ 240 V = 33.33 AMPS
- 8 AWG COPPER IS THE MINIMUM WIRE SIZE ACCORDING TO TABLE 310.15(B)(16)

THE MAXIMUM OVERCURRENT PROTECTION DEVICE WILL BE RATED AT 40 AMPS ACCORDING TO SECTIONS 210.19(A)(3), 240.4, AND 240.6.

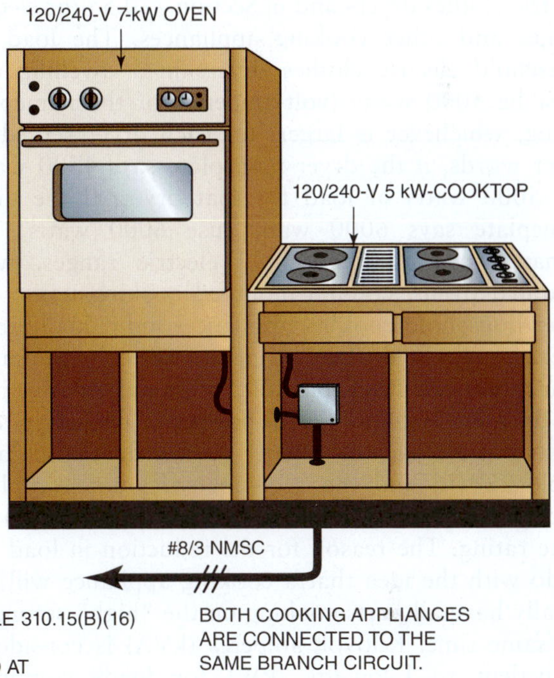

120/240-V 7-kW OVEN

120/240-V 5 kW-COOKTOP

#8/3 NMSC

BOTH COOKING APPLIANCES ARE CONNECTED TO THE SAME BRANCH CIRCUIT.

© Cengage Learning 2012.

FIGURE 6-10   A calculation for a 5 kW cooktop and a 7 kW oven. Both appliances are connected to the same branch circuit.

EXAMPLE 4

A house will have a counter-mounted cooking unit with a nameplate rating of 4 kW and a wall-mounted oven with a nameplate rating of 6 kW. A separate individual branch circuit will be run to the counter-mounted cooking unit and to the wall-mounted oven. To calculate the branch circuit load for each of the appliances, use Note 4 to Table 220.55. It states that the branch circuit load for one wall-mounted oven or one counter-mounted cooking unit must be the nameplate rating of the appliance. The branch circuit load for the countertop cooking unit is calculated as: 4000 W / 240 V = 16.67 amps. This will require a minimum branch circuit wire size of 12 AWG copper. The branch circuit overcurrent protection device will be a 20 amp two-pole circuit breaker. The branch circuit load for the wall-mounted oven is calculated as: 6000 W / 240 V = 25 amps. This will require a minimum branch circuit wire size of 10 AWG copper. The branch circuit overcurrent protection device for the wall-mounted oven will be a 30-amp, two-pole circuit breaker.

Individual branch-circuit loads for electric clothes dryers and household electric cooking appliances are based on the information found in Section 220.54 for electric clothes dryers and in Section 220.55 for electric ranges and other cooking appliances. The load for household electric clothes dryers in a dwelling unit must be 5000 watts (volt-amperes) or the nameplate rating, whichever is larger, for each dryer served. In other words, if the dryer nameplate says 4500 watts, use 5000 watts of load for that dryer. If the dryer nameplate says 6000 watts, use 6000 watts. The demand load for household electric ranges, wall-mounted ovens, counter-mounted cooking units, and other household cooking appliances individually rated in excess of 1¾ kilowatts is permitted to be computed in accordance with Table 220.55. The alternative is to use the nameplate ratings for household cooking equipment, but electricians usually do not do this because Table 220.55 and the accompanying notes apply a demand factor that reduces the load from the nameplate rating. The reason for the reduction in load has to do with the idea that a cooking appliance will not usually have all parts cooking on the "high" setting at the same time. Kilovolt-amperes (kVA) is considered equivalent to kilowatts (kW) for loads computed under Sections 220.54 and 220.55. The following examples show how to calculate common residential cooking equipment applications.

For household electric ranges and other cooking appliances, the size of the conductors must be determined by the rating of the range. The rating is calculated as shown earlier. The minimum conductor size for the range in Example 1 would be an 8 AWG copper conductor with 60°C insulation. Note that Section 210.19(A)(3) does not permit the branch-circuit rating (fuse or circuit breaker) of a circuit supplying household ranges with a nameplate rating of 8¾ kW or more to be less than 40 amperes.

One other item to consider when getting ready to calculate the total electrical load for a residential feeder or service entrance is what the *NEC®* calls "non-coincident loads." Section 220.60 states that where it is unlikely that two or more loads will be in use simultaneously, it is permissible to use only the largest load that will be used at one time in computing the total load of a feeder or service. This can be applied to the individual branch-circuit loading for air-conditioning equipment and heating equipment. For example, if a house had 10 kW of electric heating load and 8 kW of air-conditioning load, the 8 kW of air-conditioning load can be eliminated from the total service entrance load calculation because the cooling load will not be used at the same time as the heating load. Therefore, if the service entrance is sized using the larger of the two loads, there will be more than enough service capacity to supply the smaller cooling load when it is used.

DETERMINING THE AMPACITY OF A CONDUCTOR

Ampacity is defined as the current, in amperes, that a conductor can carry continuously under the conditions of use without exceeding its temperature rating. Section 210.19(A)(1) requires branch-circuit conductors to have an ampacity at least equal to the maximum electrical load to be served. This rule makes a lot of sense when you think about it. For example, if you are wiring a branch circuit that will supply electrical power to a load that will draw 16 amperes, you need to use a wire size for the branch circuit that has an ampacity of at least 16 amps. Otherwise, the conductor will heat up and cause the insulation on the conductor to melt.

Electricians use Table 310.15(B)(16) (Figure 6-11) to determine the ampacity of a conductor for use in residential wiring. You were introduced to this table in Chapter 2, but now we will look at it more closely and see how to use it. The table is set up so that the left half covers copper conductors and the right half covers aluminum and copper-clad aluminum conductors. Because the vast majority of the conductors used for branch-circuit wiring in residential applications are copper, we will concentrate on the copper half of the table. On the far-left-hand side of the table, you will see a column titled "Size AWG or kcmil." This column represents all the wire sizes you will encounter in residential wiring. The next three columns to the right are titled "60°C (140°F)," "75°C (167°F)," and "90°C (194°F)." These temperatures represent the maximum temperature ratings for the conductor insulation types listed in the columns under the temperature ratings. Temperature ratings for conductors, along with other insulation properties, are located in Table 310.104(A) of the *NEC®*. For example, a "TW" insulated conductor is rated at 60°C, a "THW" insulated conductor is rated at 75°C, and a "THHW" insulated conductor is rated at 90°C. The numbers in the table that correspond to a particular conductor size with a specific insulation type represent the ampacity of that conductor. For example, the ampacity of a 10 AWG copper conductor with a TW insulation is 30 amperes. The same conductor size with THW insulation has an ampacity of 35 amperes, and with THHW insulation it has an ampacity of 40 amperes. A conductor with a higher-temperature-rated insulation will have a higher-rated ampacity. Let's look at a few more examples:

- 12 AWG copper conductor with TW (60°C) insulation has an ampacity of 20 amps.

- 12 AWG aluminum conductor with TW (60°C) insulation has an ampacity of 15 amps.

Table 310.15(B)(16) Allowable Ampacities of Insulated Conductors Rated Up to and Including 2000 Volts, 60°C Through 90°C (140°F Through 194°F), Not More Than Three Current-Carrying Conductors in Raceway, Cable, or Earth (Directly Buried), Based on Ambient Temperature of 30°C (86°F)*

Size AWG or kcmil	Temperature Rating of Conductor [See Table 310.104(A).]						Size AWG or kcmil
	60°C (140°F)	75°C (167°F)	90°C (194°F)	60°C (140°F)	75°C (167°F)	90°C (194°F)	
	Types TW, UF	Types RHW, THHW, THW, THWN, XHHW, USE, ZW	Types TBS, SA, SIS, FEP, FEPB, MI, RHH, RHW-2, THHN, THHW, THW-2, THWN-2, USE-2, XHH, XHHW, XHHW-2, ZW-2	Types TW, UF	Types RHW, THHW, THW, THWN, XHHW, USE	Types TBS, SA, SIS, THHN, THHW, THW-2, THWN-2, RHH, RHW-2, USE-2, XHH, XHHW, XHHW-2, ZW-2	
	COPPER			ALUMINUM OR COPPER-CLAD ALUMINUM			
18	—	—	14	—	—	—	—
16	—	—	18	—	—	—	—
14**	15	20	25	—	—	—	—
12**	20	25	30	15	20	25	12**
10**	30	35	40	25	30	35	10**
8	40	50	55	35	40	45	8
6	55	65	75	40	50	55	6
4	70	85	95	55	65	75	4
3	85	100	115	65	75	85	3
2	95	115	130	75	90	100	2
1	110	130	145	85	100	115	1
1/0	125	150	170	100	120	135	1/0
2/0	145	175	195	115	135	150	2/0
3/0	165	200	225	130	155	175	3/0
4/0	195	230	260	150	180	205	4/0
250	215	255	290	170	205	230	250
300	240	285	320	195	230	260	300
350	260	310	350	210	250	280	350
400	280	335	380	225	270	305	400
500	320	380	430	260	310	350	500
600	350	420	475	285	340	385	600
700	385	460	520	315	375	425	700
750	400	475	535	320	385	435	750
800	410	490	555	330	395	445	800
900	435	520	585	355	425	480	900
1000	455	545	615	375	445	500	1000
1250	495	590	665	405	485	545	1250
1500	525	625	705	435	520	585	1500
1750	545	650	735	455	545	615	1750
2000	555	665	750	470	560	630	2000

*Refer to 310.15(B)(2) for the ampacity correction factors where the ambient temperature is other than 30°C (86°F)

**Refer to 240.4(D) for conductor overcurrent protection limitations.

FIGURE 6-11 *NEC*® Table 310.15(B)(16) is used by electricians to determine the minimum conductor size needed for a certain ampacity. *Reprinted with permission from NFPA 70*®, *National Electric Code*®, *Copyright © 2010, National Fire Protection Association, Quincy, MA. This reprinted material is not the complete and official position of the NFPA on the referenced subject, which is represented only by the standard in its entirety.*

- 8 AWG copper conductor with TW (60°C) insulation has an ampacity of 40 amps.

- 8 AWG aluminum conductor with TW (60°C) insulation has an ampacity of 35 amps.

- 6 AWG copper conductor with TW (60°C) insulation has an ampacity of 55 amps.

- 6 AWG aluminum conductor with TW (60°C) insulation has an ampacity of 40 amps.

- 4 AWG copper conductor with THW (75°C) insulation has an ampacity of 85 amps.

- 4 AWG aluminum conductor with THW (75°C) insulation has an ampacity of 65 amps.

- 4/0 copper conductor with THHN (90°C) insulation has an ampacity of 260 amps.

- 4/0 AWG aluminum conductor with THW (90°C) insulation has an ampacity of 205 amps.

CAUTION

CAUTION: Next to the 14, 12, and 10 AWG wire sizes in Table 310.15(B)(16), you will see a double asterisk (**). At the bottom of Table 310.15(B)(16), the double asterisk tells us to refer to Section 240.4(D). This section states that the overcurrent protection must not exceed 15 amperes for 14 AWG, 20 amperes for 12 AWG, and 30 amperes for 10 AWG copper, or 15 amperes for 12 AWG and 25 amperes for 10 AWG aluminum and copper-clad aluminum. So, no matter what the actual ampacity is for 14, 12, and 10 AWG conductors according to Table 310.15(B)(16), the fuse or circuit breaker size can never be larger than 15 amp for 14 AWG, 20 amp for 12 AWG, and 30 amp for 10 AWG copper conductors for residential branch-circuit applications.

Table 310.15(B)(2)(a) Ambient Temperature Correction Factors Based on 30°C (86°F)

For ambient temperatures other than 30°C (86°F), multiply the allowable ampacities specified in the ampacity tables by the appropriate correction factor shown below.

Ambient Temperature (°C)	Temperature Rating of Conductor			Ambient Temperature (°F)
	60°C	75°C	90°C	
10 or less	1.29	1.20	1.15	50 or less
11–15	1.22	1.15	1.12	51–59
16–20	1.15	1.11	1.08	60–68
21–25	1.08	1.05	1.04	69–77
26–30	1.00	1.00	1.00	78–86
31–35	0.91	0.94	0.96	87–95
36–40	0.82	0.88	0.91	96–104
41–45	0.71	0.82	0.87	105–113
46–50	0.58	0.75	0.82	114–122
51–55	0.41	0.67	0.76	123–131
56–60	—	0.58	0.71	132–140
61–65	—	0.47	0.65	141–149
66–70	—	0.33	0.58	150–158
71–75	—	—	0.50	159–167
76–80	—	—	0.41	168–176
81–85	—	—	0.29	177–185

FIGURE 6-12   Table 310.15(B)(2)(a) is used to adjust the ampacity of a conductor when an ambient temperature greater than 30°C is present. *Reprinted with permission from NFPA 70®, National Electric Code®, Copyright © 2010, National Fire Protection Association, Quincy, MA. This reprinted material is not the complete and official position of the NFPA on the referenced subject, which is represented only by the standard in its entirety.*

The ampacity of a conductor, as determined in Table 310.15(B)(16) and used in residential applications, may have to be modified slightly. As the "conditions of use" change for a conductor, the ampacity is required to be adjusted so that at no time could the temperature of the conductor exceed the temperature of the insulation on the conductor. The following adjustments may have to be made for certain residential situations:

1. The ampacities given in Table 310.15(B)(16) are based on an **ambient temperature** of no more than 30°C (86°F). When the ambient temperature that a conductor is exposed to exceeds 30°C (86°F), the ampacities from Table 310.15(B)(16) must be multiplied by the appropriate correction factor shown in Table 310.15(B)(2)(a) (Figure 6-12). You will notice that the far-left-hand column for the correction factors is the ambient temperature in degrees Celsius. The far-right-hand column has temperatures listed in degrees Fahrenheit. Let's look at an example where the ambient temperature correction factors must be applied (Figure 6-13).

A 3 AWG copper conductor with THW (75°C) insulation has an ampacity of 100 amperes according to Table 310.15(B)(16). However, if an electrician installs the conductor in an area where the ambient temperature is 100°F, the actual ampacity of the wire can be no more than 88 amperes. Here is how you would determine this. First, find the 100°F ambient temperature in the far-right-side column in Table 310.15(B)(2)(a). 100°F falls between 96 and 104, so we will use that row. Follow the row to the left until you find the number that is in the 75°C column. You will see that the correction factor is 0.88. Now multiply the original ampacity by the correction factor to get the new allowable ampacity: 100 amps × .88 correction factor = 88 amps.

2. The ampacities listed in Table 310.15(B)(16) are also based on not having any more than three current-carrying conductors in a raceway or cable. Section 310.15(B)(3)(a) states that where the number of current-carrying conductors in a raceway or cable exceeds three or where single conductors or multi-conductor cables are stacked or **bundled** longer than 24 inches (600 mm) without maintaining spacing and are not installed in raceways, the allowable ampacity of each conductor must be reduced by the adjustment factor listed in Table 310.15(B)(3)(a) (Figure 6-14). This is called "derating" and it results in a lower conductor ampacity when there are more than three current-carrying conductors in a cable or raceway. The reason for this reduction in ampacity is the fact that when you have more current-carrying conductors close to each other, there is an overall increase in temperature, which could result in damage to the conductor insulation.

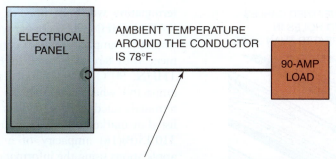

3 AWG COPPER WITH THW INSULATION INSTALLED IN AN AMBIENT TEMPERATURE NO GREATER THAN 30°C (86°F). THE AMPACITY OF THE CONDUCTOR IS 100 AMPS ACCORDING TO TABLE 310.15(B)(16). THIS CONDUCTOR SIZE WILL HANDLE THE 90-AMP LOAD.

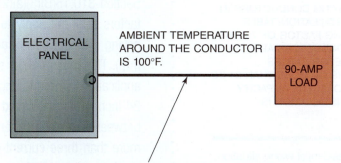

3 AWG COPPER WITH THW INSULATION INSTALLED IN AN AMBIENT TEMPERATURE THAT IS 100°F. THE AMPACITY OF THE CONDUCTOR IS NOW REQUIRED TO BE ADJUSTED. ACCORDING TO TABLE 310.15(B)(2)(a) THE ADJUSTMENT FACTOR IS .88, SO 100 AMPS X .88 = 88 AMPS. THE NEW AMPACITY OF THE CONDUCTOR IS 88 AMPS, WHICH WILL NOT HANDLE THE 90-AMP LOAD. A LARGER WIRE WILL HAVE TO BE USED.

© Cengage Learning 2012.

FIGURE 6-13 An example showing the application of the ambient temperature correction factor.

Table 310.15(B)(3)(a) Adjustment Factors for More Than Three
Current-Carrying Conductors in a Raceway or Cable

Number of Conductors[1]	Percent of Values in Tables 310.15(B)(16) through 310.15(B)(19) Adjusted for Ambient Temperature if Necessary
4–6	80
7–9	70
10–20	50
21–30	45
31–40	40
41 and above	35

[1]Number of Conductors is the total number of conductors in the raceway or cable adjusted in accordance with 310.15(B)(5) and (6).

FIGURE 6-14 *NEC*® Table 310.15(B)(3)(a). Reprinted with permission from NFPA 70®, National Electric Code®, Copyright © 2010, National Fire Protection Association, Quincy, MA. This reprinted material is not the complete and official position of the NFPA on the referenced subject, which is represented only by the standard in its entirety.

Let's look at an example of conductor derating for more than three current-carrying conductors (Figure 6-15). It is common wiring practice in residential situations to drill holes in building framing members and run several cables through the holes. If the length of these "bundled" cables exceeds 24 inches without any spacing between the cables and the total number of current-carrying conductors exceeds three, the ampacity of the conductors in the cables must be derated. Let's say that an electrician bundles four 12/2 nonmetallic sheathed cables through the drilled holes of some wall studs in a house. This installation results in having eight current-carrying conductors. According to Table 310.15(B)(16), the ampacity of a 12 AWG nonmetallic sheathed cable (using the "90°C" column as permitted by Section 334.80) is 30 amperes. Table 310.15(B)(3)(a) tells us that when seven to nine current-carrying conductors are in the same cable or raceway, a reduction to 70% of the conductor ampacity is required. This will mean that the ampacity of the 12 AWG Romex™ cables is 21 amperes: 30 amps × .70 = 21 amps.

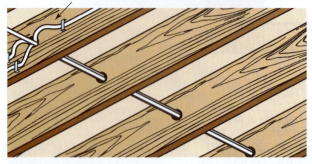

FOUR 12/2 NONMETALLIC SHEATHED CABLES INSTALLED THROUGH BORED HOLES IN JOISTS FOR A DISTANCE GREATER THAN 24 IN.

- TABLE 310.15(B)(16) SHOWS AN AMPACITY OF 30 AMPS FOR A 12 AWG NMSC (90°C COLUMN).
- THERE ARE EIGHT CURRENT-CARRYING CONDUCTORS IN THE FOUR 12/2 CABLES IN THIS INSTALLATION. TABLE 310.15(B)(3)(a) REQUIRES A DERATING FACTOR OF 70% BE APPLIED WHEN SEVEN TO NINE CONDUCTORS ARE RUN TOGETHER IN A RACEWAY, A CABLE, OR CABLES BUNDLED TOGETHER FOR LENGTHS OF MORE THAN 24 IN.
- 30 AMPS X .70 = 21 AMPS. THEREFORE, THE AMPACITY OF THE CONDUCTORS FOR THIS INSTALLATION IS 21 AMPERES.

© Cengage Learning 2012.

FIGURE 6-15   An example of a residential wiring situation where there are more than three current-carrying conductors. This wiring technique is called "bundling."

CAUTION

CAUTION: Sometimes a conductor is installed in an ambient temperature that is greater than 30°C (86°F) *and* there are more than three conductors in the same raceway or cable. If this is the case, a "double derating" will need to take place. Here is an example. Suppose an electrician installs five current-carrying conductors in a raceway. The raceway is located in an area where the ambient temperature is 100°F. If the conductor used is a 3 AWG copper with THW insulation, the normal ampacity from Table 310.15(B)(16) is 100 amperes. However, you need to apply the ambient temperature correction factor of .88 from Table 310.15(B)(2)(a) *and* the derating factor of .80 for more than three current-carrying conductors: 100 amps × .88 × .80 = 70.4 amps.

---

3. Section 110.14(C) requires an electrician to take a look at the temperature ratings of the equipment terminals that the electrical conductors will be connected to. The temperature rating (60°C, 75°C, or 90°C) associated with the ampacity of a conductor from Table 310.15(B)(16) must be selected, so the temperature produced by that conductor cannot exceed the lowest temperature rating of any connected termination, conductor, or device. If the temperature produced by a current-carrying conductor at a termination does exceed the termination temperature rating, the termination will burn up. For example, the load on an 8 AWG THHN, 90°C copper wire is limited to 40 amperes (the 60°C ampacity) where connected to electrical equipment with terminals rated at 60°C. This same 8 AWG THHN, 90°C wire is limited to 50 amperes (the 75°C ampacity) where connected to electrical equipment with terminals rated at 75°C. Unless the electrical equipment is listed or marked otherwise, you must determine the Table 310.15(B)(16) ampacity of a conductor for residential applications using the information outlined next.

from experience...

Section 310.15(B)(3)(a)(2) states that derating factors shall not apply to conductors in nipples having a length not exceeding 24 inches (600 mm). This may apply to some residential wiring applications where a length of electrical conduit 24 inches or less, called a **nipple**, is placed between two electrical enclosures. In this case, more than three current-carrying conductors can be run in a nipple, and no derating will have to be done.

CAUTION

CAUTION: Do not forget that any electrical circuit has a beginning and an end. You must always know what the termination temperature rating is at the panel and the temperature rating at the equipment (like switches and receptacles) at the end of the circuit. A typical residential circuit may have a 60°C rating at one end and a 75°C rating at the other. Conductor ampacity in this case would have to be based on the "60°C" column in Table 310.15(B)(16).

Termination provisions of equipment for circuits rated 100 amperes or less, or marked for 14 AWG through 1 AWG conductors, must be used only for one of the following:

- Conductors rated 60°C (140°F): This means that for determining the ampacity of a conductor used in circuits rated 100 amps or less or wire sizes of 14 AWG through 1 AWG, always use the ampacity in the "60°C" column of Table 310.15(B)(16) for the wire size you wish to use. The only conductor insulation shown in the "60°C" column is Type TW and Type UF. For example, it is determined that an electric dryer you have to install in a house requires a minimum 30-amp circuit. Because it is a circuit that is 100 amps or less, you go to the "60°C" column of

Table 310.15(B)(16) to find a wire size that has a minimum ampacity of 30 amps. In this case, you would use a 10 AWG insulated conductor (Figure 6-16).

- Conductors with higher temperature ratings, provided that the ampacity of such conductors is determined by the 60°C (140°F) ampacity of the conductor size used. This means that if you want to use conductors with insulations whose temperature ratings are 75°C or 90°C, you may, but you must determine their ampacity from the "60°C" column in Table 310.15(B)(16). For example, a 10 AWG copper conductor with a 75°C insulation has an ampacity of 35 amps, and

the same 10 AWG conductor with a 90°C insulation type has an ampacity of 40 amps. Both wire insulation types could be used in the electric dryer example from the previous section, but the ampacity of the wire must be determined from the "60°C" column of Table 310.15(B)(16), which is 30 amps (Figure 6-17).

- Conductors with higher temperature ratings if the equipment is listed and identified for use with such conductors: This means that if equipment terminations are rated for higher temperatures than 60°C, the ampacity for a conductor with a 75°C ampacity or 90°C ampacity can be used. For example, if the

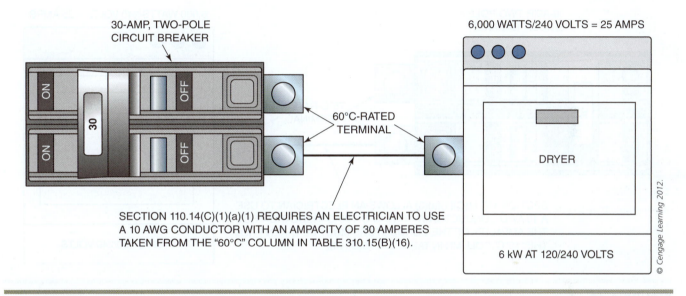

FIGURE 6-16   An example for determining the ampacity of a conductor when the circuit rating is 100 amps or less and the conductor size is no larger than 1 AWG. Section 110.14(C)(1)(a)(1).

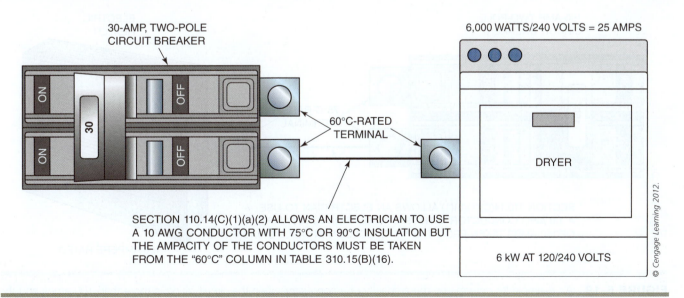

FIGURE 6-17   An example for determining the ampacity of a conductor when the circuit rating is 100 amps or less, the conductor size is no larger than 1 AWG, and the conductor insulation is 75°C or 90°C. Section 110.14 (C)(1)(a)(2).

electric dryer from the previous section has terminals marked for 75°C, you can determine the ampacity of a conductor using the "75°C" column of Table 310.15(B)(16). The conductor would have to have either a 75°C or 90°C insulation type (Figure 6-18).

Termination provisions of equipment for circuits rated over 100 amperes or marked for conductors larger than 1 AWG must be used only for one of the following:

• Conductors rated 75°C (167°F): This means that for determining the ampacity of a conductor used in

circuits rated more than 100 amps or when wire sizes larger than 1 AWG are used, always use the ampacity in the "75°C" column of Table 310.15(B)(16) for the wire size you wish to use. For example, it is determined that an electric heater you have to install in a house requires a minimum 150-amp circuit. Because it is a circuit that is more than 100 amps and the terminals are marked 75°C, you go to the "75°C" column of Table 310.15(B)(16) to find a wire size that has a minimum ampacity of 150 amps. In this case, you could use a 1/0 AWG insulated conductor (Figure 6-19).

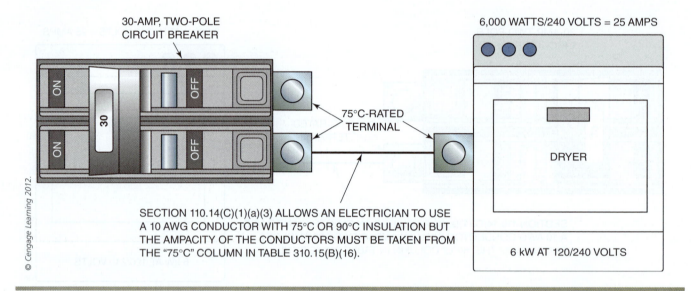

30-AMP, TWO-POLE CIRCUIT BREAKER

6,000 WATTS/240 VOLTS = 25 AMPS

75°C-RATED TERMINAL

DRYER

6 kW AT 120/240 VOLTS

SECTION 110.14(C)(1)(a)(3) ALLOWS AN ELECTRICIAN TO USE A 10 AWG CONDUCTOR WITH 75°C OR 90°C INSULATION BUT THE AMPACITY OF THE CONDUCTORS MUST BE TAKEN FROM THE "75°C" COLUMN IN TABLE 310.15(B)(16).

© Cengage Learning 2012.

FIGURE 6-18   An example for determining the ampacity of a conductor when the circuit rating is 100 amps or less, the conductor size is no larger than 1 AWG, and the termination ratings are higher than 60°C. Section 110.14(C)(1)(a)(3).

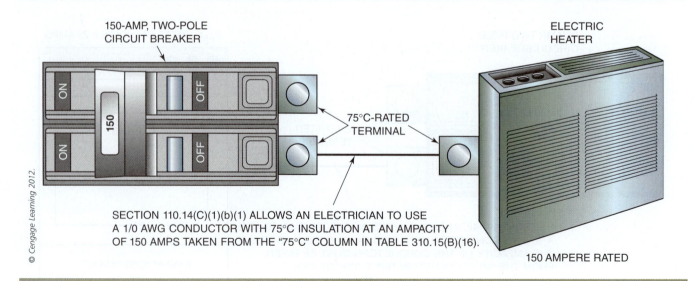

150-AMP, TWO-POLE CIRCUIT BREAKER

ELECTRIC HEATER

75°C-RATED TERMINAL

SECTION 110.14(C)(1)(b)(1) ALLOWS AN ELECTRICIAN TO USE A 1/0 AWG CONDUCTOR WITH 75°C INSULATION AT AN AMPACITY OF 150 AMPS TAKEN FROM THE "75°C" COLUMN IN TABLE 310.15(B)(16).

150 AMPERE RATED

© Cengage Learning 2012.

FIGURE 6-19   An example for determining the ampacity of a conductor when the circuit rating is more than 100 amps and the conductor size is larger that 1 AWG. Section 110.14(C)(1)(b)(1).

- Conductors with higher temperature ratings, provided that the ampacity of such conductors does not exceed the 75°C (167°F) ampacity of the conductor size used or up to their ampacity if the equipment is listed and identified for use with such conductors. This means that if you want to use a conductor with an insulation temperature rating of 90°C, you may, but you must determine the ampacity from the "75°C" column in Table 310.15(B)(16). For example, a 1 AWG copper conductor with a 90°C insulation has an ampacity of 150 amps. But the ampacity of the wire must be determined from the "75°C" column of Table 310.15(B)(16), which is 130 amps. This ampacity is less than the required circuit ampacity. A 1/0 AWG copper conductor with a 90°C insulation could be used, but the ampacity is taken from the "75°C" column in Table 310.15(B)(16) (Figure 6-20).

Separately installed pressure connectors must be used with conductors with an ampacity not exceeding the ampacity for the listed and identified temperature rating of the connector. This means that when using wire connectors such as wirenuts or split bolt connectors, the temperature rating of the connector must be equal to or greater than the temperature rating of the wire it is being used with. For example, say you are splicing wires in a junction box that has 75°C insulation on the wires and the ampacity of the wires has been determined using the "75°C" column in Table 310.15(B)(16). The wirenuts you are using must have at least a 75°C temperature rating. If they had only a 60°C temperature rating, they could burn up (Figure 6-21).

BRANCH-CIRCUIT RATINGS AND SIZING THE OVERCURRENT PROTECTION DEVICE

The branch-circuit rating is based on the size of the fuse or circuit breaker protecting the circuit according to Section 210.3 in the *NEC*®. This section also states that standard branch-circuit ratings for receptacle and lighting circuits are 15, 20, 30, 40, and 50 amp. The circuit ratings for receptacles and lighting in a residential wiring system are usually 15 or 20 ampere. Individual branch-circuit ratings can be any ampere rating, depending on the size of the load served. Standard sizes of fuses and circuit breakers are found in Section 240.6(A). There are many standard sizes available. A 200-ampere fuse or circuit breaker is usually the largest overcurrent protection device used in residential wiring. However, for very large homes, you may encounter service entrance main fuses or circuit breakers that could be larger than 200 ampere. The standard fuse or circuit breaker sizes through 200 ampere include ampere ratings of 15, 20, 25, 30, 35, 40, 45, 50, 60, 70, 80, 90, 100, 110, 125, 150, 175, and 200.

Table 210.24 (Figure 6-22) summarizes the requirements for the size of conductors and the size of the overcurrent protection for branch circuits that have two or more **outlets** or receptacles. The wire sizes listed in the table are for copper conductors.

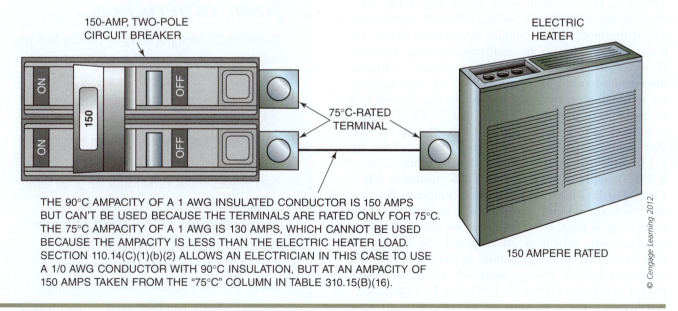

150-AMP, TWO-POLE CIRCUIT BREAKER

ELECTRIC HEATER

75°C-RATED TERMINAL

THE 90°C AMPACITY OF A 1 AWG INSULATED CONDUCTOR IS 150 AMPS BUT CAN'T BE USED BECAUSE THE TERMINALS ARE RATED ONLY FOR 75°C. THE 75°C AMPACITY OF A 1 AWG IS 130 AMPS, WHICH CANNOT BE USED BECAUSE THE AMPACITY IS LESS THAN THE ELECTRIC HEATER LOAD. SECTION 110.14(C)(1)(b)(2) ALLOWS AN ELECTRICIAN IN THIS CASE TO USE A 1/0 AWG CONDUCTOR WITH 90°C INSULATION, BUT AT AN AMPACITY OF 150 AMPS TAKEN FROM THE "75°C" COLUMN IN TABLE 310.15(B)(16).

150 AMPERE RATED

© Cengage Learning 2012.

FIGURE 6-20   An example for determining the ampacity of a conductor when the circuit rating is more than 100 amps, the conductor size is larger than 1 AWG, and the conductor insulation is 90°C. Section 110.14 (C)(1)(b)(2).

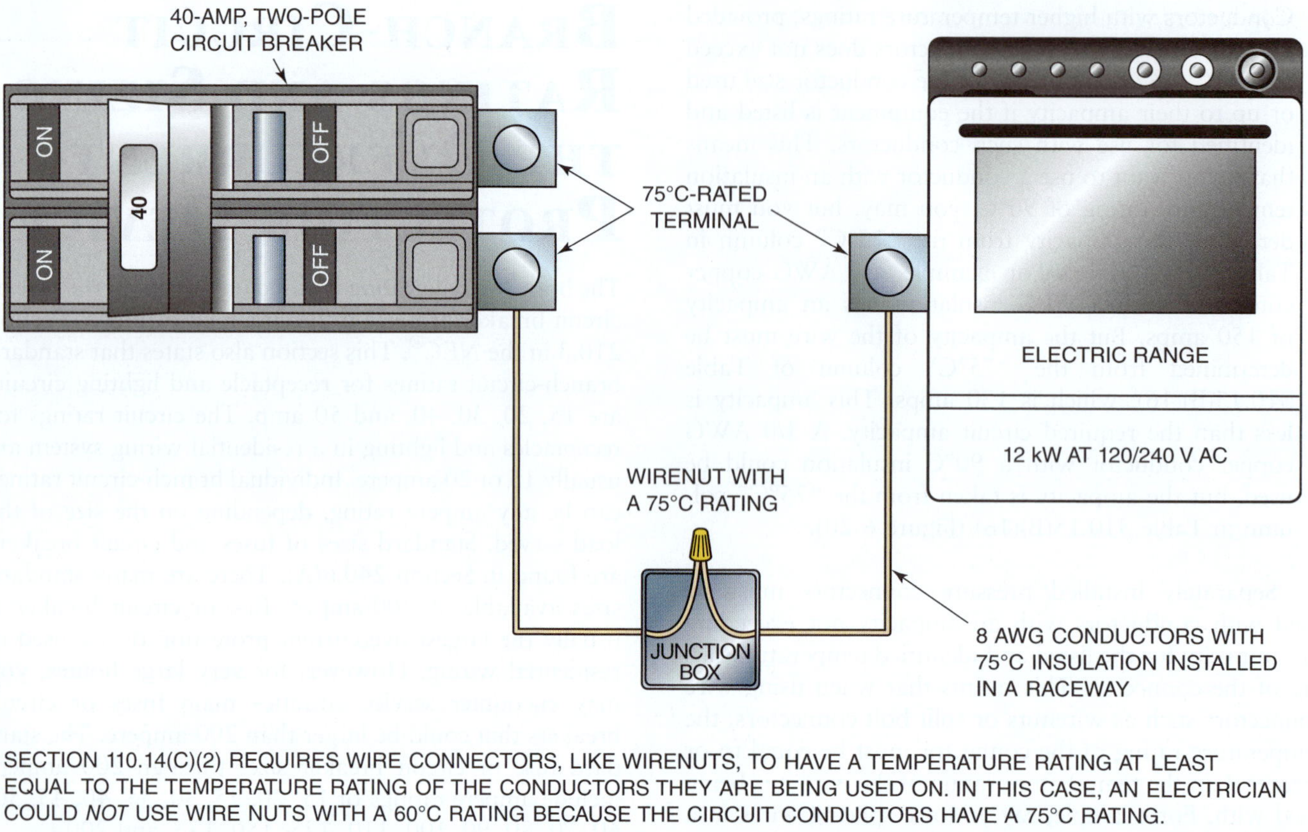

40-AMP, TWO-POLE
CIRCUIT BREAKER

75°C-RATED
TERMINAL

ELECTRIC RANGE

12 kW AT 120/240 V AC

WIRENUT WITH
A 75°C RATING

JUNCTION
BOX

8 AWG CONDUCTORS WITH
75°C INSULATION INSTALLED
IN A RACEWAY

© Cengage Learning 2012.

SECTION 110.14(C)(2) REQUIRES WIRE CONNECTORS, LIKE WIRENUTS, TO HAVE A TEMPERATURE RATING AT LEAST
EQUAL TO THE TEMPERATURE RATING OF THE CONDUCTORS THEY ARE BEING USED ON. IN THIS CASE, AN ELECTRICIAN
COULD *NOT* USE WIRE NUTS WITH A 60°C RATING BECAUSE THE CIRCUIT CONDUCTORS HAVE A 75°C RATING.

FIGURE 6-21 Wirenuts and other wire connector types must have a temperature rating that is not less than the temperature rating of the conductors they are being used with. Section 110.14(C)(2).

If the ampacity of a conductor in Table 310.15(B)(16) does not match the rating of the standard overcurrent device, Section 240.4(B) permits the use of the next larger standard overcurrent device for those circuits found in residential wiring systems. However, if the ampacity of a conductor matches the standard rating of Section 240.6(A), that conductor must be protected at the standard size device. For example, in Table 310.15(B)(16), a 3 AWG copper conductor with THW insulation has an ampacity of 100 amperes. This conductor would be protected by a 100-amp overcurrent device. On the other hand, a 6 AWG copper conductor with THW insulation has a Table 310.15(B)(16) ampacity of 65 amperes. Because there is not a 65-amp standard size of fuse or circuit breaker, it is permissible to go up to the next higher standard size, which is 70 amperes. Do not forget that based on the information in Section 240.4(D), residential branch circuits using 14, 12, or 10 AWG copper conductors must be protected by fuse or circuit breakers rated 15 amp for the 14 AWG, 20 amp for the 12 AWG, and 30 amp for the 10 AWG conductor sizes.

from experience...

In residential wiring, it is highly unlikely that you will ever use equipment that has 90°C-rated terminations. The majority of the circuits in residential wiring are 100 ampere or less, and the circuit wire sizes are smaller than 1 AWG. This means that unless you know for sure that both ends (and the middle if spliced in a junction box) of your circuit conductors are to be terminated on equipment that is rated at 75°C, always determine your conductor ampacity based on the "60°C" column of Table 310.15(B)(16). Also, nonmetallic sheathed cable, the most popular residential wiring method, always has its ampacity determined from the "60°C" column in Table 310.15(B)(16).

Table 210.24 Summary of Branch-Circuit Requirements

Circuit Rating	15 A	20 A	30 A	40 A	50 A
Conductors (min. size):					
Circuit wires[1]	14	12	10	8	6
Taps	14	14	14	12	12
Fixture wires and cords — see 240.5					
Overcurrent Protection	**15 A**	**20 A**	**30 A**	**40 A**	**50 A**
Outlet devices:					
Lampholders permitted	Any type	Any type	Heavy duty	Heavy duty	Heavy duty
Receptacle rating[2]	15 max. A	15 or 20A	30A	40 or 50 A	50A
Maximum Load	**15 A**	**20 A**	**30 A**	**40 A**	**50 A**
Permissible load	See 210.23(A)	See 210.23A	See 210.23(B)	See 210.23(C)	See 210.23(C)

[1]These gauges are for copper conductors.
[2]For receptacle rating of cord-connected electric-discharge luminaries, see **410.62(C)**.

FIGURE 6-22 *NEC*® Table 210.24 summary of branch-circuit requirements. *Reprinted with permission from NFPA 70*®, *National Electric Code*®, *Copyright © 2010, National Fire Protection Association, Quincy, MA. This reprinted material is not the complete and official position of the NFPA on the referenced subject, which is represented only by the standard in its entirety.*

SIZING THE SERVICE ENTRANCE CONDUCTORS

For dwelling units, wire sizes as listed in Table 310.15(B)(7) (Figure 6-23) are permitted as 120/240-volt, three-wire, single-phase service entrance conductors, service lateral conductors, and **feeder** conductors that serve as the main power feeder to a dwelling unit and are installed in raceway or cable. The wire sizes allowed in Table 310.15(B)(7) are smaller than the wire sizes given in Table 310.15(B)(16) for a particular ampacity. This reduction in wire size reflects the fact that residential service entrances are not typically loaded very heavily and the heavier loads that are on the service are not operated for long time periods. Also, the total residential electrical load is not usually energized at the same time. To be able to use this table for feeder sizing, the feeder between the main disconnect and the lighting and appliance branch-circuit panelboard must carry the total residential load. An example of a feeder carrying the total load would be when a service entrance is installed on the side of an attached garage. The main **service disconnect** equipment would be located in the garage. A feeder can be taken from the garage and used

to feed a loadcenter in a basement under the main part of the house. In this case, the feeder will carry the total residential load and could be sized according to Table 310.15(B)(7). The service entrance grounded conductor, or service neutral, is permitted to be smaller than the ungrounded conductors as long as it is never sized smaller than the grounding electrode conductor.

Now that you have an idea of the types of circuits found in a residential wiring system, the loading associated with those circuits, and how to size circuit conductors, the calculation of the dwelling service entrance can be covered. Two methods for calculating the minimum-size service entrance for dwellings are covered in Article 220. Article 220, Part III, covers the "standard method." Article 220, Part IV, covers the "optional method." It is easier to make the service entrance calculations when there is a series of steps to follow.

(See Procedure 6-1 on pages 216-218 for the steps used for calculating a residential service entrance using the standard method.)

(See Procedure 6-2 on pages 219-221 for the steps used for calculating a residential service entrance using the optional method.)

Table 310.15(B)(7) Conductor Types and Sizes for 120/240-Volt, Single-Phase Dwelling Services and Feeders. Conductor Types RHH, RHW, RHW-2, THHN, THHW, THW, THW-2, THWN, THWN-2, XHHW, XHHW-2, SE, USE, USE-2

Service or Feeder Rating (Amperes)	Conductor (AWG or kcmil)	
	Copper	Aluminum or Copper-Clad Aluminum
100	4	2
110	3	1
125	2	1/0
150	1	2/0
175	1/0	3/0
200	2/0	4/0
225	3/0	250
250	4/0	300
300	250	350
350	350	500
400	400	600

FIGURE 6-23 *NEC®* Table 310.15(B)(7) conductor types and sizes for 120/240-volt, single-phase dwelling services and feeders. *Reprinted with permission from NFPA 70®, National Electric Code®, Copyright © 2010, National Fire Protection Association, Quincy, MA. This reprinted material is not the complete and official position of the NFPA on the referenced subject, which is represented only by the standard in its entirety.*

SERVICE ENTRANCE CALCULATION EXAMPLES

The following examples show you how to apply the steps for calculating a residential service entrance using either the standard or the optional method. For this example, let's assume that a dwelling unit has a habitable floor area of 1800 square feet and has the electrical equipment listed here:

1.	Electric range	12 kW	120/240 volt
2.	Electric water heater	4 kW	240 volt
3.	Electric clothes dryer	6 kW	120/240 volt
4.	Garbage disposal	6.2 amps	120 volt
5.	Dishwasher	10 amps	120 volt
6.	Trash compactor	8 amps	120 volt
7.	Freezer	6.8 amps	120 volt
8.	Garage door opener	6.0 amps	120 volt
9.	Attic exhaust fan	3.5 amps	120 volt
10.	Central air conditioner	10 kW	240 volt
11.	Electric furnace	20 kW	240 volt

By following the steps outlined next, you can easily calculate the minimum-size residential service entrance. Each step has a short explanation and a reference from the *NEC®*. There is also a math formula at the end of each step to help you do the calculation. Commentary written in *italics* is included with each step and will help you understand what is done in each step.

Calculation Steps: The Standard Method for a Single-Family Dwelling

1 Calculate the square-foot area by using the outside dimensions of the dwelling. Do not include open porches, garages, and floor spaces unless they are adaptable for future living space. Include all floors of the dwelling. Refer to Section 220.12.

<u>1800</u> sq. ft.

Commentary: In this example, the habitable living area is given. Sometimes you will have to take a look at the building plans and determine the habitable living space yourself.

2 Multiply the square-foot area by the unit load per square foot as found in Table 220.12. Use 3 VA per square foot for dwelling units. Refer to Section 220.12 and Table 220.12.

<u>1800</u> sq. ft. × 3 VA/sq. ft. = <u>5400</u> VA

Commentary: The habitable living area is multiplied by 3 VA/sq. ft. to get the general lighting load in volt-amperes.

3 Add the load allowance for the required two small-appliance branch circuits and the one laundry circuit. These circuits are computed at 1500 VA each. Minimum total used is 4500 VA. Be sure to add in 1500 VA each for any additional small-appliance or laundry circuits. Refer to Section 220.52(A) and (B).

<u>5400</u> VA + 4500 VA
+ <u>0</u> VA (additional circuits) = <u>9900</u> VA

Commentary: There will always be at least two small-appliance branch circuits and one laundry circuit rated at 1500 VA each in every dwelling unit. Because the information for this example does not indicate that there are any more small-appliance branch circuits or laundry circuits, you will assume that only the minimum number required by the NEC® are to be installed.

4 Apply the demand to the total of steps 1 through 3. According to Table 220.42, the first 3000 VA is taken at 100%, 3001 to 120,000 VA is taken at 35%, and anything over 120,000 VA is taken at 25%. Refer to Table 220.42.

If Step 3 is over 120,000 VA, use

$$3000 \text{ VA} + [(\underline{} - 120{,}000 \text{ VA}) \times 25\%]$$
$$+ (117{,}000 \text{ VA} \times 35\%) = \underline{} \text{ VA}$$

If step 3 is under 120,000 VA, use

$$3000 \text{ VA} + [(\underline{9900} \text{ VA} - 3000 \text{ VA}) \times 35\%]$$
$$= \underline{5{,}415} \text{ VA}$$

Commentary: Applying the demand allows the service calculation to reflect the fact that not all the general lighting, small-appliance, or laundry circuits in a house will be fully loaded at any one time. By the way, it would be a tremendously large house that would have a 120,000 VA load at this point in the service calculation. You will use the "under 120,000" formula listed in this step for 99.9% of your residential service calculations.

5 Add in the loading for electric ranges, counter-mounted cooking units, or wall-mounted ovens. Use Table 220.55 and the Notes for cooking equipment. Refer to Table 220.55 and the Notes.

$$\underline{5415} \text{ VA} + 8000 \text{ VA (range)} = \underline{13{,}415} \text{ VA}$$

Commentary: There is one 12,000-watt range in this example. According to the title of Table 220.55, Column C must be used unless Note 3 can be applied. In this case, Note 3 does not apply because the range rating is more than 8¾ kW. Column C tells us to use 8000 watts as the maximum demand load when the rating of the range does not exceed 12,000 watts and there is only one range.

6 Add in the electric clothes dryer load, using Table 220.54. Remember to use 5000 VA or the nameplate rating of the dryer if it is greater than 5 kW. Refer to Section 220.54 and Table 220.54.

$$\underline{13{,}415} \text{ VA} + \underline{6000} \text{ VA (dryer)} = \underline{19{,}415} \text{ VA}$$

Commentary: In this example, the electric clothes dryer has a nameplate rating of 6 kW. This is greater than the minimum of 5 kW, so the 6 kW is used in the calculation.

7 Add the loading for heating (H) or the air-conditioning (AC), whichever is greater. Remember that both loads will not be operating at the same time. Electric heating and air conditioning are taken at 100% of their nameplate ratings. Refer to Sections 220.51 and 220.60.

$$\underline{19{,}415} \text{ VA} + \underline{20{,}000} \text{ VA (H or AC)} = \underline{39{,}415} \text{ VA}$$

Commentary: In this example, the air-conditioning load is 10 kW, and the electric heating load is 20 kW. Because both loads will not be used at the same time, the larger of the two loads is included in the service entrance calculation. In this case, it is the electric heating load of 20 kW.

8 Now add the loading for all other fixed appliances, such as dishwashers, garbage disposals, trash compactors, water heaters, and so on. Section 220.53 allows you to apply an additional demand of 75% to the fixed appliance load total if there are four or more of the appliances. If there are three or fewer, then take their total at 100%. Refer to Section 220.53.

APPLIANCES	VA OR W	
Electric water heater	4000 W	
Garbage disposal	744 VA	6.2 amps × 120 volt
Dishwasher	1200 W	10 amps × 120 volt
Trash compactor	960 VA	8 amps × 120 volt
Freezer	816 VA	6.8 amps × 120 volt
Garage door opener	720 VA	6.0 amps × 120 volt
Attic exhaust fan	420 VA	3.5 amps × 120 volt

Total VA $\underline{} \times 1.0 = \underline{}$ VA (if three or fewer fixed appliances)

Total VA $\underline{8860} \times 0.75 = \underline{6645}$ VA (if four or more fixed appliances)

$\underline{39{,}415}$ VA + $\underline{6645}$ Total VA (fixed appliances)
$= \underline{46{,}060}$ VA

Commentary: In this step, all the fixed-appliance loads are considered. The number and types of appliance loads vary greatly from house to house. Many appliances will have a wattage rating on their nameplate, while others will have an amperage and a voltage rating. Simply multiply the amperage rating by the voltage rating to get the volt-ampere load of the appliance. Remember, as far as service entrance calculations go, a watt and a volt-amp are considered to be the same thing. In this step, there is also a 75% demand that can be applied to the total fixed-appliance load if there are four or more of them. No demand factor is applied if there are three or fewer appliances.

9 According to Section 220.50, which refers to Section 430.24, you must add an additional 25% of the largest motor load. Use the largest motor load included in the service entrance calculation. The largest motor load is based on the highest-rated full-load current for the motor listed. Refer to Sections 220.50 and 430.24.

46,060 VA + 240 VA
(25% of the largest motor load) = 46,300 VA
(trash compactor is the largest motor load:
960 VA × .25 = 240 VA)

Commentary: Because service entrance conductors are feeding several load types, including motor loads, Section 430.24 requires us to increase the largest motor load by 25%. You choose the largest motor load based on the highest-rated amperage of a motor. In this example, the trash compactor has the largest motor based on its amperage and is multiplied by 25%. This amount is then added to the service calculation. Only include a motor load that is used in the service entrance calculation. Motors that are part of a load not used in Step 7 are not included.

10 The sum of steps 1 through 9 is called the "total computed load." Divide this figure by 240 volts to find the minimum ampacity required for the ungrounded service entrance conductors. Size them according to Table 310.15(B)(7). If the total computed load in this step is less than 100 amps, Sections 230.42(B) and 230.79(C) require both the service entrance ungrounded conductors and the service disconnect means to be rated not less than 100 amps. Refer to Table 310.15(B)(7) and Sections 230.42(B) and 230.79(C).

46,300 VA (total computed load)/240 volts
= 193 amps
Minimum service rating = 200 ampere
based on Table
310.15(B)(7)

Minimum ungrounded conductor size =

2/0 copper or
4/0 aluminum

Commentary: Remember that volt-amperes divided by voltage will give you amperage. Once the total computed load in volt-amperes is calculated, dividing by the service voltage of 240 volts will give you the minimum ampacity that the service conductors must be rated for. Table 310.15(B)(7) is used to find the minimum standard rating of the service and will also give the minimum-size copper and aluminum conductors required for that service size.

11 Calculate the minimum neutral ampacity by referring to Section 220.61. Add the volt-amp load for the 120-volt general lighting, small-appliance, and laundry circuits from step 4. Add the cooking and the drying loads and multiply the total by 70%. Now add in the loads of all appliances that operate on 120 volts. (Keep in mind that loads operating on two-wire 240 volts do not utilize a grounded neutral conductor.) Remember to

apply the additional demand of 75% to the total neutral loading of the appliances if there are four or more of them. Refer to Sections 220.61, 220.53 and 430.24.

Step 4: 120-volt circuits load = 5415 VA

Cooking load from step 5
8000 VA × .70   = 5600 VA

Drying load from step 6
6000 VA × .70 5   = 4200 VA

120-volt appliance load from step 8

APPLIANCES	VA
Garbage disposal	744 VA
Dishwasher	1200 W
Trash compactor	960 VA
Freezer	816 VA
Garage door opener	720 VA
Attic exhaust fan	420 VA

Total VA _____ × 1.00
(if three or fewer)   = _____ VA

Total VA 4860 × .75
(if four or more)   = 3645 VA

Largest 120-volt motor
load 960 VA × .25   = 240 VA

Total (neutral)   19,100 VA

Commentary: In most residential service entrance calculations, the grounded neutral conductor will be smaller than the ungrounded "hot" service conductors. This is because the house's 240-volt loads are connected across the two ungrounded service conductors. Because these load types (like an electric water heater) have no need for a connection to the service neutral conductor, their loads are not used in the calculation for finding the minimum neutral size. Electric cooking equipment and electric dryers use both 120- and 240-volt electricity. However, because they do use some 120-volt electricity (the dryer motor that turns the drum is connected to 120 volts), a portion (70%) of their load is used in the service neutral calculation.

12 Divide the total found in step 11 by 240 volts to find the minimum ampacity of the service neutral.

19,100 VA (step 11)/240 volts
= 80 amperes (ampacity of neutral)

Commentary: Once the total volt-ampere load for the service entrance neutral conductor is determined, dividing it by 240 volts will give you the total calculated neutral load in amperes. From this number, you can then determine the minimum-size conductor that has an ampacity equal to or greater than the calculated number.

13 According to Section 310.15(B)(7), the neutral may be smaller than the ungrounded service entrance conductors, but in no case can it be smaller than the grounding electrode conductor as per Section 250.24(C)(1). With this in mind, size the neutral conductor, based on the ampacity found in step 12, by using Table 310.15(B)(7) or Table 310.15(B)(16) (if the neutral load is less than 100 amps).

> Table 310.15(B)(7)
> Minimum neutral size = $\underline{4}$ copper or $\underline{2}$ aluminum
> Table 310.15(B)(16) (75°C column)
> Minimum neutral size = $\underline{4}$ copper or $\underline{2}$ aluminum

Commentary: This step can be a little tricky when it comes to choosing the minimum neutral conductor size. In this example, the calculated neutral amperage is 80 amps. Because the lowest ampacity listed in Table 310.15(B)(7) is 100 amps, you can simply use the wire sizes that are listed in the table for 100 amperes (4 AWG copper, 2 AWG aluminum). You can also take a look at Table 310.15(B)(16) to see if a smaller neutral conductor could be used. You can use the "75°C" column since service equipment terminations used for residential services usually have a 75°C temperature rating. In this example, it was determined that by using Table 310.15(B)(16), 4 AWG copper or 2 AWG aluminum can be used, which are the same sizes as permitted in Table 310.15(B)(7).

14 Calculate the minimum-size copper grounding electrode conductor by using the largest-size ungrounded service entrance conductor and Table 250.66. Refer to Table 250.66.

> Minimum grounding electrode conductor
> = $\underline{4 \text{ AWG copper}}$

Commentary: Based on the largest size service entrance conductor in this example, a 2/0 AWG copper or a 4/0 AWG aluminum, a 4 AWG copper grounding electrode conductor is the minimum size that must be used according to Table 250.66.

Shown next is the service entrance calculation for the same dwelling unit but using the optional method. Because sizing the service neutral conductor is done exactly the same way as in the standard method, only the steps used to get the minimum ungrounded service conductor size, the minimum service rating, and the grounding electrode conductor are shown.

from experience...

Some electricians do not bother to calculate the minimum service neutral size. Instead, they will automatically use a neutral conductor size that is the same size as the ungrounded "hot" service conductors. While there is nothing wrong with this practice, you should be aware that a small amount of money, and possibly time, can be saved by using a smaller neutral service conductor. Another common practice is to automatically size the service neutral at two wire sizes smaller than the ungrounded service conductors. As a matter of fact, many service entrance cables (Type SEU) will come with the neutral wire already sized two wire sizes smaller than the ungrounded wires. Make sure the service entrance you are installing can use a smaller neutral conductor. Do not assume it can always be smaller.

Calculation Steps: The Optional Method for a Single-Family Dwelling

1 Calculate the square-foot area by using the outside dimensions of the dwelling. Do not include open porches, garages, and floor spaces unless they are adaptable for future living space. Include all floors of the dwelling. Refer to Section 220.82(B)(1).

> $\underline{1800}$ sq. ft.

Commentary: Same first step as in the standard method.

2 Multiply the square-foot area by 3 VA per square foot. Refer to Section 220.82(B)(1).

> $\underline{1800}$ sq. ft. $\times$ 3 VA/sq. ft. = $\underline{5400}$ VA

Commentary: Same second step as in the standard method.

3 Add the load allowance for the required two small-appliance branch circuits and the one laundry circuit. These circuits are computed at 1500 VA each. Minimum total used is 4500 VA. Be sure to add in 1500 VA each for any additional small-appliance or laundry circuits. Refer to Section 220.82(B)(2).

5400 VA + 4500 VA + 0 VA (additional circuits)
= 9900 VA

Commentary: Same third step as in the standard method.

4   Add the nameplate ratings of all appliances that are fastened in place, permanently connected, or located to be on a specific circuit, such as electric ranges, wall-mounted ovens, counter-mounted cooktops, electric clothes dryers, trash compactors, dishwashers, and electric water heaters. Refer to Section 220.82(B)(3).

Example: 12-kW electric range = 12,000 W;
a 3-kW electric water heater = 3000 W.

9900 VA + 26,860 VA (nameplate rating total of
all appliances) = 36,760 VA

APPLIANCES	VA OR W	
Electric range	12,000 W	
Electric dryer	6000 W	
Electric water heater	4000 W	
Garbage disposal	744 VA	6.2 amps × 120 volts
Dishwasher	1200 W	10 amps × 120 volts
Trash compactor	960 VA	8 amps × 120 volts
Freezer	816 VA	6.8 amps × 120 volts
Garage door opener	720 VA	6.0 amps × 120 volts
Attic exhaust fan	420 VA	3.5 amps × 120 volts
Total VA	**26,860**	

Commentary: In the optional method, you add all the nameplate ratings of all the appliances together. Many electricians like this method because they do not have to bother with electric range or electric clothes dryer calculations.

5   Take 100% of the first 10,000 VA and 40% of the remaining load. Refer to Section 220.82(B).

10,000 VA + [(36,760 VA − 10,000 VA) × .40]
= 20,704 VA

Commentary: A demand still has to be applied in the optional method, but as you can see, it is slightly different than in the standard method. Remember, by

applying the demand, your service entrance calculation is taking into account the fact that all the loads in the previous steps are not likely to be energized at exactly the same time.

6   Add the value of the largest heating and air-conditioning load from the list included in Section 220.82(C). These loads include air conditioners, heat pumps, central electric heat (electric furnaces), and separately controlled electric space heaters, such as electric baseboard heaters. Refer to Section 220.82(C).

20,704 VA + 13,000 VA (largest from Section
220.82(C)) = 33,704 VA

Commentary: This step takes into account that both the electric heating load and the air-conditioning load will not be energized at exactly the same time. One of the tougher things to figure out in this step is what kind of electric heating system will be in the house. This example included an electric furnace. In most parts of the country, an electric furnace is not practical because of the high cost of electricity. However, there may be electric heat baseboard units or some other style of electric heating that must be considered in this step. This house has an air-conditioning load of 10 kW and an electric furnace rated at 20 kW. Section 220.82(C)(1) says to take the air-conditioning load at 100%. Section 220.82(C)(4) says to apply a factor of 65% to the electric heating load if there is only one separately controlled unit. 10 kW of air conditioning × 100% = 10 kw; 20 kW of central electric heat × 65% = 13 kW. Therefore, the largest load is the electric heating load of 13 kW.

7   The sum of steps 1 through 6 is called the "total computed load." Divide this figure by 240 volts to find the minimum ampacity required for the ungrounded service entrance conductors. Size them according to Table 310.15(B)(7). If the total computed load calculated in this step is less than 100 amps, Sections 230.42(B) and 230.79(C) require both the service entrance ungrounded conductors and the service disconnecting means to be rated not less than 100 amps. Refer to Table 310.15(B)(7) and Sections 230.42(B) and 230.79(C).

33,704 VA (total computed
load)/240 volts = 140 amps
Minimum service rating = 150 amperes
based on Table
310.15(B)(7)

Minimum ungrounded conductor size = 1 copper or
2/0 aluminum

from experience...

Doing a service entrance calculation using the optional method will almost always result in a smaller service rating and smaller service conductor size than when using the standard method. It really does not matter which method is used. Both are considered adequate for service entrance sizing. You may want to check with the authority having jurisdiction and the local electric utility to see if they have a requirement on using either the standard or the optional method.

Commentary: *Remember that volt-amperes divided by voltage will give you amperage. Once the total computed load in volt-amperes is calculated, dividing by the service voltage of 240 volts will give you the minimum ampacity that the service conductors must be rated for. Table 310.15(B)(7) is used to find the minimum standard rating of the service and will also give the minimum-size copper and aluminum conductor sizes required for that service size.*

8 The neutral size is calculated exactly the same as for the standard method. There is no optional method for calculating the minimum neutral size. The calculation for the minimum neutral size for this service entrance was done in the previous standard method example.

Commentary: *Because calculating the service entrance neutral conductor is exactly the same as for the standard method, this example will jump to step 11 of the optional service entrance calculation for determining the minimum-size grounding electrode conductor.*

11 Calculate the minimum-size grounding electrode conductor by using the largest-size ungrounded service entrance conductor and Table 250.66.

Minimum grounding electrode conductor
= 6 AWG copper

Commentary: *Based on the largest-size service entrance conductor in this example of a 1 AWG copper or a 2/0 AWG aluminum, a 6 AWG copper grounding electrode conductor is the minimum size that must be used according to Table 250.66.*

SIZING THE LOADCENTER

Once the service entrance calculation has been done and the minimum-size service entrance has been calculated, a loadcenter size can be determined. Most residential loadcenters used today are designed to accommodate circuit breakers, and since circuit breaker loadcenters are used primarily in residential wiring, this section limits the discussion for the sizing of loadcenters to those that take circuit breakers. Loadcenters used in residential wiring are often called "panels" by electricians. Chapter 2 discussed loadcenters, or panels, and you discovered that a panelboard located in a cabinet makes up what many electricians call a "loadcenter." The *NEC*® does not use the term "loadcenter" but does use the term "panelboard" and states several rules that must be followed when installing them. The installation of service entrance equipment, including a loadcenter, is covered in Chapter 8.

Panels used in residential wiring are available with a main circuit breaker already installed at the factory or in a style that has just wire **lugs** used to terminate the wires bringing electrical power to the panel. A panel with a main breaker is commonly called a "main breaker panel," and a panel with just wire lugs is called a "main lug only," or "MLO," panel (Figure 6-24). For example, a service calculation resulting in a minimum-size service entrance of 200 amperes would require a 200-amp main breaker panel. The 200-amp main breaker is the required main service disconnecting means used to disconnect all electrical power from the residential wiring system. Main lug-only panels are used primarily as "subpanels," which are discussed in greater detail in the next part of this chapter.

Section 408.30 of the *NEC*® says that all panelboards must have a rating not less than the minimum feeder capacity required for the load as computed in accordance with Article 220. The rating of a panelboard is based on the ampacity of the bus bar(s) in the panelboard. The bus bar is the part of the panelboard that is energized and distributes current to the individual circuit breakers located in the panel. In the most common style of panel used in residential wiring, the circuit breakers are installed by pushing them onto the bus bar. The circuit breakers are designed to clamp onto the bus bar once they have been pushed into place. Some panels require circuit breakers to be secured to the bus bar with screws, but this panel type is used mainly in commercial and industrial wiring systems. Panelboards must be durably marked by the manufacturer with the voltage, the current rating, and the number of phases (single phase or three phase) for which the panelboard is designed. The manufacturer's name or trademark must

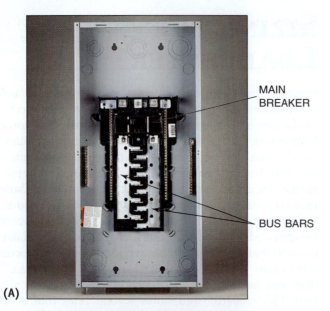

(A)

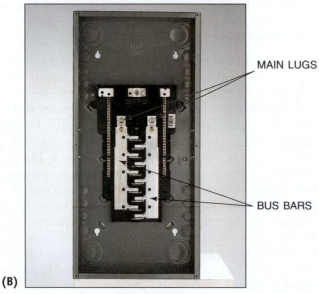

(B)

FIGURE 6-24 (A) A typical 120/240-volt, single-phase, main breaker loadcenter. *Courtesy of Square D Company.* (B) A 120/240-volt, single phase, main-lug-only (MLO) panel. *Courtesy of Square D Company, Group Schneider.*

be visible after installation. Some panels are suitable for use as service equipment and are so marked. Only those panels that are marked can be used as service entrance equipment. Listed panelboards are used with copper conductors, unless marked to indicate which terminals are suitable for use with aluminum conductors. This marking must be independent of any marking on terminal connectors and must appear on a wiring diagram or other readily visible location. If all terminals are suitable for use with aluminum conductors as well as with copper conductors, the panelboard will be marked "Use Copper or Aluminum Wire." Unless the panelboard is

marked to indicate otherwise, the termination provisions are based on the use of 60°C (140°F) ampacities for wire sizes 14 through 1 AWG and 75°C (167°F) ampacities for wire sizes 1/0 AWG and larger. As noted earlier in this chapter, most panels used in residential wiring will have terminals marked for 75°C.

In Chapter 2, you learned that there is a single-pole circuit breaker and a two-pole circuit breaker used in residential electrical work. The single-pole breaker is used to provide overcurrent protection to those circuits that operate on 120 volts. The two-pole circuit breaker is used to provide overcurrent protection to those circuits that operate on 240 volts. Once you have determined the number of circuits that will need to be installed for a house electrical system, you can choose a panel that is designed to accommodate the number of circuit breakers for the number of circuits you will install. For example, a 100-ampere-rated panel may be designed to accommodate 16, 20, or 24 circuits. A 200-ampere-rated panel may be designed to accommodate 30, 40, or 42 circuits. Some manufacturers make loadcenters that can accommodate many more circuits. One particular company offers a loadcenter suitable for residential use that can accommodate up to 60 circuits. So, based on the minimum number of circuits you plan on installing, you need to specify not only the ampere rating of the panel but also the number of circuits. For example, a 100-ampere main breaker, 120/240-volt, 24-circuit panel may be required to provide space for the minimum number of circuits needed in a house electrical system.

from experience...

It is a good idea to oversize the panel when it comes to the number of circuits for which it is rated. A good rule of thumb to follow is to never fill a panel to more than 80% of its capacity. For example, do not fill more than 16 spaces in a 20-circuit panel (20 × 80% = 16) or more than 32 spaces in a 40-circuit panel (40 × 80% = 32). This practice will help ensure that there is room in an existing panel for future expansion of the electrical system.

The number of circuits that each panel can handle is based on the 120-volt circuits. The circuit breaker for each 120-volt circuit you will be installing will take up one space in the panel. On the other hand, each 240-volt

Labels in figure A: MAIN BREAKER, BUS BARS

Labels in figure B: MAIN LUGS, BUS BARS

circuit you install will require a two-pole circuit breaker, which will take up two spaces in the panel. For example, each small-appliance branch circuit required in a residential electric system is wired with 12 AWG wire and uses a 20-amp single-pole circuit breaker for overcurrent protection of the circuit. If you only install the *NEC®* minimum number of small-appliance branch circuits, two spaces in the panel will be taken up by the two 20-amp circuit breakers. If a 240-volt electric water heater is to be installed, a common 40-gallon size will call for it to be wired with 10 AWG wire with a two-pole 30-amp circuit breaker. The circuit breaker for this water heater circuit will take up two spaces in the panel.

SIZING FEEDERS AND SUBPANELS

Some residential wiring situations may call for another loadcenter to be located in a house. This additional loadcenter is called a "subpanel." Subpanels may also be located in a detached garage. Garage wiring, including installing a subpanel in a detached garage, is covered in Chapter 15. The reason for installing a subpanel is usually to locate a loadcenter closer to an area of the house where several circuits are required. This technique will eliminate having to run long distances back to the main panel with many branch circuits. For example, a kitchen in a house usually has more required circuits than other parts of the house. If the location of the main panel is a long distance away from the kitchen area, it is a wise practice to install a subpanel closer to the kitchen and feed the kitchen with circuits that originate in the subpanel.

The wiring from the main panel to the subpanel is called a "feeder." The feeder is made up of wiring in a cable or individual wires in an electrical conduit that are large enough to feed the required electrical load that the subpanel serves. Sizing the feeder to a subpanel is done exactly the same way as for sizing service entrance conductors feeding a main service panel. You can use either the standard or the optional method. Your calculation will include only those electrical loads fed from the subpanel.

Smaller loadcenters are typically used as subpanels. Although it is permissible to have a subpanel with a main circuit breaker, most residential subpanels will be of the MLO type. The overcurrent protection device for the subpanel feeder is located at the main service entrance panel. The ampere rating of a subpanel is determined by the rating of the bus bar in the subpanel. For example, a 125-amp-rated MLO subpanel has a bus bar rating of 125 amps. This means that the subpanel could supply an electrical load of up to 125 amps. Many times, the feeder overcurrent protection device is sized less than the rating of the subpanel. For example, a 125-amp-rated MLO subpanel may be fed with an 8 AWG copper feeder with an ampacity of 40 amps. The feeder will be protected with a 40-amp circuit breaker in the main service panel.

SUMMARY

This chapter has taken a look at the types of circuits found in a residential wiring system. Sizing the circuit conductors and calculating the conductor ampacity based on the conditions of use were discussed. Both the standard and the optional methods for calculating the minimum-size service for a house were presented. The last area covered was sizing a loadcenter and a subpanel, as well as sizing the feeder supplying a subpanel. Overall, there was a great amount of material for you to become familiar with. While many of these calculations are not done by beginning electricians, you should be aware of how to do them. Electrical licensing exams always have many questions about the material covered in this chapter. Knowing how to perform the calculations covered here will help you become a more professional electrician.

PROCEDURE 6-1

Calculation Steps: The Standard Method for a Single-Family Dwelling

Steps:

1 Calculate the square foot area by using the outside dimensions of the dwelling. Do not include open porches, garages, and floor spaces unless they are adaptable for future living space. Include all floors of the dwelling. Refer to Section 220.12.

___ sq. ft.

2 Multiply the square-foot area by the unit load per square foot as found in Table 220.12. Use 3 VA per square foot for dwelling units. Refer to Section 220.12 and Table 220.12.

___ sq. ft. × 3 VA/sq. ft. = ___ VA

3 Add the load allowance for the required two small-appliance branch circuits and the one laundry circuit. These circuits are computed at 1500 VA each. Minimum total used is 4500 VA. Be sure to add in 1500 VA each for any additional small-appliance or laundry circuits. Refer to Section 220.52(A) and (B).

___ VA + 4500 VA + ___ VA (additional circuits)

= ___ VA

4 Apply the demand to the total of steps 1 through 3. According to Table 220.42, the first 3000 VA is taken at 100%, 3001 to 120,000 VA is taken at 35%, and anything over 120,000 VA is taken at 25%. Refer to Table 220.42.

If step 3 is over 120,000 VA, use

3000 VA + [(___ − 120,000 VA) × 25%] + (117,000 VA × 35%) = ___ VA

If step 3 is under 120,000 VA, use

3000 VA + [(___ VA − 3000 VA) × 35%] = ___ VA

5 Add in the loading for electric ranges, counter-mounted cooking units, or wall-mounted ovens. Use Table 220.55 and the Notes for cooking equipment. Refer to Table 220.55 and the Notes.

___ VA + ___ VA (range) = ___ VA

6 Add in the electric clothes dryer load, using Table 220.54. Remember to use 5000 VA or the nameplate rating of the dryer if it is greater than 5 kW. Refer to Section 220.54 and Table 220.54.

___ VA + ___ VA (dryer) = ___ VA

7 Add the loading for heating (H) or the air-conditioning (AC), whichever is greater. Remember that both loads will not be operating at the same time. Electric heating and air conditioning are taken at 100% of their nameplate ratings. Refer to Sections 220.51 and 220.60.

___ VA + ___ VA (H or AC) = ___ VA

8 Now add the loading for all other fixed appliances, such as dishwashers, garbage disposals, trash compactors, water heaters, and so on. Section 220.53 allows you to apply an additional demand of 75% to the fixed appliance load total if there are four or more of the appliances. If there are three or fewer, then take their total at 100%. Refer to Section 220.53.

Appliances   VA or W

_____

_____

_____

_____

_____

_____

_____

_____

**Total VA ___ × 1.0 = ___ VA (if three or fewer
fixed appliances)**

**Total VA ___ × .75 = ___ VA (if four or more fixed
appliances)**

**___ VA + ___ Total VA (fixed appliances)
= ___ VA**

9 According to Section 220.50, which refers to Section 430.24, you must add an additional 25% of the largest motor load. Use the largest motor load included in the service entrance calculation. The largest motor load is based on the highest-rated full-load current for the motor listed. Refer to Sections 220.50 and 430.24.

**___ VA + ___ VA (25% of the largest
motor load) = ___ VA**

10 The sum of steps 1 through 9 is called the "total computed load." Divide this figure by 240 volts to find the minimum ampacity required for the ungrounded service entrance conductors. Size them according to Table 310.15(B)(7). If the total computed load calculated in this step is less than 100 amps, Sections 230.42(B) and 230.79(C) require both the service entrance ungrounded conductors and the service disconnecting means to be rated not less than 100 amps. Refer to Table 310.15(B)(7) and Sections 230.42(B) and 230.79(C).

**___ VA (total computed load)/240 volts
= ___ amps**

**Minimum service rating = ___ ampere based
on Table 310.15(B)(7)**

**Minimum ungrounded conductor
size = ___ copper or**

___ aluminum

PROCEDURE 6-1

Calculation Steps: The Standard Method for a Single-Family Dwelling (Continued)

11 Calculate the minimum neutral ampacity by referring to Section 220.61. Add the volt-amp load for the 120-volt general lighting, small-appliance, and laundry circuits from step 4. Add the cooking and the drying loads and multiply the total by 70%. Now add in the loads of all appliances that operate on 120 volts. (Keep in mind that loads operating on two-wire 240 volts do not utilize a grounded neutral conductor.) Remember to apply the additional demand of 75% to the total neutral loading of the appliances if there are four or more of them. Refer to Sections 220.61, 220.53, and 430.24.

Step 4: 120-volt circuits load   **= ___ VA**

Cooking load from step 5 = ___ VA × .70 = ___ VA

Drying load from step 6   ___ VA × .70 = ___ VA

120-volt appliances from step 8

Appliances	VA
_____	
_____	
_____	
_____	
_____	
_____	
_____	
_____	
_____	
_____	

Total VA ___   × .75 = ___ VA (if four or more)

Total VA ___   × 1.00 = ___ VA (if three or fewer)

Largest 120-volt motor load ___ VA × .25 = ___ VA

Total ___ VA (neutral)

12 Divide the total found in step 11 by 240 volts to find the minimum ampacity of the service neutral.

___ VA (step 11)/240 volts = ___ ampacity of neutral

13 According to Section 310.15(B)(7), the neutral may be smaller than the ungrounded service entrance conductors, but in no case can it be smaller than the grounding electrode conductor as per Section 250.24(C)(1). With this in mind, size the neutral conductor, based on the ampacity found in step 12, by using Table 310.15(B)(7) or Table 310.15(B)(16) (if the neutral load is less than 100 amps).

(Table 310.15(B)(7))

Minimum neutral size = ___ copper or

___ aluminum

(Table 310.15(B)(16) (75°C))

Minimum neutral size = ___ copper or

___ aluminum

14 Calculate the minimum-size copper grounding electrode conductor by using the largest-size ungrounded service entrance conductor and Table 250.66. Refer to Table 250.66.

Minimum grounding electrode conductor = ___ AWG copper

PROCEDURE 6-2

Calculation Steps: The Optional Method for a Single-Family Dwelling

Steps:

1 Calculate the square-foot area by using the outside dimensions of the dwelling. Do not include open porches, garages, and floor spaces unless they are adaptable for future living space. Include all floors of the dwelling. Refer to Section 220.82(B)(1).

$$\underline{\quad} \text{ sq. ft.}$$

2 Multiply the square-foot area by 3 VA per square foot. Refer to Section 220.82(B)(1).

$$\underline{\quad} \text{ sq. ft.} \times 3 \text{ VA/sq. ft.} = \underline{\quad} \text{ VA}$$

3 Add the load allowance for the required two small-appliance branch circuits and the one laundry circuit. These circuits are computed at 1500 VA each. Minimum total used is 4500 VA. Be sure to add in 1500 VA each for any additional small-appliance or laundry circuits. Refer to Sections 220.82(B)(2).

$$\underline{\quad} \text{ VA} + 4500 \text{ VA} + \underline{\quad} \text{ VA (additional circuits)} = \underline{\quad} \text{ VA}$$

4 Add the nameplate ratings of all appliances that are fastened in place, permanently connected, or located to be on a specific circuit, such as electric ranges, wall-mounted ovens, counter-mounted cooktops, electric clothes dryers, trash compactors, dishwashers, and electric water heaters. Refer to Section 220.82(B)(3).

Example: 12 kW range = 12,000 W; a 3 kW water heater = 3000 W.

$$\underline{\quad} \text{ VA} + \underline{\quad} \text{ VA (nameplate rating total of all appliances)} = \underline{\quad} \text{ VA}$$

Appliances	VA
_____	_____
_____	_____
_____	_____
_____	_____
_____	_____
_____	_____
_____	_____
_____	_____

Total VA _____

5 Take 100% of the first 10,000 VA and 40% of the remaining load. Refer to Section 220.82(B).

$$10,000 \text{ VA} + [(\underline{\quad} \text{ VA} - 10,000 \text{ VA}) \times .40] = \underline{\quad} \text{ VA}$$

6 Add the value of the largest heating and air-conditioning load from the list included in Section 220.82(C). These loads include air conditioners, heat pumps, central electric heat (electric furnaces), and separately controlled electric space heaters, such as electric baseboard heaters. Refer to Section 220.82(C).

$$\underline{\quad} \text{ VA} + \underline{\quad} \text{ VA (largest from Section 220.82(C))} = \underline{\quad} \text{ VA}$$

PROCEDURE 6-2

Calculation Steps: The Optional Method for a Single-Family Dwelling (Continued)

7 The sum of steps 1 through 6 is called the "total computed load." Divide this figure by 240 volts to find the minimum ampacity required for the ungrounded service entrance conductors. Size them according to Table 310.15(B)(7). If the total computed load calculated in this step is less than 100 amps, Sections 230.42(B) and 230.79(C) require both the service entrance ungrounded conductors and the service disconnecting means to be rated not less than 100 amps. Refer to Table 310.15(B)(7) and Sections 230.42(B) and 230.79(C).

___ VA (total computed load)/240 volts = ___ amps

Minimum service rating = ___ ampere based on Table 310.15(B)(7)

Minimum ungrounded conductor size = ___ copper or ___ aluminum

8 The neutral size is calculated exactly the same as for the standard method. There is no optional method for calculating the minimum neutral size. Calculate the minimum neutral ampacity by referring to Section 220.61. Add the volt-amp load for the 120-volt general lighting, small-appliance, and laundry circuits from step 4. Add the cooking and the drying loads and multiply the total by 70%. Now add in the loads of all appliances that operate on 120 volts. (Keep in mind that loads operating on two-wire 240 volts do not utilize a grounded neutral conductor.) Remember to apply the additional demand of 75% to the total neutral loading of the appliances if there are four or more of them. Refer to Sections 220.61, 220.53, and 430.24.

*Step 4: 120-volt circuits load = ___ VA

*Cooking load from step 5 ___ VA × .70 = ___ VA

*Drying load from step 6 ___ VA × .70 = ___ VA

*From standard calculation steps

120-volt appliances:

Appliances	VA
_____	_____
_____	_____
_____	_____
_____	_____
_____	_____
_____	_____
_____	_____
_____	_____

Total VA ___   × .75 = ___ VA (if four or more)

Total VA ___   × 1.00 = ___ VA (if three or fewer)

Largest 120-volt motor load ___ VA × .25 = ___ VA

Total ___ VA (neutral)

9 Divide the total found in step 8 by 240 volts to find the minimum ampacity of the service neutral.

___ VA (step 8)/240 volts = ___ amps of neutral

10 According to Section 310.15(B)(7), the neutral may be smaller than the ungrounded service entrance conductors, but in no case can it be smaller than the grounding electrode conductor as per Section 250.24(C) (1). With this in mind, size the neutral conductor based on the ampacity found in step 9 by using Table 310.15(B)(7) or Table 310.15(B)(16) (if the neutral load is less than 100 amps). Refer to Tables 310.15(B)(16) and 310.15(B)(7).

(Table 310.15(B)(7))

Minimum neutral size = ___ copper

or ___ aluminum

Table 310.15(B)(16) (75°C)

Minimum neutral size = ___ copper

or ___ aluminum

11 Calculate the minimum-size grounding electrode conductor by using the largest-size ungrounded service entrance conductor and Table 250.66.

Minimum grounding electrode conductor = ___ AWG copper

REVIEW QUESTIONS

Directions: Answer the following items with clear and complete responses.

1. The minimum size of a service entrance for a single-family dwelling is 100 amperes. Name the section in the *NEC®* that states this rule.

2. The general lighting circuit unit load per square foot for a dwelling is _____ VA/sq. ft. This information is found in *NEC®* Table _____ .

3. The maximum demand for one electric range rated not over 12 kW is _____ . This information is found in *NEC®* Table _____ .

4. Calculate the computed load for an electric range that has a nameplate rating of 15 kW.

5. Name the standard sizes of circuit breakers up to and including 100 amperes. This information is found in *NEC®* Section _____ .

6. The *NEC®* states that there must be a minimum of _____ small-appliance branch circuits installed in each dwelling unit. This information is found in *NEC®* Section _____ .

7. Calculate the minimum number of 15-amp and 20-amp general lighting circuits required in a 1800-square-foot house.

8. Define a branch circuit.

9. The load for each small-appliance and laundry circuit is calculated at _____ VA each. This information is found in *NEC®* Section _____ .

10. Explain why both the electric heating load and the air-conditioning load are not used in a service entrance calculation. This information is found in *NEC®* Section _____ .

11. Name the ampacity for the following copper conductors based on Table 310.15(B)(16):
    a. 10 AWG THW _____ amps
    b. 8 AWG THWN _____ amps
    c. 1 AWG THHN _____ amps
    d. 4/0 TW _____ amps

12. Determine the ampacity of a 3 AWG THW copper conductor that is being installed in an ambient temperature of 110°F. Show all work.

13. Name the part that is used to determine the ampere rating of a panelboard.

14. Explain why the 240-volt, two-wire circuits (like an electric water heater) are not used to calculate the service neutral conductor.

15. A 1500-square-foot house will have a 12 kW electric range; a 5 kW electric clothes dryer; eight individually controlled 2 kW, 240-volt electric baseboard heaters; a 9 kW-rated air conditioner; a 3 kW electric water heater; a 6.8-amp, 120-volt garbage disposal; and a 10-amp, 120-volt freezer. Use the standard method to calculate the following:
    a. Minimum-size service entrance: _____ amp
    b. Minimum-size service conductors: _____ copper; _____ aluminum
    c. Minimum-size service neutral conductor: _____ copper; _____ aluminum
    d. Minimum-size grounding electrode conductor: _____ copper

16. List at least five types of branch circuits that are installed in a residential wiring system.

17. Define the term "dwelling unit."

18. An electrician is installing "home runs" from a loadcenter in a basement to various locations throughout a house. The wiring is being pulled through large holes drilled in the first floor joists. When the electrician is done, there are as many as five 12/2 Type NM cables bundled together through the holes. The 90°C ampacity of the cables must be derated by a factor of _____ %. The derated ampacity for this installation is _____ amps.

19. Small-appliance branch circuits and laundry branch circuits installed in a dwelling must have a rating of at least _____ amperes. This information is found in *NEC*® Section _____ .

20. At least one _____ ampere-rated bathroom circuit must be installed to supply electrical power to a residential bathroom. This information is found in *NEC*® Section _____ .

21. Calculate the minimum copper conductor size needed for a 120/240 volt electric range branch circuit. The electric range has a nameplate rating of 12 kW. Show all work.

22. Calculate the minimum copper conductor size needed for a 120/240 volt electric range branch circuit. The electric range has a nameplate rating of 15 kW. Show all work.

23. Define the term "ambient temperature".

24. According to Table 310.15(B)(7), list the copper and aluminum conductor sizes for the following service entrance sizes:

    a. 100 ampere ___ AWG CU   ___ AWG AL

    b. 150 ampere ___ AWG CU   ___ AWG AL

    c. 200 ampere ___ AWG CU   ___ AWG AL

    d. 300 ampere ___ kcmil CU   ___ kcmil AL

25. Define the term "ampacity".

Residential Service Entrances and Equipment

CHAPTER 7
Introduction to Residential Service Entrances

CHAPTER 8
Service Entrance Equipment and Installation

Introduction to Residential Service Entrances

- Demonstrate an understanding of an overhead and an underground residential service entrance.

- Define common residential service entrance terms.

- Demonstrate an understanding of *National Electrical Code* (*NEC*) requirements for residential service entrances.

- Demonstrate an understanding of grounding and bonding requirements for residential service entrances.

- List several *NEC* requirements that apply to residential service entrances.

- Demonstrate an understanding of common electric utility company requirements.

- Demonstrate an understanding of how to establish temporary and permanent power with an electric utility company.

GLOSSARY OF TERMS

accessible, readily (readily accessible) capable of being reached quickly for operation, renewal, or inspections without requiring a person to climb over or remove obstacles or to use portable ladders

bonding connected to establish electrical continuity and conductivity; the purpose of bonding is to establish an effective path for fault current that facilitates the operation of the overcurrent protective device

bonding jumper a conductor used to ensure electrical conductivity between metal parts that are required to be electrically connected

concentric knockout a series of removable metal rings that allow the knockout size to vary according to how many of the metal rings are removed; the center of the knockout hole stays the same as more rings are removed; some standard residential wiring sizes are ½, ¾, 1, 1¼, 1½, 2, and 2½ inches

drip loop an intentional loop put in service entrance conductors at the point where they extend from a weatherhead; the drip loop conducts rainwater to a lower point than the weatherhead, helping to ensure that no water will drip down the service entrance conductors and into the meter enclosure

eccentric knockout a series of removable metal rings that allow a knockout size to vary according to how many of the metal rings are removed; the center of the knockout hole changes as more metal rings are removed; common sizes are the same as for concentric knockouts

equipment a general term including material, fittings, devices, appliances, luminaires (lighting fixtures), apparatus, machinery, and other parts used in connection with an electrical installation

equipment-grounding conductor the conductor used to connect the noncurrent-carrying metal parts of equipment, raceways, and other enclosures to the system-grounded conductor, the grounding electrode conductor, or both at the service equipment

ground the earth

grounded connected to earth or to some conducting body that serves in place of the earth

grounded conductor a system or circuit conductor that is intentionally grounded

grounding conductor a conductor used to connect equipment or the grounded conductor of a wiring system to a grounding electrode or electrodes

grounding electrode a conducting object through which a direct connection to earth is established

grounding electrode conductor the conductor used to connect the system grounded conductor to a grounding electrode or to a point on the grounding electrode system

intersystem bonding termination a device that provides a means for connecting bonding conductors for communications systems to the grounding electrode system

main bonding jumper a jumper used to provide the connection between the grounded service conductor and the equipment-grounding conductor at the service

meter enclosure the weatherproof electrical enclosure that houses the kilowatt-hour meter; also called the "meter socket" or "meter trim"

neutral conductor the conductor connected to the neutral point of a system that is intended to carry current under normal conditions; the neutral point is the midpoint on a single-phase, three-wire system; the voltage from either ungrounded "hot" wire in this system to the neutral point is 120 volts

pad-mounted transformer a transformer designed to be mounted directly on a pad foundation at ground level; single-phase pad-mounted transformers are designed for underground residential distribution systems where safety, reliability, and aesthetics are important

riser a length of raceway that extends up a utility pole and encloses the service entrance conductors in an underground service entrance

service the conductors and equipment for delivering electric energy from the serving utility to the wiring system of the premises served

service entrance conductors the conductors from the service point to the service disconnecting means

service drop the overhead conductors between the utility electric supply system and the service point

service entrance cable service entrance conductors made up in the form of a cable

service entrance conductors, overhead the overhead service conductors between the service point and the first point of connection to the service conductors at the building or other structure

service entrance conductors, underground the underground service conductors between the service point and the first point of connection to the service entrance conductors inside or outside the building wall

service equipment the necessary equipment connected to the load end of the service entrance conductors supplying a building and intended to be the main control and cutoff of the supply

service head the fitting that is placed on the service drop end of service entrance cable or service entrance raceway and is designed to minimize the amount of moisture that can enter the cable or raceway; the service head is commonly referred to as a "weatherhead"

service lateral the underground conductors between the utility electric supply system and the service point

service mast a piece of rigid metal conduit or intermediate metal conduit, usually 2 or 2½ inches in diameter, that provides service conductor protection and the proper height requirements for service drops

service point the point of connection between the wiring of the electric utility and the premises wiring

service raceway the rigid metal conduit, intermediate metal conduit, electrical metallic tubing, rigid PVC conduit, or any other approved raceway that encloses the service entrance conductors

supplemental grounding electrode a grounding electrode that is used to "back up" a metal water pipe grounding electrode

transformer a piece of electrical equipment used by the electric utility to step down the high voltage of the utility system to the 120/240 volts required for a residential electrical system

triplex cable a cable type used as the service drop for a residential service entrance; it consists of a bare messenger wire that also serves as the service grounded conductor and two black insulated ungrounded conductors wrapped around the bare wire

utility pole a wooden circular column used to support electrical, video, and telecommunications utility wiring; it may also support the transformer used to transform the high utility company voltage down to the lower voltage used in a residential electrical system

One of the most important parts of a residential electrical system is the service entrance. The service entrance provides a way for the home electrical system to get electrical power from the electric utility company. This chapter will introduce you to the types of service entrances used in residential wiring. It will also discuss the most common items involved in preparing and planning for the installation of a residential service entrance, service entrance terminology, and several important *National Electrical Code*® rules that electricians must apply when installing a residential service entrance.

SERVICE ENTRANCE TYPES

The *National Electrical Code*® (*NEC*®) defines a service as the conductors and **equipment** for delivering electric energy from the serving electric utility to the wiring system of the premises served. There are two types of service entrances used to deliver electrical energy to a residential wiring system: an overhead service and an underground service. The advantages and disadvantages of each service type need to be considered in the preparation and planning stage of a residential electrical system. Residential electricians must recognize the differences between the two service types as well as the specific installation techniques required for each of them.

OVERHEAD SERVICE

The overhead service is the service type most often installed in residential wiring. It is less expensive and takes less time to install than an underground service. Electric utility companies encourage the use of overhead services because of the ready access to the service conductors if there is ever a problem. An overhead service entrance includes the service conductors between the terminals of the service equipment main disconnect and a point outside the home where they are connected to overhead wiring, which is connected to the electric utility's electrical system. The overhead wiring is placed high enough to protect it from physical

damage and to keep it away from contact with people. Figure 7-1 shows an example of a typical overhead service entrance.

UNDERGROUND SERVICE

An underground service entrance is often installed as an alternative to an overhead service. The service conductors between the terminals of the service main disconnect and the point of connection to the utility wiring are buried in the ground at a depth that protects the conductors from physical damage and also prevents accidental human contact with the conductors. Because underground services have no exposed overhead wiring, some people consider that this service type makes a

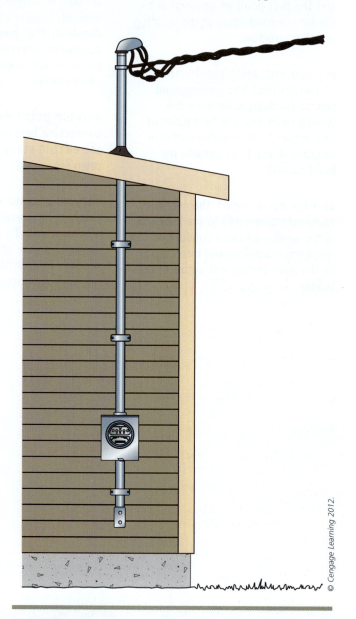

© Cengage Learning 2012.

FIGURE 7-1   A typical overhead service entrance.

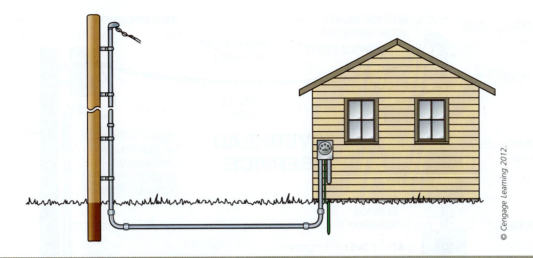

© Cengage Learning 2012.

FIGURE 7-2   A typical underground service entrance.

residence more attractive and worth the extra cost and time for the installation. If a problem arises with the underground wiring, the repair procedure usually means digging it up and fixing the problem. This is a more costly and time-consuming procedure for repair than for an overhead service. Figure 7-2 shows a typical underground service entrance.

Service Entrance Terms and Definitions

There is a lot of special terminology that applies to residential service entrances. Most of the terms commonly used are defined in Article 100 of the *NEC*®. However, there are many residential service entrance terms that the *NEC*® does not define. The following terms are those that residential electricians will find *most* important when working with residential services. Figure 7-3 illustrates the different service terms defined in the following list:

- The conductors and equipment for delivering electric energy from the electric utility to the house electrical system are called the **service**.

- The fitting that is placed on the service drop end of service entrance cable or service entrance raceway and is designed to minimize the amount of moisture that can enter the cable or raceway is called the **service head**. This part of a service entrance is usually called a "weatherhead."

- The point of connection between the wiring of the electric utility and the house wiring is called the **service point**. With an overhead service, the service

point is at the location where the service drop conductors are connected to the service conductors extending from the weatherhead. In an underground service, the service point is at the location where the service lateral conductors are connected to the electric utility wiring.

- An intentional loop put in service entrance conductors that extends from a weatherhead at the point where they connect to the service drop conductors is called a **drip loop**. The purpose of the drip loop is to conduct any rainwater to a lower point than the weatherhead. This will help ensure that no water will drip down the service entrance conductors and into the meter enclosure.

- A piece of rigid metal conduit or intermediate metal conduit, usually 2 or 2½ inches (53 or 63 Metric Designator) in diameter, that provides protection for service conductors and the proper height requirements for service drops is called a **service mast**. The mast usually extends from a meter enclosure through the overhanging portion of a house roof to a height above the roof that allows the attached service drop to have the required distance above grade. This method of installation is used on lower-roofed homes served by an overhead service so that the minimum service drop heights can be met. Some electricians refer to the service raceway that extends up from a meter enclosure but does not go above the roofline as a "mast," but most electricians use the term "service mast" only to refer to the raceway that extends above the roofline.

- The overhead service conductors from the utility pole to the point where the connection is made to the service entrance conductors at the house is called the **service drop**. In most installations, the utility company owns and installs the service drop. Aluminum **triplex cable** is normally used for the service drop.

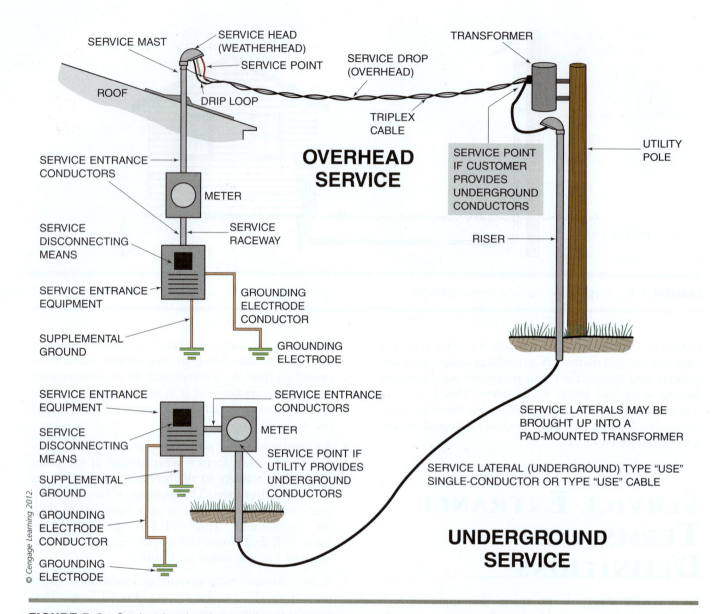

FIGURE 7-3 Overhead and underground service entrance terms.

- Service entrance conductors that are made up in the form of a cable are called **service entrance cable**. There is service entrance cable designed to be used outdoors on the side of a house (Type SEU cable) and cable designed to be buried in a trench for an underground service (Type USE cable). These cable types are discussed in more detail in Chapter 8.

- The conductors from the service point to the service disconnecting means are called the **service entrance conductors**. The service entrance conductors can be enclosed in a raceway or be part of a service entrance cable assembly.

- The service conductors between the terminals of the service equipment and a point usually outside the building where they are joined by tap or splice to the service drop are called the **overhead system service entrance conductors**.

- The service conductors between the terminals of the service equipment and the point of connection to the service lateral are called the **underground system service entrance conductors**.

- The rigid metal conduit, intermediate metal conduit, electrical metallic tubing, rigid PVC conduit, or any other approved raceway that encloses the service entrance conductors is called the **service raceway**. Installation of the raceway must meet the *NEC®* requirements that are outlined in the raceway's specific article.

- The weatherproof electrical enclosure that houses the kilowatt-hour meter is called the **meter enclosure**. Other names used to refer to a meter enclosure are "meter socket," "meter trim," and "meter base."

- The necessary equipment connected to the load end of the service conductors supplying a building and intended to be the main control and cutoff of the supply is called the **service equipment**. This equipment can consist of a fusible disconnect switch or a main breaker panel that also accommodates branch-circuit overcurrent protection devices (fuses or circuit breakers).

- A part of the service entrance that allows for the transfer of current to the earth under certain fault conditions is called the **grounding electrode**. The grounding electrode also helps limit the voltage imposed by lightning, line surges, or unintentional contact with higher-voltage lines and will stabilize the voltage to earth during normal operation. The grounding electrode is usually the metal water pipe that brings water to the home. Other types of grounding electrodes are discussed later in this chapter.

- A grounding electrode that is used to "back up" a metal water pipe grounding electrode is called a **supplemental grounding electrode**. If a metal water pipe electrode breaks and is replaced with a length of plastic plumbing pipe, grounding continuity will be lost. Because this situation occurs fairly often, the *NEC®* requires metal water pipe electrodes to be supplemented by another electrode. Electricians usually drive an 8-foot ground rod as the supplemental electrode.

- A piece of electrical equipment used by the electric utility to step down the high voltage of the utility system to the 120/240 volts required for a residential electrical system is called a **transformer**. It can be either mounted on a utility pole or placed on a concrete pad on the ground. When installed on a pad at ground level the transformer is called a **pad-mounted transformer**.

- A circular column usually made of treated wood and set in the ground for the purpose of supporting utility equipment and wiring is called a **utility pole**. Utility poles typically support transformers and electrical system wiring for electric utilities, telephone equipment and telephone wiring for communication utilities, and fiber-optic cable and coaxial cable for cable television providers.

- A length of raceway that extends up a utility pole and encloses the service entrance conductors in an underground service entrance is called a **riser**. The riser is usually made of rigid metal conduit, intermediate metal conduit, electrical metallic tubing, or rigid PVC conduit. An electrician may have to install other risers on the utility pole for such things as telephone lines and cable television lines. A piece of electrical equipment called a "standoff" allows ease of supporting multiple risers on one utility pole. Some utility companies require the use of standoffs whenever more than one riser is to be installed on the pole. Figure 7-4 shows an example of how a standoff on a utility pole is used.

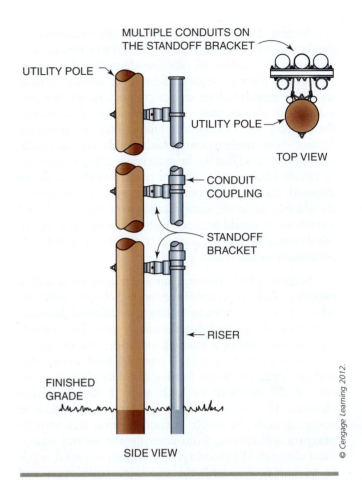

FIGURE 7-4 A standoff bracket is used to secure riser conduits on a utility pole for an underground service installation. Many electric utility companies allow electricians to secure one riser conduit directly to a pole but require the use of standoff brackets when securing multiple riser conduits to one pole.

- The underground service conductors between the electric utility transformer, including any risers at a pole or other structure, and the first point of connection to the service entrance conductors in a meter enclosure is called the **service lateral**. Electricians usually refer to the underground service conductors as simply the "lateral."

RESIDENTIAL SERVICE REQUIREMENTS (ARTICLE 230)

Article 230 of the *NEC®* covers many of the requirements for the installation of service entrances. The following discussion covers those *NEC®* rules that electricians installing residential services need to know.

Section 230.7 states that wiring other than service conductors must not be installed in the same service raceway or service cable. All other residential wiring system conductors must be separated from the service conductors. The reason is that service conductors are not provided with overcurrent protection where they receive their supply from the electric utility. They are protected against short circuits, ground faults, and overload conditions at their load end by the main service disconnect fuses or circuit breaker. The amount of current that could be imposed on the residential wiring system conductors, should they be in the same raceway or cable and should a fault occur, would be much higher than the ampacity of the wiring, and extreme damage to the wiring and unsafe conditions would result.

Section 230.8 requires an electrician to install a raceway seal in accordance with Section 300.5(G) where a service raceway enters a residential building from an underground distribution system. The sealant, such as duct seal or a bushing incorporating the physical characteristics of a seal, must be used to seal the end of service raceways. Sealing can take place at either end of the raceway or both ends if the electrician chooses. The intent of this requirement is to prevent water, usually the result of condensation due to temperature differences, from entering the service equipment through the raceway. The sealant material needs to be compatible with the conductor insulation and should not cause deterioration of the insulation over time. Any spare or unused raceways that are installed also must be sealed.

Section 300.7(A) also requires a seal any time a raceway containing electrical conductors is installed where there will be a temperature difference. The reason for sealing this type of raceway installation is to keep moisture away from electrical equipment. Again, duct seal or a bushing incorporating the physical characteristics of a seal can be used (Figure 7-5). This requirement comes into play more often in the northern parts of the United States where outside temperatures could be several degrees below zero and a home's inside temperature could be around 72°F. When a temperature difference like this occurs, a fairly large amount of moisture could accumulate because of condensation.

Service conductor clearance from building openings is covered in Section 230.9. Service conductors must have a minimum clearance of not less than 3 feet (900 mm) from windows that can be opened, doors, porches, balconies, ladders, stairs, fire escapes, or similar locations in a residential building (Figure 7-6). The intent is to protect the conductors from physical damage and protect people from accidental contact with the conductors. An *Exception* states that conductors run above the top level of a window are permitted to be closer than 3 feet (900 mm). This *Exception* permits

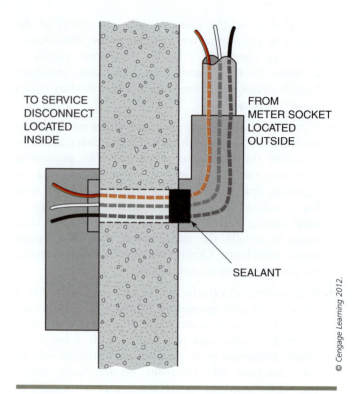

FIGURE 7-5   Section 300.7(A) requires a seal any time a raceway containing electrical conductors is installed from a cold area (outside the house) to a warmer area (inside the house).

service conductors, including drip loops and service drop conductors, to be located just above window openings because they are considered out of reach. It is important to understand that the 3-foot clearance applies only to open conductors, such as those extending from the service head, and not to the service raceway or service entrance cable that encloses the service conductors. It is a common installation practice to have the service raceway or service entrance cable assembly closer than 3 feet (900 mm) to windows or doors on the side of a residence. Section 230.9(B) requires that the vertical clearance of service conductors above or within 3 feet (900 mm) of (measured horizontally) surfaces from which they might be reached (like a deck) be at least 10 feet above the surface (Figure 7-7).

Installing an overhead service to a home in a heavily wooded area may tempt the electrician to use a tree or several trees to provide support for the overhead service conductors. Section 230.10 clearly states that vegetation such as trees must not be used to support overhead service conductors (Figure 7-8).

Section 230.22 requires individual overhead service conductors to be insulated. However, an *Exception* allows the grounded (neutral) conductor of a multi-conductor

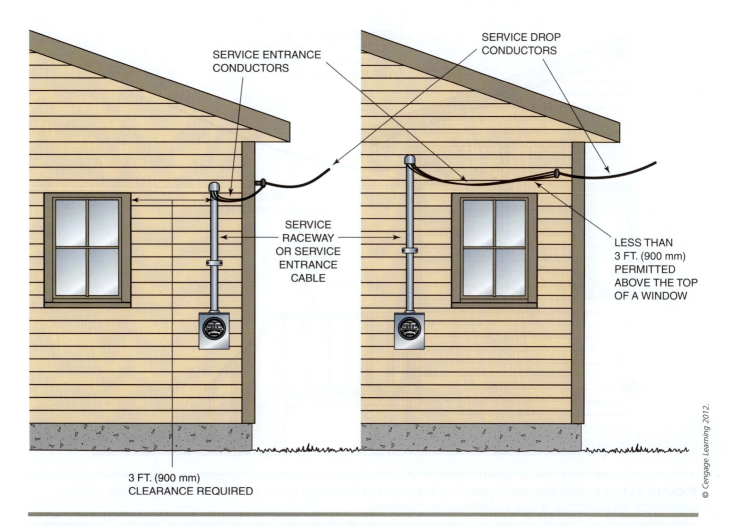

SERVICE ENTRANCE
CONDUCTORS

SERVICE DROP
CONDUCTORS

SERVICE
RACEWAY
OR SERVICE
ENTRANCE
CABLE

LESS THAN
3 FT. (900 mm)
PERMITTED
ABOVE THE TOP
OF A WINDOW

3 FT. (900 mm)
CLEARANCE REQUIRED

© Cengage Learning 2012.

FIGURE 7-6 The required dimensions for service conductors located alongside a window (left) and service conductors above the top level of a window that is designed to be opened (right).

cable to be bare. The intent of this section is to prevent problems created by weather, abrasion, and other effects that could reduce the insulating quality of the conductor insulation. The grounded conductor (neutral) of triplex service drop cable is usually bare and is used to mechanically support the two ungrounded conductors.

Section 230.23 guides us in determining the requirements for the minimum size of the overhead service conductors. The overhead service conductors must have sufficient ampacity to carry the current for the computed residential electrical load and must have adequate mechanical strength. Chapter 6 covers how to use Article 220 to calculate the electrical load for a residential wiring system.

Section 230.24 provides the dimensions for the minimum amount of vertical clearance for overhead service conductors. When installing overhead service conductors above a residential roof, conductors must have a vertical clearance of at least 8 feet (2.5 m) above the

roof. This includes overhead service conductors going over another building on the property, like a garage or storage shed. The 8-foot (2.5 m) vertical clearance above the roof must be maintained for a distance of at least 3 feet (900 mm) in all directions from the edge of the roof (Figure 7-9). *Exception No. 4* to 230.24(A) allows the requirement for maintaining the vertical clearance of 3 feet (900 mm) from the edge of the roof to not apply to the final conductor length where the overhead service drop is attached to the side of a building. *Exception No. 2* to Section 230.24(A) permits a reduction in overhead service conductor clearance above the roof from 8 feet to 3 feet if the voltage between the service conductors does not exceed 300 volts and the roof is sloped at least 4 inches vertically in 12 inches horizontally (Figure 7-10). This *Exception* is applied often to residential buildings where peaked roofs with slopes of 4/12 or greater are common. The reason for this *Exception* is that a steeply sloped roof is less likely to be walked on, and if a person

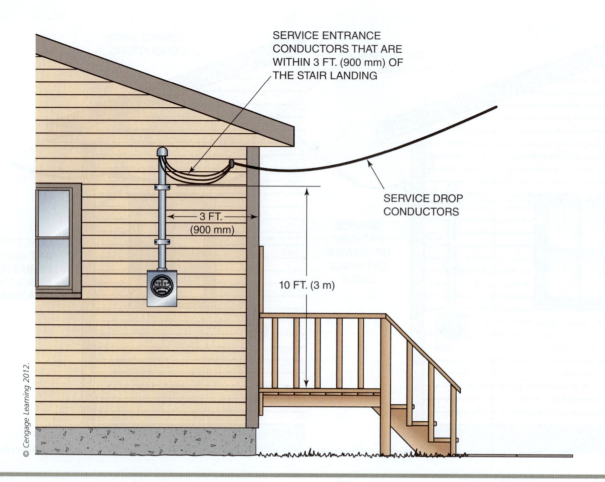

SERVICE ENTRANCE
CONDUCTORS THAT ARE
WITHIN 3 FT. (900 mm) OF
THE STAIR LANDING

SERVICE DROP
CONDUCTORS

3 FT.
(900 mm)

10 FT. (3 m)

© Cengage Learning 2012.

FIGURE 7-7 If service conductors are within 3 feet measured horizontally from a balcony, stair landing, or other platform, clearance to the platform of at least 10 feet must be maintained.

FIGURE 7-8 Trees are not allowed to support overhead service entrance conductors.

© Cengage Learning 2012.

were on the roof, he or she would be bent over and not likely to come in contact with the service drop conductors. There is no limit given for the length of the conductors over the roof. *Exception No. 5* also permits a reduction from 8 feet to 3 feet as long as the roof area is guarded or isolated and the voltage between conductors doesn't exceed 300 volts. *Exception No. 3* to Section 230.24(A) allows a reduction of overhead service conductor clearance to 18 inches (450 mm) above the roof (Figure 7-11). This reduction is designed for residential services using a service mast (through-the-roof) installation where the voltage between conductors does not exceed 300 volts and the mast is located within 4 feet (1.2 m) of the edge of the roof. This *Exception* applies to either sloped or flat roofs that are easily walked on. To use this *Exception*, not more than 6 feet of overhead service conductor is permitted to pass over the roof and attach to the service mast.

Section 230.24(B) sets the requirements for the vertical clearance from the ground for overhead service conductors (Figure 7-12). Keep in mind that the vertical

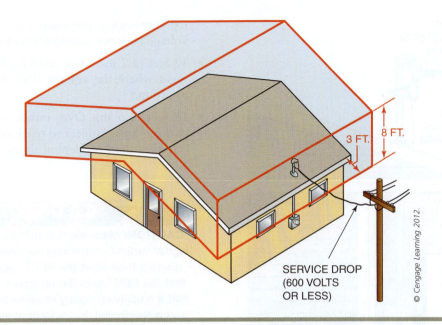

FIGURE 7-9 Section 230.24(A) applies to the vertical clearance above roofs for overhead service conductors up to 600 volts. This rule requires a vertical clearance of 8 feet above the roof, including those areas 3 feet in all directions beyond the edge of the roof. *Exception No. 4* to Section 230.24(A) exempts the final span of the overhead service conductors attached to the side of a building from the 8- and 3-foot requirements to allow the service conductors to be attached to the building.

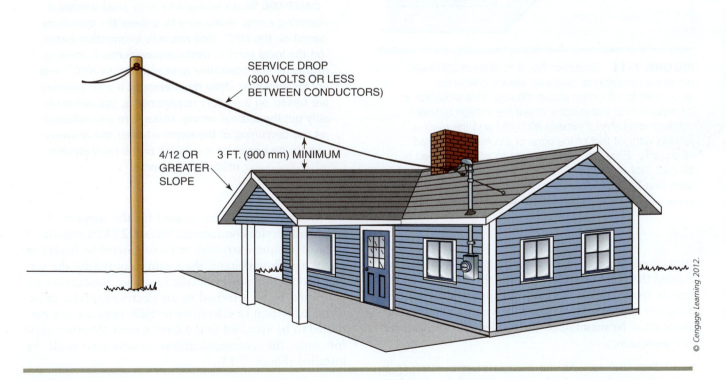

FIGURE 7-10 *Exception No. 2* to Section 230.24(A) permits a reduction in overhead service conductor clearance above the roof from 8 feet to 3 feet if the voltage between conductors does not exceed 300 volts and the roof is sloped not less than 4 inches vertically in 12 inches horizontally. Steeply sloped roofs are less likely to be walked on. There are no restrictions on the length of the conductors over the roof.

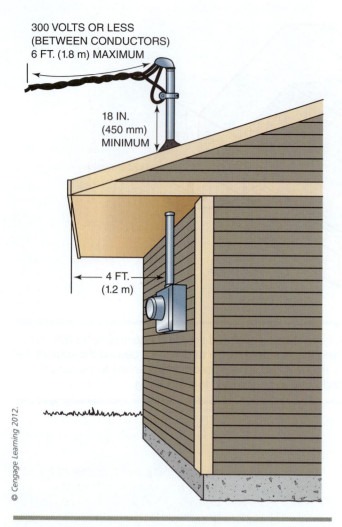

300 VOLTS OR LESS
(BETWEEN CONDUCTORS)
6 FT. (1.8 m) MAXIMUM

18 IN.
(450 mm)
MINIMUM

4 FT.
(1.2 m)

© Cengage Learning 2012.

FIGURE 7-11 *Exception No. 3* to Section 230.24(A) permits a reduction of overhead service conductor clearances to 18 inches above the roof. This reduction is for service mast installations where the voltage between conductors does not exceed 300 volts and the mast is located within 4 feet of the edge of the roof, measured horizontally. *Exception No. 3* applies to either a sloped or flat roof. Not more than 6 feet of service drop is permitted to pass over the roof.

together with a grounded bare messenger where the voltage does not exceed 150 volts to ground

- 12 feet (3.7 m): Over residential property and driveways where the voltage does not exceed 300 volts to ground

- 18 feet (5.5 m): Over public streets, alleys, roads, parking areas subject to truck traffic, and driveways on other than residential property

CAUTION

CAUTION: Many electric utility companies require higher vertical clearances for overhead service conductors than what the *NEC®* requires. Remember that the *NEC®* sets the minimum requirements and that if a utility company requires a vertical clearance over a residential driveway of at least 15 feet instead of the *NEC®* requirement of 12 feet, you must set up your service installation to meet the utility company's 15-foot requirement.

CAUTION

CAUTION: When taking a local or state electrical licensing exam, make sure to answer the questions based on the *NEC®* and not with information based on the local electric utility requirements. Licensing exams usually base their questions on the *NEC®* and not local electric utility requirements. If your answers are based on the utility requirements, you will probably get the answer wrong. Make sure you establish at the beginning of the exam whether the answers are to be based on the *NEC®* or the local electric utility requirements.

When a service mast is used for the support of the overhead service conductors, Section 230.28 requires it to be of adequate strength or be supported by braces or guy wires to withstand the strain imposed by the service drop. Only power overhead service conductors are permitted to be attached to an electrical service mast. Cable television or telephone service wires are not permitted to be attached to the service mast. Another mast for only the communications conductors must be installed (Figure 7-13).

Table 300.5 gives the minimum cover requirements for underground service conductors. The minimum depth for burying service entrance conductors or cables is 24 inches (600 mm). Service conductors or cables buried under residential driveways must be at least

clearance is dependent on how high the electrician has located the point of attachment for the service drop and that in most cases it has been preapproved by a utility company representative. The following minimum clearances must be maintained for residential overhead service conductors:

- 10 feet (3.0 m): At the electric service entrance to buildings, also at the lowest point of the drip loop of the building electric service entrance, and above areas or sidewalks accessible only to pedestrians, measured from final grade or other accessible surface for overhead service cables supported on and cabled

CLEARANCES FOR SERVICE
DROPS – RESIDENTIAL –
120/240 VOLT SINGLE PHASE:

Ⓐ = 10 FT. (3.0 m) MINIMUM
Ⓑ = 12 FT. (3.7 m) MINIMUM
Ⓒ = 10 FT. (3.0 m) MINIMUM
Ⓓ = 18 FT. (5.5 m) MINIMUM

NOTE: ELECTRIC UTILITIES FOLLOW THE *NATIONAL ELECTRICAL SAFETY CODE* (*NESC*). THE CLEARANCE REQUIREMENTS IN THE *NESC* ARE DIFFERENT THAN THOSE IN THE *NATIONAL ELECTRICAL CODE* (*NEC*®). THE DECIDING FACTOR MIGHT BE WHETHER THE INSTALLATION IS CUSTOMER INSTALLED, OWNED, AND MAINTAINED OR WHETHER THE INSTALLATION IS UTILITY INSTALLED, OWNED, AND MAINTAINED.

© Cengage Learning 2012.

FIGURE 7-12   Section 230.24(B) provides the minimum overhead service conductor vertical clearances over residential property. Be aware that some local electric utility companies may require clearances that are greater than those given in the *NEC*®.

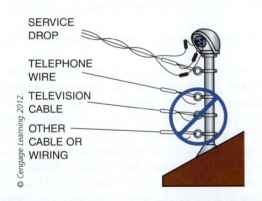

© Cengage Learning 2012.

SERVICE DROP
TELEPHONE WIRE
TELEVISION CABLE
OTHER CABLE OR WIRING

FIGURE 7-13   Only power conductors can be attached to the service mast. Section 230.28 allows no cable television wiring or telephone wiring to be attached to the service mast.

18 inches (450 mm) below grade (Figure 7-14). Remember that the local electric utility may require greater burial depths than what Table 300.5 requires.

Section 300.5 also requires underground service conductors to be protected according to the following (Figure 7-15):

- Direct-buried conductors emerging from the ground must be protected by enclosures or raceways extending from a minimum depth of 18 inches (450 mm) below grade to a point at least 8 feet (2.5 m) above finished grade.

- Conductors entering a building must be protected to the point of entrance. For example, where the conductors emerge from the ground at the side of a house, the conductors must be protected up to the first enclosure, which is usually a meter socket. The meter socket is typically mounted at approximately 5 feet (1.5 m) to the top of the meter enclosure from finished grade.

- Underground service conductors that are buried 18 inches (450 mm) or more below grade must have their location identified by a warning ribbon that is placed in the trench at least 12 inches (300 mm) above the underground installation. Providing a warning ribbon reduces the risk of an accident or electrocution during digging near underground service conductors that are not encased in concrete.

- Where the enclosure or raceway is subject to physical damage, the conductors must be installed in rigid metal conduit, intermediate metal conduit, Schedule 80 rigid PVC conduit, or a method that provides equivalent protection. Because any in-stallation outdoors is subject to physical damage, underground service conductors will always be installed in one of the raceways listed in this section.

- Any cables or insulated service conductors installed in raceways in underground installations must be listed for use in wet locations. Table 310.104(A) is

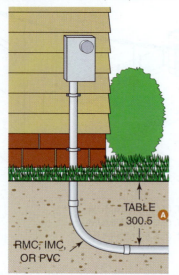

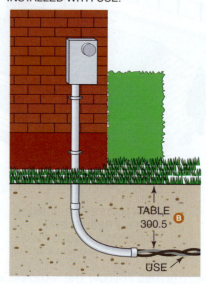

UNDERGROUND SERVICE ENTRANCE
INSTALLED WITH RMC, IMC, OR PVC.

UNDERGROUND SERVICE ENTRANCE
INSTALLED WITH USE.

Ⓐ RMC AND IMC MINIMUM BURIAL DEPTH IS 6 INCHES (150 mm); PVC MINIMUM
BURIAL DEPTH IS 18 INCHES (450 mm). UNDER DWELLING UNIT DRIVEWAYS,
THE MINIMUM BURIAL DEPTH FOR RMC, IMC, AND PVC IS 18 INCHES (450 mm).

Ⓑ THE MINIMUM BURIAL DEPTH FOR UNDERGROUND SERVICE ENTRANCE CABLE
(USE) IS 24 INCHES (600 mm). UNDER DWELLING UNIT DRIVEWAYS, THE MINIMUM
BURIAL DEPTH IS 18 INCHES (450 mm).

© Cengage Learning 2012.

FIGURE 7-14 Minimum burial depths for residential underground service conductors installed in a raceway or as a cable assembly are found in Table 300.5 of the *NEC®*.

used to determine which general wiring conductor types are permitted to be installed in wet locations. Conductor insulations with a "W" in the insulation type name are permitted in wet locations. For example, an XHHW type of insulated conductor can be installed in a raceway buried in a trench as part of an underground service entrance.

- Backfill that contains large rocks, paving materials, cinders, large or sharply angular substances, or corrosive material must not be placed in a trench on top of the service conductors, where they may damage raceways or cables. Where necessary to prevent physical damage to the raceway or cable, protection must be provided in the form of sand or other suitable material, suitable running boards, suitable sleeves, or other approved means. Electric utility companies usually will have their own requirements for the type of backfill used and the methods for backfilling an underground service trench.

- A bushing must be used at the end of a raceway that terminates underground where the conductors or cables emerge as a direct burial wiring method. A seal that provides the physical protection characteristics of a bushing is permitted to be used instead of a bushing.

- All conductors of the same circuit, including the grounded conductor and all **equipment-grounding conductors**, must be installed in the same raceway or cable. If buried underground, they must be installed close together in the same trench. This requirement is designed to make sure that the installation does not produce any inductive heating under the ground, which could produce temperatures high enough to damage the service entrance conductors.

- Where direct-buried conductors, raceways, or cables are subject to movement by settlement of the ground or frost, they must be arranged to prevent damage to the enclosed conductors or to equipment connected to the raceways. An informational note to this section suggests that "S" loops in underground direct-burial cable to raceway transitions, expansion fittings in raceway risers to fixed equipment, and flexible connections to equipment be used when subject to settlement or frost heaves. Slack must be allowed in cables, expansion joints must be used with raceways (especially rigid PVC conduit), or other measures must be taken if ground movement due to frost or settlement is anticipated.

Section 230.43 lists the wiring methods that could be used to install a residential service entrance. The various wiring methods must be installed in accordance

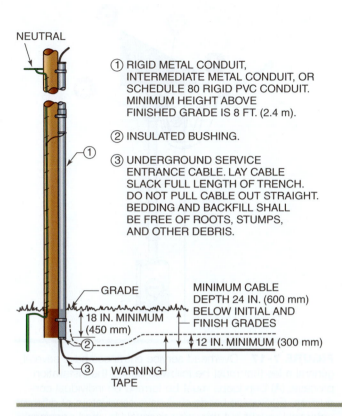

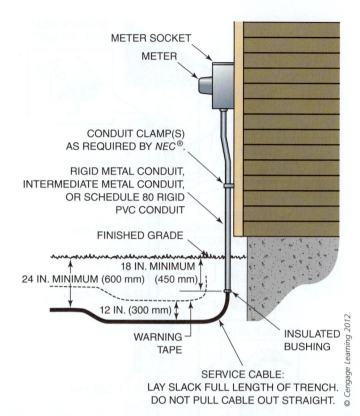

FIGURE 7-15 Protection of service conductors installed underground according to Section 300.5. The left side shows the utility pole end and the right side shows the house end of a typical underground service entrance.

with the applicable article covering that wiring method. For example, if using rigid metal conduit, you would install the conduit according to the requirements in Article 344. The following methods are the most common used for residential service installation:

- Rigid metal conduit

- Intermediate metal conduit

- Electrical metallic tubing

- Rigid PVC conduit

- Service entrance cables. According to Section 230.50(B)(1), when using service entrance cables where they are subject to physical damage, they must be protected by any of the following:

  - Rigid metal conduit

  - Intermediate metal conduit

  - Schedule 80 rigid PVC conduit

  - Electrical metallic tubing

  - Other approved means (remember that the authority having jurisdiction approves methods and materials)

Section 230.51 gives the requirements for supporting service entrance cable. Service entrance cables must be supported by straps, staples, or other approved means within 12 inches (300 mm) of the service head or any electrical enclosure and at intervals not exceeding 30 inches (750 mm) (Figure 7-16).

Section 230.54 lists some rules that apply to overhead service locations (Figure 7-17):

- Service raceways must be equipped with a rain-tight service head at the point of connection to overhead service conductors.

- Service entrance cables must be equipped with a rain-tight service head (weatherhead) that is listed for use in a wet location.

- Service heads in service entrance cables must be located above the point of attachment of the overhead service conductors to the building or other structure. An *Exception* says that where it is impracticable to locate the service head above the point of attachment, the service head location is permitted not farther than 24 inches (600 mm) from the point of attachment.

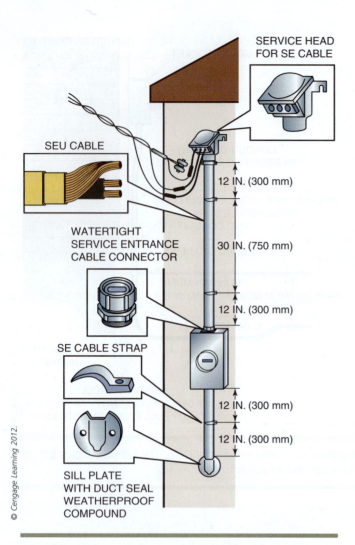

FIGURE 7-16 Service entrance cable support requirements are given in Section 230.51.

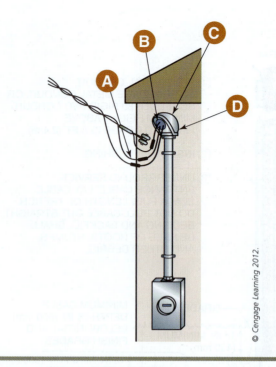

FIGURE 7-17 Overhead service entrances have several general rules that must be followed during the installation process. (A) Drip loops must be formed on individual conductors. (B) Service heads shall have conductors of different potential brought out through separately bushed openings. (C) Service heads must be located above the point of attachment of the overhead service conductors to the building or other structure. However, where it is impracticable to locate the service head above the point of attachment, the service head location is permitted not farther than 24 inches away (600 mm) from the point of the attachment. (D) Service raceways and service entrance cables must be equipped with a rain-tight service head at the point of connection to service drop conductors.

- Service entrance cables must be held securely in place.
- Weatherheads must have the service entrance conductors brought out through separately bushed openings in the weatherhead.
- Drip loops must be formed on individual conductors. To prevent the entrance of moisture, service entrance conductors must be connected to the overhead service conductors either (1) below the level of the service head or (2) below the level of the termination of the service entrance cable sheath.
- Overhead service conductors and service entrance conductors must be arranged so that water will not enter service raceway or equipment.

Section 230.70 applies to both an overhead and an underground service entrance type. A means must be provided to disconnect all conductors in a building or other structure from the service entrance conductors

(Figure 7-18). It must be installed at a **readily accessible** location either outside of a building or structure, or, if inside, as near as possible to the point of entrance of the service conductors. No maximum distance is specified from the point of entrance of service conductors to a readily accessible location for the installation of a service disconnecting means. The authority enforcing the code has the responsibility for making the decision as to how far inside the building the service entrance conductors are allowed to travel to the main disconnecting means. The length of service entrance conductors should be kept to a minimum inside buildings because electric utilities provide limited overcurrent protection, and, in the event of a fault, the service conductors could ignite nearby combustible materials (Figure 7-19). In addition, the service disconnecting means cannot be installed in bathrooms, and each service disconnect must be permanently marked to identify it as a service disconnect.

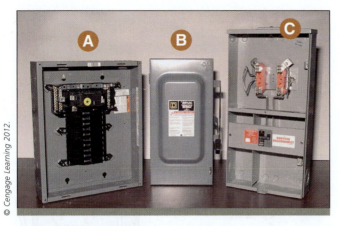

FIGURE 7-18 Section 230.70 requires a means to disconnect all conductors in a building from the service entrance conductors. This is accomplished with a main service disconnect switch. The main disconnect can consist of (A) the main circuit breaker in a loadcenter, (B) a fusible disconnect switch, or (C) a combination meter socket/disconnect switch.

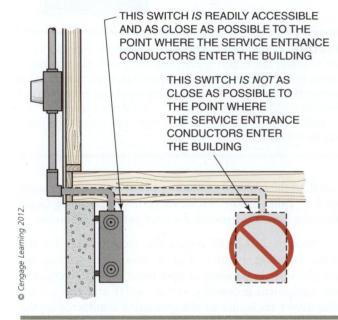

THIS SWITCH *IS* READILY ACCESSIBLE AND AS CLOSE AS POSSIBLE TO THE POINT WHERE THE SERVICE ENTRANCE CONDUCTORS ENTER THE BUILDING

THIS SWITCH *IS NOT* AS CLOSE AS POSSIBLE TO THE POINT WHERE THE SERVICE ENTRANCE CONDUCTORS ENTER THE BUILDING

FIGURE 7-19 Section 230.70(A)(1) requires the service disconnect switch to be installed in a readily accessible location as soon as the service conductors enter the building. The length of service entrance conductors should be kept to a minimum inside buildings because electric utilities provide limited overcurrent protection and, in the event of a fault, the service conductors could ignite nearby combustible materials.

CAUTION

CAUTION: Some local electric utilities have rules that allow service entrance conductors to run within the building up to a specified length to terminate at the disconnecting means. They call this "limiting the length of inside run."

Section 230.71(A) covers the maximum number of disconnects permitted as the disconnecting means for the service conductors that supply a building. One set of service entrance conductors, either overhead or underground, is permitted to supply two to six service disconnecting means in lieu of a single main disconnect. A single-occupancy building can have up to six disconnects for each set of service entrance conductors.

Section 230.79 states that the service disconnecting means must have a rating not less than the load to be carried as determined in accordance with Article 220. Section 230.79(C) states that three-wire services supplying one-family dwellings are required to have a service disconnecting means with a minimum rating of 100 amps.

Section 230.90 requires each ungrounded service conductor to have overload protection. Service entrance conductors are the supply conductors between the point of connection to the service drop or service lateral conductors and the service equipment. The service equipment is the main control and cutoff of the electrical supply to the house wiring system. In the service equipment, a circuit breaker or a fuse is installed in series with each ungrounded service conductor to provide the required overload protection.

GROUNDING AND BONDING REQUIREMENTS FOR RESIDENTIAL SERVICES (ARTICLE 250)

Article 250 of the *NEC®* requires that alternating current (AC) systems of 50 to 1000 volts that supply premises wiring systems, where the system can be **grounded** so that the maximum voltage to **ground** on the ungrounded conductors does not exceed 150 volts, must be grounded. This means that all residential electrical systems rated at 120/240 volts must be grounded. Figure 7-20 illustrates the grounding requirements for a 120-volt, single-phase, two-wire system and for a 120/240-volt, single-phase, three-wire system. Since grounding a residential electrical system starts at the service entrance, the following grounding requirements must be followed when installing a residential service.

Section 250.24(A) states that a residential wiring system supplied by a grounded AC service must have a **grounding electrode conductor** connected to the

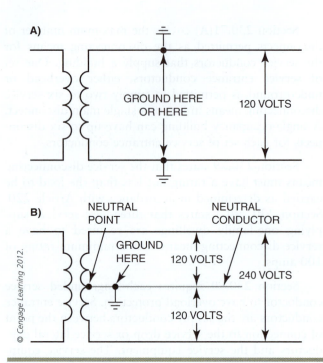

© Cengage Learning 2012.

FIGURE 7-20 Article 250 in the *NEC*® requires certain electrical system conductors to be grounded. (A) A 120-volt, single-phase, two-wire electrical system must have one conductor grounded. (B) This shows which conductor must be grounded in a 120/240-volt, single-phase, three-wire electrical system.

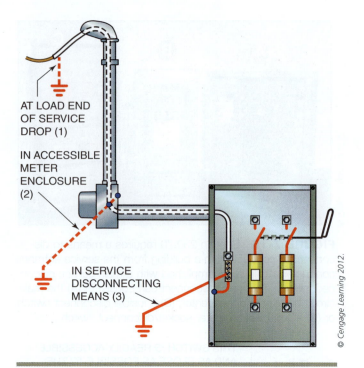

© Cengage Learning 2012.

FIGURE 7-21 A residential service entrance supplied from an overhead distribution system illustrating three possible connection points where the grounded service conductor could be connected to the grounding electrode conductor according to Section 250.24. Locations 2 and 3 are the most common locations for electricians to make the grounding electrode conductor connection.

grounded service conductor. The grounded service conductor is normally referred to as the **neutral conductor**. The connection can be made at any accessible point from the load end of the service drop or service lateral to the terminal strip to which the grounded service conductor is connected at the service disconnecting means. Figure 7-21 illustrates the two most common connection points for connecting the **grounded conductor** of the service to the grounding electrode conductor and grounding electrode for a residential installation. Section 250.4 gives information that helps explain why the grounded service conductor must be connected to a grounding electrode:

- The grounded conductor of an AC service is connected to a grounding electrode system to limit the voltage to ground imposed on the system by lightning, line surges, and (unintentional) high-voltage crossovers.

- Another reason for requiring this connection is to stabilize the voltage to ground during normal operation, including short circuits.

Section 250.24(C) requires the grounded conductor of a residential service to be run to the service disconnecting means and be bonded (attached) to the disconnecting means enclosure. The grounded conductor must be routed with the ungrounded conductors and cannot be smaller in

size than the required grounding electrode conductor specified in Table 250.66. However, it never has to be larger than the largest ungrounded service entrance conductor.

Section 250.24(B) covers the requirements for the **main bonding jumper**. For a grounded system, an unspliced main bonding jumper must be used to connect the **equipment-grounding conductor(s)** and the service disconnect enclosure to the grounded conductor of the system within the enclosure for each service disconnect. Section 250.28 lists some installation requirements for the main bonding jumper (Figure 7-22):

- Main bonding jumpers can be made of copper or other corrosion-resistant material. A main **bonding jumper** can be a wire, screw, or similar suitable conductor.

- Where a main bonding jumper is a screw, such as those used in most residential service equipment, the screw must have a green finish that shall be visible with the screw installed. The requirement for a green screw makes it possible to readily distinguish the main bonding jumper screw for inspection.

- The main bonding jumper cannot be smaller than the sizes shown in Table 250.66 for grounding electrode conductors. In residential applications, the size of the

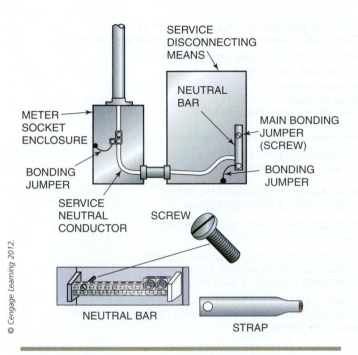

FIGURE 7-22 A main bonding jumper connects the neutral service conductor to the service entrance main disconnect enclosure and the equipment-grounding conductors.

main bonding jumper has been sized at the factory and is included with every new loadcenter. Sizing it is not something that the electrician normally has to do.

Section 250.50 covers the grounding electrode system. If present at each residential building, each item in Section 250.52(A)(1) through (A)(7) must be bonded together to form the grounding electrode system. Where none of these electrodes are present, one or more of the electrodes specified in Section 250.52(A)(4) through (A)(8) must be installed and used. Section 250.50 introduces the important concept of a "grounding electrode system," in which all electrodes are bonded together. Rather than relying totally on a single electrode to perform its function over the life of the electrical installation, the *NEC*® encourages the formation of a system of electrodes "if present at each building." There is no doubt that building a system of electrodes adds a level of reliability and helps ensure system performance over a long period of time. It is the intent of Section 250.50 that reinforcing steel, if used in a residential building footing or foundation, must be used for grounding. An *Exception* exempts reinforcing steel in existing buildings from having to be used as part of the grounding electrode system.

Section 250.52 lists the electrodes permitted for grounding:

- A metal underground water pipe in direct contact with the earth for 10 feet (3 m) or more. This is the electrode type that is most often used in residential

applications. Figure 7-23 shows the complete wiring for a typical residential service entrance with a metal water pipe being used as the grounding electrode. If a residence is located in a rural area not served by a water utility but instead has a drilled well with plastic pipe bringing water to the house, a ground rod (discussed later in this section) is often used as the grounding electrode. The metal well casing is not recognized as a grounding electrode by the *NEC*® but is required to be grounded when electrical conductors are used to supply power to a submersible pump located in the well casing. Be aware that interior metal water piping located more than 5 feet (1.52 m) from the point of entrance to the building cannot be used as a part of the grounding electrode system or as a conductor to interconnect electrodes that are part of the grounding electrode system.

- The metal frame of the building or structure, when earth connected, can be used as a grounding electrode. This electrode type is rarely used in residential applications since residential construction methods usually do not include structural metal framing.

- A concrete-encased electrode is an excellent choice (Figure 7-24). It is an electrode encased by at least 2 inches (50 mm) of concrete and is located within and near the bottom of a concrete foundation or footing that is in direct contact with the earth. This electrode type must consist of at least 20 feet (6.0 m) of one or more electrically conductive steel reinforcing rods of not less than 1/2 inch (13 mm) in diameter. It can also be at least 20 feet (6.0 m) of bare copper conductor not smaller than 4 AWG. Electricians who use concrete-encased electrodes usually choose to install at least 20 feet of 4 AWG bare copper in the footing. Obviously, the electrician will have to be at the job site when the footing is being poured. A length of the wire is left coiled up at the location where the service disconnecting means will be located for connection in the future.

- A ground ring encircling the house, in direct contact with the earth, consisting of at least 20 feet (6.0 m) of bare copper conductor at least 2 AWG. This is a great grounding electrode, but it has been the author's experience that this electrode type is not used very often.

- Rod and pipe electrodes. Rod electrodes, commonly called "ground rods," are very popular with residential electricians. They must be at least 8 feet (2.44 m) in length and must be made of the following materials:

  - Grounding electrodes of pipe or conduit cannot be smaller than 3/4-inch trade size (metric designator 21) and, if made of steel, must have their outer surface galvanized or otherwise metal-coated for corrosion protection. Pipe and conduit electrodes are rarely used in residential work.

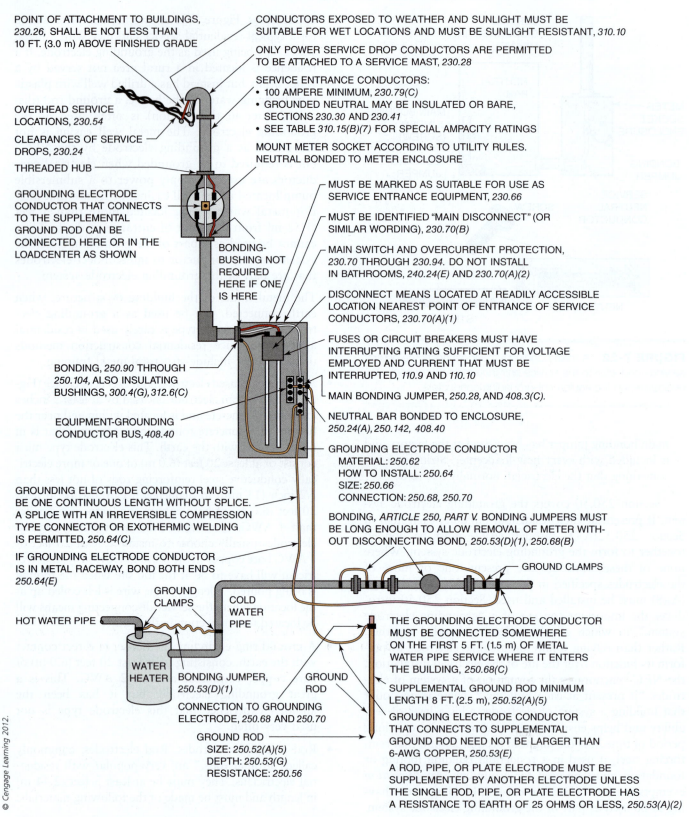

POINT OF ATTACHMENT TO BUILDINGS, *230.26,* SHALL BE NOT LESS THAN 10 FT. (3.0 m) ABOVE FINISHED GRADE

CONDUCTORS EXPOSED TO WEATHER AND SUNLIGHT MUST BE SUITABLE FOR WET LOCATIONS AND MUST BE SUNLIGHT RESISTANT, *310.10*

ONLY POWER SERVICE DROP CONDUCTORS ARE PERMITTED TO BE ATTACHED TO A SERVICE MAST, *230.28*

SERVICE ENTRANCE CONDUCTORS:
• 100 AMPERE MINIMUM, *230.79(C)*
• GROUNDED NEUTRAL MAY BE INSULATED OR BARE, SECTIONS *230.30* AND *230.41*
• SEE TABLE *310.15(B)(7)* FOR SPECIAL AMPACITY RATINGS

MOUNT METER SOCKET ACCORDING TO UTILITY RULES. NEUTRAL BONDED TO METER ENCLOSURE

OVERHEAD SERVICE LOCATIONS, *230.54*

CLEARANCES OF SERVICE DROPS, *230.24*

THREADED HUB

GROUNDING ELECTRODE CONDUCTOR THAT CONNECTS TO THE SUPPLEMENTAL GROUND ROD CAN BE CONNECTED HERE OR IN THE LOADCENTER AS SHOWN

MUST BE MARKED AS SUITABLE FOR USE AS SERVICE ENTRANCE EQUIPMENT, *230.66*

MUST BE IDENTIFIED "MAIN DISCONNECT" (OR SIMILAR WORDING), *230.70(B)*

MAIN SWITCH AND OVERCURRENT PROTECTION, *230.70* THROUGH *230.94.* DO NOT INSTALL IN BATHROOMS, *240.24(E)* AND *230.70(A)(2)*

DISCONNECT MEANS LOCATED AT READILY ACCESSIBLE LOCATION NEAREST POINT OF ENTRANCE OF SERVICE CONDUCTORS, *230.70(A)(1)*

BONDING-BUSHING NOT REQUIRED HERE IF ONE IS HERE

FUSES OR CIRCUIT BREAKERS MUST HAVE INTERRUPTING RATING SUFFICIENT FOR VOLTAGE EMPLOYED AND CURRENT THAT MUST BE INTERRUPTED, *110.9* AND *110.10*

BONDING, *250.90* THROUGH *250.104,* ALSO INSULATING BUSHINGS, *300.4(G), 312.6(C)*

MAIN BONDING JUMPER, *250.28,* AND *408.3(C).*

EQUIPMENT-GROUNDING CONDUCTOR BUS, *408.40*

NEUTRAL BAR BONDED TO ENCLOSURE, *250.24(A), 250.142, 408.40*

GROUNDING ELECTRODE CONDUCTOR
    MATERIAL: *250.62*
    HOW TO INSTALL: *250.64*
    SIZE: *250.66*
    CONNECTION: *250.68, 250.70*

GROUNDING ELECTRODE CONDUCTOR MUST BE ONE CONTINUOUS LENGTH WITHOUT SPLICE. A SPLICE WITH AN IRREVERSIBLE COMPRESSION TYPE CONNECTOR OR EXOTHERMIC WELDING IS PERMITTED, *250.64(C)*

IF GROUNDING ELECTRODE CONDUCTOR IS IN METAL RACEWAY, BOND BOTH ENDS *250.64(E)*

BONDING, *ARTICLE 250, PART V.* BONDING JUMPERS MUST BE LONG ENOUGH TO ALLOW REMOVAL OF METER WITH-OUT DISCONNECTING BOND, *250.53(D)(1), 250.68(B)*

GROUND CLAMPS

GROUND CLAMPS

COLD WATER PIPE

HOT WATER PIPE

WATER HEATER

BONDING JUMPER, *250.53(D)(1)*

CONNECTION TO GROUNDING ELECTRODE, *250.68* AND *250.70*

GROUND ROD
    SIZE: *250.52(A)(5)*
    DEPTH: *250.53(G)*
    RESISTANCE: *250.56*

GROUND ROD

THE GROUNDING ELECTRODE CONDUCTOR MUST BE CONNECTED SOMEWHERE ON THE FIRST 5 FT. (1.5 m) OF METAL WATER PIPE SERVICE WHERE IT ENTERS THE BUILDING, *250.68(C)*

SUPPLEMENTAL GROUND ROD MINIMUM LENGTH 8 FT. (2.5 m), *250.52(A)(5)*

GROUNDING ELECTRODE CONDUCTOR THAT CONNECTS TO SUPPLEMENTAL GROUND ROD NEED NOT BE LARGER THAN 6-AWG COPPER, *250.53(E)*

A ROD, PIPE, OR PLATE ELECTRODE MUST BE SUPPLEMENTED BY ANOTHER ELECTRODE UNLESS THE SINGLE ROD, PIPE, OR PLATE ELECTRODE HAS A RESISTANCE TO EARTH OF 25 OHMS OR LESS, *250.53(A)(2)*

FIGURE 7-23 A typical residential service entrance installation with a water pipe being used as the grounding electrode. Notice that the water pipe is supplemented by a ground rod. The rod electrode will "back up" the water pipe electrode in the event that the water pipe breaks and is repaired with a length of plastic pipe.

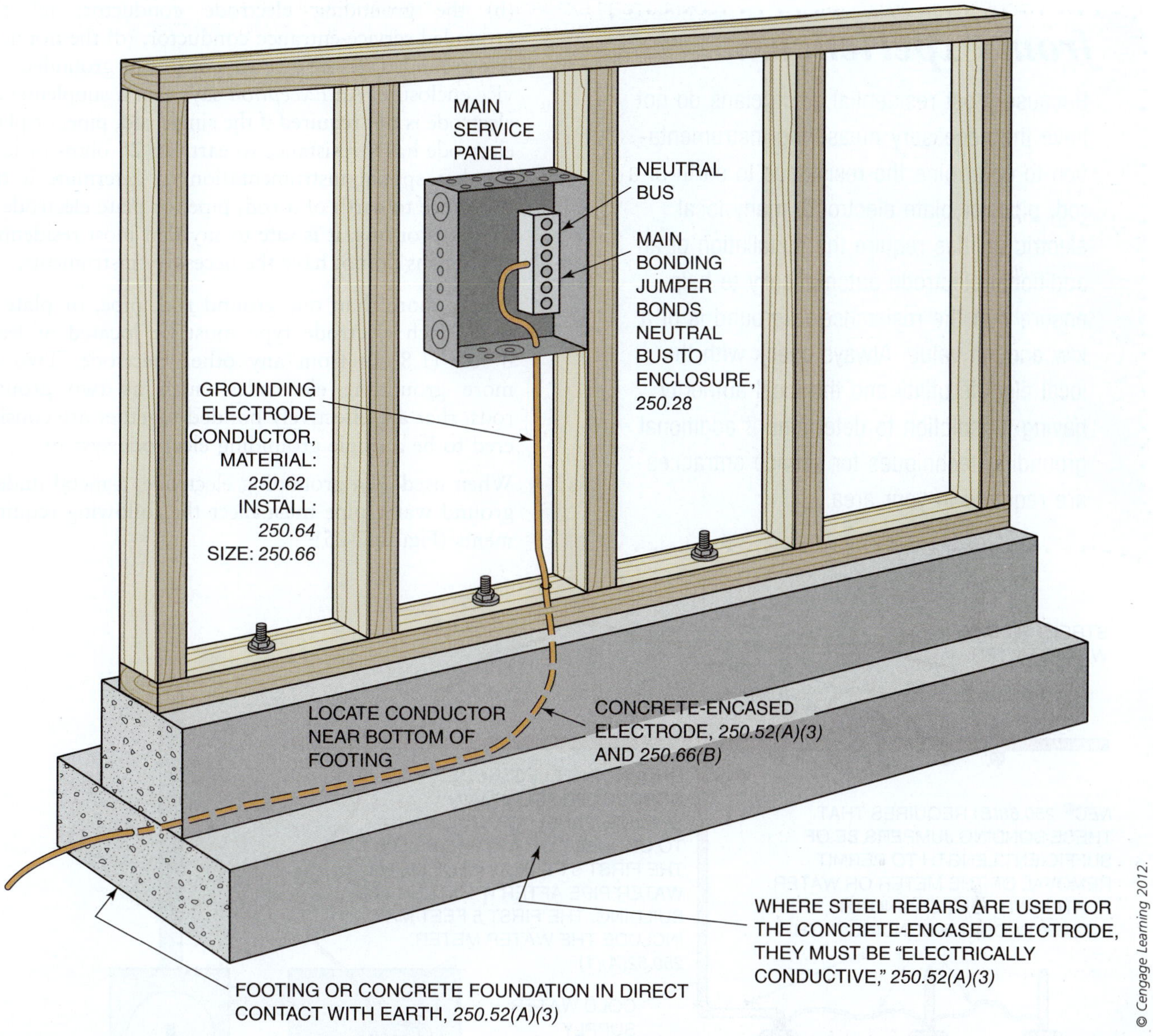

MAIN
SERVICE
PANEL

NEUTRAL
BUS

MAIN
BONDING
JUMPER
BONDS
NEUTRAL
BUS TO
ENCLOSURE,
250.28

GROUNDING
ELECTRODE
CONDUCTOR,
MATERIAL:
250.62
INSTALL:
250.64
SIZE: *250.66*

LOCATE CONDUCTOR
NEAR BOTTOM OF
FOOTING

CONCRETE-ENCASED
ELECTRODE, *250.52(A)(3)*
AND *250.66(B)*

WHERE STEEL REBARS ARE USED FOR
THE CONCRETE-ENCASED ELECTRODE,
THEY MUST BE "ELECTRICALLY
CONDUCTIVE," *250.52(A)(3)*

FOOTING OR CONCRETE FOUNDATION IN DIRECT
CONTACT WITH EARTH, *250.52(A)(3)*

© Cengage Learning 2012.

FIGURE 7-24 A concrete-encased electrode is allowed by Section 250.52(A)(3). At least 20 feet of 4 AWG or larger copper conductor can be located near the bottom of the foundation or footing as long as it is encased in at least 2 inches (51 mm) of concrete.

- Grounding electrodes of rods made of steel must be at least 5/8 inch (15.87 mm) in diameter. Copper-coated steel rods, which are very popular in residential work, or their equivalent must be listed and cannot be less than 1/2 inch (12.70 mm) in diameter.

- Plate electrodes can be used, but again, they are rarely used in residential work. Each plate electrode must have at least 2 square feet (0.186 square m^2) of surface exposed to the soil. Electrodes of iron or steel plates must be at least 1/4 inch (6.4 mm) in thickness. Electrodes of nonferrous metal, like copper, must be at least 0.06 inch (1.5 mm) in thickness.

Section 250.53 covers some installation rules for the grounding electrode system:

- Where practicable, rod, pipe, and plate electrodes must be embedded below the permanent moisture level. They must also be free from nonconductive coatings, such as paint or enamel.

- Unless a single rod, pipe or plate electrode has a resistance of 25 ohms or less to earth, it must be supplemented by an additional electrode of a type specified in Section 250.52(A)(2) through (A)(8). You are permitted to connect the supplemental electrode to one of the following: (a) the rod, pipe or plate electrode;

from experience...

Because most residential electricians do not have the necessary measuring instrumentation to determine the resistance to earth of a rod, pipe, or plate electrode, many local electric utilities require the installation of an additional electrode automatically to help ensure that the resistance to ground is of a low enough value. Always check with the local electric utility and the local authority having jurisdiction to determine if additional grounding techniques for service entrances are required in your area.

(b) the grounding electrode conductor; (c) the grounded service-entrance conductor; (d) the nonflexible grounded service raceway; or (e) any grounded service enclosure. An *Exception* says that a supplemental electrode is not required if the single rod, pipe, or plate electrode has a resistance to earth of 25 ohms or less. It takes special instrumentation to determine if the resistance to earth of a rod, pipe, or plate electrode is 25 ohms or less. It is safe to say that most residential electricians do not have the necessary instruments.

- Where more than one ground rod, pipe, or plate is used, each electrode type must be located at least 6 feet (1.8 m) from any other electrode. Two or more grounding electrodes, such as two ground rods, that are effectively bonded together are considered to be a single grounding electrode system.

- When used as a grounding electrode, a metal underground water pipe must meet the following requirements (Figure 7-25):

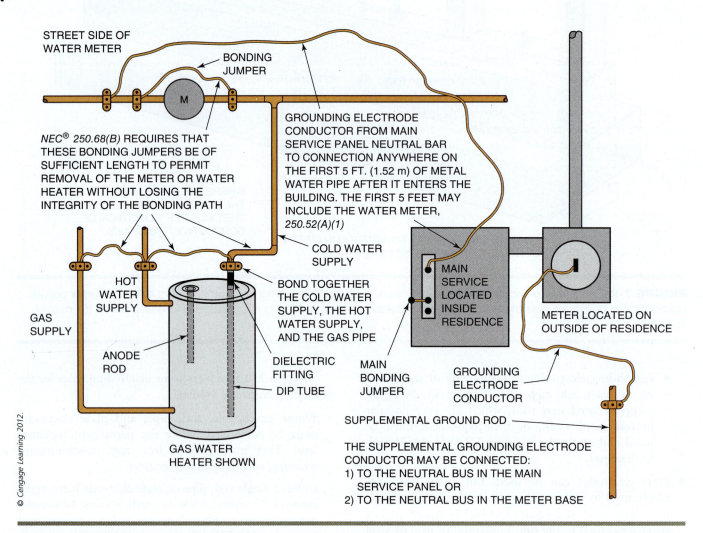

STREET SIDE OF WATER METER

BONDING JUMPER

NEC® 250.68(B) REQUIRES THAT THESE BONDING JUMPERS BE OF SUFFICIENT LENGTH TO PERMIT REMOVAL OF THE METER OR WATER HEATER WITHOUT LOSING THE INTEGRITY OF THE BONDING PATH

GROUNDING ELECTRODE CONDUCTOR FROM MAIN SERVICE PANEL NEUTRAL BAR TO CONNECTION ANYWHERE ON THE FIRST 5 FT. (1.52 m) OF METAL WATER PIPE AFTER IT ENTERS THE BUILDING. THE FIRST 5 FEET MAY INCLUDE THE WATER METER, 250.52(A)(1)

COLD WATER SUPPLY

HOT WATER SUPPLY

GAS SUPPLY

ANODE ROD

BOND TOGETHER THE COLD WATER SUPPLY, THE HOT WATER SUPPLY, AND THE GAS PIPE

DIELECTRIC FITTING

DIP TUBE

GAS WATER HEATER SHOWN

MAIN SERVICE LOCATED INSIDE RESIDENCE

MAIN BONDING JUMPER

METER LOCATED ON OUTSIDE OF RESIDENCE

GROUNDING ELECTRODE CONDUCTOR

SUPPLEMENTAL GROUND ROD

THE SUPPLEMENTAL GROUNDING ELECTRODE CONDUCTOR MAY BE CONNECTED:
1) TO THE NEUTRAL BUS IN THE MAIN SERVICE PANEL OR
2) TO THE NEUTRAL BUS IN THE METER BASE

© Cengage Learning 2012.

FIGURE 7-25 Section 250.53(D) states that when using a metal water pipe as the grounding electrode, the continuity of the grounding path cannot rely on water meters, water filters, or similar equipment. For this reason, an electrician must bond around these items.

- The continuity of the grounding path must not rely on water meters or filtering devices and similar equipment. **Bonding** around such equipment is required to ensure good grounding continuity.

- A metal underground water pipe must be supplemented by an additional electrode. The supplemental electrode is permitted to be connected to the grounding electrode conductor, the grounded service entrance conductor, the grounded service raceway, a nonflexible service raceway, or any grounded service enclosure. The requirement to supplement the metal water pipe is based on the practice of using a plastic pipe for replacement when the original metal water pipe fails. This type of replacement leaves the system without a grounding electrode unless a supplementary electrode is provided.

- Where the supplemental electrode is a rod, pipe, or plate electrode, that portion of the bonding jumper that is the sole connection to the supplemental grounding electrode is not required to be larger than 6 AWG copper wire or 4 AWG aluminum wire. For example, if a metal underground water pipe is used as the grounding electrode, Table 250.66 must be used for sizing the grounding electrode conductor, and the size may be required to be larger than a 6 AWG copper conductor. However, the size of the grounding electrode conductor for ground rod, pipe, or plate electrodes between the service equipment and the electrodes is not required to be larger than 6 AWG copper or 4 AWG aluminum when the rod, pipe, or plate is being used either as a supplemental electrode or as the only grounding electrode.

- If a ground ring is installed as the grounding electrode, it must be buried at least 30 inches (750 mm).

- Rod and pipe electrodes must be installed so that at least 8 feet (2.44 m) of length is in contact with the soil. Where large rocks or ledge is encountered and a rod or pipe cannot be driven straight down into the ground, the electrode must be either driven at not more than a 45-degree angle or buried in a 2½-foot-deep (750 mm) trench (Figure 7-26). Ground clamps used on buried electrodes must be listed for direct earth burial. Ground clamps installed aboveground must be protected where subject to physical damage. It is very common for an electrician to install two ground rods when they need to have the resistance to earth be 25 ohms or less.

- Plate electrodes must be buried at least 30 inches (750 mm) below the surface of the earth.

Section 250.64 covers the installation of the grounding electrode conductor. The following rules must be followed when installing a grounding electrode conductor:

- Bare aluminum or copper-clad aluminum grounding electrode conductors must not be used in direct

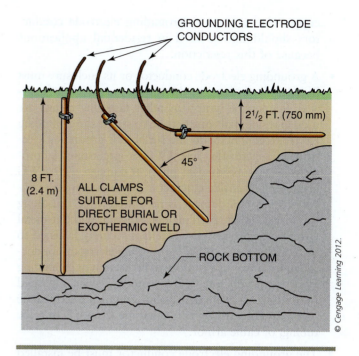

FIGURE 7-26   The installation requirements for rod and pipe electrodes as specified by Section 250.53(G).

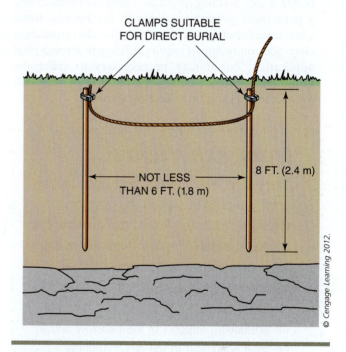

FIGURE 7-27   Electricians often install two ground rods together.

contact with masonry or the earth or where subject to corrosive conditions. Where used outside, aluminum or copper-clad aluminum grounding conductors must be terminated within 18 inches (450 mm) of the ground. For all practical purposes, aluminum or copper-clad aluminum grounding electrode conductors should not be used in residential applications because of this restriction.

- A grounding electrode conductor or its enclosure must be securely fastened to the surface on which it is carried. Staples, tie wraps, or other fastening means are used in residential construction to secure the grounding electrode conductor to the surface. A grounding electrode conductor that is 4 AWG or larger must be protected if it is exposed to physical damage. A 6 AWG grounding electrode conductor that is free from exposure to physical damage is permitted to be run along the surface of the building construction without metal covering or protection where it is securely fastened to the construction. 8 AWG grounding electrode conductors must always be protected. Protection of grounding electrode conductors, when required, is accomplished by putting them in rigid metal conduit, intermediate metal conduit, rigid PVC conduit, electrical metallic tubing, or cable armor.

- The grounding electrode conductor must be installed in one continuous length without a splice. Splicing is allowed only by irreversible compression-type connectors listed for the purpose or by the exothermic (CAD weld) welding process. These methods create a permanent connection that will not become loose after installation. It is very rare for the grounding electrode conductor to require splicing in a residential application, but it may be necessary to splice the grounding electrode conductor because of remodeling of the building or to add equipment.

from experience...

There is no *NEC*® requirement listed for how often you need to support the grounding electrode conductor. A good rule to follow is to support it within 12 inches from where it is terminated and no more than 24 inches between supports after that. If the grounding electrode conductor is placed in a raceway for protection, support of the raceway should be done according to the *NEC*® requirements located in the article that covers that raceway type.

- Metal raceways for grounding electrode conductors must be electrically continuous from the point of attachment to cabinets or equipment to the grounding electrode. They also must be securely fastened to the ground clamp or fitting. Metal raceways that are not physically continuous from cabinet or equipment to the grounding electrode must be made electrically continuous by bonding each end to the grounding electrode conductor (Figure 7-28).

Section 250.66 specifies how to determine the size of the grounding electrode conductor. Table 250.66 (Figure 7-29) is used to size the grounding electrode conductor of a grounded AC system and is based on the largest-size service entrance conductor. For example, a 200-ampere residential service entrance can use a 2/0 copper or 4/0 aluminum service entrance conductor size. (Sizing a residential service entrance is covered in Chapter 6.) Table 250.66 tells us that the minimum-size copper grounding electrode conductor would be a 4 AWG copper conductor. Special sizing for connections to rods, pipes, plates, concrete-encased electrodes, and ground rings are listed as follows:

- Where the grounding electrode conductor is connected to rod, pipe, or plate electrodes, that portion of the conductor that is the only connection to the grounding electrode is not required to be larger than 6 AWG copper wire or 4 AWG aluminum wire.

- Where the grounding electrode conductor is connected to a concrete-encased electrode, that portion of the conductor that is the only connection to the grounding electrode is not required to be larger than 4 AWG copper wire.

- Where the grounding electrode conductor is connected to a ground ring, that portion of the conductor that is the only connection to the grounding electrode is not required to be larger than the conductor used for the ground ring.

Section 250.68 covers the grounding electrode conductor connection to the grounding electrode:

- The connection of a grounding electrode conductor to a grounding electrode must be accessible. However, Section 250.68 (A) *Exception No. 1* states that an encased or buried connection to a concrete-encased, driven, or buried grounding electrode is not required to be accessible. Ground clamps and other connectors suitable for use where buried in earth or embedded in concrete must be listed for such use, either by a marking on the connector or by a tag attached to the connector. Ground clamps that are suitable for direct burial will have the words "direct burial" written on the clamp. Section 250.68(A) *Exception No. 2* states that exothermic or irreversible compression connections used at terminations are not required to be accessible.

- The connection of a grounding electrode conductor to a grounding electrode must be made in a manner that will ensure a permanent and effective grounding path.

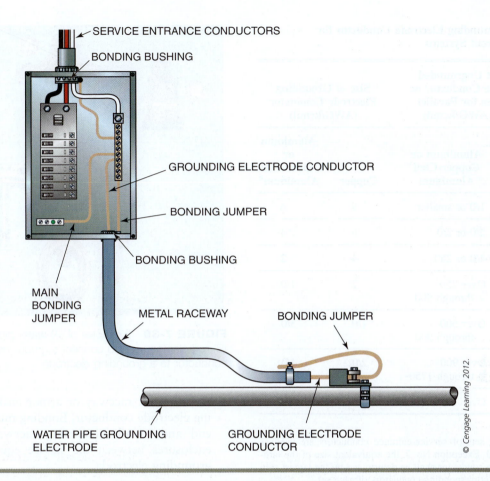

SERVICE ENTRANCE CONDUCTORS

BONDING BUSHING

GROUNDING ELECTRODE CONDUCTOR

BONDING JUMPER

BONDING BUSHING

MAIN BONDING JUMPER

METAL RACEWAY

BONDING JUMPER

WATER PIPE GROUNDING ELECTRODE

GROUNDING ELECTRODE CONDUCTOR

© Cengage Learning 2012.

FIGURE 7-28   Bonding of a metal raceway to the grounding electrode conductor at <u>both</u> ends is required by Section 250.64(E).

Where necessary to ensure the grounding path for a metal piping system used as a grounding electrode, effective bonding must be provided around insulated joints and around any equipment likely to be disconnected for repairs or replacement. Bonding jumpers must be of sufficient length to permit removal of the equipment while still maintaining the integrity of the grounding path. Examples of equipment likely to be disconnected for repairs or replacement are water meters and water filter systems.

- Interior metal water piping located not more than 5 feet (1.52 m) from the point of entrance to the building is permitted to be used as a conductor to interconnect electrodes that are part of the grounding electrode system.

Section 250.70 lists some methods of connecting the grounding conductor to an electrode. The grounding or bonding conductor must be connected to the grounding electrode by exothermic welding, listed lugs, listed pressure connectors, listed clamps, or other listed means (Figure 7-30). Connections depending on solder cannot be used. Ground clamps must be listed for the material that the grounding electrode is made of and the

grounding electrode conductor. Where used on pipe, rod, or other buried electrodes, the clamp must also be listed for direct soil burial or concrete encasement. Not more than one conductor shall be connected to the grounding electrode by a single clamp or fitting unless the clamp or fitting is listed for multiple conductors.

Section 250.80 states that metal enclosures and raceways for service conductors and equipment must be connected to the grounded (neutral) system conductor. An *Exception* allows a metal elbow that is installed in an underground installation of nonmetallic raceway and is isolated from possible contact by a minimum cover of 18 inches (450 mm) to any part of the elbow to not be required to be grounded. Metal sweep elbows are often installed in underground installations of nonmetallic conduit. The metal elbows are installed because nonmetallic elbows can be damaged by friction from the pulling ropes used during conductor installation. The elbows are isolated from physical contact by burial so that no part of the elbow is less than 18 inches below grade.

Section 250.92 requires the noncurrent-carrying metal parts of service equipment to be effectively bonded together. Figure 7-31 illustrates grounding and

Table 250.66 Grounding Electrode Conductor for Alternating-Current Systems

Size of Largest Ungrounded Service-Entrance Conductor or Equivalent Area for Parallel Conductors[a] (AWG/kcmil)		Size of Grounding Electrode Conductor (AWG/kcmil)	
Copper	Aluminum or Copper-Clad Aluminum	Copper	Aluminum or Copper-Clad Aluminum[b]
2 or smaller	1/0 or smaller	8	6
1 or 1/0	2/0 or 3/0	6	4
2/0 or 3/0	4/0 or 250	4	2
Over 3/0 through 350	Over 250 through 500	2	1/0
Over 350 through 600	Over 500 through 900	1/0	3/0
Over 600 through 1100	Over 900 through 1750	2/0	4/0
Over 1100	Over 1750	3/0	250

Notes:

1. Where multiple sets of service-entrance conductors are used as permitted in 230.40, Exception No. 2, the equivalent size of the largest service-entrance conductor shall be determined by the largest sum of the areas of the corresponding conductors of each set.

2. Where there are no service-entrance conductors, the grounding electrode conductor size shall be determined by the equivalent size of the largest service-entrance conductor required for the load to be served.

[a]This table also applies to the derived conductors of separately derived ac systems.

[b]See installation restrictions in 250.64(A).

FIGURE 7-29   Table 250.66 in the *NEC®* specifies the minimum-size grounding electrode conductor for a certain size service entrance. The minimum-size grounding electrode conductor is determined by using the largest-size ungrounded conductor of a service entrance. *Reprinted with permission from NFPA 70®, National Electric Code®, Copyright © 2010, National Fire Protection Association, Quincy, MA. This reprinted material is not the complete and official position of the NFPA on the referenced subject, which is represented only by the standard in its entirety.*

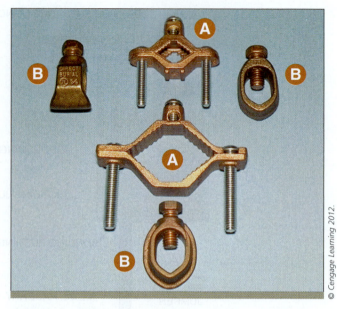

© Cengage Learning 2012.

FIGURE 7-30   Examples of (A) water pipe clamps and (B) rod clamps used to connect a grounding electrode conductor to a grounding electrode.

- Any metallic raceway or armor enclosing a grounding electrode conductor. Bonding must apply at each end and to all intervening raceways, boxes, and enclosures between the service equipment and the grounding electrode.

Section 250.92(B) lists the methods of bonding at the service. Electrical continuity at service equipment, service raceways, and service conductor enclosures must be ensured by one of the following methods:

- Bonding equipment to the grounded service conductor by exothermic welding, listed pressure connectors, listed clamps, or other listed means.

- Connections utilizing threaded couplings or threaded bosses on enclosures where made up wrenchtight.

- Threadless couplings and connectors where made up tight for metal raceways and metal-clad cables.

- Other listed devices, such as bonding-type locknuts and bushings. Standard locknuts or sealing locknuts are not acceptable as the "sole means" for bonding on the line side of service equipment. Grounding and bonding bushings for use with rigid or intermediate metal conduit are provided with means (usually one or more set screws that make positive contact with the conduit) for reliably bonding the bushing and the conduit on which it is threaded to the metal equipment enclosure or box. Grounding and bonding type bushings used with rigid or intermediate metal conduit, such as those shown in Figure 7-32, have provisions for connecting a bonding jumper or have means provided by the manufacturer for use in mounting a wire connector.

bonding at an individual service. The metal parts that require bonding together for a residential service entrance are the following:

- The service raceways.

- All service enclosures containing service conductors, including meter enclosures, boxes, or other metal electrical equipment connected to the service raceway.

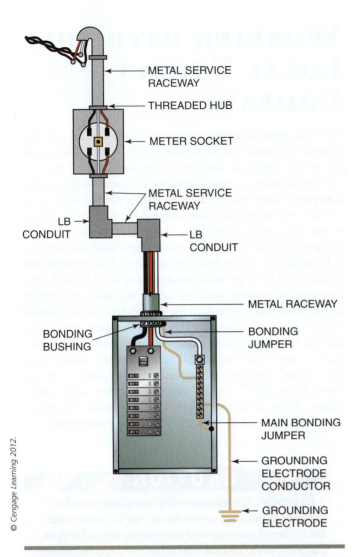

FIGURE 7-31 Grounding and bonding for a service with one disconnecting means.

FIGURE 7-32 An example of an insulated bonding-type bushing.

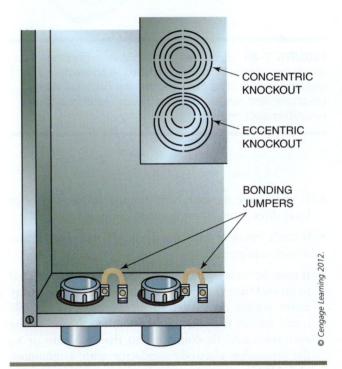

FIGURE 7-33 Bonding jumpers installed around concentric or eccentric knockouts.

- Bonding jumpers must be used around impaired connections such as reducing washers, oversized knockouts, **concentric knockouts**, or **eccentric knockouts**. Standard locknuts or bushings cannot be the only means for the bonding required by this section but can still be used to secure a raceway to a service entrance enclosure. Both concentric- and eccentric-type knockouts can reduce the electrical conductivity between the metal parts and may actually introduce unnecessary resistance into the grounding path. Installing bonding jumpers is one method often used between metal raceways and metal parts to ensure electrical conductivity. Figure 7-33 shows the difference between concentric- and eccentric-type knockouts and illustrates one method of applying bonding jumpers at these types of knockouts.

Section 250.94 requires an **intersystem bonding termination** (Figure 7-34). This termination will allow for the connection of bonding and **grounding conductors** from other systems such as cable television, telephone, and satellite TV to the electrical power system grounding electrode system. The intersystem bonding termination must comply with the following:

- Must be accessible for both making connections and inspection. Therefore, it can't be located inside a service enclosure. However, it can be located either outside or inside of a house.

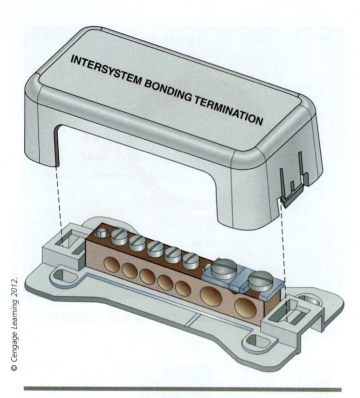

© Cengage Learning 2012.

FIGURE 7-34 Example of an intersystem bonding termination device. This device is used to connect grounding conductors from systems like cable television, telephone, and satellite TV to the house electrical system grounding electrode.

- It must have the capability for the connection of at least three intersystem bonding conductors.

- It can't interfere with the opening or closing of any service equipment.

- It must be securely mounted and electrically connected to an enclosure for the service equipment, to the utility meter socket enclosure, to an exposed nonflexible metal service raceway, or be mounted at one of the enclosures and be connected to the enclosure or to the grounding electrode conductor with a minimum of 6 AWG wire.

- It must be securely mounted and electrically connected to the house disconnecting means, or be mounted at the disconnecting means and be connected to the metal enclosure or to the grounding electrode conductor with a minimum 6 AWG copper wire.

- The terminals of the intersystem bonding termination device must be listed as grounding and bonding equipment.

WORKING WITH THE LOCAL UTILITY COMPANY

Once the type of service entrance being installed is determined, the local electric utility must be contacted, and the service type, the location of the service, and the service installation is coordinated with them. Electric utility companies have many rules governing the installation of service entrances that electricians have to follow. There is usually a publication from the utility company that is made available to electrical contractors that outlines the rules to follow when installing services. The intent of the publication is to provide information to electrical contractors, engineers, and architects in order to enable home electrical systems to be connected to the electric utility system in a safe and uniform manner. The utility publication usually states that the provisions of the publication are based on the *NEC*®. However, many times the utility requires additional requirements that go beyond the *NEC*® in the interest of safety and convenience. Some of the more common utility rules for service installation are covered in the following paragraphs.

CAUTION

CAUTION: The electric utility company rules discussed in this section are those that the author has encountered in the area of the country where he does electrical work. Be aware that the rules may be somewhat different in your area. Always be sure to check with the electric utility company in your area and the authority having jurisdiction before starting to install an overhead or underground residential service entrance.

An application for a new service connection must be initiated with the local electric utility as far in advance as possible. The location of the service entrance, the service entrance type, and the location of any transformers and poles must be reviewed and approved by the utility before any service wiring is installed. Most utilities will charge the customer whenever any wiring installed without prior company approval results in an additional expense to the utility. Electrical contractors should not start any service installation, purchase service equipment, or install wiring for additional electrical loads in existing installations until all negotiations have been completed with the utility company and it has been determined that the required service can be supplied.

A utility company representative, usually from the metering department, will work with the electrician to determine a suitable meter location, point of attachment for the service drop, and the location of the service entrance weatherhead. The meter enclosure will have to be located in a safe and readily accessible location.

from experience...

An electrical plan might call for a particular service type, but that type may not be able to be installed because of local electric utility restrictions. For example, an underground service may be called for in the building plan, but because of the existence of extensive amounts of ledge in the area where the service trench has to be dug, the minimum burial depth for the service entrance conductors cannot be met. In this case, unless an agreement can be worked out with the electric utility that allows a shallower burial depth, an overhead service will have to be installed.

The utility publication will usually tell what the electrical contractor is responsible for. For example, for an overhead service, it may say that the customer is responsible for purchasing and installing the service entrance, which includes all the wiring and parts from the weatherhead down to the main service disconnecting equipment. The service drop, which brings the electrical energy from the utility's system overhead to the house, will be installed and owned by the utility.

In municipalities where electrical inspections are required, a certificate of approval must be given to the utility company before the service can be connected to the utility system. This certificate will be obtained from the authority having jurisdiction (electrical inspector) and forwarded to the utility by the electrical contractor. In those areas that do not require an inspection certificate from an electrical inspector, utility companies will accept a certification stating that the service installation and all wiring in the house is done according to the *NEC*®. A representative of the electrical contracting company signs this certification. Many utility companies also send a representative from their company to do an inspection of the service entrance to make sure that all company installation rules have been met. Once the

service installation has been approved, the utility can be contacted to set up a time when the line crew (sometimes the metering department) can come to the house site and energize the service.

Sometimes it is necessary for an electrician to establish temporary service with a utility company during the initial construction phase of the house. Many times installing temporary service is the very first thing done by the electrical contractor. The temporary service will provide the electrical power for the various craft people, including electricians, to use power tools during the construction of the house. Electricity for temporary lighting can also be supplied by the temporary service. Most utility companies have certain requirements to be met when an electrician sets up a temporary service. The electrical contractor will need to contact the utility company to set up a temporary service and to have it inspected prior to connection. The process is very similar to what is described in the preceding paragraphs for establishing permanent service. An example of an overhead temporary service and typical utility requirements is shown in Figure 7-35.

from experience...

With the availability of powerful cordless power tools and reliable portable generators, many new residential construction sites do not require a temporary service. The author has installed many residential electrical systems where the building contractor has furnished electrical power through the use of a small electric generator.

Figure 7-36 shows an example of an underground temporary service along with some typical utility requirements.

Utility companies usually require a temporary service for a residential application to be at least 100 amps in size. The enclosure housing the temporary service disconnecting means must be rain-tight. The temporary installation is usually wired to provide the construction site with a minimum of two 125-volt, 20-amp-rated duplex receptacles (GFCI protected) and at least one 250-volt-rated receptacle. The receptacles will need to be installed with special covers, often called "bubble covers," that allow the installation to remain weatherproof even when a cord is plugged into them.

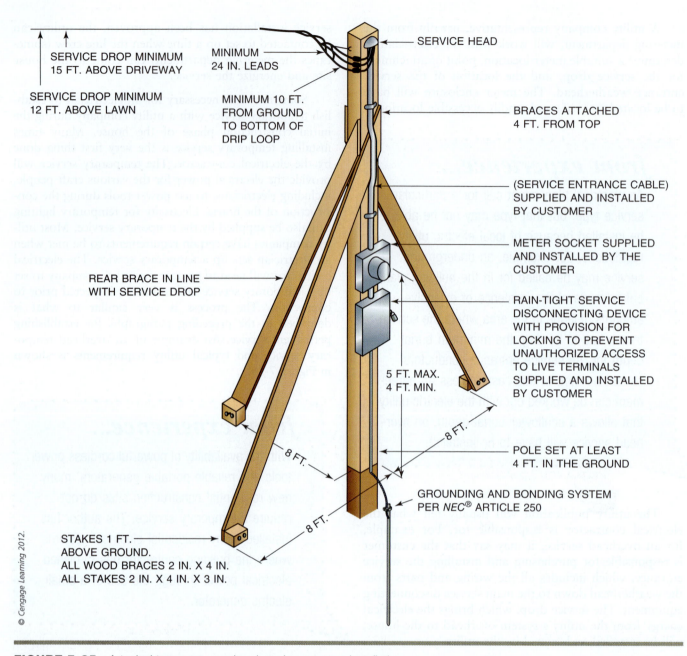

SERVICE DROP MINIMUM
15 FT. ABOVE DRIVEWAY

SERVICE DROP MINIMUM
12 FT. ABOVE LAWN

MINIMUM
24 IN. LEADS

MINIMUM 10 FT.
FROM GROUND
TO BOTTOM OF
DRIP LOOP

SERVICE HEAD

BRACES ATTACHED
4 FT. FROM TOP

(SERVICE ENTRANCE CABLE)
SUPPLIED AND INSTALLED
BY CUSTOMER

METER SOCKET SUPPLIED
AND INSTALLED BY THE
CUSTOMER

REAR BRACE IN LINE
WITH SERVICE DROP

RAIN-TIGHT SERVICE
DISCONNECTING DEVICE
WITH PROVISION FOR
LOCKING TO PREVENT
UNAUTHORIZED ACCESS
TO LIVE TERMINALS
SUPPLIED AND INSTALLED
BY CUSTOMER

5 FT. MAX.
4 FT. MIN.

8 FT.

8 FT.

POLE SET AT LEAST
4 FT. IN THE GROUND

GROUNDING AND BONDING SYSTEM
PER *NEC*® ARTICLE 250

8 FT.

STAKES 1 FT.
ABOVE GROUND.
ALL WOOD BRACES 2 IN. X 4 IN.
ALL STAKES 2 IN. X 4 IN. X 3 IN.

FIGURE 7-35   A typical temporary overhead service entrance installation.

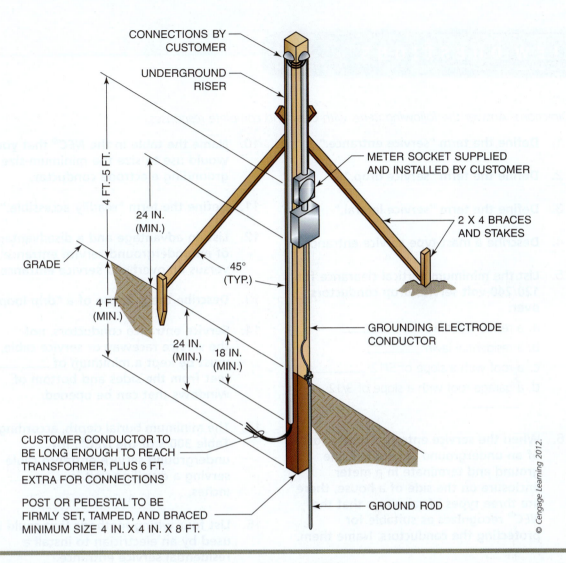

CONNECTIONS BY CUSTOMER

UNDERGROUND RISER

METER SOCKET SUPPLIED AND INSTALLED BY CUSTOMER

2 X 4 BRACES AND STAKES

4 FT.–5 FT.

24 IN. (MIN.)

GRADE

45° (TYP.)

4 FT. (MIN.)

24 IN. (MIN.)

18 IN. (MIN.)

GROUNDING ELECTRODE CONDUCTOR

CUSTOMER CONDUCTOR TO BE LONG ENOUGH TO REACH TRANSFORMER, PLUS 6 FT. EXTRA FOR CONNECTIONS

POST OR PEDESTAL TO BE FIRMLY SET, TAMPED, AND BRACED MINIMUM SIZE 4 IN. X 4 IN. X 8 FT.

GROUND ROD

© Cengage Learning 2012.

FIGURE 7-36 A typical temporary underground service entrance installation.

SUMMARY

In this chapter you were introduced to the different service entrance types used in residential wiring. Several *NEC®* rules that apply to residential service entrances were explained and illustrated. Working with the local electric utility to establish electrical service was also covered in detail. A lot of information was presented in this chapter, and all of it is important to an electrician getting ready to install the service entrance for a residential wiring system.

REVIEW QUESTIONS

Directions: Answer the following items with clear and complete responses.

1. Define the term "service entrance."

2. Define the term "service drop."

3. Define the term "service lateral."

4. Describe a mast-type service entrance.

5. List the minimum vertical clearance for 120/240-volt service drop conductors over:

   a. a residential driveway _____

   b. a residential lawn _____

   c. a roof with a slope of 5/12 _____

   d. a garage roof with a slope of 3/12 _____

6. When the service entrance conductors of an underground service exit the ground and terminate in a meter enclosure on the side of a house, there are three types of raceways that the *NEC®* recognizes as suitable for protecting the conductors. Name them.

7. Certain sizes of grounding electrode conductors that are subject to physical damage must be protected. Indicate whether the following size grounding electrode conductors need protection:

   a. 8 AWG copper grounding electrode conductor _____

   b. 6 AWG grounding electrode conductor _____

   c. 4 AWG grounding electrode conductor _____

8. List at least five items, other than a metal water pipe, that could be used as a grounding electrode for a residential service entrance.

9. Describe why it is required by the *NEC®* to install a supplemental grounding electrode when a metal water pipe is used as the main grounding electrode.

10. Name the table in the *NEC®* that you would use to size the minimum-size grounding electrode conductor.

11. Define the term "readily accessible."

12. List an advantage and a disadvantage of an underground service entrance versus an overhead service entrance.

13. Describe the purpose of a "drip loop."

14. Service entrance conductors, not the service raceway or service cable, must be kept a minimum of _____ feet from the sides and bottom of windows that can be opened.

15. The minimum burial depth, according to Table 300.5 of the *NEC®*, for underground service entrance cable serving a dwelling unit is _____ inches.

16. List four wiring methods that could be used by an electrician to install a residential service entrance.

17. Describe why the amount of "inside run" for the service entrance conductors must be as short as possible.

18. An electrician is driving an 8-foot ground rod and strikes ledge at about 3 feet. There is no way he or she can drive the rod all the way into the ground, as the *NEC®* requires. Describe what you might do if you were the electrician to meet *NEC®* requirements.

19. Explain why it is necessary to bond around reducing washers, oversized, eccentric, or concentric knockouts when bonding service entrance equipment.

20. Describe the support requirements for service entrance cable.

21. Define an "intersystem bonding termination."

22. Describe the type of service entrance that would require a pad-mounted transformer.

23. Explain why a seal must be used when a service entrance raceway is installed from a colder (outside) location to a warmer (inside) location.

24. Name the letter found on a conductor's insulation that designates it as suitable for installation in a wet location.

25. According to Table 250.66, name the minimum size grounding electrode conductor for a 200 ampere service entrance installed with 4/0 aluminum wire.

KNOW YOUR CODES

1. Identify the local electrical power utility company in your area. Contact the company and find out what an electrical contractor needs to do to establish temporary electrical power for a residential building project.

2. Compare the service entrance installation rules that your local electrical power utility company requires an electrician to follow with the *National Electrical Code* rules outlined in this chapter.

3. Contact at least three electrical contracting companies in your area who do residential wiring. Find out if they use a generator on the job site, establish a temporary service with the local electrical power utility, or only use battery powered tools and don't require any temporary power on the job site.

4. Identify the local or state electrical inspector who covers your area. Contact them and ask what they see as the most common electrical code violations for residential service entrance installations.

5. Ask your local or state electrical inspector what, if any, electrical code rules exist locally that are not in the *National Electrical Code*.

Service Entrance Equipment and Installation

OBJECTIVES

Upon completion of this chapter, the student should be able to:

- Identify common overhead service entrance equipment and materials.

- Identify common underground service entrance equipment and materials.

- Demonstrate an understanding of common installation techniques for overhead services.

- Demonstrate an understanding of common installation techniques for underground services.

- Demonstrate an understanding of voltage drop in underground service laterals.

- Demonstrate an understanding of service panel installation techniques.

- Demonstrate an understanding of subpanel installation techniques.

- Demonstrate an understanding of service entrance upgrade techniques.

GLOSSARY OF TERMS

backboard the surface on which a service panel or subpanel is mounted; it is usually made of plywood and is painted a flat black color

backfeeding a wiring technique that allows electrical power from an existing electrical panel to be fed to a new electrical panel by a short length of cable; this technique is commonly used when an electrician is upgrading an existing service entrance

bushing (insulated) a fiber or plastic fitting designed to screw onto the ends of conduit or a cable connector to provide protection to the conductors

cable hook also called an "eyebolt"; it is the part used to attach the service drop cable to the side of a house in an overhead service entrance installation

conduit "LB" a piece of electrical equipment that is connected in-line with electrical conduit to provide for a 90-degree change of direction

enclosure the case or housing of apparatus or the fence or walls surrounding an installation to prevent personnel from accidentally contacting energized parts or to

protect the equipment from physical damage

line side the location in electrical equipment where the incoming electrical power is connected; an example is the line-side lugs in a meter socket where the incoming electrical power conductors are connected

load side the location in electrical equipment where the outgoing electrical power is connected; an example is the load-side lugs in a meter socket where the outgoing electrical power conductors to the service equipment are connected

mast kit a package of additional equipment that is required for the installation of a mast-type service entrance; it can be purchased from an electrical distributor

porcelain standoff a fitting that is attached to a service entrance mast, which provides a location for the attachment of the service drop conductors in an overhead service entrance

rain-tight constructed or protected so that exposure to a beating rain will

not result in the entrance of water under specified test conditions

roof flashing/weather collar two parts of a mast-type service entrance that, when used together, will not allow water to drip down and into a house through the hole in the roof that the service mast extends through

sill plate a piece of equipment that, when installed correctly, will help keep water from entering the hole in the side of a house that the service entrance cable from the meter socket to the service panel goes through

threaded hub the piece of equipment that must be attached to the top of a meter socket so that a raceway or a cable connector can be attached to the meter socket

voltage drop (VD) the amount of voltage that is needed to "push" the house electrical load current through the wires from the utility transformer to the service panel and back to the transformer; the amount of voltage drop depends on the resistance of the wires, the length of the wires, and the actual current carried on the wires

In previous chapters, you were introduced to many of the pieces of electrical equipment and materials commonly used in residential wiring, including some items that could be used as part of a service entrance. You have learned that a service entrance is the part of a residential wiring system that provides the means for electrical energy from the electric utility to be delivered to a residential electrical system. You have also learned how to do the calculations necessary to determine the minimum-size service entrance that will supply electricity to a house electrical system. This chapter brings together all the things you have learned about service entrances from prior chapters and covers service entrance equipment and materials in detail. Common installation techniques for overhead and underground service entrances are also covered.

OVERHEAD SERVICE EQUIPMENT AND MATERIALS

Because an overhead service entrance is the most common service type installed in residential applications, we will cover the equipment and materials used to install them first. There are three different ways to install an overhead service entrance: (1) using service entrance cable installed on the side of a house, (2) using electrical conduit installed on the side of a house, and (3) using a mast-style service installation.

OVERHEAD SERVICE ENTRANCE USING SERVICE ENTRANCE CABLE

If the service will be installed with service entrance cable (Figure 8-1), the equipment will include the service drop conductors, a cable hook, a cable weatherhead, service entrance cable, clips for the service entrance cable, a meter socket with a threaded hub, a rain-tight service entrance cable connector, "regular" service entrance

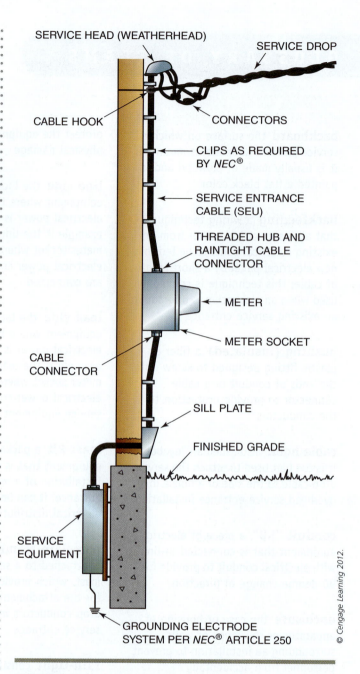

FIGURE 8-1 A typical overhead service entrance installed using Type SEU service entrance cable.

© Cengage Learning 2012.

cable connectors, a sill plate, and the service equipment. The purpose for each part is explained in the following list:

- Service drop conductors: As described in a previous chapter, the service drop conductors bring the electrical energy from the local electric utility system to the house. The overhead cable, typically called "triplex cable," is usually installed and owned by the electric utility company (Figure 8-2). So even though the service drop conductors are part of an overhead

FIGURE 8-2 The service drop for a residential overhead service entrance is usually installed with "triplex cable."
© Cengage Learning 2012.

FIGURE 8-3 A cable hook is used to attach the service drop to the side of the house in an overhead service entrance installation. This piece of equipment is also called a "service drop hook" or "lag screw eye hook." Other names may be used depending on where in the country the installation is being done.
© Cengage Learning 2012.

© Cengage Learning 2012.

FIGURE 8-4 A weatherhead is installed on the line side end of the service entrance cable (SEU) to stop water from entering into the end of the cable. This piece of equipment may also be called a "service head" or "service entrance cable cap." Other names may be used depending on where in the country the installation is being done.

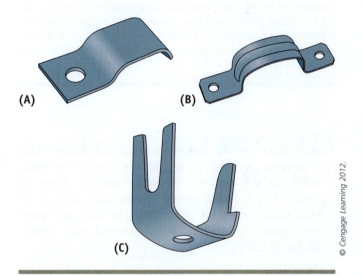

(A)   (B)

(C)

© Cengage Learning 2012.

FIGURE 8-5 Type SEU service entrance cable is typically secured and supported using (A) one-hole straps, (B) two-hole straps, or (C) fold-over clips.

service entrance, electricians typically do not have to purchase or install it. The area where you live may have different electric utility rules.

- Cable hook: The **cable hook**, or service drop hook as it is sometimes called, is used to attach the service drop cable to the side of the house (Figure 8-3). The electrician is usually required to install the cable hook even if the electric utility is installing the service drop. Many electric utilities will provide the cable hook to the electrician. The hook must be installed so that it is below the weatherhead.

- Weatherhead: The weatherhead (Figure 8-4) is used to stop the entrance of water into the end of a service entrance cable. Each service entrance conductor must exit from a separate hole in the weatherhead.

- Service entrance cable: Overhead services will usually use an SEU-type service entrance cable to bring the electricity down the side of the house from the point of attachment to the meter socket and then on into the main service equipment.

- Clips for the service entrance cable: Clips are used to provide support for the cable (Figure 8-5). Section 230.51(A) of the *National Electrical Code® (NEC®)* states that they must be located no more than 12 inches (300 mm) from the weatherhead and meter socket. Additional support is required at intervals no more than 30 inches (750 mm). Clip types that are often used include "fold-over" clips, one-hole straps, and two-hole straps.

- Meter socket and threaded hub: The meter socket is used to provide a location in the service entrance for the local electric utility to install a kilowatt-hour meter that will measure the amount of electrical energy used by the house electrical system. When you buy the meter socket, a **threaded hub** for the top of the socket must also be purchased. The hub is sized for the size of the service entrance cable connector coming into the top of the meter socket (Figure 8-6). For example, if you are using a 1¼-inch connector, use a 1¼-inch threaded hub on the top of the meter socket.

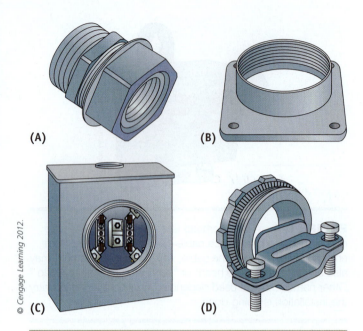

(A)   (B)

(C)   (D)

© Cengage Learning 2012.

FIGURE 8-6   (A) The rain-tight connector screws into the threaded hub. (B) The threaded hub is attached to the top of (C) the meter socket with four screws. (D) A "regular" service entrance cable connector is attached to the bottom of the meter socket and to the service equipment enclosure located inside the house.

© Cengage Learning 2012.

FIGURE 8-7   A sill plate is used to cover the hole in the side of the house where the service entrance cable enters the house. It is designed to fit over both the service cable and the hole. Remember to use a sealing compound around it and the cable to make it watertight.

- Rain-tight service entrance cable connector: This connector is used to secure the service entrance cable to the top of the meter socket. Refer to Figure 8-6. It has a special design that allows it to provide a **rain-tight** connection so that no water can enter the meter socket around the service entrance cable.

- "Regular" service entrance cable connectors: This connector type is used to connect the service entrance cable to the bottom of the meter socket. Refer to Figure 8-6. It is also the connector type used to connect the service cable to the main service equipment **enclosure** when the enclosure is located inside the house. A rain-tight connector is not required since the bottom of the meter socket and an inside-located enclosure are not subject to rain or snow.

- Sill plate: A **sill plate** is used to protect service entrance cable at the point where it enters a house through an outside wall (Figure 8-7). It is usually made of metal and comes with a couple of screws to attach it with and some duct seal to help make a good weatherproof seal at the point where the cable enters the outside wall of the house.

from experience...

Some electric utilities will provide the meter socket to the electrician. Be sure to check with the electric utility in your area to determine whether you supply the meter socket or whether the local electric utility supplies the meter socket.

- Service equipment: The service equipment contains the main service disconnecting means, usually a main circuit breaker, which is used to disconnect the supply electricity from the house electrical system. There are basically two types of service equipment commonly used in residential wiring systems:

1 A main breaker panel that is installed inside the house: It also contains the circuit breakers that provide the overcurrent protection required for the branch circuits that supply electricity to the different electrical loads in a house. This is the most common style of service entrance equipment (Figure 8-8).

2 A combination meter socket/main breaker disconnect that is installed outside the house: This style of service equipment allows easy access to the main disconnecting means in the case of an emergency. A feeder is run from this type of service equipment to a subpanel located inside the house. The subpanel will contain the circuit breakers that provide the overcurrent protection required for the branch circuits supplying electricity to the different electrical loads in a house (Figure 8-9).

from experience...

Some homes have an attached garage, and the service entrance is often installed on the side of the garage. In this case, the meter socket will be located on the outside wall of the garage. The main service disconnect may be either included with the meter socket (combination meter socket/main breaker disconnect) or located on the inside wall of the garage, usually directly behind the meter socket. In either case, it is common wiring practice to install a subpanel in the main part of the house and run the electrical system branch circuits from there.

from experience...

It is common practice in some areas of the country, where house building styles do not provide for a good location inside a house for the service equipment, to use an enclosure located on the outside of a house that contains both a main circuit breaker and the branch-circuit overcurrent protection devices. In this case, branch-circuit wiring is run from the outside-located service equipment to the different electrical loads in a house. This installation will require a NEMA Type 3R enclosure. Homes with no attached garage or a basement or crawl space under the home are candidates for this type of installation.

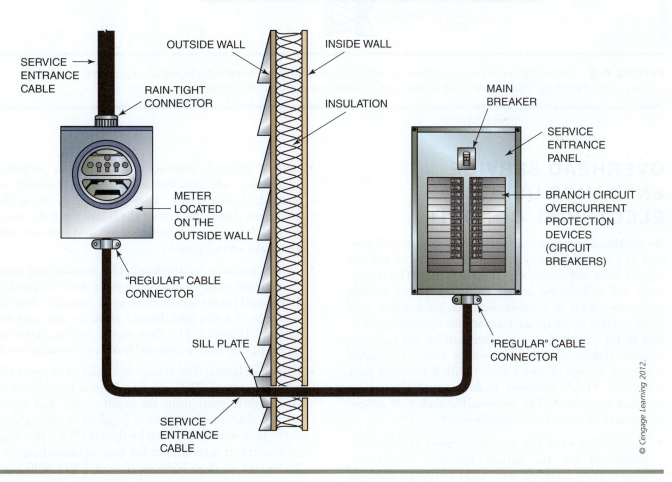

© Cengage Learning 2012.

FIGURE 8-8   A service entrance panel with a main circuit breaker located inside a house. The service panel will contain circuit breakers that protect the branch-circuit wiring.

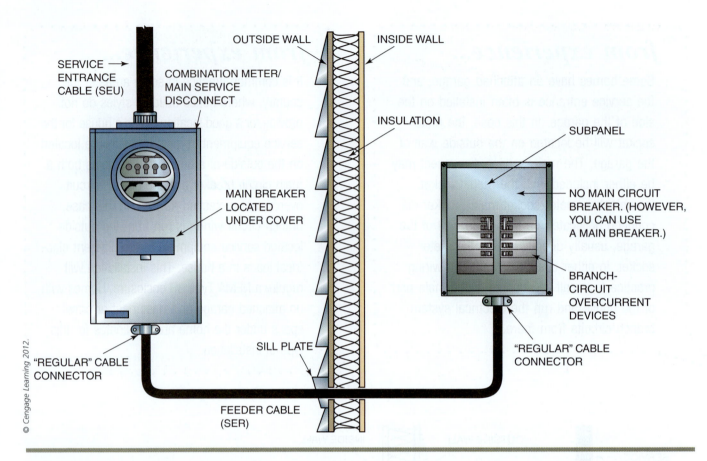

© Cengage Learning 2012.

FIGURE 8-9 A combination meter socket/main service disconnect located on the outside wall of a house and a subpanel located inside the house. The subpanel will contain circuit breakers that protect the branch-circuit wiring.

OVERHEAD SERVICE ENTRANCE USING ELECTRICAL CONDUIT

An overhead service entrance that utilizes electrical conduit as a service raceway on the side of a house will use slightly different equipment (Figure 8-10). This equipment will include the service drop conductors, a cable hook, the metal or nonmetallic electrical conduit used as the service raceway, separate service entrance conductors of the proper size, a weatherhead for the raceway, conduit clamps for service entrance raceway, conduit connector fittings, a meter socket with a threaded hub, a conduit "LB," insulated bushings, and the service entrance equipment. The purpose for each part unique to this service type is explained here:

- Service raceway: The service raceway will provide protection for the service entrance conductors. It is usually installed with rigid polyvinyl chloride conduit (PVC), rigid metal conduit (RMC), intermediate metal conduit (IMC), or electrical metallic tubing (EMT).

- Service entrance conductors: The service entrance conductors are installed in the service raceway as individual conductors. They are used to bring the electricity down the side of the house from the point of attachment to the meter socket and then on into the main service panel.

- Service raceway weatherhead: The weatherhead must be designed to fit the size and type of electrical conduit being used for the service raceway. It is used to stop the entrance of water into the end of the service entrance raceway (Figure 8-11). Each service entrance conductor must exit from a separate hole in the weatherhead.

- Conduit clamps: The clamps are used to support the service raceway. They must be located no more than 3 feet (900 mm) from the weatherhead and meter socket. Additional support is required at intervals no more than what is required by the *NEC®* for that type of conduit. It is recommended that additional support be located so there is no more than 3 feet (900 mm) between additional supports. Conduit clamp types that are often used include one-hole straps, two-hole straps, and conduit hangers (Figure 8-12).

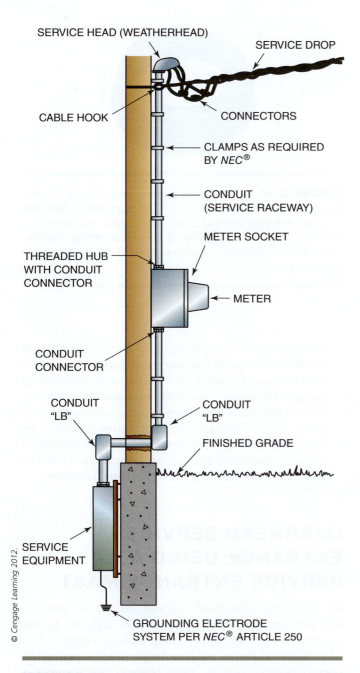

FIGURE 8-10 A typical overhead service entrance installed using a service raceway.

FIGURE 8-11 A service raceway weatherhead is installed on the top of the service raceway to stop the entrance of water into the end of the raceway. This piece of equipment may also be called a "service raceway head" or "service raceway cap." Other names may be used depending on where in the country the installation is being done. The weatherhead shown here is designed to be clamped onto the end of a raceway.

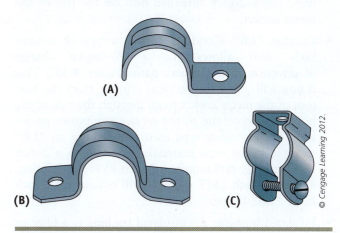

FIGURE 8-12 Service entrance raceway is typically secured and supported using (A) one-hole straps, (B) two-hole straps, or (C) conduit hangers.

• Conduit connector fittings: The conduit connectors are used to connect the service raceway to the various enclosures and conduit fittings used in the service entrance installation. If PVC is used, connector fittings are attached to the conduit using a PVC cement, which allow for attachment to the meter socket and service equipment. If EMT is used, compression-type fittings are attached to the conduit ends, allowing the EMT to be attached to the meter socket threaded hub and the service entrance equipment. Compression-type fittings that are watertight must be used for outside use of

EMT. Inside the house, setscrew-type fittings could be used. RMC and IMC raceway is threaded on each end and will be directly tightened into the meter socket threaded hub. Locknuts will hold the RMC and IMC to the bottom of the meter socket and the service equipment enclosure. Conduit connectors are covered in more detail in Chapter 12.

• Meter socket with a threaded hub: The meter socket is used to provide a location in the service entrance for the local electric utility to install a kilowatt-hour meter. When you buy the meter socket, a threaded hub for the top of the socket must also be purchased. The hub is sized for the size of the service

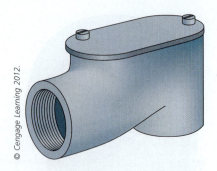

© Cengage Learning 2012.

FIGURE 8-13 A conduit "LB" is designed to allow for a 90 degree change of direction in a conduit path. For a service entrance installation that uses conduit for the entire conductor path, you will need one "LB" outside to allow the conduit to enter the side of a house and another "LB" inside to allow the conduit to enter the service panel.

© Cengage Learning 2012.

FIGURE 8-14 An insulating bushing must be used on the end of a conduit if it contains 4 AWG or larger conductors. For a service entrance installed totally with conduit, insulating bushings must be used where the conduit enters the bottom of the meter socket and where the raceway enters the service entrance panel.

raceway coming into the top of the meter socket. For example, if you are using a 2-inch service raceway, use a 2-inch threaded hub on the top of the meter socket.

- Conduit "LB": **Conduit "LB"** is a type of conduit body that allows for a 90-degree change of direction in a raceway path (Figure 8-13). This fitting will allow the raceway coming from the bottom of the meter socket to go through the side of the house and continue to the service equipment enclosure. Each conduit type used will require an "LB" made from the same material. For example, a 2-inch PVC raceway will require a 2-inch PVC "LB" fitting, and a 1½-inch EMT raceway will require a 1½-inch EMT "LB" fitting.

- Insulated bushing: An insulated **bushing** is a fiber or plastic fitting designed to screw onto the ends of conduit or a cable connector to provide protection to the conductors. Section 300.4(G) in the *NEC*® states that where raceways containing conductors 4 AWG or larger enter a cabinet, box enclosure, or raceway, the conductors must be protected by an identified fitting providing a smoothly rounded insulating surface (Figure 8-14). The *Exception* to this section says that where threaded hubs are used, such as at the top of the meter socket, an insulated bushing does not have to be used on the raceway end. For this type of service entrance, insulated bushings would have to be used where the raceway enters the bottom of the meter socket and where the raceway enters the service entrance panel. Conduit bushings constructed wholly of insulating material must not be used to secure the raceway to the enclosure. The insulating bushing must have a temperature rating not less than the insulation temperature rating of the installed conductors. The reason for this requirement for using insulated

bushings on the ends of conduits containing wires of 4 AWG or larger is that heavy conductors and cables tend to stress the conductor insulation at terminating points. Providing insulated bushings at raceway reduces the risk of insulation failure at conductor insulation "stress" points. The temperature rating of an insulating bushing must coordinate with the insulation of the conductor to ensure that the protection remains intact over the life cycle of the insulated conductor.

OVERHEAD SERVICE ENTRANCE USING A SERVICE ENTRANCE MAST

A mast-type overhead service entrance uses a few different pieces of equipment in addition to the equipment listed in the preceding section (Figure 8-15). The additional pieces include a length of raceway used as the service entrance mast, a weatherhead to fit the mast, a porcelain standoff fitting, and a weather collar and roof flashing. A **mast kit** (Figure 8-16) is often purchased through an electrical distributor and includes the required additional pieces of equipment for a mast-type service. The purpose for each part unique to this service type is explained next:

- Mast: As explained previously, a mast-type service extends above the roofline so that the minimum required service drop clearance can be achieved when bringing a service drop to a low building like a ranch-style house. The mast should be made of RMC or IMC to provide adequate strength. A minimum of a 2-inch or 2½-inch trade-size mast is usually used. Larger sizes may be needed if the service size requires conductors that are larger than what a 2½-inch conduit can

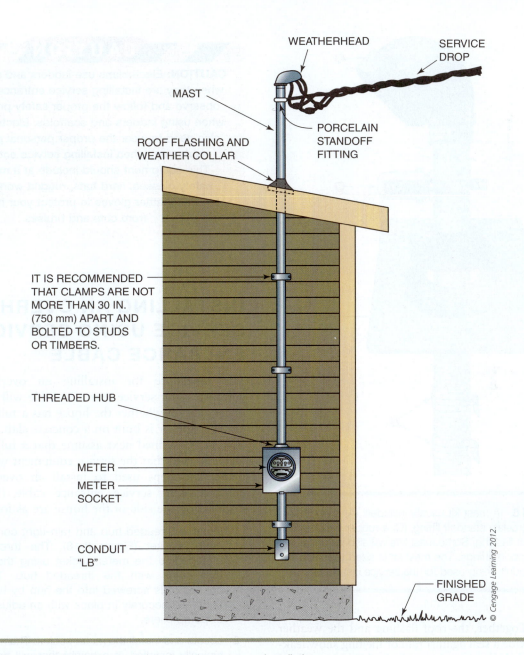

WEATHERHEAD

SERVICE DROP

MAST

PORCELAIN STANDOFF FITTING

ROOF FLASHING AND WEATHER COLLAR

IT IS RECOMMENDED THAT CLAMPS ARE NOT MORE THAN 30 IN. (750 mm) APART AND BOLTED TO STUDS OR TIMBERS.

THREADED HUB

METER

METER SOCKET

CONDUIT "LB"

FINISHED GRADE

© Cengage Learning 2012.

FIGURE 8-15 A typical mast-type overhead service entrance installation.

handle. The size and type of raceway make the mast strong enough to resist the tendency to bend from the force placed on it from the service drop.

- Weatherhead: The weatherhead must be designed to fit the size of electrical conduit being used for the service mast. As in other service entrance installation types, it is used to stop the entrance of water into the end of the service entrance mast. Each service entrance conductor must exit from a separate hole in the weatherhead.

- Porcelain standoff fitting: Because the mast extends above the roof in this service entrance style, there is not a convenient place to install the cable hook

for the attachment of the service drop. A **porcelain standoff** fitting is placed on the mast and provides a spot for the service drop to be attached.

- Weather collar and roof flashing: The **roof flashing** is placed around the hole in the roof that is made for the service mast to extend through. It is fastened in place with several screws or roofing nails. The house roofing material is eventually installed over the edges of the roof flashing. Once the mast is installed through the hole, a rubber weather collar is installed on the mast. The collar fits tightly around the mast and is placed at the point where the mast extends from the roof

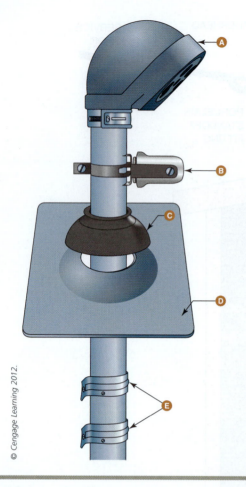

FIGURE 8-16 A mast kit usually includes: (A) a weather-head, (B) a porcelain standoff fitting, (C) a rubber weather collar, and (D) roof flashing. Some mast kits will also include a few (E) conduit support fittings. The mast kit is sized to match the size of the conduit being used for the service installation.

© Cengage Learning 2012.

flashing. Together, the roof flashing and the **weather collar** provide a seal against rain or melting snow leaking around the mast and into the building.

OVERHEAD SERVICE INSTALLATION

Earlier we covered the preparation and planning for a service entrance installation. Remember that the location of the service entrance, commonly called "spotting the meter," is a joint effort between the electrician and a representative of the local electric utility company. Once you have determined that the service will be an overhead type, verified the service location, and purchased the required service entrance equipment, all that is left is the actual installation.

INSTALLING AN OVERHEAD SERVICE USING SERVICE ENTRANCE CABLE

The procedure for installing an overhead service entrance using service entrance cable will vary slightly depending on whether the house has a full basement, a crawl space, or is built on a concrete slab. The installation steps outlined next assume that a full basement is available and that the service equipment will be located there. The steps used to install an overhead service entrance with service entrance cable (Figure 8-17) installed on the side of the house are as follows:

1   Install a threaded hub and rain-tight connector to the meter socket (Figure 8-18). The threaded hub is attached to the meter socket using the four screws that come with the threaded hub. The rain-tight connector is screwed into the hub by hand and then tightened securely in place with an adjustable wrench or pump pliers.

2   Locate and install the meter socket. The meter socket is typically installed at a height that will provide a measurement of 4 to 5 feet above ground level to the center of the kilowatt-hour meter (Figure 8-19). This is not an *NEC*® requirement but rather a requirement by the local electric utility. The meter socket is surface-mounted to the outside wall of the house using four screws inserted through the mounting holes provided in the back of the meter socket. Make sure to install the meter socket so it is level and plumb on the top, bottom, and sides. A torpedo level works well for this. Where the meter socket is located and how it is mounted provides a point of reference from which all other service entrance equipment installation measurements can be made.

3   Measure and cut the service entrance cable to the length required to go from the meter socket up to a height on the side of the house that will keep the service drop conductors at the minimum required clearance from grade. Allow 3 feet extra at the weatherhead end and 1 foot extra at the meter socket end.

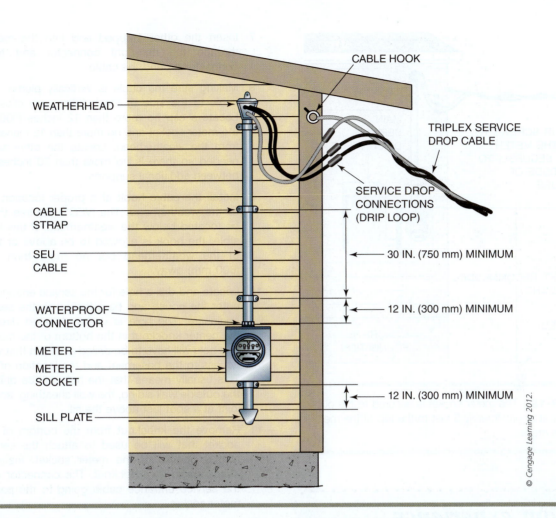

FIGURE 8-17   A service entrance installation using type SEU cable attached to the side of the house.

FIGURE 8-18   A meter socket enclosure with a threaded hub attached to the top. The rain-tight connector that secures the service entrance cable to the meter socket will be threaded into the hub. *Courtesy of Milbank Mfg. Co.*

4   Strip off approximately 3 feet of outside sheathing from one end of the SEU cable and 1 foot of outside sheathing from the other end. Twist together the individual conductors that make up the neutral conductor in Type SEU cable.

5   Install the cable weatherhead on the end that has 3 feet of free conductor. Insert the two black-colored ungrounded conductors and the bare neutral conductor through separate holes in the weatherhead so that there is approximately 3 feet of service conductor that can be formed into a drip loop and connected to the service drop conductors by the utility company. Secure the weatherhead to the cable by tightening the screws of the built-in clamping mechanism on the weatherhead.

6   Mount the assembled service entrance cable with attached weatherhead on the side of the house in a position that is directly above and vertically aligned with the rain-tight connector previously installed in the meter socket hub. A screw installed through a hole in the cable weatherhead will support the entire cable assembly at this point.

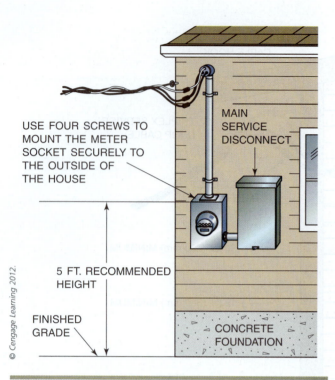

USE FOUR SCREWS TO MOUNT THE METER SOCKET SECURELY TO THE OUTSIDE OF THE HOUSE

MAIN SERVICE DISCONNECT

5 FT. RECOMMENDED HEIGHT

FINISHED GRADE

CONCRETE FOUNDATION

© Cengage Learning 2012.

FIGURE 8-19   The meter socket is secured to the outside wall surface at approximately 5 feet to the top of the meter socket from grade.

from experience...

Sometimes it is necessary to mount the meter socket on the side of a house before any outside siding has been installed by the carpenters. If this is the case, a good practice is to cut out a ½-inch or ¾-inch thick piece of moisture-resistant (or marine grade) plywood and install it on the wall where the meter socket will be located using galvanized screws. The piece of plywood should be the same size or just slightly larger than the meter socket. Once the meter socket is secured to the piece of plywood, it sits out from the wall surface so that it is easier for the siding to be installed around it and behind the service entrance parts connected to it. There are also commercially available nonmetallic meter socket backboards designed for this type of application, which are often referred to as a "meter mounting base."

7   Insert the other stripped end into the meter socket through the rain-tight connector and tighten the connector around the cable.

8   Making sure the cable is vertically plumb, secure the cable to the side of the building with clips or straps. Locate a clip no more than 12 inches (300 mm) from the meter socket and no more than 12 inches (300 mm) from the weatherhead. Locate the other supports as needed so there is no more than 30 inches (750 mm) between additional supports.

9   Install the cable hook at a proper location in relation to the weatherhead. The *NEC*® requires the hook to be located below the weatherhead. If this is not possible, the hook is allowed to be above or to the side of the weatherhead but no more than 24 inches (600 mm) away.

10 Locate and drill a hole for the service entrance cable to enter the house and be attached to the service panel enclosure. The hole is usually located directly below the center knockout in the bottom of the meter socket. The hole will have to be drilled at a spot that will provide access to the basement and the location of the panel. This usually means that the hole will be drilled though the outside wall siding, the wall sheathing, and the band joist at a spot just above the sill.

11 Remove the knockout from the bottom of the meter socket that will be used to attach the service cable to the bottom of the meter socket. Install a cable connector in the knockout. The connector will secure the service entrance cable going to the panel, to the meter socket.

12 Measure and cut the service entrance cable to the length required to go from the bottom of the meter socket to the service panel enclosure. Allow 1 foot extra at the meter socket end and enough at the panel end to allow the conductors to be easily attached to the main circuit breaker and neutral lugs. The length necessary to make connections at the panel will vary, depending on the style of the panel used.

13 Insert the 1-foot stripped end of service entrance cable through the cable connector on the bottom of the meter socket and tighten the connector onto the cable.

14 Insert the other end of the cable through the hole you previously drilled in the wall. Make sure to secure the cable within 12 inches (300 mm) of the meter socket. If additional support is required, make sure the supports are located no more than 30 inches (750 mm) apart.

15 Install a sill plate over the cable at the point where the cable goes through the hole in the side of the house. Sill plates usually come with a small amount of duct seal and two screws. The duct seal is used to help make a more watertight seal around the sill plate. If there is not enough duct seal that comes with the sill plate, additional duct seal can be purchased at your local electrical distributor.

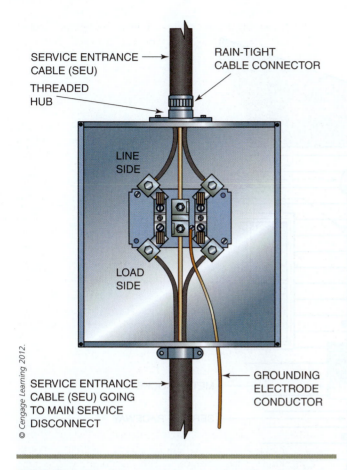

SERVICE ENTRANCE CABLE (SEU)

THREADED HUB

RAIN-TIGHT CABLE CONNECTOR

LINE SIDE

LOAD SIDE

SERVICE ENTRANCE CABLE (SEU) GOING TO MAIN SERVICE DISCONNECT

GROUNDING ELECTRODE CONDUCTOR

© Cengage Learning 2012.

FIGURE 8-20 Meter socket connections for a typical three-wire overhead service entrance installed with service entrance cable.

16 Make the meter socket connections (Figure 8-20). Connect the conductors from the weatherhead end to the **line-side** lugs in the meter socket. Connect the conductors from the service panel to the **load-side** lugs.

17 Secure the service entrance cable to the service entrance panel and make the proper connections in the service entrance panel. (Panel installation is covered later in this chapter.)

INSTALLING AN OVERHEAD SERVICE USING SERVICE ENTRANCE RACEWAY

The procedure for installing an overhead service entrance using service entrance raceway will also vary slightly, depending on whether the house has a full basement, a crawl space, or is built on a concrete slab. The installation steps outlined next assume that a full basement is available and that the service equipment will be located there. The steps used to install an overhead service entrance with service raceway (Figure 8-21) installed on the side of the house are as follows:

1 Install a threaded hub to the meter socket. The threaded hub is attached to the meter socket using the four screws that come with the threaded hub. It must be the same size as the size of the raceway being used for the service entrance conductors.

2 Locate and install the meter socket. The meter socket is typically installed at a height that will provide a measurement of 4 to 5 feet above ground level to the center of the kWH meter. This is not an *NEC®* requirement but rather a requirement by the local electric utility. It is surface-mounted to the outside wall of the house using four screws inserted through the mounting holes provided in the back of the meter socket. Make sure to install the meter socket so it is level and plumb on the top, bottom, and sides. A torpedo level works well for this. Where the meter socket is located and how it is mounted provides a point of reference from which all other service entrance equipment installation measurements can be made.

3 Measure and cut the electrical conduit to the length required to go from the meter socket up to a height on the side of the house that will keep the service drop conductors at the minimum required clearance from grade. Remember that electrical conduit will come in 10-foot lengths. In many conduit service entrance installations, a 10-foot length of conduit from the top of a meter socket located approximately 5 feet from the ground will provide all the height that is needed. If more height is needed, you will have to couple together the necessary raceway lengths.

4 Measure and cut the length of service entrance conductor wire needed. Cut enough length to leave approximately 3 feet of wire on the weatherhead end and 1 foot of wire for the meter socket end. You will need to cut three equal lengths: two lengths for the ungrounded "hot" service conductors and one length for the grounded service neutral conductor. At this time, use white electrical tape to identify each end of the service neutral conductor.

5 Insert one end of the service raceway into the meter socket hub and tighten it. Mount the service entrance raceway on the side of the house in a position that is directly above and vertically aligned with the meter socket hub. Use your torpedo level for this. One-hole straps, two-hole straps, or conduit hangers can be used to provide the support. Make sure to support the raceway within 3 feet (900 mm) of the meter socket and the weatherhead. Additional support is required, depending on the *NEC®* requirements for the type of raceway you are using.

6 Install the raceway weatherhead on the top of the raceway and then remove the top cover of the weatherhead.

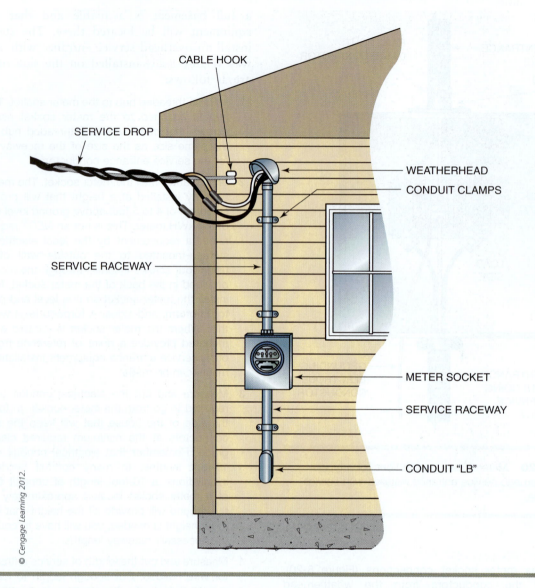

© Cengage Learning 2012.

CABLE HOOK

SERVICE DROP

WEATHERHEAD

CONDUIT CLAMPS

SERVICE RACEWAY

METER SOCKET

SERVICE RACEWAY

CONDUIT "LB"

FIGURE 8-21   An overhead service entrance with a service raceway installed on the side of the house.

7   At the meter socket end, insert all three service entrance conductors and push them up through the raceway until approximately 3 feet of the conductors come out of the top.

8   At the weatherhead end, insert the two black-colored ungrounded conductors and the white-tape-identified neutral conductor through separate holes in the weatherhead. There should be approximately 3 feet of service conductor coming out of the weatherhead for connection to the service drop conductors by the utility company. Reinstall the cover on the weatherhead.

9   Install the cable hook at a proper location in relation to the weatherhead on the service raceway. The *NEC*® requires the hook to be located below the weatherhead. If this is not possible, the hook is allowed to be

above or to the side of the weatherhead but no more than 24 inches (600 mm) away.

10   Locate and drill a hole for the service entrance raceway to enter the house and be attached to the service panel enclosure. The hole is usually located directly below the center knockout in the bottom of the meter socket. The hole will have to be drilled at a spot that will provide access to the basement and the location of the panel. This usually means that the hole will be drilled though the outside wall siding, the wall sheathing, and the band joist at a spot just above the sill. A hole saw or some other tool for making larger holes can be used.

11   Remove the knockout from the bottom of the meter socket that will be used to attach the service raceway to the bottom of the meter socket.

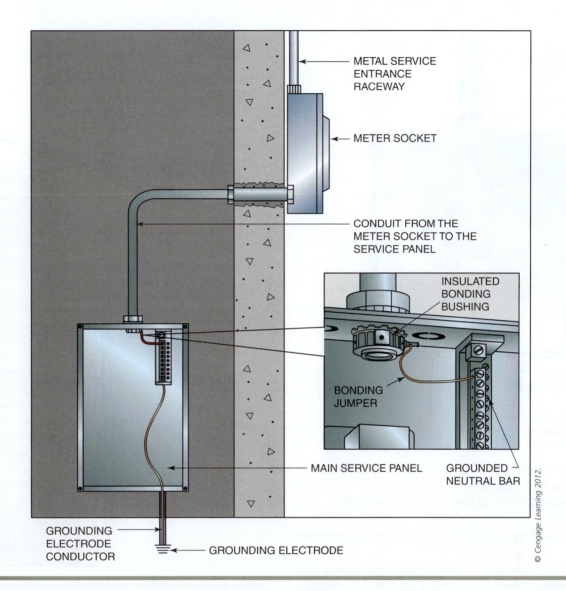

FIGURE 8-22 A bonding bushing must be installed on the end of a metal service raceway that runs from the meter socket to the service panel and encloses the service entrance conductors. The bonding bushing ensures that the metal conduit will provide a low-resistance path for current to flow in a fault condition. A bonding jumper is run from the bushing to the grounded neutral bar in the panel.

12 Install the service raceway between the meter socket and the service panel. "LB" fittings will have to be used to make a 90-degree turn through the drilled hole and into the house. The location of the service panel inside the house may allow you to run the raceway directly from the "LB" to the back of the panel. If the location of the panel is in the basement, another "LB" fitting will be needed to change direction again to bring the raceway into the top of the panel. If you are using a metal raceway, Section 250.92 states that a bonding bushing is required at the connection point where the conduit is attached to the service entrance panel *or* at the connection point where the conduit is attached to the meter socket. The bonding bushing is normally located at the service panel end. A bonding jumper bonds this bushing to the neutral bus bar in the service

panel (Figure 8-22). Make sure to install the insulated bushings required by Section 300.4(G) on the meter socket end of the raceway as well as the panel end. The bonding bushing required on one end of the metal service raceway is usually insulated; therefore, a separate insulated bushing will not have to be used.

13 Measure and cut the service entrance conductors to the length required to go from the bottom of the meter socket to the service panel enclosure. Allow 1 foot extra at the meter socket end and enough at the panel end to allow the conductors to be easily attached to the main circuit breaker and neutral lugs. The length necessary to make connections at the panel will vary, depending on the style of the panel used. Remember to identify the service neutral conductor with white tape.

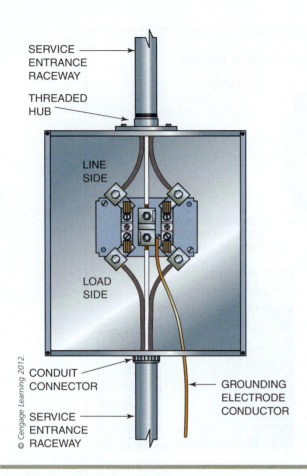

FIGURE 8-23 Meter socket connections for a typical three-wire overhead service entrance installed in a raceway.

14 Install the conductors between the meter socket and the service panel. The cover(s) on the "LB" fitting(s) will have to be removed to install the wires through them. Make sure to reinstall the "LB" covers and weatherproof gaskets after you have installed the service conductors.

15 Make the meter socket connections (Figure 8-23). Connect the conductors from the weatherhead end to the line-side lugs in the meter socket. Connect the conductors from the service panel to the load-side lugs.

16 Make the proper connections in the service entrance panel. (Panel installation is covered later in this chapter.)

INSTALLING AN OVERHEAD SERVICE USING A MAST

The installation of an overhead service entrance that utilizes a service mast (Figure 8-24) will require slightly different installation steps. The installation steps outlined next assume that a full basement is available and

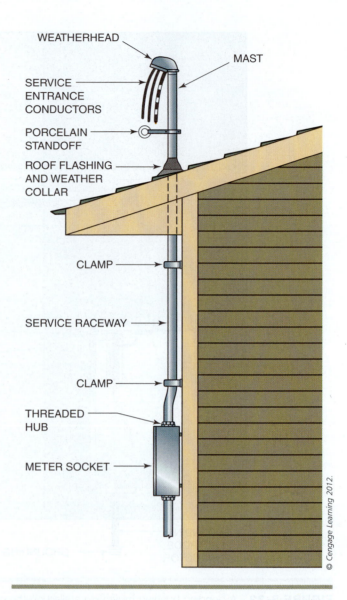

FIGURE 8-24 The parts of a mast-type service entrance.

that the service equipment will be located there. The installation steps for a typical overhead service entrance using a mast are as follows:

1 Install a threaded hub to the meter socket. The threaded hub is attached to the meter socket using the four screws that come with the threaded hub. It must be the same size as the size of the raceway being used for the service mast.

2 Locate and install the meter socket. The meter socket is typically installed at a height that will provide a measurement of 4 to 5 feet above ground level to the center of the kilowatt-hour meter. This is not an *NEC*® requirement but rather a requirement by the local electric utility. It is surface-mounted to the outside wall of the house using four screws inserted through the mounting holes provided in the back of the meter socket. Make sure to install the meter socket so it is level and plumb on the top, bottom, and sides. A torpedo level works

well for this. Where the meter socket is located and how it is mounted provides a point of reference from which all other service entrance equipment installation measurements can be made.

3 Locate and cut a hole in the roof that is aligned with the threaded hub of the meter socket. Make the hole as small as possible but big enough to accommodate the size of the mast pipe.

4 Measure and cut the electrical conduit to the length required to go from the meter socket up through the hole in the roof and to a height that will keep the service drop conductors at the minimum required clearance from grade. In most mast service entrance installations, a 10-foot length of conduit from the top of a meter socket will provide all the height that is needed.

5 Insert the mast pipe through the hole you cut in the roof and raise the mast into position. Attach it to the meter socket and secure the mast in position with the appropriate support straps or brackets.

6 Install the roof flashing and weather collar (rubber boot).

7 Measure and cut the length of service entrance conductor wire needed. Cut enough length to leave approximately 3 feet of wire on the weatherhead end and 1 foot of wire for the meter socket end. You will need to cut three equal lengths: two lengths for the ungrounded "hot" service conductors and one length for the grounded service neutral conductor. At this time, use white electrical tape to identify each end of the service neutral conductor.

8 At the meter socket end, insert all three service entrance conductors and push them up through the raceway (Figure 8-25) until approximately 3 feet of the conductors come out of the top of the mast.

9 Remove the top cover of the weatherhead and install the weatherhead body on the top of the mast. Insert the two black-colored ungrounded conductors and the white-tape-identified neutral conductor through separate holes in the weatherhead. There should be approximately 3 feet of service conductor coming out of the weatherhead for connection to the service drop conductors by the utility company. Reinstall the cover on the weatherhead (Figure 8-26).

10 Install the porcelain standoff onto the mast and align it in the direction that the service drop will be coming from. If necessary, install a guy wire to add additional support to the mast. The *NEC*® says that a guy wire *may* be

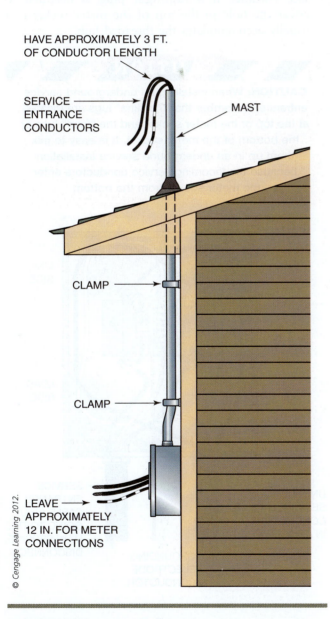

FIGURE 8-25 Approximately 3 feet of service entrance conductor must extend from the top of the service mast.

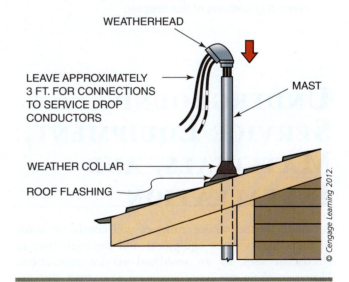

FIGURE 8-26 Once the service conductors are inserted in the proper holes of the weatherhead, the weatherhead must be secured to the top of the mast. Make sure to leave approximately 3 feet of service conductor for connection to the service drop conductors.

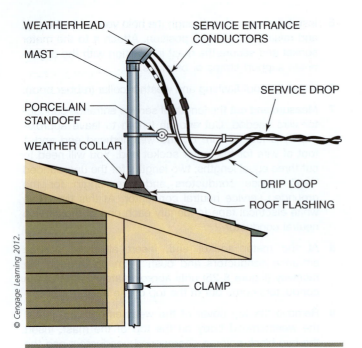

WEATHERHEAD

MAST

SERVICE ENTRANCE
CONDUCTORS

SERVICE DROP

PORCELAIN
STANDOFF

WEATHER COLLAR

DRIP LOOP

ROOF FLASHING

CLAMP

© Cengage Learning 2012.

FIGURE 8-27   The porcelain standoff provides the attachment point for the service drop to the service mast.

needed. Some electric utilities will *require* a guy wire if, for example, the point of attachment of the service drop is 30 inches or higher on the mast (Figure 8-27).

11   The installation of a mast-style service entrance from the meter socket to the service entrance panel is the same as the installation steps already outlined in the preceding sections of this chapter.

UNDERGROUND SERVICE EQUIPMENT, MATERIALS, AND INSTALLATION

An underground service entrance will consist of many of the same kinds of equipment and materials that an electrician uses in an overhead service installation. However, there are some pieces of equipment that are needed for the underground service that are not needed for an overhead service. These pieces include a meter socket designed for use in an underground service entrance, a service raceway type that conforms with Section 300.5(D)(4), bushings that conform with Section 300.5(H), underground service entrance cable, and a

marking tape that conforms with Section 300.5(D)(3). The purpose for each of these parts unique to an underground service is explained here:

- Meter socket for an underground service: The meter socket for an underground service entrance installation (Figure 8-28) must be large enough to allow a minimum of two service entrance raceways to enter the bottom of the socket: one raceway for the incoming electrical power and one raceway to come back out of the meter socket and to the service panel. Sometimes the load-side conductors exit directly out of the back of the meter socket and go directly into the service equipment. An overhead service meter socket may be used for an underground service entrance if a rain-tight plug is installed to cover the hole in the top of the meter socket that usually accommodates the threaded hub.

> **CAUTION**
>
> **CAUTION:** When installing an underground service entrance, remember that the "line" lugs are located at the top of the meter socket and the "load" lugs at the bottom of the meter socket. It is easy to mix them up in an underground service installation because the incoming service conductors enter the meter socket from the bottom.

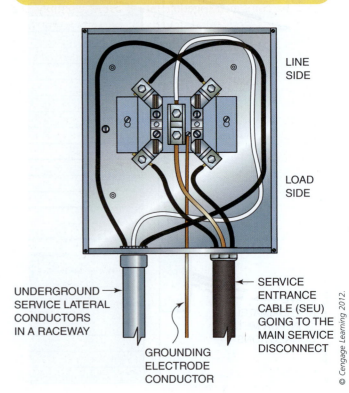

LINE
SIDE

LOAD
SIDE

UNDERGROUND
SERVICE LATERAL
CONDUCTORS
IN A RACEWAY

GROUNDING
ELECTRODE
CONDUCTOR

SERVICE
ENTRANCE
CABLE (SEU)
GOING TO THE
MAIN SERVICE
DISCONNECT

© Cengage Learning 2012.

FIGURE 8-28   The meter connections for a typical three-wire underground service entrance lateral.

- Service raceways used in an underground service: Where the underground service conductors exit the trench at the meter socket side or at the transformer side, Section 300.5(D)(4) requires the conductors to be protected from physical damage with RMC, IMC, or Schedule 80 PVC. These conduit styles must extend upward to a height of 8 feet (2.4 m) above grade at the riser end of the service lateral or at the point where the conductors enter the meter socket on the house side of the service lateral.

- Underground service conductors: The underground service entrance conductors can be installed using Type USE service entrance cable, which can be direct buried at a depth required by the local electric utility and the *NEC*®. The other way that underground service entrance conductors are installed is to run a raceway for the entire length of the service lateral and to then pull in suitable conductors. Because a conduit located underground is considered a wet location by the *National Electrical Code*, the insulation on conductors pulled into such a conduit must have a "W" as part of their insulation letter designation. A good example of a conductor insulation that could be pulled into an underground conduit is an "XHHW" insulation type.

Type USE underground service entrance cable is listed in Table 310.104(A) of the *NEC*® as a single conductor. However, three (or four) USE conductors can be combined together into a cable assembly and used in a residential underground service entrance installation. This type of cable is often referred to as URD cable. URD is an acronym for "underground residential distribution." The underground service entrance cable shown in Figure 2-22 is URD cable. In most areas of the country, when an electrician goes to an electrical distributor and asks for underground service entrance cable, URD cable will usually be in stock.

URD cable is designed for underground services operating at 600 volts or less. It is constructed of three or four insulated and stranded aluminum conductors. Three-conductor URD cable is used in an underground service entrance to a house. Four-wire URD cable, often referred to as "mobile home cable" or "trailer wire," is used as an underground feeder to mobile homes. The ungrounded "hot" conductors are black insulated while the grounded neutral conductor is black insulated with yellow identification stripes.

- Underground service raceway bushings: Section 300.5(H) requires a bushing to be placed on the end of a raceway that ends underground and provides protection for the underground service entrance conductors emerging from the ground. The purpose of the bushing is to provide a smooth opening to the raceway so that the sharp edges of the raceway will not cut the insulation of the underground service conductors.

- Marking tape: Section 300.5(D)(3) requires a warning ribbon to be used when installing an underground service. The warning tape is to be located at least 12 inches (300 mm) above the buried service entrance conductors. The purpose of the tape is to give someone a warning that they are digging in an area where live electrical wires are buried.

from experience...

Some areas of the country allow electricians to use URD cable as the service entrance conductors in an overhead service entrance installed with conduit. It is easier to cut and install a length of three-conductor URD cable than to cut, identify the neutral, and install three individual service conductors. URD cable is also less expensive to buy than three individual conductors. Check with your local authority having jurisdiction (AHJ) and the local electrical power utility before you install URD cable in an overhead service conduit to determine if it is allowed.

CAUTION

CAUTION: Electricians will have to work in trenches during the installation of an underground service entrance. OSHA Section 1926.651 sets out specific requirements for trenching and excavating. Of special interest is the requirement for sloping or shoring up the sides of trenches by means of bracing, underpinning, or piling. The work here has to be done by specialists (in engineering), but you, the electrician, have to be particularly concerned with the safety implications. Without shoring, the threat of an excavated hole caving in while you are in the hole is very real—and also very sobering. There have been many workers who have been buried alive. Do not put yourself in a position where this could happen to you.

Electrical power from the local electric utility can be delivered to the meter socket from a pole-mounted transformer or a pad-mounted transformer (Figure 8-29). Both delivery methods will require a trench to be dug to the house at the point where the meter socket is located.

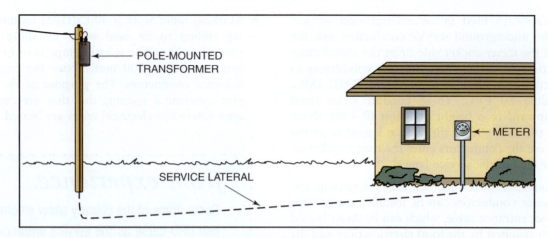

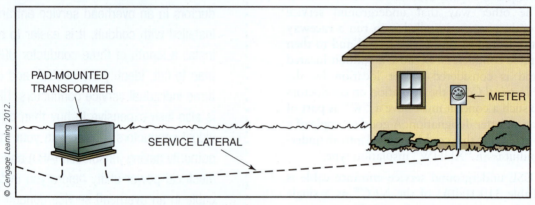

© Cengage Learning 2012.

FIGURE 8-29   Electric power from the local electric utility can be delivered underground to a house from a pole-mounted transformer or a pad-mounted transformer.

The trench must be dug to a depth that will provide the minimum burial requirements as stated in Table 300.5 of the *NEC*® or at the depth required by the local electric utility company. Figure 8-30 shows a common installation for an underground service entrance using underground service entrance (USE) cable. Figure 8-31 shows a common installation for an underground service entrance using a continuous conduit run. If you are installing an underground service entrance, always check with the local electric utility to see if there are any specific requirements that must be followed.

INTERSYSTEM BONDING TERMINATION

Introduced in Chapter 7, this device is used to provide a connection point for cable television, satellite TV, telephone, and any other system that requires being bonded

to the house electrical power system grounding electrode. Section 250.94 requires that an intersystem bonding termination be installed. The best time to install the intersystem bonding termination is when the service entrance is installed. There are various ways for the intersystem bonding termination to be installed, including connecting a termination device directly onto the meter socket. Figure 8-32 shows an example of one of the most common ways that an intersystem bonding termination is installed.

VOLTAGE DROP

If you were to take a voltage reading at the utility transformer terminals supplying a residential service entrance and compare it to the voltage at the terminals of the main circuit breaker in the service panel, they should be approximately the same. This is the case with overhead service entrances because the length of wire between the utility transformer and the service entrance main circuit breaker is normally not

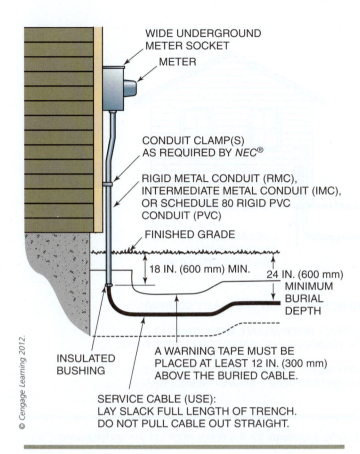

WIDE UNDERGROUND
METER SOCKET

METER

CONDUIT CLAMP(S)
AS REQUIRED BY *NEC*®

RIGID METAL CONDUIT (RMC),
INTERMEDIATE METAL CONDUIT (IMC),
OR SCHEDULE 80 RIGID PVC
CONDUIT (PVC)

FINISHED GRADE

18 IN. (600 mm) MIN.

24 IN. (600 mm)
MINIMUM
BURIAL
DEPTH

INSULATED
BUSHING

A WARNING TAPE MUST BE
PLACED AT LEAST 12 IN. (300 mm)
ABOVE THE BURIED CABLE.

SERVICE CABLE (USE):
LAY SLACK FULL LENGTH OF TRENCH.
DO NOT PULL CABLE OUT STRAIGHT.

© Cengage Learning 2012.

FIGURE 8-30 A typical underground service entrance installed using Type USE underground service cable.

excessively long. This wiring is also usually installed and owned by the utility company and has been engineered to provide the proper voltage at a customer's service panel. However, when underground service conductors are installed, they are normally customer-owned and installed by an electrical contractor. The length of the underground service conductors can be quite long, such as when a house is located on the building lot some distance from the street or road. As a result, the voltage available at the service panel in the house may be much less than the voltage at the utility transformer supplying the system located out at the street or road. This difference in voltage is called **voltage drop (VD)**, the amount of voltage that is needed to "push" the house electrical load current through the wires from the utility transformer, to the service panel, and back to the transformer. The amount of voltage drop depends on the resistance of the wires, the length of the wires, and the actual current carried on the wires.

The *NEC*® recommends that both branch circuits and feeder circuits be sized to limit the voltage drop to no more than 3 percent. There is no *NEC*® recommendation specifically for service entrance conductors but

many utility companies want no more than a 3 percent voltage drop in a service installation assuming that the loading on the service conductors is no more than 80 percent of the service size. This means that for a 100-amp-rated service entrance, the voltage drop should be no more than 7.2 volts (240 volts × 3% = 7.2 volts), assuming a maximum current load of 80 amps (100 amps × 80% = 80 amps). There is a common formula used to calculate voltage drop:

$$VD = \frac{K \times I \times L \times 2}{cm}$$

K = a constant that is 12.9 ohms for copper wire; 21.2 ohms for aluminum wire

I = load current in amps

L = the one-way length of the conductors from the source to the load

2 = the multiplier used because the length of wire goes to the load and back to the source

cm = the cross sectional area of the conductor in circular mils; this can be found in Table 8, Chapter 9 of the *NEC*®

If the formula is manipulated properly, the minimum conductor size that would need to be installed to allow a certain voltage drop can be calculated. This formula is

$$cm = \frac{K \times I \times L \times 2}{VD}$$

Two examples of how to use the voltage drop formulas are shown on page 283.

Table 8-1 shows the minimum conductor sizes needed to be installed for an underground service entrance based upon 80 percent loading and a voltage drop of no more than 3 percent. The distances given are the total length in feet from the utility transformer terminals to the customer's service entrance panel.

SERVICE EQUIPMENT INSTALLATION

The service entrance equipment provides the means of disconnecting the house electrical system from the local electric utility electrical system. It will also usually contain the branch-circuit overcurrent protection devices that provide protection for the various circuits that make up the residential electrical system. As previously discussed, the service entrance equipment can be located inside the house or, if in a NEMA Type 3R enclosure, outside the house. Because the most common location is inside a house, the discussion here on the installation of the service entrance equipment is limited to an inside location.

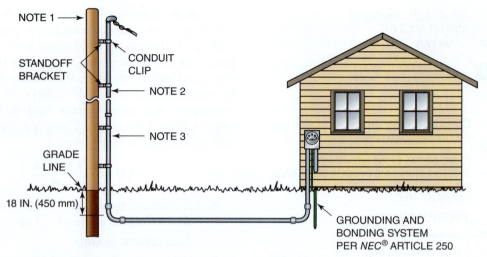

TYPICAL UTILITY COMPANY REQUIREMENTS
NOTES:
1) LOCATION AND HEIGHT OF RISER POLE TO BE SPECIFIED BY THE UTILITY COMPANY.

2) RISER SHOULD BE RIGID METAL CONDUIT (RMC), INTERMEDIATE METAL CONDUIT (IMC), OR SCHEDULE 40 RIGID PVC CONDUIT (PVC) (MINIMUM 2 IN.) RATED FOR OUTDOOR USE.

3) RIGID METAL CONDUIT (RMC), INTERMEDIATE METAL CONDUIT (IMC), OR SCHEDULE 80 RIGID PVC CONDUIT (PVC) (MINIMUM 2 IN.). MINIMUM HEIGHT ABOVE FINISHED GRADE IS 8 FT (2.4 m).

© Cengage Learning 2012.

FIGURE 8-31 A typical underground service entrance installed using a raceway for the entire length of the underground service lateral.

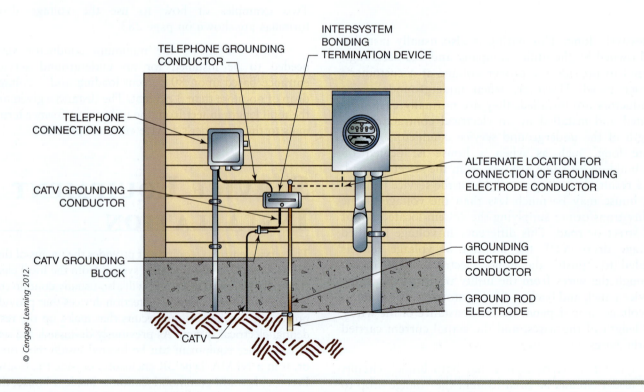

© Cengage Learning 2012.

FIGURE 8-32 A good example of how an intersystem bonding termination is often installed.

EXAMPLE 1

Calculate the voltage drop on the service entrance conductors for a 100 amp underground service entrance. The total length of the service conductors from the transformer terminals at the top of the utility pole, through the meter socket, and to the service entrance main disconnect is 100 feet. Assume a maximum load current of 80 amps. The electrician doing the installation installs 2 AWG aluminum conductors, which is the "normal" size wire for a 100-amp residential service entrance. Here is how the calculation is done:

$$\text{VD} = \frac{21.2 \text{ ohms} \times 80 \text{ amps} \times 100 \text{ feet} \times 2}{66,360 \text{ cm (from Table 8, Chapter 9)}}$$

$$= \frac{339,200}{66,360} = 5.1 \text{ volts}$$

The percent voltage drop is 2.13% ([5.1 volts/ 240 volts] × 100). This is less than the *NEC*®-recommended voltage drop of 3 percent.

EXAMPLE 2

Calculate the minimum size conductor needed for a 100-amp underground service entrance where the total length of the underground service conductors will be 250 feet. Assume a 3 percent voltage drop maximum, the use of copper conductors, and a maximum load current of 80 amps. Here is how the calculation is done:

$$\text{cm} = \frac{12.9 \text{ ohms} \times 80 \text{ amps} \times 250 \text{ feet} \times 2}{7.2 \text{ volts (3\% of 240 volts)}}$$

$$= \frac{516,000}{7.2} = 71,667 \text{ cm}$$

A minimum wire size of 1 AWG copper must be used according to Table 8 in Chapter 9 of the *NEC*®, based on a minimum circular mil requirement of 71,667 cm.

WORKING SPACES AND CLEARANCES

Article 110 of the *NEC*® provides specifications for working space around electrical equipment. Specifications are outlined for two situations: (1) 600 V or less (110.26) and (2) over 600 V (110.30). In residential work, we will be concerned only with the requirements for 600 V or less. These requirements will have to be observed when installing an electrical panel in a house.

Section 110.26 states that sufficient access and working space must be provided and maintained about all electric equipment to permit ready and safe operation and maintenance of such equipment. Storage of materials that block access to electrical equipment or prevent safe work practices when an electrician is working on a piece of electrical equipment must be avoided at all times. Section 110.26(A)(1) tells us that the depth of the working space in front of residential electrical equipment must not be less than that specified in Table 110.26(A)(1), which specifies the space requirements for working space in front of electrical equipment. Distances are measured from the face of the panel enclosure. Because residential electrical systems never exceed 150 volts to ground, the 0–150-volt row is used in the table. The dimensions given in the table are based on Conditions 1, 2, or 3. For all three conditions in the 0–150-volt row, the minimum free space that must be maintained in front of residential electrical equipment is 3 feet (900 mm). Notice that Conditions 1, 2, and 3 are explained under the table, but for residential wiring it really does not make any difference which condition we are under because they all have the same clearance for the 0–150-volt locations. Section 110.26(A)(2) states that the width of the working space in front of the electric equipment must be the width of the equipment or 30 inches (750 mm), whichever is greater. Residential electrical panels are usually no more than 14½ inches wide so that they can fit between wall studs that are installed at 16 inches center to center. In all cases, the work space must permit at least a 90-degree opening of equipment doors on the panels.

TABLE 8-1 This Table Shows the Maximum Allowable Length in Feet for Common Underground Service Entrance Sizes. The Lengths Given Were Calculated Based on a 3 Percent Voltage Drop and Conductor Loading of 80 Percent.

CONDUCTOR SIZE	100 AMP SERVICE ALUMINUM	100 AMP SERVICE COPPER	150 AMP SERVICE ALUMINUM	150 AMP SERVICE COPPER	200 AMP SERVICE ALUMINUM	200 AMP SERVICE COPPER	400 AMP SERVICE ALUMINUM	400 AMP SERVICE COPPER	MINIMUM CONDUIT SIZE
2	150'	235'							2"
1	190'	290'		195'					2"
1/0	235'	350'		235'					2"
2/0	295'	435'	195'	290'		215'			2"
3/0	360'	530'	240'	355'		265'			2"
4/0	440'	650'	295'	435'	220'	325'			2½"
250 kcmil	500'	725'	340'	485'	250'	360'			2½"
350 kcmil	660'	910'	440'	605'	350'	455'			3"
500 kcmil	875'	1150'	580'	765'	435'	575'		290'	4"
2–4/0	885'	1300'	585'	865'	440'	650'		325'	4"
2–250	1,015'	1450'	675'	965'	506'	725'	225'	365'	4"
2–350	1,320'	1835'	885'	1225'	660'	915'	330'	460'	4"

from experience...

Even though the *NEC®* recommends that the voltage drop of a branch circuit or a feeder circuit be no more than 3 percent, this is not something that electricians worry much about when sizing branch circuits or feeder circuits in a residential installation. The reason is that the branch circuit or feeder runs are relatively short in house wiring and the load on a branch circuit or feeder circuit is rarely close to the maximum amperage allowed on the conductor. For example, a 120-volt branch circuit installed with a 12/2 Type NM cable can be run 50 feet from the circuit breaker panel, have a load of 16 amps (20 amps × 80% = 16 amps), and still have a voltage drop of less than 3 percent. However, a good rule of thumb is that if you are installing a branch circuit or feeder that will be 100 feet or more in length, you should determine if the voltage drop is acceptable by using the formula given in this chapter. It is much easier to install a larger wire initially to compensate for voltage drop on an excessively long run than to replace previously installed wiring that was too small to begin with.

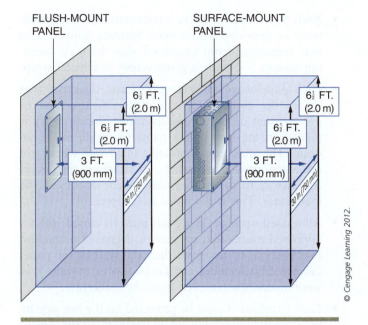

FIGURE 8-33 *Minimum working space clearances in front of service entrance equipment.*

must extend from the floor to a height of 6½ feet (2.0 m) (Figure 8-33). Within the 6½ foot (2.0 m) height requirement, only equipment associated with the electrical installation can be placed above or below the electrical equipment but it cannot extend more than 6 inches (150 mm) beyond the front of the equipment. *Exception No. 2* does allow kWh meters installed in meter sockets to extend beyond the front of the electrical equipment.

In Articles 312 and 408, there are a few rules for installing panelboards and their cabinets. Because panelboards are usually used as the service equipment in a house electrical system, it is appropriate to review these installation rules here. The rules that affect residential installations are summarized as follows:

- In a damp or wet location, a surface-mounted panel must be placed or made so that moisture is kept from entering and accumulating in it. This is why a NEMA 3R cabinet is used outdoors.

- When mounting a panel in a damp or wet location, a ¼-inch (6 mm) airspace must be maintained between it and the wall. If the panel is made of a nonmetallic material, the ¼-inch (6 mm) air gap is not required.

- If a panel is installed in a finished wall, there must be no more than a ⅛-inch (3 mm) gap around the edge of the panel in the wall.

- When mounting the panel in a noncombustible wall, the front edge must be flush or set back no more than ¼-inch (6 mm) from the surface. In a combustible wall, the panel edge must always be flush or project out from the wall surface.

So regardless of the width of the electrical equipment, the working space cannot be less than 30 inches (750 mm) wide. This allows an individual to have at least shoulder-width space in front of the equipment. This 30-inch (750 mm) measurement can be based on the center of the panel or made from either the left or the right edge of the equipment. Section 110.26(A)(3) addresses the required headroom. The minimum headroom of working spaces about electrical service equipment is 6½ feet (2.0 m) or the height of the equipment, whichever is greater. *Exception No. 1* to this section states that in existing dwelling units, service equipment or panelboards that do not exceed 200 amperes are permitted in spaces where the headroom is less than 6½ feet (2.0 m). This *Exception* was put in the *NEC®* to allow service upgrades that include a new panel in older basements that have less than a 6½-foot ceiling height. So, when you install an electrical panel for a residential service entrance, there must be at least 3 feet (900 mm) of free space in front of it, the space must be at least 30 inches (750 mm) wide, and the space

- Each panelboard used as service entrance equipment must be provided with a main bonding jumper. You may remember from Chapter 7 that the main bonding jumper is usually a green screw or a metal strap.

- Each panelboard must have a circuit directory that shows the clear, evident, and specific purpose of every circuit or circuit modification. The directory is placed on the face of the panel or inside the panel door. The directory is usually filled out when you are trimming out the panel and attaching the circuit conductors to the proper overcurrent devices in the loadcenter. This is discussed in Chapter 19.

- All unused openings in a panel must be closed with an approved means that provides a degree of protection that is equal to or better than the wall of the panel cabinet. This includes any knockouts or pryouts that were mistakenly removed from the loadcenter.

- Each panelboard must be protected on the line side by no more than two main circuit breakers (or two sets of fuses) that have a combined rating equal to or less than the panelboard's ampacity rating. In residential wiring, there is usually just one main circuit breaker or one main set of fuses in the panelboard. This protection is not required if the panelboard feeder has a circuit breaker protecting it that has a rating equal to or less than the rating of the panelboard. This is the case when a subpanel is used in a house and is supplied by a feeder from the main panel.

- Panelboard metal cabinets must be grounded. This is done by terminating a grounding conductor to a grounding bar that is mechanically attached to the metal cabinet.

The preceding sections in this chapter outlined the steps needed to install the service entrance up to the service panel. Now let's take a look at the actual installation of the panel. The service panel (Figure 8-34) should be installed before any branch-circuit installations are made since all the branch-circuit wiring originates at the service panel. The following steps should be followed when installing the service panel:

1. Fasten the service panel in place. If it is a flush-mount installation, the panel is designed to fit between two studs that are 16 inches on center. Make sure to mount the panel enclosure so the front edge of the enclosure will be flush with the finished wall surface. Secure the panel enclosure to the studs with at least two screws per side. If it will be a surface installation, install a mounting backboard for the service entrance panel on the wall where you wish to locate the panel. Mounting **backboards** are often made of 3/4-inch plywood and are painted black. If the backboard is to be mounted on a masonry wall in a basement, an air gap of at least 1/4 inch behind the backboard should be maintained to allow for moisture to dry out. The backboard must

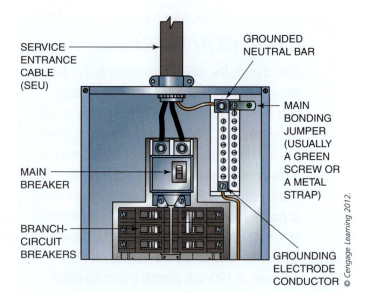

FIGURE 8-34 The connections for a 120/240-volt, single-phase, main circuit breaker entrance panel.

be large enough for the panel and any other items installed on it, such as a surface-mount box with a duplex receptacle.

CAUTION

CAUTION: Section 230.70(A)(1) requires the service conductors to terminate at the main disconnecting means as soon as the conductors enter the house from the outside meter socket. Make sure that you install the service panel at a location that gives you the least amount of "inside run" of the service entrance conductors. Failure to do this will usually result in the electrical inspector turning down your installation.

CAUTION

CAUTION: Make sure to install the service entrance panel so that the requirements of Section 110.26 are met. Remember that a working space must be provided in front of the panel that is at least 30 inches (750 mm) wide, extends at least 3 feet (900 mm) in front of the panel, and extends from the floor up to a height of 6½ feet (2.0 m) or more.

2. Attach the service conductors from the meter socket to the service panel. Make sure there is enough "free" conductor to make the necessary connections. Remember, service entrance cable or a service raceway will be coming to the panel from the meter

socket. The cable will require a cable connector to secure it to the enclosure, and the raceway will require some type of conduit connector to secure it to the enclosure.

3 Connect both ungrounded ("hot") service entrance conductors to the main circuit breaker terminals: 100-ampere-rated main circuit breakers have slotted set-screws that will need to be tightened, and 150- and 200-amp-rated and larger main circuit breakers will have hexagonal countersunk screws that will require an Allen wrench or a ratchet wrench with a hexagonal socket to tighten them. Remember that the proper torque must be applied to these terminals with a torque wrench or torque screwdriver. The proper torque amount can be found, along with other information about the panel, inside the panel enclosure on a sticker installed by the panel manufacturer.

> **CAUTION**
>
> **CAUTION:** If aluminum wire is being used as service conductors, make sure to use an antioxidant and prepare the aluminum wires according to the instructions that come with the antioxidant.

4 Connect the grounded neutral service entrance conductor to the proper terminal on the neutral bus bar.

5 Bond the enclosure to the equipment grounding bus bar and the neutral bus bar using a main bonding jumper. The main bonding jumper can be a green screw, a short length of wire, or a strap. Most new residential service panels use a green screw as the main bonding jumper.

6 Connect the grounding electrode conductor to the proper terminal on the neutral bus. Route this conductor to the grounding electrode. If a metal water pipe bringing water to the house is available, it must be used as the grounding electrode. Attach the grounding electrode conductor to the water pipe at a point that is 5 feet (1.5 m) or less from where the pipe enters the house. Use a water pipe grounding clamp for the attachment (Figure 8-35).

> **CAUTION**
>
> **CAUTION:** Remember that if you use a metal water pipe as the grounding electrode, you must supplement it with another electrode. Usually, a ground rod type of electrode is used as the supplement and is connected at the meter socket as shown in Figure 8-36. Also, remember that a rod, pipe, or plate electrode must be supplemented by another electrode unless a single rod, pipe, or plate electrode has a resistance to earth of less than 25 ohms. If installing another ground rod, make sure it is driven at least 6 feet away.

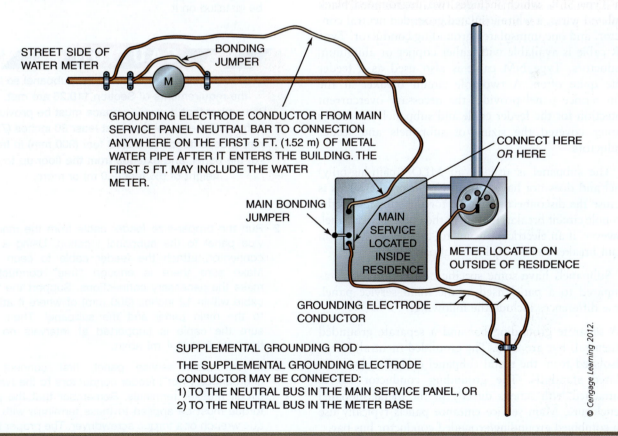

STREET SIDE OF WATER METER

BONDING JUMPER

M

GROUNDING ELECTRODE CONDUCTOR FROM MAIN SERVICE PANEL NEUTRAL BAR TO CONNECTION ANYWHERE ON THE FIRST 5 FT. (1.52 m) OF METAL WATER PIPE AFTER IT ENTERS THE BUILDING. THE FIRST 5 FT. MAY INCLUDE THE WATER METER.

MAIN BONDING JUMPER

MAIN SERVICE LOCATED INSIDE RESIDENCE

CONNECT HERE *OR* HERE

METER LOCATED ON OUTSIDE OF RESIDENCE

GROUNDING ELECTRODE CONDUCTOR

SUPPLEMENTAL GROUNDING ROD

THE SUPPLEMENTAL GROUNDING ELECTRODE CONDUCTOR MAY BE CONNECTED:
1) TO THE NEUTRAL BUS IN THE MAIN SERVICE PANEL, OR
2) TO THE NEUTRAL BUS IN THE METER BASE

© Cengage Learning 2012.

FIGURE 8-35 The installation of the grounding electrode conductor from the main service panel to a water pipe electrode.

SUBPANEL INSTALLATION

Subpanels are often used to eliminate the need for long branch-circuit runs from the main service entrance panel to the various loads of a residential electrical system. They also will provide an opportunity for the installation of more circuits than what the main service panel is designed to supply. The main service entrance panel is usually located inside the house, but as we learned previously, it can be located outside in a proper weatherproof enclosure if there is not a suitable location to install the panel inside. Subpanels will be installed inside the house at a location that is near the area it is intended to serve. If you are installing a subpanel solely for the extra circuits that it provides, you can install it alongside the main service entrance panel.

Electrical power is supplied to a subpanel using a feeder cable run from the main service panel. An electrical raceway, such as EMT, can also be used to run feeder conductors from the service panel to a subpanel. Most residential wiring systems use a feeder cable to bring power to a subpanel, so the discussion on the installation of a subpanel in this section assumes the use of a feeder cable. The feeder cable most often used is a Type SER, which includes two ungrounded black insulated wires, a white insulated grounded neutral conductor, and one uninsulated grounding conductor. Type SER cable is available with either copper or aluminum conductors. Type NM cable is also used as a feeder cable quite often. A two-pole circuit breaker at the main service panel provides the necessary overcurrent protection for the feeder cable and subpanel. An earlier chapter covered the sizing of subpanels and feeder conductors.

The subpanel is usually an MLO (main-lug-only) panel and does not have a main circuit breaker. This is because the disconnecting means for the subpanel is the two-pole circuit breaker located in the main service panel. However, if an electrician wants to, a panel with a main circuit breaker can be used as a subpanel.

Subpanels have some specific design differences as compared to a panel used as the main service panel. These differences include the following:

- A separate grounding bar and a separate grounded (neutral) bar are used. The grounded (neutral) bar is isolated from the metal subpanel enclosure by insulated standoffs. The grounding conductor bar is attached with screws directly to the metal subpanel enclosure. Main service entrance panels typically use a combined grounding/grounded conductor bus bar.

- A four-conductor feeder cable is used to provide electrical power to a subpanel. A three-wire service entrance cable or three wires in a raceway supply the electrical power to a main service entrance panel.

- Subpanels do not normally have a main circuit breaker. The ungrounded feeder conductors are terminated at the panel's main lugs.

The installation procedures for a subpanel are similar to the procedures for installing a main service panel; however, there are a few differences that a residential electrician must be aware of. The following steps should be followed when installing a subpanel:

1 Fasten the subpanel in place. If it is a flush-mount installation, the panel is designed to fit between two studs that are 16 inches on center. Make sure to mount the panel enclosure so the front edge of the enclosure will be flush with the finished wall surface. Secure the panel enclosure to the studs with at least two screws per side. If it will be a surface installation, install a mounting backboard for the subpanel on the wall where you wish to locate it. Mounting backboards are often made of 3/4-inch plywood and are painted black. If the backboard is to be mounted on a masonry wall in a basement, an air gap of at least 1/4 inch behind the backboard should be maintained to allow for moisture to dry out. The backboard must be large enough for the panel and any other items that need to be installed on it.

> **CAUTION**
>
> **CAUTION:** Make sure to install the subpanel so that the requirements of Section 110.26 are met. Remember that a working space must be provided in front of the panel that is at least 30 inches (750 mm) wide, extends at least 3 feet (900 mm) in front of the panel, and extends from the floor up to a height of 6½ feet (2.0 m) or more.

2 Run the proper-size feeder cable from the main service panel to the subpanel location. Using a cable connector, attach the feeder cable to each panel. Make sure there is enough "free" conductor to make the necessary connections. Support the feeder cable within 12 inches (300 mm) of where it attaches to the main panel and the subpanel. Then, make sure the cable is supported at intervals no more than 4½ feet (1.4 m) apart.

3 In the main service panel, first connect both ungrounded ("hot") feeder conductors to the two-pole circuit breaker terminals. Remember that the proper torque must be applied to these terminals with a torque wrench or a torque screwdriver. The proper torque

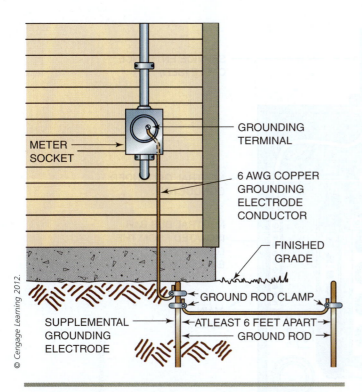

© Cengage Learning 2012.

METER SOCKET

GROUNDING TERMINAL

6 AWG COPPER GROUNDING ELECTRODE CONDUCTOR

FINISHED GRADE

GROUND ROD CLAMP

SUPPLEMENTAL GROUNDING ELECTRODE

ATLEAST 6 FEET APART

GROUND ROD

FIGURE 8-36   A water pipe electrode must be supplemented by another electrode. The most common wiring practice is to use a driven ground rod that is then conducted to the meter socket with a 6 AWG copper grounding electrode conductor. Remember that a rod, pipe, or plate electrode must be supplemented by another electrode unless a single rod, pipe, or plate electrode has a resistance to earth of less than 25 ohms. If installing another ground rod make sure it is driven at least 6 feet away.

amount can be found, along with other information about the panel, inside the panel enclosure on a sticker installed by the panel manufacturer. Then, connect *both* the feeder grounded neutral conductor and the feeder grounding (bare) conductor to the grounded neutral bus bar.

CAUTION

CAUTION: If aluminum wire is being used as the feeder conductors, make sure to use an antioxidant and prepare the aluminum wires according to the instructions that come with the antioxidant.

4   In the subpanel (Figure 8-37), connect both ungrounded ("hot") feeder conductors to the main lug terminals. Remember that the proper torque must be applied to

these terminals with a torque wrench or a torque screwdriver. The proper torque amount can be found, along with other information about the panel, inside the panel enclosure on a sticker installed by the panel manufacturer. Next, connect the grounded neutral feeder conductor to the neutral bus bar. Finally, connect the grounding (bare) conductor of the feeder to the equipment-grounding bar.

from experience...

Most panels used as a subpanel do not come from the factory with an equipment-grounding conductor bar already installed in the panel. The electrician must buy a separate grounding bar kit that is designed for the subpanel and install it. The installation of the grounding bar kit can actually take place at any time during the installation process.

CAUTION

CAUTION: Do not bond the subpanel enclosure to the neutral bus bar using a main bonding jumper like you do with a main service panel. If a main bonding jumper (a green screw or a strap) is included with the subpanel, *throw it away!* Section 250.142(B) clearly states that the grounded circuit conductor cannot be used to ground non-current-carrying metal parts on the *load* side of the service entrance main disconnect. This is exactly what the situation is when using a sub-panel and is the reason why the grounded neutral bar and the equipment-grounding bar must be separated in a subpanel. Failure to follow this *NEC*® requirement will result in an electrical safety hazard to the occupants of the house.

SERVICE ENTRANCE UPGRADING

The preceding sections of this chapter discussed the installation of service entrances and equipment for new residential applications. However, many service entrance installations involve upgrading an existing service. The reason for the upgrade is usually to increase

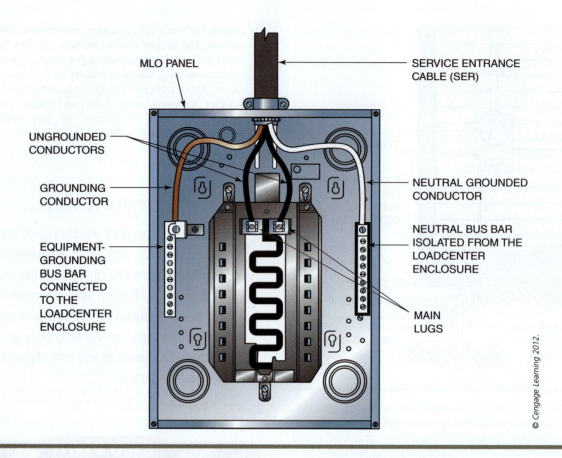

MLO PANEL

SERVICE ENTRANCE
CABLE (SER)

UNGROUNDED
CONDUCTORS

GROUNDING
CONDUCTOR

NEUTRAL GROUNDED
CONDUCTOR

EQUIPMENT-
GROUNDING
BUS BAR
CONNECTED
TO THE
LOADCENTER
ENCLOSURE

NEUTRAL BUS BAR
ISOLATED FROM THE
LOADCENTER
ENCLOSURE

MAIN
LUGS

© Cengage Learning 2012.

FIGURE 8-37   The connections for a 120/240-volt, single-phase, main-lug-only (MLO) subpanel.

the service size because of an increase in the load being served by the service. Another reason for an upgrade may be because the condition of the service entrance equipment has deteriorated over time to a point where it is not safe to continue to use the service equipment in its present condition. Whatever the reason, there are some special installation procedures to consider that are not part of the installation of a service entrance in a new residential building.

There are three different service upgrade scenarios:

• A house where nobody lives, electricity has been disconnected, and the house is being remodeled for new occupants: In this case, the old service entrance and equipment can be taken out and a new service entrance installed by following the same procedures as used for installing a service entrance in a new house.

• A house that is occupied, and the service entrance needs to be upgraded at a new location on the house: Some modification to the procedures for a new house service installation is needed.

• A house that is occupied, and the service entrance needs to be upgraded in the same location: Again, some modification to the procedures for a new house service installation is needed.

Upgrading an existing service entrance will require an electrician to work very closely with the local electric utility. If the upgrade requires a new location for the service, contact must be made with the local electric utility to establish the new location for the meter socket and other service equipment. Once the new location for the service has been established, the electrician will install the new service equipment. If the house is occupied, the electrician will typically need to continue electrical service to the house until the switch-over from the old service to the new service is accomplished. This is done by running a jumper cable from a load-side circuit breaker in the old service entrance main panel to a load-side circuit breaker in the new service entrance main panel. The new service entrance main disconnect must remain in the OFF or open position. This procedure is commonly called **backfeeding** and should be done only by an experienced electrician. Once the upgraded service is installed, an inspection will be required. Following approval of the installation, the electric utility will install the new service drop (assuming an overhead installation) and meter in the new meter socket. The utility workers will then de-energize the old service and energize the new service. The electrician will disconnect the jumper cable and then close the main disconnect for the new service panel. The electrician will then switch the circuit breakers in the new panel ON and energize the house branch circuits. The old service entrance and equipment can now be taken out.

If the service entrance upgrade requires an installation of the new service in the same location as the existing service, contact must still be made with the local electric utility. The local utility will "float" the meter socket so that the electrician can install a new meter socket in the same location on the building. The electrician will then install a new meter socket, new service panel, and new service conductors from the meter socket up the side of the house to the new point of attachment. If the house is occupied, the electrician will typically need to continue electrical service to the house until the switch-over from the old service to the new service is accomplished. This is done by running a jumper cable from a load-side circuit breaker in the old service entrance main panel to a load-side circuit breaker in the new service entrance main panel. The new service entrance main disconnect must remain in the OFF or open position. This procedure is commonly called "backfeeding" and should be done only by an experienced electrician. The electrician will then make an appointment with the electric utility company so that the utility can disconnect the old service drop while the electrician is removing the old service and

installing new service conductors and associated equipment from the new meter socket to the new main panel through the same hole that the old service conductors used. The utility workers will then install a new meter in the meter socket and energize the new service. The electrician will disconnect the jumper cable and then turn on the main disconnect for the new service panel. The electrician will then switch the circuit breakers in the new panel ON and energize the house branch circuits.

SUMMARY

This chapter covered residential service entrance equipment in detail and the procedures that are commonly used to install a service entrance. You learned that overhead and underground service entrances have many similarities and use many of the same parts. However, the few differences between the two service types greatly affect how each is installed. Installation of a service panel and a subpanel was presented in detail. Because many electricians find themselves doing electrical upgrade work in an existing house, upgrading an existing service was also covered.

REVIEW QUESTIONS

Directions: Answer the following items with clear and complete responses.

1. Describe the purpose of a weatherhead.

2. Describe the purpose of the cable hook used in an overhead service entrance.

3. Name the part that is installed on a meter socket (with four screws) that provides a connection point to the meter socket for the service entrance conductors. Assume an overhead service entrance installation.

4. Explain the difference between Types SEU, SER, and USE service entrance cable. Be as specific as possible.

5. Describe the purpose of a sill plate.

6. Name four common raceway types used where the service entrance conductors are installed in a raceway installed on the side of a house. Assume an overhead service entrance installation.

7. Describe the purpose of an "LB" conduit body used in a residential service entrance installation that uses electrical conduit.

8. Name two types of electrical conduit that are suitable as a service entrance mast.

9. A mast-type service entrance will need a porcelain standoff fitting and a weather collar/roof flashing. Describe the purpose of each item.

10. The *NEC®* requires the cable hook to be located below the weatherhead in an overhead service installation. If this is not possible, the *NEC®* allows the hook to be above or to the side of the weatherhead, but no more than _____ away. Name the *NEC®* section that states this requirement.

11. Describe the difference between a surface-mount and a flush-mount installation of a service entrance panel.

12. Explain why backfeeding a panel is sometimes done during a service entrance upgrade.

13. Explain the difference between the load-side lugs and the line-side lugs in a meter socket.

14. If Type SER service entrance cable is used by an electrician to feed a subpanel from a main panel, the cable must be secured within _____ of the panel enclosures and at intervals not exceeding _____. Name the *NEC®* section that states this requirement.

15. An MLO panel is usually used as a subpanel in residential wiring systems. Explain what an MLO panel is.

16. URD cable is made up of three or four USE conductors and is often the preferred wiring method for the installation of a residential underground service entrance. What do the letters URD stand for?

17. Calculate the voltage drop of a 120/240-volt, three-wire underground residential service installed with 2 AWG aluminum URD cable if the load is 80 amps and the one-way length of the underground service conductors is 100 feet. *Hint:* Use the voltage drop formula presented in this chapter.

18. The minimum working space width in front of service equipment installed in a house must be _____ inches wide or the width of the equipment, whichever is greater. The minimum amount of clear space in front of service equipment must be at least _____ feet. This space must be maintained to a height from the floor of _____ feet. Name the *NEC*® section that covers these requirements.

19. When a house is built on a concrete slab and there is no full basement under it, the service entrance panel is often installed outdoors next to the meter socket. What NEMA enclosure rating is necessary for the panel in this outdoor installation?

20. There is no *NEC*® rule that states how high a meter socket must be mounted on the side of a house, but they are typically installed at a height of _____ feet to the center of the meter from finished grade.

KNOW YOUR CODES

1. Check with your local electric utility to see if they supply and install the service drop cable.

2. Check with your local electric utility to see if they supply the service entrance meter socket.

3. Check with your local electric utility company to see if a guy wire is required on a mast type service entrance.

4. Check with your local electric utility to see if there are any specific requirements that must be followed when installing an underground service entrance.

Residential Electrical System Rough-In

CHAPTER 9
General Requirements for Rough-In Wiring

CHAPTER 10
Electrical Box Installation

CHAPTER 11
Cable Installation

CHAPTER 12
Raceway Installation

CHAPTER 13
Switching Circuit Installation

CHAPTER 14
Branch-Circuit Installation

CHAPTER 15
Special Residential Wiring Situations

CHAPTER 16
Video, Voice, and Data Wiring Installation

General Requirements for Rough-In Wiring

OBJECTIVES

Upon completion of this chapter, the student should be able to:

- Discuss the selection of appropriate wiring methods, conductor types, and electrical boxes for a residential electrical system rough-in.

- Demonstrate an understanding of general requirements for wiring as they apply to residential rough-in wiring.

- Demonstrate an understanding of general requirements for conductors as they apply to residential rough-in wiring.

- Demonstrate an understanding of general requirements for electrical box installation as they apply to residential rough-in wiring.

- List several general requirements that apply to wiring methods, conductors, and electrical boxes installed during the rough-in stage of a residential wiring system.

GLOSSARY OF TERMS

deteriorating agents a gas, fume, vapor, liquid, or any other item that can cause damage to electrical equipment

foyer an entranceway or transitional space from the exterior to the interior of a house

furring strip long, thin strips of wood (or metal) used to make backing surfaces to support the finished surfaces in a room; furring refers to the backing surface, the process of installing it, or may also refer to the strips themselves; furring strips typically measure $1'' \times 2''$ or $1'' \times 3''$ and are typically laid out perpendicular to studs or joists and nailed to them, or set vertically against an existing wall surface

inductive heating the heating of a conducting material in an expanding and collapsing magnetic field; inductive heating will occur when current-carrying conductors of a circuit are brought through separate holes in a metal electrical box or enclosure

knockout plug a piece of electrical equipment used to fill unused openings in boxes, cabinets, or other electrical equipment

rough-in the stage in an electrical installation when the raceways, cable, boxes, and other electrical equipment are installed; this electrical work must be completed before any construction work can be done that covers wall and ceiling surfaces

sheetrock a popular building material used to finish off walls and ceilings in residential and commercial construction; it is available in standard sizes, such as 4 by 8 feet, and is constructed of gypsum sandwiched between a paper front and back; often referred to as wallboard

wiring a term used by electricians to describe the process of installing a residential electrical system

Wiring is a term that electricians use to describe the process of installing an electrical system. The **rough-in** stage of **wiring** a residential electrical system involves mounting boxes and installing the circuit conductors using an appropriate wiring method. There are many *National Electrical Code®* (*NEC®*) requirements that need to be followed during the rough-in stage. This chapter looks at several general requirements that an electrician must consider when installing the rough-in wiring.

GENERAL WIRING REQUIREMENTS

Chapter 2 covered several types of hardware and material used to install a residential electrical system. It included an introduction to the different cable and raceway types used in residential wiring. Conductor types and sizes were also presented. Based on the *NEC®* and local electrical code requirements, an electrician must determine an appropriate wiring method and conductor type to be used for installing the residential electrical system. Residential wiring systems are usually installed using a cable-type wiring method. In most areas of the country, electricians use nonmetallic sheathed cable (Type NM), commonly called Romex™, to install residential electrical systems. In certain areas of the country, the authority having jurisdiction does not allow Type NM cable to be used in residential construction. If this is the case, armored-clad cable (Type AC) or metal-clad cable (Type MC) are great alternative wiring methods to Romex™. Rarely are residential electrical systems installed using a raceway wiring method. However, some local electrical codes require dwelling units to be wired using a raceway like electrical metallic tubing (EMT). We discuss raceway installation in a later chapter. In an effort to keep this book easier to understand for the new electrician, and because the vast majority of residential electrical system wiring is done using nonmetallic sheathed cable (Type NM), this chapter and subsequent chapters assume the use of Type NM cable as further discussion of the rough-in stage is presented.

REQUIREMENTS FOR ELECTRICAL INSTALLATIONS

Article 110 of the *NEC®* includes several requirements that apply to rough-in wiring. These requirements will need to be followed by electricians during the rough-in stage of a residential wiring job. The following paragraphs discuss in detail the Article 110 sections that must be considered.

Section 110.3(B) states that listed or labeled equipment must be installed and used in accordance with any instructions included in the listing or labeling. Manufacturers usually supply installation instructions with equipment for use by electrical contractors, electrical inspectors, and others concerned with an installation. It is important to follow the listing or labeling installation instructions when installing electrical equipment in the rough-in stage.

> **CAUTION**
>
> **CAUTION:** Never throw away the instructions that come with any piece of electrical equipment. According to Section 110.3(B), all electrical equipment must be installed according to the instructions that are included by the manufacturer of the equipment.

Section 110.7 addresses wiring integrity. It states that completed wiring installations must be free from short circuits and ground faults. Insulation is the material that prevents the short circuits and faults to ground. Failure of the insulation system is one of the most common causes of problems in residential electrical systems. The installing electrician must take care not to damage the conductor insulation in any way during the rough-in stage.

Section 110.11 covers **deteriorating agents** and states that, unless identified for use in the operating environment, no conductors or equipment can be located in damp or wet locations; be exposed to gases, fumes, vapors, liquids, or other agents that have a deteriorating effect on the conductors or equipment; or be exposed to excessive temperatures. When choosing a cable or conductor type, an electrical box style, or any other piece of electrical equipment used in a residential electrical system, an electrician must make sure that it is suitable for the location where you wish to install it. Otherwise, damage to the electrical equipment can occur that will cause problems—if not immediately, then down the road. Informational Note 2 of Section 110.11 tells us that some cleaning and lubricating compounds can cause severe deterioration of plastic materials used for insulating and structural applications in electrical equipment. Equipment identified as "dry locations," "NEMA Type 1," or "indoor use only" must be protected against permanent damage from the weather during building construction. This last sentence requires electricians to cover and protect any electrical equipment that is being used during the rough-in stage and that might be subject to rain or snow damage. This is especially true when electricians start to install the rough-in wiring and the doors and windows (and sometimes even the roof!) have not been installed yet.

Section 110.12 states that all electrical equipment must be installed in a neat and workmanlike manner. The "neat and workmanlike" installation requirement has appeared in the *NEC®* for more than 50 years. It stands as a basis for pride in one's work and helps make electrical work a profession and not just a "job." Electrical inspectors have cited many *NEC®* violations based on their interpretation of "neat and workmanlike manner." Many electrical inspectors use their own experience or common wiring practice in their local areas as the basis for their judgments. Examples of installations that do not qualify as "neat and workmanlike" include exposed runs of cables or raceways that are not properly supported and result in sagging between supports; field-bent and kinked, flattened, or poorly measured electrical conduit; or electrical boxes and enclosures that are not level or not properly secured.

Section 110.12(A) covers unused openings and states that any unused opening other than those for the operation of equipment or for mounting purposes must be closed. The material being used to cover the opening must be at least as strong as the material the electrical box is made of. A common piece of equipment used to meet this requirement is called a **knockout plug** and is inserted into any knockout opening that is open and not used (Figure 9-1).

Section 110.12(B) requires that any internal parts of electrical equipment not be damaged or contaminated by foreign materials such as paint, plaster, cleaners, abrasives, or corrosive residues. There must be no damaged parts that may adversely affect safe operation or mechanical strength of the equipment, such as parts that are broken, bent, cut, or deteriorated by corrosion, chemical action, or overheating. This rule will mean that after the rough-in wiring has been done, an electrician must look at the installed electrical equipment and determine if any of the equipment may have to be covered so that when the house finish work begins, no paint, plaster, or anything else could contaminate the insides of the electrical equipment. For example, a flush-mounted service panel located in an area of the house that will be finished off may require covering the enclosure so that contamination of the inside of the panel cannot take place by paint that is being applied to the walls with a spray gun.

FIGURE 9-1   Section 110.12(A) requires unused cable or raceway openings in electrical boxes to be effectively closed.

© Cengage Learning 2012.

WIRING METHODS

Article 300 of the *NEC®* contains several requirements for the wiring methods used by an electrician when installing a residential electrical system. Several of the more important requirements that electricians need to be aware of are covered in the following paragraphs.

Section 300.3(A) states that single conductors with an insulation type that is listed in *NEC®* Table 310.104(A) can be installed only as part of an *NEC®*-recognized wiring method. In other words, individual insulated conductors with, for example, a THHN insulation cannot be used in a wiring method that is not listed in the *NEC®*. Here is an example where an electrician probably wishes this rule was not in the *NEC®*. An electrician installs a switching circuit in a house and, after everything is installed, discovers that a two-wire cable was installed instead of the required three-wire cable. Some electricians believe that they are allowed to simply run one more individual insulated conductor in this situation to fix the problem. They are wrong! Section 300.3(A) does not allow one individual conductor to be run unless it is part of a recognized wiring method, like a Romex™ cable. In this case, a new three-wire cable would have to be installed.

Section 300.3(B) states that all conductors of the same circuit and, where used, the grounded conductor and all equipment-grounding conductors and bonding conductors must be contained within the same raceway, trench, cable, or cord. This is designed to eliminate **inductive heating**. By keeping all circuit conductors of an individual circuit grouped together, the magnetic fields around the conductors cancel each other out. This means that an expanding and collapsing magnetic field will not be present and will not cause the molecules in the metal enclosure to move around and produce heat, which could damage the conductor insulation (Figure 9-2).

Section 300.4 states that where a wiring method is subject to physical damage, conductors must be adequately protected. If a cable or raceway is going to be installed through a wood framing member, the following rules must be followed (Figure 9-3):

• Bored holes: Section 300.4(A)(1) states that in both exposed (like exposed studs in a garage) and concealed (wall studs inside with **sheetrock** on them)

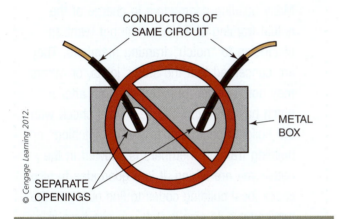

FIGURE 9-2 Section 300.3(B) requires all circuit conductors of an individual circuit to be grouped together and run through the same box opening to reduce inductive heating and to avoid increases in overall circuit impedance.

locations, where a cable or raceway type wiring method is installed through bored holes in joists, rafters, or wood members, holes must be bored so that the edge of the hole is not less than 1¼ inches (32 mm) from the nearest edge of the wood member. Where this distance cannot be maintained, the cable or raceway must be protected from penetration by screws or nails by a steel plate or bushing, at least ¹⁄₁₆ inch (1.6 mm) thick, and of appropriate length and width installed to cover the area of the wiring. *Exception No. 1* states that steel plates are not required to protect rigid metal conduit (RMC), intermediate metal conduit (IMC), rigid polyvinyl chloride conduit (PVC), or electrical metallic tubing (EMT) that is installed through a wood framing member. *Exception No. 2* to Section 300.4(A)(1) allows a listed and marked steel plate to be less than ¹⁄₁₆ inch (1.6 mm) thick if it still provides equal or greater protection against nail or screw penetration.

• Notches in wood: Section 300.4(A)(2) states that where there is no objection because of weakening the building structure, in both exposed and concealed locations, cables or raceways are permitted to be laid in notches in wood studs, joists, rafters, or other wood members where the cable or raceway at those points is protected against nails or screws by a steel plate at least ¹⁄₁₆ inch (1.6 mm) thick installed before the building finish is applied. *Exception No. 1* states that steel plates are not required to protect RMC, IMC, PVC, or EMT when these raceways are installed in a notch. *Exception No. 2* to Section 300.4(A)(2) allows a listed and marked steel plate

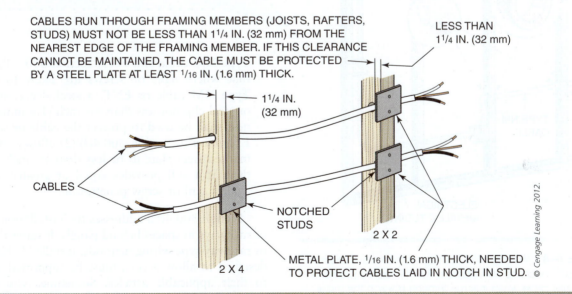

FIGURE 9-3 Cables or raceways installed through a wood framing member must be protected according to Section 300.4(A). The intent of this section is to prevent nails and screws from being driven into cables and raceways. An *Exception* to this section permits RMC, IMC, PVC, or EMT to be installed in wood framing members without additional protection.

to be less than 1/16 inch (1.6 mm) thick if it still provides equal or greater protection against nail or screw penetration.

The intent of Section 300.4(A)(1) is to prevent nails and screws from being driven into cables and raceways. Keeping the edge of a drilled hole 1¼ inches (32 mm) from the nearest edge of a stud should prevent nails from penetrating the wooden framing member far enough to injure a cable. Building codes limit the maximum size of bored or notched holes in studs, and Section 300.4(A)(2) indicates that consideration should be given to the size of notches in studs so they do not affect the strength of the structure. Most electricians will bore a hole in the framing member rather than notch it.

Sometimes, metal framing members are encountered in residential construction. Section 300.4(B) covers the use of Type NM cable and electrical nonmetallic tubing (ENT) through metal framing members. If a Type NM cable or ENT raceway is going to be installed through a metal framing member, the following rules must be followed (Figure 9-4):

- Type NM cable: Section 300.4(B)(1) states that in both exposed and concealed locations where Type NM cable passes through either factory- or field-punched, -cut, or -drilled slots or holes in metal members, the cable must be protected by listed bushings

or listed grommets covering all metal edges that are securely fastened in the opening prior to installation of the cable. The listed grommets or listed bushings must completely encircle Type NM cables as they pass through holes in metal studs. This requirement affords physical protection for Type NM as the cables are "pulled" through the openings in metal studs. Fastening the listed grommet or listed bushing in place prior to installing cable is mandatory.

> ### *from experience...*
> Many building contractors in charge of the actual framing of a house will not want an electrician to "notch" framing members. They are concerned that the studs, joists, or rafters may not have the required strength after a notch has been cut into it. Always check with the building contractor to see if "notching" of building framing members is allowed in the house you are wiring. It is a good idea to also check local building codes to find out what the maximum hole size is that you can bore in a building framing member. Be aware that some pre-manufactured roof truss systems are designed and engineered so that drilling them for cable runs is not allowed.

- Type NM cable and ENT: Section 300.4(B)(2) states that where nails or screws are likely to penetrate Type NM cable or ENT, a steel sleeve, steel plate, or steel clip not less than ¹⁄₁₆ inch (1.6 mm) in thickness shall be used to protect the cable or tubing. An *Exception* to Section 300.4(B)(2) allows a listed and marked steel plate to be less than ¹⁄₁₆ inch (1.6 mm) thick if it still provides equal or greater protection against nail or screw penetration.

Section 300.4(C) addresses the installation of cables or raceways in spaces behind panels. It states that cables or raceway-type wiring methods, installed behind panels designed to allow access, must be supported according to their applicable articles. Sometimes you will find yourself installing wiring above a suspended ceiling. This ceiling type is sometimes called a "dropped ceiling." Any cable- or raceway-type wiring methods installed above suspended ceilings with lift-up panels must not be allowed to lay on the suspended ceiling

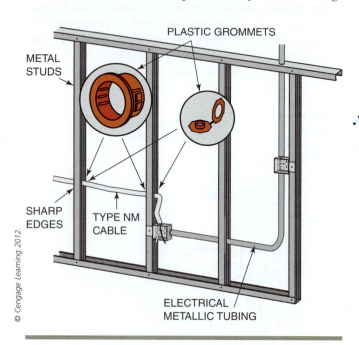

PLASTIC GROMMETS

METAL STUDS

SHARP EDGES

TYPE NM CABLE

ELECTRICAL METALLIC TUBING

© Cengage Learning 2012.

FIGURE 9-4 Where metal framing members are encountered, *NEC®* Sections 300.4(B) and 334.17 require protection for Type NM cable when it is run through holes in the metal framing members. This protection is provided by using listed bushings, or grommets, in the holes that are field punched or drilled by an electrician or holes that are provided by the manufacturer. The grommet must cover all metal edges of the hole.

panels or grid system. They are required to be supported according to Sections 300.11(A) and 300.23 and the requirements of the article applicable to the wiring method involved. This also applies to the installation of low-voltage cable for chime or thermostat wiring, telephone wiring, cable television wiring, or home computer network wiring. They are not permitted to block access to equipment above the suspended ceiling.

Section 300.4(D) covers the requirements for installing cables and raceways parallel to framing members and furring strips. In both exposed and concealed locations, where a cable- or raceway-type wiring method is installed along framing members (such as joists, rafters, studs, or furring strips), the cable or raceway must be installed and supported so that the nearest outside edge of the cable or raceway is not less than 1¼ inches (32 mm) from the nearest edge of the framing member or **furring strip** where nails or screws are likely to penetrate. Where this distance cannot be maintained, the cable or raceway shall be protected from penetration by nails or screws by a steel plate, sleeve, or equivalent at least ¹⁄₁₆ inch (1.6 mm) thick (Figure 9-5). *Exception No. 1* states that steel plates or sleeves are not required to protect RMC, IMC, PVC, or EMT where they have an edge that is less than 1¼ inches (32 mm) to the edge of the framing member. The intent of Section 300.4(D) is to prevent mechanical damage to cables and raceways from nails and screws. One way to do this is to fasten the cable or raceway so that it is at least 1¼ inches (32 mm) from the edge of the framing member, as illustrated in Figure 9-6. This requirement generally applies to exposed and concealed work. *Exception No. 1* permits the cable or raceway to be installed closer than 1¼ inches (32 mm) from the edge of the framing member if physical protection, such as a steel plate or a sleeve, is provided (a steel plate is illustrated in Figure 9-7). *Exception No. 2* states that for concealed work in a finished existing building, it is permissible to fish the cables between access points and not have the cable or raceway be at least 1¼ inches (32 mm) from the edge of a framing member. *Exception No. 3* to Section 300.4(D) allows a listed and marked steel plate to be less than ¹⁄₁₆ inch (1.6 mm) thick if it still provides equal or greater protection against nail or screw penetration.

Section 300.4(F) covers cables and raceways installed in shallow grooves. Cable- or raceway-type wiring methods installed in a groove and to be covered by wallboard, siding, paneling, carpeting, or similar finish must be protected by a steel plate or sleeve ¹⁄₁₆ inch (1.6 mm) thick or by not less than 1¼ inches (32 mm) free space for the full length of the groove in which the cable or raceway is installed. For example, an installation may require an electrician to groove out a solid wooden beam so that a Romex™ cable can be laid in it and run to a certain location (Figure 9-8). Before a wall

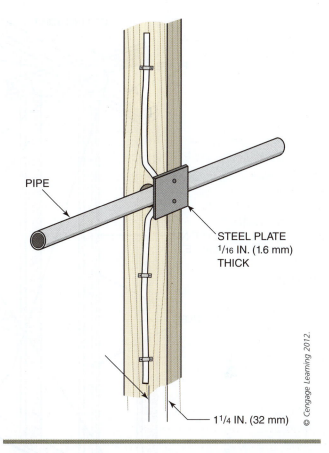

PIPE

STEEL PLATE
¹⁄₁₆ IN. (1.6 mm)
THICK

1¼ IN. (32 mm)

© Cengage Learning 2012.

FIGURE 9-5 Cables run parallel to framing members must have a clearance of at least 1¼ inches (32 mm) from the cable to the edge of the framing member. If it is not possible to maintain this clearance, a steel plate at least ¹⁄₁₆ inch (1.6 mm) thick must be installed.

covering can be placed over the beam with the cable in the groove, the groove must be covered with a metal plate at least ¹⁄₁₆ inch thick (1.6 mm). If the electrician makes the groove so that the cable sits at least 1¼ inches (32 mm) down into it, no steel plate would have to be used. *Exception No. 1* says that steel plates or sleeves are not required to protect RMC, IMC, PVC, or EMT used in a groove. *Exception No. 2* to Section 300.4(F) allows a listed and marked steel plate to be less than ¹⁄₁₆ inch (1.6 mm) thick if it still provides equal or greater protection against nail or screw penetration.

Section 300.10 covers the electrical continuity of metal raceways and enclosures. It states that metal raceways, cable armor, and other metal enclosures for conductors must be metallically joined together into a continuous electric conductor and must be connected to all boxes, fittings, and cabinets to provide effective electrical continuity. Unless specifically permitted elsewhere in the *NEC®*, raceways and cable assemblies must be mechanically secured to boxes, fittings, cabinets, and other enclosures. Section 250.4(A) states

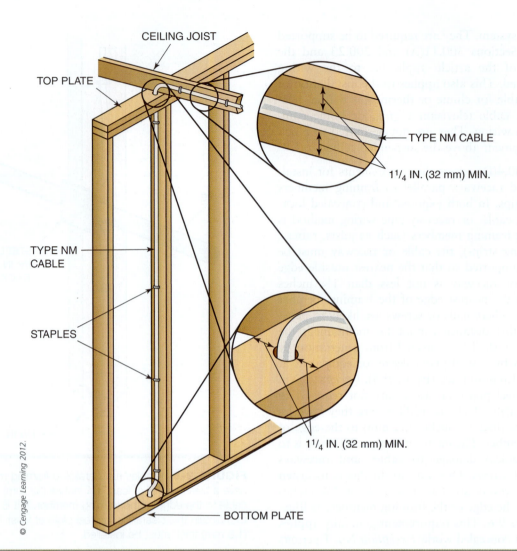

CEILING JOIST

TOP PLATE

TYPE NM CABLE

$1\frac{1}{4}$ IN. (32 mm) MIN.

TYPE NM
CABLE

STAPLES

$1\frac{1}{4}$ IN. (32 mm) MIN.

BOTTOM PLATE

© Cengage Learning 2012.

FIGURE 9-6 The intent of Section 300.4(D) is to prevent mechanical damage to cables and raceways from nails and screws. Cable is fastened to framing members so that it is at least 1¼ inches (32 mm) from the edge of the framing member. If the cable is installed closer than 1¼ inches (32 mm) from the edge of the framing member, physical protection, such as a steel plate or a sleeve, must be provided. An *Exception* to this section says that this requirement does not apply to RMC, PVC, IMC, or EMT wiring methods because these methods provide physical protection for the conductors.

what must be accomplished by grounding and bonding the metal parts of the electrical system. These metal parts must form an effective low-resistance path to ground in order to safely conduct any fault current and facilitate the operation of overcurrent devices protecting the enclosed circuit conductors. If an electrician is installing metal boxes during the rough-in stage, a wiring method that can provide a way to ground all the metal boxes must be used. Type NM cable with a bare grounding wire connected to each box with a green grounding screw is very common and works well to meet the intent of this section.

Section 300.11(A) states that raceways, cables, boxes, cabinets, and fittings must be securely fastened in place (Figure 9-9). The specific *NEC®* article that

covers a raceway or cable type will state the support requirements for that type of wiring method.

Section 300.11(B) states that raceways may be used as a means of support for other raceways, cables, or non-electric equipment only under the following conditions:

1 Where the raceway or means of support is identified for the purpose.

2 Where the raceway contains power supply conductors for electrically controlled equipment and is used to support Class 2 circuit conductors or cables that are solely for the purpose of connection to the equipment control circuits.

3 Where the raceway is used to support boxes or conduit bodies in accordance with Section 314.23 or to support luminaires (fixtures) in accordance with Article 410.

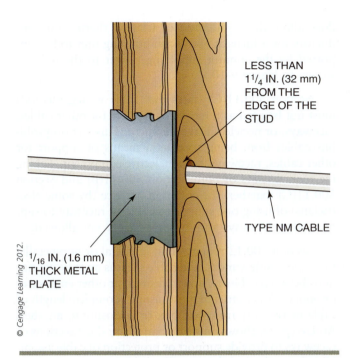

LESS THAN
1¼ IN. (32 mm)
FROM THE
EDGE OF THE
STUD

TYPE NM CABLE

1/16 IN. (1.6 mm)
THICK METAL
PLATE

FIGURE 9-7   A steel plate used to protect Type NM cable within 1¼ inches (32 mm) of the edge of a wood stud. A listed and marked steel plate less than 1/16 inch (1.6 mm) thick may be used if it still provides equal or better protection against nail or screw penetration.

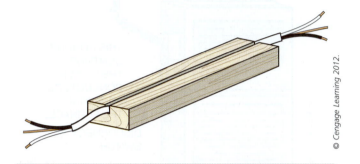

FIGURE 9-8   A groove can be cut in a solid wood framing member. After the cable has been laid in the groove, a metal plate not less than 1/16 inch (1.6 mm) thick must be installed over the groove to protect the cable. A listed and marked steel plate less than 1/16 inch (1.6 mm) thick may be used if it still provides equal or better protection against nail or screw penetration.

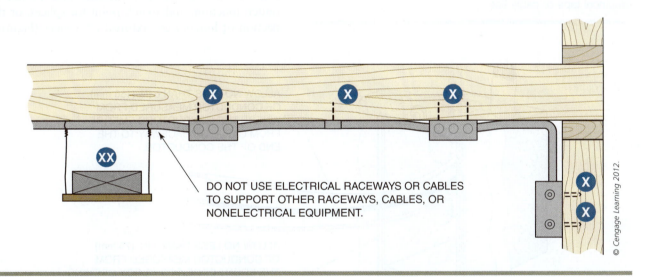

DO NOT USE ELECTRICAL RACEWAYS OR CABLES
TO SUPPORT OTHER RACEWAYS, CABLES, OR
NONELECTRICAL EQUIPMENT.

FIGURE 9-9   Section 300.11(A) requires all electrical boxes, raceways, and cable assemblies to be securely fastened in place (points X). Section 300.11(B) and (C) do not allow raceways or cables to support other raceways or cables, electrical boxes, or nonelectrical equipment (point XX).

The purpose of this section is to prevent cables from being attached to the exterior of a raceway. Electrical, telephone, and computer cables wrapped around a raceway can prevent dissipation of heat from the raceway and affect the temperature of the conductors inside.

This section also prohibits the use of a raceway as a means of support for nonelectric equipment, such as suspended ceilings, water pipes, nonelectric signs, and the like, which could cause a mechanical failure of the raceway (Figure 9-9). However, Section 300.11(B)(2)

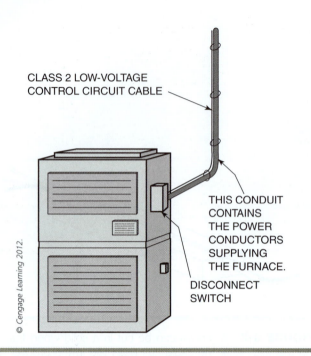

CLASS 2 LOW-VOLTAGE
CONTROL CIRCUIT CABLE

THIS CONDUIT
CONTAINS
THE POWER
CONDUCTORS
SUPPLYING
THE FURNACE.

DISCONNECT
SWITCH

© Cengage Learning 2012.

FIGURE 9-10   Section 300.11(B)(2) allows Class 2 control wiring to be supported by the raceway that supplies the power to the piece of equipment that the control wiring is associated with. A common residential wiring practice is to use the piece of EMT conduit that supplies a furnace to also provide support to the Class 2 thermostat control cable. The thermostat cable is usually secured to the raceway with electrical tape or cable ties.

does allow the installation of Class 2 thermostat conductors for a furnace or air-conditioning unit to be supported by the conduit supplying power to the unit, as shown in Figure 9-10.

Section 300.11(C) states that cable wiring methods must not be used as a means of support for other cables, raceways, or nonelectrical equipment. This section prohibits cables from being used as a means of support for other cables, raceways, or nonelectric equipment. Taking the requirements of both Section 300.11(B) and Section 300.11(C) together, the common practice (by some electricians) of using one supported cable or raceway to support other raceways and cables is clearly not allowed.

Section 300.12 requires that all metal or nonmetallic raceways, cable armors, and cable sheaths must be continuous between cabinets, boxes, fittings, or other enclosures or outlets. An electrician must install a complete length of cable or raceway from one box or enclosure to another. An *Exception* allows an electrician to use short sections of raceways to provide support or protection of cable assemblies from physical damage. Used this way, a raceway is not required to be mechanically continuous.

Section 300.14 covers a very important requirement: the length of free conductors at outlets, junctions, and switch points. At least 6 inches (150 mm) of free conductor, measured from the point in the box where it emerges from its raceway or cable sheath, must be left at each outlet, junction, and switch point for splices, or the connection of luminaires (fixtures) or devices (Figure 9-11).

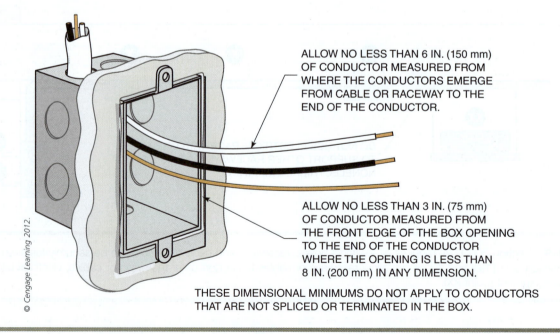

ALLOW NO LESS THAN 6 IN. (150 mm)
OF CONDUCTOR MEASURED FROM
WHERE THE CONDUCTORS EMERGE
FROM CABLE OR RACEWAY TO THE
END OF THE CONDUCTOR.

ALLOW NO LESS THAN 3 IN. (75 mm)
OF CONDUCTOR MEASURED FROM
THE FRONT EDGE OF THE BOX OPENING
TO THE END OF THE CONDUCTOR
WHERE THE OPENING IS LESS THAN
8 IN. (200 mm) IN ANY DIMENSION.

THESE DIMENSIONAL MINIMUMS DO NOT APPLY TO CONDUCTORS
THAT ARE NOT SPLICED OR TERMINATED IN THE BOX.

© Cengage Learning 2012.

FIGURE 9-11   When roughing in the wiring to electrical boxes mounted in walls and ceilings, Section 300.14 requires a minimum of 6 inches (150 mm) of conductor length at each location. No maximum length is stated, but good wiring practice dictates that electricians leave no more than 8 inches (200 mm). Too much conductor left in a box makes it more difficult to place connected receptacles and switches into the box.

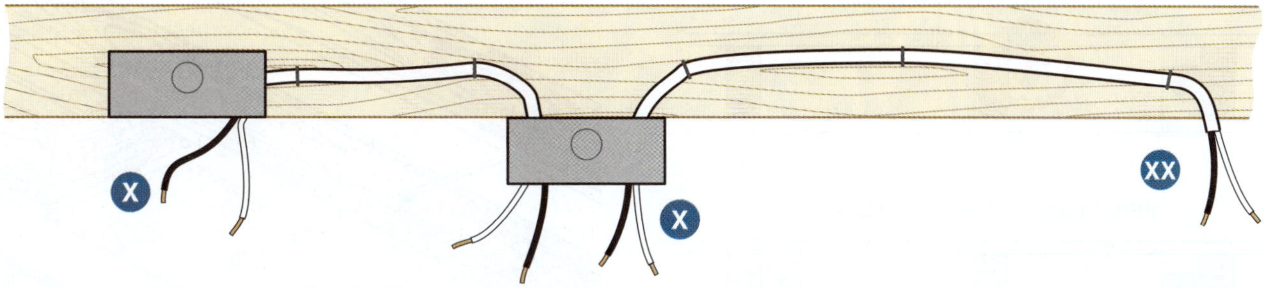

FIGURE 9-12 Section 300.15 requires that where the wiring method is conduit or cable, an electrical box must be installed at each conductor splice point, outlet point, switch point, junction point, or termination point as shown at points X. Point XX would be a code violation if no box were used.

Where the opening to an outlet, junction, or switch point is less than 8 inches (200 mm) in any dimension, each conductor must be long enough to extend at least 3 inches (75 mm) outside the opening. An *Exception* states that conductors that are not spliced or terminated at the outlet, junction, or switch point are not required to have the minimum 6 inches (150 mm) of conductor. This is the case when conductors installed in a raceway may go straight through one box to get to another box. This section is very specific about the amount of free conductor length required at each splice point or device outlet.

Section 300.15 states that where the wiring method is conduit, tubing, Type AC cable, Type MC cable, or Type NM cable, a box or conduit body must be installed at each conductor splice point, outlet point, switch point, junction point, termination point, or pull point. There are some wiring methods used where a box is not required because the wiring method provides interior access to the wires by design or a built-in box is provided. For all practical purposes, all receptacle outlets, switch locations, or junction locations in residential wiring will require that an electrical box be installed (Figure 9-12).

Section 300.21 covers the spread of fire or products of combustion. It states that electrical installations in hollow spaces, vertical shafts, and ventilation or air-handling ducts must be made so that the possible spread of fire or products of combustion (i.e., smoke) will not be substantially increased. Openings around electrical penetrations through fire-resistant-rated walls, partitions, floors, or ceilings must be fire-stopped using approved methods to maintain the fire-resistance rating (Figure 9-13). The intent of Section 300.21 is that cables and raceways must be installed through fire-rated walls, floors, or ceilings in such a manner that they do not contribute to the spread of fire or smoke. NFPA 221, Standard for Fire Walls and Fire Barrier Walls, defines fire-resistance rating as

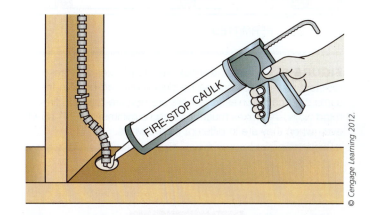

FIGURE 9-13 Section 300.21 requires all openings made to route cables or raceways through fire-rated walls and ceilings to be fire-stopped. This is to maintain the proper fire-resistance rating. Some state and local jurisdictions require that fire-rated as well as non-fire-rated penetrations be fire-stopped.

"the time, in minutes or hours, that materials or assemblies have withstood a fire exposure as established in accordance with the test procedures of NFPA 251, Standard Methods of Tests of Fire Endurance of Building Construction and Materials." The Informational Note to Section 300.21 points out that directories of electrical construction materials published by qualified testing laboratories contain many listing installation restrictions necessary to maintain the fire-resistive rating of assemblies where penetrations or openings are made. Building codes also contain restrictions on penetrations on opposite sides of a fire-resistant-rated wall assembly. An example is the 24 inches (600 mm) of minimum horizontal separation that applies between boxes installed on opposite sides of the wall (Figure 9-14).

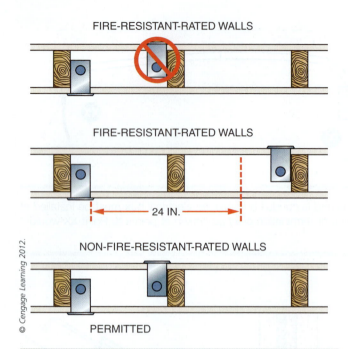

FIRE-RESISTANT-RATED WALLS

FIRE-RESISTANT-RATED WALLS

← 24 IN. →

NON-FIRE-RESISTANT-RATED WALLS

PERMITTED

© Cengage Learning 2012.

FIGURE 9-14 Back-to-back electrical boxes in a fire-resistant wall. Boxes cannot be placed back-to-back in the same stud cavity unless the walls are non-fire-rated. In fire-rated walls, the boxes must be placed a minimum of 24″ apart, even when they are in different stud cavities.

> **CAUTION**
>
> **CAUTION:** It is a good idea to always check with the authority having jurisdiction to determine which wiring method penetrations you have made will require fire-stopping. Some state and local inspectors will require both non-fire-rated and fire-rated penetrations to be fire-stopped.

Section 300.22 covers wiring in ducts, plenums, and other air-handling spaces. Usually, the information contained in this section is something that electricians doing commercial electrical work are more concerned with. However, Section 300.22(C) addresses wiring in spaces used for environmental air handling, such as those found in dwelling unit forced-hot-air furnace systems. A common practice of heating system installers in a house is to install sheet metal across the bottom of two joists in a basement and use this newly created space as a cold-air return to the furnace. The *Exception* to Section 300.22(C) states that this section does not apply to the joist or stud spaces of dwelling units where the wiring passes through such spaces perpendicular to the long dimension of such spaces. This permits cable to pass

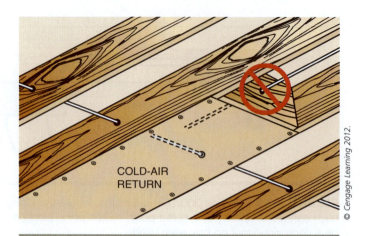

COLD-AIR RETURN

© Cengage Learning 2012.

FIGURE 9-15 A Section 300.22(C) *Exception* allows wiring to pass through in a direction that is perpendicular to the long dimension of air-handling spaces. This situation could present itself if the space between two joists is used as a cold-air return.

through joist or stud spaces of a dwelling unit, as illustrated in Figure 9-15. Equipment such as a junction box or device box is not permitted in this location.

GENERAL REQUIREMENTS FOR CONDUCTORS

Article 310 of the *NEC*® contains several sections that cover general requirements for the conductors that an electrician will be installing during the rough-in stage. It is essential that a residential electrician understand these requirements. The most important requirements are presented in the following paragraphs.

Section 310.106(A) addresses the minimum size of conductors. The minimum size of conductors is shown in *NEC*® Table 310.106(A). This table tells us that in the voltage range of 0 to 2000 volts, the minimum-size conductor allowed is 14 AWG copper and 12 AWG aluminum. Because dwelling unit voltages (120/240 volts) fall into this range, we can say that the smallest wire size allowed in residential wiring is 14 AWG copper or 12 AWG aluminum. Other sections of the *NEC*® allow the use of 16 or 18 AWG in residential wiring situations for such things as chime circuits, furnace thermostat circuits, lighting fixture wires, and flexible cords.

Section 310.10(F) states that conductors used for direct burial applications must be of a type identified for such use. If any of your rough-in wiring will involve installing conductors underground, the wiring method must be listed as suitable for use in an underground location. For example, if you are installing wiring for an outside-located pole light, Type NM cable could not be installed underground to feed the fixture. A wiring method listed for such use would have to be used. Type UF cable would be a good choice.

Section 310.10(D) addresses those locations exposed to direct sunlight. It states that insulated conductors and cables used where exposed to the direct rays of the sun must be of a type listed for sunlight resistance or listed and marked "sunlight resistant." Section 310.10(D)(2) allows tape or sleeving listed or marked as being sunlight resistant to be used on conductors and cables that are not sunlight resistant to make them so. Sometimes an electrician is required to install conductors outside a house where the wiring method is exposed to the direct rays of the sun; Type NM cable could not be used in this situation since it is not marked as sunlight resistant. Too much exposure to the direct rays of the sun will cause the insulation on Romex™ cable to deteriorate quickly. Again, a Type UF cable marked sunlight resistant would be a good choice in this application.

Section 310.15(A)(3) covers the temperature limitation of conductors. No conductor can be used in such a manner that its operating temperature exceeds that designated for the type of insulated conductor involved. In no case shall conductors be associated together in such a way with respect to type of circuit, the wiring method employed, or the number of conductors that the limiting temperature of any conductor is exceeded. Residential terminations are normally designed for 60°C (140°F) or 75°C (167°F) maximum temperatures. Therefore, the higher-rated ampacities for conductors of 90°C (194°F) cannot be used unless the terminals at which the conductors terminate have 90°C (194°F) ratings. Table 310.15(B)(2)(a) has ampacity correction factors for ambient temperatures above 30°C (86°F). The ambient temperature correction factor is applied to the ampacity found for a conductor with a specific insulation in Table 310.15(B)(16). The correction factor is also applied in addition to any adjustment factor, such as in Section 310.15(B)(3)(a). The information presented in this section was covered in Chapter 6.

from experience...

An electrician will often use Type NM cable from a partially used roll. Typically, the package that the cable originally came in, and that has information about the cable written on it, has been discarded. The conductor size, maximum rated voltage, and letter type for the cable will be clearly written on the sheathing.

Section 310.120(A) states that all conductors and cables must be marked to indicate the following information:

- The maximum rated voltage
- The proper type letter or letters for the type of wire or cable
- The manufacturer's name, trademark, or other distinctive marking by which the organization responsible for the product can be readily identified
- The AWG size or circular mil area
- Cable assemblies where the neutral conductor is smaller than the ungrounded conductors

This information is marked on the surface of the following conductors and cables: single-conductor (solid or stranded) insulated wire, nonmetallic sheathed cable, service entrance cable, and underground feeder and branch-circuit cable. The information is on a printed tag attached to the coil, reel, or carton for the following conductors and cables: Type AC cable and Type MC cable.

Section 310.110(A) states that insulated grounded conductors must be identified in accordance with Section 200.6. Section 200.6(A) requires a grounded conductor of 6 AWG or smaller to be identified by a white or gray color along its entire length. An alternative method of identification is described as "three continuous white stripes on other than green insulation along the conductor's entire length." An insulated grounded conductor larger than 6 AWG must be identified either by a continuous white or gray outer finish, by three continuous white stripes on other than green insulation along its entire length, or, at the time of installation, by a distinctive white marking at its terminations. This marking must encircle the conductor or insulation. The general rule of Section 200.6(B) requires the insulated conductors to be white or gray for their entire length or to be identified by three continuous white stripes along the entire length of the insulated conductor. Another permitted method for these larger conductors is applying a distinctive white marking in the field, such as white electrical tape. The tape is applied at the time of installation at all the grounded conductor termination points. If field-applied, the white marking must completely encircle the conductor in order to be clearly visible. This method of identification is shown in Figure 9-16.

Section 310.110(B) states that equipment-grounding conductor identification must be in accordance with Section 250.119. Equipment-grounding conductors are permitted to be bare or insulated. Individually insulated equipment-grounding conductors of 6 AWG or smaller must have a continuous outer finish that is either green or green with one or more yellow stripes. For equipment-grounding conductors larger than 6 AWG, it is permitted, at the time of installation, to be permanently identified as an equipment-grounding conductor at each

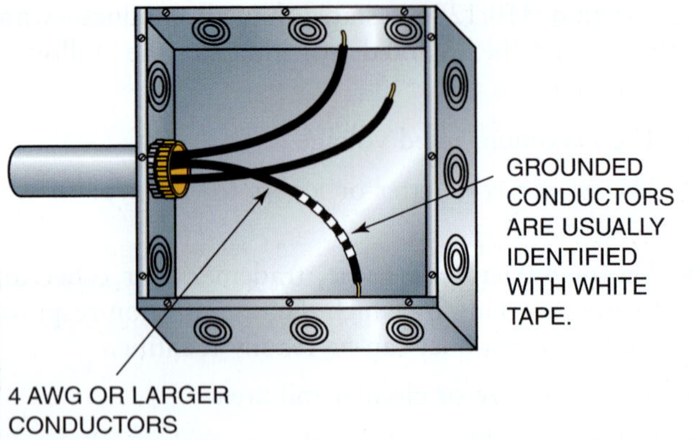

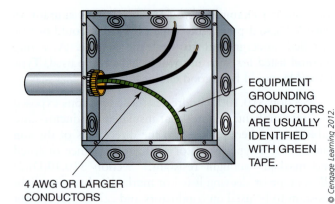

GROUNDED CONDUCTORS ARE USUALLY IDENTIFIED WITH WHITE TAPE.

4 AWG OR LARGER CONDUCTORS

FIGURE 9-16 The general rule of Section 200.6(B) requires insulated grounded conductors to be white or gray for their entire length or to be identified by three continuous white stripes along the entire length of the insulated conductor. The most often used method to identify grounded conductors larger than 6 AWG having an insulation that is not white or gray in color is to field-apply white marking tape at the time of installation at all the conductor termination points. If field applied, the white marking tape must completely encircle the conductor in order to be clearly visible.

EQUIPMENT GROUNDING CONDUCTORS ARE USUALLY IDENTIFIED WITH GREEN TAPE.

4 AWG OR LARGER CONDUCTORS

FIGURE 9-17 For equipment-grounding conductors larger than 6 AWG, it is permitted to be permanently identified as an equipment-grounding conductor at each end and at every point where the conductor is accessible. Identification must encircle the conductor. The usual way electricians reidentify is to use green tape that covers the entire exposed length of conductor.

end and at every point where the conductor is accessible. Identification must encircle the conductor and must be accomplished by one of the following (Figure 9-17):

- Stripping the insulation or covering from the entire exposed length
- Coloring the exposed insulation or covering green at the termination
- Marking the exposed insulation or covering with green tape at the termination

Conductors that are intended for use as ungrounded or "hot" conductors, whether used as a single insulated conductor or in a multi-conductor cable like Romex™, must be colored to be clearly distinguishable from grounded and grounding conductors. In other words, the ungrounded conductors can be identified with any color other than white, gray, or green. An *Exception* to Section 310.110(C) refers you to Section 200.7. Section 200.7(C)(1) states that ungrounded conductors with white or gray insulation in a cable are permitted if the conductors are permanently reidentified at termination points and if the conductor is visible and accessible. The normal method of reidentification is to use black-colored tape (Figure 9-18). Other applications where white conductors are permitted include flexible cords and circuits less than 50 volts. A white conductor used in single-pole, three-way, and four-way switch loops also requires reidentification (a color other than white, gray, or green) if it is used as an ungrounded conductor. Switching circuits are covered in Chapter 13.

TYPE NM-B

BLACK PLASTIC TAPE

WHITE WIRE REIDENTIFIED AS A "HOT" CONDUCTOR

FIGURE 9-18 A white insulated wire in a cable assembly can be reidentified for use as an ungrounded "hot" conductor by marking it with a piece of black electrical tape.

from experience...

Electricians typically use green marking tape to identify a grounding conductor. Make sure that the entire exposed length of conductor is reidentified with the tape. This requirement is slightly different than the reidentification of a grounded conductor in that as little as one wrap of white tape satisfactorily reidentifies a grounded conductor, while a grounding conductor must have its entire exposed length covered with green tape.

Table 310.104(A), Conductor Applications and Insulations, lists and describes the insulation types recognized by the *NEC®*. These conductors are permitted for use in any of the wiring methods recognized in Chapter 3 of the *NEC®*. This table also includes conductor applications and maximum operating temperatures. Some conductors have dual ratings. For example, Type XHHW is rated 90°C (194°F) for dry and damp locations and 75°C (167°F) for wet locations; Type THW is rated 75°C (167°F) for dry and wet locations, and 90°C (194°F) for special applications within electric-discharge (fluorescent) lighting equipment. Types RHW-2, XHHW-2, and other types identified by the suffix "-2" are rated 90°C (194°F) for wet locations as well as dry and damp locations.

Section 310.15 covers the ampacities for conductors rated 0 to 2000 volts. Table 310.15(B)(16) is referenced as the place to look for determining the ampacity of a conductor used in residential wiring. A detailed description of ampacity and how to determine a conductor's ampacity using Table 310.15(B)(16) was presented in Chapter 6. At this time, you should review the material in Chapter 6 that covers determining the ampacity of a conductor.

GENERAL REQUIREMENTS FOR ELECTRICAL BOX INSTALLATION

As we discussed earlier, installing electrical boxes is part of the rough-in stage for the wiring of a residential electrical system. Article 314 of the *NEC®* covers several requirements that an electrician must comply with when installing electrical boxes. Article 210 has several requirements for the actual location of receptacle outlets, lighting outlets, and switching locations. It is important for an electrician to understand the following sections that apply to electrical boxes.

INSTALLATION AND USE OF BOXES USED AS OUTLET, DEVICE, OR JUNCTION BOXES

Section 314.16 of the *NEC®* provides the guidelines for the calculation of the maximum number of conductors in outlet, device, and junction boxes. It states that boxes must be of sufficient size to provide free space for all enclosed conductors. In no case can the volume of the box, as calculated in Section 314.16(A), be less than the fill calculation, as calculated in Section 314.16(B). Calculating the maximum number of conductors in an electrical box is covered in detail in Chapter 10.

Section 314.17 states that any conductors entering electrical boxes must be protected from abrasion. This is true whether the box is metal or made of a nonmetallic material. Protection from abrasion can be accomplished by using bushings on sharp raceway ends, using connectors that are designed and built with a smooth opening for the conductors to go through, or simply by making sure that a short section of cable sheathing extends past the clamping mechanism of a cable clamp.

Section 314.17(B) applies to metal boxes. It states that where a raceway or cable is installed with metal boxes, the raceway or cable must be secured to such boxes. This is accomplished by using the proper cable or raceway connector. The connector may be an internal clamp or an external type of connector. Figure 9-19 shows internal cable clamps in both a metal device box and a metal octagon box.

CAUTION

CAUTION: Never install a Romex™ cable to a metal box by simply taking out the knockout and pushing the cable through the hole. Remember, there must always be a connector to secure the cable to the box.

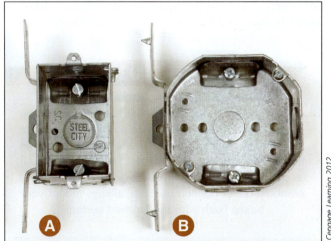

© Cengage Learning 2012.

FIGURE 9-19   (A) A metal device box and (B) octagon box with internal clamps. Section 314.17(B) requires cables and raceways to be attached to all metal electrical boxes. External or internal clamps can be used.

Section 314.17(C) applies to nonmetallic boxes. It says that nonmetallic boxes must be suitable for the lowest temperature-rated conductor entering the box. Where Type NM or Type UF cable is used, the sheath must extend not less than ¼ inch (6 mm) inside the box and beyond any cable clamp. In all instances, all permitted wiring methods must be secured to the boxes. The *Exception* to Section 314.17(C) states that where Type NM or Type UF cable is used with single-gang boxes not larger than 2¼ by 4 inches (57 by 100 mm) mounted in walls or ceilings and where the cable is fastened within 8 inches (200 mm) of the box measured along the sheath and where the sheath extends through a cable knockout not less than ¼ inch (6 mm), securing the cable directly to the box is not required. Multiple cable entries are permitted in a single-cable knockout opening (Figure 9-20). For nonmetallic boxes that are larger than 2¼ by 4 inches (57 by 100 mm), some type of cable-securing means is required (Figure 9-21). The requirement is based on the width of the box and the likelihood that the cable will be pushed back out of the box when the conductors and device, if any, are folded back into the box during installation of receptacles and switches.

Section 314.20 requires boxes that are installed in walls or ceilings with a surface of concrete, tile, gypsum (sheetrock), plaster, or other noncombustible material to be installed so that the front edge of the box will not be set back of the finished surface more than ¼ inch (6 mm). In walls and ceilings constructed of wood or other combustible surface material, boxes must be installed flush with the finished surface (Figure 9-22). For example, a wall constructed of gypsum board fastened to the face of wood studs is permitted to contain boxes set back or recessed not more than ¼ inch. Another example is a wall constructed of wood paneling fastened to the face of wood (or metal) studs; this requires that installed electrical boxes be mounted flush with the combustible finish.

> **CAUTION**
>
> **CAUTION:** Electrical inspectors may consider certain sheetrock, such as ⅜-inch and ½-inch, to be combustible. Check with the local inspector to make sure that it is okay for you to install electrical boxes that set back ¼ inch from the finished surface. When in doubt as to whether the wall or ceiling surface is combustible or noncombustible, always mount your boxes flush with the finished surface.

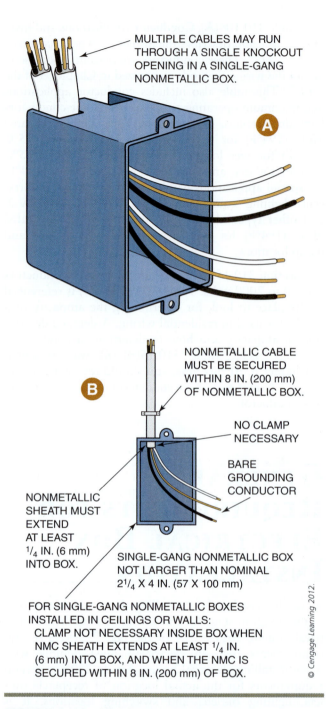

MULTIPLE CABLES MAY RUN THROUGH A SINGLE KNOCKOUT OPENING IN A SINGLE-GANG NONMETALLIC BOX.

A

B

NONMETALLIC CABLE MUST BE SECURED WITHIN 8 IN. (200 mm) OF NONMETALLIC BOX.

NO CLAMP NECESSARY

BARE GROUNDING CONDUCTOR

NONMETALLIC SHEATH MUST EXTEND AT LEAST ¼ IN. (6 mm) INTO BOX.

SINGLE-GANG NONMETALLIC BOX NOT LARGER THAN NOMINAL 2¼ X 4 IN. (57 X 100 mm)

FOR SINGLE-GANG NONMETALLIC BOXES INSTALLED IN CEILINGS OR WALLS: CLAMP NOT NECESSARY INSIDE BOX WHEN NMC SHEATH EXTENDS AT LEAST ¼ IN. (6 mm) INTO BOX, AND WHEN THE NMC IS SECURED WITHIN 8 IN. (200 mm) OF BOX.

© Cengage Learning 2012.

FIGURE 9-20 (A) A Section 314.17(C) *Exception* allows single-gang nonmetallic electrical boxes to have more than one cable installed in one knockout opening. (B) Nonmetallic sheathed cable must be secured within 8 inches (200 mm) of a single-gang nonmetallic box, and the sheathing must extend into the box at least ¼ inch (6 mm).

Section 314.23 addresses support of electrical boxes. 314.23(A) states that an electrical box mounted on a building or other surface must be rigidly and securely fastened in place. If the surface does not provide rigid and secure support, additional support must be provided. Although there is no *NEC®* rule to address it, it is a common wiring practice to use at least two screws, nails, or other fastening means to properly secure a box to any surface.

Section 314.23(B) says that a box supported from a structural member of a building must be rigidly supported either directly or by using a metal, polymeric, or wood brace (Figure 9-23). Section 314.23(B)(1) says that if nails or screws are used as a fastening means, side-mounting brackets on the outside of the electrical box should be used. Some electricians still use nails that are driven through holes inside an electrical box. This practice is allowed as long as the nails are within ¼ inch (6 mm) of the back, top, or bottom of the box (Figure 9-24). This requirement prevents the nails from interfering with the installation of switches and receptacles. If screws are used and pass through the box, some approved means must be used to cover the screw threads to protect the conductors from abrasion.

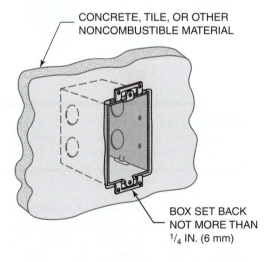

FIGURE 9-21   Nonmetallic electrical boxes, other than single gang, must have a way to secure a cable to the box. These boxes come from the manufacturer with internal clamps. Nonmetallic sheathed cable must be secured within 12 inches (300 mm) of this box type.

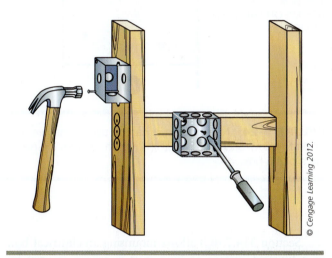

FIGURE 9-23   A building structural framing member must support electrical boxes. Attaching the box directly to the framing member or using a brace can accomplish this.

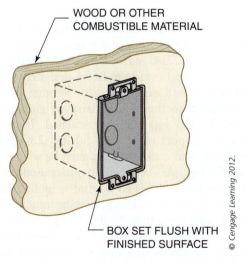

FIGURE 9-22   Section 314.20 requires electrical boxes installed in walls or ceilings with a surface of concrete, tile, gypsum, plaster, or other noncombustible material, to be installed so that the front edge of the box will not be set back of the finished surface more than ¼ inch (6 mm). In walls and ceilings constructed of wood or other combustible surface material, boxes must be installed flush with the finished surface.

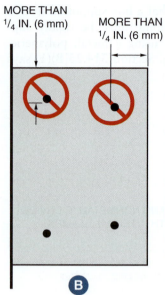

MAXIMUM ¼ IN. (6 mm)   MAXIMUM ¼ IN. (6 mm)

MORE THAN ¼ IN. (6 mm)   MORE THAN ¼ IN. (6 mm)

(A) CORRECT

THE NAILS OR SCREWS ARE KEPT WITHIN ¼ IN. (6 mm) FROM THE BACK AND/OR ENDS OF THE BOX. THIS ALLOWS ROOM FOR THE WIRING DEVICE.

(B) VIOLATION

THE NAILS OR SCREWS HAVE BEEN INSTALLED MORE THAN ¼ IN. (6 mm) FROM THE BACK AND/OR ENDS OF THE BOX. THESE NAILS OR SCREWS WILL INTERFERE WITH THE WIRING DEVICE. IT MAY ALSO BE IMPOSSIBLE TO INSTALL A WIRING DEVICE DUE TO THE INTERFERENCE OF THESE NAILS OR SCREWS.

A   B

© Cengage Learning 2012.

FIGURE 9-24 Nails or screws can be used to secure an electrical box to a building framing member. However, care must be taken to follow Section 314.23(B)(1), which requires the nails or screws to be located at least ¼ inch (6 mm) from the back and ends of the box.

Section 314.23(C) allows mounting an electrical box in a finished surface as long as it is rigidly secured by clamps, anchors, or fittings identified for the application. This wiring practice is used in remodel work (old work) where boxes are cut into existing walls. Figure 9-25 shows one example of an acceptable mounting method. More information on old-work wiring is included in later chapters.

Section 314.27 has some requirements that must be observed when installing lighting outlet boxes in walls or ceilings. Section 314.27(A) states that boxes used at luminaire (lighting fixture) outlets must be designed for the purpose. At every wall outlet used for lighting, the box must be designed so that a luminaire weighing a minimum of 50 lbs (23 kg) may be attached. If the box is designed to support more than 50 lbs (23 kg), the maximum weight that it can support must be written on the inside of the box. At every ceiling outlet used for lighting, the box must be designed so that a luminaire weighing a minimum of 50 lbs (23 kg) may be attached. If the luminaire weighs more than 50 lbs (23 kg), it must be independently supported unless the box is listed and marked for the maximum weight that it is designed to support. Metal octagon boxes and nonmetallic round nail-on ceiling boxes are used most often in residential wiring at light fixture locations. These boxes are designed so that the 8-32 size screws used with these box types will allow attachment of a lighting fixture to the box (Figure 9-26). Device boxes, such as a 3- by

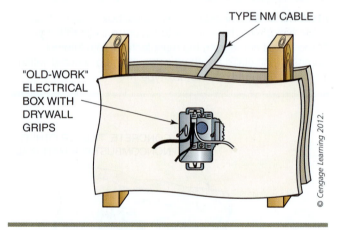

TYPE NM CABLE

"OLD-WORK" ELECTRICAL BOX WITH DRYWALL GRIPS

© Cengage Learning 2012.

FIGURE 9-25 Section 314.23(C) allows an electrical box to be mounted in an existing wall or ceiling by using clamps, anchors, or other fittings identified for the application.

2- by 3½-inch metal box or a single-gang plastic nail-on box, are designed to have only devices like switches or receptacles attached to them. However, Section 314.27(A)(1) *Exception* does allow a wall-mounted luminaire weighing not more than 6 pounds (3 kg) to be permitted to be supported on a device box (Figure 9-27) or on plaster rings that are secured to other boxes (like a 4-inch-square box), provided the luminaire or its supporting yoke is secured to the box with no fewer than two 6-32 or larger screws.

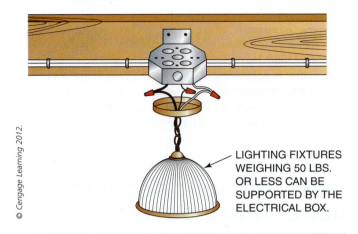

© Cengage Learning 2012.

LIGHTING FIXTURES WEIGHING 50 LBS. OR LESS CAN BE SUPPORTED BY THE ELECTRICAL BOX.

FIGURE 9-26 Section 314.27 states that electrical boxes used to support a luminaire must be designed specifically for that purpose. A lighting fixture that weighs more than 50 pounds (23 kg) cannot rely on just the electrical box for support and must be independently supported.

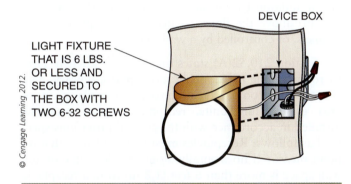

DEVICE BOX

LIGHT FIXTURE THAT IS 6 LBS. OR LESS AND SECURED TO THE BOX WITH TWO 6-32 SCREWS

© Cengage Learning 2012.

FIGURE 9-27 Device boxes are not specifically designed to support luminaires. However, a luminaire of no more than 6 pounds (3 kg) can be supported by a device box when at least two 6-32 screws are used.

CAUTION

CAUTION: Make sure to follow the installation instructions that come with a lighting fixture, and support the fixture in such a way that the weight of the fixture is adequately supported.

Section 314.27(B) states that boxes listed specifically for floor installation must be used for receptacles located in the floor. No other box type can be

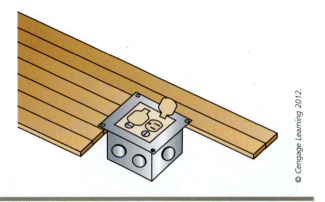

© Cengage Learning 2012.

FIGURE 9-28 Floor boxes must be designed for the purpose. Regular device boxes, square boxes, or octagon boxes cannot be installed in a floor of a dwelling unit.

used in a floor installation. Make sure to install only boxes that are specifically designed for floor installation (Figure 9-28).

CAUTION

CAUTION: Regular electrical device boxes are not suitable for installation in a floor. Only boxes that are listed for floor installation can be used. Floor boxes tend to be expensive and harder to install than a "regular" box. Try to avoid installing floor receptacles that require a special box whenever possible. Always try to install the box in a wall.

Section 314.27(C) tells us that where a box is used as the sole support for a ceiling-suspended paddle fan, the box must be listed for the application, be marked by the manufacturer as suitable for this purpose, and must not support paddle fans that weigh more than 70 pounds (32 kg). If an outlet box is designed to support a paddle fan that weighs more than 35 pounds (16 kg), the box must have the maximum weight it can support marked on it. Outlet boxes specifically listed to adequately support ceiling-mounted paddle fans are available, as are several alternative and retrofit methods that can provide suitable support for a paddle fan (Figure 9-29). Ceiling-suspended paddle fan installation is covered in greater detail later in this textbook.

The last section covered here is Section 314.29, which states that boxes must be installed so that the wiring contained in them can be rendered accessible without removing any part of the building. A box is permitted to be used at any point for the connection of conduit, tubing, or cable, provided it is not rendered inaccessible. See Article 100 for the definition of "accessible" (as applied to wiring methods).

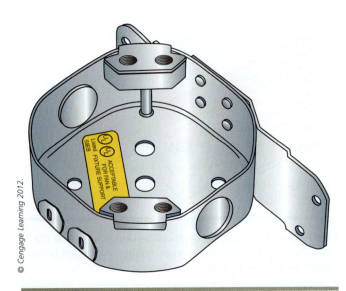

© Cengage Learning 2012.

FIGURE 9-29 A box that is designed for support of a ceiling-suspended paddle fan. Special "beefed-up" mounting brackets to attach the box to a building framing member and larger tapped holes for attaching a ceiling-suspended paddle fan to the box are included.

CAUTION

CAUTION: Never install an electrical box in a location that is accessible during the rough-in stage but is rendered inaccessible once wall and ceiling materials have been installed. Remember, all electrical boxes that contain conductors must be accessible.

DWELLING UNIT REQUIRED RECEPTACLE OUTLETS

Section 210.52 of the *NEC*® tells an electrician where receptacle outlets must be installed in a dwelling unit. This information is very important for the electrician to know so that electrical boxes installed during the rough-in stage are located to meet or exceed the requirements of this section. The requirements of Section 210.52 apply to dwelling unit receptacles that are rated 125-volts and 15- or 20-amperes and that are not part of a luminaire or an appliance. These receptacles are normally used to supply lighting and general-purpose electrical equipment and are in addition to the ones that are more than 5½ feet (1.7 m) above the floor, located within cupboards and cabinets, or controlled by a wall switch.

Section 210.52(A) states that in every kitchen, family room, dining room, living room, parlor, library, den, sunroom, bedroom, recreation room, or similar room or area of dwelling units, receptacle outlets must be installed in accordance with the general provisions specified as follows: Receptacles must be installed so that no point measured horizontally along the floor line in any wall space is more than 6 feet (1.8 m) from a receptacle outlet. This means that electrical boxes for receptacles must be installed during the rough-in stage so that no point in any wall space is more than 6 feet (1.8 m) from a receptacle. This rule means that an appliance or lamp with a flexible cord attached may be placed anywhere in the room near a wall and be within 6 feet (1.8 m) of a receptacle (Figure 9-30). This required placement of receptacles will eliminate the need for long extension cords running all over the place.

As used in this section, a wall space includes the following:

- Any space 2 feet (600 mm) or more in width (including space measured around corners) and unbroken along the floor line by doorways, fireplaces, and similar openings (Figure 9-31): Isolated, individual wall spaces 2 feet (600 mm) or more in width are considered usable for the location of a lamp or appliance, and a receptacle outlet is required to be provided.

- The space occupied by fixed panels in exterior walls, excluding sliding panels: Fixed panels in exterior walls, such as the fixed glass section of a sliding

from experience...

Another thing to consider is that homeowners often replace an existing ceiling lighting fixture with a combination ceiling-suspended paddle fan/lighting fixture. When they do this, separate wall switching of the paddle fan and the lighting fixture is desirable. In anticipation of this happening, many electricians install the necessary wiring for separate switching from a switch box to the paddle fan/lighting fixture box during the rough-in stage. Section 314.27(C) requires that whenever wiring is roughed-in to a ceiling lighting outlet box to provide multiple switching in case a ceiling suspended paddle fan/lighting fixture is someday installed at that location, the outlet box should be a type listed for the sole support of a ceiling-suspended paddle fan. This is true whether a paddle fan is ever actually installed or not. If a listed ceiling-suspended paddle fan box was not installed and a homeowner attached a ceiling-suspended paddle fan/lighting fixture to a "regular" box, the whole thing could come crashing down and seriously injure someone.

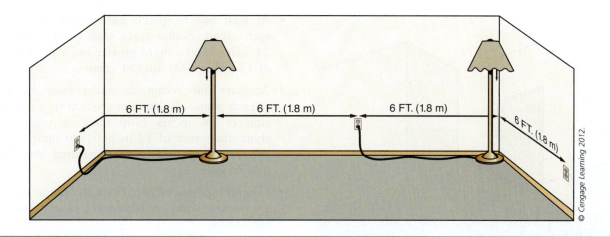

FIGURE 9-30 A good way to understand the placement of dwelling unit receptacles along a wall is to make sure that any piece of electrical equipment with a power cord that is 6 feet (1.8 m) long can be placed anywhere in a room and still be able to reach a receptacle.

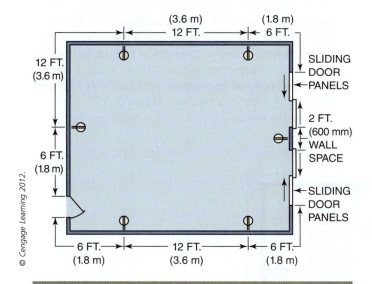

FIGURE 9-31 A typical receptacle layout in a dwelling unit room that meets the requirements of Section 210.52(A).

FIGURE 9-32 Fixed panels, like the part of a sliding glass door that does not move, are considered wall space.

glass door, are counted as regular wall space. A floor-type receptacle installed no more than 18 inches (450 mm) from the wall can be used if the spacing requirements might require a receptacle at the location of a glass fixed panel (Figure 9-32).

- The space afforded by fixed room dividers, such as freestanding bar-type counters or railings: Fixed room dividers, such as bar-type counters and railings, are to be included in the 6-foot (1.8 m) measurement (Figure 9-33).

Section 210.52(C) covers the required location of receptacles at countertop locations in a dwelling unit kitchen or dining room. This information is extremely important for the electrician because electrical boxes for the receptacles will have to be installed and wire run to them before any of the kitchen cabinets and countertop have been installed. The correct placement of the electrical boxes during the rough-in stage is imperative so that electrical boxes do not end up hidden behind cabinets or other kitchen equipment. The following requirements must be met when installing boxes for receptacle outlets to serve countertops in a kitchen or dining room (Figure 9-34):

- A receptacle outlet must be installed at each wall counter space that is 12 inches (300 mm) or wider. Receptacle outlets must be installed so that no point along the wall line is more than 24 inches (600 mm) measured horizontally from a receptacle outlet in that space.

FIGURE 9-33 Fixed room dividers, like the railing shown here, are considered to be wall space and must meet the requirements of Section 210.52(A).

- At least one receptacle outlet must be installed at each island counter space with a long dimension of 24 inches (600 mm) or greater and a short dimension of 12 inches (300 mm) or greater.

- At least one receptacle outlet must be installed at each peninsular counter space with a long dimension of 24 inches (600 mm) or greater and a short dimension of 12 inches (300 mm) or greater. A peninsular countertop is measured from the connecting edge.

- Countertop spaces separated by range tops, refrigerators, or sinks are considered as separate countertop spaces.

- Receptacle outlets must be located on or above, but not more than 20 inches (500 mm) above, the countertop. On island and peninsular countertops where the countertop is flat across its entire surface (no backsplashes, dividers, and so on) and there is no way to mount a receptacle within 20 inches (500 mm) above the countertop, receptacle outlets are permitted to be mounted not more than 12 inches (300 mm) below the countertop. However, receptacles mounted below a countertop cannot be located where the countertop extends more than 6 inches (150 mm) beyond its support base, such as at a bar-type eating area in a kitchen (Figure 9-35).

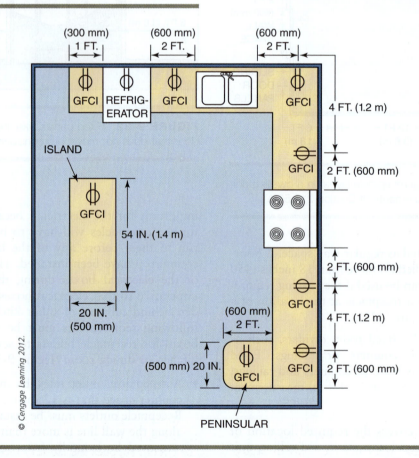

FIGURE 9-34 Dwelling unit receptacles serving countertop spaces in a kitchen and installed in accordance with Section 210.52(C).

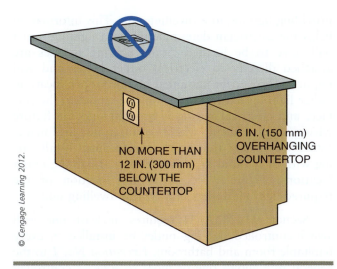

FIGURE 9-35 Receptacles cannot be installed face-up in a countertop. However, the receptacles required by Section 210.52(C) can be mounted below countertops as long as they are located no lower than 12 inches (300 mm). If the countertop has an overhanging portion that is more than 6 inches (300 mm), receptacles are not allowed under that area.

CAUTION

CAUTION: According to Section 406.5(E), receptacles cannot be installed in a face-up position in a countertop. Receptacles installed in a face-up position could collect crumbs, liquids, and other debris, resulting in a potential fire or shock hazard. However, a new addition to Sections 210.52(C)(5) and 210.52(D) in the 2011 NEC states that receptacle outlet assemblies that are listed for the application are permitted to be installed in a countertop.

Section 210.52(D) requires one wall receptacle in each bathroom of a dwelling unit to be installed adjacent and within 36 inches (900 mm) of the outside edge of the basin (Figure 9-36). This receptacle is required in addition to any receptacle that may be part of any luminaire or medicine cabinet. If there is more than one basin, a receptacle outlet is required adjacent to each basin location. If the basins are in close proximity, one duplex receptacle outlet installed between the two basins will satisfy this requirement (Figure 9-37). The receptacle is allowed to be installed no more than 12 inches (300 mm) below the countertop on the side or face of the basin cabinet instead of on the wall adjacent to the basin. It can also be installed on the countertop.

Section 210.52(E) states that for a one-family dwelling and each unit of a two-family dwelling that is at grade level, at least one receptacle outlet accessible at grade level and not more than 6.5 feet (2.0 m) above

FIGURE 9-36 At least one receptacle must be installed within 36 inches (900 mm) of a bathroom basin.

FIGURE 9-37 If a bathroom has more than one basin, receptacles can be placed within 36 inches (900 mm) of each basin or one receptacle may be placed so that it is within 36 inches (900 mm) of either basin.

grade must be installed at the front and back of the dwelling (Figure 9-38). Balconies, decks, and porches that are attached to a house and are accessible from inside the house need to have at least one receptacle outlet installed that is accessible from the balcony, deck, or porch. This rule helps eliminate the use of extension cords running through doors or windows to provide power to appliances or decorations located on a balcony, deck, or porch. The receptacle must be located no more than 6½ feet (2.0 m) above the balcony, deck, or porch.

Section 210.52(F) requires at least one receptacle outlet to be installed for the laundry. Remember that a 20-ampere branch circuit, which can have no other outlets on the circuit, supplies the laundry receptacle outlet(s).

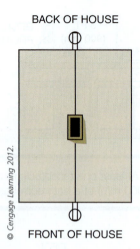

BACK OF HOUSE

FRONT OF HOUSE

© Cengage Learning 2012.

FIGURE 9-38 Receptacles are required on the front and back of a one-family dwelling unit as well as each unit of a two-family dwelling. They must be located no more than 6½ feet (2 m) above grade.

Section 210.52(G) states that for a one-family dwelling, at least one receptacle outlet, in addition to any provided for specific equipment (like laundry equipment), must be installed in each basement. At least one receptacle is also required to be installed in each attached garage and in each detached garage or accessory building (like a storage shed) with electric power. Where a portion of the basement is finished into one or more habitable rooms, each separate unfinished portion must have a receptacle outlet installed in accordance with this section.

Section 210.52(H) requires that in dwelling units, hallways of 10 feet (3.0 m) or more in length must have at least one receptacle outlet. In determining the hallway length, use the measured length along the centerline of the hall without passing through a doorway.

Section 210.52(I) requires a receptacle be installed in each wall space 3 feet (900 mm) or more in width in a **foyer** that has an area of more than 60 square feet. If the foyer is part of a hallway, then Section 210.52(H) applies. Very few houses have a foyer. It is simply an entranceway or transitional space from the exterior to the interior of a house.

DWELLING UNIT REQUIRED LIGHTING OUTLETS

During the rough-in stage, electrical boxes will have to be installed for the lighting outlets required in a house. Section 210.70 contains the minimum requirements for

providing lighting in a dwelling unit. This information helps the electrician determine where a lighting outlet will have to be located. Some lighting fixtures are attached directly to electrical outlet boxes and will require a box to be mounted at the proper location. Other lighting fixtures are simply mounted to the surface, and electrical wiring is brought into the fixture wiring compartment where connections are made. Either way means that an electrician must install wiring with or without an electrical box at lighting outlet locations. Figure 9-39 shows the location of the required lighting outlets in a typical dwelling unit.

Section 210.70(A)(1) requires at least one wall switch-controlled lighting outlet be installed in every habitable room and bathroom. *Exception No. 1* to the general rule allows one or more receptacles controlled by a wall switch to be permitted instead of lighting outlets, but only in areas other than kitchens and bathrooms. A wall switch-controlled lighting outlet is required in the kitchen and bathroom. A receptacle outlet controlled by a wall switch is not permitted to serve as a lighting outlet in these rooms. *Exception No. 2* allows lighting outlets to be controlled by occupancy sensors that are (1) in addition to wall switches or (2) located at a customary wall switch location and equipped with a manual override that will allow the sensor to function as a wall switch.

Section 210.70(A)(2) lists three additional locations where lighting outlets need to be installed:

- At least one wall switch-controlled lighting outlet must be installed in hallways, stairways, attached garages, and detached garages with electric power.

- In attached garages and detached garages with electric power, at least one wall switch-controlled lighting outlet must be installed to provide illumination on the exterior side of outdoor entrances or exits with grade-level access. A vehicle door in a garage is not considered as an outdoor entrance or exit.

- Where one or more lighting outlet(s) are installed for interior stairways, there must be a wall switch at each floor level and at each landing level that includes an entryway to control the lighting outlet(s) where the stairway between floor levels has six risers or more.

An *Exception* states that in hallways, stairways, and at outdoor entrances, remote, central, or automatic control of lighting is permitted.

Section 210.70(A)(3) addresses storage or equipment spaces. It says that for attics, crawl spaces, utility rooms, and basements, at least one lighting outlet containing a switch or controlled by a wall switch must be

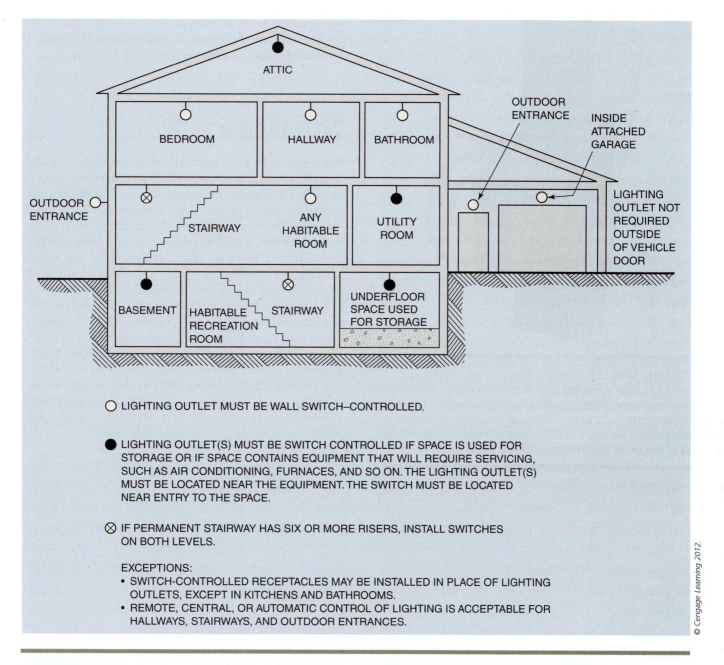

ATTIC

BEDROOM   HALLWAY   BATHROOM

OUTDOOR
ENTRANCE

INSIDE
ATTACHED
GARAGE

OUTDOOR
ENTRANCE

STAIRWAY

ANY
HABITABLE
ROOM

UTILITY
ROOM

LIGHTING
OUTLET NOT
REQUIRED
OUTSIDE
OF VEHICLE
DOOR

BASEMENT   HABITABLE
RECREATION
ROOM

STAIRWAY

UNDERFLOOR
SPACE USED
FOR STORAGE

○ LIGHTING OUTLET MUST BE WALL SWITCH–CONTROLLED.

● LIGHTING OUTLET(S) MUST BE SWITCH CONTROLLED IF SPACE IS USED FOR
STORAGE OR IF SPACE CONTAINS EQUIPMENT THAT WILL REQUIRE SERVICING,
SUCH AS AIR CONDITIONING, FURNACES, AND SO ON. THE LIGHTING OUTLET(S)
MUST BE LOCATED NEAR THE EQUIPMENT. THE SWITCH MUST BE LOCATED
NEAR ENTRY TO THE SPACE.

⊗ IF PERMANENT STAIRWAY HAS SIX OR MORE RISERS, INSTALL SWITCHES
ON BOTH LEVELS.

EXCEPTIONS:
• SWITCH-CONTROLLED RECEPTACLES MAY BE INSTALLED IN PLACE OF LIGHTING
OUTLETS, EXCEPT IN KITCHENS AND BATHROOMS.
• REMOTE, CENTRAL, OR AUTOMATIC CONTROL OF LIGHTING IS ACCEPTABLE FOR
HALLWAYS, STAIRWAYS, AND OUTDOOR ENTRANCES.

FIGURE 9-39   The required lighting outlets in a dwelling unit according to Section 210.70(A).

installed where these spaces are used for storage or contains equipment requiring servicing. At least one point of control must be at the door to these spaces. The lighting outlet must be provided at or near equipment, like furnaces or other heating, ventilating, and air-conditioning equipment that requires servicing. Installation of lighting outlets in attics, crawl spaces, utility rooms, and basements is required when these spaces are used for storage of items such as holiday decorations or luggage.

COMMUNICATIONS OUTLET

Section 800.156 requires a minimum of one communications outlet (Figure 9-40) to be installed in all newly constructed dwellings units. The communication cabling must be run to the location where the telephone company brings in service from the street. This is called the

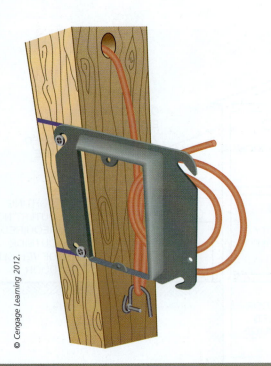

FIGURE 9-40 At least one communication outlet is required inside a dwelling unit. A common wiring practice is to secure a raised plaster ring to a framing member and run the communication cable to it. Because of the low amount of electrical power available on a communication circuit, an electrical box is not required but could be used.

demarcation point and is usually at a weatherproof box on the side of the house. Communications wiring is covered in detail in Chapter 16.

Summary

Before an electrician starts the rough-in stage of a residential wiring job, certain *NEC*® requirements must be considered. During the rough-in stage, these requirements are then applied. After completing the rough-in stage, an electrical inspector will determine if you correctly met the *NEC*® requirements with your installation. At this point in the installation of the wiring system, no ceiling or wall coverings have been installed, and the inspector will have easy access to the boxes and wire installed by the electrician. This chapter covered several common *NEC*® general requirements that must be considered when wiring a house. Also discussed and presented were several requirements that pertain to the conductors and electrical boxes installed during the rough-in stage. Understanding and following the *NEC*® requirements presented in this chapter will result in the rough-in stage wiring being free of code violations.

REVIEW QUESTIONS

Directions: Answer the following items with clear and complete responses.

1. Define the term "wiring" as it is used by electricians doing residential work.

2. Describe the "rough-in" stage of a residential wiring installation.

3. Name the most common wiring method used in residential wiring and which *NEC®* article covers this wiring method.

4. The *NEC®* requires electricians to install the electrical system in a "neat and workmanlike manner." Describe what this means and give an example of something not installed in a neat and workmanlike manner.

5. Name the table in the *NEC®* that lists the insulation types for conductors used in residential wiring.

6. When boring holes in a wooden framing stud for the installation of Romex™ cable, electricians must make sure that the distance from the edge of the hole to the face of the stud is no less than _____. If this distance cannot be maintained, a metal plate at least _____ thick must be used to protect the cable. Name the *NEC®* section that covers this wiring situation.

7. The minimum amount of free conductor required to be available at an electrical box used in residential wiring for termination purposes is _____. As long as the box has no dimension larger than 8 inches (200 mm), there must be at least _____ of free conductor extending out from the box opening. Name the *NEC®* section that covers this wiring practice.

8. Openings around electrical penetrations through fire-resistant-rated walls, partitions, floors, or ceilings must be fire-stopped using approved methods to maintain the fire-resistance rating. Explain why fire-stopping is required.

9. The minimum size of wire used in residential wiring is _____ AWG. (general rule)

10. The *NEC®* recognizes two wire sizes that are smaller than the answer to question 9. List the two sizes and give an example of where they might be used in a residential wiring system.

11. Name a cable wiring method that would be a good choice for an installation that would result in the cable being located outdoor where the direct rays of the sun could shine on it.

12. Name four items that must be written on the sheathing of a nonmetallic sheathed cable, according to the *NEC®*.

13. Describe how a 6 AWG or smaller insulated grounded conductor must be identified.

14. Describe how an insulated grounded conductor larger than 6 AWG is identified.

15. Describe how a white insulated conductor in a cable assembly is reidentified as a "hot" conductor.

16. Describe how a 6 AWG or smaller grounding conductor is identified.

17. Describe how a conductor that is larger than 6 AWG can be identified as a grounding conductor.

18. Conductors that are intended for use as ungrounded or "hot" conductors, whether used as a single insulated conductor or in a multi-conductor cable like Romex™, must be colored to be clearly distinguishable from grounded and grounding conductors. Describe how the identification is accomplished.

19. Explain what is required by Section 110.3(B) of the *NEC*®.

20. Name the common piece of electrical equipment that electricians use to insert into any knockout opening that is open and not used. Name the *NEC*® section that covers this wiring practice.

21. In single-gang nonmetallic boxes, the sheathing of a Romex™ cable must extend at least _____ into the box and be visible. Romex™ must also be secured no more than _____ from this same single-gang nonmetallic box.

22. A wall constructed of sheetrock fastened to the face of wood studs is permitted to contain boxes set back or recessed not more than _____. Name the *NEC*® section that covers this requirement.

23. Some electricians use nails that are driven through holes inside an electrical box to attach the box to a wall stud. This practice is allowed as long as the nails are within _____ of the back, top, or bottom of the box. Name the *NEC*® section that covers this requirement.

24. A wall-mounted luminaire (fixture) weighing not more than _____ is permitted to be supported on a device box provided the luminaire or its supporting yoke is secured to the box with no fewer than two 6-32 or larger screws. Name the *NEC*® section that covers this requirement.

25. The 8-32 screws used with a lighting outlet box can support a lighting fixture to the box as long as the fixture weighs no more than _____. Name the *NEC*® section that covers this requirement.

26. Switch-controlled receptacles, instead of lighting outlets, can be used in all areas of a dwelling unit except _____ and _____.

27. Receptacles located more than _____ above the floor are not counted in the required number of receptacles along a wall.

28. The maximum distance between wall receptacles in a dwelling unit is _____.

29. The maximum distance between countertop receptacles in a dwelling unit kitchen is _____.

30. Interior stairway lighting must be controlled by a wall switch placed at the top and bottom of the stairway when the stairway has _____ or more steps.

CHAPTER 10

Electrical Box Installation

OBJECTIVES

Upon completion of this chapter, the student should be able to:

- Select an appropriate electrical box type for a residential application.

- Size electrical boxes according to the *NEC®*.

- Demonstrate an understanding of the installation of metal and nonmetallic electrical device boxes in residential wiring situations.

- Demonstrate an understanding of the installation of lighting outlet and junction boxes in residential wiring situations.

- Demonstrate an understanding of the installation of electrical boxes in existing walls and ceilings.

GLOSSARY OF TERMS

accessible (as applied to wiring methods) capable of being removed or exposed without damaging the building structure or finish, or not permanently closed in by the structure or finish of the building

box fill the total space taken up in an electrical box by devices, conductors, and fittings; box fill is measured in cubic inches

lighting outlet an outlet intended for the direct connection of a lamp holder, a luminaire (lighting fixture), or a pendant cord terminating in a lamp holder

Madison hold-its thin metal straps that are used to hold "old-work" electrical boxes securely to an existing wall or ceiling

pigtail a short length of wire used in an electrical box to make connections to device terminals

setout the distance that the face of an electrical box protrudes out from the face of a building framing member; this distance is dependent on the thickness of the wall or ceiling finish material

Electrical boxes are required at each location on an electrical circuit where a device or lighting fixture is connected to the circuit. An electrical box, called a junction box, is also required at all locations where circuit conductors are simply spliced together. During the rough-in stage, an electrician will install electrical boxes that are designed for the specific wiring application indicated on the building electrical plans. Switches and receptacles will require a device box to be installed, lighting fixtures will require an outlet box designed to allow the fixture to be secured to it, and junction box locations will require a box designed to accommodate a certain number of spliced conductors. This chapter covers how to select the appropriate electrical box type and how to properly size the electrical box according to the *National Electrical Code*® (*NEC*®). Installation methods for various electrical box types in both new construction and remodeling situations are covered in detail.

SELECTING THE APPROPRIATE ELECTRICAL BOX TYPE

Earlier you were introduced to several different electrical box types. You learned that there are metal and nonmetallic electrical boxes available for residential use. You also learned that electrical boxes used in residential wiring are often categorized as device boxes, outlet boxes, and junction boxes. Device boxes are designed to allow a device, like a switch or receptacle, to be installed in them. Device boxes come in specific sizes, such as a 3″ × 2″ × 3½″ metal device box or a 20-cubic-inch plastic nail-on device box. Outlet boxes are designed to have lighting fixtures or ceiling-suspended paddle fans attached to them. This box style is a square or octagon metal box or a round nonmetallic box.

Junction boxes are designed to safely contain conductors that are spliced together and are usually just outlet boxes with blank covers on them. Sometimes conductors are spliced together in a box that also has a device in it. In this case, the box serves as both a device box and a junction box.

Selecting the proper electrical box requires an electrician to check the electrical plans to determine what is being installed at a certain location. If the symbol for a duplex receptacle is shown on the plans, a device box will need to be installed. If the symbol for a ceiling-mounted LED lighting fixture is shown on the plans, an outlet box designed to accommodate a lighting fixture will have to be installed. If the symbol for a junction box is shown on the plans, a box type that is designed to accommodate the proper number of conductors coming into that box will have to be installed. The bottom line is that the electrician will have to install a box type that allows the item shown in the plans to be installed.

The determination of whether the boxes will be nonmetallic or metal will normally be left for the electrical contractor to decide. Sometimes a set of residential building plans will have specifications that call for the use of only metal boxes or only nonmetallic boxes. In today's residential construction market, most new houses are wired using nonmetallic boxes at those locations in the house where they are appropriate. A house may be wired using all nonmetallic electrical boxes or all metal electrical boxes, but usually a house is wired using a combination of these.

CAUTION

CAUTION: If you are not absolutely sure, always check with the authority having jurisdiction in your area to make sure that nonmetallic electrical boxes are allowed. For the most part, nonmetallic boxes are only used with nonmetallic sheathed cable in residential wiring. If the authority having jurisdiction in your area does not allow houses to be wired with Type NM cable (Romex™) but rather an armored cable or metal electrical conduit, then you will have to use metal boxes.

An important item to consider when selecting an electrical device box is whether it must be a single gang, two gang, three gang, or more (Figure 10-1). This information is found on the electrical plan. For example, if you see an indication by the electrical symbols used that two single-pole switches are to be located at a specific spot, a two-gang box will need to be used to accommodate the two switches. If only one

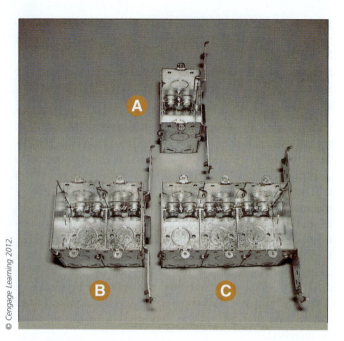

© Cengage Learning 2012.

FIGURE 10-1   Metal device boxes in (A) a one-, (B) two-, and (C) three-gang configuration.

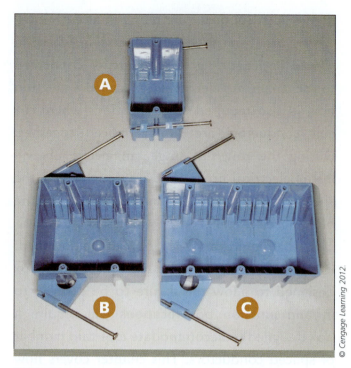

© Cengage Learning 2012.

FIGURE 10-2   Plastic nonmetallic device boxes in (A) a one-, (B) two-, and (C) three-gang configuration.

symbol for a switch or receptacle is shown on the plans, only a single-gang box will need to be installed. Remember, if you are using metal device boxes that are gangable, you may have to take the sides off the boxes and configure them in a way that gives you an electrical box that will accommodate the required number of devices at that location. The procedure for ganging together metal device boxes was presented in Chapter 2. Nonmetallic electrical boxes are manufactured to accommodate a specific number of devices. For example, a location that calls for two switches will require you to install a nonmetallic electrical box that comes from the manufacturer already configured as a two-gang box (Figure 10-2).

Another thing to consider at this time in the proper selection of an electrical box is whether it will be used in "new work" or "old work" (Figure 10-3). If the house being wired is new construction, new-work boxes will be used. These boxes have no "ears" on them and are designed to be attached directly to building framing members like studs and joists. If the electrical wiring is to be done in an existing house, it is called "old work" or "remodel work" and will require old-work electrical boxes that have ears on them. In old work, electrical boxes are often secured directly in a wall or ceiling material such as sheetrock and require special mounting accessories to hold them in place. We discuss installing

boxes in existing walls and ceilings at the end of this chapter. For now, we concentrate on the installation of electrical boxes in new construction.

The last thing to consider has to do with choosing a box that will not lower the overall energy efficiency of the house. The air barrier of a house is one of the most important elements of the energy efficiency of a green home. It blocks air from moving between the inside of the house and the outside, and holds the conditioned (heated or cooled) air in the house. The air barrier may be the drywall covering the inside walls and ceiling, the exterior wall sheathing, the housewrap, the floor sheathing, and other materials that block air movement. The air tightness of a house is achieved through meticulous air sealing at all penetrations that go from inside the house to the outside through the air barrier. The best practice is to avoid making holes in the air barrier of the house whenever possible. The electrician should discuss the steps for air sealing penetrations with the building team during the planning stage. An air-sealing specialist or the insulation contractor often handles the sealing task; but sometimes the building team assigns the electrician the responsibility for sealing the penetrations he or she makes. Even when another trade contractor handles the air sealing work, the electrician may still have to supply and install special boxes to simplify the air sealing process.

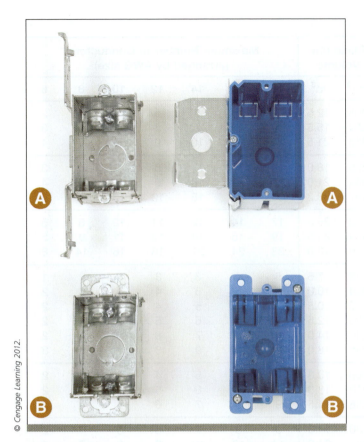

© Cengage Learning 2012.

FIGURE 10-3   (A) New-work versus (B) old-work metal and nonmetallic electrical device boxes. Notice the plaster ears on the old-work boxes.

Green Tip

Electric boxes that penetrate the air barrier should be special airtight models. The airtight boxes are used for switches, receptacles, lights, paddle fans, or other fixtures mounted in exterior walls, floors with a basement or crawl space beneath or ceilings with attic above. These boxes have gaskets where the wires enter and exit; and they have gasketed flanges that seal to the drywall or exterior wall sheathing.

SIZING ELECTRICAL BOXES

Once an electrician has selected the proper box type for a specific wiring application, he or she must make sure that the box can accommodate the required number of conductors and other electrical equipment that will be placed in the box. Understanding how to properly size electrical boxes is very important. An electrical box that is sized too small makes it harder to install electrical devices and other items, such as **pigtails** and wirenuts, in the box. It also presents a greater safety hazard. Article 314 of the *NEC®* shows us the proper procedures to follow to determine a minimum box size for the number of electrical conductors and other items going in the box. Tables 314.16(A) and 314.16(B) should become very familiar to an electrician (Figure 10-4).

Section 314.16 provides the requirements and identifies the allowances for the number of conductors permitted in a box. This section requires that the total box volume be equal to or greater than the total **box fill**. The total box volume is determined by adding the individual volumes of the box components. The components include the box itself plus any attachments to it, such as a plaster ring, an extension ring, or a lighting fixture dome cover. Figure 10-5 shows an example of a 4″ × 1½″ square box and a ¾-inch raised plaster ring being used together. The total volume results from adding the volume of the box and the raised plaster ring together. The volume of each box component is determined either from the volume marking on the box itself for nonmetallic boxes or from the standard volumes listed in Table 314.16(A) for metal boxes. If a box is marked with a larger volume than listed in Table 314.16(A), the larger volume can be used instead of the table value. Adding all the volume allowances for all items contributing to box fill determines the total box fill. The volume allowance for each item is based on the volume listed in Table 314.16(B) for the conductor size indicated. Table 10-1 summarizes the items that contribute to box fill and is based on the requirements of Section 314.16. Figure 10-6 is a selection guide to determine the maximum number of conductors allowed for the more common metal box styles and sizes used in residential wiring.

Table 314.16(A) Metal Boxes

Box Trade Size			Minimum Volume		Maximum Number of Conductors* (arranged by AWG size)						
mm	in.		cm³	in³	18	16	14	12	10	8	6
100 × 32	(4 × 1¼)	round/octagonal	205	12.5	8	7	6	5	5	5	2
100 × 38	(4 × 1½)	round/octagonal	254	15.5	10	8	7	6	6	5	3
100 × 54	(4 × 2⅛)	round/octagonal	353	21.5	14	12	10	9	8	7	4
100 × 32	(4 × 1¼)	square	295	18.0	12	10	9	8	7	6	3
100 × 38	(4 × 1½)	square	344	21.0	14	12	10	9	8	7	4
100 × 54	(4 × 2⅛)	square	497	30.3	20	17	15	13	12	10	6
120 × 32	(4¹¹⁄₁₆ × 1¼)	square	418	25.5	17	14	12	11	10	8	5
120 × 38	(4¹¹⁄₁₆ × 1½)	square	484	29.5	19	16	14	13	11	9	5
120 × 54	(4¹¹⁄₁₆ × 1⅛)	square	689	42.0	28	24	21	18	16	14	8
75 × 50 × 38	(3 × 2 × 1½)	device	123	7.5	5	4	3	3	3	2	1
75 × 50 × 50	(3 × 2 × 2)	device	164	10.0	6	5	5	4	4	3	2
75 × 50 × 57	(3 × 2 × 2¼)	device	172	10.5	7	6	5	4	4	3	2
75 × 50 × 65	(3 × 2 × 2½)	device	205	12.5	8	7	6	5	5	4	2
75 × 50 × 70	(3 × 2 × 2¾)	device	230	14.0	9	8	7	6	5	4	2
75 × 50 × 90	(3 × 2 × 3½)	device	295	18.0	12	10	9	8	7	6	3
100 × 54 × 38	(4 × 2⅛ × 1½)	device	169	10.3	6	5	5	4	4	3	2
100 × 54 × 48	(4 × 2⅛ × 1⅞)	device	213	13.0	8	7	6	5	5	4	2
100 × 54 × 54	(4 × 2⅛ × 1⅛)	device	238	14.5	9	8	7	6	5	4	2
95 × 50 × 65	(3¾ × 2 × 2½)	masonry box/gang	230	14.0	9	8	7	6	5	4	2
95 × 50 × 90	(3¾ × 2 × 3½)	masonry box/gang	344	21.0	14	12	10	9	8	7	4
min. 44.5 depth	FS—single cover/gang (1¾)		221	13.5	9	7	6	6	5	4	2
min. 60.3 depth	FD—single cover/gang (2⅜)		295	18.0	12	10	9	8	7	6	3
min. 44.5 depth	FS—single cover/gang (1¾)		295	18.0	12	10	9	8	7	6	3
min. 60.3 depth	FD—single cover/gang (2⅜)		395	24.0	16	13	12	10	9	8	4

*Where no volume allowances are required by 314.16(B)(2) through (B)(5)

Table 314.16(B) Volume Allowance Required per Conductor

Size of Conductor (AWG)	Free Space Within Box for Each Conductor	
	cm³	in.³
18	24.6	1.50
16	28.7	1.75
14	32.8	2.00
12	36.9	2.25
10	41.0	2.50
8	49.2	3.00
6	81.9	5.00

FIGURE 10-4 *NEC® Tables 314.16(A) and 314.16(B). Reprinted with permission from NFPA 70®, National Electric Code®, Copyright © 2010, National Fire Protection Association, Quincy, MA. This reprinted material is not the complete and official position of the NFPA on the referenced subject, which is represented only by the standard in its entirety.*

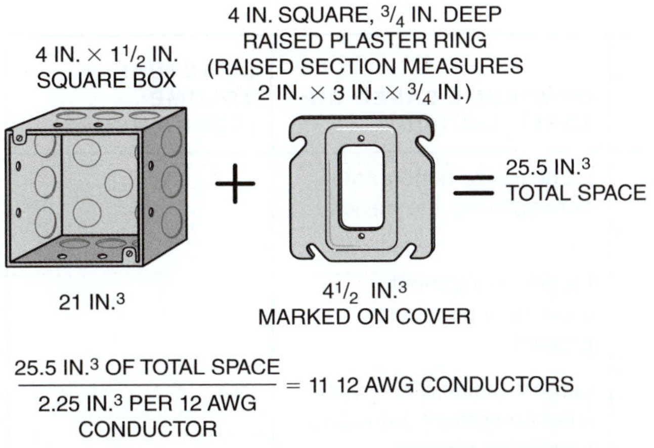

4 IN. × 1½ IN.
SQUARE BOX

4 IN. SQUARE, ¾ IN. DEEP
RAISED PLASTER RING
(RAISED SECTION MEASURES
2 IN. × 3 IN. × ¾ IN.)

= 25.5 IN.³
TOTAL SPACE

21 IN.³

4½ IN.³
MARKED ON COVER

$$\frac{25.5 \text{ IN.}^3 \text{ OF TOTAL SPACE}}{2.25 \text{ IN.}^3 \text{ PER 12 AWG CONDUCTOR}} = 11 \text{ 12 AWG CONDUCTORS}$$

© Cengage Learning 2012.

FIGURE 10-5 Total volume in an electrical box includes the box itself and any raised cover that is attached to it. However, in order for the volume of the cover to be included, the cubic inch capacity of the cover must be marked on it.

The following examples illustrate the applicable requirements of Section 314.16 and Tables 314.16(A) and 314.16(B):

EXAMPLE 1

An electrician has a case of 3″ × 2″ × 3½″ metal device boxes with side-mounting brackets, but no internal clamps. He would like to know the maximum number of 12 AWG conductors that can be installed in this box. This information is very easy to find. The electrician simply needs to look at Table 314.16(A) and find the row for a 3″ × 2″ × 3½″ device box. Once the proper row is found, find the maximum number of 12 AWG conductors by locating the number that is in the same row but under the column for 12 AWG conductors. In this case, eight 12 AWG conductors is the maximum allowed.

TABLE 10-1 This Table Summarizes the Box Fill Rules From Section 314.16 in the *NEC®*. Use the Table to Calculate the Actual Box Fill for a Specific Electrical Box Situation and then Choose a Box with a Volume in Cubic Inches that Meets or Exceeds the Actual Box Fill.

(1) ITEMS IN THE ELECTRICAL BOX	(2) VOLUME ALLOWANCE	(3) VOLUME BASED ON TABLE 314.16(B)	(4) ACTUAL VOLUME* (CUBIC INCHES)
Conductors that originate outside the box	One for each conductor	Actual conductor size	
Conductors that pass through the box without splice or connection	One for each conductor	Actual conductor size	
A looped or coiled, unbroken length of conductor at least 12 inches (300 mm) long that passes through a box without splice or connection	Two for each conductor	Actual conductor size	
Conductors that originate within the box and do not leave the box (pigtails and jumpers)	None (These conductors are not counted.)		
Light fixture conductors (four or fewer that are smaller than #14 awg)	None (These conductors are not counted.)		
Internal cable clamps (one or more)	One only	Largest size conductor present	
Support fittings (such as fixture studs, hickeys)	One for each type of support fitting	Largest size conductor present	
Devices (such as receptacles, switches)	Two for each yoke or mounting strap	Largest size conductor connected to the device	

TABLE 10-1 *(continued)*

(1) ITEMS IN THE ELECTRICAL BOX	(2) VOLUME ALLOWANCE	(3) VOLUME BASED ON TABLE 314.16(B)	(4) ACTUAL VOLUME* (CUBIC INCHES)
Wide devices (such as a range or dryer receptacle)	Two for each gang of the box required for mounting the device	Largest size conductor connected to the device	
Equipment grounding conductor (one or more)	One only	Largest equipment grounding conductor present	
Isolated equipment grounding conductor (one or more)	One only	Largest isolated and insulated equipment grounding conductor present	
			Total Volume

*Actual volume is found by multiplying the volume allowance in column #2 by the volume per conductor in column #3.

EXAMPLE 2

An electrician needs to select a standard-size 4-inch square box for use as a junction box where all the conductors are the same size and the box does not contain any cable clamps, support fittings, devices, or equipment-grounding conductors (Figure 10-7). To determine the number of conductors permitted in the standard 4-inch square box, count the conductors in the box and compare the total to the maximum number of conductors permitted by Table 314.16(A). Each unspliced conductor running through the box is counted as one conductor, and every other conductor that originates outside the box is counted as one conductor. Therefore, the total conductor count for this box is nine. Now look at Table 314.16(A) to find a 4-inch square box that will accommodate nine 12 AWG conductors. A 4″ × 1½″ (21.0-cubic-inch) square box is the minimum size box that will meet the requirements for this application.

EXAMPLE 3

A common method for determining proper box size is to find the total box volume in Table 314.16(A) and then subtract the actual total box fill from that amount. If the result is zero or some number greater than zero, the box is adequate for the application. Using this method, refer to Figure 10-8 and determine whether the box is adequately sized. For a standard 3″ × 2″ × 3½″ device box (18 cubic inches), Table 314.16(A) allows up to a maximum of nine 14 AWG conductors. Using Table 10-1 to calculate the total cubic inch volume of the box fill, we come up with 16 cubic inches (Figure 10-9). Therefore, because the total box fill of 16 cubic inches is less than the 18 cubic inch total box volume permitted, the box is adequately sized.

BOX SELECTION GUIDE
FOR METAL BOXES GENERALLY USED FOR RESIDENTIAL WIRING

DEVICE BOXES	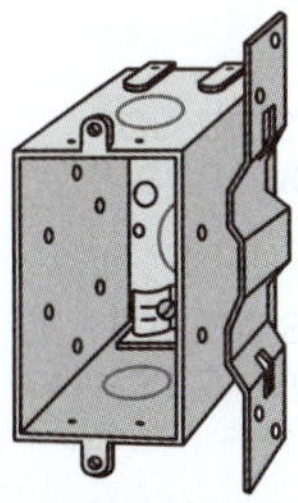	WIRE SIZE	$3x2x2^1/_2$ (12.5 IN.^{3})	$3x2x2^3/_4$ (14 IN.^{3})	$3x2x3^1/_2$ (18 IN.^{3})	
		14 AWG	6	7	9	
		12 AWG	5	6	8	
		10 AWG	5	5	7	

SQUARE BOXES	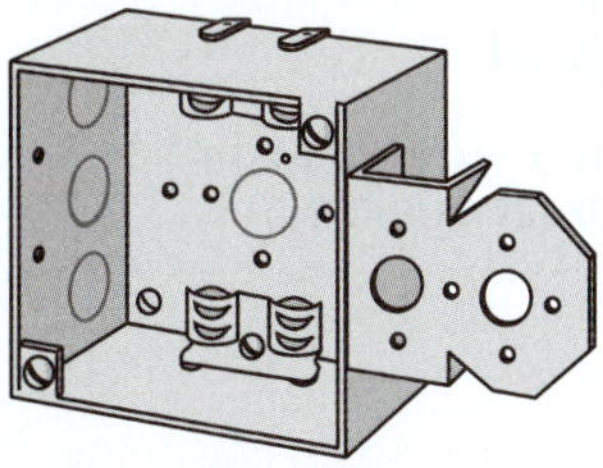	WIRE SIZE	$4x4x1^1/_2$ (21 IN.^{3})	$4x4x2^1/_8$ (30.3 IN.^{3})	
		14 AWG	10	15	
		12 AWG	9	13	
		10 AWG	8	12	

OCTAGON BOXES	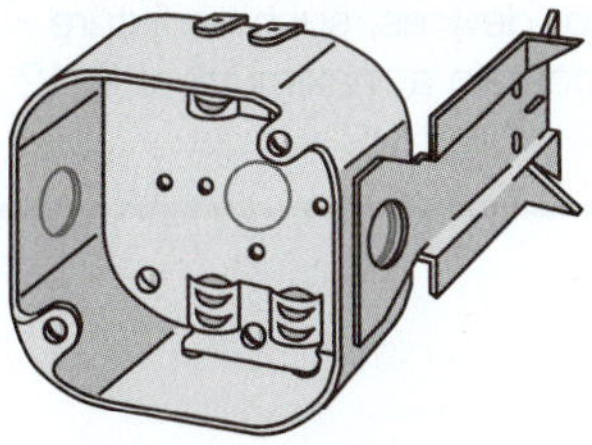	WIRE SIZE	$4x1^1/_2$ (15.5 IN.^{3})	$4x2^1/_8$ (21.5 IN.^{3})	
		14 AWG	7	10	
		12 AWG	6	9	
		10 AWG	6	8	

HANDY BOXES	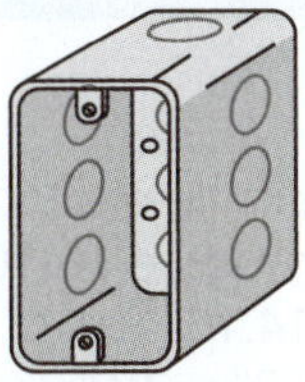	WIRE SIZE	$4x2^1/_8x1^1/_2$ (10.3 IN.^{3})	$4x2^1/_8x1^7/_8$ (13 IN.^{3})	$4x2^1/_8x2^1/_8$ (14.5 IN.^{3})	
		14 AWG	5	6	7	
		12 AWG	4	5	6	
		10 AWG	4	5	5	

RAISED COVERS

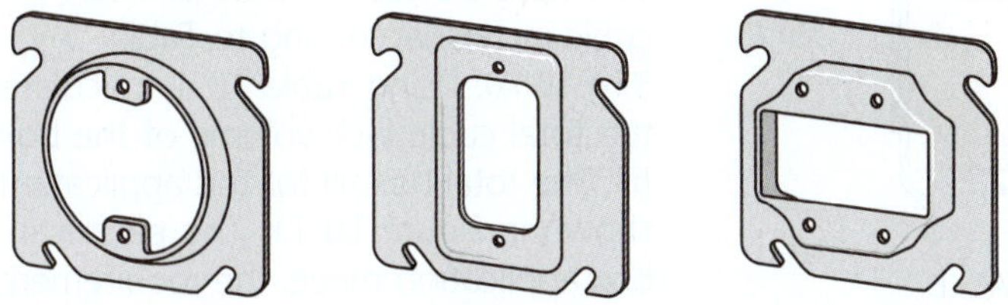

WHERE RAISED COVERS ARE MARKED WITH THEIR VOLUME IN CUBIC INCHES, THAT VOLUME MAY BE ADDED TO THE BOX VOLUME TO DETERMINE MAXIMUM NUMBER OF CONDUCTORS IN THE COMBINED BOX AND RAISED COVER.

NOTE: BE SURE TO MAKE DEDUCTIONS FROM THE ABOVE MAXIMUM NUMBER OF CONDUCTORS PERMITTED FOR WIRING DEVICES, CABLE CLAMPS, FIXTURE STUDS, AND GROUNDING CONDUCTORS. THE CUBIC INCH (IN.^{3}) VOLUME IS TAKEN DIRECTLY FROM *TABLE 314.16(A)* OF THE *NEC*.® NONMETALLIC BOXES ARE MARKED WITH THEIR CUBIC INCH CAPACITY.

© Cengage Learning 2012.

FIGURE 10-6   Metal electrical box selection guide.

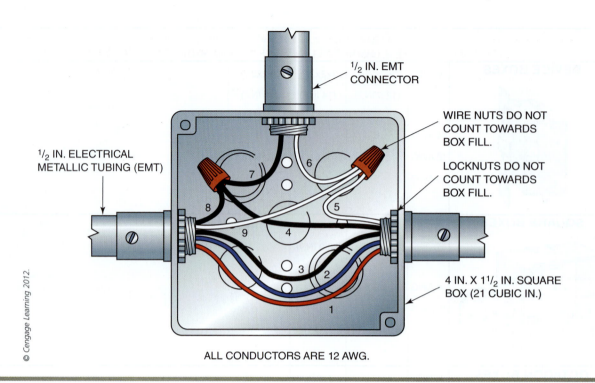

ALL CONDUCTORS ARE 12 AWG.

FIGURE 10-7 A standard-size 4″ × 1½″ square box (21 cubic inches) containing no fittings or devices, such as fixture studs, cable clamps, switches, receptacles, or equipment-grounding conductors. This box can contain a maximum of 9–12 AWG conductors.

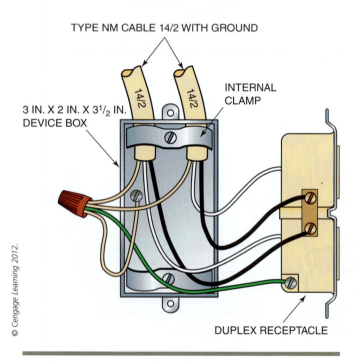

FIGURE 10-8 A 3″ × 2″ × 3½″ device box with a device and 14 AWG conductors.

EXAMPLE 4

Using the same method we used in Example 3, determine if the electrical box application shown in Figure 10-10 complies with Section 314.16. This application has two 3″ × 2″ × 3½″ device boxes ganged together to make a two-gang electrical box. The total box volume is 36 cubic inches (2 × 18 cubic inches) according to Table 314.16(A). Using Table 10-1, calculate the total cubic inch volume of the box fill. The total box fill for this application is shown in Figure 10-11. This electrical box application meets the requirements of the *NEC®* because only 26 cubic inches of the total allowed volume of 36 cubic inches is used.

(1) Items in the Electrical Box	(2) Volume Allowance	(3) Volume Based on Table 314.16(B)	(4) Actual Volume* (Cubic Inches)
Conductors that originate outside the box	4	2.0 cubic inches	8.0 cubic inches
Internal cable clamps (one or more)	1	2.0 cubic inches	2.0 cubic inches
Devices (such as receptacles and switches)	2	2.0 cubic inches	4.0 cubic inches
Equipment grounding conductor (one or more)	1	2.0 cubic inches	2.0 cubic inches
		Total Volume	16.0 cubic inches

*Actual volume is found by multiplying the volume allowance in column 2 by the volume per conductor in column 3.

© Cengage Learning 2012.

FIGURE 10-9 The total calculated box fill for the box used in Example 3. The calculation proves that this box is adequately sized.

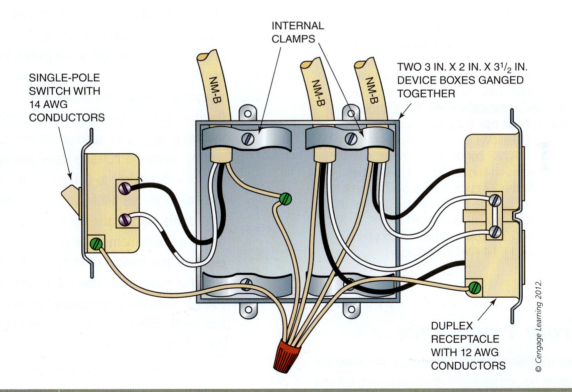

FIGURE 10-10 Two 3″ × 2″ × 3½″ device boxes ganged together and containing conductors of different sizes. A switch and receptacle are to be installed in the box.

(1) Items in the Electrical Box	(2) Volume Allowance	(3) Volume Based on Table 314.16(B)	(4) Actual Volume* (Cubic Inches)
Conductors that originate outside the box	Two 14 AWG Four 12 AWG	2.0 cubic inches 2.25 cubic inches	4.0 cubic inches 9.0 cubic inches
Internal cable clamps (one or more)	One 12 AWG	2.25 cubic inches	2.25 cubic inches
Devices (such as receptacles and switches)	Two 14 AWG Two 12 AWG	2.0 cubic inches 2.25 cubic inches	4.0 cubic inches 4.5 cubic inches
Equipment grounding conductor (one or more)	One 12 AWG	2.25 cubic inches	2.25 cubic inches
		Total Volume	26.0 cubic inches

© Cengage Learning 2012.

*Actual volume is found by multiplying the volume allowance in column 2 by the volume per conductor in column 3.

FIGURE 10-11 The total calculated box fill for the box used in Example 4. The calculation proves that this box is adequately sized.

EXAMPLE 5

An electrician is installing some single-gang plastic nail-on boxes and is not sure how many 14 AWG conductors can be installed in the box. Nonmetallic boxes that are used in residential wiring will have their cubic inch volume marked on the box. A typical volume for a single-gang plastic box is 22 cubic inches. Table 314.16(B) allows 2.0 cubic inches for each 14 AWG conductor in a box, so all the electrician has to do in this situation is divide the 22 cubic inches by 2.0 cubic inches to get the maximum number of 14 AWG conductors allowed in this size box: 22.0 cubic inches / 2.0 cubic inches = 11 14 AWG conductors.

from experience...

Nonmetallic electrical boxes used in residential wiring usually have both their cubic inch volume marked on the box and the maximum number of 14, 12, and 10 AWG conductors that can be contained in the box. The box manufacturer has done the calculations for the electrician, which can save valuable time during the installation process.

from experience...

There is no *NEC*® rule for the height of electrical boxes in residential wiring. A common height for switch boxes is 48 inches to the center of the box. A common height for receptacle boxes is 16 inches to the center of the box. Table 10-2 shows some common mounting heights for various box locations in a house. Check the building plans and the electrical supervisor to determine the required box height for the house you are working on. Also, some electricians like to make their box height measurement to the bottom or top of an electrical box instead of to the center. It really does not make any difference as long as you are consistent throughout the house so that some boxes are not higher or lower than others.

TABLE 10-2 Common box mounting heights for receptacles, switches, and wall-mounted lighting fixtures in residential wiring.

SWITCHES	
Regular	48 inches (1.12 m)
Between counter and kitchen cabinets—depends on backsplash	45 inches (1.125 m)
RECEPTACLE OUTLETS	
Regular	16 inches (400 mm)
Between counter and kitchen cabinets—depends on backsplash	45 inches (1.125 m)
In garages	48 inches (1.2 m), minimum 18 inches (450 mm)
In unfinished basements	48 inches (1.2 m)
In finished basements	16 inches (400 mm)
Outdoors (above grade or deck)	18 inches (450 mm)
WALL LUMINAIRE (FIXTURE) OUTLETS	
Outside entrances—depends on luminaire (fixture)	66 inches (1.7 m). If luminaire is "upward" from box on luminaire, mount wall box lower. If luminaire is "downward" from box on luminaire, mount wall box higher.
Inside wall brackets	5 feet (1.5 m)
Side of or above medicine cabinet or mirror	You need to know the measurement of the medicine cabinet. Check the rough-in opening of medicine cabinet measurement of mirror. Mount electrical wall box approximately 6 inches (150 mm) to center above rough-in opening or mirror. Medicine cabinets that come complete with luminaires have a wiring compartment with a knockout(s) in which the supply cable or conduit is secured using the appropriate fitting.
	Many strip luminaires have a backplate with a conduit knockout in which the supply cable or conduit is secured using the appropriate fitting. Where to bring in the cable or conduit takes careful planning.

Note: All dimensions are from finished floor to center of the electrical box. If possible, try to mount wall boxes for luminaires based on the type of luminaire to be installed. Verify all dimensions before "roughing in." If wiring for physically handicapped, the above heights may need to be lowered in the case of switches and raised in the case of receptacles.

INSTALLING NONMETALLIC DEVICE BOXES

Because most device boxes used in residential rough-in wiring are nonmetallic, we will cover their installation first. The most common method for installing nonmetallic boxes is to simply nail them directly to a wood framing member (Figure 10-12). Nonmetallic device boxes usually come equipped with nails and are ready for installation as soon as an electrician takes it out of the cardboard box they come in. Some styles of nonmetallic boxes come with a side-mounting bracket that is nailed or screwed to a framing member (Figure 10-13). Many nonmetallic box manufacturers provide indicators on the boxes (Figure 10-14) that an electrician can use to set the box out from the framing member at the proper depth, depending on what is being used as the wall covering.

CAUTION

CAUTION: Remember that Section 314.20 requires electrical boxes to be flush with the finished surface in combustible materials. Boxes may be set back no more than ¼ inch in noncombustible materials.

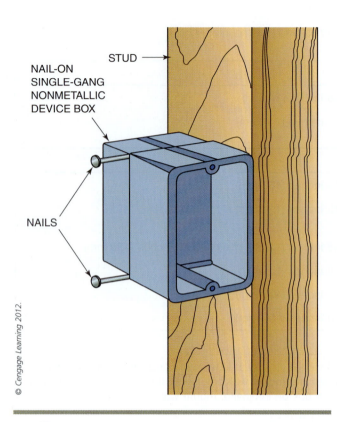

FIGURE 10-12   A nonmetallic device box nailed directly to a stud.

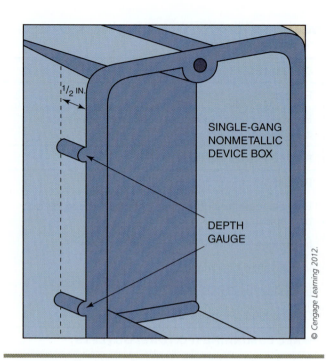

FIGURE 10-14   Most nonmetallic device boxes have a depth gauge that helps the electrician mount the box with the proper setout.

(See Procedure 10-1 on pages 347–349 for step-by-step instructions on installing nonmetallic device boxes in new construction.)

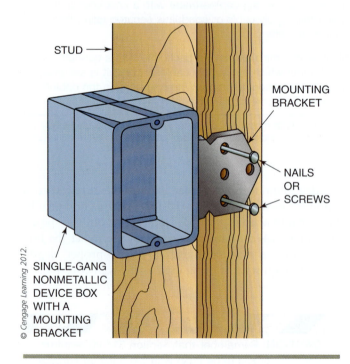

FIGURE 10-13   A nonmetallic device box with a side-mounting bracket attached to a stud.

INSTALLING METAL DEVICE BOXES

Like nonmetallic electrical device boxes, metal device boxes are usually nailed to building framing members. Sometimes screws, such as sheetrock screws, are used to attach them to studs, joists, or rafters. The best way to attach a metal device box with nails is to keep the nails outside the box. The box shown in Figure 10-15 is a common style of metal device box that comes from the manufacturer already equipped with nails. Many electricians use metal device boxes that are attached to framing members using a side-mounting bracket as shown in Figure 10-16. Also, like nonmetallic boxes, metal device boxes are available with depth gauge markings (Figure 10-17) on the side of the box to help electricians determine the proper setout for the box.

(See Procedure 10-1 on pages 347–349 for step-by-step instructions on installing metal device boxes in new construction.)

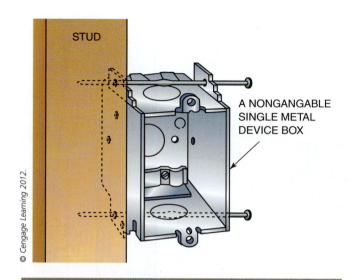

FIGURE 10-15 A common metal device box style with manufacturer-equipped nails.

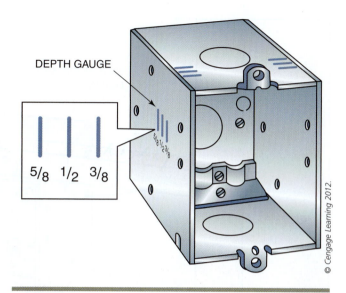

FIGURE 10-17 Some metal device boxes have depth gauge markings to help the electrician determine the proper box setout.

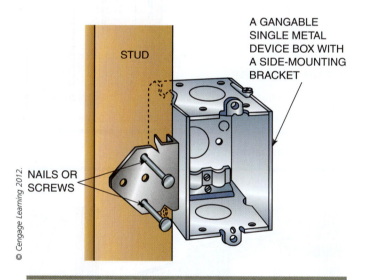

FIGURE 10-16 A metal device box with a side-mounting bracket.

There are times when installing a box directly to a stud, joist, or rafter does not put it in a location that is called for in the building electrical plan. Once in a while, a box has to be mounted so that it is actually positioned between two studs, joists, or rafters. When this mounting situation is encountered, wood or metal strips can be placed between the framing members, and the electrical box can be secured to them (Figure 10-18). Wood straps can usually be made from some scrap wood pieces found at the building site. The carpenters at the job site will be more than willing to help you out and cut the wood pieces to fit. If you decide to cut

from experience...

When there is a need to mount electrical boxes between two studs and you don't have a factory-made mounting bracket to do the job, many electricians will simply nail a short length of 2-by-4-inch lumber to the side of the stud at the location where the box needs to be installed. This will set the box mounting location about 1½ inches closer to the center of the two studs. Additional short lengths of 2-by-4-inch lumber are nailed on if the box needs to be moved closer to the center. There are often short lengths of 2-by-4-inch lumber on the building site that have been sawed off longer pieces by the building carpenters. They usually have no problem with you using the pieces to relocate a box mounting location, but you should ask them first. This technique can also be used when mounting electrical boxes between joists or rafters in a house.

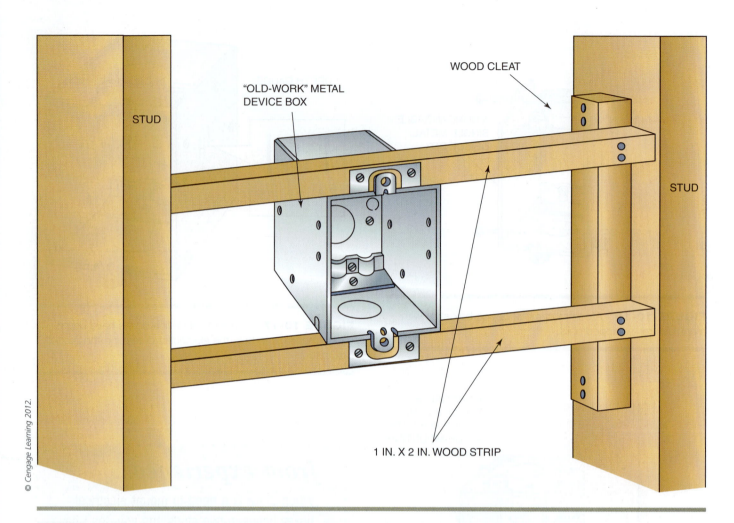

FIGURE 10-18   Wood cleats and strips can be used to mount electrical boxes between studs.

the pieces of wood yourself, use proper safety equipment and follow proper safety procedures when using power saws. Metal mounting brackets are readily available at the local electrical supplier (Figure 10-19). They are designed to fit directly between standard widths of studs, joists, or rafters. Metal mounting brackets may cost more initially but they are a lot faster to install than finding and cutting to length wood box mounting pieces.

The last item that we cover in this section is the mounting of handy boxes, or utility boxes, as they are often called. You may remember that this box type is classified by the *NEC*® as a device box and is normally mounted directly on the surface of a framing member or existing wall. They are often used in places where a finished wall surface is not going to be used, like an unfinished basement or unfinished garage.

(See Procedure 10-2 on pages 350–351 for step-by-step instructions on installing a handy box on a wood or masonry surface.)

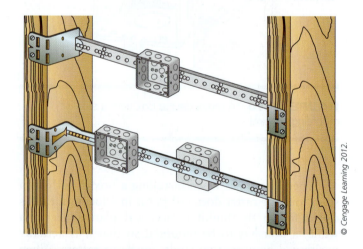

FIGURE 10-19   Metal box mounting brackets are available that allow easy placement of metal electrical boxes between building framing members. Remember that a raised plaster ring would need to be used with the 4-inch square boxes shown if a switch or receptacle is going to be installed. These mounting brackets can be used on wood or metal framing members.

INSTALLING LIGHTING OUTLET AND JUNCTION BOXES

Electrical boxes used in walls or ceilings for lighting outlets in residential electrical systems must be designed for the purpose and are round or octagonal in shape (Figure 10-20). The round lighting outlet boxes are usually nonmetallic and made of plastic but can be made of other nonmetallic materials. They are sized according to their cubic-inch capacity, which is plainly marked on each box. The octagon boxes are made of metal and are used in residential wiring in two common sizes: $4'' \times 1\frac{1}{2}''$ and $4'' \times 2\frac{1}{8}''$. Both of these boxes are listed in Table 314.16(A). Some wiring schemes will result in lighting outlet boxes also being used as junction boxes. This is allowed by the *NEC®* because the wires are accessible by simply removing the lighting fixture. Boxes used as purely junction boxes are usually square, although some electricians find octagon boxes may work just as well. Boxes used only as junction boxes will be covered with a flat blank cover. The two most common square box sizes are $4'' \times 1\frac{1}{2}''$ and $4'' \times 2\frac{1}{8}''$ (Figure 10-21). Like the octagon boxes, these square boxes are also listed in Table 314.16(A). Junction boxes that are to be surface-mounted are installed the same way as handy boxes. Junction boxes that are to be mounted flush with a finished surface are installed the same way as lighting outlet boxes.

Lighting outlet boxes are mounted to the building framing members with side-mounting brackets or bar hangers that are adjustable (Figure 10-22). Installing outlet boxes with side-mounting brackets requires the box to be mounted directly to the stud, joist, or rafter.

Using an adjustable bar hanger allows an outlet box to be properly positioned between two studs, rafters, or joists. Both nonmetallic and metal outlet boxes are available with adjustable bar hangers.

CAUTION

CAUTION: Houses are sometimes built with a metal roof. Section 300.4(E) addresses installation of boxes under metal-corrugated sheet roofing. It requires boxes to be installed and supported so there is at least 1½ inches (38 mm) from the lowest surface of the roof material to the top of the box. This rule is designed to minimize the chance of long roofing screws penetrating through a box and damaging the conductors inside.

(See Procedure 10-3 on page 352 for step-by-step instructions on installing outlet boxes with a side-mounting bracket.)

© Cengage Learning 2012.

FIGURE 10-21   Square boxes like the ones shown are usually used as junction boxes. Common sizes are (A) 4 by 1½ inches and 4 by 2⅛ inches square. Junction boxes are required to have a (B) flat blank cover installed.

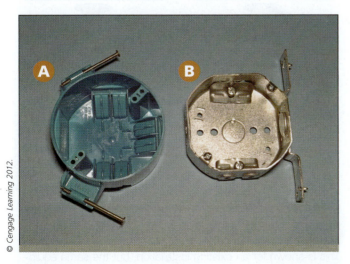

© Cengage Learning 2012.

FIGURE 10-20   Lighting outlet boxes are usually (A) round and nonmetallic or (B) octagon and made of metal.

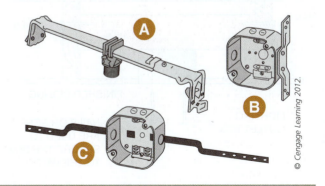

© Cengage Learning 2012.

FIGURE 10-22   (A) Adjustable bar hanger with a fixture stud; (B) an octagon box with a side-mounting bracket; and (C) an octagon box with an offset-style bar hanger.

(See Procedure 10-4 on pages 353–354 for step-by-step instructions on installing outlet boxes with an adjustable bar hanger.)

If an electrician needs to position an outlet box between two framing members and there is no adjustable bar hanger available, an alternative method is to use some scrap wood from the construction site and fabricate your own bar hanger. Short lengths of 2-by-4-inch or 2-by-6-inch pieces of lumber will work well. Simply cut the piece of lumber so it will fit tightly between the two studs, joists, or rafters where you want to locate the outlet box. Place the wood box hanger between the framing members and nail it in place at a spot that, when the outlet box is attached to the hanger, results in an installation where the box face is flush with the finished surface (Figure 10-23).

Lighting fixtures that weigh more than 50 pounds (23 kilograms) cannot be supported by just the electrical outlet box, according to Section 314.27 of the *NEC*®. A lighting fixture weighing more than 50 pounds (23 kg) must be supported independently of the outlet box unless the box is listed and marked at or above the weight of the fixture. Section 314.27 also requires boxes used as the sole support of a ceiling-suspended paddle fan to be listed and marked by the manufacturer as suitable for the application and may not support paddle fans weighing more than 70 pounds (32 kg). When paddle fans weighing more than 35 pounds (16 kg) are to be supported by a box, the box must have the maximum weight it can support marked on it. For these special heavy-load installations, boxes that are rated for ceiling-suspended paddle fan and heavy-load installations are available in both metal and nonmetallic boxes. Figure 10-24 shows a common box type used to hang a heavy lighting fixture or a ceiling-suspended paddle fan. This box is usually nonmetallic but can be found without the bracket in a round metal version that is screwed directly to the underside of a joist. This shallow box type is often referred to as a "pancake" box. Figure 10-25 shows how a listed ceiling fan hanger could be installed in an existing ceiling or a new construction ceiling.

Electricians will sometimes encounter a suspended ceiling (often called a "dropped ceiling") while doing the rough-in wiring in a house. Suspended ceilings are often used in a home's basement where a finished family room or recreation room is to be located. The suspended-ceiling panels can be removed to provide access to electrical, plumbing, heating, and other system equipment installed in the ceiling. Section 300.11(A) of the *NEC*® states that suspended-ceiling support wires that do not provide secure support must not be permitted as the only support for boxes. However, support wires and associated fittings that provide secure support

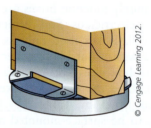

FIGURE 10-24   A common box type used to hang a heavy lighting fixture or a ceiling-suspended paddle fan. This box is usually nonmetallic but can be found without the bracket in a round metal version. Often called a "pancake" box.

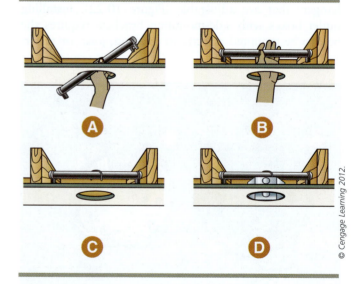

FIGURE 10-25   (A–D) A listed ceiling-fan hanger and box assembly installed through a properly sized hole in a ceiling. This is shown for an existing installation, but the same equipment can be used in new construction as well.

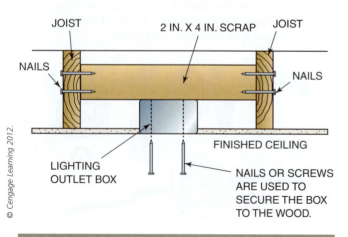

FIGURE 10-23   Scrap wood can be used to fabricate an electrical box hanger when factory-made hangers are not available.

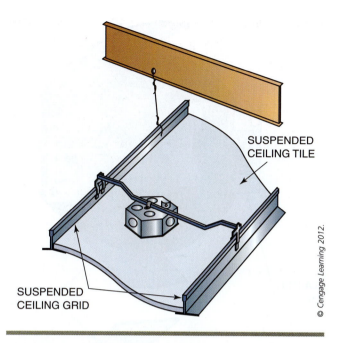

SUSPENDED CEILING TILE

SUSPENDED CEILING GRID

© Cengage Learning 2012.

FIGURE 10-26 Attaching an electrical box to the grid of a suspended ceiling must be done using mechanical means identified for the purpose. This illustration shows an electrical box attached to the suspended ceiling grid with a special box hanger.

and that are installed *in addition* to the ceiling grid support wires are permitted as electrical box support. Where independent support wires are used, they must be secured at both ends. They must also be distinguishable from the ceiling grid support wires by color, tagging, or some other method. If an electrical box is to be mounted to the suspended ceiling structural frame or to wires installed specifically for box support, Section 314.23(D) requires it to be not more than 100 cubic inches (1650 cm^3) in size and be securely fastened in place by either:

- Fastening to the framing members by mechanical means such as bolts, screws, or rivets, or by the use of clips or other securing means identified for use with the type of ceiling framing member(s) and enclosure(s) employed (Figure 10-26). The framing members must be adequately supported and securely fastened to each other and to the building structure.

- Using methods identified for the purpose, to ceiling support wire(s), including any additional support wire(s) installed for that purpose. Support wire(s) used for enclosure support must be fastened at each end so as to be taut within the ceiling cavity. Some manufacturers make specially designed fittings that attach to a support wire and provide a way to secure an electrical box or cable to the support wire.

INSTALLING BOXES IN EXISTING WALLS AND CEILINGS

Up until now, we have discussed box installation in new construction. However, most remodel electrical jobs will require mounting boxes in existing walls and ceilings.

Newer homes have existing walls and ceilings made of wood or metal framing with a drywall surface. This wall or ceiling type is relatively easy to cut in order to install an electrical box. Older homes typically have wood lath and plaster walls and ceilings, which are much harder to cut for the installation of an electrical box. Old-work boxes will have to be used when installing electrical boxes in an existing wall or ceiling. The old-work boxes are available in metal or nonmetal, but both have one common characteristic: They have plaster ears that help hold the box in the wall or ceiling.

To install old-work boxes, a hole must be cut into the wall or ceiling at the location where you wish to mount the box. The plaster ears on the old-work boxes keep the boxes from going too far into the hole, but there has to be some other device that will keep the box from coming back out of the hole. There are many ways this is accomplished.

Figure 10-27 shows nonmetallic old-work boxes with metal straps that, along with the plaster ears, hold the box in the wall or ceiling opening. Once this box type is inserted into the cutout box opening, the metal straps spread open and do not allow the box to come back out of the hole. The screws are tightened, and the box is held into the box opening by the plaster ears on the outside and the metal strap on the inside. The box shown in Figure 10-28 is a common style of old-work nonmetallic device box. It has ears like other

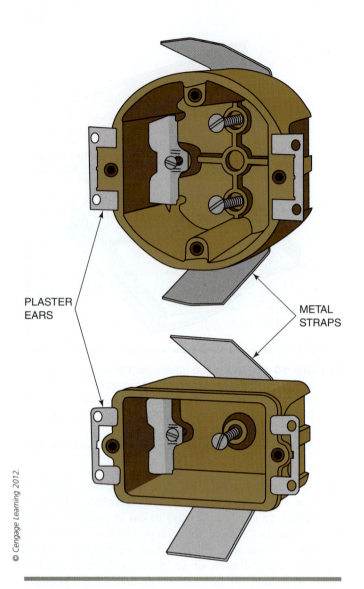

PLASTER
EARS

METAL
STRAPS

© Cengage Learning 2012.

FIGURE 10-27 Two types of nonmetallic old-work boxes.

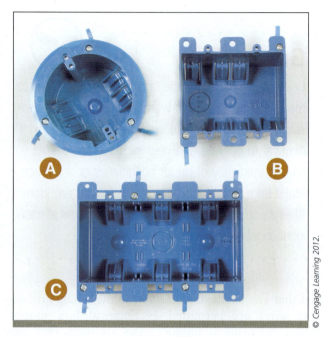

© Cengage Learning 2012.

FIGURE 10-28 A common style of old-work box. This box has brackets that swing up behind the wall or ceiling to help hold the box in place. (A) is an old-work box typically used in a ceiling to install a lighting fixture, (B) is a two-gang old-work device box, and (C) is a three-gang old-work device box.

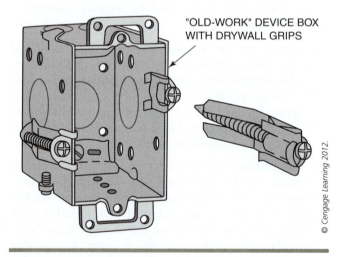

"OLD-WORK" DEVICE BOX WITH DRYWALL GRIPS

© Cengage Learning 2012.

FIGURE 10-29 This box style is commonly used in remodel work. The box has drywall grips that, when properly tightened, will hold the box in a wall.

old-work boxes and has brackets that swing up and into place once the box has been inserted into a wall. A Phillips screwdriver is used to turn the screws that swing the brackets into the correct holding position.

The box shown in Figure 10-29 shows another common old-work box with sheetrock grips. Once a hole has been cut, the box is pushed into the hole. The plaster ears keep the box from going into the hole too far, and the sheetrock grips are tightened to hold the box firmly in the wall or ceiling.

Another very common method for holding old-work electrical boxes in a wall or ceiling is by using Madison hold-its (Figure 10-30). Electricians commonly refer to this type of device as a "Madison strap" or "Madison hanger." These thin metal straps are installed on each side of an old-work box and provide a secure attachment of the box to the wall or ceiling material.

(See Procedure 10-5 on pages 355–356 for step-by-step instructions on installing old-work electrical boxes in a wood lath and plaster wal, or ceiling.)

(See Procedure 10-6 on page 357 for step-by-step instructions on installing old-work metal electrical boxes in a sheetrock wall or ceiling.)

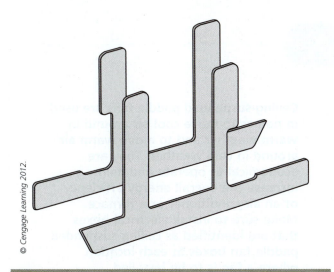

FIGURE 10-30   These thin metal old-work box supports are called Madison hold-its or Madison straps. They are used to help hold an old-work electrical box in a wall.

(See Procedure 10-7 on pages 358–359 for step-by-step instructions on installing old-work nonmetallic electrical boxes in a sheetrock wall or ceiling.)

SUMMARY

There was a lot of material covered in this chapter that has to do with the installation of electrical boxes during the rough-in stage of installing a residential electrical wiring system. The type of electrical equipment to be installed at a particular location in a house will determine what kind of electrical box will have to be installed at that location. Device boxes are used where switches or receptacles will be located and can be metal or nonmetal. Lighting outlet locations usually require a round nonmetallic box or an octagon metal box to be installed. Square metal boxes are normally installed in those areas where a junction box is required. All electrical boxes installed during the rough-in stage will have to be sized according to Section 314.16 so that adequate room will be available in the boxes for devices, conductors, and other items, such as cable clamps. This chapter also presented many installation techniques for mounting electrical boxes in both new construction and existing homes where remodeling work is being done. A good rough-in job starts with the proper installation of the electrical boxes. By following the information presented in this chapter, electricians should have a good understanding of electrical box selection, sizing, and mounting techniques.

GREEN CHECKLIST

☐ Electric boxes that penetrate the air barrier should be special airtight models. The airtight boxes are used for switches, receptacles, lights, paddle fans, or other fixtures mounted in exterior walls, floors with a basement or crawl space beneath or ceilings with attic above. These boxes have gaskets where the wires enter and exit; and they have gasketed flanges that seal to the drywall or exterior wall sheathing.

☐ Ceiling-suspended paddle fans are used in homes to move cool air around in warm weather and to move warm air around in cold weather. They are economical to operate and help increase the overall energy efficiency of an air conditioner or a furnace. Make sure to install electrical boxes that are identified as ceiling-suspended paddle fan boxes at each location where a fan is to be installed.

PROCEDURE 10-1

Installing Device Boxes in New Construction

- Put on safety glasses and follow all applicable safety rules.

- Determine the depth at which the box is to be mounted. Remember, the depth will depend on the material thickness that is used for the finished wall surface.

 Determine the height from the finished floor that the box is to be mounted and make a mark using a permanent marker or pencil on the framing member at the proper mounting height.

CAUTION

CAUTION: Most boxes in a residential application are mounted with a setout from the framing member of ½ inch, assuming the use of ½-inch sheetrock. However, always check the building plans and ask the construction supervisor to find out for sure what the required setout will need to be for the electrical boxes.

© Cengage Learning 2012.

PROCEDURE 10-1

Installing Device Boxes in New Construction (Continued)

B If installing a plastic nail-on device box to a wood framing member, nail the box to the framing member at the required height and with the proper box setout.

B

© Cengage Learning 2012.

C If installing a metal device box to a wood framing member, nail or screw the box to the framing member at the required height and with the proper box setout.

C

© Cengage Learning 2012.

D If installing a metal device box to a metal framing member, screw the box to the framing member at the required height and with the proper box setout.

D

© Cengage Learning 2012.

E If installing a metal device box to a metal framing member using a box mounting bracket, screw the mounting bracket to the framing member and then attach the box and raised ring to the mounting bracket.

- Clean up the work area and put all tools and materials away.

E

© Cengage Learning 2012.

PROCEDURE 10-2

Installing a Handy Box on a Wood or Masonry Surface

- Put on safety glasses and follow all applicable safety rules.

Wood Surface

- Determine the height from the finished floor that the box is to be mounted and make a mark using a permanent marker or pencil on the wood at the proper mounting height.

(A) Nail or screw the box to the wood surface at the required height.

- Clean up the work area and put all tools and materials away.

Masonry Surface

- Put on safety glasses and follow all applicable safety rules.

- Determine the height from the finished floor that the box is to be mounted and make a mark on the masonry at the proper mounting height.

(A)

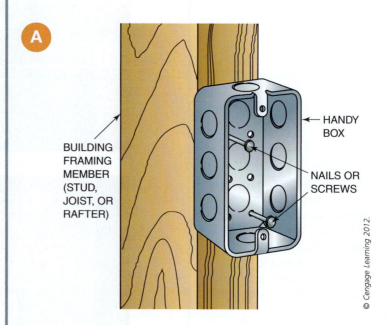

BUILDING FRAMING MEMBER (STUD, JOIST, OR RAFTER)

HANDY BOX

NAILS OR SCREWS

© Cengage Learning 2012.

- Hold the handy box at the location you want to mount it and, using a pencil or permanent marker, mark the masonry through the two mounting holes in the bottom of the box that you wish to use for mounting.

- Using a masonry bit, drill into the masonry at the locations marked in the previous step and install masonry anchors.

B Position the box at the proper location and secure the box to the masonry with the proper screws.

- Clean up the work area and put all tools and materials away.

B

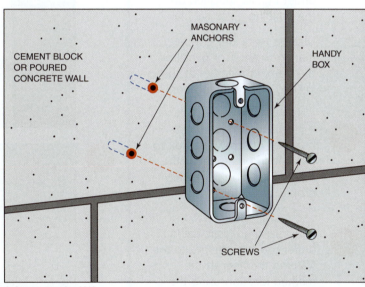

CEMENT BLOCK OR POURED CONCRETE WALL

MASONARY ANCHORS

HANDY BOX

SCREWS

© *Cengage Learning 2012.*

PROCEDURE 10-3

Installing Outlet Boxes with a Side-Mounting Bracket

- Put on safety glasses and follow all applicable safety rules.

- Determine the depth at which the box is to be mounted. Remember the depth will depend on the material thickness that is used for the finished wall or ceiling surface.

- Determine which building framing member will provide the correct location for the outlet box as specified on the building electrical plan.

- Make a mark using a permanent marker or pencil on the framing member at the desired mounting location.

A If installing a metal outlet box, nail or screw the box to the framing member at the marked location using the proper box setout.

B If installing a plastic nail-on outlet box, nail the box to the framing member at the marked location using the proper box setout.

- Clean up the work area and put all tools and materials away.

A

© Cengage Learning 2012.

B

© Cengage Learning 2012.

PROCEDURE 10-4

Installing Outlet Boxes with an Adjustable Bar Hanger

- Put on safety glasses and follow all applicable safety rules.

- Determine the depth at which the box is to be mounted. Remember, the depth will depend on the material thickness that is used for the finished wall or ceiling surface.

- Determine which building framing members will provide the correct location for the outlet box as specified on the building electrical plan.

- Make a mark using a permanent marker or pencil on the framing members at the desired mounting location. Because the bar hanger is secured between two different framing members, be sure to mark the location on both of the studs, joists, or rafters.

 Attach the outlet box to the bar hanger with the screw(s) and fitting supplied with the bar hanger. Do not completely tighten the screw(s) that hold the box to the bar at this time. This will allow the bar to be adjusted back and forth so it fits between the two framing members.

- Adjust the bar so it fits tightly between the two framing members and align the bar hanger with the marks at the mounting location.

© Cengage Learning 2012.

PROCEDURE 10-4

Installing Outlet Boxes with an Adjustable Bar Hanger (Continued)

B While making sure that the box depth will result in a flush box position with the finished surface, nail or screw the bar hanger in place.

B

© Cengage Learning 2012.

C Move the box into the desired position between the framing members and tighten the screw(s) that hold the box on to the hanger.

• Clean up the work area and put all tools and materials away.

C

© Cengage Learning 2012.

PROCEDURE 10-5

Installing Old-Work Electrical Boxes in a Wood Lath and Plaster Wall or Ceiling

- Put on safety glasses and follow all applicable safety rules.

- Determine the location where you want to mount the box and make a mark. Make sure there are no studs or joists directly behind where you want to install the box. Also, make sure that you are able to get a cable to this location.

(A) Make a small cut through the plaster at the location you marked and carefully cut the plaster away until you have exposed one full lath.

(A)

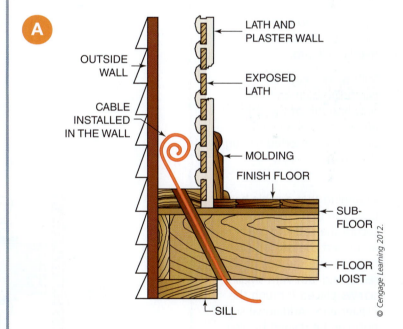

LATH AND PLASTER WALL

EXPOSED LATH

OUTSIDE WALL

CABLE INSTALLED IN THE WALL

MOLDING

FINISH FLOOR

SUB-FLOOR

FLOOR JOIST

SILL

© Cengage Learning 2012.

PROCEDURE 10-5

Installing Old-Work Electrical Boxes in a Wood Lath and Plaster Wall or Ceiling (Continued)

B Turn the old-work box you are installing backward and place it at the mounting location so the center of the box is centered on the full lath. Trace around the box with a pencil. Do not trace around the plaster ears.

- Now, using a sharp tool such as a wood chisel, carefully remove all the plaster from inside the lines of the box tracing.

- Using a keyhole saw, carefully cut out the middle full lath.

- Still using a keyhole saw, carefully notch and remove only enough of the lath above and below so the box can be inserted through the hole.

C Assuming that a cable has been run to the box opening, secure the cable to the box and insert the box into the hole. Secure the box to the remaining lath with small screws placed through the plaster ears. Additional support can be gained by also using Madison hold-its.

- Clean up the work area and put all tools and materials away.

B

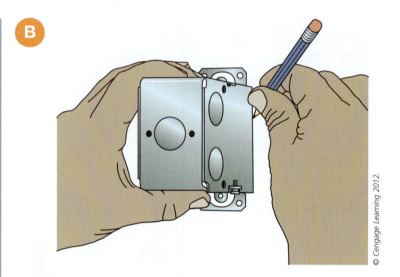

© Cengage Learning 2012.

C

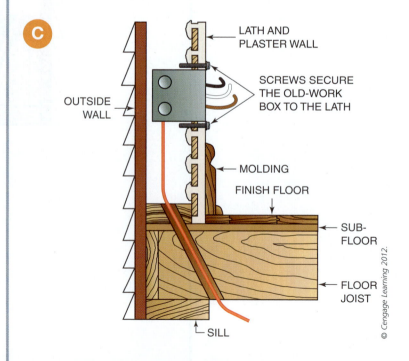

LATH AND PLASTER WALL

SCREWS SECURE THE OLD-WORK BOX TO THE LATH

OUTSIDE WALL

MOLDING

FINISH FLOOR

SUB-FLOOR

FLOOR JOIST

SILL

© Cengage Learning 2012.

PROCEDURE 10-6

Installing Old-Work Metal Electrical Boxes in a Sheetrock Wall or Ceiling

- Put on safety glasses and follow all applicable safety rules.

- Determine the location where you want to mount the box and make a mark. Make sure there are no studs or joists directly behind where you want to install the box.

- Turn the old-work box you are installing backward and place it at the mounting location so the center of the box is centered on the box location mark. Trace around the box with a pencil. Do not trace around the plaster ears.

- Using a keyhole saw, carefully cut out the outline of the box. There are two ways to get the cut started. One way is to use a drill and a flat blade bit (say, a ½-inch size), drill out the corners, and start cutting in one of the corners. The other way that is used often is to simply put the tip of the keyhole saw at a good starting location and, with the heel of your hand, hit the keyhole saw handle with enough force to cause the blade to go through the sheetrock. It usually does not require much force to start the cut this way.

A Assuming that a cable has been run to the box opening, secure the cable to the box and insert the box into the hole. Secure the box to the wall or ceiling surface with Madison hold-its or use a metal or nonmetallic box with built-in drywall grips.

- Clean up the work area and put all tools and materials away.

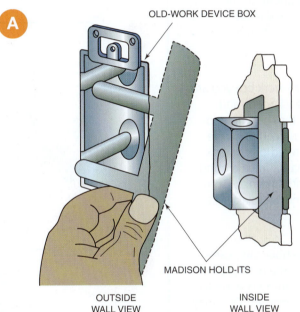

A

OLD-WORK DEVICE BOX

MADISON HOLD-ITS

OUTSIDE WALL VIEW

INSIDE WALL VIEW

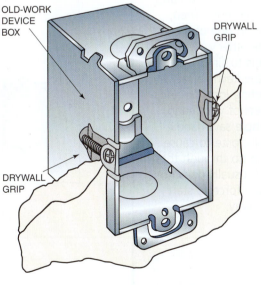

OLD-WORK DEVICE BOX

DRYWALL GRIP

DRYWALL GRIP

© Cengage Learning 2012.

Installing Old-Work Nonmetallic Electrical Boxes in a Sheetrock Wall or Ceiling

- Put on safety glasses and follow all applicable safety rules.

- Determine the location where you want to mount the box and make a mark. Make sure there are no studs or joists directly behind where you want to install the box.

A Turn the old-work box you are installing backwards and place it at the mounting location so the center of the box is centered on the box location mark. Trace around the box with a pencil. Do not trace around the plaster ears.

B Using a keyhole saw, carefully cut out the outline of the box. There are two ways to get the cut started. One way is to use a drill and a flat blade bit (say, a ½″ size) and drill out the corners. Then, start cutting in one of the corners. The other way that is often used is to simply put the tip of the keyhole saw at a good starting location and, with the heel of your hand, hit the keyhole saw handle with enough force to cause the blade to go through the sheetrock. It usually does not require much force to start the cut this way.

A

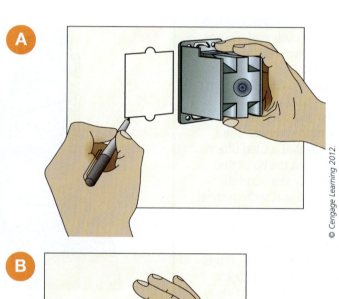

© Cengage Learning 2012.

B

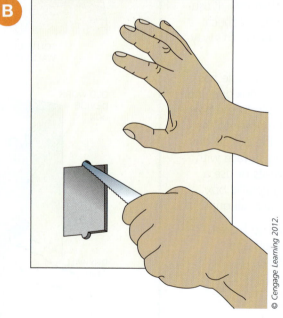

© Cengage Learning 2012.

C Assuming a cable has been run to the box opening, secure the cable to the box and insert the box into the hole.

C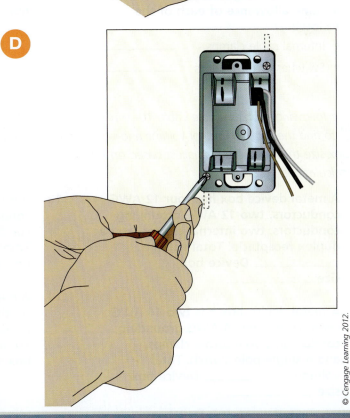

© Cengage Learning 2012.

D Using a Phillips head screwdriver, tighten the screws at both the top and bottom corners of the box. Tightening the screws will cause the holding bracket to rise behind the sheetrock. Continue to tighten the screws until the box is secured in the opening.

- Clean up the work area and put all tools and materials away.

D

© Cengage Learning 2012.

REVIEW QUESTIONS

Directions: Answer the following items with clear and complete responses.

1. Who is it that determines whether metal or nonmetallic electrical boxes will be used in a house?

2. Explain the difference between "new-work" and "old-work" electrical boxes.

3. What does "ganging" electrical boxes mean? Also, why would electricians need a "ganged" electrical box?

4. Describe why an electrician would use an adjustable bar hanger to mount a ceiling electrical box.

5. What is the size opening of a standard metal device box used to mount a switch or receptacle in a wall?

6. Explain why is it important to have the proper "setout" when installing electrical boxes.

7. When calculating box fill, what is the volume allowance of each of the following?

   • Internal cable clamps _____

   • Switches or receptacles _____

   • Pigtails and jumpers _____

   • Wirenuts _____

8. A luminaire weighing more than _____ pounds (_____ kg) must be supported independently of the outlet box unless the box is listed at or above the weight of the fixture.

9. An outlet box used for the support of a ceiling-suspended paddle fan must be listed and marked as suitable for the application and can't support paddle fans weighing more than _____ pounds (_____ kg). When ceiling-suspended paddle fans weighing more than _____ pounds (_____ kg) are supported by an outlet box, the box must have the maximum weight it can support marked on it.

10. Describe a wiring situation where an electrician would be using Madison hold-its.

For the following problems, use Table 10–1 in this book and Table 314.16(A) or Table 314.16(B) of the NEC® to find the total conductor volume and correct box size for the situations listed in the following. List the box size that will be the smallest based on the volume of the box. Show all work in making the box size calculations.

11. A metal device box has four 12 AWG conductors, two 12 AWG grounding conductors, two internal clamps, and a duplex receptacle. Total volume _____ Device box size _____

12. A metal device box has two 14 AWG conductors, one 14 AWG grounding conductor, two internal clamps, and a single-pole switch. Total volume _____ Device box size _____

13. A metal octagon box has seven 14 AWG conductors, three 14 AWG grounding conductors, and two internal clamps. Total volume _____ Octagon box size _____

14. A metal octagon box has eight 14 AWG conductors, two 14 AWG grounding conductors, and two external clamps. Total volume _____ Octagon box size _____

15. A metal 4-inch square box has ten 12 AWG conductors, four 12 AWG grounding conductors, and two internal clamps. Total volume _____ Square box size _____

16. A metal 4-inch square box has six 12 AWG conductors, two 12 AWG grounding conductors, and two internal clamps. Total volume _____ Square box size _____

17. A plastic single-gang device box has a capacity of 25 cubic inches. Calculate the minimum number of 12 AWG conductors allowed. _____ 12 AWG conductors

18. A plastic ceiling box has a capacity of 32 cubic inches. Calculate the minimum number of 10 AWG conductors allowed. _____ 10 AWG conductors

19. A $4'' \times 2\frac{1}{8}''$ square box has a 31-cubic-inch capacity. A $4'' \times \frac{3}{4}''$ raised plaster ring is used in conjunction with the square box and has a capacity of 4.5 cubic inches. Calculate the maximum number of 12 AWG conductors allowed in this box-and-cover combination. _____ 12 AWG conductors

20. Determine the number of 8 AWG conductors allowed in a $4\frac{11}{16}'' \times 2\frac{1}{8}''$ square box. _____ 8 AWG conductors

KNOW YOUR CODES

1. Check with the local or state electrical inspector to find out if there are any specific box mounting heights in your area.

2. Check with the local or state electrical inspector to find out if there are any rules in your area that specify the type of electrical box that must be installed.

3. Check with the local or state electrical inspector to find out if there are any special requirements for supporting ceiling-suspended paddle fans in your area.

4. Check with the local or state electrical inspector to find out if there are any special requirements for supporting lighting fixtures on the wall or ceiling in your area.

Cable Installation

OBJECTIVES *Upon completion of this chapter, the student should be able to:*

- Select an appropriate cable type for a residential application.

- State several *NEC®* requirements for the installation of the common cable types used in residential wiring.

- Demonstrate an understanding of the proper techniques for preparing, starting, and supporting a cable run in a residential wiring application.

- Demonstrate an understanding of the proper installation techniques for securing the cable to an electrical box and preparing the cable for termination in the box.

- Demonstrate an understanding of the common installation techniques for installing cable in existing walls and ceilings.

GLOSSARY OF TERMS

fish the process of installing cables in an existing wall or ceiling

home run the part of the branch-circuit wiring that originates in the loadcenter and provides electrical power to the first electrical box in the circuit

jug handles a term used to describe the type of bend that must be made with certain cable types; bending cable too tightly will result in damage to the cable and conductor insulation; bending them in the shape similar to a "handle" on a "jug" will help satisfy *NEC*® bending requirements

pulling in the process of installing cables through the framework of a house

redhead sometimes called a "red devil"; an insulating fitting required to be installed in the ends of a Type AC cable to protect the wires from abrasion

reel a drum having flanges on each end; reels are used for wire or cable storage; also called a "spool"

running boards pieces of board lumber nailed or screwed to the joists in an attic or basement; the purpose of using running boards is to have a place to secure cables during the rough-in stage of installing a residential electrical system

secured (as applied to electrical cables) fastened in place so the cable cannot move; common securing methods include staples, tie wraps, and straps

supported (as applied to electrical cables) held in place so the cable is not easily moved; common supporting methods include running cables horizontally through holes or notches in framing members, or using staples, tie wraps, or straps after a cable has been properly secured close to a box according to the *NEC*®

n previous chapters, you were introduced to the different cable types used in residential wiring. Now it is time to take a look at how to properly install cables during the rough-in stage. An electrician must have a planned route for the cable before he or she starts to install it. A cable run starts at the service entrance panel (or subpanel if one is installed) and is then run from box to box to power lighting and receptacle outlets on the circuit. Switching locations on a circuit will require cabling to be installed between the switches and the outlets they control. Cables are installed along or through the framing members of a house. In this chapter, we take a look at common methods of installation for electrical cables in new residential construction as well as in existing homes.

SELECTING THE APPROPRIATE CABLE TYPE

The common cable types used in residential wiring were introduced earlier in this text. While nonmetallic sheathed cable (Type NM) is the most often used cable, other types, such as underground feeder cable (Type UF), armored-clad cable (Type AC), metal-clad cable (Type MC), service entrance cable (Type SEU, SER, or USE), and even low-voltage cable for door chime and thermostat wiring, are used in residential wiring. Selecting the appropriate cable type depends on: (1) the cable type allowed by the authority having jurisdiction in your area and (2) the specific wiring application. Many residential building plans will state either on the electrical plan or in the electrical specifications what the required cable type is. In those situations that do not have a specific type listed in the building plans, the electrical contractor will decide the cable type to be used on the basis of the two previously mentioned criteria.

REQUIREMENTS FOR CABLE INSTALLATION

No matter which cable type is used, there are certain installation requirements for that cable type that will have to be followed during the installation of the cable in the rough-in stage. These requirements are outlined in the specific article for that cable type in the *National Electrical Code®* (*NEC®*). Let's take a look at the most important installation requirements that residential electricians need to know.

NONMETALLIC SHEATHED CABLE (TYPE NM)

Article 334 of the *NEC®* covers the installation requirements for nonmetallic sheathed cable (Romex™). Section 334.10(1) states that Type NM cable is allowed to be used in one- and two-family homes, including their attached or detached garages and storage buildings. The following paragraphs summarize the most important installation requirements.

> **⚠ CAUTION**
>
> **CAUTION:** Check with the authority having jurisdiction to make sure Type NM cable is permitted in one- and two-family homes in your area of the country.

Section 334.15 states that when Type NM cable is run exposed and not concealed in a wall or ceiling, the cable must be installed as follows:

- The cable must closely follow the surface of the building finish or be **secured** to **running boards**.

> ### *from experience...*
> Manufacturers of nonmetallic sheathed cable are now color-coding their cables. The cable colors are white for 14/2 and 14/3 Type NM, yellow for 12/2 and 12/3 Type NM, and orange for 10/2 and 10/3 Type NM. This allows ready identification of the cable size, especially during the rough-in stage. An electrician (or electrical inspector) can look around a house that has been roughed-in with Romex™ and easily determine where the 14 AWG, 12 AWG, or 10 AWG cables have been installed.

- The cable must be protected from physical damage where necessary by rigid metal conduit, electrical metallic tubing, Schedule 80 PVC rigid conduit, guard strips, or other approved means. Where passing through a floor, the cable must be enclosed in

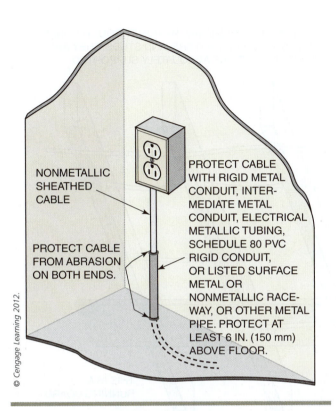

NONMETALLIC SHEATHED CABLE

PROTECT CABLE FROM ABRASION ON BOTH ENDS.

PROTECT CABLE WITH RIGID METAL CONDUIT, INTERMEDIATE METAL CONDUIT, ELECTRICAL METALLIC TUBING, SCHEDULE 80 PVC RIGID CONDUIT, OR LISTED SURFACE METAL OR NONMETALLIC RACEWAY, OR OTHER METAL PIPE. PROTECT AT LEAST 6 IN. (150 mm) ABOVE FLOOR.

© Cengage Learning 2012.

FIGURE 11-1 When Type NM cable is installed exposed where it passes through a floor, the cable must be protected for the entire exposed length or for at least 6 inches (150 mm) from where it comes through the floor.

rigid metal conduit, intermediate metal conduit, electrical metallic tubing, or Schedule 80 PVC rigid conduit extending at least 6 inches (150 mm) above the floor (Figure 11-1).

- Where the cable is run across the bottom of joists in unfinished basements and crawl spaces, it is permissible to secure cables not smaller than two 6 AWG conductors (such as a 6/2 Type NM cable) or three 8 AWG conductors (such as 8/3 Type NM cable) directly to the bottom edge of the joists. Smaller cables must be run either through bored holes in joists or on running boards. Figure 11-2 shows how nonmetallic sheathed cables are installed in an unfinished basement or crawl space through holes drilled in joists, attached to the side or face of joists or beams, and on running boards. Section 300.4(D) requires cables that are run along the side of framing members to be installed at least 1¼ inches (32 mm) from the nearest edge of studs, joists, or rafters.

Type NM cable used on a masonry wall of an unfinished basement is permitted to be installed in a listed conduit or tubing. A plastic bushing or adapter must be used on the end where the cable enters the raceway (Figure 11-3). In some areas of the country, a running board is attached to a masonry wall in an unfinished basement or crawl space and then Type

NM cable is secured to it with staples (Figure 11-4). This wiring practice may not be allowed by some electrical inspectors because they feel that the Type NM cable needs to be protected from physical damage by sleeving it through conduit or tubing.

Section 334.17 covers installing Type NM cable through or parallel to framing members. It states that Type NM cable must be protected in accordance with Section 300.4 where installed through or parallel to framing members. Section 300.4 requirements were covered in Chapter 9, and you should review that section of the text at this time for a more detailed explanation. Only a short overview of Section 300.4 is given in the following:

- If Type NM cable is run through a hole that has an edge closer than 1¼ inches (32 mm) to the edge of the framing member, a metal plate at least ¹⁄₁₆ inch (1.6 mm) thick must be used to protect that spot from nails or screws that could penetrate the cable sheathing.

- If the cable is placed in a notch in wood studs, joists, rafters, or other wood members, a steel plate must protect the cable and be at least ¹⁄₁₆ inch (1.6 mm) thick.

- If the cable is run along a framing member, where the cable is closer than 1¼ inches (32 mm) to the edge, metal protection plates at least ¹⁄₁₆ inch (1.6 mm) thick must be used to protect the cable.

Installing Romex™ in accessible attics is covered in Section 334.23. The installation of Type NM cable in accessible attics or roof spaces must comply with Section 320.23, which also applies to armored-clad cable. This section tells us that if Romex™ is run across the top of floor joists, or within 7 feet (2.1 m) of an attic floor or floor joists, the cable must be protected by guard strips that are at least as high as the cable (Figure 11-5). The 7-foot (2.1 m) rule is also applied when Romex™ is run across the face of rafters or studding in an attic. Where the attic space is not accessible by permanent stairs or ladders, protection is only required within 6 feet (1.8 m) of the nearest edge of a scuttle hole or attic entrance (Figure 11-6). Where the cable is installed along the sides of rafters, studs, or floor joists in an attic, neither guard strips nor running boards are required, but the cable must be kept at least 1¼ inches (32 mm) from the edge of the framing member.

Section 334.24 states that bends in nonmetallic sheathed cable must be made so that the cable will not be damaged. The radius of the curve of the inner edge of any bend during or after installation must not be less than five times the diameter of the cable. This means that when a Romex™ cable is brought through a hole in a building framing member, the cable must be formed in a **jug handle** to ensure that the cable is not bent too sharply. Bending the cable too much will weaken the outer cable sheath as well as the insulation on the individual conductors (Figure 11-7).

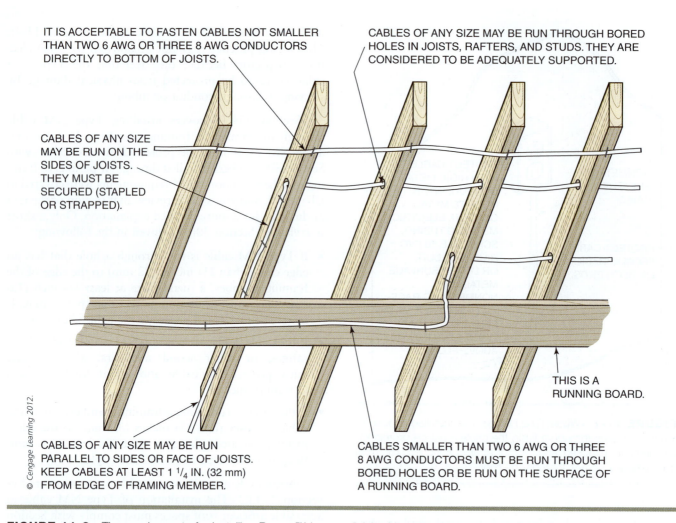

IT IS ACCEPTABLE TO FASTEN CABLES NOT SMALLER THAN TWO 6 AWG OR THREE 8 AWG CONDUCTORS DIRECTLY TO BOTTOM OF JOISTS.

CABLES OF ANY SIZE MAY BE RUN THROUGH BORED HOLES IN JOISTS, RAFTERS, AND STUDS. THEY ARE CONSIDERED TO BE ADEQUATELY SUPPORTED.

CABLES OF ANY SIZE MAY BE RUN ON THE SIDES OF JOISTS. THEY MUST BE SECURED (STAPLED OR STRAPPED).

THIS IS A RUNNING BOARD.

CABLES OF ANY SIZE MAY BE RUN PARALLEL TO SIDES OR FACE OF JOISTS. KEEP CABLES AT LEAST 1 1/4 IN. (32 mm) FROM EDGE OF FRAMING MEMBER.

© Cengage Learning 2012.

CABLES SMALLER THAN TWO 6 AWG OR THREE 8 AWG CONDUCTORS MUST BE RUN THROUGH BORED HOLES OR BE RUN ON THE SURFACE OF A RUNNING BOARD.

FIGURE 11-2 The requirements for installing Romex™ in an unfinished basement.

Securing and supporting Type NM cable is addressed in Section 334.30. It requires nonmetallic sheathed cable to be secured by staples, cable ties, straps, hangers, or similar fittings designed and installed so as not to damage the cable at intervals not exceeding 4½ feet (1.4 m) and within 12 inches (300 mm) of every cabinet, box, or fitting (Figure 11-8). The general requirement of Section 334.30 requires that the cable be secured. Simply draping the cable over air ducts, joists, pipes, and ceiling grid members is not permitted (Figure 11-9).

Section 334.30 also says that flat cables, like 14/2, 12/2, or 10/2 Romex™, must not be stapled on edge. The intent of this section is to prohibit the cable from being installed with its short dimension against a wood framing member. When stapled in this manner, two cables are placed side by side under the staple. If the staple is driven too far into the stud, damage to the insulation and conductors can occur. Flat cables can be installed on top of each other as long as the "flat side" of the cable is against the framing member and the next cable (Figure 11-10). There is no *NEC*® rule against

installing round cables (like a 14/3 or 12/3 Romex™) on top of each other, although this is generally considered not the best way to install these cables.

Section 334.30(A) states that Type NM cables that run horizontally through framing members that are spaced less than 4½ feet (1.4 m) apart and that pass through bored or punched holes in framing members without additional securing are considered *supported* by the framing members. Cable ties or staples are not required as the cable passes through these members. However, the Romex™ cable must be *secured* within 12 inches (300 mm) of any electrical box. Where the cable terminates at a single-gang nonmetallic electrical box that does not contain cable clamps, the cable must be secured within 8 inches (200 mm) of the box, according to an *Exception* to Section 314.17(C) (Figure 11-11).

Section 334.30(B) allows Type NM cables to *not* be **supported** in these situations:

- Where the cable is fished between access points or concealed in finished buildings and supporting is impractical. This is the situation with remodel electrical work.

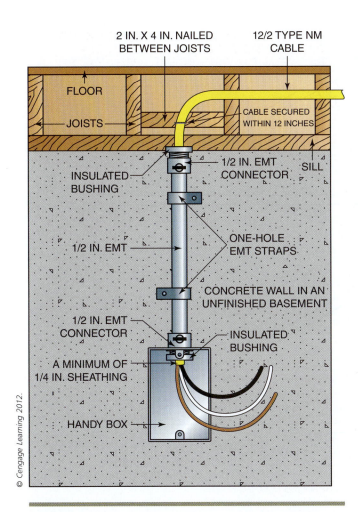

FIGURE 11-3 It is common wiring practice to install Type NM cable through a ½-inch EMT raceway when roughing-in switch or receptacle locations on a wall in an unfinished basement.

- Where the cable is not more than 4½ feet (1.4 m) from the last point of support for connections within an accessible ceiling to luminaires (lighting fixture). This allows short, unsupported lengths of Type NM cable for luminaire connections.

UNDERGROUND FEEDER CABLE (TYPE UF)

Article 340 of the *NEC*® allows Type UF cable to be used in residential wiring for direct burial applications or for installation in the same wiring situations in which you would use nonmetallic sheathed cable. For underground installation requirements, Section 300.5 must be followed. Underground residential branch-circuit installation is discussed in Chapter 15. When installed as nonmetallic sheathed cable, the installation and conductor requirements must comply with the provisions of Article 334.

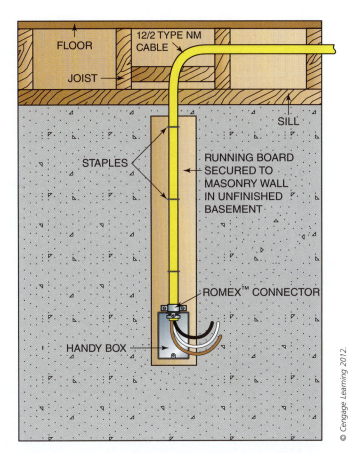

FIGURE 11-4 Electricians often install a running board on a masonry wall and then secure Type NM cable to it with staples. This wiring practice can be used to rough-in wiring to a receptacle or switch location in an unfinished basement or crawl space. Check with the authority having jurisdiction (AHJ) in your area to be sure this wiring practice is allowed. The AHJ may require exposed Type NM cable to be installed in a raceway that will provide protection from physical damage.

Section 340.24 states that bends in Type UF cable must be made so that the cable is not damaged. The radius of the curve of the inner edge of any bend must not be less than five times the diameter of the cable. This is the same rule as for nonmetallic sheathed cable and is intended to prevent damage to the cable and individual conductor insulation.

ARMORED-CLAD CABLE (TYPE AC)

Armored-clad cable, sometimes referred to as "BX" cable by electricians, can be installed exposed or concealed in a house. Section 320.15 requires exposed runs of Type AC cable to closely follow the surface of the building finish or be secured to running boards.

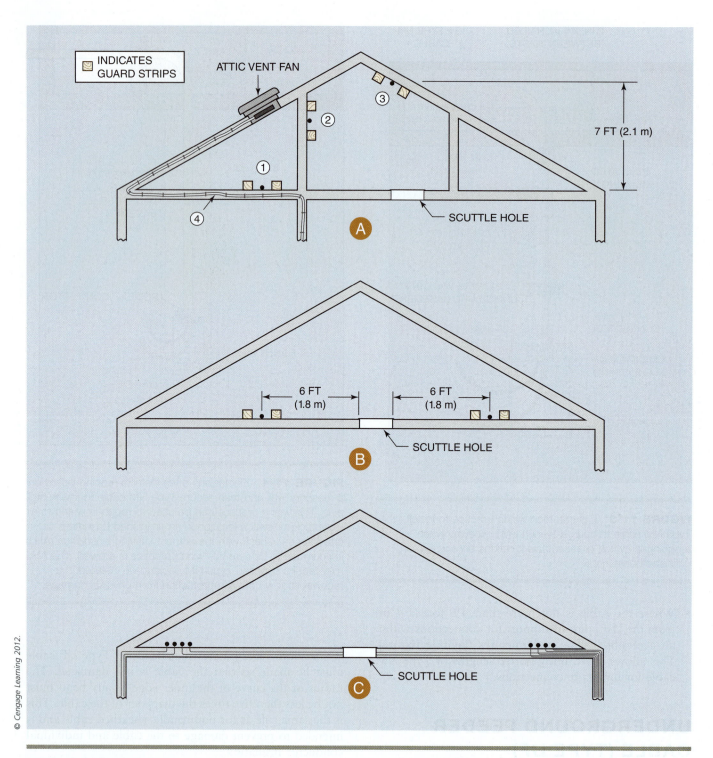

FIGURE 11-5   (A) In accessible attics, cables must be protected by guard strips when: (1) they are run across the top of joists, (2) they are run across the face of studs and are within 7 feet (2.1 m) of the floor or floor joists, or (3) they are run across the face of rafters and are within 7 feet (2.1 m) of the floor or floor joists. Also, (4) guard strips are not required when the cable is run along the sides of joists or rafters in the attic. (B) In attics that are accessible only through a scuttle hole, protection of the cable is required only within 6 feet (1.8 m) of the scuttle hole. (C) An installation of cable in an attic where the cables are located close to the "eaves" is considered safe because it is unlikely that items can ever be stored on top of the cables located in this position.

GUARD STRIPS REQUIRED WHEN CABLES
ARE RUN ACROSS THE TOP OF JOISTS.

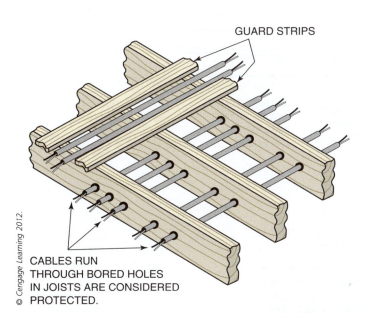

GUARD STRIPS

CABLES RUN
THROUGH BORED HOLES
IN JOISTS ARE CONSIDERED
PROTECTED.

© Cengage Learning 2012.

FIGURE 11-6 Two methods for protecting cables in an accessible attic are to (1) use suitable guard strips and (2) install the cables through holes that are bored into the joists. Remember that there must be at least 1¼ inches (32 mm) from the edge of the bored holes to the edge of the joists.

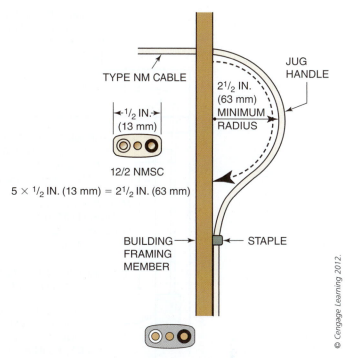

TYPE NM CABLE

½ IN.
(13 mm)

12/2 NMSC

5 × ½ IN. (13 mm) = 2½ IN. (63 mm)

2½ IN.
(63 mm)
MINIMUM
RADIUS

JUG
HANDLE

BUILDING
FRAMING
MEMBER

STAPLE

© Cengage Learning 2012.

FIGURE 11-7 Type NM cable must be bent so that the radius of the bend is not less than five times the cable's diameter. As shown in this illustration, the cable will have to be installed using jug handles when it is bent to avoid damaging the cable.

Exposed runs are permitted to be installed on the underside of joists, like in a basement, where they are supported at each joist and located so as not to be subject to physical damage (Figure 11-12).

Section 320.17 allows Type AC cable to be run through or along framing members in a house. When Type AC cable is run this way, it must be protected in accordance with Section 300.4. In other words, if Type AC cable is run through a hole that has an edge closer than 1¼ inches (32 mm) to the edge of the framing member, a metal plate at least 1/16 inch (1.6 mm) thick must be used to protect that spot from nails or screws that could penetrate the cable sheathing. If the cable is placed in a notch in wood studs, joists, rafters, or other wood members, a steel plate must protect the cable and be at least 1/16 inch (1.6 mm) thick. Also, if the cable is run along a framing member, where the cable is closer than 1¼ inches (32 mm) to the edge, metal protection plates at least 1/16 inch (1.6 mm) thick must be used to protect the cable. A listed and marked steel plate can be less than 1/16 inch (1.6 mm) thick if it still provides equal or greater protection against nail or screw penetration. These requirements are the same as for Type NM cable.

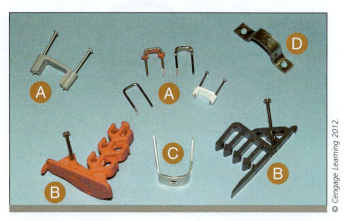

© Cengage Learning 2012.

FIGURE 11-8 Examples of common cable items used to support and secure cables in residential wiring. (A) Staples are the most common type of supporting and securing method; (B) Cable stackers are used to keep multiple cables in the center of a stud; (C) Fold over clip for Type SEU cable; and (D) Two-hole strap for Type SEU and larger Type NM cable.

CAUTION

CAUTION: Armored-clad cable must be protected in the same manner as nonmetallic sheathed cable. Many electricians believe that because Type AC cable has an armored sheathing, no additional protection for the cable is required. They are wrong.

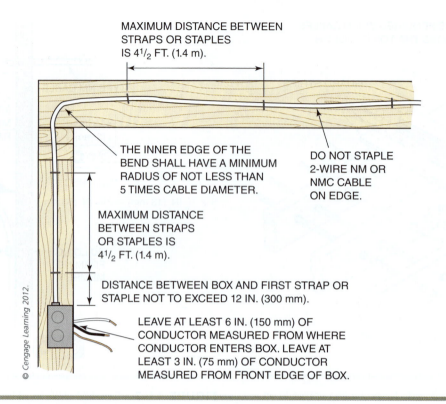

MAXIMUM DISTANCE BETWEEN STRAPS OR STAPLES IS 4¹/₂ FT. (1.4 m).

THE INNER EDGE OF THE BEND SHALL HAVE A MINIMUM RADIUS OF NOT LESS THAN 5 TIMES CABLE DIAMETER.

DO NOT STAPLE 2-WIRE NM OR NMC CABLE ON EDGE.

MAXIMUM DISTANCE BETWEEN STRAPS OR STAPLES IS 4¹/₂ FT. (1.4 m).

DISTANCE BETWEEN BOX AND FIRST STRAP OR STAPLE NOT TO EXCEED 12 IN. (300 mm).

LEAVE AT LEAST 6 IN. (150 mm) OF CONDUCTOR MEASURED FROM WHERE CONDUCTOR ENTERS BOX. LEAVE AT LEAST 3 IN. (75 mm) OF CONDUCTOR MEASURED FROM FRONT EDGE OF BOX.

© Cengage Learning 2012.

FIGURE 11-9   Securing and supporting requirements for nonmetallic sheathed cable.

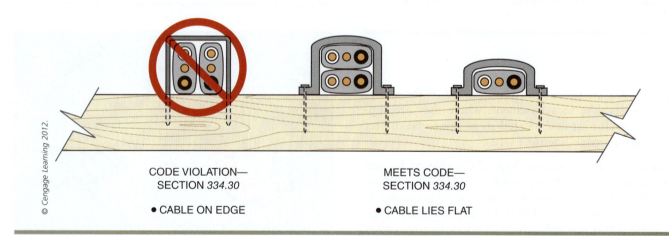

CODE VIOLATION— SECTION *334.30*
● CABLE ON EDGE

MEETS CODE— SECTION *334.30*
● CABLE LIES FLAT

© Cengage Learning 2012.

FIGURE 11-10   Flat-type nonmetallic sheathed cable cannot be stapled on edge.

If installed in accessible attics, Section 320.23 requires Type AC cable to be installed exactly the same way as nonmetallic sheathed cable in that situation. The installation requirements for Type NM cable installed in an accessible attic were presented earlier in this chapter.

Also, like Type NM cable, Type AC cable must not be bent too much. Section 320.24 states that bends in Type AC cable must be made so that the cable will not be damaged. The radius of the curve

of the inner edge of any bend must not be less than five times the diameter of the Type AC cable. Again, the intent of this requirement is to prevent damage to the cable sheathing and the individual conductor insulation (Figure 11-13).

Type AC cable must be supported and secured by staples, cable ties, straps, hangers, or similar fittings designed and installed so as not to damage the cable at intervals not exceeding 4½ feet (1.4 m) and within 12 inches (300 mm) of every outlet box, junction box,

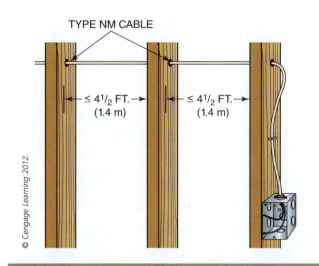

FIGURE 11-11 Horizontal (or diagonal) runs of Type NM cable through framing members are considered supported and secured as long as the distance between the framing members is not more than the minimum interval for support of 4½ feet (1.4 m).

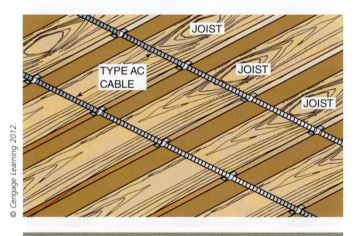

FIGURE 11-12 Exposed runs of Type AC cable on the underside of joists are allowed as long as the cable is secured at each joist.

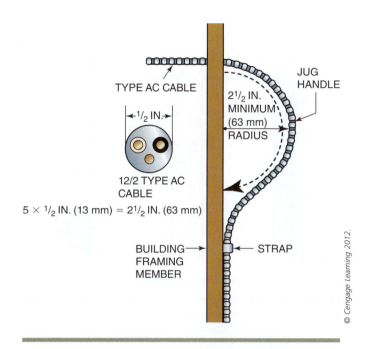

FIGURE 11-13 Type AC cable must be bent so that the radius of the bend is not less than five times the cable's diameter. As shown in this illustration, the cable will have to be installed using jug handles when it is bent to avoid damaging the cable.

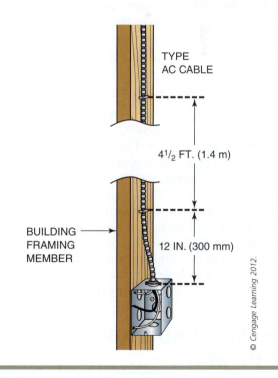

FIGURE 11-14 Type AC cable securing and supporting requirements.

cabinet, or fitting according to Section 320.30. These are the same securing and supporting requirements as for Type NM cable (Figure 11-14).

Section 320.30(C) addresses Type AC being run horizontally through holes and notches in framing members. In other than vertical runs, cables installed in accordance with Section 300.4 are to be considered supported and secured where the support does not exceed 4½-foot (1.4-m) intervals and the armored cable is securely fastened in place by an approved means within 12 inches (300 mm) of each box, cabinet, conduit body, or other armored cable termination (Figure 11-15).

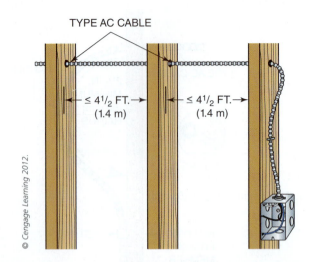

FIGURE 11-15 Horizontal runs of Type AC cable through framing members are considered supported as long as the distance between the framing members is not more than the minimum interval for support of 4½ feet (1.4 m).

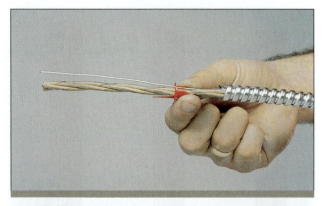

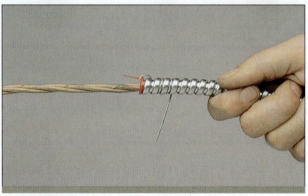

FIGURE 11-16 Anti short bushings used with Type AC cable, commonly called "redheads," prevent the cable conductors from being cut by the sharp edges of the metal armor sheathing. *Courtesy of AFC Cable Systems, Inc.*

According to 320.30(D), Type AC cable is permitted to be unsupported in these situations:

- Where the cable is fished between access points or concealed in finished buildings or structures and supporting is impractical such as in remodel work in existing houses

- Where the cable is not more than 2 feet (600 mm) in length at terminals where flexibility is necessary, like a connection to an electric motor

- Where the cable is not more than 6 feet (1.8 m) from the last point of support for connections within an accessible ceiling to a luminaire

The last installation requirement to be looked at for Type AC cable is in Section 320.40. It states that at all points where the armor of Type AC cable terminates, a fitting must be provided to protect wires from abrasion, unless the design of the outlet boxes or fittings provides equivalent protection. The required insulating fitting is called many names by electricians in the field, including "**redhead**" or "red devil." It gets its name from the red color that the fitting has. The connector or clamp by which the Type AC cable is fastened to boxes or cabinets must be of such design that the insulating bushing (redhead) will be visible for inspection (Figure 11-16).

METAL-CLAD CABLE (TYPE MC)

Metal-clad cable is another cable type that some electricians use to wire a house. This cable type is very popular in some parts of the country. It has gotten to a point that many local electrical distributors are stocking only

Type MC cable and not Type AC cable any longer. Article 330 of the *NEC*® provides some installation requirements for Type MC cable.

> **CAUTION**
>
> **CAUTION:** While there is no *NEC*® requirement that redheads be used with Type MC cable, it is a good idea to do so. Rolls of Type MC cable come from the manufacturer with small bags that contain a few insulated fittings.

Section 330.17 requires Type MC cable to be protected in accordance with Section 300.4 where installed through or parallel to framing members. If Type MC cable is run through a hole that has an edge closer than 1¼ inches (32 mm) to the edge of the framing member, a metal plate at least ¹⁄₁₆ inch (1.6 mm) thick must be used to protect that spot from nails or screws that could penetrate the cable sheathing. If the cable is placed in a notch in wood studs, joists, rafters, or other wood members, a steel plate must protect the cable and be at least ¹⁄₁₆ inch (1.6 mm) thick. Also, if the cable is run along a framing member where the cable is closer than 1¼ inches (32 mm) to the edge, metal protection plates at least ¹⁄₁₆ inch (1.6 mm) thick must be used to protect the cable. A listed and marked steel plate can be less than ¹⁄₁₆ inch (1.6 mm) thick if it still provides equal or greater protection against nail or screw penetration. These requirements are the same as for Type NM cable and Type AC cable.

Section 330.23 requires the installation of Type MC cable in accessible attics or roof spaces to comply with Section 320.23. In accessible attics, Type MC cable installed across the top of floor joists or within 7 feet (2.1 m) of the floor or floor joists must be protected by guard strips. Where the attic is not accessible by a permanent ladder or stairs, guard strips are required only within 6 feet (1.8 m) of the scuttle hole or opening. This requirement is the same as for Type NM and Type AC cable.

Bends in Type MC cable must be made so that the cable will not be damaged. Section 330.24(B) states that the radius of the curve of the inner edge of any bend must not be less than seven times the external diameter of the metallic sheath for the style of Type MC used by electricians in residential wiring. This style has an interlocking-type armor or corrugated sheathing and looks very similar on the outside to Type AC cable. This requirement simply means that when installing Type MC cable, bigger turns and jug handles will have to be used.

Securing and supporting Type MC cable is addressed in Section 330.30. The general rule states that Type MC cable must be supported and secured by staples, cable ties, straps, or similar means and installed in a way that won't damage the cable. Section 330.30(C) covers horizontal installation through framing members and allows Type MC cables installed in accordance with Section 300.4 to be considered supported and secured where the framing members are no more than 6 feet (1.8 m) apart. Cable ties or other securing methods are not required as the cable passes through these members. According to Section 330.30(B), Type MC cable containing four or fewer conductors of 10 AWG or smaller is required to be secured within 12 inches (300 mm) from every box, cabinet, or fitting and at intervals not exceeding 6 feet (18 m). This means that in residential work, Type MC for branch circuits of 14, 12, and 10 AWG must be secured within 12 inches (300 mm) of each box and then no longer than every 6 feet (1.8 m) (Figure 11-17). Section 330.30(D) allows Type MC cable to be installed unsupported in these situations:

- When it is fished between access points or concealed in finished buildings or structures and supporting is impracticable (Figure 11-18)

- When it is not more than 6 feet (1.8 m) from the last point of support for connections within an accessible ceiling to a luminaire

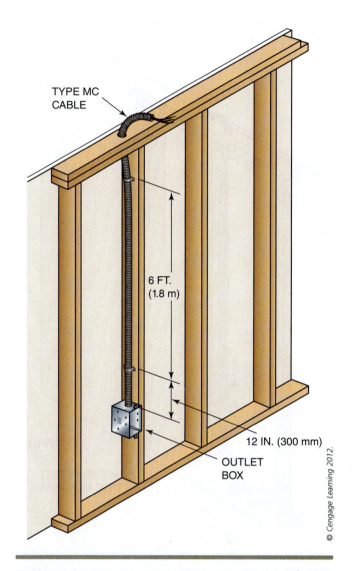

FIGURE 11-17 Type MC cable must be supported and secured at intervals not exceeding 6 feet (1.8 meters) and within 12 inches (300 mm) of an electrical box.

Fittings used for connecting Type MC cable to boxes, cabinets, or other equipment must be listed and identified for such use according to Section 330.40. Connectors should be selected in accordance with the size and type of cable for which they are designated. Some Type AC cable connectors are also acceptable for use with Type MC cable when specifically indicated on the device or the shipping carton.

SERVICE ENTRANCE CABLE (TYPE SEU AND SER)

Service entrance cable installation, when used as a service entrance wiring method, was covered in Chapter 8. However, there are certain situations in residential wiring when service entrance cable is used as the branch-

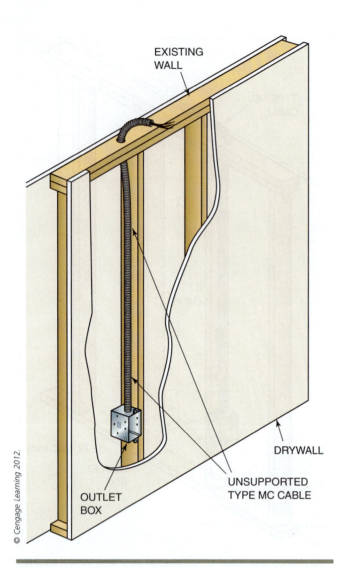

EXISTING WALL

DRYWALL

UNSUPPORTED TYPE MC CABLE

OUTLET BOX

© Cengage Learning 2012.

FIGURE 11-18 The *NEC*® permits Type NM, Type AC, and Type MC cable to be fished in walls, floors, or ceilings of existing buildings without support. This illustration shows a Type MC cable fished into the wall of an existing building. It is not required to be secured and supported.

counter-mounted cooking units, or electric clothes dryers. Type SER cable would have to be installed for these situations. The *Exception* to this section does permit a bare neutral service entrance cable for existing installations only and is coordinated with Sections 250.32 and 250.140 of the *NEC*®. Prior to the 1996 *NEC*®, it was permissible to use Type SEU with a bare grounded (neutral) conductor as the wiring method used to supply electric ranges and electric clothes dryers.

Section 338.10(B)(4) says that, in addition to the provisions of Article 338, Type SE, cable used for interior wiring must comply with the installation requirements of Article 334 for nonmetallic sheathed cable. This means that when using Type SE for branch-circuit or feeder runs inside a dwelling unit, the electrician must install the Type SEU or Type SER service entrance cable the same way as for nonmetallic sheathed cable. When used inside a house and installed in thermal installation, the ampacity of Type SE cable is found in the 60 degree column of Table 310.15(B)(16).

PREPARING THE CABLE FOR INSTALLATION

The first thing that needs to be done in preparation for installing branch-circuit and feeder cables during the rough-in stage is to simply get them ready for installation. It is important that a residential electrician know the correct procedure for setting up and then unrolling cables from a box, roll, spool, or **reel**. In this section, the terms "spool" and "reel" mean the same thing. When electrical contractors purchase nonmetallic sheathed cable (Type NM) or underground feeder cable (Type UF) from the local electrical supplier, they usually get it in one of two ways: (1) as a 250-foot (75 m) roll packaged in plastic or in a cardboard box or (2) on a 1000-foot (300 m) reel. Type AC and Type MC cables are purchased in much the same way. Each is available in rolls of 250 feet (75 m) but do not come in a plastic wrapping or in a cardboard box. Rolls of Type AC or Type MC cable are usually held together with wire ties that must be cut away before the cables may be unrolled. Type AC and Type MC cable is also available on reels of 1000 feet (300 m), and many electrical contractors choose to purchase these cable types on reels. Service entrance cable (Type SE) is usually purchased from the electrical supplier in the lengths needed for the service installation of the house being wired. However, there are many electrical contractors who purchase larger reels of service

circuit wiring method for individual appliances, such as electric furnaces, electric ranges, and electric clothes dryers, or as the wiring method for a feeder going to a subpanel. Section 338.10(B) of the *NEC*® allows us to use service entrance cable as branch-circuit or feeder conductors. Branch circuits using service entrance cable as a wiring method are permitted only if all circuit conductors within the cable are fully insulated with an insulation recognized by the *NEC*®. The equipment-grounding conductor is the only conductor permitted to be bare within service entrance cable used for branch circuits. Service entrance cable containing an uninsulated conductor (Type SEU) is not permitted for new installations where it is used as a branch circuit to supply appliances such as ranges, wall-mounted ovens,

FIGURE 11-19   When unrolling cable from a roll, the best way is to pull the cable out of the roll from the inside. In this picture, an electrician is pulling Type NM cable from the inside of a 250-foot roll.

FIGURE 11-20   Unrolling cable from a roll is very easy when using a device commonly called a "spinner." This picture shows an electrician pulling a length of Type NM cable from a spinner attached to a wall stud.

FIGURE 11-21   Spinners can also be located on the floor.

entrance cable, especially if they have many homes to wire that all require service entrances installed with service entrance cable.

When cables are used from a roll, the best way to begin unrolling them is to start from the inside of the roll. If the roll was purchased in a box, many manufacturers provide a round cutout on the box that is used to indicate the starting point for unrolling the cable. If the cable is not unrolled properly, twists and kinks of the cable will develop that could damage the cable insulation (Figure 11-19).

Rolls of cable can also be unrolled easily with a device commonly called a "spinner." The spinner is usually attached to a wall stud in the building being wired (Figure 11-20) or the spinner can be set on the floor (Figure 11-21). A roll of cable is placed on the spinner, and the electrician simply pulls the cable off the roll from the outside. As the cable is pulled, the spinner spins, allowing the cable to be unrolled in a manner that does not produce twists or kinks.

For cable that comes on reels, a commercial reel holder helps simplify unrolling cable from the reel. These reel holders are readily available from the local electrical supplier. Some electricians build their own by

the cables are pulled in, the job will be done in a safe, neat, and workmanlike manner. The installation must also meet or exceed the installation requirements of the *NEC®*. The following items should always be considered by an electrician when installing a cable run:

- Plan the cable route ahead of time and visually line it all up in your mind.

- If holes have to be drilled in the building framing members, drill them all, making sure to maintain a minimum clearance of 1¼ inches (32 mm) from the edge of the hole to the edge of the framing member.

- When drilling holes through consecutive framing members for longer cable runs, align the holes as closely as possible. This will result in making it easier to pull cables through the holes and will make the installation neat and organized.

© Cengage Learning 2012.

FIGURE 11-22 Large cable sizes and long lengths of smaller cable are usually purchased on reels.

simply using a length of pipe placed between two framing members or between two step ladders. The reel is placed on the pipe, and the cable is unrolled from the reel. Either using a commercial reel holder or a "homemade" reel holder will result in cable that is unrolled without twists or kinks (Figure 11-22).

INSTALLING THE CABLE RUNS

Installing the cable in the framework of a house is often referred to as **pulling in** the cable. The cables must be routed through or along studs, joists, and rafters to all of the receptacle, switch, or lighting outlet locations that make up a residential circuit. There are a few common installation techniques and procedures that electricians need to become familiar with so that when

Green Tip

During the rough wiring phase, an electrician looks for the shortest practical route for feeders and branch circuit cable runs. Sometimes routing cable through a floor or ceiling is shorter than running it through perimeter walls; other times the walls provide the shortest route. A few minutes spent considering alternative cable routes and selecting the most efficient will save materials and often labor as well.

from experience...

Many electricians have found that the best way for them to unroll cable from a roll is to bend over and pick up the roll and, with their hands and wrists placed though the opening in the roll, rotate the roll and play out the cable as they walk across a room. This technique will allow two-wire "flat" cable, like 14/2, 12/2, and 10/2 Romex™, to be unrolled without twists or kinks. The cable is then ready for installation. This technique also works well with three-wire Romex™ and armored cables (Figure 11-23).

© Cengage Learning 2012.

FIGURE 11-23 This picture is showing an electrician unrolling Type NM from the outside of the roll. This technique will result in very few kinks or twists in the cable.

Green Tip

Reducing the amount of materials needed to build or remodel a house is a green building practice. The one material that electricians use more of than any other to wire a house is cable. Every green home project should have a material recycling and waste management plan. The job site recycling program minimizes the amount of waste that ends up in landfills and saves money in trash hauling and dumping fees. The electrician can incorporate waste from rough and finish wiring phases into the job site recycling system.

- Once the cable has been pulled to all the electrical boxes on the circuit run, double check your work for the following:

  - All locations on the circuit have the correct cable installed.

  - The cable is correctly secured and supported according to the *NEC®*.

  - There are no twists or kinks in the cable.

  - Cable turns have not been bent too tightly.

  - There is enough cable at each electrical box location for the required connections.

> ### CAUTION
>
> **CAUTION:** Houses are sometimes built with a metal roof. Section 300.4(E) addresses installation of a cable under metal-corrugated sheet roofing. It requires a cable to be installed and supported so there is at least 1½ inches (38 mm) from the lowest surface of the roof material to the top of the cable. This rule is designed to minimize the chance of long roofing screws penetrating through a cable sheathing and damaging the conductors inside.

INSTALLING CABLES THROUGH STUDS AND JOISTS

There are several ways for electricians to install the cable runs through the wall studs of a house; however, there are four methods commonly used:

1 Drilling holes that are all a certain distance from the floor and in line. This method will allow the easiest pulling of the cables.

2 When there is a need to make changes in elevation along the cable run, there are two ways to accomplish it:

- Drill holes parallel to the floor at a certain height, keeping the holes in line, and then repeat the drilling process at a new elevation from the floor. The cable will be run with small turns to get from one elevation to the next (Figure 11-24).

- Drill holes that are gradually higher from one stud to the next (Figure 11-25).

3 Drilling holes close to the bottom of the studs on outside walls. This is a technique that keeps the cable run close to the bottom plate of a wall and allows for a better and easier fit for insulation that will be placed in the wall between the studs. (Figure 11-26).

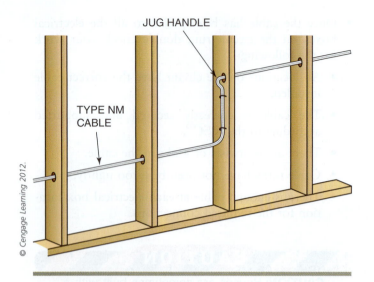

FIGURE 11-24   A cable run through studs with a change in elevation.

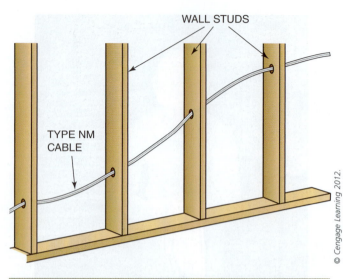

FIGURE 11-25   A cable run through studs with a gradual elevation change.

> ## *from experience...*
>
> Drawing a cable diagram for the circuits you are running in a house helps keep the cable installation organized. Use the common electrical symbols discussed in Chapter 5 when doing a cable diagram. The cable diagram will quickly tell a new electrician where to run a cable, as well as how many conductors must be in each cable run. As each cable run is installed, an electrician can cross off the completed runs on the cable diagram. This will help ensure that all the wiring has been roughed in that is supposed to be roughed in. It is a very bad feeling when the walls and ceilings are installed and an electrician discovers that some cables were not run.

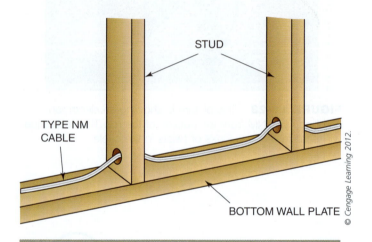

FIGURE 11-26   Drilling the holes for installing cables at the bottom of wall studs is a good practice. It allows the cables to be placed along the bottom of the wall and results in an easier installation of thermal insulation placed in the wall spaces between the studs.

Installing cables through ceiling or floor joists is similar to installing cables through wall studs, and some of the methods previously described will work just fine. However, some cable installations in joists will require the use of one of the installation methods discussed here:

- Drill holes through the joists. This is similar to drilling holes through wall studs. Try to line up the holes as you drill them so the cables will pull through easier (Figure 11-28).

4 Some electricians notch the studs instead of drilling them. The cable is simply laid into the notches, and then a 1/16-inch (1.6 mm) or thicker metal plate is placed over the notch to protect the cables from screws and nails and also to keep the cable from coming out of the notch. Not all areas allow notching of studs. Check with the local electrical inspector to find out for sure if notching studs is allowed in your area (Figure 11-27).

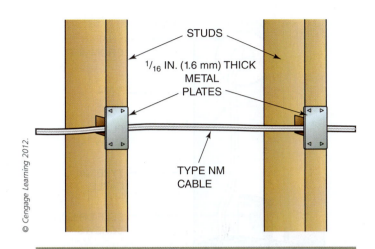

FIGURE 11-27 The *NEC®* allows cutting notches in wall studs for cable installation, but not all local codes allow it.

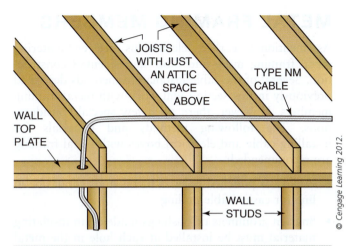

FIGURE 11-29 Running cables over the tops of joists.

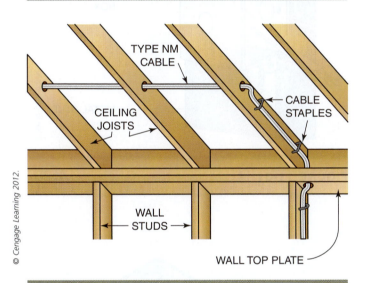

FIGURE 11-28 Drilling holes in joists and routing cable through the holes.

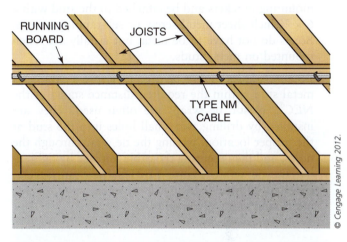

FIGURE 11-30 Running boards can be used to route cables along the undersides of joists and rafters.

- Running the cables over the joists. This is accomplished by drilling a hole through the top plate of a wall section and running the cable above the ceiling joists. This technique will work when the joists are in an attic and no flooring is to be put on top of them. No additional support is required in this case since the distance between the joists will be much less than the maximum distance required between supports of all cables used in residential wiring (Figure 11-29).

- Running boards are sometimes used in a basement. The running boards are secured to the bottom of the joists in a basement in an end-to-end manner for the full length of the cable run. The cable is then stapled to the running boards at intervals consistent with *NEC®* requirements. This technique eliminates the need to drill through joists (Figure 11-30).

Green Tip

An electrician generates a lot of cable cut-offs when installing cables for branch and feeder circuits. Left-over materials often end up being tossed into the trash can, even though they are still good materials that can be used. Following green practices, the electrician should save and sort lengths of different size cable and either use the cable on the current job or save them for the next job. Cable cut-offs 6 inches or longer can still be used for pigtails, short box to box runs, switch legs, and other short runs.

METAL FRAMING MEMBERS

As mentioned earlier, electricians are encountering metal framing more and more in residential construction. Several of the cable installation methods described previously in this section will work with metal framing members. However, there are a few special considerations. The following methods and materials for installing cable and electrical boxes with metal framing are recommended:

- Drilled or punched holes in metal studs should be in line for easier cable pulling.

- Snap-in grommets or bushings made of an insulating material must be installed at each hole in the metal framing member. They are required to protect the cable sheathing from the sharp edges of metal.

- Electrical device boxes should be mounted with side-mounting brackets and be attached to the stud with a 6 × ¾-inch sheet metal screw or self-tapping screws. Boxes do not have to be metal but usually are when mounted on metal studs.

- Because staples cannot be used to secure a cable to a metal stud within the required distance stated in the *NEC®*, plastic tie wraps are often used. They are installed by drilling two small holes into the stud at the proper location, passing the tie wrap through the holes, and then securing the cable in place by tightening the tie wrap (Figure 11-31). There are several types of commercially available cable supports for metal stud work (Figure 11-32).

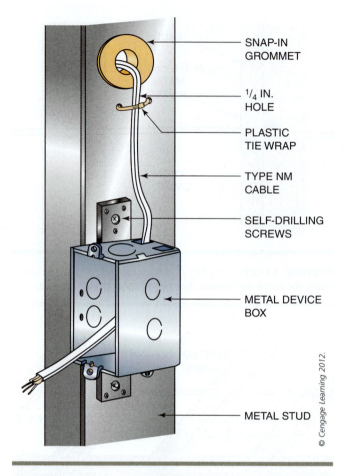

SNAP-IN GROMMET

¼ IN. HOLE

PLASTIC TIE WRAP

TYPE NM CABLE

SELF-DRILLING SCREWS

METAL DEVICE BOX

METAL STUD

© Cengage Learning 2012.

FIGURE 11-31   A cable installed in a metal stud application.

STARTING THE CABLE RUN

Although it is not a specific requirement, the best place to start a cable run for a branch circuit is at the service panel or subpanel. This is considered to be the starting point of a circuit because it is the location on the circuit where electrical power is provided. It is also where the circuit overcurrent protection devices are located. The branch-circuit cable run that goes from the service panel or subpanel to the first electrical box of the circuit is called a **home run**. Once an electrician has more experience and understands the wiring process better, starting the circuit run at other locations is possible. However, for now we will assume the starting point to be at the loadcenter. The procedures discussed in this section assume the use of nonmetallic sheathed cable. Other wiring methods may require some modification of the procedures shown.

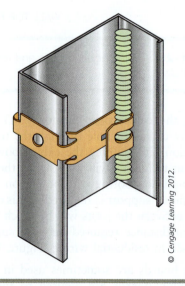

© Cengage Learning 2012.

FIGURE 11-32   A common style of metal stud cable support. This type of support can be used with metal clad cables or Type NM cables.

(See Procedure 11-1 on pages 387–390 for step-by-step instructions on starting a cable run from a loadcenter.)

SECURING AND SUPPORTING THE CABLE RUN

Once the cable has been run to each electrical outlet box on the circuit, an electrician must make sure that the cable is properly secured and supported. Some electricians secure and support the cable as they pull the cable into place. Other electricians prefer to make the full cable run for the circuit and then go back and properly secure and support the run. There is no right or wrong way. It is up to the electrician to decide which technique to use.

AT THE DEVICE AND OUTLET BOX

The *NEC*® requirements for securing and supporting cables attached to nonmetallic or metal electrical boxes were discussed earlier in this book. This section provides a quick review of common securing techniques.

If you are using single-gang nonmetallic electrical device boxes, at least 1/4 inch (6 mm) of the cable sheathing must extend into the box. This technique provides extra protection for the cable conductors from cuts to the insulation that could occur from the sharp edges of the box (Figure 11-33). Also, make sure to secure the cable to the building framing member no more than 8 inches (200 mm) from the box with a staple or some other approved method (Figure 11-34).

If you are using nonmetallic round ceiling boxes, nonmetallic device boxes of two-gang or more, or any metal boxes, you must use an approved connector to secure the cable to the box. Built-in cable clamps or a separate cable connector can be used. Although it is not an *NEC*® rule, approximately 1/4 inch (6 mm) of cable sheathing should be exposed inside the box (Figure 11-35). Again, the technique of having the cable sheath extend into the electrical box a short distance will help keep the conductor insulation from being cut from sharp edges on the box. The cable must be secured with staples or other approved method within 12 inches (300 mm) of any of these box types (Figure 11-36).

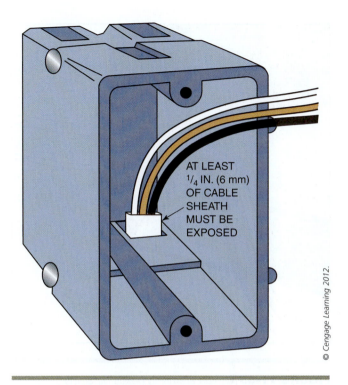

FIGURE 11-33 The Type NM cable must extend at least ¼ inch into a single-gang nonmetallic device box.

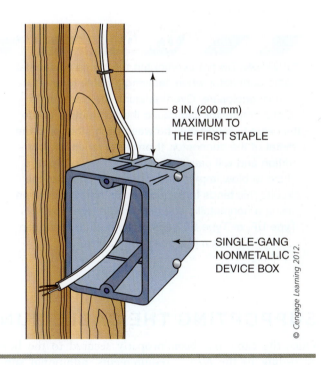

FIGURE 11-34 The cable must be secured within 8 inches (200 mm) of a single-gang nonmetallic device box.

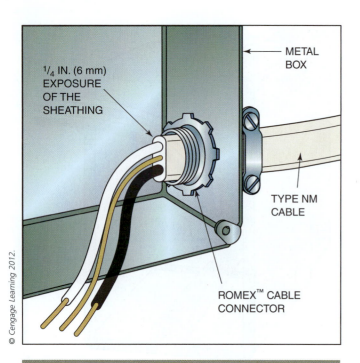

© Cengage Learning 2012.

1/4 IN. (6 mm) EXPOSURE OF THE SHEATHING

METAL BOX

TYPE NM CABLE

ROMEX™ CABLE CONNECTOR

FIGURE 11-35   If you are using boxes other than single-gang nonmetallic device boxes, it is recommended that you extend at least ¼ inch (6 mm) of Type NM cable into the box. This illustration shows a metal box with an external cable clamp securing the cable to the box. Notice that ¼ inch (6 mm) of cable is exposed in the box.

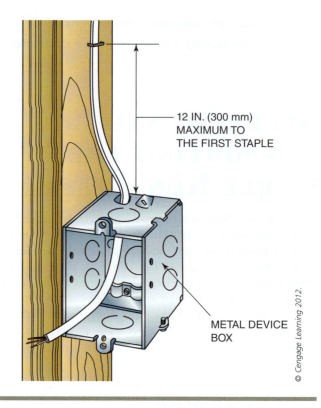

12 IN. (300 mm) MAXIMUM TO THE FIRST STAPLE

METAL DEVICE BOX

© Cengage Learning 2012.

FIGURE 11-36   The cable must be secured within 12 inches (300 mm) of all metal boxes types and all nonmetallic boxes other than single-gang nonmetallic device boxes.

CAUTION

CAUTION: Do not overtighten the clamp of a metal cable connector when securing a cable to an electrical enclosure. Overtightening the clamp on a metal connector can cause the conductors inside the cable to "short out" between each other or to the metal of the connector. It presents an unsafe condition and will cause a circuit breaker to trip or a fuse to blow immediately when you energize the circuit. This tends to happen much more often when using a nonmetallic sheathed cable like Type NM, Type UF, or Type SE cable, but could also happen with metal sheathed cables.

SUPPORTING THE CABLE RUN

Once the cable has been properly secured to the box according to the *NEC*® requirements, additional support of the cable must be done. Horizontal runs of cable through holes in building framing members is adequately supported as long as the distance between framing members does not exceed the *NEC*® support requirements for a particular wiring method. For example, a Romex™ cable installed through holes drilled in studs that are placed 16 inches (400 mm) O.C. (on center) means that the cable is adequately supported because the distance between supports is less than the 4½-foot (1.4 meters) requirement for supporting Type NM cable. When the cable is run along the sides of a building framing member, it must be supported according to the *NEC*® requirements. For example, Romex™ cable running along the side of a stud must be supported with staples or other approved means not more than every 4½ feet (1.4 meters) (Figure 11-37).

Section 300.11(A) states that suspended-ceiling support wires and associated fittings that provide secure support and that are installed *in addition to* the ceiling grid support wires are permitted as the cable support. Where independent support wires are used, they must be secured at both ends. Suspended-ceiling grids cannot support cables. Wiring methods of any type and all

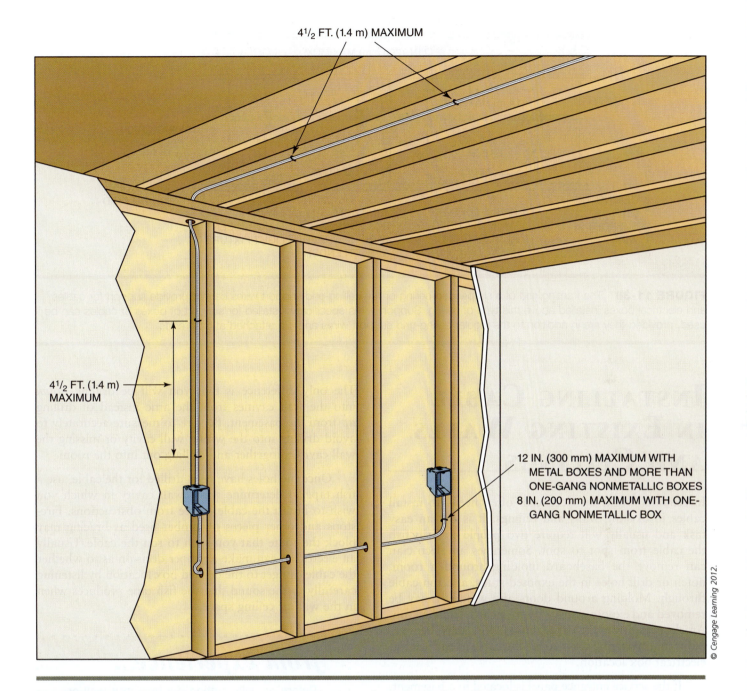

4¹/₂ FT. (1.4 m) MAXIMUM

4¹/₂ FT. (1.4 m) MAXIMUM

12 IN. (300 mm) MAXIMUM WITH METAL BOXES AND MORE THAN ONE-GANG NONMETALLIC BOXES
8 IN. (200 mm) MAXIMUM WITH ONE-GANG NONMETALLIC BOX

© Cengage Learning 2012.

FIGURE 11-37 Type NM cable must be supported at intervals no greater than 4½ feet (1.8 m) when it is run along studs, joists, and rafters.

luminaires are not allowed to be supported or secured to the support wires or "T-bars" of a suspended-ceiling assembly unless the assembly has been tested and listed for that use. If support wires are selected as the supporting means for cables within the ceiling cavity, they must be distinguishable from the ceiling support wires and secured at both ends (Figure 11-38). A common method used by electricians to distinguish a cable support wire from a ceiling support wire is to color-code the cable support wire with tape or paint.

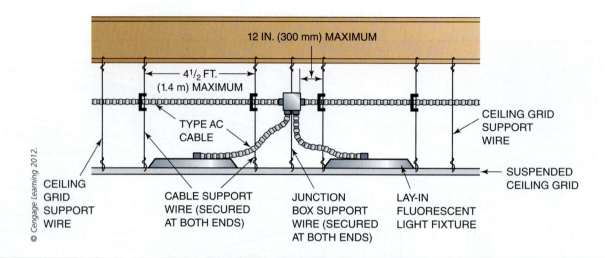

12 IN. (300 mm) MAXIMUM

4½ FT. (1.4 m) MAXIMUM

TYPE AC CABLE

CEILING GRID SUPPORT WIRE

CEILING GRID SUPPORT WIRE

CABLE SUPPORT WIRE (SECURED AT BOTH ENDS)

JUNCTION BOX SUPPORT WIRE (SECURED AT BOTH ENDS)

LAY-IN FLUORESCENT LIGHT FIXTURE

SUSPENDED CEILING GRID

© Cengage Learning 2012.

FIGURE 11-38 The framing grid of a suspended-ceiling or the ceiling-grid support wires cannot provide support for cables and electrical boxes installed above this type of ceiling. Support wires specifically installed for support of boxes or cables can be used, provided they are in addition to the regular ceiling-grid support wires and are attached at both ends.

INSTALLING CABLE IN EXISTING WALLS AND CEILINGS

In remodel work, an electrician often needs to install cables in existing walls and ceilings. It is not an easy task and usually will require two people to help **fish** the cable from spot to spot. Sometimes an electrician can remove the baseboard molding around a room, notch or drill holes in the exposed studs, and run cable through. Molding around doors and windows may be removed and expose a space where a cable could be run. An electrician running cables in old work must be ingenious in discovering ways to get the cable to each circuit electrical box location.

If the service entrance panel is located in a basement, new cable runs can start at the panel and be run along the exposed basement framing members until the cable is under the wall section in which you want to install an outlet. Drill a hole from the basement up and into the wall cavity as shown in Figure 11-39. A long auger bit works well for this operation. The drilling must be done at the proper angle to get from the basement into the wall. If you are drilling from the basement up through the floor and into an inside wall partition, there is no need to drill at an angle. Be prepared to drill through several items, including flooring, bottom wall plates, studs, joists, and so on. This same drilling technique can be used when running cable from an unfinished attic space into the wall cavities of the rooms below.

The only difference is that you will be drilling *down* into the wall cavities from the attic instead of drilling *up* from the basement. Be sure to measure accurately to avoid drilling into the wrong wall cavity or missing the wall cavity altogether and drilling out into the room.

Once the holes have been drilled for the cable, use a fish tape to determine if the wall cavity in which you wish to install the cable is free from obstructions. Fire-stops and other pieces of lumber used as bracing may block the route that you wish to run the cable. Usually an electrician can make a proper decision as to whether the cable can get to the desired box location by listening carefully to the sound that the fish tape produces when in the wall or ceiling space.

from experience...

Before actually cutting the hole in a wall or ceiling for the installation of an old-work electrical box, be sure you can get the wiring method you are using to the box location. Many electricians have cut holes in existing walls and ceilings and then found out that there is no available route for the circuit wiring to get to the box. The result of this poor planning is a hole in a wall or ceiling that will now have to be patched.

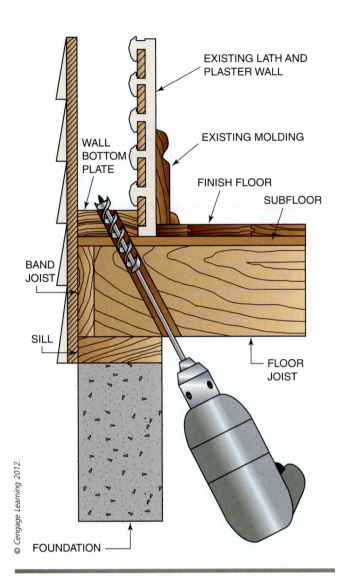

EXISTING LATH AND
PLASTER WALL

EXISTING MOLDING

WALL
BOTTOM
PLATE

FINISH FLOOR

SUBFLOOR

BAND
JOIST

SILL

FLOOR
JOIST

FOUNDATION

© Cengage Learning 2012.

FIGURE 11-39 Drilling into a wall from a basement in an existing dwelling often involves angling the drill bit so you can get into the proper wall cavity.

Once the electrical outlet box hole has been cut, the cable can be installed through the walls or ceiling to the box location. This is usually a two-person job. One electrician will push a fish tape through the bored holes toward the electrical box location. Another electrician will look for the fish tape at the electrical box location and catch the end of the fish tape when it gets to the box hole. The electrician at the box hole will attach a cable to the end of the fish tape and feed it carefully through the box hole as the electrician at the other end carefully pulls the fish tape with the attached cable back into the basement or attic space. The installed cable can then be attached to a junction box or to the service entrance panel.

Sometimes a cable will have to be installed in a location that does not allow one end of a fish tape to be easily seen and caught at the electrical box location. Two fish tapes with hooks formed on the ends are required for this installation. One electrician will feed fish tape 1 into the wall or ceiling space from one end of the desired cable run. At the other end of the run, another electrician will feed fish tape 2 into the same area. By twisting and moving the fish tapes around, an electrician can feel and hear when contact is made between the two fish tapes. Once contact is established, the electricians will slowly pull back on one end or the other until the two hooks "catch" each other. This may take several attempts before the ends are hooked together. While keeping steady pressure on the fish tapes so the hooked ends do not come apart, one electrician will pull fish tape 1 and the "hooked" fish tape 2 back to them. A cable is then attached to the fish tape 2 end, and the other electrician simply pulls fish tape 2 and the newly attached cable back to the new location, completing the cable installation between the two points.

SUMMARY

Once an electrician has determined the routing for the various branch circuits in a house, the first step is to locate and install the appropriate electrical boxes. The next step is to install the appropriate circuit conductors. In this chapter, we took a look at common installation techniques for cables. Selecting the appropriate cable for the job and installing it according to the proper *NEC*® requirements was discussed. Preparing the cable for installation was covered and included a discussion of how to properly unroll cables from reels and other packaging methods. Installation techniques discussed included how to actually start a cable run, how to properly support a cable run, how to properly secure a cable to an electrical box, and how to prepare the cable for connection to electrical equipment at each electrical box. The last item covered in the chapter was a discussion of cable installation techniques in existing homes. Following the cable installation techniques presented in this chapter and the *NEC*® installation requirements for the cable type you are installing in a house will result in a wiring job that will be safe and reliable.

GREEN CHECKLIST

☐ During the rough wiring phase, an electrician looks for the shortest practical route for feeders and branch circuit cable runs. A few minutes spent considering alternative cable routes and selecting the most efficient will save materials and often labor as well.

☐ Every green home project should have a material recycling and waste management plan. The one material that electricians use more of than any other to wire a house is cable. The electrician can incorporate waste from rough and finish wiring phases into the job site recycling system.

☐ An electrician generates a lot of cable cut-offs when installing cables for branch and feeder circuits. Cable cut-offs 6 inches or longer can still be used for pigtails, short box to box runs, switch legs, and other short runs.

PROCEDURE 11-1

Starting a Cable Run from a Loadcenter

- Put on safety glasses and observe all applicable safety rules.

- Remove a knockout in the loadcenter cabinet and install a cable connector that is appropriate for the size and type of cable you are using.

- Place one end of the cable for the circuit you are installing through the connector and into the loadcenter. Leave enough cable in the loadcenter so the wires in the cable will have plenty of length to get to the circuit breaker, grounded neutral bar, and grounding bar connection points. It is up to each individual electrician whether the cable sheathing is stripped away at this time or during the trim-out stage of the loadcenter. The trimming out of a loadcenter is when the circuit breakers are installed and the circuit wires are properly terminated. We discuss trimming out the loadcenter in Chapter 19.

- Secure the cable to the cabinet by tightening the cable connector onto the cable.

from experience...

There is a good rule to follow for determining how much cable to leave in the loadcenter so you will have enough conductor length to make all the necessary connections. Measure the height and width of the cabinet, add them together, and use that dimension for the minimum amount of cable to be left in the cabinet. For example, if a loadcenter is 24 inches (600 mm) high and 14.5 inches (363 mm) wide, the minimum length of cable that should be left in the cabinet would be 24 inches (600 mm) + 14.5 inches (363 mm) = 38.5 inches (963 mm).

from experience...

The cables can enter the loadcenter from the top, bottom, or sides. Most cables are brought into the loadcenter from the top or bottom. If the loadcenter is located in a basement, the cables are almost always brought into the loadcenter through the top so the cables can easily be routed to the floors above. If a loadcenter is flush-mounted in a wall of a house that has no basement, cables are commonly brought into the loadcenter from both the top and the bottom. Manufacturers of loadcenters provide knockouts in the back of the panel cabinet as well. Sometimes it is necessary to bring your home-run cable through the back of the loadcenter. An example of this situation would be when the loadcenter is a weatherproof type located on the outside of a house and electricians route the circuit cables from the back of the loadcenter into the house and to the various outlets.

PROCEDURE 11-1

Starting a Cable Run from a Loadcenter (Continued)

- Identify each circuit cable in the loadcenter cabinet with the circuit number, type, and location as it is shown on the wiring plan. This will help eliminate any confusion about which cable goes to which circuit breaker during the trim-out stage of the loadcenter.

(A) Once the cable run is started from the loadcenter, it is continued along or through building framing members until you reach the first electrical box on the circuit.

from experience...

A permanent marker may be used to write directly on the cable sheathing. Other methods of marking include using tags or marker tape.

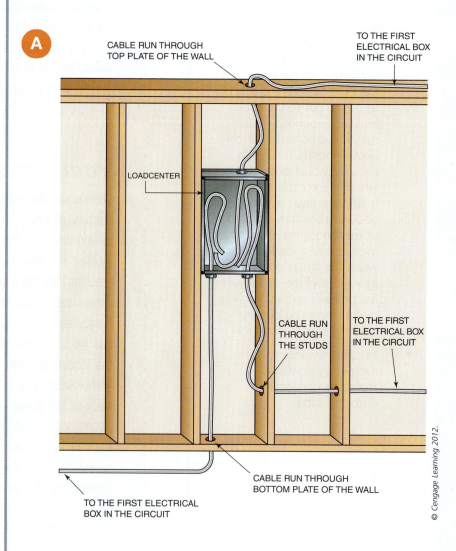

(A)

CABLE RUN THROUGH
TOP PLATE OF THE WALL

TO THE FIRST
ELECTRICAL BOX
IN THE CIRCUIT

LOADCENTER

CABLE RUN
THROUGH
THE STUDS

TO THE FIRST
ELECTRICAL BOX
IN THE CIRCUIT

CABLE RUN THROUGH
BOTTOM PLATE OF THE WALL

TO THE FIRST ELECTRICAL
BOX IN THE CIRCUIT

© Cengage Learning 2012.

B At each outlet box, leave enough cable so that approximately 8 inches (200 mm) of conductor will be available in each box for connection purposes. Remember that the *NEC*® requires at least 6 inches (150 mm) of free conductor in electrical boxes for connection to devices. At least 3 inches (75 mm) of conductor must extend from the front of each box.

C Install another cable into the electrical box and continue the cable run to the next box in the circuit. Again, be sure to leave approximately 8 inches (200 mm) of cable (or conductor) in the box.

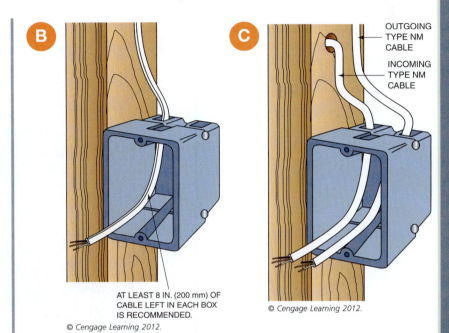

B

AT LEAST 8 IN. (200 mm) OF CABLE LEFT IN EACH BOX IS RECOMMENDED.

© *Cengage Learning 2012.*

C

OUTGOING TYPE NM CABLE

INCOMING TYPE NM CABLE

© *Cengage Learning 2012.*

from experience...

Some electricians install the cable into the electrical outlet box without stripping away the outside sheathing first. This technique will require the sheathing to be stripped later. Stripping the outside sheathing from a cable while it is in an electrical box can be tricky, especially with deeper boxes, and many electricians (including the author) prefer to strip the outside sheathing before the cable is secured to the electrical box. There is no absolutely right or wrong way to do this. The best way to do it is the way your instructor or supervisor suggests.

PROCEDURE 11-1

Starting a Cable Run from a Loadcenter (Continued)

D Fold the cable (or conductors if the cable is stripped) back into the box and continue to pull the cable to the rest of the boxes in the circuit.

D

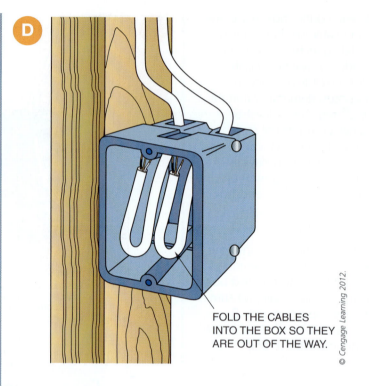

FOLD THE CABLES INTO THE BOX SO THEY ARE OUT OF THE WAY.

© Cengage Learning 2012.

REVIEW QUESTIONS

Directions: Answer the following items with clear and complete responses.

1. Name five common cable types used in residential wiring and the *NEC®* article that covers each type.

2. Nonmetallic sheathed cable must be secured within _____ of a two-gang plastic nail-on device box.

3. Type AC cable must be secured within 12 inches (300 mm) of each electrical box and then supported no more than every _____.

4. The minimum bending radius for both Type NM and Type AC cable is _____ times the cable diameter.

5. When a Type NM cable is installed through a bored hole in a wood stud that is less than 1¼ inches (32 mm) from the edge of the stud, it must be protected by a metal plate at least _____ thick.

6. A 12/2 Type MC cable must be secured within _____ of a metal device box and then supported at intervals not exceeding _____.

7. When Type NM or Type AC cable is installed in an accessible attic space used for storage, the cable must be protected within _____ of the attic scuttle hole.

8. The maximum distance permitted between supports for a run of Type NM cable is _____.

9. When Type AC cable is used, the *NEC®* requires a special fitting to be used in the ends. State the common trade name for the fitting and describe its purpose.

10. Service entrance cable can be installed as an interior wiring method, such as for the branch-circuit wiring to an

electric range. Name the article in the *NEC®* that covers the installation requirements for this situation.

11. Horizontal runs of Type NM cable through holes in building framing members are adequately supported as long as the distance between framing members does not exceed the *NEC®* support requirements. True or False (Circle the correct response.)

12. If you are using single-gang nonmetallic electrical device boxes, at least _____ of the Romex™ cable sheathing must extend into the box.

13. Although it is not a specific requirement, most electricians feel that the best place to start a cable run for a branch circuit is at the service panel or subpanel. Explain why this is so.

14. When electricians use Romex™ cable that comes in a plastic wrapping or in a box, it usually is bought in _____ lengths. When Romex™ cable is bought on reels, it usually comes in _____ lengths.

15. Flat cable, like 14/2 or 12/2 Romex™, cannot be stapled on edge. Explain why this is true and state the *NEC®* section that prohibits it.

16. Manufacturers of nonmetallic sheathed cable are now color-coding their cables. List the color-coding used on the outside sheathing for 14, 12, and 10 AWG Type NM cables.

17. Electricians sometimes drill holes and run cables at the bottom of wall studs. Discuss the benefit of drilling holes and routing cables in this manner.

18. When drilling holes through consecutive framing members for longer cable runs, electricians align the holes as close as possible. Explain why this is done.

19. Discuss the advantages of having a cable diagram of the circuits that you have to install.

20. Name the two main criteria for selecting an appropriate cable type to be used in house wiring.

KNOW YOUR CODES

1. Check with the authority having jurisdiction to make sure Type NM cable is permitted in one- and two-family homes in your area of the country.

2. Check with the authority having jurisdiction to find out if Type NM cable can be installed exposed in areas like unfinished basements and garages without physical protection.

Raceway Installation

OBJECTIVES

Upon completion of this chapter, the student should be able to:

- Select an appropriate raceway size and type for a residential application.

- Demonstrate an understanding of the proper techniques for cutting, threading, and bending electrical conduit for residential applications.

- Demonstrate an understanding of the proper installation techniques for common raceway types used in residential wiring.

- Demonstrate an understanding of the common installation techniques for installing conductors in an installed raceway system.

GLOSSARY OF TERMS

back-to-back bend a type of conduit bend that is formed by two 90-degree bends with a straight length of conduit between the two bends

box offset bend a type of conduit bend that uses two equal bends to cause a slight change of direction for a conduit at the point where it is attached to an electrical box

conduit a raceway with a circular cross section, such as electrical metallic tubing, rigid metal conduit, or intermediate metal conduit

conduit body a separate portion of a conduit system that provides access through a removable cover to the inside of the conduit body; a conduit body is used to connect two or more sections of a conduit system together; sometimes referred to as a "condulet"

field bend any bend or offset made by installers, using proper tools and equipment, during the installation of conduit systems

offset bend a type of conduit bend that is made with two equal-degree bends in such a way that the conduit changes elevation and avoids an obstruction

raceway an enclosed channel of metal or nonmetallic materials designed expressly for holding wires or cables; raceways used in residential wiring include rigid metal conduit, rigid nonmetallic conduit, intermediate metal conduit, liquid tight flexible conduit, flexible metal conduit, electrical nonmetallic tubing, and electrical metallic tubing

saddle bend a type of conduit bend that results in a conduit run going over an object that is blocking the path of the run; there are two styles: a three-point saddle and a four-point saddle

stub-up a type of conduit bend that results in a 90-degree change of direction

take-up the amount that must be subtracted from a desired stub-up height so the bend will come out right; the take-up is different for each conduit size

thinwall a trade name often used for electrical metallic tubing

Raceway is not used very often as the primary wiring method in residential wiring. However, certain service entrance installations use raceway, and there are some local electrical codes that require all wiring in houses to be installed in a raceway. When a raceway is used in residential wiring, it has a circular cross section and is called "conduit." Conduit may be made of metal or nonmetallic materials, and it may be rigid or flexible. It is up to the electrician to select the proper type of conduit for the wiring application that is encountered. This chapter discusses the common raceway types used in residential wiring. The selection of the proper raceway type and size is covered, as are the applicable *National Electrical Code*® *(NEC*®*)* installation requirements for each raceway type. An introduction to proper cutting, threading, and bending techniques is presented. The last area covered is the actual installation of electrical conductors in a completely installed raceway system.

SELECTING THE APPROPRIATE RACEWAY TYPE AND SIZE

The raceway types used in residential wiring installations were introduced in Chapter 2. While service entrance installations using conduit of some type are very common, branch-circuit installation using a raceway wiring method is seldom used. There is no question that cable wiring methods are easier to install than raceway wiring methods, and this is the main reason why most houses are wired using as little conduit as possible. However, as we have mentioned before, some areas of the country require that all wiring in a house be installed in a raceway wiring method.

When installing a circuit in a raceway wiring method, individual conductors can be used. If the raceway type is made of metal, the raceway itself may serve as the equipment-grounding conductor so there is no need to run a grounding wire in the raceway. However, it is common wiring practice for many electricians to install a green insulated equipment-grounding conductor in every raceway. Those circuits that operate on 120 volts will require a black insulated ungrounded conductor and a white insulated grounded conductor to be installed in the raceway. Those circuits that operate on 240 volts will require two black insulated ungrounded conductors or possibly one black and one red insulated ungrounded conductor. Circuits that operate on 120/240 volts will need one white insulated grounded conductor and two insulated ungrounded conductors of any color other than white, gray, or green. As mentioned previously, a green insulated equipment-grounding conductor can be installed with the regular circuit conductors, or an electrician can use the raceway itself as the grounding conductor, provided that the raceway is a listed grounding conductor in Section 250.118 of the *NEC*®. Although it is usually done only when added physical protection is desired, cables are permitted to be installed in raceways as long as the article covering that raceway does not prohibit it. For example, Article 358 does not prohibit Type NM cable from being installed in electrical metallic tubing (EMT); therefore, an electrician can install Type NM cable in EMT if he or she chooses to do so.

Selecting the type of raceway to be used in a residential wiring installation will depend on the actual wiring application, an electrician's personal preference of raceway types, and what any local electrical codes will or will not allow. As mentioned in Chapter 8, rigid metal conduit (RMC) or intermediate metal conduit (IMC) are the raceway types used for a mast-style service entrance installation. Service entrance installations that use a raceway for the installation of service conductors that do not require the raceway to extend above a roof (a mast-type service entrance) will be able to use electrical metallic tubing (EMT) or rigid polyvinyl chloride conduit (PVC) in addition to RMC and IMC. The most common raceway type used for branch-circuit installation is EMT, mainly because of its relatively easy bending and connection techniques. It is also much less expensive than other metal raceways. Installations where flexibility is desired, such as connections to air-conditioning systems, often require a flexible raceway like flexible metal conduit (FMC) for use indoors or liquid tight flexible metal conduit (LFMC) for use outdoors.

Installation requirements for the various raceways are located in the appropriate articles of the *NEC*®. The following paragraphs cover the most important installation requirements for the raceway types used most often in residential wiring.

RIGID METAL CONDUIT: TYPE RMC

Article 344 covers the installation requirements for RMC. Section 344.2 defines RMC as a threadable raceway of circular cross section designed for the physical protection and routing of conductors and cables, and for use as an equipment-grounding conductor when installed with appropriate fittings (Figure 12-1). RMC is generally made of steel with a protective galvanized coating and can be used in all atmospheric conditions and occupancies. It can also be made of aluminum, red brass, or stainless steel.

Section 344.22 states that the number of conductors or cables allowed in RMC must not exceed that permitted by the percentage fill specified in Table 1 in Chapter 9 of the *NEC*® (Figure 12-2). Table 1 in Chapter 9 specifies the maximum fill percentage of a conduit or tubing. It says that for an application where only two conductors are installed, the conduit cannot be filled to more than 31 percent of the conduit's cross-sectional area. If three or more conductors are to be installed (a common practice), the conduit cannot be filled to more than 40 percent of its cross-sectional area. If the conductors being installed are all of the same wire size and have the same insulation type, Informative Annex C is used. Informative Annex C is located in the back of the *NEC*® and indicates the maximum number of conductors

Table 1 Percent of Cross Section of Conduit and Tubing for Conductors

Number of Conductors	All Conductor Types
1	53
2	31
Over 2	40

FIGURE 12-2 Table 1 in Chapter 9 of the *NEC*® specifies the maximum fill percentage for conduit and tubing. *Reprinted with permission from NFPA 70*®, *National Electric Code*®, *Copyright © 2010, National Fire Protection Association, Quincy, MA. This reprinted material is not the complete and official position of the NFPA on the referenced subject, which is represented only by the standard in its entirety.*

permitted in a conduit or tubing. The maximum number of conductors allowed in a conduit according to Informative Annex C takes into account the fill percentage requirements of Table 1 in Chapter 9. No additional calculations are necessary.

from experience...

Rigid metal conduit (RMC) is available in trade sizes ½″ to 6″ (Metric Designator 16–155). Intermediate metal conduit (IMC) is available in trade sizes ½″ to 4″ (Metric Designator 16–103). Manufacturers of these two conduit types put different color plastic caps on the end opposite the coupling end. The caps do two things: (1) they protect the threaded ends from damage, and (2) they identify the size of conduit based on the color of the cap. A black cap designates ½″ trade sizes (½″, 1½″, 2½″, and 3½″), a red cap designates ¼″ trade sizes (¾″ and 1¼″), and a blue cap designates inch trade sizes (1″, 2″, 3″, 4″, 5″, and 6″).

For example, to select the proper trade size of RMC for an application calling for the installation of sixteen 12 AWG conductors with THHN insulation, the following procedure should be followed:

1 Go to Table C8 in Informative Annex C (Figure 12-3).

2 Look in the far-left-hand column and locate the THHN insulation type.

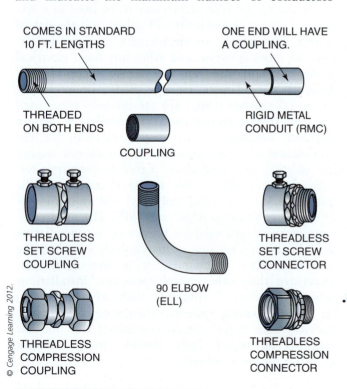

COMES IN STANDARD 10 FT. LENGTHS

ONE END WILL HAVE A COUPLING.

THREADED ON BOTH ENDS

RIGID METAL CONDUIT (RMC)

COUPLING

THREADLESS SET SCREW COUPLING

90 ELBOW (ELL)

THREADLESS SET SCREW CONNECTOR

THREADLESS COMPRESSION COUPLING

THREADLESS COMPRESSION CONNECTOR

FIGURE 12-1 Rigid metal conduit (RMC) and associated fittings.

CONDUCTORS

Type	Conductor Size (AWG/ kcmil)	16 (½)	21 (¾)	27 (1)	35 (1¼)	41 (1½)	53 (2)	63 (2½)	78 (3)	91 (3½)	103 (4)	129 (5)	155 (6)
RHH*,	6	1	3	5	8	11	18	27	41	55	71	111	160
RHW*,	4	1	1	3	6	8	14	20	31	41	53	83	120
RHW-2*,	3	1	1	3	5	7	12	17	26	35	45	71	103
TW,	2	1	1	2	4	6	10	14	22	30	38	60	87
THW,	1	1	1	1	3	4	7	10	15	21	27	42	61
THHW,	1/0	0	1	1	2	3	6	8	13	18	23	36	52
THW-2	2/0	0	1	1	2	3	5	7	11	15	19	31	44
	3/0	0	1	1	1	2	4	6	9	13	16	26	37
	4/0	0	0	1	1	1	3	5	8	10	14	21	31
	250	0	0	1	1	1	3	4	6	8	11	17	25
	300	0	0	1	1	1	2	3	5	7	9	15	22
	350	0	0	0	1	1	1	3	5	6	8	13	19
	400	0	0	0	1	1	1	3	4	6	7	12	17
	500	0	0	0	1	1	1	2	3	5	6	10	14
	600	0	0	0	1	1	1	1	3	4	5	8	12
	700	0	0	0	0	1	1	1	2	3	4	7	10
	750	0	0	0	0	1	1	1	2	3	4	7	10
	800	0	0	0	0	1	1	1	2	3	4	6	9
	900	0	0	0	0	1	1	1	1	3	4	6	8
	1000	0	0	0	0	0	1	1	1	2	3	5	8
	1250	0	0	0	0	0	1	1	1	1	2	4	6
	1500	0	0	0	0	0	1	1	1	1	2	3	5
	1750	0	0	0	0	0	0	1	1	1	1	3	4
	2000	0	0	0	0	0	0	1	1	1	1	3	4
THHN,	14	13	22	36	63	85	140	200	309	412	531	833	1202
THWN,	12	9	16	26	46	62	102	146	225	301	387	608	877
THWN-2	10	6	10	17	29	39	64	92	142	189	244	383	552
	8	3	6	9	16	22	37	53	82	109	140	221	318
	6	2	4	7	12	16	27	38	59	79	101	159	230
	4	1	2	4	7	10	16	23	36	48	62	98	141
	3	1	1	3	6	8	14	20	31	41	53	83	120
	2	1	1	3	5	7	11	17	26	34	44	70	100
	1	1	1	1	4	5	8	12	19	25	33	51	74
	1/0	1	1	1	3	4	7	10	16	21	27	43	63
	2/0	0	1	1	2	3	6	8	13	18	23	36	52
	3/0	0	1	1	1	3	5	7	11	15	19	30	43
	4/0	0	1	1	1	2	4	6	9	12	16	25	36
	250	0	0	1	1	1	3	5	7	10	13	20	29
	300	0	0	1	1	1	3	4	6	8	11	17	25
	350	0	0	1	1	1	2	3	5	7	10	15	22
	400	0	0	1	1	1	2	3	5	7	8	13	20
	500	0	0	0	1	1	1	2	4	5	7	11	16
	600	0	0	0	1	1	1	1	3	4	6	9	13
	700	0	0	0	1	1	1	1	3	4	5	8	11
	750	0	0	0	0	1	1	1	3	4	5	7	11
	800	0	0	0	0	1	1	1	2	3	4	7	10
	900	0	0	0	0	1	1	1	2	3	4	6	9
	1000	0	0	0	0	1	1	1	1	3	4	6	8
FEP,	14	12	22	35	61	83	136	194	300	400	515	808	1166
FEPB,	12	9	16	26	44	60	99	142	219	292	376	590	851
PFA,	10	6	11	18	32	43	71	102	157	209	269	423	610
PFAH,	8	3	6	10	18	25	41	58	90	120	154	242	350
TFE	6	2	4	7	13	17	29	41	64	85	110	172	249
	4	1	3	5	9	12	20	29	44	59	77	120	174
	3	1	2	4	7	10	17	24	97	50	64	100	145
	2	1	1	3	6	8	14	20	31	41	53	83	120

FIGURE 12-3 Table C8 in Informative Annex C of the *NEC*® lists the maximum number of conductors allowed in RMC that are the same size and have the same insulation. *Reprinted with permission from NFPA 70*®, *National Electric Code*®, *Copyright © 2010, National Fire Protection Association, Quincy, MA. This reprinted material is not the complete and official position of the NFPA on the referenced subject, which is represented only by the standard in its entirety.*

from experience...

from experience...

Informative Annex C is used to determine the maximum number of conductors that are all the same size and have the same insulation type for *all* of the conduit types typically used in residential wiring.

from experience...

There is an index at the beginning of Informative Annex C that will tell you which table applies to the specific type of conduit you are using.

3 Look in the next column to the right to find the conductor size, which in this example is 12 AWG.

4 Continue to look to the right until you find the number of conductors you will be installing *or* the next higher number to the number of conductors you are installing. In this case, the number of installed conductors is 16.

5 Look at the top of the table for the conduit size. It will be at the top of the column where you found the number of installed conductors in the previous step. In this example, you will find that a ¾-inch (21 metric designator) minimum trade size is required for this installation.

When determining the maximum number of conductors of different sizes and of different insulations in a conduit, Table 4 in Chapter 9 will provide the usable area within the selected conduit or tubing, and Table 5 in Chapter 9 will provide the required area for each conductor. The calculation procedure is a little complicated but, once explained, is easily applied by an electrician. Let's look at an example. An electrician needs to determine the minimum-size RMC that can be used for the installation of eight 14 AWG THHN conductors, five 12 AWG THW conductors, and two 8 AWG THWN conductors. (This is not a common situation in residential wiring, but it will help you understand conduit sizing when different wire sizes with different insulation types are used.) The procedure for calculating

the minimum size of RMC for this application is as follows:

1 From Table 5 in Chapter 9, determine the cross-sectional area (in.²) of each conductor size with its insulation type (Figure 12-4):

- 14 AWG THHN = 0.0097 in.²
- 12 AWG THW = 0.0181 in.²
- 8 AWG THWN = 0.0366 in.²

2 Multiply each cross-sectional area by the number of conductors to be installed in the conduit:

- 14 AWG THHN = 0.0097 in.² × 8 = 0.0776 in.²
- 12 AWG THW = 0.0181 in.² × 5 = 0.0905 in.²
- 8 AWG THWN = 0.0366 in.² × 2 = 0.0732 in.²

3 Add all the conductor cross-sectional areas together to get a total cross-sectional area for all the conductors to be installed:

- 0.0776 in.² + 0.0905 in.² + 0.0732 in.² = 0.2413 in.²

4 Using Table 4 in Chapter 9, go to the section that covers RMC and find the minimum conduit size that will work (Figure 12-5). Use the "Over 2 Wires 40%" column and look for 0.2413 in.² *or* the number that is the next higher listed number. In this example, you will not find 0.2413, but you will find the next higher listed number, which is 0.355. Now look to the left to find the conduit size that corresponds with this number. It is 1 inch, and this is the minimum-size RMC that could be used for this application.

Section 344.24 requires that when bending RMC, the bends must be made so that the conduit is not damaged and the internal diameter of the conduit is not effectively reduced. Section 344.26 limits the number of bends in one conduit run from one box to another to no more than 360 degrees total (Figure 12-6). Limiting the number of bends in a conduit run will reduce the pulling tension on conductors and help ensure easy insertion or removal of conductors.

When cutting RMC, Section 344.28 requires all cut ends to be reamed or otherwise finished to remove rough edges. Conduit is cut using a saw or a pipe cutter. Care should be taken to ensure a straight cut. After the cut is made, the conduit must be reamed. Proper reaming removes any burrs from the interior of the cut conduit so that as wires and cables are pulled through the conduit, no chafing of the insulation or cable jacket can occur. Finally, the conduit is threaded. Where conduit is threaded in the field, a standard cutting die with a 1 in 16 taper (¾-inch taper per foot) must be used.

RMC must be installed as a complete system as required in Section 300.18(A) (Figure 12-7) and must be securely fastened in place and supported in

Type	Size (AWG or kcmil)	Approximate Diameter		Approximate Area	
		mm	in.	mm²	in.²
TW, THHW, THW, THW-2	12	3.861	0.152	11.68	0.0181
	10	4.470	0.176	15.68	0.0243
	8	5.994	0.236	28.19	0.0437
RHW*, RHW*, RHW-2*	14	4.140	0.163	13.48	0.0209
RHH*, RHW*, RHW-2*, XF, XFF	12	4.623	0.182	16.77	0.0260
Type: RHH*, RHW*, RHW-2*, THHN, THHW, THW, THW-2, TFN, TFFN, THWN-2, XF, XFF					
RHH,* RHW,* RHW-2,* XF, XFF	10	5.232	0.206	21.48	0.0333
RHH*, RHW*, RHW-2*	8	6.756	0.266	35.87	0.0556
TW, THW, THHW, THW-2, RHH*, RHW*, RHW-2*	6	7.722	0.304	46.84	0.0726
	4	8.941	0.352	62.77	0.0973
	3	9.652	0.380	73.16	0.1134
	2	10.46	0.412	86.00	0.1333
	1	12.50	0.492	122.6	0.1901
	1/0	13.51	0.532	143.4	0.2223
	2/0	14.68	0.578	169.3	0.2624
	3/0	16.00	0.630	201.1	0.3117
	4/0	17.48	0.688	239.9	0.3718
	250	19.43	0.765	296.5	0.4596
	300	20.83	0.820	340.7	0.5281
	350	22.12	0.871	384.4	0.5958
	400	23.32	0.918	427.0	0.6619
	500	25.48	1.003	509.7	0.7901
	600	28.27	1.113	627.7	0.9729
	700	30.07	1.184	710.3	1.1010
	750	30.94	1.218	751.7	1.1652
	800	31.75	1.250	791.7	1.2272
	900	33.38	1.314	874.9	1.3561
	1000	34.85	1.372	953.8	1.4784
	1250	39.09	1.539	1200	1.8602
	1500	42.21	1.662	1400	2.1695
	1750	45.11	1.776	1598	2.4773
	2000	47.80	1.882	1795	2.7818
TFN, TFFN	18	2.134	0.084	3.548	0.0055
	16	2.438	0.096	4.645	0.0072
THHN, THWN, THWN-2	14	2.819	0.111	6.258	0.0097
	12	3.302	0.130	8.581	0.0133
	10	4.166	0.164	13.61	0.0211
	8	5.486	0.216	23.61	0.0366
	6	6.452	0.254	32.71	0.0507

FIGURE 12-4 Table 5 in Chapter 9 of the *NEC®* lists the dimensions of insulated conductors. *Reprinted with permission from NFPA 70®, National Electric Code®, Copyright © 2010, National Fire Protection Association, Quincy, MA. This reprinted material is not the complete and official position of the NFPA on the referenced subject, which is represented only by the standard in its entirety.*

Type	Size (AWG or kcmil)	Approximate Diameter		Approximate Area	
		mm	in.	mm²	in.²
THHN, THWN, THWN-2	4	8.230	0.324	53.16	0.0824
	3	8.941	0.352	62.77	0.0973
	2	9.754	0.384	74.71	0.1158
	1	11.33	0.446	100.8	0.1562
	1/0	12.34	0.486	119.7	0.1855
	2/0	13.51	0.532	143.4	0.2223
	3/0	14.83	0.584	172.8	0.2679
	4/0	16.31	0.642	208.8	0.3237
	250	18.06	0.711	256.1	0.3970
	300	19.46	0.766	297.3	0.4608

Type: FEP, FEPB, PAF, PAFF, PF, PFA, PFAH, PFF, PGF, PGFF, PTF, PTFF, TFE, THHN, THWN-2, Z, ZF, ZFF

Type	Size (AWG or kcmil)	Approximate Diameter		Approximate Area	
		mm	in.	mm²	in.²
THHN, THWN, THWN-2	350	20.75	0.817	338.2	0.5242
	400	21.95	0.864	378.3	0.5863
	500	24.10	0.949	456.3	0.7073
	600	26.70	1.051	559.7	0.8676
	700	28.50	1.122	637.9	0.9887
	750	29.36	1.156	677.2	1.0496
	800	30.18	1.188	715.2	1.1085
	900	31.80	1.252	794.3	1.2311
	1000	33.27	1.310	869.5	1.3478
PF, PGFF, PGF, PFF, PTF, PAF, PTFF, PAFF	18	2.184	0.086	3.742	0.0058
	16	2.489	0.098	4.839	0.0075
PF, PGFF, PGF, PFF, PTF, PAF, PTFF, PAFF, TFE, FEP, PFA, FEPB, PFAH	14	2.870	0.113	6.452	0.0100
TFE, FEP, PFA, FEPB, PFAH	12	3.353	0.132	8.839	0.0137
	10	3.962	0.156	12.32	0.0191
	8	5.232	0.206	21.48	0.0333
	6	6.198	0.244	30.19	0.0468
	4	7.417	0.292	43.23	0.0670
	3	8.128	0.320	51.87	0.0804
	2	8.941	0.352	62.77	0.0973
TFE, PFAH	1	10.72	0.422	90.26	0.1399
TFE, PFA, PFAH, Z	1/0	11.73	0.462	108.1	0.1676
	2/0	12.90	0.508	130.8	0.2027
	3/0	14.22	0.560	158.9	0.2463
	4/0	15.70	0.618	193.5	0.3000
ZF, ZFF	18	1.930	0.076	2.903	0.0045
	16	2.235	0.088	3.935	0.0061

FIGURE 12-4 *(Continued)*

accordance with Section 344.30(A) and (B). Straight runs of RMC in ½- and ¾-inch trade sizes are required to be securely fastened at intervals of no more than 10 feet (3 m) (Figure 12-8). As the trade size of RMC gets larger, Table 344.30(B)(2) allows support distances of more than 10 feet (3 m). Secure fastening is also required within 3 feet (900 mm) of outlet boxes, junction boxes, cabinets, and conduit bodies (Figure 12-9).

Article 344—Rigid Metal Conduit (RMC)

Metric Designator	Trade Size	Nominal Internal Diameter		Total Area 100%		60%		1 wire 53%		2 wires 31%		Over 2 Wires 40%	
		mm	in.	mm²	in.²	mm²	in.²	mm²	in.²	mm²	in.²	mm²	in.²
12	⅜	—	—	—	—	—	—	—	—	—	—	—	—
16	½	16.1	0.632	204	0.314	122	0.188	108	0.166	63	0.097	81	0.125
21	¾	21.2	0.836	353	0.549	212	0.329	187	0.291	109	0.170	141	0.220
27	1	27.0	1.063	573	0.887	344	0.532	303	0.470	177	0.275	229	0.355
35	1¼	35.4	1.394	984	1.526	591	0.916	522	0.809	305	0.473	394	0.610
41	1½	41.2	1.624	1333	2.071	800	1.243	707	1.098	413	0.642	533	0.829
53	2	52.9	2.083	2198	3.408	1319	2.045	1165	1.806	681	1.056	879	1.363
63	2½	63.2	2.489	3137	4.866	1882	2.919	1663	2.579	972	1.508	1255	1.946
78	3	78.5	3.090	4840	7.499	2904	4.499	2565	3.974	1500	2.325	1936	3.000
91	3½	90.7	3.570	6461	10.010	3877	6.006	3424	5.305	2003	3.103	2584	4.004
103	4	102.9	4.050	8316	12.882	4990	7.729	4408	6.828	2578	3.994	3326	5.153
129	5	128.9	5.073	13050	20.212	7830	12.127	6916	10.713	4045	6.266	5220	8.085
155	6	154.8	6.093	18821	29.158	11292	17.495	9975	15.454	5834	9.039	7528	11.663

FIGURE 12-5 Table 4 in Chapter 9 of the *NEC*® lists the dimensions and percent area of conduit and tubing. Each conduit or tubing type has its own section in Table 4. *Reprinted with permission from NFPA 70*®*, National Electric Code*®*, Copyright © 2010, National Fire Protection Association, Quincy, MA. This reprinted material is not the complete and official position of the NFPA on the referenced subject, which is represented only by the standard in its entirety.*

However, where structural support members do not permit fastening within 3 feet (900 mm), secure fastening may be located up to 5 feet (1.5 m) away (Figure 12-10).

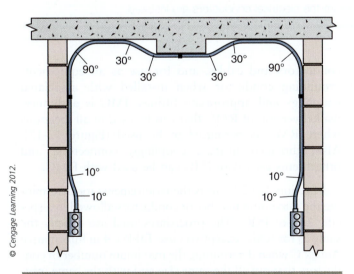

© Cengage Learning 2012.

FIGURE 12-6 No more than 360 degrees of bends are allowed in a conduit run (Section 344.26). Because the total bends for this application is only 340 degrees, it meets the requirements of the *NEC*®.

Section 344.30(B)(4) permits lengths of RMC to be supported by framing members at 10-foot (3 m) intervals, provided the RMC is secured and supported at least 3 feet (900 mm) from the box or enclosure. This would apply when RMC is run through holes drilled in framing members of a residential structure.

Section 344.46 requires a bushing to be installed on the end of an RMC conduit where it enters a box, fitting, or other enclosure. The purpose of the bushing is to prevent any wires installed in the RMC from abrasion when they are being pulled through the RMC. Section 300.4(G) requires an insulated bushing be used for conductors 4 AWG and larger. When 6 AWG and smaller conductors are installed, a metal bushing could be used but most electricians use an insulated (usually plastic) bushing on RMC no matter what the conductor size is (Figure 12-11).

The last installation requirement mentioned here for RMC is that according to Section 344.60, RMC is permitted as an equipment-grounding conductor. This means that if the RMC is attached correctly at each electrical enclosure and any couplings used to connect lengths of RMC together are properly tightened, the conduit itself can be the equipment-grounding conductor, and there is no need to run an additional grounding conductor in the conduit along with the regular circuit conductors.

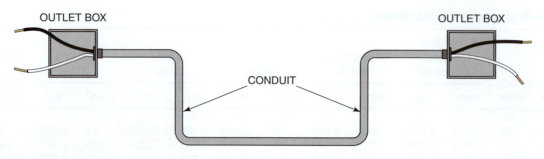

OUTLET BOX

OUTLET BOX

CONDUIT

INSTALL CONDUIT COMPLETELY BETWEEN BOXES BEFORE PULLING IN CONDUCTORS.

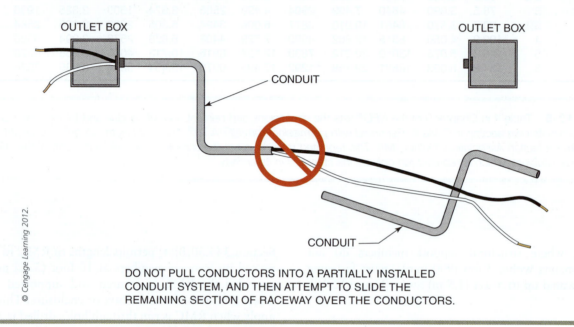

OUTLET BOX

OUTLET BOX

CONDUIT

CONDUIT

© Cengage Learning 2012.

DO NOT PULL CONDUCTORS INTO A PARTIALLY INSTALLED CONDUIT SYSTEM, AND THEN ATTEMPT TO SLIDE THE REMAINING SECTION OF RACEWAY OVER THE CONDUCTORS.

FIGURE 12-7 Conduit systems must be completely installed before the electrical conductors are installed.

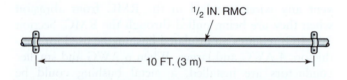

¹/₂ IN. RMC

10 FT. (3 m)

FIGURE 12-8 The maximum support interval for RMC is 10 feet (3 m) in accordance with Section 344.30(B)(1). However, Section 344.30(B)(2) and Table 344.30(B)(2) allow greater distances between supports for straight runs of conduit made up with threaded couplings.

© Cengage Learning 2012.

INTERMEDIATE METAL CONDUIT: TYPE IMC

IMC is covered in Article 342. Section 342.2 defines IMC as a steel threadable raceway of circular cross section designed for the physical protection and routing of conductors and cables, and for use as an equipment-grounding conductor when installed with associated couplings and appropriate fittings. IMC is a thinner-walled version of RMC that can be used in all locations where RMC is permitted to be used (Figure 12-12). Also, threaded fittings, couplings, connectors, and other items used with IMC can be used with RMC.

Section 342.22 covers the requirements for determining the maximum number of conductors allowed in a specific size of IMC. The procedures used are exactly the same as for RMC except you use Table C4 in Informative Annex C when determining the maximum number of conductors that are all the same size and with the same insulation type. Also, you must use the IMC section of Table 4 in Chapter 9 when determining the minimum-size IMC for conductors of different sizes that have different insulation types. All other *NEC*® installation requirements for IMC, including support, are the same as for RMC and were presented earlier in this chapter.

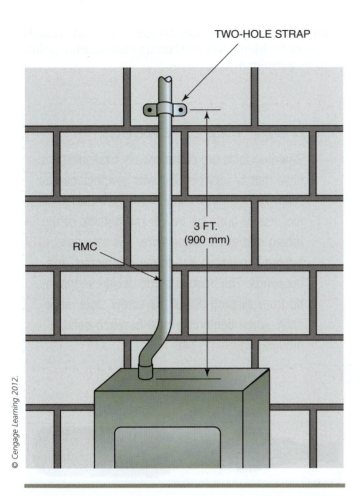

TWO-HOLE STRAP

RMC

3 FT.
(900 mm)

© Cengage Learning 2012.

FIGURE 12-9 RMC must be securely fastened within 3 feet (900 mm) of each conduit termination point.

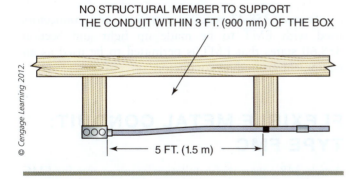

NO STRUCTURAL MEMBER TO SUPPORT
THE CONDUIT WITHIN 3 FT. (900 mm) OF THE BOX

5 FT. (1.5 m)

© Cengage Learning 2012.

FIGURE 12-10 If structural members are not available for securing RMC within 3 feet (900 mm), a distance of 5 feet (1.5 m) is acceptable.

ELECTRICAL METALLIC TUBING: TYPE EMT

Article 358 covers the installation requirements for EMT. Section 358.2 defines EMT as an unthreaded **thinwall** raceway of circular cross section designed for

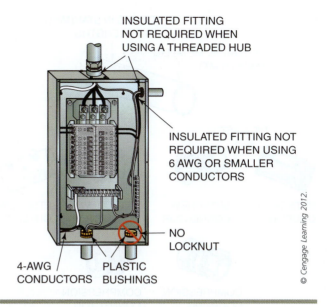

INSULATED FITTING
NOT REQUIRED WHEN
USING A THREADED HUB

INSULATED FITTING NOT
REQUIRED WHEN USING
6 AWG OR SMALLER
CONDUCTORS

NO
LOCKNUT

4-AWG
CONDUCTORS

PLASTIC
BUSHINGS

© Cengage Learning 2012.

FIGURE 12-11 Bushings are required with RMC to protect the wires from abrasion. Insulated bushings are required when the conductors are 4 AWG or larger. At no time can an insulated bushing be used to secure a conduit to a box.

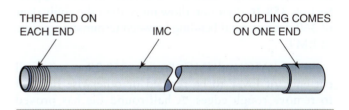

THREADED ON
EACH END

IMC

COUPLING COMES
ON ONE END

FIGURE 12-12 IMC is lighter in weight than RMC and has slightly thinner walls. It can be used with the same fittings as RMC.
© Cengage Learning 2012.

the physical protection and routing of conductors and cables, and for use as an equipment-grounding conductor when installed utilizing appropriate fittings (Figure 12-13). EMT is generally made of steel.

Section 358.22 covers the requirements for determining the maximum number of conductors allowed in a specific size of EMT. The procedures used are exactly the same as for RMC and IMC except you use Table C1 in Informative Annex C when determining the maximum number of conductors that are all the same size and with the same insulation type. Also, you must use the EMT section of Table 4 in Chapter 9 when determining the minimum-size EMT for conductors of different sizes that have different insulation types.

Section 358.24 requires bends in EMT to be made so that the tubing is not damaged and the internal diameter of the tubing is not effectively reduced. Like all

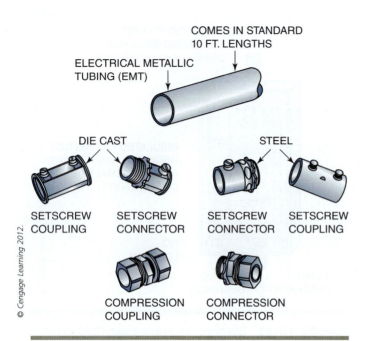

FIGURE 12-13 EMT and associated fittings.

© Cengage Learning 2012.

intervals not greater than 10 feet (3 m) and securely fastened within 3 feet (900 mm) of termination points to be permitted.

from experience...

Reaming tools are commercially available. One style that is very popular with electricians can ream ½-, ¾-, and 1-inch trade-size EMT. This tool reams both the inside and outside of the tubing end at the same time. It is also used to tighten screws on set-screw connectors and couplings. The hooded-blade design keeps the tip from slipping out of the screw slots, especially when tightening hard-to-reach conduit fittings (Figure 12-14).

FIGURE 12-14 A commercially available reaming tool. *Courtesy of Klein Tools Inc.*

the other circular raceways discussed in this chapter, Section 358.26 does not allow more than the equivalent of 360 degrees total bending between termination points of EMT.

Like RMC and IMC, Section 358.28 requires all cut ends of EMT to be reamed or otherwise finished to remove rough edges. A half-round file has proved practical for removing rough edges. However, many electricians find that the nose of lineman pliers, the nose of diagonal cutting pliers, or an electrician's knife can be an effective reaming tool on the smaller sizes of EMT.

Securing and supporting requirements of EMT is similar to both RMC and IMC. Section 358.30 requires EMT to be installed as a complete system and to be securely fastened in place and supported in accordance with the following:

- EMT must be securely fastened in place at least every 10 feet (3 m).

- In addition, each EMT run between termination points must be securely fastened within 3 feet (900 mm) of each outlet box, junction box, device box, cabinet, conduit body, or other tubing termination.

- *Exception No. 2* to Section 358.30(A) states that for concealed work in finished buildings or prefinished wall panels where such securing is impractical, unbroken lengths (without a coupling) of EMT are permitted to be fished.

- Section 358.30(B) allows horizontal runs of EMT supported by openings through framing members at

Section 358.42 requires couplings and connectors used with EMT to be made up tight and Section 358.60 states that EMT is permitted to be used as an equipment-grounding conductor.

FLEXIBLE METAL CONDUIT: TYPE FMC

Article 348 covers the installation requirements for FMC (Figure 12-15). Many electricians refer to this raceway type as "Greenfield." FMC is appropriate for use indoors where a need for flexibility at the connection points is required. Section 348.2 defines FMC as a raceway of circular cross section made of helically wound, formed, interlocked metal strip. On the outside, it looks a lot like Type AC cable or Type MC cable. However, unlike the armored cables, FMC requires an electrician to install electrical wires in the raceway. Although FMC can be used in longer lengths if desired, electricians usually use it in lengths of 6 feet (1.8 m) or less where a flexible connection is necessary (Figure 12-16).

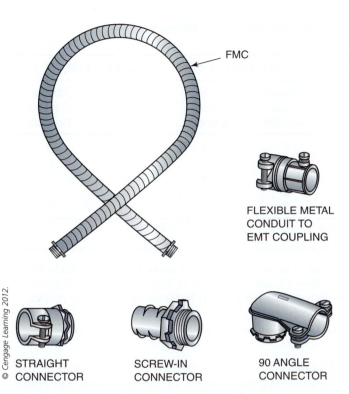

FIGURE 12-15 FMC and associated fittings.

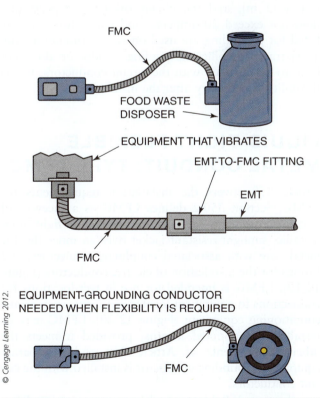

FIGURE 12-16 Some common FMC applications in residential wiring.

Section 348.22 covers the requirements for determining the maximum number of conductors allowed in a specific size of FMC. The procedures used are exactly the same as for the raceways we have previously discussed, except you use Table C3 in Informative Annex C when determining the maximum number of conductors that are all the same size and with the same insulation type. Also, you must use the FMC section of Table 4 in Chapter 9 when determining the minimum-size FMC for conductors of different sizes that have different insulation types. There is one other difference, and that is the use of Table 348.22 when determining the maximum number of conductors in a ⅜-inch trade-size FMC. This table must be used because the ⅜-inch size is not included in Table C3 of Annex C (Figure 12-17).

For example, according to Table 348.22, the maximum number of 14 AWG conductors with THHN insulation that you could install in a ⅜-inch trade-size FMC is three if the connector fittings are inside or four if the connector fittings are outside. The title of the table has an asterisk at the end which refers to a note at the bottom of the table. The note allows one equipment grounding conductor of the same size to be installed as well. So in this example, if you are installing an equipment grounding conductor, you are permitted to actually install four total 14 AWG THHN conductors if inside connector fittings are used or five total 14 AWG THHN conductors if outside connector fittings are used.

As with other raceways, a run of flexible metal conduit installed between boxes, conduit bodies, and other electrical equipment is not permitted to contain more than the equivalent of 360 degrees total. Proper shaping and support of this flexible wiring method will ensure that conductors can be easily installed or taken out at any time. These requirements are found in Sections 348.24 and 348.26.

Section 348.28 requires all cut ends to be trimmed or otherwise finished to remove rough edges, except where fittings that thread into the convolutions (so-called inside fittings) are used. Many electricians believe that an antishort bushing similar to the "redhead" style used with Type AC cable must be used with FMC. While this is not an *NEC®* requirement, it is a good idea to use a redhead in each end of FMC.

The securing and supporting requirements are given in Section 348.30. They are exactly the same as for Type AC cable and Type NM cable. FMC must be securely fastened in place and supported by an approved means within 12 inches (300 mm) of each box, cabinet, conduit body, or other conduit termination and be supported and secured at intervals not to exceed 4½ feet (1.4 m). The supporting and securing rules do not have to be followed when FMC is fished in a wall or ceiling. The rules also do not apply at terminals where flexibility is

Table 348.22 Maximum Number of Insulated Conductors in Metric Designator 12 (Trade Size ⅜) Flexible Metal Conduit*

Size (AWG)	Types RFH-2, SF-2		Types TF, XHHW, TW		Types TFN, THHN, THWN		Types FEP, FEBP, PF, PGF	
	Fittings Inside Conduit	Fittings Outside Conduit	Fittings Inside Conduit	Fittings Outside Conduit	Fittings Inside Conduit	Fittings Outside Conduit	Fittings Inside Conduit	Fittings Outside Conduit
18	2	3	3	5	5	8	5	8
16	1	2	3	4	4	6	4	6
14	1	2	2	3	3	4	3	4
12	—	—	1	2	2	3	2	3
10	—	—	1	1	1	1	1	2

*In addition, one insulated, covered, or bare equipment grounding conductor of the same size shall be permitted.

FIGURE 12-17   Table 348.22 in the *NEC*® lists the maximum number of insulated conductors allowed in 3/8-inch trade-size FMC and LFMC. *Reprinted with permission from NFPA 70*®, *National Electric Code*®, *Copyright © 2010, National Fire Protection Association, Quincy, MA. This reprinted material is not the complete and official position of the NFPA on the referenced subject, which is represented only by the standard in its entirety.*

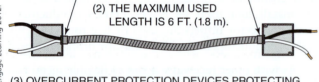

SECTION 250.118 (5) ALLOWS LISTED FMC TO BE USED AS AN EQUIPMENT GROUNDING CONDUCTOR WHEN:

(1) CONNECTORS ARE LISTED FOR GROUNDING.

(2) THE MAXIMUM USED LENGTH IS 6 FT. (1.8 m).

(3) OVERCURRENT PROTECTION DEVICES PROTECTING THE ENCLOSED CONDUCTORS ARE 20 AMPS OR LESS.

FIGURE 12-18   FMC used as a grounding path.

required when lengths do not exceed 3 feet (900 mm) for trade sizes ½ inch–1¼ inches; 4 feet (1200 mm) for trade sizes 1½ inch–2 inches; and 5 feet (1500 mm) for trade sizes 2½ inches and larger. Horizontal runs of FMC are considered supported by openings through framing members when the framing members are located at intervals not greater than 4½ feet (1.4 m).

FMC can be used as a grounding means, but there are a few requirements in Section 348.60 that have to be met first. According to Underwriters Laboratories, FMC longer than 6 feet (2 m) has not been judged to be suitable for grounding purposes. Therefore, any installation of FMC over 6 feet (2 m) in length will require an installed grounding conductor (Figure 12-18). The general rules for permitting or not permitting FMC for grounding purposes are found in Section 250.118(5). One specific *Exception* is where FMC is used for flexibility. An additional equipment-grounding conductor is *always* required where FMC is used for flexibility.

Examples of such installations include using FMC to minimize the transmission of vibration from equipment such as motors or to provide flexibility for floodlights, spotlights, or other equipment that requires adjustment. According to Section 250.118(5), where the length of the total ground-fault return path does not exceed 6 feet (2 m), and the circuit overcurrent protection does not exceed 20 amperes, and connectors that are listed for grounding are used on each termination end, a separate equipment-grounding conductor does not have to be installed with the circuit conductors, unless flexibility is necessary after the installation.

LIQUID TIGHT FLEXIBLE METAL CONDUIT: TYPE LFMC

Article 350 covers the installation requirements for LFMC. Section 350.2 defines LFMC as a raceway of circular cross section having an outer liquid tight, nonmetallic, sunlight-resistant jacket over an inner flexible metal core with associated couplings, connectors, and fittings for the installation of electric conductors (Figure 12-19). LFMC is intended for use in wet locations for connections to equipment located outdoors, such as air-conditioning equipment (Figure 12-20). LFMC may be installed in unlimited lengths, provided it meets the other requirements of Article 350 and a separate equipment-grounding conductor is installed with the circuit conductors.

Section 350.22 covers the requirements for determining the maximum number of conductors allowed in a specific size of LFMC. The procedures used are exactly the same as for the raceways we have previously

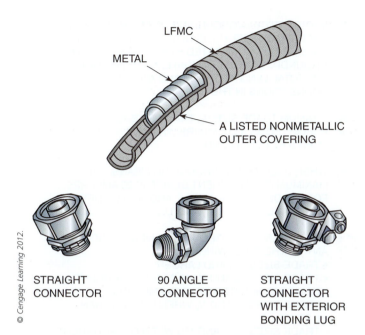

LFMC

METAL

A LISTED NONMETALLIC
OUTER COVERING

STRAIGHT
CONNECTOR

90 ANGLE
CONNECTOR

STRAIGHT
CONNECTOR
WITH EXTERIOR
BONDING LUG

© Cengage Learning 2012.

FIGURE 12-19 LFMC and associated fittings.

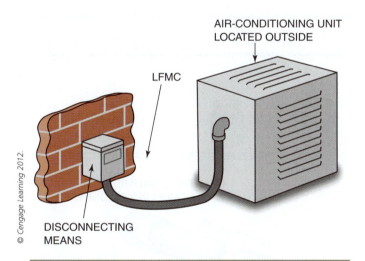

AIR-CONDITIONING UNIT
LOCATED OUTSIDE

LFMC

DISCONNECTING
MEANS

© Cengage Learning 2012.

FIGURE 12-20 A common LFMC application in residential wiring.

determining the maximum number of conductors in a ⅜-inch trade-size LFMC. This table must be used because the ⅜-inch size is not included in Table C7 of Informative Annex C. For LFMC, use the "Fittings Outside Conduit" columns in Table 348.22.

As with FMC, a run of LFMC installed between boxes, conduit bodies, and other electrical equipment is not permitted to contain more than the equivalent of 360 degrees total. Proper shaping and support of this flexible wiring method will ensure that conductors can be easily installed or taken out at any time. These requirements are found in Sections 350.24 and 350.26.

The securing and supporting requirements are given in Section 350.30. LFMC must be securely fastened in place and supported by an approved means within 12 inches (300 mm) of each box, cabinet, conduit body, or other conduit termination and be supported and secured at intervals not to exceed 4½ feet (1.4 m). The supporting and securing rules do not have to be followed when LFMC is fished in a wall or ceiling. The rules also do not apply at terminals where flexibility is required when lengths do not exceed 3 feet (900 mm) for trade sizes ½ inch–1¼ inches; 4 feet (1200 mm) for trade sizes 1½ inches–2 inches; and 5 feet (1500 mm) for trade sizes ½ inches and larger. Horizontal runs of LFMC are considered supported by openings through framing members when the framing members are located at intervals not greater than 4½ feet (1.4 m).

Section 350.60 allows LFMC to be used as a grounding means, but there are a few requirements that have to be met first (Figure 12-21). Like FMC, Underwriters Laboratories has determined that lengths of LFMC longer than 6 feet (2 m) have not been judged to be suitable for grounding. Therefore, any installation of LFMC over 6 feet (2 m) in length will require an installed grounding conductor. The general rules for permitting or not permitting FMC for grounding purposes are found in Section 250.118(6). One specific *Exception* is where LFMC is used for flexibility. An additional equipment-grounding conductor is *always* required where LFMC is used for flexibility. According to Section 250.118(6)(b), where the LFMC is size ⅜ through ½ inch and the circuit overcurrent protection does not exceed 20 amperes, the raceway itself is allowed to be the equipment-grounding conductor, provided the connectors used on each end are listed for grounding. According to Section 250.118(6)(c), where the LFMC is size ¾ through 1¼ inches, and the circuit overcurrent protection does not exceed 60 amperes, the raceway itself is allowed to be the equipment-grounding conductor, provided the connectors used on each end are listed for grounding.

discussed, except you use Table C7 in Informative Annex C when determining the maximum number of conductors that are all the same size and with the same insulation type. Also, you must use the LFMC section of Table 4 in Chapter 9 when determining the minimum-size LFMC for conductors of different sizes that have different insulation types. There is one other difference, and that is the use of Table 348.22 when

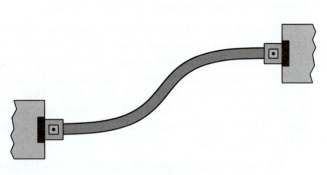

A PERMITTED AS A GROUNDING MEANS IF IT IS NOT OVER TRADE SIZE 1¹/₄, IS NOT OVER 6 FT. (1.8 m) LONG, AND IS CONNECTED BY FITTINGS LISTED FOR GROUNDING. THE 6 FT. (1.8 m) LENGTH INCLUDES THE TOTAL LENGTH OF ANY AND ALL FLEXIBLE CONNECTIONS IN THE RUN.

B MINIMUM TRADE SIZE ¹/₂. TRADE-SIZE ³/₈ PERMITTED AS A LUMINAIRE WHIP NOT OVER 6 FT. (1.8 m) LONG.

C WHEN USED AS THE GROUNDING MEANS, THE MAXIMUM OVERCURRENT DEVICE IS 20 AMPERES FOR TRADE-SIZES ³/₈ AND ¹/₂, AND 60 AMPERES FOR TRADE-SIZES ³/₄, 1, AND 1¹/₄.

B NOT SUITABLE AS A GROUNDING MEANS UNDER ANY OF THE FOLLOWING CONDITIONS:
• TRADE-SIZES 1¹/₂ AND LARGER
• TRADE-SIZES ³/₈ AND ¹/₂ WHEN OVERCURRENT DEVICE IS GREATER THAN 20 AMPERES
• TRADE-SIZES ³/₈ AND ¹/₂ WHEN LONGER THAN 6 FT. (1.8 m)
• TRADE-SIZES ³/₄, 1 AND 1¹/₄ WHEN OVERCURRENT DEVICE IS GREATER THAN 60 AMPERES
• TRADE-SIZES ³/₄, 1 AND 1¹/₄ WHEN LONGER THAN 6 FT. (1.8 m)
FOR THESE CONDITIONS, INSTALL A SEPARATE EQUIPMENT-GROUNDING CONDUCTOR SIZED PER *TABLE 250.122.*

© Cengage Learning 2012.

FIGURE 12-21   LFMC used as a grounding means.

RIGID POLYVINYL CHLORIDE CONDUIT: TYPE PVC

Article 352 covers the installation requirements for rigid polyvinyl chloride conduit (PVC). Section 352.2 defines PVC as a rigid nonmetallic conduit of circular cross section, with integral or associated couplings, connectors, and fittings for the installation of electrical conductors and cables (Figure 12-22). Two types are commonly used in residential wiring. Underwriters Laboratories recognizes the two types as: (1) Schedule 40 and (2) Schedule 80 PVC conduit. Schedule 40 PVC conduit is suitable for underground use by direct burial or encasement in concrete. Unless marked "Underground Use Only" or equivalent wording, Schedule 40 PVC conduit is also suitable for aboveground use indoors or outdoors and exposed to sunlight and weather (where not subject to physical damage). Schedule 80 PVC conduit is suitable for use wherever Schedule 40 PVC conduit may be used. The marking "Schedule 80" identifies the conduit as suitable for use where exposed to physical damage. Unless marked for a higher temperature, PVC is intended for use with wires rated 75°C or less. PVC conduit is designed for connection to couplings, fittings, and boxes by the use of a suitable solvent-type cement.

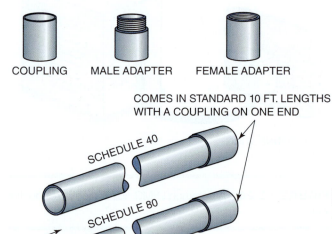

COUPLING   MALE ADAPTER   FEMALE ADAPTER

COMES IN STANDARD 10 FT. LENGTHS WITH A COUPLING ON ONE END

SCHEDULE 40

SCHEDULE 80

PVC

© Cengage Learning 2012.

FIGURE 12-22   PVC conduit and associated fittings.

Instructions supplied by the solvent-type cement manufacturer describe the method of assembly and precautions to be followed.

Section 352.22 covers the requirements for determining the maximum number of conductors allowed in a specific size of PVC. The procedures used are exactly the same as for the other solid-length conduits discussed in this chapter except you use Table C9 in Informative Annex C when determining the maximum number of conductors that are all the same size and with the same insulation type. Also, you must use the PVC section of Table 4 in Chapter 9 when determining the minimum-size PVC for conductors of different sizes that have different insulation types.

Section 352.24 requires that when bending PVC, the bends must be made so that the conduit is not damaged and the internal diameter of the conduit is not effectively reduced. **Field bends** must be made only with bending equipment identified for the purpose (Figure 12-23).

Section 352.26 limits the number of bends in one conduit run from one box to another to no more than 360 degrees total. Limiting the number of bends in a conduit run will reduce the pulling tension on conductors and help ensure easy insertion or removal of conductors.

When cutting PVC, Section 352.28 requires all cut ends to be reamed or otherwise finished to remove rough edges. The rough edges should be removed on both the inside and outside of the PVC.

Section 352.30 requires PVC to be installed as a complete system and to be fastened so that movement from thermal expansion or contraction is permitted. Expansion and contraction caused by temperature changes can cause damage to the raceway, its supports, and the electrical boxes the PVC is attached to. Expansion fittings should be used, and the supports must be installed to allow expansion or contraction cycles without damage. PVC must be securely fastened and supported in accordance with the following (Figure 12-24):

from experience...

When trimming and reaming the ends of a cut piece of Schedule 40 or Schedule 80 PVC conduit, most electricians use their knives. Files and other tools used to ream the cut ends of metal conduits don't work very well on PVC. Be careful when using your knife to trim and ream the conduit. Leather or Kevlar gloves should be worn when doing this so that your hands are protected against knife cuts.

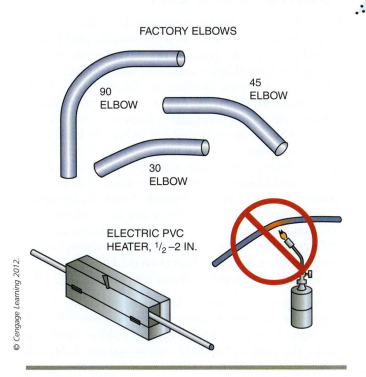

FIGURE 12-23 When electricians make field bends with PVC conduit, they must use equipment that is designed for the purpose. Some electricians find it easier to simply buy factory elbows that are already bent.

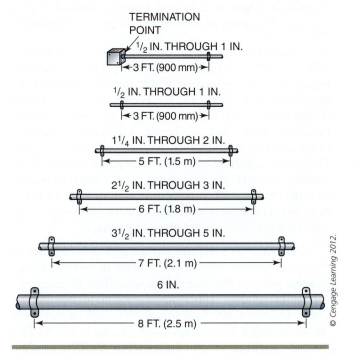

FIGURE 12-24 Support requirements for PVC conduit based on Table 352.30.

Table 352.30 Support of Rigid Polyvinyl Chloride Conduit (PVC)

Conduit Size		Maximum Spacing Between Supports	
Metric Designator	Trade Size	mm or m	ft
16–27	½–1	900 mm	3
35–53	1¼–2	1.5 m	5
63–78	2½–3	1.8 m	6
91–129	3½–5	2.1 m	7
155	6	2.5 m	8

FIGURE 12-25 Table 352.30 in the NEC lists the maximum distances between supports for different sizes of PVC conduit. *Reprinted with permission from NFPA 70®, National Electric Code®, Copyright © 2010, National Fire Protection Association, Quincy, MA. This reprinted material is not the complete and official position of the NFPA on the referenced subject, which is represented only by the standard in its entirety.*

- PVC must be securely fastened within 3 feet (900 mm) of each outlet box, junction box, device box, conduit body, or other conduit termination.

- PVC conduit must be supported as required in Table 352.30 (Figure 12-25).

- Horizontal runs of PVC supported by openings through framing members at intervals not exceeding those in Table 352.30 and securely fastened within 3 feet (900 mm) of termination points are permitted.

Expansion fittings for PVC are covered in Section 352.44 and are required to compensate for thermal expansion and contraction where the length change, in accordance with Table 352.44, is expected to be ¼ inch (6 mm) or greater in a straight run between securely-mounted items such as boxes, cabinets, elbows, or other conduit terminations. Expansion fittings (Figure 12-26) are generally provided in exposed runs of PVC conduit where (1) the run is long, (2) the run is subjected to large temperature variations during or after installation, or (3) expansion and contraction measures are provided for the building or other structures.

PVC exhibits a greater change in length per degree change in temperature than do metal raceway systems. In some parts of the United States, outdoor temperature variations of over 100°F are common. According to Table 352.44, a 100-foot run of PVC will change 4.06 inches in length if the temperature change is 100°F. The normal expansion range of most large size PVC conduit expansion fittings is generally 6 inches. Information concerning installation and application of expansion fittings is found in the manufacturer's instructions.

EXPANSION FITTING

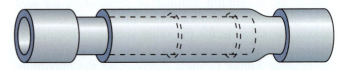

FIGURE 12-26 Expansion fittings must be used when there will be thermal expansion or contraction of ¼ inch or more in a run of PVC conduit.
© *Cengage Learning 2012.*

Where PVC enters a box, fitting, or other enclosure, a bushing must be provided to protect the wire from abrasion unless the design of the box, fitting, or enclosure is such as to afford equivalent protection, according to Section 352.46. Section 300.4(G) requires that an insulated bushing (like plastic) be used for the protection of conductors sizes 4 AWG and larger that is installed in conduit.

Since PVC is made of a nonconductive material, Section 352.60 requires a separate equipment-grounding conductor to always be installed in the conduit.

ELECTRICAL NONMETALLIC TUBING: TYPE ENT

Article 362 covers electrical nonmetallic tubing (ENT). Section 362.2 defines ENT as a nonmetallic pliable corrugated raceway of circular cross section with integral or associated couplings, connectors, and fittings for the installation of electric conductors (Figure 12-27). ENT is composed of a material that is resistant to moisture and chemical atmospheres and is flame-retardant. A pliable raceway is a raceway that can be bent by hand with a reasonable force and does not require a special bending tool. Because of the corrugations, ENT can be bent by hand and has some degree of flexibility. The outside diameter of ENT for ½- through 2-inch trade-size is made so that standard couplings and other fittings for PVC conduit can be used if an electrician so desires. ENT is suitable for the installation of conductors having a temperature rating as indicated on the ENT. However, the maximum allowable ambient temperature is 122°F.

Section 362.22 covers the requirements for determining the maximum number of conductors allowed in a specific size of ENT. The procedures used are exactly the same as for EMT except you use Table C2 in Informative Annex C when determining the maximum number of conductors that are all the same size and with the same insulation type. Also, you must use the ENT section of Table 4 in Chapter 9 when determining the minimum-size ENT for conductors of different sizes that have different insulation types.

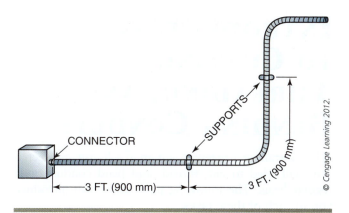

FIGURE 12-28 Securing and support requirements for ENT.

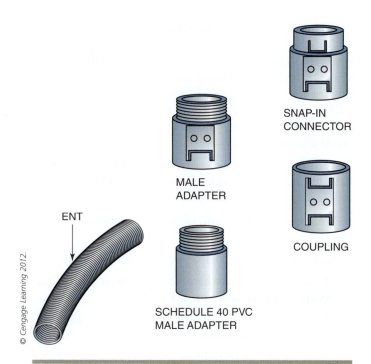

FIGURE 12-27 ENT with associated fittings.

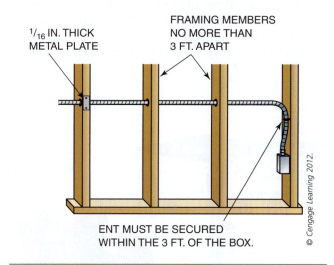

FIGURE 12-29 ENT run horizontally through framing members is considered supported as long as the framing members are no more than 3 feet (900 mm) apart. Section 300.4(A)(1) requires a ¹⁄₁₆-inch-thick metal plate if the ENT is run through a bored hole that is less than 1¼ inches (31 mm) from the edge of the framing member.

Section 362.24 requires bends in ENT to be made so that the tubing is not damaged and the internal diameter of the tubing is not effectively reduced, and, like all the other circular raceways discussed in this chapter, Section 362.26 does not allow more than the equivalent of 360 degrees total bending between termination points of ENT.

Like the other raceway types discussed in this chapter, Section 362.28 requires all cut ends of ENT to be reamed or otherwise finished to remove rough edges.

Section 362.30 covers securing and supporting of ENT. It states that ENT must be installed as a complete system and be securely fastened in place and supported according to the following:

- ENT must be securely fastened at intervals not exceeding 3 feet (900 mm). In addition, ENT must be securely fastened in place within 3 feet (900 mm) of each outlet box, device box, junction box, cabinet, or fitting where it terminates. Where ENT is run on the surface of framing members, it is required to be fastened to the framing member every 3 feet (900 mm) and within 3 feet (900 mm) of every box (Figure 12-28).

- Horizontal runs of ENT supported by openings in framing members at intervals not exceeding 3 feet (900 mm) and securely fastened within 3 feet (900 mm) of termination points is permitted (Figure 12-29).

- An unbroken length (without couplings) of ENT is permitted to be fished in finished walls of houses without having to be secured and supported.

Where ENT enters a box, fitting, or other enclosure, a bushing must be provided to protect the wire from abrasion unless the design of the box, fitting, or enclosure is such as to afford equivalent protection, according to Section 362.46. Section 300.4(G) requires that an insulated bushing (like plastic) be used for the protection of conductors sizes 4 AWG and larger that is installed in conduit.

Because electrical nonmetallic tubing (ENT) is made of a nonconductive material, Section 362.60 requires a separate equipment-grounding conductor to always be installed in the conduit.

INTRODUCTION TO CUTTING, THREADING, AND BENDING CONDUIT

When installing a raceway system in a house, an electrician will need to cut, thread, and bend conduit on a regular basis. The following paragraphs will introduce you to each of these tasks.

CUTTING CONDUIT

Electricians will need to know how to cut to length the various conduit types used in residential wiring. Solid-length metal conduit (like RMC, IMC, and EMT) is cut to length using a pipe cutter, a tubing cutter, a portable band saw, or simply a hacksaw. The ends of cut lengths

of RMC and IMC will also have to be threaded. Figure 12-30 shows some of the tools used to cut and thread RMC and IMC. PVC conduit can be cut using a hacksaw or with saws equipped with special blades for PVC (Figure 12-31). FMC and LFMC are usually cut with a hacksaw. ENT can also be cut with a hacksaw, but special nonmetallic tubing cutters that look like big scissors are available.

No matter what raceway type you encounter and have to cut, the following procedures should always be followed:

• Wear safety glasses and observe all applicable safety rules.

• Secure the raceway before you attempt to cut it.

• Mark the locations where you wish to make the cut clearly and accurately on the raceway.

• Choose a cutting tool that is appropriate for the raceway type you are using.

• Make the cut as straight as possible.

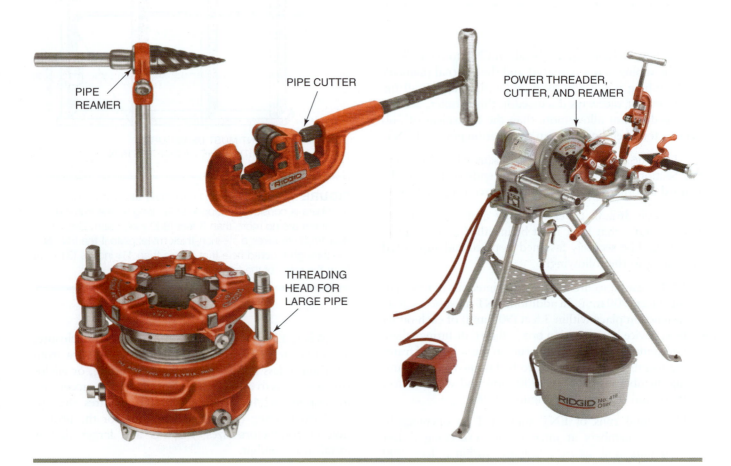

FIGURE 12-30   RMC and IMC cutting, threading, and reaming tools. *Courtesy of Ridgid/Emerson.*

FIGURE 12-31   PVC conduit is cut with special PVC cutting saws (left) or with a hacksaw (right). *Courtesy of Greenlee, a Textron Company.*

- File or ream the ends of the raceway after making the cut to remove any sharp or rough edges that could damage the conductor insulation.

(See Procedure 12-1 on pages 422–423 for the proper procedure to follow for cutting conduit with a hacksaw.)

> ## *from experience...*
>
> Some electricians cut FMC by intentionally bending it so tightly that it actually breaks open. A pair of diagonal cutting pliers is then used to complete cutting off the length needed. This method is not recommended because of the damage it can cause the raceway. Check with your supervisor to see if this is an acceptable method for cutting FMC in your area.

THREADING CONDUIT

It does not happen very often in residential wiring, but once in a while an electrician will have to cut to length a piece of RMC or IMC and then thread the ends. Either hand threaders or power threading equipment like those shown in Figure 12-30 can be used. When cutting new threads on a conduit end, the following items should always be observed:

- Wear safety glasses as well as proper foot and hand protection. Observe all applicable safety rules.
- Secure the conduit in a pipe vise before you attempt to thread it.
- Choose the proper threading die for the size of conduit you are threading.
- Always use plenty of cutting oil during the threading process.

CAUTION

CAUTION: Always use plenty of cutting oil when threading a raceway. The cutting oil will allow the cutting die to cut the threads into the pipe easier. It extends the life of the cutting die and, when applied properly, will flush away the metal shavings produced during the threading process. Cutting oil can also help cool down the pipe end that is being threaded.

(See Procedure 12-2 on pages 424–426 for the proper procedure to follow for cutting and threading RMC conduit with a hand threader.)

BENDING CONDUIT

Bending conduit is definitely a skill that improves with practice. The more you bend conduit, the better conduit bender you become. The most common electrical conduit installed in houses is EMT, and for this reason, the discussion that follows focuses on EMT. However, the bending techniques described also apply to the other types of circular metal raceway, such as RMC and IMC.

There are a few common bends that electricians will need to know. The most common bend is a 90-degree bend, commonly called a **stub-up** (Figure 12-32). When electricians refer to "stubbing up" a pipe, they mean that a 90-degree bend is to be made. A **back-to-back bend** is a type of bend where the distance is measured between the outside diameters of two sections of the pipe (Figure 12-33). This type of bend consists of putting a 90-degree bend on each end of one length of pipe, resulting in a straight section of pipe between the two 90-degree bends. This bending technique allows the conduit to fit precisely between two objects. An **offset bend** requires two equal bends in a conduit that results in the direction of the conduit being changed so that it can avoid an obstruction blocking the conduit run (Figure 12-34). A **saddle bend** is similar to an offset bend in that it results

in a conduit run going around an object that is blocking the path of the run (Figure 12-35). The difference with a saddle bend is that it actually goes *over* the obstruction rather than *around* it. There are two styles of saddle bends: a three-point saddle and a four-point saddle. Later in this section, we discuss the actual bending of a three-point saddle because it is considered to be a little easier to do than the four-point saddle. The last type of bend we discuss is one that is used when conduits enter a box or other electrical enclosure that is surface-mounted. It is called a **box offset bend** and is really just a smaller version of the regular offset bend (Figure 12-36). An installation where the electrician has used box offsets at each box location is considered to be a neater and more professional installation.

EMT is bent in the field using either a hand bender, hydraulic bender, or an electric bender. Sizes from ½ inch through 1¼ inches can be bent with a hand bender

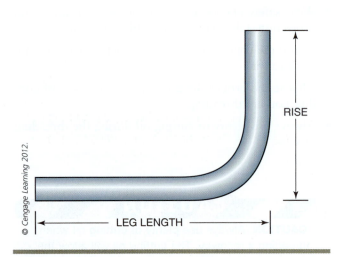

© Cengage Learning 2012.

FIGURE 12-32   A stub-up bend.

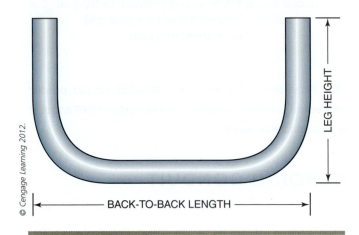

© Cengage Learning 2012.

FIGURE 12-33   A back-to-back bend.

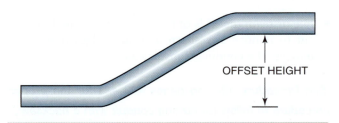

FIGURE 12-34   An offset bend.
© Cengage Learning 2012.

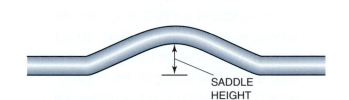

FIGURE 12-35   A three-point saddle bend.
© Cengage Learning 2012.

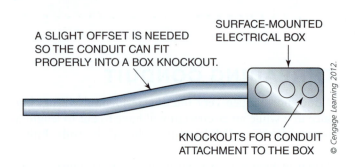

A SLIGHT OFFSET IS NEEDED SO THE CONDUIT CAN FIT PROPERLY INTO A BOX KNOCKOUT.

SURFACE-MOUNTED ELECTRICAL BOX

KNOCKOUTS FOR CONDUIT ATTACHMENT TO THE BOX

© Cengage Learning 2012.

FIGURE 12-36   A box offset bend.

(Figure 12-37). However, electricians normally don't bend 1¼ inch trade size EMT with a hand bender because of how hard it is to bend. Larger sizes will require the use of a hydraulic (Figure 12-38) or electric power bender (Figure 12-39). Since most EMT installed in houses will be ½-, ¾-, or 1-inch trade sizes, we will keep our discussion focused on bending with a hand bender.

FIGURE 12-39   A popular model of electric bender. The bending shoes are changed when bending EMT or RMC/IMC. *Courtesy of Greenlee, a Textron Company.*

FIGURE 12-37   An example of a hand bender that can be used for EMT, RMC, and IMC. A handle is used with each bender shown. *Courtesy of Greenlee, a Textron Company.*

- STRONG PORTABLE PIN-ASSEMBLED ALUMINUM COMPONENTS
- EASY-TO-USE RAM TRAVEL SCALE AND BENDING CHARTS
- CONDUIT SUPPORTS INDEX TO SUIT ALL SIZES
- MODEL AVAILABLE TO BEND PVC-COATED RIGID CONDUIT

from experience...

In some parts of the country, the wiring method required by the authority having jurisdiction to be used in houses is RMC or IMC. Trade sizes ½-inch, ¾-inch, and 1-inch RMC or IMC can be bent with a hand bender but to make the job easier many electricians use a mechanical bender. This type of bender is often referred to as a "Chicago Bender" (Figure 12-40).

FIGURE 12-40   A mechanical bender. It is used to bend ½-, ¾-, and 1-inch trade size RMC. It is often called a "Chicago Bender."

FIGURE 12-38   A typical hydraulic bender and associated parts. *Courtesy of Greenlee, a Textron Company.*

© Cengage Learning 2012.

An EMT bender has many features (Figure 12-41). A hook at the front of the bender is used to help secure and correctly place the bender onto a piece of conduit. The opposite end of the bender will have a foot pedal that is designed to allow an electrician to exert foot pressure on the bender while making a bend. The foot pressure does two things: It helps keep the bender seated against the conduit, and it actually causes the bend to be made. A handle is attached to the bender and is used more for allowing the hands and arms to guide the bender during the bending operation than it is for bending the pipe. An EMT bender also has many markings on it, and an electrician will need to know what each of the markings indicates. The bender will usually have a degree scale marked on the bender in 22-, 30-, 45-, 60-, and 90-degree increments. The degree indicator will help an electrician determine the amount of bend put into a piece of EMT. There is also an arrow mark on the bender that is located toward the hook end. The arrow is aligned with bending marks placed on the conduit at the locations needed for a specific type of bend, like a 90-degree stub bend. There will be a mark that indicates the location of the back of a bend. On most

EMT benders, this mark is a star with one of the star points being longer than the other points. The longer star point is aligned with the marking on the conduit where the back of a bend will be. The last marking on an EMT bender that you need to be aware of is commonly called a "rim notch." This mark is aligned with the marking on a conduit that represents the center of a three-point saddle bend. The amount of **take-up** for the bender is also marked on it. Take-up is the amount that must be subtracted from a desired stub-up height so the bend will come out right. Take-up must be taken into account because in a 90-degree bend, the actual conduit bend is in an arc and not a true right angle. The arc, or sweep, of the bend is due to the shape of the bender. The take-up for a ½-inch EMT hand bender is 5 inches and for a ¾-inch hand bender is 6 inches.

When actually making a bend using a hand bender, the following procedures should always be followed:

- Wear safety glasses and observe all applicable safety rules.

- Bend on a flat surface that is not slippery.

- Mark the locations on the conduit where you wish to make the bends clearly and accurately.

- Apply heavy foot pressure on the foot pedal to keep the conduit tightly in the bender.

- When making multiple bends on the same pipe length, keep all bends in the same plane.

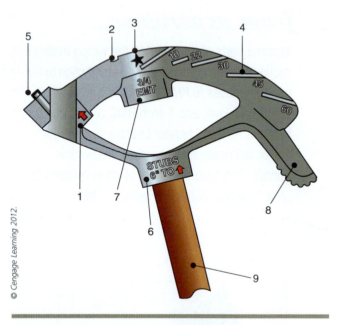

© Cengage Learning 2012.

FIGURE 12-41 EMT hand bender features: (1) Arrow: used with stub-up, offset, and the outer marks of saddle bends; (2) Rim notch: used to bend the center bend of a three-point saddle; (3) Star point: indicates the back of a 90-degree bend; (4) Degree scale: used to indicate the amount of angle in the bend; (5) Hook: used to help secure the bender to the pipe being bent; (6) Take-up: indicates the amount of take-up for the bender head; (7) Pipe size: indicates the sizes of conduit the bender can bend; (8) Foot pedal; and (9) Handle.

(See Procedure 12-3 on pages 427–428 for the proper procedure to follow for bending a stub-up in a length of ½-inch EMT.)

(See Procedure 12-4 on pages 429–430 for the proper procedure to follow for bending a back-to-back bend in a length of ½-inch EMT.)

(See Procedure 12-5 on pages 431–433 for the proper procedure to follow for bending an offset bend in a length of ½-inch EMT.)

(See Procedure 12-6 on pages 434–436 for the proper procedure to follow for bending a three-point saddle in a length of ½-inch EMT.)

(See Procedure 12-7 on pages 437–439 for the proper procedure to follow for bending box offsets in a length of ½-inch EMT.)

INSTALLATION OF RACEWAY IN A RESIDENTIAL WIRING SYSTEM

In this section, we discuss the installation of raceway in a residential wiring system. The raceway will contain the conductors for the various branch circuits or feeder circuits used throughout the house. As we mentioned earlier, EMT is the most common raceway type used in residential work, and for this reason we limit our discussion in this section to the installation of EMT.

When EMT is installed, metal outlet and metal device boxes are used. The EMT will be connected to the boxes with approved fittings, called "connectors" (Figure 12-42). The setscrew type of connector is used most often because of its ease of installation and lower cost. Some electricians like to use the compression type of connector, which is a little more expensive to buy. Either connector type will provide a solid and secure connection of the conduit to an electrical box. When lengths of EMT need to be coupled together, only approved couplings may be used. Just like the connectors, couplings are available in a setscrew or a compression type (Figure 12-43). In locations where EMT is used outdoors, the compression connectors and couplings must be used for a watertight connection. Setscrew connectors and couplings are not allowed in wet locations.

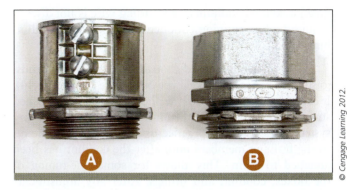

FIGURE 12-42   (A) Setscrew and (B) compression-type EMT connectors.

© Cengage Learning 2012.

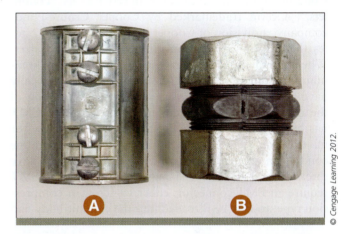

FIGURE 12-43   (A) Setscrew and (B) compression-type EMT couplings.

© Cengage Learning 2012.

from experience...

Not all compression-type EMT connectors and couplings are listed as being watertight. Always check the listing of the fitting to make sure you can use it in an outdoor (or indoor) wet location. Manufacturers include some type of rubber sealing ring in these fittings to make them watertight. They also typically provide some distinguishing characteristic that can be seen once the fittings are installed so that an electrical inspector can tell quickly whether watertight connectors or couplings have been installed.

As we learned earlier, EMT is an approved grounding method and, as such, does not always require an equipment-grounding conductor to be run in the raceway with the other circuit conductors. However, most electricians will run a green insulated grounding conductor in the raceway. Always check with your local electrical inspector to determine if an insulated equipment-grounding conductor is always required or if it is left up to the electrician to decide. Electricians choose the conductor insulation color for the conductors they install in the raceway. Usually, the same color coding found in cables is used. For a 120-volt branch circuit, an electrician will use a white insulated wire and a black insulated wire. For a straight 240-volt circuit (like an electric water heater), two black conductors or a black and a red conductor will usually be used. If the circuit is a 120/240-volt circuit (like an electric clothes dryer), a white insulated wire, a black insulated wire, and a red insulated wire will be run. Remember, a green insulated equipment-grounding wire is not always required when running conductors in an EMT conduit, but it is recommended.

EMT can be installed though drilled holes in studs, joists, and rafters just like Type NM cable (Figure 12-44). It can also be installed on a wall or on a ceiling's finished surface (Figure 12-45). When installing an EMT raceway system, it is common practice to not fully tighten boxes to the framing members until after the conduit has been attached to them. This allows for some fine-tuning of the boxes, conduit, and fittings. After any adjustments have been made, the boxes are then fully secured to the framing members.

Electricians often use a **conduit body** to connect two or more conduits together as part of a conduit system. Conduit bodies are commonly referred to as "condulets," a term trademarked by Cooper Crouse-Hinds company, a division of Cooper Industries. Conduit bodies (Figure 12-46) come in different shapes and sizes. When used with EMT they are usually made of aluminum. Condulets designed for use with EMT typically have set-screw connections built-in at each opening of the conduit body. Once the end of a length of EMT has been inserted into the condulet end, a set screw (or multiple screws in larger sizes) is tightened to secure the conduit in place. Some conduit bodies simply have threaded holes at the openings and when using EMT require an EMT connector, either set-screw or compression type, to first be threaded into the conduit body end

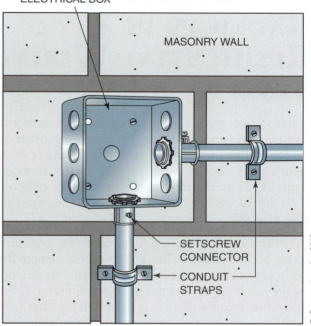

SURFACE-MOUNTED ELECTRICAL BOX

MASONRY WALL

SETSCREW CONNECTOR

CONDUIT STRAPS

© Cengage Learning 2012.

FIGURE 12-45 EMT run on the surface.

and then the EMT is secured to the conduit body by tightening the EMT connector. The most common types of conduit bodies include:

- L-shaped bodies ("Ells") include the LB, LL, and LR, where one opening is in line with the access cover and the other is on the back (LB), left (LL) and right (LR), respectively. In addition to providing access to wires for pulling, "L" fittings allow a 90 degree turn in conduit where there is insufficient space for a full-radius 90 degree sweep of a conduit section.

- T-shaped bodies ("Tees") feature an opening in line with the access cover and openings to both the cover's left and right.

- C-shaped bodies ("Cees") have identical openings on both sides of the access cover, and are used to pull conductors in straight runs because they make no turn between the openings.

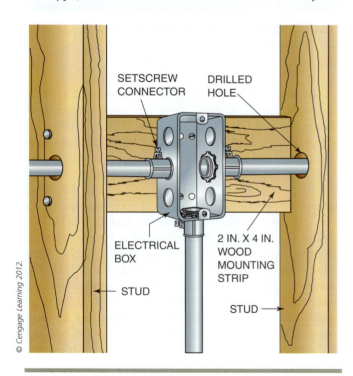

SETSCREW CONNECTOR

DRILLED HOLE

ELECTRICAL BOX

2 IN. X 4 IN. WOOD MOUNTING STRIP

STUD

STUD

© Cengage Learning 2012.

FIGURE 12-44 EMT installed through holes in building framing members.

> **CAUTION**
>
> **CAUTION:** Section 300.9 states that when installing conduits above grade in a wet location (like outdoors on the side of a house), the interior of the conduit is considered a wet location. This means that insulated conductors installed in the conduit must have an insulation type that is permitted in wet locations like THWN, XHHW, or THW. A conductor with a THHN insulation could not be used.

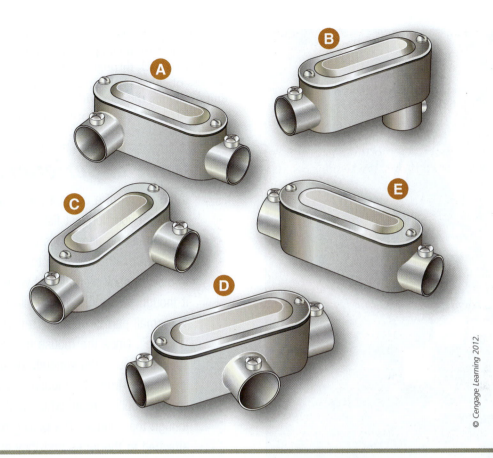

FIGURE 12-46 Common conduit body styles include (A) LR; (B) LB; (C) LL; (D) T; and (E) C.

© Cengage Learning 2012.

CAUTION

CAUTION: Houses are sometimes built with a metal roof. Section 300.4(E) addresses installation of a raceway under metal-corrugated sheet roofing. It requires EMT to be installed and supported so there is at least 1½ inches (38 mm) from the lowest surface of the roof material to the top of the EMT conduit. This rule is designed to minimize the chance of long roofing screws penetrating through the EMT and damaging the conductors inside.

RACEWAY CONDUCTOR INSTALLATION

Once the raceway system has been installed, the electrician must install the various branch-circuit or feeder conductors in the conduit. The conductors are usually pulled into the conduit, but in shorter runs between electrical boxes, the conductors may be pushed through the raceway.

The conductors will be taken off spools. The electrician will need to set up the pull by arranging the spools of the required circuit conductors in such a way that the conductors on the spools can be easily pulled off without becoming tangled with each other. One of the easiest ways to do this is to use a commercially available wire cart that allows several spools of wire to be put on them at one time (Figure 12-47). Once the electrician has determined how many wires must be installed in the conduit run, a few feet of each wire are pulled from the spools on the wire cart, and the ends are taped together. The taped end of the wires is then inserted into the raceway and pushed to the electrical boxes. If the length of conduit between boxes is fairly long, a fish tape must be used (the fish tape was introduced in Chapter 3). It is made of a flexible metal or nonmetallic tape that is enclosed in a metal or plastic enclosure. The tape can be pulled out of the enclosure and inserted into a raceway and pushed through it until it comes out at a box location. The fish tape will have a hook on the end of it, and the conductors are attached by the electrician to the fish tape end. While one

electrician pulls the conductors slowly off the spools, another electrician will pull the fish tape with the attached conductors back through the raceway (Figure 12-48). This process of either pushing the conductors through the conduit or pulling them in with a fish tape is repeated until all the required conductors are installed in the raceway system.

from experience...

If the length of a conduit run is longer than the length of your longest fish tape, another technique must be used. This technique uses a powerful machine similar to a vacuum cleaner to blow or suck a small part with a string tied to it though the length of conduit (Figure 12-49). The small part is normally referred to as a "mouse". Once the mouse has been blown or sucked through the conduit, the attached string is removed from the mouse and tied to a stronger pulling rope, which is then pulled though the conduit. The pulling rope is then attached to the wires and they are pulled into the conduit. This technique is used much more often in commercial and industrial wiring because the conduit runs in house wiring are usually not longer than the length of a fish tape.

© Cengage Learning 2012.

FIGURE 12-47   A commercially available wire cart that can hold several spools of wire at one time. The cart allows the wires to be easily pulled off the spools without tangling.

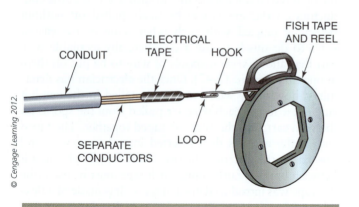

© Cengage Learning 2012.

FIGURE 12-48   A fish tape is used to pull wires through a length of conduit. Once the conductor is looped through the hook, tape it up to prevent the loop separating from the hook during installation.

FIGURE 12-49   A tool used to blow or suck a pull string into a conduit run that is too long to use a fish tape. *Courtesy of Greenlee, a Textron Company.*

from experience...

Most residential electrical contractors do not encounter raceway wiring very often. As such, they probably do not have a commercially available wire cart. If this is the case, an electrician can cut a length of ½-inch EMT conduit, put the pipe through the holes in the wire spools, and then use a step ladder (or two step ladders) to support the pipe with the spools on it. The electrician can then pull the wires from the spools as needed. Another common practice is to support the pipe length with the spools on it by drilling holes in two studs and inserting the pipe through the holes.

CAUTION

CAUTION: Do not use a metal fish tape to pull or push conductors in a raceway that is connected to an energized loadcenter. Nonconductive fish tapes are available for this wiring situation.

SUMMARY

In this chapter, we have taken a look at the different raceway types that are used in residential wiring. The *NEC®* installation requirements for these raceways were covered in detail, and although there are many similarities in the raceway installation requirements, it is important for a residential electrician to recognize the specific rules for the type of conduit being installed. Because EMT is the most often used raceway system in a residential wiring installation, the proper bending and installation techniques for it were explained and examples given. The last thing covered in this chapter was some common installation techniques for installing the conductors in a completed raceway system.

Even though many electricians who do residential wiring seldom have the opportunity to install raceway, they must still be aware of the common raceway types and installation practices for those times that they do encounter it. State and local electrical licensing exams also have many questions about raceways and their installation requirements. For electricians who live and work in an area of the country where they have to install a residential electrical system using a raceway wiring method or for residential electricians who work with raceways only once in a while, following the information presented in this chapter will make for a professionally installed and *NEC®*-compliant raceway installation.

PROCEDURE 12-1

Cutting Conduit with a Hacksaw

Follow this procedure for cutting a length of ½-inch EMT. **Note:** *This procedure can also be used for cutting other conduit types with a hacksaw.*

- Wear safety glasses and observe all applicable safety rules.

 A Secure a length of ½-inch EMT in a pipe vise. Other securing methods can be used.

A

© Cengage Learning 2012.

B Measure and clearly mark the pipe at the location you wish to make the cut.

B

© Cengage Learning 2012.

C Prepare to cut the conduit using a hacksaw. Place the hacksaw blade at the cutting mark and using smooth strokes start to cut the conduit. Remember that the forward stroke is used for cutting.

D Continue to cut using smooth strokes until you have cut completely through the conduit.

E Using a reaming tool, ream the conduit end. If necessary, use a file to eliminate all burrs on the inside and outside of the conduit ends.

• Clean up your work area and put all tools and materials away.

C

© Cengage Learning 2012.

D

© Cengage Learning 2012.

E

© Cengage Learning 2012.

PROCEDURE 12-2

Cutting and Threading RMC Conduit

Follow this procedure for cutting and threading a length of ½-inch RMC with a hand threader. **Note:** *This procedure can also be used to cut and thread IMC.*

- Wear safety glasses and observe all applicable safety rules.

A Secure a length of ½-inch RMC in a pipe vise.

A

© Cengage Learning 2012.

B Measure and clearly mark the pipe at the location you wish to make the cut.

B

© Cengage Learning 2012.

C Prepare to cut the conduit using a pipe cutter. Do this by turning the handle to open up the cutter so it will easily fit around the pipe. Place the cutter wheel at the cut mark and then tighten the handle of the pipe cutter until the cutting wheel is tight against the pipe.

C

© Cengage Learning 2012.

- Rotate the pipe cutter around the conduit. Tighten the pipe cutter a little more each time you make a complete rotation around the pipe.

D Continue to alternately rotate and tighten the pipe cutter until you have completely cut off the end of the conduit.

 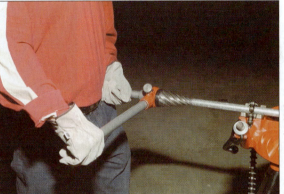

E Using a reaming tool, ream the cut ends of the conduit. If necessary, use a file to eliminate any burrs on the outside and inside of the conduit ends.

- Inspect the hand threader before using. Replace cutting dies or any other parts that show damage or wear.

F Using the hand threader with the proper-size die, place the cutting die on the end of the conduit. Most hand threader handles are of the ratchet type. Make sure it is set to turn in the right direction.

© Cengage Learning 2012.

Cutting and Threading RMC Conduit (Continued)

G To get it started, use one hand to exert some force on the end of the die head and at the same time use your other hand to rotate the threader handle in a clockwise direction. Once the cutting die starts to cut the threads on the end of the conduit, there will be no need to continue pushing the die onto the conduit end.

- When hand-threading, your weight should be above the handle. This ensures maximum leverage. If possible, don't do all the work with your arms, use your weight. Keep proper footing and balance to maintain control.

- Continue to cut the new threads onto the conduit until the length of the cut thread is the same as the length of the cutting threads of the die. Do not forget to apply a good amount of cutting oil during the threading process.

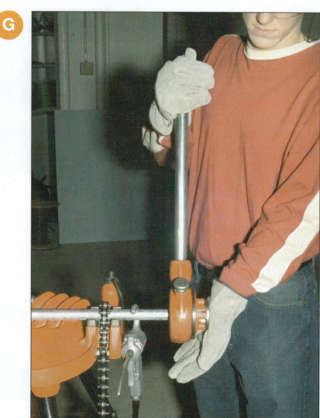

© Cengage Learning 2012.

H Now change the direction of the ratcheting handle and slowly remove the cutting die from the end of the conduit. Be careful, as the newly cut threads are sharp.

- Clean up your work area and put all tools and materials away.

© Cengage Learning 2012.

PROCEDURE **12-3**

Bending a 90-Degree Stub-Up

- Wear safety glasses and observe all applicable safety rules.

- Subtract the "take-up" from the required finished stub height. The take-up for a ½-inch bender is 5 inches, so 12 inches − 5 inches = 7 inches.

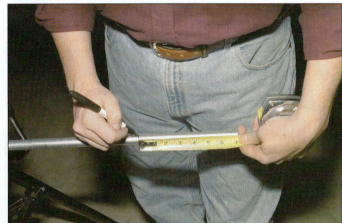

A Measure back from the end of the EMT a distance of 7 inches. Mark this dimension clearly on the conduit.

- Place the conduit on a flat surface, such as the floor.

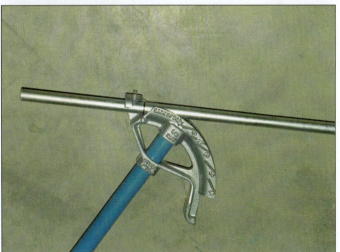

B Line up the arrow on the bender head with the mark on the conduit.

from experience...

It is a good idea is to use a pencil or a permanent marker for marking conduit. Be sure to wrap your mark all the way around the pipe in case a pipe is not bent correctly and the bender has to be placed back on the conduit for additional bending. This is called "fine-tuning" by some electricians. Adding or taking away some angle from a bend must be done at exactly the same spot as the original bend, or damage to the conduit will result.

PROCEDURE 12-3

Bending a 90-Degree Stub-Up (Continued)

C While applying constant foot pressure to the bender, bend the conduit to 90 degrees.

D Measure the stub height. You should have a stub that is 12 inches high.

• Clean up your work area and put all tools and materials away.

© Cengage Learning 2012.

© Cengage Learning 2012.

from experience...

You can use the degree indicator on the bender to tell if you have a 90-degree bend, but a torpedo level placed on the side of the pipe works well. Remember, a torpedo level used in electrical work will have one side magnetized so that it will stick to the side of a metal conduit made of steel.

from experience...

Do not be surprised if you come out a little over or a little under your target measurement when using a bender. They are fairly accurate but are not precise. Also, each bender will have its own characteristic that you will have to get used to. For example, if a bender keeps making bends that are consistently ¼ inch *under* the target measurement, *add* ¼ inch to the measurement that is marked on the conduit and then make the bend.

PROCEDURE 12-4

Bending a Back-to-Back Bend

Follow this procedure for bending a back-to-back bend in a length of ½-inch EMT with a hand bender. For this example, leg 1 of the bend will be 25 inches high and leg 2 will be 30 inches high. The actual length of the bend from the outside of one leg to the outside of the other leg will be 48 inches.

- Wear safety glasses and observe all applicable safety rules.

- Subtract the "take-up" from the finished stub height for leg 1. Because the first leg is to be 25 inches high, 25 inches − 5 inches (½-inch EMT bender take-up) = 20 inches.

- Measure 20 inches back from one end of the conduit and mark this dimension clearly on the conduit.

- Place the conduit on a flat surface, such as the floor.

A   Line up the arrow on the EMT bender with the mark on the conduit.

B   Apply constant foot pressure to the bender and bend the conduit to 90 degrees. You should now have leg 1 bent to a 90-degree angle with a height of 25 inches.

A

© Cengage Learning 2012.

B

© Cengage Learning 2012.

PROCEDURE 12-4

Bending a Back-to-Back Bend (Continued)

C Now lay the conduit on a flat surface and measure 48 inches from the outside of leg 1 to the point where the back of the second bend is to be. Mark the conduit.

C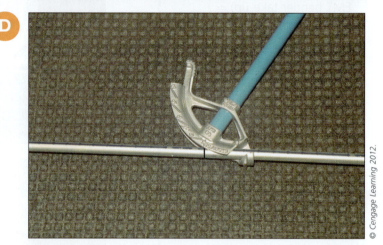

D Align this mark on the conduit with the starpoint (or the diamond) of the bender and bend to 90 degrees.

- Leg 2 must be 30 inches high. Simply cut off any extra conduit until the leg is the correct length.

D

E Measure the leg heights and the distance from the back of one leg to the back of the other leg. You should have a back-to-back bend with one leg that is 25 inches high, the other leg 30 inches high, and the back-to-back distance should be 48 inches. Adjust the pipe back-to-back bend as needed.

- Clean up your work area and put all tools and materials away.

E

PROCEDURE 12-5

Bending an Offset Bend

- Wear safety glasses and observe all applicable safety rules.

A Cut the conduit to the proper length.

A

B Make sure to ream the end of the conduit after cutting it to length.

B

PROCEDURE 12-5

Bending an Offset Bend (Continued)

C Determine the center of the pipe length and mark it.

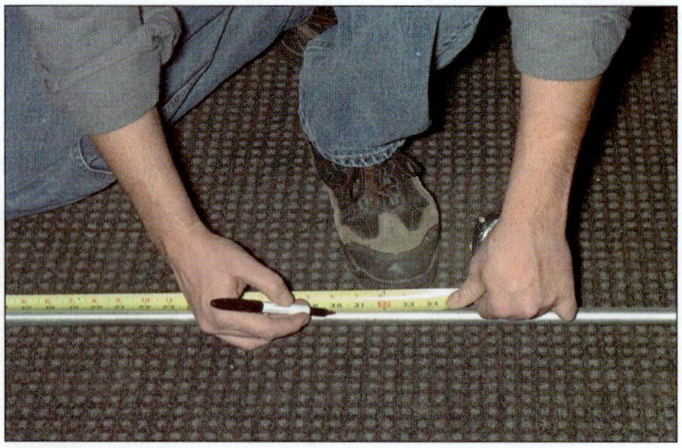

C

© Cengage Learning 2012.

D Determine the distance between the two bends of the offset. Consult an offset table like the one shown. In this example, the distance is 8 inches for an offset bend of 4 inches.

D

Offset Bending Chart

Offset Depth in Inches	Degree of Offset Bends			
	2.5°	**30°**	**45°**	**60°**
2″	5¼″			
3″	7¾″	6″		
4″	10½″	**8″**		
5″	13″	10″	7″	
6″	15½″	12″	8½″	7¼″
7″	18¼″	14″	9¾″	8⅜″
8″	20¾″	16″	11¼″	9⅝″
9″	23½″	18″	12½″	10⅞″
10″	26″	20″	14″	12″

© Cengage Learning 2012.

E Mark the pipe with the dimensions determined from the offset table on the conduit. Do this by making a mark an equal distance on each side of the center mark. The distance from the center of each mark is found by dividing the dimension from the offset table in half. For this example, 8 inches ÷ 2 = 4 inches.

• Place the conduit on a flat surface, such as the floor.

• Line up the arrow on the bender with the first mark on the conduit.

E

© Cengage Learning 2012.

F Apply heavy foot pressure to the bender and bend the conduit to 30 degrees. Note that when the bender handle is straight up, you have a 30-degree bend (this is true for most but not all hand benders).

G While keeping the conduit in the bender, invert the bender and place the handle end on the floor. Rotate the conduit 180 degrees, slide it ahead in the bender head, and align the arrow with the next mark. Then, while standing up, bend the conduit to 30 degrees. Use the degree scale on the bender to determine when you have bent to 30 degrees.

• Make sure to keep the bends in the same plane. If they are not, fine-tune the offset as necessary.

H Check to be sure that the offset amount is correct. If it is not enough or is too much, add or subtract some angle from the bends. When adding or taking away angle from the bends, make sure to always bend on exactly the same spot on the conduit. Proper marking of the pipe at the beginning of the process is extremely important, especially if you need to make bending adjustments.

• Clean up your work area and put all tools and materials away.

© Cengage Learning 2012.

PROCEDURE 12-6

Bending a Three-Point Saddle

Follow this procedure for bending a three-point saddle bend in a 5-foot length of ½-inch EMT with a hand bender. The saddle must go over a 4-inch obstruction and the saddle will be bent using a 45-degree center bend and two 22.5-degree side bends. This is considered the most common three-point saddle type.

- Wear safety glasses and observe all applicable safety rules.

- Cut the conduit to the proper length.

- Make sure to ream the end of the conduit after cutting it to length.

- Determine the mark spacing for the 4-inch three-point saddle. There will be three marks made on the conduit: a center mark and a mark placed on either side of the center.

A First, determine the location on the conduit where you want the center of the three-point bend and mark it.

B Next, determine the dimension of the two outer marks from the center mark. The multiplier is 2.5 for all pipe sizes when using a 45-degree center bend and two 22.5-degree outer bends or a 60-degree center bend and two 30-degree outer bends. Figure B shows actual mark spacing for saddle heights of 1 inch through 6 inches. In this example, 10 inches is the distance from the center mark for each outside mark. This is found by multiplying the desired saddle height by 2.5: 4 inches × 2.5 = 10 inches.

A

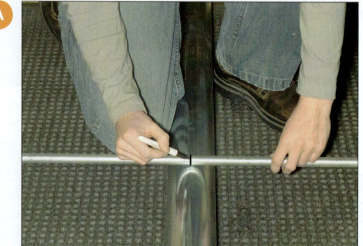

© Cengage Learning 2012.

B **Saddle Bending Marks for a 45° Center Bend and 22.5° Outer Bends**

If the obstruction to be saddled over is:	Make the outside marks on either side of the center mark:
1″	2½″
2″	5″
3″	7½″
4″	10″
5″	12½″
6″	15″

© Cengage Learning 2012.

C Mark this dimension on the conduit by making a mark 10 inches on each side of the center mark.

- Place the conduit on a flat surface, such as the floor.

D Line up the *rim notch* or *saddle mark* on the bender with the center mark on the conduit.

- Keeping the rim notch and the center mark aligned, place the conduit on the floor. Apply heavy foot pressure to the bender and bend the conduit to 45 degrees.

E While keeping the conduit in the bender head, invert the bender so the handle is now on the floor and rotate the conduit 180 degrees.

PROCEDURE 12-6

Bending a Three-Point Saddle (Continued)

F Slide the conduit ahead and align the *arrow* with the first outer mark.

G While standing up, bend the conduit to 22.5 degrees.

- Remove the conduit from the bender and reverse it end to end.

H Insert the other end of the conduit into the bender. Align the remaining outside mark at the arrow and make another 22.5-degree bend.

I Check the saddle bend and fine-tune as necessary. Be sure to have all bends in the same plane.

- Clean up your work area and put all tools and materials away.

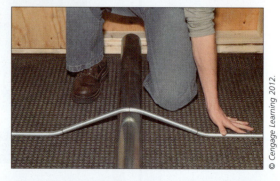

© Cengage Learning 2012.

PROCEDURE 12-7

Box Offsets

Follow this procedure for bending a box offset at the end of a ½-inch EMT with a hand bender.

- Wear safety glasses and observe all applicable safety rules.

- Cut the conduit to the proper length for attachment to an electrical box or enclosure.

- Make sure to ream the end of the conduit.

A Mark the pipe at 2 inches and 8 inches from the end of the pipe.

B Invert the bender by placing the handle on the floor and placing the 2-inch mark at the arrow; make a small bend (approximately 5 degrees).

© Cengage Learning 2012.

© Cengage Learning 2012.

PROCEDURE 12-7

Box Offsets (Continued)

C Rotate the pipe 180 degrees and, keeping the bender in the same position, slide the pipe ahead so that the 8-inch mark now lines up with the arrow.

• Make a small bend that is equal to the first bend made.

C

© Cengage Learning 2012.

from experience...

After electricians have installed box offsets enough times, they find that there is no need to make marks on the conduit for the location of the small bends. However, it is a good idea for electricians to always mark the pipe for the box offset bends so that if some adjustment has to be made, the pipe can be aligned in the bender at the same spots as the original bends. This will help eliminate kinking at the bends.

D Check the box offset by placing the conduit at the knockout opening where you want to secure it and see if it will slide into the opening without hitting the sides of the box. If it does not, fine-tune the pipe by adding or taking away offset height until the pipe easily fits into the knockout opening.

• Clean up your work area and put all tools and materials away.

D

© Cengage Learning 2012.

from experience...

Some electricians use a commercially available mechanical box offset bending tool. It works on ½- and ¾-inch trade-size EMT. The conduit is simply inserted in the bender, and the bender handle is depressed, which causes a perfect box offset to be formed in the pipe.

© Cengage Learning 2012.

REVIEW QUESTIONS

Directions: Answer the following items with clear and complete responses.

1. Name the article in the *NEC®* that covers RMC.

2. What does the *NEC®* say about reaming and threading of RMC?

3. Describe the securing and support requirements of RMC as outlined in the *NEC®*.

4. Discuss the reasons for using cutting oil when threading RMC.

5. List the minimum trade size of EMT for each of the following applications. Assume the use of conductors with THHN insulation.
   a. Six 14 AWG _____
   b. Six 12 AWG _____
   c. Three 10 AWG _____
   d. Four 8 AWG _____

6. Name the article of the *NEC®* that covers EMT.

7. List the minimum trade size EMT for each of the following applications.
   a. Three 14 AWG THHN, four 12 AWG THHN _____
   b. Three 12 AWG THHN, four 8 AWG TW _____
   c. Five 12 AWG THW, three 10 AWG THWN, three 8 AWG THHN _____

8. Describe the securing and support requirements for EMT as outlined in the *NEC®*.

9. Name the article in the *NEC®* that covers IMC.

10. IMC can be used in all the locations where RMC can be used. (Circle the correct response.) True or False

11. Describe the securing and support requirements for IMC.

12. Name the article in the *NEC®* that covers PVC conduit.

13. Describe the securing and support requirements for PVC conduit as outlined in the *NEC®*.

14. Name the article in the *NEC®* that covers ENT.

15. Describe the securing and support requirements for ENT as outlined in the *NEC®*.

16. Name the article in the *NEC®* that covers FMC.

17. Describe the securing and support requirements for FMC as outlined in the *NEC®*.

18. Name and describe the four basic conduit bend types used in residential wiring.

19. An electrician must install branch-circuit conductors in a metal raceway that will supply a 120/240-volt electric range. The raceway will not be used as the equipment-grounding conductor. How many conductors must the electrician install? What would be the colors for the required conductors?

20. Name the article in the *NEC®* that covers LFMC.

21. Describe the securing and support requirements for LFMC.

22. When installing more than two conductors in a conduit or tubing, what is the maximum percentage of fill that is allowed by the *NEC*®?

23. An electrician must always install an equipment grounding wire along with the regular circuit conductors in RMC, IMC, and EMT. (Circle the correct response.) True or False

24. Expansion fittings are required for PVC conduit to compensate for thermal expansion and contraction when there is expected to be a ____ inch or greater length change in a straight run of the conduit.

25. A nonconductive fish tape should be used when pulling conductors though an installed conduit system and into an energized loadcenter. (Circle the correct response.) True or False

KNOW YOUR CODES

1. Check with your local or state electrical inspector and determine whether a raceway wiring system must be used when wiring a house in your area.

2. If a raceway wiring method must be used to wire a house in your area, check with your local or state electrical inspector to determine which raceway type must be used.

Switching Circuit Installation

OBJECTIVES

Upon completion of this chapter, the student should be able to:

- Select an appropriate switch type for a specific residential switching situation.

- Select a switch with the proper rating for a specific switching application.

- List several *National Electrical Code*® (*NEC*®) requirements that apply to switches.

- Demonstrate an understanding of the proper installation techniques for single-pole, three-way, and four-way switches.

- Demonstrate an understanding of the proper installation techniques for switched duplex receptacles, combination switches, and double-pole switches.

- Demonstrate an understanding of the proper installation techniques for single-pole and three-way dimmer switches.

- Demonstrate an understanding of the proper installation techniques for ceiling-suspended paddle fan/light switches.

GLOSSARY OF TERMS

combination switch a device with more than one switch type on the same strap or yoke

dimmer switch a switch type that raises or lowers the lamp brightness of a lighting fixture

double-pole switch a switch type used to control two separate 120-volt circuits or one 240-volt circuit from one location

four-way switch a switch type that, when used in conjunction with two three-way switches, will allow control of a 120-volt lighting load from more than two locations

motion sensor a device that upon detecting movement in a specific area will switch on a lighting load; once all movement stops and a short amount of time goes by the sensor device switches off the electricity to the lighting load; often used for outdoor lighting control in residential applications

occupancy sensor a device that detects when a person has entered a room and switches on the room lighting; when everybody leaves a room and a short amount of time goes by the sensor device switches the room lighting off; often used for automatic control of room lighting in residential applications

single-pole switch a switch type used to control a 120-volt lighting load from one location

switch loop a switching arrangement where the feed is brought to the lighting outlet first and a two-wire loop run from the lighting outlet to the switch

three-way switch a switch type used to control a 120-volt lighting load from two locations

timer a device that controls the flow of electricity on a circuit for a certain amount of time; used in residential applications to switch off a bathroom fan or ceiling-suspended paddle fan after a specific length of time has gone by

n residential wiring systems, switches are used to control various electrical loads. Understanding the common connection techniques for each switch type is very important for residential electricians because they must know how many conductors to include in the wiring method installed during the rough-in stage between switches and the loads they control. This chapter presents an overview of the different types of switches commonly used in residential wiring as well as the common connection techniques for each switch type. Lighting outlets are the most common loads controlled by switches in a residential wiring system. Because this is the case, the 120-volt switching circuits covered in this chapter are limited to common switching circuits with lighting outlets as the load. Double-pole switching of 240-volt loads is also covered.

SELECTING THE APPROPRIATE SWITCH TYPE

Chapter 9 covered all of the locations in a house where switched lighting outlets are required. An electrician must install boxes, wiring, and switches so that the locations specified in Section 210.70 of the National Electrical Code have the required lighting outlets with the proper switch control. You may want to review this information in Chapter 9 at this time. In Chapter 2, you were introduced to the different types of switches used in residential wiring. Single-pole, double-pole, three-way, and four-way switches were all discussed. Specialty switches, such as dimmer switches and combination switches, were also discussed. A quick review of the switch types introduced in Chapter 2 is appropriate at this time.

The most common switch type used in residential wiring is a **single-pole switch** (Figure 13-1). This switch type is used in 120-volt residential circuits to control

FIGURE 13-1 A single-pole switch is used to control a lighting load from one location. *Courtesy of Hubbell Incorporated.*

FIGURE 13-2 A double-pole switch is sometimes used in residential wiring to control a 240-volt load. This switch is shown in the decorator style. *Courtesy of Hubbell Incorporated.*

FIGURE 13-3 Three-way switches are used in pairs to control lighting loads from two separate locations. *Courtesy of Hubbell Incorporated.*

FIGURE 13-4 Four-way switches, in combination with three-way switches, can control a lighting load from three or more locations. *Courtesy of Hubbell Incorporated.*

a lighting outlet load from one specific location. A **double-pole switch** (Figure 13-2) is used to control a 240-volt load in a residential wiring system, such as an electric water heater. Like the single-pole switch, the double-pole switch can control the load from only one specific location. **Three-way switches** (Figure 13-3) are used to control lighting outlet loads from two separate locations, like at the top and bottom of a stairway. In combination with three-way switches, **four-way switches** (Figure 13-4) allow for control of lighting outlet loads from three or more locations, such as in a room with three doorways and the wiring plan calls for a switch controlling the room lighting to be located at each doorway. A **dimmer switch** (Figure 13-5) can be found in both a single-pole and a three-way configuration and is used to brighten or dim a lamp or lamps in a lighting fixture. **Combination switch** (Figure 13-6) devices consist of two switches on one strap or yoke. This allows the placement of multiple switches in a single-gang electrical box. For example, the wiring plans may call for two switches—a single-pole switch and a three-way switch—at a location next to a doorway. However, when the electrician attempts to install a two-gang electrical box for the two required switches at that location, it is found that there is not enough space between the building framing members to attach the electrical box. An alternative installation would be to use a single-gang box and install a combination switch

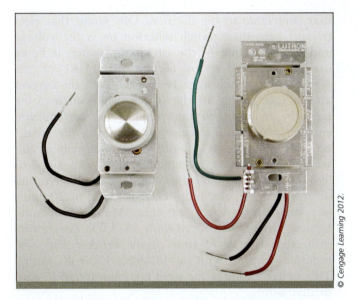

FIGURE 13-5 A dimmer switch is used to raise or lower the brightness of the light in a specific area of a house. Notice that dimmer switches have insulated pigtail wires rather than terminal screws.

that has a single-pole switch and a three-way switch on the same strap.

When selecting the proper switch type for a residential lighting application, there are many factors

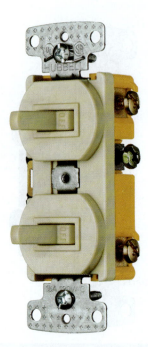

FIGURE 13-6 A combination switch consists of two switches on one strap or yoke. *Courtesy of Hubbell Incorporated.*

from experience...

In addition to single-pole, three-way, four-way, and double-pole switches, electricians install dimmer switches, **timers**, **motion sensors**, and **occupancy sensors** in residential electrical systems. These devices are not only convenient and provide added safety, they also reduce electricity consumption. There is another big reason to use lighting controls in house wiring and that is because energy efficiency codes are increasingly becoming the law of the land and these devices need to be installed to help meet the energy code requirements. Fortunately for us, these devices connect into a lighting circuit much the same way that "regular" switches do. Always read the manufacturers' installation instructions that are included with each type of device. More information on the use and installation of timers, motion sensors, and occupancy sensors are included later in this textbook.

that contribute to the decision. One factor that electricians base their switch selection on is the voltage and current rating of the circuit the switch is being used in. According to Section 404.15(A) of the *NEC®*, each switch must be marked with the current and voltage rating for the switch. The switch rating must be matched to the voltage and current you encounter with the circuit you are using the switch on. For example, a 20-amp-rated, 120-volt circuit wired with 12 AWG conductors having 16 amps of lighting load must have a switch with a voltage rating of at least 120 volts and a current rating of 20 amps (a standard switch rating). Many residential lighting circuits are wired with 14 AWG conductors protected with a 15-amp circuit breaker and will require switches with a 15-amp, 120-volt rating. This switch rating is the most common found in residential wiring.

Another factor to consider is the number of switching locations for the lighting load on the circuit. Knowing this information will allow the electrician to choose a single-pole switch when the load is controlled from one location, two three-way switches when the lighting load is controlled from two separate locations, or two three-way switches and the required number of four-way switches when the load is controlled from three or more locations.

Section 404.2(A) of the *NEC®* requires three- and four-way switches to be wired so that all switching is done in the ungrounded circuit conductor. No switching in the grounded conductor is allowed. Single-pole switches are also wired to switch the ungrounded circuit conductor. The switching techniques described later in this chapter will show common connections where only the "hot" ungrounded circuit conductor is switched. Section 404.2(B) actually says that switches must not disconnect the grounded conductor of a circuit. There is no need to ever connect a white insulated grounded conductor to any switch in a residential switching circuit (Figure 13-7).

The *NEC®* and Underwriters Laboratories refer to the switches used to control lighting outlets in residential applications as "snap switches." However, most electricians refer to them as "toggle switches" or simply as "switches." Switches that are used to control lighting circuits are classified as "general-use snap switches," and the requirements for these switches are the same

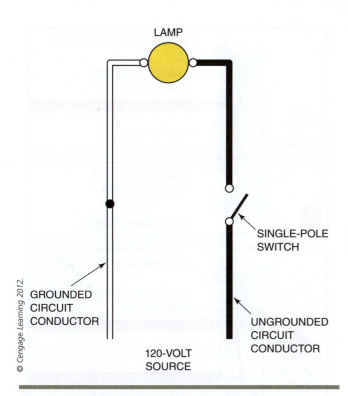

LAMP

SINGLE-POLE SWITCH

GROUNDED CIRCUIT CONDUCTOR

UNGROUNDED CIRCUIT CONDUCTOR

120-VOLT SOURCE

© Cengage Learning 2012.

FIGURE 13-7   Switches are to be wired so that all switching is done in the ungrounded circuit connector.

from experience...

Electricians doing residential wiring will only be using the alternating current general-use snap switch. When switches are purchased at the electrical equipment distributor, they will be for use only on alternating current circuits unless an electrician specifically asks for an AC/DC general-use snap switch.

for both the *NEC®* and Underwriters Laboratories. The requirements are found in Article 404 of the *NEC®*. Section 404.14(A) and (B) covers the requirements and recognizes two distinct switch categories. Section 404.14(A) categorizes one switch type as an "alternating current general-use snap switch" and states that it is suitable for use only on alternating current circuits for controlling the following:

- Resistive and inductive loads, including fluorescent lamps, not exceeding the ampere rating of the switch at the voltage involved

- Tungsten-filament lamp loads (incandescent lamps) not exceeding the ampere rating of the switch at 120 volts

- Motor loads not exceeding 80 percent of the ampere rating of the switch at its rated voltage

Section 404.14(B) categorizes the other switch type as an "alternating current or direct current general-use snap switch" and states that it can be used on either alternating current or direct current circuits for controlling the following:

- Resistive loads not exceeding the ampere rating of the switch at the voltage applied.

- Inductive loads not exceeding 50 percent of the ampere rating of the switch at the applied voltage. Switches rated in horsepower are suitable for controlling motor loads within their rating at the voltage applied.

- Tungsten-filament lamp loads not exceeding the ampere rating of the switch at the applied voltage if T-rated.

After an electrician has selected the appropriate switch type with the proper current and voltage rating for the switching circuit encountered, installing the wiring required for the actual switching connections is the next step. Installing a cable wiring system was presented in Chapter 11, and installing a raceway wiring system was presented in Chapter 12. Knowing and understanding common switch connection techniques will allow an electrician to install the appropriate number of conductors between switches and the loads they control during the rough-in stage. Once the wiring method has been installed, the switching connections are made.

To help electricians understand switching circuit connections, there are three things to always remember when making the switch and load connections:

1 First, at all electrical box locations in the switching circuit, connect the circuit-grounding conductors to:

- Metal electrical boxes with a green grounding screw

- The green screw on the switching devices

- The lighting fixture ground screw or grounding conductor

2 Next, at the lighting outlet locations, connect the white insulated grounded conductor to the silver screw terminal or wire identified as the grounded wire on the lighting fixture. Then, connect the "hot" ungrounded conductor at the lighting fixture to the brass screw or the wire identified as the ungrounded conductor on the lighting fixture.

3 Finally, at the switch locations, determine the conductor connections to the switch, depending on the specific switching situation, and make those connections.

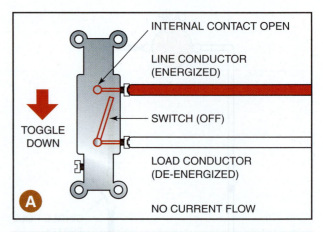

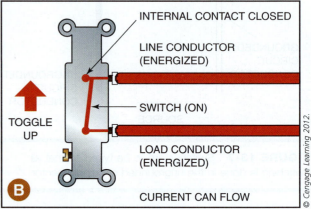

from experience...

When installing switches in a metal or non-metallic switch box, a connection must be made between the switch and the branch-circuit equipment-grounding conductor. Electricians usually do this by connecting the branch-circuit equipment-grounding conductor in the switch box to the green grounding screw on the switch. This is the recommended procedure, and all switching illustrations throughout this textbook will show this connection. However, Section 404.9(B) does allow a switch mounted to a metal box with metal screws to be considered grounded. This method, though rarely used, does not require an equipment-grounding conductor connection to the green grounding screw on the switch. Always check with your supervisor to see if this wiring technique is acceptable.

FIGURE 13-8   How a single-pole switch works. (A) The toggle is in the OFF (down) position. Current cannot flow from one terminal to the other. (B) The toggle is in the ON (up) position. Current can flow from one terminal to the other.

INSTALLING SINGLE-POLE SWITCHES

Because single-pole switching is the most common switching found in residential wiring, we start our discussion of switching connections with single-pole switches. On a single-pole switch, two wires will be connected to the two terminal screws on the switch. Both of these wires are considered "hot" ungrounded conductors. Figure 13-8 shows how a single-pole switch works. When the toggle is in the ON position, the internal contacts are closed, and current can flow through the switch and energize the lighting outlet. When the toggle is in the OFF position, the internal contacts are open, and current cannot flow, resulting in a de-energized lighting outlet. The following three single-pole switching applications assume the use of nonmetallic sheathed cable with nonmetallic boxes. Nonmetallic sheathed cable (Type NM) and nonmetallic boxes are shown in the accompanying single-pole switching illustrations because they are used the most in house wiring. The only adjustment you will need to make if metal boxes are used is to make sure to attach the circuit equipment-grounding conductor to each metal box with a green grounding screw.

SINGLE-POLE SWITCHING CIRCUIT 1

The most common single-pole switching wiring arrangement is when the electrical power feed is brought to the switch and then the cable continues on to the lighting load. Follow these steps when installing a single-pole switch for a lighting load with the power source feeding the switch (Figure 13-9):

1 Always wear safety glasses and observe all applicable safety rules.

2 At the lighting fixture box, connect the bare grounding conductors to the grounding connection of the lighting fixture. If the fixture does not have any metal parts that must be grounded, simply fold the grounding conductor into the back of the box. Do not cut it off. Remember that a nonmetallic box does not require the circuit-grounding conductor to be connected to it.

3 At the lighting fixture box, connect the white insulated grounded conductor to the silver screw terminal or wire identified as the grounded wire on the lighting fixture.

4 At the lighting fixture box, connect the black "hot" ungrounded conductor to the brass screw or a wire identified as the ungrounded conductor on the lighting fixture.

5 At the switch box, connect the bare grounding conductors together and, using a bare pigtail, connect all the grounding conductors to the green grounding screw on the switch. Remember that a nonmetallic box does not require the circuit-grounding conductor to be connected to it.

6 At the switch box, connect the white insulated grounded conductors together using a wirenut.

7 At the switch box, connect the black "hot" conductor in the incoming power source cable to one of the terminal screws on the single-pole switch.

8 At the switch box, connect the black "hot" conductor in the two-wire cable going to the lighting outlet to the other terminal screw on the single-pole switch.

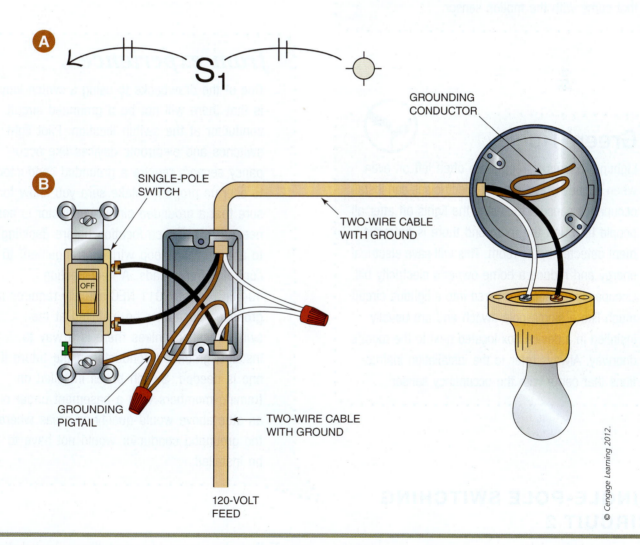

FIGURE 13-9 (A) A cabling diagram and (B) a wiring diagram for a switching circuit that has a single-pole switch controlling a lighting load. The power source is feeding the switch.

Green Tip

Outside lighting fixtures are often switched on for extended periods of time. They are also often left on during the daylight hours when they are not needed. Motion sensors can be installed to control outdoor lighting fixtures so that they operate only when needed. This will result in significant energy saving for a home owner. Motion sensors are wired into a lighting circuit similar to a single-pole switch and are usually located right at the outside lighting fixture. Always refer to the installation instructions that come with the motion sensor.

Green Tip

Lighting fixtures in rooms are often left on even when nobody is in the room. Electricians can install occupancy sensors to switch the lights off after all people have left the room and there is no movement detected in the room. This will save electrical energy and reduce a home owner's electricity bill. Occupancy sensors are wired into a lighting circuit much like a single-pole switch and are usually installed in a device box located next to the room's doorway. Always refer to the installation instructions that come with the occupancy sensor.

available with only a white insulated conductor and a black insulated conductor. The switch loop arrangement will require the white wire in the cable to be used as a "hot" ungrounded conductor. When this situation arises, the white wire must be reidentified as a "hot" conductor. This is usually accomplished by wrapping some black electrical tape on the conductor, although another color tape (like red) may also be used. A permanent black marker can also be used to reidentify a white insulated conductor as a "hot" conductor in a cable. The *NEC®* requires that any reidentification marking completely encircle the conductor. This requirement is easily taken care of when you wrap a piece of black electrical tape around a conductor, but you have to be careful when using a marker to make sure that the mark goes all the way around the conductor insulation.

from experience...

One of the drawbacks to using a switch loop is that there will not be a grounded circuit conductor at the switch location. Pilot light switches and electronic devices like occupancy sensors require a grounded conductor to operate properly. Make sure you know for sure that a grounded circuit conductor is not needed at a switch location before deciding to use a switch loop wiring arrangement to control lighting loads. A new Section 404.2(C) in the 2011 NEC actually requires a grounded conductor be installed at the switch location unless there is a way to install a grounded conductor in the future if one is needed. Switch boxes installed on framing members with a basement under or an attic above would qualify as areas where the grounded conductor would not have to be installed.

SINGLE-POLE SWITCHING CIRCUIT 2

It is very common for residential electricians to run the power source to the lighting outlet first and to then run a two-wire cable to the single-pole switching location. This is called a **switch loop**. When wiring a house with nonmetallic sheathed cable, a two-wire cable is

Follow these steps when installing a (switch loop) single-pole switch for a lighting load with the power source feeding the lighting outlet first (Figure 13-10):

1 Always wear safety glasses and observe all applicable safety rules.

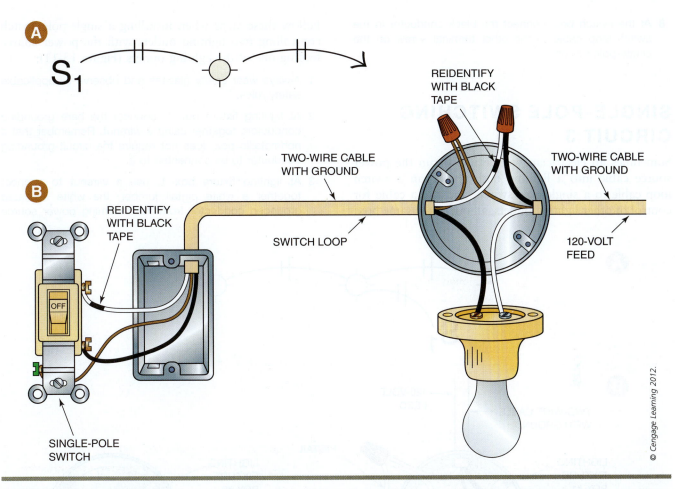

FIGURE 13-10   (A) A cabling diagram and (B) a wiring diagram for a switching circuit that has a single-pole switch controlling a lighting load with the power source feeding the lighting outlet. This switching arrangement is called a "switch loop."

from experience...

When wiring a house with a conduit wiring method, a white insulated wire cannot be used in a switch loop switching arrangement as a "hot" conductor. Section 200.7(C)(1) of the *NEC*® allows the reidentification of a white insulated conductor as a "hot" conductor only in a cable assembly, such as nonmetallic sheathed cable. Use two black insulated conductors or one black and one red insulated conductor as the switch loop wiring in a conduit system.

2 At the lighting fixture box, connect the bare grounding conductors together using a wirenut. Remember that a nonmetallic box does not require the circuit-grounding conductor to be connected to it.

3 At the lighting fixture box, connect the white insulated grounded conductor from the incoming power source cable to the silver screw terminal or wire identified as the grounded wire on the lighting fixture.

4 At the lighting fixture box, use a wirenut to connect the black "hot" conductor from the incoming power source cable to the white insulated conductor of the two-wire cable going to the single-pole switch. Reidentify the white wire as a "hot" conductor with a piece of black electrical tape.

5 At the lighting fixture box, connect the black conductor from the two-wire switch loop cable to the brass screw or a wire identified as the ungrounded conductor on the lighting fixture.

6 At the switch box, connect the bare grounding conductor to the green grounding screw on the switch. Remember that a nonmetallic box does not require the circuit-grounding conductor to be connected to it.

7 At the switch box, reidentify the incoming white insulated conductor as a "hot" conductor with a piece of black tape and connect it to one of the terminal screws on the single-pole switch.

8 At the switch box, connect the black conductor in the switch loop cable to the other terminal screw on the single-pole switch.

SINGLE-POLE SWITCHING CIRCUIT 3

Sometimes an electrician will decide to run the power source cable into a lighting outlet box, run a switch loop cable to a single-pole switch, and run a cable for control of other lighting outlets, all out of the same box.

Follow these steps when installing a single-pole switch controlling two lighting outlets with the power source feeding one of the lighting outlets (Figure 13-11):

1 Always wear safety glasses and observe all applicable safety rules.

2 At lighting fixture box 1, connect the bare grounding conductors together using a wirenut. Remember that a nonmetallic box does not require the circuit-grounding conductor to be connected to it.

3 At lighting fixture box 1, use a wirenut to connect together a white pigtail jumper, the white insulated grounded conductor from the incoming power source

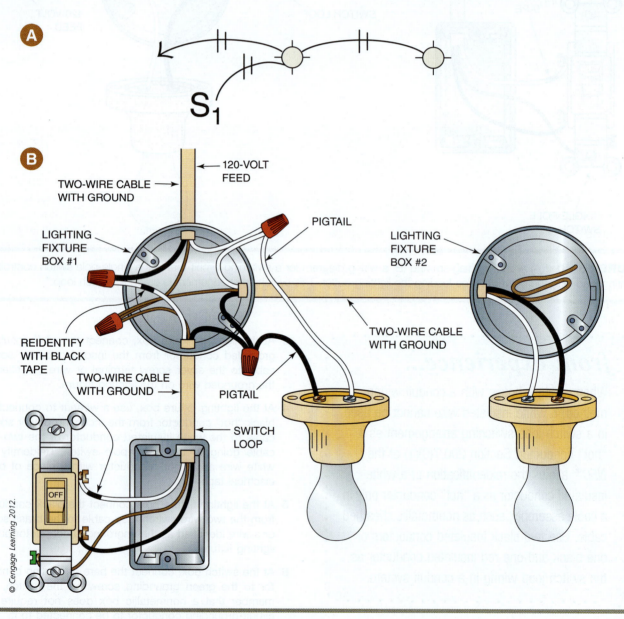

FIGURE 13-11 (A) A cabling diagram and (B) a wiring diagram for a switching circuit that has a single-pole switch controlling two lighting outlets. The power source is feeding one of the lighting outlets.

cable, and the white insulated conductor of the cable feeding light fixture box 2. Then, connect the white pigtail jumper to the silver screw or the wire identified as the grounded conductor on the lighting fixture.

4 At lighting fixture box 1, use a wirenut to connect the black "hot" conductor from the incoming power source cable to the white insulated conductor of the two-wire cable going to the single-pole switch. Reidentify the white wire as a "hot" conductor with a piece of black electrical tape.

5 At lighting fixture box 1, use a wirenut to connect together a black pigtail jumper, the black conductor from the two-wire switch loop cable, and the black conductor from the cable feeding light box 2. Then, connect the black pigtail jumper to the brass screw or the wire identified as the ungrounded conductor on the lighting fixture.

6 At lighting fixture box 2, fold the grounding conductor into the back of the box, connect the white insulated grounded conductor to the silver screw terminal or wire identified as the grounded wire on the lighting fixture, and connect the black "hot" ungrounded conductor to the brass screw or a wire identified as the ungrounded conductor on the lighting fixture.

7 At the switch box, connect the bare grounding conductor to the green grounding screw on the switch, reidentify the incoming white insulated conductor as a "hot" conductor with a piece of black tape and connect it to one of the terminal screws on the single-pole switch, and connect the black conductor in the switch loop cable to the other terminal screw on the single-pole switch.

INSTALLING THREE-WAY SWITCHES

Three-way switches are used to control a lighting load from two different locations. You may remember from Chapter 2 that three-way switches have three terminal screws on them. One of the screws is called the "common" terminal and is usually colored black. The other two terminal screws are the same color, usually brass or bronze, and are called the "traveler terminals." Beginning electricians often find the connections for three-way switches confusing. As a matter of fact, even some experienced electricians find themselves confused when wiring three-way switches after they have not had to wire them for a while. Learning some common rules will make the process much easier whether you wire three-way switches all the time or

only once in a while. Common rules to keep in mind when wiring switching circuits with three-way switches include the following:

- Three-way switches must always be installed in pairs. There is no such thing as an installation with only one three-way switch.

- A three-wire cable must always be installed between the two three-way switches. If you are wiring with conduit, three separate wires must be pulled into the conduit between the two three-way switches.

- The black-colored "common" terminal on a three-way switch should always have a black insulated wire attached to it. One three-way switch will have a black "hot" feed conductor attached to it, and the other three-way switch will have the black insulated conductor that will be going to the lighting load attached to it.

- Assuming the use of a nonmetallic sheathed cable, when the power source feed is brought to the first three-way switch, the traveler wires that interconnect the traveler terminals of both switches will be black and red in color.

- Assuming the use of a nonmetallic sheathed cable, when the power source feed is brought to the lighting outlet first, the traveler wires will be red and white in color. The white traveler conductors will need to be reidentified with black tape at each switch location.

- There is no marking for the ON or OFF position of the toggle on a three-way switch, so it does not make any difference which way it is positioned in the electrical device box.

Three-way switches work a little differently than single-pole switches. Single-pole switches either let current flow through the switch and to the load or do not. The internal configuration of a three-way switch always allows current to flow from the common terminal to either of the two traveler terminals (Figure 13-12). The three-way switch is sometimes called a single-pole double-throw switch because of this characteristic. When the three-way switch is mounted in a device box and the toggle is placed in the "down" position, contact is made between one of the traveler terminals and the common terminal. When the toggle is moved to the "up" position, contact is made between the common and the other traveler terminal. There is always one "set" of contacts in a three-way switch that are closed.

The following applications will assume the use of nonmetallic sheathed cable with nonmetallic boxes. Nonmetallic sheathed cable (Type NM) and nonmetallic boxes are shown in the accompanying three-way switching illustrations because they are used the most in house

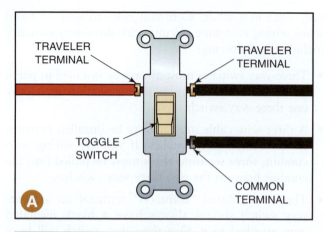

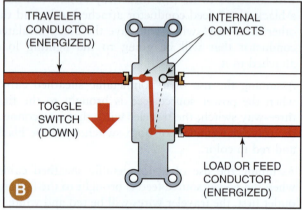

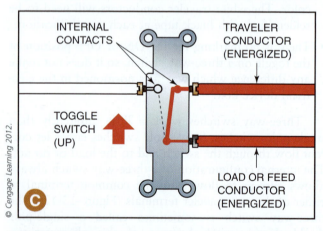

FIGURE 13-12 How a three-way switch works. (A) Three-way switches have two traveler terminals and one common terminal. (B) When the switch toggle is moved, a connection is made from the common terminal to one of the traveler terminals. (C) When the switch toggle is moved in the opposite direction, a connection is made from the common terminal to the other traveler terminal.

wiring. The only adjustment you will need to make if metal boxes are used is to make sure to attach the circuit equipment-grounding conductor to each metal box with a green grounding screw.

THREE-WAY SWITCHING CIRCUIT 1

Follow these steps when installing a three-way switching circuit with the power source feeding the first three-way switch location (Figure 13-13):

1 Always wear safety glasses and observe all applicable safety rules.

2 At the lighting fixture box, connect the bare grounding conductors to the grounding connection of the lighting fixture. If the fixture does not have any metal parts that must be grounded, simply fold the grounding conductor into the back of the box. Remember that a nonmetallic box does not require the circuit-grounding conductor to be connected to it.

3 At the lighting fixture box, connect the white insulated grounded conductor to the silver screw terminal or wire identified as the grounded wire on the lighting fixture.

4 At the lighting fixture box, connect the black ungrounded conductor to the brass screw or a wire identified as the ungrounded conductor on the lighting fixture.

5 At the first three-way switch location, connect the bare grounding conductors together and, using a bare pigtail, connect all the grounding conductors to the green grounding screw on the switch.

6 At the first three-way switch box, connect the white insulated grounded conductors together using a wirenut.

7 At the first three-way switch box, connect the black "hot" conductor in the incoming power source cable to the common terminal screw on three-way switch 1.

8 At the first three-way switch box, connect the black and red traveler wires in the three-wire cable going to three-way switch 2 to the traveler terminal screws on the three-way switch.

9 At the second three-way switch box, connect the bare grounding conductors together and, using a bare pigtail, connect all the grounding conductors to the green grounding screw on the switch.

10 At the second three-way switch box, connect the white insulated grounded conductors together using a wirenut.

11 At the second three-way switch box, connect the black conductor in the two-wire cable going to the lighting outlet to the common terminal screw on three-way switch 2.

12 At the second three-way switch box, connect the black and red traveler wires in the three-wire cable coming from three-way switch 1 to the traveler terminal screws on three-way switch 2.

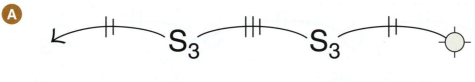

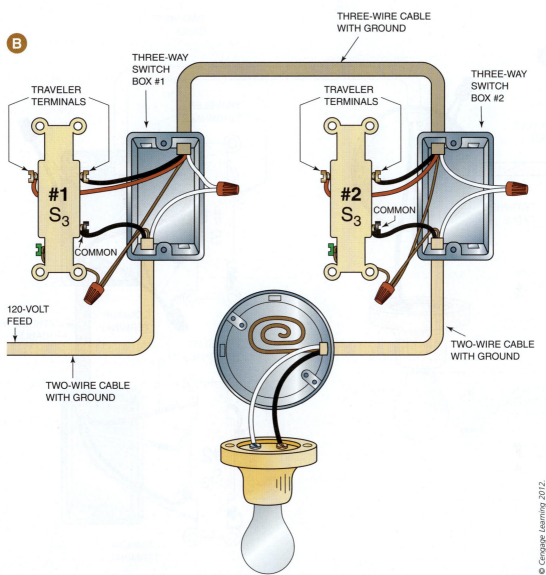

FIGURE 13-13   (A) A cabling diagram and (B) a wiring diagram for a switching circuit that has two three-way switches controlling a lighting load. The power source is feeding the first three-way switch location.

THREE-WAY SWITCHING CIRCUIT 2

Follow these steps when installing a three-way switching circuit with the power source feeding the lighting outlet location (Figure 13-14):

1 Always wear safety glasses and observe all applicable safety rules.

2 At the lighting fixture box, connect the bare grounding conductors to the grounding connection of the lighting fixture. If the fixture does not have any metal parts that must be grounded, simply use a wirenut to connect the grounding conductors together and place them in the back of the box.

3 At the lighting fixture box, connect the white insulated grounded conductor of the incoming power feed cable to the silver screw terminal or wire identified as the grounded wire on the lighting fixture.

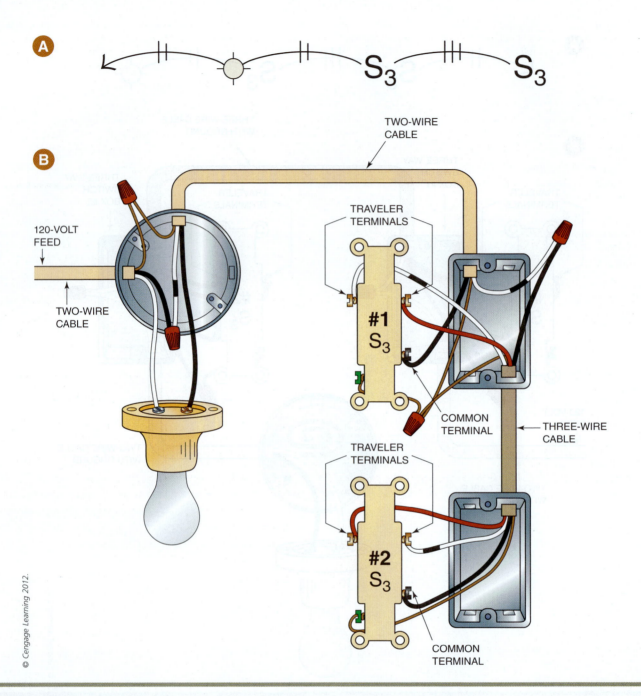

A

B

120-VOLT
FEED

TWO-WIRE
CABLE

TWO-WIRE
CABLE

TRAVELER
TERMINALS

TRAVELER
TERMINALS

#1
S₃

#2
S₃

COMMON
TERMINAL

COMMON
TERMINAL

THREE-WIRE
CABLE

FIGURE 13-14   (A) A cabling diagram and (B) a wiring diagram for a switching circuit that has two three-way switches controlling a lighting load. The power source is feeding the lighting outlet location.

4   At the lighting fixture box, use a wirenut to connect the black "hot" conductor from the incoming power source cable to the white insulated conductor of the two-wire cable going to three-way switch 1. Reidentify the white wire as a "hot" conductor with a piece of black electrical tape.

5   At the lighting fixture box, connect the black conductor from the switch loop cable going to three-way switch 1 to the brass screw or a wire identified as the ungrounded conductor on the lighting fixture.

6   At the first three-way switch location, connect the bare grounding conductors together and, using a bare pigtail, connect all the grounding conductors to the green grounding screw on the switch.

7   At the first three-way switch box, use a wirenut to connect the white insulated conductor from the two-wire cable coming from the lighting outlet to the black insulated conductor in the three-wire cable going to three-way switch 2. Be sure to reidentify the white insulated conductor with a piece of black electrical tape.

8 At the first three-way switch box, connect the black conductor in the two-wire cable from the lighting outlet to the common terminal screw on three-way switch 1.

9 At the first three-way switch box, connect the red and white traveler wires in the three-wire cable going to three-way switch 2 to the traveler terminal screws on three-way switch 1. Be sure to reidentify the white insulated conductor with a piece of black electrical tape.

10 At the second three-way switch box, connect the bare grounding conductor to the green grounding screw on the switch.

11 At the second three-way switch box, connect the black conductor in the three-wire cable coming from three-way switch 1 to the common terminal on three-way switch 2.

12 At the second three-way switch box, connect the red and white traveler wires in the three-wire cable coming from three-way switch 1 to the traveler terminal screws on three-way switch 2. Be sure to reidentify the white insulated conductor with a piece of black electrical tape.

THREE-WAY SWITCHING CIRCUIT 3

Follow these steps when installing a three-way switching circuit with the power source feeding the lighting outlet location and a three-wire cable is run from the lighting outlet to each of the three-way switches (Figure 13-15):

1 Always wear safety glasses and observe all applicable safety rules.

2 At the lighting fixture box, connect the bare grounding conductors to the grounding connection of the lighting fixture. If the fixture does not have any metal parts that must be grounded, simply use a wirenut to connect the grounding conductors together and place them in the back of the box.

3 At the lighting fixture box, connect the white insulated grounded conductor from the incoming power source cable to the silver screw terminal or wire identified as the grounded wire on the lighting fixture.

4 At the lighting fixture box, use a wirenut to connect the black "hot" conductor from the incoming power source cable to a black insulated conductor in the three-wire cable going to three-way switch 1.

5 At the lighting fixture box, connect the two red traveler wires together and the two white traveler wires together using wirenuts. Remember to reidentify the white traveler wires with black tape.

6 At the lighting fixture box, connect the black conductor from the three-wire cable going to three-way switch 2 to the brass screw or a wire identified as the ungrounded conductor on the lighting fixture.

7 At three-way switch 1, connect the bare grounding conductor to the green grounding screw on the switch.

8 At three-way switch 1, connect the black conductor in the three-wire cable coming from the lighting outlet to the common terminal on three-way switch 1.

9 At three-way switch 1, connect the red and white traveler wires in the three-wire cable coming from the lighting outlet to the traveler terminal screws on three-way switch 1. Be sure to reidentify the white insulated conductor with a piece of black electrical tape.

10 At three-way switch 2, connect the bare grounding conductor to the green grounding screw on the switch.

11 At three-way switch 2, connect the black conductor in the three-wire cable coming from the lighting outlet to the common terminal on three-way switch 2.

12 At three-way switch 2, connect the red and white traveler wires in the three-wire cable coming from the lighting outlet to the traveler terminal screws on three-way switch 2. Be sure to reidentify the white insulated conductor with a piece of black electrical tape.

INSTALLING FOUR-WAY SWITCHES

Four-way switches are used in conjunction with three-way switches to control a lighting load from more than two different locations. Chapter 2 introduced you to four-way switches, and you learned that they have four terminal screws on them. All four terminal screws are called traveler terminals. The four screws are divided into two pairs. Each pair is of the same color. One pair is usually brass or bronze, and the other pair is some other color like black. Once you understand the common three-way switching connections, four-way connections should be relatively easy to understand. Four-way switches are simply inserted into the wiring between three-way switches. Only traveler wires are connected to the four-way switch screw terminals. Learning some common rules will make the process much easier. Common rules to keep in mind when wiring switching circuits with four-way switches include the following:

• Four-way switches must always be installed between two three-way switches. There is no such thing as an installation with only one four-way switch or a switching circuit that has only four-way switches. For example, if you wire a switching situation that requires four switching locations for the same lighting load, you would need two three-way switches and two four-way switches.

• A three-wire cable must always be installed between all four-way and three-way switches. If you are wiring with conduit, three separate wires must be pulled into the conduit between all the four-ways and three-ways.

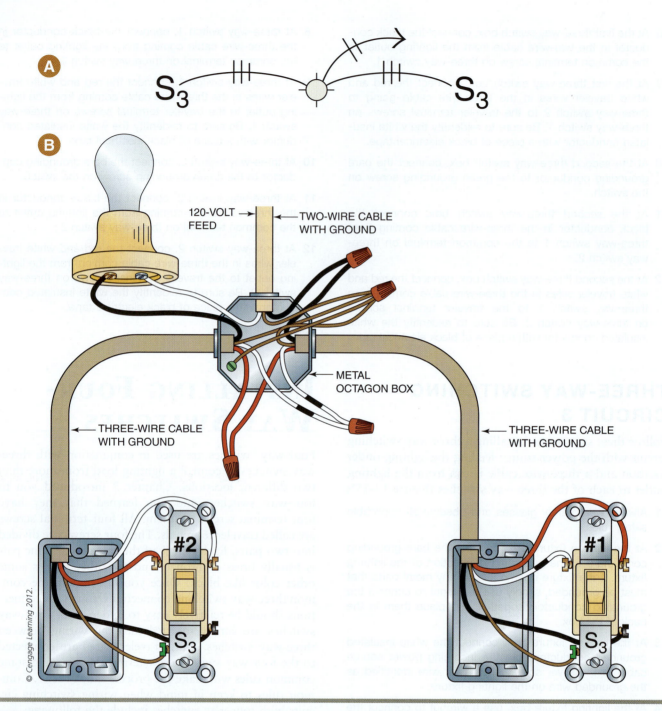

FIGURE 13-15 (A) A cabling diagram and (B) a wiring diagram for a switching circuit that has two three-way switches controlling a lighting load. The power source is feeding the lighting outlet location, and a three-wire cable is run from the lighting outlet to each of the three-way switches.

- Assuming the use of a nonmetallic sheathed cable, when the power source feed is brought to the first three-way switch in the circuit, the traveler wires that interconnect the traveler terminals of all four-way and three-way switches will be black and red in color.

- Assuming the use of a nonmetallic sheathed cable, when the power source feed is brought to the lighting outlet first, the traveler wires will be red and

white in color. The white traveler conductors will need to be reidentified with black tape at each switch location.

- Most four-way switch traveler terminals are vertically configured. This means that when the four-way switch is positioned in a vertical position, the top two screws have the same color and are a traveler terminal pair. The bottom two screws are the other traveler terminal pair and have the same color. Remember that the color of each traveler pair is different.

• Like a three-way switch, there is no marking for the ON or OFF position of the toggle on the four-way switch, so it does not make any difference which way it is positioned in the electrical device box.

Like single-pole and three-way switches, four-way switches work by having their internal contacts open or closed. Four-ways have four sets of internal contacts. The contacts are closed or opened in a sequence determined by the position of the toggle. Figure 13-16 shows the internal configuration of the contacts when the toggle is in the up or down position.

The following applications will assume the use of nonmetallic sheathed cable with nonmetallic boxes. Nonmetallic sheathed cable (Type NM) and nonmetallic boxes are shown in the accompanying four-way switching illustrations because they are used the most in house wiring. The only adjustment you will need to make if metal boxes are used is to make sure to attach the circuit equipment-grounding conductor to each metal box with a green grounding screw.

FOUR-WAY SWITCHING CIRCUIT 1

Follow these steps when installing a four-way switching circuit with the power source feeding the first three-way switch location (Figure 13-17):

1 Always wear safety glasses and observe all applicable safety rules.

2 At the lighting fixture box, connect the bare grounding conductors to the grounding connection of the lighting fixture. If the fixture does not have any metal parts that

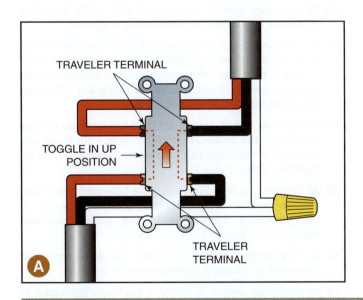

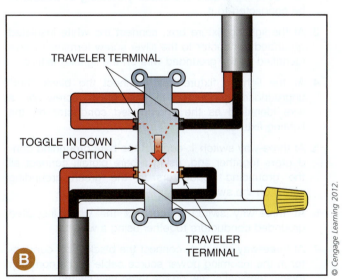

FIGURE 13-16 How a four-way switch works. (A) Connections are made vertically from the traveler terminals on the bottom of the switch to the traveler terminals on the top of the switch when the toggle switch is placed in one position. (B) Connections are made diagonally from the traveler terminals on the bottom to the traveler terminals on the top of the switch when the switch toggle is placed in the opposite position.

© Cengage Learning 2012.

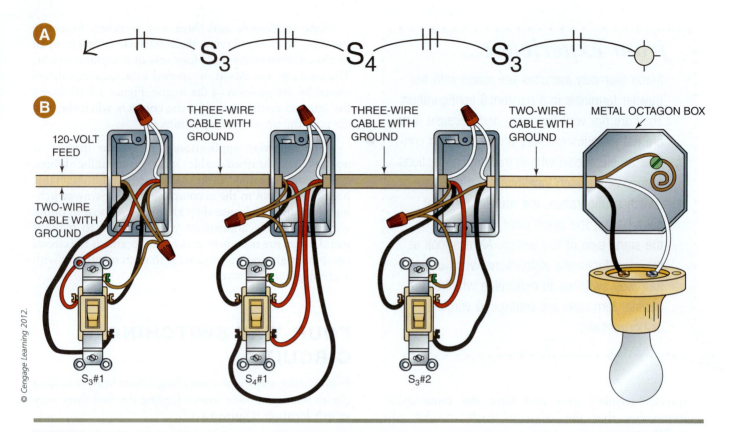

FIGURE 13-17 (A) A cabling diagram and (B) a wiring diagram for a switching circuit that has a four-way switch and two three-way switches controlling a lighting load. The power source is feeding the first three-way switch location.

must be grounded, simply fold the grounding conductor into the back of the box. Remember that a nonmetallic box does not require the circuit-grounding conductor to be connected to it.

3 At the lighting fixture box, connect the white insulated grounded conductor to the silver screw terminal or wire identified as the grounded wire on the lighting fixture.

4 At the lighting fixture box, connect the black "hot" ungrounded conductor to the brass screw or a wire identified as the ungrounded conductor on the lighting fixture.

5 At three-way switch 1, connect the bare grounding conductors together and, using a bare pigtail, connect all the grounding conductors to the green grounding screw on the switch.

6 At three-way switch 1, connect the white insulated grounded conductors together using a wirenut.

7 At three-way switch 1, connect the black "hot" conductor in the incoming power source cable to the common (black) terminal screw on the three-way switch.

8 At three-way switch 1, connect the black and red traveler wires in the three-wire cable going to four-way switch 1 to the traveler terminal screws on three-way switch 1.

9 At four-way switch 1, connect the bare grounding conductors together and, using a bare pigtail, connect all

the grounding conductors to the green grounding screw on the switch.

10 At four-way switch 1, connect the white insulated grounded conductors together using a wirenut.

11 At four-way switch 1, connect the black and red traveler wires in the three-wire cable coming from three-way switch 1 to a pair of traveler terminal screws on the four-way switch. It really does not make any difference if it is the top pair or the bottom pair.

12 At four-way switch 1, connect the black and red traveler wires from the three-wire cable coming from three-way switch 2 to the other pair of traveler terminal screws on four-way switch 1.

13 At three-way switch 2, connect the bare grounding conductors together and, using a bare pigtail, connect all the grounding conductors to the green grounding screw on the switch.

14 At three-way switch 2, connect the white insulated grounded conductors together using a wirenut.

15 At three-way switch 2, connect the black conductor in the two-wire cable going to the lighting outlet to the common (black) terminal screw on three-way switch 2.

16 At three-way switch 2, connect the black and red traveler wires in the cable coming from four-way switch 1 to the traveler terminal screws on three-way switch 2.

FOUR-WAY SWITCHING CIRCUIT 2

Follow these steps when installing a four-way switching circuit with the power source feeding the lighting outlet (Figure 13-18):

1 Always wear safety glasses and observe all applicable safety rules.

2 At the lighting fixture box, connect the bare grounding conductors to the grounding connection of the lighting fixture. If the fixture does not have any metal parts that must be grounded, simply use a wirenut to connect the grounding conductors together and place them in the back of the box.

3 At the lighting fixture box, connect the white insulated grounded conductor from the incoming power source cable to the silver screw terminal or wire identified as the grounded wire on the lighting fixture.

4 At the lighting fixture box, use a wirenut to connect the black "hot" conductor from the incoming power source cable to the white insulated conductor in the three-wire cable going to three-way switch 1. Reidentify the white wire as a "hot" conductor with a piece of black electrical tape.

5 At the lighting fixture box, connect the black "hot" conductor from the three-wire cable going to three-way switch 1 to the brass screw or a wire identified as the ungrounded conductor on the lighting fixture.

6 At three-way switch 1, connect the bare grounding conductors together and, using a bare pigtail, connect all the grounding conductors to the green grounding screw on the switch.

7 At three-way switch 1, use a wirenut to connect the white insulated conductor from the two-wire cable coming from the lighting outlet to the black insulated conductor in the three-wire cable going to four-way switch 1. Be sure to reidentify the white insulated conductor with a piece of black electrical tape.

8 At three-way switch 1, connect the black conductor in the two-wire cable from the lighting outlet to the common (black) terminal screw on three-way switch 1.

9 At three-way switch 1, connect the red and white traveler wires in the three-wire cable going to four-way switch 1 to the traveler terminal screws on three-way switch 1. Be sure to reidentify the white insulated conductor with a piece of black electrical tape.

10 At four-way switch 1, connect the bare grounding conductors together and, using a bare pigtail, connect all the grounding conductors to the green grounding screw on the switch.

11 At four-way switch 1, connect the black insulated conductors together using a wirenut.

12 At four-way switch 1, connect the red and white traveler wires in the three-wire cable coming from three-way switch 1 to a pair of traveler terminal screws on the four-way switch. It really does not make any difference if it is the top pair or bottom pair. Reidentify the white traveler wire with black tape.

13 At four-way switch 1, connect the red and white traveler wires from the three-wire cable going to three-way switch 2 to the other pair of traveler terminal screws on the four-way switch. Reidentify the white traveler wire with black tape.

14 At three-way switch 2, connect the bare grounding conductor to the green grounding screw on the switch.

15 At three-way switch 2, connect the black conductor in the three-wire cable coming from four-way switch 1 to the common (black) terminal on three-way switch 2.

16 At three-way switch 2, connect the red and white traveler wires in the three-wire cable coming from four-way switch 1 to the traveler terminal screws on three-way switch 2. Be sure to reidentify the white insulated conductor with a piece of black electrical tape.

INSTALLING SWITCHED DUPLEX RECEPTACLES

Switched receptacles are often found in areas such as bedrooms, living rooms, and family rooms of homes. In these areas, lamps are often the primary lighting source, and electricians place switches next to the doorways of these areas for control of the receptacles that the lamps are plugged into. The switch on the lamp is left in the ON position, so when a switch is activated and energizes the receptacle that the lamp is plugged into, the lamp comes on. Switching of receptacles can be done so the whole receptacle is switched on or off, or the wiring can be installed so that half a duplex receptacle is energized with the switch while the other half remains "hot" at all times. This is accomplished by "splitting" a duplex receptacle. Splitting a duplex receptacle means removing the tab between the two brass screw terminals on the ungrounded side of a duplex receptacle. It is common wiring practice to switch the bottom half of the receptacle and leave the top half "hot" at all times because having a lamp cord plugged into the bottom half of the split-duplex receptacle allows the top half to be clear for plugging in another piece of electrical equipment. On the other hand, if the top half were switched and a lamp cord plugged into it, the cord would hang down in front of the "hot" half of the duplex receptacle and be in the way when another piece of electrical equipment needed to be plugged into that half. A lamp can be plugged into the bottom half of the duplex receptacle and an electrical load that needs to be energized at all

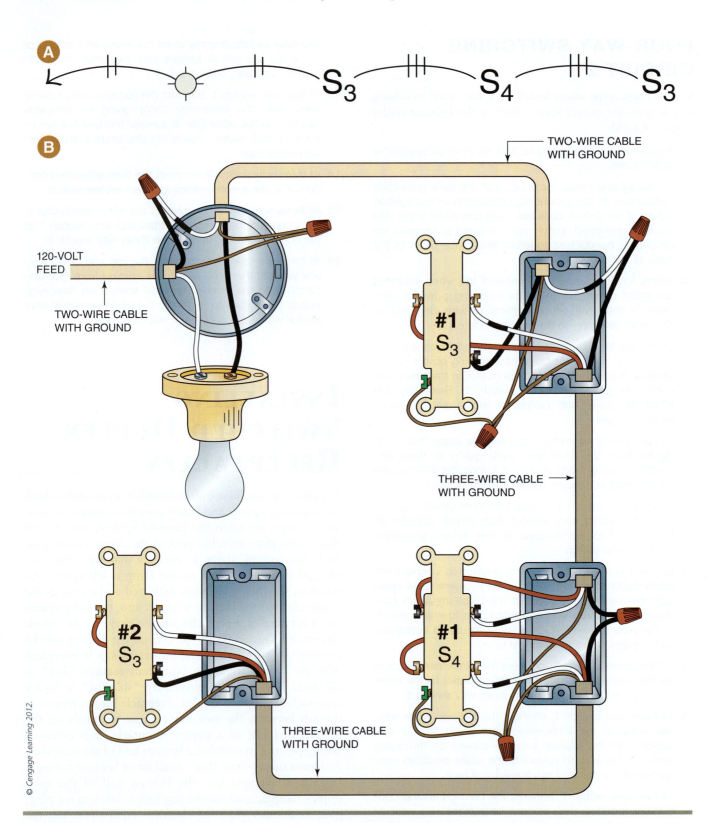

FIGURE 13-18 (A) A cabling diagram and (B) a wiring diagram for a switching circuit that has a four-way switch and two three-way switches controlling a lighting load. The power source is feeding the lighting outlet.

times (like a television) can be plugged into the top half. Wiring connections for both of these wiring applications are presented in this section.

The following applications will assume the use of nonmetallic sheathed cable with nonmetallic boxes. Nonmetallic sheathed cable (Type NM) and nonmetallic boxes are shown in the accompanying receptacle switching illustrations because they are used the most in house wiring. The only adjustment you will need to make if metal boxes are used is to make sure to attach the circuit equipment-grounding conductor to each metal box with a green grounding screw.

SWITCHED DUPLEX RECEPTACLE CIRCUIT 1

Follow these steps when installing a duplex receptacle that is controlled by a single-pole switch with the power source feeding the switch location (Figure 13-19):

1 Always wear safety glasses and observe all applicable safety rules.

2 At the single-pole switch location, connect the bare grounding conductors together and, using a bare pigtail, connect all the grounding conductors to the green grounding screw on the switch. Remember that a non-metallic box does not require the circuit-grounding conductor to be connected to it.

3 At the single-pole switch location, connect the white insulated grounded conductors together using a wirenut.

4 At the single-pole switch location, connect the black "hot" conductor in the incoming power source cable to one of the terminal screws on the single-pole switch.

5 At the single-pole switch location, connect the black conductor in the two-wire cable going to the duplex receptacle outlet to the other terminal screw on the single-pole switch.

6 At the duplex receptacle location, connect the bare grounding conductor to the green grounding screw on the duplex receptacle.

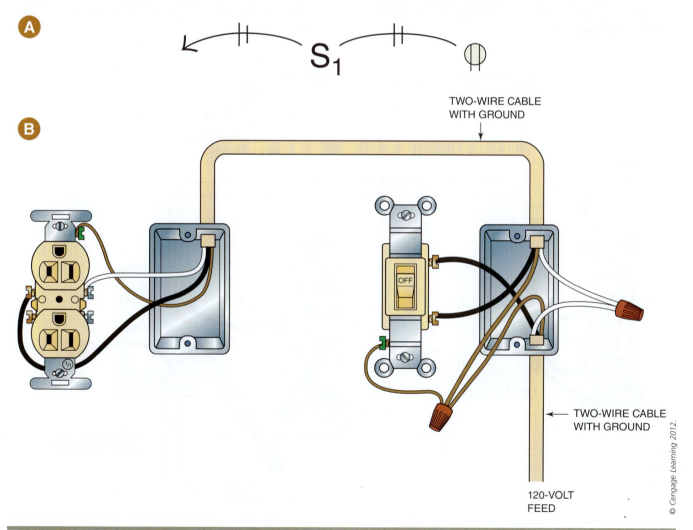

FIGURE 13-19 (A) A cabling diagram and (B) a wiring diagram for a switching circuit that has a duplex receptacle controlled by a single-pole switch. The power source is feeding the switch location.

7 At the duplex receptacle location, connect the white grounded conductor to one of the silver terminal screws on the duplex receptacle. It does not make a difference which of the two silver screws you use.

8 At the duplex receptacle location, connect the black conductor to one of the brass terminal screws on the duplex receptacle. Again, it does not make a difference which of the two brass screws you use.

SWITCHED DUPLEX RECEPTACLE CIRCUIT 2

Follow these steps when installing a duplex receptacle that is controlled by a single-pole switch with the power source feeding the receptacle location (a switch loop situation) (Figure 13-20):

1 Always wear safety glasses and observe all applicable safety rules.

2 At the single-pole switch location, connect the bare grounding conductor to the green grounding screw on the single-pole switch.

3 At the single-pole switch location, connect the white conductor to one of the terminal screws on the single-pole switch. It does not make a difference which of the two screws you use. Make sure to reidentify the white wire with black electrical tape.

4 At the single-pole switch location, connect the black conductor to the other terminal screw on the single-pole switch.

5 At the duplex receptacle location, connect the bare grounding conductors together and, using a bare pigtail, connect all the grounding conductors to the green grounding screw on the receptacle.

6 At the duplex receptacle location, connect the white grounded conductor of the incoming power source cable to one of the silver terminal screws on the duplex receptacle. It does not make a difference which of the two silver screws you use.

7 At the duplex receptacle location, connect the black "hot" conductor of the incoming power source cable to the white insulated conductor in the two-wire cable going to the single-pole switch with a wirenut. Reidentify the white conductor with a piece of black electrical tape.

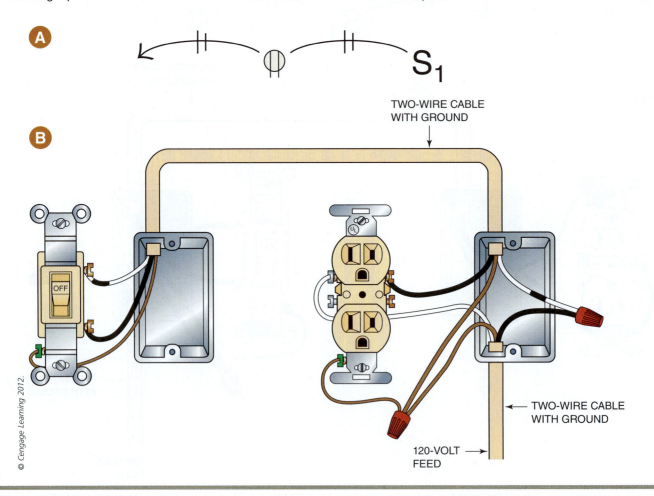

© Cengage Learning 2012.

FIGURE 13-20 (A) A cabling diagram and (B) a wiring diagram for a switching circuit that has a duplex receptacle controlled by a single-pole switch. The power source is feeding the receptacle location. This is a switch-loop wiring situation.

8 At the duplex receptacle, connect the black insulated wire from the two-wire cable coming from the switch to one of the brass terminal screws.

SWITCHED DUPLEX RECEPTACLE CIRCUIT 3

Follow these steps when installing a split-duplex receptacle so that the bottom half is controlled by a single-pole switch and the top half is "hot" at all times. The power source feeds the switch location (Figure 13-21):

1 Always wear safety glasses and observe all applicable safety rules.

2 At the single-pole switch location, connect the bare grounding conductors together and, using a bare pigtail, connect all the grounding conductors to the green grounding screw on the switch.

3 At the single-pole switch location, connect the white insulated grounded conductors together using a wirenut.

4 At the single-pole switch location, using a wirenut, connect together a black pigtail, the black conductor of the incoming power source cable, and the black conductor of the three-wire cable going to the split-duplex receptacle. Connect the black pigtail end to one of the terminal screws on the single-pole switch.

5 At the single-pole switch location, connect the red conductor in the three-wire cable going to the split-duplex receptacle outlet to the other terminal screw on the single-pole switch.

6 Split the duplex receptacle by using a pair of pliers to take off the tab between the two brass terminals on the ungrounded side of the receptacle.

7 At the split-duplex receptacle location, connect the bare grounding conductor to the green grounding screw on the duplex receptacle.

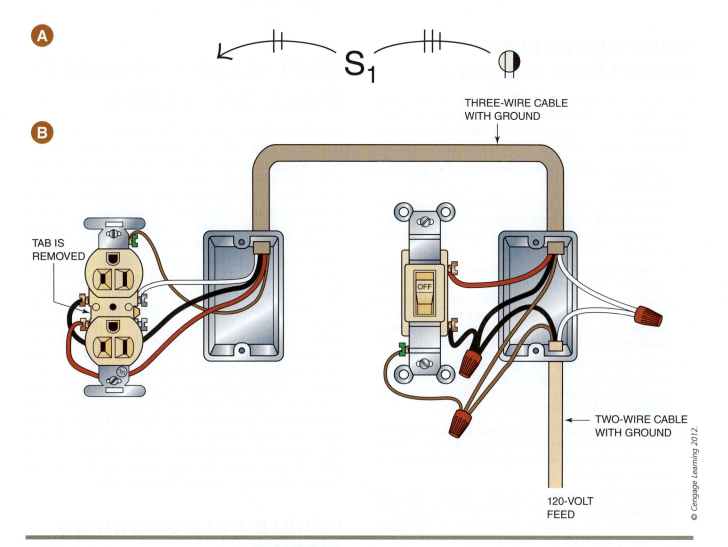

FIGURE 13-21 (A) A cabling diagram and (B) a wiring diagram for a switching circuit that has a split-duplex receptacle wired so that the bottom half is controlled by a single-pole switch and the top half is "hot" at all times. The power source feeds the switch location.

8 At the split-duplex receptacle location, connect the white grounded conductor to one of the silver terminal screws on the duplex receptacle. It does not make a difference which of the two silver screws you use.

9 At the split-duplex receptacle location, connect the black conductor from the three-wire cable to the top brass terminal screw.

10 At the split-duplex receptacle location, connect the red conductor from the three-wire cable to the bottom brass terminal screw.

In Chapter 11, you learned that in a residential wiring system it is common practice to install wiring from outlet to outlet. Sometimes it is desired to continue an unswitched part of a circuit "downstream" of a switched lighting outlet. To accomplish this, a three-wire cable is used between the switch location and the lighting outlet. A two-wire cable is then continued from the lighting outlet to other nonswitched loads, such as receptacle outlets.

SWITCHED DUPLEX RECEPTACLE CIRCUIT 4

Follow these steps when installing a single-pole switch controlling a lighting load with a continuously "hot" receptacle located downstream of the lighting outlet. The power source will feed the switch (Figure 13-22):

1 Always wear safety glasses and observe all applicable safety rules.

2 At the lighting fixture box, connect the bare grounding conductors together using a wirenut. Remember that a nonmetallic box does not require the circuit-grounding conductor to be connected to it.

3 At the lighting fixture box, use a wirenut and connect together a white pigtail, the white conductor of the incoming three-wire cable, and the white conductor of the two-wire cable going to the duplex receptacle. Connect the white pigtail end to the silver screw terminal or wire identified as the grounded wire on the lighting fixture.

4 At the lighting fixture box, use a wirenut and connect together the black conductor from the incoming three-wire cable to the black conductor of the two-wire cable going to the duplex receptacle.

5 At the lighting fixture box, connect the red conductor from the three-wire incoming cable to the brass screw or a wire identified as the ungrounded conductor on the lighting fixture.

6 At the single-pole switch location, connect the bare grounding conductors together and, using a bare pigtail, connect all the grounding conductors to the green grounding screw on the switch.

7 At the single-pole switch location, connect the white insulated grounded conductors together using a wirenut.

8 At the single-pole switch location, using a wirenut, connect together a black pigtail, the black conductor of the incoming power source cable, and the black conductor

of the three-wire cable going to the lighting outlet. Connect the black pigtail end to one of the terminal screws on the single-pole switch.

9 At the single-pole switch location, connect the red conductor in the three-wire cable going to the lighting outlet to the other terminal screw on the single-pole switch.

10 At the duplex receptacle location, connect the bare grounding conductor to the green grounding screw on the duplex receptacle.

11 At the duplex receptacle location, connect the white grounded conductor to one of the silver terminal screws on the grounded side of the duplex receptacle. It does not make a difference which of the two silver screws you use.

12 At the duplex receptacle location, connect the black "hot" conductor to one of the brass terminal screws on the ungrounded side of the duplex receptacle. Again, it does not make a difference which of the two brass screws you use.

INSTALLING DOUBLE-POLE SWITCHES

As you have learned, electrical equipment in a house can operate on 120 volts, 120/240 volts, or 240 volts. For those wiring situations when a switch is needed to control a 240-volt load, double-pole (also called two-pole) switches are used. The double-pole switch has four terminals on it, and, at first glance, it looks like a four-way switch. However, unlike a four-way (or three-way) switch, the toggle on the double-pole switch has the words ON and OFF written on it. This means that like a single-pole switch, there is a correct mounting position for the switch, so when it is ON, the toggle will indicate it. The double-pole switch also has markings that usually indicate the "load" and the "line" sides of the switch. Usually the "line" set of terminals is colored black and the "load" set of terminals is colored a brass or bronze color to help distinguish them from each other. Many times a double-pole switch is used to control a 240-volt-rated receptacle. These receptacle types have special configurations (covered in Chapter 2) for the specific current and voltage rating of the receptacle. The 240-volt piece of electrical equipment will be plugged into the 240-volt-rated receptacle using a plug that has the same current and voltage rating as the receptacle.

DOUBLE-POLE SWITCHING CIRCUIT 1

Follow these steps when installing a double-pole switch controlling a 240-volt receptacle. The power source will feed the switch. For this circuit, assume the use of nonmetallic boxes and nonmetallic sheathed cable (Figure 13-23):

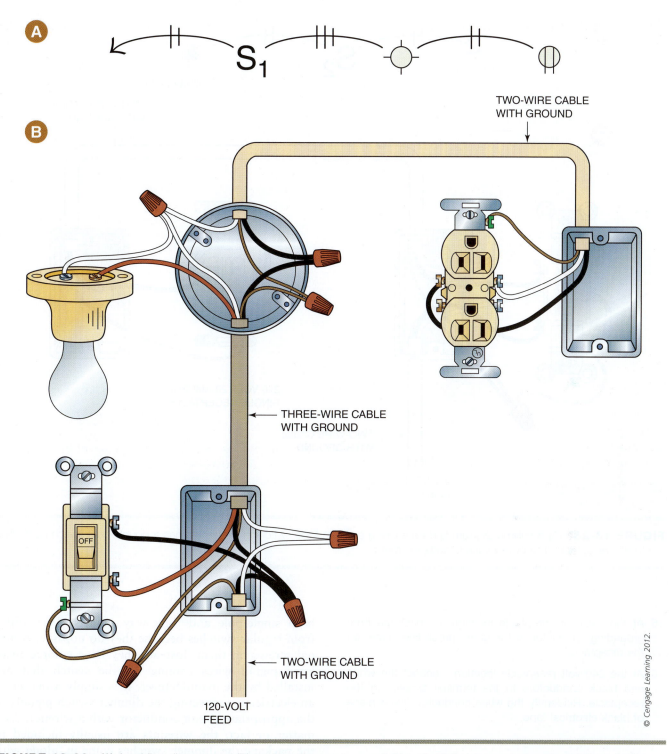

Ⓐ

Ⓑ

TWO-WIRE CABLE
WITH GROUND

THREE-WIRE CABLE
WITH GROUND

TWO-WIRE CABLE
WITH GROUND

120-VOLT
FEED

FIGURE 13-22   (A) A cabling diagram and (B) a wiring diagram for a switching circuit that has a single-pole switch controlling a lighting load with a continuously "hot" receptacle located downstream of the lighting outlet. The power source is feeding the switch.

1   Always wear safety glasses and observe all applicable safety rules.

2   At the double-pole switch location, connect the bare grounding conductors together and, using a bare pigtail, connect all the grounding conductors to the green grounding screw on the switch.

3   At the double-pole switch location, connect the white and black insulated conductors of the incoming two-wire power source feed cable to the line-side terminal screws on the switch. Reidentify the white conductor with a piece of black electrical tape.

4   At the double-pole switch location, connect the white and black conductors of the two-wire cable going to the 240-volt receptacle to the load-side terminal screws on the double-pole switch. Reidentify the white conductor with a piece of black electrical tape.

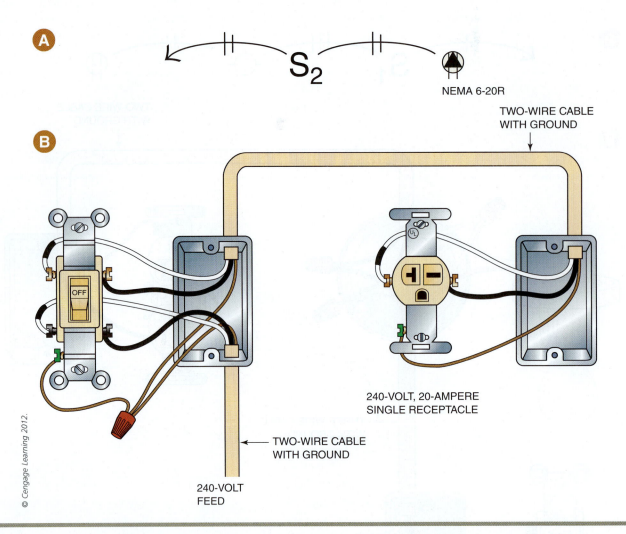

A

S_2

NEMA 6-20R

TWO-WIRE CABLE
WITH GROUND

B

OFF

240-VOLT, 20-AMPERE
SINGLE RECEPTACLE

TWO-WIRE CABLE
WITH GROUND

240-VOLT
FEED

© Cengage Learning 2012.

FIGURE 13-23 (A) A cabling diagram and (B) a wiring diagram for a switching circuit that has a double-pole switch controlling a 240-volt receptacle. The power source feeds the switch.

5 At the 240-volt receptacle location, connect the bare grounding conductor to the green grounding screw on the receptacle.

6 At the 240-volt receptacle location, connect the white and black conductors to the terminal screws on the receptacle. Reidentify the white conductor with a piece of black electrical tape.

INSTALLING DIMMER SWITCHES

Dimmer switches are used in residential wiring applications to provide control for the brightness of the lighting in a specific area of a house. They are available in both single-pole and three-way models. They differ from regular switches because they do not have terminal screws on them. Instead, they have colored insulated pigtail wires coming off the switch that are installed by the manufacturer. It is simply a matter of an electrician connecting the dimmer switch pigtails to the appropriate circuit conductor with a wirenut. As a matter of fact, the wirenuts are usually included in the package the dimmer switches come in, along with the 6-32 screws that secure the dimmer switch to the device box.

Dimmer switches are larger in size than regular single-pole and three-way switches and take up more room in an electrical box. They also tend to generate more heat and should be installed in electrical device boxes that do not have many other circuit wires in them.

Green Tip

Using dimmer switches can be a big part of a home's energy management system. According to a leading manufacturer of dimmer switches, a dimmer that is set to reduce the amount of light that a lighting fixture produces by 25% will result in a 20% reduction in energy costs. It also will result in an increase in the life expectancy of the fixture lamps of up to four times.

Single-pole or three-way dimmer switches are connected in a switching circuit exactly the same way as regular single-pole and three-way switches (Figure 13-24).

Follow these rules when connecting dimmer switches:

- Single-pole dimmers have two black insulated conductors coming off them. Like the two brass screw terminals on a regular single-pole toggle switch, it does not make any difference which of the two wires is connected to the incoming "hot" feed wire or the outgoing switch leg.

- Three-way dimmer switches typically come from the manufacturer with two red insulated pigtails and one black insulated pigtail. The two red conductors are the "traveler" connections, and the black conductor is the "common" connection.

- When using three-way dimmers, it is common wiring practice to use one three-way dimmer and one three-way regular toggle switch when controlling and dimming a lighting load from two locations. The reason for this is that, if you use two three-way dimmers, the *highest* level the lighting load will be able to be brightened is whatever the *lowest* level is on the other three-way dimmer switch. Be aware that some dimmer switch manufacturers make "digital" dimmers that

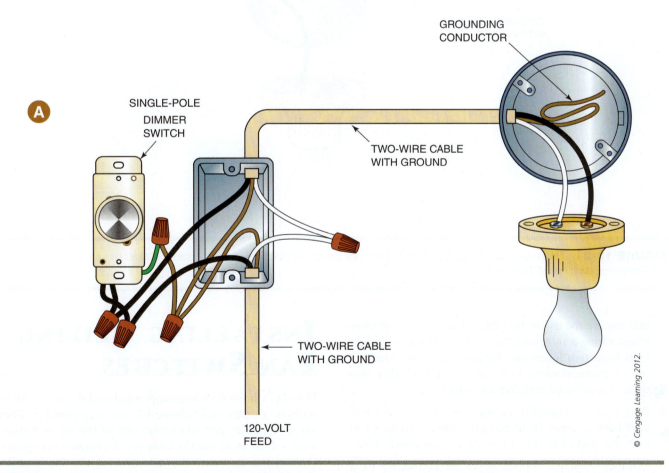

FIGURE 13-24   Single-pole and three-way dimmer switches are connected in a circuit the same way as regular (A) single-pole.

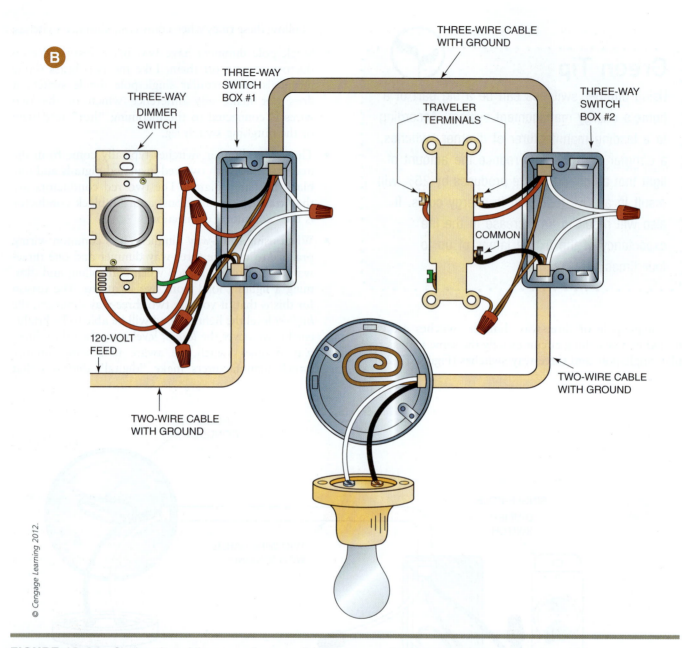

B

THREE-WAY
DIMMER
SWITCH

THREE-WAY
SWITCH
BOX #1

THREE-WIRE CABLE
WITH GROUND

TRAVELER
TERMINALS

THREE-WAY
SWITCH
BOX #2

COMMON

120-VOLT
FEED

TWO-WIRE CABLE
WITH GROUND

TWO-WIRE CABLE
WITH GROUND

FIGURE 13-24   Single-pole and three-way dimmer switches are connected in a circuit the same way as regular
(B) Three-way switches. (Continued)

will require the use of two three-way dimmer switches when wiring a switching circuit that controls a lighting load from two locations. Always check the manufacturer's instructions to find out for sure whether two three-way dimmer switches must be used.

- Both single-pole and three-way dimmer switches will also have a green insulated grounding pigtail. Connect this pigtail to the circuit-grounding conductor.

- Dimmer switches should never be connected to a "live" electrical circuit. The solid-state electronics that allow a dimmer switch to operate will be severely damaged. Always de-energize the electrical circuit first before installing a dimmer switch.

INSTALLING CEILING FAN SWITCHES

In today's homes, ceiling-suspended paddle fans, with or without an attached lighting kit, are very popular. They are designed to provide movement of the air in a room so it will be more comfortable for the room's occupants. During the summer, the direction of the fan paddles should direct the air down. During the winter, the fan paddle direction should be reversed so the air is pulled upward. There is a switch on the paddle fan to switch the direction of rotation. The switching to provide electrical power to the paddle fan and any light fixture attached to it is normally placed in a wall-mounted switch box for easy access by the home owner. In this section, we look at the switching connections needed to control a ceiling-suspended paddle fan/light.

There are several styles of paddle fan control switches available for an electrician to install. A common switch style is shown in Figure 13-25. The rotary switch is used to control the paddle fan. It has three specific speed settings: low, medium, and high. The single-pole switch provides control of the attached lighting fixture. A two-gang electrical device box must be installed for this switch combination. Another common switching installation to control a ceiling-suspended paddle fan/light is to use two single-pole switches. One switch controls the paddle fan, and the other switch controls the lighting fixture. The speed of the paddle fan will need to be set by a switch located on the fan itself in this switch arrangement. Like the other switching circuits described in this chapter, the power source feed for a ceiling-suspended paddle fan switching circuit can originate at the paddle fan itself or at the switching location.

CEILING-SUSPENDED PADDLE FAN/LIGHT CIRCUIT 1

Follow these steps when installing individual single-pole switch control for a ceiling-suspended paddle fan with an attached light fixture. The power source will feed the switch location. For this circuit, assume the use of non-metallic boxes and nonmetallic sheathed cable. The

outlet box used for the paddle fan/light must be listed and identified as suitable for the purpose (Figure 13-26). Nonmetallic sheathed cable (Type NM) and nonmetallic boxes are shown because they are used the most in house wiring. The only adjustment you will need to make if metal boxes are used is to make sure to attach the circuit equipment-grounding conductor to each metal box with a green grounding screw. The actual installation of a ceiling-suspended paddle fan is covered in the "Trim-Out" section of this textbook.

Green Tip

It is easy for a home owner to turn a ceiling-suspended paddle fan on and forget about it. Timers can be installed in a ceiling-suspended paddle fan circuit so that the fan does not end up running constantly. This will result in a lower use of electrical energy. Timers can also be installed to control bathroom exhaust fans so that they get switched off after a reasonable amount of time.

1 Always wear safety glasses and observe all applicable safety rules.

2 At the switch location, using a wirenut, connect the bare grounding conductors together and, using two bare pigtails, connect all the grounding conductors to the green grounding screws on the two single-pole switches. Remember that a nonmetallic box does not require the circuit-grounding conductor to be connected to it.

3 At the switch location, connect the white insulated grounded conductors together with a wirenut.

4 At the switch location, using a wirenut, connect the black "hot" conductor of the incoming power source cable to one end of two black pigtails. Connect the other end of the black pigtails to a terminal screw on each single-pole switch. It does not matter which screw is used on each switch.

5 At the switch location, connect the black conductor in the three-wire cable going to the paddle fan/light to the other terminal screw on single-pole switch 2.

6 At the switch location, connect the red conductor in the three-wire cable going to the paddle fan/light to the other terminal screw on single-pole switch 1.

7 At the paddle fan/light location, use a wirenut to connect the bare grounding conductor to the green insulated grounding conductor on the paddle fan/light. Typically, the paddle fan/light mounting bracket also has a green insulated grounding pigtail that must be connected to the incoming grounding conductor.

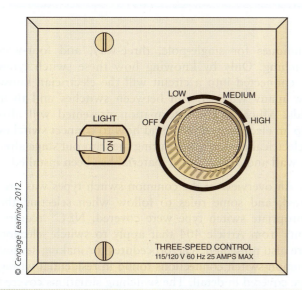

© Cengage Learning 2012.

LIGHT   ON/OFF

OFF   LOW   MEDIUM   HIGH

THREE-SPEED CONTROL
115/120 V 60 Hz 25 AMPS MAX

FIGURE 13-25   A common switch style for a ceiling-suspended paddle fan/light. The single-pole switch controls the lighting fixture attached to the paddle fan, and the rotary switch provides three-speed control of the paddle fan.

A

B

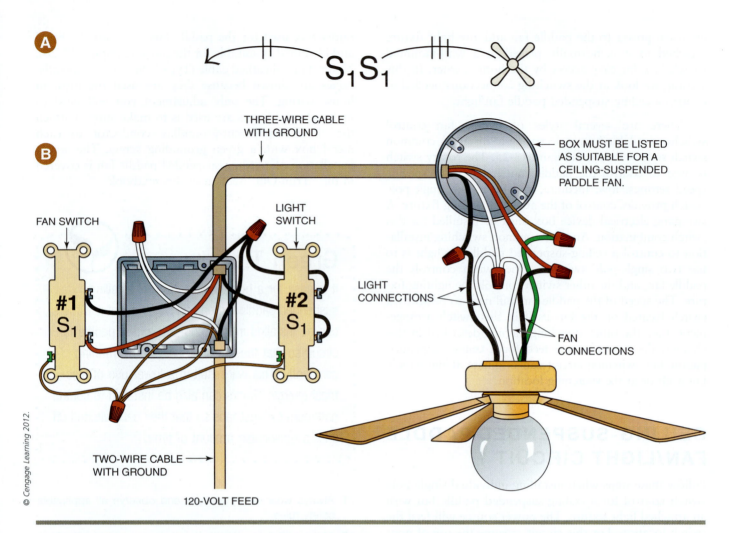

THREE-WIRE CABLE
WITH GROUND

BOX MUST BE LISTED
AS SUITABLE FOR A
CEILING-SUSPENDED
PADDLE FAN.

FAN SWITCH

LIGHT
SWITCH

#1
S_1

#2
S_1

LIGHT
CONNECTIONS

FAN
CONNECTIONS

TWO-WIRE CABLE
WITH GROUND

120-VOLT FEED

© Cengage Learning 2012.

FIGURE 13-26   (A) A cabling diagram and (B) a wiring diagram for a switching circuit that has individual single-pole switch control for a ceiling-suspended paddle fan with an attached light fixture. The power source will feed the switch location.

8   At the paddle fan/light location, using a wirenut, connect the white insulated grounded conductor of the three-wire cable to the white insulated grounded conductor of the paddle fan/light.

9   At the paddle fan/light location, using a wirenut, connect the black insulated conductor of the three-wire cable to the insulated conductor of the paddle fan/light that is labeled "Light Fixture."

10   At the paddle fan/light location, using a wirenut, connect the red insulated conductor of the three-wire cable to the insulated conductor of the paddle fan/light that is labeled "Paddle Fan."

SUMMARY

This chapter presented the information that residential electricians need to know when they are roughing-in the wiring for the switching circuits found in residential wiring. An electrician must know the common connection techniques for single-pole, three-way, and four-way switching. Only by knowing how these switch types are connected into a circuit will the electrician know how many wires to run between switches and their loads. Much of the information presented will also help an electrician understand how to connect switches in electrical device boxes during the trim-out stage after the wall and ceiling finish material has been installed.

An overview of the common switch types was presented, and some rules to follow when selecting the appropriate switch type were covered. *NEC*® requirements from Article 404 that apply to switch selection were also presented. The procedures for making several common switch connections found in residential wiring were covered in detail. The switching situations covered in this chapter cover most of the switching situations that a residential electrician will encounter. A good understanding of the switching connection techniques described in this chapter will allow an electrician to correctly wire any switch situation that may arise.

GREEN CHECKLIST

☐ Outside lighting outlets are often switched on for extended periods of time. They are also often left on during the daylight hours when they are not needed. Motion sensors can be installed to control outdoor lighting fixtures so that they operate only when needed. This will result in significant energy saving for a home owner.

☐ Lighting fixtures in rooms are often left on when everybody has left the room. Electricians can install occupancy sensors to turn the lights off after all people have left the room and there is no movement detected in the room. This will save electrical energy and reduce a home owner's electricity bill.

☐ Using dimmer switches can be a big part of a home's energy management system. According to a leading manufacturer of dimmer switches, a dimmer set to reduce the amount of light that a lighting fixture produces by 25% will result in a 20% reduction in energy costs. It also will result in an increase in the life expectancy of the fixture lamps of up to four times.

☐ It is easy for a home owner to turn a ceiling-suspended paddle fan on and forget about it. Timers can be installed in a ceiling-suspended paddle fan circuit so that the fan does not end up running constantly. This will result in a lower use of electrical energy. Timers can also be installed to control bathroom exhaust fans so that they get turned off after a reasonable amount of time.

REVIEW QUESTIONS

Directions: Answer the following items with clear and complete responses.

1. Explain how the *NEC*® and Underwriters Laboratories categorize general-use snap switches.

2. Determine the switch type and rating for a single-pole switch controlling eight 100-watt incandescent lamps in a 120-volt residential switching circuit.

3. Describe the type of wiring situation that would require the use of a single-pole switch.

4. Describe the type of wiring situation that would require the use of three-way switches.

5. The *NEC*® requires switching to be done in the _____ circuit conductor.

6. Draw a line from the switch type to the correct number of terminal screws for the switch.

    single-pole switch   four terminal screws

    three-way switch   three terminal screws

    four-way switch   two terminal screws

7. The number of conductors required to be installed between three-way and four-way switches is _____.

8. Explain why it is common wiring practice to switch the bottom half of a split-duplex receptacle.

9. Describe a situation where you would need to use a combination single-pole/three-way switch.

10. A switching circuit used to control a lighting load from six locations would require _____ three-way switches and _____ four-way switches.

11. The number of conductors required to be installed between four-way switches is _____.

12. The number of conductors required to be installed from a switch location to a ceiling-suspended paddle fan with a lighting fixture is _____ when separate switching is desired.

13. A _____ _____ is the name given to a switching circuit when the feed is brought to the lighting load first and then a two-wire cable (or two conductors in a raceway) is run to the switch location.

14. Describe a wiring situation where a two-pole switch would be used.

15. Explain why it is common wiring practice to install one three-way dimmer switch and one regular three-way switch when it is desired to have a lighting load be controlled and dimmed from two locations.

16. Complete the connections in the following diagram. Assume the use of nonmetallic sheathed cable and nonmetallic electrical boxes. A 120-volt feed is brought to the switch box, and the single-pole switch will control the lighting outlet.

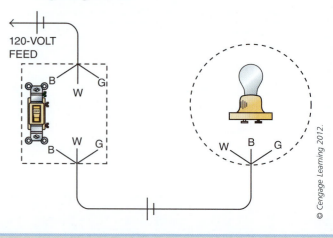

120-VOLT FEED

17. Complete the connections in the following diagram. Assume the use of nonmetallic sheathed cable and nonmetallic electrical boxes. The single-pole switch will control the lighting outlet, but notice that the power source feed is brought to the lighting outlet. This is a switch loop wiring situation.

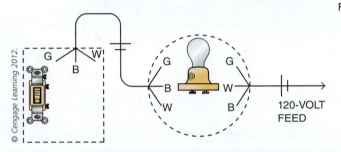

18. Complete the connections in the following diagram. Assume the use of nonmetallic sheathed cable and nonmetallic electrical boxes. Either three-way switch will control the lighting outlet. The 120-volt feed is brought to the first three-way switch.

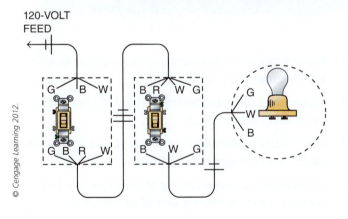

19. Complete the connections in the following diagram. Assume the use of nonmetallic sheathed cable and nonmetallic electrical boxes. The single-pole switch controls the lighting outlet but not the duplex receptacle. The duplex receptacle is "hot" at all times.

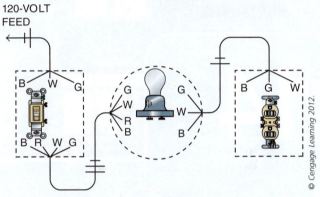

20. Complete the connections in the following diagram. Assume the use of nonmetallic sheathed cable and nonmetallic electrical boxes. The 120-volt feed is brought to the lighting outlet. All three switches will control the lighting outlet.

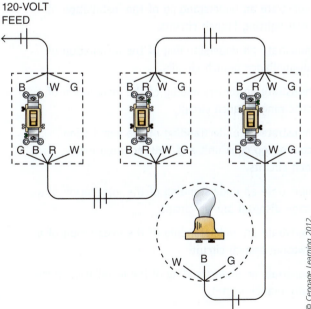

Branch-Circuit Installation

OBJECTIVES

Upon completion of this chapter, the student should be able to:

- Demonstrate an understanding of the installation of general lighting branch circuits.

- Demonstrate an understanding of the installation of small-appliance branch circuits.

- Demonstrate an understanding of the installation of electric range branch circuits.

- Demonstrate an understanding of the installation of countertop cook unit and wall-mounted oven branch circuits.

- Demonstrate an understanding of the installation of a garbage disposal branch circuit.

- Demonstrate an understanding of the installation of a dishwasher branch circuit.

- Demonstrate an understanding of the installation of the laundry branch circuit.

- Demonstrate an understanding of the installation of an electric clothes dryer branch circuit.

- Demonstrate an understanding of the installation of branch circuits in a bathroom.

- Demonstrate an understanding of the installation of a water pump branch circuit.

- Demonstrate an understanding of the installation of an electric water heater branch circuit.

- Demonstrate an understanding of the installation of branch circuits for heating and air-conditioning.

- Demonstrate an understanding of the installation of branch circuits for electric heating.

- Demonstrate an understanding of the installation of a branch circuit for smoke detectors.

- Demonstrate an understanding of the installation of a branch circuit for carbon monoxide detectors.

- Demonstrate an understanding of the installation of a low-voltage chime circuit.

GLOSSARY OF TERMS

carbon monoxide detector a device that detects the presence of carbon monoxide (CO) gas in order to prevent carbon monoxide poisoning; CO is a colorless and odorless compound produced by incomplete combustion

cord-and-plug connection an installation technique in which electrical appliances are connected to a branch circuit with a flexible cord with an attachment plug; the attachment plug end is plugged into a receptacle of the proper size and type

counter-mounted cooktop a cooking appliance that is installed in the top of a kitchen countertop; it contains surface cooking elements, usually two large elements and two small elements

dual-element time-delay fuses a fuse type that has a time-delay feature built into it; this fuse type is used most often as an overcurrent protection device for motor circuits

electric range a stand-alone electric cooking appliance that typically has four cooking elements on top and an oven in the bottom of the appliance

hardwired an installation technique in which the circuit conductors are brought directly to an electrical appliance and terminated at the appliance

heat pump a reversible air-conditioning system that will heat a house in cool weather and cool a house in warm weather

hydronic system a term used when referring to a hot water heating system

interconnecting the process of connecting together smoke detectors or carbon monoxide detectors so that if one is activated, they all will be activated

inverse time circuit breaker a type of circuit breaker that has a trip time that gets faster as the fault current flowing through it gets larger; this is the circuit breaker type used in house wiring

jet pump a type of water pump used in home water systems; the pump and electric motor are separate items that are located away from the well in a basement, garage, crawl space, or other similar area

nameplate the label located on an appliance that contains information such as amperage and voltage ratings, wattage ratings, frequency, and other information needed for the correct installation of the appliance

smoke detector a safety device that detects airborne smoke and issues an audible alarm, thereby alerting nearby people to the danger of fire; most smoke detectors work either by optical detection or by ionization, but some of them use both detection methods to increase sensitivity to smoke

submersible pump a type of water pump used in home water systems; the pump and electric motor are enclosed in the same housing and are lowered down a well casing to a level that is below the water line

thermostat a device used with a heating or cooling system to establish a set temperature for the system to achieve; they are available as line voltage models or low-voltage models; some thermostats can also be programmed to keep a home at a specific temperature during the day and another temperature during the night

transformer the electrical device that steps down the 120-volt house electrical system voltage to the 16 volts a chime system needs to operate correctly

wall-mounted oven a cooking appliance that is installed in a cabinet or wall and is separated from the counter-mounted cooktop

A residential electrical system branch circuit is run from the main service panel or a subpanel to the various lighting and power outlets on the circuit. It consists of the conductors and other necessary electrical equipment needed to supply and control electrical power to the branch circuit's electrical loads. Branch circuits are installed using a variety of methods to achieve the goal of getting electrical power to the electrical loads. Earlier in this book, you learned that there are several types of branch circuits found in a residential electrical system. These branch circuits include general lighting branch circuits, small-appliance branch circuits, laundry branch circuits, bathroom branch circuits, and individual branch circuits. In this chapter, we take a close look at common installation methods and requirements for the types of branch circuits found in residential wiring.

INSTALLING GENERAL LIGHTING BRANCH CIRCUITS

Because general lighting branch circuits make up the largest portion of the branch circuits installed in a residential wiring job, we start our discussion with them. A residential general lighting branch circuit will include the ceiling- and wall-mounted lighting outlets as well as most of the receptacles in a house. The only receptacles not found on a general lighting branch circuit will be those receptacles on small-appliance branch circuits, laundry circuits, a bathroom circuit, or those located on an individual branch circuit for a specific appliance.

General lighting branch circuits will be wired with either a 14 AWG copper conductor and protected with a 15-ampere fuse or circuit breaker or a 12 AWG copper conductor protected with a 20-ampere fuse or circuit breaker. The general lighting branch circuit will originate in the service entrance panel or a subpanel and will be routed to the various lighting outlets, switching points for the lighting outlets, and receptacle outlets on the circuit.

In Chapter 6, you were taught how to determine the minimum number of general lighting branch circuits for a specific size of house. It is a good idea to review that

information at this time. Because there is no maximum number, as long as you meet the *National Electrical Code®* (*NEC®*) requirements for the minimum number installed, you can install as many general lighting branch circuits as you want. It is common practice to install more general lighting branch circuits than the minimum number required by the *NEC®*. It is up to the electrician installing the residential electrical system just how many general lighting branch circuits are ultimately installed.

It is also up to the electrician installing a general lighting branch circuit to determine how many lighting outlets and how many receptacle outlets will be on each circuit. Electricians should try to divide the number of lighting and receptacle outlets on each circuit evenly among all the general lighting branch circuits. This is not always practical, but the effort should be made. It is common practice for some general lighting branch circuits to have more (or less) lighting and receptacle outlets than on other general lighting branch circuits. As we mentioned before, most general lighting branch circuits will have both lighting outlets and receptacle outlets on the circuit; however, sometimes a general lighting branch circuit will have only lighting outlets *or* receptacle outlets. While there is no limit to the number of receptacles that can be connected to a general lighting branch circuit in a residential installation, a good rule of thumb is to assign each receptacle outlet a value of 1.5 amperes. Using this assumption, the maximum number of receptacles on a 15-amp-rated general lighting branch circuit would be 10 (15 amps/1.5 amps for each receptacle = 10). The maximum number of receptacles on a 20-amp-rated general lighting branch circuit would be 13 (20 amps/1.5 amps for each receptacle = 13). For those general lighting branch circuits that have both lighting outlets and receptacle outlets, it is recommended that not more than a total of 8 outlets (either lighting or receptacle) be put on one 15-amp-rated general lighting branch circuit and not more than 10 outlets (either lighting or receptacle) on a 20-amp-rated general lighting branch circuit. These numbers take into account the possibility of multiple lamp lighting fixtures being installed or the installation of high-wattage lamps in the lighting fixtures.

CAUTION

CAUTION: The *NEC®* permits loading an overcurrent protection device to 100 percent of its rating only if the device has a listing that allows a loading of 100 percent. Underwriters Laboratories (UL) has no 100 percent listing for molded-case circuit breakers of the type used in residential wiring applications. Normally, they are listed only for loading of up to 80 percent of their ratings. Fuses, on the other hand, can be loaded to 100 percent of their ratings if an electrician so chooses.

from experience...

It is really up to the electrical contractor whether 15- or 20-amp-rated general lighting branch circuits are to be installed. Also, some building plan electrical specifications may require all wiring in a house to be a minimum of 12 AWG copper with 20-amp-rated over-current protection devices.

from experience...

It is a good idea for electricians planning and installing general lighting branch circuits to include some lighting and receptacle outlets in a room on different circuits. This practice can help balance the load more evenly and may result in using less cable or conduit when wiring the installation. Another benefit to this practice is that should a circuit breaker trip on one circuit, the other outlets protected by a circuit breaker on another circuit will still provide some light and electrical power in the room.

When roughing in general lighting branch circuits, there are a few steps to always follow:

1 Determine whether you will be installing 15- or 20-amp-rated general lighting branch circuits. Remember to use 14 AWG wire for 15-amp-rated branch circuits and 12 AWG wire for 20-amp-rated branch circuits.

2 Determine the number of lighting and receptacle outlets to be included on the general lighting branch circuit (Figure 14-1).

3 If they have not been installed previously, install the required lighting outlet boxes, switching location boxes, and receptacle outlet boxes at the appropriate locations.

4 Starting at the electrical panel where the circuit overcurrent protection device is located, install the wiring to the first electrical box on the circuit. Be sure to mark the cable or other wiring method with the circuit number and type and the area served at the panel. For example, general lighting branch circuit 1 serving the master bedroom can be marked "GL1MBR," or general lighting branch circuit 3 serving bedroom number two can be marked "GL3BR2."

5 Continue roughing in the wiring for the rest of the circuit, following the installation practices described in previous chapters. Make sure to tuck the wiring in the back of all electrical boxes on the circuit.

6 Double-check to make sure all circuit wiring for the general lighting branch circuit you are installing has been roughed in.

from experience...

Make sure to tuck the conductors all the way into the back of electrical boxes when roughing in branch circuit wiring. It is common practice for sheetrock installers to cover all electrical boxes with sheetrock and to then cut out the electrical box openings. They usually cut out the openings by using what is often called a roto-zip with a zip bit. A roto-zip and zip bit is simply a high-speed drill with a sheetrock cutting bit. If the conductors are not placed in the back of an electrical box, the roto-zip bit may cut through the conductor insulation or even completely through a conductor. This can present problems for the electricians when they go back to the house to install switches and receptacles in the trim-out stage and find that (even though they left at least 6 inches of free conductor with which to make device connections) the wires have been cut, making it very difficult to terminate the conductors. If this happens to you, use electrical tape to re-insulate a conductor whose insulation has been damaged by the roto-zip and, if a conductor gets cut off by the roto-zip, wirenut a 6–8-inch pigtail to it and then make the device connection.

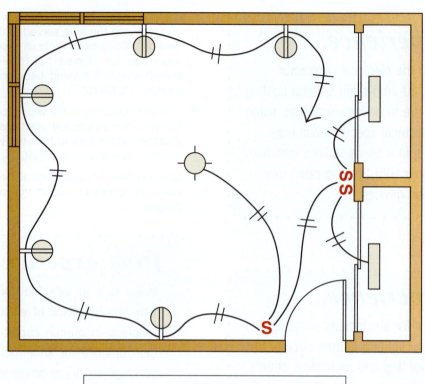

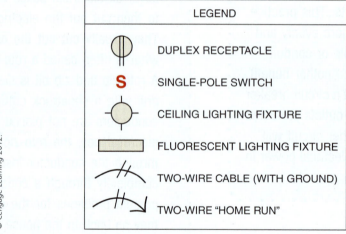

© Cengage Learning 2012.

LEGEND	
⏻	DUPLEX RECEPTACLE
S	SINGLE-POLE SWITCH
⊕	CEILING LIGHTING FIXTURE
▭	FLUORESCENT LIGHTING FIXTURE
⌒⧸⧸	TWO-WIRE CABLE (WITH GROUND)
⌒⧸⧸	TWO-WIRE "HOME RUN"

FIGURE 14-1   A partial electrical plan showing the cabling diagram of a general lighting circuit serving this area of the house. Notice that both lighting outlets and receptacle outlets are included in this branch circuit.

INSTALLING SMALL-APPLIANCE BRANCH CIRCUITS

Small-appliance branch circuits are the circuits that supply electrical power to receptacles in kitchens, dining rooms, pantries, and other similar areas of a house. The *NEC®* requires that they must be protected with 20-ampere overcurrent protection devices. The *NEC®* also requires a minimum of two small-appliance branch circuits be installed in each house. However, in larger homes, it is normal wiring practice to install more than two small-appliance branch circuits to adequately feed all the small-appliance loads. There is no maximum number that may be installed. When small-appliance branch circuits are used to feed the countertop receptacles in a kitchen or dining area, at least two different small-appliance branch circuits must be used. It is common wiring practice to split the countertop receptacles as evenly as possible on two different small-appliance branch circuits. The *NEC®* allows no other outlets (lighting or receptacle) on a small-appliance branch circuit except a clock receptacle and/or the receptacle providing power for gas-fired ranges, ovens, or counter-mounted cooking units. Therefore, ceiling lighting outlets, under-cabinet lighting, hood fan/lights, outside receptacles, garbage disposals, dishwashers, trash compactors, or any other electrical load in a kitchen may not be connected to a small-appliance branch circuit.

from experience...

Section 210.12 requires all 120-volt, 15- and 20-amp branch circuits supplying outlets in a dining room to be arc fault circuit interrupter (AFCI) protected. It does not require AFCI protection for outlets in kitchens. For this reason, many electricians install one or more small-appliance branch circuits that serve receptacles in just the dining room area and protect these circuits with AFCI circuit breakers in the electrical panel. Two or more additional small-appliance branch circuits are then installed to serve the receptacles in the kitchen. They are protected with a "regular" circuit breaker or a ground fault circuit interrupter (GFCI) circuit breaker in the electrical panel. If "regular" circuit breakers are used to protect the kitchen area small-appliance branch-circuit wiring then GFCI receptacles will be used to provide the necessary GFCI protection for the receptacles serving the kitchen countertop.

Electricians wire small-appliance branch circuits with 12 AWG copper conductors, although a larger-size wire, such as a 10 AWG, could be used to compensate for voltage drop problems when the distance back to the electrical panel is very long. Like general lighting branch circuits, the small-appliance branch circuits originate in the service entrance panel or a subpanel and are routed to the various receptacle outlet boxes on the circuit. Also, like general lighting branch circuits, there is no maximum number of receptacles that can exist on a small-appliance branch circuit, but again it is good wiring practice to limit the number of receptacles on them to 13 (20 amps/1.5 amps per receptacle = 13) or less.

When roughing in small-appliance branch circuits, there are a few steps to always follow:

1 Use 12 AWG wire and 20-amp overcurrent protection devices for all small-appliance branch circuits.

2 Always install a minimum of two small-appliance branch circuits.

3 Determine the number of receptacle outlets to be included on each small-appliance branch circuit (Figure 14-2).

4 If they have not been installed previously, install the required receptacle outlet boxes at the appropriate locations.

5 Starting at the electrical panel where the overcurrent protection device is located, install the wiring to the first electrical box on the circuit. Be sure to mark the cable or other wiring method with the circuit number, type, and the area served at the panel. For example, small-appliance branch circuit 1 serving the kitchen countertop can be marked "SA1KCT."

6 Continue roughing in the wiring for the rest of the circuit, following the installation practices described in previous chapters. Make sure to tuck the wiring in the back of all electrical boxes on the circuit.

7 Double-check to make sure all circuit wiring for the small-appliance branch circuit you are installing has been roughed in.

INSTALLING ELECTRIC RANGE BRANCH CIRCUITS

In addition to the small-appliance branch circuits and the general lighting branch circuits serving the kitchen area in a house, there are several other branch-circuit types required. **Electric ranges**, counter-mounted cook-tops, built-in ovens, garbage disposals, dishwashers, and other appliances are commonly found in residential kitchens and must have branch-circuit wiring installed with the proper voltage and ampere ratings. In this section, a detailed look at the installation requirements for electric ranges is presented.

To install the branch circuit for an electric range, an electrician must first determine the minimum size of the wire and maximum overcurrent protection device required. In Chapter 6, you were shown how to determine this information (see Figures 6-8 and 6-9 for a review of the procedure). The branch circuit voltage will be 120/240 volts and it will need a conductor size that will handle, at a minimum, the calculated load. Electric ranges are usually connected to the electrical system in a house through a **cord-and-plug connection**. An 8/3 copper cable with a grounding conductor protected

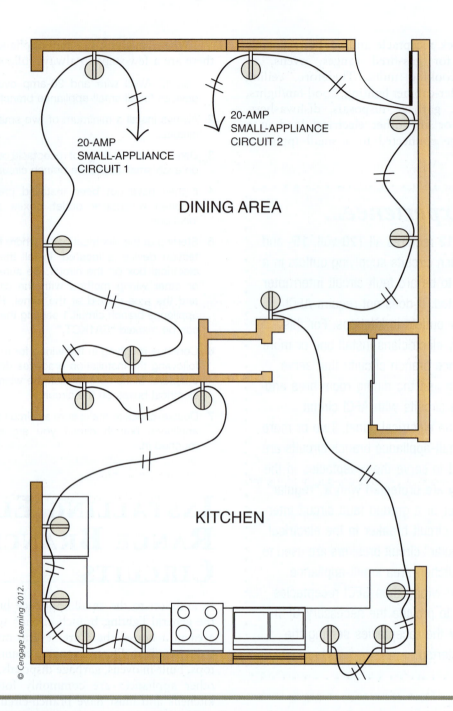

20-AMP
SMALL-APPLIANCE
CIRCUIT 1

20-AMP
SMALL-APPLIANCE
CIRCUIT 2

DINING AREA

KITCHEN

© Cengage Learning 2012.

FIGURE 14-2   A partial electrical plan showing the cabling diagram of two small-appliance branch circuits serving a kitchen and dining room area. No lighting outlets are allowed on these circuits. Remember that AFCI protection is needed for the receptacle outlets in the dining area and GFCI protection is required for the receptacles in the kitchen that serve the countertop.

by a 40-ampere circuit breaker or a 6/3 copper cable with a grounding conductor protected by a 50-ampere circuit breaker is often used. Nonmetallic sheathed cable, Type AC cable, Type MC cable, or Type SER service entrance cable may be used for the branch-circuit wiring method. Three conductors (with a fourth grounding conductor) of the proper size installed in a raceway may also be used. An electrician will install the wiring method between the electrical panel and the location of

the receptacle for the electric range. The receptacle will be either surface-mounted or flush-mounted. Flush mounting the receptacle will require an electrical box. The receptacle location is usually close to the floor at the rear of the final position of the electric range. Access to the receptacle is by removing the bottom drawer of the electric range or simply by pulling the electric range out from the wall. The wiring method is attached to either the electrical box for flush mounting or the receptacle

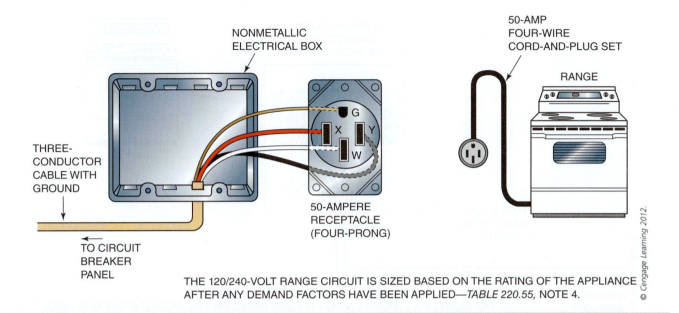

THE 120/240-VOLT RANGE CIRCUIT IS SIZED BASED ON THE RATING OF THE APPLIANCE AFTER ANY DEMAND FACTORS HAVE BEEN APPLIED—*TABLE 220.55,* NOTE 4.

FIGURE 14-3 A four-wire range installation using a cord-and-plug assembly.

body itself for surface mounting. Assuming the use of a cable wiring method, the wires of the branch circuit will be connected to the receptacle terminals as follows (Figure 14-3):

1 Connect the white grounded neutral conductor to the terminal marked "W."

2 Connect the black ungrounded conductor to the terminal marked "Y."

3 Connect the red ungrounded conductor to the terminal marked "X."

4 Connect the green or bare equipment-grounding conductor to the terminal marked "G."

The cord that connects the electric range to the receptacle is often not installed on the electric range by the appliance dealer and an electrician will have to install it. Make sure that the cord is a four-wire cord and has a plug that will properly match the receptacle that it will be plugged into. Follow the manufacturer's instructions, often found on the back of an electric range, for the correct wiring connections for the cord.

Sometimes the electrical installation calls for the connection to an electric range to be **hardwired**. This is done by bringing the branch-circuit wiring method directly to the back of an electric range and making the terminal connections on the terminal block on the electric range. Assuming a cable wiring method, the connections are made as follows (Figure 14-4):

1 Connect the white grounded neutral conductor to the silver screw terminal on the electric range terminal block.

2 Connect the black ungrounded conductor to the brass screw on the electric range terminal block.

3 Connect the red ungrounded conductor to the other brass screw on the electric range terminal block.

4 Connect the green or bare equipment-grounding conductor to the green grounding screw on the electric range terminal block.

Although three-conductor electric range installations have not been allowed by the *NEC®* in new residential construction since 1996, electricians may run into this wiring situation when installing new electric ranges in homes built prior to 1996. If this is the case, the *NEC®* allows the existing three-wire branch circuit to stay in place, and a three-wire cord-and-plug connection will have to be made to the electric range. The connections for attaching the three-wire cord to the electric range (or electric clothes dryer) are made as follows (Figure 14-5):

1 Connect the black (or one of the outside conductors of the cord assembly) to one of the brass screws on the terminal board.

2 Connect the red (or the other outside conductor of the cord assembly) to the other brass terminal screw on the terminal board.

3 Connect the white (or middle wire on the three-wire cord assembly) to the silver screw on the terminal board.

4 Connect a jumper, bare or green, from the silver screw grounded neutral terminal on the terminal board to the green screw grounding terminal. Check to make sure that the green screw location is bonded (connected) to the frame of the electric range. *Note 1:* This jumper is usually installed at the factory, but if it is not, you will have to install one. *Note 2:* This jumper must be removed and discarded if this electric range is installed on a four-wire system in a new installation and the jumper was installed at the factory.

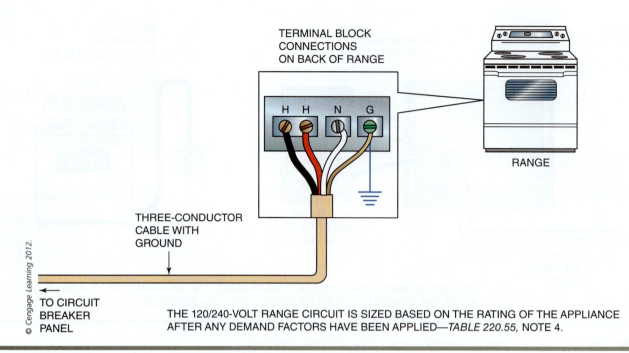

TERMINAL BLOCK
CONNECTIONS
ON BACK OF RANGE

H H N G

RANGE

THREE-CONDUCTOR
CABLE WITH
GROUND

TO CIRCUIT
BREAKER
PANEL

© Cengage Learning 2012.

THE 120/240-VOLT RANGE CIRCUIT IS SIZED BASED ON THE RATING OF THE APPLIANCE
AFTER ANY DEMAND FACTORS HAVE BEEN APPLIED—*TABLE 220.55,* NOTE 4.

FIGURE 14-4   A four-wire range installation using the hardwired method.

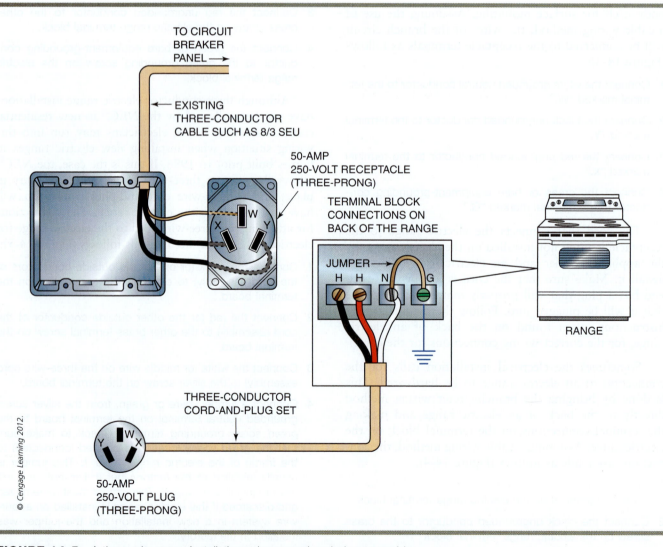

TO CIRCUIT
BREAKER
PANEL →

EXISTING
THREE-CONDUCTOR
CABLE SUCH AS 8/3 SEU

50-AMP
250-VOLT RECEPTACLE
(THREE-PRONG)

X W Y

TERMINAL BLOCK
CONNECTIONS ON
BACK OF THE RANGE

JUMPER →
H H N G

RANGE

THREE-CONDUCTOR
CORD-AND-PLUG SET

Y W X

© Cengage Learning 2012.

50-AMP
250-VOLT PLUG
(THREE-PRONG)

FIGURE 14-5   A three-wire range installation using a cord-and-plug assembly.

INSTALLING THE BRANCH CIRCUIT FOR COUNTER-MOUNTED COOKTOPS AND WALL-MOUNTED OVENS

Many home owners choose a separate **counter-mounted cooktop** and either one or two **wall-mounted ovens** instead of a single stand-alone electric range. Installing a branch circuit to these appliances is very similar to installing a branch circuit to a stand-alone electric range. Each countertop cook unit and wall-mounted oven normally comes from the factory equipped with a short length of flexible metal conduit enclosing the feed wires for the appliance. The feed wires are already connected to the terminal block in the appliance, and the only electrical connection that an electrician will need to make is the connection of the feed wires to the branch-circuit wiring method brought from the electrical panel to the appliance.

There are three common methods used to install the branch circuit and make the necessary connections to counter-mounted cooktops and wall-mounted ovens. The first method involves running a separate branch circuit from the electrical panel to both the countertop cook unit and the wall-mounted oven (Figure 14-6). The branch-circuit wiring method is brought to a metal junction box located near the cooking appliance. The flexible metal conduit with the appliance feed wires is installed into the junction box, and the proper connections are made using wirenuts. A blank cover is then put on the junction box. In the second method (Figure 14-7), an electrician runs one larger conductor-size branch circuit from the electrical panel to a metal junction box located close to both appliances. The flexible metal conduit-enclosed appliance wires from both appliance types are connected to the junction box, and the proper connections to the branch circuit wiring are made in the junction box using wirenuts. In the third method (Figure 14-8), individual branch-circuit wiring is installed to a receptacle outlet box located in a cabinet near the cooking appliance. The flexible metal conduit-enclosed wiring installed at the factory is disconnected, and a properly sized cord-and-plug assembly is attached to each unit. A receptacle with the proper voltage and amperage rating, which correctly matches the plug on the end of the newly installed appliance cord, will be installed in the electrical box. Each appliance will then be plugged into the receptacle to get electrical power. One advantage to this last method is the fact that if one of the

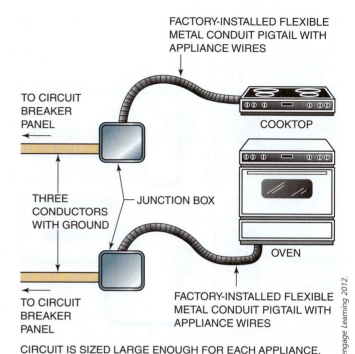

CIRCUIT IS SIZED LARGE ENOUGH FOR EACH APPLIANCE. USE THE NAMEPLATE RATING—*TABLE 220.55*, NOTE 4.

© Cengage Learning 2012.

FIGURE 14-6   A counter-mounted cooktop and wall-mounted oven installation using separate branch circuits to each unit.

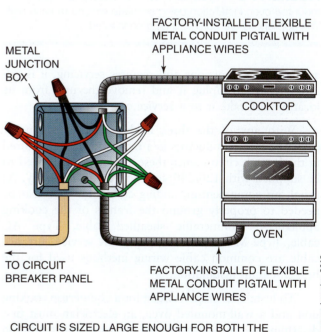

CIRCUIT IS SIZED LARGE ENOUGH FOR BOTH THE COOKTOP AND THE OVEN—*TABLE 220.55*, NOTE 4.

© Cengage Learning 2012.

FIGURE 14-7   A counter-mounted cooktop and wall-mounted oven installation with both appliances fed from the same branch circuit.

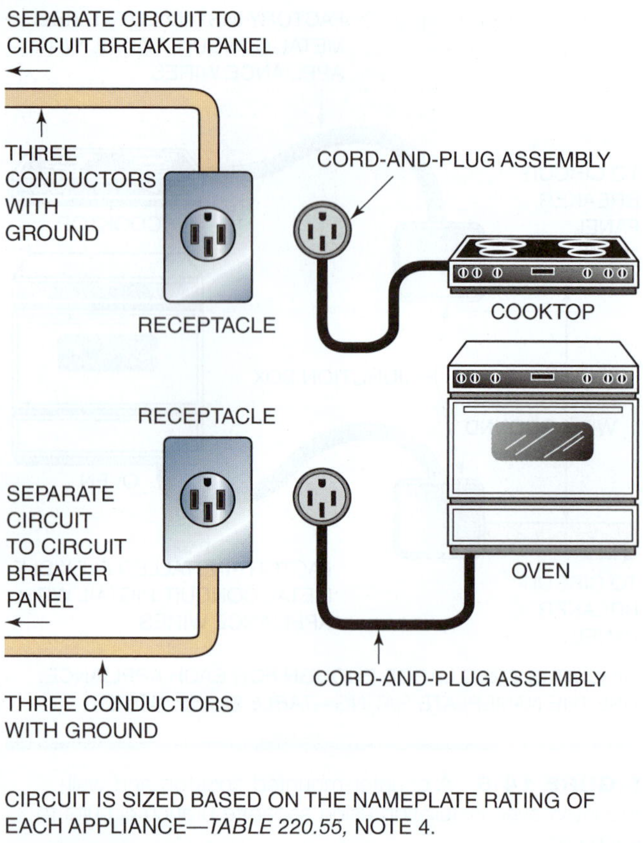

SEPARATE CIRCUIT TO
CIRCUIT BREAKER PANEL
←

THREE
CONDUCTORS
WITH
GROUND

CORD-AND-PLUG ASSEMBLY

RECEPTACLE

COOKTOP

RECEPTACLE

SEPARATE
CIRCUIT
TO CIRCUIT
BREAKER
PANEL

OVEN

CORD-AND-PLUG ASSEMBLY

THREE CONDUCTORS
WITH GROUND

CIRCUIT IS SIZED BASED ON THE NAMEPLATE RATING OF
EACH APPLIANCE—*TABLE 220.55*, NOTE 4.

© Cengage Learning 2012.

FIGURE 14-8 A counter-mounted cooktop and wall-mounted oven installation using separate circuits to each unit. The appliances are cord-and-plug connected.

appliances malfunctions and requires servicing, it is relatively easy to unplug it and remove the unit from its location and take it to a service center for repair.

With any of the three methods, three conductor cables or three conductors in a raceway will be required in the branch circuit since these appliances will need to be supplied with 120/240 volts to operate correctly. As usual, an equipment-grounding conductor will also be needed to properly ground the frames of the cooking appliances. Nonmetallic sheathed cable, Type AC cable, Type MC cable, and Type SER service entrance cable are common cable wiring methods used for the branch-circuit installation.

To install the branch circuit for a countertop cooking unit and a wall-mounted oven, an electrician must first determine the minimum size of the wire and maximum overcurrent protection device required for the installation method used. In Chapter 6, you were shown how to determine this information if one larger branch circuit is sized to feed both a countertop cooking unit and one or two wall-mounted ovens (see Figure 6-10 for a review of the calculation). When individual branch circuits are going to be installed for each cooking appliance,

Table 220.55, Note 4, of the *NEC®* tells us that the nameplate wattage rating of each appliance is used. This rating is divided by 240 volts to get the current rating of the appliance. It is this amperage that is used to size the branch-circuit conductors and the branch-circuit overcurrent protection device. Here is an example: A countertop cook unit with a nameplate rating of 6.6 kilowatts and a wall-mounted oven with a nameplate rating of 7.5 kilowatts is to be installed in a house with separate branch circuits run to each unit.

- Cooktop load in amps: 6600 watts/240 volts = 27.5 amps.

- Cooktop wire size: The minimum conductor size according to Table 310.15(B)(16) would be a 10 AWG copper.

- Cooktop fuse or circuit breaker size: The overcurrent protection device would be a 30-amp-rated fuse or circuit breaker.

- Wall-mounted oven load in amps: 7500 watts/240 volts = 31.25 amps.

- Wall-mounted oven wire size: The minimum conductor size according to Table 310.15(B)(16) would be an 8 AWG copper.

- Wall-mounted oven fuse or circuit breaker size: The overcurrent protection device would be a 40-amp-rated fuse or circuit breaker.

INSTALLING THE GARBAGE DISPOSAL BRANCH CIRCUIT

Most new and existing homes have a garbage disposal installed under the kitchen sink. It is up to the electrical contractor to install a branch circuit to feed the disposal unit and to make the electrical connections at the unit itself. It is normally the plumbing contractor who actually installs the garbage disposal under the sink. The branch circuit size and the overcurrent protection device are sized based on the nameplate rating of the disposal. Residential garbage disposals are rated at 120 volts and are normally installed on a separate 20-amp-rated branch circuit. The branch-circuit wire size will be 12 AWG copper.

CAUTION

CAUTION: Remember that appliances such as garbage disposals may not be connected to a small-appliance branch circuit supplying receptacles in a kitchen.

Two installation methods are used by electricians to connect electrical power to the garbage disposal. The first method involves running the wiring to a single-pole switch and then hardwiring directly to the garbage disposal. The switch is usually located above the kitchen countertop near the kitchen sink (Figure 14-9). The second method involves installing a receptacle outlet under the kitchen sink during the rough-in stage that will be controlled by a single-pole switch located above the countertop and near the sink (Figure 14-10).

A cord-and-plug assembly, which according to Section 422.16(B)(1) cannot be shorter than 18 inches (450 mm) and not longer than 3 feet (900 mm), is connected by the electrician to the disposal unit. The cord is plugged into the receptacle, and when the switch is ON, electrical power is supplied to the disposal (Figure 14-11). This method has one major advantage over the hardwired method in that if a garbage disposal malfunctions and requires servicing, it is a simple job to unplug it and remove the appliance for servicing.

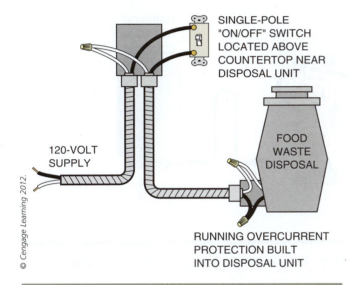

© Cengage Learning 2012.

FIGURE 14-9   The branch-circuit wiring for a hardwired garbage disposal operated by a separate switch located above the countertop and near the sink. The wiring method shown is Type AC cable (BX).

CAUTION

CAUTION: Electricians often field install flexible power cords with an attachment plug to garbage disposals, trash compactors, and dishwashers. They usually use a length of flexible cord that they buy from the local electrical supplier along with a NEMA 5-15P attachment plug. While this will work fine, it may not be allowed by the authority having jurisdiction because Section 422.16(B) states that the cord-and-plug assembly must be identified as suitable for the purpose *in the installation instructions that come with the appliance.* Most inspectors interpret this to mean that the only cord-and-plug assembly that can be used is one that is supplied by the appliance manufacturer either with the appliance or in a separate connection kit package.

Figure 14-12 shows a typical branch-circuit installation and switch connection for a garbage disposal. The electrical connections at the garbage disposal are not

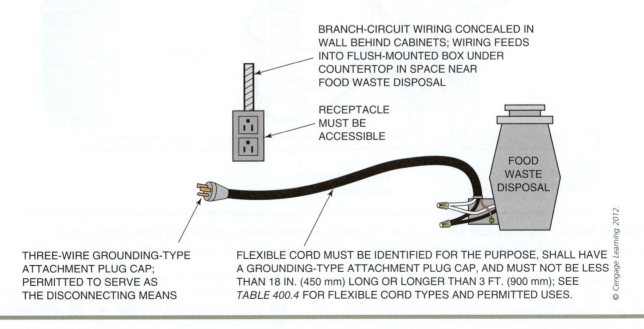

© Cengage Learning 2012.

FIGURE 14-10   A typical garbage disposal installation using a cord-and-plug assembly.

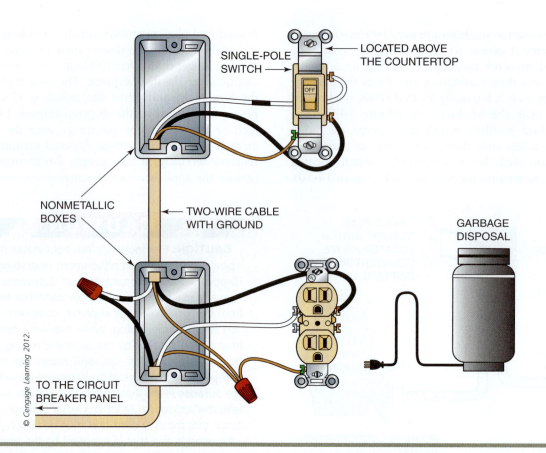

© Cengage Learning 2012.

FIGURE 14-11   The receptacle is controlled by the single-pole switch. The garbage disposal is connected to the receptacle with a cord-and-plug assembly.

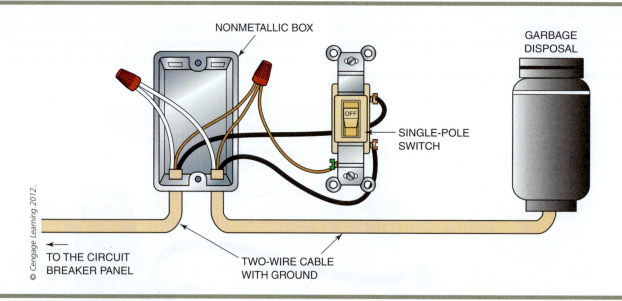

© Cengage Learning 2012.

FIGURE 14-12   The single-pole switch controls the garbage disposal, which is hardwired into the branch circuit.

very complicated. Assuming the use of a cable wiring method, the wiring connections in the disposal junction box are done as follows:

1 Loosen the screw that secures the blank cover to the junction box and take off the cover.

2 Strip the outside sheathing back about 6 inches and attach the cable to the disposal junction box through the ½-inch knockout located on the unit using an approved connector.

3 Connect the incoming equipment-grounding conductor to the green screw or green grounding jumper.

4 Wire-nut the white incoming grounded conductor to the white pigtail.

5 Wire-nut the black incoming ungrounded wire to the black pigtail.

6 Carefully push the wire connections into the junction box area and put the blank cover back on.

INSTALLING THE DISHWASHER BRANCH CIRCUIT

There are very few homes today that do not have an automatic built-in dishwasher installed. The plumbing contractor usually installs the dishwasher along with the necessary water and drain connections. It is up to the electrical contractor to supply and install a branch circuit to the dishwasher and make the necessary electrical connections at the dishwasher itself. Also, like a garbage disposal, there are two common ways to electrically connect a dishwasher: (1) hardwire it in the branch circuit or (2) use a cord-and-plug connection. It is this author's experience that hardwiring dishwashers is the most often used method, but you may find in your area that cord-and-plug connections are normally done.

from experience...

Trash compactors are often installed in homes. The branch-circuit installation and electrical connection procedure is the same for a trash compactor as it is for a dishwasher.

The branch-circuit size and the overcurrent protection device are sized based on the nameplate rating of the dishwasher. Residential dishwashers are rated at 120 volts and are normally installed on a separate 20-amp-rated branch circuit. The branch-circuit wire size will be 12 AWG copper. Nonmetallic sheathed cable, Type AC cable, or Type MC cable are common cable wiring methods used for the branch-circuit wiring.

When wiring the dishwasher, an electrician will install the branch-circuit wiring from the electrical panel to the dishwasher location. The wiring method used will be brought to the dishwasher location from underneath (through the basement) or from behind (through a wall). Each dishwasher will have a junction box on the appliance where (for the hardwired method) the branch-circuit wiring is terminated. During the rough-in stage of the electrical system installation, enough cable needs to be left at the dishwasher location so that when the kitchen cabinets are finally installed and the dishwasher is moved into its final position, there will be enough wire to make the necessary connections. The electrical connections will be made in the dishwasher junction box. This junction box is normally located on the right side at the front of the appliance. The bottom "skirt" on the dishwasher will have to be removed to gain access to the junction box. Assuming the use of a cable wiring method, the wiring connections are done as follows (Figure 14-13):

1 Take off the junction box cover and take out the ½-inch knockout on the junction box.

2 Attach the wiring method used to the dishwasher junction box through the ½-inch knockout with an approved connector.

3 Connect the incoming equipment-grounding conductor to the green screw or green grounding jumper.

4 Wire-nut the white incoming grounded conductor to the white pigtail.

5 Wire-nut the black incoming ungrounded wire to the black pigtail.

6 Carefully push the wire connections into the junction box area and put the blank cover back on.

If a dishwasher is to be cord-and-plug connected, the installation is similar to that of a cord-and-plug-connected garbage disposal. The difference is that there is no need to switch the receptacle that the dishwasher cord will be plugged into (Figure 14-14). In this method, an electrician will install branch-circuit wiring to a receptacle outlet box at a location that is adjacent to the location of the dishwasher. It is not a good idea to locate the receptacle behind the dishwasher location since it will be virtually impossible to access it once the dishwasher is installed. Also, locating the receptacle in a basement below the dishwasher location is not allowed because the cord-and-plug connection is the disconnecting means for

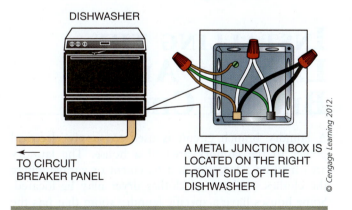

DISHWASHER

TO CIRCUIT BREAKER PANEL

A METAL JUNCTION BOX IS LOCATED ON THE RIGHT FRONT SIDE OF THE DISHWASHER

© Cengage Learning 2012.

FIGURE 14-13 A hardwired dishwasher branch circuit.

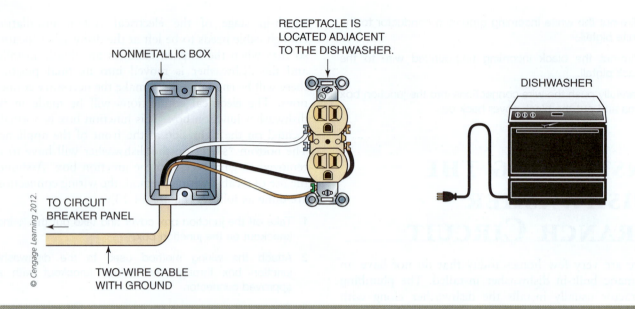

NONMETALLIC BOX

RECEPTACLE IS
LOCATED ADJACENT
TO THE DISHWASHER.

DISHWASHER

TO CIRCUIT
BREAKER PANEL

TWO-WIRE CABLE
WITH GROUND

© Cengage Learning 2012.

FIGURE 14-14   A cord-and-plug-connected dishwasher branch circuit.

the dishwasher and is not located within sight of the appliance. The cord-and-plug assembly must be at least 3 feet (900 mm) long and no more than 4 feet (1.2 m) long according to Section 422.16(B)(2).

> ## from experience...
>
> Receptacles located under the kitchen cabinets for cord-and-plug connections for garbage disposals or dishwashers do not need to be GFCI (ground fault circuit interrupter) protected because they do not serve the kitchen countertop.

INSTALLING THE LAUNDRY AREA BRANCH CIRCUITS

A laundry branch circuit is used to supply electrical power to the laundry area of a house. The laundry area may be a location in a basement or garage where the clothes washer and clothes dryer may be located. Some houses have a specific laundry room that has the clothes washer, clothes dryer, and other appliances that assist the home owner when doing laundry.

Section 210.11(C)(2) of the *NEC®* requires the laundry branch circuit to be rated at 20 amps, which results in wiring it with 12 AWG wire. The laundry circuit may consist of only one receptacle outlet, such as when the clothes washer is located in the basement of a home and requires a receptacle to plug into for power. It can also consist of several receptacles located in a specific laundry room (Figure 14-15). In either case, no other outlets (lighting or receptacle) can be connected to the 20-amp laundry circuit.

> ## CAUTION
>
> **CAUTION:** Many electricians install a single receptacle at the outlet box when they install a 20-amp laundry circuit and the only thing served is a clothes washer. Because the branch-circuit rating is 20 amps, the rating of the single receptacle must be a 20-amp-rated receptacle (NEMA 5-20R). It is interesting to note that if the electrician chooses to install a duplex receptacle at the outlet box, either a 15-amp (NEMA 5-15R) or a 20-amp (NEMA 5-20R) duplex receptacle can be used. These rules also apply when 20-amp individual branch circuits are installed for other appliance types such as dishwashers, trash compactors, and garbage disposals.

An electrician will usually start the rough-in wiring of a laundry circuit from the main panel or subpanel and then route the wiring to the laundry area. Section 210.50(C) of the *NEC®* requires that the receptacle for the clothes washer be located no more than 6 feet

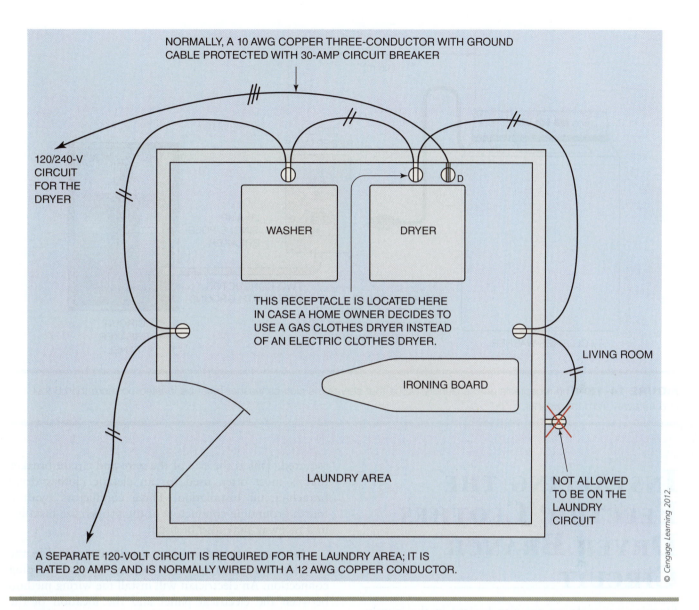

NORMALLY, A 10 AWG COPPER THREE-CONDUCTOR WITH GROUND CABLE PROTECTED WITH 30-AMP CIRCUIT BREAKER

120/240-V CIRCUIT FOR THE DRYER

WASHER

DRYER

THIS RECEPTACLE IS LOCATED HERE IN CASE A HOME OWNER DECIDES TO USE A GAS CLOTHES DRYER INSTEAD OF AN ELECTRIC CLOTHES DRYER.

IRONING BOARD

LIVING ROOM

LAUNDRY AREA

NOT ALLOWED TO BE ON THE LAUNDRY CIRCUIT

A SEPARATE 120-VOLT CIRCUIT IS REQUIRED FOR THE LAUNDRY AREA; IT IS RATED 20 AMPS AND IS NORMALLY WIRED WITH A 12 AWG COPPER CONDUCTOR.

© Cengage Learning 2012.

FIGURE 14-15 The laundry area in a house must be served by at least one 20-amp-rated laundry circuit. Only receptacles located in the laundry area can be connected to the circuit. An electric clothes dryer requires a separate 120/240-volt branch circuit.

(1.8 m) from the appliance because most clothes washer cord-and-plug assemblies are no more than 6 feet (1.8 m) long (Figure 14-16). If the clothes washer is located in a basement or garage area and is the only laundry appliance to be plugged into a receptacle, a surface-mounted box is often installed a little higher and just behind the location of the clothes washer, and the wiring method used is brought to that box. A single receptacle is recommended for this situation. If a laundry room is used, a determination of how many receptacle outlets required in the room is done and the wiring is roughed-in from box to box in that room. Remember that in a laundry room, regular

receptacle spacing is not required. In other words, a laundry room can have receptacle outlets located so there is more than 12 feet between them if an electrician so chooses.

> ### CAUTION
>
> **CAUTION:** If the laundry area is located in an unfinished basement or a bathroom and a receptacle is used to serve the clothes washing machine, it must be GFCI protected.

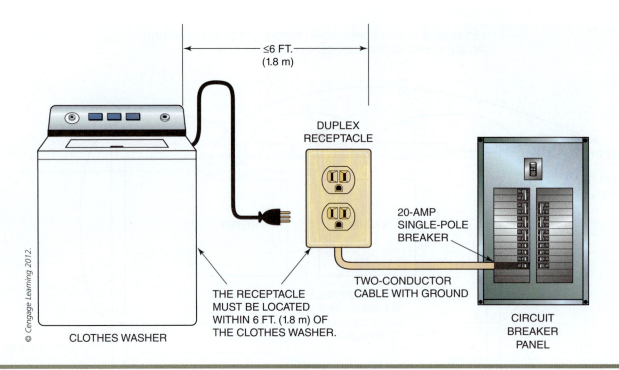

≤6 FT.
(1.8 m)

DUPLEX
RECEPTACLE

20-AMP
SINGLE-POLE
BREAKER

TWO-CONDUCTOR
CABLE WITH GROUND

CIRCUIT
BREAKER
PANEL

THE RECEPTACLE
MUST BE LOCATED
WITHIN 6 FT. (1.8 m) OF
THE CLOTHES WASHER.

CLOTHES WASHER

© Cengage Learning 2012.

FIGURE 14-16 The receptacle on the laundry circuit that serves the clothes washer must be located no more than 6 feet (1.8 m) away from the washer.

INSTALLING THE ELECTRIC CLOTHES DRYER BRANCH CIRCUIT

Electric clothes dryers are also located in the laundry area of a house. However, they are not supplied by the 20-ampere laundry circuit discussed in the previous section. Instead, they are supplied by an individual branch circuit (see Figure 14-15) that is installed similar to the branch circuit for an electric range.

To install the branch circuit for an electric clothes dryer, an electrician must first determine the minimum size of the wire and maximum overcurrent protection device required. The branch circuit will be rated at 120/240 volts and will need a conductor size that will handle, at a minimum, the calculated load. The nameplate of an electric clothes dryer will give the load in watts. An electrician will simply need to divide the wattage rating by 240 volts to get the electric clothes dryer load in amperes. For example, an electric clothes dryer has a nameplate that says it is rated at 5.5 kilowatts. Therefore, 5500 watts/240 volts = 23 amps. In this example, a 10/3 copper cable with a grounding conductor and protected by a 30-ampere circuit breaker is

required. This is the size of the wire and circuit breaker that is most often used for an electric clothes dryer branch-circuit installation. Three conductors (with a fourth grounding conductor) of the proper size installed in a raceway may also be used.

Electric clothes dryers are usually connected to the electrical system in a house through a cord-and-plug connection. An electrician will install the wiring method between the electrical panel and the location of the receptacle for the electric clothes dryer. The receptacle location is usually behind the electric clothes dryer and slightly higher than the top of the electric clothes dryer for easy access. The receptacle will either be surface-mounted or flush-mounted. Flush mounting the receptacle will require an electrical box. The wiring method is attached to either the electrical box for flush mounting or the receptacle body itself for surface mounting. Assuming the use of a cable wiring method, the wires of the branch circuit will be connected to the receptacle terminals as follows (Figure 14-17):

1 Connect the white grounded neutral conductor to the terminal marked "W."

2 Connect the black ungrounded conductor to the terminal marked "Y."

3 Connect the red ungrounded conductor to the terminal marked "X."

4 Connect the green or bare equipment-grounding conductor to the terminal marked "G."

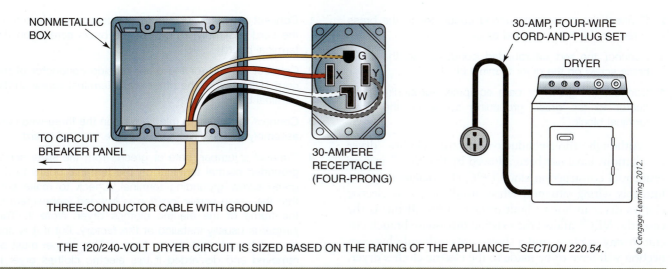

THE 120/240-VOLT DRYER CIRCUIT IS SIZED BASED ON THE RATING OF THE APPLIANCE—*SECTION 220.54.*

© Cengage Learning 2012.

FIGURE 14-17   A cord-and-plug-connected clothes dryer branch-circuit installation. A four-wire connection is shown.

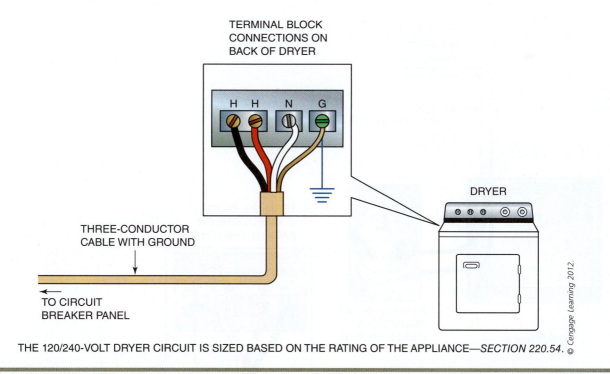

THE 120/240-VOLT DRYER CIRCUIT IS SIZED BASED ON THE RATING OF THE APPLIANCE—*SECTION 220.54.*

© Cengage Learning 2012.

FIGURE 14-18   A hardwired clothes dryer branch-circuit installation. A four-wire connection is shown.

The cord-and-plug assembly that connects the electric clothes dryer to the receptacle is often not installed by the appliance dealer. If this is the case, an electrician will have to install it. Make sure that the cord is a four-wire cord and has a plug that will properly match the receptacle that it is to be plugged into. Follow the manufacturer's instructions, often found on the back of the electric clothes dryer, for the correct wiring connections for the cord to the terminal block at the back of the electric clothes dryer.

Sometimes the electrical installation calls for the connection to an electric clothes dryer to be hardwired. This is done by bringing the branch-circuit wiring method directly to the back of the electric clothes dryer and making the terminal connections on the terminal block on the electric clothes dryer. Assuming the use of a 10/3 cable wiring method, the connections are made as follows (Figure 14-18):

1 Connect the white grounded neutral conductor to the silver screw terminal on the dryer terminal block.

2 Connect the black ungrounded conductor to the brass screw on the dryer terminal block.

3 Connect the red ungrounded conductor to the other brass screw on the dryer terminal block.

4 Connect the green or bare equipment-grounding conductor to the green grounding screw on the dryer terminal block.

Although three-conductor electric clothes dryer installations have not been allowed by the *NEC®* in new residential construction since 1996, electricians may run into this wiring situation when installing a new electric clothes dryer in homes built prior to 1996. If this is the case, the *NEC®* allows the existing three-wire branch circuit to stay in place, and a three-wire cord-and-plug connection will have to be made to the electric clothes dryer. The connections for attaching the three-wire cord to the electric clothes dryer are made as follows (Figure 14-19):

1 Connect the black (or one of the outside conductors of the cord assembly) to one of the brass screws on the terminal board.

2 Connect the red (or the other outside conductor of the cord assembly) to the other brass terminal screw on the terminal board.

3 Connect the white (or middle wire on the three-wire cord assembly) to the silver screw on the terminal board.

4 Connect a jumper, bare or green, from the silver screw grounded neutral terminal on the terminal board to the green screw grounding terminal. Check to make sure that the green screw location is bonded (connected) to the frame of the electric clothes dryer. *Note 1:* This jumper is usually installed at the factory, but if it is not, you will have to install one. *Note 2:* This jumper must be removed and discarded if this electric clothes dryer is installed on a four-wire system in a new installation and the jumper was installed at the factory.

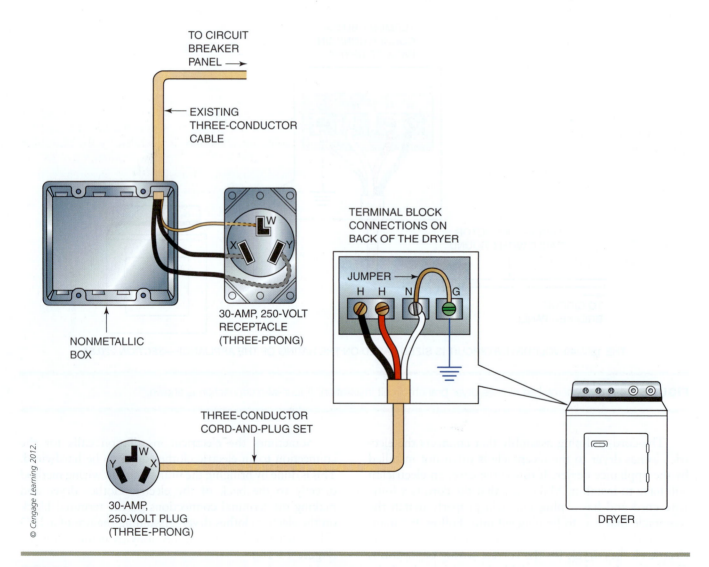

TO CIRCUIT BREAKER PANEL →

EXISTING THREE-CONDUCTOR CABLE

NONMETALLIC BOX

30-AMP, 250-VOLT RECEPTACLE (THREE-PRONG)

TERMINAL BLOCK CONNECTIONS ON BACK OF THE DRYER

JUMPER →

THREE-CONDUCTOR CORD-AND-PLUG SET

30-AMP, 250-VOLT PLUG (THREE-PRONG)

DRYER

© Cengage Learning 2012.

FIGURE 14-19 A three-wire dryer branch-circuit installation.

INSTALLING THE BATHROOM BRANCH CIRCUIT

A bathroom is defined by the *NEC®* as an area that has a basin and one or more toilets, urinals, tubs, foot baths, bidets, or showers. A bathroom circuit is a circuit that supplies electrical power to this area in a residential installation. The *NEC®* requires at least one 20-amp-rated branch circuit be installed to supply the bathroom receptacle or receptacles. You will remember that the *NEC®* requires only one receptacle in a bathroom and that it must be located no more than 3 feet (900 mm) from the edge of the basin. Instead of installing the receptacle on a wall or partition adjacent to the basin, it may be installed on the side or face of the basin cabinet, as long as it is located no more than 12 inches (300 mm) below the countertop. Additionally, each 15- and 20-amp 125-volt-rated receptacle in a bathroom must be GFCI protected (Figure 14-20).

The *NEC®* does allow one 20-amp-rated branch circuit to supply two (or more) bathrooms if certain conditions are met. These conditions were covered in Chapter 6 (see Figure 6-6 for a review). When roughing in bathroom branch circuits, there are a few steps to always follow:

1 Use 12 AWG wire and 20-amp overcurrent protection devices for all bathroom branch circuits.

2 Install at least one bathroom branch circuit for each bathroom.

3 Determine the number of receptacle outlets to be included on each bathroom branch circuit.

4 If they have not been installed previously, install the required lighting outlet and receptacle outlet boxes at the appropriate locations in the bathroom.

5 Starting at the electrical panel where the circuit overcurrent protection device is located, install the wiring to the first electrical box on the circuit. Be sure to mark the cable or other wiring method with the circuit number and type and the area served at the panel. For example, bathroom branch circuit 1 serving the first-floor bathroom can be marked "BATH1FF."

6 Continue roughing in the wiring for the rest of the circuit, making sure to follow the installation practices described in previous chapters. Make sure to tuck the wiring in the back of all electrical boxes on the circuit.

7 Double check to make sure all circuit wiring for the bathroom branch circuit you are installing has been roughed in.

INSTALLING A WATER PUMP BRANCH CIRCUIT

Houses built in a town or city generally get their supply of water from a municipal water system. The water-supply piping for the house is installed by the plumbing contractor, and the electricians have no circuitry to run for the water system. However, in rural areas where water system piping is not available, a well is used to supply water to the house. Getting the water from the well to the house will require a pump. It is usually up to the electricians wiring the house to supply a correctly sized water pump circuit to the control equipment for the pump. It is usually the plumbing contractor who, along with installing the necessary water piping throughout the house, will actually install the pump (with attached electric motor) and the control equipment used with the pump and water system. There are two types of water pumps commonly used: the deep-well jet pump and the submersible pump. In this section, we take a look at some common characteristics of each and what kind of electrical wiring must be done for these pumps.

JET PUMP

The jet pump and the electric motor that turns it are usually located in a basement or crawl-space area, although other locations may be used. The electric motors are usually around 1 horsepower, single phase, capacitor-start, and operate on either 115 or 230 volts. (*Note:* Electric motors use the voltage designations of 115 and 230 volts instead of 120 and 240 volts.)

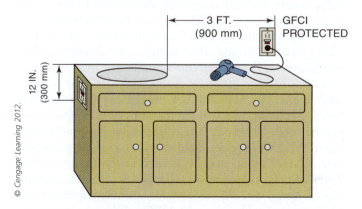

FIGURE 14-20 A bathroom receptacle must be located adjacent to the basin and no more than 3 feet (900 mm) away. It could also be installed in the front or side of the cabinet as long as it is no more than 12 inches (300 mm) below the countertop.

© Cengage Learning 2012.

Capacitor-start electrical motors are designed to produce a high starting torque that will be enough to get the pump motor started. A 115/230-volt motor is called a dual-voltage electric motor and can operate at the same speed and produce the same horsepower at either voltage. When the motor is connected at the higher voltage (230 volts), the motor draws half as much current as when it is connected to operate on the lower voltage (115 volts). For this reason, it is common wiring practice to connect water pump motors to the higher voltage. You should remember that at the higher voltage of 230 volts (actually 240 volts from the circuit breaker panel), a two-pole circuit breaker is required.

The size of the circuit breaker and the circuit conductors is determined by the current draw of the motor. Article 430 of the *NEC®* tells us how to size the circuit conductors, the circuit breaker or fuses protecting the branch circuit, and the overload protection for the motor itself. Figure 14-21 shows a typical electrical circuit for a jet pump. The illustration shows the circuit wiring originating in a circuit breaker panel, going to the control equipment for the pump motor and then to the pump motor itself. In this example, the pump motor will be a 1-horsepower, single-phase motor connected at

Table 430.248 Full-Load Currents in Amperes, Single-Phase Alternating-Current Motors

The following values of full-load currents are for motors running at usual speeds and motors with normal torque characteristics. The voltages listed are rated motor voltages. The currents listed shall be permitted for system voltage ranges of 110 to 120 and 220 to 240 volts.

Horsepower	115 Volts	200 Volts	208 Volts	230 Volts
1/6	4.4	2.5	2.4	2.2
1/4	5.8	3.3	3.2	2.9
1/3	7.2	4.1	4.0	3.6
1/2	9.8	5.6	5.4	4.9
3/4	13.8	7.9	7.6	6.9
1	16	9.2	8.8	8.0
1 1/2	20	11.5	11.0	10
2	24	13.8	13.2	12
3	34	19.6	18.7	17
5	56	32.2	30.8	28
7 1/2	80	46.0	44.0	40
10	100	57.5	55.0	50

FIGURE 14-22   Table 430.248 of the *NEC®*. *Reprinted with permission from NFPA 70®, National Electric Code®, Copyright © 2010, National Fire Protection Association, Quincy, MA. This reprinted material is not the complete and official position of the NFPA on the referenced subject, which is represented only by the standard in its entirety.*

230 volts. When calculating the size of branch-circuit conductors and the size of the fuse or circuit breaker for the circuit, Section 430.6 requires electricians to look at Table 430.248 (Figure 14-22). This table lets us determine the full-load current of a single-phase motor. For a 1-horsepower motor connected at 230 volts, Table 430.248 says that the full-load current is 8 amperes. Section 430.22 requires electricians to now multiply the full-load current of the motor by 125 percent and use that number to determine the minimum-size branch-circuit conductor. In this case, 8 amps × 125% = 10 amps. Table 310.15(B)(16) is now used to size the minimum-size branch circuit. Using the "60°C" column, you will see that 14 AWG copper wire can be used. However, it is common wiring practice to install 12 AWG copper conductors to a water pump in case a larger pump may be needed in the future as the water supply requirements change. So, an electrician would install a 12/2 nonmetallic sheathed cable or a 12/2 armored cable or run 12 AWG conductors in a raceway system to the pump motor. The size of the circuit breaker or fuse is determined by following Section 430.52. This section tells us to take a look at Table 430.52 (Figure 14-23) and to multiply the full-load current that we got from Table 430.248 by the percentage found in Table 430.52 for a specific type of overcurrent

120/240-VOLT
CIRCUIT BREAKER
PANEL

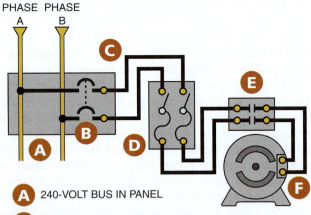

A   240-VOLT BUS IN PANEL

B   20-AMPERE, TWO-POLE CIRCUIT BREAKER

C   12 AWG CONDUCTORS

D   TWO-POLE DISCONNECT SWITCH AND MOTOR OVERLOAD PROTECTION

E   TWO-POLE PRESSURE SWITCH

F   1-HORSEPOWER, 230-VOLT, SINGLE-PHASE MOTOR CONNECTED TO THE PUMP

© Cengage Learning 2012.

FIGURE 14-21   A typical electrical installation for a jet pump.

Table 430.52 Maximum Rating or Setting of Motor Branch-Circuit Short-Circuit and Ground-Fault Protective Devices

Type of Motor	Percentage of Full-Load Current			
	Nontime Delay Fuse[1]	Dual Element (Time-Delay) Fuse[1]	Instantaneous Trip Breaker	Inverse Time Breaker[2]
Single-phase motors	300	175	800	250
AC polyphase motors other than wound-motor	300	175	800	250
Squirrel cage — other than Design B energy-efficient	300	175	800	250
Design B energy-efficient	300	175	1100	250
Synchronous[3]	300	175	800	250
Wound rotor	150	150	800	150
Direct current (constant voltage)	150	150	250	150

Note: For certain exceptions to the values specified, see 430.54.
[1]The values in the Nontime Delay Fuse column apply to Time-Delay Class CC fuses.
[2]The values given in the last column also cover the ratings of nonadjustable inverse time types of circuit breakers that may be modified as in 430.53(C) (1), Exception No. 1 and No. 2.
[3]Synchronous motors of the low-torque, low-speed type (usually 450 rpm or lower), such as are used to drive reciprocating compressors, pumps, and so forth, that start unloaded, do not require a fuse rating or circuit-breaker setting in excess of 200 percent of full-load current.

FIGURE 14-23 Table 430.52 of the *NEC®*. *Reprinted with permission from NFPA 70®, National Electric Code®, Copyright © 2010, National Fire Protection Association, Quincy, MA. This reprinted material is not the complete and official position of the NFPA on the referenced subject, which is represented only by the standard in its entirety.*

protection device. We will limit our discussion to using either **dual-element time-delay fuses** or **inverse time circuit breakers** (the type of circuit breaker used in residential electrical panels), as these two overcurrent device types are commonly used in residential electrical system installation. The calculations are done as follows:

- Dual-element time-delay fuses: 175% × 8 amperes (from Table 430.248) = 14 amps.

- Because 14 amps is not a standard fuse size [see Section 240.6(A)], the maximum-size dual-element time-delay fuse we can use is 15 amps.

and

- Inverse time circuit breaker: 250% × 8 amps (from Table 430.248) = 20 amps.

- Because 20 amps is a standard size of circuit breaker [see Section 240.6(A)], the maximum-size inverse time circuit breaker we can use is 20 amps.

Overload protection for an electric motor is required by Article 430 so that the motor will not overheat from dangerous current overload levels and burn up. The overload protection can be thermal overloads (commonly called "heaters"), electronic sensing devices, thermal devices built into the motor, or time-delay fuses. Section 430.32 requires the vast majority of motor applications to be protected by overload devices sized not more than 125 percent of the motor's nameplate full-load ampere rating, not the full-load current rating from Table 430.248. In this example, we assume that the full-load ampere rating listed on the motor's nameplate is the same as the full-load current rating found in Table 430.248. Therefore, 125% × 8 amps (full-load ampere rating from the nameplate) = 10 amps. This is the maximum size in amperes of the motor overload protection device. In the "real world," the overload protection devices normally are sized at the factory and are included in the controller equipment that comes with the pump.

Provided that the pump is in sight of the enclosure that contains the fuse or circuit breaker protecting the water pump branch circuit, no additional disconnecting means is required next to the pump. An electrician will install a branch circuit from the electrical panel to the controller or, if the controller has already been connected to the pump circuit by the pump installer, to the two-pole pressure switch. If the pump motor is located in another room, or is more than 50 feet (15 m) away from the fuse or circuit breaker enclosure, or is just not in sight of the enclosure, a disconnecting means must be installed and located next to the pump.

SUBMERSIBLE PUMP

A submersible pump is a pump type where both the electric motor and the pump are enclosed in the same housing. The housing is lowered into a well casing and placed below the water level in the well. The standard parts of a submersible pump system are shown in Figure 14-24.

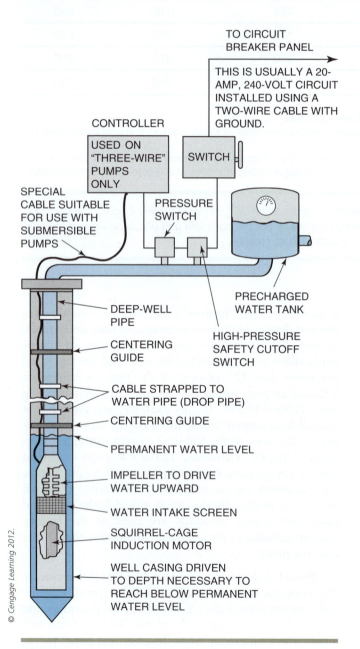

TO CIRCUIT BREAKER PANEL

THIS IS USUALLY A 20-AMP, 240-VOLT CIRCUIT INSTALLED USING A TWO-WIRE CABLE WITH GROUND.

CONTROLLER

USED ON "THREE-WIRE" PUMPS ONLY

SWITCH

SPECIAL CABLE SUITABLE FOR USE WITH SUBMERSIBLE PUMPS

PRESSURE SWITCH

DEEP-WELL PIPE

CENTERING GUIDE

CABLE STRAPPED TO WATER PIPE (DROP PIPE)

CENTERING GUIDE

PERMANENT WATER LEVEL

IMPELLER TO DRIVE WATER UPWARD

WATER INTAKE SCREEN

SQUIRREL-CAGE INDUCTION MOTOR

WELL CASING DRIVEN TO DEPTH NECESSARY TO REACH BELOW PERMANENT WATER LEVEL

PRECHARGED WATER TANK

HIGH-PRESSURE SAFETY CUTOFF SWITCH

© Cengage Learning 2012.

FIGURE 14-24 A typical submersible pump installation.

Many submersible pumps are of the "two-wire" type, and the housing contains all the parts of a motor starter and overload protection. This means there will be no need to have an aboveground controller box. Other submersible pump types, called "three-wire" pumps, will require a controller to be installed above the ground next to the pre-charged water tank. The controller will contain items such as a starting relay, overload protection, starting and running capacitors, and the necessary connection terminals for connection of the pump branch-circuit wiring.

The calculations for sizing the branch-circuit conductor size, the fuse or circuit breaker size, and the overload protection size are the same as for the jet pump. One of the differences in wiring a submersible pump versus a jet pump is that a special submersible pump cable is used. It is buried in the ground, usually in the same trench with the incoming water pipe from the well. The cable is attached to the submersible pump, and electrical connections are made, then the pump is lowered down the well casing and into the water. It is the submersible pump installer who makes the connection of the submersible pump cable to the submersible pump and lowers it all into the well casing. The other end of the submersible pump cable is then left at the location of the pressure switch or the controller for the proper electrical connections to be made later. The size of the submersible pump cable is often larger than the calculated minimum size to overcome voltage drop problems that may be encountered because of the long distances from the house to the actual location of the submersible pump. It is not uncommon for the pump installer to do most, if not all, of the wiring to the controller (if there is one) and to the load side of the pressure switch. The electrician often is required only to bring a branch circuit from the fuse or circuit breaker enclosure to the pressure switch, where the proper connections are made on the line side. A submersible pump usually operates on 240 volts, and, as a result, an electrician will install a two-wire cable or run two wires in a raceway to the controller or pressure switch location. The conductor size is often 10 AWG with a two-pole, 30-amp circuit breaker, but some installations can get by with a 12 AWG conductor and a two-pole, 20-amp circuit breaker.

Similar to a jet pump, if the controller is in sight of the enclosure that contains the fuse or circuit breaker protecting the water pump branch circuit, no additional disconnecting means is required next to the pump. If the pump controller motor is located in another room, or is more than 50 feet (15 m) away from the fuse or circuit breaker enclosure, or is just not in sight of the enclosure, a disconnecting means must be installed and located next to the controller location.

INSTALLING AN ELECTRIC WATER HEATER BRANCH CIRCUIT

Electric water heaters are quite common and are used to supply a house with domestic hot water. Residential electric hot water heaters come in several sizes, including 30, 40, 50, 80, 100, and 120 gallons. Electric water heaters used in homes usually operate on 240 volts and normally will require a 10 AWG conductor size with a 30-ampere overcurrent protection device. Some smaller electrical water heaters may require 120 volts and will be wired with a 12 AWG conductor with a 20-ampere overcurrent protection device.

Section 422.13 of the *NEC*® states that a fixed-storage-type electric water heater of 120 gallons or less is considered to be a continuous load. Section 422.10(A) requires a continuously loaded appliance to have a branch-circuit rating not less than 125 percent of the nameplate rating. You will remember that the rating of the branch circuit is based on the size of the overcurrent protection device. To calculate the size of the overcurrent protection device, refer to Section 422.11(E), which says to size the overcurrent device no larger than the protective device rating on the nameplate. If the water heater has no size listed on the nameplate, use the following:

• If the water heater current rating does not exceed 13.3 amps, use a 20-amp overcurrent protective device.

• If the water heater draws more than 13.3 amps, use 150 percent of the nameplate rating. If this rating does not correspond with a standard rating shown in Section 240.6(A), go to the next higher standard size.

Let us look at an example: A water heater nameplate shows a rating of 4800 watts at 240 volts. What minimum-size conductor and maximum-size overcurrent protection device are required?

1 Use the formula $I = P/E$, or 4800 watts/240 volts = 20 amps.

2 Section 422.10(A) requires the branch-circuit rating to not be less than 125 percent of the water heater's nameplate rating. So, 20 amps × 125% = 25 amps.

3 Therefore, a 25-amp fuse or circuit breaker [Section 240.6(A)] would be the minimum size used and results in a 25-amp rating for the water heater branch circuit.

4 Because the water heater draws more than 13.3 amps, the maximum size overcurrent protective device would be 20 amps × 150% = 30 amps.

5 Using Table 310.15(B)(16), we would use a 10 AWG conductor and a 30-amp fuse or circuit breaker for the installation of this electric water heater.

The individual branch circuit to an electric water heater is usually made directly from the overcurrent protection device in an electrical panel to the junction box or terminal block on the appliance. The overcurrent protection device in the panel will satisfy Section 422.30, which requires a disconnecting means for the water heater appliance. However, if the overcurrent protection device in the panel is not visible from the location of the water heater or is more than 50 feet (15 m) away, a separate disconnect must be installed within 50 feet (15 m) and within sight of the water heater. This could be a two-pole toggle switch with the proper voltage and ampere rating or a non-fusible safety switch type of disconnect. Assuming the use of a 10/2 nonmetallic sheathed cable wiring method, the wiring connections are done as follows (Figure 14-25):

1 Take off the junction box cover and take out the knock-out on the junction box.

2 Attach the wiring method used to the water heater junction box through the knockout with an approved connector.

3 Connect the incoming equipment-grounding conductor to the green screw or green grounding jumper.

4 Wire-nut the white incoming conductor to one of the black pigtail wires in the junction box. Reidentify the white wire as a "hot" conductor with black tape.

5 Wire-nut the black incoming ungrounded wire to the other black pigtail.

6 Carefully push the wire connections into the junction box area and put the blank cover back on.

Note: Some electric water heaters may have a terminal block that is used to terminate the branch-circuit conductors.

Section 400.7(A)(8) and Section 422.16(A) of the *NEC*® describe the uses for flexible cords. Flexible cords can be used only where the appliance is designed for ready removal for maintenance or repair and the appliance is identified for use with flexible cord connections. Electric water heaters are usually not allowed to be cord-and-plug connected because they are not considered portable and cannot be easily removed for maintenance and repair. However, if an electric water heater has been connected to the water piping system with connections that are designed so the water heater can be quickly and easily removed, a flexible cord-and-plug electrical connection may be allowed. Check with the authority having jurisdiction (AHJ) in your area to make sure that a flexible cord-and-plug connection to an electric water heater is allowed. It is this author's experience that electric water heaters are normally hard-wired and are very seldom connected with a flexible cord and plug.

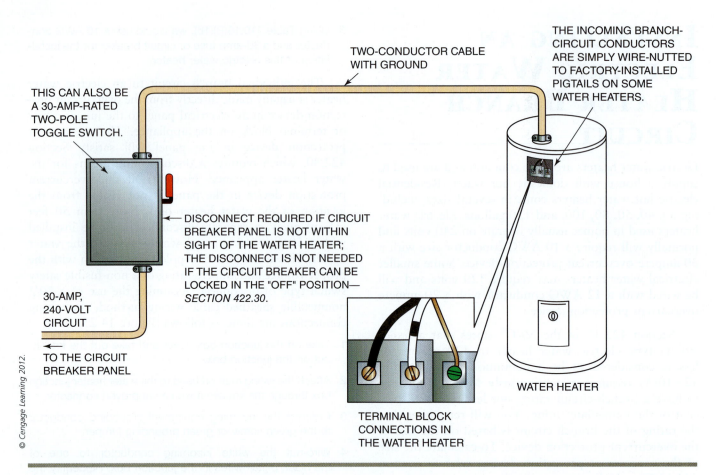

THE INCOMING BRANCH-CIRCUIT CONDUCTORS ARE SIMPLY WIRE-NUTTED TO FACTORY-INSTALLED PIGTAILS ON SOME WATER HEATERS.

TWO-CONDUCTOR CABLE WITH GROUND

THIS CAN ALSO BE A 30-AMP-RATED TWO-POLE TOGGLE SWITCH.

DISCONNECT REQUIRED IF CIRCUIT BREAKER PANEL IS NOT WITHIN SIGHT OF THE WATER HEATER; THE DISCONNECT IS NOT NEEDED IF THE CIRCUIT BREAKER CAN BE LOCKED IN THE "OFF" POSITION— *SECTION 422.30.*

30-AMP, 240-VOLT CIRCUIT

TO THE CIRCUIT BREAKER PANEL

TERMINAL BLOCK CONNECTIONS IN THE WATER HEATER

WATER HEATER

© Cengage Learning 2012.

FIGURE 14-25 A typical water heater branch-circuit installation.

from experience...

Heating hot water only when it is needed with a tankless electric hot water heater can lower household energy bills by as much as 40%. While tankless units may require a larger initial investment than traditional tank units, the higher initial cost is offset by long-term benefits and cost savings. Tankless electric hot water heaters require relatively large amounts of electrical power to operate. Most installations will require wiring and circuit breakers that will allow from 50- to 150-amps at 240-volts, depending on the size of the tankless electric hot water heater. Always consult the manufacturer's installation manual when installing a tankless electric hot water heater.

INSTALLING BRANCH CIRCUITS FOR ELECTRIC HEATING

There are many different styles of electric heating available to a home owner, and electricians should be familiar with the most common types and installation techniques. Electric furnaces used to heat an entire house, individually controlled baseboard electric heaters used to heat specific rooms, and unit heaters used to heat a specific area like a basement or garage are the most common types installed. Other types of electric heating are beyond the scope of this book.

Electric heat has many advantages. It is very convenient to control because thermostats are easily installed in each room. It is considered safer than using fossil fuels like oil or gas because there is no storage of fuel on the premises and the explosion and fire hazard is certainly not as great. Electric heat is very quiet since there are few if any moving parts. It is relatively inexpensive and easy to install. Once it is installed, there is little maintenance required to keep it working. One last advantage to consider is that no chimney is required to exhaust the combustion gasses into the outside air. This results in a more economical heating installation because the home owner does not have to pay for a chimney installation; it is also an environmentally cleaner heating system. You would think that with all the advantages of electric heat, the result would be electric heating installations in all homes. However, one big disadvantage of heating with electricity is its relatively high cost to operate, which severely limits the electric heating installations done by electricians. In some parts of the country that have lower electricity rates, you will see an increase in the number of installations of electric heat. In those areas of the country where electricity rates are high, you will not see as many electric heat installations. For example, in the northeastern area of the country, electricians almost never install electric heating. In the Pacific Northwest, electric heating is very popular, and electricians install electric heating systems on a regular basis.

from experience...

There are several ways to calculate the amount of electric heating you need to install. The simplest way is to use 10 watts per square foot for each heated area in a house. This calculation assumes a ceiling height of 8 feet. For example, a living room that is 20 feet × 20 feet has a total square foot area of 400 square feet. 400 square feet × 10 watts per square foot results in the need for 4000 watts of electric heating for that room. If the electric baseboard heater units you are installing are rated for 2000 watts each, you would need to install two units in this room (4000 watts/2000 watts per unit = 2 baseboard units).

When making an electric furnace installation (Figure 14-26), an electrician needs to do the following:

- Install the electric furnace on a separate circuit. Section 422.12 specifies that a central heating unit must be supplied by a separate branch circuit.

- Determine the size and type of the individual branch circuit required. Section 422.11(A) requires the overcurrent protective device to be sized

according to the nameplate on the furnace. The nameplate will also list the minimum-size branch-circuit conductor required. Section 422.62(B)(1) states that the nameplate must specify the minimum size of the supply conductor's ampacity and the maximum size of the branch-circuit overcurrent protection device. In other words, always install the electric furnace according to the manufacturer's instructions and specifications.

- Determine the size and location of the electric furnace disconnecting means. Section 422.30 requires a disconnecting means. Section 422.31(B) requires it to be within sight of the furnace. A common wiring practice is to mount the disconnecting means on the side of the electric furnace or on a wall space adjacent to the furnace.

- Determine the location of the thermostat and install the Class 2 wiring from the furnace controller box to the thermostat.

When making an electric baseboard heater installation (Figure 14-27), an electrician needs to do the following:

1 Determine whether a line-voltage thermostat or a low-voltage thermostat will be used to control the electric baseboard heating (Figure 14-28). The most common thermostat type used is the line-voltage type. This thermostat works at the full 240 volts required for the baseboard heating units and is connected directly in the electric baseboard heater branch circuit. Line-voltage thermostats are rated in watts. Some common sizes include 2500, 3000, and 5000 watts. If a 5000-watt line-voltage thermostat is used, the total connected load of electric baseboard heat on it must not exceed 20.8 amps (5000 watts/240 volts = 20.8 amperes). A low-voltage thermostat is used in conjunction with a relay when the load that a thermostat needs to control is more than what the line-voltage thermostat is rated for (Figure 14-29).

2 Determine the size and type of the individual branch circuit required. Most electric baseboard heating installations are done on 240-volt branch circuits with a 20-amp-rated, two-pole circuit breaker as the overcurrent protection device. This means that an electrician would use 12 AWG conductors to wire the electric baseboard branch circuits. Larger groups of electric baseboard heaters may require 10 AWG conductors with 30-amp, two-pole circuit breakers. The wattage rating of electric baseboard heaters is on the nameplate of each unit. By comparing the wattage ratings of each baseboard unit with a line-voltage thermostat's wattage rating, a determination can be made as to the proper size of a line-voltage thermostat for a specific electric baseboard heating load. For example, three 1500-watt electric baseboard units (a total load of 4500 watts) can be connected to one circuit and controlled by a line-voltage thermostat with a rating of 4500 watts or greater.

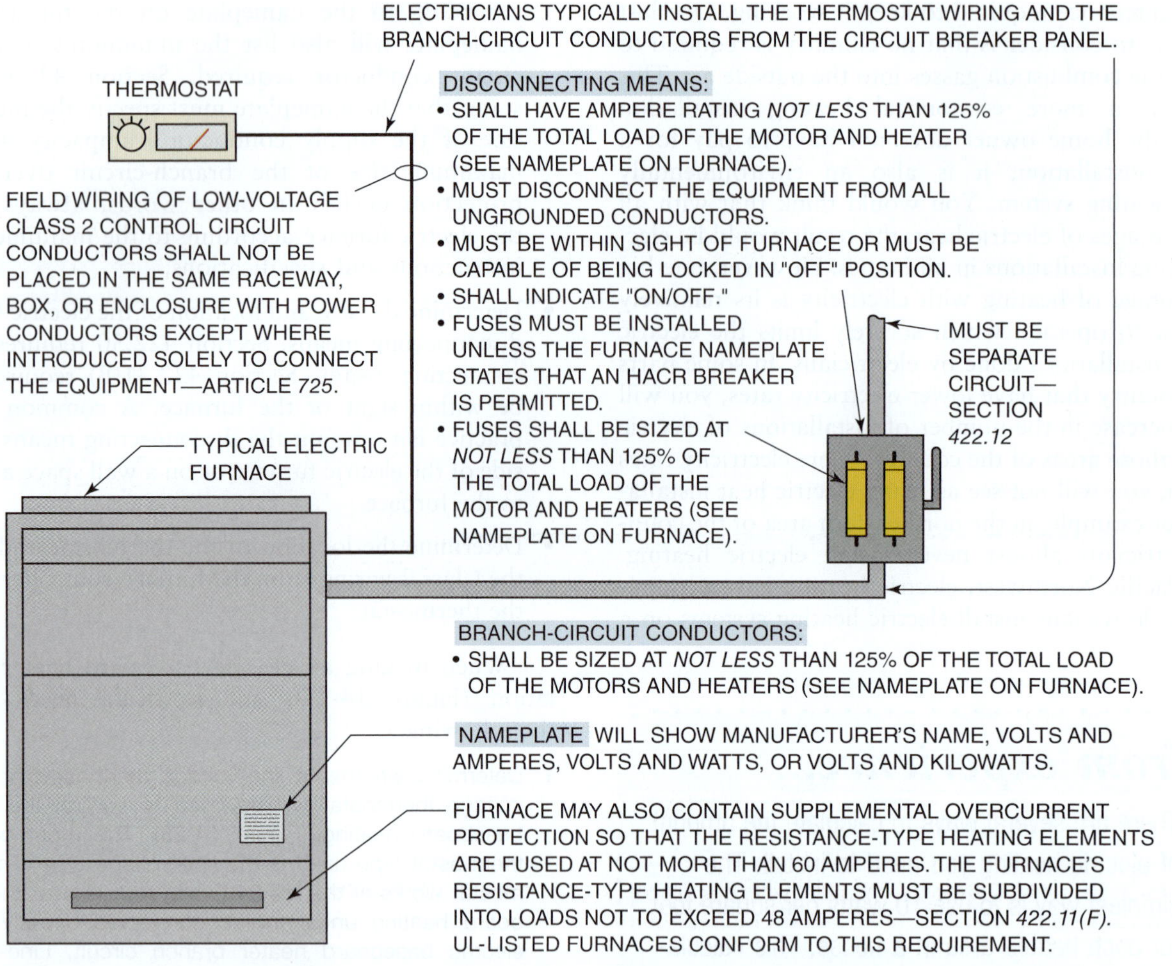

ELECTRICIANS TYPICALLY INSTALL THE THERMOSTAT WIRING AND THE BRANCH-CIRCUIT CONDUCTORS FROM THE CIRCUIT BREAKER PANEL.

THERMOSTAT

FIELD WIRING OF LOW-VOLTAGE CLASS 2 CONTROL CIRCUIT CONDUCTORS SHALL NOT BE PLACED IN THE SAME RACEWAY, BOX, OR ENCLOSURE WITH POWER CONDUCTORS EXCEPT WHERE INTRODUCED SOLELY TO CONNECT THE EQUIPMENT—ARTICLE *725*.

TYPICAL ELECTRIC FURNACE

DISCONNECTING MEANS:
- SHALL HAVE AMPERE RATING *NOT LESS* THAN 125% OF THE TOTAL LOAD OF THE MOTOR AND HEATER (SEE NAMEPLATE ON FURNACE).
- MUST DISCONNECT THE EQUIPMENT FROM ALL UNGROUNDED CONDUCTORS.
- MUST BE WITHIN SIGHT OF FURNACE OR MUST BE CAPABLE OF BEING LOCKED IN "OFF" POSITION.
- SHALL INDICATE "ON/OFF."
- FUSES MUST BE INSTALLED UNLESS THE FURNACE NAMEPLATE STATES THAT AN HACR BREAKER IS PERMITTED.
- FUSES SHALL BE SIZED AT *NOT LESS* THAN 125% OF THE TOTAL LOAD OF THE MOTOR AND HEATERS (SEE NAMEPLATE ON FURNACE).

MUST BE SEPARATE CIRCUIT— SECTION *422.12*

BRANCH-CIRCUIT CONDUCTORS:
- SHALL BE SIZED AT *NOT LESS* THAN 125% OF THE TOTAL LOAD OF THE MOTORS AND HEATERS (SEE NAMEPLATE ON FURNACE).

NAMEPLATE WILL SHOW MANUFACTURER'S NAME, VOLTS AND AMPERES, VOLTS AND WATTS, OR VOLTS AND KILOWATTS.

FURNACE MAY ALSO CONTAIN SUPPLEMENTAL OVERCURRENT PROTECTION SO THAT THE RESISTANCE-TYPE HEATING ELEMENTS ARE FUSED AT NOT MORE THAN 60 AMPERES. THE FURNACE'S RESISTANCE-TYPE HEATING ELEMENTS MUST BE SUBDIVIDED INTO LOADS NOT TO EXCEED 48 AMPERES—SECTION *422.11(F)*. UL-LISTED FURNACES CONFORM TO THIS REQUIREMENT.

EXAMPLE: WHAT SIZE COPPER CONDUCTORS (THHN), FUSES, AND DISCONNECT SWITCH ARE REQUIRED FOR A FURNACE MARKED 79 AMPERES, 240 VOLT, SINGLE PHASE, 60 CYCLES? TERMINALS ON FURNACE AND SWITCH MARKED 75°C. TO SELECT THE PROPER AMPACITY OF THE CONDUCTORS IN ACCORDANCE WITH SECTION *110.14(C)*, BE SURE TO USE THE "75°C" AMPACITY COLUMN IN *TABLE 310.15(B)(16)*.

ANSWER:

CONDUCTOR SIZE: 79 × 1.25 = 98.75 AMPERES
FROM *TABLE 310.15(B)(16)*, SELECT 3 AWG THHN (100 AMPERES AT 75°C)

FUSE SIZE: 79 × 1.25 = 98.75 AMPERES
INSTALL 100-AMPERE FUSES

SWITCH: 100-AMPERE SWITCH

© Cengage Learning 2012.

FIGURE 14-26 An electric furnace installation. The furnace components are pre-wired at the factory so an electrician usually only has to provide the branch-circuit wiring to the unit. The thermostat wiring may also be installed by an electrician.

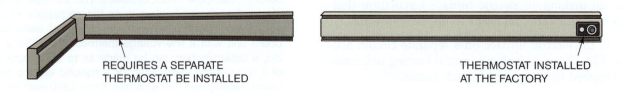

REQUIRES A SEPARATE THERMOSTAT BE INSTALLED

THERMOSTAT INSTALLED AT THE FACTORY

FIGURE 14-27 Typical electric baseboard heating units.

© Cengage Learning 2012.

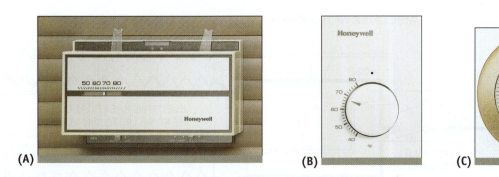

FIGURE 14-28   Thermostats for electric heating systems. (A) and (C) Low-voltage thermostats; (B) Line-voltage thermostat.

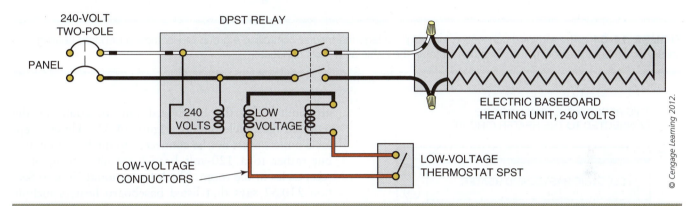

FIGURE 14-29   Typical wiring for electric baseboard heating controlled by a low-voltage thermostat and relay. This wiring arrangement is used when the electrical load of the electric heating is greater than the rating of a high-voltage thermostat. The relay contacts are sized to switch "ON" and "OFF" the heavy current drawn by the heating units.

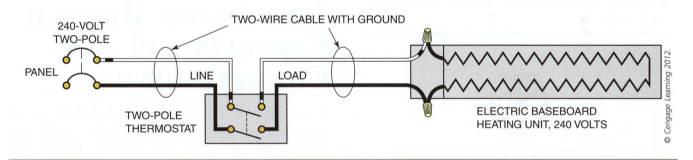

FIGURE 14-30   The electrical connection for a high-voltage thermostat controlling one section of electric baseboard heating. Remember to reidentify the white conductor in the cable as a "hot" conductor.

3   Starting at the location where the branch-circuit overcurrent protection device is located, install the wiring to the electrical box on the circuit that will contain the line-voltage thermostat. Be sure to mark the cable or other wiring method with the circuit number and type and the area served at the panel. For example, electric heating circuit 1 serving the master bedroom can be marked "EH1MBR." Continue roughing in the wiring to the electric heating baseboard unit (Figure 14-30) or to a group of baseboard units (Figure 14-31), making sure to follow the installation practices described in previous chapters.

CAUTION

CAUTION: When using a cable wiring method like nonmetallic sheathed cable, make sure you reidentify the white insulated conductor in the cable as a "hot" conductor by wrapping some black electrical tape on the conductor wherever it is visible, such as in the circuit breaker panel, line-voltage thermostat box, and at the electric baseboard heater itself.

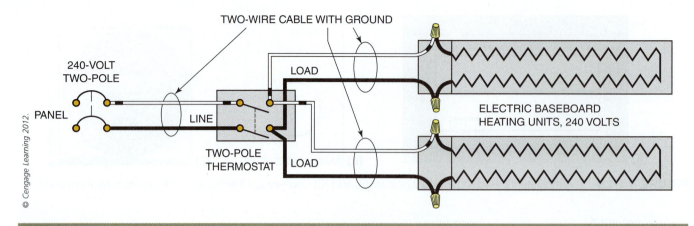

FIGURE 14-31 The electrical connection for a high-voltage thermostat controlling a group of electrical baseboard heating units. Remember to reidentify the white conductor in the cable as a "hot" conductor.

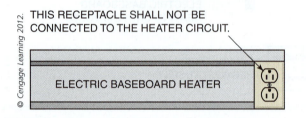

FIGURE 14-32 A length of electric baseboard heater with a factory-installed receptacle outlet. The receptacle will meet Section 210.52 requirements for receptacle placement in a house. The receptacle(s) may not be connected to the electric baseboard heating branch circuit. A separate 120-volt circuit must be used.

FIGURE 14-33 Unless the manufacturer's installation instructions specifically permit the electrical baseboard to be installed directly below receptacles, it is a violation to do so.

4 Double check to make sure all circuit wiring for the electric baseboard heating branch circuit you are installing has been roughed in.

When installing electric baseboard heating units along a wall space that requires receptacle outlets, Section 210.52 allows factory-installed receptacles that

are built in to the baseboard unit to count as the required receptacle outlets (Figure 14-32). These receptacle outlets must not be connected to the heater circuits but rather to a 120-volt branch circuit serving other receptacles in the room. The Informational Note to Section 210.52 says that listed baseboard heaters include instructions that may not permit their installation below receptacle outlets (Figure 14-33). The correct positioning of electric baseboard heating units is shown in Figure 14-34.

INSTALLING BRANCH CIRCUITS FOR AIR-CONDITIONING

Air-conditioning a house is accomplished in one of two ways: (1) a central air conditioner system that blows cooled air from one centrally located unit into various areas of a house through a series of ducts or (2) individual room air conditioners located in different rooms throughout a house. Like a central-heating system, when installing the wiring for a central air-conditioning system, an electrician usually just provides the rough-in wiring, and an air-conditioning technician who works for the heating and refrigeration contractor will make the final connections. If room air conditioners are going to be used in a house, the electrician may have to install specific branch circuits that will supply the proper amperage and voltage to the location where the air-conditioning unit will be located. Some homes are heated *and* cooled with a **heat pump**. During cooler weather, the heat pump supplies heat to a house. In warmer weather, the heat pump is simply run in reverse and, as a result, can supply cool air to the house as well.

POSITION ELECTRIC BASEBOARD HEATING UNITS SO THEY WILL *NOT* BE DIRECTLY BELOW A WALL RECEPTACLE OUTLET.

IF INSTALLED AS SHOWN, ELECTRICAL CORDS COULD COME IN CONTACT WITH THE BASEBOARD UNIT, SUBJECTING THIS CORD TO RUBBING (ABRASION) AND HEAT, WHICH MIGHT RESULT IN FAILURE OF THE INSULATION OF THE CORD, A POTENTIAL FIRE AND SHOCK HAZARD.

© Cengage Learning 2012.

FIGURE 14-34 Correct positioning of electric baseboard heating units.

A heat pump is installed similarly to a central air-conditioning system, and the wiring required to be installed by an electrician is basically the same. In this section, we look at the branch-circuit installation for both central air conditioner and room air conditioner wiring situations.

CENTRAL AIR CONDITIONERS

If the system is to be a central air-conditioning system, the nameplate on the unit will give the branch-circuit ratings, the minimum conductor size, the overcurrent protection device size and type, and what the amperage and voltage ratings are for the unit. The manufacturer's instruction booklet will contain more information concerning the installation and should be consulted by an electrician installing the supply wiring to the system. Figure 14-35 shows a typical central air-conditioning system installation. An electrician will need to install branch-circuit wiring from the electrical panel to the main air-handling unit, which is located inside the home. Another branch circuit will need to be installed from the electrical panel to the disconnect switch of the compressor unit, which is located outside the house. The air-handling unit contains the fan that blows the cool air

throughout a house and will need either a 120-volt or a 240-volt circuit. Be sure to consult the manufacturer's instructions to verify which voltage and amperage must be supplied to the air-handling unit. The circuit supplying the compressor unit will require a 240-volt, two-wire circuit with a grounding conductor. If the disconnect located outside next to the compressor contains fuses or a circuit breaker, the circuit supplying it is a "feeder." If the disconnect has no overcurrent protection devices in it and is a non-fusible disconnect, the circuit supplying it is considered a "branch circuit." Article 440 of the *NEC®* contains most of the electrical installation requirements for air-conditioning systems. Figure 14-36 shows the basic circuit requirements for the wiring for the compressor unit.

When selecting the type of overcurrent protection device for an air-conditioning circuit (or heating circuit), the nameplate on the unit must be consulted (Figure 14-37). If it says "maximum size fuse," the overcurrent protection device *must* be a fuse. If the nameplate says "maximum size fuse or HACR-type circuit breaker," a fuse *or* a HACR circuit breaker could be used. (HACR is an acronym for "heating, air-conditioning, and refrigeration.") Always look for the HACR label on a circuit breaker before using it to

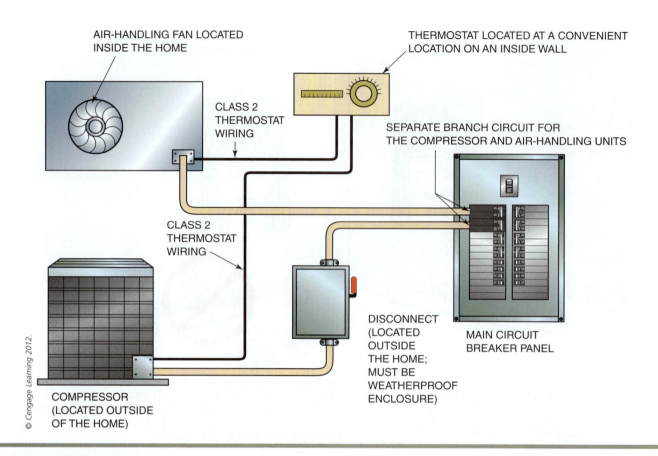

AIR-HANDLING FAN LOCATED
INSIDE THE HOME

THERMOSTAT LOCATED AT A CONVENIENT
LOCATION ON AN INSIDE WALL

CLASS 2
THERMOSTAT
WIRING

SEPARATE BRANCH CIRCUIT FOR
THE COMPRESSOR AND AIR-HANDLING UNITS

CLASS 2
THERMOSTAT
WIRING

DISCONNECT
(LOCATED
OUTSIDE
THE HOME;
MUST BE
WEATHERPROOF
ENCLOSURE)

MAIN CIRCUIT
BREAKER PANEL

COMPRESSOR
(LOCATED OUTSIDE
OF THE HOME)

© Cengage Learning 2012.

FIGURE 14-35 A typical electrical installation of a central air conditioner.

protect a heating or air-conditioning circuit. (Figures 14-38, 14-39, and 14-40). The disconnecting means must be within sight of the unit it disconnects (Section 440.14), and it must be readily accessible. The disconnecting means enclosure is normally mounted outside on the side of the house and in a spot that is on the side of and not directly behind the air-conditioning unit. Remember that you have to follow Section 110.26 working space requirements, so install the disconnect so that a working space of at least 30 inches (750 mm) in width, 3 feet (900 mm) out from the face of the disconnect enclosure, and a height of 6.5 feet (2 m) is maintained.

A thermostat is required for the system control. It allows home owners to set the temperature to a value they want, and the air-conditioning system will automatically adjust the temperature in the area of the thermostat to the desired set temperature. Thermostat wire will need to be run from the compressor and the air handler to the thermostat. This low-voltage wiring is called a Class 2 circuit. Article 725 covers Class 2 wiring. Some information electricians need to be familiar with concerning Class 2 wiring includes the following:

- Class 2 circuits are power limited because of the **transformers** that supply the low voltage they operate on. As such, they are considered to be safe from

electric shock and do not present much, if any, of a fire hazard. The transformers used to supply the low voltage for a Class 2 circuit in residential areas are normally marked "Class 2 transformer."

- The Class 2 conductors do not need separate fuses or circuit breakers to protect them, because they are inherently current limited by the transformers they originate from.

- Class 2 wiring can be run exposed or, if an electrician so chooses, in a raceway for added protection. All Class 2 wiring must be installed in a neat and workmanlike manner. If cables are installed exposed, they need to be supported by the building structure with straps, staples, hangers, or other similar fittings in a way that will not damage the cable.

- Class 2 wiring cannot be run in the same raceway, cable, compartment, or electrical box with regular light and power conductors.

Thermostat wire comes in a cable form and is usually 16 or 18 AWG solid copper wire. Some installations may require the electrician to install the thermostat cable to and from the proper locations in the system; however, most central air-conditioning system thermostats and thermostat wiring is done by the air-conditioning technician.

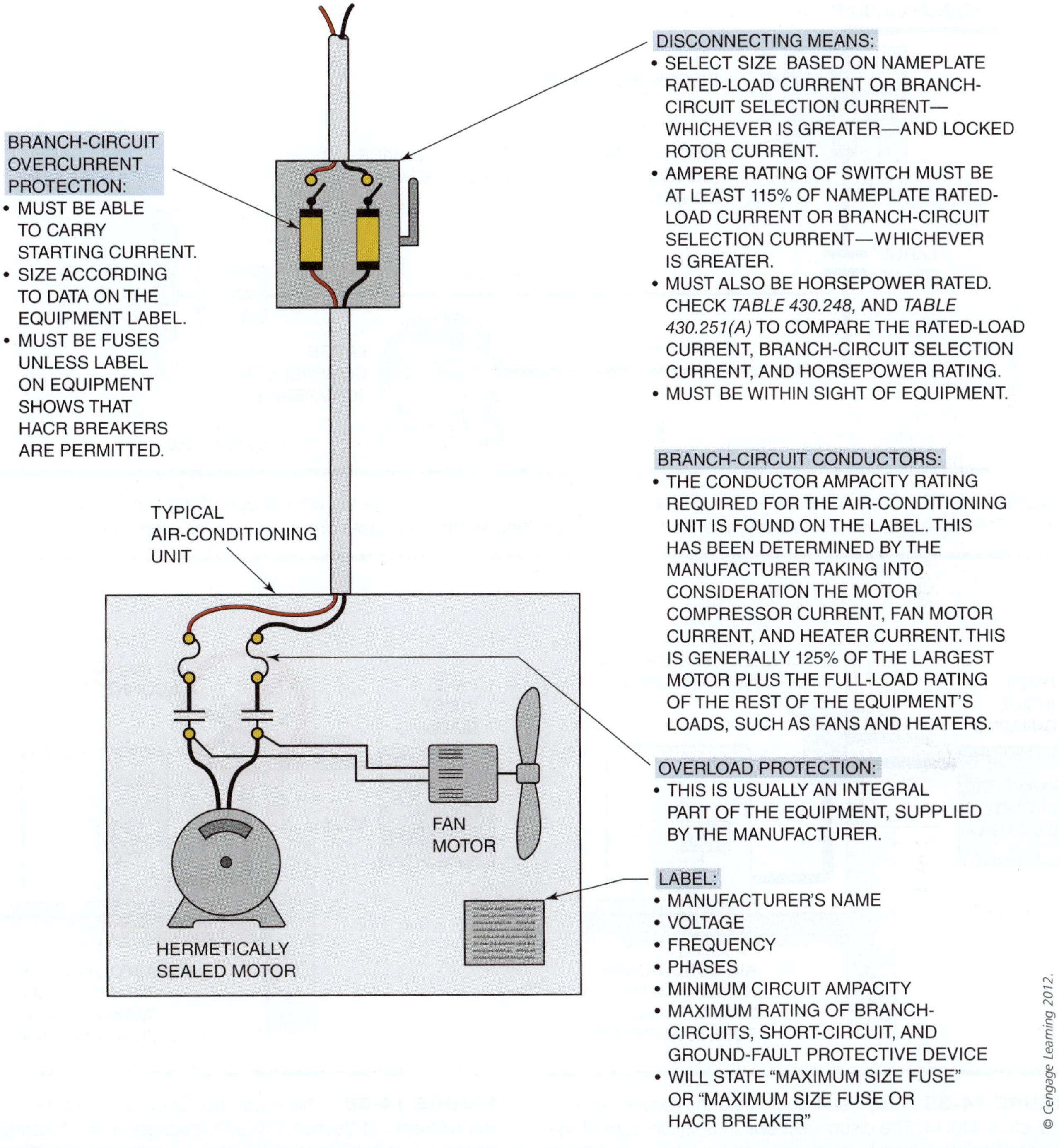

BRANCH-CIRCUIT OVERCURRENT PROTECTION:
- MUST BE ABLE TO CARRY STARTING CURRENT.
- SIZE ACCORDING TO DATA ON THE EQUIPMENT LABEL.
- MUST BE FUSES UNLESS LABEL ON EQUIPMENT SHOWS THAT HACR BREAKERS ARE PERMITTED.

TYPICAL AIR-CONDITIONING UNIT

HERMETICALLY SEALED MOTOR

FAN MOTOR

DISCONNECTING MEANS:
- SELECT SIZE BASED ON NAMEPLATE RATED-LOAD CURRENT OR BRANCH-CIRCUIT SELECTION CURRENT—WHICHEVER IS GREATER—AND LOCKED ROTOR CURRENT.
- AMPERE RATING OF SWITCH MUST BE AT LEAST 115% OF NAMEPLATE RATED-LOAD CURRENT OR BRANCH-CIRCUIT SELECTION CURRENT—WHICHEVER IS GREATER.
- MUST ALSO BE HORSEPOWER RATED. CHECK *TABLE 430.248,* AND *TABLE 430.251(A)* TO COMPARE THE RATED-LOAD CURRENT, BRANCH-CIRCUIT SELECTION CURRENT, AND HORSEPOWER RATING.
- MUST BE WITHIN SIGHT OF EQUIPMENT.

BRANCH-CIRCUIT CONDUCTORS:
- THE CONDUCTOR AMPACITY RATING REQUIRED FOR THE AIR-CONDITIONING UNIT IS FOUND ON THE LABEL. THIS HAS BEEN DETERMINED BY THE MANUFACTURER TAKING INTO CONSIDERATION THE MOTOR COMPRESSOR CURRENT, FAN MOTOR CURRENT, AND HEATER CURRENT. THIS IS GENERALLY 125% OF THE LARGEST MOTOR PLUS THE FULL-LOAD RATING OF THE REST OF THE EQUIPMENT'S LOADS, SUCH AS FANS AND HEATERS.

OVERLOAD PROTECTION:
- THIS IS USUALLY AN INTEGRAL PART OF THE EQUIPMENT, SUPPLIED BY THE MANUFACTURER.

LABEL:
- MANUFACTURER'S NAME
- VOLTAGE
- FREQUENCY
- PHASES
- MINIMUM CIRCUIT AMPACITY
- MAXIMUM RATING OF BRANCH-CIRCUITS, SHORT-CIRCUIT, AND GROUND-FAULT PROTECTIVE DEVICE
- WILL STATE "MAXIMUM SIZE FUSE" OR "MAXIMUM SIZE FUSE OR HACR BREAKER"

© Cengage Learning 2012.

FIGURE 14-36 The basic requirements for the wiring of an air conditioner compressor unit. Both the compressor unit and the disconnect switch are typically located outside the house.

ROOM AIR CONDITIONERS

When homes have no central air-conditioning but home owners still want cooler temperatures inside a house during days when the outside temperature is very hot, room air conditioners are used. They are available in 120-volt and 240-volt models and typically fit into a window opening or are installed in a more permanent way directly in an outside wall. These air conditioners are cord-and-plug connected. The circuits that feed them must be sized and installed according to the installation

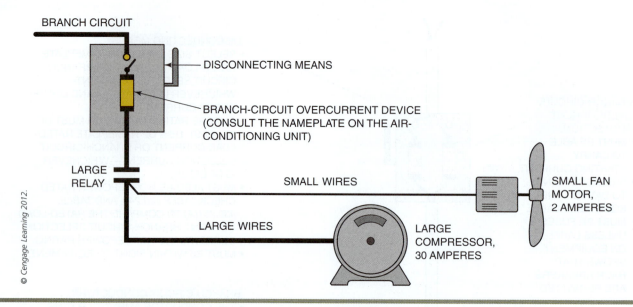

FIGURE 14-37 Consult the nameplate on the air-conditioning unit to determine the correct size of the overcurrent protection device. The nameplate will also specify whether the overcurrent device can be a fuse, circuit breaker, or either one.

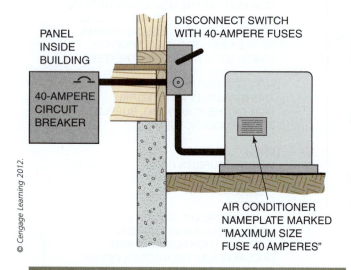

FIGURE 14-38 This installation meets the requirements of Section 440.14. The disconnect switch is within sight of the air-conditioning unit and contains the 40-amp fuses that the nameplate specified.

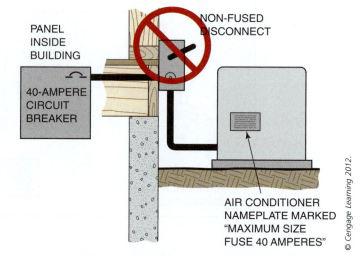

FIGURE 14-39 This installation does not meet the requirements of Section 110.3(B), because a non-fused disconnect switch is used as the disconnecting means. The nameplate specifically calls for 40-amp *fuses* to be used as the overcurrent protection device, but an electrician has installed a 40-amp circuit breaker in the electrical panel.

requirements as outlined in Article 440. Sections 440.60 through 440.64 cover room air conditioners. These code rules are summarized as follows:

- The air conditioners must be grounded and connected with a cord-and-plug set.

- The air conditioner may not have a rating that is greater than 40 amps at 250 volts.

- The rating of the branch circuit overcurrent protection device must not exceed the branch-circuit conductor rating or the receptacle rating, whichever is less.

- On an individual branch-circuit where no other items are served by the circuit, the air conditioner load cannot exceed 80 percent of the branch-circuit ampacity.

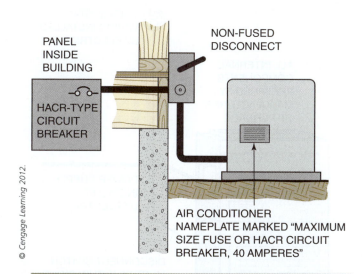

PANEL INSIDE BUILDING

HACR-TYPE CIRCUIT BREAKER

NON-FUSED DISCONNECT

AIR CONDITIONER NAMEPLATE MARKED "MAXIMUM SIZE FUSE OR HACR CIRCUIT BREAKER, 40 AMPERES"

© Cengage Learning 2012.

FIGURE 14-40 This installation meets the requirements of the *NEC*® because even though a non-fused disconnect switch has been installed, a 40-amp HACR circuit breaker has also been installed in the electrical panel. This satisfies the nameplate requirement that fuses *or* an HACR circuit breaker must be used to provide the overcurrent protection for the air-conditioning unit.

- On a branch circuit where other loads are also served, the air conditioner load cannot exceed 50 percent of the branch-circuit ampacity.

- The plug on the end of the cord can serve as the air conditioner's disconnecting means.

- The maximum length of the cord on a 120-volt room air conditioner is 10 feet (3 m).

- The maximum length of the cord on a 240-volt room air conditioner is 6 feet (1.8 m).

If a room air conditioner is rated at 120 volts and is purchased by a home owner after the electrical system has been installed, the room air conditioner is typically just plugged into a receptacle located next to the window or wall location where the air conditioner is placed. This receptacle will most likely be on a branch circuit that supplies electricity to other lighting and receptacle outlets in that particular room. If an electrician knows where room air conditioners will be located during the installation of the house electrical system, they can rough in the wiring to a receptacle location that is dedicated to a specific room air conditioner. The electrician will need to know the amperage and voltage rating of the room air conditioner so that the proper wire size and overcurrent protection device can be installed. Knowing the specific amperage and voltage rating of the unit will also enable the electrician to install the proper receptacle for the cord-and-plug connection of the air conditioner. In earlier chapters, we discussed different receptacle configurations based on

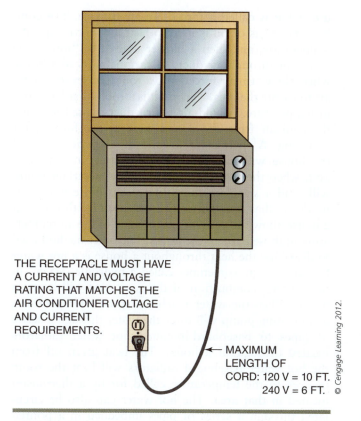

THE RECEPTACLE MUST HAVE A CURRENT AND VOLTAGE RATING THAT MATCHES THE AIR CONDITIONER VOLTAGE AND CURRENT REQUIREMENTS.

MAXIMUM LENGTH OF CORD: 120 V = 10 FT. 240 V = 6 FT.

© Cengage Learning 2012.

FIGURE 14-41 A typical installation of a room air conditioner.

specific voltage and amperage ratings. Care must be taken by the electrician to install a receptacle that has the proper configuration for the plug on the end of the room air conditioner cord. Figure 14-41 shows a typical installation of a room air conditioner.

INSTALLING THE BRANCH CIRCUIT FOR GAS AND OIL CENTRAL HEATING SYSTEMS

Not all residential heating systems use electricity to produce the heat. In fact, the majority of central heating systems installed in houses use gas- or oil-fired furnaces. There are two methods for the heat to be transferred throughout a house. The first and most popular method uses what is called a forced-hot-air furnace. The combustion of gas or oil is used to heat air up to a specific temperature, and then a blower is used to "force" the warm air through air ducts to various parts of a house where the warm air forced into a room causes the colder air to vacate the area. The result is a room that has a rise in temperature. A cold-air return duct is used to bring the cool air from the rooms back to the furnace for reheating. A thermostat located in a specific area of the house will monitor the temperature in that area and, when the temperature gets below a certain point, will send a signal to the furnace that more hot air is needed in that area. Different zones installed throughout a house allow for more precise control of the air temperature in those parts of the house. The other method used to distribute the heat throughout a house is a hot water heating system, sometimes called a **hydronic system**. In this system, combustion of the gas or oil will heat up water. When the water reaches a specific temperature, a circulating pump will force the water through a series of pipes to baseboard-mounted hot water radiators located throughout a house. The heat given off from the hot water baseboard radiators will heat the room or area to the temperature called for by a thermostat located in that area. The hot water can also be circulated through a series of pipes imbedded in a poured concrete floor. The concrete floor itself heats up and radiates heat into the area above it. This type of radiant heating is popular in homes that are one story and built on a concrete slab.

It is usually a heating technician who sets up and wires the controls for the heating equipment used in a home. New furnaces come from the factory with most of their components pre-wired. Electricians will install the separate branch circuit required for a central heating system. The wiring will normally be run from the electrical panel to a point over the location of the gas- or oil-fired furnace, leaving enough of the wiring method used for the heating technician to make the final connection to the furnace. The electrician will also make the necessary connections to the fuses or circuit breaker in the electrical panel. This circuit most often consists of a 12 AWG conductor size with a 20-amp-rated overcurrent protection device but could be a 14 AWG conductor with a 15-amp overcurrent protection device. Electricians may also install the Class 2 control wiring from a furnace controller box to the thermostat location (Figure 14-42). Electricians installing the Class 2 thermostat wiring must make sure to leave enough extra wire for the heating technician to make the necessary connections at the furnace controller box. Because electricians usually do not make the actual electrical connections for the wiring associated with a central heating system furnace, this book will not cover the specific wiring for a gas- or oil-fired furnace.

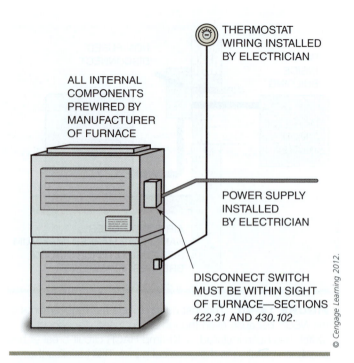

THERMOSTAT WIRING INSTALLED BY ELECTRICIAN

ALL INTERNAL COMPONENTS PREWIRED BY MANUFACTURER OF FURNACE

POWER SUPPLY INSTALLED BY ELECTRICIAN

DISCONNECT SWITCH MUST BE WITHIN SIGHT OF FURNACE—SECTIONS *422.31* AND *430.102*.

© Cengage Learning 2012.

FIGURE 14-42 An electrician typically installs the branch circuit power supply to the furnace, a disconnect switch, and the thermostat wiring. The other components of a furnace are usually pre-wired at the factory.

INSTALLING THE SMOKE DETECTOR BRANCH CIRCUIT

Because fire in homes is the third-leading cause of accidental death, some type of fire warning system is required in all new home construction. The National Fire Protection Association (NFPA) publishes the National Fire Alarm Code, called NFPA 72. (You might remember that NFPA 70 is actually the *NEC®*.) NFPA 72 outlines the minimum requirement for the selection, installation, and maintenance of fire alarm equipment. Chapter 2 of NFPA 72 covers residential fire alarm systems. It defines a household fire alarm system as a system of devices that produces an alarm signal in the house for the purpose of notifying the occupants of the house of a fire so that they will evacuate the house. NFPA 72 is usually adopted by the building codes that must be followed when building a house and, as such, must be followed by electricians when installing a residential electrical system. The most common fire warning device used in a house is a **smoke detector** (Figure 14-43). In this section, we take a look at the installation requirements and methods for smoke detectors.

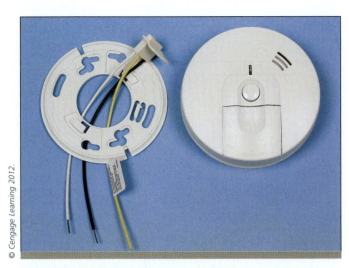

FIGURE 14-43   A typical smoke detector.

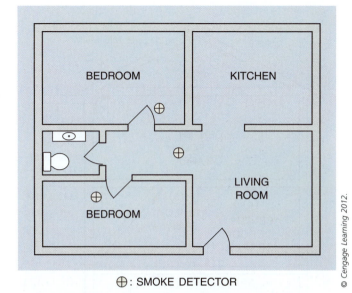

⊕ : SMOKE DETECTOR

FIGURE 14-44   Recommended locations for smoke detectors in a house. Make sure a smoke detector is located in each bedroom of any new house.

---

from experience...

Local oil burner codes often require a safety switch to be installed at the head of the basement stairway or at the entrance to a room or crawl space that houses an oil-fired furnace or water heater. This allows a home owner to shut off the electrical power to the furnace or water heater should a fire break out. This will stop the oil burner pump from continuing to pump oil and feed the fire with fuel. An electrician installing the branch-circuit wiring for the oil-fired furnace or water heater will route the wiring to the switch location and then to a location above the furnace or water heater for future connections by an oil burner technician. Once the single-pole switch is installed, a red cover must be used on the switch that identifies it as the oil burner safety switch. Check with the authority having jurisdiction in your area to see if this requirement applies to you.

When wiring the smoke detector circuit for a new house, a detector unit should be placed in each bedroom and in the area just outside the bedroom areas (Figure 14-44). They also should be installed on each level of a house (Figure 14-45). The smoke detectors must be installed in new residential construction so that when one detector is operated, all other detectors in the house will also operate. This is called **interconnecting**. The following installation requirements should always be followed when installing smoke detectors in a house:

- Install the smoke detectors on the ceiling at a location where there is no "dead" airspace (Figure 14-46).

- Install smoke detectors at the top of a stairway instead of the bottom because smoke will rise. An exception is in the basement, where it is better to install a smoke detector on the ceiling but close to the stairway to the first floor.

- Install smoke detectors in new houses that are hard-wired directly to a 120-volt circuit. Dual-powered smoke detectors that have a battery backup when the electrical power is off are a good idea but are not normally required. Wire the smoke detectors so that they are interconnected—when one goes off, they all go off.

- Do not install smoke detectors to branch-circuit wiring that is controlled by a wall switch. The electrical power to the smoke detectors must be on at all times.

- Do not install smoke detectors on circuits that are GFCI protected. If the GFCI trips off, you have lost power to the smoke detectors.

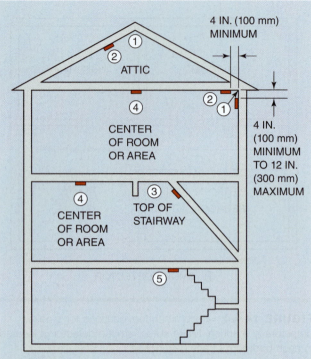

NOTES:

① DO *NOT* INSTALL DETECTORS IN *DEAD* AIRSPACES.

② MOUNT DETECTORS ON THE BOTTOM EDGE OF JOISTS OR BEAMS. THE SPACE BETWEEN THESE JOISTS AND BEAMS IS CONSIDERED TO BE *DEAD* AIRSPACE.

③ DO *NOT* MOUNT DETECTORS IN *DEAD* AIRSPACE AT THE TOP OF A STAIRWAY IF THERE IS A DOOR AT THE TOP OF THE STAIRWAY THAT CAN BE CLOSED. DETECTORS *SHOULD* BE MOUNTED AT THE TOP OF AN OPEN STAIRWAY BECAUSE HEAT AND SMOKE TRAVEL UPWARD.

④ MOUNT DETECTORS IN THE CENTER OF A ROOM OR AREA.

⑤ BASEMENT SMOKE DETECTORS MUST BE LOCATED IN CLOSE PROXIMITY TO THE STAIRWAY LEADING TO THE FLOOR ABOVE.

© Cengage Learning 2012.

FIGURE 14-45   Smoke detectors are required on each level of a house. Locate them carefully so they will be sure to work correctly.

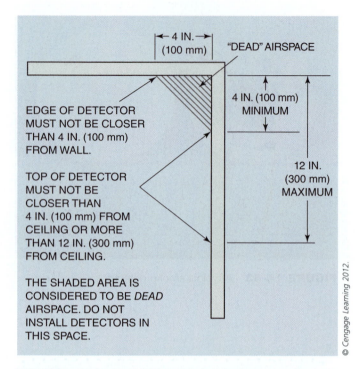

EDGE OF DETECTOR MUST NOT BE CLOSER THAN 4 IN. (100 mm) FROM WALL.

TOP OF DETECTOR MUST NOT BE CLOSER THAN 4 IN. (100 mm) FROM CEILING OR MORE THAN 12 IN. (300 mm) FROM CEILING.

THE SHADED AREA IS CONSIDERED TO BE *DEAD* AIRSPACE. DO NOT INSTALL DETECTORS IN THIS SPACE.

© Cengage Learning 2012.

FIGURE 14-46   Never install smoke detectors in the "dead" airspace where a wall meets the ceiling.

connect to the red wire in a three-wire cable. The maximum number of interconnected smoke detectors varies by manufacturer, but a good rule of thumb is to not connect more than 10 smoke detectors on a circuit. See Figure 14-47 for a typical smoke detector branch-circuit installation.

INSTALLING THE CARBON MONOXIDE DETECTOR BRANCH CIRCUIT

Statistics indicate that the leading cause of accidental death in homes is from carbon monoxide (CO) poisoning. CO is an invisible, odorless, tasteless, and non-irritating gas that is completely undetectable to your senses. Inside a home, appliances used for heating and cooking are the most likely sources of CO. Vehicles running in attached garages can also produce dangerous levels of CO. CO can be produced when burning any fuel, such as gasoline, propane, natural gas, oil and wood. It can be produced by any fuel-burning appliance that is malfunctioning, improperly installed, or not

When wiring smoke detectors, electricians will often run a two-wire cable (or two conductors in a raceway) to the first smoke detector location. A three-wire cable (or three wires in a raceway) is then run to each of the other smoke detector locations. The black and white conductors in the circuit are used to provide 120 volts to each smoke detector. The red conductor (or third wire in a raceway) is used as the interconnection between all the smoke detectors. Each smoke detector usually has a yellow wire that is used to

SMOKE DETECTORS

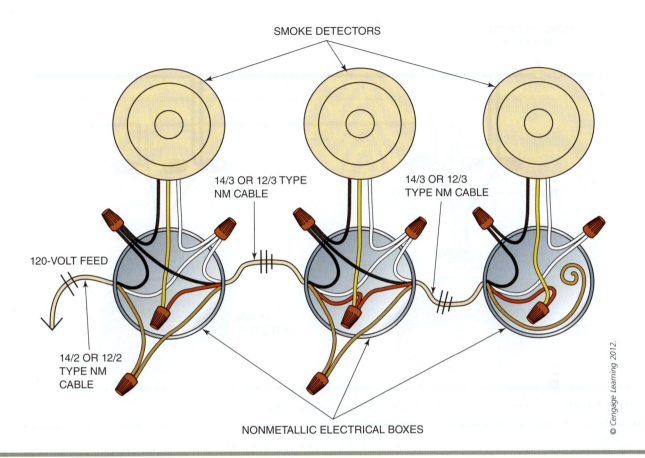

14/3 OR 12/3 TYPE NM CABLE

14/3 OR 12/3 TYPE NM CABLE

120-VOLT FEED

14/2 OR 12/2 TYPE NM CABLE

NONMETALLIC ELECTRICAL BOXES

© Cengage Learning 2012.

FIGURE 14-47   A typical interconnected smoke detector installation.

ventilated correctly, such as automobiles, furnaces, gas ranges/stoves, gas clothes dryers, water heaters, portable fuel-burning space heaters and generators, fireplaces, wood-burning stoves and certain swimming pool heaters.

Carbon monoxide detectors (Figure 14-48) are now required in most areas of the country to alert people living in the home in case there is a dangerously high level of CO. Even if your area does not require them to be installed—they should be. Like smoke detectors, CO detectors monitor the air in your home and sound a loud alarm to warn you of trouble. Also, like smoke detectors they can be installed so that only one detector is on a circuit or they can be interconnected so that when one device detects high levels of CO all the other detectors in the house will sound an alarm.

CO detectors should be mounted in or near bedrooms and living areas. It is recommended that you install a CO detector on each level of a home. When choosing the installation locations, make sure the home occupants will be able to hear the alarm from all sleeping areas. If you install only one CO detector in the home, install it near bedrooms, not in the basement or

© Cengage Learning 2012.

FIGURE 14-48   A typical carbon monoxide (CO) detector.

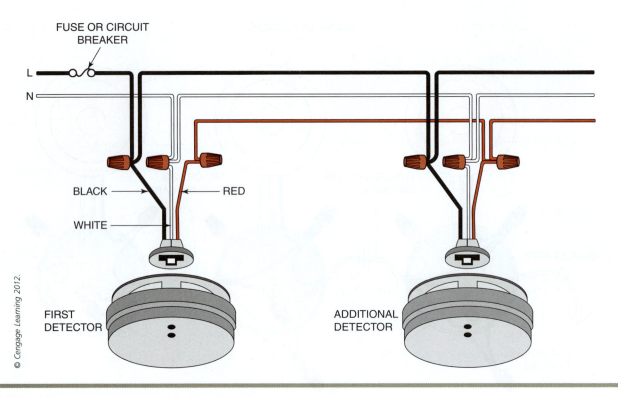

© Cengage Learning 2012.

FIGURE 14-49 An interconnected carbon monoxide detector installation.

furnace room. When wall mounting, place the detector high enough so it is out of reach of children. To avoid causing damage to the unit, to provide optimum performance, and to prevent unnecessary nuisance alarms:

- Do not install in kitchens, garages or furnace rooms that may expose the sensor to substances that could damage or contaminate it.

- Do not install in areas where the temperature is colder than 40°F (4.4°C) or hotter than 100°F (37.8°C) such as crawl spaces, attics, porches and garages.

- Do not install within 15 feet of heating or cooking appliances.

- Do not install near vents, flues, chimneys or any forced/unforced air ventilation openings.

- Do not install near ceiling fans, doors, windows or areas directly exposed to the weather.

- Do not install in dead air spaces, such as peaks of vaulted ceilings or gabled roofs, where CO may not reach the sensor in time to provide early warning (see Figures 14-45 and 14-46).

- Do not install on a switched or ground fault circuit interrupter protected branch circuit.

Carbon monoxide detectors are hardwired into a 120-volt branch circuit. There should be no other outlets on the branch circuit. Like smoke detectors,

CO detectors are available with a battery backup so that if the power should ever go off for an extended period of time the CO detector will still be working. Install the wiring for interconnected CO detectors similar to the way interconnected smoke detectors are wired (Figure 14-49). It is recommended that 12 AWG conductors be used with a 20-amp overcurrent device for the CO detector branch circuit but 14 AWG conductors with a 15-amp overcurrent device could be used. Combination smoke/carbon monoxide detectors are available and are a good way to satisfy both smoke detector and CO detector requirements at the same time.

INSTALLING THE LOW-VOLTAGE CHIME CIRCUIT

A chime system is used in homes to signal when someone is at the front or rear door. A chime system will consist of the chime, the momentary contact switch buttons, a transformer, and the wire used to connect the system together. This system is often referred to as the "doorbell system" even though bells (and buzzers) have not been used in homes for many years. However, the term "doorbell" has stayed with us, and many electricians still refer to this part of the electrical system installation as "installing the doorbell." A chime system sounds a musical note or a series of notes when a button located next to an outside entrance or exit door is pushed. The signal from the door buttons is delivered to the chime itself through low-voltage Class 2 wiring. In this section, we look at common installation practices for a chime circuit.

The chime should be located in an area of the house where once it "chimes" and sounds a tone, the people in the house can hear it. Sometimes chimes are located on each floor of a larger home to make sure that everyone in the house will hear the chime when a tone is sounded. Chimes are available in a variety of styles (Figure 14-50). If a chime is to be surface-mounted, it is common wiring practice to just bring the low-voltage wiring into the back of the chime enclosure and use hollow wall anchors to hold the chime onto the wall. Sometimes a flush-mounted chime installation is needed, and the installation requires adequate backing and an electrical box for attachment of the chime enclosure to the wall (Figure 14-51).

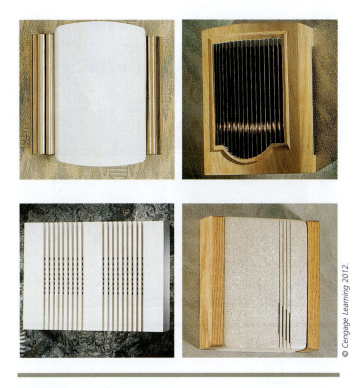

© Cengage Learning 2012.

FIGURE 14-50 Chimes are available in a variety of styles.

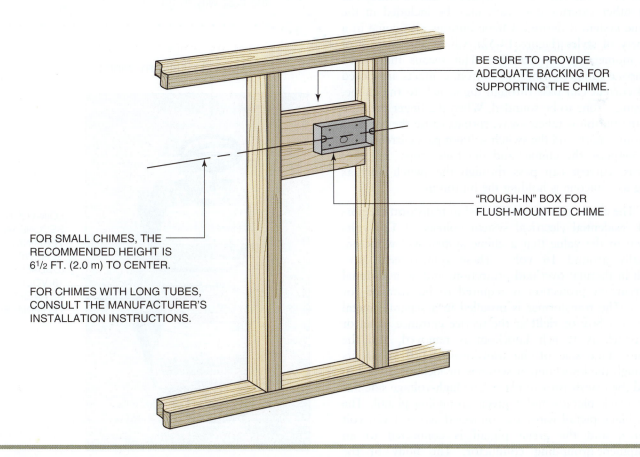

BE SURE TO PROVIDE ADEQUATE BACKING FOR SUPPORTING THE CHIME.

"ROUGH-IN" BOX FOR FLUSH-MOUNTED CHIME

FOR SMALL CHIMES, THE RECOMMENDED HEIGHT IS 6½ FT. (2.0 m) TO CENTER.

FOR CHIMES WITH LONG TUBES, CONSULT THE MANUFACTURER'S INSTALLATION INSTRUCTIONS.

FIGURE 14-51 The rough-in for a flush-mounted chime installation. *Courtesy of NuTone, Inc.*

FIGURE 14-52 Various styles of chime buttons. *Courtesy of NuTone, Inc.*

A chime button will need to be located next to the front door and rear door of a house. Chime buttons at any other exterior doorways may be included in the chime system if desired. Chime buttons also come in a variety of styles (Figure 14-52). Chime buttons are of the momentary contact type. This means that when someone's finger pushes the button, contacts are closed and current can pass through the switch to the chime, causing a tone to be sounded. When the finger pressure on the button is taken away, springs cause the contacts to come apart, and the switch will not pass current. This de-energizes the chime, and the tone stops. In other words, current can pass through the switch only as long as someone is holding the button in.

The chime transformer is used to transform the normal residential electrical system voltage of 120 volts down to the value that a chime system will work on, usually around 16 volts. These transformers have built-in thermal overload protection, and no additional overcurrent protection is required to be installed for them. The transformer is installed in a separate metal electrical box or right at the service entrance panel or subpanel. A ½-inch knockout is removed, and the high-voltage side of the transformer (120 volts) fits through the knockout. A setscrew or locknut is used to hold the transformer in place. The high-voltage side has two black pigtails and a green grounding pigtail. The two black pigtail wires are connected across a 120-volt source, and the green pigtail is connected to an equipment-grounding conductor. The body of the transformer with the low-voltage side is located on the outside of the electrical box or panel. Two screws

on the low-voltage side (16 volts) of the transformer are used to connect the chime wiring to the transformer (Figure 14-53).

from experience...

Wireless chimes are being installed more and more. Some models plug into any available receptacle outlet to provide power to the chime. Push-buttons are then placed at entrances and when activated will send a signal through the air that will cause the chime to sound a tone. The big drawback with this type of chime is that you are limited to only receptacle locations for chime placement. Other wireless chime models are battery-powered and can be placed anywhere. The big drawback to this type of chime is that when the batteries die there is no door chime. It may be old-fashioned, but in my opinion a hardwired door chime system is still the best way to go.

120-VOLT PRIMARY WIRES

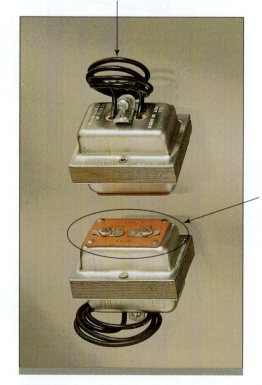

LOW-VOLTAGE SECONDARY TERMINAL SCREWS (USUALLY 16 VOLTS BETWEEN TERMINALS)

FIGURE 14-53 Chime transformers. *Courtesy of NuTone, Inc.*

The wire used to connect the buttons, chime, and transformer is often called "bell wire" or simply "thermostat wire" since it is the same type as that used to wire heating and cooling system thermostats. It is usually 16 or 18 AWG solid copper with an insulation type that limits it to use on 30-volt-or-less circuits. It comes in a cable assembly with two, three, or more single conductors covered with a protective outer sheathing. Each conductor is color-coded in the cable assembly. The cable is run through bored holes in the building framing members or on the surface. It is supported with small insulated staples or cleats. It can also be installed in a raceway for added protection from physical damage. When running the cable, keep the following (Article 725) items in mind:

- Do not install Class 2 wiring in the same raceway or electrical box as the regular power and lighting conductors.

- Keep the Class 2 bell wire at least 2 inches (50 mm) from light and power wiring. However, when the wiring method is nonmetallic sheathed cable, Type AC cable, Type MC cable, or a raceway wiring method, there is no problem with running the bell wire right next to these wiring methods.

- Try not to run the chime wiring through the same bored holes as regular light and power wiring methods. While this is not against the *NEC®*, it will ensure that there will not be any physical damage to the chime wiring.

- Do not strap or otherwise support the chime wiring to other electrical raceways or cables.

- The cable insulation on bell wire is very thin. Be very careful when installing this cable type so as not to damage it.

There are two common wiring schemes for a chime circuit. The first is to run a two-wire thermostat cable from each doorbell button location and from the transformer location to the chime. This means that at the chime location, there will be three two-wire cables for a total of six conductors. The wiring connections for this arrangement are shown in Figure 14-54. The second scheme that electricians often use is to install a two-wire thermostat cable from each doorbell button location to the transformer location. Next, install a three-wire thermostat cable from the chime location to the transformer location. Make the connections as shown in Figure 14-55. The advantage of the second method is that there will be only three wires at the chime, making connections easier. The disadvantage is that there will be several wires at the transformer, making the connection more complicated. Another disadvantage is that an electrician will need to keep both a spool of two-wire and three-wire thermostat cable. Only a spool of two-wire thermostat cable is needed for the first method.

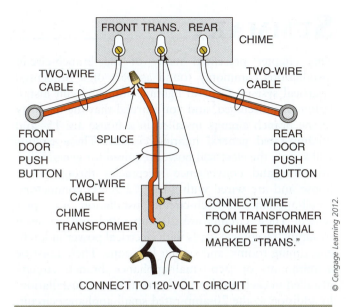

FIGURE 14-54 A typical chime installation showing the wiring connections when a two-wire cable is brought from the transformer and each button to the door chime.

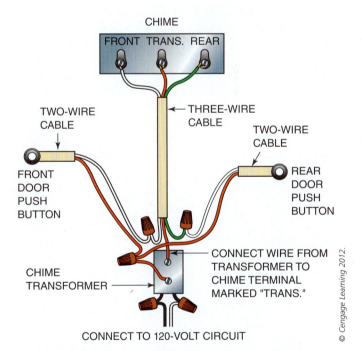

FIGURE 14-55 A typical chime installation showing the wiring connections when a two-wire cable is brought from each door chime button to the transformer and a three-wire cable is brought from the transformer to the door chime.

SUMMARY

This chapter presented many of the branch-circuit installations commonly found as part of a residential electrical system. General lighting branch-circuit installation was discussed, and you learned that the majority of the branch circuits installed in a house are 15- and 20-amp-rated general lighting circuits. These circuits will supply the electrical power required for general illumination and convenience receptacles throughout a house and are wired with 14 or 12 AWG conductors. Small-appliance branch-circuit installation was presented. These circuits are 20-amp rated and are used to supply receptacle loads with electrical power in kitchens, dining rooms, and similar locations. There must be a minimum of two small-appliance branch circuits installed in each residential electrical system installation. In addition to the 20-amp-rated small-appliance circuit, at least one 20-amp-rated laundry circuit must also be installed. The main purpose of the laundry circuit is to supply electrical power to appliances in the laundry area, specifically a clothes washer. Bathroom circuits are required to be 20-amp rated and are installed much the same way as small-appliance and laundry

branch circuits. The conductor size used for both the laundry branch circuit and the bathroom branch circuits is 12 AWG. This chapter also presented branch-circuit installation information for a variety of individual branch circuits to appliances commonly found in kitchens, such as electric ranges, counter-mounted cooktops, wall-mounted ovens, dishwashers, and garbage disposals. The wire sizes and overcurrent protection sizes are determined from the nameplate ratings and applying the proper *NEC*® section information. Other individual branch-circuit installations presented were electric clothes dryers, electric water heaters, electric heaters, air-conditioning systems, and gas or oil central heating systems. The wire sizes and fuse or circuit breaker sizes were also determined by the nameplate ratings and *NEC*® requirements. Installation of smoke detector circuits, carbon monoxide detectors, and chime systems were the last areas covered in this chapter.

By following the recommended installation practices for the branch circuits covered in this chapter, electricians will be able to determine and install the required branch circuits of a residential electrical system in a way that both meets the *NEC*® minimum requirements and adequately serves the home owner's needs for electrical power.

REVIEW QUESTIONS

Directions: Answer the following items with clear and complete responses.

1. Explain the difference between a "two-wire" submersible pump and a "three-wire" submersible pump.

2. Determine the full-load current for a 2-horsepower, 230-volt, single-phase motor. *Hint*: Use Table 430.248.

3. The *NEC®* requires electric water heaters that are 120 gallons (450 L) or less to be considered continuous duty and as such must have a circuit rating of not less than _____ percent of the water heater rating.

4. Name and describe the three installation methods for a countertop cook unit and a wall-mounted oven.

5. A wall-mounted oven has a nameplate rating of 8 kilowatts at 240 volts. The rated load is _____ watts and _____ amps.

6. Name two common installation methods for garbage disposal units.

7. A receptacle located under a kitchen sink so that a cord-and-plug-connected garbage disposal can be plugged into it must be GFCI protected. True or False

8. Name the section of the *NEC®* that limits the length of the attachment cord to a dishwasher. Specify the minimum and maximum length of the cord.

9. List several advantages of electric heating.

10. Most electric heating equipment operates on 240 volts. When the equipment is fed using a 10/2 nonmetallic sheathed cable, what has to be done to the white conductor in the cable at the connection points?

11. The total load of a single-room air conditioner on an individual branch circuit cannot exceed _____ percent of the circuit rating.

12. The total load of a single-room air conditioner on a branch circuit that also supplies lighting in a room cannot exceed _____ percent of the circuit rating.

13. The nameplate on an air conditioner unit states "maximum size fuse or HACR circuit breaker." Describe what an HACR circuit breaker is.

14. The disconnecting means for an air conditioner or heat pump must be installed _____ of the unit.

15. Name the *NEC®* article that prohibits the use of low-voltage Class 2 thermostat wires from being in the same raceway or electrical enclosure with power and lighting circuit conductors.

16. Name the *NEC®* section that requires a separate branch circuit for a central heating system furnace.

17. The low-voltage thermostat wiring between a furnace and a thermostat is considered to be Class 1 wiring. (Circle the correct response.) True or False

18. Doorbell buttons used with a chime system are called "momentary contact switches." Describe the operation of this kind of switch.

19. Describe the purpose of the transformer as used in a residential chime system.

20. Assuming the use of a cable wiring method, describe the reason for running a three-wire cable between each smoke detector location in a house.

21. Describe carbon monoxide.

22. Discuss the placement of carbon monoxide alarms in a house.

23. Name one disadvantage of using a wireless battery-powered door chime system.

24. When determining the amount of electric heating to be installed in a house the rule of thumb is to use _____ watts per square foot.

25. Smoke detectors can be interconnected with each other so that when one is activated and sounds an alarm, they all sound an alarm. True or False

KNOW YOUR CODES

1. Check with the local or state authority having jurisdiction in your area and find out if carbon monoxide alarms are required in homes. If they are, where must they be located?

2. Check with the local or state authority having jurisdiction in your area and find out where smoke detectors are required to be located in homes.

Special Residential Wiring Situations

OBJECTIVES *Upon completion of this chapter, the student should be able to:*

- Demonstrate an understanding of the installation of garage feeders and branch circuits.

- Demonstrate an understanding of the installation of branch circuits for a swimming pool.

- Demonstrate an understanding of the installation of branch circuits in outdoor situations.

- Demonstrate an understanding of the installation of a standby power system.

GLOSSARY OF TERMS

critical loads the electrical loads that are determined to require electrical power from a standby power generator when electrical power from the local electric utility company is interrupted

dry-niche luminaire a lighting fixture intended for installation in the wall of a pool that goes in a niche; it has a fixed lens that seals against water entering the niche and surrounding the lighting fixture

exothermic welding a process for making bonding connections on the bonding grid for a permanently installed swimming pool using specially designed connectors, a form, a metal disk, and explosive powder; this process is sometimes called "Cad-Weld"

forming shell the support structure designed and used with a wet-niche lighting fixture; it is installed in the wall of a pool

generator a rotating machine used to convert mechanical energy into electrical energy

hydromassage bathtub a permanently installed bathtub with re-circulating piping, pump, and associated equipment; it is designed to accept, circulate, and discharge water each time it is used

multiwire branch circuit a branch circuit that consists of two or more ungrounded conductors that have a voltage between them and a grounded conductor that has an equal voltage between it and each ungrounded conductor

no-niche luminaire is a lighting fixture intended for installation above or below the water level; it does not have a forming shell that it fits into but rather sits on the surface of the pool wall

permanently installed swimming pool a swimming pool constructed totally or partially in the ground with a water depth capacity of greater than 42 inches (1 m); all pools, regardless of depth, installed in a building are considered permanent

photocell a light-sensing device used to control lighting fixtures in response to detected light levels; no sunlight detected causes it to switch on a lighting fixture; sunlight detected causes it to switch off a lighting fixture

pool cover, electrically operated a motor-driven piece of equipment designed to cover and uncover the water surface of a pool

self-contained spa or hot tub a factory-fabricated unit consisting of a spa or hot tub vessel having integrated water-circulating, heating, and control equipment

spa or hot tub a hydromassage pool or tub designed for immersion of users; usually has a filter, heater, and motor-driven pump; it can be installed indoors or outdoors, on the ground, or in a supporting structure; a spa or hot tub is not designed to be drained after each use

standby power system a backup electrical power system that consists of a generator, transfer switch, and associated electrical equipment; its purpose is to provide electrical power to critical branch circuits when the electrical power from the utility company is not available

storable swimming pool a swimming pool constructed on or above the ground with a maximum water-depth capacity of 42 inches (1 m) or a pool with nonmetallic, molded polymeric walls (or inflatable fabric walls) regardless of size or water-depth capacity

transfer switch a switching device for transferring one or more load conductor connections from one power source to another

twistlock receptacle a type of receptacle that requires the attachment plug to be inserted and then turned slightly in a clockwise direction to lock the plug in place; the attachment plug must be turned slightly counterclockwise to release the plug so it can be removed from the receptacle

wet location installations underground or in concrete slabs or masonry in direct contact with the earth in locations subject to saturation with water or other liquids, such as in unprotected areas exposed to the weather

wet-niche luminaire a type of lighting fixture intended for installation in a wall of a pool; it is accessible by removing the lens from the forming shell; this luminaire type is designed so that water completely surrounds the fixture inside the forming shell

There are many special wiring situations that come up during the installation of a residential electrical system. Today, most new homes are built with either an attached or a detached garage that will require electrical wiring. Many homes have swimming pools, hot tubs, or hydromassage bathtubs, which are installed either at the time of construction or sometime after the house is built. All these items require an electrician to install electrical wiring for associated equipment such as pumps and pool lighting. Outside wiring may be required for items such as outdoor receptacles and outdoor lighting. Some homes are located in an area where the electrical power supply from the local electric utility may not be dependable, and a portable generator system may need to be installed. This chapter looks at several special wiring situations that electricians encounter on a regular basis.

INSTALLING GARAGE FEEDERS AND BRANCH CIRCUITS

Garages are either attached to the main house during the construction process or are built detached at some distance away from the main house. They can be designed for a single vehicle or for multiple vehicles. Garages also are typically used to store things like lawnmowers, snowblowers, and gardening tools. In warmer climates, a clothes washer and clothes dryer may be located in the garage. You might also find refrigerators and freezers in a garage. In this section, we will take a look at the electrical wiring that is typically installed in a garage.

ATTACHED GARAGES

If the garage is attached, Section 210.52(G) of the *National Electrical Code®* (*NEC®*) requires at least one 15- or 20-amp, 125-volt receptacle be located there, and Section 210.70(A)(2)(a) requires at least one

wall switch-controlled lighting outlet in the garage (Figure 15-1). While these minimum requirements may be satisfactory for a small attached garage, most garages are large enough to require several lighting outlets and several receptacle outlets. There is no *NEC®* requirement concerning the maximum distance between receptacle outlets in a garage. Therefore, it is up to the electrician as to how many and how far apart the receptacles are located. The branch circuits supplying the receptacle(s) and lighting outlet(s) may be on the same branch circuit and can be rated either 15 or 20 amp. A good wiring practice is to install the lighting outlets on a separate 15- or 20-amp branch circuit and to install the receptacle(s) on one or more separate 20-amp-rated branch circuits. Section 210.8(A)(2) requires all the receptacles installed in a garage to be GFCI protected.

CAUTION

CAUTION: Be aware that previous editions of the *National Electrical Code* (*NEC®*) permitted 15- and 20-amp, 125-volt-rated receptacles in garages and unfinished basements to *not* require GFCI protection if they were *not* readily accessible. An example of this was a receptacle located in a ceiling of a garage into which a garage door opener would be plugged in. It was usually considered not readily accessible and did not need to be GFCI protected. Another receptacle location that was exempt from having GFCI protection was any receptacle that served cord-and-plug-connected appliances, which were not easily moved, like a refrigerator or freezer. For this exemption, the receptacle needed to be located behind the appliance. One last situation that did not require GFCI protection was when an appliance was plugged into a single receptacle. An example of this was when a washing machine located in an unfinished basement or garage was plugged into a single receptacle. Starting with the 2008 *NEC®* and continuing with the 2011 *NEC®*, it is required that all 15- and 20-amp, 125-volt-rated receptacles (whether they are readily accessible, single, or duplex) located in a garage or unfinished basement be GFCI protected. An *Exception* permits a receptacle installed to supply a permanently installed fire alarm or security system to not be GFCI protected.

It is important for an electrician to make sure that an adequate amount of light is installed in a garage. Many electricians install lighting outlets directly over a vehicle. This location will light up the roof of a car or truck but does not properly put the light where it is needed most—along the sides of the vehicles where the

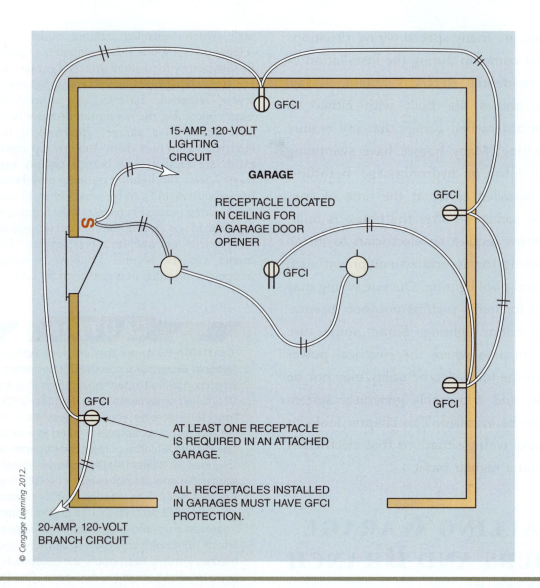

15-AMP, 120-VOLT
LIGHTING
CIRCUIT

GARAGE

RECEPTACLE LOCATED
IN CEILING FOR
A GARAGE DOOR
OPENER

GFCI

GFCI

GFCI

GFCI

GFCI

GFCI

AT LEAST ONE RECEPTACLE
IS REQUIRED IN AN ATTACHED
GARAGE.

ALL RECEPTACLES INSTALLED
IN GARAGES MUST HAVE GFCI
PROTECTION.

20-AMP, 120-VOLT
BRANCH CIRCUIT

© Cengage Learning 2012.

FIGURE 15-1 An electrical plan for a garage that meets the requirements of Sections 210.52(G) and 210.70(A)(2)(a).

doors open and people enter and exit them. Good wiring practice requires electricians to install lighting outlets as follows (Figure 15-2):

- Two lighting outlets minimum in a one-car garage located on each side of the vehicle parking area

- Three lighting outlets minimum in a two-car garage located in the middle and to the outside of each vehicle parking location

- Four lighting outlets minimum in a three-car garage located so light shines down into the areas on each side of the vehicle parking areas

Locate the lighting outlets at a location that is a little bit off of center and toward the front of the vehicle parking area so that if work is needed to be done on a car or truck and the hood is up, enough light will be

available to see. Also, by locating the lighting outlets more toward the back of the vehicle parking areas, a fully opened overhead garage door may block the light and defeat the whole purpose of having adequate lighting outlets to begin with.

In many areas of the country, garage lighting is often done with 75- or 100-watt incandescent lamps used with porcelain or plastic lamp-holders (Figure 15-3). (*Note:* Common lamp types used in residential wiring are discussed in Chapter 17.) In some areas, especially those where the outside temperature is not likely to fall below 50°F, fluorescent lamps are used as the lighting source. Without special ballasts, a fluorescent lamp will usually not operate when exposed to temperatures below 50°F. Of course, if the home has a heated garage, fluorescent lamps may be used even in the coldest of climates.

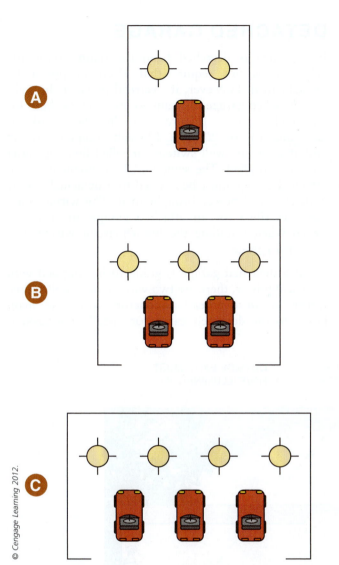

FIGURE 15-2 Suggested lighting outlet locations in a garage. (A) One-car garage, (B) two-car garage, and (C) three-car garage.

© Cengage Learning 2012.

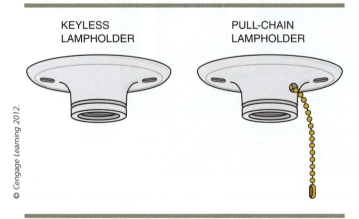

KEYLESS LAMPHOLDER   PULL-CHAIN LAMPHOLDER

© Cengage Learning 2012.

FIGURE 15-3 Typical incandescent lamp-holders. They can be made of porcelain or plastic and are used in garages, attics, unfinished basements, and crawl spaces.

Green Tip

Especially in warmer areas of the country, incandescent lamps used for lighting in a garage can be replaced with compact fluorescent (CFL) lamps. CFLs will use less electricity to operate and they tend to last for a long time. Monthly electrical bills will be less when using CFLs. It is very easy to replace incandescent lamps with CFLs because they will screw directly into the existing lamp-holders.

The lighting outlets for the incandescent lamp-holders are installed using a plastic round ceiling box or metal 4-inch octagon box. The wiring method used is brought to the outlet boxes, and the proper connections are made to the lighting fixtures. If fluorescent lighting is used, the wiring method is normally attached directly to surface-mounted fluorescent lighting fixtures, and all the wiring connections are done inside the fixtures.

When installing the electrical boxes for receptacle outlets in a garage, the usual mounting height is around 48 inches (1.2 m) to center from the finished garage floor. Remember that there is no *NEC®* requirement for the mounting height, and it is really up to the electrician as to how high the receptacle outlets are located.

The branch-circuit wiring is brought from the electrical panel to the various switching points, lighting outlets, and receptacle outlets that the electrician decides should be on each circuit. Local building codes for houses with attached garages require that the wall that separates the main house from the garage have a specific fire-resistance rating. We learned earlier that if holes are drilled and wiring methods are installed through a fire-resistance-rated wall, floor, or ceiling, the holes must be filled with a material that meets or exceeds the required fire rating. Therefore, if the garage branch circuits originate at an electrical panel located in the main part of the house and the circuit wiring penetrates the fire-resistant separating wall, be sure to plug the holes with an approved method.

Some interiors of a garage are finished off. That is, sheetrock or some other wall covering is installed on the wall studs and ceiling joists. When a garage is finished off in this manner, the lighting outlet and receptacle outlet boxes are installed in the same manner as inside the house. The proper box setout must be maintained so that the box front edge is flush with the finished wall surface. Most garages are unfinished, and proper box setout from the studs and joists is not as important.

However, in case the garage walls and ceiling might be finished off in the future, it is a good wiring practice to install the electrical boxes with a ½-inch (13 mm) setout.

CAUTION

CAUTION: Garages that are unfinished result in exposed runs of the wiring method used to install the branch-circuit wiring. Some electrical inspectors may not allow wiring methods such as nonmetallic sheathed cable to be used in a garage because it is considered to be exposed to potential physical damage. In these instances, a metal-clad cable or conduit wiring system may have to be installed in an unfinished garage. Always check with the authority having jurisdiction (AHJ) to make sure the wiring method you are using is allowed in an unfinished garage.

DETACHED GARAGE

If a garage is detached from the main house, the *NEC*® does not require that electrical power be brought to it. However, if electrical power is brought to a detached garage, the same rules apply as those for an attached garage (Figure 15-4). In other words, at least one 15- or 20-amp, 125-volt receptacle outlet and at least one wall switch-controlled lighting outlet must be installed. The same GFCI requirement for an attached garage must be applied to a detached garage with electrical power brought to it. The wiring practices described for an attached garage are also followed when installing the branch-circuit wiring in a detached garage.

If a detached garage is going to be supplied with electrical power, there are two ways to accomplish this: Either install overhead conductors from the main house to the detached garage or install underground

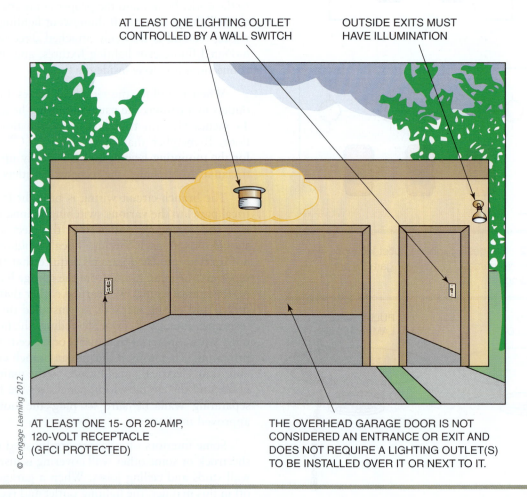

AT LEAST ONE LIGHTING OUTLET CONTROLLED BY A WALL SWITCH

OUTSIDE EXITS MUST HAVE ILLUMINATION

AT LEAST ONE 15- OR 20-AMP, 120-VOLT RECEPTACLE (GFCI PROTECTED)

THE OVERHEAD GARAGE DOOR IS NOT CONSIDERED AN ENTRANCE OR EXIT AND DOES NOT REQUIRE A LIGHTING OUTLET(S) TO BE INSTALLED OVER IT OR NEXT TO IT.

© Cengage Learning 2012.

FIGURE 15-4 A detached garage that meets the minimum *NEC*® requirements when electrical power is brought to it. Remember, a detached garage is not required to have electrical power brought to it, but if it is, the *NEC*® requirements are the same as those for an attached garage.

wiring from the main house to the detached garage. If there is not much of an electrical load requirement in a detached garage, one circuit can be installed to serve the required receptacle outlet and switched lighting outlet. Electricians can install this single branch circuit with a Type UF cable containing an equipment-grounding conductor. It will need to be buried according to the requirements of Table 300.5 of the *NEC*®, which tells us that the minimum burial depth for this type of circuit installation is 24 inches (600 mm). Another way to install a single circuit would be to bury a PVC conduit from the main house to the detached garage and then pull in the required circuit conductors, including a green grounding conductor. One last method of installation worth mentioning would be to bury a metal raceway system, like rigid metal conduit, between the main house and the detached garage. The circuit conductors would then be pulled into the metal raceway, and if approved fittings are used to attach the raceway at each end, Section 250.118 would allow us to use the metal raceway as a grounding means and an additional green grounding conductor would not have to be installed. The Section 250.32(A) *Exception* says that if a detached garage is supplied by only one branch circuit with an equipment-grounding conductor, there is no requirement to establish a grounding electrode system or to connect to one if one exists. For this situation, a **multiwire branch circuit** is considered to be a single branch circuit. A multiwire branch circuit could be installed using a three-wire Type UF cable or three wires and an equipment-grounding conductor in a raceway. This would supply two 120-volt branch circuits with a shared neutral conductor. The multiwire branch circuit is typically brought to an electrical box and then the wiring is split up from there. For example, the black wire and the shared neutral could supply lighting outlets and the red wire and the shared neutral could supply receptacle outlets. The overcurrent protection device protecting the wiring would be located back at the main house electrical panel. A two-pole circuit breaker would be required.

> ### CAUTION
> **CAUTION:** Whenever burying a cable or raceway underground, consult Table 300.5 for the minimum burial depths.

Usually the electrical loads required to be served in a detached garage require more electrical power than a single branch circuit can supply. A common wiring practice is to install feeder wiring from the main service entrance panel to a subpanel located in the garage. Several different branch circuits can originate in the garage subpanel and supply the electrical load requirements of the garage. The feeder can be brought to the detached garage as either a cable, typically Type UF cable, or a raceway wiring method, typically PVC conduit.

When a feeder supplies a detached garage, Section 250.32(A) requires that a grounding electrode system be established, unless one already exists (Figure 15-5). The most common method of establishing a grounding electrode system at the detached garage is to simply drive an 8-foot (2.5-m) ground rod. The equipment-grounding bus bar in the subpanel located in the detached garage must be bonded to the grounding electrode system. The disconnecting means enclosure must also be bonded to the grounding electrode system. Additionally, Section 250.32(B)(1) requires a feeder supplying a detached garage from the service of the main house to have an equipment-grounding conductor sized according to Table 250.122 run with the feeder. The grounded conductor (neutral) is not permitted to be connected to the equipment-grounding conductor or to the grounding electrode system (Figure 15-6). The grounding electrode conductor in the detached garage is sized according to Table 250.66, the same way the grounding electrode conductor for the service entrance in the main house is sized.

Section 250.32(B)(1) *Exception* says that for buildings wired in compliance with previous *NEC*® editions that permitted such connections, when the feeder to the detached garage does not have an equipment-grounding conductor run with the feeder to the garage and there are no continuous metallic paths bonded to the grounding system in both the main house and the detached garage (like metal water pipes or metal raceways), the grounded neutral circuit conductor run with the supply to the detached garage must be connected to the subpanel enclosure and to the grounding electrode (Figure 15-7). In this case, the connections in the garage subpanel are the same as for the main service panel in the main house. That is, use the main bonding jumper and connect the equipment-grounding bus bar and the grounded neutral bus bars together. The size of the grounded neutral conductor serving the detached garage must not be smaller than the size required by Table 250.122 or the size required for the calculated neutral load according to Section 220.61. Common wiring practice calls for the grounded neutral conductor to be the same size as the ungrounded "hot" circuit conductors feeding the garage subpanel. Remember, this method can only be used on an existing building that was wired according to a previous *NEC*® edition.

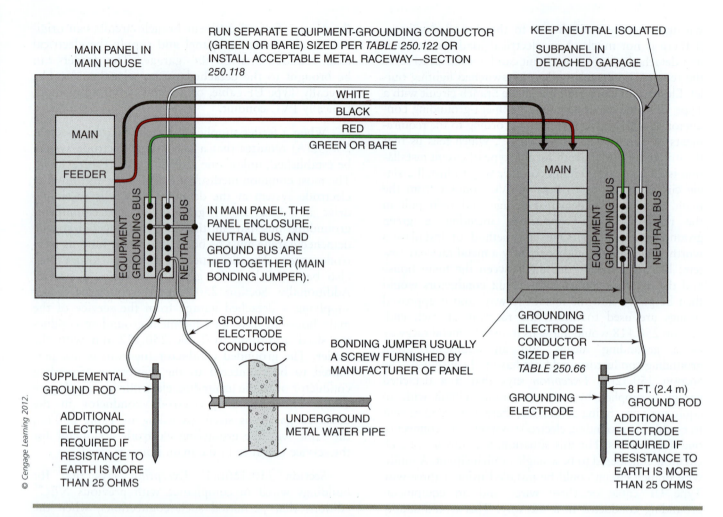

FIGURE 15-5 Proper grounding connections at a detached garage when a grounding conductor is included with the feeder from the main house to a detached garage, Section 250.32(B).

from experience...

If you are required to install wiring for a storage shed or other accessory building, install the wiring the same as for a detached garage. There is no requirement that electrical power be brought to a storage shed or accessory building, but if it is, at least one wall switch-controlled lighting outlet and at least one receptacle outlet must be installed. The GFCI requirements for receptacles as outlined in the *NEC*® for garages must be followed.

INSTALLING BRANCH-CIRCUIT WIRING FOR A SWIMMING POOL

Swimming pools are often installed at the same time a new home is built. Many owners of existing homes have swimming pools installed after the house has been built for a few years. Whether it is during the initial installation of the electrical system or sometime after, electricians find themselves installing the branch-circuit wiring required for a swimming pool quite often. In this section, we look at the most important *NEC*® rules and common wiring practices for swimming pools and other similar equipment, such as hot tubs and hydromassage baths.

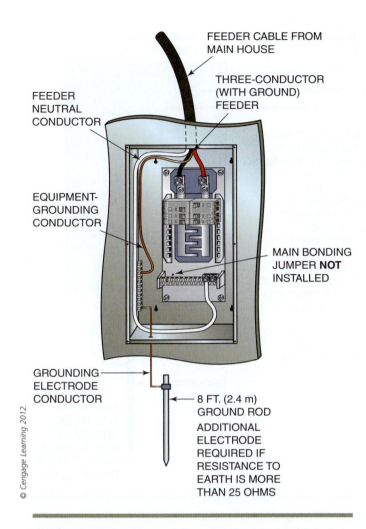

FEEDER CABLE FROM MAIN HOUSE

THREE-CONDUCTOR (WITH GROUND) FEEDER

FEEDER NEUTRAL CONDUCTOR

EQUIPMENT-GROUNDING CONDUCTOR

MAIN BONDING JUMPER **NOT** INSTALLED

GROUNDING ELECTRODE CONDUCTOR

8 FT. (2.4 m) GROUND ROD ADDITIONAL ELECTRODE REQUIRED IF RESISTANCE TO EARTH IS MORE THAN 25 OHMS

© Cengage Learning 2012.

FIGURE 15-6 The neutral conductor to the feeder is not allowed to be connected to the equipment-grounding bar or to the grounding electrode system in the panel when an equipment-grounding conductor is brought to the detached garage with the feeder from the main house.

Article 680 covers swimming pools, hydromassage bathtubs, hot tubs, and spas. This article applies to both permanent and storable equipment and includes rules for auxiliary equipment such as pumps and filters. These are all items that an electrician may encounter during a residential electrical system installation.

Section 680.2 has many terms and definitions that relate specifically to the items covered in Article 680. The following are considered the most important to know for residential wiring:

- A **spa or hot tub** is a hydromassage pool or tub designed for immersion of users and usually has a filter, heater, and motor-driven pump. It can be installed indoors or outdoors on the ground or in a supporting structure. Generally, a spa or hot tub is not designed to be drained after each use (Figure 15-8).

- A **self-contained spa or hot tub** is a factory-fabricated unit consisting of a spa or hot tub vessel having integrated water-circulating, heating, and control equipment (Figure 15-9).

- A **hydromassage bathtub** is a permanently installed bathtub with re-circulating piping, pump, and associated equipment. It is designed to accept, circulate, and discharge water on each use (Figure 15-10).

- A **wet-niche luminaire** is a type of lighting fixture intended for installation in a wall of a pool. It is accessible by removing the lens from the forming shell. This luminaire type is designed so that water completely surrounds the fixture inside the forming shell (Figure 15-11).

- A **dry-niche luminaire** is a lighting fixture intended for installation in the wall of a pool that goes in a niche. However, unlike the wet-niche luminaire, the dry-niche luminaire has a fixed lens that seals against water entering the niche and surrounding the lighting fixture (Figure 15-12).

- A **no-niche luminaire** is a lighting fixture intended for installation above or below the water level. It does not have a forming shell that it fits into. It simply sits on the surface of the pool wall (Figure 15-13).

- A **permanently installed swimming pool** is one constructed totally or partially in the ground with a water-depth capacity of greater than 42 inches (1 m). All pools installed in a building, regardless of depth, are considered permanent (Figure 15-14).

- A **storable swimming pool** is one constructed on or above the ground and has a maximum water-depth capacity of 42 inches (1 m) or a pool with nonmetallic, molded polymeric walls (or inflatable fabric walls) regardless of size or water depth capacity (Figure 15-15).

- A **pool cover, electrically operated**, is a motor-driven piece of equipment designed to cover and uncover the water surface of a pool (Figure 15-16).

- A **forming shell** is the support structure designed and used with a wet-niche lighting fixture. It is installed in the wall of a pool (Figure 15-17).

PERMANENTLY INSTALLED POOLS

The *NEC*® rules for permanently installed pools are located in Part II of Article 680. Part II provides installation rules for clearances of overhead lighting, underwater lighting, receptacles, switching, associated equipment, bonding of metal parts, grounding, and electric heaters.

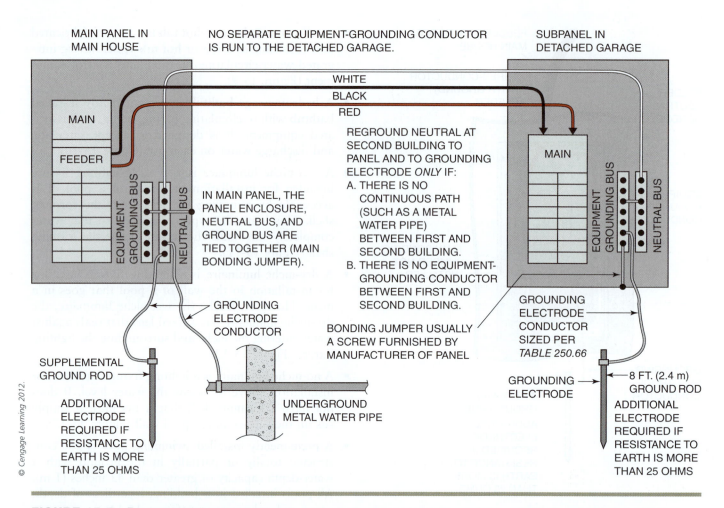

MAIN PANEL IN
MAIN HOUSE

NO SEPARATE EQUIPMENT-GROUNDING CONDUCTOR
IS RUN TO THE DETACHED GARAGE.

SUBPANEL IN
DETACHED GARAGE

WHITE

BLACK

RED

MAIN

FEEDER

EQUIPMENT GROUNDING BUS

NEUTRAL BUS

IN MAIN PANEL, THE
PANEL ENCLOSURE,
NEUTRAL BUS, AND
GROUND BUS ARE
TIED TOGETHER (MAIN
BONDING JUMPER).

REGROUND NEUTRAL AT
SECOND BUILDING TO
PANEL AND TO GROUNDING
ELECTRODE *ONLY* IF:
A. THERE IS NO
   CONTINUOUS PATH
   (SUCH AS A METAL
   WATER PIPE)
   BETWEEN FIRST AND
   SECOND BUILDING.
B. THERE IS NO EQUIPMENT-
   GROUNDING CONDUCTOR
   BETWEEN FIRST AND
   SECOND BUILDING.

MAIN

EQUIPMENT GROUNDING BUS

NEUTRAL BUS

GROUNDING
ELECTRODE
CONDUCTOR

BONDING JUMPER USUALLY
A SCREW FURNISHED BY
MANUFACTURER OF PANEL

GROUNDING
ELECTRODE
CONDUCTOR
SIZED PER
TABLE 250.66

SUPPLEMENTAL
GROUND ROD

GROUNDING
ELECTRODE

8 FT. (2.4 m)
GROUND ROD

UNDERGROUND
METAL WATER PIPE

ADDITIONAL
ELECTRODE
REQUIRED IF
RESISTANCE TO
EARTH IS MORE
THAN 25 OHMS

ADDITIONAL
ELECTRODE
REQUIRED IF
RESISTANCE TO
EARTH IS MORE
THAN 25 OHMS

© Cengage Learning 2012.

FIGURE 15-7 Proper grounding connections at a detached garage when a grounding conductor is not included with the feeder from the main house to a detached garage, Section 250.32(B). This method can only be used on existing building wiring systems.

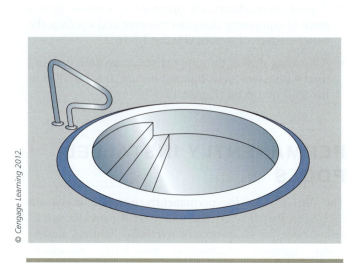

© Cengage Learning 2012.

FIGURE 15-8 A spa or hot tub.

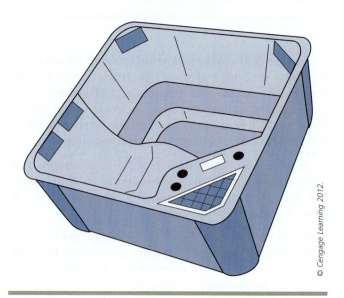

© Cengage Learning 2012.

FIGURE 15-9 A self-contained spa or hot tub.

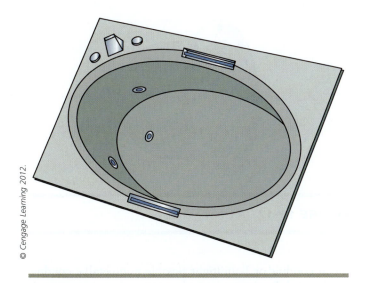

FIGURE 15-10 A hydromassage bathtub.

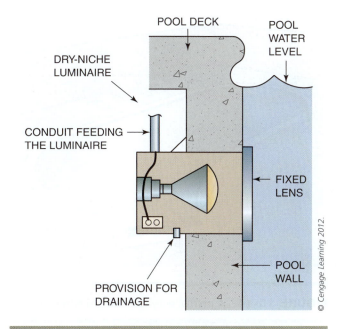

FIGURE 15-12 A dry-niche luminaire.

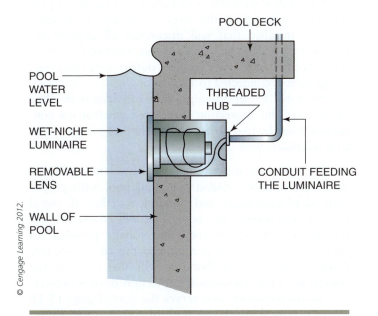

FIGURE 15-11 A wet-niche luminaire.

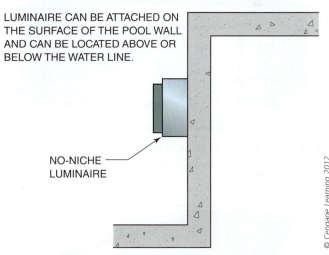

FIGURE 15-13 A no-niche luminaire.

The electrical panel in a house or a subpanel in a garage or other building located on the premises can supply the electrical power to the pool. Branch-circuit wiring will be brought from the appropriate electrical panel to serve the various swimming pool loads. When running branch-circuit wiring from an electrical panel inside a house or any other associated building, like a garage, to pool-associated motors, the wiring method can be any of the regular wiring methods recognized in Chapter 3 of the *NEC*®. If a raceway is used, whether it is nonmetallic or metal, an insulated equipment-grounding conductor must be used. If a cable assembly is used, like a nonmetallic sheathed cable, an uninsulated

equipment-grounding conductor can be used, but it must be enclosed within the cable's outer sheathing. Many permanently installed swimming pool pump motors are installed using a cord-and-plug connection. If so, the pump motor must have an approved double-insulation system and must provide a grounding means for the pump's internal and inaccessible noncurrent-carrying metal parts. Section 680.21(A) requires the branch-circuit wiring for pool-associated motors to be installed in rigid metal conduit, intermediate metal conduit, PVC conduit, or metal-clad (Type MC) cable listed for the location-specific conditions encountered. Any wiring method used must contain a copper equipment-grounding

FIGURE 15-14 A permanently installed swimming pool.

FIGURE 15-15 A storable swimming pool.

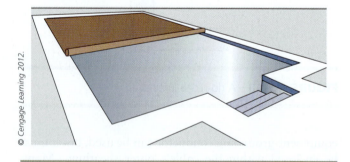

FIGURE 15-16 A pool cover, electrically operated.

conductor that is sized according to Table 250.122, but in no case can it be smaller than 12 AWG. Electrical metallic tubing is allowed to be used when the wiring will be installed on or within a building. PVC conduit is often used by electricians to install the wiring for pool motors and other pool-associated equipment. If a flexible connection to the pool motor is necessary, liquidtight flexible

FIGURE 15-17 A forming shell.

metal conduit or liquidtight flexible nonmetallic conduit can be used. Pool motors can be cord-and-plug connected. If they are, the cord cannot be more than 3 feet (900 mm) long and must contain an equipment-grounding conductor sized according to Table 250.122. The attachment plug will have to be a grounding type. A disconnecting means must be installed that will disconnect all ungrounded conductors. It must be located within sight of the equipment. Section 680.21(C) requires outlets supplying pool pump motors connected to 120- through 240-volt branch circuits, rated 15- or 20-amps, to be provided with GFCI protection. This protection is required whether the connection is hard-wired to the pump or the pump is connected to a receptacle with a cord and plug.

Probably the most important installation practice for a permanently installed swimming pool is the proper method of grounding and bonding together all metal parts in and around the pool. This will ensure that all the metal parts are at the same voltage potential to ground, which will help minimize the shock hazard. Proper grounding and bonding will also facilitate the operation of the overcurrent protection devices protecting the circuit wiring that serves the pool. Figure 15-18 shows a typical grounding path of common items used with a permanently installed swimming pool. Pertinent *NEC*® requirements and the proper section location are also shown in the illustration.

Section 680.26 requires an electrician to properly bond together all the metal parts in and around the permanently installed swimming pool. The bonding conductor is not required to be connected to a grounding electrode, to any part of the house service entrance, or to any other part of the building electrical system. It is simply used to tie all the metal parts together. Figure 15-19 shows an example of what items should be connected together. An 8 AWG or larger *solid* copper conductor must be used to make the bonding connections. The bonding conductor can be bare or insulated, and the connection to the metal parts must be made with **exothermic welding** or by pressure connectors labeled for the purpose

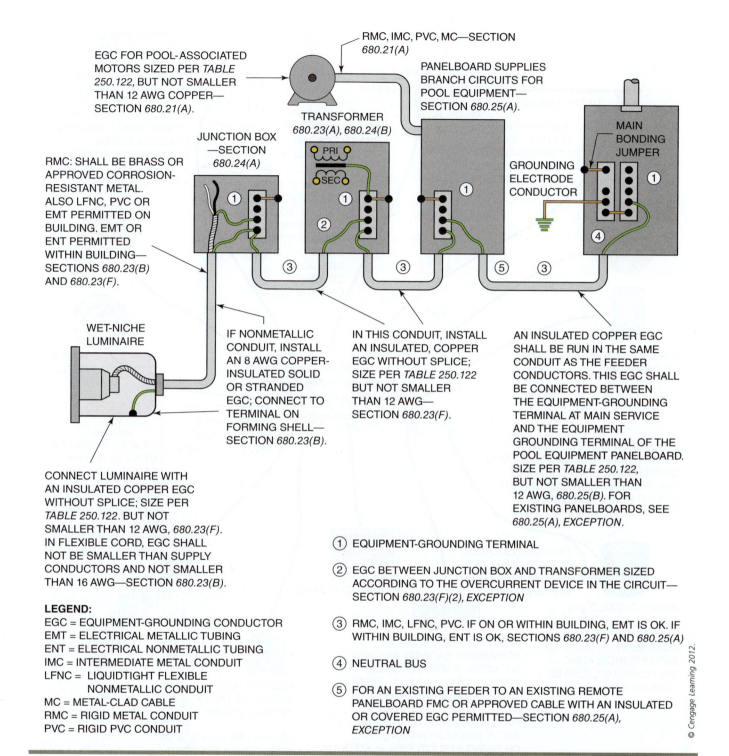

EGC FOR POOL-ASSOCIATED MOTORS SIZED PER *TABLE 250.122*, BUT NOT SMALLER THAN 12 AWG COPPER—SECTION *680.21(A)*.

RMC, IMC, PVC, MC—SECTION *680.21(A)*

PANELBOARD SUPPLIES BRANCH CIRCUITS FOR POOL EQUIPMENT—SECTION *680.25(A)*.

TRANSFORMER *680.23(A), 680.24(B)*

MAIN BONDING JUMPER

JUNCTION BOX —SECTION *680.24(A)*

RMC: SHALL BE BRASS OR APPROVED CORROSION-RESISTANT METAL. ALSO LFNC, PVC OR EMT PERMITTED ON BUILDING. EMT OR ENT PERMITTED WITHIN BUILDING—SECTIONS *680.23(B)* AND *680.23(F)*.

GROUNDING ELECTRODE CONDUCTOR

WET-NICHE LUMINAIRE

IF NONMETALLIC CONDUIT, INSTALL AN 8 AWG COPPER-INSULATED SOLID OR STRANDED EGC; CONNECT TO TERMINAL ON FORMING SHELL—SECTION *680.23(B)*.

IN THIS CONDUIT, INSTALL AN INSULATED, COPPER EGC WITHOUT SPLICE; SIZE PER *TABLE 250.122* BUT NOT SMALLER THAN 12 AWG—SECTION *680.23(F)*.

AN INSULATED COPPER EGC SHALL BE RUN IN THE SAME CONDUIT AS THE FEEDER CONDUCTORS. THIS EGC SHALL BE CONNECTED BETWEEN THE EQUIPMENT-GROUNDING TERMINAL AT MAIN SERVICE AND THE EQUIPMENT GROUNDING TERMINAL OF THE POOL EQUIPMENT PANELBOARD. SIZE PER *TABLE 250.122*, BUT NOT SMALLER THAN 12 AWG, *680.25(B)*. FOR EXISTING PANELBOARDS, SEE *680.25(A), EXCEPTION*.

CONNECT LUMINAIRE WITH AN INSULATED COPPER EGC WITHOUT SPLICE; SIZE PER *TABLE 250.122*. BUT NOT SMALLER THAN 12 AWG, *680.23(F)*. IN FLEXIBLE CORD, EGC SHALL NOT BE SMALLER THAN SUPPLY CONDUCTORS AND NOT SMALLER THAN 16 AWG—SECTION *680.23(B)*.

① EQUIPMENT-GROUNDING TERMINAL

② EGC BETWEEN JUNCTION BOX AND TRANSFORMER SIZED ACCORDING TO THE OVERCURRENT DEVICE IN THE CIRCUIT—SECTION *680.23(F)(2), EXCEPTION*

LEGEND:
EGC = EQUIPMENT-GROUNDING CONDUCTOR
EMT = ELECTRICAL METALLIC TUBING
ENT = ELECTRICAL NONMETALLIC TUBING
IMC = INTERMEDIATE METAL CONDUIT
LFNC = LIQUIDTIGHT FLEXIBLE NONMETALLIC CONDUIT
MC = METAL-CLAD CABLE
RMC = RIGID METAL CONDUIT
PVC = RIGID PVC CONDUIT

③ RMC, IMC, LFNC, PVC. IF ON OR WITHIN BUILDING, EMT IS OK. IF WITHIN BUILDING, ENT IS OK, SECTIONS *680.23(F)* AND *680.25(A)*

④ NEUTRAL BUS

⑤ FOR AN EXISTING FEEDER TO AN EXISTING REMOTE PANELBOARD FMC OR APPROVED CABLE WITH AN INSULATED OR COVERED EGC PERMITTED—SECTION *680.25(A), EXCEPTION*

FIGURE 15-18 An overview of the grounding requirements for a typical permanently installed swimming pool installation.

and made of stainless steel, brass, copper, or a copper alloy. Exothermic welding is a process for making grounding and bonding connections using specially designed connectors, a form, a metal disk, and explosive powder. The copper bonding conductor is placed into the connector, and a form is placed around the connector and the part that the conductor is being connected to, like another piece of grounding conductor. The electrician puts the metal disk and explosive powder into the form and then ignites it. The violent chemical reaction that takes place "welds" the copper conductor to the other conductor. This process is sometimes called "Cad-Weld." Other methods acceptable to the *NEC*® for bonding the swimming pool metal parts include the following:

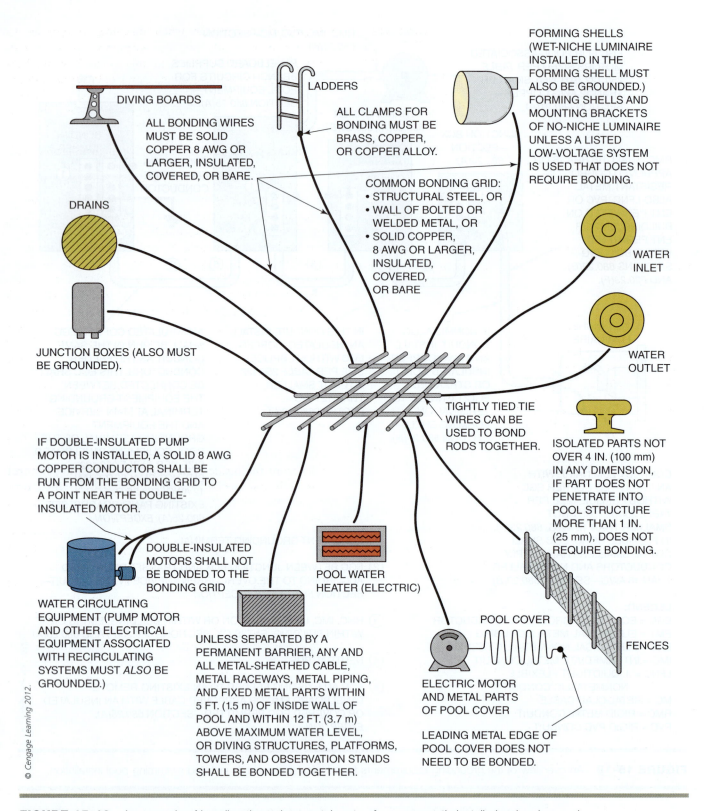

DIVING BOARDS

ALL BONDING WIRES MUST BE SOLID COPPER 8 AWG OR LARGER, INSULATED, COVERED, OR BARE.

LADDERS

ALL CLAMPS FOR BONDING MUST BE BRASS, COPPER, OR COPPER ALLOY.

FORMING SHELLS (WET-NICHE LUMINAIRE INSTALLED IN THE FORMING SHELL MUST ALSO BE GROUNDED.) FORMING SHELLS AND MOUNTING BRACKETS OF NO-NICHE LUMINAIRE UNLESS A LISTED LOW-VOLTAGE SYSTEM IS USED THAT DOES NOT REQUIRE BONDING.

DRAINS

COMMON BONDING GRID:
• STRUCTURAL STEEL, OR
• WALL OF BOLTED OR WELDED METAL, OR
• SOLID COPPER, 8 AWG OR LARGER, INSULATED, COVERED, OR BARE

WATER INLET

WATER OUTLET

JUNCTION BOXES (ALSO MUST BE GROUNDED).

TIGHTLY TIED TIE WIRES CAN BE USED TO BOND RODS TOGETHER.

ISOLATED PARTS NOT OVER 4 IN. (100 mm) IN ANY DIMENSION, IF PART DOES NOT PENETRATE INTO POOL STRUCTURE MORE THAN 1 IN. (25 mm), DOES NOT REQUIRE BONDING.

IF DOUBLE-INSULATED PUMP MOTOR IS INSTALLED, A SOLID 8 AWG COPPER CONDUCTOR SHALL BE RUN FROM THE BONDING GRID TO A POINT NEAR THE DOUBLE-INSULATED MOTOR.

DOUBLE-INSULATED MOTORS SHALL NOT BE BONDED TO THE BONDING GRID

POOL WATER HEATER (ELECTRIC)

WATER CIRCULATING EQUIPMENT (PUMP MOTOR AND OTHER ELECTRICAL EQUIPMENT ASSOCIATED WITH RECIRCULATING SYSTEMS MUST *ALSO* BE GROUNDED.)

UNLESS SEPARATED BY A PERMANENT BARRIER, ANY AND ALL METAL-SHEATHED CABLE, METAL RACEWAYS, METAL PIPING, AND FIXED METAL PARTS WITHIN 5 FT. (1.5 m) OF INSIDE WALL OF POOL AND WITHIN 12 FT. (3.7 m) ABOVE MAXIMUM WATER LEVEL, OR DIVING STRUCTURES, PLATFORMS, TOWERS, AND OBSERVATION STANDS SHALL BE BONDED TOGETHER.

ELECTRIC MOTOR AND METAL PARTS OF POOL COVER

POOL COVER

FENCES

LEADING METAL EDGE OF POOL COVER DOES NOT NEED TO BE BONDED.

© Cengage Learning 2012.

FIGURE 15-19   An example of bonding the various metal parts of a permanently installed swimming pool.

- The structural reinforcing steel of a concrete pool where the rods are bonded together by steel tie wires or the equivalent

- The walls of a bolted together or welded together metal pool

- Brass rigid metal conduit or other approved corrosion-resistant metal conduit

Swimming pool electrical installation usually includes installing one or more receptacles. Section 680.22(A) includes the installation requirements for

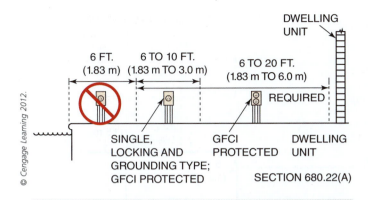

FIGURE 15-20 Required receptacles and their locations for a permanently installed swimming pool.

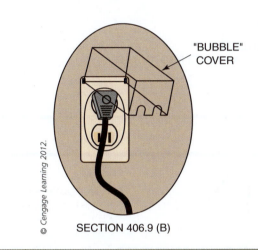

"BUBBLE" COVER

SECTION 406.9 (B)

FIGURE 15-21 Any 15- or 20-amp, 125- through 250-volt receptacle installed in a wet location must have a type of cover or enclosure that keeps it weatherproof at all times, whether or not an attachment plug is inserted.

receptacles and must be followed closely by an electrician (Figure 15-20). Receptacles installed for a permanently installed swimming pool's water pump motor or other loads directly related to the circulation and sanitation system for the pool can be located between 6 feet (1.83 m) and 10 feet (3 m) from the inside walls of the pool. If this receptacle is used and located accordingly, it must be a single receptacle of the proper voltage and amperage rating, a locking and grounding type, and GFCI protected. A permanently installed swimming pool must have at least one 125-volt, 15- or 20-ampere-rated receptacle located a minimum of 6 feet (1.83 m), but no more than 20 feet (6 m), from the pool's inside wall installed by an electrician. This receptacle cannot be located more than 6 feet, 6 inches (2 m) above the floor, platform, or grade level serving the pool. Any 15- or 20-amp, 125-volt receptacle located within 20 feet (6 m) of the pool's inside wall must be GFCI protected. One last thing for an

electrician to consider when installing receptacles to serve a permanently installed swimming pool is that, according to Section 406.9(B), any 15- or 20-amp, 125- or 250-volt receptacle installed in a **wet location** (indoors or outdoors) must have a type of cover or enclosure that keeps it weatherproof at all times whether a plug is inserted in it or not (Figure 15-21).

Some permanently installed pool installations will require luminaires or ceiling-suspended paddle fans to be installed. Section 680.22(B) lists the installation requirements for luminaires over a pool. In an outdoor pool area (Figure 15-22), luminaires, ceiling-suspended paddle fans, or lighting outlets cannot be installed over the pool or over the area extending 5 feet (1.5 m) horizontally from the inside wall of the pool. This rule does not need to be followed if the lighting fixtures are located 12 feet (3.7 m) or more above the maximum water level in the pool. If there are existing lighting fixtures (such as when a pool is installed sometime after the house has been built and there are outside lighting fixtures on the house) and they are located less than 5 feet (1.5 m) measured horizontally from the inside wall of the pool, they must be at least 5 feet (1.5 m) above the maximum water level of the pool, rigidly attached to the existing structure they are on, and GFCI protected. In the area that is more than 5 feet (1.5 m) but no more than 10 feet (3 m) measured horizontally from the inside wall of the pool, lighting fixtures and ceiling-suspended paddle fans must be GFCI protected. No GFCI protection is required if the luminaires or paddle fans are located more than 5 feet (1.5 m) above the maximum water level of the pool and are rigidly attached to an adjacent structure. Section 680.22(C) states that any switching device used for luminaires or ceiling-suspended paddle fans in or above a pool must be located at least 5 feet (1.5 m) horizontally from the inside wall of the pool, unless the switch location is separated from the pool by a solid fence, wall, or other permanent barrier. If the pool is to be installed indoors (Figure 15-23), Section 680.22(B)(2) does not require electricians to follow the rules described above if the following are true:

- The luminaires are of a totally enclosed type.

- Ceiling-suspended paddle fans are identified for use beneath ceiling structures, such as provided on porches or patios.

- The branch circuit supplying the equipment is GFCI protected.

- The bottom of the luminaire or ceiling-suspended paddle fan is at least 7 feet, 6 inches (2.3 m) above the maximum water level of the pool.

Some swimming pools will require the installation of luminaires underwater in the walls of the pool (Figure 15-24). Section 680.23(A) requires lighting

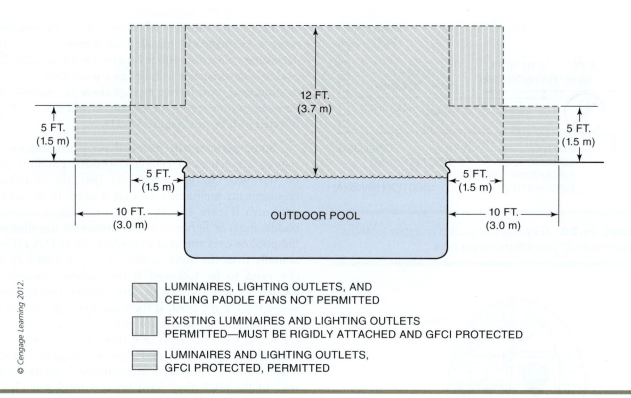

LUMINAIRES, LIGHTING OUTLETS, AND CEILING PADDLE FANS NOT PERMITTED

EXISTING LUMINAIRES AND LIGHTING OUTLETS PERMITTED—MUST BE RIGIDLY ATTACHED AND GFCI PROTECTED

LUMINAIRES AND LIGHTING OUTLETS, GFCI PROTECTED, PERMITTED

© Cengage Learning 2012.

FIGURE 15-22   Outdoor pool lighting fixture locations.

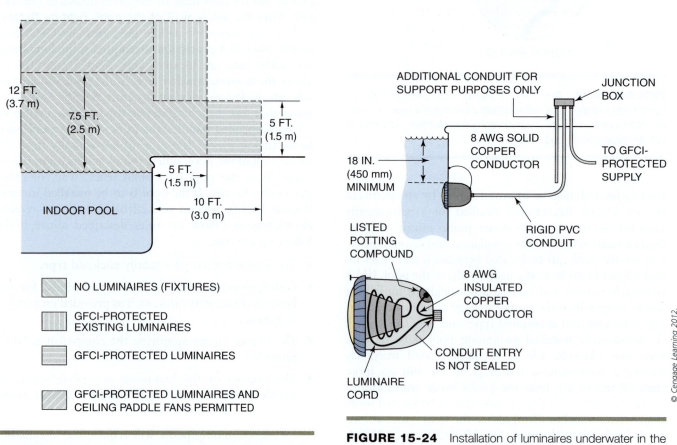

NO LUMINAIRES (FIXTURES)

GFCI-PROTECTED EXISTING LUMINAIRES

GFCI-PROTECTED LUMINAIRES

GFCI-PROTECTED LUMINAIRES AND CEILING PADDLE FANS PERMITTED

© Cengage Learning 2012.

© Cengage Learning 2012.

FIGURE 15-23   Indoor pool lighting fixture locations.

FIGURE 15-24   Installation of luminaires underwater in the walls of a permanently installed swimming pool.

fixtures that are installed in the walls of pools to be located so that the top of the luminaire (fixture) lens is at least 18 inches (450 mm) below the pool's normal water level, unless the fixture is listed and identified for use at a depth of not less than 4 inches (100 mm) below the pool's normal water depth. Section 680.23(B) requires that in nonmetallic conduit the termination point for the 8 AWG solid copper (or larger) bonding conductor located inside the forming shell for the luminaire must be covered with a listed potting compound or covered in some other way so as to protect the connection from the deteriorating effects of the pool water. The wet-niche, dry-niche, or no-niche luminaires must be connected to a 12 AWG or larger equipment-grounding conductor sized according to Table 250.122 as stated in Section 680.23(F). Section 680.26(B) requires that the metal forming shells for the luminaires be bonded to the common bonding grid with a solid copper conductor no smaller than 8 AWG. These lighting fixtures are connected through a junction box, and Section 680.23(B) requires that the metal conduit connection from the junction box to the forming shell must be made with a brass rigid metal conduit or other identified corrosion resistant metal. Nonmetallic conduit can also be used and typically PVC conduit is what most electricians install. An insulated copper equipment-grounding conductor must be run along with the circuit conductors in the conduit. Section 680.23(B) requires that when using PVC conduit, an 8 AWG insulated copper equipment-grounding conductor be installed in the conduit and connected to the forming shell, junction box, or other enclosure in the circuit. This requirement is not necessary if a listed low-voltage lighting system is used that does not require grounding. The supply conductors brought to the junction box by the electrician must be GFCI protected if the voltage is above 15 volts.

There are a few specific rules to follow from Section 680.24(A) when installing the junction boxes used with underwater lighting. They are summarized as follows (Figure 15-25):

- Locate the junction box at least 4 feet (1.2 m) from the pool's inside wall, unless separated from the pool by a solid fence, wall, or other permanent barrier.

- The junction box must be made of copper, brass, suitable plastic, or other approved corrosion-resistant material.

- Measured from the inside of the bottom of the junction box, the box must be located at least 8 inches (200 mm) above the maximum water level of the pool.

- The junction box must be equipped with threaded entries or a specifically listed nonmetallic hub.

- Locate the junction box no less than 4 inches (100 mm) above the ground level or pool deck. This is measured from the inside of the bottom of the junction box.

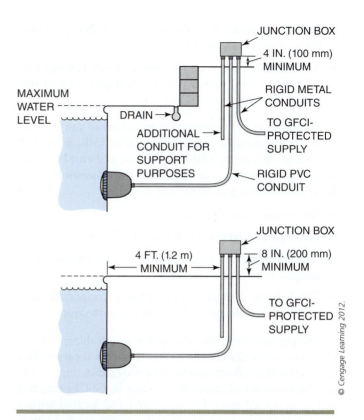

FIGURE 15-25 Installation of junction boxes used in underwater lighting.

- The junction box support must comply with Section 314.23(E), which requires the box to be supported by at least two rigid metal conduits, or intermediate metal conduits, threaded into the box. Each conduit must be secured within 18 inches (450 mm) of the enclosure. PVC conduit is not permitted for junction box support, so if it is used, additional support will be required.

STORAGE POOLS

So-called aboveground storage pools are very popular with home owners. The electrical requirements and subsequent circuit installation for a storage pool is not as complicated as it is for a permanently installed swimming pool. In this section, we look at the most common wiring requirements for storage pools.

Most storable pools have no lighting fixtures installed in or on them. However, Section 680.33(A) says that if lighting fixtures are installed, they must be cord-and-plug connected and be a "made at the factory" assembly. The lighting fixture assembly must be properly listed for use with storage pools and must have the following construction features:

- No exposed metal parts

- A lamp that operates at no more than 15 volts

- An impact-resistant polymeric lens, lighting fixture body, and transformer enclosure

- A transformer, to drop the voltage from 120 volts to 15 volts or less, with a primary voltage rating of not over 150 volts

Lighting fixtures without a transformer can also be used with a storable pool. Section 680.33(B) says that they must operate at 150 volts or less and can be cord-and-plug connected. The lighting fixture assembly must have the following construction features:

- No exposed metal parts

- An impact-resistant polymeric lens and fixture body

- A GFCI with open neutral conductor protection as an integral part of the assembly

- The lighting fixture is permanently connected to a GFCI with open neutral conductor protection

When installing the wiring for a storable pool, an electrician typically installs a receptacle in a location that allows the filter system pump to be plugged in (Figure 15-26). Section 680.32 states that it must be GFCI protected if it is 15- or 20-amp, 125-volt rated and located within 20 feet (6.0 m) of the inside walls of the pool. Section 406.9(B) requires the receptacle to have a cover or enclosure that maintains its weatherproof capability, whether a pump cord is plugged in or not. The *NEC®* also requires that all 15- and 20-amp, 125- and 250-volt non-locking receptacles must be a listed weather-resistant type when installed in damp and wet locations. Section 680.31 states that the cord-and-plug-connected pool filter pump must

incorporate an approved double-insulated system and must have a means for grounding the appliance's internal and non-accessible noncurrent-carrying metal parts. The grounding means must consist of an equipment-grounding conductor (run with the power supply conductors) in the flexible cord that properly terminates in a grounding-type attachment plug with a fixed grounding contact member. Cord-connected pool filter pumps must be provided with a ground fault circuit interrupter that is built-in to the attachment plug or is located in the power supply cord within 12 inches (300 mm) of the attachment plug.

SPAS AND HOT TUBS

Electrical installation requirements for spas and hot tubs are found in Part IV of Article 680. A hot tub is made of wood, typically redwood, teak, or oak. A spa is made of plastic, fiberglass, concrete, tile, or some other man-made product. However, even though they are made from different materials and do not look exactly the same, both have certain electrical installation rules that apply to both of them.

Section 680.42 specifies the wiring methods that are allowed for outdoor installations. They include flexible connections using flexible raceway or cord-and-plug connections. For a one-family dwelling unit, a spa or hot tub assembly can be connected using regular wiring methods recognized in Chapter 3 of the *NEC®*, as long as the wiring method has a copper equipment-grounding conductor not smaller than 12 AWG.

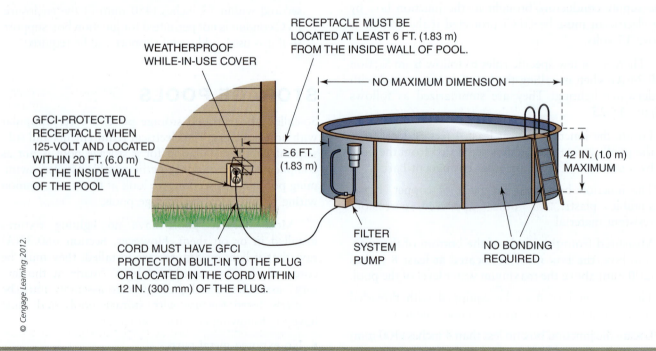

© Cengage Learning 2012.

FIGURE 15-26   Installation of a receptacle for a storable pool. The receptacle will serve the pump motor.

Section 680.43 covers indoor installation requirements. The following installation requirements apply to receptacle installation (Figure 15-27):

- At least one 125-volt, 15- or 20-amp-rated receptacle connected to a general-purpose branch circuit must be installed at least 6 feet (1.83 m) but not more than 10 feet (3 m) from the inside wall of the spa or hot tub.

- All other 125-volt, 15- or 20-amp-rated receptacles located in the area of the spa or hot tub must be located at least 6 feet (1.83 m) measured horizontally from the inside wall of the spa or hot tub.

- All 125-volt, 15-, 20-, or 30-amp-rated receptacles located within 10 feet (3 m) of the inside walls of the spa or hot tub must be GFCI protected.

- The receptacle that supplies power to the spa or hot tub must be GFCI protected.

The following installation requirements apply to luminaire installations or ceiling-suspended paddle fans installation around a spa or hot tub (see Figure 15-27):

- They must not be installed less than 12 feet (3.7 m) above the spa or hot tub unless GFCI protected.

- They must not be installed less than 7 feet, 6 inches (2.3 m) above the spa or hot tub, even when GFCI protected.

- If located less than 7 feet, 6 inches (2.3 m) above the spa or hot tub, they must be suitable for a damp location and be a recessed luminaire with a glass or plastic lens and a nonmetallic or isolated metal trim and be a surface-mounted luminaire with a glass or plastic globe, a nonmetallic body, or a metal body that is isolated from contact.

- If underwater lighting is installed for a spa or hot tub, the same rules for underwater lighting in a swimming pool apply.

- When installing wall switches for the lighting fixtures or ceiling-suspended paddle fans, they must be kept a minimum of 5 feet (1.5 m) from the inside edge of the spa or hot tub.

All electrical equipment located within 5 feet (1.5 m) of the inside walls of a spa or hot tub and all electrical equipment associated with the water circulating system must be grounded. Also, the following parts must be bonded together:

- All metal fittings within or attached to the spa or hot tub structure

- Metal parts of electrical equipment associated with the spa or hot tub water-circulating system, including pump motors

- Metal raceways, metal piping, or any metal surfaces that are within 5 feet (1.5 m) of the inside walls of the spa or hot tub and that are not separated from the spa or hot tub with a permanent barrier like a wall

The bonding can be done by the interconnection of threaded metal piping and fittings, metal-to-metal mounting on a common frame, or by installing a copper

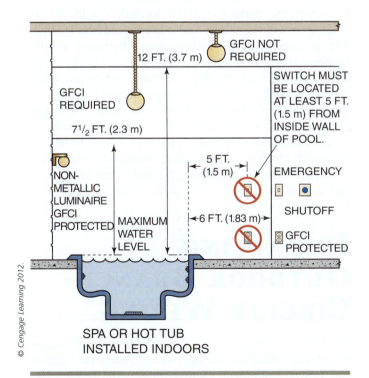

FIGURE 15-27 Installation requirements that apply to receptacles installed at indoor spa or hot tub locations.

solid bonding jumper of 8 AWG or larger. If the spa or hot tub is field-assembled, an electrician will need to make sure the metal parts are properly bonded together. If the spa or hot tub is listed by a Nationally Recognized Testing Laboratory (NRTL) like Underwriters Laboratories (UL) and comes as a pre-assembled unit from the factory, the grounding and bonding of the proper parts will have been done at the factory and all an electrician will need to do is connect the incoming electrical supply.

HYDROMASSAGE BATHTUBS

A **hydromassage bathtub** is intended to be filled, used, and then drained after each use. Some people refer to this type of unit as a "whirlpool bath." Part VII of Article 680 covers the installation requirements for hydromassage bathtubs (Figure 15-28). Hydromassage bathtubs are usually installed in a bathroom of a home. Remember that Section 210.8(A)(1) requires all bathroom receptacles (125-volt, 15- or 20-amp rated) to be GFCI protected. Additionally, Section 680.71 requires all 125-volt receptacles that are 15-, 20-, or 30-amp rated to be GFCI protected when they are located within 6 feet (1.83 m) measured horizontally of the inside walls of the hydromassage tub. This is true whether the hydromassage tub is installed in a bathroom area or some other area of the house. Section 680.71 also states that all hydromassage bathtubs must be on a dedicated branch circuit and any associated electrical equipment must be protected by a readily accessible GFCI.

Section 680.74 requires all metal piping, electrical equipment metal parts, and pump motors associated with the hydromassage bathtub to be bonded together using an 8 AWG or larger solid copper conductor. The bonding wire can be bare or insulated and is connected to the terminal on the circulating pump that is intended for this purpose. It does not have to be connected to a double-insulated circulating pump motor. Additionally, the bonding jumper does not have to be extended or attached to any panelboard, service equipment, or grounding electrode. The bonding jumper does have to be long enough to terminate on a replacement non-double-insulated pump motor and must be terminated to the equipment-grounding conductor of the branch circuit of the motor when a double-insulated circulating pump motor is used.

The hydromassage bathtub electrical equipment must be accessible without causing damage to any part of the building, according to Section 680.73. This means that there will have to be a panel of some type that can be easily removed for access to the electrical equipment. Where the hydromassage tub is cord-and-plug connected, the receptacle must be located no more than 1 foot (300 mm) from the opening of a service access opening.

Electricians usually run a separate circuit to a hydromassge bathtub. It is normally either a 15-amp circuit using a 14 AWG conductor in a cable or raceway or a 20-amp circuit using a 12 AWG conductor. Remember that the circuit must be protected by a readily accessible GFCI, which will mean the use of a GFCI circuit breaker at the electrical panel. The circuit wiring is run during the rough-in stage to a location that will be near to where the pump and control/junction box for the hydromassage pump will be located. A junction box is installed as close to this location as possible. Consult the building plans and the plumbing contractor to determine exactly where the hydromassage bathtub will be located, specifically at what spot will the pump and control panel be located. The manufacturers of hydromassage bathtubs usually supply a length of liquidtight flexible raceway that contains a black ungrounded conductor, a white grounded conductor, and a green equipment-grounding conductor. The length supplied is around 3 feet (900 mm). The length of liquidtight flex is brought to the junction box, and the proper connections are made (Figure 15-29). If the hydromassage bathtub is cord-and-plug connected, a receptacle (GFCI protected) is installed at the end of the branch circuit at a location that is close to the pump location.

INSTALLING OUTDOOR BRANCH-CIRCUIT WIRING

Outdoor electrical wiring in residential situations includes installing the wiring and equipment for lighting and power equipment located outside the house. The wiring may be installed overhead or underground. In this section, we take a look at the installation of outdoor receptacles and outdoor lighting applications.

FIGURE 15-28 The installation requirements for hydromassage bathtubs.

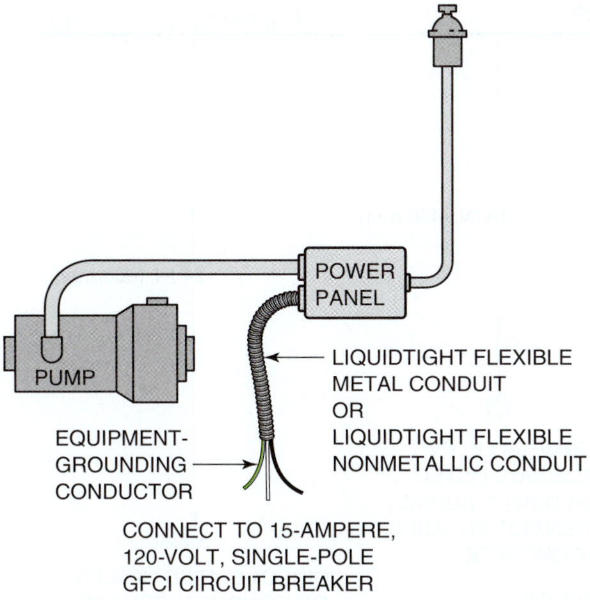

FIGURE 15-29 Electrical connections for a hydromassage bathtub. The motor, power panel, and the electrical supply leads are shown.

UNDERGROUND WIRING

Because the majority of underground receptacle and lighting circuits installed in residential wiring is done using Type UF cable, a review of this wiring method is appropriate at this time. Article 340 covers Type UF cable. When determining the size of Type UF cable to use for a particular wiring situation, the ampacity of the cable is found using the "60°C" column in Table 310.15(B)(16), just like nonmetallic sheathed cable. According to Article 340, the following apply to Type UF cable:

• Must be marked as underground feeder cable

• Is available from 14 AWG through 4/0 copper and from 12 AWG through 4/0 aluminum

• Can be used outdoors in direct exposure to the sun only if it is listed as being sunlight resistant and has a sunlight-resistant marking on the cable sheathing

• Can be used with the same fittings as is used with nonmetallic sheathed cable

• Can be buried directly in the ground and installed according to Section 300.5 and Table 300.5

• When used as an interior wiring method, must be installed according to the same rules as for nonmetallic sheathed cable

• Contains an equipment-grounding conductor (bare) that is used to ground equipment fed by the Type UF cable

See Procedure 15-1 on pages 553–554 for a procedure for stripping the insulation from a Type UF cable.

Some electricians prefer to use a conduit wiring method when installing underground conductors. All underground installations are considered a wet location by definition, meaning that only certain types of conduit will be able to be used. Also, any wiring installed in the underground conduits must have a "W" in their insulation designation, such as "THWN" or "XHHW." The "W" means that the conductor insulation is suitable for installation in a wet location (refer to Table 310.104(A) of the *NEC*®). The conduits installed underground must be of a type that is resistant to corrosion. The manufacturer of the conduit and Underwriters Laboratories (UL) listings will give you information that will let you determine if the conduit type you wish to install underground is suitable for that location. Additional protection to a metal conduit can be provided by applying a protective coating to the conduit (Figure 15-30). Most electricians will use PVC conduit for underground conduit installation. A separate equipment-grounding conductor will have to be installed, and it can either be bare or green insulated. It will be sized according to Table 250.122 of the *NEC*®.

Is Supplemental Corrosion Protection Required?

	In Concrete Above Grade?	In Concrete Below Grade?	In Direct Contact with Soil?
Rigid metal conduit[1]	No	No	No[2]
Intermediate metal conduit[1]	No	No	No[2]
Electrical metallic tubing[1]	No	Yes	Yes[3]

[1]Severe corrosion can be expected where ferrous metal conduits come out of concrete and enter the soil. Here again, some electrical inspectors and consulting engineers might specify the application of some sort of supplemental nonmetallic corrosion protection.
[2]Unless subject to severe corrosive effects. Different soils have different corrosive characteristics.
[3]In most instances, electrical metallic tubing is not permitted to be installed underground in direct contact with the soil because of corrosion problems.

FIGURE 15-30 Supplemental corrosion protection may be required for the metal raceways you are installing underground.

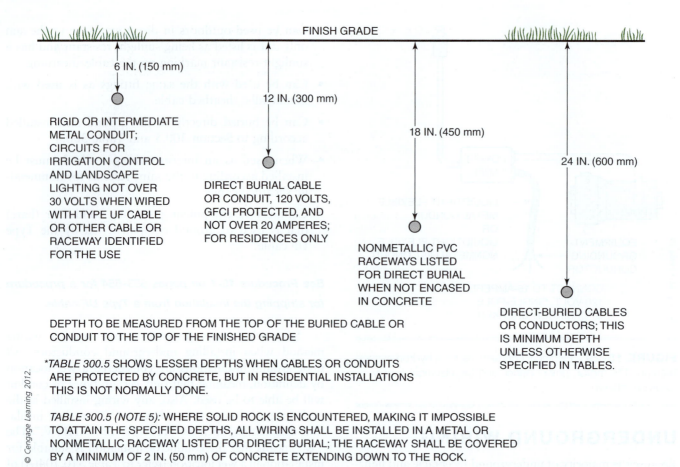

FINISH GRADE

6 IN. (150 mm)

RIGID OR INTERMEDIATE
METAL CONDUIT;
CIRCUITS FOR
IRRIGATION CONTROL
AND LANDSCAPE
LIGHTING NOT OVER
30 VOLTS WHEN WIRED
WITH TYPE UF CABLE
OR OTHER CABLE OR
RACEWAY IDENTIFIED
FOR THE USE

12 IN. (300 mm)

DIRECT BURIAL CABLE
OR CONDUIT, 120 VOLTS,
GFCI PROTECTED, AND
NOT OVER 20 AMPERES;
FOR RESIDENCES ONLY

18 IN. (450 mm)

NONMETALLIC PVC
RACEWAYS LISTED
FOR DIRECT BURIAL
WHEN NOT ENCASED
IN CONCRETE

24 IN. (600 mm)

DIRECT-BURIED CABLES
OR CONDUCTORS; THIS
IS MINIMUM DEPTH
UNLESS OTHERWISE
SPECIFIED IN TABLES.

DEPTH TO BE MEASURED FROM THE TOP OF THE BURIED CABLE OR
CONDUIT TO THE TOP OF THE FINISHED GRADE

TABLE 300.5 SHOWS LESSER DEPTHS WHEN CABLES OR CONDUITS
ARE PROTECTED BY CONCRETE, BUT IN RESIDENTIAL INSTALLATIONS
THIS IS NOT NORMALLY DONE.

TABLE 300.5 (NOTE 5): WHERE SOLID ROCK IS ENCOUNTERED, MAKING IT IMPOSSIBLE
TO ATTAIN THE SPECIFIED DEPTHS, ALL WIRING SHALL BE INSTALLED IN A METAL OR
NONMETALLIC RACEWAY LISTED FOR DIRECT BURIAL; THE RACEWAY SHALL BE COVERED
BY A MINIMUM OF 2 IN. (50 mm) OF CONCRETE EXTENDING DOWN TO THE ROCK.

© Cengage Learning 2012.

FIGURE 15-31   Minimum burial depths for cables and conduits installed underground.

OUTDOOR RECEPTACLES

Receptacle outlets located outdoors must be installed in weatherproof enclosures. The electrical boxes are usually made of metal and are of the "FS" or "FD" type (Figure 15-33). This box type is often called a "bell box" by electricians working in the field and has threaded openings or hubs that allow attachment to the box with conduit or a cable connector. Each of these boxes comes from the factory with a few "threaded plugs" that are used in any unused threaded openings to make the box truly weatherproof. These boxes can be mounted on the surface of an outside wall or on some other structural support, such as a wooden post driven into the ground. They are often installed with underground wiring and supported by conduits coming up out of the ground. Section 314.23(E) and (F) covers the support rules for electrical boxes when they are fed using an underground wiring method and supported by conduit. Basically, the rules state that the box must be supported by at least two conduits threaded wrench-tight into the box. For boxes that contain a receptacle device, or if the box is just a junction box with both conduits threaded into the same side, the maximum distance from where the conduit exits the ground to the box is 18 inches (450 mm) (Figure 15-34).

When a receptacle is installed outdoors, the enclosure and cover combination must maintain its weatherproof characteristics, whether a cord plug is inserted into the receptacle or not. This is required by Section 406.9(B) of the *NEC*®. This requirement is met by installing a self-closing cover that is deep enough to also cover the attached plug cap on a cord (Figure 15-35). The *NEC*®

The minimum burial depths for both Type UF cable and for any of the conduit wiring methods are shown in Table 300.5 of the *NEC*®. When installing the underground conductors across a lawn or field, use the minimum depths as shown in Figure 15-31. When installing a cable or conduit system under a residential driveway or parking lot, use the minimum depths as shown in Figure 15-32.

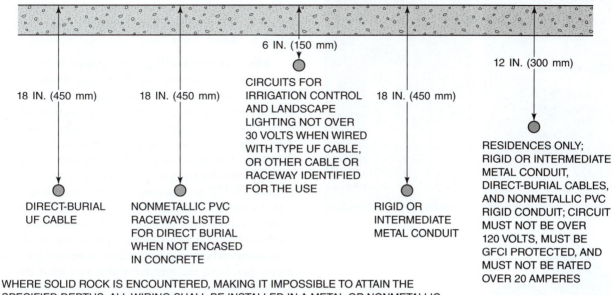

SURFACE OF DRIVEWAY OR PARKING AREA

6 IN. (150 mm)

12 IN. (300 mm)

18 IN. (450 mm)

18 IN. (450 mm)

CIRCUITS FOR
IRRIGATION CONTROL
AND LANDSCAPE
LIGHTING NOT OVER
30 VOLTS WHEN WIRED
WITH TYPE UF CABLE,
OR OTHER CABLE OR
RACEWAY IDENTIFIED
FOR THE USE

18 IN. (450 mm)

RESIDENCES ONLY;
RIGID OR INTERMEDIATE
METAL CONDUIT,
DIRECT-BURIAL CABLES,
AND NONMETALLIC PVC
RIGID CONDUIT; CIRCUIT
MUST NOT BE OVER
120 VOLTS, MUST BE
GFCI PROTECTED, AND
MUST NOT BE RATED
OVER 20 AMPERES

DIRECT-BURIAL
UF CABLE

NONMETALLIC PVC
RACEWAYS LISTED
FOR DIRECT BURIAL
WHEN NOT ENCASED
IN CONCRETE

RIGID OR
INTERMEDIATE
METAL CONDUIT

WHERE SOLID ROCK IS ENCOUNTERED, MAKING IT IMPOSSIBLE TO ATTAIN THE
SPECIFIED DEPTHS, ALL WIRING SHALL BE INSTALLED IN A METAL OR NONMETALLIC
PVC RACEWAY LISTED FOR DIRECT BURIAL; THE RACEWAY SHALL BE COVERED BY
A MINIMUM OF 2 IN. (50 mm) OF CONCRETE EXTENDING DOWN TO THE ROCK.

DEPTH TO BE MEASURED FROM THE TOP OF THE BURIED CABLE OR
CONDUIT TO THE TOP OF THE FINISHED DRIVEWAY.

© Cengage Learning 2012.

FIGURE 15-32 Minimum burial depths for cables and raceways under residential driveways or parking lots.

© Cengage Learning 2012.

FIGURE 15-33 FS or FD boxes are used to install surface-mounted receptacles outdoors in residential applications. The FS box is a shallow box, while the FD box is a deep box. These boxes are often called "bell boxes."

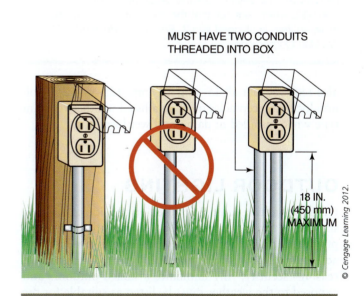

MUST HAVE TWO CONDUITS
THREADED INTO BOX

18 IN.
(450 mm)
MAXIMUM

© Cengage Learning 2012.

FIGURE 15-34 Section 314.23(F) covers the support rules for electrical boxes with a receptacle when they are fed using an underground wiring method and supported by conduit.

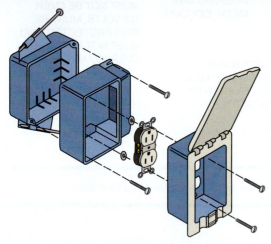

FIGURE 15-35 A recessed type of weatherproof receptacle. This style meets the requirements of Section 406.9(B), which requires that the enclosure be weatherproof while an item is plugged into it. *Photos courtesy of TayMac Corporation.*

also requires that all 15- and 20-amp, 125- and 250-volt non-locking receptacles must be a listed weather-resistant type when installed in damp and wet locations.

OUTDOOR LIGHTING

Outdoor lighting can be mounted on the side of building structures, on poles, or even on trees. The lighting provides illumination on a home owner's property for activities after dark and for one other very important reason—security. Any luminaire installed outdoors and exposed to the weather must be listed as suitable for the location. A label with a marking that states "suitable for wet locations" must be found on the fixture. If a luminaire is to be installed under a canopy or under an open

porch, it is considered a "damp" location and the fixture only needs a label that states "suitable for damp locations." A wet-location luminaire may be used in either a "damp" or a "wet" outdoor location.

Most wiring installed for outdoor lighting is done underground, but overhead wiring from the main house to the location of the outdoor lighting may be done. Contrary to popular belief, Section 410.36(G) allows outdoor lighting fixtures to be mounted on trees (Figure 15-36). However, Sections 225.26 and 590.4(J) state that overhead conductor spans cannot be supported by trees or other living or dead vegetation (Figure 15-37). So, if an electrician is installing wiring to a tree-mounted lighting fixture, it will need to be by an underground wiring method. Make sure to use a wiring method that provides adequate protection to the circuit conductors when installing wiring from the point where it emerges from the ground and is run up a tree to the lighting fixture.

FIGURE 15-36 Locating lighting fixtures on trees is not prohibited by the *NEC®*. A recognized wiring method that adequately protects the conductors must be used to carry the conductors up to the tree trunk to the fixture.

© Cengage Learning 2012.

FIGURE 15-37   The *NEC*® does not allow trees and other vegetation, either dead or alive, to support overhead spans of permanently installed conductors.

Outside lighting can be manually controlled from switch locations inside a house or they can be automatically controlled with timers or sunlight sensing devices. Photoelectric sensors, often called a "photocell" (Figure 15-38), can be used to switch outside lighting fixtures on when it gets dark outside and then switch them off when the sun comes up in the morning. The photocells can be a built-in part of an outside lighting fixture or an electrician can install a separate photocell and have it control as many lighting outlets as it is rated for. Always follow the installation instructions that come with a photocell but most of the time the wiring diagram shown in Figure 15-39 will work. Photocells are typically rated in watts and can control an electrical load up to that specific amount. For example, a 600-watt photocell could control six 100-watt incandescent lamps or up to ten 60-watt incandescent lamps.

A common outdoor wiring installation for electricians in residential wiring is a pole light. Figure 15-40 shows a typical pole light installation.

Green Tip

Motion sensors can be used to control outside lighting so that they will be switched on when movement is detected. During the sunlight hours, a photocell built-in to the motion sensor keeps the lighting fixture from switching on and wasting electrical power even if there is motion in the area. A motion sensor with a built in photoelectric sensor controlling outside lighting can eliminate the lighting being left on for long periods of time when not needed. This type of lighting control will cut down on the amount of electricity used to light the outside of a home and will result in electric bill savings.

© Cengage Learning 2012.

FIGURE 15-38 A common style of photocell. The cap covering the "eye" can be adjusted to let more or less sunlight hit the photocell. This has a direct effect on how late at night or early in the morning the lighting outlets controlled by the photocell are switched on or off.

INSTALLING THE WIRING FOR A STANDBY POWER SYSTEM

In many areas, the reliability of electric service to a home is not very good. Ice storms, high winds, heavy snows, and other severe weather conditions can cause power outages. Many home owners have opted to have the electrical contractor install a standby power system so they will not need to be without electrical power for prolonged periods of time. In this section,

we take a look at a common installation of standby power systems for residential customers. It is beyond the scope of this book to cover all possible standby power installations for houses.

There are two basic types of residential generators that can be used in a standby power system. Portable generators are designed primarily to supply electricity to individual appliances with extension cords. Permanent generators are designed to supply electricity directly to a house electrical system. Typically, portable generators cannot provide enough power to meet a home's entire electrical load, unless it is relatively small. When portable generators are used, they are most often connected to power-critical loads like refrigerators, lighting, and a few selected receptacles. When a home owner wants to be able to use all of the house electrical system capacity, a permanent generator system will need to be installed. An air-cooled permanent generator can provide up to 16,000 watts of electricity, usually enough to power a good-sized home until regular power is restored. Liquid-cooled larger permanent generators can provide from around 15,000 watts up to approximately 40,000 watts. Permanent generator systems are becoming increasingly popular with home owners, and more electrical contractors are installing complete permanent generator systems. However, because they are the most economical systems to purchase and install, portable generator systems are still the most popular and are the focus of this section of the textbook. The portable **generator** is a packaged unit that has a gasoline-powered engine turning a generator that produces alternating current electricity. These manufactured generator units are usually started manually, like a lawnmower, but many of the larger units offer electric starting. Portable generators are rated in watts and can range from around 3000 watts up to 12,000 watts or more. Each generator comes factory equipped with one or more 15- or 20-amp-rated receptacles for direct connection of extension cords for powering small loads. A larger 120/240-volt polarized **twistlock receptacle** is installed on the unit so the generator can be connected to power larger loads.

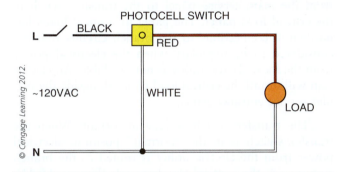

FIGURE 15-39 A wiring diagram showing how a photocell is connected in a circuit.

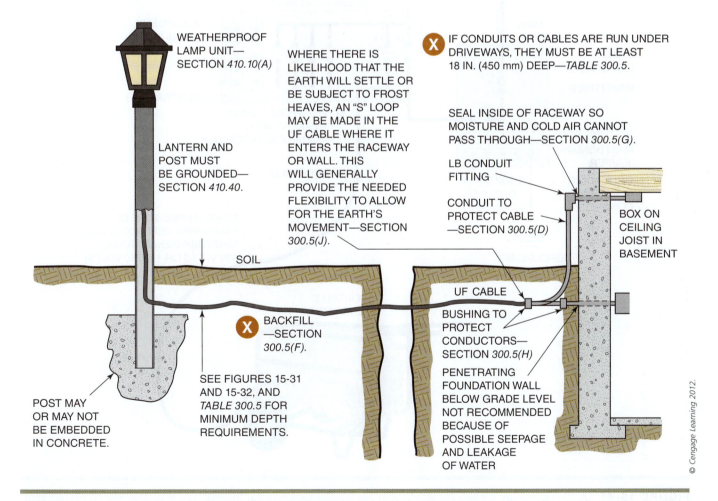

FIGURE 15-40 A typical pole light installation.

Before installing the standby power system, a home owner must determine which electrical loads are critical and which are not. On the basis of this information, the correct size of generator can be determined. Some generator manufacturers suggest that after all the **critical loads** are added up, another 20 percent is added for any future loads that may be included. Generators can be purchased that can supply the electrical power needs

for a whole house, or, as is more often the case, generators are purchased that can supply only a portion of the house electrical loads.

The installation for the standby power system presented in this section includes a portable generator that is cord-and-plug connected to a **transfer switch** located in a critical load panel. There is also wiring installed from the main service panel to the transfer switch in the critical load panel. The panel for critical loads has hardwired connections to the branch circuits that are considered to be necessary when the electrical power from the local electric utility is not available. An electrician will install the critical load panel alongside the regular service entrance panel.

The transfer switch is very important. When the transfer switch is in the "normal" position, electrical power from the electric utility is routed to the branch circuits in the critical load panel (Figure 15-41). When the transfer switch is in the "generator" position

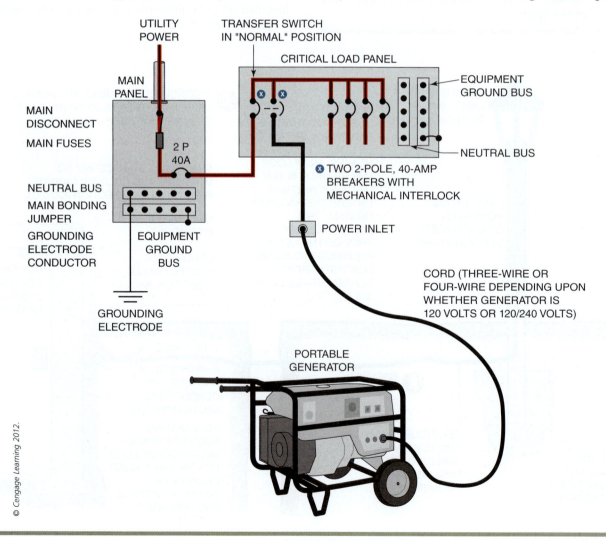

FIGURE 15-41 A standby power system installation with the transfer switch in the "normal" position. The branch circuits identified as being critical are being fed from the service entrance conductors.

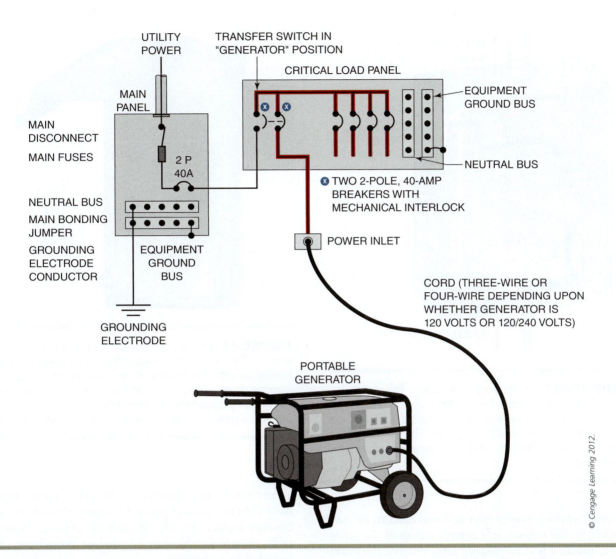

UTILITY POWER

TRANSFER SWITCH IN "GENERATOR" POSITION

CRITICAL LOAD PANEL

MAIN PANEL

MAIN DISCONNECT

MAIN FUSES

2 P 40A

EQUIPMENT GROUND BUS

NEUTRAL BUS

⊗ TWO 2-POLE, 40-AMP BREAKERS WITH MECHANICAL INTERLOCK

NEUTRAL BUS

MAIN BONDING JUMPER

GROUNDING ELECTRODE CONDUCTOR

EQUIPMENT GROUND BUS

POWER INLET

CORD (THREE-WIRE OR FOUR-WIRE DEPENDING UPON WHETHER GENERATOR IS 120 VOLTS OR 120/240 VOLTS)

GROUNDING ELECTRODE

PORTABLE GENERATOR

© Cengage Learning 2012.

FIGURE 15-42 A standby power system installation with the transfer switch in the "generator" position. When the transfer switch is in this position, the critical load branch circuits are being fed by the generator power. Notice that with the transfer switch in this position, there is no way that electrical power from the generator can be "backfed" to the electrical utility power lines.

(Figure 15-42), it disconnects the critical load branch circuits from the incoming electric utility power. Now, only the electrical power from the generator can be delivered to the critical load panel branch circuits. The transfer switch eliminates the possibility of the generator voltage being applied to the secondary of the electric utility's supply transformer, resulting in a very high voltage being put back on the "dead" electric utility wiring from the primary of the home supply transformer. This situation is commonly called "backfeeding" and could be a very dangerous situation when utility linemen working on the power lines believe them to be de-energized but they are really "live" with a high voltage developed from the home owner's generator. Many utility linemen have been severely injured or even killed by a home electrical system standby generator not being properly installed with an approved transfer switch. Transfer switches are rated in amperes and for residential applications range from 40 to 200 amps (Figure 15-43). They are a double-pole, double-throw (DPDT) type of switch. If the transfer switch is connected to the line side of the main service disconnecting means (that is, between the utility meter and the main circuit breaker), the transfer switch must be listed as suitable for use as service entrance equipment. The installation discussed in this section assumes the transfer switch to be located on the load side of the service disconnecting means and therefore does not need to be listed as suitable for service entrance equipment.

The connection of the generator to the transfer switch is generally done by a flexible cord with cord-and-plug connections. There will be a polarized twist-lock receptacle on the generator that is installed by the generator manufacturer. The receptacle installed for the connection to the transfer switch will need to be a

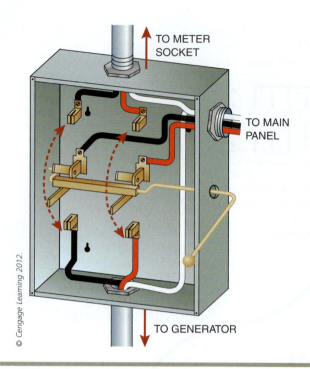

TO METER SOCKET

TO MAIN PANEL

TO GENERATOR

© Cengage Learning 2012.

FIGURE 15-43 A manual transfer switch. This switch is really a double-pole, double-throw (DPDT) switch and is necessary so that the generator voltage cannot be "backfed" out onto the utility company electrical system. The transfer switch must be listed as suitable for service entrance equipment when used, as shown here.

© Cengage Learning 2012.

FIGURE 15-44 An outdoor power inlet for portable generators. They are used in conjunction with a generator panel, loadcenter with an interlock kit, or a manual transfer switch for standby power. The power inlet enclosure allows the generator cord to be plugged into it with the cover closed. This provides maximum protection against weather and normal field use.

polarized twistlock power inlet and will need to be sized according to the voltage and current rating of the generator receptacle it will be plugged into. Because the portable generator is gasoline-powered and will need to be located outdoors, the location of the receptacle that the generator is plugged into may be located outdoors on an outside wall or possibly just inside the overhead door of an attached garage (Figure 15-44). The size of the conductors must be sized according to Section 445.13, which requires them to be at least 115 percent of the generator's nameplate current rating. For example, if you were installing an 8-kilowatt, 120/240-volt generator, the minimum conductor size from the generator to the power inlet and then to the transfer switch would be found as follows:

- 8000 watts/240 volts = 33.33 amps
- 33.33 amps × 115% = 38.33 amps
- Minimum wire size Type NM cable would be an 8 AWG copper conductor according to Table 310.15(B)(16) (the "60°C" column)
- Minimum-size Type SO flexible cord would be 8 AWG copper according to Table 400.5(A)(1)

The connection from the main service panel to the transfer switch is also installed as a 120/240-volt system with a two-pole circuit breaker in the main panel sized

according to the number of branch circuits supplied from the critical load panel. Common sizes of the circuit breaker include 30, 40, 50, and 60 amperes. A nonmetallic sheathed cable can be used or the wiring can be installed in a raceway.

from experience...

To save money, some electrical contractors use two two-pole circuit breakers that are mechanically interlocked as the transfer switch and are located in the main panel. The mechanical interlock keeps them from both being "on" at the same time. One of the two-pole breakers is the main breaker for the loadcenter and is connected to the incoming service entrance conductors. The other two-pole breaker is connected to the standby power system generator (Figure 15-45).

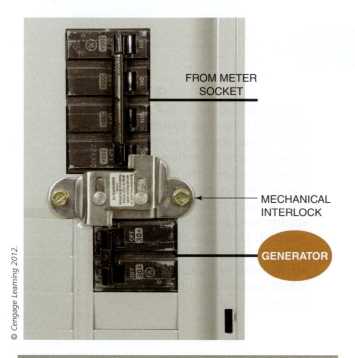

FROM METER SOCKET

MECHANICAL INTERLOCK

GENERATOR

© Cengage Learning 2012.

FIGURE 15-45 A transfer switch in a loadcenter made up with two double-pole circuit breakers and a mechanical interlock that will only let one of the breakers be turned "on" at a time. The mechanical interlock comes in a kit that is available for most brands of main circuit breaker panels.

(See Procedure 15-2 on page 555 for the recommended procedure for safely connecting a generator's electrical power to the critical load branch circuits.)

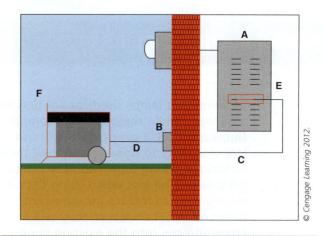

A

F

B

E

D

C

© Cengage Learning 2012.

FIGURE 15-46 A connection to a generator-ready load-center. (A) Generator-ready loadcenter; (B) Outdoor power inlet box; (C) Interior wiring; (D) Power cord from generator to power inlet; (E) Mechanical interlock for transfer switch in loadcenter; (F) Portable generator.

from experience...

Many manufacturers are now making generator-ready loadcenters (Figure 15-46). These new loadcenters are only slightly more expensive than standard loadcenters but can save the home owner a lot of money in future generator installation expense. The loadcenter is provided with two interiors, both being powered by the utility power supply. Critical circuits are placed in the lower interior, which is backed up by a generator in the event of a power outage. Providing back up to only the critical circuits in a house reduces the size and expense of the generator required.

SUMMARY

When an electrician installs the electrical wiring system for a house, there are many special wiring applications that may be encountered. Garages, either attached or detached, are often found in residential construction and must be wired as necessary. Many of the common wiring practices that relate to both detached and attached garages were covered in this chapter. More and more homes have swimming pools, spas or hot tubs, or a hydromassage bathtub installed. All these items require some type of electrical wiring that must be installed by an electrician. This chapter presented the basic installation requirements for these specific installations. Outdoor wiring was also presented with the emphasis being on outdoor receptacles and outdoor luminaires. The last special wiring situation discussed in this chapter was standby electrical power systems. Portable generators were described, and the proper installation for a typical standby power system with a portable generator was presented. Not all residential wiring systems will require an electrician to deal with the areas covered in this chapter. However, when these wiring situations are encountered, you will now have a better idea of the electrical wiring requirements and installation techniques needed for these applications.

GREEN CHECKLIST

☐ Incandescent lamps used for lighting in a garage can be replaced with compact fluorescent (CFL) lamps. CFLs will use less electricity to operate and they tend to last for a long time. Monthly electrical bills will be less when using CFLs. It is very easy to replace incandescent lamps with CFLs because they will screw directly into the existing lampholders.

☐ Motion sensors can be used to control outside lighting so that they will be switched on when movement is detected. During the sunlight hours, a photocell built-in to the motion sensor keeps the lighting fixture from switching on and wasting electrical power even if there is motion in the area. A motion sensor with a built in photoelectric sensor controlling outside lighting can eliminate the lighting being left on for long periods of time when not needed. This type of lighting control will cut down on the amount of electricity used to light the outside of a home and will result in electric bill savings.

PROCEDURE 15-1

Stripping the Outside Sheathing of Type UF Cable

The outside sheathing of a Type UF cable is much more difficult to remove than the outside sheathing on a nonmetallic sheathed cable. In a Type UF cable, the outer sheathing is molded around each of the circuit conductors. This manufacturing process helps make the cable suitable for underground installation but does make the cable rather difficult to strip. While every electrician eventually finds a way to strip away the UF cable sheathing, the following method has worked well for this author.

- Put on your safety glasses and observe all applicable safety rules. Wear gloves during this procedure to help prevent hand injuries.

A At the end of the cable you wish to strip, use a knife to cut a 1-inch slit in the cable above the bare grounding conductor.

- Using your fingers, open up the cable sheathing and grab the end of the bare grounding conductor with your long-nose pliers.

A

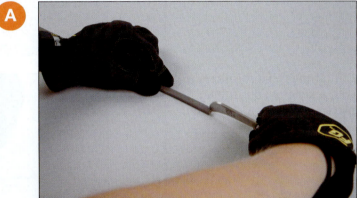
© Cengage Learning 2012.

B While holding on to the cable end with one hand, pull the grounding conductor back about 8 inches. This action will slit the cable sheathing as you are pulling the grounding conductor.

B

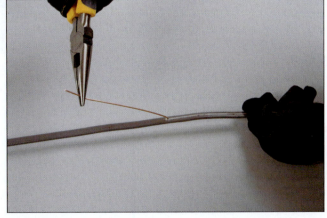
© Cengage Learning 2012.

PROCEDURE 15-1

Stripping the Outside Sheathing of Type UF Cable (Continued)

C Now, using your long-nose pliers, grab the end of the other conductors in the cable and pull them back to the same point as the grounding conductor.

C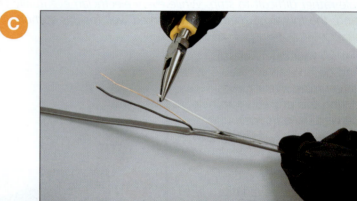

© Cengage Learning 2012.

D Cut off the cable sheathing with a knife or your cutting pliers.

- Clean up the work area and put all tools and materials away.

D

© Cengage Learning 2012.

PROCEDURE 15-2

Connecting a Generator's Electrical Power to the Critical Load Branch Circuits

When a standby power system includes a portable generator that is cord-and-plug connected to the transfer switch, the following procedure should be followed to safely connect the generator's power to the critical load branch circuits when electrical power from the utility is lost:

1 Plug in the four-wire flexible cord to the polarized twistlock receptacle on the generator.

2 Plug the other end into the polarized twistlock power inlet that is connected to the transfer switch.

3 Switch the transfer switch to the "generator" position.

4 Making sure that the generator is located outdoors, start the generator.

5 When electrical power from the utility is restored, switch the transfer switch to its "normal" position.

6 Shut down the generator.

7 Unplug the power cord from the transfer switch end and then the generator end.

8 Put everything back in its proper storage area.

REVIEW QUESTIONS

Directions: Answer the following items with clear and complete responses.

1. Describe what is meant by the phrase "standby power."

2. Describe the function of a transfer switch in a standby power system.

3. A typical transfer switch used in a residential standby power system is a:

   a. Double-pole, single-throw switch (DPST)

   b. Double-pole, double-throw switch (DPDT)

   c. Single-pole, double-throw switch (SPDT)

   d. Single-pole, single-throw switch (SPST)

4. The conductors running from a standby system generator to the transfer switch must not be less than _____ percent of the generator's output rating. The *NEC*® section that states this is _____.

5. Article _____ covers the requirements for swimming pools, hot tubs, and hydromassage bathtubs.

6. The *NEC*® requires all metal parts of a permanently installed swimming pool to be bonded together. The bonding grid must be connected to a grounding electrode. (Circle the correct response.) True or False. Name the *NEC*® section that backs up your answer. Section _____.

7. Receptacles located within 20 feet (6 m) from the inside edge of a swimming pool must be _____ protected.

8. Luminaires not GFCI protected and installed over a swimming pool must be located no less than _____ feet above the maximum water level.

9. All 125-volt, 15-, 20-, and 30-amp receptacles that are located within 10 feet (3 m) of the inside walls of an indoor hot tub must be GFCI protected. (Circle the correct response.) True or False. Name the *NEC*® Section that backs up your answer. Section _____.

10. Wall switches must be located at least 5 feet (1.5 m) from the edge of an indoor hot tub. (Circle the correct response.) True or False. Name the *NEC*® section that backs up your answer. Section _____.

11. When installing underwater luminaires in a swimming pool, they must be installed so that the top of the lens is not less than _____ inches below the normal water level, unless they are designed for a lesser depth.

12. Describe the difference between a hot tub and a hydromassage bathtub.

13. All 125-volt, 15-, 20-, and 30-amp receptacles installed within _____ of the hydromassage bathtub must be GFCI protected.

14. State the minimum burial depth for a Type UF cable feeding a pole light in a residential application. The circuit is rated 20 amperes and *is not* GFCI protected.

15. State the minimum burial depth for a Type UF cable feeding a pole light in a residential application. The circuit is rated 20 amperes and *is* GFCI protected.

16. State the minimum burial depth for a PVC conduit feeding a detached garage and installed across a lawn in a residential application.

17. When receptacles are installed outdoors and are intended to serve cord-and-plug-connected items, the cover must provide a weatherproof environment whether the plug is inserted into the receptacle or not. (Circle the correct response.) True or False. Name the *NEC*® section that backs up your answer. Section _____.

18. Describe what kind of electrical hazard can occur when a standby system generator is used, no transfer switch has been installed, and the service entrance main circuit breaker is not turned off.

19. Define the following:

    a. Dry-niche luminaire

    b. Wet-niche luminaire

    c. No-niche luminaire

20. A _____ or larger solid copper conductor must be used to make the bonding connections in the bonding grid for a permanently installed swimming pool. Name the *NEC*® section that backs up your answer. Section _____.

21. Name the two types of residential generators used in a standby power system.

22. An electrician is installing a 10 kW, 120/240-volt portable generator system for a home owner. Calculate the minimum-size copper conductors that will need to be installed from the power inlet to the transfer switch.

23. Lighting fixtures can be mounted on trees. (Circle the correct response.) True or False

24. Explain why Type UF cable is so much harder to strip than Type NM cable.

25. The *NEC*® requires at least one 15- or 20-amp, 125-volt rated receptacle be installed in an attached garage or a detached garage with electrical power; however, more receptacles can be installed if you choose. Explain the GFCI-protection requirements for receptacles installed in a garage.

KNOW YOUR CODES

1. Check with the authority having jurisdiction (AHJ) in your area to determine if nonmetallic-sheathed cable is allowed in an unfinished garage or accessory building.

2. Check with the authority having jurisdiction (AHJ) in your area to determine if there are any local wiring requirements for swimming pools.

3. Check with your local electric utility company to determine if there are any special rules to follow when setting up a standby generator system for a house.

CHAPTER

16

Video, Voice, and Data Wiring Installation

OBJECTIVES   *Upon completion of this chapter, the student should be able to:*

- List several common terms and definitions used in video, voice, and data cable installations.

- Demonstrate an understanding of EIA/TIA 570-B standards for the installation of video, voice, and data wiring in residential applications.

- Identify common materials and equipment used in video, voice, and data wiring.

- Demonstrate an understanding of the installation of video, voice, and data wiring in residential applications.

- Install crimp-on and compression style F-Type coaxial cable connectors.

- Install RJ-45 jacks and plugs on Category 5e and Category 6 unshielded twisted pair cable.

GLOSSARY OF TERMS

attenuation the decrease in the power of a signal as it passes through a cable; it is measured in decibels; often increases with frequency, cable length, and the number of connections in a circuit

bandwidth identifies the amount of data that can be sent on a given cable; it is measured in hertz (Hz) or megahertz (MHz); a higher frequency means higher data-sending capacity

category ratings on the bandwidth performance of UTP cable; categories include 3, 4, 5, 5e, and 6; Category 5e is rated to 100 MHz and is often installed in houses

coaxial cable a cable in which the center signal-carrying conductor is centered within an outer shield and separated from the conductor by a dielectric; used to deliver video signals in residential structured cabling installations

EIA/TIA an acronym for the Electronic Industry Association and Telecommunications Industry Association; these organizations create and publish compatibility standards for the products made by member companies

EIA/TIA 570-B the newest standard document for structured cabling installations in residential applications

F-type connector a 75-ohm coaxial cable connector that can fit RG-6 and RG-59 cables and is used for terminating video system cables in residential wiring applications; the most common styles used are crimp-on and compression

horizontal cabling the connection from the distribution center to the work area outlets

IDC an acronym for "insulation displacement connection"; a type of termination where the wire is "punched down" into a metal holder with a punch-down tool; no prior stripping of the wire is required

jack the receptacle for an RJ-45 plug

jacketed cable a voice/data cable that has a nonmetallic polymeric protective covering placed over the conductors

megabits per second (Mbps) refers to the rate that digital bits (1s and 0s) are sent between two pieces of equipment

megahertz (MHz) refers to the upper frequency band on the ratings of a cabling system

near end crosstalk (NEXT) electrical noise coupled from one pair of wires to another within a multi-pair voice or data cable

patch cord a short length of cable with an RJ-45 plug on either end; used to connect hardware to the work area outlet or to connect cables in a distribution panel

punch-down block the connecting block that terminates cables directly; 110 blocks are most popular for residential situations

RG-59 a type of coaxial cable typically used for residential video applications; EIA/TIA 570A

recommends that RG-6 coaxial cable be used instead of RG-59 because of the better performance characteristics of the RG-6 cable

RG-6 (Series 6) a type of coaxial cable that is "quad shielded" and is used in residential structured cabling systems to carry video signals such as cable and satellite television

ring one of the two wires needed to set up a telephone connection; it is connected to the negative side of a battery at the telephone company; it is the telephone industry's equivalent to an ungrounded "hot" conductor in a normal electrical circuit

RJ-11 a popular name for a six-position UTP connector

RJ-45 the popular name for the modular eight-pin connector used to terminate Category 5e and 6 UTP cable

service center the hub of a structured wiring system with telecommunications, video, and data communications installed; it is usually located in the basement next to the electrical service panel or in a garage; sometimes called a "distribution center"

STP an acronym for "shielded twisted pair" cable; it resembles UTP but has a foil shield over all four pairs of copper conductors and is used for better high-frequency performance and less electromagnetic interference

structured cabling an architecture for communications cabling specified by the EIA/TIA TR41.8 committee and used as a voluntary standard by manufacturers to ensure compatibility

tip the first wire in a pair; a conductor in a telephone cable pair that is usually connected to the positive side of a battery at the telephone company's central office; it is the telephone industry's equivalent to a grounded conductor in a normal electrical circuit

UTP an acronym for "unshielded twisted pair" cable; it is composed of four pairs of copper conductors and graded for bandwidth as "categories" by EIA/TIA 568; each pair of wires is twisted together

work area outlet the jack on the wall that is connected to the desktop computer by a patch cord

According to the Consumer Electronics Association, approximately two-thirds of all new homes are being built with a structured cabling system. Today's residential electricians are often required to install a low-voltage structured cabling system that provides video, voice, and data signals to locations throughout a house. A residential structured cabling system can also include an audio system that allows music from a variety of sources to be heard and controlled from any room in the house. An intercom system can be installed to let you talk to visitors and see them on your TV when they are at the front or rear doors. It will also allow communication with family members from any room in the house. Another popular option is a home monitoring system, where home owners can view various locations both inside and outside the home on any computer or television monitor connected to the system. While an audio system, an intercom system, and a home monitoring system are great additions to a residential structured cabling system, most new homes have their structured cabling system limited to video, voice, and data. It is for this reason that this chapter concentrates on the materials, installation methods, and termination techniques of copper structured cabling systems for video, voice, and data.

INTRODUCTION TO EIA/TIA 570-B STANDARDS

In today's homes, there is often a need for a "structured cabling" system to be installed. This system is completely separate from the residential electrical power system that we have discussed up to now. The structured cabling system includes the wiring and other necessary components for providing video, voice, and data signals throughout a house. It is installed using special cables and associated equipment by the same electricians who install the regular electrical power system. These systems offer multiple signal outlets for faster Internet connections and the ability to network home computers and peripheral devices like printers. They also provide for a number of conveniently located signal outlets for all the video devices (televisions, DVD players, Blue-Ray players, digital recorders, and so on) and telephones that are used in a modern home.

These systems are wired with special cables that include unshielded twisted pair cable (UTP), shielded twisted pair cable (STP), and coaxial cables. Sometimes optical fiber cables (FO) are used in the installation of a structured cabling system. However, because optical fiber cable is normally used only in commercial and industrial structured cabling applications, the installation of optical fiber cable is beyond the scope of this book and is not covered here.

In an effort to standardize the installation of a structured cabling system, the Telecommunications Industry Association (TIA) and the Electronic Industry Association (EIA) standards development committee was formed to develop installation standards for structured cabling systems in residential applications. EIA/TIA 570-A was developed by the committee and published in 1999 to establish a standard for a generic cabling system that can accommodate many different applications. In 2004, **EIA/TIA 570-B** was published and included updates to the original 570-A standard. In 2008, EIA/TIA 570-B-1 was published with an addendum that included some new guidelines for coaxial cable broadband installations. It was determined that proper installation and connector termination are critical to a video, voice, and data system's overall performance. By following EIA/TIA 570-B, a structured cabling system will perform properly.

To be able to truly understand the installation requirements as outlined in EIA/TIA 570-B, an electrician needs to become familiar with several terms that are used in the area of structured cabling. These terms are not normally used when referring to the regular electrical power system in a house, and not understanding the meaning of these terms can cause a lot of confusion for an electrician installing a structured cabling system. The following structured cabling system terms and definitions are used often and should become familiar to you:

- **Structured cabling** is a system for video, voice, and data communications cabling specified by EIA/TIA and used as a voluntary standard by manufacturers to ensure compatibility. Notice that the word "voluntary" is used. These installation guidelines are not mandatory. Interestingly enough, the only mandatory installation rules for installing a structured cabling system are those referenced in the *National Electrical Code® (NEC®)*. We will look at the *NEC®* rules later in this chapter.

- **EIA/TIA** is the acronym for the Electronic Industry Association and the Telecommunications Industry Association. Both organizations create compatibility standards for the products made by their member companies.

- **Bandwidth** is the term that identifies the amount of data that can be sent on a given cable. It is measured in hertz (Hz) or megahertz (MHz). The higher the frequency, the higher the data-sending capacity.

- **Insulation displacement connection (IDC)** is a type of termination where the wire is "punched down" into a metal holder with a punch-down tool (Figure 16-1). No prior stripping of the wire is required.

- **Unshielded twisted pair (UTP)** (Figure 16-2) is the type of cable normally used to install voice and data communication wiring in a house. It is composed of four pairs of copper conductors and graded for bandwidth as "categories."

- A **category** is the rating, based on the bandwidth performance, of UTP cable.

- **Shielded twisted pair (STP)** is a cable that resembles UTP but has a foil shield over all four pairs of copper conductors and is used for better high-frequency performance and less electromagnetic interference (EMI).

- **Coaxial cable** is a cable in which the center signal-carrying conductor is centered within an outer shield and separated from the conductor by a dielectric. This cable type is used to install video signal wiring in residential applications. It can also be used to carry a high-speed Internet signal.

FIGURE 16-1 A punch-down tool with a 110 blade. This tool terminates voice and data wires to a 110 punch-down block. It also cuts the wires to length. The impact absorbing cushioned grip reduces operator stress for high-volume users. The actuation mechanism can be adjusted, depending on the operator's preference. *Courtesy of IDEAL INDUSTRIES, INC.*

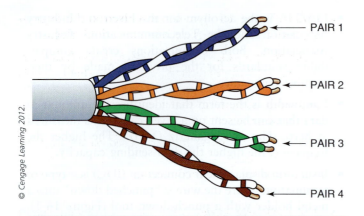

FIGURE 16-2 An unshielded twisted pair (UTP) cable. This cable example has four twisted pairs.

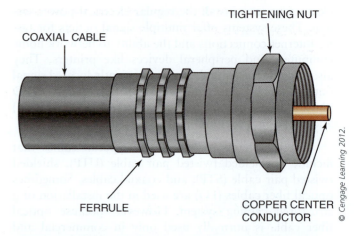

FIGURE 16-4 A crimp-on style F-type coaxial cable connector. This connector is threaded onto a female adapter at a television outlet.

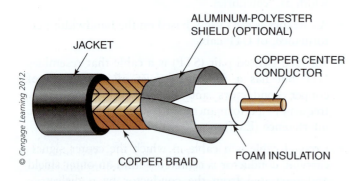

FIGURE 16-3 An RG-6 coaxial cable.

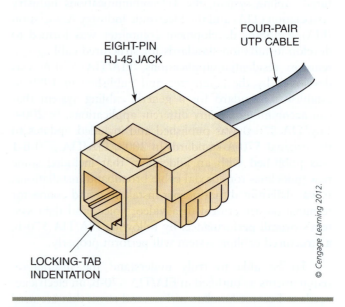

FIGURE 16-5 An eight-pin connector or jack used to terminate UTP cable. It is commonly called an RJ-45 jack.

- **RG-6** (Figure 16-3) is a type of coaxial cable that is used in residential structured cabling systems to carry video signals such as cable and satellite television. The letters "RG" stand for "radio guide."

- **RG-59** is a type of coaxial cable typically used for residential video applications. However, EIA/TIA-570-B recommends that RG-6 coaxial cable be used instead of RG-59 because of the better performance characteristics of the RG-6 cable. Do not use RG-59 coaxial cable in any new video signal wiring installation.

- An **F-type connector** (Figure 16-4) is a 75-ohm coaxial cable connector that can fit RG-6 and RG-59 cables and is used for terminating video system cables. It is a connector type that is threaded onto the television outlet termination point for a good, solid connection. It can be terminated on the end of a coaxial cable in a variety of ways. It is highly recommended that F-type connectors be either crimped or compressed on to the end of a coaxial cable with a proper tool. The procedures for doing this are shown later in this chapter.

(See Procedure 16-1 on page 579 for the proper procedure for installing a crimp-on style F-type connector on an RG-6 coaxial cable.)

(See Procedure 16-2 on page 580 for the proper procedure for installing a compression style F-Type connector on an RG-6 coaxial cable.)

- **RJ-11** is popular name given to a six-position connector or jack. The letters "RJ" stand for "registered jack."

- **RJ-45** is the popular name given to an eight-pin connector or jack used to terminate UTP cable (Figure 16-5).

(See Procedure 16-3 on pages 581–583 for the proper procedure for installing an RJ-45 jack on a four-pair UTP Category 5e cable.)

(See Procedure 16-4 on pages 584–585 for the proper procedure for installing an RJ-45 jack on a four-pair UTP Category 6 cable.)

from experience...

In many areas of the country, compression F-type coaxial connectors (Figure 16-6) have become the coaxial cable connector of choice. They are different from a crimp-on F-type connector in that they use a radial crimp method that compresses uniformly around the connector body for better audio/video signal quality. They install easily and also provide a watertight connection that makes them especially useful when making coaxial cable connections outside. A disadvantage to using the compression F-type coaxial cable connector instead of the crimp-on style is that they are more expensive and require a special installation tool.

© Cengage Learning 2012.

FIGURE 16-6 A compression type coaxial cable connector.

- A **jack** (Figure 16-7) is the term given to the receptacle device that accepts an RJ-11 or RJ-45 plug.

- A **punch-down block** is the connecting block that terminates UTP cables directly; 110 blocks (Figure 16-8) are most popular for residential applications and require a 110-block punch-down tool for making the terminations.

- **Horizontal cabling** is the term used to identify the cables run from a service center that serves as the "hub" for the structured cabling system to the work area outlet.

- The **work area outlet** is the jack on the wall that is connected to a desktop computer or to a telephone by a patch cord plugged in to the appropriate jack (Figure 16-9).

- A **patch cord** is the short length of cable with an RJ-45 plug on either end used to connect a home computer or telephone to the work area outlet. A patch cord can also be used to interconnect various punch-down blocks in the service center.

(See Procedure 16-5 on pages 586–588 for the proper procedure for assembling a patch cord with RJ-45 plugs and Category 5e UTP cable.)

- The **service center** (Figure 16-10) is the hub of a structured wiring system with telecommunications, video, and data communications installed. It is usually located in the basement next to the electrical service panel or in a garage.

- **Megahertz (MHz)** refers to the upper frequency band on the ratings of a cabling system. "Mega" is the prefix for "one million." "Hertz" is the term used for the number of cycles per second of the specific signal.

- **Megabits per second (Mbps)** refers to the rate that digital bits (1s and 0s) are sent between two pieces of digital electronic equipment. A bit is the smallest unit of measurement in the binary system. The expression "bit" comes from the term "binary digit."

The cables used for voice and data signal transmission have been categorized by the EIA and TIA. All these cables are tested and listed according to Underwriters Laboratories Standard 444. The cable is categorized as follows (Figure 16-11):

- Category 1
  - Four wires, not twisted
  - Called "quad wire" by electricians
  - Okay for audio and low-speed data transmission
  - Not recommended for use in residential applications

SURFACE-MOUNTED JACK

WALL-MOUNTED JACK

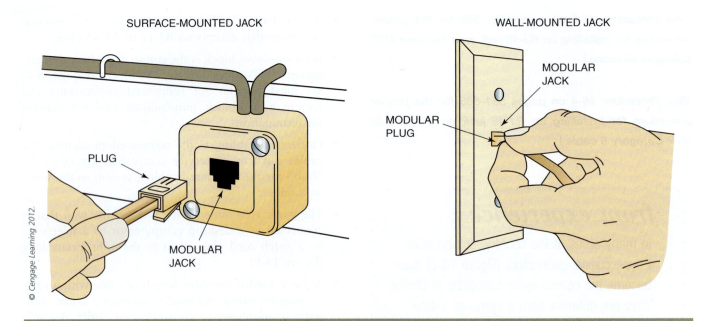

© Cengage Learning 2012.

FIGURE 16-7 A "jack" is the receptacle that the "plug" on the end of voice and data cables fits into. The jack can be surface-mounted or flush-mounted.

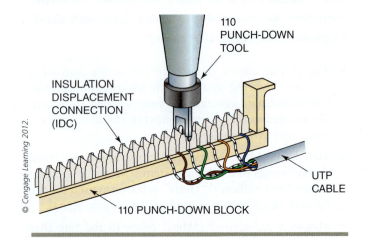

© Cengage Learning 2012.

FIGURE 16-8 A 110-style punch-down block. A punch-down tool with a 110 blade is used to terminate the wires of a UTP cable to the block.

© Cengage Learning 2012.

FIGURE 16-9 A work area outlet may consist of one or more jacks for video, voice, or data connections.

- Category 2
  - Four pairs with a slight twist to each pair
  - Okay for audio and low-speed data
  - Not recommended for use in residential applications
- Category 3
  - 16-MHz bandwidth
  - Supports applications up to 10 Mbps
  - 100-ohm UTP-rated Category 3

- Declining in popularity but often still used in residential telephone only applications
- Category 4
  - 20-MHz bandwidth
  - Supports applications up to 16 Mbps

- 100-ohm UTP-rated Category 4
- Basically obsolete
- Category 5
  - 100-MHz bandwidth
  - Supports applications up to 100 Mbps
  - 100-ohm UTP-rated Category 5
  - Often used for residential voice and data applications
- Category 5e
  - 100-MHz bandwidth
  - Supports applications up to 100 Mbps
  - 100-ohm UTP-rated Category 5e
  - Higher performance over a minimally compliant Category 5 installation by following Category 5e Technical Specifications
  - The preferred cable type for both voice and data in residential applications
- Category 6
  - 250-MHz bandwidth
  - Supports applications over 100 Mbps
  - 100-ohm UTP-rated Category 6
  - The favorite in commercial applications and is recommended for use in residential applications where the high bandwidth of Cat 6 is needed to provide the speed that many newer applications require

FIGURE 16-10 A service-center, sometimes called a distribution center, is the origination point for the video, voice, and data circuits installed throughout a house. It is located in a basement, garage, or some other utility area.

© Cengage Learning 2012.

Category 1:	Four-pair nontwisted cable. Referred to as "quad wire." Okay for audio and low-speed data. Not suitable for modern audio, data, and imaging applications. This is the old-fashioned plain old telephone service (POTS) cable.
Category 2:	Usually four pairs with a slight twist to each pair. OK for audio and low-speed data.
Category 3:	UTP. Data networks up to 16 MHz.
Category 4:	UTP. Data networks up to 20 MHz.
Category 5.	UTP. Data networks up to 100 MHz. By far the most popular, having four pairs with 24 AWG copper connectors, solid (for structured wiring) and stranded (for patch cords). Cable has an outer PVC jacket.
Category 5e:	UTP. Basically an enhanced Category 5 manufactured to tighter tolerances for higher performance. Data networks up to 350 MHz. Will eventually replace Category 5 for new installations.
Category 6:	UTP. Four twisted pairs. Data networks up to 250 MHz. Has a spline between the four pairs to minimize interference between the pairs. Installed in most new commercial applications.
Category 7:	Four twisted pairs surrounded by a metal shield. Data networks up to 750 MHz. Not used in residential or commercial applications at this time.

© Cengage Learning 2012.

FIGURE 16-11 Cables are rated in "categories." This figure shows the various category types and has a short description of each.

- Category 7
  - 750-MHz bandwidth
  - A metal shield around the conductors
  - Not currently used in residential applications

The voice and data cable recommended by EIA/TIA 570-B for use in residential applications has the following characteristics:

- It carries both voice and data.
- It is normally 22 or 24 AWG copper.
- It is always described and connected in pairs.
- The wire pairs should be twisted together to preserve signal quality.

Each pair of wires in a voice/data structured cabling system cable consists of a **tip** and a **ring** wire. This is a carryover from the old days in the telephone industry. When an operator was asked to connect a call to another party, a cable was plugged into a connection panel in front of the operator. This cable had a connector on the end of it with a tip and a ring that made the proper connection so that the call could go through. Because direct current is used by telephone systems, polarity must be maintained, so the standard wiring practice is to use the tip wire as the positive (+) conductor and the ring as the negative (−) conductor.

Each pair of conductors in a UTP or STP cable is twisted together in a certain way. The reason for the twisted pairs is that it helps prevent interference like induction and "crosstalk" from other pairs in the same cable and from other sources like electrical power circuits and electric motors. You may have experienced "crosstalk" when you have been using a phone and heard another conversation going on in the background at the same time. **Jacketed cable** with four twisted conductor pairs is recommended for all inside residential voice and data wiring.

⚠ CAUTION

CAUTION: The old "quad wire" telephone cable that has been used for years to wire telephone jacks in residential applications is not recommended for use by EIA/TIA 570-B. Use a four-pair UTP cable like Cat 5e instead.

Each pair of conductors in voice/data cable has a specific color coding so an electrician will know which pair of conductors connects to the proper terminals at a voice/data jack or in the service center (Figure 16-12). The standard color coding for a four-pair UTP cable is as follows:

- Pair 1: Tip is white/blue; ring is blue.
- Pair 2: Tip is white/orange; ring is orange.
- Pair 3: Tip is white/green; ring is green.
- Pair 4: Tip is white/brown; ring is brown.

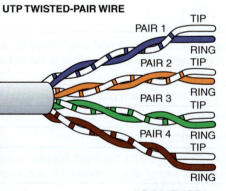

STANDARD FOUR-PAIR UTP COLOR CODES		
PAIR 1	T	WHITE/BLUE STRIPE
	R	BLUE
PAIR 2	T	WHITE/ORANGE STRIPE
	R	ORANGE
PAIR 3	T	WHITE/GREEN STRIPE
	R	GREEN
PAIR 4	T	WHITE/BROWN STRIPE
	R	BROWN

NOTE: FOR SIX-WIRE JACKS USE PAIR 1, 2, AND 3 COLOR CODES. FOR FOUR-WIRE JACKS, USE PAIR 1 AND 2 COLOR CODES.

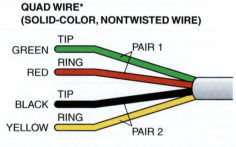

*CAUTION

QUAD WIRE IS NO LONGER RECOMMENDED FOR INSTALLATION. IF ENCOUNTERED DURING A RETROFIT, QUAD WIRE SHOULD BE REPLACED WITH 100-Ohm UTP IF POSSIBLE. CONNECTING NEW QUAD TO INSTALLED QUAD WILL ONLY AMPLIFY EXISTING PROBLEMS AND LIMITATIONS ASSOCIATED WITH QUAD WIRE; LEAVING EXISTING QUAD IN PLACE AND CONNECTING 100-Ohm UTP TO IT MAY ALSO BE INEFFECTIVE, AS THE QUAD WIRE MAY NEGATE THE DESIRED EFFECT OF THE UTP.

FIGURE 16-12 The TIA standard color coding for residential voice and data wiring.

from experience...

An easy way to remember the color coding for a four-pair UTP cable is to use the acronym "BLOGB." "BL" stands for blue, "O" stands for orange, "G" stands for green, and "B" stands for brown.

The standard color coding for a quad wire (solid-color, non-twisted pair) is as follows:

- Pair 1: Tip is green; ring is red.
- Pair 2: Tip is black; ring is yellow.

INSTALLING RESIDENTIAL VIDEO, VOICE, AND DATA CIRCUITS

In today's residential structured cabling systems, it is common practice to install a service center that serves as the origination point for all video, voice, and data systems in a house. Install the service center in any well-lighted area that will give a technician clear access to it such as in a basement, garage, or utility closet. It should not be located close to the main service panel or a subpanel. If possible, maintain a minimum separation of 5 feet from any electrical power panel. Do not locate the service center above major appliances like a clothes washer and avoid moist locations such as in a bathroom or over a sink. Try to choose a location that is central to the rooms being served to keep all cable runs as short as possible so that there will be better system performance. EIA/TIA 570-B recommends that you install a 15- or 20-amp, 125-volt-rated receptacle next to the service center so electrical power is available for service center modules that require 120 volts alternating current to operate. It also recommends a location that is no more than 5 feet from the house electrical power system's grounding electrode. Installation instructions that come with a service center will contain instructions for proper grounding of the metal service center enclosure. The instructions typically require an electrician to run a minimum of 10 AWG copper wire to the nearest grounded metal water pipe location. So that the service center can be easily accessed and worked on, mount it on or in a wall

so the top of the panel is approximately 6 feet from the floor. In new residential construction, the "home runs" from the service center are run in concealed pathways, such as in wall or ceiling cavities. Some electricians choose to install the cable wiring in conduit so that it will be easier to add or take away cables in the future as customer needs change. In existing homes, the cable wiring is concealed as much as possible in attics, basements, and crawl spaces. When the wiring cannot be concealed, it is run on the surface using a surface metal or nonmetallic raceway.

VOICE WIRING SYSTEM

Figure 16-13 shows a suggested wiring layout for the voice system in a typical house. Use four-pair 100-ohm UTP cable. Remember to not use the old quad cable. A Category 3 cable is the absolute minimum performance category that should be installed. It is recommended that at least a Category 5e be used in all new voice installations. At each wall outlet location, an eight-position RJ-45 jack with T568A wiring should be used. T568B wiring of the jack may also be used (Figure 16-14). Whichever wiring scheme is used, make sure that all jacks in the house are wired the same way. There should be a minimum of one voice jack per outlet location. Each voice outlet location should have a separate home run back to the service center. The home runs are terminated at the service center by punching down the cable to the proper 110 terminal blocks. Once the system has been completely installed, connection of individual telephones to the wall jacks is accomplished by simply plugging the phone into any jack in the house.

> **CAUTION**
>
> **CAUTION:** Even though many people use a cell phone as their main telephone communication device and do not have traditional telephone service installed in their home, Section 800.156 requires that at least one communications outlet be installed in each new home and that proper wiring be brought to the telephone company's demarcation point. It is common practice to install a wall-mounted telephone jack in the kitchen area of a house and then to install additional telephone jacks and associated wiring throughout the house. It will be up to the home owner whether the local telephone company connects to the house hard-wired communication system.

The telephone company will bring telephone service to the house by either an overhead wiring method or an underground wiring method. Either way, the telephone

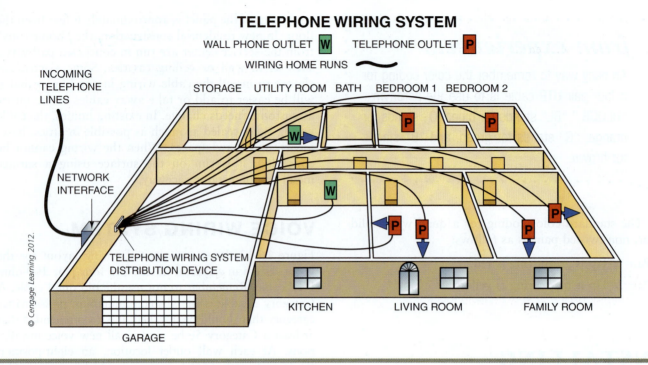

FIGURE 16-13 A typical residential telephone wiring system installation from a central distribution panel.

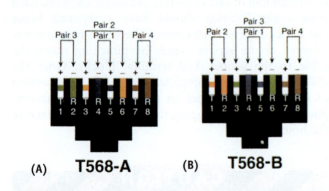

FIGURE 16-14 The wiring configurations for a T568-A (A) and a T568-B (B) scheme. Modular plugs are shown in this illustration, but the wiring configuration for the two schemes is the same for jacks and punch-down blocks as well. *Courtesy of IDEAL INDUSTRIES, INC.*

company will terminate its wiring at a telephone network interface box mounted on the outside of the house. This enclosure is weatherproof and is suitable for mounting on the outside of the house. If the telephone company terminates its wiring inside a house, a standard network interface box is used. The point where the telephone company's wiring ends and the home owner's interior wiring begins is called the "demarcation point." As an electrician installing the voice system in a home, you will need to install telephone wiring from the demarcation point to the service center.

Even though a service center is recommended to be installed for all new residential telephone installations, many home owners will choose to have the telephone system installed in a more traditional manner. In most cases, when a home owner chooses the traditional method over the newer EIA/TIA method described previously, it is based on the fact that the older method costs less money to install. The older method, even though it is not EIA/TIA recommended, is still the most often used technique for installing a telephone system in a house. It requires an electrician to install a telephone junction box (Figure 16-15) at a convenient location. It is usually located inside a home and is often placed next to the service entrance panel. A connection is made by the electrician from the demarcation point to the junction box with a suitable length of telephone cable. From the telephone junction box, individual runs of telephone cable (Category 3 or higher) are wired to the various telephone outlet locations throughout a house. Figure 16-16 shows a typical telephone system installation using the more traditional method.

DATA WIRING SYSTEM

Figure 16-17 shows a suggested wiring layout for the data wiring system in a typical house. Use a four-pair 100-ohm UTP cable. It is recommended that a

Category 5e be used as the minimum category rated cable for this type of installation. Category 6 cable is often specified in anticipation of faster data

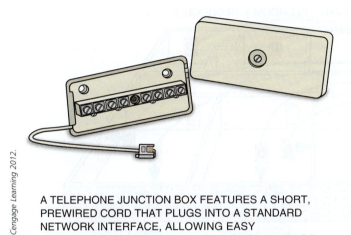

A TELEPHONE JUNCTION BOX FEATURES A SHORT, PREWIRED CORD THAT PLUGS INTO A STANDARD NETWORK INTERFACE, ALLOWING EASY CONNECTION OF ADDITIONAL TELEPHONE CABLES.

© Cengage Learning 2012.

FIGURE 16-15   A telephone junction box.

from experience...

Category 5e can more than adequately handle the data speeds required in most home networks. It is also easier and less costly to install than Category 6 UTP. For these reasons, Category 5e is still the preferred cable category for voice and data wiring in a residential structured cabling system. If possible, check with the home owner to find out if their home network will require the high transmission speeds of Cat 6 UTP cable. If so, you will need to install Cat 6 and not a Cat 5e data cabling system.

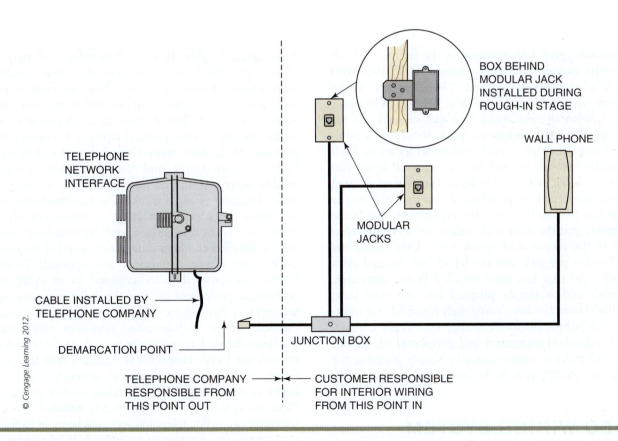

© Cengage Learning 2012.

FIGURE 16-16   A typical residential telephone installation using a more traditional method. The telephone company installs and protects the incoming telephone service cable. It can be installed underground or overhead.

DATA WIRING SYSTEM

DATA OUTLET D   WIRING HOME RUNS ~

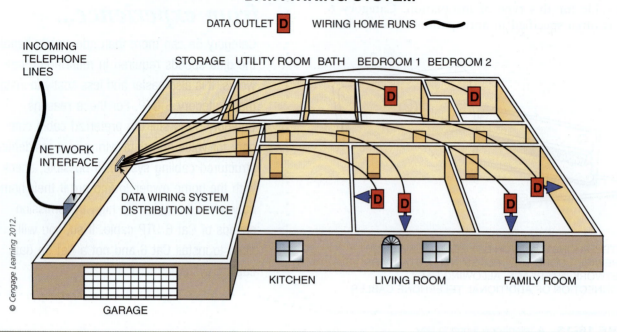

INCOMING TELEPHONE LINES

STORAGE   UTILITY ROOM   BATH   BEDROOM 1   BEDROOM 2

NETWORK INTERFACE

DATA WIRING SYSTEM DISTRIBUTION DEVICE

KITCHEN   LIVING ROOM   FAMILY ROOM

GARAGE

© Cengage Learning 2012.

FIGURE 16-17   A typical residential data wiring system installation from a central distribution center.

transmission speed requirements in the future. At each wall outlet location, an eight-position RJ-45 jack with T568-A wiring should be used. T568-B wiring of the jack may also be used (see Figure 16-14). Whichever wiring scheme is used, make sure that all jacks in the house are wired the same way. There should be a minimum of one data jack at each wall outlet location. Each data outlet location should have a separate home run back to the service center. The home runs are terminated at the service center by punching down the cable to the proper 110 terminal blocks. The equipment, like home computers, that the data wall outlets are serving is connected to the system with patch cords. One end of the patch cord is plugged into the RJ-45 jack located on a network card that has been installed in the computer. The other end is simply plugged into the wall jack. Once the system has been completely installed and Internet service is brought to the service center, the connection of individual computers and peripheral equipment to the wall jacks is accomplished by simply plugging the equipment into any jack in the house with a patch cord.

VIDEO WIRING SYSTEM

Figure 16-18 shows a suggested wiring layout for the video wiring system in a typical house. Use a 75-ohm

RG-6 coaxial cable. It is recommended that two runs of coaxial cable be installed from the service center to each television outlet location. The two runs of cable will allow for video distribution from any video source as well as distribution to a television at each outlet location. Also, coaxial cable is becoming more popular as a data transmission line for high-speed Internet connections, and the extra cable could be used to serve a computer located close to the television outlet location. Two runs are also recommended to be run from the service center to a convenient attic or basement location in case a satellite television system will be installed at a later date. At each end of the coaxial cable, male F-type connectors are installed. Threaded F-type connectors that are crimped on or compressed on using a special tool are recommended to reduce signal interference. Push-on or screw-on fittings are not recommended. At the wall-mounted television outlet end, a female-to-female F-type coupler is used. At the service center end, the F-type connector is threaded onto the proper fitting. At the television outlet end, connection to a video device is done using a 75-ohm RG-6 coaxial patch cord. Once the system has been completely installed and cable television or satellite television service is brought to the service center, the connection of individual televisions and other video devices to the wall jacks is accomplished by simply connecting them to any video outlet in the house.

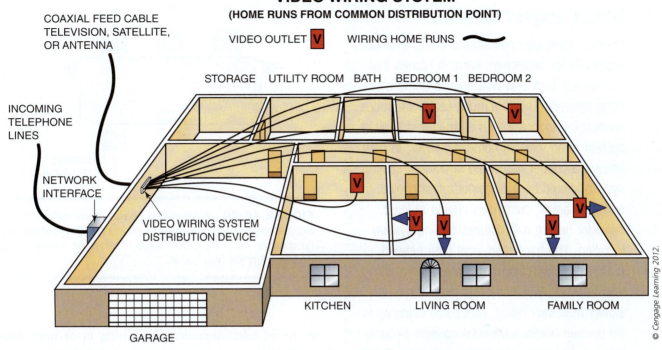

FIGURE 16-18   A typical residential video wiring system installation from a central distribution center. EIA/TIA-70-B recommends two RG6 cables be run to each video outlet.

from experience...

Use 75-ohm F-type termination caps in each unused television outlet and at each unused splitter port throughout the house. The termination cap will maintain the 75-ohm impedance of the system and help keep the signal to the other video devices strong and reliable (Figure 16-19).

FIGURE 16-19   A 75-ohm F-Type dtermination cap.

Like the voice (telephone) system described earlier, a more traditional method of installing the coaxial cable throughout a house is often used. Again, it is usually the higher cost associated with the EIA/TIA recommended method that causes home owners to choose the more traditional coaxial wiring technique. In this method, television service from a cable television company, a television antenna, or a satellite dish is brought to a distribution point in the house. The television service wiring is terminated at a splitter (Figure 16-20). The splitter "splits up" the incoming signal so it can be sent on individual RG-6 coaxial cables from the splitter to different television outlets throughout the house.

COMMON VIDEO, VOICE, AND DATA INSTALLATION SAFETY CONSIDERATIONS

The following safety items should be considered when installing cabling and jacks for a video, voice, and data application in a house:

- Never install or connect telephone wiring during an electrical storm. Telephone wiring can carry a fatal lightning surge for many miles.

Wireless computer networks are very popular, especially for customers living in homes that are already built and who want a home computer network. They are much easier and more economical to install than a structured cabling system in an existing home. However, in new homes, a copper hardwired structured cabling system should be a customer's first choice. In existing homes, customers should at least consider having a structured cabling system installed. The hardwired local area network (LAN) speed typically runs faster than a wireless network. Wireless LANs usually run slower than their rating, with some being up to 50 percent below advertised speeds! Security is a huge concern on wireless networks. Great improvements have been made in encryption, but for the average home owner, managing and setting up wireless security is beyond their expertise. With a hardwired network, a customer does not have to worry about someone in their neighborhood being able to use their wireless Internet connection. I often hear students at my school say that they do not pay for Internet access, they simply log on to their neighbor's wireless network. Another common practice out there is what is called "network sniffing," or searching the neighborhood for open networks at night in vehicles. Who knows what kinds of things are being downloaded and could be traced to your computer's IP address! So, even though a wireless network can be very convenient, a hardwired structured cabling system is the best way to go because of its speed, security, and reliability.

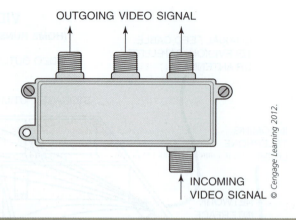

FIGURE 16-20 A three-way coaxial cable splitter. One signal-carrying cable comes into the splitter. Three-signal carrying cables leave the splitter to go to television outlets located in different areas of a house.

- Avoid telecommunications wiring in or near damp locations.
- Never place telephone wires near bare power wires or lightning rods.
- Never place voice and data wiring in any conduit, box, or other enclosure that contains power conductors.
- Always maintain adequate separation between voice and data wiring, and electrical wiring according to the *NEC®*.
- 50 to 60 volts direct current is normally present on an idle telephone tip and ring pair. An incoming call consists of 90 volts alternating current. This can cause a shock under the right conditions.

> **CAUTION**
>
> **CAUTION:** Many electricians are under the impression that you cannot get an electrical shock from a telephone line. This is not true. An incoming telephone call consists of 90 volts alternating current between the tip and ring conductors. If an electrician happens to be working on a telephone circuit when a call comes in, a 90-volt shock is possible. This author can tell you from personal experience that a 90-volt telephone circuit shock hurts.

- Jacks should never be installed where a person could use a telephone (hardwired) while in a bathtub, hot tub, spa, or swimming pool.
- Do not run open communications wiring between structures where it may be exposed to lightning.

- Always disconnect the dial-tone service from the house when working on an existing phone system. If you cannot disconnect, simply take the receiver off the hook. The direct-current value will drop, and the 90-volt alternating-current ring will not be available.

- Use caution when running telephone wiring on or near metal siding. Always check for stray voltages present on any metal surface.

- Be careful to not cut through or drill into concealed wiring. Always inspect the area before making a cut or starting to drill.

COMMON VIDEO, VOICE, AND DATA INSTALLATION PRACTICES

The "Star" wiring method is the recommended way to install the wiring. This is an easy method for electricians because it is the same as running home runs in electrical work. The video, voice, and data cables are run from the service center to each outlet location. If it is more convenient, the cables can be run from the various video, voice, and data outlets back to the service center.

Unlike the regular electrical power system wiring, video, voice, and data wiring is considered low voltage and does not require any kind of outlet box. However, something needs to be installed that will allow the video, voice, or data connection device to be installed. While both metal and nonmetallic boxes can be installed at each outlet location, it is recommended that mud rings (also called plaster rings) be used. Mud rings are typically installed on studs (Figure 16-21) during the rough-in stage, or they can be installed in existing walls (Figure 16-22) during the trim-out stage. When roughing in a mud ring, install it at the same height from the floor as the regular 120-volt receptacle outlets in the room. A wall-mounted telephone jack will require a mud ring (or box) to be installed at approximately 5 feet from the floor. The advantage of a mud ring over a regular electrical box is that the mud ring allows more room for space-consuming devices used in a video, voice, or data system, and they allow the installer to leave a long cable loop inside the stud cavity. It is recommended that you leave a 24-inch loop for any cable used in a video, voice, or data structured cabling system. This leaves you plenty of spare cable to correct wiring errors and for future expansion should the device at that location need changing. The mud ring provides a sturdy surface for attaching a device or cover plate during the trim-out stage.

The following items should be followed by an electrician when installing the structured wiring cables:

- Keep the cable runs as short as possible.

- Do not splice wires on the cable runs. Run the cables as one continuous length.

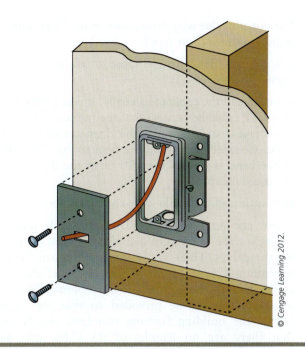

FIGURE 16-21   One style of mud ring being used in a new home. The raised part of the mud ring must match the thickness of the wall covering. For example, if ½-inch sheetrock is being used, a ½-inch raised mud ring will be needed.

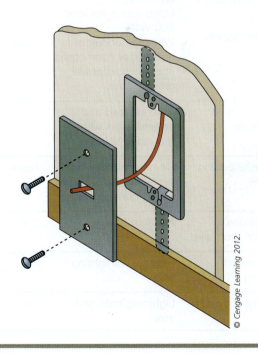

FIGURE 16-22   This style of mud ring is used in an existing wall. Cutting the hole in the wall for the mud ring is similar to cutting the hole in a wall for the installation of an old-work electrical box.

- Do not pull the wire with more than 25 pounds of pulling tension (four pair).

- Do not run the wire too close to electrical power wiring (Figure 16-23).

- Do not bend the cable too sharply. It is recommended that UTP cable be bent so that the radius of the bend is no less than four times the diameter of the UTP cable. Coaxial cable should be bent so that the radius of the bend is no less than 10 times the diameter of the coaxial cable.

- Do not install the cable with kinks or knots.

- There are no *NEC®* or EIA/TIA 570-B requirements for the maximum distance between supports for the cables. There is also no requirement for the minimum distance from a box or panel that the cables have to be secured. It is recommended that adequate support be provided so that the cables follow the building framing members closely and that there are no lengths of the cable that sag excessively. A good rule of thumb is to secure all video, voice, and data cables within 12 inches (300 mm) of an enclosure and then support them so that there is a maximum distance between supports of 4.5 feet (1.4 m).

- Use J-hooks or similar cable supports instead of staples for supporting cables along joists (Figure 16-24).

- Use insulated rounded or depth-stop plastic staples to secure the cables to the building framing members.

- Use tie wraps (secured loosely), Velcro, or cable straps when you have a bundle of several cables to support.

- Maintain polarity and match color coding throughout the house.

- To provide compatibility with two-line telephones, which are common in many homes, be sure to wire up the two inner pairs of an RJ-45 jack. Home telephones are typically connected to a telephone jack with a four-wire flat cable that has four-pin plugs on each end. When a four-pin plug is connected to an eight-pin RJ-45 jack, it makes connections with only the four middle pins of the jack. The four middle pins correspond to pair 1 and pair 2 of the voice cabling system.

Purpose	Type of Wire Involved	Minimum Separation
Electric supply	Bare light or power of any voltage	5 ft.
	Open wiring not over 300 volts	2 in.
	Wires in conduit or in armored or nonmetallic sheath cable/power ground wires	None
Radio and television	Antenna lead and ground wires without grounded shield	4 in.
Cable television cables	Community television systems coaxial cables with grounded shield	None
Telephone service drop wire	Aerial or buried	2 in.
Fluorescent lighting	Fluorescent lighting wire	5 in.
Lightning system	Lightning rods and wires	6 ft.

© Cengage Learning 2012.

FIGURE 16-23 This table shows the minimum recommended separations between residential video, voice, and data wiring and the electrical power system conductors.

CAUTION

CAUTION: Always use a T568-A wiring scheme when wiring two-line telephone jacks. This wiring scheme will result in both incoming lines being available at each telephone jack. If a T568-B wiring scheme is used, both pair 1 and pair 3 are used as the two innermost pairs, and the pair 2 telephone line will not be available when a four-pin plug is connected to the RJ-45 jack.

© Cengage Learning 2012.

FIGURE 16-24 A J-hook is used to support multiple structured wiring system cables run along joists. It is nailed or screwed to the joist at regular intervals of about 3 feet. A cable-tie is usually used to secure the cables in the J-hook after all of the cables have been placed in it. The cable-tie goes through the holes in the front and back of the J-hook.

from experience...

A common wiring practice used to support bundles of several video, voice, or data cables is to drive a regular Romex™ cable staple into a wooden framing member until it is almost all the way in. Then insert a tie wrap through the area between the staple and the framing member, place the tie wrap around the bundle of cables, and loosely tighten the tie wrap. This will hold the bundle of cables in place. Remember to not tighten the tie wrap too tightly.

- If conduit is installed, leave a pull string so that the video, voice, and data cable can be easily pulled in later.
- Never run video, voice, or data wiring in the same conduit with power wires.
- Use inner structural walls instead of outer walls for cable runs whenever possible.
- Do not run the cables through bored holes with power wires (Figure 16-25).

USE INSULATED NM-TYPE STAPLES AND LEAVE WIRE LOOSE INSIDE STAPLE.

DO NOT SHARE BORED HOLES WITH POWER WIRES.

IF POWER MUST BE CROSSED, CROSS AT 90.

WALL PHONE OUTLET BOXES ARE 54 IN. TO 60 IN. (1.3 m–1.5 m) FROM FLOOR.

© Cengage Learning 2012.

FIGURE 16-25 The suggested installation practice for running video, voice, and data cables through bored holes in building framing members.

- Keep the cables away from heat sources, such as hot water pipes and furnaces.
- Avoid running exposed cables whenever possible.
- Leave about 24 inches of wire at outlets and connection points. This will ensure enough wire to properly make the necessary terminations. Push any extra cabling back into the wall or ceiling in case future work will need to be done on the system.
- Always check for shorts, opens, and grounds when the rough-in is complete.
- It is recommended that a separate 15- or 20-amp branch circuit be run to the service center.

The following items should be considered when terminating the cables:

- Binding posts—still commonly used for residential applications (Figure 16-26):
  - Be careful to not nick the inner conductors when stripping the outer jacket of the cable.
  - Wrap the conductor in a clockwise direction between two washers.

FRONT

BACK

BINDING POST TERMINALS

FIGURE 16-26 Binding post connections are found on the back of some residential telephone wall jacks. This type of jack is fine for voice-only applications. However, it is no longer recommended for use in residential installations. *Courtesy of Hubbell Incorporated.*

from experience...

When roughing in the wiring for a residential structured cabling system in a house that has more than one floor, it is a good idea to install a couple of empty conduits from the basement, or from wherever else the distribution center is located, to the attic. The conduits should be a minimum of ¾-inch trade size and be installed using EMT or ENT. This will allow for easy expansion of the structured cabling system if and when new cables for video, voice, data, audio, or home monitoring need to be installed. Simply pull the new cables through the conduits, into the attic, and then to the proper locations in the rooms on the top floor. A pull string should be installed in each conduit run.

- Be sure the wire does not get caught in the screw threads—it may break.

- Trim off any excess exposed bare wire.

- Use no more than two or three wires under a single screw.

- Always leave plenty of spare wire at the connection point.

- Do not overtighten the binding posts with more than 7 inch-pounds of torque.

- Insulation displacement connectors (IDCs) (see Figure 16-8):

  - Do not untwist any more of the wire pairs than absolutely necessary to make a termination. Make sure that the twists are no more than ½ inch away from where the wires are terminated.

  - IDCs are faster and more reliable than binding post terminations.

  - IDCs require a special punch-down tool.

  - The most common IDC type for residential applications is a 110 clip.

- Wiring jacks for residential applications:

  - EIA/TIA 570-B recommends eight conductor RJ-45 jacks.

  - Six (RJ-11) or four conductor jacks (binding post type) are still widespread but are not recommended to be installed for any new residential installation or system upgrade.

- Category 5e wiring and devices are recommended for all residential applications. Category 6 wiring and devices may be needed depending on the home network speed requirements.

- Wire one or two jacks per room. *Warning*: Do not use a screwdriver blade to terminate IDCs. Always use a 110-clip punch-down tool.

- Use standard wiring pattern T568-A or T568-B.

- Install the jacks at the same height as electrical receptacle outlets.

- Cover unused wall boxes with blank wall plates.

VIDEO, VOICE, AND DATA INSTALLATIONS AND THE *NEC*®

The installation of a structured cabling system must conform to these Article 800 and Article 820 rules:

- Access to equipment must not be denied by an accumulation of communication wires and cables or coaxial cables that prevents removal of panels, including suspended-ceiling panels. Structured cabling system wiring must always be supported. It cannot just lie on top of panels so that access to other electrical equipment is hindered.

- Communications circuits and equipment and coaxial cable must be installed in a neat and workmanlike manner. Communications cables installed exposed on the outer surface of ceilings and walls must be supported by the structural components of the building structure in such a manner that the cable cannot be damaged by normal building use. The cables must be attached to structural components by straps, staples, cable ties, hangers, or similar fittings designed and installed so as not to damage the cable. The installation must also conform with Section 300.4(D) and Section 300.11. Coaxial cables must be attached to or supported by the structure with staples, straps, cable ties, clamps, hangers, and the like. The installation method must not damage the cable. In addition, the location of the cable should be carefully evaluated to ensure that activities and processes within the building do not cause damage to the cable.

- Communications wires and cables installed as wiring within buildings must be listed as being suitable for the purpose, and must be marked in accordance with Table 800.179. Examples:

  - CMP = communications plenum cable

  - CMR = communications riser cable

  - CMG = communications general-purpose cable

  - CM = communications general-purpose cable

- CMX = communications cable, limited use
- CMUC = undercarpet communications wire and cable
- Coaxial cables in a building must be listed as being suitable for the purpose, and must be marked in accordance to Table 820.179. Examples:
  - CATVP = plenum-rated coaxial
  - CATVR = riser-rated coaxial
  - CATV = general-purpose coaxial
  - CATVX = limited use for dwelling units or in raceways
- Communications wires and cables and coaxial cable must be separated at least 2 inches (50 mm) from conductors of any electric light, power, Class 1, nonpower-limited fire alarm, or medium-power network-powered broadband communications circuits. However, where either (1) all the conductors of the electric light and power circuits are in a metal-sheathed, metal-clad, nonmetallic-sheathed, Type AC, or Type UF cable or (2) all the conductors of communications circuits are in a raceway, the 2-inch (50 mm) clearance does not have to be adhered to. This *Exception* allows electricians to install video, voice, and data wiring directly alongside of the power and lighting wiring, depending on what wiring method is being used. This is the case in residential wiring.
- Raceways must be used for their intended purpose. Communications cables or coaxial cable must not be strapped, taped, or attached by any means to the exterior of any conduit or raceway as a means of support.

TESTING A RESIDENTIAL STRUCTURED CABLING SYSTEM

In commercial building structured cabling systems, standards are well-established for the testing and certification of structured wiring systems. These video, voice, and data systems are "certified" to meet certain standards. Certification refers to the process of making measurements and then comparing the results obtained to pre-defined standards, so that a pass/fail determination can be made.

Rather than being "certified," most home voice, video, and data wiring systems are "verified." Verification ensures that basic continuity and correct terminations have been applied, but does not attempt to measure the information-carrying capacity of the cable runs like you would do in a commercial installation. This is because

home cabling runs are considerably shorter than commercial wiring runs. Because the runs are shorter, they do not suffer nearly as much from **attenuation** losses, and because the signal is typically much stronger, problems such as **near end crosstalk (NEXT)** are much less of a concern.

In verifying residential cabling, the most important measurement is often called wiremapping. Wiremapping ensures proper pin-to-pin connectivity between both ends of a cable run. In a home, a cable could be cut or shorted out by a nail or staple. It could be incorrectly terminated or simply wired incorrectly. It could also be damaged when drywall was put on the walls and ceilings. A good field tester (Figure 16-27) will quickly find any breaks, shorts, or incorrect wire connections. Correct wiring and common errors are shown in Figure 16-28, while a typical fault displayed in a field tester is shown in Figure 16-29.

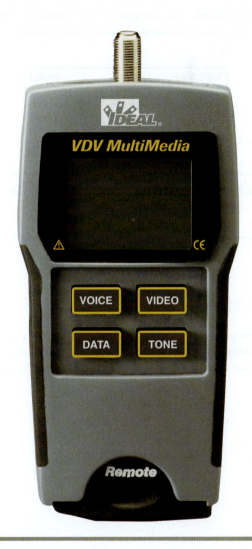

FIGURE 16-27 A good example of a voice, video, and data cable tester. It tests for opens, shorts, miswires, reversals, and split pairs. It also indicates a correct wiremap. *Courtesy of IDEAL INDUSTRIES, INC.*

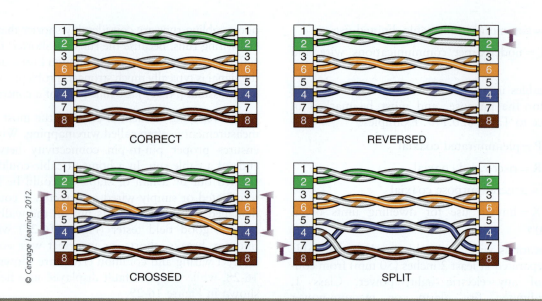

© Cengage Learning 2012.

FIGURE 16-28   An example of correct wiring and examples of crossed, reversed, and split errors.

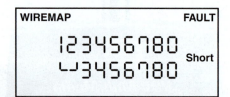

FIGURE 16-29   An example of a tester display indicating a short between wires 1 and 2.
© *Cengage Learning 2012.*

SUMMARY

This chapter gave you a good introduction to the parts and materials, as well as the common installation practices, for a structured cabling system in a residential application. The installation of a structured cabling system by electricians has become quite common throughout the country, and knowing as much as possible about structured cabling is necessary for any electrician doing residential wiring. As these systems get more complex and new materials and installation techniques become available, you will need to consult Web sites, read trade magazines, and attend upgrade classes so that you can stay current in this ever changing area of electrical wiring.

Hardwired structured cabling installation was covered in this chapter, but be aware that wireless technology is available that may someday be the preferred way that the video, audio, voice, and data signals are sent from device to device in a house. However, for the time being and for the foreseeable future, electricians will be asked to install UTP cable, coaxial cable, and the various jacks and connectors so that video, voice, and data signals can be distributed throughout a home.

PROCEDURE 16-1

Installing a Crimp-On Style F-Type Connector on an RG-6 Coaxial Cable

- Put on safety glasses and observe all applicable safety rules.

- Measure and mark ¾ inch from the end of the cable.

A Using a coaxial cable stripper, remove all ¾ inch of the outside sheathing of the cable. Note: Some coaxial cable strippers have two blades, an inside blade that cuts the outer cable sheathing and an outside blade that cuts the white dielectric that surrounds the center conductor.

B Using a coaxial stripper, remove ½ inch of the white dielectric insulation from the center conductor. Note: This step may already be done if you are using a two bladed coaxial cable stripper.

C Fold the copper braiding back over the outer sheathing and then push the RG-6 F-type connector onto the end of the cable until the white dielectric insulation is even with the hole on the inside of the connector.

D If necessary, install the proper size crimping die into the crimping tool. Then, using the crimping tool, crimp the connector ferrule onto the cable. Make sure the crimp is made toward the cable side of the connector. Too close to the ferrule head will pinch the crimp.

- Clean up the work area and put all tools and materials away.

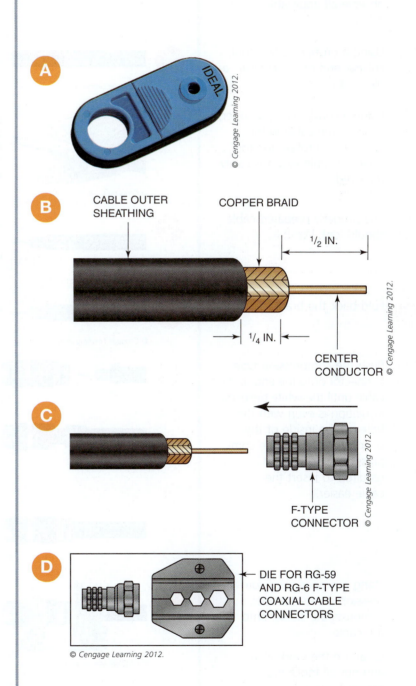

A

B CABLE OUTER SHEATHING — COPPER BRAID — ½ IN. — ¼ IN. — CENTER CONDUCTOR

C F-TYPE CONNECTOR

D DIE FOR RG-59 AND RG-6 F-TYPE COAXIAL CABLE CONNECTORS

© Cengage Learning 2012.

PROCEDURE 16-2

Installing a Compression Style F-Type Connector on an RG-6 Coaxial Cable

- Put on safety glasses and observe all applicable safety rules.

A Using a proper cutting tool, cut the end of the coaxial cable so it is straight.

B Using a coaxial cable stripper, position the cable in the stripper and rotate the tool until the cable jacket is easily removed.

C The properly prepared cable should look like this.

D Fold back the braid.

E Push the compression type connector onto the end of the cable until the white dielectric insulation is even with the hole on the inside of the connector. Detach the compression sleeve only if required to insert the cable easier.

F Using a proper-sized compression connector tool, compress the sleeve onto the cable.

- Clean up the work area and put all tools and materials away.

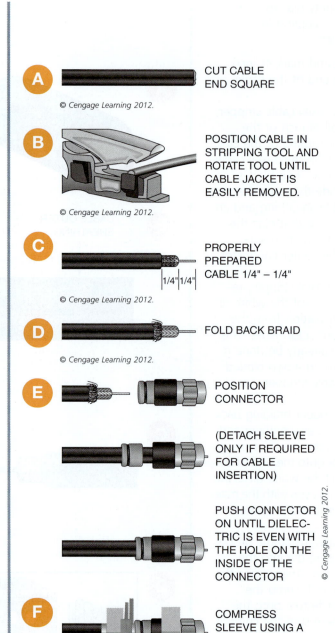

A CUT CABLE END SQUARE

© Cengage Learning 2012.

B POSITION CABLE IN STRIPPING TOOL AND ROTATE TOOL UNTIL CABLE JACKET IS EASILY REMOVED.

© Cengage Learning 2012.

C PROPERLY PREPARED CABLE 1/4" – 1/4"

1/4" 1/4"

© Cengage Learning 2012.

D FOLD BACK BRAID

© Cengage Learning 2012.

E POSITION CONNECTOR

(DETACH SLEEVE ONLY IF REQUIRED FOR CABLE INSERTION)

PUSH CONNECTOR ON UNTIL DIELECTRIC IS EVEN WITH THE HOLE ON THE INSIDE OF THE CONNECTOR

© Cengage Learning 2012.

F COMPRESS SLEEVE USING A COMPRESSION CONNECTOR TOOL

© Cengage Learning 2012.

Installing an RJ-45 Jack on a Four-Pair UTP Category 5e Cable

- Put on safety glasses and observe all applicable safety rules.

A Using a UTP cable jacket stripping tool, remove about 2 inches of jacket from the cable end.

B Determine which wiring scheme you will be using (T568-A or T568-B) and note the color coding and pin numbers on the jack. Note: Most manufacturers of RJ-45 jacks and plugs will have color coding and pin numbering available on their products.

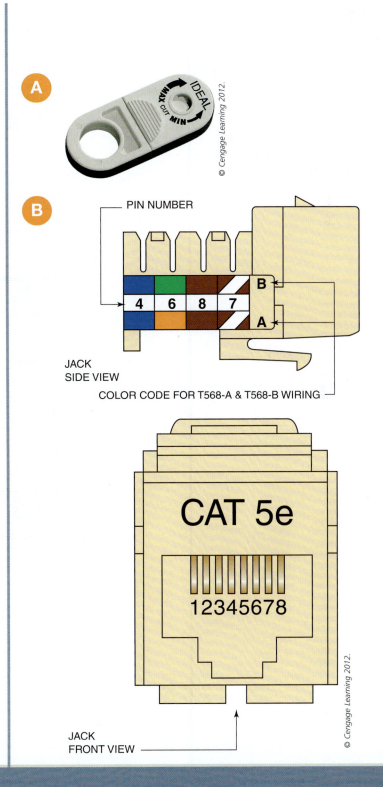

A

© Cengage Learning 2012.

B

PIN NUMBER

| 4 | 6 | 8 | 7 |

B

A

JACK
SIDE VIEW

COLOR CODE FOR T568-A & T568-B WIRING

CAT 5e

12345678

JACK
FRONT VIEW

© Cengage Learning 2012.

PROCEDURE 16-3

Installing an RJ-45 Jack on a Four-Pair UTP Category 5e Cable (Continued)

 Route the conductors for termination. Terminate one pair at a time starting from the rear of the jack. Terminating each pair after placement will prevent crushing the inside pairs with the punch-down tool. Note: The cable should be placed so that the cable jacket touches the rear of the jack housing as shown.

C

T568-A		
①	5	WHITE/BLUE
	4	BLUE
②	3	WHITE/ORANGE
	6	ORANGE
③	1	WHITE/GREEN
	2	GREEN
④	7	WHITE/BROWN
	8	BROWN

T568-B		
①	5	WHITE/BLUE
	4	BLUE
②	3	WHITE/GREEN
	6	GREEN
③	1	WHITE/ORANGE
	2	ORANGE
④	7	WHITE/BROWN
	8	BROWN

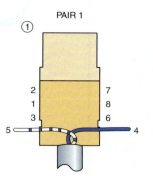
PAIR 1

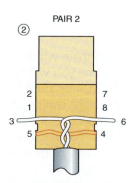
PAIR 2

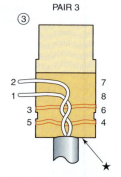
PAIR 3

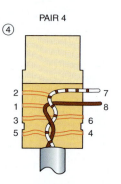
PAIR 4

★ KEEP THE CABLE OUTER SHEATHING TIGHT AGAINST THE BODY OF THE RJ-45 JACK.

© Cengage Learning 2012.

D Using a 110-style punch-down tool, seat each conductor into the proper IDC slot. Be sure to keep the twists within ½ inch of the IDC slot. The punch-down tool, if used properly, will trim any excess wire off flush with the device body.

E Place the cap that comes with the RJ-45 device over the terminated wires and press it into place. This cap will provide a more secure connection of the wires to the IDC slots as well as provide some additional strain relief.

- The jack can now be inserted into a wall plate assembly and secured to an electrical box or mounting plate placed at the desired location.

- Clean up the work area and put all tools and materials away.

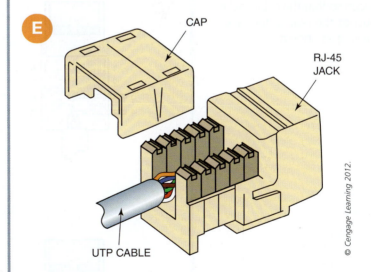

D

110 PUNCH-DOWN TOOL

RJ-45 JACK

CUT

UTP CABLE

© Cengage Learning 2012.

E

CAP

RJ-45 JACK

UTP CABLE

© Cengage Learning 2012.

PROCEDURE 16-4

Installing an RJ-45 Jack on a Four-Pair UTP Category 6 Cable

- Put on safety glasses and observe all applicable safety rules.

A Using a UTP cable jacket stripping tool, remove about 2 inches of jacket from the cable end.

B Route the wires for termination following the T568A or T568B wiring standard.

C Terminate one pair at a time starting from the rear of the jack. Terminating each pair after placement will prevent crushing the inside pairs with the punch-down tool. Note: The cable should be placed so that the cable jacket touches the rear of the jack housing as shown.

A

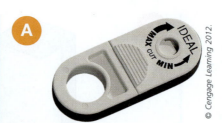

© Cengage Learning 2012.

B

T568A Wiring Standard	T568B Wiring Standard
1 White/Green	1 White/Orange
2 Green	2 Orange
3 White/Orange	3 White/Green
4 Blue	4 Blue
5 White/Blue	5 White/Blue
6 Orange	6 Green
7 White/Brown	7 White/Brown
8 Brown	8 Brown

© Cengage Learning 2012.

C

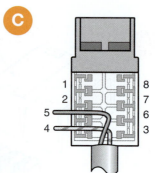

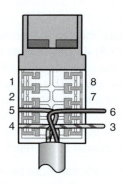

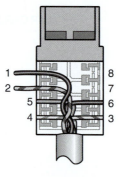

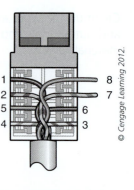

© Cengage Learning 2012.

D Using a 110-style punch-down tool, seat each conductor into the proper IDC slot. Be sure to keep the twists as close as possible to the IDC slot. The punch-down tool, if used properly, will trim any excess wire off flush with the device body.

- Place the cap that comes with the RJ-45 device over the terminated wires and press it into place. This cap will provide a more secure connection of the wires to the IDC slots as well as provide some additional strain relief.

- The jack can now be inserted into a wall plate assembly and secured to an electrical box or mounting plate placed at the desired location.

- Clean up the work area and put all tools and materials away.

D

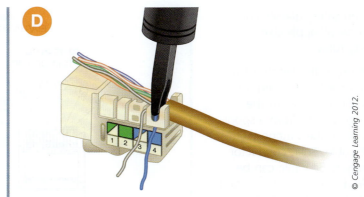

© Cengage Learning 2012.

PROCEDURE 16-5

Assembling a Patch Cord with RJ-45 Plugs and Category 5e UTP Cable

- • Put on safety glasses and observe all applicable safety rules.

- • Determine the length of the patch cord and, using a pair of cable cutters, cut the desired length from a spool or box of the Category 5e UTP cable. Solid conductor Category 5e cable can be used, but it is recommended that stranded conductor Category 5e cable be used for added flexibility.

- • Using a proper UTP cable stripping tool, strip the cable jacket back about 3/4 inch from each end of the cable.

- • Determine whether the connection will be done to the EIA/TIA 568-A color scheme or to the EIA/TIA 568-B color scheme.

A On one end of the patch cord, sort the pairs out so they fit into the plug in the order shown for the T568-A scheme or for the T568-B scheme.

- • Insert the pairs into the plug.

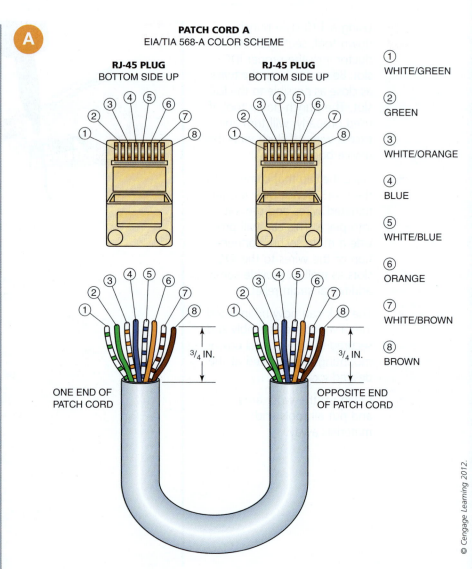

A

PATCH CORD A
EIA/TIA 568-A COLOR SCHEME

RJ-45 PLUG BOTTOM SIDE UP

RJ-45 PLUG BOTTOM SIDE UP

ONE END OF PATCH CORD

OPPOSITE END OF PATCH CORD

3/4 IN.

3/4 IN.

① WHITE/GREEN

② GREEN

③ WHITE/ORANGE

④ BLUE

⑤ WHITE/BLUE

⑥ ORANGE

⑦ WHITE/BROWN

⑧ BROWN

© Cengage Learning 2012.

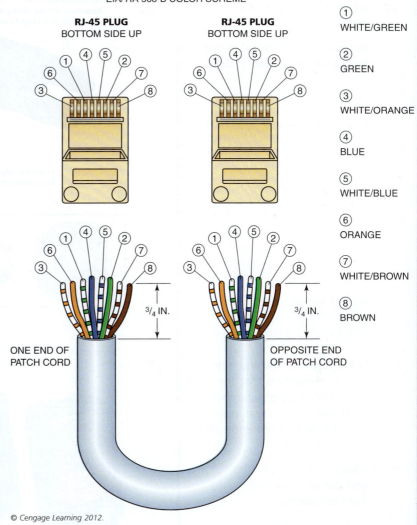

PATCH CORD B
EIA/TIA 568-B COLOR SCHEME

① WHITE/GREEN

② GREEN

③ WHITE/ORANGE

④ BLUE

⑤ WHITE/BLUE

⑥ ORANGE

⑦ WHITE/BROWN

⑧ BROWN

© Cengage Learning 2012.

PROCEDURE 16-5

Assembling a Patch Cord with RJ-45 Plugs and Category 5e UTP Cable (Continued)

B Using a proper crimping tool, crimp the pins with the crimping tool.

- Install a plug on the other end following the preceding steps.

B

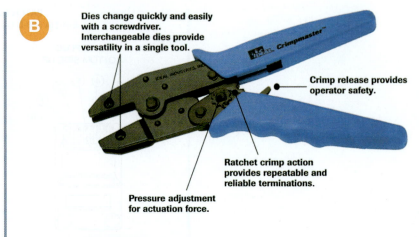

Dies change quickly and easily with a screwdriver. Interchangeable dies provide versatility in a single tool.

Crimp release provides operator safety.

Ratchet crimp action provides repeatable and reliable terminations.

Pressure adjustment for actuation force.

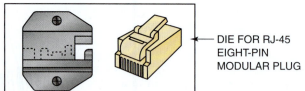

DIE FOR RJ-45 EIGHT-PIN MODULAR PLUG

© Cengage Learning 2012.

C Using a tester, test the patch cord to determine that the correct color scheme has been used and that there is continuity between both ends of the patch cord.

- Clean up the work area and put all tools and materials away.

C

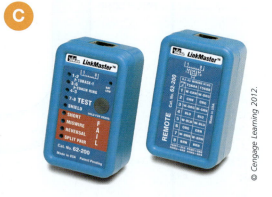

© Cengage Learning 2012.

REVIEW QUESTIONS

Directions: Answer the following items with clear and complete responses.

1. The most often used category-rated UTP cable for wiring from various outlets back to the service center in a residential application is _____.

2. The minimum bending radius for a Category 5e-rated cable is recommended to be no less than _____ times the diameter of the cable.

3. The minimum bending radius for a coaxial cable is recommended to be no less than _____ times the diameter of the cable.

4. Define the term "structured cabling."

5. Name the EIA/TIA standard that is used as a guide for the installation of a structured cabling system in a residential application.

6. Describe why each pair of wires in the voice and data cables used in a residential structured cabling system is twisted.

7. Describe the difference between a 568-A and a 568-B connection.

8. What do the letters "IDC" stand for?

9. What is the maximum amount of pulling force that an electrician can exert on a Category 5e cable?

10. Name the document that contains the installation requirements for a structured cabling system that must be followed.

11. A _____ is the rating, based on the bandwidth performance, of UTP cable.

12. A type of coaxial cable that is "quad shielded" and is used in residential structured cabling systems to carry video signals such as cable and satellite television is called _____.

13. The popular name given to an eight-pin connector or jack used to terminate UTP cable is _____.

14. A _____ _____ is the short length of cable with an RJ-45 plug on either end and is used to connect a home computer to the work area outlet.

15. The standard color coding for a four-pair UTP cable is:
    - Pair 1: tip is _____; ring is _____.
    - Pair 2: tip is _____; ring is _____.
    - Pair 3: tip is _____; ring is _____.
    - Pair 4: tip is _____; ring is _____.

16. Explain why mud rings rather than regular electrical boxes are recommended to be used at video, voice, and data outlet locations.

17. Explain what the demarcation point is for a typical residential telephone installation.

18. What is the purpose of a 75-ohm, F-type termination cap? Where are they installed?

19. An incoming telephone call has _____ volts AC between the tip and ring conductors. This amount of voltage is more than enough to cause an electrical shock.

20. A coaxial cable you are installing has the letters "CATVR" written on it. What do the letters mean?

Residential Electrical System Trim-Out

CHAPTER 17
Lighting Fixture Installation

CHAPTER 18
Device Installation

CHAPTER 19
Service Panel Trim-Out

CHAPTER 17

Lighting Fixture Installation

OBJECTIVES

Upon completion of this chapter, the student should be able to:

- Demonstrate an understanding of lighting basics.

- Demonstrate an understanding of common lamp and lighting fixture terminology.

- Demonstrate an understanding of the four different lamp types used in residential wiring applications: incandescent, LED, fluorescent, and high-intensity discharge.

- Select a lighting fixture for a specific residential living area.

- Demonstrate an understanding of the installation of common residential lighting fixtures.

- Demonstrate an understanding of the installation of ceiling-suspended paddle fans.

GLOSSARY OF TERMS

ballast a component in a fluorescent lighting fixture that controls the voltage and current flow to the lamp

ceiling-suspended paddle fan a type of fan installed in a ceiling, usually having four or five blades, that is used for air circulation in residential applications; it can be installed close to a ceiling or hung from a ceiling with a metal rod; the fan often has a lighting fixture attached to the bottom of the unit; a special outlet box that is listed as being suitable for paddle-fan support is required

Class P ballast a ballast with a thermal protection unit built in by the manufacturer; this unit opens the lighting electrical circuit if the ballast temperature exceeds a specified level

color rendition a measure of a lamp's ability to show colors accurately; the color rendering index (CRI) is a scale that compares color rendering for different light sources; the scale ranges from 1 (low-pressure sodium) to 100 (the sun); a CRI of 85 is considered to be very good

color temperature a measure of the color appearance of a light source that helps describe the apparent "warmth" (yellowish) or "coolness" (bluish) of that light source; light sources below 3200 K are considered

"warm" while those above 4000 K are considered "cool" light sources; the letter, K, stands for Kelvin

fluorescent lamp a gaseous discharge light source; light is produced when the phosphor coating on the inside of a sealed glass tube is struck by energized mercury vapor

high-intensity discharge (HID) lamp another type of gaseous discharge lamp, except the light is produced without the use of a phosphor coating

incandescent lamp the original electric lamp; light is produced when an electric current is passed through a filament; the filament is usually made of tungsten

lamp efficacy a measure used to compare light output to energy consumption; it is measured in lumens per watt; a 100-watt light source producing 1750 lumens of light has an efficacy (efficiency) of 17.5 lumens per watt (L/W)

LED lamp a light-emitting-diode lamp is a solid-state lamp that uses light-emitting diodes (LEDs) as the source of light; because the light output of individual LEDs is small compared to incandescent and

compact fluorescent lamps, multiple diodes are used together

lumen the unit of light energy emitted from a light source

luminaire a complete lighting unit consisting of a lamp or lamps together with the parts designed to distribute the light, to position and protect the lamps and ballast (where applicable), and to connect the lamps to the power supply

sconce a wall-mounted lighting fixture

track lighting a type of lighting fixture that consists of a surface-mounted or suspended track with several lighting heads attached to it; the lighting heads can be adjusted easily to point their light output at specific items or areas

troffer a term commonly used by electricians to refer to a fluorescent lighting fixture installed in the grid of a suspended-ceiling

Type IC a light fixture designation that allows the fixture to be completely covered by thermal insulation

Type Non-IC a light fixture that is required to be kept at least 3 inches from thermal insulation

nstalling the lighting fixtures in a house is one of the last things an electrician does to complete the residential electrical system. The ceilings and walls will have their final coat of paint or other finish already applied. Choosing the lighting fixture types to be used throughout a house is usually done during the initial planning stages for the installation of the electrical system. Planning early allows the electrician to install the proper rough-in wiring and electrical boxes required for a specific lighting fixture type. The electrical connections for most types of lighting fixtures used in residential wiring are very similar. However, the method of mounting the fixtures to a wall or ceiling varies with the type of fixture used. Manufacturers of lighting fixtures enclose installation instructions with each lighting fixture, and these should always be consulted during the installation. This chapter introduces you to basic lighting fundamentals, lamp types, and common lighting fixture types used in residential lighting systems. The methods most often used for installing lighting fixtures are also covered.

LIGHTING BASICS

Today the lighting industry produces many different styles of lighting fixtures and lamps to provide the home owner with an overwhelming amount of flexibility when planning their lighting system. The *National Electrical Code® (NEC®)* uses the term "luminaire" when referring to lighting fixtures. A **luminaire** is defined as a complete lighting unit consisting of a lamp or lamps together with the parts designed to distribute the light, to position and protect the lamps and ballast (where applicable), and to connect the lamps to the power supply.

The overall performance of a lighting system is a combination of the quantity of light the lamps produce and the quality of that light. Light is the visible portion of the electromagnetic spectrum. Figure 17-1 shows the relationship of the various visible wavelengths. The shortest visible wavelength appears as violet, while the longest visible wavelength appears red. The other colors—blue, green, yellow, and orange—are intermediate wavelengths and are between violet and red.

We tend to think that each object has a fixed color. In reality, an object's appearance results from the way it reflects the light that falls on it. For example, a red sports car appears red to you because it reflects light in the red part of the spectrum and absorbs the light of the other wavelengths. The quality of the red wavelength striking the sports car determines the shade of red we see.

White light is formed by nearly equal parts of all the visible wavelengths. The balance does not have to be precise for the light to appear white. Red, blue, and green are the primary colors, and they can be combined to make any other color. This means that a light source that contains a good balance of red, blue, and green will provide excellent color appearance for objects it illuminates.

Lamp manufacturers are concerned with three main factors: color temperature, color rendering, and lamp efficacy. The **color temperature** of a light source is a measurement of its color appearance and is measured in degrees Kelvin (K) (Figure 17-2). It is based on the premise that any object heated to a high enough temperature will emit light and that as the temperature rises, the color will shift from red to orange to yellow to white and then to blue. One of the more confusing aspects of lighting is that light from the higher-temperature wavelengths, blue and white, is referred to as "cool," while the light from the lower-temperature wavelengths is referred to as "warm." For example, a lamp with a color temperature rating of 2000 K is a "warm" yellow light, and a lamp with a color temperature rating of 6000 K puts off a light that is a "cool" white color. These descriptions have nothing to do with temperature but rather with the way the colors are perceived. Warm light sources are ideal for residential applications because they make colors appear more natural and vibrant. In residential settings, earth tones usually dominate the color scheme, and warm light sources enhance their appearance.

The standard for determining the quality of light is based on the sunlight that strikes an object at high noon. This is when light has the best combination of the primary colors and provides the best **color rendition**. Lamp manufacturers, attempting to reproduce this effect, have used various combinations of materials with varying degrees of success. These combinations produce light with slightly different mixtures of the primary color wavelengths, which causes us to see colors in different shades. This means that the red sports car discussed earlier may be seen in different shades of red, depending on the type of lamp that is illuminating it.

Many electricians think that a higher-wattage lamp will produce more light. They are actually confusing light output with the amount of electricity a lamp uses. Light output is measured in "lumens," while the amount of electrical power used by a lamp type is measured in watts. A **lumen** is defined as the unit of light that is emitted from

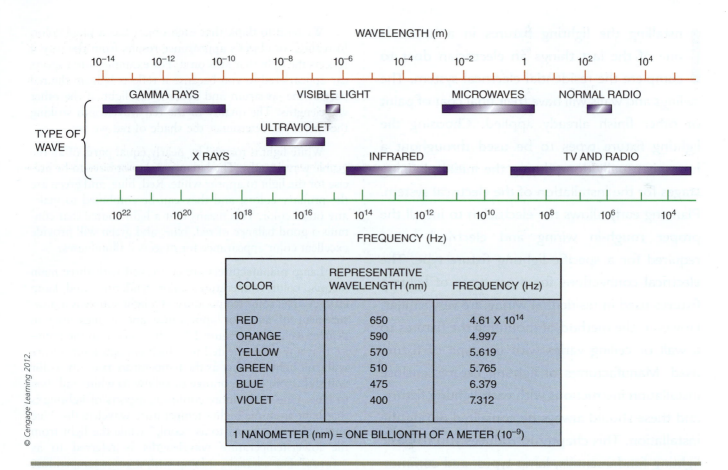

© Cengage Learning 2012.

FIGURE 17-1   The electromagnetic spectrum. Notice that visible light falls between ultraviolet light and infrared light. Red has the longest wavelength and violet the shortest.

a light source. For example, a 20-watt compact fluorescent lamp will produce as much usable light as a 75-watt incandescent lamp and use much less electricity. In today's energy-conscious world, the ability to get sufficient light for a lower price is very important. The best indicator of a lamp's performance, called its **lamp efficacy**, is its LPW (lumens per watt) rating. This rating is a ratio of the number of lumens a lamp produces to each watt of power it uses. The higher the LPW of a light source, the better job it does at changing electrical energy to light energy.

OVERVIEW OF LAMP TYPES FOUND IN RESIDENTIAL LIGHTING

There are four main lamp types found in residential wiring. They are incandescent lamps, LED lamps, fluorescent lamps, and high-intensity discharge lamps. In this section we will take a close look at each lamp type.

INCANDESCENT LAMPS

Incandescent lamps were the first type of electric lamp. They have used the same basic technology for over 100 years. Light is produced when a tungsten filament, placed inside a glass enclosure, has an electric current passed through it. The resistance of the filament causes it to heat up, giving off light. These lamps come in a variety of sizes and shapes to meet various lighting needs. Incandescent lamp bases also come in several sizes to meet residential lighting needs. Figure 17-3 shows additional information for incandescent lamp and base types used in residential wiring.

There have been several improvements to the incandescent lamp over the years. The straight filament has been replaced by a coiled filament to increase the life of the lamp. The sealed lamp is now filled with an inert gas, such as argon. Early incandescent lamps had the air pumped out to create a vacuum around the filament. This was done to keep oxygen away from the filament. As the filament heated up, oxygen would cause it to burn out almost immediately. Because a true vacuum could not be created inside the lamp, the remaining oxygen would work to shorten the life of the lamp.

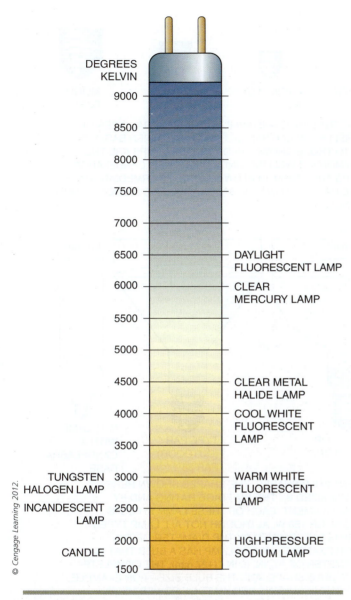

FIGURE 17-2 Color temperature scale. This scale assigns a numeric value to the color appearance of a light source ranging from yellow (warm light) to white, and then to blue (cool light).

The latest improvement is the tungsten halogen lamp. In standard incandescent lamps, the tungsten from the filament is burned off over time. This causes the filament to become thinner and eventually break. As the tungsten is burned off, it is deposited on the inside of the lamp surface, giving it a black coating, which causes the lamp to dim. By using a thicker filament and filling the lamp with halogen gas, lamp efficiency increases by at least 20 percent. This occurs because the tungsten that is burned off the filament combines with the halogen gas and is redeposited back onto the filament, extending the life of the filament and the lamp.

CAUTION

CAUTION: Halogen lamps give off extreme amounts of heat and have been associated with several fires. An electrician must carefully follow all manufacturers' installation instructions when installing halogen lamps. These lamps are required to be enclosed by a glass lens to prevent contact with flammable materials.

Incandescent lamps and their associated fixtures are used in every aspect of residential lighting. They are the most common lamp type used in residential systems for several reasons:

- Come in a wide variety of sizes and styles
- Have the lowest initial cost of any lamp type
- Produce a warm light that produces excellent color tones
- Can be easily controlled with dimmers

However, they are very inefficient because they produce light by heating a solid object, and most of the energy they consume is released as heat, not light.

from experience...

By the end of 2014, the old style (non-halogen) tungsten filament incandescent lamps as we know them will be phased out in the United States because of their low efficiency. 100-watt incandescent lamps will be the first to be phased out in 2012 and 40-watt incandescent lamps will be the last size to be phased out in 2014. Higher-efficiency halogen incandescent lamps are being developed and are now available. The intent of the phase out is to encourage the use of more compact fluorescent (CFL) and light-emitting diode (LED) lamps at traditional incandescent lamp locations in a home as a way to save energy. Many other countries in the world have placed an outright ban on incandescent lamps, including Canada and the United Kingdom.

GLASS BULB

CONDUCTOR

GROUNDED
CIRCUIT
CONDUCTOR
CONTACT

FILAMENT

CONDUCTOR

GLASS STEM

PHASE
CONDUCTOR
CONTACT

CANDELABRA
BASE

INTERMEDIATE
BASE

MEDIUM
BASE

MOGUL
BASE

INCANDESCENT LAMPS ARE AVAILABLE IN SEVERAL SIZES OF
SCREW SHELLS, CALLED BASES. THERE ARE OTHER BASES IN
ADDITION TO THOSE SHOWN. MEDIUM-BASE LAMPS ARE THE
MOST COMMON IN DWELLING-UNIT LUMINAIRES. CANDELABRA-
BASE LAMPS ARE COMMON IN DWELLINGS. INTERMEDIATE-BASE
LAMPS ARE RARE, AND MOGUL-BASE LAMPS ARE SELDOM FOUND
IN DWELLINGS.

19/8 IN.
(60 mm)

30/8 IN.
(95 mm)

40/8 IN.
(127 mm)

38/8 IN.
(121 mm)

10/8 IN.
(32 mm)

TYPE A LAMP
(STANDARD)
A-19

TYPE R LAMP
(REFLECTOR)
R-30

TYPE R LAMP
(REFLECTOR)
R-40

TYPE PAR
(OUTDOOR)
PAR 38

G-10 LAMP
WITH A
CANDELABRA
BASE

INCANDESCENT LAMPS ARE AVAILABLE IN SEVERAL DIFFERENT BULB SHAPES. THE SHAPES SHOWN HERE
ARE THE ONES MOST OFTEN FOUND IN DWELLINGS. LAMPS ARE SIZED BY THE WATTAGE RATING AND BY
THE SIZE AND SHAPE OF THE GLASS BULB SURROUNDING THE FILAMENT. EACH OF THESE LAMP TYPES IS
AVAILABLE IN SEVERAL WATT (W) RATINGS, SUCH AS 60 W, 75 W, OR 150 W, ALTHOUGH NOT ALL LAMP TYPES
ARE AVAILABLE IN ALL WATT RATINGS. THE SIZE OF THE LAMP IS DETERMINED BY THE DIAMETER OF THE
GLASS BULB MEASURED IN 1/8 IN. (3.175 mm) INCREMENTS. FOR EXAMPLE, AN A-19 LAMP HAS A BULB THAT IS
19/8 IN. (2 3/8 IN.) (60 mm) IN DIAMETER. AN R-40 LAMP HAS A DIAMETER OF 40/8 IN. (5 IN.) (127 mm). THE COMPLETE
DESCRIPTION OF THE LAMP INCLUDES THE WATT RATING, THE BULB SHAPE, AND THE BULB SIZE—FOR EXAMPLE,
75 W R-40 OR 40 W G-10. INCANDESCENT LAMPS PRODUCE ABOUT 20 LUMENS PER WATT OF POWER USED.

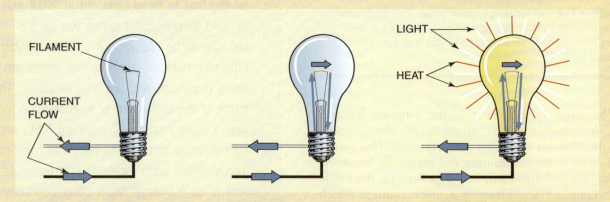

FILAMENT

CURRENT
FLOW

LIGHT

HEAT

INCANDESCENT LAMPS PRODUCE LIGHT BECAUSE THE FILAMENT IS HEATED BY THE CURRENT UNTIL IT IS SO HOT
THAT IT GLOWS. UNFORTUNATELY, THIS METHOD OF PRODUCING LIGHT PRODUCES A LOT OF HEAT. THIS WASTES
ELECTRICITY AND REDUCES THE LIFE OF THE LAMP. ONLY ABOUT 11 PERCENT OF THE APPLIED ELECTRICITY IS
TRANSFORMED INTO LIGHT. AS A RESULT, INCANDESCENT LAMPS HAVE THE LOWEST LAMP EFFICACY.

FIGURE 17-3 A guide to incandescent lamp types used in residential wiring.

LED LAMPS

Not too many years ago light-emitting diode (LED) lamps were available only in small sizes that could not really produce enough light to be useful as a light source in homes. Today that has all changed. LEDs are a viable light source for homes and can produce usable amounts of light using very small amounts of electricity and with very little or no production of heat when in use. The money savings because of the reduced amount of electricity used to light a home is significant.

LED lamps (Figure 17-4) have many advantages over traditional incandescent lamps, including:

- Better energy efficiency with an energy saving of 80 percent to 90 percent

- A very long operating life (up to 100,000 hours)

- An ability to operate at many different voltages, including 12V AC/DC or 85- to 240-volt AC

- A cool light output with little or no heat energy produced

- Available with narrow, medium, and wide angle lenses for a variety of light directions

- Available in a variety of lighting colors, including: Warm White (2700K to 4300K), Daylight white (5000K to 6500K), Cool white (up to 8000K), Amber, Orange, Pink, Red, Cyan, Blue, Royal Blue, and Green

The lighting efficiency of the new high-power LED lamps is more than eight times that of incandescent lamps, and twice as high as compact fluorescent lamps. LED bulbs also emit a much higher percentage of light in the desired direction. This makes them even more efficient compared to either incandescent or fluorescent lamps for task lighting, desk lamps, reading lights, spotlights, flood lights, and track lighting. LED lamps generate very little unwanted heat. The energy savings may be doubled in air-conditioned environments where each watt of incandescent lighting can add another watt or more to the power needed for air-conditioning.

LED lighting instantly achieves full brightness with no warm up time. Fluorescent lamps are dim when first turned on, and get brighter as they warm up. Depending on the temperature and the age of the lamp, they can take a long time to warm up. If it is cold enough, they will not warm up at all. LED lamps always start at full brightness. Fluorescent lamps contain mercury and must be treated as hazardous waste. LED lighting contains no mercury or other dangerous substances. LED lamps emit no damaging ultraviolet (UV) light, so they will not cause fading and aging of artwork or other sensitive materials. Fluorescent and halogen lamps can cause significant damage over time.

LED bulbs can operate for up to 100,000 hours or more. One LED light bulb can easily outlast 30 incandescent bulbs, or 6 compact fluorescents. Operating 8 hours per day, LED light bulbs can last 10 years or more. LED light bulbs are less sensitive to shock, vibration, and the extreme temperature changes that can quickly ruin fragile incandescent bulbs. Unlike fluorescent bulbs which wear out much faster if they are frequently turned on and off, LED bulbs are not affected by frequent on-off switching. The long life of LED light bulbs reduce the time, effort, and cost of replacement. The high reliability of LED lights increases safety and security. LED bulbs light up instantly to full brightness, even in the coldest weather. LED bulbs operate at much lower temperatures. Halogen and incandescent lights are hot enough to cause fire, and they frequently do.

One thing that has not been mentioned yet is the initial cost. LED lamps are expensive to buy because of their long life and low operating cost, but will cost less than incandescent or fluorescent lamps in the long run. LED lamps are presently available in sizes that produce an equivalent amount of light to a 60-watt incandescent lamp. In the near future, LEDs will be developed that will produce greater amounts of light and will be able to replace 100-watt incandescent lamps.

LED lighting will be used more and more in residential lighting. If you have not been exposed to this type of lighting yet … you will be.

© Cengage Learning 2012.

FIGURE 17-4 An example of an LED lamp. This lamp is designed to produce the same amount of light as a 40-watt incandescent lamp but with using a much smaller amount of electricity.

FLUORESCENT LAMPS AND BALLASTS

Fluorescent lamps are referred to as "electric discharge lamps" by the *NEC*®. They have been around for many years and are an excellent light source. Light is produced when an electric current is passed through tungsten cathodes at each end of a sealed glass tube. The tube is filled with an inert gas, such as argon or krypton. A very small amount of mercury is also in the tube. Electrons are emitted from the cathodes and strike particles of mercury vapor. This results in the production of ultraviolet radiation, which causes a phosphor coating on the inside of the glass tube to glow. Figure 17-5 shows additional information about fluorescent lamps used in residential wiring applications.

Today's fluorescent lamps offer more options in light quality than any other type of lamp. Refinements in the composition of phosphor coatings and better control over the generation of primary color light wavelengths allow fluorescent lamps to provide excellent color rendition of virtually all colors. While their color rendering is not quite as good as incandescent lamps, fluorescent lamps offer much better efficacy. Strip luminaires (fixtures without lenses) are usually found in garages, basements, closets, and work areas. Wraparound fixtures (those with lenses) are often used in living areas of a residence. The availability of a wide range of lens types provide for better appearance and better light distribution in living areas that require a large amount of general lighting.

Fluorescent lamps have two electrical requirements. First, a high-voltage source is needed to start the lamp. Second, once the lamp is started, the mercury vapor offers a decreasing amount of resistance to the current. To prevent the lamp from drawing more and more current and quickly burning itself out, the current flow must be regulated. The **ballast** was developed to fill these needs (Figure 17-6).

Ballasts

A ballast is a device that provides both the high voltage needed to start the lamp and the current control that allows the lamp to operate efficiently. There are two types available today: the magnetic ballast and the electronic ballast. While both perform the same functions, the electronic ballast offers several advantages over the magnetic ballast. When compared to a magnetic ballast, electronic ballasts produce their full light output using 25 percent to 40 percent less energy, are more reliable, last longer, and produce a constant flicker-free light. However, a disadvantage is that they are more expensive.

Section 410.130(E) of the *NEC*® requires that all fluorescent ballasts installed indoors have thermal protection built into the unit by the manufacturer for both new and replacement installations. This protection causes the electrical circuit to the fixture to open when the temperature level inside the unit exceeds a preset level. Heat, caused by electrical shorts, grounds, or a lack of air circulation, could cause a fire. Ballasts with built-in thermal protection are called **Class P ballasts**.

Another factor to consider with fluorescent ballasts is their location. Standard ballasts, either magnetic or electronic, do not perform well in cold temperatures. A special ballast has been developed to overcome this problem. This unit is expensive and has had some reliability problems. Many electricians recommend that incandescent luminaires be used in these situations.

Fluorescent ballasts have three types of circuitry. The first is called "preheat" (Figure 17-7). They are easily identified because they have a "starter." The most common starter looks like a silver button on the fixture. When the ballast receives power, the starter causes the cathodes to glow, or preheat, for a few seconds before the arc is established to produce light. Lamps used with preheat ballasts have two pins (bipin) on each end of the lamp (Figure 17-8). Preheat lamps and ballasts cannot be used with dimmers.

The most common type of ballast circuitry is the "rapid start" (Figure 17-9). This type of starting circuitry does not require a starter. Instead, the filaments remain energized by a low-voltage circuit from the ballast. This allows the lamps to come to full power in less than 1 second. To ensure the lamps start properly, ballast manufacturers recommend the ballast and the fluorescent fixture case be grounded to the supply circuit-grounding conductor. Lamps used with rapid-start ballasts are bipin and may be dimmed with a special dimming ballast. Figure 17-10 shows examples of bipin sockets that are placed at each end of a fluorescent lighting fixture and provide a way for bipin lamps to be connected to the ballast circuitry.

The third type of ballast circuitry is the "instant start" (Figure 17-11). These ballasts provide a high-voltage surge to start the lamp instantly. They require special lamps that do not require preheating of the cathodes. The main disadvantage of this system is that lamp life is shortened by 20 percent to 40 percent. Most instant-start lamps have a single pin on each end of the lamp and cannot be used with dimmers. Figure 17-12 shows an example of single-pin sockets that are placed at each end of a fluorescent lighting fixture and provide a way for single-pin lamps to be connected to the ballast circuitry.

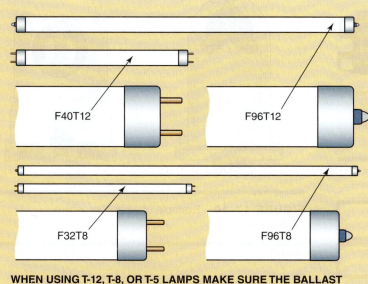

FLUORESCENT TUBES COME IN SEVERAL STANDARD SIZES, LENGTHS, AND SHAPES. THE MOST COMMON LENGTHS OF STRAIGHT TUBES ARE 18 IN. (457 mm), 24 IN. (610 mm), 48 IN. (1.22 m), AND 96 IN. (2.44 m). THEY ARE ALSO AVAILABLE IN A U-SHAPED TUBE AND A CIRCULAR TUBE. 96 IN. (2.44 m) TUBES HAVE A SINGLE-PIN CONNECTION TO THE FIXTURE. OTHER SIZE TUBES HAVE A 2-PIN CONNECTION METHOD.

FLUORESCENT TUBES ARE ALSO AVAILABLE IN DIFFERENT DIAMETERS. LIKE INCANDESCENT LAMPS, FLUORESCENT LAMPS ARE MEASURED IN $1/8$ IN. (3 mm) UNITS AND CARRY THE PREFIX "T" IN THE PART NUMBER. THE LAMPS USED IN DWELLINGS WILL BE EITHER T-12 [$^{12}/_{8}$ IN. (38 mm)] OR T-8 [$^{8}/_{8}$ IN. (25 mm)]. NEW T-5 LAMPS [$^{5}/_{8}$ IN. (16 mm)] ARE BEGINNING TO BE USED IN RESIDENTIAL APPLICATIONS.

WHEN USING T-12, T-8, OR T-5 LAMPS MAKE SURE THE BALLAST IS DESIGNED FOR THE LAMP SIZE YOU ARE USING.
USING THE WRONG LAMP OR THE WRONG BALLAST WILL SHORTEN THE LIFE OF BOTH COMPONENTS.

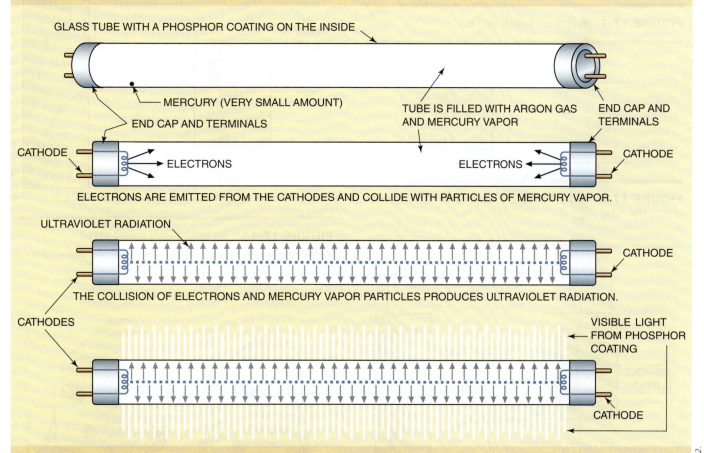

GLASS TUBE WITH A PHOSPHOR COATING ON THE INSIDE

MERCURY (VERY SMALL AMOUNT)

TUBE IS FILLED WITH ARGON GAS AND MERCURY VAPOR

END CAP AND TERMINALS

END CAP AND TERMINALS

CATHODE

ELECTRONS

ELECTRONS

CATHODE

ELECTRONS ARE EMITTED FROM THE CATHODES AND COLLIDE WITH PARTICLES OF MERCURY VAPOR.

ULTRAVIOLET RADIATION

CATHODE

THE COLLISION OF ELECTRONS AND MERCURY VAPOR PARTICLES PRODUCES ULTRAVIOLET RADIATION.

CATHODES

VISIBLE LIGHT FROM PHOSPHOR COATING

CATHODE

THE ULTRAVIOLET RADIATION CAUSES THE PHOSPHOR MATERIAL TO PRODUCE VISIBLE LIGHT. BY CONTROLLING THE MATERIAL COMPOSITION OF THE COATING, THE COLOR OF THE LIGHT CAN BE CONTROLLED TO SOME DEGREE. THIS METHOD OF LIGHTING PRODUCES MORE LIGHT AND LESS HEAT FOR EACH WATT OF ENERGY USED. AN INCANDESCENT LAMP WILL PRODUCE ABOUT 20 LUMENS PER WATT OF POWER USED, A FLUORESCENT LAMP PROVIDES ABOUT 80 TO 100 LUMENS PER WATT OF POWER USED.

© Cengage Learning 2012.

FIGURE 17-5 A guide to fluorescent lamps used in residential wiring.

FIGURE 17-6 An electronic Class P ballast that is thermally protected when used indoors as required in Section 410.130(E) of the *NEC®. Courtesy of Motorola Lighting.*

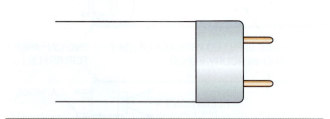

FIGURE 17-7 A preheat ballast circuit.

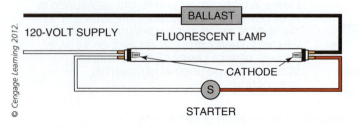

FIGURE 17-8 Many fluorescent lamps have bipins at each end of the tube.
© Cengage Learning 2012.

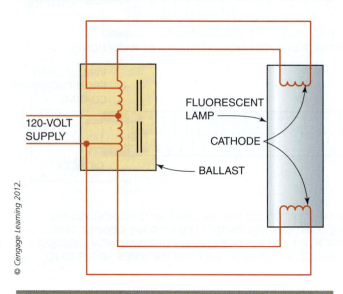

FIGURE 17-9 A rapid-start ballast circuit.

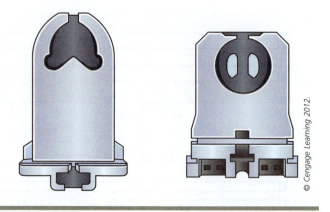

FIGURE 17-10 Example of bipin fluorescent lamp sockets.

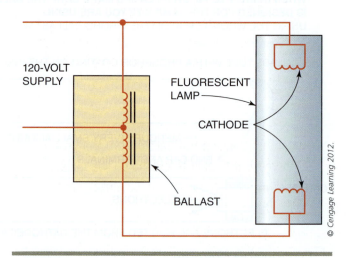

FIGURE 17-11 An instant-start ballast circuit.

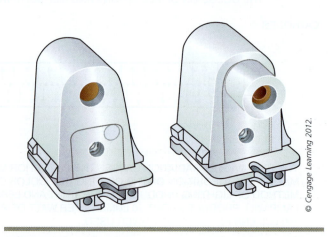

FIGURE 17-12 Example of single-pin fluorescent lamp sockets.

from experience...

Many residential customers request that fluorescent lighting be installed in basements, garages, and in some types of outdoor buildings, such as a workshop or storage shed. While there is no reason that fluorescent lighting cannot be used in these locations, it may not be the best choice because in colder temperatures the lamps may not come on, and in warmer temperatures they may cycle on and off. Regular fluorescent lamp ballasts are designed to start the lamps in temperatures above 50°F. Cold weather ballasts that can start the lamps at temperatures below 0°F should be specified for applications where the ambient temperature is expected to be below 50°F. In areas experiencing high ambient temperatures, the ballast can overheat. Class P ballasts contain a thermal protection device that disconnects the ballast from the power source if it begins to overheat. The ballast then cools until an automatic switch reconnects it to the power supply. If the overheating is not corrected, the process, called "cycling," will repeat itself. In addition, if the temperatures are high enough, ballast life could be shortened and light output reduced. Fixtures in areas with high ambient temperatures should be well-ventilated. Always check the nameplate on a ballast to verify what its temperature rating is.

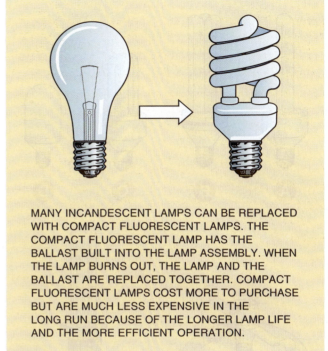

MANY INCANDESCENT LAMPS CAN BE REPLACED WITH COMPACT FLUORESCENT LAMPS. THE COMPACT FLUORESCENT LAMP HAS THE BALLAST BUILT INTO THE LAMP ASSEMBLY. WHEN THE LAMP BURNS OUT, THE LAMP AND THE BALLAST ARE REPLACED TOGETHER. COMPACT FLUORESCENT LAMPS COST MORE TO PURCHASE BUT ARE MUCH LESS EXPENSIVE IN THE LONG RUN BECAUSE OF THE LONGER LAMP LIFE AND THE MORE EFFICIENT OPERATION.

FIGURE 17-13 A compact fluorescent lamp is a great replacement for an incandescent lamp.

Compact Fluorescent Lamps (CFLs)

The fastest-growing application of fluorescent lighting is in CFLs. These lamps are often designed to replace incandescent lamps (Figure 17-13). CFLs provide the same light output as an incandescent lamp, but typically use only a quarter of the energy that an incandescent would. CFLs also produce only a fraction of the heat that incandescent lamps produce. This saves even more on electric bills in homes with air-conditioning and makes homes without air-conditioning more comfortable in the summer. Besides saving so much energy, a CFL lasts about five to fifteen times longer than a standard incandescent bulb. Therefore, by using CFLs, a home owner can save money in heating and cooling costs *and* on the amount spent on purchasing light bulbs. Figure 17-14 shows some of the common sizes and shapes of CFLs.

CFLs come in three basic types: self-ballasted, adapter kits, and new fixtures. A self-ballasted (screw-in) CFL is a one-piece fluorescent lamp and ballast combination. The ballast end of the unit screws directly into an ordinary incandescent light socket just like an incandescent lamp. Because the lamp is not separate from the unit, when it burns out the entire unit is disposed of and replaced. This type of CFL is the easiest to install and replace. It is very well suited for retrofits in homes. These lamps are available in many different shapes that give them a wide variety of applications.

Adapter kit CFLs involve two pieces. One piece is the ballast, which screws directly into an ordinary incandescent light socket just like an incandescent lamp. This piece is called the adapter. The second piece is the fluorescent lamp, which plugs into the ballast. This allows the ballast to be used through the life span of more than one lamp. This CFL is also very good

- Replace existing 75- and 65-watt incandescent bulbs with a 20-watt CFL.

- Replace existing 60-, 55-, 52-, and 50-watt incandescent bulbs with a 15-watt CFL.

HIGH-INTENSITY DISCHARGE LAMPS

High-intensity discharge (HID) lamps are another type of gaseous discharge lamp (Figure 17-15). The method used to produce light is similar to fluorescent lighting. An arc is established between two cathodes in a glass tube filled with a metallic gas, such as mercury, halide, or sodium. The arc causes the metallic gas to produce radiant energy, resulting in light. The electrodes are only a few inches (or less) apart in the lamp arc tube, not at opposite ends of a glass tube like a fluorescent lamp. This causes the arc to generate extremely high temperatures, allowing the metallic elements of the gas to release large amounts of visible energy. Because the energy is released as light energy, no phosphors are needed in HID lighting.

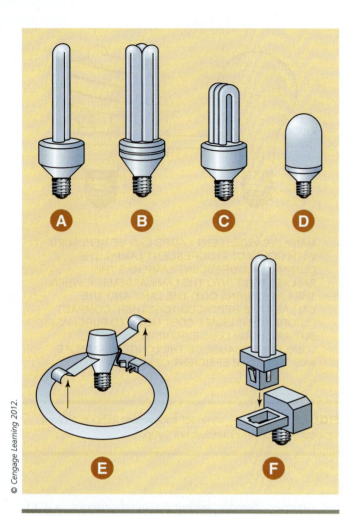

© Cengage Learning 2012.

FIGURE 17-14 Compact fluorescent lamps (CFLs) come in a variety of sizes and shapes including (A) self-ballasted twin-tube, (B and C) self-ballasted triple-tube, (D) a self-ballasted with a cover that reduces glare, (E) adapter kit circline and ballast, and (F) adapter kit quad-tube. Other CFL styles are available.

for most retrofits. It is especially useful for retrofits in homes where some dedicated ballast CFLs are already being used.

New fixture CFLs require a new lighting fixture that includes a CFL ballast. The lamps are separate from the fixture and simply plug in, making lamp changes easy. These lights use the same type of lamps as the adapter kits. This CFL is an excellent choice for new construction, building additions, and remodeling.

When replacing incandescent lamps with CFLs, the following guide will help you determine which CFL lamp size to use:

- Replace existing 100-, 95- and 90-watt incandescent bulbs with a 25-watt CFL.

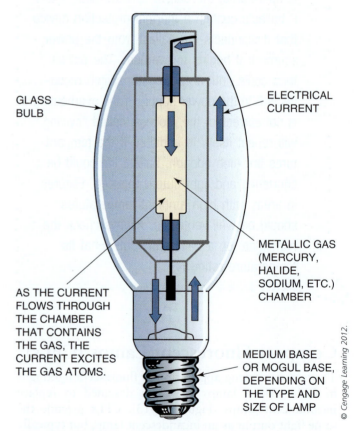

GLASS BULB

ELECTRICAL CURRENT

METALLIC GAS (MERCURY, HALIDE, SODIUM, ETC.) CHAMBER

AS THE CURRENT FLOWS THROUGH THE CHAMBER THAT CONTAINS THE GAS, THE CURRENT EXCITES THE GAS ATOMS.

MEDIUM BASE OR MOGUL BASE, DEPENDING ON THE TYPE AND SIZE OF LAMP

© Cengage Learning 2012.

FIGURE 17-15 A typical HID lamp and an explanation of how it produces light.

Like other electric discharge lighting, HID lamps require the use of a ballast to operate properly. It is extremely important that the ballast be designed for the lamp type and wattage being used. All HID lamps require a warm-up period before they are able to provide full light output. The main disadvantage of HID lighting is that even a very brief power interruption can cause the lamp to restart its arc and have to warm up again—a process that usually takes several minutes. For this reason, HID lighting is not generally used inside residences. It is used for outdoor security and area lighting applications.

Metal halide and sodium vapor are the two main types of HID lamps used today. Their names reflect the material inside the arc chamber. Mercury vapor is the oldest HID lamp and is obsolete as a light source. Even though the mercury vapor lamp was very efficient and had an extremely long life, it is not used in new residential applications but may be encountered in existing dwellings as a light source for outside area lighting. Metal halide lamps are the most energy efficient source of white light on the market today. These lamps offer excellent color rendition and long life. They are widely used for large areas that require good illumination, such as highways, parking lots, and the interiors of large commercial buildings. In residential systems, they are limited to large outdoor applications. Sodium vapor lamps provide the highest efficacy of any lamp type. These lamps are produced in both high- and low-pressure versions. While both are very efficient, their poor color rendition limits their use to areas where high-quality lighting is not an important factor. These lamps are seldom found in any residential lighting system but can be used to illuminate residential driveways and other outdoor areas.

Table 17-1 lists the different types of lamps discussed in this section and lists each lamp's characteristics.

SELECTING THE APPROPRIATE LIGHTING FIXTURE

Today, the lighting industry produces many different types of lighting fixtures to satisfy almost every need. Residential luminaires are readily available and easy to install. They are usually either incandescent or fluorescent and can be surface-mounted, recessed, or mounted in suspended-ceilings.

Green Tip

One of the best choices for energy efficient lighting is fluorescent hardwired Energy Star qualified fixtures. These fixtures generally have electronic ballasts and accept pin style fluorescent lamps. The advantage of pin style fixtures is that home owners can use fluorescent replacement lamps and not less efficient incandescent lamps. Fluorescent lamps use only one quarter of the electricity that incandescent lamps use. They also last 5 to 10 times longer.

TABLE 17-1 Characteristics of Lamp Types Used in Residential Wiring

	INCANDESCENT	LED	FLUORESCENT	CFL	METAL HALIDE	HIGH-PRESSURE SODIUM
Lumens per watt	6–23	15–25	25–100	25–100	65–120	75–140
Wattage range	40–1500	2–12	4–215	3–32	175–1500	35–1000
Life (hours)	750–8000	25,000–100,000	9000–20,000	10,000	5000–15,000	20,000–24,000
Color temperature (K)	2400–3100	2500–3500	2700–7500	2700	3000–5000	2000
Color Rendition Index	90–100	60–90	50–100	80–90	60–70	20–25
Potential for good color rendition	High	Very Good	High	High	Good	Fair
Lamp cost	Low	High	Moderate	Moderate	High	High
Operational cost	High	Low	Good	Good	Moderate	Low

Note: The values given are generic for general service lamps. A survey of lamp manufacturers' catalogs should be made before specifying or purchasing any lamp.

Green Tip

LED lighting fixtures are even more energy efficient than fluorescent lamps and last even longer. Today, they cost more than fluorescent fixtures and there are limited fixture choices so LEDs may not be the first choice for lighting. However, LED fixture choices are expected to increase over the next decade and the costs will go down so it will not be long before LED fixtures and lamps are the preferred lighting for green homes.

Installation requirements for light fixtures are contained in Article 410 of the *NEC®*. This article also covers fixture mounting, supporting, clearances, and construction requirements. Each fixture comes with specific installation instructions provided by the manufacturer. It is very important for the electrician to read, understand, and follow these instructions. Each fixture also comes with labeling that informs the installer of any installation restrictions pertaining to location and wiring methods. Some common information found on a label includes the following:

- For wall mount only
- Ceiling mount only
- Maximum lamp wattage: _____ watts
- Lamp type
- Suitable for operation in an ambient temperature not exceeding _____°F (°C)
- Suitable for use in suspended ceilings
- Suitable for damp locations
- Suitable for wet locations
- Suitable for mounting on low-density cellulose fiberboard
- For supply connections, use wire rated at least _____°F (°C)
- Thermally protected
- Type IC
- Type Non-IC

Each light fixture and lamp installed in any residential lighting system should carry the label of a nationally recognized testing laboratory. This assures the installer that the luminaire has met a minimum set of performance standards. The three major testing laboratories are Underwriters Laboratories (UL), Canadian Standards Association (CSA), and Electrical Testing Laboratories (ETL). These testing labs were discussed in Chapter 2. If an electrician finds a fixture that does not carry one of these labels, he or she should check with the local inspector to be sure the luminaire is approved for installation.

Because of the very personal nature of lighting and the vast number of choices available, selecting the "perfect" light fixture for a particular location can be a very daunting task. There is no set rule explaining who will choose the luminaires for a residence. If a building contractor builds a house to sell (a "spec house"), the contractor will usually choose the fixtures. If a building contractor builds a house for a specific individual, the customer will have the final say on the type of fixtures installed.

Residential lighting can be divided into four separate groups: general lighting, accent lighting, task lighting, and security lighting. The requirements of each group are unique and require different approaches. The electrician must work closely with the home owner or the builder to ensure that the lighting they install produces the desired effect. A description of each lighting group follows:

- General lighting provides a room or other area with adequate overall lighting. It can range from very basic, like a plastic keyless fixture in a garage, to very ornate, such as a chandelier hanging over the home foyer. General lighting is often used with either accent lighting or task lighting in bedrooms, kitchens, living rooms, and garages or workshops.

- Accent lighting is used to focus your attention on a particular area or object, such as a fireplace or work of art. Recessed spotlights and track lighting are two popular fixture types used for accent lighting. To be effective, accent lighting should be at least five times brighter than the room general lighting.

- Task lighting provides an adequate light source for tasks to be performed without distracting glare or shadows. It is important that the task lighting complement the general lighting of the room. To avoid too much contrast, task lighting should be no more than three times brighter than the general room lighting.

- Security lighting is designed to meet safety and security concerns around the exterior of the house. When properly designed and installed, security lighting will enhance the appearance of the residence and increase its value.

There are four factors that should be considered when selecting light fixtures for a particular location:

- Match both the lamp and the fixture to the desired application. Is the light source appropriate for the type of lighting required? For example, you would not want to use floodlights to accent a wall-hung painting. Does the light fixture complement the light fixtures in the rest of the room? Lighting manufacturers will provide a wealth of information to simplify this decision. They have pamphlets and booklets that give information on the types of fixtures that will work in most residential situations.

- How important is color rendition to the lighting application? Color rendition is a term used to describe how well a particular light source allows us to see an object's true color. All lamps are assigned a number reflecting how their light compares to sunlight at noon. The higher the assigned number, the better the color rendition of the lamp. Most incandescent lamps have a color rendition number of 95, and fluorescent lamps have a color rendition of 80 to 85. In residential lighting, color rendition is very important for indoor applications. For unfinished basements, garages, and outdoor locations, the color rendition may not be as important.

- The energy efficiency of the lamp and fixture must be considered. While the most light output for the lowest cost (purchase price plus operating cost) is certainly important, in some cases a less efficient luminaire will provide a better lighting result and must be considered. The goal is find the right light source with the highest efficiency for the lowest total cost.

- How long will a particular lamp last? In residential lighting, this is usually not a high priority because most lamps are relatively inexpensive and are easy to replace. Lamp life becomes an issue in areas where lamps are difficult to reach. Lamps located in cathedral ceilings or in chandelier fixtures hung in open foyers are examples of areas where lamp life is important.

LIGHTING FIXTURES INSTALLED IN CLOTHES CLOSETS

Lighting fixtures are not required by the *NEC®* to be installed in clothes closets. However, in many new homes the building plans call for luminaires to be installed in all or some of the home's clothes closets. Article 410 has some special rules to follow when installing luminaires in clothes closets because of the fire hazard from clothing, boxes, and other things that people typically store in them. A clothes closet is defined in the *NEC®* as a non-habitable room or space intended primarily for storage of garments and apparel. Only certain types of luminaires can be installed in a clothes closet and there are rules on placing them certain distances from closet storage space. Section 410.2 has defined what is considered to be storage space. It is the space that extends from the bottom of the closet to a height of 6 feet (1.8 m) or to the highest clothes hanging rod. This space is measured in 24 inches (600 mm) from the closet sides and back wall. Above 6 feet (1.8 m) or the highest hanging rod, the storage space continues up to the ceiling but is now measured in 12 inches (300 mm) from the sides and back wall of the closet

or the width of the storage shelf, whichever is greater. Figure 17-16 shows the storage space of a typical clothes closet.

Section 410.16(A) lists the luminaires that *are* permitted to be installed in clothes closets. They are: a surface-mounted or recessed incandescent or LED fixture with a completely enclosed lamp(s); a surface-mounted or recessed fluorescent fixture; and a surface-mounted fluorescent or LED luminaire identified as suitable for installation in the storage area. Section 410.16(B) lists the luminaires that *are not* permitted in a clothes closet. They are: incandescent light fixtures with open or partially enclosed lamps and pendant light fixtures or lampholders. Figure 17-17 illustrates the types of fixtures permitted and not permitted in clothes closets.

When installing luminaires on the wall above the closet door or on the closet ceiling, Section 410.16(C) limits their location to the following (Figure 17-18):

- Surface-mounted incandescent or LED fixtures with a completely enclosed light source must have a minimum clearance of 12 inches (300 mm) from the nearest point of storage space.

- Surface-mounted fluorescent fixtures must have a minimum clearance of 6 inches (150 mm) from the nearest point of storage space.

- Recessed incandescent or LED fixtures with a completely enclosed lamp(s) must have a minimum clearance of 6 inches (150 mm) from the nearest point of storage space.

- Recessed fluorescent fixtures must have a minimum clearance of 6 inches (150 mm) from the nearest point of storage space.

from experience...

When considering the different lighting fixture types that are permitted to be installed in a clothes closet, unless the choice has already been made and is indicated on the building plans, most electricians will install a surface-mounted fluorescent strip. It is an economical way to provide lighting in a clothes closet and even the shallowest clothes closets in today's homes should have enough space on the ceiling or on the wall above the door to allow at least 6 inches (150 mm) from the edge of a surface-mounted fluorescent strip to the closet storage space. Common strip sizes include 18-inch and 24-inch fixtures, but longer strips can be used if there is room.

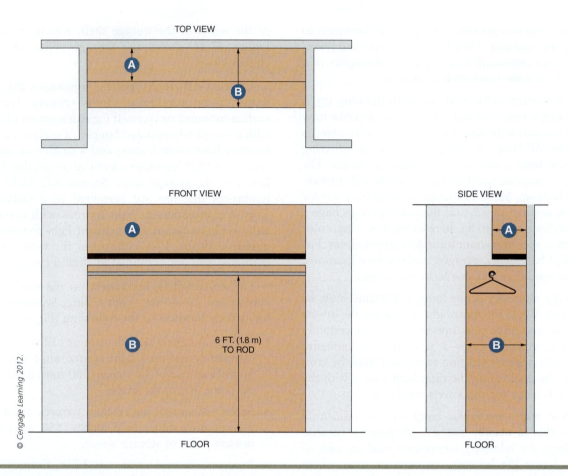

© Cengage Learning 2012.

FIGURE 17-16   The storage space of a typical closet that has a clothes hanging rod and a storage shelf. Shaded area A is the width of the shelf or 12 inches (300 mm), whichever is greater. Shaded area B is 24 inches (600 mm) from the sides and back wall up to a height of 6 feet (1.8 m) or the height of the rod, whichever is greater.

LIGHTING FIXTURES INSTALLED IN BATHTUB AND SHOWER AREAS

Section 410.10(D) requires an electrician to follow some specific installation rules when installing lighting fixtures in a bathtub or shower area. A luminaire that is cord-connected or chain-, cable-, or cord-suspended cannot have any part of the fixture within a zone measuring 3 feet (900 mm) horizontally and 8 feet (2.5 m) vertically from the top of the bathtub rim or from a shower stall threshold. Track lighting, pendant lighting fixtures, and ceiling-suspended paddle fans must follow the same rule. The zone consists of the area directly over the tub or shower stall. Any luminaire that is legally installed in this zone must be listed for damp locations or listed for wet locations if it is going to be subject to shower spray (Figure 17-19).

RECESSED LUMINAIRES

Recessed luminaires (Figure 17-20) are fixtures installed above the ceiling. They are designed so that little or none of the fixture, lamp, lens, or trim extends below the level of the sheetrock or drop ceiling tile. Fluorescent, incandescent, and LED recessed fixtures are available.

In residential lighting systems, recessed fluorescent lighting is not used that often. When it is used, it is usually in areas of a house such as kitchens, laundry rooms, finished basements, or recreation rooms. This type of fixture produces a softer and more even light distribution. These fixtures are available in two-, three-, and four-lamp models. The two-lamp fixtures require one ballast, while the three- and four-lamp fixtures usually require two ballasts. Recessed fluorescent fixtures have lenses to protect the lamps and to disperse the light in a particular pattern. Each lens is designed to provide maximum light output for that particular light fixture.

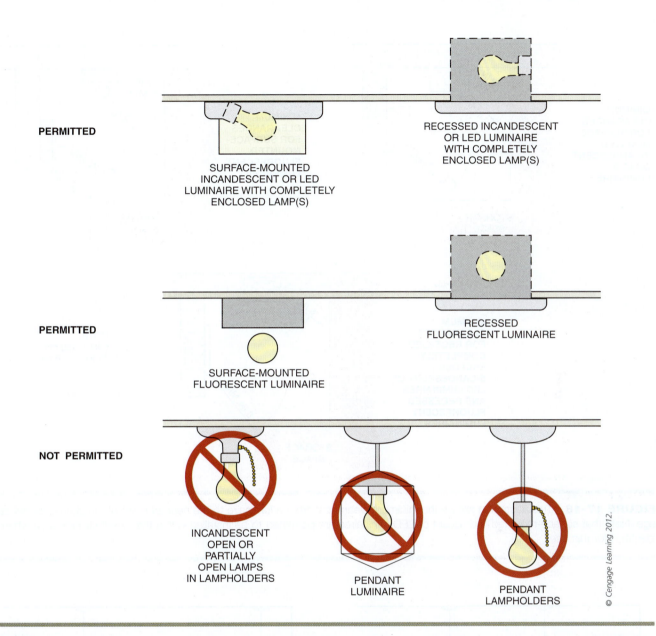

PERMITTED

SURFACE-MOUNTED
INCANDESCENT OR LED
LUMINAIRE WITH COMPLETELY
ENCLOSED LAMP(S)

RECESSED INCANDESCENT
OR LED LUMINAIRE
WITH COMPLETELY
ENCLOSED LAMP(S)

PERMITTED

SURFACE-MOUNTED
FLUORESCENT LUMINAIRE

RECESSED
FLUORESCENT LUMINAIRE

NOT PERMITTED

INCANDESCENT
OPEN OR
PARTIALLY
OPEN LAMPS
IN LAMPHOLDERS

PENDANT
LUMINAIRE

PENDANT
LAMPHOLDERS

© Cengage Learning 2012.

FIGURE 17-17   This illustration shows the types of lighting fixtures permitted to be installed in a clothes closet. It also shows the types that are not permitted. Note that surface-mounted fluorescent or LED luminaires identified as suitable for installation within the closet storage area are also permitted.

Recessed incandescent or LED lighting is a very versatile type of lighting available for residential lighting and, as a result, is used often in residential lighting systems. The light fixture comes in two separate pieces: the recessed fixture rough-in frame (Figure 17-21) and the trim. Each recessed fixture, commonly called a "can" or "high hat," is available with trim rings that will fit only that fixture. The variety of trim rings available allow the fixture to be used as general lighting, accent lighting, task lighting, or security lighting.

from experience...

Recessed incandescent and LED light fixture rough-in frames and trim rings are sold separately. Be sure to get both parts when purchasing the light fixtures.

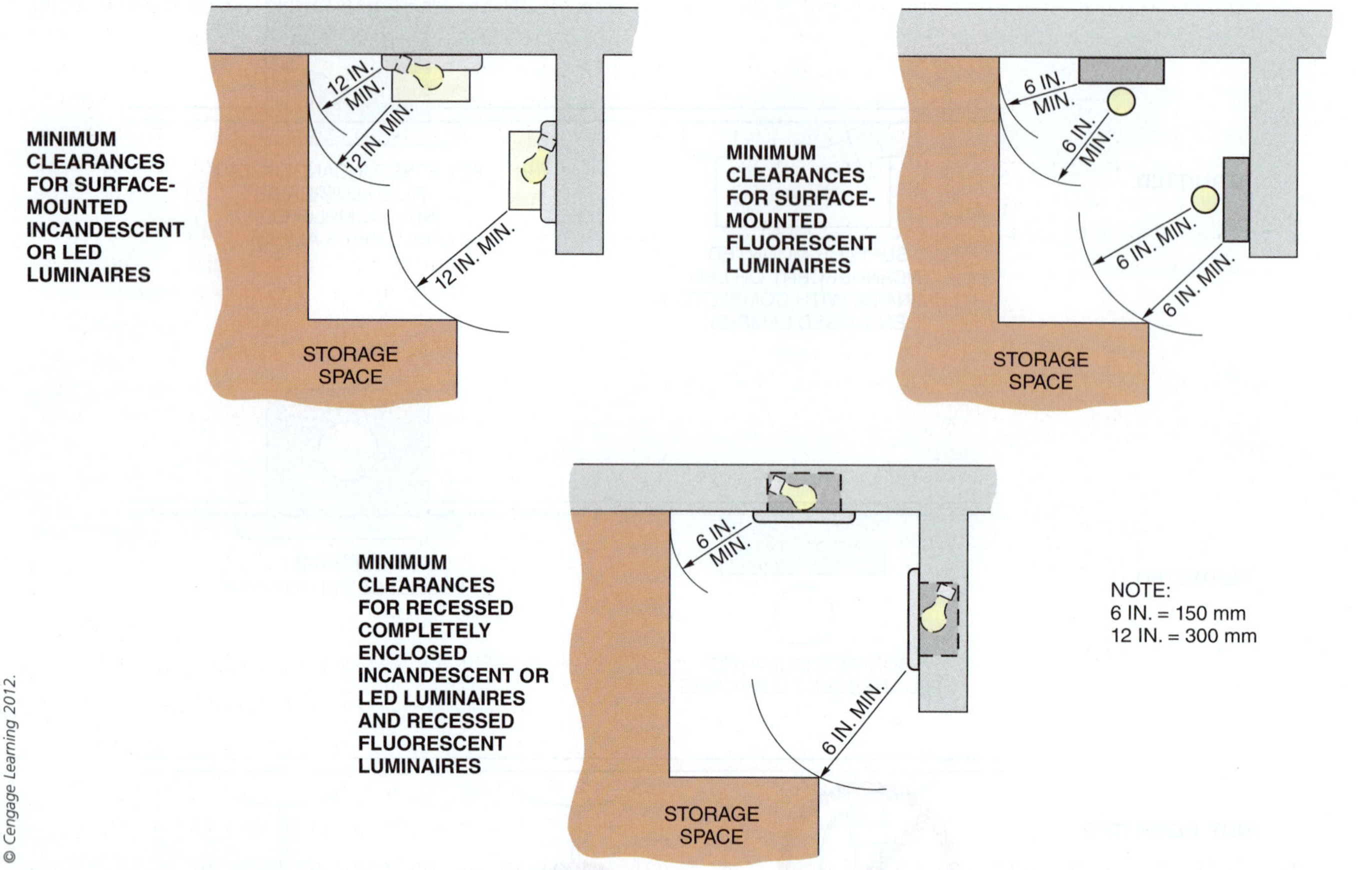

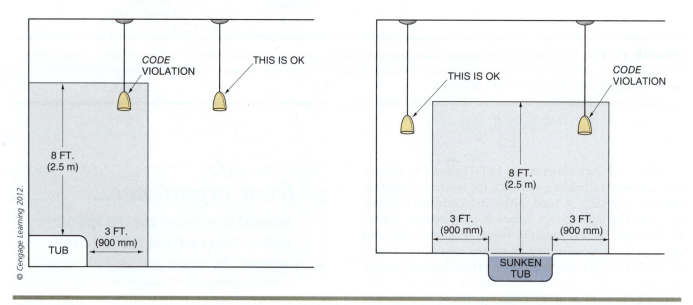

FIGURE 17-18   This illustration shows the minimum clearances from a luminaire to the nearest point of the clothes closet storage. Note that surface-mounted fluorescent or LED luminaires are permitted to be installed within the closet storage area where identified for this use.

FIGURE 17-19   In a bathtub or shower area, no part of a hanging lighting fixture or ceiling-suspended paddle fan can be in the shaded area shown in this illustration.

The rough-in frame is made of metal and holds the light fixture and a junction box. This framework is installed in the ceiling joists or trusses, and the fixture is connected to the electrical system by means of the junction box before the ceiling is installed. The trim rings and lamp are installed after the ceiling is painted. It is important the electrician know the trim ring to be installed so that the proper fixture may be installed during the rough-in phase.

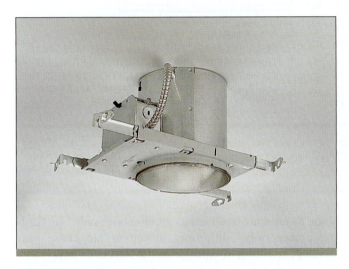

FIGURE 17-20   A recessed luminaire rough-in frame.
Photo courtesy of Progress Lighting.

The electrician must also know the insulation rating of the fixture housing. Recessed fixtures are rated either **Type IC** (insulated ceiling) or **Type Non-IC**. Type IC fixtures are designed to be in direct contact with thermal insulation. Type Non-IC fixtures must be installed so that the insulation is no closer than 3 inches (75 mm) to any part of the fixture. Figure 17-22 shows both types of recessed lighting fixtures. This allows any heat generated by the lighting fixture to dissipate away from the fixture, preventing damage to the fixture and possibly fire. This is covered in Section 410.116 of the *NEC®*. Each recessed incandescent fixture must be equipped with a built-in thermal protection unit, as stated in Section 410.115(C) of the *NEC®*. This is to lessen the threat of fire from heat buildup.

from experience...

The fish-eye trim is not a good choice for high ceilings or cathedral ceilings. Changing lamps can be extremely difficult unless you have the proper ladder. Telescoping tools designed for changing lamps will not work with this style of trim ring. They simply push the fish-eye part of the trim ring back up into the can.

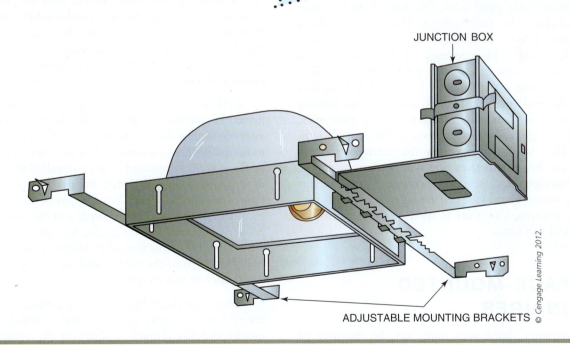

JUNCTION BOX

© Cengage Learning 2012.

ADJUSTABLE MOUNTING BRACKETS

FIGURE 17-21   A recessed lighting fixture rough-in frame. The adjustable mounting brackets allow the frame to be easily installed between two ceiling joists. The cable feeding the fixture is terminated in the attached junction box. The *NEC®'s* installation requirements for recessed fixtures are found in Article 410.

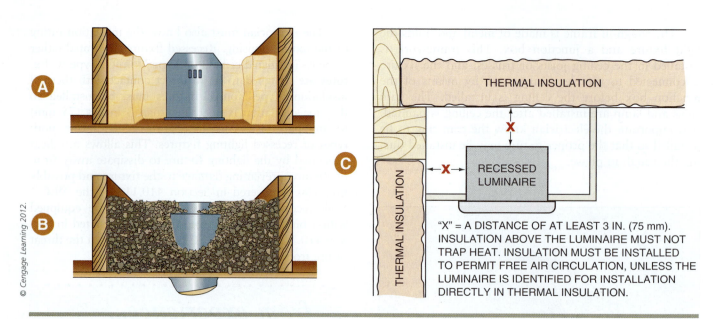

© Cengage Learning 2012.

FIGURE 17-22 (A) A Type Non-IC recessed fixture installed with the thermal insulation kept away from the fixture. (B) A Type IC fixture that may be completely covered with thermal insulation. (C) The required clearances from thermal insulation for a Type Non-IC recessed fixture according to Section 410.116 of the *NEC®*.

There are four main types of trim rings available for recessed incandescent or LED light fixtures (Figure 17-23). They are held in place with springs or clips. The open baffle is used to provide general lighting. This trim type allows for a softer illumination of a wide area. This trim is available in various colors, with either smooth or stepped sides. The down-light trim ring is a good choice for task lighting. This trim provides a more focused beam of light for the task at hand. The trim rings come with openings of different sizes to accommodate lamps of different size and wattage. The fish-eye trim ring is used mainly for accent lighting but also works well in task lighting and general lighting situations. This trim can be rotated along two axes, allowing the light to be aimed to best fit the desired application. The fourth trim type is the waterproof trim. Basically, a down-light trim ring with a glass or acrylic lens, it is designed for use in wet locations. Outdoors, this trim ring is used to provide general lighting on decks and porches. Indoors, this trim is used to light bathroom shower and tub units.

SURFACE-MOUNTED LUMINAIRES

Surface-mounted luminaires make up the majority of light fixtures installed in a residence. There are a multitude of styles and sizes available, so you are limited only by your imagination. Surface-mounted luminaires can be either incandescent, LED, or fluorescent and are designed for either wall or ceiling mounting. However, you should always check the manufacturer's instructions because

some fixtures are designed as "wall mount only" or "ceiling mount only." Improper mounting can cause heat buildup, resulting in damage to the fixture or fire.

Wall-mounted lighting fixtures (Figure 17-24), also called **sconces**, are usually connected to either metallic or nonmetallic lighting outlet boxes after the sheetrock is installed and finished. Sconces can be used in any room in the house and are used for general lighting or accent lighting. There are many styles and shapes of sconces to meet almost any lighting need. Units generally have one or two lamps with either a candelabra or a medium base.

Ceiling-mounted fixtures (Figure 17-25) are also connected to metallic or nonmetallic lighting outlet boxes after the ceiling has been installed and finished. Ceiling-mounted fixtures are designed to use incandescent, LED, or fluorescent lamps and may be installed on the bottom of dropped ceilings if listed for that purpose. For this type of mounting, special brackets, available in most electrical supply stores, must be used, and the fixture must be properly supported to the ceiling framework. Installation requirements for this type of installation are found in Section 410.36(B) of the *NEC®* and are discussed in greater detail later in this chapter.

There are three basic types of ceiling-mounted fixtures: direct mount, pendant, and chandelier. Direct-mount fixtures (Figure 17-26) are the most common type used in residential lighting applications. Some manufacturers refer to this type of mounting as "close to ceiling." They can be used to provide general lighting in any room in a house. Incandescent and LED models usually have one, two, or three lamps, while fluorescent models may have two, three, or four lamps.

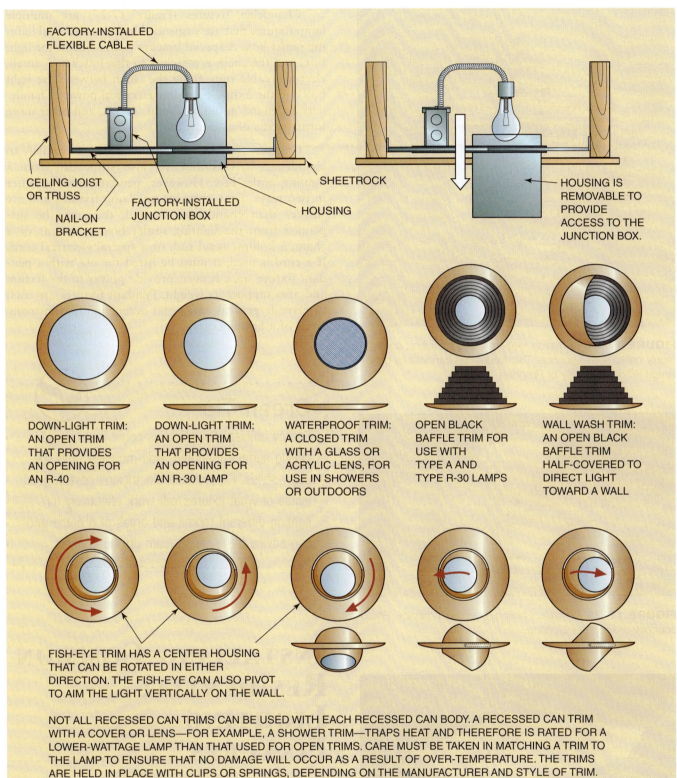

FACTORY-INSTALLED
FLEXIBLE CABLE

CEILING JOIST
OR TRUSS

NAIL-ON
BRACKET

FACTORY-INSTALLED
JUNCTION BOX

SHEETROCK

HOUSING

HOUSING IS
REMOVABLE TO
PROVIDE
ACCESS TO THE
JUNCTION BOX.

DOWN-LIGHT TRIM:
AN OPEN TRIM
THAT PROVIDES
AN OPENING FOR
AN R-40

DOWN-LIGHT TRIM:
AN OPEN TRIM
THAT PROVIDES
AN OPENING FOR
AN R-30 LAMP

WATERPROOF TRIM:
A CLOSED TRIM
WITH A GLASS OR
ACRYLIC LENS, FOR
USE IN SHOWERS
OR OUTDOORS

OPEN BLACK
BAFFLE TRIM FOR
USE WITH
TYPE A AND
TYPE R-30 LAMPS

WALL WASH TRIM:
AN OPEN BLACK
BAFFLE TRIM
HALF-COVERED TO
DIRECT LIGHT
TOWARD A WALL

FISH-EYE TRIM HAS A CENTER HOUSING
THAT CAN BE ROTATED IN EITHER
DIRECTION. THE FISH-EYE CAN ALSO PIVOT
TO AIM THE LIGHT VERTICALLY ON THE WALL.

NOT ALL RECESSED CAN TRIMS CAN BE USED WITH EACH RECESSED CAN BODY. A RECESSED CAN TRIM
WITH A COVER OR LENS—FOR EXAMPLE, A SHOWER TRIM—TRAPS HEAT AND THEREFORE IS RATED FOR A
LOWER-WATTAGE LAMP THAN THAT USED FOR OPEN TRIMS. CARE MUST BE TAKEN IN MATCHING A TRIM TO
THE LAMP TO ENSURE THAT NO DAMAGE WILL OCCUR AS A RESULT OF OVER-TEMPERATURE. THE TRIMS
ARE HELD IN PLACE WITH CLIPS OR SPRINGS, DEPENDING ON THE MANUFACTURER AND STYLE OF TRIM.

FIGURE 17-23 A guide to recessed-lighting fixture trims used in residential applications.

FIGURE 17-24   A typical wall-mounted luminaire, commonly called a wall sconce. This fixture has candelabra base lamps. *Photo courtesy of Progress Lighting.*

FIGURE 17-25   A typical ceiling-mounted luminaire. *Photo courtesy of Progress Lighting.*

FIGURE 17-26   A direct-mount lighting fixture. The lighting fixture shown is mounted directly to the ceiling in a room like a kitchen. This fixture is designed to take two 40-watt, U-shaped fluorescent tubes. *Photo courtesy of Progress Lighting.*

Chandelier fixtures (Figure 17-27) are multiple lamp fixtures that are suspended from an electrical lighting outlet box. A special bracket is attached to the light box, and the chain is attached to this bracket. A small, electrical cable runs along the chain between the light box and the fixture providing electrical power. Chandeliers are found in dining rooms, foyers, and other more formal areas of a residence.

Pendant fixtures (Figure 17-28) are similar to chandeliers in that they are also suspended from a lighting outlet box. However, pendant fixtures differ in two ways. First, they are one-lamp fixtures and are smaller than chandeliers. Second, they may be suspended from the lighting outlet box by means of a chain, a hollow metal rod, or a special electrical cord. If a cord is used, it must be listed for use with a pendant fixture since it must provide power to the fixture and also support its weight. Pendant fixtures are used to provide general, task, and accent lighting in many areas of a house.

Green Tip

Energy efficient light fixtures come in recessed, ceiling surface-mounted, wall-mounted, and specialty styles. Experienced electricians have a good sense of what fixtures will work effectively for lighting different rooms and areas of a home and can advise the building team on the choices.

INSTALLING COMMON RESIDENTIAL LIGHTING FIXTURES

Installing lighting fixtures in a house is one of the last things an electrician does during the installation of the electrical system. Before an electrician actually installs the lighting fixtures, there are three factors that need to be considered:

- You must remember that light fixtures are very fragile and are easily damaged or broken. It is important to handle and store these fixtures with care.

FIGURE 17-27   Several examples of chandelier-type lighting fixtures. They are commonly installed over a dining room table. *Photos courtesy of Progress Lighting.*

FIGURE 17-28 An example of a pendant-type lighting fixture that could be installed over a small table in the eating area of a kitchen. Other pendant-type lighting fixture styles are popular for installation in areas of a house with high ceilings. *Photo courtesy of Progress Lighting.*

CAUTION

CAUTION: Make sure the power to the circuit is disconnected before starting to install a lighting fixture. Many electricians are needlessly shocked because they *assumed* the power had been turned off. Always check the circuit with a voltage tester before beginning and make sure the circuit is locked out.

from experience...

On some lighting fixtures, especially the ones hung with chain, the fixture wiring is not color-coded, and at first glance the grounded wire and the ungrounded wire are not distinguishable from each other. However, you will find that one of the wires has a smooth surface and that one has a ribbed surface. The wire with the ribbed surface is the grounded conductor, and the wire with the smooth surface is the ungrounded conductor.

• Read the manufacturer's instructions that come with each lighting fixture. The instructions can make the installation process much easier if they are read *before* you start the installation. Nothing is more frustrating than to get most of the installation complete and then realize that you did not install a part in the proper location—now you have to take the fixture completely apart and start over. This can be avoided by simply reading the instructions. Remember, instructions are not a means of last resort.

• When installing lighting fixtures, an electrician must ensure that they are connected to the electrical system with the proper polarity. Section 410.50 of the *NEC*® states that the grounded conductor must be connected to the silver screw shell. The ungrounded, or "hot," conductor is connected to the brass contact in the bottom center of the screw shell. If the fixture has both a black and a white conductor, and the polarity has been determined, you may proceed with the installation. If the fixture has two wires of the same color, usually black, a continuity tester can be used to determine the proper polarity. One test lead of the tester is placed against the silver screw shell, and the other test lead is placed against the bare end of one of the fixture conductors. If continuity is indicated, that wire is the grounded conductor and should be marked with white tape. If continuity is not indicated, the wire must be the ungrounded conductor and should be marked accordingly.

DIRECT CONNECTION TO A LIGHTING OUTLET BOX

There are not many lighting fixtures that are mounted directly to the outlet box. The two main types of fixtures in this category are keyless fixtures and pull-chain fixtures (Figure 17-29). Keyless fixtures are typically made of porcelain or plastic and are often used in basements,

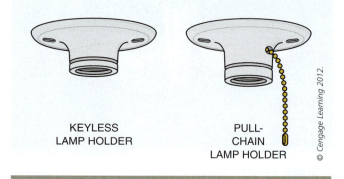

KEYLESS LAMP HOLDER

PULL-CHAIN LAMP HOLDER

© Cengage Learning 2012.

FIGURE 17-29 Keyless and pull-chain direct-connection fixtures. They may be made of porcelain, plastic, or fiberglass.

crawl spaces, attics, and garages. Because the fixture is made from nonmetallic materials, it has no provisions for grounding.

(See Procedure 17-1 on page 625 for installation steps for installing a light fixture directly to an outlet box.)

DIRECT CONNECTION TO THE CEILING

Surface-mounted fluorescent fixtures (Figure 17-30) are installed and connected to the electrical system in a slightly different manner. Attaching the fixture to the ceiling requires a different approach because the fixture is not attached directly to the lighting outlet box. Instead, the fixture is attached directly to the ceiling, and the electrical wiring is brought into the fixture. Surface-mounted fluorescent fixtures have wiring compartments where the lighting branch-circuit wiring is connected to the fixture wiring. The electrical connections are very basic: white wire to white wire, black wire to black wire; and the grounding wire to the fixture's metal body.

There are two ways that a surface-mounted fluorescent fixture can be connected to the lighting branch-circuit wiring. The first way is to have the wiring method used, such as a Type NM cable, connected directly to the lighting fixture. In this scenario, a length of cable is installed at the location of the fixture(s) during the rough-in stage and left long enough to hang down from the ceiling a short distance. It is then coiled up out of the way and placed between the ceiling joists. When the ceiling workers put up the ceiling, they will put a small hole in the ceiling material and bring the cable down through the hole so that a short length hangs down below the ceiling. During the installation of the surface-mounted fixture, the electrician attaches this cable to the lighting fixture with a cable clamp, secures the fixture to the ceiling, and makes the necessary electrical connections. Manufacturers provide knockouts on their surface-mounted fluorescent fixtures for this installation method. A good example of this type of connection is in the installation of a surface-mounted fluorescent strip in a clothes closet. In this case, the cable may be hanging from a spot on the wall above the closet door or from the ceiling, depending on where the electrician has left the cable in anticipation of installing the fluorescent strip.

The second wiring method is similar to the one where the lighting fixture is attached directly to a lighting outlet box. During the rough-in stage, a lighting outlet box is installed at the location for the surface-mounted fluorescent fixture, and the lighting circuit wiring is run to the box. When a surface-mounted fluorescent fixture is installed, it will cover the lighting outlet box. The wiring is brought from the lighting outlet box down and into the wiring compartment of the surface-mounted fluorescent fixture. When using this method, make sure to follow Section 410.24(B) of the *NEC*®, which requires surface-mounted fixtures to allow access to the wiring in the lighting outlet box *after* the surface-mounted fixture has been installed. This means that a large opening will have to be knocked out in the surface-mounted fluorescent fixture. The hole is then aligned with the lighting outlet box that the fixture is covering (Figure 17-31). The large hole allows easy access to the wiring in the lighting outlet box.

(See Procedure 17-2 on page 626 for installation steps for installing a cable-connected surface-mounted fluorescent lighting fixture directly to a ceiling.)

STRAP TO LIGHTING OUTLET BOX

This light fixture installation process is very similar to the direct connection method discussed previously. The main difference is that a metal strap is connected to the lighting outlet box and the fixture is attached to the metal strap with headless bolts and decorative nuts. The metal strap must be connected to the system-grounding conductor. For this type of installation, it is important to read, understand, and follow the manufacturer's instructions. Figure 17-32 shows an example of a lighting fixture that is installed using the strap to lighting outlet box method.

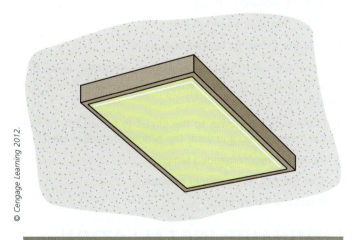

© Cengage Learning 2012.

FIGURE 17-30   A surface-mounted fluorescent fixture. If they are to be mounted on low-density cellulose fiberboard, they must be marked with a label that says "Suitable for Surface Mounting on Low-Density Cellulose Fiberboard."

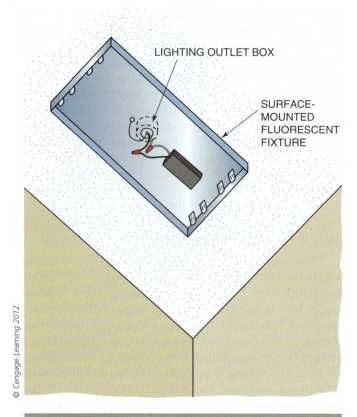

FIGURE 17-31 According to Section 410.24(B) of the *NEC*®, there must be access to the lighting outlet box even when a surface-mounted fluorescent lighting fixture is installed over the box location.

FIGURE 17-32 Examples of fixtures that use the strap to outlet box mounting method. *Photos courtesy of Progress Lighting.*

(See Procedure 17-3 on page 627 for installation steps for the installation of a strap to lighting outlet box lighting fixture.)

STUD AND STRAP CONNECTION TO A LIGHTING OUTLET BOX

There are larger and heavier types of light fixtures that use the stud and strap method of installation. Hanging fixtures, like a chandelier or pendant fixture, often require extra mounting support when compared to smaller light fixtures. The electrical connections for each installation are the same as discussed previously, so we will not spend a lot of time on them here. It is extremely important to read and follow the manufacturer's instructions because there are some slight variations with this type of installation.

(See Procedure 17-4 on page 628 for installation steps for installing a chandelier-type light fixture using the stud and strap connection method to a lighting outlet box.)

RECESSED LUMINAIRE INSTALLATION

At this stage of the process, recessed fixtures are very easy to finish out. The recessed fixture rough-in frame has already been installed and connected to the electrical system (Figure 17-33). All that remains to be done is to install the lamps and the trim rings or lenses.

For recessed incandescent or LED lighting fixtures, make sure the power to the circuit is turned off, review the instructions, and wear the proper safety gear:

1 Install the appropriate trim rings with the provided springs or clips. Make sure the trim ring is listed for use with the installed can light.

2 Install the recommended lamp.

3 Test the fixture for proper operation.

For recessed fluorescent lighting fixtures, keep the following items in mind when you are completing the installation:

1 Make sure the power to the circuit is turned off, review the instructions, and wear the proper protective gear.

2 Install the recommended lamps.

3 Test the fixture for proper operation.

4 Install the fixture lens.

LUMINAIRE INSTALLATION IN DROPPED CEILINGS

When installing luminaires in dropped ceilings, also called "suspended ceilings," it is important to make sure the fixture is listed for that type of installation. The

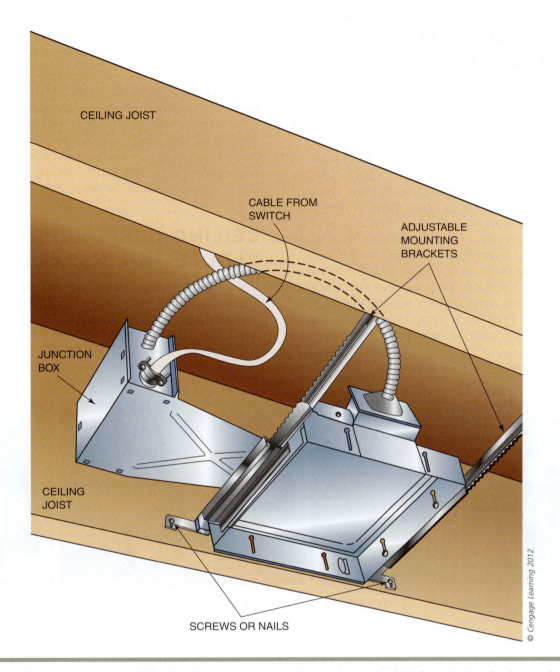

CEILING JOIST

CABLE FROM
SWITCH

ADJUSTABLE
MOUNTING
BRACKETS

JUNCTION
BOX

CEILING
JOIST

© Cengage Learning 2012.

SCREWS OR NAILS

FIGURE 17-33   Recessed fixture installation. The rough-in frame is pre-wired with wiring from the junction box to the lamp socket. The electrician runs a cable to the junction box and connects the fixture to the lighting branch circuit by making the necessary electrical connections in the junction box. The fixture is secured to the ceiling joists with screws or nails.

two listings to look for are "suitable for use in suspended ceilings" and "suitable for mounting on low-density cellulose fiberboard." The second listing is important because most drop ceiling tiles are made of that material. These fixtures are not considered recessed because there is usually a great deal of room above them and easy access to the electrical outlet box on each fixture.

There are several types of light fixtures that may be used in a suspended ceiling. They include recessed incandescent, LED, and fluorescent fixtures and any surface-mounted fixture that can be connected to a metallic or nonmetallic outlet box and is listed as suitable for installation in a dropped ceiling. It is important to note that the lighting outlet boxes must be connected to the ceiling support grid by means of an approved mounting bar. Fluorescent lighting fixtures installed in a dropped ceiling grid are often referred to by electricians as a **troffer**.

(See Procedure 17-5 on pages 629–630 for installation steps for installing a fluorescent fixture in a dropped ceiling.)

OUTSIDE LUMINAIRE INSTALLATION

When installing light fixtures outside a residence, you can use any of the methods that have been discussed previously in this chapter. Review the method that is appropriate for the fixture you are installing.

There are several factors that must be considered when installing light fixtures outdoors. One is whether the fixture is listed for the desired application. Outdoor fixtures are listed for use in either damp or wet locations. A damp location is any area that is subject to some degree of moisture but not saturation from water or other liquids. Examples of a damp location are fixtures installed in the eaves of a house or in the ceiling of an open porch. A wet location is any area that is subject to saturation from water or other liquids. Wet location fixtures may be installed in damp locations, but damp location fixtures cannot be installed in wet locations.

Figure 17-34 shows several examples of outdoor lighting fixtures that are suitable for installation in a wet (or damp) location.

Always check with the local code enforcement office for regulations concerning the placement of outdoor fixtures. Many communities have restrictions limiting the amount of light that may be seen from neighboring properties. There may also be regulations that limit the brightness of outdoor fixtures and the amount of light that is directed up into the sky.

CEILING-SUSPENDED PADDLE-FAN INSTALLATION

Earlier chapters in this textbook covered the rough-in wiring for a ceiling-suspended paddle fan (Figure 17-35), including the special electrical box required to support them. You may want to review that information at this

FIGURE 17-34   Examples of outdoor lighting fixtures. These fixtures are designed for wall mounting. *Photos courtesy of Progress Lighting.*

© Cengage Learning 2012.

FIGURE 17-35   A ceiling-suspended paddle fan with an installed light kit.

time. Most ceiling-suspended paddle fans installed in homes have a light kit attached to the bottom of the fan. In this section, the explanation of how to install a ceiling-suspended paddle fan will assume that a light kit is included. When the height of a ceiling is low, paddle fans need to be mounted close to the ceiling surface so that the fan blades are a minimum of 7 feet above the floor. For high or sloped ceilings, extension rods are used with paddle fans that allow them to hang down into the room, but no lower than 7 feet from the floor (Figure 17-36). Paddle fans need to be installed so the

tips of the blades are at least 18 inches from the wall in order to provide sufficient clearance. Use extra caution when installing a paddle fan on a sloped ceiling to make sure that after the fan is installed the blades will not hit the ceiling.

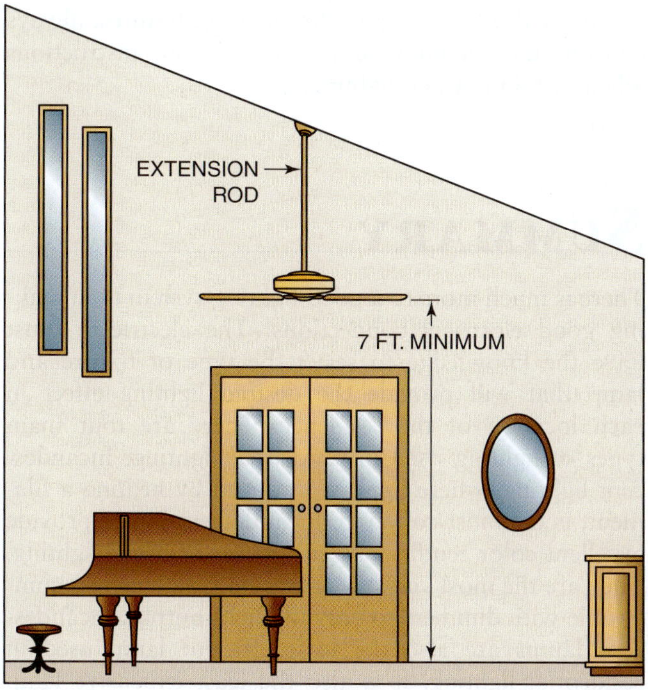

Before you start to install a ceiling-suspended paddle fan you should make sure that:

- You open the paddle-fan carton, locate the manufacturer's installation instructions, and follow them.

- You inspect the contents of the paddle fan carton for possible shipping or handling damage to the contents.

- All wiring to the fan has been installed in accordance with the *NEC*® and the appropriate local electrical codes.

- The outlet box is UL-listed for ceiling-suspended paddle fan installation.

- The outlet box is supplied by a grounded electrical circuit of 120 VAC, 60 Hz with a minimum 15-amp rating.

(See Procedure 17-6 on pages 632–633 for installation steps for installing a ceiling-suspended paddle fan with a light kit.)

TRACK LIGHTING INSTALLATION

Track lighting (Figure 17-37) is often installed in new homes or in older homes as a replacement for an existing lighting fixture. It is used to illuminate a specific work area or to highlight special items like a painting hanging on a wall. Track lighting is available in 12-volt or 120-volt models and consists of a track (Figure 17-38) with a number of lighting "heads" (Figure 17-39) attached to it. The lighting heads can be adjusted to direct their light wherever a home owner may want it.

© Cengage Learning 2012.

FIGURE 17-36 Extension rods are used to allow a ceiling-suspended paddle fan to hang down lower in the room. However, the blades must not be lower than 7 feet from the floor.

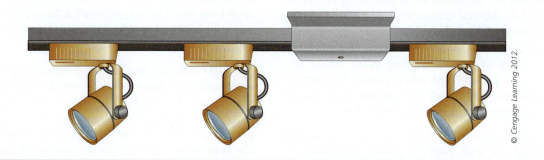

© Cengage Learning 2012.

FIGURE 17-37 A typical track light.

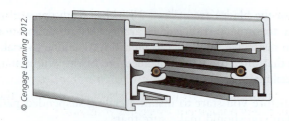

© Cengage Learning 2012.

FIGURE 17-38 A closer look at a lighting track.

© Cengage Learning 2012.

FIGURE 17-39 A typical lighting head for track lighting.

The most common lighting track is often called "basic track" and is the most user-friendly. It is installed directly to a ceiling with screws, but can be dropped from the ceiling with the use of suspensions. Lighting heads are installed by sliding them onto the track. The track is then finished with end caps for a smooth look. To power a basic track system you can use a number of different power feeds. A floating canopy connector can be installed anywhere along the track and covers a junction box located in the ceiling while providing power to the system track. An end connector can be used if you are running wiring (like a Type NM cable) directly into the track without using a junction box. You can also power the system at any connecting point where the connector has a power feed entry point. A junction box can also be used at any of these power entry points and can be covered with a canopy cover plate.

If the track lighting system you are installing is low voltage, it will have a power feed with a built-in transformer, sometimes called a "surface

transformer." You can also use a remote transformer. Remote transformers provide a cleaner look since they are not enclosed in a large canopy where they can be seen. They can also reduce any audible dimming hum when they are placed in a remote location. It is recommended that for low-voltage track lighting system a remote transformer be used whenever possible. Add the total watts for all lighting heads and make sure that the transformer has a wattage that is equal to or higher than the total watts of all of the heads. If you think lighting heads may be added to the track in the future, a higher-wattage transformer is needed.

As with installing all other lighting fixtures, always consult the manufacturer's installation instructions when installing track lighting.

SUMMARY

There is much more to a good lighting system than making good electrical connections. The electrician must have the knowledge to select the type of fixture and lamp that will provide the desired lighting effect in each location of the residence. There are four main types of lighting used in residential lighting. Incandescent lighting, where light is produced by heating a filament, is the most common. Incandescent lamps provide excellent color rendition in all phases of home lighting. They are the most versatile lamps because they are compatible with dimmers to vary the light output. Incandescent lamps are also the least efficient lamp used in residential lighting. It is also the least expensive lamp type to purchase.

The newest lamp type being used in residential wiring is the LED lamp. It is initially expensive to purchase but because of its long life and very high efficiency will cost very little to use in the long run.

Fluorescent lighting gives the best result when used for general lighting in kitchens, laundry rooms, basements, and work areas. It is a very efficient light source with long life. The major drawbacks to fluorescent lighting are poor cold temperature operation and incompatibility with dimmers.

HID lighting produces the whitest light and is the most efficient lamp used in residential lighting. The major drawback to HID lighting is the warm-up period before full light output is achieved. This warm-up period can last several minutes and occurs every time the lamp is lit and after power interruptions. This limits HID lighting to outdoor applications.

It is extremely important to install luminaires according to the manufacturer's instructions. Even similar-looking fixtures from different manufacturers can have different installation procedures. Reading the instructions before you start the installation will result in a smoother, faster, and correct installation. The installation steps provided in this chapter are meant to be a guide to the learner. They are not meant to replace the manufacturer's instructions.

Above all, be safe. Always make sure the power to the circuit is off before you start the installation, and test the circuit before you touch the conductors. Many electricians have been shocked because they thought that simply having the switch controlling the fixture turned off was adequate protection. Also, always use the proper safety equipment to ensure your safety.

GREEN CHECKLIST

☐ One of the best choices for energy efficient lighting is fluorescent hardwired Energy Star qualified fixtures. These fixtures generally have electronic ballasts and accept pin style fluorescent lamps. The advantage of pin style fixtures is that home owners can use fluorescent replacement lamps and not less efficient incandescent lamps. Fluorescent lamps use only one-quarter of the electricity that incandescent lamps use. They also last 5 to 10 times longer.

☐ LED lighting fixtures are even more energy efficient than fluorescent lamps and last even longer. Today, they cost more than fluorescent fixtures and there are limited fixture choices so LEDs may not be the first choice for lighting. However, LED fixture choices are expected to increase over the next decade and the costs will go down so it will not be long before LED fixtures and lamps are the preferred lighting for green homes.

☐ Energy-efficient light fixtures come in recessed, ceiling surface-mounted, wall-mounted, and specialty styles. Experienced electricians have a good sense of what fixtures will work effectively for lighting different rooms and areas of a home and can advise the building team on the choices.

PROCEDURE 17-1

Installing a Light Fixture Directly to an Outlet Box

- Put on safety glasses and observe all applicable safety rules.

- Using a voltage tester, verify that there is no electrical power at the lighting outlet where the fixture will be installed. If electrical power is present, turn off the power and lock out the circuit.

(A) Locate and identify the ungrounded, grounded, and grounding conductors in the lighting outlet box and connect them as explained below.

- The grounding conductor will not be connected to this fixture. If there is a grounding conductor in a nonmetallic box, simply coil it up and push it to the back or bottom of the electrical box. Do not cut it off, as it may be needed if another type of light fixture is installed at that location. If there are two or more grounding conductors in a nonmetallic box, connect them together with a wirenut and push them to the back or bottom of the lighting outlet box. If there is a grounding conductor in a metal outlet box, it must be connected to the outlet box by means of a listed grounding screw or clip. If there are two or more grounding conductors in a metal box, use a wirenut to connect them together along with a grounding pigtail. Attach the grounding pigtail to the metal box with a grounding screw identified for this purpose.

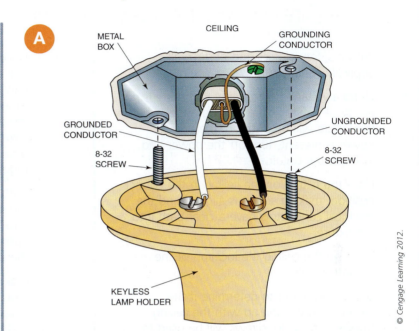

A

CEILING

METAL BOX

GROUNDING CONDUCTOR

GROUNDED CONDUCTOR

UNGROUNDED CONDUCTOR

8-32 SCREW

8-32 SCREW

KEYLESS LAMP HOLDER

© Cengage Learning 2012.

- Connect the white grounded conductor(s) to the silver screw or white fixture pigtail. If there is one grounded conductor in the box, strip approximately 3/4 inch (19 mm) of insulation from the end of the conductor and form a loop at the end of the conductor using an approved tool, such as a T-stripper. Once the loop is made, slide it around the silver terminal screw on the fixture so the end is pointing in a clockwise direction. Hold the conductor in place and tighten the screw. If there are two or more grounded conductors in the box, strip the ends as described previously and use a wirenut to connect them and a white pigtail together. Attach the pigtail to the silver grounded screw as described previously.

- Connect the black ungrounded conductor(s) to the brass-colored terminal screw or black pigtail on the fixture. The connection procedure is the same as for the grounded conductor(s).

- Now the fixture is ready to be attached to the lighting outlet box. Make sure the grounding conductors are positioned so they will not come in contact with the grounded or ungrounded screw terminals. Align the mounting holes on the fixture with the mounting holes on the lighting outlet box. Insert the 8-32 screws that are usually equipped with the fixture through the fixture holes. Then, thread them into the mounting holes in the outlet box.

- Tighten the screws until the fixture makes contact with the ceiling or wall. Be careful not to overtighten the screws, as you may damage the fixture.

- Install the proper lamp, remembering not to exceed the recommended wattage.

- Turn on the power and test the light fixture.

PROCEDURE 17-2

Installing a Cable-Connected Fluorescent Lighting Fixture Directly to the Ceiling

- Put on safety glasses and observe all applicable safety rules.

- Using a voltage tester, verify that there is no electrical power at the lighting outlet where the fixture will be installed. If electrical power is present, turn off the power and lock out the circuit.

- Place the fixture on the ceiling in the correct position, making sure it is aligned and the electrical conductors have a clear path into the fixture.

- Mark on the ceiling the location of the mounting holes.

- Use a stud finder to determine if the mounting holes line up with the ceiling trusses. If they do, screws will be used to mount the fixture. If they do not, toggle bolts will be necessary. For some installations, a combination of screws and toggle bolts will be required.

- If screws are used, drill holes into the ceiling using a drill bit that has a smaller diameter than the screws to be used. This will make installing the screws easier. If toggle bolts are to be used, use a flat-bladed screwdriver to punch a hole in the sheetrock only large enough for the toggle to fit through.

- Remove a knockout from the fixture where you wish the conductors to come through. Install a cable connector in the knockout hole.

- Place the fixture in its correct position and pull the cable through the connector and into the fixture. Tighten the cable connector to secure the cable to the fixture. This part of the process may require the assistance of a coworker. If using toggle bolts, put the bolt through the mounting hole and start the toggle on the end of the bolt.

- With a coworker holding the fixture, install the mounting screws or push the toggle through the hole until the wings spring open. This will hold the fixture in place until the fixture is secured to the ceiling.

- Make the necessary electrical connections. The grounding conductor should be properly wrapped around the fixture grounding screw and the screw tightened. The white grounded conductor is connected to the white conductor lead, then the black ungrounded conductor is connected to the black fixture conductor.

- Install the wiring cover by placing one side in the mounting clips, squeezing it, and then snapping the other side into its mounting clips.

- Install the recommended lamp(s). Usually, they have two contact pins on each end of the lamp. Align the pins vertically, slide them up into the lamp-holders at each end of the fixture, and rotate the lamp until it snaps into place.

- Test the fixture and lamps for proper operation.

- Install the fixture lens cover if there is one.

PROCEDURE 17-3

Installing a Lighting Fixture to an Outlet Box with a Strap

- Put on safety glasses and observe all applicable safety rules.

- Using a voltage tester, verify that there is no electrical power at the lighting outlet where the fixture will be installed. If electrical power is present, turn off the power and lock out the circuit.

- Before starting the installation process, read and understand the manufacturer's instructions.

A Mount the strap to the outlet box using the slots in the strap. With metal boxes, the screws are provided with the box. With nonmetallic boxes, you must provide your own 8-32 mounting screws. Put the 8-32 screws through the slot and thread them into the mounting holes on the outlet box. Tighten the screws to secure the strap to the box.

- Identify the proper threaded holes on the strap and install the fixture-mounting headless bolts in the holes so the end of the screw will point down.

- Make the necessary electrical connections. Make sure that all metal parts (including the outlet box), the strap, and the fixture are properly connected to the grounding conductor in the power feed cable.

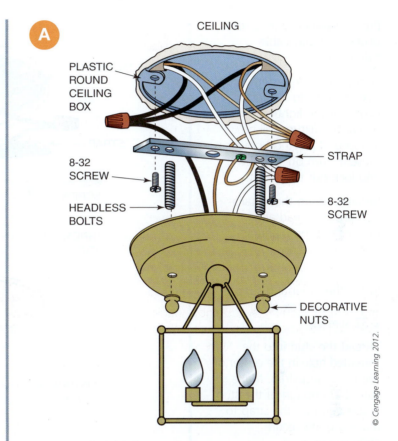

A CEILING

PLASTIC ROUND CEILING BOX

8-32 SCREW

HEADLESS BOLTS

STRAP

8-32 SCREW

DECORATIVE NUTS

© Cengage Learning 2012.

- Neatly fold the conductors into the outlet box. Align the headless bolts with the mounting holes on the fixture. Slide the fixture over the headless bolts until the screws stick out through the holes. Do not be alarmed if the mounting screws seem to be too long. Thread the provided decorative nuts onto the headless bolts. Keep turning the nuts until the fixture is secure to the ceiling or wall.

- Install the recommended lamp and test the fixture operation.

- Install any provided lens or globe. They are usually held in place by three screws that thread into the fixture. Start the screws into the threaded holes, position the lens or globe so it touches the fixture, and tighten the screws until the globe or lens is snug. Do not overtighten the screws. You may return the next day and find the globe or lens cracked or broken.

PROCEDURE 17-4

Installing a Lighting Fixture to a Lighting Outlet Box Using a Stud and Strap

- Put on safety glasses and observe all applicable safety rules.

- Using a voltage tester, verify that there is no electrical power at the lighting outlet where the fixture will be installed. If electrical power is present, turn off the power and lock out the circuit.

- Before starting the installation process, read and understand the manufacturer's instructions.

A Install the mounting strap to the outlet box using 8-32 screws.

- Thread the stud into the threaded hole in the center of the mounting strap. Make sure that enough of the stud is screwed into the strap to make a good secure connection.

- Measure the chandelier chain for the proper length, remove any unneeded links, and install one end to the light fixture.

- Thread the light fixture's chain-mounting bracket onto the stud. Remove the holding nut and slide it over the chain.

- Slide the canopy over the chain.

- Attach the free end of the chain to the chain-mounting bracket.

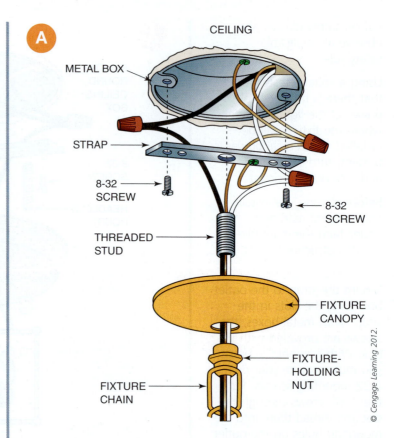

- Weave the fixture wires and the grounding conductor up through the chain links, being careful to keep the chain links straight. Section 410.56(F) of the *NEC*® states that the conductors must not bear the weight of the fixture. As long as the chain is straight and the conductors make all the bends, the chain will support the fixture properly.

- Now run the fixture wires up through the fixture stud and into the lighting outlet box.

- Make all necessary electrical connections.

- Slide the canopy up the chain until it is in the proper position. Slide the nut up the chain and thread it on to the chain-mounting bracket until the canopy is secure.

- Install the recommended lamp and test the fixture for proper operation.

PROCEDURE 17-5

Installing a Fluorescent Fixture (Troffer) in a Dropped Ceiling

- Put on safety glasses and observe all applicable safety rules.

- Before starting the installation process, read and understand the manufacturer's instructions.

- During the rough-in stage, mark the location of the fixtures on the ceiling.

- Using standard wiring methods, place lighting outlet boxes on the ceiling near the marked fixture locations and connect them to the lighting branch circuit.

- Once the dropped ceiling grid has been installed by the ceiling contractor, install the fluorescent light fixtures in the ceiling grid at the proper locations. Some electricians refer to this action as "laying in" the fixture. Once the fixture is installed, some electricians refer to the fixtures as being "laid in."

 Support the fixture according to *NEC®* requirements. Section 410.36(B) requires that all framing members used to support the ceiling grid be securely fastened to each other and to the building itself. The fixtures themselves must be securely fastened to the grid by an approved means, such as bolts, screws, rivets, or clips. This is to prevent the fixture from falling and injuring someone.

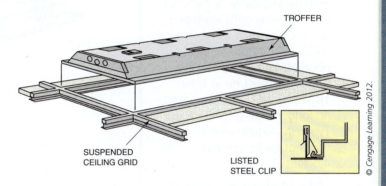

A **IMPORTANT:** TO PREVENT THE LUMINAIRE FROM INADVERTENTLY FALLING, *410.36(B)* OF THE *NEC ®* REQUIRES THAT (1) SUSPENDED CEILING FRAMING MEMBERS THAT SUPPORT RECESSED LUMINAIRES MUST BE SECURELY FASTENED TO EACH OTHER, AND MUST BE SECURELY ATTACHED TO THE BUILDING STRUCTURE AT APPROPRIATE INTERVALS, AND (2) RECESSED LUMINAIRES MUST BE SECURELY FASTENED TO THE SUSPENDED CEILING FRAMING MEMBERS BY BOLTS, SCREWS, RIVETS, OR SPECIAL LISTED CLIPS PROVIDED BY THE MANUFACTURER OF THE LUMINAIRE FOR THE PURPOSE OF ATTACHING THE LUMINAIRE TO THE FRAMING MEMBER.

TROFFER

SUSPENDED CEILING GRID

LISTED STEEL CLIP

© Cengage Learning 2012.

- Using a voltage tester, verify that there is no electrical power at the lighting outlet where the fixture will be installed. If electrical power is present, turn off the power and lock out the circuit.

PROCEDURE **17-5**

Installing a Fluorescent Fixture (Troffer) in a Dropped Ceiling (Continued)

B Connect the fixture to the electrical system. This is done by means of a "fixture whip." A fixture whip is often a length of Type NM, Type AC, or Type MC cable. It can also be a raceway with approved conductors, such as flexible metal conduit or electrical nonmetallic tubing. The fixture whip must be at least 18 inches (450 mm) long and no longer than 6 feet (1.8 m). Type NM fixture whips may not be longer than 4½ feet (1.4 m)

- Make all necessary electrical connections. The fixture whip should already be connected to the outlet box mounted in the ceiling. Using an approved connector, connect the cable or raceway to the fixture outlet box and run the conductors into the outlet box. Make sure that all metal parts are properly connected to the grounding system. Connect the white grounded conductors together and then the black ungrounded conductors together. Close the connection box.

- Install the recommended lamps and test the fixture for proper operation.

- Install the lens on the fixture.

B

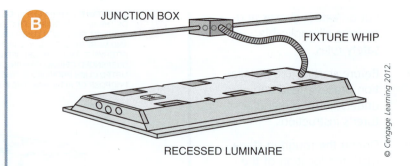

JUNCTION BOX

FIXTURE WHIP

RECESSED LUMINAIRE

© Cengage Learning 2012.

PROCEDURE **17-6**

Installing a Ceiling-Suspended Paddle Fan with an Extension Rod and a Light Kit

- Put on safety glasses and observe all applicable safety rules.

- Open the paddle fan carton and locate the instruction manual. Remove all parts from the carton and check them against the parts listed in the manual. Make certain that all parts have been removed.

- Using a voltage tester, verify that there is no electrical power at the outlet where the paddle fan is to be installed.

- Locate and identify the circuit conductors in the outlet box.

(A) Route the wires from the ceiling outlet box through the center hole of the paddle fan mounting bracket. Attach the bracket, with ground wire down, to the ceiling fixture outlet box with the special screws supplied with the listed paddle fan box. Tighten the screws firmly, being careful not to bend the bracket by overtightening.

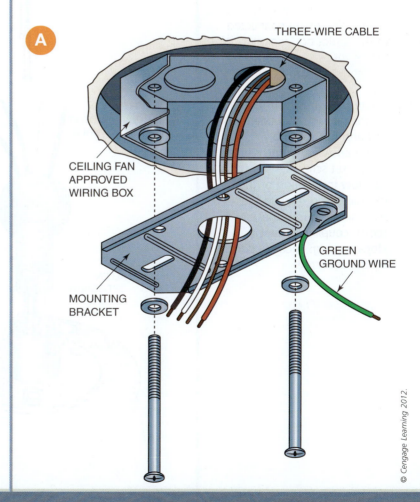

THREE-WIRE CABLE

CEILING FAN APPROVED WIRING BOX

GREEN GROUND WIRE

MOUNTING BRACKET

© Cengage Learning 2012.

Installing a Ceiling-Suspended Paddle Fan with an Extension Rod and a Light Kit (Continued)

B Attach the canopy to the mounting bracket. Tighten the screws firmly.

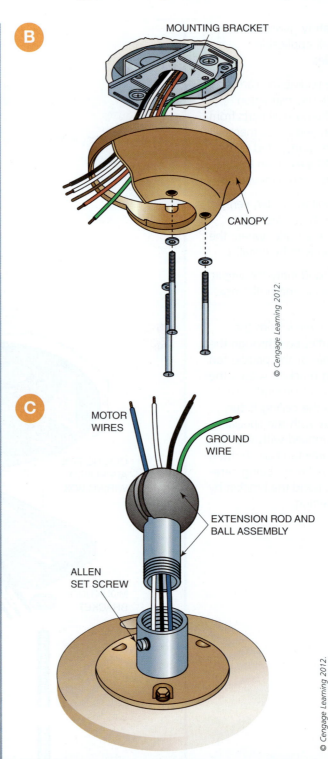

B MOUNTING BRACKET

CANOPY

© Cengage Learning 2012.

C Route the wires from the fan motor through the extension rod and ball assembly. Using an Allen wrench, loosen the set screw several turns to allow installation of the extension rod. Thread the extension rod into the motor coupling until it stops turning. Securely tighten the set screw.

C MOTOR WIRES

GROUND WIRE

EXTENSION ROD AND BALL ASSEMBLY

ALLEN SET SCREW

© Cengage Learning 2012.

D Hang the fan body in the canopy by holding the fan body firmly and insert the ball into the canopy opening. Check that no wires were pinched. Rotate the fan body until the slot in the nylon ball fits into the pin opposite the canopy opening.

• Make the electrical connections following the paddle fan manufacturer's instructions. Push the wiring back into the canopy and install the canopy hatch cover. The connections will vary depending on what kind of switching is desired. See Figure 13-26 in this textbook for the connections when there will be one switch controlling the fan and another switch controlling the lighting fixture.

E Attach the blades to the blade holders with three blade screws and flat washers. Tighten securely. Install the assembled blade and blade holder to the fan motor. Tighten securely. Repeat for each blade assembly.

F Remove the bottom of the switch housing and install the light kit according to the manufacturer's instructions.

• Install the proper size and type of lamp in the light kit.

• Turn on the power and test the fan and light fixture.

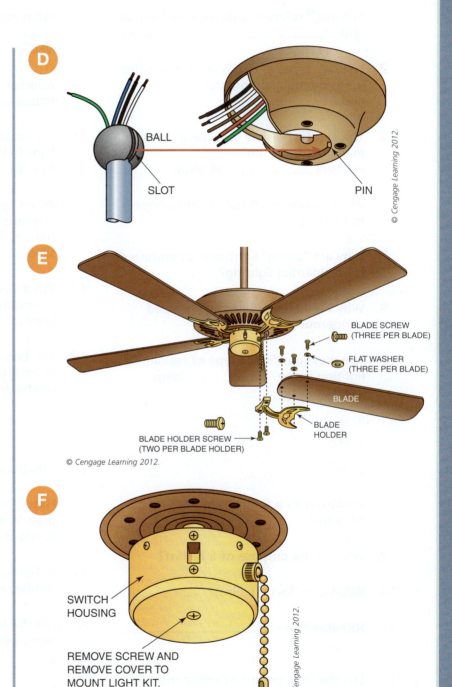

D

BALL

SLOT

PIN

© Cengage Learning 2012.

E

BLADE SCREW (THREE PER BLADE)

FLAT WASHER (THREE PER BLADE)

BLADE

BLADE HOLDER

BLADE HOLDER SCREW (TWO PER BLADE HOLDER)

© Cengage Learning 2012.

F

SWITCH HOUSING

REMOVE SCREW AND REMOVE COVER TO MOUNT LIGHT KIT.

© Cengage Learning 2012.

REVIEW QUESTIONS

Directions: Answer the following items with clear and complete responses.

1. The *NEC®* refers to a fluorescent lamp as a(n) _____ _____ lamp.

2. The two types of lamps used most often in residential lighting are _____ and _____.

3. The shortest visible wavelength in the magnetic spectrum appears _____ in color.

4. What causes a red sports car to appear red to you?

5. Why are "warm" light sources preferred for residential lighting?

6. Why are halogen lamps considered dangerous?

7. The most inefficient type of residential lighting is the _____ lamp.
   a. fluorescent
   b. incandescent
   c. LED
   d. HID

8. What causes a fluorescent lamp to give off light?

9. What is the purpose of a ballast?

10. What is a Class P ballast?

11. HID stands for _____ _____ _____.

12. List the four groups of residential lighting.
   1.
   2.
   3.
   4.

13. What is a Type Non-IC recessed light fixture, and where is it covered in the *NEC®*?

14. Another name for a wall-mounted light fixture is a _____ _____.

15. The requirements for dropped ceiling light fixture installations are covered in Article _____ of the *NEC®*.

16. When connecting the grounded conductor to an incandescent light fixture, it is always connected to the silver _____ _____.

17. What size screws are used to attach the mounting strap to a lighting outlet box? _____.

18. The two listings you should look for before installing a fixture in a dropped ceiling are:
   1.
   2.

19. What is a fixture whip?

20. Fixtures rated for damp locations can be used in wet locations. True or False. (Circle the correct response.)

21. A T-8 fluorescent lamp is _____ inch(es) in diameter.

22. List the four types of incandescent lamp bases presented in this chapter.
   1.
   2.
   3.
   4.

23. List three shapes that fluorescent lamps are available in.

1.

2.

3.

24. Define the term "luminaire."

25. A fluorescent lighting fixture that is "laid in" to the grid of a suspended ceiling is commonly called a _____ by many electricians.

26. What do the letters "LED" stand for?

27. Describe a track lighting system.

28. List the types of lighting fixtures that are *not* allowed in clothes closets.

Device Installation

OBJECTIVES

Upon completion of this chapter, the student should be able to:

- Demonstrate an understanding of the proper way to terminate circuit conductors to a switch or receptacle device.

- Select the proper receptacle for a specific residential application.

- Demonstrate an understanding of the proper installation techniques for receptacles.

- Select the proper switch type for a specific residential application.

- Demonstrate an understanding of the proper installation techniques for switches.

- Demonstrate an understanding of GFCI receptacle installation.

- Demonstrate an understanding of AFCI devices.

- Demonstrate an understanding of TVSS devices.

GLOSSARY OF TERMS

arc fault circuit interrupter (AFCI) a device intended to provide protection from the effects of arc faults by recognizing characteristics unique to arcing and by functioning to de-energize the circuit when an arc is detected

duplex receptacle the most common receptacle type used in residential wiring; it has two receptacles on one strap; each receptacle is capable of providing power to a cord-and-plug-connected electrical load

ground fault circuit interrupter (GFCI) receptacle a receptacle device that protects people from dangerous amounts of current; it provides protection at the receptacle outlet where it is installed and can also provide GFCI protection to other

"regular" receptacle devices on the same branch circuit; it is designed to trip when fault current levels are in the range of 4 to 6 milliamps or more

impulse a type of transient voltage that originates outside the home and is usually caused by utility company switching or lightning strikes

receptacle a device installed in an electrical box for the connection of an attachment plug

ring wave a type of transient voltage that originates inside the home and is usually caused by home office photocopiers, computer printers, the cycling on and off of heating, ventilating, and air-conditioning equipment, and spark igniters on gas appliances like furnaces and ranges

single receptacle a single contact device with no other contact device on the same strap (yoke)

split-wired receptacle a duplex receptacle wired so that the top outlet is "hot" all the time and the bottom outlet is switch-controlled

strap (yoke) the metal frame that a receptacle or switch is built around; it is also used to mount a switch or receptacle to a device box

transient voltage surge suppressor (TVSS) (receptacle or strip) an electrical device designed to protect sensitive electronic circuit boards from voltage surges

The trim-out stage of the residential wiring system installation is when the required receptacles and switches are installed by the electrician. In order to have the branch circuits installed during the rough-in stage work properly, the conductor connections to the various receptacles and switches on the circuits must be made properly. This means that all conductor terminations to the devices must be done in such a way that they function properly for many years of trouble-free service. It is imperative that the devices be installed in a professional, high-quality manner. At best, a device not installed properly simply does not work correctly. At worst, a device not installed properly could cause a fire. The fire could result in serious damage to the home or, worse yet, serious personal injury to the home's occupants. Good workmanship needs to be practiced at all times but is especially important when connecting devices to a branch circuit. This chapter presents the proper selection and installation procedures for the receptacle and switch types required on the branch circuits that make up a residential electrical system. It also presents a discussion on installing ground fault circuit interrupter receptacles and transient voltage surge suppressors.

SELECTING THE APPROPRIATE RECEPTACLE TYPE

Chapter 2 presented an overview of receptacle devices. It is appropriate at this time for you to review that information. During the rough-in stage, the electrical boxes and a wiring method with the required number of conductors are installed at each receptacle outlet location. Surface-mounted receptacles, like those sometimes used for an electric range or electric clothes dryer, will not need separate electrical boxes. During the trim-out stage, an electrician selects the proper

receptacle type for a specific location and installs it. The installation for regular flush-mount wall receptacles will consist of making the proper electrical connections to the terminal screws on the receptacle and then securely mounting the device to the electrical box. A plastic or metal cover is then attached to the receptacle, which finishes off the installation. A surface-mounted receptacle installation involves attaching the receptacle base to a wall or floor surface with screws, making the proper terminations to the terminal screws (or lugs), and securing the surface mount cover on the receptacle to complete the installation. A short review of receptacle devices is presented in the following paragraphs.

A receptacle is an electrical device that allows the home owner to access the electrical system with cord-and-plug-connected equipment, such as lamps, computers, kitchen appliances, and tools. The *National Electrical Code*® (*NEC*®) defines a receptacle as a contact device installed at an outlet for the connection of an attachment plug. A **single receptacle** is a single contact device with no other contact device on the same strap (yoke). A **duplex receptacle** has two contact devices on the same strap (yoke). A multiple receptacle has more than two contact devices on the same strap (yoke) (Figure 18-1). In residential applications, general-use receptacles are usually 125-volt, 15- or 20-amp devices and have specific slot configurations (Figure 18-2) and are found in regular and decorator styles (Figure 18-3). Standard colors available from the factory are brown, ivory, and white, but receptacles in other colors are available. For appliances that require just 240 volts, like a room air conditioner, receptacles are made with special blade configurations to meet the requirements of the appliance (Figure 18-4). For appliances that require 120/240 volts, such as electric

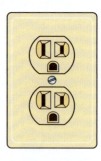
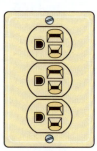

SINGLE RECEPTACLE

DUPLEX RECEPTACLE

MULTIPLE RECEPTACLE (TRIPLEX)

© Cengage Learning 2012.

FIGURE 18-1 Single, duplex, and triplex receptacle devices.

A 15-AMPERE, 125-VOLT, NEMA 5-15R RECEPTACLE, USED ON MAXIMUM 15-AMPERE BRANCH CIRCUITS. THE WIDE SLOT IS THE GROUNDED CONDUCTOR SLOT.

A 20-AMPERE, 125-VOLT, NEMA 5-20R RECEPTACLE, USED ON 15- OR 20-AMPERE BRANCH CIRCUITS. THE "T" SLOT IS THE GROUNDED CONDUCTOR SLOT.

© Cengage Learning 2012.

FIGURE 18-2   125-volt, 15- or 20-amp devices have specific slot configurations.

A 15-AMPERE, 250-VOLT, NEMA 6-15R RECEPTACLE, USED ON MAXIMUM 15-AMPERE BRANCH CIRCUITS. NOTE THE SLOT ARRANGEMENT SO STANDARD 125-VOLT ATTACHMENT PLUG CAPS WILL NOT FIT; COULD BE USED FOR A 240-VOLT WINDOW AIR CONDITIONER.

A 20-AMPERE, 250-VOLT, NEMA 6-20R RECEPTACLE, USED ON 15- OR 20-AMPERE BRANCH CIRCUITS. NOTE THE SLOT ARRANGEMENT SO STANDARD 125-VOLT ATTACHMENT PLUG CAPS WILL NOT FIT; COULD BE USED FOR A 240-VOLT WINDOW AIR CONDITIONER.

© Cengage Learning 2012.

FIGURE 18-4   250-volt rated receptacles with special blade configurations.

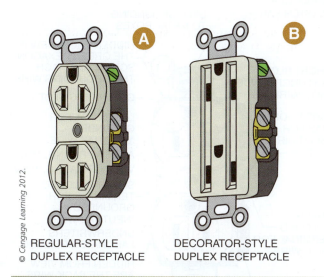

REGULAR-STYLE DUPLEX RECEPTACLE

DECORATOR-STYLE DUPLEX RECEPTACLE

© Cengage Learning 2012.

FIGURE 18-3   (A) Regular-style and (B) decorator-style duplex receptacles.

50-AMPERE
3-POLE, 3-WIRE
125/250-VOLT
NEMA 10-50R
PERMITTED
PRIOR TO 1996 *NEC*®

50-AMPERE
4-POLE, 4-WIRE
125/250-VOLT
NEMA 14-50R REQUIRED
SINCE THE 1996 *NEC*®
FOR NEW INSTALLATIONS

30-AMPERE
3-POLE, 3-WIRE
125/250-VOLT
NEMA 10-30R
PERMITTED
PRIOR TO 1996 *NEC*®

30-AMPERE
4-POLE, 4-WIRE
125/250-VOLT
NEMA 14-30R REQUIRED
SINCE THE 1996 *NEC*®
FOR NEW INSTALLATIONS

© Cengage Learning 2012.

FIGURE 18-5   125/250-volt rated range (50-amp) and dryer (30-amp) receptacles have special blade configurations to meet the requirements of the appliance.

clothes dryers and electric ranges, special blade configurations to meet the requirements of the appliance are also available (Figure 18-5).

General-use receptacles are made from high-strength plastic with a metal **strap (yoke)** running through it or behind it. The metal strap provides a means for the grounding blade of a male plug to be attached to the grounding system and for the receptacle to be attached to the device box. Section 406.4(A) of the *NEC®* requires receptacles installed on 15- and 20-amp branch circuits to be of the grounding type. Grounding type receptacles must be installed on branch circuits that have a voltage and current rating that matches the rating of the receptacle, except as permitted in Tables 210.21(B)(2) and (3). These tables apply to circuits with two or more receptacles. A duplex receptacle is considered to be two <u>single</u> receptacles so it would fall under these tables.

Table 210.21(B)(2) lists the maximum cord-and-plug connected loads to a receptacle.

- For a 15-amp rated receptacle the maximum load is 12 amps.
- For a 20-amp rated receptacle the maximum load is 16 amps.
- For a 30-amp rated receptacle the maximum load is 24 amps.

Table 210.21(B)(3) lists the maximum receptacle ratings for various sizes of circuits. Note: The circuit rating is based on the fuse or circuit breaker size protecting the circuit.

- For a circuit rating of 15 amps the receptacle rating is 15 amps.
- For a circuit rating of 20 amps the receptacle rating can be 15 or 20 amps.
- For a circuit rating of 30 amps the receptacle rating is 30 amps.
- For a circuit rating of 40 amps the receptacle rating can be 40 or 50 amps.
- For a circuit rating of 50 amps the receptacle rating is 50 amps.

For single receptacles on an individual branch circuit, Section 210.21(B)(1) states that the single receptacle must have a rating that is <u>not less</u> than the rating of the branch circuit. On a 20-amp individual branch circuit a 20-amp single receptacle must be used. A 15-amp-rated single receptacle may not be used on a 20-amp-rated individual branch circuit. On a 15-amp individual branch circuit a 15-amp-rated single receptacle or a 20-amp-rated single receptacle can be used.

Section 406.4(B) requires receptacles that have equipment-grounding conductor contacts to be connected to an equipment-grounding conductor. This is accomplished by attaching a branch-circuit grounding conductor to a green grounding screw, which is connected to the receptacle strap.

from experience...

When houses are roughed-in with metal device boxes, self-grounding receptacles are often used. The self-grounding feature (see Figure 18-6) bonds the receptacle yoke to the metal electrical box through the 6-32 screws used to secure the device to the box. All the electrician has to do in this situation is attach the branch-circuit equipment-grounding conductor to the metal device box with a green grounding screw. There is no need to bring out a grounding jumper to the green screw on the receptacle. Self-grounding receptacles cost more than "regular" receptacles, but money can be saved from the reduced time that it takes to install the self-grounding receptacle.

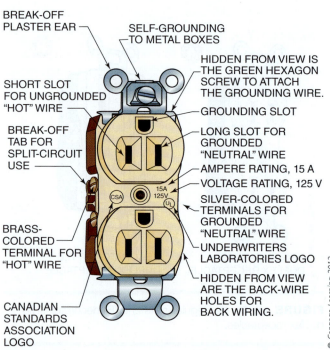

BREAK-OFF PLASTER EAR

SELF-GROUNDING TO METAL BOXES

SHORT SLOT FOR UNGROUNDED "HOT" WIRE

BREAK-OFF TAB FOR SPLIT-CIRCUIT USE

BRASS-COLORED TERMINAL FOR "HOT" WIRE

CANADIAN STANDARDS ASSOCIATION LOGO

HIDDEN FROM VIEW IS THE GREEN HEXAGON SCREW TO ATTACH THE GROUNDING WIRE.

GROUNDING SLOT

LONG SLOT FOR GROUNDED "NEUTRAL" WIRE

AMPERE RATING, 15 A

VOLTAGE RATING, 125 V

SILVER-COLORED TERMINALS FOR GROUNDED "NEUTRAL" WIRE

UNDERWRITERS LABORATORIES LOGO

HIDDEN FROM VIEW ARE THE BACK-WIRE HOLES FOR BACK WIRING.

© Cengage Learning 2012.

FIGURE 18-6 125-volt, 15- or 20-amp receptacles used in residential wiring come with a long slot for the grounded plug blade of the attachment plug, a short slot for the ungrounded plug blade of the attachment plug, and a U-shaped slot for the attachment plug grounding prong.

Receptacles rated 125 volts and 15 or 20 amperes come with a long slot for the grounded plug blade of the attachment plug, a short slot for the ungrounded plug blade of the attachment plug, and a U-shaped slot for the attachment plug grounding prong (Figure 18-6). There are terminal screws on each side of the receptacle, brass colored for the ungrounded conductors and silver colored for the grounded (neutral) conductors. In between the screw terminals of a 125-volt duplex receptacle is a removable tab used for **split-wired receptacles**. There is also a green grounding screw on each receptacle for grounding.

> ### *from experience...*
>
> Some wiring situations call for a split receptacle. This wiring technique will allow the bottom half of a duplex receptacle to be controlled by a switch, and the top half will be "hot" at all times. A load like a lamp can be plugged into the bottom half of the receptacle and be controlled by a single-pole switch. An electrician creates a split-wired receptacle by removing the connecting tab between the brass terminal screws on the "hot" side of a duplex receptacle (Figure 18-7).

Many receptacles also have holes in the back, which are used for "quick wiring" or "back wiring." Back wiring can be done in one of two ways: screw-connected or push-in. The screw-connected method is by far the most recommended. This connection feature is found on higher-quality (and higher-priced) receptacles. However, the time saved in making this type of connection can more than pay for the extra cost of the receptacle. In this method, the conductor is stripped back according to a strip gauge located on the receptacle and inserted into the proper hole. The terminal screw is then tightened, securing the conductor to the receptacle. In the push-in method, the wire is inserted into the proper hole (called a "push-in terminal") and held in place by a thin copper strip that pushes against the conductor (Figure 18-8). The drawback to this method is that as current flows through the copper strip, the strip can weaken, allowing the conductor to pull free. This can cause open circuits, short circuits, ground faults, or fires. The push-in method is UL listed for use with only 14 AWG solid copper conductors. Aluminum conductors, stranded conductors, or conductors sized 12 AWG and larger may not be used. The holes on the back of push-in receptacles are sized so that 12 AWG conductors will not fit easily into them.

Even though "back wiring" of receptacles is quick and easy, most electricians use terminal loops to attach the circuit conductors to the screws on a receptacle. Terminal loop connections are discussed in detail later in this chapter.

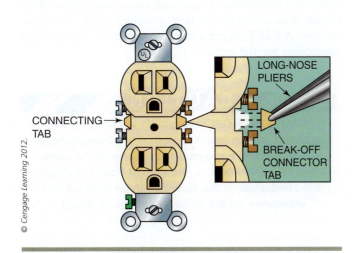

CONNECTING TAB

LONG-NOSE PLIERS

BREAK-OFF CONNECTOR TAB

© Cengage Learning 2012.

FIGURE 18-7 An electrician creates a split-wired receptacle by removing the connecting tab between the brass terminal screws on the "hot" side of a duplex receptacle.

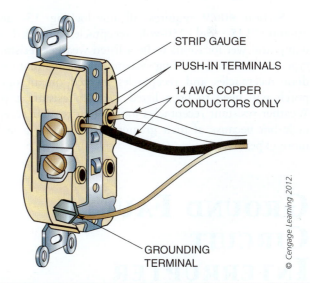

STRIP GAUGE

PUSH-IN TERMINALS

14 AWG COPPER CONDUCTORS ONLY

GROUNDING TERMINAL

© Cengage Learning 2012.

FIGURE 18-8 A duplex receptacle may be back wired by using the "push-in terminals" on the back of the device.

Section 406.12 requires all of the 15- and 20-amp, 125-volt receptacles required by Section 210.52, including GFCI receptacles, to be tamper-resistant. There are a few exceptions including: receptacles located more than 5½ feet (1.7 m) above the floor; receptacles that are part of a light fixture or appliance; a single receptacle for one appliance or duplex receptacle for two appliances located in dedicated space with appliances that are unlikely to be unplugged and moved. The most common tamper-resistant receptacle is one that has internal shutters that are closed behind the receptacle slots when there is no attachment plug inserted. The shutters open when a plug is inserted and allow the plug to become energized. It is connected to the branch-circuit wiring at a receptacle outlet box in exactly the same way as a regular receptacle. The purpose of a tamper-resistant receptacle is to prevent children from inserting foreign objects into the slots of receptacles and getting injured or even electrocuted. Tamper-resistant receptacles are a little more expensive than regular receptacles. Refer to Figure 2-36 in Chapter 2 for a look at a tamper-resistant receptacle.

Section 406.9 requires all non-locking 15- and 20-amp, 125- and 250-volt receptacles installed in damp and wet locations to be a listed weather-resistant type. For residential wiring systems this means that outdoor receptacles and receptacles located on an open porch will need to be a listed weather-resistant type. Weather-resistant receptacles look like a regular receptacle but because of their greater durability are a little more expensive.

GROUND FAULT CIRCUIT INTERRUPTER RECEPTACLES

Ground fault circuit interrupter (GFCI) receptacles (Figure 18-9) were developed to protect individuals from electric shock hazards. GFCI receptacles protect the locations where they are installed and can also protect regular duplex receptacles installed farther along the circuit. Ground-fault protection is also available with a GFCI circuit breaker. The breaker provides ground-fault protection for the entire circuit. GFCI breakers are covered in greater detail in Chapter 19. Shock hazards exist because of wear and tear on conductor insulation, defective materials, or misuse of equipment. This can result in contact with "hot" conductors or the metal parts of equipment that are in contact with "hot" conductors. Now would be a good time to review the shock hazards discussed in Chapter 1. Remember that the effect of the electric shock on the body is determined by the amount of current flowing through the body, the path the current takes through the body, and the length of time the current flows through the body.

A GFCI receptacle operates by monitoring the current flow on both the "hot" ungrounded circuit conductor and the grounded conductor. Whenever the GFCI receptacle senses a current on the "hot" ungrounded conductor that is 6 milliamps or greater than the current flowing on the grounded conductor, it trips the circuit off. This imbalance means that some of the current is taking a path other than the normal return path on the grounded conductor (Figure 18-10). Class A GFCI receptacle devices are designed to not trip when the current to ground is less than 4 milliamps according to UL 943, Standard for Ground Fault Circuit Interrupters.

A GFCI receptacle is designed to protect against ground faults only. It does not provide protection against short circuits, which occur when two "hot" ungrounded conductors touch each other or when a "hot" ungrounded conductor comes in contact with a grounded conductor. The circuit overcurrent protection device (fuse or circuit breaker) provides this type of protection. A GFCI receptacle will not provide overcurrent protection for the circuit.

GFCI protection is required in many areas, both inside and outside the house. These requirements are listed in Section 210.8 of the *NEC*® and were covered in earlier chapters. Remember that a GFCI protection device must be installed in a readily accessible location so do not install one behind something like a refrigerator. Now is a good time to review these requirements.

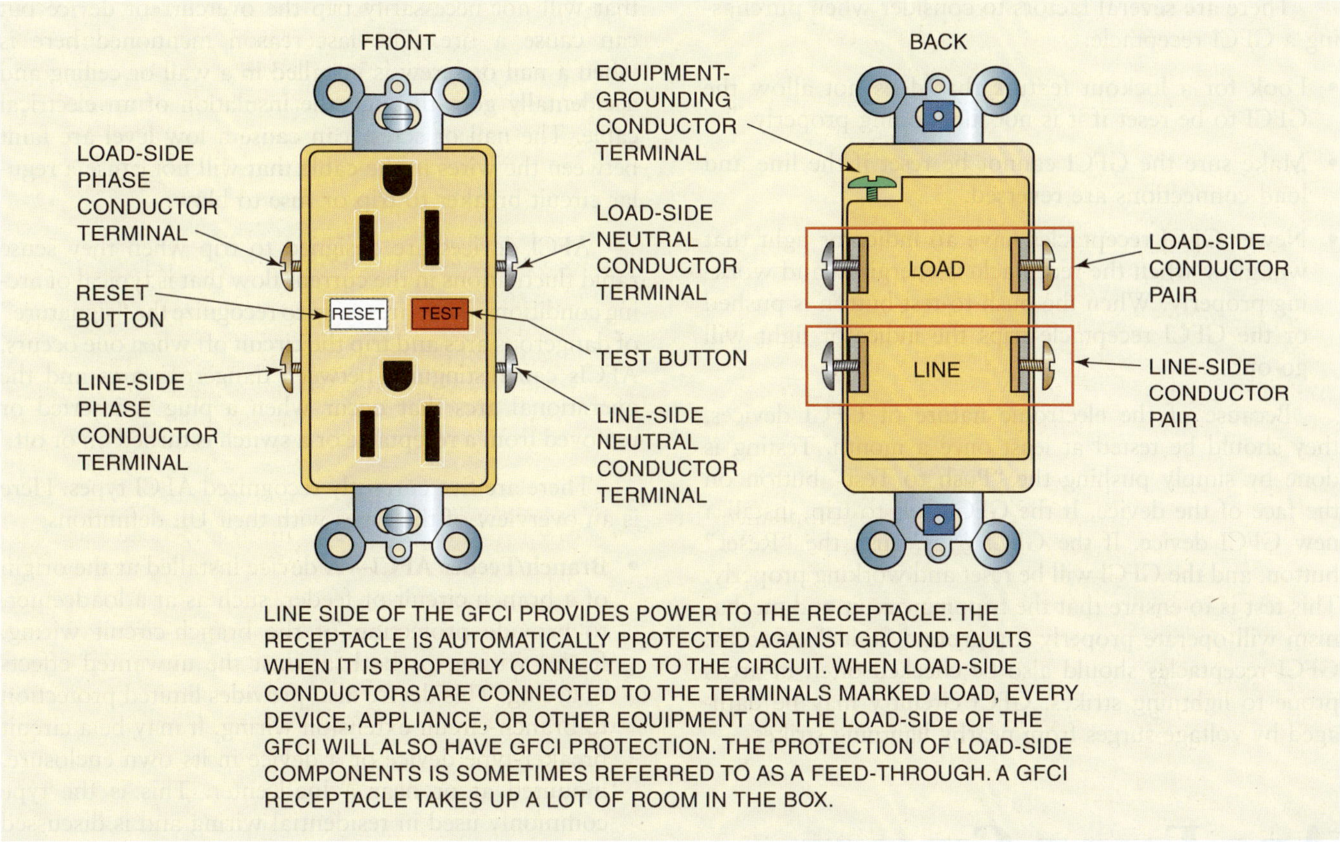

FRONT

LOAD-SIDE PHASE CONDUCTOR TERMINAL

RESET BUTTON

LINE-SIDE PHASE CONDUCTOR TERMINAL

RESET   TEST

EQUIPMENT-GROUNDING CONDUCTOR TERMINAL

LOAD-SIDE NEUTRAL CONDUCTOR TERMINAL

TEST BUTTON

LINE-SIDE NEUTRAL CONDUCTOR TERMINAL

BACK

LOAD

LINE

LOAD-SIDE CONDUCTOR PAIR

LINE-SIDE CONDUCTOR PAIR

LINE SIDE OF THE GFCI PROVIDES POWER TO THE RECEPTACLE. THE RECEPTACLE IS AUTOMATICALLY PROTECTED AGAINST GROUND FAULTS WHEN IT IS PROPERLY CONNECTED TO THE CIRCUIT. WHEN LOAD-SIDE CONDUCTORS ARE CONNECTED TO THE TERMINALS MARKED LOAD, EVERY DEVICE, APPLIANCE, OR OTHER EQUIPMENT ON THE LOAD-SIDE OF THE GFCI WILL ALSO HAVE GFCI PROTECTION. THE PROTECTION OF LOAD-SIDE COMPONENTS IS SOMETIMES REFERRED TO AS A FEED-THROUGH. A GFCI RECEPTACLE TAKES UP A LOT OF ROOM IN THE BOX.

© Cengage Learning 2012.

FIGURE 18-9   Ground fault circuit interrupter receptacles.

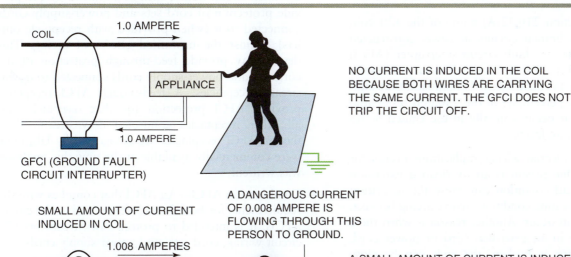

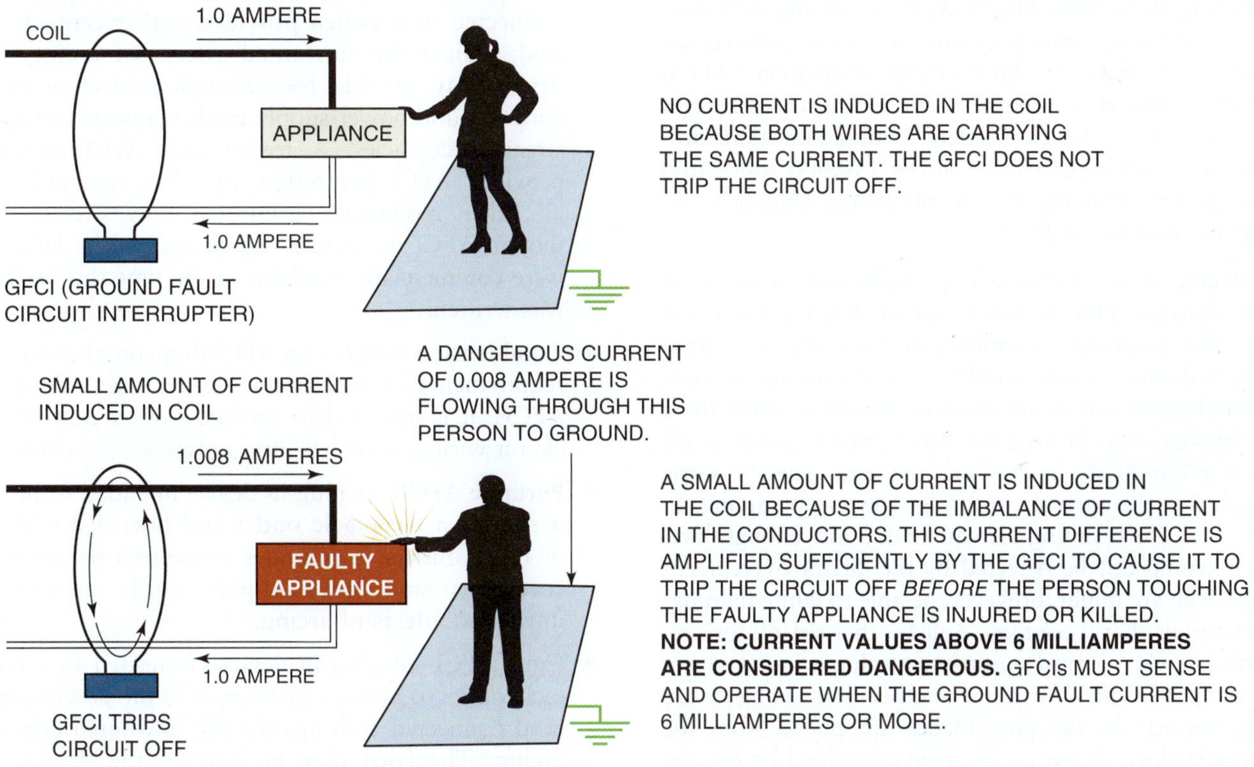

COIL

1.0 AMPERE →

APPLIANCE

← 1.0 AMPERE

GFCI (GROUND FAULT CIRCUIT INTERRUPTER)

NO CURRENT IS INDUCED IN THE COIL BECAUSE BOTH WIRES ARE CARRYING THE SAME CURRENT. THE GFCI DOES NOT TRIP THE CIRCUIT OFF.

SMALL AMOUNT OF CURRENT INDUCED IN COIL

1.008 AMPERES

FAULTY APPLIANCE

← 1.0 AMPERE

GFCI TRIPS CIRCUIT OFF

A DANGEROUS CURRENT OF 0.008 AMPERE IS FLOWING THROUGH THIS PERSON TO GROUND.

A SMALL AMOUNT OF CURRENT IS INDUCED IN THE COIL BECAUSE OF THE IMBALANCE OF CURRENT IN THE CONDUCTORS. THIS CURRENT DIFFERENCE IS AMPLIFIED SUFFICIENTLY BY THE GFCI TO CAUSE IT TO TRIP THE CIRCUIT OFF *BEFORE* THE PERSON TOUCHING THE FAULTY APPLIANCE IS INJURED OR KILLED. **NOTE: CURRENT VALUES ABOVE 6 MILLIAMPERES ARE CONSIDERED DANGEROUS.** GFCIs MUST SENSE AND OPERATE WHEN THE GROUND FAULT CURRENT IS 6 MILLIAMPERES OR MORE.

© Cengage Learning 2012.

FIGURE 18-10   GFCI operation.

There are several factors to consider when purchasing a GFCI receptacle:

- Look for a lockout feature that does not allow the GFCI to be reset if it is not functioning properly.

- Make sure the GFCI cannot be reset if the line and load connections are reversed.

- Newer GFCI receptacles have an indicator light that will glow when the receptacle is energized and working properly. When the push-to-test button is pushed or the GFCI receptacle trips the indicator light will go out.

Because of the electronic nature of GFCI devices, they should be tested at least once a month. Testing is done by simply pushing the "Push to Test" button on the face of the device. If the GFCI fails to trip, install a new GFCI device. If the GFCI trips, push the "Reset" button, and the GFCI will be reset and working properly. This test is to ensure that the tripping circuit and mechanism will operate properly if a ground fault does occur. GFCI receptacles should also be checked often in areas prone to lightning strikes. GFCI circuitry may be damaged by voltage surges from nearby lightning strikes.

ARC FAULT CIRCUIT INTERRUPTER DEVICES

According to Section 210.12(A) many of the 120 volt, 15- and 20-amp branch circuits in newly constructed dwellings must be arc fault circuit interrupter (AFCI) protected. This comes as the result of a series of studies by the Consumer Product Safety Commission that indicated that a high percentage of the 150,000 residential electrical fires that occur annually in the United States occur because of arc faults.

Arcing in residential wiring applications occurs for many reasons. One reason is an overloaded extension cord. The overload condition can cause the insulation to break down, creating conditions where arcing between the conductors can occur. Another reason is when there is a broken wire in an extension cord or power cord. When a conductor is broken, the ends can be close enough that the current can bridge the gap by arcing. In either case, the resulting arc can be at a low enough level that it would not be detected by the circuit overcurrent device. The arc might, however, be able to ignite flammable materials that it contacts. Still another reason for arcing occurs when a cord's attachment plug is knocked out of a receptacle when furniture or other similar objects are being moved. As the plug blades are pulled from the receptacle slots, arcing occurs. The more load on the circuit, the greater the arc. Again, this is a low-level arcing

that will not necessarily trip the overcurrent device but can cause a fire. The last reason mentioned here is when a nail or screw is installed in a wall or ceiling and accidentally goes through the insulation of an electrical cable. The nail or screw can cause a low level arc fault between the wires in the cable that will not cause a regular circuit breaker to trip or fuse to blow.

AFCI devices are designed to trip when they sense rapid fluctuations in the current flow that is typical of arcing conditions. They are set up to recognize the "signature" of dangerous arcs and trip the circuit off when one occurs. AFCIs can distinguish between dangerous arcs and the operational arcs that occur when a plug is inserted or removed from a receptacle or a switch is turned on or off.

There are five currently recognized AFCI types. Here is an overview of the types with their UL definitions:

- **Branch/Feeder AFCI**—A device installed at the origin of a branch circuit or feeder, such as at a loadcenter, to provide protection of the branch-circuit wiring, feeder wiring, or both against the unwanted effects of arcing. This device also provides limited protection to branch-circuit extension wiring. It may be a circuit breaker-type device or a device in its own enclosure, mounted at or near a loadcenter. This is the type commonly used in residential wiring and is discussed in more detail in Chapter 19.

- **Outlet Circuit AFCI**—A receptacle device installed at a branch-circuit outlet, such as an outlet box, to provide protection of cord sets and power-supply cords connected to it (when provided with receptacle outlets) against the unwanted effects of arcing. This device may provide feed-through protection of the cord sets and power-supply cords connected to downstream receptacles. A feed-though AFCI receptacle provides AFCI protection for that receptacle and all other receptacles connected downstream. Even though AFCI receptacles are recognized by UL, none were commercially available at the time this textbook was written.

- **Combination AFCI**—An AFCI that complies with the requirements for both branch feeder and outlet circuit AFCIs. It is intended to protect downstream branch-circuit wiring, cord sets, and power-supply cords.

- **Portable AFCI**—A plug-in device intended to be connected to a receptacle outlet and provided with one or more outlets. It provides protection to connected cord sets and power-supply cords against the unwanted effects of arcing.

- **Cord AFCI**—A plug-in device connected to a receptacle outlet to provide protection to the power-supply cord connected to it against the unwanted effects of arcing. The cord may be part of the device. The device has no additional outlets.

CAUTION

CAUTION: Section 210.12(A) in the *NEC®* requires all of the 15- and 20-amp, 120-volt branch circuits installed in dwelling unit family rooms, dining rooms, living rooms, parlors, libraries, dens, bedrooms, sun rooms, recreation rooms, closets, hallways, or similar rooms or areas to be AFCI protected. Be aware that with the types of AFCI devices that are commercially available at this time, all 15- and 20-amp, 120-volt branch circuits feeding the areas listed above will require combination AFCI circuit breakers.

TRANSIENT VOLTAGE SURGE SUPRESSORS

Many of the appliances used in homes today contain sensitive electronic circuitry, circuit boards, or microprocessors. These components are susceptible to transient voltages, also known as voltage surges or voltage spikes. Voltage spikes can destroy components or cause microprocessors to malfunction.

Line surges can be line to line, line to neutral, or line to ground. Transient voltages are grouped into two categories. The first, called an **impulse**, is when the transient voltage starts outside the residence. They are generally caused by power company equipment. The second category is the **ring wave**. This line surge begins inside the residence and is caused by inductive loads, such as the spark igniters on gas clothes dryers, gas ranges, or gas water heaters, electric motors, and computers.

Transient voltage surge suppressors (TVSS) are used to minimize voltage surges and spikes. A TVSS device works by clamping the high-surge voltage to a level that a piece of equipment like a computer can withstand (Figure 18-11). A TVSS reacts to a voltage surge in less than 1 nanosecond (one-billionth of a second) and permits only a safe amount of energy to enter the connected load.

TVSS devices are also known as Surge Protection Devices (SPDs). The *NEC®* covers these devices in Article 285 and refers to TVSS devices as Type 2 and Type 3 SPDs. The American Standards Institute (ANSI) and the Institute for Electrical and Electronic Engineers (IEEE) recognize three tiers of TVSS protection: Class A, Class B, and Class C (Figure 18-12). Class C protection is located at the service entrance main disconnect or meter socket; Class B serves sub-panels and power strips; and Class A provides protection at the receptacle outlets and is often referred to as point-of-use (POU) devices.

TVSS devices are available in several different styles that require an electrician to install them. One such style is a device that looks very much like a GFCI receptacle and mounts in a device box (Figure 18-13). This is a Class A POU type of device and would be installed by an electrician at an outlet designed to provide power to electronic equipment like a high-definition (HD) television. It installs like a regular receptacle but always be sure to follow the manufacturer's installation instructions when installing this device.

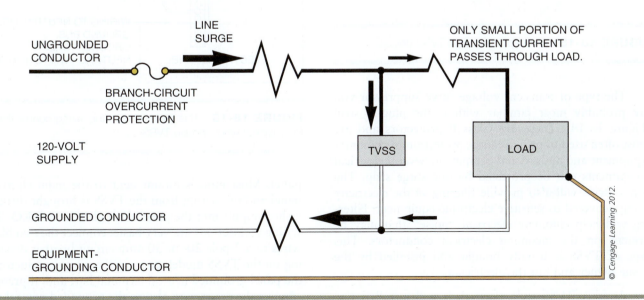

FIGURE 18-11 A TVSS device works by clamping the high-surge voltage to a level that a piece of equipment like a computer can withstand.

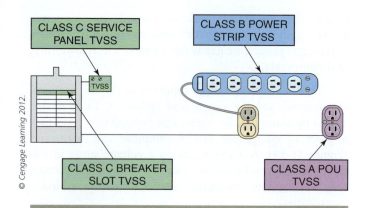

FIGURE 18-12 The American Standards Institute (ANSI) and the Institute for Electrical and Electronic Engineers (IEEE) recognize three tiers of TVSS protection: Class A, Class B, and Class C.

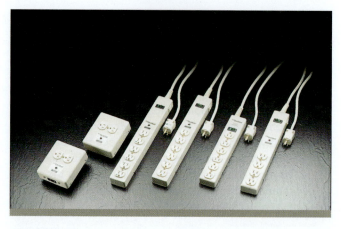

FIGURE 18-14 Plug-in TVSS strips. The strips allow multiple loads to be protected from voltage surges at the same time. *Courtesy of Hubbell, Incorporated.*

FIGURE 18-13 A receptacle model TVSS device.

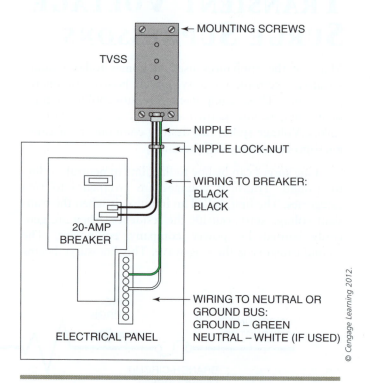

FIGURE 18-15 This figure shows the wiring connections for a typical whole house TVSS.

The type of transient voltage surge suppressor you are probably most familiar with is the plug-in strip (Figure 18-14). These are Class B protectors and are most often used to provide surge protection to computer equipment and audiovisual equipment. Several electrical components can be protected by one surge strip. The better strips will also provide filtering of the electricity being delivered to sensitive electronic equipment. Filtering will help eliminate electrical "noise" that may be present on the incoming electrical conductors. This type of TVSS is usually bought and installed by the home owner and not the electrician.

Another TVSS style electricians will install is one unit that protects the entire house from transient voltages. These units are considered Class C protectors and are hardwired directly into the main electrical

panel. Most models mount next to the main electrical panel and the wiring from the TVSS is brought through a short nipple into the panel through a ½-inch KO. The wiring connections are very basic: connect the two black wires to a 2-pole 20- or 30-amp circuit breaker depending on the TVSS model; connect the white conductor to the panel grounded neutral terminal bar; and, if present, connect the green wire also to the grounded neutral terminal bar (Figure 18-15). Not all whole house TVSS devices are installed the same so always be sure to follow the manufacturer's installation instructions when installing this device. This type of whole house protection can also be found in a circuit breaker style that fits into the house service panel and takes up two circuit breaker spaces.

INSTALLING RECEPTACLES

The amount of work required to install the receptacles in the trim-out phase is directly proportional to the amount of work that is put into the rough-in phase of the installation process. The more circuit testing and wire marking that occurs during the rough-in stage, the easier receptacle installation will be during the trim-out stage. Circuits should always be checked before the sheetrock or other wall covering is installed and finished. Circuit tracing and, if necessary, cable or raceway replacement is much easier during the rough-in stage.

Before the actual installation begins, check to make sure that no receptacle boxes are buried under the sheetrock. Drywall installers are very conscientious about exposing all the receptacle boxes, but on occasion they will miss one. Marking receptacle and switch locations on the floor and having an accurate circuit layout will make finding buried receptacle boxes much easier. Once a buried box location is determined, a keyhole saw is used to carefully cut the drywall to expose the device box.

PIGTAILS

The listing instructions for receptacles and switches normally allow only one wire to be terminated to each terminal screw. However, many electrical boxes containing receptacles (or switches) will have many circuit conductors requiring connections to the device terminal screws. The best way to make the necessary connections so that only one conductor gets connected to a terminal screw is to use a "pigtail." Figure 18-16 shows the correct method for connecting circuit conductors to a device with a pigtail splice. The usual method of making pigtail splices in an electrical box involves using a wirenut to connect the necessary wires together. Wirenuts were first introduced in Chapter 2. Other connector types were also discussed. Conductor splices made by following the proper installation procedures for wirenuts will result in a device installation that is safe and trouble free.

TERMINAL LOOPS

The preferred method for connecting circuit conductors to a receptacle (or switch) is to form terminal loops in the wire, put the loop under a terminal screw, and

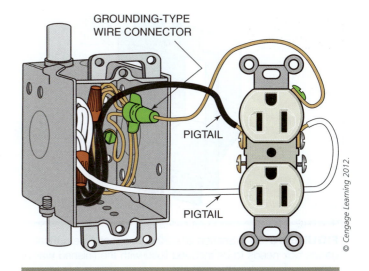

FIGURE 18-16 This connection shows the use of a white and a black pigtail. Notice the use of a green grounding wirenut used to provide a grounding pigtail that is terminated to the green grounding screw on the receptacle.

tighten the screw the proper amount. This sounds like a simple wiring practice, but a terminal screw that is not tightened properly or a wire that is not looped properly around a screw will typically be the cause of future problems. An electrician must practice good workmanship when connecting wires to any device so that the installation will be trouble free.

(See Procedure 18-1 on pages 656–657 for the correct procedure for using terminal loops to connect circuit conductors to terminal screws on a receptacle or switch.)

RECEPTACLE INSTALLATION

The first step in connecting a receptacle to the branch-circuit wiring is to locate the proper conductors and pull them out from the device box. Once all the circuit conductors have been properly identified, the conductors are ready to be connected to the receptacle. For regular duplex receptacles, there should be three wires: an ungrounded black conductor, a grounded white conductor, and a bare grounding conductor. Remember that the hot conductor is connected to the brass-colored terminal, the grounded conductor is connected to the silver screw terminal, and the grounding conductor is connected to the green grounding screw.

The next step is to take a close look at the position of the device box and determine if it is set back too far from the wall or ceiling surface. You may remember that according to Section 314.20, electrical device boxes need to be flush with the finished wall or ceiling

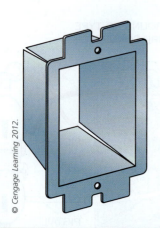

© Cengage Learning 2012.

FIGURE 18-17   Section 314.20 states that an electrical device box needs to be installed flush with the finished wall or ceiling surface when the material is combustible but is allowed to be set back no more than ¼ inch (6 mm) when the material is noncombustible. A box extender like the one illustrated here is used when an electrical device box has been installed too far behind the surface of a wall or ceiling.

surface when the material is combustible but are allowed to be set back no more than ¼ inch (6 mm) when the material is considered to be noncombustible. If the device box is set back more than what it should be, a box extender will need to be used (Figure 18-17). Box extenders are available in both metal and nonmetallic styles, but the nonmetallic box extender is used most often. A box extender:

- Levels and supports the wiring device when the device box is set back too far from the wall or ceiling surface

- Works in single-gang nonmetallic device boxes, 3-inch × 2-inch metal device boxes, handy boxes, and 4-inch square electrical boxes with plaster rings

- For multiple gang device boxes, trim off the sidewall for each nonmetallic extender where that sidewall would otherwise be between wiring devices

- Can typically be adjusted and used in boxes that were incorrectly installed up to 1½ inches behind the wall or ceiling surface

(See Procedure 18-2 on pages 658–660 for the correct procedure for installing a duplex receptacle in a nonmetallic electrical device box with one cable.)

(See Procedure 18-3 on pages 661–662 for the correct procedure for installing a duplex receptacle in a metal electrical device box with one cable.)

(See Procedure 18-4 on pages 663–665 for the correct procedure for installing a duplex receptacle in a nonmetallic electrical device box with two cables.)

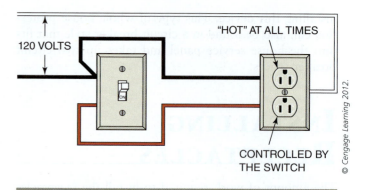

120 VOLTS   "HOT" AT ALL TIMES

CONTROLLED BY THE SWITCH

© Cengage Learning 2012.

FIGURE 18-18   A split-wired duplex receptacle has a switch controlling the bottom half of the receptacle. The other half is "hot" at all times.

A split-wired duplex receptacle usually has a switch controlling half of the receptacle, and the other half is "hot" all the time (Figure 18-18). This method of wiring is generally used when a lamp is providing the general room lighting instead of an overhead light fixture. Before installing this type of duplex receptacle, the tab connecting the terminal screws on the "hot" side of the receptacle must be removed. Do not remove the tab connecting the silver terminal screws. The connections for this wiring situation were presented in Chapter 13 (see Figure 13-21).

GFCI receptacles are connected to the electrical system in much the same manner as with regular duplex receptacles. However, on the back of GFCI receptacles, one of the brass screw terminals and one of the silver screw terminals are marked for the "Line," or incoming power conductors. The other set of screw terminals are marked as "Load" terminals, or the outgoing power conductors. On new GFCI receptacles, the "Load" terminals are covered with a yellow tape. If there is only one cable in the device box, the connection is quite simple. The grounding conductor is connected to the green grounding screw. The white grounded conductor is connected to the silver line terminal screw, and the black ungrounded conductor is connected to the brass line terminal screw. The conductors can either be wrapped around the screw terminals or back wired as previously discussed.

from experience...

Arc fault circuit interrupter (AFCI) receptacles were not commercially available at the time this third edition of *House Wiring* was in production, but are expected to be in the future. When they become available, it is expected that they will install the same way as a ground fault circuit interrupter (GFCI) receptacle.

If the GFCI is a feed-through receptacle, the connection will be slightly different. The grounding conductors need to be connected together with a grounding pigtail, and the pigtail will be connected to the green grounding screw. The conductors from the cable supplying power to the receptacle will be connected to the "Line" terminals, while the cable running to the receptacles downstream from the receptacle will be connected to the "Load" terminals. Remember, the "Load" terminals will have the yellow tape over them that will need to be removed. Now all the receptacles connected to the load side of the receptacle will be protected by the feed-through GFCI receptacle. If a ground fault occurs, the receptacle will trip off, stopping the current flow to the receptacle and all the other receptacles connected to it.

(See Procedure 18-5 on page 666 for the correct procedure for installing feed-through GFCI duplex receptacles in nonmetallic electrical device boxes.)

Once the receptacle you are installing is connected to the electrical system, it is time to secure it to the device box. Make sure the conductors inside the box are pushed to the back of the device box, leaving enough room to install the receptacle. Next, position the receptacle so the grounding slot is on the top (Figure 18-19). While this is not an *NEC*® requirement, it is the recommended positioning of a receptacle. This will lessen the chances of a piece of metal coming in contact with both the "hot" and grounded conductors, causing arcing and a short circuit. Carefully push the receptacle into the

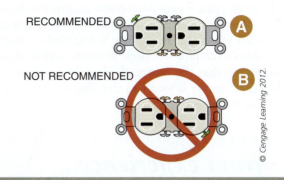

FIGURE 18-20 When a receptacle is installed in a horizontal position (A), the grounded slot should be on the top. Although the *NEC*® does not prohibit the grounded slot from being on the bottom (B), this practice is not recommended.

device box, checking that the ears on the top and bottom of the receptacle yoke will rest against the sheetrock when the receptacle is installed. Once you have determined that the receptacle ears will rest against the drywall properly, the receptacle can be attached to the device box using the provided 6-32 machine screws.

On occasion, a receptacle will need to be installed in the horizontal position. In this situation, the recommended receptacle position is to have the grounded conductor slot, the long slot, on the top (Figure 18-20). Again, this is to help keep metal objects away from the "hot" conductor plug blade.

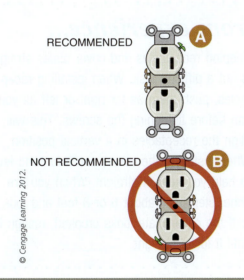

FIGURE 18-19 Position the receptacle so the grounding slot is on the top (A). Although the *NEC*® does not prohibit the grounding slot from being on the bottom (B), this practice is not recommended.

from experience...

Be careful not to strip out the threads in the device box mounting holes when installing the receptacles, especially when you are working with nonmetallic boxes. This creates many problems trying to get the receptacle to mount securely to the device box. When installing the receptacle, turn it approximately 30 degrees in both directions and push it into the box. Pull it back out, align the receptacle with the mounting holes, and then install the receptacle. This takes all the pressure off the threads while you tighten the screws. If you do encounter a stripped mounting hole, use your triple-tap tool (described in Chapter 3) to rethread the hole.

If the receptacle ears do not rest against the drywall, use spacers to properly position the receptacle. Spacers are needed when the device box is installed so that the front edge of the box sets back from the wall surface. It is important that the receptacle is positioned properly with respect to the drywall so that the receptacle cover plate will fit correctly.

from experience...

There are many common items used as spacers by electricians to place on the 6-32 screw between a wall box and a receptacle or switch yoke. For example, some electricians use a length of bare 14 AWG grounding conductor wrapped several times around the shank of a No. 2 Phillips screwdriver. The coils of wire can be cut off to provide the needed spacers. Many electricians use small flat washers, and some electricians use plastic tubing cut to the desired length. Still other electricians remove and save the mounting ears from receptacles and switches installed in previous installations that do not require the mounting ears.

Now the receptacle cover, or plate, can be installed. Receptacle covers come in two types (Figure 18-21). The most common is the regular receptacle cover. This cover comes in the same colors that receptacles do. It is important to purchase the covers from the same company as the receptacles to ensure a proper color match. The second cover has a rectangular hole in it to fit GFCI, TVSS, and decorator-style receptacles. In most cases, the cover is provided with the GFCI, and the TVSS receptacles. For the decorator-style receptacles and regular receptacles, the covers are purchased separately.

The cover is attached to a regular duplex receptacle with a single 6-32 screw in the center of the cover plate. For cover plates with rectangular openings, two 6-32 screws are needed. It is important to make sure the receptacle is flush with the wall and straight. This allows the cover to be straight and flush with the receptacle. The device and its cover are the part of the installation the home owner and everyone who enters the house will see. The quality of this part of the installation will determine, to a large extent, the impression

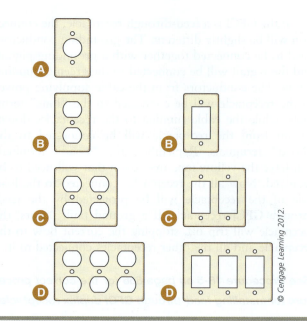

FIGURE 18-21   Receptacle wall covers come in two styles: regular and decorator. (A) shows a single receptacle cover; (B) shows a regular and decorator duplex receptacle cover; (C) shows a two-gang regular and decorator duplex receptacle cover; and (D) shows a three-gang regular and decorator duplex cover. Receptacle covers come in many colors but ivory and brown are the two most common.

that others will have concerning your work. Another nice touch is to align the cover screw slots to face in the same direction.

from experience...

Keeping receptacles and cover plates straight is not a difficult task. When installing receptacles, push them as far right or left as you can before tightening the screws. This will align the receptacles in a vertical position. When installing the cover, use a torpedo level to help you keep it straight. When you are done, step back about 6 or 8 feet and look at it. If the cover plate looks crooked, realign it until it looks straight.

When connecting a receptacle to a multiwire circuit, it is important that the conductors be pigtailed when there are two cables in the device box. Section 300.13

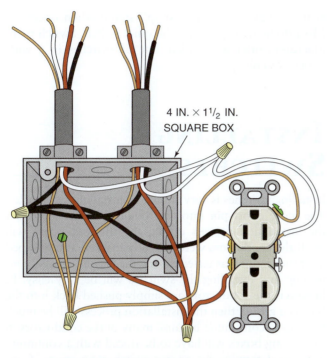

A PROPER WAY TO CONNECT GROUNDED NEUTRAL CONDUCTORS IN A MULTIWIRE BRANCH CIRCUIT

4 IN. × 1½ IN. SQUARE BOX

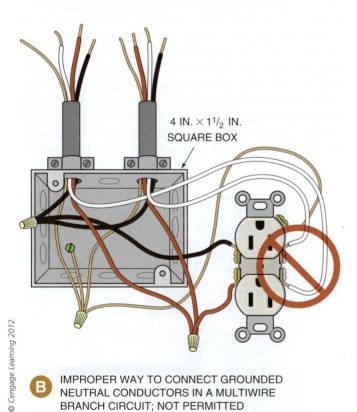

B IMPROPER WAY TO CONNECT GROUNDED NEUTRAL CONDUCTORS IN A MULTIWIRE BRANCH CIRCUIT; NOT PERMITTED

4 IN. × 1½ IN. SQUARE BOX

© Cengage Learning 2012.

FIGURE 18-22 Connecting the grounded neutral conductors to a split-wired receptacle on a multiwire branch circuit.

(B) of the *NEC*® requires that the continuity of the grounded conductor not depend on device connections to receptacles, lamp-holders, and so on. This means that you cannot connect one of the grounded conductors to one of the silver terminal screws and the second grounded conductor to the other silver terminal screw (Figure 18-22). If a problem developed with the receptacle, the "hot" conductor could be unaffected while the grounded conductor would be open. Any receptacle downstream would have power to the "hot" side of the receptacle but there would be no return path through the electrical system. This would create a hazard for anyone using any of those receptacles. This is one reason it is a good wiring practice to pigtail all conductors whenever possible. In this way, you ensure a complete circuit no matter what happens to one receptacle.

SELECTING THE APPROPRIATE SWITCH

The primary factor in selecting the appropriate switch is having the proper switch, or combination of switches, control the desired luminaire (light fixture) or receptacle. Chapter 2 presented a very detailed look at the different switch types used in residential wiring. Chapter 13 presented several different switching circuits and how the various switch types are connected into those circuits. It also showed how each switch actually works. A review of the information presented in Chapters 2 and 13 is appropriate at this time.

Single-pole, three-way, and four-way switches are the main switch types used in residential wiring. Single-pole switches are used to control one or more light fixtures or receptacles from a single location. Three-way switches are used in pairs to control one or more light fixtures or receptacles from two locations. Four-way switches are used with three-way switches to control one or more light fixtures from three or more locations. Both single-pole and three-way switches are available as dimmer switches. Double-pole switches are used to control 240-volt loads, such as electric water heaters. The last type of switch you may have to use in a residence is a low-voltage switch, such as the type needed for controlling a door chime. Residential switches are generally rated at 15 amps, 120/277 volts. Higher-ampere-rated switches are available. They can be wired to screw terminals or may be back wired. Remember that push-in back wiring is not a recommended method of installation.

Once the appropriate switch has been determined for a particular location on the wiring system, the color and style of the switch must be decided. Switch color is usually determined by the room color and varies throughout the house. Ivory and white are the two most popular colors installed in residences. Brown is the color that is preferred in rooms with darker-colored walls. The two styles of switches used most are the toggle switch and the rocker switch, also referred to as a decorator switch (Figure 18-23). Toggle switches are the standard switch used. Rocker switches present a cleaner line, and some people find them easier to operate than toggle switches. Toggle switches are available as snap switches or quiet switches. The designation refers to the sound the switch makes when it is operated. Snap

switches have a definite "snap" sound when operated. Most home owners prefer the quiet switch. Switch manufacturers offer a complete line of switches in a wide variety of colors.

INSTALLING SWITCHES

Installing switches is very similar to installing receptacles. The more effort put into the rough-in phase, the easier the installation process during the trim-out phase will be. If the conductors were properly marked and grouped after the circuit was tested during the rough-in stage, connecting and installing the switches will be very easy. If, however, the conductors are simply pushed back into the box as a group, then the installation process will be much more time-consuming because many of the conductors in multi-gang boxes will have to be traced with a continuity tester to determine the proper switch connections. Connecting the conductors to the switch is done basically in the same manner as connecting conductors to receptacles. The insulation is stripped from the conductor, and a loop is made in the end of the conductor. The loop is placed around the proper terminal in a clockwise manner and the terminal screw tightened (Figure 18-24). If the switch is back wired, the insulation is stripped from the conductor, the end of the conductor is inserted into the proper hole, and the terminal screw is tightened (Figure 18-25). Push-in terminations (Figure 18-26) are found on many devices. As we discussed earlier in this chapter, only 14 AWG copper wire is suitable for these terminations

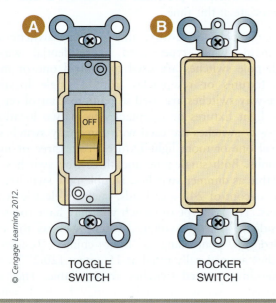

FIGURE 18-23 The two styles of switches used in residential wiring are the (A) toggle switch and the (B) rocker switch.

FIGURE 18-24 Terminal loop connections to the screws are a reliable way to terminate the circuit conductors to a switch or receptacle device. *Courtesy of Hubbell Wiring Systems.*

FIGURE 18-25 Back-wired connections are another reliable way to terminate the circuit conductors to a switch or receptacle device. This type of termination can save time during the installation of switches and receptacles because a terminal loop is not required to be made. However, devices with this type of termination are more expensive. *Courtesy of Hubbell Wiring Systems.*

FIGURE 18-26 Push-in terminations are available on most wiring devices, but they do not provide as good a termination as a terminal loop or back-wired termination. Always check with your supervisor before using push-in terminations on switches or receptacles. *Courtesy of Hubbell Wiring Systems.*

and most electrical contractors do not want their employees to use the push-in terminals because of the high frequency of problems that exist when using them. Always check with your supervisor before using push-in terminations on devices.

For single-pole switches, two conductors (usually black) will be connected to the switch. One is the incoming power wire, and the other wire runs to the light fixture or receptacle. Be sure to ground the switch as required by Section 404.9(B) (see Figure 13-9). Also, be sure the switch is set up so it will read "OFF" when the toggle is in the down position. If the two conductors to be connected are black and white, you will be using a connection called a "switch loop" (see Figure 13-10). On this connection, the white wire is used as an ungrounded conductor and must be identified as such. Section 200.7(C) requires that the white conductor be identified at both ends as a "hot" conductor. Most electricians use black electrical tape to mark the conductor.

(See Procedure 18-6 on pages 667–668 for the correct procedure for installing single-pole switches in a single-gang nonmetallic box.)

(See Procedure 18-7 on page 669 for the correct procedure for installing single-pole switches in a multi-gang nonmetallic box.)

For three-way switches, three conductors will be connected to the switch. The two traveler conductors are connected to the two brass-colored terminals, and the conductor that provides power or the conductor that takes power to the light fixture is connected to the third, dark-colored terminal. If the conductors are not marked, looking into the device box can be an easy way to identify the conductors. If the conductors come from two different cables, the black and red wires in the three-wire cable are the travelers. The black conductor in the two-wire cable will be connected to the common terminal. If all three conductors come from the same cable, a continuity tester is used to identify the conductors. Section 200.7(C) requires the white conductor to be reidentified if it is used as an ungrounded conductor. Remember to ground the switch (see Figures 13-13, 13-14, and 13-15).

Four-way switches have four conductors connected to them. Two conductors will come from one three-wire cable, and two conductors will come from a second three-wire cable. The conductors that come from one cable will be connected to the two screw terminals that are the same color. The remaining two conductors will be connected to the two screw terminals that are a different color. Remember to properly ground the switch (see Figures 13-17 and 13-18).

Once the switches are connected to the electrical system, they are ready to install in the device box. Again, this procedure is very similar to the receptacle installation procedure. The main difference is the number of switches that are installed in multi-gang boxes. In residential installations, two- and three-gang switch boxes are common.

In these installations, particular care must be taken to ensure there is enough room in the device box for all the conductors and switches. Installing a switch in the device box is easily done by tilting the switch and pushing it into the device box, pulling it back out and aligning the mounting holes, and then installing the 6-32 mounting screws and tightening them. Make sure the ears on the yoke of the switch are mounted flush with the drywall. Use the spacer method described in the receptacle installation section of this chapter to obtain the desired result. It may be necessary to use box extenders.

Installing faceplates on switches is done in just about the same manner as installing cover plates on receptacles. However, instead of one screw, switch plates (Figure 18-27) require two screws to attach them to a switch. Because there are so many multi-gang switch boxes, it is important to make sure all the switches are level so the faceplate will be level when it is installed. The mounting holes are large enough to allow enough movement to keep the switches level. Using the short, colored 6-32 screws that come with a faceplate,

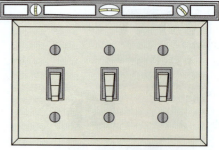

ALIGN SCREW SLOTS
LIKE THIS,

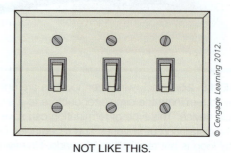

NOT LIKE THIS.

FIGURE 18-28 When installing faceplates, always align the screw slots in one direction, usually vertical, to provide a more finished look.

attach the faceplate to the switches and the wall surface. Do not overtighten the screws on a plastic faceplate, or you may crack it and have to replace it. Align the screw slots in one direction, usually vertical, to provide a more finished look (Figure 18-28).

If metal faceplates are installed, be sure to provide a means to ground them to the grounding system, as required in Section 404.9(B). The metal screws provided with the switch cover are an acceptable means of grounding these faceplates. If installing decorator switch covers, make sure the provided screws are metal and not nylon. Many manufacturers of decorator switch covers provide nylon screws to keep from damaging the faceplate finish. These screws do not meet the requirements of Section 404.9(B). You will need to install metal screws to properly ground these faceplates.

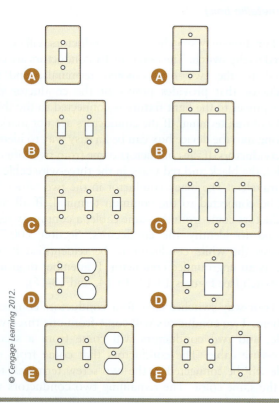

FIGURE 18-27 Switch covers are available in metal or nonmetallic materials. They come in a variety of configurations and colors. (A) shows a single-gang regular and decorator switch cover; (B) shows a two-gang regular and decorator switch cover; (C) shows a three-gang regular and decorator switch cover; (D) shows a two-gang switch/receptacle cover in both regular and decorator styles; (E) shows a three-gang switch/receptacle cover in both regular and decorator styles.

from experience...

When installing switches in multi-gang device boxes, a good practice is to push all the switches either all the way to the left or all the way to the right, then tighten the mounting screws. This will space the switches so that the faceplate will fit the first time.

SUMMARY

In this chapter, we have looked at how to install the many different types of switches and receptacles used in residential wiring. Grounding-type duplex receptacles are the primary means by which a home owner accesses the home electrical system. GFCI receptacles are specialized devices that help prevent people from receiving electrical shock. The *NEC*® requires that GFCIs be installed in specific locations, such as kitchens, bathrooms, garages, and unfinished basements. AFCI devices are installed to control dangerous arc faults that can cause fires. TVSSs are devices designed to protect sensitive electronic circuits found in many of today's appliances and audiovisual equipment from voltage surges.

Switches are devices used to control light fixtures from one, two, or more locations. Switches can also control all or part of receptacles. Dimming switches are used on incandescent and special fluorescent light fixtures to control the light output. Dimmers can be used to control light fixtures from one location or in circuits that control light fixtures from two or more locations.

When installing receptacles or switches, it is important that the conductors be securely attached to the device. If the side screw terminals are used, be sure the conductor loop is wrapped around the screw terminal in a clockwise direction and then properly tightened. If the device is back wired, make sure the conductor is inserted into the proper hole and then secured to the device. All devices installed in new construction are required by the *NEC*® to be grounded to the house grounding system.

The installation of devices, either receptacles or switches, and their cover plates are a very important part of the electrical wiring process. This part of the installation is the most visible part of the electrical system, and many people form their impression of the quality of your work by how well the switches, receptacles, and their covers look. Switches and receptacles installed so they are straight and flush with the wall surface, along with covers that are level and flush with the wall surface, demonstrate your commitment to quality workmanship.

PROCEDURE 18-1

Using Terminal Loops to Connect Circuit Conductors to Terminal Screws on a Receptacle or Switch

- Wear safety glasses and observe all applicable safety rules.

- Using a wire stripper, remove approximately ¾ inch of insulation from the end of the wire.

A Using the wire strippers (or possibly long-nosed pliers), make a terminal loop at the end of the wire.

A

© Cengage Learning 2012.

B Place the loop around the terminal screw so that the loop is going in the clockwise direction. This will cause the loop to close around the screw post. If the loop is put on so it is going in the counterclockwise direction, the loop will open up and actually become loose as the terminal screw is turned clockwise to tighten it.

B

© Cengage Learning 2012.

Using a screwdriver, tighten the terminal screw until it is snug. Then, tighten it approximately one-quarter turn more. Note: Electricians should use a torque screwdriver, set to the proper torque for the screw you are tightening, to tighten the terminal screws the proper amount.

© Cengage Learning 2012.

PROCEDURE 18-2

Installing a Duplex Receptacle in a Nonmetallic Electrical Device Box with One Cable

- Wear safety glasses and observe all applicable safety rules.

A Using a wire stripper, remove approximately ¾ inch of insulation from the end of the insulated wires.

B Using the wire strippers (or possibly long-nosed pliers), make a terminal loop at the end of each of the wires.

C Place the terminal loop on the end of the bare grounding wire around the green terminal screw of the receptacle so that the loop is going in the clockwise direction. While pulling the loop snug around the screw terminal, tighten the screw to the proper amount with a screwdriver.

- Place the terminal loop on the black wire around a receptacle brass terminal screw so that the loop is going in the clockwise direction. While pulling the loop snug around the screw terminal, tighten the screw to the proper amount with a screwdriver.

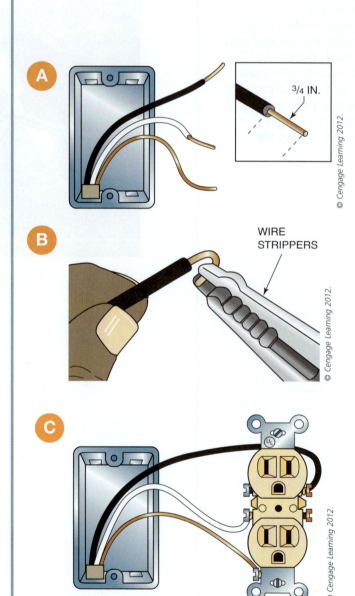

A

¾ IN.

© Cengage Learning 2012.

B

WIRE STRIPPERS

© Cengage Learning 2012.

C

© Cengage Learning 2012.

- Place the terminal loop on the white wire around a receptacle silver terminal screw so that the loop is going in the clockwise direction. While pulling the loop snug around the screw terminal, tighten the screw to the proper amount with a screwdriver.

D Place the receptacle into the outlet box by carefully folding the conductors back into the device box.

D

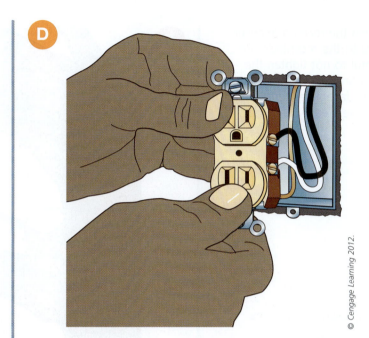

© Cengage Learning 2012.

E Secure the receptacle to the device box using the 6-32 screws. Mount the receptacle so it is vertically aligned. Use a box extender or spacers to align the receptacle with the wall surface if the box is set back too far.

E

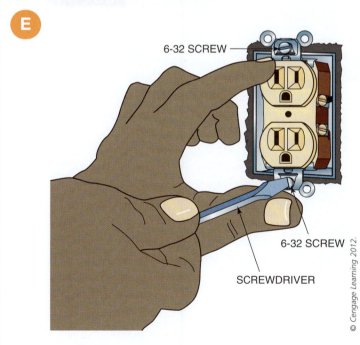

6-32 SCREW

6-32 SCREW

SCREWDRIVER

© Cengage Learning 2012.

PROCEDURE 18-2

Installing a Duplex Receptacle in a Nonmetallic Electrical Device Box with One Cable (Continued)

F Attach the receptacle cover plate to the receptacle. Be careful to not tighten the mounting screw(s) too much. Plastic faceplates tend to crack very easily.

F

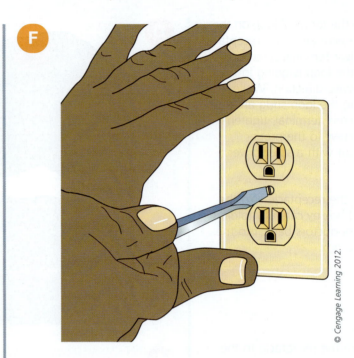

© Cengage Learning 2012.

PROCEDURE 18-3

Installing a Duplex Receptacle in a Metal Electrical Device Box with One Cable

- Wear safety glasses and observe all applicable safety rules.

A Attach an 8-inch-long equipment bonding jumper (pigtail) to the metal electrical outlet box with a 10-32 green grounding screw. The equipment bonding jumper can be a bare or green insulated conductor and will be the same size as the conductors used to wire the receptacle.

Note: There is an optional way to connect the equipment-grounding conductor (EGC) to a metal box and receptacle when using Type NM cable. The first thing is to strip the cable so the bare EGC is at least 8 inches long in the box. Then, using your fingers or long-nosed pliers, grab the EGC as close to where the cable enters the box and wrap it around the green grounding screw in the metal box in a clockwise direction. Taking the long end of the EGC, put a terminal loop in it, and connect it to the green grounding screw on the receptacle. This method saves a little time and money.

B Attach another 8-inch-long equipment bonding jumper (pigtail) to the green screw on the receptacle.

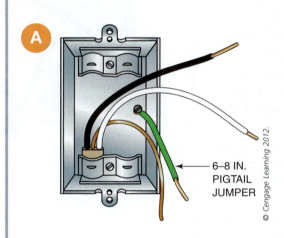

A

6–8 IN.
PIGTAIL
JUMPER

© Cengage Learning 2012.

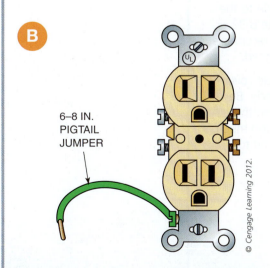

B

6–8 IN.
PIGTAIL
JUMPER

© Cengage Learning 2012.

PROCEDURE 18-3

Installing a Duplex Receptacle in a Metal Electrical Device Box with One Cable (Continued)

C Using a wirenut, connect the branch-circuit grounding conductor(s), the equipment bonding jumper attached to the box, and the equipment bonding jumper attached to the receptacle together.

- Using a wire stripper, remove approximately 3/4 inch of insulation from the end of the insulated wires.

- Using wire strippers (or possibly long-nosed pliers), make a terminal loop at the end of each of the wires.

- Place the terminal loop on the black wire around a brass terminal screw and the terminal loop on the white wire around a silver terminal screw so that the loops are going in the clockwise direction. Tighten the screws to the proper amount with a screwdriver.

- Place the receptacle into the outlet box by carefully folding the conductors back into the device box.

- Secure the receptacle to the device box using the 6-32 screws. Mount the receptacle so it is vertically aligned.

- Attach the receptacle cover plate to the receptacle. Be careful to not tighten the mounting screw(s) too much. Plastic faceplates tend to crack very easily.

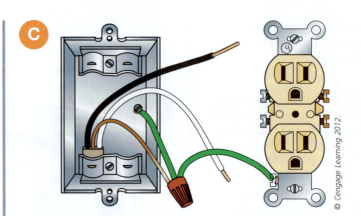

© Cengage Learning 2012.

PROCEDURE 18-4

Installing a Duplex Receptacle in a Nonmetallic Electrical Device Box with Two Cables

- Wear safety glasses and observe all applicable safety rules.

- Using a wire stripper, remove approximately ¾ inch of insulation from the end of the insulated wires.

- Using the wire strippers (or possibly long-nosed pliers), make a terminal loop at the end of each of the black and white wires.

A Using at least an 8 inch length of bare copper wire as an equipment bonding jumper wirenut one end together with the two other equipment-grounding conductors in the box.

- Using the wire strippers (or possibly long-nosed pliers), make a terminal loop in the end of the equipment bonding jumper.

A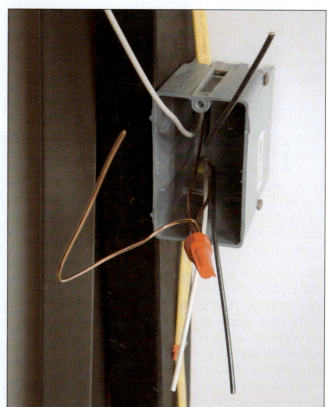

© Cengage Learning 2012.

PROCEDURE 18-4

Installing a Duplex Receptacle in a Nonmetallic Electrical Device Box with Two Cables (Continued)

 B Place this terminal loop around the green terminal screw of the receptacle so that the terminal loop is going in the clockwise direction. While pulling the loop snug around the screw terminal, tighten the screw to the proper amount with a screwdriver.

• Place the terminal loop of one of the black wires around a brass terminal screw so that the terminal loop is going in the clockwise direction. While pulling the loop snug around the screw terminal, tighten the screw to the proper amount with a screwdriver.

 C Place the terminal loop of the other black wire around a brass terminal screw so that the terminal loop is going in the clockwise direction. Tighten the screw to the proper amount with a screwdriver.

• Place the terminal loop of one of the white wires around a silver terminal screw so that the terminal loop is going in the clockwise direction. While pulling the loop snug around the screw terminal, tighten the screw to the proper amount with a screwdriver.

B

© Cengage Learning 2012.

C

© Cengage Learning 2012.

D Place the terminal loop of the other white wire around a silver terminal screw so that the terminal loop is going in the clockwise direction. While pulling the loop snug around the screw terminal, tighten the screw to the proper amount with a screwdriver.

- Place the receptacle into the outlet box by carefully folding the conductors back into the device box.

- Secure the receptacle to the device box using the 6-32 screws. Mount the receptacle so it is vertically aligned. Use a box extender or spacers to align the receptacle with the wall surface if the box is set back too far.

- Attach the receptacle cover plate to the receptacle. Be careful to not tighten the mounting screw(s) too much. Plastic faceplates tend to crack very easily.

D

© Cengage Learning 2012.

PROCEDURE 18-5

Installing Feed-Through GFCI Duplex Receptacles in Nonmetallic Electrical Device Boxes

- Wear safety glasses and observe all applicable safety rules.

A At the electrical outlet box containing the GFCI feed-through receptacle, use a wirenut to connect the branch-circuit grounding conductors and a grounding pigtail together. Connect the grounding pigtail to the receptacle's green grounding screw.

- At the electrical outlet box containing the GFCI feed-through receptacle, identify the incoming power conductors and connect the white grounded wire to the line-side silver screw and the incoming black ungrounded wire to the line-side brass screw.

- At the electrical outlet box containing the GFCI feed-through receptacle, identify the outgoing conductors and connect the white grounded wire to the load-side silver screw and the outgoing black ungrounded wire to the load-side brass screw.

- Secure the GFCI receptacle to the electrical box with the 6-32 screws provided by the manufacturer.

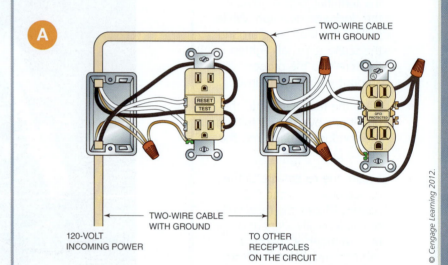

A

TWO-WIRE CABLE WITH GROUND

RESET
TEST

GFCI PROTECTED

TWO-WIRE CABLE WITH GROUND

120-VOLT INCOMING POWER

TO OTHER RECEPTACLES ON THE CIRCUIT

© Cengage Learning 2012.

- A proper GFCI cover is provided by the device manufacturer; attach it to the receptacle with the short 6-32 screws also provided.

- At the next "downstream" electrical outlet box containing a regular duplex receptacle, connect the white grounded wire(s) to the silver screw(s) and the black ungrounded wire(s) to the brass screw(s) in the usual way. Place a label on the receptacle that states "GFCI Protected." These labels are provided by the manufacturer.

- Continue to connect and label any other "downstream" duplex receptacles as outlined in the previous step.

PROCEDURE 18-6

Installing Single-Pole Switches in a Single-Gang Nonmetallic Box

- Wear safety glasses and observe all applicable safety rules.

- Using a wire stripper, remove approximately ¾ inch of insulation from the end of the insulated wires.

(A) Assuming that one cable is entering the box and it is a switch loop situation, use wire strippers (or possibly long-nosed pliers) and make a terminal loop at the end of each of the wires.

- Place the terminal loop on the bare grounding wire around the green terminal screw so that the terminal loop is going in the clockwise direction. While pulling the loop snug around the screw terminal, tighten the screw to the proper amount with a screwdriver.

- Place the terminal loop on the black wire around a terminal screw so that the terminal loop is going in the clockwise direction. While pulling the loop snug around the screw terminal, tighten the screw to the proper amount with a screwdriver.

- Reidentify the white conductor with black tape and then place the terminal loop on the white wire around the other terminal screw so that the terminal loop is going in the clockwise direction. While pulling the loop snug around the screw terminal, tighten the screw to the proper amount with a screwdriver.

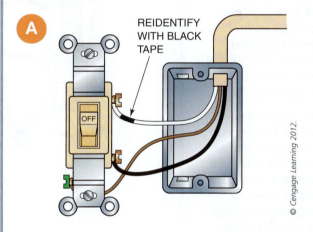

(A) REIDENTIFY WITH BLACK TAPE

© Cengage Learning 2012.

PROCEDURE 18-6

Installing Single-Pole Switches in a Single-Gang Nonmetallic Box (Continued)

B Assuming that two cables are entering the box, use wire strippers (or possibly long-nosed pliers) and make a terminal loop at the end of each of the black wires.

- Connect the two cable equipment-grounding conductors and an 8-inch-long bare copper equipment bonding jumper together with a wirenut. Put a terminal loop on the end of the equipment bonding jumper.

- Place the terminal loop on the bare equipment bonding jumper around the green terminal screw of the switch so that the terminal loop is going in the clockwise direction. While pulling the loop snug around the screw terminal, tighten the screw to the proper amount with a screwdriver.

- Connect the two white conductors together with a wirenut and place the connection in the back of the device box.

- Place the terminal loop on one of the black wires around a terminal screw so that the terminal loop is going in the clockwise direction. While pulling the loop snug around the screw terminal, tighten the screw to the proper amount with a screwdriver.

B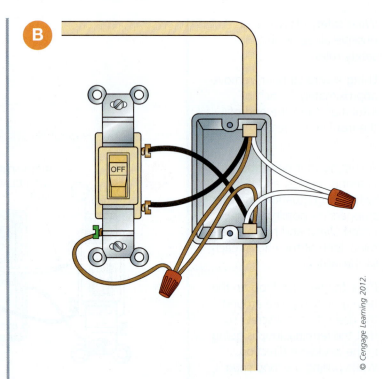

© Cengage Learning 2012.

- Place the terminal loop of the other black wire around the other terminal screw so that the terminal loop is going in the clockwise direction. While pulling the loop snug around the screw terminal, tighten the screw to the proper amount with a screwdriver.

- Place the switch into the device box by carefully folding the conductors back into the device box.

- Secure the switch to the device box using the 6-32 screws. Mount the switch so it is vertically aligned. Use a box extender or spacers to align the switch with the wall surface if the box is set back too far.

- Attach the switch cover plate to the switch. Be careful to not tighten the mounting screw(s) too much. Plastic faceplates tend to crack very easily.

Installing Single-Pole Switches in a Multi-Ganged Nonmetallic Box

- Wear safety glasses and observe all applicable safety rules.

- **A** Assuming that two cables (a 3-wire and a 2-wire Type NM cable) are entering a two-gang box containing two single-pole switches, use wire strippers to remove approximately ¾ inch of insulation from the end of the insulated wires.

- Connect the two cable equipment-grounding conductors and two 8-inch-long bare copper equipment bonding jumpers together with a wirenut. Put a terminal loop on the end of the two equipment bonding jumpers.

- Connect the two white conductors together with a wirenut and place the connection in the back of the device box.

- Place the terminal loop at the end of each bare equipment bonding jumper around the green terminal screw of each switch so that the terminal loop is going in the clockwise direction. While pulling the loop snug around the screw terminal, tighten the screw to the proper amount with a screwdriver.

- Connect the black conductor from the 2-wire feed cable and two 8-inch-long black pigtails together with a wirenut and place the connection in the back of the device box.

- Place the terminal loop of one of the black pigtails around a terminal screw on switch #1

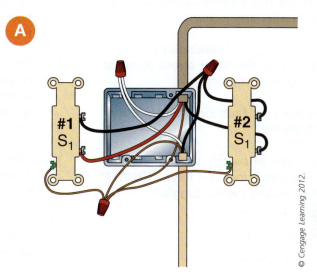

© Cengage Learning 2012.

so that the terminal loop is going in the clockwise direction. While pulling the loop snug around the screw terminal, tighten the screw to the proper amount with a screwdriver.

- Place the terminal loop of the other black pigtail around a terminal screw on switch #2 so that the terminal loop is going in the clockwise direction. While pulling the loop snug around the screw terminal, tighten the screw to the proper amount with a screwdriver.

- Place the terminal loop of the black wire from the 3-wire cable around the other terminal screw on switch #2 so that the terminal loop is going in the clockwise direction. While pulling the loop snug around the screw terminal, tighten the screw to the proper amount with a screwdriver.

- Place the terminal loop of the red wire from the 3-wire cable around the other terminal screw on switch #1 so that the terminal loop is going in the clockwise direction. While pulling the loop snug around the screw terminal, tighten the screw to the proper amount with a screwdriver.

- Place the switches into the device box by carefully folding the conductors back into the device box.

- Secure the switches to the device box using the 6-32 screws. Mount the switches so they are vertically aligned. Use a box extender or spacers to align the switches with the wall surface if the box is set back too far.

- Attach the switch cover plate to the switches. Be careful to not tighten the mounting screw(s) too much. Plastic faceplates tend to crack very easily.

REVIEW QUESTIONS

Directions: Answer the following items with clear and complete responses.

1. Section _____ of the *NEC®* covers the required locations for GFCIs in residential applications.

2. The _____ conductor is connected to the silver screw terminal on a duplex receptacle.

3. What is an arc fault circuit interrupter (AFCI) device?

4. Which type of switch has four screw terminals and does not have "ON/OFF" marked on the switch toggle?

   a. Three-way switch

   b. Four-way switch

   c. Double-pole switch

   d. Single-pole switch

5. When a receptacle is installed horizontally in a device box, it is recommended that the _____ slot be on the top.

6. To GFCI protect receptacles that are downstream from a GFCI receptacle, the wiring going to those receptacles is connected to the _____ terminals on the GFCI receptacle.

7. A GFCI is designed to protect you if you touch both the "hot" and the grounded conductors of a circuit at the same time. (Circle the correct response.) True or False

8. The two categories of transients that a TVSS device can protect sensitive electronic equipment from are:

   1.

   2.

9. A GFCI is designed to protect you if you come in contact with a circuit "hot" conductor and the metal frame of an appliance. (Circle the correct response.) True or False

10. When installed in noncombustible materials, a device box cannot be set back from the wall surface more than _____.

11. When wiring a split-circuit receptacle, it is recommended that the part of the duplex receptacle controlled by a switch be the _____ half and the part that is "hot" at all times is the _____ half.

12. Three-way switches are used to control a luminaire from three locations. (Circle the correct response.) True or False

13. Section _____ of the *NEC®* states that all switches must be effectively grounded, except when _____.

14. If nonmetallic device boxes are used, how are metal cover plates grounded?

15. Conductor reidentification, required when connecting a switch loop, is covered in Section _____ of the *NEC®*.

16. A GFCI will trip when it senses a current imbalance of _____ or more.

17. A 15-amp duplex receptacle may be installed on a 20-amp-rated circuit. (Circle the correct response.) True or False

18. A 15-amp single receptacle may be installed on a 20-amp-rated branch circuit. (Circle the correct response.) True or False

19. The metal frame of a switch or receptacle is called a(n) _____ or _____.

20. The push-in method of connecting conductors to a switch or receptacle is UL listed for use with only _____ solid copper conductors. Aluminum conductors, stranded conductors, or conductors sized _____ and larger may not be used.

CHAPTER 19

Service Panel Trim-Out

OBJECTIVES

Upon completion of this chapter, the student should be able to:

- Select the proper overcurrent protection device for a specific residential branch circuit or feeder circuit.

- Demonstrate an understanding of common fuses and circuit breakers used in residential wiring.

- Demonstrate an understanding of installing circuit breakers or fuses in an electrical panel.

- Demonstrate an understanding of the common techniques for trimming out a residential electrical panel.

GLOSSARY OF TERMS

arc fault circuit interrupter (AFCI) circuit breaker a device intended to provide protection from the effects of arc faults by recognizing characteristics unique to arcing and by functioning to de-energize the circuit when an arc fault is detected

cartridge fuse a fuse enclosed in an insulating tube that confines the arc when the fuse blows; this fuse may be either a ferrule or a blade type

Edison-base plug fuse a fuse type that uses the same standard screw base as an ordinary lightbulb; different fuse sizes are interchangeable with each other; this fuse type may only be used as a replacement for an existing Edison-base plug fuse

interrupting rating the highest current at rated voltage that a device is intended to interrupt under standard test conditions

overcurrent protection device (OCPD) a fuse or circuit breaker used to protect an electrical circuit from an overload, a short circuit, or a ground fault

stab a term used to identify the location on a loadcenter's ungrounded bus bar where a circuit breaker is snapped on

Type S plug fuse a fuse type that uses different fuse bases for each fuse size; an adapter that matches each Type S fuse size is required; this fuse type is not interchangeable with other Type S fuse sizes

An electrical panel serves as the origination point for the branch-circuit and feeder-circuit wiring. It contains the fuses or circuit breakers that provide overcurrent protection to the circuits that make up a residential wiring system. During the rough-in stage, the various cable or raceway wiring methods used to wire the electrical system's circuits were installed. Enough conductor length is left at each electrical panel so that at some future time an electrician can connect the circuit wiring to the overcurrent protection devices in the panel. Some electricians make these connections at the same time they rough-in the circuit wiring, but most electricians make the connections during the trim-out stage. They refer to this process as "trimming out the panel." In this chapter, we look at trimming out the electrical panel. An overview of the types of overcurrent protection devices used in electrical panels will be presented. Both ground fault circuit interrupter (GFCI) circuit breakers and arc fault circuit interrupter (AFCI) circuit breakers are discussed. Because most electrical panels in today's residential electrical systems use circuit breakers, an overview of common methods of circuit breaker installation and panel trim-out procedures is also presented.

UNDERSTANDING RESIDENTIAL OVERCURRENT PROTECTION DEVICES

Chapter 2 presented a good overview of the different types of overcurrent protection devices (OCPDs) used in residential wiring. Both circuit breakers and fuses are used to provide overcurrent protection in residential

wiring, but there is no question that between the two, circuit breakers are used the most. Circuit breakers are available as a single-pole device for 120-volt applications, as a two-pole device for 240-volt applications, and as a twin (dual) device for 120-volt applications when there is a limited number of spaces available in an electrical panel (Figure 19-1). Ground fault circuit interrupter (GFCI) and arc fault circuit interrupter (AFCI) circuit breakers are also available (Figure 19-2). Fuses are available in two basic styles: plug and cartridge. The plug-type fuses are available as an Edison-base model and as a Type S model (Figure 19-3). Cartridge fuses are available as a ferrule model or a blade-type model (Figure 19-4). Fuses may be found with time delay or non-time delay characteristics. A review of the OCPD information in Chapter 2 is appropriate at this time.

> ## ⚠ CAUTION
> **CAUTION:** Use time delay fuses to protect circuits that supply motor loads. Remember that electric motors draw approximately six to eight times their normal running current at start-up. A time delay fuse will allow the large inrush of current to get the motor started. Using a non-time delay fuse will result in having the fuse blow each time the motor starts. Circuit breakers used in residential wiring are called inverse time circuit breakers and have built-in time delay characteristics.

Article 240 of the *National Electrical Code®* (*NEC®*) provides many installation requirements for OCPDs. A discussion of the more important sections

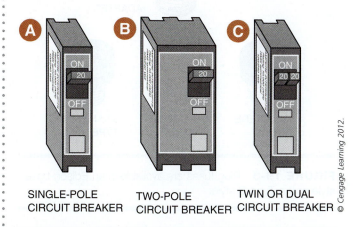

© Cengage Learning 2012.

SINGLE-POLE CIRCUIT BREAKER   TWO-POLE CIRCUIT BREAKER   TWIN OR DUAL CIRCUIT BREAKER

FIGURE 19-1 Circuit breakers are available (A) as a single-pole device for 120-volt applications and (B) as a two-pole device for 240-volt applications. (C) Twin or dual circuit breakers can be used for two 120-volt branch circuits.

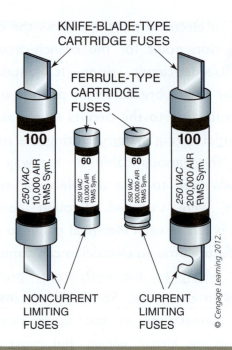

FIGURE 19-2 (A) A GFCI circuit breaker and (B) an AFCI circuit breaker.

FIGURE 19-4 Cartridge fuses are available as a ferrule model or a blade-type model.

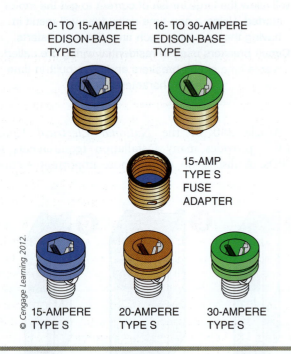

FIGURE 19-3 Plug fuses are available in an Edison-base style and a Type S style.

of Article 240 that apply specifically to overcurrent protection devices used in residential wiring is presented in the next few paragraphs. A discussion on OCPD interrupting ratings versus equipment short-circuit ratings is also presented.

GENERAL REQUIREMENTS FOR OCPDs

Section 240.4 states that circuit conductors must be protected against overcurrent in accordance with their ampacity as specified in Table 310.15(B)(16). Section 240.6(A) lists the standard sizes of fuses and circuit breakers. Common standard sizes used in residential wiring include 15, 20, 25, 30, 35, 40, 45, 50, 60, 70, 80, 90, and 100 amperes. Other standard sizes are available for larger electrical loads. Because all the ampacities listed in Table 310.15(B)(16) do not correspond with a standard fuse or circuit breaker size, Section 240.4(B) allows the next higher standard overcurrent device rating to be used, provided that all the following conditions are met:

- The conductors being protected are not part of a multi-outlet branch circuit supplying receptacles for cord-and-plug-connected portable loads.

- The ampacity of the conductors does not correspond with the standard ampere rating of a fuse or a circuit breaker.

- The next higher standard rating selected does not exceed 800 amperes.

Section 240.4(B) applies in most residential wiring situations, but there are a couple of specific requirements that must be observed first:

- Table 210.24 (Figure 19-5) summarizes the requirements for the size of conductors and the size of the overcurrent protection for branch circuits where two

Table 210.24 Summary of Branch-Circuit Requirements

Circuit Rating	15 A	20 A	30 A	40 A	50 A
Conductors (min. size):					
Circuit wires[1]	14	12	10	8	6
Taps	14	14	14	12	12
Fixture wires and cords — see 240.5					
Overcurrent Protection	**15 A**	**20 A**	**30 A**	**40 A**	**50 A**
Outlet devices:					
Lampholders permitted	Any type	Any type	Heavy duty	Heavy duty	Heavy duty
Receptacle rating[2]	15 max. A	15 or 20 A	30 A	40 or 50 A	50 A
Maximum Load	**15 A**	**20 A**	**30 A**	**40 A**	**50 A**
Permissible load	See 210.23(A)	See 210.23(A)	See 210.23(B)	See 210.23(C)	See 210.23(C)

[1]These gauges are for copper conductors.
[2]For receptacle rating of cord-connected electric-discharge luminaires, see **410.62(C)**.

FIGURE 19-5 Table 210.24 of the *NEC®* summarizes the requirements for the size of conductors and the size of the overcurrent protection for branch circuits where two or more receptacle outlets are required. *Reprinted with permission from NFPA 70®, National Electric Code®, Copyright © 2010, National Fire Protection Association, Quincy, MA. This reprinted material is not the complete and official position of the NFPA on the referenced subject, which is represented only by the standard in its entirety.*

or more receptacle outlets are required. The first footnote also indicates that the wire sizes are for copper conductors.

- Section 210.3 indicates that branch-circuit conductors for other than individual branch circuits must be rated 15, 20, 30, 40, and 50 amperes. For example, a branch circuit wired with 12 AWG wire and protected with a 20-amp fuse or circuit breaker will have a 20-amp circuit rating.

In summary, Table 310.15(B)(16) lists the ampacities of conductors used in residential wiring. Section 240.6(A) lists the standard ratings of overcurrent devices. If the ampacity of the conductor does not match the rating of the standard overcurrent device, Section 240.4(B) permits the use of the next larger standard overcurrent device. However, if the ampacity of a conductor matches the standard rating of Section 240.6(A), that conductor must be protected by the standard-size device. For example, in Table 310.15(B)(16), a 3 AWG, 75°C copper, Type THWN is listed as having an ampacity of 100 amperes. That conductor would be protected by a 100-ampere overcurrent device. Remember that Section 310.15(B)(7) allows an increase in the ampacity of the conductor and the overcurrent device for a single-phase, 120/240-volt residential service entrance.

Section 240.4(D) requires the OCPD to not exceed 15 amperes for 14 AWG, 20 amperes for 12 AWG, and 30 amperes for 10 AWG copper, or 15 amperes for

12 AWG and 25 amperes for 10 AWG aluminum and copper-clad aluminum after any correction factors for ambient temperature and number of conductors have been applied. Because these wire sizes are common in residential wiring, circuits installed with 14 AWG copper conductors will have a 15-amp OCPD, circuits installed using 12 AWG copper conductors will have a 20-amp or lower OCPD, and circuits installed with 10 AWG copper conductors will have 30-amp or lower OCPDs. This rule has also been explained in previous chapters of this book.

Section 240.24 addresses the location of the OCPDs in the house. Section 240.24(A) requires that OCPDs be readily accessible. Section 240.24(C) requires the OCPDs to be located where they will not be exposed to physical damage. Section 240.24(D) prohibits the OCPDs from being located in the vicinity of easily ignitable material. Examples of locations where combustible materials may be stored are linen closets, paper storage closets, and clothes closets. Finally, Section 240.24(E) says that in dwelling units, OCPDs cannot be located in bathrooms and Section 240.24(F) does not allow OCPDs to be located over the steps of a stairway.

No discussion of the general requirements for OCPDs is complete without a discussion of interrupting ratings. Section 110.9 states that all fuses and circuit breakers intended to interrupt the circuit at fault levels must have an adequate interrupting rating wherever they are used in the electrical system. Fuses or circuit breakers that do not have adequate interrupting ratings could rupture while attempting to clear a short circuit. The *NEC®* defines the term **interrupting rating** as the highest current at rated voltage that a device is intended to interrupt under standard test conditions. It is important that the test conditions match the actual installation needs. Interrupting ratings should not be confused with short-circuit current ratings.

Short-circuit current ratings are marked on the equipment. This marking appears on many pieces of equipment, such as panelboards. Remember, the basic purpose of overcurrent protection is to open the circuit before conductors or conductor insulation is damaged when an overcurrent condition occurs. An overcurrent condition can be the result of an overload, a ground fault, or a short circuit and must be eliminated before the conductor insulation damage point is reached. Fuses and circuit breakers must be selected to ensure that the short-circuit current rating of any electrical system component is not exceeded should a short circuit or a high-level ground fault occur. Electrical system components include wire, bus structures in load-centers, switches, disconnect switches, and other electrical distribution equipment, all of which have limited short-circuit ratings and would be damaged or destroyed if those short-circuit ratings were exceeded. Merely providing OCPDs with sufficient interrupting ratings does not ensure adequate short-circuit protection for the system components. When the available short-circuit current exceeds the short-circuit current rating of an electrical component, the OCPD must limit the let-through current to within the rating of that electrical component. Utility companies can determine and provide information on available short-circuit current levels at the service equipment. Calculating the available short-circuit current is beyond the scope of this book.

> **CAUTION**
>
> **CAUTION:** Fuses and circuit breakers that are subjected to fault currents that exceed their interrupting ratings may actually explode like a bomb. Damage to people and equipment can result. It is extremely important for overcurrent protection devices to have an interrupting rating high enough to handle any available fault current.

The interrupting rating of most circuit breakers is either 5000 or 10,000 amps. The interrupting rating of fuses can be as high as 300,000 amps. For circuit breakers, if the interrupting rating is other than 5000 amps, it must be marked on the breaker. The interrupting rating for all fuses must be marked on the fuse. The interrupting rating of a fuse is defined as the maximum amount of current the device can handle without failing. The interrupting rating of a device must be at least as high

as the available fault current. The available fault current is the maximum amount of current that can flow through the circuit if a fault occurs. It depends on the size of the utility company transformer that is supplying the electrical power to the residential wiring system.

> **CAUTION**
>
> **CAUTION:** When replacing fuses, never use a fuse that has a lower interrupting rating than the one that is being replaced. This could cause severe damage to the circuit components.

GENERAL REQUIREMENTS FOR PLUG FUSES

Section 240.50 applies to plug fuses of both the Edison-base type and Type S type. This section limits the voltage on the circuits using this fuse type to the following:

- Circuits not exceeding 125 volts between conductors. All 120-volt residential branch circuits meet this requirement.

- Circuits supplied by a system having a grounded neutral where the line-to-neutral voltage does not exceed 150 volts. All residential 120/240-volt circuits meet this requirement. An example would be a 120/240-volt electric clothes dryer branch circuit. A fusible disconnect that uses plug fuses could be used with a 120/240-volt dryer branch circuit.

Section 240.50(B) requires each fuse, fuseholder, and adapter to be marked with its ampere rating. Section 240.50(C) requires plug fuses of 15-ampere and lower rating to be identified by a hexagonal configuration of the window, cap, or other prominent part to distinguish them from fuses of higher ampere ratings. Figure 19-6 shows some examples of Edison-base plug fuses and Type S fuses. Note the hexagonal feature on the 10- and 15-ampere fuses.

SPECIFIC REQUIREMENTS FOR EDISON-BASE PLUG FUSES

Section 240.51 applies specifically to Edison-base plug fuses and requires the following:

- Plug fuses of the Edison-base type must be classified at not over 125 volts and 30 amperes or below.

- Plug fuses of the Edison-base type can be used only for replacements in existing installations where there is no evidence of overfusing or tampering.

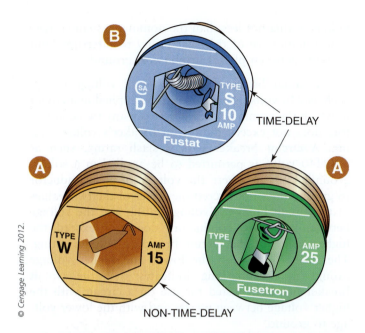

FIGURE 19-6 Some examples of (A) Edison-base plug fuses and (B) a Type S fuse. Note the hexagonal feature on the 10- and 15-ampere fuses. The Fustat Type S fuse (B) and the Fusetron Edison-base fuse (right A) are time delay fuses. The Edison-base (left A) fuse is a non-time-delay fuse.

CAUTION

CAUTION: Edison-base plug fuses may *not* be used as OCPDs in any new electrical installation. If an electrician chooses to use plug fuses in a new electrical installation, they must be Type S plug fuses.

SPECIFIC REQUIREMENTS FOR TYPE S PLUG FUSES

Section 240.53 applies specifically to Type S plug fuses and requires the following:

- Type S fuses are classified at not over 125 volts and 0 to 15 amperes, 16 to 20 amperes, and 21 to 30 amperes. Type S plug fuses are found in three sizes for residential work: 15-amp-rated, which is blue in color; 20-amp rated, which is orange in color; and 30-amp-rated, which is green in color.

- Type S fuses of a higher ampere rating are not interchangeable with a lower-ampere-rated fuse. For example, a 20-amp Type S plug fuse will not work if inserted into a 15-amp adapter. This feature prevents overfusing of branch-circuit conductors.

- They must be designed so that they cannot be used in any fuseholder other than a Type S fuseholder or a fuseholder with a Type S adapter inserted.

Section 240.54 applies specifically to Type S fuses, adapters, and fuseholders, and requires the following:

- Type S adapters must fit Edison-base fuseholders.

- Type S fuseholders and adapters must be designed so that either the fuseholder itself or the fuseholder with a Type S adapter inserted cannot be used for any fuse other than a Type S fuse.

from experience...

Once a Type S fuse adapter has been installed in an Edison-base screw shell, it is virtually impossible to remove. Therefore, make sure that you put the correct adapter for the size of Type S fuse you wish to use. Color coding makes this choice fairly simple. The bottom inside of the Type S fuse adapter is blue to match the 15-amp Type S fuse, orange to match the 20-amp Type S fuse, and green to match the 30-amp Type S fuse. Anytime this author has tried to remove a Type S fuse adapter once it has been completely installed has always resulted in the destruction of the adapter.

- Type S adapters must be designed so that once inserted in a fuseholder, they cannot be removed.

- Type S fuses, fuseholders, and adapters must be designed so that tampering or shunting (bridging) would be difficult.

- Dimensions of Type S fuses, fuseholders, and adapters must be standardized to permit interchangeability regardless of the manufacturer.

SPECIFIC REQUIREMENTS FOR CARTRIDGE FUSES

Sections 240.60 and 240.61 apply specifically to cartridge fuses and state the following:

- Cartridge fuses and fuseholders of the 300-volt type are permitted to be used in circuits not exceeding 300 volts between conductors.

- Fuses rated 600 volts or less are permitted to be used for voltages at or below their ratings.

- Fuseholders must be designed so that it will be difficult to put a fuse of any given class into a fuseholder that is designed for a current lower or a voltage higher than that of the class to which the fuse belongs. Fuseholders for current-limiting fuses cannot permit insertion of fuses that are not current limiting.

- Fuses must be plainly marked, either by printing on the fuse barrel or by a label attached to the barrel showing the following: ampere rating, voltage rating, interrupting rating where other than 10,000 amperes, current limiting where applicable, and the name or trademark of the manufacturer.

- Cartridge fuses and fuseholders are to be classified according to their voltage and amperage ranges.

SPECIFIC REQUIREMENTS FOR CIRCUIT BREAKERS

Section 240.81 states that circuit breakers must clearly indicate whether they are in the open OFF or closed ON position. Where circuit breaker handles are operated vertically, the "up" position of the handle must be the ON position. Also, circuit breakers must be designed so that any fault must be cleared before the circuit breaker can be reset. This means that even if the handle is held in the ON position, the circuit breaker will remain tripped as long as there is a trip-rated fault on the circuit.

Section 240.83 requires that circuit breakers be marked with their ampere rating in a manner that will be durable and visible after installation. This marking is permitted to be made visible by removal of a trim or cover. Section 240.83(B) says that circuit breakers rated at 100 amperes or less and 600 volts or less must have the ampere rating molded, stamped, etched, or similarly marked on their handles. Section 240.83(C) requires every circuit breaker having an interrupting rating other than 5000 amperes to have its interrupting rating shown on the circuit breaker. Section 240.83(D) requires circuit breakers used as switches in 120- and 277-volt fluorescent lighting circuits to be listed and marked SWD or HID. Circuit breakers used as switches in high-intensity discharge lighting circuits must be listed and marked as HID. Circuit breakers marked SWD are 15- or 20-ampere breakers that have been subjected to additional endurance and temperature testing. If high-intensity discharge (HID) lighting such as high-pressure sodium or metal halide lighting is used, the breaker used for switching must be marked HID. Section 240.83(E) says that circuit breakers must be marked with a

voltage rating not less than the nominal system voltage that is indicative of their capability to interrupt fault currents between phases or phase to ground.

A circuit breaker with a straight voltage rating, such as 240 volts, is permitted to be applied in a circuit in which the nominal voltage between any two conductors does not exceed the circuit breaker's voltage rating. A circuit breaker with a slash rating, such as 120/240 volts, is permitted to be applied in a solidly grounded circuit where the voltage of any conductor to ground does not exceed the lower of the two values of the circuit breaker's voltage rating and the voltage between any two conductors does not exceed the higher value of the circuit breaker's voltage rating. The slash (/) between the lower and higher voltage ratings in the marking indicates that the circuit breaker has been tested for use on a circuit with the higher voltage between phases and with the lower voltage to ground.

GFCI AND AFCI CIRCUIT BREAKERS

There are several receptacle outlet locations that require ground fault circuit interrupter (GFCI) protection in a house. In previous chapters these locations were discussed and in Chapter 18 GFCI receptacle installation was covered in detail. Previous chapters also introduced you to arc fault circuit interrupter (AFCI) protection and the locations that require AFCI protection in a house. In this section we will take a close look at providing GFCI and AFCI protection using circuit breakers.

GFCI CIRCUIT BREAKERS

There are two ways that an electrician can provide GFCI protection to the receptacles that require it. One way is to use GFCI receptacles. The other way is to use a GFCI circuit breaker. An advantage to the GFCI circuit breaker is that the whole circuit becomes protected. However, with the GFCI receptacle, only that receptacle location and any "downstream" receptacles are GFCI protected. A disadvantage is the relatively high cost of the GFCI breaker as compared to the GFCI receptacle.

The GFCI circuit breaker looks very similar to a "regular" circuit breaker (Figure 19-7). However, there are two differences that you should be aware of. The first is that the GFCI breaker has a white pigtail attached to it. It is all curled up (like a pig's tail) when the GFCI breaker is first taken out of the box that it comes in. The second difference is the "Push to Test"

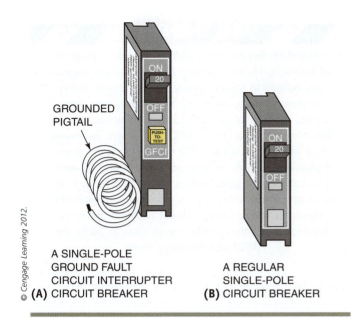

GROUNDED
PIGTAIL

ON
20

OFF

PUSH-
TO-
TEST

GFCI

ON
20

OFF

A SINGLE-POLE
GROUND FAULT
CIRCUIT INTERRUPTER
(A) CIRCUIT BREAKER

A REGULAR
SINGLE-POLE
(B) CIRCUIT BREAKER

© *Cengage Learning 2012.*

FIGURE 19-7   A GFCI circuit breaker has a white pigtail attached to it. It also has a "Push to Test" button for testing the ground-fault interrupting capabilities of the breaker.

button that is located on the front of the breaker. This allows the breaker to be tested for correct operation once it has been installed and energized.

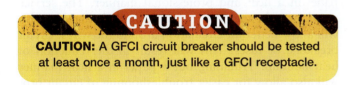

CAUTION

CAUTION: A GFCI circuit breaker should be tested at least once a month, just like a GFCI receptacle.

AFCI CIRCUIT BREAKERS

Arc fault circuit interrupter (AFCI) devices were introduced in Chapter 18. In this section, AFCI circuit breakers are covered. The *NEC®* defines an AFCI as a device intended to provide protection from the effects of arc faults by recognizing characteristics unique to arcing and by functioning to de-energize the circuit when an arc fault is detected. Arcing faults can occur in homes when insulation around cords, wires, or cables is damaged or deteriorated. The arc faults often happen where they cannot be seen, like behind walls and between floors. They can create up to a 10,000°F arc that will ignite nearby combustible materials. Arc faults are likely to occur during the following situations:

- When new cabling is being installed during the rough-in stage of new construction or when remodeling, the cable can be pinched or punctured by being

pressed against a nail or by being stapled too tightly against a stud.

- In older homes, wire insulation can deteriorate from age and arcing can occur between conductors.

- When pictures or other items are hung on a wall in a house, the nails or screws can sometimes puncture the cable insulation and cause arcing.

- Cords running in front of hot air registers or just sitting in sunlight for long amounts of time can deteriorate the cord insulation.

- Loose connections on switches and receptacles can cause arcing.

- Furniture pushed up against or resting on a cord can cause arcing.

Any of the arcing faults produced by these situations will not be detected by conventional circuit breakers.

There are currently five types of AFCIs recognized by Underwriters Laboratories but only three of the types have significance when discussing AFCI circuit breakers. The **branch/feeder AFCI** type is a device installed at the origin of a branch circuit or feeder, such as in a circuit breaker panel, which provides protection of the branch-circuit wiring, feeder wiring, or both against the unwanted effects of arcing. It is typically a circuit breaker device but can be a device in its own enclosure mounted at or near an electrical panel. The **outlet circuit AFCI** type is a receptacle device installed at a branch-circuit outlet to provide protection of cord sets and power-supply cords connected to it (when provided with receptacle outlets) against the unwanted effects of arcing. This device may provide feed-through protection to downstream receptacles. Even though these AFCI receptacles are recognized by UL, they are not commercially available at this time. The **combination AFCI** is designed to comply with the requirements for both **branch/feeder** and **outlet circuit AFCIs**. It is intended to protect downstream branch-circuit wiring, cord sets, and power-supply cords. The *NEC®* requires all AFCI protection to be a **combination AFCI** type.

Section 210.12(A) requires that AFCI protection be provided on *all* 120-volt, 15- and 20-amp branch circuits supplying outlets in family rooms, dining rooms, living rooms, parlors, libraries, dens, bedrooms, sun rooms, recreation rooms, closets, hallways, or similar rooms or areas in dwelling units. This means that *all* 15- or 20-amp, 120-volt receptacle outlets, lighting outlets, or any other power outlet in these areas of a house must be AFCI protected. Areas in a dwelling unit that use different types of electrical loads are exempt from this AFCI requirement. These exempt areas include unfinished basements, garages, bathrooms, outdoor applications, and kitchens.

AFCIs are evaluated to UL 1699, Safety Standard for Arc-Fault Circuit Interrupters, using testing methods that create or simulate arcing conditions to determine the product's ability to detect and interrupt arcing faults. These devices are also tested to verify that arc detection is not made any less effective by the presence of loads and circuit characteristics that may mask the hazardous arcing condition. In addition, these devices are evaluated to determine resistance to unwanted tripping due to the presence of arcing that occurs in control and utilization equipment under normal operating conditions or to a loading condition that closely mimics an arcing fault, such as a solid-state electronic ballast or a dimmed lighting load.

The *NEC®* is clear that the objective is to provide protection of the entire branch circuit, including any cords plugged into receptacles on the circuit. For instance, a cord AFCI could not be used to comply with the requirement of Section 210.12 to protect the entire branch circuit. This means that for an electrician to comply with Section 210.12, a combination-type AFCI circuit breaker will need to be used so that the entire branch circuit, from the overcurrent protective device to the last outlet installed on the circuit, and any cords plugged into the circuit, are AFCI protected.

AFCI circuit breakers look very similar to GFCI circuit breakers (Figure 19-8). The AFCI breaker has a "Push to Test" button, but it is typically a different color than the "Push to Test" button on a GFCI breaker.

CAUTION

CAUTION: Section 210.12(A) in the *NEC®* requires all 120-volt, 15-and 20-amp branch circuits supplying outlets in family rooms, dining rooms, living rooms, parlors, libraries, dens, bedrooms, sun rooms, recreation rooms, closets, hallways, or similar rooms or areas to be AFCI protected. Be aware that with the types of AFCI devices that are commercially available at this time, all 15- and 20-amp, 120-volt branch circuits in a house that have to be AFCI protected will require a combination-type AFCI circuit breaker. Always check with the authority having jurisdiction (AHJ) in your area to verify the AFCI requirements there.

INSTALLING CIRCUIT BREAKERS IN A PANEL

At this point in the installation of the electrical system, the branch and feeder circuit wiring has been installed into the service entrance panel or subpanels, and it is time to install the OCPD for each circuit.

It is important for an electrician to trim out the panel in a neat and professional manner. The service panel or subpanel will not only look better but also will allow better air circulation around the wires, circuit breakers, or fuses. This will result in the reduction of heat buildup. Trimming out a panel in a neat manner will also allow for easier troubleshooting of circuit problems should they arise. Figure 19-9 shows what a well trimmed out electrical panel should look like.

from experience...

It is a good idea to provide the home owner with at least two extra circuit breakers to be used as spares. This allows the home owner to add to the electrical system without having to replace the existing panel or add a subpanel. Do not forget to identify the circuit breakers as spares on the circuit directory.

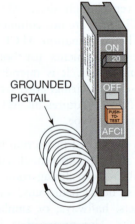

GROUNDED PIGTAIL

(A) A SINGLE-POLE ARC FAULT CIRCUIT INTERRUPTER CIRCUIT BREAKER

(B) A REGULAR SINGLE-POLE CIRCUIT BREAKER

© Cengage Learning 2012.

FIGURE 19-8 An AFCI circuit breaker looks very similar to a GFCI circuit breaker.

In this section, we look at safely installing circuit breakers in electrical panels. We do not cover fuse installation since the majority of OCPDs used in today's homes are circuit breakers.

FIGURE 19-9   A good example of a neatly trimmed-out electrical panel.

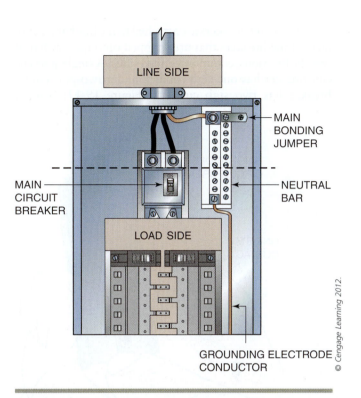

FIGURE 19-10   The line side versus the load side in an electrical panel. Remember that the line side will still be "hot" even when the main circuit breaker is off. Only the load side will be de-energized.

Before installing circuit breakers, some factors must be considered. Make sure that the circuit breakers being used are compatible with the loadcenter already installed. Circuit breakers are not always interchangeable with different manufacturers. Most 100-amp-rated electrical panels used in residential wiring are designed for 20 or 24 over-current devices. Most 200-amp-rated electrical panels are designed for 40 circuits. Electrical panels designed for over 42 circuits are now allowed by the *NEC®*. Be sure that your panel will handle the number of circuits the electrical system requires.

SAFETY RULES

When installing or removing circuit breakers from an electrical panel in a house where the service entrance conductors have been connected to the local utility company system and a kilowatt-hour meter has been installed in the meter socket, there a couple of safety rules to follow:

- Always turn off the electrical power at the main circuit breaker when working in an energized main breaker panel. This will disconnect the power from the load side of the panel, but be aware that the line side of the panel will still be energized (Figure 19-10).

If you are working on an energized subpanel, find the circuit breaker in the service panel, turn it off, and lock it in the OFF position.

- Test the panel you are working on with a voltage tester to verify that the electrical power is indeed off. Never assume that the panel is de-energized.

> **CAUTION**
>
> **CAUTION:** When installing or removing circuit breakers from an electrical panel, make sure that the panel has been de-energized and the electrical power supplying the panel has been locked out.

CIRCUIT BREAKER INSTALLATION

Circuit breakers are installed in an electrical panel by attaching them to the bus bar assembly in the panel. The bus bar assembly is connected to the incoming service entrance conductors or, in the case of a subpanel, to the incoming feeder conductors. The bus bar distributes the electrical power to each of the circuit breakers located in the panel.

The circuit breakers are attached to the bus bar by contacts in the breakers that are snapped onto the bus bar at specific locations, commonly called **stabs**. A single-pole circuit breaker has one stab contact, and a two-pole circuit breaker has two stab contacts (Figure 19-11). Circuit breakers are constructed in such a way that the end opposite the stab contacts has a slot that fits onto a retainer clip or mounting rail in the panel. Between the retainer clip or mounting rail and the circuit breaker contacts being snapped onto the bus bar stabs, the breaker is held in place. In panels typically used in commercial and industrial locations, the circuit breakers are more likely to be bolted onto the bus bar rather than just snapping onto them. Once all the breakers have been installed and the wiring has been correctly attached to them, a cover is put on the panel. The cover also helps hold the circuit breakers in place. Figure 19-12 shows how both single-pole circuit breakers and double-pole circuit breakers are attached to a bus bar in a main breaker panel.

> **CAUTION**
>
> **CAUTION:** Only use identified handle ties to connect two single-pole circuit breakers together as a double-pole breaker.

Most branch circuits in a residential wiring system are 120-volt circuits. They will be wired with 14 AWG or 12 AWG copper conductors. Remember that 14 AWG conductors require a single-pole circuit breaker that is 15-amp rated and that 12 AWG conductors require a single-pole circuit breaker rated at 20 amps.

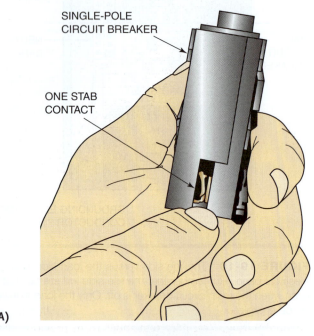

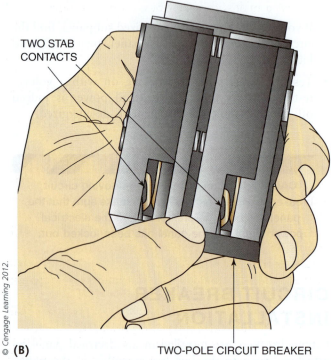

© Cengage Learning 2012.

FIGURE 19-11 (A) A single-pole circuit breaker has one stab contact, and (B) a two-pole circuit breaker has two stab contacts.

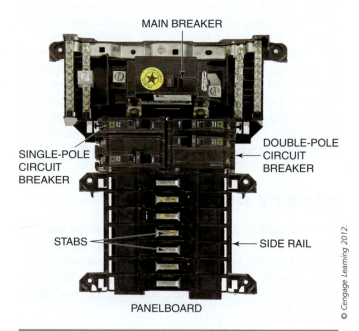

© Cengage Learning 2012.

FIGURE 19-12 A single-pole circuit breaker and a double-pole circuit breaker attached to a bus bar in a main breaker panel.

(See Procedure 19-1 on pages 685–686 for the correct procedure for installing a single-pole circuit breaker.)

(See Procedure 19-2 on pages 687–688 for the correct procedure for installing a single-pole GFCI circuit breaker.)

(See Procedure 19-3 on pages 689–690 for the correct procedure for installing a single-pole AFCI circuit breaker.)

from experience...

The covers used for electrical panels come from the factory with several partially punched-out rectangular pieces of metal that will need to be removed by an electrician. The number and location of the pieces to be removed is determined by the number of circuit breakers used in the panel and their actual location. Once the correct number of pieces is removed, the cover will fit correctly onto the panel and over the installed circuit breakers. Be careful to not take out too many of the pieces. If you do, a plug must be inserted in the spot to prevent the "hot" ungrounded panel bus bar from being exposed (Figure 19-13).

© Cengage Learning 2012.

FIGURE 19-13 Remove only those "twistouts" that match the locations of the installed circuit breakers.

Many residential branch circuits serve appliances like electric water heaters, air conditioners, and electric heating units. These loads require 240 volts to operate properly. Like the 120-volt branch circuits, the 240-volt branch circuits require a 15-amp circuit breaker when wired with 14 AWG wire, a 20-amp circuit breaker when wired with 12 AWG wire, and a 30-amp circuit breaker when wired with 10 AWG wire. However, because it is a 240-volt circuit, a two-pole circuit breaker will be needed.

(See Procedure 19-4 on pages 691–692 for the correct procedure for installing a two-pole circuit breaker for a 240-volt branch circuit.)

Branch circuits in residential wiring can also supply 120/240 volts to appliances such as electric clothes dryers and electric ranges. This branch-circuit installation requires a two-pole circuit breaker, just like the 240-volt-only application described in the preceding paragraph. The big difference with this type of circuit is that a three-wire cable with a grounding conductor is used.

Multiwire branch circuits are often installed in residential wiring. You may remember from previous chapters that a multiwire circuit consists of two ungrounded conductors (usually a black and red wire) that have 240 volts between them and a grounded conductor (the white neutral conductor) that has 120 volts between it and either of the two ungrounded conductors. An advantage to installing a multiwire circuit in residential wiring is that two 120-volt circuits can be run using just three wires instead of four. An example of where a multiwire branch circuit may be installed would be when an electrician runs a 12/3 Type NM cable from the main electrical panel to a junction box located in an accessible attic. From the junction box two 2-wire 120-volt branch circuits can be run to feed various receptacle or lighting outlets on the second floor of the house. Section 210.4 has a couple of rules to consider when trimming out an electrical panel. Section 210.4(B) states that each multiwire branch circuit must be provided with a means that will simultaneously disconnect all ungrounded conductors at the point where the multiwire branch circuit originates. This means you will need to install a two-pole circuit breaker in the loadcenter to protect the multiwire branch-circuit wiring. However, Section 240.15(B)(1) permits two individual single-pole circuit breakers with identified handle ties to be used instead of a two-pole breaker. Section 210.4(D) states that ungrounded and grounded conductors of each multiwire branch circuit must be grouped together by wire ties or some other means in at least one location in the electrical panel.

This means that in the electrical panel you will need to tie wrap the conductors together at one or more places for each multiwire branch circuit. An *Exception* says that you do not have to group the conductors together if the multiwire branch circuit comes from a cable or raceway that makes the grouping obvious.

Feeder circuits from a main electrical panel to one or more subpanels located throughout a house supply 120/240 volts and will also need to be protected with two-pole circuit breakers. The size of the feeder wiring is typically larger than branch-circuit wiring and that will mean having larger ampere rated circuit breakers for protection.

(See Procedure 19-5 on pages 693–694 for the correct procedure for installing a two-pole circuit breaker for a 120/240-volt branch circuit.)

SUMMARY

The main service panel and subpanels contain the OCPDs for the various branch circuits that make up a residential wiring system. The OCPDs may consist of fuses or circuit breakers. It is during the trim-out stage that an electrician will "trim out" the electrical panel(s) used in a house. OCPDs are installed in the panel(s), and the proper wiring terminations are made to them one branch circuit at a time.

This chapter presented an overview of the types of OCPDs used in residential wiring. Edison-base plug fuses, Type S plug fuses, and cartridge fuses were presented. Because circuit breakers are used the most as overcurrent protection in residential installations, a thorough discussion of the different types of circuit breakers was also presented. This included GFCI and AFCI circuit breakers. Circuit breaker installation was covered in detail, and several actual installation procedures were presented.

Trimming out the panel for a residential wiring system is a very important part of the trim-out stage. It must be done in a neat and professional manner. Making sure that the OCPD is installed properly and that the correct size has been chosen will allow the branch circuits to supply electricity to all the loads located throughout a house in a safe and trouble-free manner.

PROCEDURE 19-1

Installing a Single-Pole Circuit Breaker

- Put on safety glasses and observe all applicable safety rules.

- Using a voltage tester, make sure that the panel is de-energized and that the electrical power source for the panel has been locked out.

A Locate a free space in the panel for the circuit breaker. Insert the slot on the back of the circuit breaker onto the retainer clip or mounting rail. Then, using firm pressure, snap the single-pole circuit breaker onto the bus bar stab.

- Determine the location on the equipment-grounding bar where you want to terminate the bare or green equipment-grounding conductor. Loosen the termination screw. (*Note:* In a main service panel the equipment-grounding conductor is terminated on the grounded neutral terminal bar.)

B Insert the end of the bare or green equipment-grounding conductor into the termination hole on the equipment-grounding bar and tighten the screw according to the torque requirements listed on the panel enclosure. (*Note:* It is recommended that you do not put more than three equipment-grounding wires in each termination hole.)

- Determine the location on the grounded neutral terminal bar where you want to terminate the white grounded circuit conductor. Loosen the termination screw.

© Cengage Learning 2012.

© Cengage Learning 2012.

PROCEDURE 19-1

Installing a Single-Pole Circuit Breaker (Continued)

C Using a wire stripper, remove approximately ½ inch (13 mm) from the end of the white grounded circuit conductor and insert the tip of the white conductor into the termination hole on the grounded terminal bar. Tighten the screw according to the torque requirements listed on the panel enclosure. (*Note:* The *NEC®* and UL allow only one grounded conductor to be installed in each termination hole. Do not install two or more grounded conductors in one hole or combine a grounded conductor with a bare or green equipment-grounding conductor in one hole.)

C

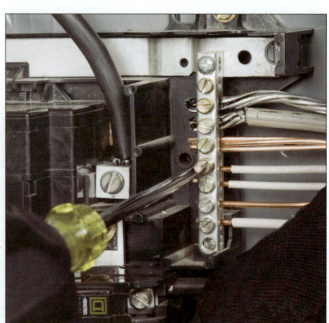

© Cengage Learning 2012.

D Using a wire stripper, remove approximately ½ inch (13 mm) from the end of the black ungrounded circuit conductor and insert the tip of the black wire under the terminal screw of the single-pole circuit breaker. (*Note:* You may have to loosen the terminal screw on the circuit breaker.) Tighten the screw according to the torque requirements listed for the breaker.

D

© Cengage Learning 2012.

PROCEDURE 19-2

Installing a Single-Pole GFCI Circuit Breaker

- Put on safety glasses and observe all applicable safety rules.

- Using a voltage tester, make sure that the panel is de-energized and that the electrical power source for the panel has been locked out.

Ⓐ Locate a free space in the panel for the circuit breaker. Insert the slot on the back of the circuit breaker onto the retainer clip or mounting rail. Then, using firm pressure, snap the single-pole GFCI circuit breaker onto the bus bar stab.

- Determine the location on the equipment-grounding bar where you want to terminate the bare or green equipment-grounding conductor. Loosen the termination screw. (*Note:* In a main service panel the equipment-grounding conductor is terminated on the grounded neutral terminal bar.)

- Insert the end of the bare or green equipment-grounding conductor into the termination hole on the equipment-grounding bar and tighten the screw according to the torque requirements listed on the panel enclosure.

Ⓑ Connect the white grounded pigtail of the GFCI circuit breaker to the grounded terminal bar in the panel. The pigtail usually comes with its end stripped back at the factory.

Ⓐ

© Cengage Learning 2012.

Ⓑ

© Cengage Learning 2012.

PROCEDURE 19-2

Installing a Single-Pole GFCI Circuit Breaker (Continued)

C Using a wire stripper, remove approximately ½ inch (13 mm) from the end of the white grounded branch-circuit conductor and insert the tip of the white conductor into the terminal on the GFCI breaker marked "load neutral." Tighten the screw according to the torque requirements listed for the breaker. (*Note:* Because the "load neutral" and "load power" terminals are hard to reach when the breaker is installed in the panel, some electricians will make the connections with the GFCI breaker in their hand and then install the breaker in the panel.)

C
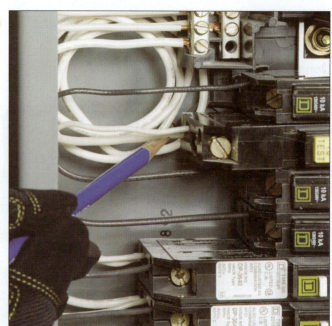
© Cengage Learning 2012.

D Using a wire stripper, remove approximately ½ inch (13 mm) from the end of the black ungrounded circuit conductor and insert the tip of the black conductor into the terminal on the GFCI breaker marked "load." Tighten the screw according to the torque requirements listed for the breaker.

D

© Cengage Learning 2012.

PROCEDURE 19-3

Installing a Single-Pole AFCI Circuit Breaker

- Put on safety glasses and observe all applicable safety rules.

- Using a voltage tester, make sure that the panel is de-energized and that the electrical power source for the panel has been locked out.

A Locate a free space in the panel for the circuit breaker. Insert the slot on the back of the circuit breaker onto the retainer clip or mounting rail. Then, using firm pressure, snap the single-pole AFCI circuit breaker onto the bus bar stab.

- Determine the location on the equipment-grounding bar where you want to terminate the bare or green equipment-grounding conductor. Loosen the termination screw. (*Note:* In a main service panel the equipment-grounding conductor is terminated on the grounded neutral terminal bar.)

- Insert the end of the bare or green equipment-grounding conductor into the termination hole on the equipment-grounding bar and tighten the screw according to the torque requirements listed on the panel enclosure.

B Connect the white grounded pigtail of the AFCI circuit breaker to the grounded terminal bar in the panel. The pigtail usually comes with its end stripped back at the factory.

A

© Cengage Learning 2012.

B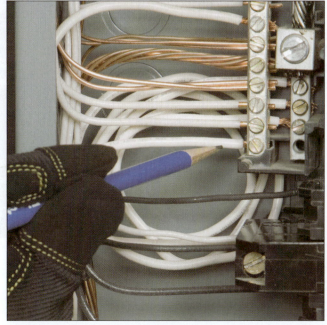

© Cengage Learning 2012.

PROCEDURE 19-3

Installing a Single-Pole AFCI Circuit Breaker (Continued)

C Using a wire stripper, remove approximately ½ inch (13 mm) from the end of the white grounded branch-circuit conductor and insert the tip of the white conductor into the terminal on the AFCI breaker marked "load neutral." Tighten the screw according to the torque requirements listed for the breaker. (*Note:* Because the "load neutral" and "load power" terminals are hard to reach when the breaker is installed in the panel, some electricians will make the connections with the AFCI breaker in their hand and then install the breaker in the panel.)

© Cengage Learning 2012.

D Using a wire stripper, remove approximately ½ inch (13 mm) from the end of the black ungrounded branch-circuit conductor and insert the tip of the black conductor into the terminal on the AFCI breaker marked "load." Tighten the screw according to the torque requirements listed for the breaker.

© Cengage Learning 2012.

PROCEDURE 19-4

Installing a Two-Pole Circuit Breaker for a 240-Volt Branch Circuit

- Put on safety glasses and observe all applicable safety rules.

- Using a voltage tester, make sure that the panel is de-energized and that the electrical power source for the panel has been locked out.

A Locate a free space in the panel for the two-pole circuit breaker. Remember that this breaker takes up two regular spaces. Insert the slots on the back of the circuit breaker onto the retainer clips or mounting rail. Then, using firm pressure, snap the double-pole circuit breaker onto the bus bar stabs.

- Determine the location on the equipment-grounding bar where you want to terminate the bare or green equipment-grounding conductor. Loosen the termination screw. (*Note:* In a main service panel the equipment-grounding conductor is terminated on the grounded neutral terminal bar.)

B Insert the end of the bare or green equipment-grounding conductor into the termination hole on the equipment-grounding bar and tighten the screw according to the torque requirements listed on the panel enclosure.

© Cengage Learning 2012.

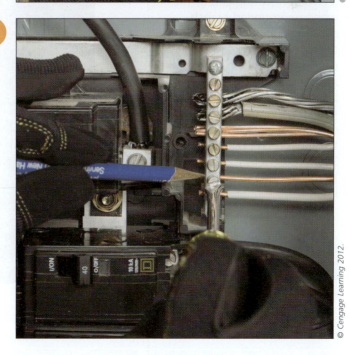

© Cengage Learning 2012.

PROCEDURE 19-4

Installing a Two-Pole Circuit Breaker for a 240-Volt Branch Circuit (Continued)

C Using a wire stripper, remove approximately ½ inch (13 mm) from the end of the black ungrounded branch-circuit conductor and insert the tip of the black wire under one of the terminal screws of the two-pole circuit breaker. (*Note:* You may have to loosen the terminal screws on the circuit breaker.) Tighten the screw according to the torque requirements listed for the breaker.

C

© Cengage Learning 2012.

D Using a wire stripper, remove approximately ½ inch (13 mm) from the end of the white ungrounded branch-circuit conductor. Reidentify this white wire as a "hot" conductor by wrapping it with some black electrical tape. Insert the tip of the reidentified white wire under the other terminal screw of the two-pole circuit breaker. Tighten the screw according to the torque requirements listed for the breaker.

D

© Cengage Learning 2012.

PROCEDURE **19-5**

Installing a Two-Pole Circuit Breaker for a 120/240-Volt Branch Circuit

- Put on safety glasses and observe all applicable safety rules.

- Using a voltage tester, make sure that the panel is de-energized and that the electrical power source for the panel has been locked out.

 Locate a free space in the panel for the two-pole circuit breaker. Remember that this breaker takes up two regular spaces. Insert the slots on the back of the circuit breaker onto the retainer clips or mounting rail. Then, using firm pressure, snap the double-pole circuit breaker onto the bus bar stabs.

- Determine the location on the equipment-grounding bar where you want to terminate the bare or green equipment-grounding conductor. Loosen the termination screw. (*Note:* In a main service panel the equipment-grounding conductor is terminated on the grounded neutral terminal bar.)

 Insert the end of the bare or green equipment-grounding conductor into the termination hole on the equipment-grounding bar and tighten the screw according to the torque requirements listed on the panel enclosure.

- Determine the location on the grounded terminal bar where you want to terminate the white grounded circuit conductor. Loosen the termination screw.

© Cengage Learning 2012.

© Cengage Learning 2012.

PROCEDURE 19-5

Installing a Two-Pole Circuit Breaker for a 120/240-Volt Branch Circuit (Continued)

C Using a wire stripper, remove approximately ½ inch (13 mm) from the end of the white grounded branch-circuit conductor and insert the tip of the white conductor into the termination hole on the grounded terminal bar. Tighten the screw according to the torque requirements listed on the panel enclosure.

D Using a wire stripper, remove approximately ½ inch (13 mm) from the end of the black ungrounded branch-circuit conductor and insert the tip of the black wire under one of the terminal screws of the two-pole circuit breaker. Tighten the screw according to the torque requirements listed for the breaker.

E Using a wire stripper, remove approximately ½ inch (13 mm) from the end of the red ungrounded branch-circuit conductor. Insert the tip of the red wire under the other terminal screw of the two-pole circuit breaker. Tighten the screw according to the torque requirements listed for the breaker. (*Note:* If a raceway wiring method was used for the circuit, there may be two black wires instead of a black and red as found in cable wiring methods.)

C

© Cengage Learning 2012.

D

© Cengage Learning 2012.

E
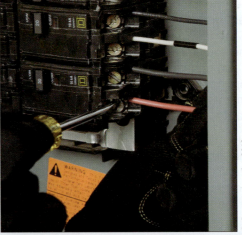
© Cengage Learning 2012.

REVIEW QUESTIONS

Directions: Answer the following items with clear and complete responses.

1. Describe a Type S fuse.

2. The maximum voltage between conductors on circuits protected with plug fuses is _____ volts.

3. List the standard sizes of fuses and circuit breakers up to and including 100 amperes.

4. A 240-volt electric water heater branch circuit will require a (single-pole or double-pole) circuit breaker. (Circle the correct response.)

5. A 120-volt dishwasher branch circuit will require a (single-pole or double-pole) circuit breaker. (Circle the correct response.)

6. Overcurrent protection devices may not be installed in clothes closets. The *NEC*® section that states this is _____. (Circle the correct response.) True or False

7. Overcurrent protection devices may be installed in bathrooms. The *NEC*® section that states this is _____. (Circle the correct response.) True or False

8. Overcurrent protection devices must be readily accessible. The *NEC*® section that states this is _____. (Circle the correct response.) True or False

9. Define an arc fault circuit interrupter (AFCI) circuit breaker.

10. Explain why it is important for an electrician to trim out the electrical panel in a neat and professional manner.

11. Name the two types of cartridge fuses found in residential wiring.

12. Name the maximum overcurrent protective devices allowed for the following residential branch-circuit conductor sizes:
    1) AWG copper _____
    2) 12 AWG copper _____
    3) 10 AWG copper _____

13. What does the hexagonal configuration on a plug fuse indicate?

14. A circuit breaker used to switch on or off fluorescent lighting must have the letters _____ written on the breaker. The *NEC*® section that states this is _____.

15. Name an advantage to using a GFCI circuit breaker instead of a GFCI receptacle to provide the *NEC*®-required ground-fault protection.

16. Edison-base plug fuses can be used for replacement purposes only and may not be used to provide overcurrent protection for a new branch circuit. (Circle the correct response.) True or False

17. An OCPD must be sized according to the ampacity of a conductor. If the ampacity of a conductor is not more than 800 amps and does not correspond with a standard size of OCPD, the next standard size higher may be used. (Circle the correct response.) True or False

18. Describe the danger that exists if an OCPD is subjected to a fault current that exceeds its interrupting rating.

19. Two 20-amp single-pole circuit breakers are sitting next to each other on a table. One is a regular breaker and the other is a GFCI breaker. Based on the breaker's physical appearance, describe how you would be able to tell which one is the GFCI circuit breaker.

20. Describe the difference between a branch/feeder AFCI, an outlet circuit AFCI, and a combination AFCI.

21. Describe a multiwire branch circuit as used in residential wiring.

22. Name an advantage for installing multiwire branch circuits in residential wiring.

23. Explain why a two-pole circuit breaker is used to protect a multiwire branch circuit.

24. Explain what must be done if an electrician mistakenly takes out too many rectangular pieces from an electrical panel cover while getting the cover ready for installation on an electrical panel.

25. Explain why time delay fuses are used to protect circuits that have an electric motor on them.

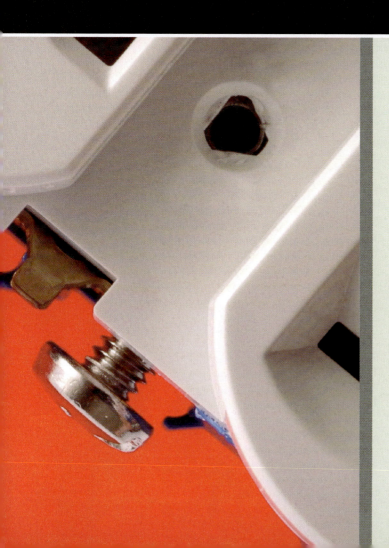

Maintaining and Troubleshooting a Residential Electrical Wiring System

CHAPTER 20
Checking Out and Troubleshooting Electrical Wiring Systems

Checking Out and Troubleshooting Electrical Wiring Systems

OBJECTIVES

Upon completion of this chapter, the student should be able to:

- Follow a checklist to determine if the basic requirements of the *NEC®* were met in the electrical system installation.

- Demonstrate an understanding of how to test for current and voltage in an energized circuit.

- Demonstrate an understanding of how to test for continuity in existing branch-circuit wiring and wiring devices.

- Demonstrate an understanding of the common testing techniques to determine whether a circuit has a short circuit, a ground fault, or an open circuit.

- Troubleshoot common residential electrical system problems.

- Demonstrate an understanding of how to use a circuit tracer.

- Demonstrate an understanding of how to perform a successful service call.

GLOSSARY OF TERMS

buried box an electrical box that has been covered over with sheetrock or some other building material

checkout the process of determining if all parts of a recently installed electrical system are functioning properly

circuit tracer a type of tester that is used to determine which circuit breaker in an electrical panel protects a specific circuit; used when a circuit directory does not properly indicate what is protected by a specific breaker

ground fault an unintended low-resistance path in an electrical circuit through which some current flows to ground using a pathway other than the intended pathway; it results when an ungrounded "hot" conductor unintentionally touches a grounded surface or grounded conductor

open circuit a circuit that is energized but does not allow useful current to flow on the circuit because of a break in the current path

receptacle polarity when the white-colored grounded conductor is attached to the silver screw of the receptacle and the "hot" ungrounded circuit conductor is attached to the brass screw on the receptacle; this results in having the long slot always being the grounded slot and the short slot always being the ungrounded slot on 125-volt, 15- or 20-amp-rated receptacles used in residential wiring

short circuit an unintended low-resistance path through which current flows around rather than along a circuit's intended current path; it results when two circuit conductors come in contact with each other unintentionally

troubleshooting the process of determining the cause of a malfunctioning part of a residential electrical system

The installation of a residential electrical system must meet the minimum installation requirements of the *National Electrical Code*® (*NEC*®). An electrician who follows the installation information presented in the previous chapters of this book should have an electrical installation that is essentially safe and free from defects. The installed electrical system should provide home owners with the electrical service they desire. However, during the installation process, mistakes can happen. Some electrical equipment may be faulty. Other trades people, such as carpenters and plumbers working in the house, may inadvertently damage a part of the electrical system. Electricians can mix up traveler wires on three-way and four-way switching circuits. Wirenut connections may not be properly tightened. In other words, a just completed residential electrical installation may have problems that are not readily apparent without a thorough **checkout** of all parts of the system. When problems are found, **troubleshooting** the problem will have to be done, and then the identified problem will need to be fixed. This chapter looks at common checkout, troubleshooting, and repair procedures with which all electricians should be familiar.

DETERMINING IF ALL APPLICABLE NEC® INSTALLATION REQUIREMENTS ARE MET

It is common practice for an electrical inspector to inspect the electrical installation at the end of the electrical system rough-in and at the end of the electrical system trim-out to make sure that the system has been installed according to the requirements of national, state, and local electrical codes. At the end of the rough-in stage and on completion of the residential electrical system trim-out stage, an electrician should go over the following checklist (Table 20-1) to make sure the basic installation requirements of the *NEC*® have been met. State and local electrical codes are beyond the scope of this book but should always be considered by the electrician doing the residential electrical system installation. Making sure that these guidelines are met will result in an electrical system installation that will very likely pass an inspection by the electrical inspector. As an electrician goes over the checklist, they need to make note of those areas that were not installed according to the *NEC*®. Any areas of the installation that are found to be deficient will need to be updated to meet the minimum *NEC*® standards.

DETERMINING IF THE ELECTRICAL SYSTEM IS WORKING PROPERLY

Once an electrician has determined that all applicable *NEC*® installation requirements have been met, a check of the electrical system must be done to determine if everything is working properly. The check of the system is usually done after an electrical inspector has approved the installation and the local electric utility has connected the house service entrance to the utility electrical system. At this time, the main service disconnecting means is turned on, which energizes the fuse or circuit breaker panel. Each circuit is then energized one by one, and each receptacle and lighting outlet on the circuit is checked for proper voltage, proper polarity of the connections, and proper switch control.

The electrical plans used as a guide for the initial installation of the electrical system can be used during the checkout. As discussed earlier in this book, it is a good idea to draw out a cabling diagram on the building electrical plan so that an electrician will know exactly which receptacle and lighting outlets, as well as which switch locations, are on a particular circuit. By comparing the outlets that are included on the electrical plan cabling diagram to the actual outlets being checked, an electrician is able to determine which outlets are working properly and which outlets are not. If a lighting outlet or power outlet is found to be not working as intended, a troubleshooting procedure will need to be initiated to find the problem. Once the cause of the problem is found, the electrician will need to correct the problem and then verify that the correction has caused the circuit and its components to operate properly.

TABLE 20-1   2011 National Electrical Code Requirements Checklist for House Wiring

GENERAL CIRCUITRY

NEC SECTION	REQUIREMENT	COMPLETED YES, <u>N</u>O, OR <u>N</u>OT APPLICABLE
Section 210.11(C) and Section 422.12	In addition to the branch circuits installed to supply general illumination and receptacle outlets in dwelling units, there are a *minimum* of: • Two 20-amp circuits for the kitchen receptacles. • One 20-amp circuit for the laundry receptacle(s). • One 20-amp circuit for the bathroom receptacles. • One separate, individual branch circuit for central heating equipment.	
Section 210.52(B)(3)	The receptacles installed in the kitchen to serve countertop surfaces are supplied by at least two separate small-appliance branch circuits.	
Section 300.3(B)	All conductors of the same circuit, including grounding conductors, are contained in the same raceway, cable, or trench.	
Section 408.4	All circuits and circuit modifications are legibly identified as to purpose or use on a directory located on the face or inside of the electrical panel doors.	
Section 406.4	All receptacle outlets are of the grounding type, effectively grounded, and are wired to have the proper polarity.	
Section 406.9(A) and Section 406.9(B)	All 15- and 20-amp, 125- and 250-volt receptacles installed in a wet location are installed in an enclosure that maintains its weatherproofing whether a cord attachment plug is inserted or not. All nonlocking 15- and 20-amp, 125- and 250-volt receptacles installed in damp or wet locations are a weather-resistant type.	
Section 210.52	Receptacle outlets in every kitchen, family room, dining room, living room, parlor, library, den, sunroom, bedroom, recreation room, or similar room or area are installed so that no point measured horizontally along the floor line in any wall space is more than 6 feet (1.8 m) from a receptacle outlet. A receptacle is installed in each wall space 2 feet (600 mm) or more in width. At kitchen countertops, receptacle outlets are installed so that no point along the wall line is more than 24 inches (600 mm) measured horizontally from a receptacle outlet in that space. A receptacle outlet is installed at each counter space that is 12 inches (300 mm) or wider. A receptacle outlet is installed at each island counter or peninsular space 24 inches (600 mm) by 12 inches (300 mm) or larger. At least one receptacle outlet is installed in bathrooms within 3 feet (900 mm) of the outside edge of each basin. Outdoor receptacles, accessible at grade level and no more than 6.5 feet (2 m) above grade, are installed at the front and back of the house. Balconies, decks, and porches that are accessible from inside the dwelling unit have at least one receptacle outlet installed within the perimeter of the balcony, deck, or porch. At least one receptacle outlet is installed for the laundry. At least one receptacle outlet, in addition to those for specific equipment, is installed in each basement, in each attached garage, and in each detached garage or accessory building with electric power.	

TABLE 20-1 (continued)

GENERAL CIRCUITRY		
NEC SECTION	**REQUIREMENT**	**COMPLETED YES, NO, OR NOT APPLICABLE**
	Hallways of 10 feet (3.0 m) or more in length have at least one receptacle outlet.	
	Foyers that have an area greater than 60 square feet have a receptacle(s) located in each wall space that is 2 feet (600 mm) or wider.	
Section 406.12	Tamper-resistant receptacles are installed for all locations of 15- and 20-amp, 125-volt nonlocking receptacles required by Section 210.52. Exceptions include receptacles more than 5½ feet (1.7 m) above the floor and receptacles located within dedicated space for appliances that in normal use are not easily moved from one place to another and that are cord-and-plug connected.	
Section 210.12	All branch circuits supplying 120-volt, 15- and 20-ampere outlets in dwelling unit family rooms, dining rooms, living rooms, parlors, libraries, dens, bedrooms, sun rooms, recreation rooms, closets, hallways, or similar rooms or areas are protected by a listed combination arc fault circuit interrupter device (AFCI).	
REQUIRED GROUND FAULT CIRCUIT INTERRUPTER PROTECTION		
Section 210.8	Ground fault circuit interrupter (GFCI) protection is provided for all 15- and 20-amp, 125-volt receptacle outlets installed in: • Bathrooms, • Garages, • Crawl spaces, • Unfinished basements, • Locations within 6 feet (1.8 m) of the outside edge of sinks in other than kitchens, • In boathouses and for boat hoists, • Outdoors, and • Receptacles serving the countertop in kitchens.	
WIRING METHODS		
Section 314.23	All electrical boxes are securely supported by the building structure.	
Section 314.27(C)	Boxes used for ceiling-suspended paddle fans are listed and labeled for such use.	
Section 334.30	Type NM cable is supported and secured at intervals not exceeding 4½ feet (1.4 m) and within 12 inches (300 mm) of each electrical box and within 8 inches (200 mm) of each single-gang nonmetallic electrical box.	
Section 314.17(C) *Exception*	The outer jacket of Type NM cable extends into all single-gang nonmetallic electrical boxes a minimum of ¼ inch (6 mm).	
Section 300.14	The minimum length of conductors at all boxes is at least 6 inches (150 mm). At least 3 inches (75 mm) extends outside the box.	
Section 300.4(A)	Where cables are installed through bored holes in joists, rafters, or other wood framing members, the holes are bored so that the edge of the hole is not less than 1¼ inches (32 mm) from the nearest edge of the wood member.	

TABLE 20-1 (continued)

WIRING METHODS		
NEC SECTION	**REQUIREMENT**	**COMPLETED YES, NO, OR NOT APPLICABLE**
	Where this distance cannot be maintained or where screws or nails are likely to penetrate the cable, they are protected by a steel plate at least ¹⁄₁₆ inch (1.6 mm) thick and of appropriate length and width.	
Section 300.22(C) *Exception*	Type NM cable is not installed in spaces used for environmental air; however, Type NM cable is permitted to pass through perpendicular to the long dimension of such spaces.	
Sections 250.134, 314.4, and 404.9(B)	All metal electrical equipment including boxes, cover plates, and plaster rings are grounded. All switches, including dimmer switches, are grounded.	
Sections 110.12(A) and 314.17(A)	Unused openings in boxes, other than those intended for the operation of equipment, those intended for mounting purposes, or those permitted as part of the design for listed equipment, are effectively closed. When openings in nonmetallic boxes are broken out and not used, the entire box has been replaced.	
Sections 110.14(A) and 110.14(B)	Only one conductor is installed under a terminal screw unless it is identified for more than one. Conductors are spliced or joined with splicing devices identified for the use like a wirenut. All splices and joints and the free ends of conductors are covered with an insulation equivalent to that of the conductors or with an insulating device identified for the purpose.	
Section 250.148	In boxes with more than one grounding wire, the grounding wires are spliced with an approved mechanical connector and then, using a "jumper" or a "pigtail," are attached to the grounding terminal screw of the device. A connection is made between the one or more equipment-grounding conductors and a metal box by means of a green grounding screw that is used for no other purpose, equipment listed for grounding, or a listed grounding device. One or more equipment-grounding conductors brought into a non-metallic outlet box are arranged so that a connection can be made to any fitting or device in that box that requires grounding.	
Section 300.15	When splicing underground conductors, the method and items used are identified for such use.	
Sections 314.25 and 410.22	In a completed installation, all outlet boxes have a cover, lamp-holder, canopy for a luminaire, and an appropriate cover plate for switches and receptacles.	
Section 314.29	Junction boxes are installed so that the wiring contained in them can be rendered accessible without removing any part of the building.	
Section 314.16	The volume of electrical boxes is sufficient for the number of conductors, devices, and cable clamps contained within the box.	

TABLE 20-1 (continued)

WIRING METHODS		
NEC SECTION	**REQUIREMENT**	**COMPLETED YES, NO, OR NOT APPLICABLE**
Section 410.2 and 410.16	Luminaires installed in clothes closets have the following minimum clearances from the defined storage area: • 12 inches (300 mm) for surface incandescent or LED fixtures with completely enclosed lamps • 6 inches (150 mm) for recessed incandescent or LED fixtures with completely enclosed lamps or fluorescent fixtures • 6 inches (150 mm) for surface fluorescent fixtures Incandescent luminaires with open or partially enclosed lamps and pendant fixtures or lamp-holders are not used in clothes closets.	
Section 410.116	Recessed lighting fixtures installed in insulated ceilings or installed within ½ inch (13 mm) of combustible material are approved for insulation contact and labeled Type IC.	
EQUIPMENT LISTING AND LABELING		
Section 110.3(B)	All electrical equipment is installed and used in accordance with the listing requirements and manufacturer's instructions. All electrical equipment, including luminaires, devices, and appliances, are *listed* and *labeled* by a nationally recognized testing laboratory (NRTL) as having been tested and found suitable for a specific purpose.	
SERVICE ENTRANCES		
Section 310.15(B)(7)	Service entrance conductor sizes for 120/240-volt residential services are not smaller than those given in Table 310.15(B)(7).	
Section 110.14	Conductors of dissimilar metals are not intermixed in a terminal or splicing device unless the device is listed for the purpose. Listed antioxidant compound is used on all aluminum conductor terminations, unless information from the device manufacturer specifically states that it is not required.	
Section 300.7(A)	Portions of raceways and sleeves subject to different temperatures (where passing from the interior to the exterior of a building) are sealed with an approved material to prevent condensation from entering the service equipment.	
Section 230.54	Where exposed to weather, service entrance conductors are enclosed in rain-tight enclosures and arranged in a drip loop to drain.	
Section 300.4(G)	Where raceways containing ungrounded conductors 4 AWG or larger enter a cabinet, box, or electrical enclosure, the conductors are protected by an insulated bushing providing a smoothly rounded insulating surface.	
Section 230.70(A)(1)	The electrical service disconnecting means is installed at a readily accessible location either outside a house or inside at a location that is nearest to the point of entrance of the service entrance conductors. There is no excess "inside run."	
Sections 230.70(A)(2) and 240.24(C)(D)(E)	Electrical panels containing fuses or circuit breakers are readily accessible and are not located in bathrooms or in the vicinity of easily ignitable materials such as clothes closets.	

TABLE 20-1 (continued)

SERVICE ENTRANCES		
NEC SECTION	**REQUIREMENT**	**COMPLETED YES, NO, OR NOT APPLICABLE**
Section 110.26	Sufficient working space is provided around electrical equipment. When the voltage to ground does not exceed 150 volts, the depth of that space in the direction of access to live parts is a minimum of 3 feet (900 mm). The minimum width of that space in front of the electrical equipment is the width of the equipment or 30 inches (750 mm), whichever is greater. The work space is clear and extends from the floor to a height of 6.5 feet (2 m). The work spaces are provided with illumination.	
GROUNDING AND BONDING		
Section 250.50, 250.52, and 250.53	The house electrical service is connected to a grounding electrode system consisting of a metal underground water pipe in direct contact with earth for 10 feet (3.0 m) or more. If a metal water pipe is not available as the grounding electrode, any other electrode as specified in Section 250.52 is allowed. An additional electrode supplements the water pipe electrode. If the metal water pipe is used as part of the grounding system, a bonding jumper is placed around the water meter. Rod, pipe, or plate electrodes are supplemented by an additional electrode providing the single rod, pipe, or plate electrode does not have a resistance to earth of 25 ohms or less.	
Sections 250.64(C) and 250.66	The grounding electrode conductor is unspliced and its size is determined, using the size of the service entrance conductors, by Table 250.66.	
Section 250.28	A main bonding jumper or the green bonding screw provided by the panel manufacturer is installed in the service panel to electrically bond the grounded service conductor and the equipment-grounding conductors to the service enclosure	
Section 250.104(A)(1)	The interior metal water piping and other metal piping that may become energized is bonded to the service equipment with a bonding jumper sized the same as the grounding electrode conductor.	
UNDERGROUND WIRING		
Section 300.5	All direct-buried cable or conduit or other raceways meets the minimum cover requirements shown in Table 300.5. Underground service laterals have their location identified by a warning ribbon placed in the trench at least 12 inches (300 mm) above the underground conductors. Where subject to movement, direct-buried cables or raceways are arranged to prevent damage to the enclosed conductors or connected equipment. Conductors emerging from underground are installed in rigid metal conduit, intermediate metal conduit, or Schedule 80 PVC conduit to provide protection from physical damage. This protection extends from 18 inches (450 mm) below grade or the minimum cover distance to a height of 8 feet (2.4 m) above finished grade or to the point of termination aboveground.	

TESTING RECEPTACLE OUTLETS

In Chapter 4, you were introduced to the various types of test and measurement instruments commonly used in residential wiring. You were also instructed on how to properly take a voltage reading using either a voltage tester or a voltmeter. One of the most important parts of a residential electrical system to check is the various receptacle outlets. An electrician will need to check each receptacle outlet to determine if the proper voltage is available and that the wiring connections have resulted in the proper receptacle polarity. Proper **receptacle polarity** simply means that the white-colored grounded conductor is attached to the silver screw of the receptacle and the "hot" ungrounded circuit conductor is attached to the brass screw on the receptacle. Receptacles wired with the proper polarity will ensure that electrical equipment that is cord-and-plug connected will have the "hot" ungrounded conductor going where it is supposed to go and the grounded conductor going where it is supposed to go.

CAUTION

CAUTION: Remember to always wear safety glasses and observe proper safety procedures when using measurement instruments to test a circuit.

Receptacle outlets can be checked with a voltage tester, a voltmeter, or a plug tester, such as the one shown in Figure 20-1. As mentioned in Chapter 4, a solenoid voltage tester is probably the voltage tester most often used in residential wiring. It can be used to check receptacles to determine if they are energized with the correct voltage, to determine if the receptacles are wired for the proper polarity, and to determine if the receptacle is properly grounded.

(See Procedure 20-1 on pages 716–717 for the correct procedure for testing 120-volt receptacles with a voltage tester to determine proper voltage, polarity, and grounding.)

FIGURE 20-1 A plug-in receptacle tester. This device tests for correct wiring, an open ground, reversed polarity, an open "hot," the "hot" on the grounded side, and the "hot" and ground reversed. Neon lamps glow in a certain configuration to indicate the result of the test. *Courtesy of IDEAL Industries, Inc.*

(See Procedure 20-2 on pages 718–721 for the correct procedure for testing 120/240-volt, range and dryer receptacles with a voltage tester to determine proper voltage, polarity, and grounding.)

TESTING LIGHTING OUTLETS

Luminaires are not usually taken apart and tested individually. Electricians usually test a luminaire by energizing the lighting circuit that the luminaire is on and activating the switch or switches that are supposed to control the lighting outlet. It is assumed that the luminaire has been properly wired if the lamp(s) light in the luminaire when the proper switches are activated.

from experience...

Sometimes during the installation of a residential electrical system, an incandescent lamp gets broken off while still screwed into the socket of a lighting fixture. The following steps can be used to safely remove a broken incandescent lamp. First, make sure the electric power is off and put on safety glasses. It is also a good idea to wear leather gloves so you will not cut your hands. There are two ways to take out the bulb's base. In the first method:

- Using both hands, insert long-nose pliers as far into the broken base as you can.
- Spread the handles apart, exerting force against the sides of the bulb base with the tips of the pliers, and rotate the pliers counterclockwise.
- Continue turning until the base is out. If you meet resistance, turn the base back in slightly and then back out.

If this method does not work, try this:

- Carefully insert a small screwdriver between the bulb base and the socket. Bend the bulb base slightly inward enough to allow the long-nose pliers to get a grip.
- Hold the pliers firmly and begin to turn the base out, counterclockwise. You will probably meet some resistance. When you do, turn the base back in slightly, then out again. The trick is to work the base out, not break the fixture.

Because the trim-out of lighting fixtures is very straightforward, there are not a lot of things that can go wrong during their installation. It is basically a "white to white" and a "black to black" wiring connection at each fixture. The grounding connection is also very straightforward and is done by wirenutting the circuit-grounding conductor to the fixture's grounding conductor or attaching the circuit-grounding conductor to the metal body of the lighting fixture with a proper screw or clip. Thus, if the lamps in the lighting fixture do not come on when the switches are activated, troubleshooting the problem is usually limited to the following:

- Check to see if the circuit is turned off. If it is, turn it on.
- Make sure that the lamps installed in the lighting fixture are not "burned out."
- Check to see that all circuit conductor connections are tight and that they all have continuity.
- Check for damage to the circuit wiring.
- Check to see if the lighting outlet is being fed from a **buried box** that has been sheetrocked over and never trimmed out.
- Make sure the luminaire is properly grounded, especially if it is a fluorescent fixture.
- Check out the switching scheme for the proper connections. See the next section for ideas about switch checking.

If you are testing a lighting circuit that has fluorescent fixtures, there are a few things to consider if the lamps do not light or they flicker. Lamps that do not light can be caused by lack of electrical power, a bad ballast, a bad starter if one is used, poor connections at the lamp sockets, or simply bad bulbs. Follow these steps in order:

- Check for power.
- Adjust the lamps so they are seated properly in the lamp sockets.
- Replace the starter (if applicable).
- Determine if the bulbs need to be replaced. To test fluorescent lamps, look at the bulbs and if they appear to be very dark near either end, the bulb is bad or close to failure. Replace them if there is a darkening at either end. The most reliable way to test a fluorescent bulb is to simply install it into a known working fixture.
- If everything else checks out, replace the ballast.

When flickering is the issue, you still must do the same sort of troubleshooting, as all the same problems that can cause a lamp to not work can also cause flickering. Remember that regular fluorescent ballasts are designed to light fluorescent lamps in an ambient temperature of 50°F or higher. If your lamps will not light or they flicker excessively, it may because the ambient temperature around the fluorescent lighting fixture is too cold for a regular ballast. Replace the regular ballast with a cold-temperature model, if appropriate. In many fluorescent fixtures, power is sent through a pair of bulbs. If either bulb is bad, they may both flicker, or one may flicker and the other may show no life. It makes sense in this case to always replace both bulbs. However, if both tubes are functional, the problem is with the ballast or, if it has one, the starter. The starter is replaced first, and if that does not solve the problem, the ballast should be replaced. Remember that only the pre-heat ballast circuit has a starter and it is very rare for today's fluorescent fixtures to be the pre-heat type, so you might not find one if your fixture is less than 15 to 20 years old.

TESTING FOR PROPER SWITCH CONNECTIONS

There are a few procedures to follow when checking out the various switches used to control luminaires in a house. The easiest switch type to check out is a single-pole switch. Remember that they are used to control a lighting load from one location. When a lighting circuit is energized, a single-pole switch can be checked out by simply opening and closing the switch and observing if the lighting fixture(s) it is controlling also turns on and off. If it does, an electrician can rightly assume that the switch is connected properly.

Three-way and four-way switching arrangements are not as simple to check out since there are several different switching combinations. The problem most often found with three-way and four-way switches is having an electrician mix up the traveler wires when the switches were installed as part of the trim-out process. This can be as simple as having a traveler wire attached to a common (black) terminal on a three-way switch or having all traveler wires mixed up with each other on a four-way switch.

(See Procedure 20-3 on pages 722–723 for the correct procedure for testing a standard three-way switching arrangement.)

The following example shows what can happen when a traveler conductor and a common conductor in a three-way switching circuit are mixed up (Figure 20-2):

- When three-way switch A is toggled, the lighting load is turned on and then off (Figure 20-3).

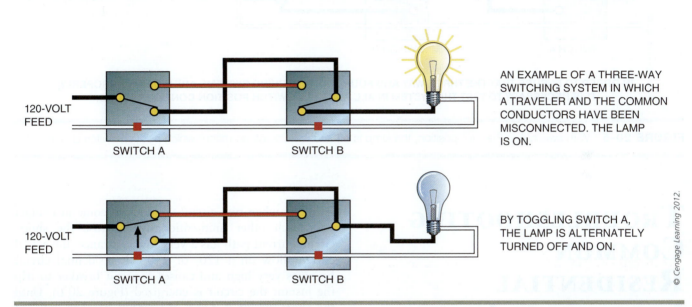

AN EXAMPLE OF THE WIRING FOR A NORMAL THREE-WAY SWITCHING SYSTEM. NOTICE THAT THE TOGGLING OF EITHER SWITCH WILL REVERSE THE ON/OFF STATUS OF THE LAMP.

120-VOLT FEED

SWITCH A   SWITCH B

FIGURE 20-2 A three-way switching connection that is properly wired. Toggling of either switch will turn the lamp on or off.

AN EXAMPLE OF A THREE-WAY SWITCHING SYSTEM IN WHICH A TRAVELER AND THE COMMON CONDUCTORS HAVE BEEN MISCONNECTED. THE LAMP IS ON.

120-VOLT FEED

SWITCH A   SWITCH B

BY TOGGLING SWITCH A, THE LAMP IS ALTERNATELY TURNED OFF AND ON.

120-VOLT FEED

SWITCH A   SWITCH B

FIGURE 20-3 A three-way switching connection in which a traveler wire and the common wire have been misconnected. Toggling switch A will cause the lamp to go on or off.

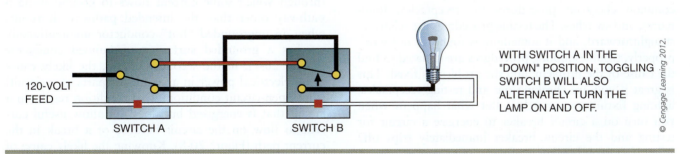

WITH SWITCH A IN THE "DOWN" POSITION, TOGGLING SWITCH B WILL ALSO ALTERNATELY TURN THE LAMP ON AND OFF.

120-VOLT FEED

SWITCH A   SWITCH B

FIGURE 20-4 When switch A is left in the "down" position, toggling switch B will also cause the lamp to go on or off.

- When three-way switch A is left in the "down" position, toggling three-way switch B will also turn the lighting load on and off (Figure 20-4).
- However, with three-way switch A in the "up" position, the lighting load cannot be turned on regardless of the position of three-way switch B (Figure 20-5).

CAUTION

CAUTION: In order to ensure that three-way and four-way switching systems are working properly, it is necessary to test the switches in all possible configurations.

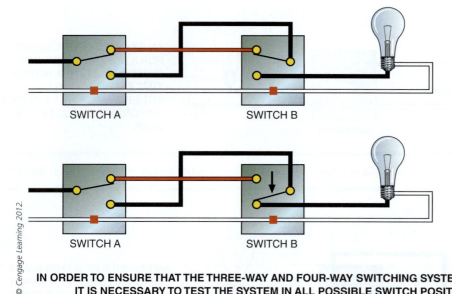

SWITCH A   SWITCH B

WITH SWITCH A IN THE "UP"
POSITION, THE LAMP IS OFF
AND WILL REMAIN OFF
REGARDLESS OF THE
POSITION OF SWITCH B.

SWITCH A   SWITCH B

© Cengage Learning 2012.

**IN ORDER TO ENSURE THAT THE THREE-WAY AND FOUR-WAY SWITCHING SYSTEMS ARE WORKING PROPERLY,
IT IS NECESSARY TO TEST THE SYSTEM IN ALL POSSIBLE SWITCH POSITION CONFIGURATIONS.**

FIGURE 20-5   With switch A in the "up" position, the lamp is off and will stay off no matter which position switch B is in.

TROUBLESHOOTING COMMON RESIDENTIAL ELECTRICAL CIRCUIT PROBLEMS

The previous sections in this chapter presented some common checkout procedures for receptacles, luminaires, and switches. The testing procedures are relatively straightforward, and if something is found to not work properly, troubleshooting procedures are initiated to find the cause of the problem, and the problem is fixed. This is great if you are checking out and testing receptacles, lighting fixtures, or switches. But what happens when you turn on a circuit breaker to energize a circuit for testing and the circuit breaker immediately trips off? What if a circuit breaker is turned on to energize a circuit for testing, and even though you have turned the breaker on, there is no electrical power on the circuit? These situations are common and require good troubleshooting techniques to locate the problem and fix it.

In the first scenario described in the previous paragraph, the likely cause of the circuit breaker tripping (or fuse blowing) as soon as the circuit is energized is a **short-circuit** condition. This condition is an unintended low-resistance path through which current flows around rather than along a circuit's intended current path.

It results when two circuit conductors come in contact with each other unintentionally. Because the short-circuit current path has a very low resistance, the actual current flow on a 120- or 240-volt residential circuit becomes very high and causes a circuit breaker to trip the instant the circuit is energized (Figure 20-6). Until the short circuit is fixed, the circuit breaker will not stay on. Another possible cause for having a circuit breaker trip (or fuse blow) as soon as a circuit is energized is a **ground fault** condition. A ground fault is an unintended low-resistance path in an electrical circuit through which some current flows to ground using a pathway other than the intended pathway. It results when an ungrounded "hot" conductor unintentionally touches a grounded surface or grounded conductor (Figure 20-7). In the second scenario, the likely cause of no electrical power in a circuit (or parts of a circuit) is an **open-circuit** condition. This condition results in a circuit that is energized but does not allow useful current to flow on the circuit because of a break in the current path (Figure 20-8). Knowing the likely cause of a problem in a circuit is great, but finding the location of the problem in the circuit will require troubleshooting the circuit.

The following troubleshooting guide will help locate a short circuit or ground fault in a circuit that causes a fuse to blow or a circuit breaker to trip. Start tracking down the problem by:

- Looking for black smudge marks on switch or receptacle cover plates. The smudge marks indicate the location of a short circuit or a fault to ground.

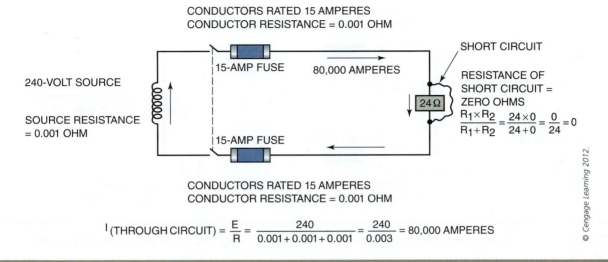

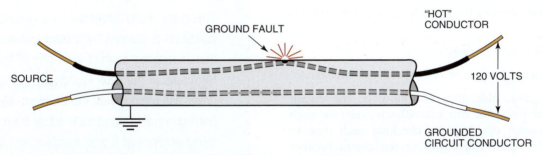

FIGURE 20-6 An example of a short circuit. The circuit conductors will get extremely hot because of the high current flow. The conductor insulation will melt off, and the conductors themselves will melt unless the fuses open the circuit in a very short amount of time.

"HOT" CONDUCTOR COMES IN CONTACT WITH METAL RACEWAY OR OTHER METAL OBJECT. IF THE RETURN GROUND PATH HAS LOW RESISTANCE (IMPEDANCE), THE OVERCURRENT DEVICE PROTECTING THE CIRCUIT WILL CLEAR THE FAULT. IF THE RETURN GROUND PATH HAS HIGH RESISTANCE (IMPEDANCE), THE OVERCURRENT DEVICE WILL NOT CLEAR THE FAULT. THE METAL OBJECT WILL THEN HAVE A VOLTAGE TO GROUND THE SAME AS THE "HOT" CONDUCTOR HAS TO GROUND. IN HOUSE WIRING, THIS VOLTAGE TO GROUND IS 120 VOLTS. PROPER GROUNDING AND GROUND-FAULT CIRCUIT INTERRUPTER PROTECTION IS DISCUSSED ELSEWHERE IN THIS TEXT. THE CALCULATION PROCEDURE FOR A GROUND FAULT IS THE SAME AS FOR A SHORT CIRCUIT; HOWEVER, THE VALUES OF "R" CAN VARY GREATLY BECAUSE OF THE UNKNOWN IMPEDANCE OF THE GROUND RETURN PATH. LOOSE LOCKNUTS, BUSHINGS, SET SCREWS ON CONNECTORS AND COUPLINGS, POOR TERMINATIONS, RUST, AND SO ON. ALL CONTRIBUTE TO THE RESISTANCE OF THE RETURN GROUND PATH, MAKING IT EXTREMELY DIFFICULT TO DETERMINE THE ACTUAL GROUND-FAULT CURRENT VALUES.

FIGURE 20-7 An example of a ground fault. A section of the "hot" circuit conductor that has had its insulation scraped off during the installation process is touching a grounded surface.

If any marks are found, take the cover off and fix any places where two "hot" conductors are touching each other or where a "hot" conductor is touching a grounded surface.

• Then, replace the fuse or reset the circuit breaker.

• If the fuse does not blow or the circuit breaker does not trip, you have found the problem.

• If the fuse blows or the circuit breaker trips immediately, a short circuit still exists in the circuit. Continue to look for locations on the circuit that could be the cause of the problem.

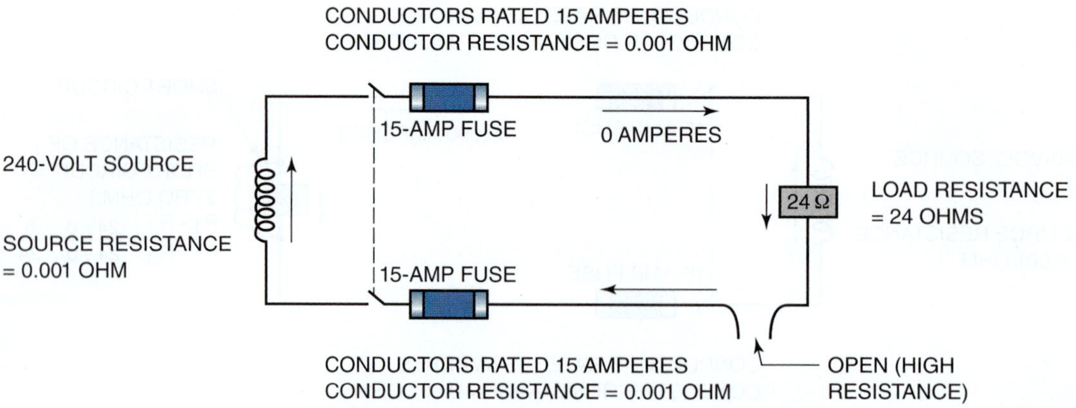

CONDUCTORS RATED 15 AMPERES
CONDUCTOR RESISTANCE = 0.001 OHM

15-AMP FUSE   0 AMPERES

240-VOLT SOURCE

SOURCE RESISTANCE
= 0.001 OHM

15-AMP FUSE

24 Ω   LOAD RESISTANCE
= 24 OHMS

CONDUCTORS RATED 15 AMPERES
CONDUCTOR RESISTANCE = 0.001 OHM   OPEN (HIGH RESISTANCE)

© Cengage Learning 2012.

FIGURE 20-8 An example of an open circuit. There is no current flow through the circuit when there is an "open" in the circuit.

- With the circuit de-energized, remove each cover plate and inspect the wiring of each device or lighting fixture on the circuit. Look for charred insulation, a wire shorted against the metal box, or a device that is defective.

- Fix any defective wiring.

- Then, replace the fuse or reset the circuit breaker.

  - If the new fuse does not blow or the circuit breaker does not trip immediately, turn on each wall switch one by one, checking each time to see if the fuse has blown or the circuit breaker has tripped.

  - If turning on a wall switch causes a fuse to blow or the breaker to trip, there is a short circuit in a light fixture or receptacle controlled by that switch, or there is a short circuit in the switch wiring.

- With the circuit de-energized, inspect the fixture, receptacle, and switch for charred wire insulation or faulty connections.

- Replace any faulty switch or fixture.

- Then, replace the fuse or reset the circuit breaker.

  - If the fuse does not blow or the circuit breaker does not trip, you have found the problem.

  - If the fuse blows or the circuit breaker trips immediately, a short circuit still exists in the circuit. Continue to look for locations on the circuit that could be the cause of the problem. Eventually you will find it!

(See Procedure 20-4 on pages 724–725 for the correct procedure for using a continuity tester to troubleshoot a receptacle circuit that has a ground fault.)

from experience...

There are many times when a ground-fault condition is caused by a bare grounding conductor making contact with one of the screw terminals in the box. Removing the switch cover will often reveal the problem. Moving the bare grounding conductor away from the terminal screw that it was touching will take care of the problem.

When performing troubleshooting tests with a continuity tester, electricians need to be aware of "feed-through," which occurs when a load, like a lightbulb, is still connected in a branch circuit that is being tested. The current from the continuity tester can flow through the load and cause a false reading. When testing lighting branch circuits with a continuity tester, remove all lamps from lighting fixtures. When using a continuity tester to test a receptacle circuit, make sure there are no portable items plugged into any of the receptacles on the circuit. Figure 20-9 shows some testing situations on a lighting circuit with the load still connected in the circuit.

Each electrical panel is supposed to have a legible directory that will tell an electrician what each circuit breaker or fuse protects. However, there will be times when it is necessary to work on a circuit but for some reason the directory does not show clearly which circuit breaker protects the circuit you need to work on. In this

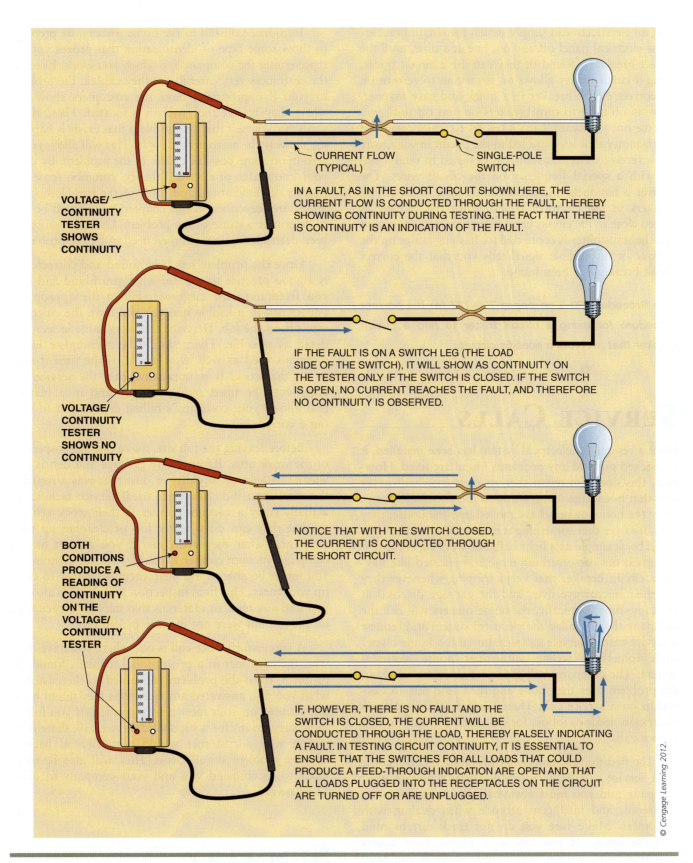

VOLTAGE/
CONTINUITY
TESTER
SHOWS
CONTINUITY

CURRENT FLOW
(TYPICAL)

SINGLE-POLE
SWITCH

IN A FAULT, AS IN THE SHORT CIRCUIT SHOWN HERE, THE
CURRENT FLOW IS CONDUCTED THROUGH THE FAULT, THEREBY
SHOWING CONTINUITY DURING TESTING. THE FACT THAT THERE
IS CONTINUITY IS AN INDICATION OF THE FAULT.

VOLTAGE/
CONTINUITY
TESTER
SHOWS NO
CONTINUITY

IF THE FAULT IS ON A SWITCH LEG (THE LOAD
SIDE OF THE SWITCH), IT WILL SHOW AS CONTINUITY ON
THE TESTER ONLY IF THE SWITCH IS CLOSED. IF THE SWITCH
IS OPEN, NO CURRENT REACHES THE FAULT, AND THEREFORE
NO CONTINUITY IS OBSERVED.

BOTH
CONDITIONS
PRODUCE A
READING OF
CONTINUITY
ON THE
VOLTAGE/
CONTINUITY
TESTER

NOTICE THAT WITH THE SWITCH CLOSED,
THE CURRENT IS CONDUCTED THROUGH
THE SHORT CIRCUIT.

IF, HOWEVER, THERE IS NO FAULT AND THE
SWITCH IS CLOSED, THE CURRENT WILL BE
CONDUCTED THROUGH THE LOAD, THEREBY FALSELY INDICATING
A FAULT. IN TESTING CIRCUIT CONTINUITY, IT IS ESSENTIAL TO
ENSURE THAT THE SWITCHES FOR ALL LOADS THAT COULD
PRODUCE A FEED-THROUGH INDICATION ARE OPEN AND THAT
ALL LOADS PLUGGED INTO THE RECEPTACLES ON THE CIRCUIT
ARE TURNED OFF OR ARE UNPLUGGED.

© Cengage Learning 2012.

FIGURE 20-9   A feed-through condition may cause false readings during testing. When a load is connected in the circuit being
tested, the current from the continuity tester may flow through the load and indicate a fault between the "hot" ungrounded conductor
and the grounded conductor. When using a continuity tester to troubleshoot a circuit, always disconnect the load from the circuit.

case, an electrician can simply switch the circuit breakers in the electrical panel off and on, one at a time, until the correct breaker is found or they can use a circuit tracer. Using a circuit tracer allows an electrician to zero in on the correct circuit breaker in a quick and easy manner. There is no need to turn breakers on and off until you find the one you want. Circuit tracers typically consist of a transmitter that is connected to the circuit in some way and a receiver. The transmitter is designed to send a signal with a special frequency on the circuit wires. The receiver is handheld and is positioned close to the circuit breakers in the electrical panel. When the receiver is moved close to the circuit breaker that protects the circuit that the transmitter is connected to, flashing lamps on the receiver or an audible signal tells you that the correct circuit breaker has been found.

(See Procedure 20-5 on pages 726–727 for the correct procedure for using a circuit tracer to find a circuit breaker that protects a specific circuit.)

SERVICE CALLS

Once a residential electrical system has been installed, it is checked out and any problems found are fixed. However, this does not mean that an electrician will never visit that house again and work on the electrical system. After the home owners have moved into the house, they may discover that something is not working correctly. It could be as simple as a light bulb not working because it is burned out or something more complicated like having a circuit breaker that keeps tripping whenever they use their microwave oven and the garbage disposal at the same time. Typically, the home owners will call the company that installed the electrical system and request a service call. A service call is required if a home owner finds something wrong with their house electrical system. The company will get some information about the problem over the phone and then send out an electrician on a service call. There are a few steps that an electrician needs to follow for successful completion of a service call.

The first thing to consider when you go into a home on a service call is to realize that you represent your company and you must always display a professional, courteous, and intelligent attitude when dealing with customers. Make sure you do not track dirt or mud into the house. If possible, have plastic boot covers available to be put on over muddy shoes if you need to walk through a house to the location where the problem exists. Be very careful so that none of your tools or other equipment leave scratches on walls, floors, or furniture.

Introduce yourself to the home owners. Be prepared to show some type of identification that proves you are representing the company for which you work. Listen to the customers very carefully as they explain the problem to you. If appropriate, have the customers show you where they believe the problem is located. Then, try to verify for yourself that the problem that exists is happening the way the customers say it is. This will allow you to zero-in on some possible causes of the problem. By using your knowledge of electrical theory, common sense, the process of elimination, and some of the troubleshooting procedures described in this chapter, you should be able to determine a cause of the problem. Once the cause has been determined, you will then need to fix the problem.

Once the problem has been found and corrected, it is a good practice to explain what you found and how you fixed it to the customer. Fill in the appropriate paperwork in a legible manner and have the customer sign off on the job. Describe the work done in as much detail as possible. There should be a complete list of materials used as well as a record of the time it took to do the work. It is imperative that the service call paperwork be filled out accurately and in a manner that allows your company's billing department to send out a correct bill.

Before leaving the job site, make one last inspection of the work area. Remove any garbage and debris that was left. Clean up any dirt or dust that was a result of the work you did. Many electrical service technicians actually carry a vacuum cleaner in their work vehicles so that they can do a better job of cleaning up their work area after they are done. If you do not have a company vacuum cleaner available, ask the customer if you might be able to use their vacuum cleaner to clean up your mess. This final inspection and cleanup also lets you find any tools or materials that may have been used on the job but were not initially picked up.

A successful service call is one where you have dealt with the customer in a professional manner, found the problem, fixed the problem, explained to the customer what you did, answered any questions they might have, and cleaned up your mess before you left. If you follow these suggestions for a successful service call, there is no reason why a customer would not call you back if another problem should arise. They will also be more likely to recommend you and your company to other home owners.

SUMMARY

An important stage in the installation of a residential electrical system is making sure that all *NEC®*, local, and state electrical requirements have been met. This chapter presented a checklist of common *NEC*® requirements that need to be included in virtually all residential wiring installations. If an electrician finds that any of the items listed are not included in the electrical system, he or she must bring the installation up to code.

The other very important part of this last stage is checking out each circuit to verify that everything is working the way it is supposed to before you leave this job and go on to the next one. Several testing procedures were presented in this chapter. If an electrical circuit is found to be working improperly or not at all, troubleshooting procedures will need to be started so that the reason for the problem can be found and

fixed. This chapter presented some common troubleshooting procedures, including the "halving the circuit" method for finding problems in a circuit.

As we learned earlier in this book, the *NEC*® requires electricians to install a residential electrical system in a neat and workmanlike manner. While many people interpret this to mean different things, there is one interpretation that all electrical people agree on: The electrical installation must be installed in a professional manner and must meet the minimum standards of the *NEC*®. This will result in an electrical system that should give reliable and safe service to the home owner for a long, long time.

PROCEDURE 20-1

Testing 120-Volt Receptacles with a Voltage Tester to Determine Proper Voltage, Polarity, and Grounding

- Wear safety glasses and observe all applicable safety rules.

A Test between the ungrounded conductor connection (short slot) and the grounded conductor connection (long slot). The meter should indicate 120 volts. *Note: With most tamper-resistant receptacles you will need to insert both test leads in the slots at the same time.*

B Test between the ungrounded conductor connection (short slot) and the grounding conductor connection (U-shaped grounding slot). The meter should indicate 120 volts. *Note: This step cannot be done with most tamper-resistant receptacles because you cannot insert only one test lead in the short slot. As an alternative, very carefully touch a test lead to one of the brass terminal screws on the side of the receptacle and insert the other test lead into the U-shaped grounding slot.*

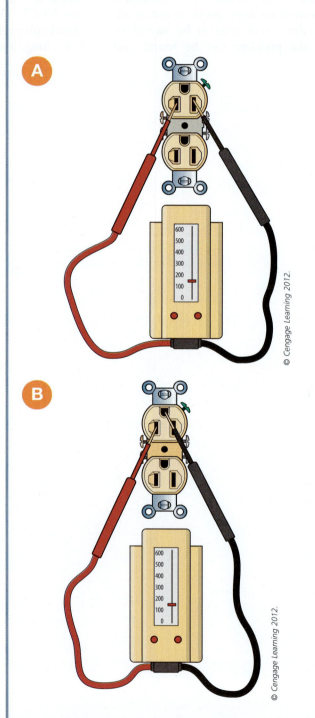

© Cengage Learning 2012.

C Test between the grounded conductor connection (long slot) and the grounding conductor connection (U-shaped grounding slot). The meter should indicate 0 volts. *Note: This step cannot be done with most tamper-resistant receptacles because you cannot insert only one test lead in the long slot. As an alternative, very carefully touch a test lead to one of the silver terminal screws on the side of the receptacle and insert the other test lead into the U-shaped grounding slot.*

- If the tests produce the proper voltage, the receptacle is wired properly, and you can go on to test the next receptacle on the circuit.

- If the tests in the previous steps do not produce the proper voltage, there is a problem with the circuit wiring, and the following troubleshooting procedures should be followed:

  - Check to see if the circuit is turned off.

  - Check to see that all circuit conductor connections are tight and that they all have continuity.

  - Check for damage to the circuit wiring.

C

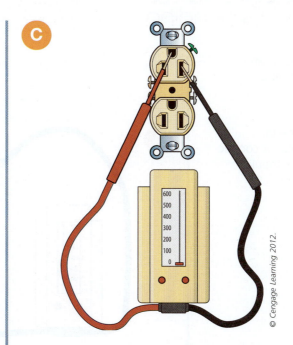

© Cengage Learning 2012.

- Check to see if the receptacle is being fed from a "buried box" that has been sheetrocked over and never trimmed out.

- Correct the problem(s) found during troubleshooting and retest the receptacle for proper voltage and grounding.

- If the test in step C produces a voltage of 120 volts, receptacle polarity is reversed, and the following troubleshooting procedures should be followed:

  - Turn off the circuit power.

  - Remove the receptacle from the outlet box.

  - Reverse the wire connections so that the "white" grounded conductor is attached to a silver screw and the "hot" ungrounded conductor is attached to a brass screw.

  - Turn on the power and retest the receptacle for proper polarity.

PROCEDURE 20-2

Testing 120/240-Volt Range and Dryer Receptacles with a Voltage Tester to Determine Proper Voltage, Polarity, and Grounding

- Wear safety glasses and observe all applicable safety rules.

A Test between the un-grounded conductor connections. The meter should indicate 240 volts.

A

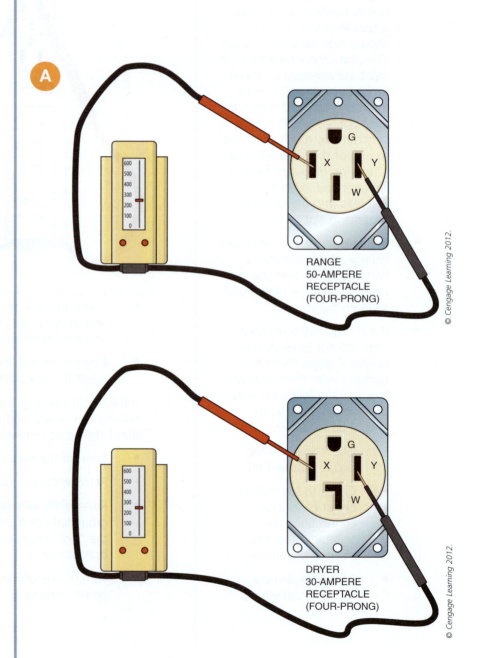

RANGE
50-AMPERE
RECEPTACLE
(FOUR-PRONG)

DRYER
30-AMPERE
RECEPTACLE
(FOUR-PRONG)

© Cengage Learning 2012.

B Test between one of the ungrounded conductor connections and the grounded conductor connection. The meter should indicate 120 volts.

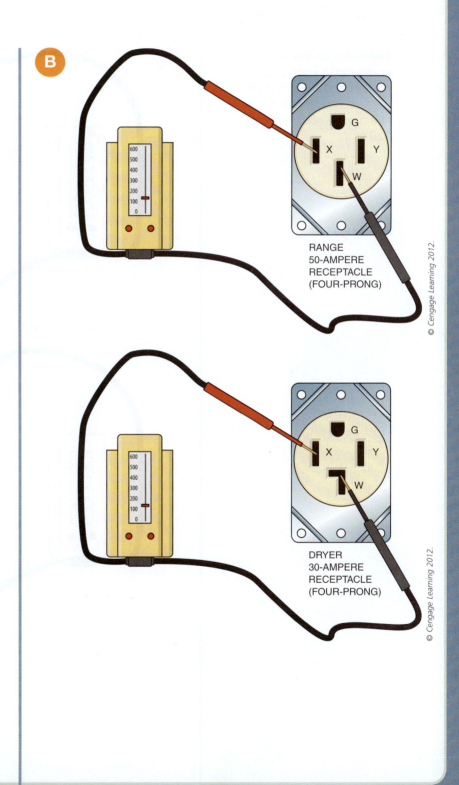

RANGE
50-AMPERE
RECEPTACLE
(FOUR-PRONG)

© Cengage Learning 2012.

DRYER
30-AMPERE
RECEPTACLE
(FOUR-PRONG)

© Cengage Learning 2012.

PROCEDURE 20-2

Testing 120/240-Volt Range and Dryer Receptacles with a Voltage Tester to Determine Proper Voltage, Polarity, and Grounding (Continued)

C Test between the other ungrounded conductor connection and the grounded conductor connection. The meter should indicate 120 volts.

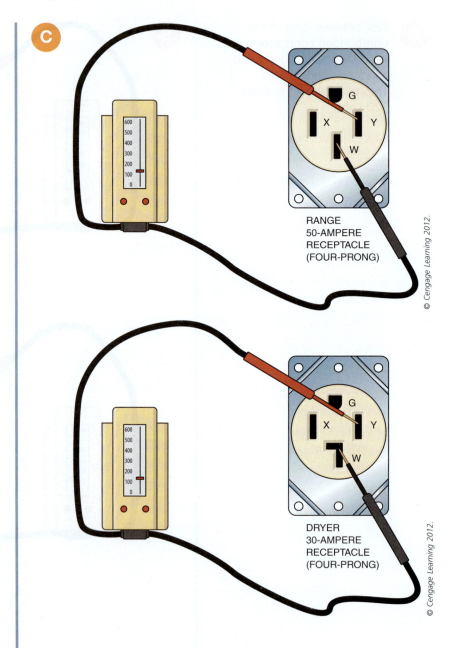

RANGE
50-AMPERE
RECEPTACLE
(FOUR-PRONG)

© Cengage Learning 2012.

DRYER
30-AMPERE
RECEPTACLE
(FOUR-PRONG)

© Cengage Learning 2012.

D Test between one of the ungrounded conductor connections and the grounding conductor connection. The meter should indicate 120 volts.

- If the tests in the previous steps produce the proper voltage, the receptacle is wired properly.

- If the tests in the previous steps do not produce the proper voltage, there is a problem with the circuit wiring, and troubleshooting procedures will need to begin. More than likely, it is simply a matter of the feed wiring connected to the wrong terminals.

- Correct the problem by rewiring the receptacle so the correct wires go to the correct terminals.

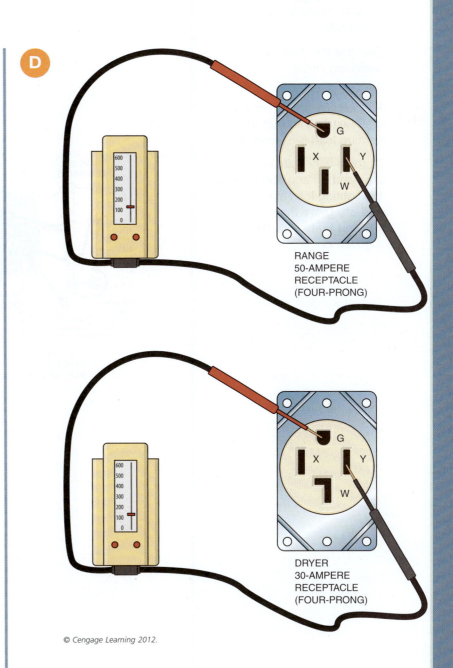

D

RANGE
50-AMPERE
RECEPTACLE
(FOUR-PRONG)

DRYER
30-AMPERE
RECEPTACLE
(FOUR-PRONG)

© *Cengage Learning 2012.*

PROCEDURE 20-3

Testing a Standard Three-Way Switching Arrangement

- Wear safety glasses and observe all applicable safety rules.

- Energize the three-way switching circuit.

A Position the toggle on three-way switch A so that the lighting load is energized.

- At three-way switch A, toggle it on and off and observe that the lighting load also goes on and off. Leave the switch in a position that keeps the lighting load on.

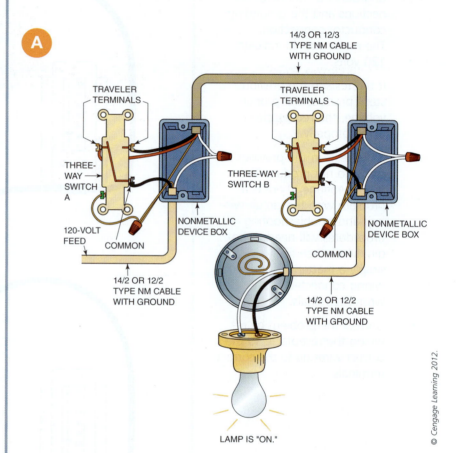

A

TRAVELER TERMINALS

14/3 OR 12/3 TYPE NM CABLE WITH GROUND

TRAVELER TERMINALS

THREE-WAY SWITCH A

THREE-WAY SWITCH B

NONMETALLIC DEVICE BOX

NONMETALLIC DEVICE BOX

120-VOLT FEED

COMMON

COMMON

14/2 OR 12/2 TYPE NM CABLE WITH GROUND

14/2 OR 12/2 TYPE NM CABLE WITH GROUND

LAMP IS "ON."

© Cengage Learning 2012.

B Go to three-way switch B and toggle it so the lighting load goes on and off. Leave this switch so that the lighting load is off.

- Finally, go back to three-way switch A and toggle it so the lighting load comes back on.

- If the three-way switches turn the lighting load on and off as required, the switching system was installed properly, and nothing more needs to be done.

- If the three-way switches do not turn the lighting load on and off as the test requires, troubleshooting procedures should be initiated. The most likely problem will be an improper connection of the traveler wires on one or both of the three-way switches.

- Turn off the electrical power to the circuit before initiating troubleshooting procedures. Lock out the power.

- Verify that the traveler wires are correctly wired to the traveler terminals of the three-way switch.

- If the traveler wires are incorrectly connected, rewire the circuit, turn the electrical power on, and retest the circuit. The switches should now work properly.

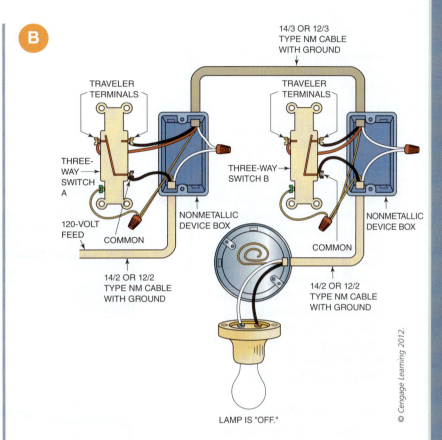

B

14/3 OR 12/3 TYPE NM CABLE WITH GROUND

TRAVELER TERMINALS

TRAVELER TERMINALS

THREE-WAY SWITCH A

THREE-WAY SWITCH B

NONMETALLIC DEVICE BOX

NONMETALLIC DEVICE BOX

120-VOLT FEED

COMMON

COMMON

14/2 OR 12/2 TYPE NM CABLE WITH GROUND

14/2 OR 12/2 TYPE NM CABLE WITH GROUND

LAMP IS "OFF."

© Cengage Learning 2012.

PROCEDURE 20-4

Using a Continuity Tester to Troubleshoot a Receptacle Circuit That has a Ground Fault

- Wear safety glasses and observe all applicable safety rules.

- Make sure no electrical equipment is plugged into any receptacle on the circuit.

A In the electrical panel where the circuit originates, turn off the circuit breaker and disconnect the circuit ungrounded conductor (the black wire) from the breaker, the circuit-grounded conductor (the white wire) from the neutral terminal bar, and the circuit equipment-grounding conductor (the bare or green wire) from the grounding terminal.

B At the electrical panel, use a continuity tester and test between the ungrounded circuit conductor (black) of the circuit and the grounded (white) circuit conductor. If there is continuity, then the problem is a short circuit. Then, test between the ungrounded circuit conductor (black) and the circuit-grounding conductor (bare or green). If there is continuity, then the problem is a ground fault.

- To find out where either the short circuit or ground fault is located on the circuit, locate a receptacle outlet box that is approximately midway on the circuit. Consulting a cabling diagram of the branch circuit will help with this determination.

A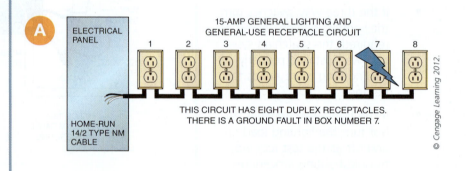

ELECTRICAL PANEL

15-AMP GENERAL LIGHTING AND GENERAL-USE RECEPTACLE CIRCUIT

1 2 3 4 5 6 7 8

THIS CIRCUIT HAS EIGHT DUPLEX RECEPTACLES. THERE IS A GROUND FAULT IN BOX NUMBER 7.

HOME-RUN 14/2 TYPE NM CABLE

© Cengage Learning 2012.

B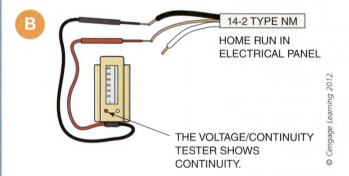

14-2 TYPE NM

HOME RUN IN ELECTRICAL PANEL

THE VOLTAGE/CONTINUITY TESTER SHOWS CONTINUITY.

© Cengage Learning 2012.

- At the midpoint receptacle outlet box, remove the circuit conductors. This will separate the circuit into two halves: the line side and the load side.

C   At the receptacle outlet box, perform the same tests with the continuity tester as described previously. This will allow you to determine if the problem is in the wiring from the electrical panel to your location (the line side) or in the wiring from your location to the end of the circuit (the load side).

D   Once you determine which half of the branch circuit contains the problem, split the part of the circuit you are working with in half and repeat the continuity testing.

E   Continue to split the part of the circuit you are working with in half and test for continuity until you locate the problem.

- Once the problem is located, make the necessary repairs, reconnect the circuit components, reconnect the wiring in the electrical panel, and retest the circuit.

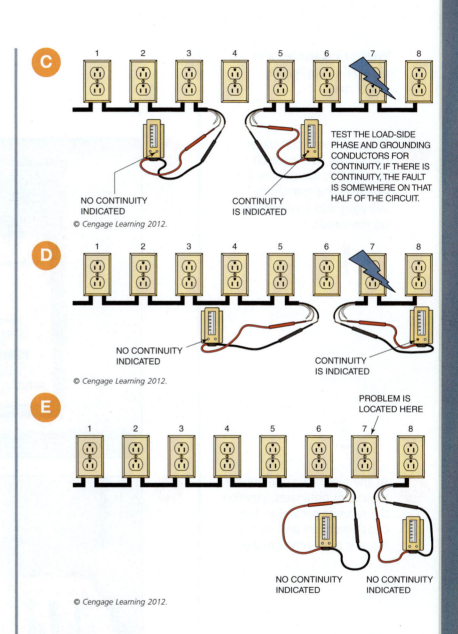

C

NO CONTINUITY INDICATED

CONTINUITY IS INDICATED

TEST THE LOAD-SIDE PHASE AND GROUNDING CONDUCTORS FOR CONTINUITY. IF THERE IS CONTINUITY, THE FAULT IS SOMEWHERE ON THAT HALF OF THE CIRCUIT.

© Cengage Learning 2012.

D

NO CONTINUITY INDICATED

CONTINUITY IS INDICATED

© Cengage Learning 2012.

E

PROBLEM IS LOCATED HERE

NO CONTINUITY INDICATED

NO CONTINUITY INDICATED

© Cengage Learning 2012.

PROCEDURE 20-5

Using a Circuit Tracer to Find a Circuit Breaker That Protects a Specific Circuit

• Wear safety glasses and observe all applicable safety rules.

 Connect the circuit tracer transmitter to the circuit you wish to identify. In this example, the transmitter will plug into a receptacle on the circuit.

© Cengage Learning 2012.

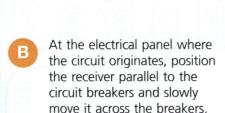 At the electrical panel where the circuit originates, position the receiver parallel to the circuit breakers and slowly move it across the breakers.

© Cengage Learning 2012.

C Once the receiver has indicated the correct circuit breaker with flashing lamps or an audible signal, turn off the breaker.

C

© Cengage Learning 2012.

D Disconnect the transmitter and verify that the electrical power is off by using a voltage tester at the point where the transmitter had been connected to the circuit.

- Perform a lock out/tag out procedure at the electrical panel and complete the work on the circuit as needed.

- Once the work is done, take off the lock out/tag out equipment and energize the circuit by turning the circuit breaker on.

D

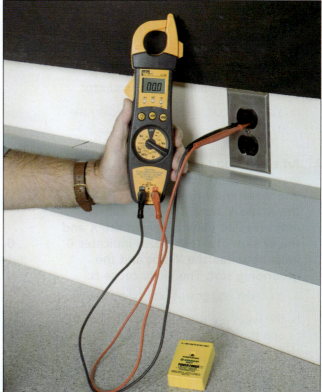

© Cengage Learning 2012.

REVIEW QUESTIONS

Directions: Answer the following items by circling the best response.

1. **An electrician tests a 15-amp, 125-volt-rated duplex receptacle with a voltage tester. The tester indicates 120 volts between the ungrounded (short) slot and the grounded (long) slot. The tester also indicates 120 volts between the grounded (long) slot and the grounding slot (U-shaped). The likely cause is:**

   a. an open equipment-grounding path.

   b. reversed polarity at the receptacle.

   c. an open grounded circuit conductor.

   d. a short circuit in the cable feeding the receptacle.

2. **An electrician tests a 15-amp, 125-volt-rated duplex receptacle with a voltage tester. The tester indicates 120 volts between the ungrounded (short) slot and the grounded (long) slot. The tester also indicates 0 volts between the ungrounded (short) slot and the grounding slot (U-shaped). The likely cause is:**

   a. an open equipment-grounding path.

   b. reversed polarity at the receptacle.

   c. an open grounded circuit conductor.

   d. a short circuit in the cable feeding the receptacle.

3. **An electrician tests a 30-amp, 250-volt dryer receptacle with a voltage tester. The tester indicates 120 volts between the X and Y slots. The tester also indicates 240 volts between the X and the W slot. The tester also indicates 0 volts between the Y slot and the grounding slot. The likely cause is:**

   a. an open equipment-grounding path.

   b. the wires are reversed between the W and the Y slots.

   c. the wires are reversed between the X and Y slots.

   d. an open grounded circuit conductor.

4. **An electrician tests a three-way switching circuit. The lighting load can be turned OFF and then ON at each of the three-way switch locations. However, when one three-way switch toggle is left in the up position, toggling the other three-way switch does not cause the lighting load to come on. The likely cause is:**

   a. an open equipment-grounding path.

   b. an open grounded conductor.

   c. the traveler wires are incorrectly connected to the traveler terminals on the switch.

   d. there is no electrical power at that switch.

5. **An electrician is testing a circuit where a single-pole switch controls an incandescent lighting fixture in a bedroom. When the single-pole switch is toggled ON, the lighting fixture does not come on. Which of the following items could *not* be a cause for this?**

   a. There is an open equipment-grounding path.

   b. The circuit breaker protecting the circuit is off.

   c. There are no lamps installed in the lighting fixture controlled by the single-pole switch.

   d. The connections in the lighting outlet box are loose.

6. **A(n) _____ is an unintended low-resistance path through which current flows around rather than along a circuit's intended current path.**

   a. open

   b. overload

   c. ground fault

   d. short circuit

7. A(n) _____ is an unintended low-resistance path in an electrical circuit through which current flows to ground using a pathway other than the one intended.

   a. open

   b. overload

   c. ground fault

   d. short circuit

8. A(n) _____ results in no current flow in a circuit, even though the circuit is energized with the proper voltage.

   a. open

   b. overload

   c. ground fault

   d. short circuit

9. A feed-through condition may cause false readings during circuit testing and can happen when a load is still connected to the circuit being tested. (Circle the correct response.) True or False

10. When testing a "live" 120-volt receptacle using a voltage tester, you should get a 120-volt reading when measuring from the grounded conductor connection (long slot) to the grounding conductor connection (U-shaped slot). (Circle the correct response.) True or False

11. When testing a "live" 120-volt receptacle using a voltage tester, you should get a 120-volt reading when measuring from the ungrounded conductor connection (short slot) to the grounding conductor connection (U-shaped slot). (Circle the correct response.) True or False

12. When testing a "live" 50-amp, 120/240-volt range receptacle using a voltage tester, you should get a 240-volt reading when measuring from the grounded conductor connection (W slot) to either ungrounded conductor connection (X or Y slot). (Circle the correct response.) True or False

13. When checking out the electrical circuits for correct operation, an electrician should turn them all on at the same time. (Circle the correct response.) True or False

14. When troubleshooting for short circuits or ground faults, the circuit being tested should be energized with electrical power. (Circle the correct response.) True or False

15. A continuity tester can be used to find short circuits, ground faults, and open circuits in de-energized circuits. (Circle the correct response.) True or False

Green House Wiring Techniques

CHAPTER 21
Green Wiring Practices

CHAPTER 22
Alternative Energy System Installation

Green Wiring Practices

OBJECTIVES

Upon completion of this chapter, the student should be able to:

- Demonstrate an understanding of how to advise a house building team about energy efficient wiring practices.

- Demonstrate an understanding of how to advise the building team about durability and water management when installing the electrical system.

- Demonstrate an understanding of how to advise the building team about selecting green products whenever they are available.

- Demonstrate an understanding of how to advise the building team about reducing material use and waste when installing the house electrical system.

- Demonstrate an understanding of how to advise the building team about what electrical system items to include in a home owner education and reference manual.

GLOSSARY OF TERMS

dioxins a family of chlorinated chemicals; toxic under certain exposure conditions; emitted when combustion of carbon compounds is inefficient; associated with cancer of the stomach, sinus lining, liver, and lymph system

Energy Star an international standard for energy efficient consumer products

flashing sheets of metal or other material used to weatherproof locations on exterior surfaces, like a roof, where penetrations have been made

green the process of designing and building a home that minimizes its impact on the environment both during construction and over its useful life

micro-hydro electric system an alternative energy system that uses water pressure to move a turbine, which then drives a generator to produce electricity

phantom load a load that consumes electricity even when the load is not performing its intended function; examples are cable TV boxes, internet routers, microwave ovens, and televisions

photovoltaic the name used to describe the effect of sunlight striking a specially prepared surface and producing electricity

sones an internationally recognized unit of loudness, measured in decibels (dB); doubling the sone value is equivalent to doubling the loudness; one sone is roughly equivalent to the sound of a quiet refrigerator

residential electrician has an important role in the process of building a green home. The role starts during the planning phase by helping the building team (the designer, the building contractor, and the home owner) make green choices about the entire electrical system. During construction the electrician works closely with the general contractor and other trade contractors to ensure the green goals of the project are met. Finally, the electrician shows the home owner how to operate and maintain the electrical system for peak energy efficiency. Green Tips have been used in previous chapters to tell you about the advantages of installing a green electrical system. This chapter covers the material that every electrician should know so that the home they are wiring has the most up-to-date and efficient green electrical system possible.

ENERGY EFFICIENCY

Energy efficiency is one of the most important features of a **green** home.

Electricity accounts for about half of the energy used in an average home. Reducing the electrical energy consumption of both new and existing homes will have a big impact on overall energy use in this country. Homes that use less electricity to operate save home owners a lot of money.

The electrician can influence the electrical efficiency of a home in a couple of ways. The first way is advising the building team on the selection of energy efficient lighting, appliances, and equipment. The second way is recommending and installing electronic control equipment that reduces the electricity used in a home. Lights, electronic devices, and appliances are often left on even when not being used. Automatic or easy-to-use controls such as timers or wall-mounted switches will reduce electricity waste.

SELECTING AND INSTALLING ENERGY EFFICIENT ELECTRICAL EQUIPMENT

There are many options when selecting lighting, appliances, and other electrical equipment for a house. Newer models are much more energy efficient than older ones; but even among new models there is a range of energy efficiency. The electrical efficiency of many products is listed. The electrician can advise the building team about what is available and what the best choices are that fit within the project budget.

LIGHTING

One of the best choices for energy efficient lighting is fluorescent hardwired **Energy Star** (Figure 21-1) qualified lighting fixtures. These light fixtures generally have electronic ballasts and accept pin style fluorescent lamps. The advantage of pin style fixtures is that home owners can only use fluorescent replacement lamps and not less efficient incandescent lamps. Fluorescent lamps only use one-quarter of the electricity that incandescent lamps use and they last up to ten times longer.

Light emitting diode (LED) lighting fixtures are another option. LEDs are even more energy efficient than fluorescent lighting fixtures and they last longer. However, they cost more than fluorescent fixtures and there are limited fixture choices. LED light fixture choices are expected to increase over the next decade and as the costs go down it will not be long before LED fixtures and lamps are the preferred lighting type for green homes.

Energy efficient light fixtures come in recessed, ceiling surface-mounted, wall-mounted, and specialty styles. Experienced electricians have a good sense of

FIGURE 21-1 This logo indicates that an electrical item is Energy Star qualified. *US Environmental Protection Agency, ENERGY STAR program.*

what fixtures will work effectively for lighting different rooms and areas of a home and can advise the building team on the choices.

APPLIANCES

There are many types of electric appliances used in homes: refrigerators, freezers, dishwashers, electric cook tops, electric ovens, electric ranges, microwave ovens, clothes washing machines, and electric clothes dryers. Energy Star rated models (Figure 21-2) that use less energy than their ordinary counterparts are available for most appliances. The electrician should advise the building team to select Energy Star models when available. Energy efficient appliances may cost more than less efficient models; but the reduced operation expense more

than offsets the upfront cost over the life of the equipment. For example, an Energy Star energy efficient model electric clothes dryer may cost $600 compared to a less energy efficient model that sells for $550. However, the energy efficient model's electrical energy cost for a year may be around $70, while the less energy efficient model's yearly electrical energy cost may be $80 or more. It is easy to see that after five years of use the extra money that the more energy efficient model costs new is made up in electrical energy savings. As the years go by the electrical energy savings will amount to a large share of the energy efficient appliance's original cost. In some cases, the entire cost of the energy efficient electric clothes dryer may be offset by the energy savings. Of course this will depend upon the number of years the appliance is used and the actual amount of electrical energy savings.

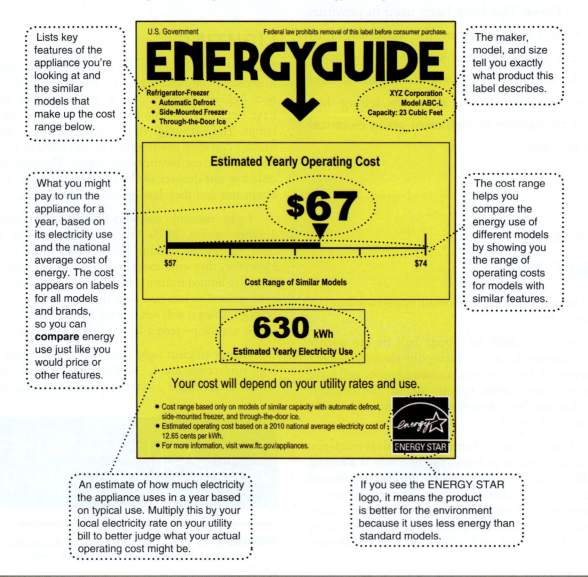

FIGURE 21-2 An example of an Energy Guide label found on appliances. This one is for a refrigerator/freezer. *Federal Trade Commission, www.ftc.gov.*

EQUIPMENT

Electric water heaters, kitchen exhaust fans, bathroom fans, heat pumps, air-conditioning condensers and air handlers, and other electrical equipment are hardwired into homes. There are energy efficient models available for each. The electrician along with the plumber and heating, ventilation, and air-conditioning (HVAC) contractor can advise the building team on equipment choices and system layout.

In many cases, the electrician will work cooperatively with another trade to supply and install equipment so both parties will need to understand installation requirements for peak performance and energy efficiency. Let's look at a couple of special cases and how equipment selection impacts green wiring practices.

Electric Water Heaters

Electric water heaters supply domestic hot water and come in a couple of different types:

- A remotely located tank type that supplies the hot water needs of an entire house;

- A tankless type that is located near the fixture(s) it supplies.

Common tank type electric water heaters use electric resistance coils to heat the water (Figure 21-3). Tank sizes range from 20 to 120 gallons. Tank type water heaters are located in a basement, garage, or utility area often far from the kitchen or bathrooms where the hot water is used. Long plumbing runs lead to some heat loss of the water in the pipes. The tanks are insulated to reduce energy loss. The more energy efficient tank type water

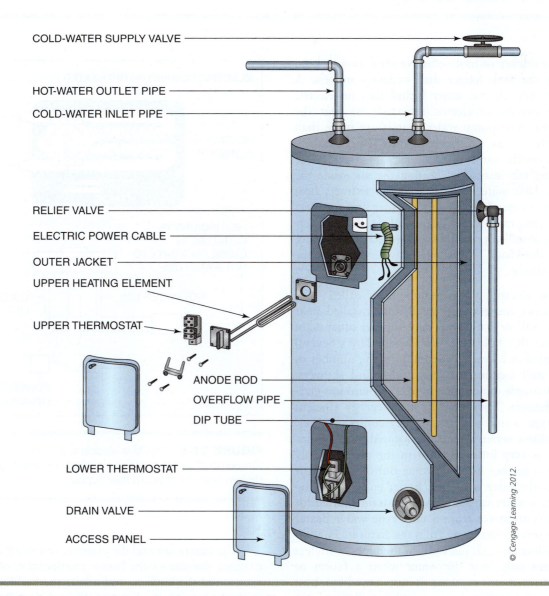

© Cengage Learning 2012.

FIGURE 21-3   A common tank type electric water heater with resistive heating elements.

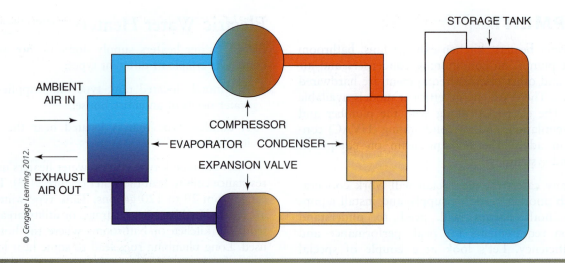

FIGURE 21-4 An electric heat pump water heater. The heat pump extracts heat from the air surrounding the water heater and transfers it to the storage tank water.

heaters have thicker or more effective insulation to keep the heat in the tank longer than ordinary models. A newer tank type electric water heater uses an electric heat pump instead of electric resistance to heat the water. Electric heat pump water heaters are more than twice as efficient as electric resistance water heaters. They work on the same principle as a heat pump for space heating but instead of heating the air inside the house, they heat water. The heat pump extracts heat from the air surrounding the water heater and transfers it to the storage tank water (Figure 21-4). Retrofit versions can be installed onto an existing electric hot water heater. The drawback of this type of water heater is the greater initial cost.

Tankless electric water heaters (also called "on-demand" water heaters) are small and installed near the kitchen sink or bathroom they supply; often right in the cabinet that holds the sink (Figure 21-5). Several may be needed in a house depending on the number of bathrooms and location of sinks, tubs, and showers. There are a couple of advantages to using tankless electric water heaters. They may be more energy efficient than tank type water heaters and they conserve water. Because tankless water heaters are located near the fixtures, there is very little heat lost in the pipes. It only takes seconds for hot water to come out of a faucet or shower once the valve is opened. Faster hot water delivery means less water wasted down the drain waiting for the hot water to arrive from a tank type water heater that can be several feet away from the fixtures. In addition, heat is lost through the walls of the pipes. Tankless water heaters only heat the water when a faucet or shower is turned on so there is not 'stand-by' heat loss. Tank type water heaters—even well insulated ones—lose heat constantly.

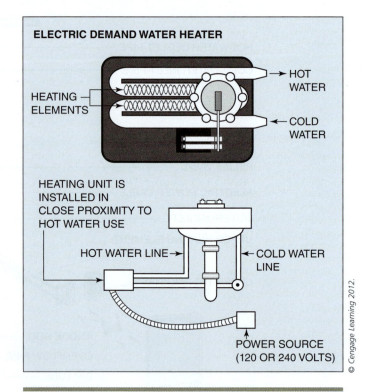

FIGURE 21-5 Tankless electric water heaters (also called "on-demand" water heaters) are small and installed near the kitchen sink or bathroom they supply.

The electrician and the plumber can work together to evaluate the size of the house and location of the bathrooms and the kitchen, and advise the building team on the most energy- and water-efficient electrically heated hot water system for a green house.

Exhaust Fans and Fresh Air Ventilation Fans

Kitchen and bathroom exhaust fans and fresh air ventilation systems are important equipment in a green home. The electrician often supplies and installs exhaust fans and can advise the building team on model selection and installation location. The electrician also usually has a big hand in the installation of the control system for a fresh air ventilation system used in a green home.

The airtight nature of an energy efficient home calls for mechanical ventilation equipment to exhaust excess moisture from the bathrooms and kitchen. Without mechanical ventilation, the humidity level inside the house will rise to an unhealthy level and the durability of the structure may suffer as well.

The simplest mechanical ventilation system uses exhaust fans located in the kitchen and bathrooms. There are three features of exhaust fans to consider when selecting one: electrical efficiency, noise level, and power. Energy Star qualified exhaust fans are efficient, quiet, and powerful which helps making the selection easier. Quiet fans are more likely to be used by owners than noisy fans, so the noise level is important. Noise level is measured in **sones** and a quiet model will be rated 1.5 sones or lower. Fan power is measured in cubic feet per minute (CFM).

Bathroom exhaust fan power is matched to the size of a bathroom and whether there is a shower or tub (Figure 21-6). Large bathrooms with a shower require more powerful fans than a small bath without a shower. The large bathroom has a greater volume of air inside to ventilate and the shower generates more moisture vapor into the air that needs exhausting than in a bathroom without one.

Special electrical controls are used to operate bathroom fans in a green home. Timers are used to continue exhaust fan operation for a period of time after a user leaves the bathroom. Dehumidistats (Figure 21-7) sense the humidity level in a bathroom and automatically run exhaust fans when moisture vapor reaches a high level. 24-hour timers can be installed to operate exhaust fans periodically throughout the day and night to provide fresh air ventilation for the house. Fan speed controls can be installed to adjust the CFM output of the fans.

A variety of controls are available. Some controls combine several functions like time delay, fan speed, and 24-hour timing together in one unit. The electrician needs to be familiar with the fan control options available in order to wire an exhaust fan system to meet the needs of a green home. Often, the controls require more conductors run between the control and the exhaust fan, so following the manufacturer's installation instructions is very important.

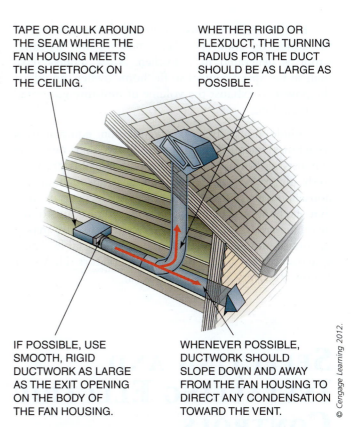

TAPE OR CAULK AROUND THE SEAM WHERE THE FAN HOUSING MEETS THE SHEETROCK ON THE CEILING.

WHETHER RIGID OR FLEXDUCT, THE TURNING RADIUS FOR THE DUCT SHOULD BE AS LARGE AS POSSIBLE.

IF POSSIBLE, USE SMOOTH, RIGID DUCTWORK AS LARGE AS THE EXIT OPENING ON THE BODY OF THE FAN HOUSING.

WHENEVER POSSIBLE, DUCTWORK SHOULD SLOPE DOWN AND AWAY FROM THE FAN HOUSING TO DIRECT ANY CONDENSATION TOWARD THE VENT.

© Cengage Learning 2012.

FIGURE 21-6 An exhaust fan in a bathroom can be ducted through the roof or through an outside wall.

© Cengage Learning 2012.

FIGURE 21-7 Dehumidistats sense the humidity level in a bathroom and automatically run exhaust fans when moisture vapor reaches a high level. The numbers on the dehumidistat represent the percent of relative humidity desired.

Kitchen exhaust fans are installed over the top of the range or cooktop where most of the moisture vapor is generated in a kitchen. Most models have built-in fan speed control so the home owner can adjust the power to exhaust the volume of moisture generated when cooking.

Whole house fresh air ventilation equipment is sometimes installed in a green home that not only exhausts indoor air from bathrooms but also draws in fresh air from outside and ducts it throughout the house. Some systems include an energy recovery feature that exchanges the energy in the exhaust air to the incoming fresh air. The energy recovery feature makes the house more energy efficient and comfortable. These systems are called Heat Recovery Ventilators (HRVs) or Energy Recovery Ventilators (ERVs).

SELECTING AND INSTALLING ELECTRIC CONTROLS

Automatic controls make life easier and can save energy at the same time. Traditional electrical controls like light switches require home owners to operate them. When someone forgets to turn off a switch, the lights stay on, using electricity unnecessarily. Lighting controls, HVAC system controls, and controls for many of the devices people have in their homes can be installed for convenience and to improve energy efficiency.

LIGHTING CONTROL EQUIPMENT

One way to eliminate lights being left on when either they are not needed or when no one is in a room is to install a motion detector (also called an occupancy sensor) to turn lights on and off automatically. A motion detector controls lighting by sensing when someone is in a room or hallway and turning the lights on. Then, after a brief time delay, if the person is no longer sensed, the detector turns the light off. Automatic lighting controls can reduce electricity by only turning on the lights when they are needed. Other controls sense the level of light (photo cells) and can automatically turn on lights inside or outside the house when the natural light fades. Timer controls can automatically turn lights on and off at prescribed periods of the day. These are commonly used for exterior lighting.

PROGRAMMABLE THERMOSTAT

A programmable thermostat (Figure 21-8) regulates the heating and air-conditioning systems. It lowers and raises the temperature inside a house periodically throughout the day and night to save energy. The programmable thermostat can be set to lower the temperature inside the house at night while home owners are sleeping and when they are away during the day at work or school. Lowering the inside temperature of a house when it is not occupied or when the occupants are asleep reduces the energy used to heat or cool the house. A programmable thermostat automatically adjusts the temperature when the family is home and awake early in the morning and again in the evening so they are comfortable. It is estimated that a properly programmed thermostat can reduce the energy to heat and cool a home by up to 15%.

CONTROLS FOR PHANTOM ELECTRIC LOADS

Some appliances and electric devices consume electricity even when they are not performing their intended function. Cable TV boxes, Internet routers, microwave ovens, and televisions are just a few examples of devices that use electricity even when turned 'off.' Televisions, video recorders, and stereo equipment consume electricity in 'standby mode' waiting for a remote control single to turn them on. Battery chargers for cell phones and laptop computers often use electricity even if the gadgets they power are fully charged or not even connected. Power converters for an Internet router and cordless phones are other examples of devices that consume electricity even when they are not being used. The total

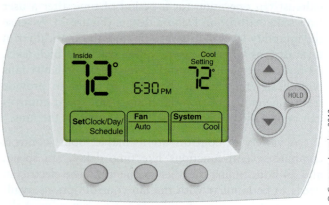

FIGURE 21-8   A programmable thermostat regulates the heating and air-conditioning systems automatically.

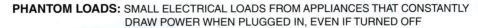

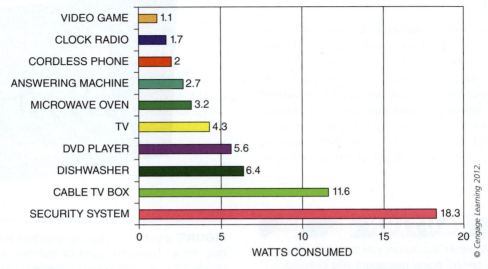

PHANTOM LOADS: SMALL ELECTRICAL LOADS FROM APPLIANCES THAT CONSTANTLY DRAW POWER WHEN PLUGGED IN, EVEN IF TURNED OFF

Appliance	Watts
VIDEO GAME	1.1
CLOCK RADIO	1.7
CORDLESS PHONE	2
ANSWERING MACHINE	2.7
MICROWAVE OVEN	3.2
TV	4.3
DVD PLAYER	5.6
DISHWASHER	6.4
CABLE TV BOX	11.6
SECURITY SYSTEM	18.3

WATTS CONSUMED

© Cengage Learning 2012.

FIGURE 21-9   This illustration shows several common phantom loads found in a home and the approximate number of watts consumed.

electrical load of all these devices can add up to 5% or more of a monthly electric bill. This load is called the '**phantom load**' because it is not obvious the devices are using electricity (Figure 21-9).

The electrician can help the building team locate and install special circuits to reduce electrical loads from sucking electricity needlessly. Special circuits with switch or timer control allow the home owner to completely turn off electricity to the loads either at will or automatically when the loads are not needed.

from experience...

A charging station for battery-operated loads can be designed and installed in a house. Several receptacles are ganged together in a closet or cabinet and controlled by an automatic timer. All the battery chargers can be left plugged into the station, but the timer will only supply electricity for a short period of time during the night to do the charging.

Small appliances like microwave ovens, DVD players, stereo systems, and televisions can be plugged into switched outlets so they can truly be 'turned off' when they are not being used.

Computer monitors, Internet routers, and other devices that have an ON/OFF switch are often left on all the time. Ideally, the home owner should turn them off when not in use but in the event they do not, a simple control system can be put to use. These loads can be plugged into either a plug-in style timer or a timer-controlled receptacle. The timer can be set to power the devices only when they are likely to be used, such as the early morning and the evening when the owners are active.

MAINTAINING THE INTEGRITY OF THE AIR BARRIER AND INSULATION

The air barrier of a house is one of the most important elements of the energy efficiency of a green home. It blocks air from moving between the inside of the house and the outside and holds the conditioned (heated or cooled) air in the house. The air barrier may be the drywall covering the inside walls and ceiling, the exterior wall sheathing, the housewrap (sometimes called Tyvek), the floor sheathing, and other materials that block air movement. The air tightness of a house is achieved through meticulous air sealing at all penetrations that go from inside the house to the outside through the air barrier.

AVOID HOLES IN THE AIR BARRIER

The best practice is to avoid making holes in the air barrier of the house whenever possible. Recessed lighting fixtures are one example of air barrier holes that can be avoided. The recessed fixtures penetrate the ceiling drywall air barrier and leak air around the perimeter and through the lighting fixtures themselves. The electrician should always advise the building team against planning recessed light fixtures in insulated ceilings. Surface-mounted ceiling or wall lighting fixtures can be used as an alternative.

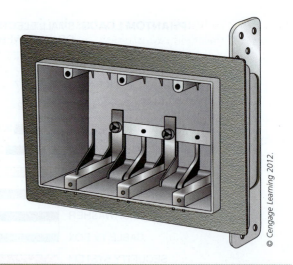

© Cengage Learning 2012.

FIGURE 21-10   A three-gang airtight nonmetallic electrical box. Airtight boxes are used for switches, receptacles, lights, paddle fans, or other fixtures mounted in exterior walls, floors with a basement or crawl space beneath, or ceilings with an attic above in green homes.

> **CAUTION**
>
> **CAUTION:** Whether the house you are wiring is a green home or not, some designers and building contractors do not want lighting fixtures installed in a ceiling because of the amount of air that can escape around them. Always check with the building contractor to find out if lighting fixtures can be ceiling installed.

When recessed lighting fixtures must be installed in an insulated ceiling, Type IC airtight rated models should be used. Airtight recessed lighting fixtures are not truly "air-tight" but they are sealed better than ordinary Type IC fixtures.

Many holes the electrician makes to run cables and equipment penetrate the air barrier of the house. Here is a list of common penetrations made by the electrician through the air barriers of a house:

- Cables to exterior lights
- Electrical boxes for exterior lights
- Recessed light fixtures
- Ceiling light fixture electrical boxes
- Cables or conduits run into an attic, basement, or crawl space
- Service entrance cable
- Exterior receptacle outlets
- Interior receptacle and switch boxes on exterior walls
- Wiring routed from an interior partition wall into an exterior wall

The electrician should discuss the steps for air sealing penetrations with the building team during the planning stage. An air-sealing specialist or the insulation contractor often handles the sealing task; but sometimes the building team assigns the electrician the responsibility for sealing the penetrations they make.

Even when another trade contractor handles the air sealing work, the electrician may still have to supply and install special equipment to simplify the air sealing process. Electrical boxes that penetrate the air barrier should be special airtight models. The airtight boxes are used for switches, receptacles, lights, paddle fans, or other fixtures mounted in exterior walls, floors with a basement or crawl space beneath, or ceilings with an attic above. These boxes have gaskets where the wires enter and exit; and they have gasketed flanges that seal to the drywall or exterior wall sheathing (Figure 21-10).

ON-SITE ELECTRIC POWER GENERATION

Some green homes incorporate on-site electric power generation from alternative energy resources such as **photovoltaic** (PV) systems (also called solar electric systems), wind turbine generators, and micro-hydro electric generators. Alternative energy systems convert energy from the sun, wind, or moving water into electricity. These systems are becoming more popular with home owners as energy costs rise, government subsidies increase, and equipment costs decrease.

Until fairly recently, alternative energy system installation has been more of an electric specialty and few electricians installed these systems. However, improvements and simplifications in the systems make them easier to install and that has resulted in the opening up of the field to properly trained electricians. Local community colleges and technical schools as well as equipment manufacturers and distributors usually offer the specialized training in alternative energy system installation.

The electrician has an important role advising a building team considering an on-site power generation system. There are many options including the type of system (solar, wind, or water), capacity of the system, and whether the system is tied to the electric grid or independent. A knowledgeable electrician can help guide the building team.

There are a variety of photovoltaic systems (Figure 21-11). Some consist of separate solar panels mounted to a roof or ground level frame, and others are integrated into roof shingles, tiles, or metal roof panels. Since most PV systems are mounted onto roofs, the electrician has to work closely with roofers who install the roofing integrated PV or install and flash mounting hardware to support a PV panel frame. Not all houses or building sites are suited to PV systems. First, the house must be in a region that receives enough sunny

days to make a PV system cost effective. Regions that have long periods of cloudy days may not be suitable for PV systems. Ideally, the roof of the house should face south so a roof-mounted PV system receives the best exposure. There should not be any trees or structures that shade the PV panels during the day, otherwise the electricity generation will be reduced. These are all site conditions that the electrician can evaluate to determine how feasible a PV system is for a specific installation. Photovoltaic system installation is presented in more detail in Chapter 22.

There are many residential sized wind turbine systems available now (Figure 21-12). They come in several different blade configuration styles, tower styles, and electric output. Wind turbines require a minimum constant wind speed in order to work, so a lot of home sites are not suitable for installation. Some regions just do not get enough wind to make them cost effective. Wind turbines need to be located in an open area away from obstructions like trees and buildings that can slow or deflect the wind. Even a small wind turbine requires sturdy mounting equipment that is usually installed by a specialty contractor. The electrician

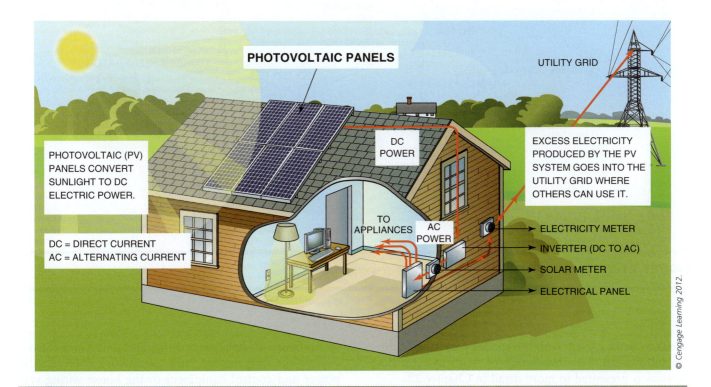

FIGURE 21-11 A typical green home photovoltaic system. This one is connected to the utility grid so any excess electricity produced by the system can be put on the grid for other people to use.

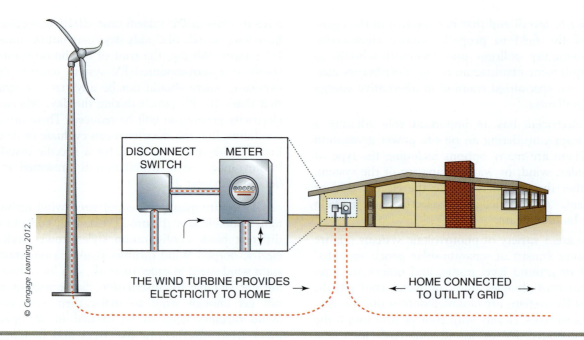

© Cengage Learning 2012.

DISCONNECT SWITCH   METER

THE WIND TURBINE PROVIDES ELECTRICITY TO HOME →

← HOME CONNECTED TO UTILITY GRID →

FIGURE 21-12 An example of a residential wind turbine system. This one is connected to the utility grid so any excess electricity produced by the system can be put on the grid for other people to use.

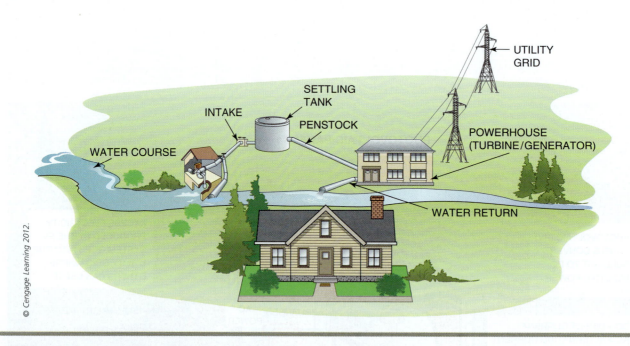

© Cengage Learning 2012.

UTILITY GRID

SETTLING TANK

INTAKE

PENSTOCK

POWERHOUSE (TURBINE/GENERATOR)

WATER COURSE

WATER RETURN

FIGURE 21-13 An example of a micro-hydro alternative energy system.

needs to work closely with the specialty contractor during installation. The installation of small wind turbine systems is presented in more detail in Chapter 22.

Because most building sites do not have streams running nearby, residential **micro-hydro electric systems** (Figure 21-13) are rare. Specialty contractors install micro-hydro electric generators. The residential electrician works with the specialty contractor to connect the power and control equipment to the house wiring system. There are many state and local rules that apply so that the environment is not damaged when a micro-hydro system is installed.

DURABILITY AND WATER MANAGEMENT

A green home is one that is built to be durable and last a long time. A home's biggest enemy is water. Water that leaks into the building can lead to rot and costly repairs. So it is important than any penetrations through the exterior walls and roof are flashed and sealed to prevent water from entering.

Electricians cut and drill a lot of holes throughout a house to run cable and mount fixtures. The electrician should review penetrations with the building team so the weather-tight details can be designed and the responsibility for flashing each one can be assigned. The responsibility may be taken on by the electrician or by another trade contractor such as the roofer, siding person, or carpenter.

During construction, the electrician should keep the builder or site supervisor aware of the penetrations made during the rough-in wiring phase so the other trade contractors who are **flashing** them can do so.

SEAL ROOFING AND SIDING PENETRATIONS

Penetrations through the roofing or siding must be sealed against water leaks. Caulking or duct seal may be part of the sealing process but the primary weather block is flashing. There are some specially made flashings for electrical boxes and cables but when they are not available, custom flashing systems will have to be made. Here are some of the common electric penetrations that need to be flashed and sealed on a house:

- Service Entrance Conductors
- Overhead Service Mast
- Exterior Receptacles
- Air Conditioner/Heat Pump Disconnect
- Exterior-mounted lighting

Wall Flashing and Sealing

Pre-made wall flashings for electrical boxes are available. They are made from pliable plastic molded to fit specific box sizes and brands. The flashing flange is integrated with the weather-resistive barrier (WRB) (also called housewrap) so water does not get into the wall. The top of the flashing tucks under the WRB and the bottom laps on top of the WRB. Water that gets behind the siding and reaches the WRB is prevented from reaching the wall by the flashing. The pre-made flashings come in one piece and two piece designs and may be installed by the plumber or the WRB installer.

Custom flashing can be made from butyl adhesive flexible flashing tape used to flash window sills. Small and odd shaped penetrations like an LB or service entrance cable are examples. After the cable or conduit is installed through the wall, a piece of flexible flashing tape can be shaped to seal the penetration to the WRB. After the siding or trim is installed, a bead of caulk can seal the surface.

Roof Flashing

Special flashings are used to flash the service mast to the roofing. They are similar to preformed wall penetration flashings but made from a rubber or soft plastic gasket (weather collar) that seals to the pipe and a plastic or metal roof flashing. Ideally, the roof flashing should be located and installed by the roofer before the mast is installed. The electrician can then insert the mast through the flashing without climbing on the roof. To facilitate precise flashing placement, the electrician can determine where the mast will rise on the wall of the house and locate the roof exit on the house plans before construction begins. Chapter 8 in this textbook shows how a mast type service entrance is installed.

GREEN PRODUCT SELECTION

Green homes are built with many green products. An electrician should learn about green electrical products and advise the building team on which ones to incorporate into a green home. An electrician can recommend and install:

- Products and materials manufactured with low environmental impact.
- Products and materials made with recycled content.
- Materials made locally reduce energy to transport.
- Materials that are more environmentally friendly.

SELECT ELECTRIC PRODUCTS AND MATERIALS MANUFACTURED WITH LOW ENVIRONMENTAL IMPACT

Today, many building components are manufactured in a way that minimizes their impact to the environment. Companies use materials like steel, glass, and plastic that have been produced in an environmentally sound manner. The companies use energy and water efficiently in their manufacturing processes. They minimize the amount of pollution they emit. All of these practices reduce the environmental impact of a product and help to make it green.

SELECT PRODUCTS AND MATERIALS MADE WITH RECYCLED CONTENT

Steel, copper, and aluminum are three of the most easily recycled materials. Many electrical products (cable, electrical enclosures, light fixtures) are made of these materials. The electrician can recommend that the building team selects electrical products that contain recycled materials. The electrician can also choose and install products that are made from materials that can be easily recycled at the end of their useful life.

SELECT LOCALLY MADE MATERIALS AND FIXTURES

When selecting electrical materials and equipment, the electrician should consider how far away from the building site they are manufactured. The farther a product or material has to be shipped, the more energy is used to deliver it. Try to limit the distance from manufacturing plant to the job site to 500 miles or less whenever possible.

SELECT AND INSTALL ENVIRONMENTALLY FRIENDLY PRODUCTS

More and more 'green' electrical products are becoming available every year. Presently, most nonmetallic sheathed cables and cable insulation is made from polyvinyl chloride (PVC) plastic. Some of the byproduct compounds of PVC manufacturing are toxic, and when PVC burns it releases toxic dioxins. Because of this many people do not consider PVC to be a green product. While there are not any practical alternative cable products to the PVC sheathed and insulated ones available today, perhaps there will be in the near future. Electricians need to keep informed about new green electrical products as they are introduced by manufacturers so they can be incorporated into green homes.

REDUCE MATERIAL USE AND RECYCLE WASTE

Reducing the amount of materials needed to build or remodel a house is a green building practice. Every green home project should have a material recycling and waste management plan. The job site recycling program minimizes the amount of waste that ends up in landfills and saves money in trash hauling and dumping fees. The electrician can incorporate waste from rough-in and trim-out wiring phases into the job site recycling system.

DESIGN A MATERIAL-EFFICIENT BRANCH CIRCUIT LAYOUT

At the planning phase, the electrician reviews the proposed electrical layout for the building and looks for ways to minimize the number or length of feeders and branch circuits. For instance, relocating the service panel from the garage to a centrally located utility area within the house or basement may save hundreds of feet of cable.

During the rough-in wiring phase, the electrician looks for the shortest practical route for feeders and branch circuit cable runs. Sometimes routing cable through a floor or ceiling is shorter than running it through perimeter walls; other times the walls provide the shortest route. A few minutes spent considering alternative cable routes and selecting the most efficient will save materials and often labor as well.

USE LEFTOVER MATERIALS

A good term for a lot of electrical scrap and waste is "leftovers." For example, an electrician generates a lot of short cable cut-offs when wiring circuits. Leftover materials often end up being tossed into the trash can even though they could still be used. Following green practices, the electrician should save and sort lengths of different size cable and either use the cable on the current job or save them for the next job. Cable cut-offs 6 inches or longer can still be used for things like pigtails, short box to box runs, switch legs, and other short runs.

Other common leftovers are nuts, bolts, washers, screws, wirenuts, crimps, and other odds and ends. They may be extra parts supplied with a light fixture, the contents of a work pouch at the end of the day or items saved when removing old wiring and fixtures. Rather than throwing these leftovers away, the electrician can sort, store, and use the parts on a future job.

INTEGRATE ELECTRICAL WASTE INTO THE JOB SITE RECYCLING PROGRAM

Every green home project has a material recycling and waste management plan. There is a wide range of scrap and waste materials that the electrician generates when wiring a house from cable cut-offs to cardboard boxes. Almost all of it can be recycled when a plan is implemented.

A lot of the scrap metal that cannot be reused can be recycled. Copper and aluminum cable are valuable and can be sold to a recycling company. The recycling company mechanically removes the plastic jacket so the metals can be manufactured into new products. Steel can be recycled as well. There are lots of small waste pieces such as bent nails, broken screws, knock-outs from junction boxes and electrical panels, side panels from multi-gang metal boxes, cut-outs from steel studs, and many others. A job site recycling station will have separate containers for each metal to make processing at the recycling company easier.

Cardboard, paper, and plastic packaging materials account for the largest percentage of waste on a typical new home building project. Lots of electrical system components from switches and receptacles to panels and light fixtures come boxed in cardboard, paper, or plastic boxes. Cardboard and paper boxes can be flattened and separated for recycling. Plastics can be separated by number for easy processing at a recycling company.

The job site recycling program minimizes the amount of waste that ends up in landfills and saves money in trash hauling and dumping fees. The electrician can advise the building team on what recyclable materials will be generated when wiring a house and how best to sort and handle them.

HOME OWNER EDUCATION AND REFERENCE MANUAL

The electrician and the building team can design and install an energy efficient electrical system throughout a house. Efficient lighting, appliances, and equipment coupled with automatic and user-operated controls can cut electricity use to half what an ordinary house uses. Renewable energy generation may be installed to power the home. But unless the home owners are trained to operate the electrical system properly for peak performance, the green goal of energy efficiency will not be met.

The electrician and building team must instruct the home owners how all of the components of the electrical system work and how to use them. A reference manual of the entire electrical system should be compiled so owners can refresh their memories and so the information is carried forward to future owners. Printed documentation from all the electrical fixtures and equipment should be saved and organized for the home owners. Information about the electrical system should be written by the electrician and included with the documentation so home owners have a permanent reference manual for the house. Things to include are:

- Locations of the main electrical panel, sub-panels, and disconnects.
- Low voltage transformer locations for door chimes, lighting, alarm, and telecommunications equipment.
- System service checklist.
- Photos and drawings of the rough wiring within walls, floors, and ceilings.
- How to operate AFCI circuit breakers.
- How to operate GFCI circuit breakers and receptacles.
- How to operate lighting, ventilation, HVAC, and other electrical control equipment.
- How to reset a circuit breaker.
- How to turn off the main breaker and other disconnects.

Save and include manufacturers' printed installation instructions, information and warranty forms for the fixtures, appliances, and equipment in the reference manual. The documentation will be a reference for the present and future owners.

Teach the owners how their behavior affects electrical energy efficiency of the house and train them on practices that reduce electricity use such as:

- Turn off lights and electrical devices when not in use.
- Adjust automatic lighting controls with each season change.
- Replace incandescent light bulbs with compact florescent or LEDs.
- Use fans or open windows instead of running air-conditioning.
- Use a microwave to heat food instead of an electric range or cooktop.
- Use a clothes line instead of a clothes dryer.

Summary

This chapter covered the major areas that each electrician should know when installing an electrical system in a green home. Energy efficiency is one of the most important features of a green home. Electricity accounts for about half of the energy used in an average home. Reducing the electrical energy consumption of both new and existing homes will have a big impact on overall energy use in the country. Homes that use less electricity to operate save home owners a lot of money.

A home's biggest enemy is water. Water that leaks into the building can lead to rot and costly repairs. So it is important than any penetrations through the exterior walls and roof are flashed and sealed to prevent water from entering.

Green homes are built with many green products. An electrician should learn about green electrical products and advise the building team on which ones to incorporate into a green home. An electrician can recommend and install products and materials manufactured with low environmental impact; products and materials made with recycled content; materials made locally, which reduces energy to transport; and materials that are more environmentally friendly.

Every green home project should have a material recycling and waste management plan. The job site recycling program minimizes the amount of waste that ends up in landfills and saves money in trash hauling and dumping fees. The electrician can incorporate waste from rough-in and trim-out wiring phases into the job site recycling system.

The electrician and building team must instruct the home owners how all of the components of the electrical system work and how to use them. A reference manual of the entire electrical system should be compiled so owners can refresh their memories and so the information is carried forward to future owners. Printed documentation from all the electrical fixtures and equipment should be saved and organized for the home owners. Information about the electrical system should be written down by the electrician and included with the documentation so home owners have a permanent reference manual for the house.

By following the recommendations in this chapter, the installed electrical system will provide the peak efficiency needed to use as little electrical energy as possible. You can also be confident that the materials and the manner in which they were installed will result in a green electrical system that was not wasteful in any way and did not affect the environment negatively.

REVIEW QUESTIONS

Directions: Answer the following items with clear and complete responses.

1. Explain why an electrician should advise the building team to select Energy Star appliances when available.

2. Discuss some of the advantages of using a tankless electric water heater.

3. Noise level for exhaust fans is measured in _____. Exhaust fan power is measured in _____.

4. Explain why a timer is often installed for a bathroom exhaust fan in a green home.

5. Name one way to eliminate lights being left on when either they are not needed or when no one is in a room.

6. It is estimated that a programmable thermostat can reduce the energy to heat and cool a home by up to _____ percent.

7. Describe a phantom electrical load.

8. Name five common penetrations made by an electrician through the air barriers of a house that will need to be sealed.

9. In a green home wiring system, electrical boxes that penetrate the air barrier must be special airtight models. Describe an airtight electrical box.

10. Name three on-site electrical power generation alternative energy sources that could be used in a green home.

11. Name four of the common electrical penetrations in roofs and siding that need to be flashed and sealed on a green house.

12. Reducing the amount of materials needed to build or remodel a house is a green building practice. True or False (Circle the correct response).

13. Explain why during the rough-in wiring phase, an electrician should look for the shortest practical route for feeders and branch circuit cable runs.

14. Name five things that should be in a home owner's reference manual.

15. Name five things that you should teach home owners to do to save money on their monthly electricity bill.

Alternative Energy System Installation

OBJECTIVES   *Upon completion of this chapter, the student should be able to:*

- Demonstrate an understanding of the different types of photovoltaic systems used in residential wiring.

- Demonstrate an understanding of the components that make up a photovoltaic system installation.

- List the system components that make up a typical stand-alone PV system.

- List the system components that make up a typical interactive (grid-tie) PV system.

- Demonstrate an understanding of how a typical photo-voltaic system is installed.

- List several *National Electric Code*® requirements that apply to photovoltaic system installation.

- Demonstrate an understanding of small wind turbine system installation.

- List the components that make up a small wind turbine system.

- List several *National Electric Code*® requirements that apply to a small wind turbine system installation.

GLOSSARY OF TERMS

amorphous having no definite form or distinct shape

anemometer a device used to measure wind speed

array a mechanically integrated assembly of modules or panels with a support structure and foundation, tracker, and other components, as required, to form a direct-current power-producing unit

azimuth the sun's apparent location in the sky east or west of true south

charge controller equipment that controls dc voltage or dc current, or both, used to charge a battery

diode a semiconductor device that allows current to pass through in only one direction

fuel cell an electrochemical system that consumes a fuel like natural gas or LP gas to produce electricity; the consumption of the fuel gas is through an electrochemical process rather than a combustion process

guy a cable that mechanically supports a wind turbine tower

hybrid system a system comprised of multiple power sources; the power sources may include photovoltaic, wind, micro-hydro generators, or engine-driven generators

insolation the term used for the measure of solar radiation striking the earth's surface at a particular time and place

irradiance a measure of how much solar power is striking a specific location

inverter equipment that is used to change voltage level or waveform, or both, of electrical energy; a device that changes dc input to an ac output; may also function as battery chargers that use alternating current from another source and convert it into direct current for charging batteries

magnetic declination the deviation of magnetic south from true south

maximum power point (MPP) indicates the maximum output of the module and is the result of the maximum voltage (Vmp) multiplied by the maximum current (Imp)

micro-hydro a small hydroelectric alternative energy system that uses water flow to turn a generator that produces electricity; the water flow can come from a stream or a small reservoir

module a complete, environmentally protected unit consisting of solar cells, optics, and other components, exclusive of

tracker, designed to generate dc power when exposed to sunlight

nacelle an enclosure housing the alternator and other parts of a wind turbine

open circuit voltage (Voc) the maximum voltage when no current is being drawn from the module

panel a collection of modules mechanically fastened together, wired, and designed to provide a field-installable unit

short circuit current (Isc) the maximum current output of a module under conditions of a circuit with no resistance (short circuit)

solar cell the basic photovoltaic device that generates electricity when exposed to light

thermals rising currents of warm air that go up and over land during sunny daylight hours

tower a pole or other structure that supports a wind turbine

wind turbine a mechanical device that converts wind energy to electrical energy

wind turbine system a small wind electric generating system

According to the U.S. Energy Information Administration (EIA), renewable generated electricity, including sources such as wind and solar, will account for almost 20 percent of total U.S. electricity generation in 2030, up from about 10 percent in 2010. More and more alternative energy systems will be installed in houses and it will be qualified electricians who will be the primary installers of these systems.

The 2011 *National Electric Code*® has included wording which clarifies that only qualified installers can install a wind or solar electricity alternative energy system. The concerns are that contractors and installers engaged in this work need to have the necessary training and qualifications in the electrical field. Solar and wind electricity systems produce electrical power and are often interconnected to the utility grid. This is not a job for an untrained person, and these systems are far more than plug-and-play. Designing a photovoltaic or wind turbine alternative energy system requires a high level of knowledge and experience. Because beginning residential electricians do not have the level of knowledge and experience necessary to design systems, this chapter is limited to covering the information needed for a good understanding of how a solar or wind electricity producing system is installed.

INTRODUCTION TO ALTERNATIVE ENERGY SYSTEMS

The two types of alternative energy systems commonly installed in house wiring are the solar electric (photovoltaic) system and the **wind turbine system**. The term "photovoltaic" can be broken down into two words. "Photo" meaning the sun and "voltaic" meaning electricity. Photovoltaic (PV) systems are solar energy systems that produce electricity directly from sunlight.

Wind turbine systems produce electricity when wind hits the blades of a turbine and spins a generator that produces electricity. PV and wind turbine power systems are simply energy converters. They convert one form of energy (sunlight or wind) into another form of energy (electricity). Although wind is not technically a form of energy, for this chapter we will consider it an energy form in order to more easily understand the material in the chapter.

There are other types of renewable energy systems like **micro-hydro** and **fuel cell** systems but at the present time they are seldom installed in homes. Because they are the system types that electricians will most likely find themselves installing in a residential environment, this chapter will concentrate on photovoltaic and small wind turbine systems.

INTRODUCTION TO PHOTOVOLTAIC SYSTEMS

Solar cells that produce electricity are not new. In 1873, scientists first noticed that selenium was sensitive to light and, in 1880, the first selenium-based solar electric cell was developed. In 1905, a fellow by the name of Albert Einstein gave his explanation of the photoelectric effect and his theories led to a greater understanding of how to generate electricity from sunlight. Research continued through the 1930s and 40s on the selenium solar cell. In the early 1950s, scientists at Bell Labs discovered that silicon, when treated with certain impurities, generated a substantial voltage when exposed to sunlight and, in 1954, Bell Labs developed a silicon-based photovoltaic cell. NASA used PV cells to power the first U.S. satellite, and by the early 1960s PV systems were used routinely on satellites and other spacecraft. The international space station that orbits the earth today primarily gets its electricity from a large number of PV panels arrayed in what look like wings on the space station. PV technology has continued to improve and today millions of homes in the world use PV technology. The world PV market is estimated to be more than 10 billion dollars!

PHOTOVOLTAIC SYSTEM ADVANTAGES AND DISADVANTAGES

The use of fossil fuel sources to produce electricity has created many problems. Some of the problems include: global warming, acid rain, smog, water pollution, destruction of natural habitat from fuel spills, and loss of natural resources. The 2010 oil spill in the Gulf of Mexico is a good reminder to us of what some of the disadvantages of using fossil fuels to produce electricity can be. I am sure that as you read this the recovery effort in the Gulf of Mexico, especially along the Louisiana shoreline, is still going on and will for years to come.

Advantages of Photovoltaic Systems

There are many advantages to using a photovoltaic system. Perhaps the biggest advantage of all is that a PV system is an investment in the future. It will produce electricity at the fixed price of the original installation for the life of a PV system. The cost of producing electricity by a utility company will continue to rise but the electricity that a PV system produces will be the same price as when it was installed. Because PV systems will last for years and years, the cost of electricity from a PV system will be the same for a very long time.

Additional advantages include:

- PV systems are very reliable; there is little to go wrong and very rarely do they breakdown.

- PV systems are very durable. They are strong and can easily withstand outside weather conditions.

- PV systems have low maintenance costs. There are few parts that ever need replacing.

- PV systems have no fuel costs so there is no need to burn fuels like oil and gas to make electricity.

- PV systems have little sound pollution. They are very quiet and make no noise when working.

- PV systems have modularity. Parts can be added easily to a system to increase the electrical output of a PV system.

- PV systems are very safe. There are no exposed electrical or moving parts that could harm anyone.

- PV systems allow for energy independence. They have the ability to make as much of your own electricity as you want.

Disadvantages of Photovoltaic Systems

Even though there are many advantages with a PV system, they are not perfect. Disadvantages include:

- PV systems are expensive. The initial cost of a PV system is expensive and many home owners cannot afford such a system.

- Solar radiation amounts vary across the country. Some areas get more sun hours than others and are better locations for a PV system installation.

- Energy storage is another disadvantage. At this time batteries are the only way to store electricity produced by the sun for use at night or during very cloudy and stormy weather. Batteries cost a lot of money and they have to be replaced every few years.

PHOTOVOLTAIC SYSTEM COMPONENTS

The photovoltaic solar cell (Figure 22-1) is the basic building block of PV technology. A typical PV solar cell is about 4 inches across and produces about 1 watt of power in full sunlight at about .5 volts DC. A **module** (Figure 22-2), often called a **panel**, is a configuration of PV cells laminated between a clear outer superstrate (glazing) and an encapsulating inner substrate. Solar cells are wired together to produce a PV module. Modules (Panels) are connected together to form a PV array (Figure 22-3). An **array** is wired to deliver a specific voltage. A **charge controller** (Figure 22-4) regulates the battery voltage and makes sure that the PV system batteries (if used) are charged properly. Batteries (Figure 22-5) are used to chemically store direct current (DC) electrical energy in a PV system. PV modules are connected to an **inverter** (Figure 22-6) that "converts" the DC electricity produced by most solar arrays to the AC electricity commonly used in a house.

TYPES OF PHOTOVOLTAIC SYSTEMS

Day use only systems (Figure 22-7) are the simplest and least expensive PV system. They consist of modules wired directly to DC loads. This system type has no electrical storage capabilities. Examples are remote water pumping; fans, blowers, or circulators for solar water heating or ventilation systems. These systems are not used in homes because electrical loads in homes must also be powered at night.

A direct current system with storage batteries (Figure 22-8) is a PV system where loads can be powered day or night. They can supply the extra surge current needed for starting electric motors. Because batteries are used, a charge controller is needed. This system is not used in very many homes because almost all home electrical loads use alternating current and not direct current.

Direct current systems powering alternating current loads (Figure 22-9) is a PV system that must use an inverter to convert the DC electricity to AC electricity. This system is commonly used in residential applications. A newer type of PV system used to supply AC

FIGURE 22-1 A photovoltaic solar cell.

FIGURE 22-2 A PV module. Modules are often referred to as panels.

FIGURE 22-3 Modules are wired together to form a PV array.

loads is available where individual micro-inverters are installed at each PV panel (Figure 22-10) so that each panel produces AC electricity. This system is becoming very popular and is relatively easy for electricians to install.

Another type of PV system is called a hybrid system (Figure 22-11). This system type incorporates a gas or diesel powered generator along with a PV system. A generator can provide the extra power needed during cloudy weather and during times of heavy electrical use. The generator can also charge the system batteries whenever it is running. Many hybrid systems also include a small wind turbine. Small wind turbines are covered later in this chapter.

PV systems can either be stand-alone or grid-tied. A stand-alone system has no connection to the local electric utility's grid system and is used when a home owner wants to be totally independent from the electric utility (Figure 22-12). A stand-alone PV system may also be installed if the location of a home is far away from a main road and the utility grid and it could cost several thousand dollars for a utility company to construct a power line to the home's location. A Utility Grid Interconnected System, often called a

FIGURE 22-4 A typical charge controller used in a PV system. Schneider Electric Xantrex XW MPPT Solar Charge Controller. *Courtesy of Schneider Electric.*

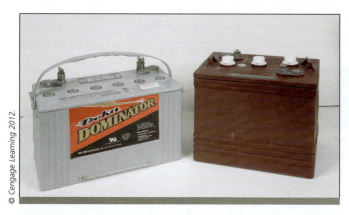

FIGURE 22-5 Batteries are used to chemically store DC electrical energy in a PV system.

FIGURE 22-6 Inverters "convert" the DC electricity produced by solar arrays to AC electricity.

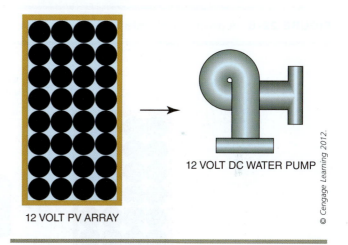

12 VOLT PV ARRAY

12 VOLT DC WATER PUMP

FIGURE 22-7 Day use only systems are the simplest and least expensive type of PV system.

Utility-Connected, Grid-Tie, Intertie, or Line-Tie system, is connected to the utility grid (Figure 22-13). They do not need battery storage because the utility grid is the power reserve. The system will automatically shut down if the utility grid goes down so electricity cannot backfeed the grid. With this PV system type the home owner "sells" excess energy to the utility company. This is the PV system type that is most often installed across the country.

PHOTOVOLTAIC GRID-TIE SYSTEMS AND NET METERING

Most locations in the United States offer some type of net metering agreement for home owners who have installed a grid-tie PV system. The net meter moves forward when electricity is flowing from the utility into the house and the meter moves backward when excess solar energy flows back to the utility grid (Figure 22-14). At the end of the month, the customer is billed only for net consumption, the amount of electricity consumed, less the amount of electricity produced. In most parts of the country an interconnection agreement with the

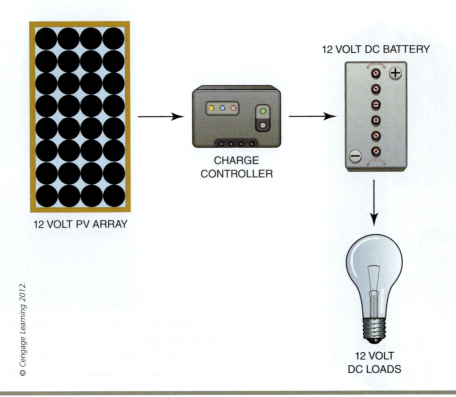

© Cengage Learning 2012.

FIGURE 22-8   A direct current system with storage batteries is a PV system where loads can be powered day or night.

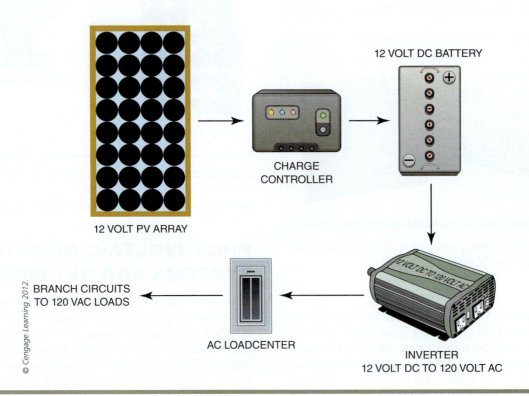

© Cengage Learning 2012.

FIGURE 22-9   Direct current systems powering alternating current loads are PV systems that must use an inverter to convert the DC electricity to AC electricity.

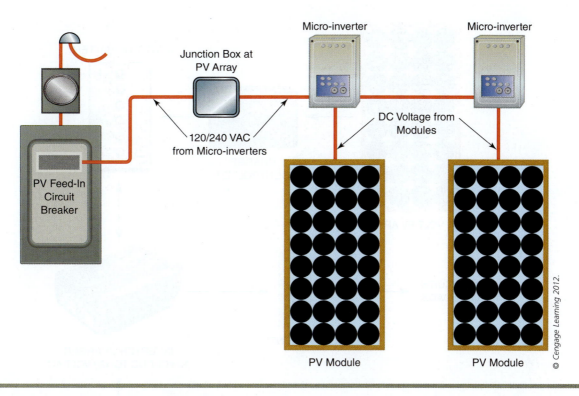

Micro-inverter   Micro-inverter

Junction Box at
PV Array

PV Feed-In
Circuit
Breaker

120/240 VAC
from Micro-inverters

DC Voltage from
Modules

PV Module

PV Module

© Cengage Learning 2012.

FIGURE 22-10   A newer type of PV system uses individual micro-inverters to produce AC electricity at each panel.

local power utility must be signed by a home owner. This agreement is a legal document that outlines certain requirements that must be met so that the customer's grid-tie PV system can be connected to the utility's power grid.

UNDERSTANDING PHOTOVOLTAIC SYSTEM ELECTRICITY BASICS

It is assumed that students studying house wiring and using this textbook already have a good understanding of both direct current (DC) and alternating current (AC) electrical theory. Therefore, a short review is all that is necessary.

An electrical circuit is the continuous path of electron flow from a voltage source through a conductor, to a load, and back to the source. A switch in the circuit controls the continuity of current flow. If the switch is in the "OFF" position, there is an open circuit and no current can flow. If the switch is in the "ON" position, there is a closed circuit and current can flow in the circuit.

Electricity is the flow of electrons through a circuit. The force that causes the electrons to move is called voltage and the unit of measure for voltage is the volt. The letter "V" or "E" is used to represent voltage and the letter "V" is used to represent volts. The rate of flow of the electrons is called current or amperage. The unit of measure for current is the ampere (or amp). The letter "I" is used to represent current and the letter "A" is used to represent amps. One amp of current flowing for one hour is equal to one amp-hour (Ah). For example, 10 amps flowing for 1 hour is 10 Ah, or 5 amps flowing for 2 hours is 10 Ah. The rate at which an electrical load uses electrical energy or the rate at which electrical energy is produced is called power. The unit of measure for electrical power is the watt. The letter "W" is used to represent watts and the letter "P" is used to represent power. The amount of electrical power (watts) is equal to volts multiplied by the amps. For example, 100 volts × 10 amps equal 1000 watts of electrical power. The amount of electrical energy used by a consumer is measured in watt-hours. The number of watt-hours equal the electrical power consumed (watts) multiplied by the length of time the load has been operated (hours). Because the amount of watt-hours is typically quite large, the unit of measurement is in kilowatt-hours (kWh). For example, 1000 watt-hours equal 1 kilowatt-hour (kWh). Another way to think of this is to consider that if ten 100-watt

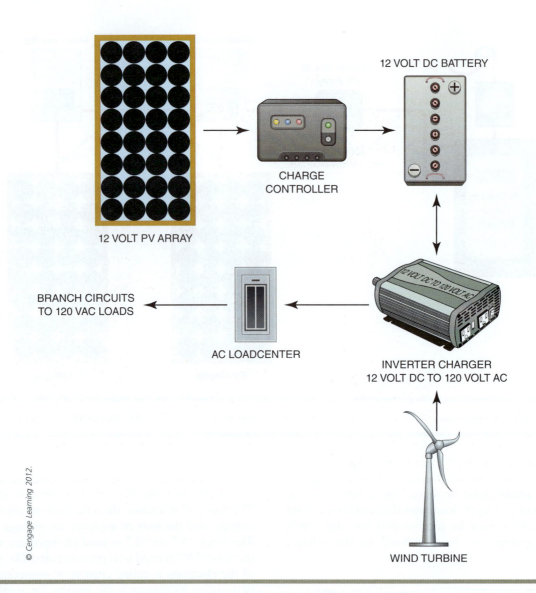

12 VOLT DC BATTERY

CHARGE
CONTROLLER

12 VOLT PV ARRAY

BRANCH CIRCUITS
TO 120 VAC LOADS

AC LOADCENTER

INVERTER CHARGER
12 VOLT DC TO 120 VOLT AC

WIND TURBINE

© Cengage Learning 2012.

FIGURE 22-11   A hybrid system uses another energy source like a small wind turbine to supplement a PV system.

incandescent light bulbs were left on for one hour it would equal 1000 watt-hours (10 bulbs × 100 watts each × 1 hour) or 1 kilowatt-hour (kWh) of energy. If the cost of electricity is 15 cents per kWh, the cost of electricity for having the ten 100-watt light bulbs operating for 1 hour is only 15 cents.

PV modules or batteries can be wired together in series, parallel, or series-parallel, to get a desired system voltage and current. Voltage sources connected in series result in the total voltage increasing but the current flow remains the same. With DC sources like PV modules and batteries, connect them positive (+) to negative (−) for a series connection (Figure 22-15). Voltage sources connected in parallel result in the total current increasing but the voltage remains the same. With DC sources like PV modules and batteries, connect them positive (+) to positive (+) and negative (−) to negative (−) for a

parallel connection (Figure 22-16). PV systems may use a mix of series and parallel connections to obtain the required voltage and amperage (Figure 22-17).

SOLAR FUNDAMENTALS

Entry level photovoltaic system installers like the students using this textbook will not yet have enough knowledge and experience to design a PV system from the ground up. However, they do need to learn a few things now about PV system design in order to better understand how to install a system. One area where beginning installers need to have some basic knowledge is called solar fundamentals.

DC CIRCUITS
(OPTIONAL)

LOW VOLTAGE
DISCONNECT

AC CIRCUITS

CHARGE
CONTROLLER

STAND-ALONE
INVERTER

GENERATOR BACKUP

BATTERY BANK

© Cengage Learning 2012.

FIGURE 22-12 A typical stand-alone PV system.

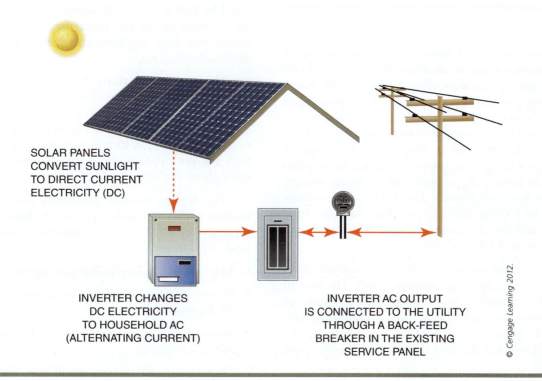

SOLAR PANELS
CONVERT SUNLIGHT
TO DIRECT CURRENT
ELECTRICITY (DC)

INVERTER CHANGES
DC ELECTRICITY
TO HOUSEHOLD AC
(ALTERNATING CURRENT)

INVERTER AC OUTPUT
IS CONNECTED TO THE UTILITY
THROUGH A BACK-FEED
BREAKER IN THE EXISTING
SERVICE PANEL

© Cengage Learning 2012.

FIGURE 22-13 A typical interactive (grid-tie) PV system.

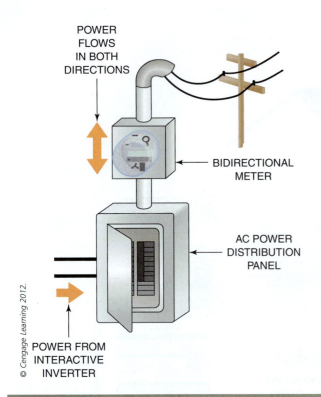

POWER
FLOWS
IN BOTH
DIRECTIONS

BIDIRECTIONAL
METER

AC POWER
DISTRIBUTION
PANEL

POWER FROM
INTERACTIVE
INVERTER

© Cengage Learning 2012.

FIGURE 22-14 In a net-metering system the net meter moves forward when electricity is flowing from the utility into the house and the meter moves backward when excess solar energy flows back to the utility grid. At the end of the month, the customer is billed for the amount of electricity consumed less the amount of electricity produced.

Solar **irradiance** is a measure of how much solar power is striking a specific location. This irradiance varies throughout the year depending on the seasons. It also varies throughout the day, depending on the position of the sun in the sky, and the weather. Solar **insolation** is a measure of solar irradiance over of period of time, typically over the period of a single day. Insolation is expressed as a number of watts per square meter when described as power. It is usually given as an average daily value for each month. The values are generally expressed in $kWh/m^2/day$. On a clear day, the total insolation striking the earth is about 1000 watts per square meter or $1\ kWh/m^2/day$.

Solar radiation received at the earth's surface varies because of atmospheric attenuation (loss). The loss of light can be caused by air molecules, water vapor, or dust scattering the light. Ozone, water vapor, and carbon dioxide absorb the light. Peak sun hours are the number of hours per day that the solar insolation equals $1000\ w/m^2$.

The earth's distance from the sun and its tilt affect how much solar energy is available. The tilt is approximately 23.5 degrees from the vertical. The northern

hemisphere is tilted towards the sun from June through August so there is more available solar energy in summer than winter. In the northern hemisphere, photovoltaic modules should point towards the southern sky to collect the maximum amount of solar energy.

A location's latitude (the distance north or south of the equator) determines whether the sun appears to travel in the northern or southern sky. For example, the location in Maine where the author of this textbook lives is approximately 44.5 degrees north latitude and the sun appears to move across the southern sky. Use this website to get the exact latitude for your area: http://www.ngdc. noaa.gov/geomagmodels/Declination.jsp.

The sun appears to rise and set at different points on the horizon throughout the year because of the 23.5 degree tilt of the earth (Figure 22-18). The sun's apparent location in the sky east or west of true south is called the **azimuth**. It is measured in degrees east or west of true south. South on a compass is not the same as true south. Compass needles align with the earth's magnetic field. The earth's magnetic field does not align with the earth's rotational axis. The deviation of magnetic south from true south is called **magnetic declination** (Figure 22-19). PV modules should be facing true south (0 degrees azimuth) for maximum efficiency. For example: a site in central Maine has a magnetic declination of 20 degrees west. This means that true south is 20 degrees west of magnetic south. On a compass with the north needle pointing at 360 (also 0) degrees, true south is in the direction indicated by 200 degrees (180 degrees plus 20 degrees). Use this website to get the exact magnetic declination for where you live: http://www.ngdc.noaa.gov/geomagmodels/Declination.jsp.

The sun's height above the horizon is called altitude (Figure 22-20). It is measured in the number of degrees above the horizon. Solar noon is when the sun is true south at 0 degrees azimuth. This is the sun's highest altitude for that day. A location's latitude determines how high the sun appears above the horizon at solar noon throughout the year.

The highest average insolation will fall on a PV module with a tilt angle equal to the latitude. The best tilt angle for a PV array for different seasonal electrical loads is:

- Year-round electrical loads: tilt angle equals latitude

- Winter electrical loads: tilt angle equals latitude plus 15 degrees

- Summer electrical loads: tilt angle equals latitude minus 15 degrees

Remember, PV arrays work best when the sun's rays shine perpendicular (90 degrees) to the cells because of the effect of tilt angle (Figure 22-21). Since grid-tie systems have the utility grid as a backup, there is

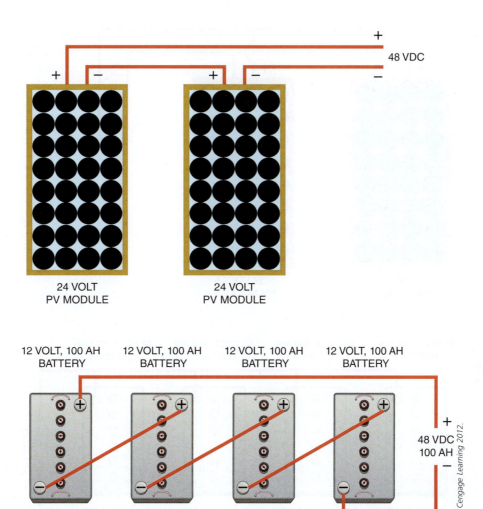

FIGURE 22-15   PV modules and batteries in a series configuration.

more flexibility in siting the PV array. A roof mount PV array probably will not result in the best tilt angle. In this case, the PV array will produce a percentage of the total available energy and the grid will provide the rest.

EVALUATING A SITE FOR PV INSTALLATION

When evaluating a site to locate where the PV array will be installed, care must be taken to identify shading obstacles. Shading greatly affects a PV array's performance and as a general rule a PV array should be free of shade from 9:00 am to 3:00 pm. This is called the "solar window" and is considered to be the peak sun harvesting hours. Shading is usually a greater problem in the winter because of the sun's low altitude and the shadows are longer. Use December 21, the shortest day of the year, for worst case shadow calculations in the northern hemisphere.

One method for finding a site with good year-round exposure is through long-term observation. Because of the amount of time involved this not very practical. A more practical way is to use sun charts to determine specific sun altitudes (Figure 22-22). The sun charts will help you determine if the PV array will be shaded during critical times of the year. There are also commercially available solar site selector devices with sun charts built-in (Figure 22-23).

PV SYSTEM COMPONENTS

The previous sections introduced you to photovoltaic system components and solar fundamentals. This section will cover the major components of a PV system in more detail.

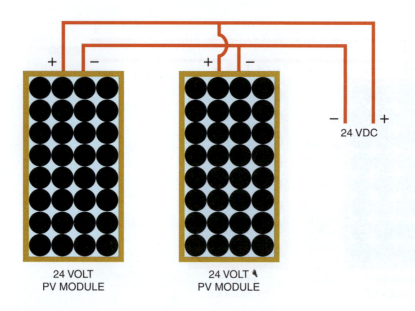

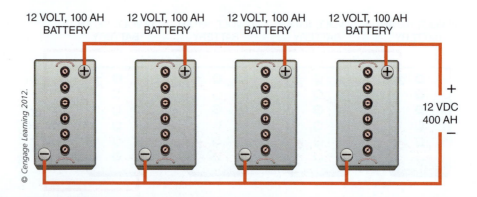

FIGURE 22-16 PV modules and batteries in a parallel configuration.

PV MODULES

The basic unit of a PV system is the photovoltaic cell. PV cells convert sunlight into DC electricity. They are about 1/100th of an inch thick and produce about .5-volt DC. Their lifespan is 25 years or more.

PV modules are assemblies of PV cells wired in series, parallel, or series/parallel to produce a desired voltage and current. The current output depends on the surface area of the cells in the modules. PV cells are encapsulated within the module framework to protect them from the weather. Modules used in residential PV systems typically produce from 100 to 300 watts. The terms "Panel" and "Module" are used interchangeably but according to the *National Electric Code®*, a panel is technically a group of modules wired together to get a desired voltage. An array is a group of panels wired together to get a desired voltage and current.

A module or array is not that efficient when it comes to changing sunlight to electricity. Many modules and arrays convert only about 10% to 15% of the available solar radiation to electrical energy. Module efficiency has continued to improve and some modules today are approaching a 20 percent efficiency rating.

The Photovoltaic Reaction

The voltage production in a PV cell is called "photovoltaic reaction." To understand photovoltaic reaction you must first understand a little about how most PV cells are made. The manufacturing process of a single crystalline cell starts with silicon being purified and grown into a crystalline structure. Silicon in its pure form is called a "semiconductor." Impurities are added through a process called "doping." The silicon is grown into a cylindrical shape and then sliced into very thin wafers. The wafers are doped with either boron or phosphorous. With boron, which has an electron deficiency, it creates a positively charged material (P-type). With phosphorous, which has an excess of electrons, it creates a negatively charged material (N-type). The region in the cell created by the positive

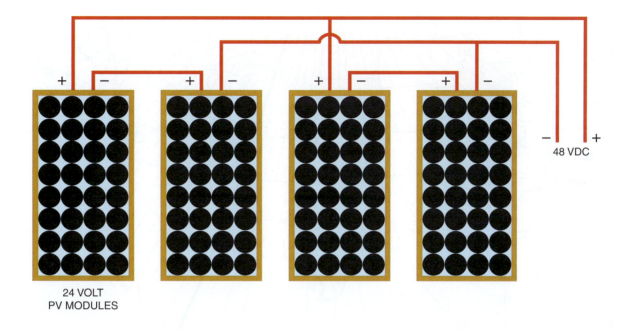

24 VOLT
PV MODULES

48 VDC

12 VOLT, 100 AH
BATTERIES

24 VDC
200 AH

© Cengage Learning 2012.

FIGURE 22-17   PV modules and batteries in a series-parallel configuration.

and negative layers is called the P-N junction. A voltage in the PV cell is produced when sunlight "knocks" loosely held electrons from the silicon layer. These electrons are attracted to the positively charged boron layer. An electrical force is developed at the P-N junction and electrons begin to flow through metal contacts built into the cell. The contacts in a cell are connected from the front of one cell to the back of another cell in the module. This circuitry enables the electrons to flow through P-N junctions of each cell, building a voltage since each cell is wired in series. The voltage at each cell's P-N junction is approximately .5 volt. If a module is made up of 72 cells all wired in series, the total voltage produced by the module would be 36 volts. The cells are covered with an anti-reflective coating to enhance sunlight absorption. The cells are then placed on a backing and wired together to get the desired voltage and current rating of the module. It is all then framed and encapsulated.

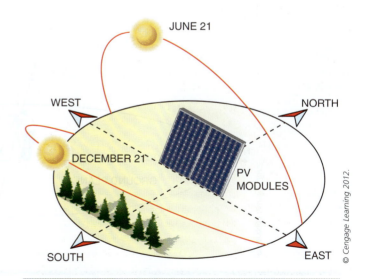

JUNE 21

WEST

NORTH

DECEMBER 21

PV
MODULES

SOUTH

EAST

© Cengage Learning 2012.

FIGURE 22-18   The sun appears to rise and set at different points on the horizon throughout the year because of the 23.5 degree tilt of the earth.

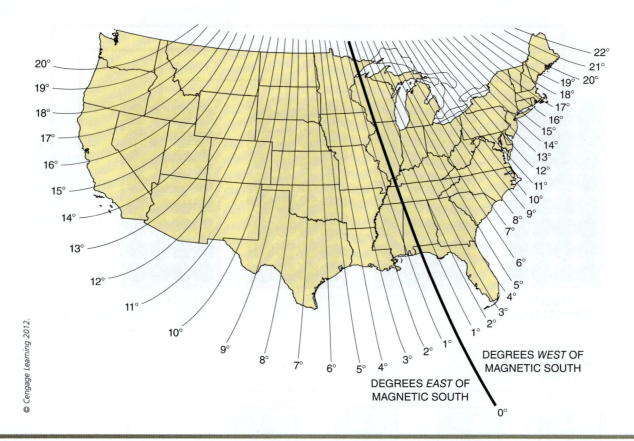

FIGURE 22-19 The deviation of magnetic south from true south is called magnetic declination and varies depending on where in the country you live.

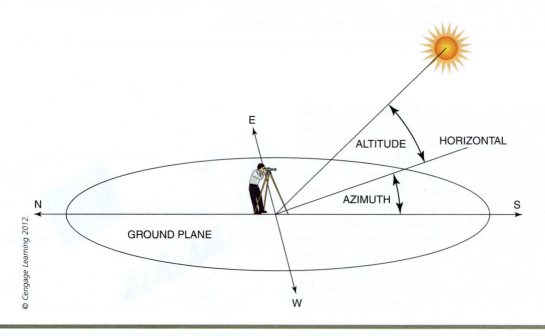

FIGURE 22-20 The sun's height above the horizon is called "altitude" and is measured in the number of degrees above the horizon.

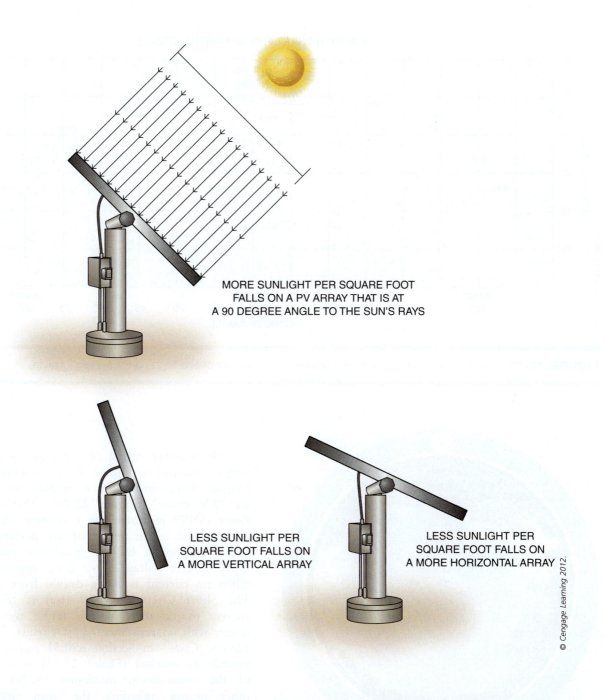

MORE SUNLIGHT PER SQUARE FOOT
FALLS ON A PV ARRAY THAT IS AT
A 90 DEGREE ANGLE TO THE SUN'S RAYS

LESS SUNLIGHT PER
SQUARE FOOT FALLS ON
A MORE VERTICAL ARRAY

LESS SUNLIGHT PER
SQUARE FOOT FALLS ON
A MORE HORIZONTAL ARRAY

© Cengage Learning 2012.

FIGURE 22-21 PV arrays work best when the sun's rays shine perpendicular (90 degrees) to the cells because of the effect of tilt angle.

PV Module Types and Characteristics

Modules can have different cell material, glazing material, and electrical connections. The type of module typically used today is determined primarily by the composition of the silicon crystalline structure. If it is grown as a single crystal, it is called single-crystalline. If it is cast into an ingot of multiple crystals, it is called poly-crystalline (also called multi-crystalline). If it is deposited as a thin film, it is called **amorphous** silicon or just thin-film. Amorphous means that there is no definite shape that the module comes in. Single-crystalline cells are slightly more efficient than poly-crystalline cells. Amorphous silicon is less expensive to manufacture but is only about half as efficient as single-crystalline cells. Concentrator modules incorporate small solar cells surrounded by some type of reflective

SUN PATH CHART FOR 40° NORTH LATITUDE

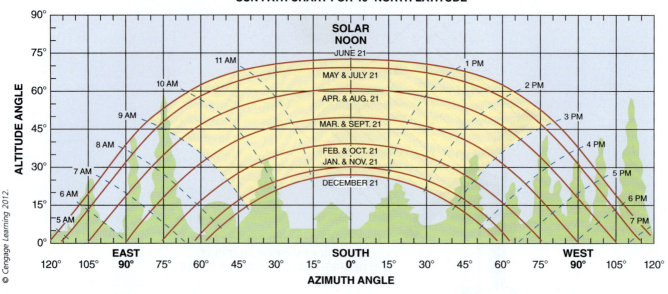

FIGURE 22-22 Sun charts can be used to help determine the amount of shading at a PV system location.

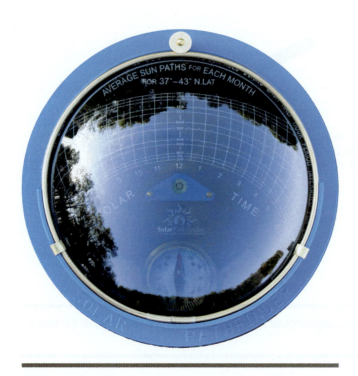

FIGURE 22-23 A commercially available solar site selector device. This one is called a Solar Pathfinder. *Courtesy of The Solar Pathfinder Company.*

material that concentrates sunlight onto the cells. They are more efficient than regular modules. Even though silicon cells are still by far the most common type of cell, other types are available including copper indium selenium (CIS) and cadmium telluride.

The wattage rating of a module is equal to its output voltage multiplied by its operating current. Output characteristics for a module can be found in an "I-V curve" (Figure 22-24). **Maximum Power Point (MPP)** indicates the maximum output of the module and is the result of the maximum voltage (Vmp) multiplied by the maximum current (Imp). **Open Circuit Voltage (Voc)** is the maximum voltage when no current is being drawn from the module. The **Short Circuit Current (Isc)** is the maximum current output of a module under conditions of a circuit with no resistance (short circuit). The *National Electric Code®* in Section 690.51 requires that each module be marked (Figure 22-25) with the polarity of the connections, maximum OCPD rating, and other ratings including: the open circuit voltage; operating voltage; maximum permissible system voltage; operating current; short circuit current; and maximum power. This information can be used to determine such things as the conductor sizes used to wire the PV system and the fuse or circuit breaker sizes needed.

PV Module Mounting

There are various systems available for mounting a PV array. Choose a mounting system based upon the orientation of the house; shading at the site; weather considerations; roof material; ground or roof-bearing capacity; and the system application.

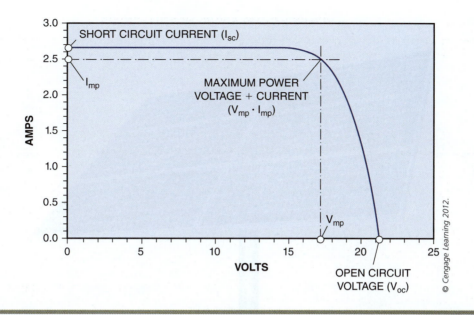

FIGURE 22-24   An IV curve for a common size of PV module.

SHARP
SOLAR MODULE
NT–S5E1U

_c(UL)_{US} **LISTED**
2P89
PHOTOVOLTAIC MODULE
E160673

THE ELECTRICAL CHARACTERISTICS ARE WITHIN $\pm$ 10 PERCENT OF THE INDICATED VALUES OF I_{SC}, V_{OC}, AND P_{MAX} UNDER STANDARD TEST CONDITIONS

(IRRADIANCE OF 1000W/m² , AM1.5 SPECTRUM AND CELL TEMPERATURE OF 25°C)

MAXIMUM POWER	(P_{MAX})	185.0 W
OPEN-CIRCUIT VOLTAGE	(V_{OC})	44.9 V
SHORT-CIRCUIT CURRENT	(I_{SC})	5.75 A
MAXIMUM POWER VOLTAGE	(V_{PMAX})	36.21 V
MAXIMUM POWER CURRENT	(I_{PMAX})	5.11 A
MAXIMUM SYSTEM VOLTAGE		600 V
FUSE RATING		10 A
FIRE RATING		CLASS C
FIELD WIRING		COPPER ONLY 14 AWG MIN. INSULATED FOR 90°C MIN.
SERIAL No.		0 3 4 0 9 0 2 7 3

FIGURE 22-25   The *National Electric Code®* in Section 690.51 requires that each PV module be marked with the correct polarity of the leads, the maximum overcurrent protection device size, and several ratings.

© Cengage Learning 2012.

FIGURE 22-26   A pole-mounted array.

POLE MOUNT SYSTEMS

Pole mount systems (Figure 22-26) allow modules to be mounted on a hardware system that is then attached to a vertical pole placed permanently and securely in the ground. This is an excellent choice but a little more expensive than simply roof or ground mounting. A big advantage to this mounting system is that the array can easily be adjusted (either manually or automatically) to optimize the array's performance. Another advantage is that regular maintenance of the array can be done easily compared to a roof-mounted array. A disadvantage is that a pole-mounted rack is usually located many feet away from the house and other PV system components. The wire run from the pole mount array can be long and extensive trenching may need to be done. Also, the long wire run may result in a very large size conductor being used to compensate for voltage drop losses.

GROUND MOUNT SYSTEMS

Ground mount systems (Figure 22-27) use a support structure that is attached directly to prepared footings in the ground. There are many standard sizes commercially available including 4, 8, or 12 module structures. Like pole-mounted arrays they may be adjusted manually to optimize system performance. Metal that is mounted directly to a concrete footing should be galvanized or otherwise protected since the lime content of the concrete can corrode aluminum and the nuts, bolts, and washers should be stainless steel. Ground mounting requires level foundations and the installation must be strong enough to avoid having the structure sink into the ground. It must also be strong enough to withstand strong winds.

ROOF MOUNT SYSTEMS

Roof mount systems are the most popular but are not necessarily the best way to mount an array. Direct-mounted PV modules are mounted directly to the roof (Figure 22-28). This eliminates the need for a frame and mounting rails. Because the modules lay so close to the roof surface there is little air circulation and access to the electrical wiring is limited after installation. Direct mounting modules to a roof is not recommended. Rack mounts installed on a roof have the modules mounted in a metal framework and set at a predetermined angle. The rack's structural members are bolted directly to the roof's structural members (Figure 22-29). The array's attachment to the rack actually places it above the roof surface. Roof mounting increases the weight load on a roof and can pose wind loading problems. The roof racks do allow air to circulate around

FIGURE 22-27   A ground-mounted array.

FIGURE 22-28   Direct-mounted PV modules are mounted directly to a roof.

the modules better and because the modules are placed above the roof surface, the electrical connections are easier to access. Some roof racks are adjustable.

Integrated mounting is when PV modules are attached directly to the roof's rafters and actually replace the conventional roof covering. This type of mounting is designed for new construction (Figure 22-30). The access to the module electrical wiring is in the attic. Another type of integrated mounting is PV shingles (Figure 22-31). The shingles are installed over the entire roof area or only a certain section of the roof. The wiring is typically connected in the attic. The shingles act

© Cengage Learning 2012.

FIGURE 22-29   Rack mounts installed on a roof have the modules mounted in a metal framework and set at a predetermined angle.

© Cengage Learning 2012.

FIGURE 22-30   An example of an integrated roof-mounted PV array.

like regular roofing shingles to keep out rain and snow but when the sun shines on them they produce electricity.

Track mount systems (Figure 22-32) are used so that the PV modules can track the sun in its daily path to maximize system performance. There are two types of trackers available: passive trackers and active trackers. Passive tracking units have no motors, controls, or gears. They use the changing weight of a gaseous refrigerant within a sealed frame. When the sunlight activates the refrigerant it causes the frame assembly to move by gravity. Active trackers use motors powered by small PV modules to move the array. There are single-axis trackers that follow the sun's azimuth but not its altitude. There are also dual-axis trackers that follow both the sun's azimuth and its altitude. Dual axis trackers are more expensive than single-axis trackers. Both types of trackers are usually not cost effective on residential installations.

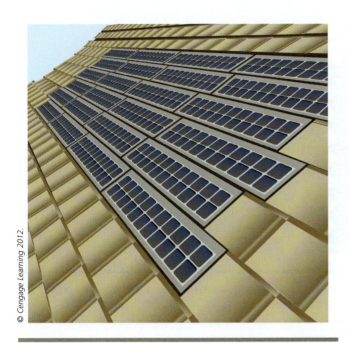

© Cengage Learning 2012.

FIGURE 22-31 PV shingles are another way to integrate a PV array into a building roof.

PV Modules and Diodes

One last item to cover about PV modules is diodes. A **diode** is a semiconductor device that allows current to pass through in only one direction. Blocking diodes (Figure 22-33) are placed in the positive line between modules and the battery to prevent battery current from reversing its flow from the batteries to the array at night or during cloudy weather. Most charge controllers already have diodes or can perform this function in some other way so an electrician does not often have to install blocking diodes. Bypass diodes (Figure 22-34) are wired in parallel with a module to divert current around the module in the event of too much shading. They are usually pre-wired in modules at the factory but you may need to install them. If you install them yourself, they should be sized to handle at least twice the maximum current they are expected to carry and sized at least twice the voltage they will have applied to them.

© Cengage Learning 2012.

FIGURE 22-32 An example of a track-mounted PV array. This mounting system type allows tracking of the sun as it moves across the sky.

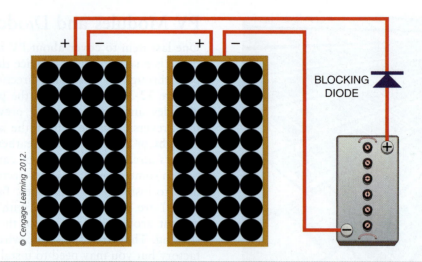

FIGURE 22-33 Blocking diodes are placed in the positive line between modules and the battery to prevent battery current from reversing its flow from the batteries to the array at night or during cloudy weather.

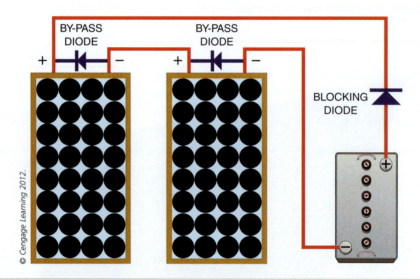

FIGURE 22-34 Bypass diodes are wired in parallel with a module to divert current around the module in the event of too much shading.

BATTERIES

Batteries store DC electricity in chemical form for later use. This stored electricity is used in a PV system at night and during cloudy weather. Batteries can also power loads when a PV array is disconnected for repair or maintenance and can supply the surge current needed by electric motor loads to get started. Not all PV systems use batteries. For example, day use PV systems do not require batteries. Grid-tie systems do not require batteries but sometimes are used as an emergency back-up. Stand-alone systems are where you typically will be installing batteries.

A battery is charging when electricity is put in, discharging when electricity is taken out. A "cycle" is one charge-discharge sequence. Batteries commonly used in PV systems are lead-acid and can either be a liquid vented style (Figure 22-35) or a sealed style (Figure 22-36). Batteries can also be alkaline, nickel-cadmium, or nickel-iron. PV systems require batteries to discharge small amounts of current over a long time and to be recharged under irregular conditions. Automotive batteries are designed to put out large amounts of current for short times, like when a car is started. Therefore, automotive style batteries are not recommended for PV systems. When properly sized and maintained, batteries will last from 3 to 10 years.

FIGURE 22-35 A liquid-vented lead-acid PV battery.

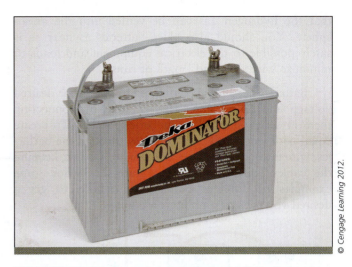

FIGURE 22-36 A sealed style lead-acid PV battery.

Liquid vented lead-acid batteries are built with lead and lead alloy positive and negative plates placed in an electrolyte of sulfuric acid and water. As the battery nears full charge, hydrogen gas is produced and vented out. Water is lost when waste gasses are vented and the battery must be refilled periodically. Some batteries have recombinator cell caps that return the vented gasses as water but it is this author's experience that recombinator caps do not work all that well. Batteries last longer if they are not completely discharged. To help them last longer in a PV system, most charge controllers have a low voltage disconnect (LVD) that shuts down the system if battery voltage gets too low. Batteries have less capacity when they are cold but can last longer. Higher temperatures mean more capacity but a shorter battery life.

> **CAUTION**
>
> **CAUTION:** Hydrogen gas is very explosive! Make sure to properly vent the area where lead-acid batteries used in a PV system are present.

Sealed lead-acid batteries have no caps so there is no access to the electrolyte. They have a valve to allow excess pressure from overcharging to escape and are called a valve-regulated lead-acid battery (VRLA). Sealed batteries are considered maintenance free. There are two types of sealed batteries used in PV systems. The gel cell, where the electrolyte is gelled with silica gel, and the absorbed glass mat (AGM) type, which uses a fibrous silica glass mat to suspend the electrolyte. Sealed batteries are spill proof.

Lead-acid batteries used in a PV system need a charge controller to prevent overcharging and discharging. Use the manufactures' specifications when setting charge termination voltages and low voltage cutoffs with the charge controller.

FIGURE 22-37 Batteries should be placed in a sturdy enclosure like this.

There are several things to consider when installing batteries as part of a PV system. Regardless of temperature concerns, batteries should be placed in a sturdy enclosure (Figure 22-37). The enclosure or area where the batteries are located needs to be well ventilated so that the hydrogen gas can be vented into the atmosphere. You should also place batteries as near as is safely possible to the electrical equipment and loads to minimize wire sizes and length of runs.

Batteries need to be configured to obtain the desired voltage and amp-hours. This is done by connecting them in series, parallel, or in series/parallel. To create equal path lengths for electron flow through the batteries, wire into opposite sides of the battery bank, keeping the cables equal length. See Figure 22-38 for some examples of different battery connections to get a desired voltage and amp-hour amount.

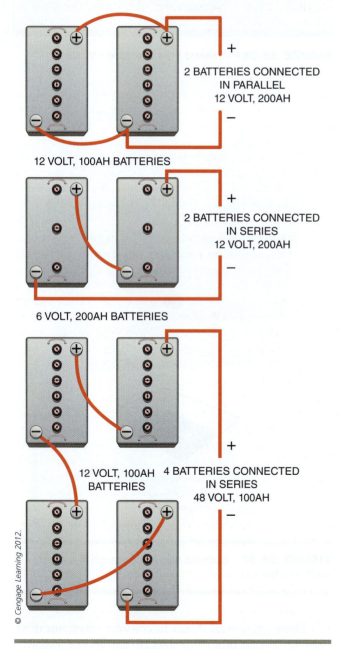

FIGURE 22-38 Examples of different battery connections to get a desired voltage and amp-hour amount.

CHARGE CONTROLLERS

A PV system charge controller is really a voltage regulator. Its primary function is to prevent the batteries from being overcharged by the array. It also protects batteries from being over discharged by the electrical load. Charge controllers come in many sizes from a few amps to 80 amps and more. If a higher current rating is needed, use two or more charge controllers in parallel and divide the array into sub-arrays.

Charge controllers must have a voltage rating that matches the DC system voltage. For example, a charge controller used in a 48 VDC system must have a voltage rating of 48 volts. It must also be capable of handling the maximum PV array short circuit (Isc) current. Use the PV array short circuit current rating plus a 25% safety margin. The *NEC®* in Section 690.8(A)(1) requires the PV array current to be calculated as the sum of the short circuit current ratings of all the parallel-connected modules multiplied by 125%. The *NEC®* defines this circuit as the PV Output Circuit in Section 690.2. It is defined as the circuit conductors between the PV source circuits (the PV array) and the DC utilization equipment. The charge controller must also be capable of handling the maximum DC load current that will pass through it.

Here is an example of how to make sure you have the correct size charge controller. Assume that a customer is installing a 24-volt PV system and has a maximum DC electrical load of 1200 watts that must be powered at any one time. The PV array consists of ten PV modules wired in parallel. Each module has a peak current of 5.95 amps and a short circuit current rating of 6.28 amps. First, calculate the PV array short circuit current: 10 modules × 6.28 amps = 62.8 amps. Then, multiply this number by 125% as the *NEC®* requires: 62.8 amps × 125% = 78.5 amps. The charge controller must be able to handle at least 78.5 amps from the PV array. Next, calculate the maximum DC load amps: 1200 watts/24 volts = 50 amps. The charge controller will also need to be able to handle at least 50 amps of DC electrical load.

INVERTERS

The type of PV module most commonly used and discussed in this chapter produces DC electricity. Batteries can be used to store the DC electricity for use at night and on dark and stormy days. The primary purpose of an inverter is to change the DC supplied from PV modules and batteries to AC when AC loads need to be supplied. Inverters can also feed AC electricity back into the utility grid in a grid-tie PV system.

There are two categories of inverters: grid-tie inverters (Figure 22-39) and stand-alone inverters (Figure 22-40). Some inverters may have both capabilities built-in for

FIGURE 22-39 A typical grid-tie inverter. Schneider Electric Conext Grid Tie Inverter. *Courtesy of Schneider Electric.*

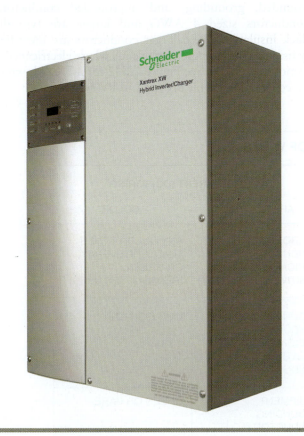

FIGURE 22-40 A typical stand-alone inverter. Schneider Electric Xantrex XW Inverter/Charger. *Courtesy of Schneider Electric.*

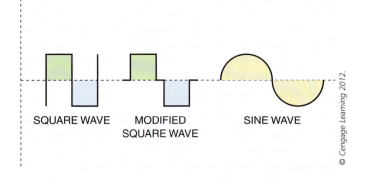

SQUARE WAVE   MODIFIED SQUARE WAVE   SINE WAVE

© Cengage Learning 2012.

FIGURE 22-41 The three most common inverter AC wave forms are the square wave, modified square wave, and the pure sine wave.

future utility connection. Inverters also are classified for the type of AC waveform they produce. The three most common are the square wave, modified square wave, and the pure sine wave (Figure 22-41). A square wave is suitable for small resistive heating loads, some small appliances, and incandescent lighting. This type of inverter is relatively inexpensive but has the disadvantage of possibly burning up electric motors in certain equipment. A modified square wave inverter is suitable for operating a wide variety of loads, including motors, lights, TVs, and other electronic equipment. However, some electronic equipment may pick up inverter noise. One disadvantage is that clocks and appliances/equipment that use digital timekeepers run either fast or slow. Another disadvantage is that you should never charge tool battery packs on a modified square wave inverter unless you know for sure that the battery charger is designed to work correctly. Many electricians and other trades people have burned up their battery chargers on a construction site when they plugged them into an on-site generator that produces a modified square wave. A sine wave inverter is suitable for sensitive electronic equipment. It produces an output with very little distortion. It has high surge capabilities and can feed electricity back into the utility grid with no problems. It is the first choice for inverters if it fits the system budget. It is the only choice if the system you are installing is a grid-tie system. Because of the need for the grid-tie inverter sine wave to be as pure as possible, the inverter must meet Institute of Electrical and Electronic Engineers (IEEE) Standard 1547 and Underwriters Laboratories (UL) 1741. The nameplate on a grid-tie inverter will have this information.

There are a few things to consider when specifying a grid-tie inverter. The AC wattage output rating must be large enough so that the inverter can handle the AC electrical load. For example, if the total AC load that needs to be operated at the same time is 3600 watts you would need an inverter with an output wattage rating of at least 3600 watts. A standard size inverter made by many manufacturers is 4000 watts. Another consideration is the input voltage. Grid-tie inverters have a voltage range that your PV array voltage must fall in. Most require a DC input of 75–600 volts. The output voltage will be 120 VAC or 240 VAC single-phase at a 60 Hz frequency.

Considerations for a stand-alone system inverter are very similar. They include the AC output wattage, the DC voltage from the batteries, the output voltage, and the frequency. In addition, you will need to consider the surge capacity and the waveform type of a stand-alone system inverter.

PV SYSTEM WIRING AND THE *NEC*®

PV system wiring must be in compliance with the *National Electric Code*®. The *NEC*® material referenced in this chapter is based on the 2011 *NEC*®. Article 690 covers Solar Photovoltaic (PV) Systems.

PV SYSTEM CONDUCTORS

Copper wire should be used when wiring PV systems. Aluminum wire can be used where allowed but most PV system installers still use copper wire. Remember

that there are drawbacks to using aluminum wire, especially because it oxidizes quickly and if not terminated properly can cause connection problems. Solid wire or stranded wire can be used but installers usually use stranded wire because of the flexibility that it has. If using flexible, fine-stranded cables they must be terminated only with terminals, lugs, devices, or connectors that are identified and listed for such use according to Section 690.31(F). Terminations that are not designed for the fine-stranded wire can burn up.

PV system installers must use conductors with insulation that is suitable for where the conductor is used. Many sections of a PV system installation, such as on the roof of a house or underground in a trench, are considered wet locations. These areas will require a conductor that has a "W" in its letter designation like THWN-2. THHN insulation cannot be used in wet locations. Table 310.104(A) is used to determine whether a specific conductor insulation can be used in a dry, damp, or wet location and should be consulted by a PV system installer.

Conductor insulation is color coded to designate its function and use. Color coding is used to identify the grounded, grounding, and ungrounded conductors. Conductors sized 4 AWG and larger are typically black insulated. These larger conductors are identified by using a colored tape on the ends where electrical connections are made. See Figure 22-42 for a breakdown of the color coding typically used in PV systems.

COLOR CODING OF WIRES			
ALTERNATING CURRENT (AC) WIRING		**DIRECT CURRENT (DC) WIRING**	
APPLICATION	COLOR	APPLICATION	COLOR
UNGROUNDED CONDUCTOR	ANY COLOR OTHER THAN WHITE, GRAY, OR GREEN	UNGROUNDED CONDUCTOR (TYPICALLY THE POSITIVE CONDUCTOR)	ANY COLOR OTHER THAN WHITE, GRAY, OR GREEN (RED IS OFTEN USED)
GROUNDED CONDUCTOR	WHITE OR GRAY	GROUNDED CONDUCTOR (TYPICALLY THE NEGATIVE CONDUCTOR)	WHITE OR GRAY
EQUIPMENT GROUNDING CONDUCTOR	GREEN OR BARE	EQUIPMENT GROUNDING CONDUCTOR	GREEN OR BARE

© Cengage Learning 2012.

FIGURE 22-42 Typical color coding for the PV system wiring.

PV system conductors are installed as a cable assembly or in a conduit system. Cable types are the same as those introduced to you in Chapter 2 and discussed throughout this textbook. They include Type NM cable, Type UF cable, Type AC cable, Type MC cable, and Type USE cable. These cables are installed in a PV system according to the requirements found in the *NEC®* articles that cover each cable type. For PV module interconnections, you can use USE-2 or any listed and labeled photovoltaic (PV) wire according to Section 690.31(B). For PV modules that do not have factory installed connection leads, many installers choose to install single conductors in conduit. Single conductors like THWN-2 are typically used. In this type of installation, a flexible raceway like liquidtight flexible nonmetallic conduit (LFNC) is used to connect module junction boxes and the THWN-2 conductors are then installed in the LFNC. Most manufacturers now make PV modules with lengths of interconnection cables permanently connected to the modules. The terminations are made with MC (multi-contact) connections (Figure 22-43). When the connectors are plugged together a water-tight connection is made. This type of module connection makes connecting the modules in a series, parallel, or series-parallel configuration very easy. Connectors that are readily accessible and that are used in circuits operating at over 30 volts require a tool for opening.

If conductors are installed in conduit located outside the house or underground in a trench, you need to use 90°C, wet rated conductors. The following conductors can be used: RHW-2, THW-2, THWN-2, and XHHW-2. Use the *NEC®* to determine the maximum number of conductors that can be installed in a conduit type. Informational Annex C and Tables 4 and 5 in Chapter 9 contain the information needed. Chapter 12 in this textbook covered how to determine the maximum number of conductors allowed in a conduit. If you are not sure how to determine the maximum number of conductors in a conduit, you should review that material.

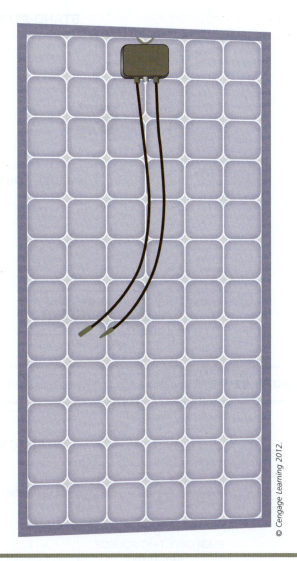

FIGURE 22-43 Most manufacturers make PV modules with lengths of interconnection cables permanently connected to the modules. The terminations are made with MC (multi-contact) connections. This figure shows the back side of a module.

CONDUCTOR SIZING

Conductor sizing for a PV system requires an installer to consider two criteria: ampacity and voltage drop. Ampacity refers to the current-carrying ability of a conductor. Tables 310.15(B)(16) and 310.15(B)(17) in the *NEC®* are used to determine the conductor ampacity for PV installations. Conductor ampacity must be derated for high ambient temperatures. Chapter 6 in this textbook presented material on how to calculate the maximum ampacity of a conductor when the ambient temperature around the conductor is higher than the base temperature rating of 30°C (86°F) for Tables 310.15(B)(16) and 310.15(B)(17). You should review that material at this time. Conductor ampacity must

also be derated if the conductors are installed in conduit on rooftops. See Section 310.15(B)(3)(c) and Table 310.15(B)(3)(c). As an installer you will need to calculate the conductor size using both the ampacity and voltage drop criteria and then choose the wire size that will have an ampacity equal to or greater than the maximum circuit amperage <u>and</u> allow a voltage drop no higher than the system's maximum allowed.

Sizing the conductors based on ampacity is found according to Section 690.8. To understand this material an electrician needs to understand the locations of the various circuit types in a PV system. Figure 22-44 shows where the various PV circuit types are located in a typical stand-alone system and should be consulted as you study the following conductor sizing sections.

STAND-ALONE PV SYSTEM

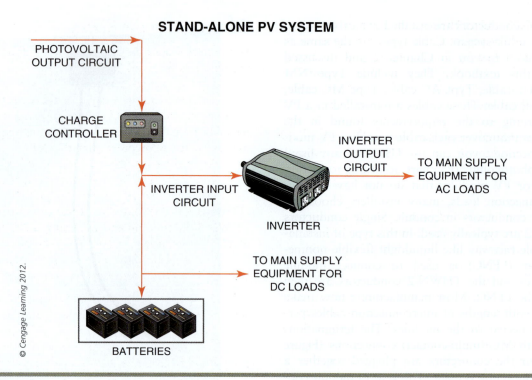

© *Cengage Learning 2012.*

FIGURE 22-44   It is important to know where the various PV circuit types are located in a typical stand-alone system.

INTERACTIVE (GRID-TIE) PV SYSTEM

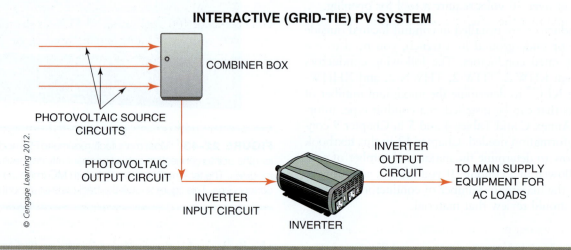

© *Cengage Learning 2012.*

FIGURE 22-45   It is important to know where the various PV circuit types are located in a typical utility-interactive (grid-tie) system.

Figure 22-45 shows where the various PV circuit types are located in a typical utility-interactive (grid-tie) system and should be consulted as you study the following conductor sizing sections.

The first thing you will need to do is find the maximum circuit current that will be on the conductors. Section 690.8(A) contains the requirements for calculating the maximum circuit current. Section 690.8(B) states that all PV system currents are considered continuous

and Section 690.8(B)(1)(a) states that circuit conductors must be sized to carry no less than 125% of the maximum circuit current.

PV Output Circuit Sizing

Section 690.8(A)(1) and (2) requires the maximum current for the PV output circuit to be found by multiplying the sum of the parallel-connected module's short circuit currents by 125%. Once the maximum circuit current is found for either PV source circuits or PV output circuit it must be multiplied by 125% because they are also considered to be a continuous load. For example, a PV output circuit is fed by an array that is made up of 4 series strings of modules with a short circuit current (Isc) of 5.25 amps each. The 4 strings of PV modules are connected in parallel with each other. The maximum circuit current used to size the PV output circuit conductors is 32.8 amps (4 strings × 5.25 I_{sc} × 1.25 × 1.25). Another way to say this is that for the wire run from the PV array to the inverter in a grid-tie system, or to the charge controller in a stand-alone system, you use the short circuit current of each module multiplied by the number of panels in parallel and then multiply by 125% and then by another 125%. Once the maximum load current is found, refer to Table 310.15(B)(16) for the minimum conductor size that will handle that amperage. Assuming that all terminations are rated 75°C and based on Table 310.15(B)(16), using THWN-2 in conduit requires at least 10 AWG conductors. Remember that if the ambient temperature around these conductors is greater than 86°F the conductor size will most likely need to be adjusted to a larger size. There will also need to be an additional adjustment to the ambient temperature if the conductors are run in a conduit exposed to sunlight on a rooftop. For example, let us assume that the conductors in the example above are installed in conduit from a rooftop PV array combiner box to a disconnect switch located on the outside wall of a house. The conduit on the roof portion of the installation is installed so that there is 2 inches from the roof to the bottom of the conduit. According to Table 310.15(B)(3)(c), a temperature adder of 40°F is added to the ambient temperature around the conductors. If the ambient temperature is determined to be 90°F, the total temperature used for ambient temperature compensation would be 130°F (40° + 90°). According to Table 310.15(B)(2)(a), the ambient temperature adjustment factor would be .76 for THWN-2 insulated conductors. This ambient temperature compensation adjustment would mean that at least 8 AWG conductors would need to be installed instead of the 10 AWG conductors. The THWN-2 conductor has an ampacity of 55 amps in the 90°C column. Remember that the NEC allows us to apply the ampacity adjustment factors to the actual ampacity based upon the conductor's insulation temperature rating. Therefore, 55 amps multiplied by the correction factor of .76 equals 41.8 amps. This adjusted ampacity will have room for the maximum circuit current of 32.8 amps calculated above. The 10 AWG with THWN-2 insulation has a 90°C ampacity of 40 amps and when multiplied by the adjustment factor of .76 results in an ampacity allowed of 30.4 amps. Since 30.4 amps is smaller than the 32.8 amps needed you must install the 8 AWG THWN-2 conductors.

Battery Circuit Sizing

For batteries used as part of a stand-alone PV system, the wire run from the batteries to the charge controller is usually the same size as the PV output circuit conductors. The conductor size used for the wire run from the batteries (or charge controller) to a DC load distribution panel is based on the total amperage requirement of the DC load. Remember that the *NEC®* requires you to multiply the current by 125% on all wire runs in a PV system. For example, if the DC load is 22 amps the maximum circuit current would be 27.5 amps (22 amps × 1.25). Based on Table 310.15(B)(16), using THWN-2 in conduit and assuming 75°C terminations would require at least a 10 AWG conductor. Remember that you may need to adjust the conductor size up because of ambient temperature compensation.

For battery interconnection wires there is no *NEC®* rule on how to size them. Because batteries can "pump out" a lot of amps in a short amount of time, you need to make sure that the battery interconnection wires are large enough to handle the high currents. A recommendation is to use 4 AWG copper wire if you have a 500-watt or smaller inverter, 2 AWG copper wire for an 800-watt inverter, and 2/0 or larger for 1000-watt and larger inverters. Another way to size the interconnection wires is to use 4/0 AWG cables for circuits protected with a 250-amp circuit breaker or 400-amp fuses, 2/0 AWG cables for circuits protected with a 175-amp circuit breaker or 200-amp fuses, and 2 AWG cables for circuits protected with 110-amp or smaller circuit breakers or fuses. If you are still not sure how to size the battery interconnection cables, consult the vendor who supplied the PV system batteries.

Because battery plates and terminals are often made of soft lead and lead alloys encased in plastic, larger sized copper interconnection cables with a low number of strands can cause distortion when they are attached to the battery terminals. The use of flexible cables will reduce the possibility of such distortions. Listed cables with the appropriate physical and chemical resistant properties should be used. Welding and "battery" cables are not allowed or described in the *NEC®* for this use. Section 690.74 states that flexible, fine-stranded cables can only be used with terminals, lugs, devices, and connectors that are listed and marked for such use. Flexible, fine-stranded battery connection cables cannot be connected to terminals identified for use with conductors using the larger strands of conductors. Flexible "building wire" type cables are available and suitable for this use.

Inverter Output Circuit Sizing

The wire size for the inverter output circuit that runs from an inverter to an AC loadcenter is based on the inverter's continuous output current rating. Again, remember that the *NEC*® requires you to multiply the current by 125% on all wire runs in a PV system. For example, if an inverter has a rating of 4500 watts, and the output voltage is 240 VAC, the continuous output current would be 23.4 amps [(4500 watts / 240 volts) × 1.25)]. Based on Table 310.15(B)(16), using THWN-2 in conduit and assuming 75°C terminations would require at least a 12 AWG conductor. Remember that you may need to adjust the conductor size up because of ambient temperature compensation.

Inverter Input Circuit Sizing for Stand-Alone Systems

Section 690.8(A)(4) states that the maximum current is the stand-alone continuous inverter input current rating when the inverter is producing rated power at the lowest input voltage. Stand-alone inverters are nearly constant output power devices so when the input voltage from the batteries decreases, the battery current increases to keep the wattage output constant. The input current for these inverters is calculated by taking the rated wattage output of the inverter and dividing it by the lowest operating battery voltage and then by the rated efficiency of the inverter under those operating conditions. The simplest way to calculate the maximum current for an inverter input circuit in a stand-alone PV system is to take the inverter rated wattage output and divide it by the DC input circuit voltage multiplied by the percent efficiency of the inverter. The percent efficiency is on the nameplate but if it is not there assume an efficiency of 90%. For example, for a 4000 watt, 24-volt inverter that is 90% efficient at 22 volts, the input current is 202 amps [4000 watts / (22 volts × .90)]. Based on Table 310.15(B)(16), using THWN-2 in conduit and assuming 75°C terminations would require at least a 4/0 AWG conductor. Remember that you may need to adjust the conductor size up because of ambient temperature compensation.

Conductor Sizing Based on Voltage Drop

In Chapter 8 of this textbook you were introduced to voltage drop. In that chapter voltage drop was considered for long runs of underground service entrance conductors. In this chapter there is a need to consider voltage drop when installing long runs of wiring as part of a PV system. Remember that voltage drop is the amount of the system voltage that is used to "push" the current through one circuit conductor from the source to a load, through that load, and back on another conductor to the source. The amount of voltage drop depends on the resistance of the wires, the length of the wires, and the actual current carried on the wires. Allowing too much voltage drop will not leave enough circuit voltage for the load to operate properly. The shorter runs of wire in a PV system

used to connect certain components may be so close together that voltage drop is not a concern. Nevertheless, you should always determine what the voltage drop is going to be in any PV system circuit so that the conductors can be sized properly.

The *NEC*® recommends that both branch circuits and feeder circuits be sized to limit the voltage drop to no more than 3 percent. Remember, this is not a requirement, only a suggestion. There is also no *NEC*® recommendation specifically for PV system circuits but many PV system installers and designers want no more than a 2 or 3 percent voltage drop in the system wiring.

There is a common formula used to calculate voltage drop:

$$VD = \frac{K \times I \times L \times 2}{cm}$$

K = a constant that is 12.9 ohms for copper wire; 21.2 ohms for aluminum wire

I = load current in amps

L = the one-way length of the conductors from the source to the load

2 = the multiplier used because the length of wire goes to the load and back to the source

cm = the cross sectional area of the conductor in circular mils; this can be found in Table 8, Chapter 9 of the *NEC*®

If the formula is manipulated properly, the minimum conductor size that would need to be installed to allow a certain voltage drop can be calculated. This formula is

$$cm = \frac{K \times I \times L \times 2}{VD}$$

Two examples of how to use the voltage drop formulas are shown on the next page.

OVERCURRENT PROTECTION

Every circuit must be protected from an electrical current that exceeds the conductor's ampacity. There are two types of overcurrent devices used in PV systems: fuses and circuit breakers. Circuit breakers must be UL listed and have a DC rating if used on DC circuits. AC-only circuit breakers will have their contacts burn up quickly if used on DC circuits. Fuses must be UL listed and be DC rated if used in a DC circuit.

Overcurrent Protection Placement

The *NEC*® requires every ungrounded conductor to be protected by an OCPD (Section 240.20). This includes conductors used in a PV system. The ungrounded conductor is the positive (+) conductor in a DC system

EXAMPLE 1

Calculate the voltage drop on the PV output circuit conductors from a roof-mounted PV array to a charge controller in a stand-alone system. Use the same PV output circuit information from above that resulted in a 10 AWG copper conductor sized on only the amperage flowing on the conductors. A PV output circuit is fed by an array that is made up of 4 series strings of modules with a short circuit current of 5.25 amps each. The 4 strings of PV modules are connected in parallel with each other. The maximum circuit current used to size the PV output circuit conductors considering voltage drop is 21 amps (4 strings × 5.25 I_{sc}). There is no need to use the 125% safety rules from Section 690.8 when calculating conductor size considering voltage drop because voltage drop will only depend on the actual amount of current flowing on the conductors.

The array voltage in this example is 24 volts DC. The charge controller in this example is located in the basement of the home and the total distance of the wire run is 80 feet. Here is how the calculation is done:

$$VD = \frac{12.9 \text{ ohms} \times 21 \text{ amps} \times 80 \text{ feet} \times 2}{10,380 \text{ cm (from Table 8, Chapter 9)}}$$
$$= \frac{43,344}{10,380} = 4.2 \text{ volts}$$

The percent voltage drop is 17.5% [(4.2 volts/24 volts) × 100]. This is a very high voltage drop and when you subtract the 4.2 volts from the 24-volt array voltage it will only leave 19.8 volts at the charge controller. This is not enough voltage for the charge controller to do its job. Obviously a larger conductor size is needed to keep the voltage drop in the 2–3 percent range.

EXAMPLE 2

Calculate the minimum size conductor needed for the PV output circuit conductors so that there will not be a voltage drop of more than 2 percent. Here is how the calculation is done:

$$cm = \frac{12.9 \text{ ohms} \times 21 \text{ amps} \times 800 \text{ feet} \times 2}{.48 \text{ volts (2% of 24 volts)}}$$
$$= \frac{43,344}{.48} = 90,300 \text{ cm}$$

Find the minimum conductor size in Table 8 in Chapter 9 of the *NEC®* based on a minimum

circular mil requirement of 90,300 cm. A minimum wire size of 1/0 AWG copper must be used.

Calculating the PV output circuit size using only amperage gave us a 10 AWG conductor size. Calculating the PV output circuit size based on no more than a 2% voltage drop gave us a 1/0 AWG conductor size. In a real installation you would use the 1/0 AWG size because it will be more than able to handle the circuit current and will keep the voltage drop to 2% or lower.

(Caution: Some systems may have the positive conductor grounded but we will limit our discussion to systems with the positive conductor being ungrounded.) In a PV system with multiple sources of power (e.g, panels, batteries, generators, etc.), the OCPD must protect the conductor coming from any power source [Section 690.9(A)]. Refer to Figure 22-46 for the proper overcurrent protection device placement in a typical PV system.

Overcurrent Protection Sizing

The general rule to follow is that the rating of the OCPD must be less than or equal to the ampacity of the wire. Standard overcurrent protection device sizes are listed in

Section 240.6(A). Remember that Section 240.4(B) allows the next standard higher OCPD to be used if the ampacity of the conductor does not match a standard size fuse or circuit breaker. Also, Section 240.4(D) requires an OCPD no larger than 15A for 14 AWG; 20A for 12 AWG; and 30A for 10 AWG copper conductors.

Section 690.8(B)(1) covers sizing for the minimum size OCPD. It requires the use of the amperage of the power source or the actual current draw, including safety factors. The maximum size OCPD allowed is based on the ampacity of the actual wire size used. This value could be a lot higher than needed because the wire may have been upsized because of voltage drop considerations.

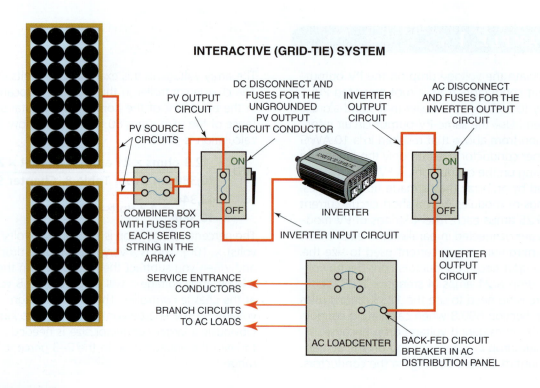

INTERACTIVE (GRID-TIE) SYSTEM

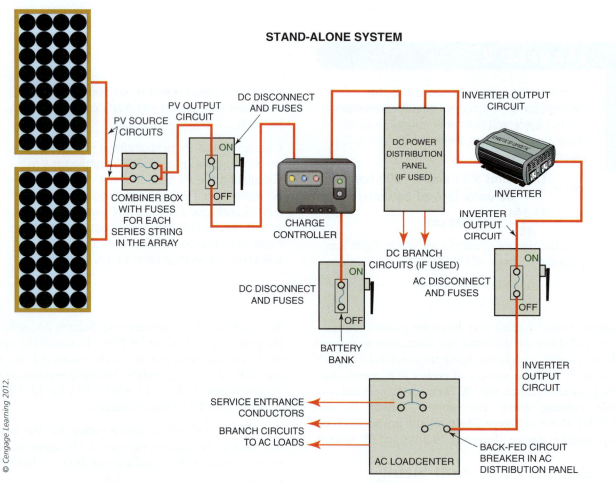

STAND-ALONE SYSTEM

FIGURE 22-46   An example of the proper overcurrent protection device placement in a PV system.

For example, let us determine the <u>minimum</u> size circuit breaker needed to protect the PV output circuit conductors that we sized in the previous section. The PV array has a maximum circuit current of 32.8 amps (4 strings × 5.25 I_{sc} × 1.25 × 1.25). Therefore, a 35-amp OCPD is the minimum size allowed. To determine the <u>maximum</u> size circuit breaker needed to protect the wire, you consider the wire size needed based on a minimum voltage drop of 2%. A 1/0 AWG wire is needed for a 2% voltage drop maximum and a 1/0 AWG conductor with THWN-2 insulation has an ampacity of 150 amps, assuming termination temperatures are all 75°C. According to Section 240.6(A), there is a standard size 150-amp circuit breaker so that is the maximum size allowed. Therefore, for this example you could choose a 35-amp circuit breaker or any size up to and including 150 amps. Some factors that will influence your choice are the circuit breaker cost, the size of the circuit breaker terminations (smaller size breakers will not be able to handle the larger size wire needed because of voltage drop), any future expansion of the PV system, and the availability of the circuit breaker size you want to use. It is recommended that you choose an overcurrent protection device that is close to the minimum size for the best circuit protection.

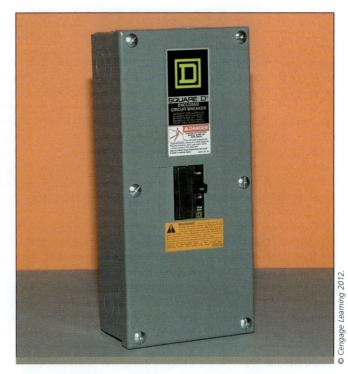

FIGURE 22-47   Circuit breakers can be used as disconnects.

DISCONNECTS

Section 690.13 states that all current-carrying conductors of the PV power source must have a means to disconnect them from other conductors in a building. The grounded conductor must not be disconnected. Section 690.15 requires that each piece of equipment in a PV system, such as inverters, batteries, and charge controllers, must be able to be disconnected from all sources of power. Circuit breakers can be used as disconnects and are used often (Figure 22-47). Fuses can only be used as a disconnecting means if they are part of a switch like a fusible safety switch (Figure 22-48). Each PV system disconnecting means must be permanently marked to identify it as a PV system disconnect according to Section 690.14(C)(2). Section 690.17 also states that where all terminals of the disconnecting means may be energized in the open position, a warning sign must be placed on or next to the disconnecting means (Figure 22-49).

According to Section 690.17, the disconnecting means must be manually operated switches or circuit breakers that meet the following criteria:

- Must be accessible
- Must not have any exposed live parts
- Must clearly indicate when they are in the open or closed positions
- Must be rated for the voltage and available current

FIGURE 22-48   Fuses can only be used as a disconnecting means if they are part of a switch like a fusible safety switch.

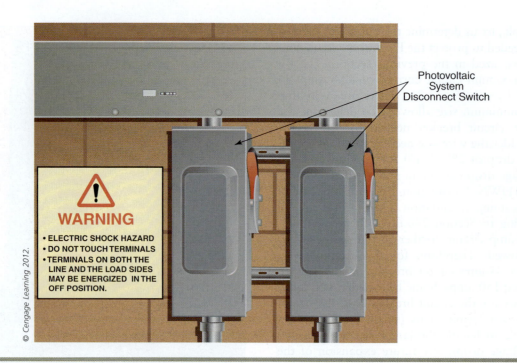

Photovoltaic
System
Disconnect Switch

WARNING

• ELECTRIC SHOCK HAZARD
• DO NOT TOUCH TERMINALS
• TERMINALS ON BOTH THE
  LINE AND THE LOAD SIDES
  MAY BE ENERGIZED IN THE
  OFF POSITION.

© Cengage Learning 2012.

FIGURE 22-49 Where all terminals of the disconnecting means may be energized in the open position, a warning sign must be placed on or next to the disconnecting means.

A disconnecting means to disconnect all conductors in a building from the PV system conductors must be installed according to Section 690.14(C). When installing the disconnect you will need to have six or fewer disconnecting devices to shut off all sources of power according to Section 690.14(C)(4). In addition, Section 690.14(C)(5) requires that the disconnects be grouped together. The PV disconnect needs to be in a readily accessible location outside or inside at the first point of building penetration of the PV conductors according to Section 690.14(C)(1). The PV conductors from the array should be installed on the outside of a building until reaching the first readily accessible disconnect. Sections 690.14(C)(1) *Exception* and 690.31(E) state that where direct-current photovoltaic source or output circuits of a grid-tie inverter from a photovoltaic system are run inside a building or structure, they must be contained in metal raceways, Type MC cable that has an equipment ground as listed in Section 250.118(10), or metal enclosures, from the point of penetration of the surface of the building or structure to the first readily accessible disconnecting means. This means that you could install a metal raceway inside a house that runs from a roof-mounted PV array to a utility room or basement where the other PV system components are located. If the PV conductors are installed in a metal raceway or you use Type MC cable through the interior of a house, the disconnect switch is permitted (but not required) to be located at some point away from the point of entry of the PV system conductors. For example, if a metal raceway is installed from a roof down to a basement where the PV system inverter is located for a grid-tie system, the disconnect switch does not need to be located at the point where the metal raceway enters the basement. You could continue the metal raceway run to a location some distance away and install the disconnecting means on a backboard with the grid-tie inverter.

There are some other *NEC*® items to consider when installing PV source and output circuit wiring inside a building. Section 690.31(E)(1) states that the wiring method cannot be installed within 10 inches (25 cm) of the roof decking or roof sheathing except directly below the roof where the PV modules are mounted. According to the Informational Note, this requirement is to prevent accidental damage from the saws used by firemen when ventilating a roof during a fire. Section 690.31(E)(2) requires the installer to protect flexible metal conduit (Type FMC is a metal raceway) smaller than ¾" trade size or Type MC cable smaller than 1 inch diameter with guard strips that are at least as high as the wiring method when run across ceilings or floor joists. Section 690.31(E)(3) requires marking exposed raceways and other wiring methods, the covers on pull or junction boxes, and conduit bodies where any of the available openings are unused with the wording "Photovoltaic Power Source." You can use labels or permanent markers. Section 690.31(E)(4) requires the labels or markings to be visible after installation. The labels or markings need to appear on every section of the wiring that is separated by enclosures, walls, partitions, ceilings, or floors. There is not supposed to be anymore than 10 feet between the labels or markings.

GROUNDING

There are two types of grounding in a PV system: system grounding and equipment grounding. Section 250.4 outlines why systems and equipment, including PV systems, are grounded. Some of the main reasons include:

- Limiting voltages due to lightning, line surges, or unintentional contact with higher voltage lines.

- Stabilizing voltages and providing a common reference point—the earth.

- Providing a low resistance current path in order to facilitate the operation of the overcurrent protection devices during faults to ground.

Equipment Grounding

Equipment grounding can provide protection from shock caused by a ground fault in PV system components by causing the overcurrent protection device to activate quickly. An equipment grounding conductor (EGC) is continuous and bonds all metal parts of the installation together. A final connection to the earth will be made with a grounding electrode conductor and a grounding electrode. Section 690.43 covers equipment grounding. Section 690.43(A) states that all exposed noncurrent-carrying metal parts of module frames, equipment, and conductor enclosures must be grounded regardless of voltage.

> **CAUTION**
>
> **CAUTION:** In addition to the requirements found in the *NEC*®, make sure each module is grounded using the supplied (or specified) hardware and use the manufacturer's grounding instructions. The grounding point on the module must be identified by the manufacturer. The designated point marked on the module (Figure 22-50) must be used since this is the only point tested and evaluated by UL for use as a long-term grounding point.

Module and Rack Grounding

The PV module equipment grounding conductor must be installed so that the removal of any module from the circuit will not interrupt the grounding of any of the other modules. Section 690.43(C) states that devices listed and identified for grounding metal frames of PV

FIGURE 22-50 The grounding point on a module must be identified by the manufacturer.

© Cengage Learning 2012.

modules can be used to bond the frames to grounded metal mounting structures. Mounting racks (structures) are not typically listed as equipment grounding conductors so just attaching the modules to a rack that is grounded does not automatically ground the modules. Section 690.43(D) requires racks and associated components that are used to ground the attached modules to be identified as being suitable for the purpose of grounding PV modules. Metal mounting racks are considered subject to being energized and as such must be grounded according to Section 250.110. Also, if the metal racks are not identified for grounding they must have bonding jumpers or other devices connected between the separate sections and they need to be bonded to the grounding system. Section 690.43(E) states that devices identified and listed for bonding the metal frames of PV modules are permitted to bond the metal frames of one module to another adjacent module. Section 690.43(F) states that equipment grounding conductors for the PV array and mounting rack must be routed in the same raceway or cable containing the PV array circuit conductors or otherwise run with the PV array circuit conductors when those conductors leave the vicinity of the PV array.

System Grounding

System grounding for a PV system is addressed in Section 690.41. It states that for a photovoltaic power source, one conductor of a 2-wire system with a photovoltaic system voltage over 50 volts must be solidly grounded. System grounding is basically taking one system conductor and intentionally connecting it to the earth. It is required for all PV systems over 50 volts. This means in the DC part of a PV system the negative conductor is connected to earth at some point in the system. You will need to locate this system ground connection point as close as possible to the PV source to better protect the system from voltage surges due to lightning according to the Informational Note following Section 690.42. Section 240.21 requires that when a PV system under 50 volts is not system grounded, both conductors will need overcurrent protection. As a result, PV systems (including those with 12-, 24-, and 48-volt DC circuits) are usually system grounded so that only the ungrounded conductor will need an overcurrent protection device. According to Section 690.42, the DC circuit-grounding connection must be made at any single point on the photovoltaic output circuit. An *Exception* to this requirement states that PV systems requiring ground-fault protection devices are permitted to have the single point grounding connection made inside the ground-fault protection equipment or inside the grid-tie inverter and additional external bonding connections are not permitted. Connections are to be made in accordance with markings on the equipment or in the installation instructions. Ground-fault protection in PV systems is covered in the next section.

Figure 22-51 shows an example of both equipment grounding and system grounding in a stand-alone PV system installation. Figure 22-52 shows an example of both equipment grounding and system grounding in a grid-tie PV system installation.

Ground-Fault Protection

Section 690.5 contains the requirements for ground-fault protection in a PV system. Grounded DC PV arrays must be provided with ground-fault protection (GFP) to reduce fire hazards. Do not confuse GFP with

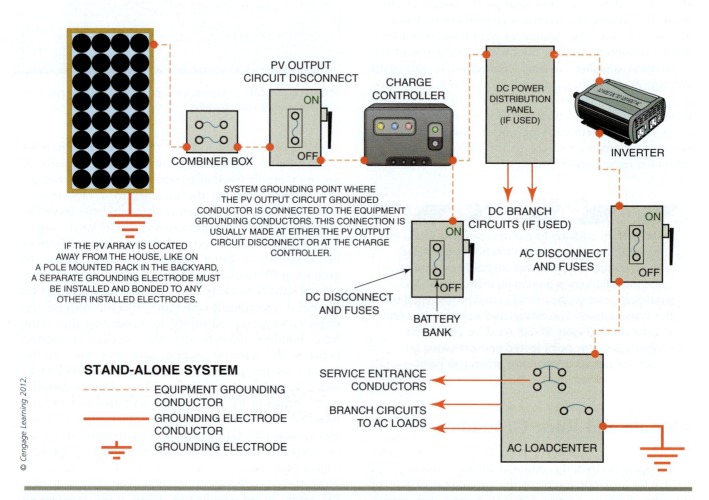

FIGURE 22-51 An example of both equipment grounding and system grounding in a stand-alone PV system installation.

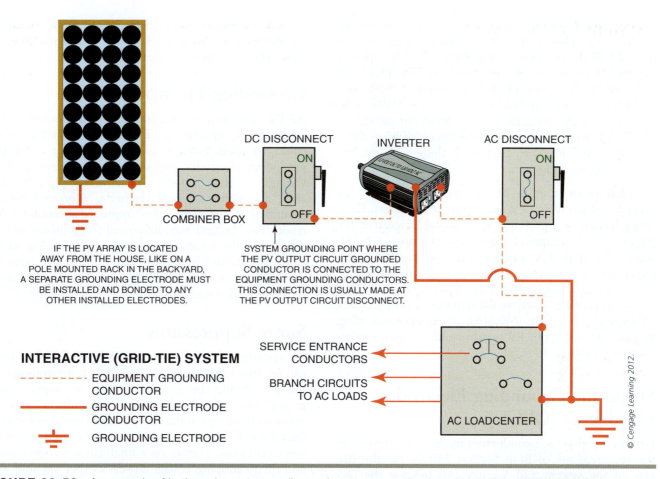

IF THE PV ARRAY IS LOCATED
AWAY FROM THE HOUSE, LIKE ON A
POLE MOUNTED RACK IN THE BACKYARD,
A SEPARATE GROUNDING ELECTRODE MUST
BE INSTALLED AND BONDED TO ANY
OTHER INSTALLED ELECTRODES.

SYSTEM GROUNDING POINT WHERE
THE PV OUTPUT CIRCUIT GROUNDED
CONDUCTOR IS CONNECTED TO THE
EQUIPMENT GROUNDING CONDUCTORS.
THIS CONNECTION IS USUALLY MADE AT
THE PV OUTPUT CIRCUIT DISCONNECT.

INTERACTIVE (GRID-TIE) SYSTEM

– – – – – – – EQUIPMENT GROUNDING
CONDUCTOR

——————— GROUNDING ELECTRODE
CONDUCTOR

GROUNDING ELECTRODE

SERVICE ENTRANCE
CONDUCTORS

BRANCH CIRCUITS
TO AC LOADS

AC LOADCENTER

© Cengage Learning 2012.

FIGURE 22-52   An example of both equipment grounding and system grounding in a grid-tie PV system installation.

GFCI protection. GFP is designed to prevent fires in DC PV circuits due to ground faults while a GFCI device is designed to protect people in single-phase AC systems. *Exception No. 1* to Section 690.5 states that ground-mounted or pole-mounted PV arrays with no more than two paralleled source circuits and with all DC source and DC output circuits isolated from buildings are permitted without ground-fault protection. So, unless a PV array that is ground-or pole-mounted has three or more paralleled strings of modules, GFP is only required on dwelling unit roof-mounted PV arrays. These devices sense DC ground faults anywhere on the DC side of a PV system and can be mounted anywhere in that system. However, GFP devices are usually installed inside the grid-tie inverter or in the DC power center in a stand-alone PV system.

The ground-fault protection device must be capable of detecting a ground-fault current, interrupting the flow of fault current, and providing an indication of the fault. Ground-fault protection must be able to isolate the grounded conductor (normally the negative wire) from ground-fault conditions. It also disconnects the ungrounded conductor (normally the positive wire) at the same time.

A warning label must be placed on the grid-tie inverter or be applied by the installer near the ground-fault indicator at a visible location. When the PV system also has batteries, the same warning label must be applied by the installer in a visible location at the batteries. The label needs to say the following:

WARNING

ELECTRIC SHOCK HAZARD IF A GROUND FAULT IS INDICATED, NORMALLY GROUNDED CONDUCTORS MAY BE UNGROUNDED AND ENERGIZED.

Sizing the Equipment Grounding Conductor (EGC)

Section 690.45(A) covers how to size the equipment grounding conductor for photovoltaic source and photovoltaic output circuits in a PV system. The equipment grounding conductor can be as large as the current-carrying conductors but not smaller than specified in Table 250.122. Where no overcurrent protective device is used in the circuit, an assumed overcurrent device rated at the photovoltaic rated short circuit current must be used. The EGC is not required to be upsized because of voltage drop considerations. The EGC can never be smaller than 14 AWG. Sections 690.46 and 250.120(C) require an EGC smaller than 6 AWG to be run in a raceway or cable armor unless not subject to physical damage or where protected from physical damage. The EGC needs to be routed along with the PV output circuit conductors.

Here is an example of how to size the EGC. An array has 4 paralleled PV source circuits (strings). Each string has an Isc of 6.0 amps. The total Isc current for the array is 30 amps [(4 strings × 6.0 amps) × 1.25]. Remember that this number is found based on the requirements of Section 690.8(A)(1) & (2). The size OCPD would be 40 amps: (30 amps × 1.25 = 37.5 amps). This size is based on the requirements of Sections 690.8(B)(2); 240.4(B); 240.6(A). Therefore, according to Table 250.122, this PV system would require a minimum size EGC of 10 AWG.

Sizing the Grounding Electrode Conductor (GEC)

Grounding electrode conductors must be sized to meet the requirements of both Section 250.66 for the AC side of the system and Section 250.166 for the DC side of the system. The GEC on the AC side would have already been sized and installed when the PV system is a grid-tie system because in a grid-tie PV system the premises grounding system serves as the PV system AC side grounding system. Remember that on the AC side the GEC is sized according to Table 250.66 and is based on the largest size service entrance conductor used. This was covered in detail in Chapter 7. You may want to review that material at this time. In a stand-alone system, the GEC on the AC side (assuming there is an AC side) is based on the equivalent size of the largest service entrance conductor required for the load to be served according to Note 2 of Table 250.66. This usually results in a GEC size of 8 AWG but it could be bigger.

The GEC on the DC side of a PV system must not be smaller than the largest conductor supplied by the system, and cannot be smaller than 8 AWG CU (even if the DC system conductors are smaller than 8 AWG) according to Section 250.166(B). Where connected to a ground rod electrode, that portion of the grounding electrode conductor that is the sole connection to the grounding electrode does not have to be larger than 6 AWG copper. Therefore, most installers use 6 AWG CU since most grounding electrodes in a PV system are ground rods and Section 250.166(C) requires a GEC not larger than 6 AWG in these situations.

Grounding Electrodes

All PV systems must be connected to a grounding electrode. An 8' ground rod is most common but others could be used. Section 250.52 lists other acceptable grounding electrodes and these were covered back in Chapter 7. They include a metal water pipe, building steel, a concrete encased electrode, a ground ring, a ground plate, or some other approved grounding electrode. Two ground rods may need to be driven at least 6' apart if the resistance is not 25 ohms or less to the earth. If two different electrodes are installed, one for the AC side and one for the DC side, the two electrodes must be bonded together according to Section 690.47(C).

Surge Suppression

PV systems mounted in the open or on building roofs can act like lightning rods. Array frame grounding conductors should be routed directly to supplemental ground rods. Metal conduits can add inductance to the array-to-building conductor runs and slow down surges. Metal oxide varistors (MOVs) can be used but have some problems. They draw a small current continuously and their clamping voltage lowers as they age and may reach the open circuit voltage of the PV system. They may also catch on fire when they fail! Silicon oxide surge arrestors offer some advantages such as no current draw when off and they rarely catch on fire.

They are easy to install and are usually placed so that a DC surge suppressor is installed on the DC side and an AC surge suppressor is installed on the AC side. Figure 22-53 shows where they are located in a grid-tie PV system.

Plaque Requirements

According to Section 690.56(A), a stand-alone PV system must have a plaque or directory permanently installed in a visible area on the exterior of the building and acceptable to the authority having jurisdiction. It must indicate the location of the system disconnecting means and that the house contains a stand-alone electrical power system.

Section 690.56(B) requires houses with a grid-tie PV system to have a permanent plaque or directory providing the location of the service disconnecting means and the photovoltaic system disconnecting means if not located at the same location.

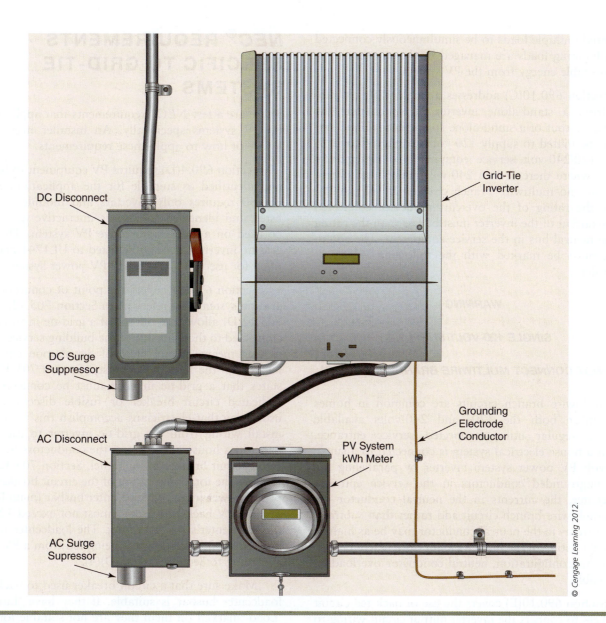

DC Disconnect

DC Surge Suppressor

AC Disconnect

AC Surge Supressor

Grid-Tie Inverter

Grounding Electrode Conductor

PV System kWh Meter

© Cengage Learning 2012.

FIGURE 22-53 An example of where surge suppressors are located in a grid-tie PV system.

NEC® REQUIREMENTS SPECIFIC TO STAND-ALONE SYSTEMS

There are a few *NEC*® requirements that apply to stand-alone PV systems specifically. An installer needs to be aware of how to apply these requirements.

Section 690.10(A) covers the inverter output. A stand-alone PV installation may have an AC output and be connected to a building wired in full compliance with all articles of the *NEC*®. Even though such an installation may have service-entrance equipment rated at 100 or 200 amperes at 120/240 volts, there is no requirement that the PV source provide either the rated full current or the dual voltages of the service equipment. While safety requirements dictate full compliance with the AC wiring sections of the *NEC*®, a PV installation is usually designed so that the actual AC demands on the system are sized to the output rating of the PV system. The inverter output is required to have sufficient capacity to power the largest single piece of utilization equipment to be supplied by the PV system, but the inverter output does not have to be sized for the potential multiple loads to be simultaneously connected to it. Lighting loads are managed by the user based on the available energy from the PV system.

Section 690.10(C) addresses a single 120-volt supply from a stand-alone inverter. It states that the inverter output of a stand-alone solar photovoltaic system is permitted to supply 120 volts to single-phase, 3-wire, 120/240-volt service equipment or distribution panels where there are no 240-volt outlets and where there are no multiwire branch circuits. In all installations, the rating of the overcurrent device connected to the output of the inverter must be less than the rating of the neutral bus in the service equipment. The equipment must be marked with the following words or equivalent:

WARNING

SINGLE 120-VOLT SUPPLY.

DO NOT CONNECT MULTIWIRE BRANCH CIRCUITS!

Multiwire branch circuits are common in homes and utilize both the 120- and 240-volts available from a regular utility connected service entrance. When a house electrical system is connected to a single 120-volt PV power system inverter by paralleling the two ungrounded conductors in the service entrance loadcenter, the currents in the neutral conductor for each multiwire branch circuit add rather than subtract. The currents in the neutral conductor may be as high as twice the rating of the branch-circuit overcurrent device. With this configuration, neutral conductor overloading is possible.

Section 690.10(E) covers the use of back fed circuit breakers to connect the inverter output circuit wiring to a circuit breaker panel. It states that plug-in type circuit breakers that are back-fed and installed in a panel must be secured according to Section 408.36(D). This section requires such circuit breakers to be secured in place by an additional fastener that requires other than a pull to release the breaker from the mounting means on the panel.

A new Section 690.11 in the 2011 *NEC®* requires DC arc-fault circuit protection for some installations. PV systems with DC source or output circuits from an array that are located on or penetrating a house and operating at 80 VDC or more must be protected by a listed DC arc-fault protection device. Other system components that can provide equivalent arc-fault protection can also be used.

NEC® REQUIREMENTS SPECIFIC TO GRID-TIE SYSTEMS

There are a few *NEC®* requirements that apply to grid-tie PV systems specifically. An installer needs to be aware of how to apply these requirements.

Section 690.4(D) requires PV equipment to be listed and identified as suitable for the application. Section 690.60 requires only inverters and modules that are listed and identified for use as interactive (grid-tie) to be used on a utility interactive PV system. Therefore, grid-tie inverters need to be listed to UL1741 and identified for use in an interactive PV power system.

Section 690.64 covers the point of connection in a grid-tie system and refers us to Section 705.12. Section 705.12(D) allows the output of a grid-tie inverter to be connected to the load side of the building service disconnecting means at any piece of distribution equipment located on the building premises. Section 705.12(D)(1) states that a grid-tie inverter must be connected to a dedicated circuit breaker or fusible disconnect. The usual way that electricians accomplish this is to simply install wiring from the grid-tie inverter to the service panel in a house and connect the conductors to a dedicated circuit breaker in the panel. Section 705.12(D)(2) requires the total amp rating of the circuit breakers supplying power to an AC loadcenter busbar (main breaker and the PV backfed breaker) must not exceed 120% of the loadcenter's busbar rating. The loadcenter must be marked to indicate that it is being fed from a PV source according to Section 705.12(D)(4).

Make sure that a circuit breaker used to backfeed a loadcenter busbar is suitable. If they have 'Line' and 'Load' marked on them they are not suitable for back-feeding according to Section 705.12(D)(5) and the Informational Note that follows. Section 705.12(D)(6) states that when back-feeding an AC loadcenter with a circuit breaker, the breaker does not need to be individually clamped to the panel busbar as required in 408.36(D). The back-fed circuit breaker is required to be located at the opposite end from the main breaker in the loadcenter according to Section 705.12(D)(7) and a permanent warning label must be applied stating:

WARNING

INVERTER OUTPUT CONNECTION

DO NOT RELOCATE THE OVERCURRENT DEVICE.

Grid-tie inverters can be mounted in not-readily accessible locations like on roofs or other exterior areas according to Section 690.14(D). To do this the DC side disconnecting means must be mounted within sight of or in the inverter; the AC side disconnecting means must be mounted within sight of or in the inverter; the AC output conductors from the inverter must have a disconnecting means installed at a readily accessible location either on the outside of the building or inside at the nearest point of entrance of the conductors; and a plaque must be installed at the service entrance which notifies someone that there is a PV power source in addition to the regular AC service from the utility.

PV System Installation Guidelines

There are several installation guidelines to follow when installing either a stand-alone or grid-tie PV system. The most common installation guidelines are presented in the following paragraphs.

PV ARRAY INSTALLATION GUIDELINES

The installer will need to select the PV array location first and then choose the appropriate mounting system. Always try to locate the PV array as close as possible to the charge controller/batteries/inverter to minimize voltage drop. Mount the array so it is easy to maintain and install it so that there is as much security as possible. For example, you can use specialized screws with unique heads so that it will not be easy for someone to take the array apart and steal the expensive PV modules.

PV module support structures should provide a simple, strong and durable mounting system. That is why most systems use an extruded aluminum frame. The mounting system should be made of weather-resistant and corrosion free materials like anodized extruded aluminum. Galvanized steel and stainless steel are excellent choices as well. Wood is not a great choice because of the extra maintenance required.

BATTERY INSTALLATION GUIDELINES

If batteries are used in the PV system good installation guidelines start with the shipping of the batteries. Remember that the batteries are heavy and prone to leakage so check the batteries when they arrive for any damage. A boost charge may be necessary when you get the batteries. When you charge them make sure to keep the batteries away from open flames and sparks and make sure the charging area is well ventilated. When transporting batteries to a PV system installation work site make sure to pack them properly to avoid spills and short circuits. A good idea is to cover the battery terminals with an electrical insulator like tape in case something should fall between the terminals and cause a short circuit.

Batteries are often housed in a ventilated box located for periodic maintenance. The battery box is usually insulated and built from materials that are corrosion resistant. There are several companies who sell battery enclosures. Installers often build a battery enclosure. Do not place the battery enclosure below anything that could fall on the batteries. Protect batteries from freezing temperatures. Vent the battery enclosure directly to the outdoors and use something to prevent animals or insects from blocking the vent. The battery enclosure should be lockable to keep untrained people away from the batteries. It is a good idea to store the maintenance log and manufacturer's information in a resealable plastic bag inside the battery enclosure. The enclosure should be made of wood, plastic, or some other corrosion resistant painted material.

When installing the batteries you should use the connector bolt torque specified by the battery manufacturer when connecting batteries to each other in a series or parallel configuration. Be sure to adequately size all battery wiring and fuses as outlined in previous sections of this chapter. Use wire connectors or grommets to protect the wiring. Remember that an OCPD must be placed on the positive (+) conductor leaving the battery box at some point. Always use stainless steel nuts, bolts, and washers on battery terminals. Coat the battery terminals with battery terminal coating, petroleum jelly, oxidation protection material, or high temperature grease to protect them from oxidizing. When placing the batteries next to each other always leave at least a 1/4" air space between all batteries so that air can circulate around them to help keep them cool.

CHARGE CONTROLLER AND INVERTER INSTALLATION GUIDELINES

Make sure the charge controller and inverter will work for your specific PV system. When installing these items make sure to protect them from excess dust, dirt, overheating, and rough handling. Try to locate away from potential hazards. Cover the array with an opaque material when installing the controller and inverter so that the conductors you are connecting to the charge controller or inverter are not energized. Use the correct size wire and proper terminal fasteners. Use a controller with field adjustable set points because of battery differences. Use an enclosure made for electrical wiring. Use OCPD and switching in or near the control enclosure to protect the controller and provide a way to disconnect the array.

PV SYSTEM WIRING INSTALLATION GUIDELINES

Use the following checklist to make sure you have installed the wiring for the PV system properly:

- All wiring has adequate ampacity for the loads
- Wiring has been sized to keep the voltage drop no more than 2%
- The fuse or circuit breaker size does not exceed a wire's ampacity
- The wiring methods used are allowed by the *NEC*® in PV systems
- All conduits used in the PV system are correctly sized
- Electrical boxes are sized correctly, covered, and accessible
- All electrical connections are accessible
- Electrical connections are protected from moisture
- All disconnect switches are rated correctly
- Equipment grounding conductors are installed correctly with bare or green wire
- Metal conduit used as an equipment grounding conductor is installed properly
- All conduits and cables are supported correctly

SAFETY AND PHOTOVOLTAIC SYSTEM INSTALLATION

Safety is a full time job and is the responsibility of everyone working with PV equipment, whether in the design, installation, maintenance, or use of the system. To work safely you must have:

- Good work habits
- A clean and orderly work area
- Proper training in the equipment and its use
- A knowledge of potential hazards and how to avoid them
- Periodic reviews of safety procedures
- Instruction in first aid and CPR

PV SYSTEM SAFETY

PV systems generate electricity and should always be considered "live." Placing a cover over the modules does not necessarily mean that it is safe because some light could still reach the panel or the cover could come off somehow. Batteries are always "live" and cannot be turned off. When working with PV modules and other system components, you should be familiar with the following safety precautions:

- Be alert, check everything, and work carefully
- Do not work alone
- Understand the system completely before you attempt the installation
- Review safety, testing, and the installation with everybody involved with the installation BEFORE you start
- Make sure that tools and test equipment are in proper working condition
- Check out your test equipment prior to starting the job
- Wear appropriate clothing, including:
  - Hard hat
  - Eye protection
  - Leather gloves
- Remove all jewelry
- Measure everything with a high quality multimeter
- Expect the unexpected
- Make safety the number one concern during the installation

Physical Hazards

You should be aware of potential hazards when installing a PV system. Physical hazards are always present. Physical hazards can cause you to trip and fall. Physical hazards include:

- Fall hazards: Lifting modules up on a roof and moving around to place them in a roof rack. Wearing a fall protection harness and being tied off while

© Cengage Learning 2012.

FIGURE 22-54   Wearing a fall protection harness and being tied off while working on a roof is the best way to protect you from falling and seriously hurting yourself.

working on a roof (Figure 22-54) is the best way to protect you from falling off the roof and seriously hurting yourself. Chapter 1 in this textbook covered fall protection and you should review it at this time.

- Environmental Hazards: Exposure to the sun for long periods of time during an installation can be dangerous. Be sure to apply sunscreen to protect your skin from the harmful effects of too much exposure to sunlight.

  - Insects, snakes and other creatures are often found at a PV system installation site so be prepared for the possibility of bites or stings.

- Other Physical Hazards: Cuts and bruises from handling the PV system equipment is always a possibility. Use caution when handling the components, especially the ones with sharp edges.

  - Falls, sprains and strains are other things to watch out for.

- Burn and Electrical Hazards: Thermal burns can happen from the PV system parts getting very hot from being exposed to the sun for a long period of time. The roof surface itself will also get hot enough to cause a thermal burn.

  - Electrical hazards are something that a PV system installer must always protect themselves against.

- Chemical Hazards: Systems that use batteries will also have chemical hazards to watch out for. These include acid burns from the leaking electrolyte used in batteries.

- Explosion Hazards: The hydrogen gas produced around lead acid batteries can also cause explosions and fire if the batteries are not properly vented and a spark of some kind causes the hydrogen to ignite.

Personal Safety Rules

There are some personal safety rules that you should follow. They include the following:

- Always work with a partner, never work alone.

- Have a good understanding of safety practices and emergency procedures.

- Always wear a proper type of hard hat.

- Always wear eye protection.

- Always wear gloves to protect your hands from cuts and abrasions.

- Wear a rubber apron and rubber gloves for battery work.

- Use a body harness for working on roofs or other elevated heights and make sure it is tied to an anchor point that is adequate to hold your weight in an emergency.

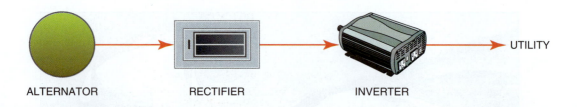

ALTERNATOR   RECTIFIER   INVERTER   UTILITY

FIGURE 22-55   The components of an interactive (grid-tie) small wind turbine system.
© *Cengage Learning 2012.*

- Use high voltage rated gloves and tools when working on "live" parts of a PV system.

Job Site Safety Rules

There are many job site safety resources that an installer should have. They include the following:

- A safety plan that has been covered with all installers so that everybody involved with a PV system installation will know what to do in the case of an emergency.

- A good first aid kit always needs to be available.

- At least one fire extinguisher should be on the job site. A good choice would be a dry chemical extinguisher that can handle all of the fire types that could typically be started at a PV system installation site.

- If a PV system installation includes batteries it is a good idea to have some distilled water available to wash off any spilled battery electrolyte from a person's body. Baking soda should also be available to neutralize the electrolyte acid in case there is a spill.

- Proper ladders are a must. Make sure the ladders are strong and in good shape and are made of a nonconductor of electricity like fiberglass.

- Proper lifting equipment should be available. This could include equipment to help lift the PV modules from the ground up to a roof. This equipment could also be used to move heavy batteries around it they are part of the PV system.

- Proper labeling of system components is a requirement of the *NEC®*. Having the correct equipment to make a permanent label for the required marking is important.

SMALL WIND TURBINE SYSTEMS

A **wind turbine system** includes the wind machine, tower, and all associated equipment. Small **wind turbine** electric systems consist of one or more wind electric generators

with individual systems up to and including 100 kW. These systems can include generators, alternators, inverters, and controllers. They can be interactive with other electrical power production sources (Figure 22-55) or may be stand-alone systems (Figure 22-56). Wind turbine systems can have AC or DC output and can be used with or without batteries. The *NEC®* limits the voltage of a small wind turbine system to 600 volts in one- and two-family homes.

INTRODUCTION TO WIND POWER

Wind power is quite reliable. The wind does not blow all of the time, and it does not always blow at the same speed, but no matter where you are there will always be some amount of wind power available. The big question will be whether there is enough wind available to adequately move the blades of a wind turbine. Wind power is economical. It happens in nature and is free for us to use. Sailing ships have taken advantage of free wind power for centuries. It makes environmental sense. The use of wind power to produce electricity can certainly lessen our dependence on the fossil fuels that are presently used to produce most of our electrical energy. Wind power is here now; wind machines are not tomorrow's technology.

To use wind power successfully you will need a good site. Not all locations are suitable for a wind turbine system. Once you find the right location you will need to select the right machine. For the purposes of this chapter we are focusing on small wind turbine systems of 100 kW or less. You will also need to have courage because wind turbine systems are not cheap!

There is a basic difference between a wind mill and a wind turbine. A lot of electricians refer to a system as a wind mill when it is really a wind turbine. Wind mills are basically water pumping machines. Farms have used them for years to pump water. Unlike a wind mill, wind turbines produce electricity. There are three main components to a wind turbine (Figure 22-57): the rotor, the alternator, and the tower.

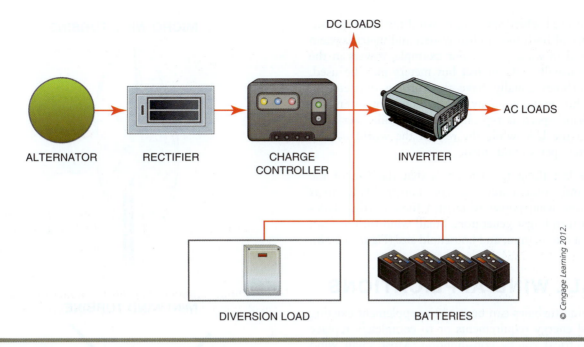

FIGURE 22-56   The components of a stand-alone small wind turbine system.

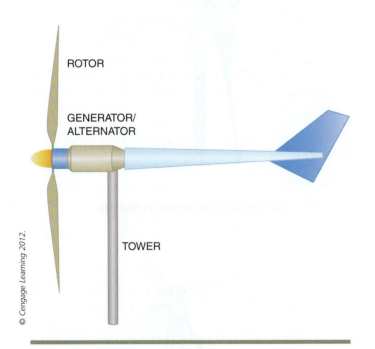

FIGURE 22-57   The three main components of a small wind turbine system are the rotor, the alternator, and the tower.

The wind turbine blades are attached to the rotor. The wind that strikes the blades will produce a turning force (torque) that turns the rotor. An alternator (generator) is connected to the rotor shaft, either directly or indirectly through some type of gearbox, and is made to rotate along with the rotor. The rotating alternator produces electricity. The **nacelle** is the enclosure housing the alternator and other parts of a wind turbine. On small wind turbine systems the nacelle is removed so that work on the individual system components can be done. On larger wind farm scale wind turbine systems the nacelle is large enough for workers to actually work inside of them. The **tower** is a pole or other structure that supports a wind turbine and places the wind turbine high enough in the air to make sure that an adequate amount of wind will strike the blades. It is suggested that the tower height used in small wind turbine installation be no less than 80 feet. Higher towers can be used but many manufacturers suggest that there is not really much (if any) advantage to having a small wind turbine placed higher than 120 feet. Shorter towers can be used and often are. However, towers shorter than 80 feet will most likely result in a wind turbine system that is not producing anything close to its rated output.

Power is the rate at which work is performed. It can also be the rate at which energy is transformed from one form to another. The rate at which the energy in the wind passes through an area can also be a type of power. Power is measured in watts (W) or kilowatts (kW) and the amount of electrical energy produced by a wind turbine is measured in kilowatt-hours (kWh).

Torque is the rotational force applied to a wind turbine's rotor. There is more torque with a longer blade. Wind speed is very important and is typically presented in three significant figures like 10.5 miles per hour (mph) or 8.25 meters per second (m/s). Metric units are often used when referring to wind but there is usually a mix of both the English system and metric system being used in wind power. For example, towers in the U.S. are usually sold in feet but meters may be used. Wind turbines usually have the turbine diameter in metrics, but again, you may see the dimension in English units. Wind speed is typically listed in miles per hour (mph) in the U.S. while the rest of the world usually uses meters per second (m/s).

One last thing to mention is that the size of the blades and rotor really matters. Larger blade areas allow more wind power to exert a force on the blades that can turn larger generators. Small wind turbines can come in a micro, mini, and household size (Figure 22-58).

SMALL WIND APPLICATIONS

Small wind turbines can be used to supplement existing electrical energy requirements or to completely replace your electrical energy consumption. Most small wind turbine systems installed in residential applications in this country are used to supplement existing electrical energy requirements.

Generating power at a remote home site is expensive but it may end up being less expensive than having the local power utility company construct a power line from the closest road to the home's location. A good rule of thumb to follow is that when a home is located more than ½ mile from existing utility service you will find it cheaper to install an independent power system. A large number of small wind turbines are used by home owners determined to produce their own power even though they could easily be connected to the utility grid. They do this to be energy independent. Remember that wind is not always available in the amounts needed to turn the rotor of a wind turbine fast enough to produce a usable amount of electricity. This means a form of electricity storage is needed, and just like in a PV system batteries are used. A gasoline, diesel, or propane fueled generator can also be connected to the small wind turbine system to provide electricity when the wind is not blowing.

Most residential wind turbine systems are interconnected with the utility. Wind turbine systems can produce electricity that is at constant voltage and at a constant 60 Hz frequency that can be easily supplied to the utility electrical grid. They are grid-tie connected in a manner similar to the way a PV system can be connected to the utility electrical grid. There are thousands of small wind turbines connected to the utility grid throughout the world. This is the small wind system type that a residential electrician is most likely to install.

There are two types of small wind turbines suitable for producing utility compatible power. Some manufacturers build their turbines so they produce an AC

MICRO WIND TURBINE

MINI WIND TURBINE

HOUSEHOLD WIND TURBINE

© Cengage Learning 2012.

FIGURE 22-58 Small wind turbines can come in a micro, mini, and household size.

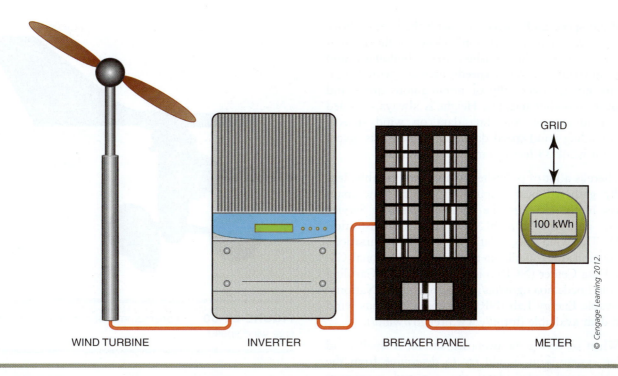

FIGURE 22-59 A small wind turbine system's output voltage is connected to the circuit breaker panel in a house.

voltage with a proper 60 Hz waveform that can be directly connected to the utility grid. Many other manufacturers build their turbines so they produce a voltage and frequency that cannot be directly connected to the grid. If this is the case an inverter is used to change the wind turbine output to a voltage with a smooth 60 Hz waveform that can be connected to the utility grid. In both system types the wiring from the wind turbine is connected to a circuit breaker in the home's service panel (Figure 22-59).

MEASURING THE WIND

Wind turbines with little wind are like hydroelectric dams with little water. We know that there is wind everywhere but we need to realize that every location does not have enough wind to allow a wind turbine to work properly. Before a wind turbine is ever installed for a home owner, you must determine whether their location is suitable for a wind turbine. Measuring the wind at that site is an important part of determining the suitability of a particular site.

The atmosphere is a huge, solar-fired engine that transfers heat from one area to another. Large scale convection currents carry heat from lower latitudes to northern latitudes. These "rivers of air" are what we call wind. **Thermals** are rising currents of warm air that go up and over land during sunny daylight hours.

Winds are stronger and steadier along shores of lakes and oceans because of the differential heating between the land and the water.

It was mentioned before that wind speed is measured in mph in the U.S. but many meteorologists use m/s. Sailors use nautical miles per hour (knots) for wind speed over the water. Some may even use kilometers per hour (km/h) to describe wind speeds. Here are some conversions that may help you.

- 1 knot = 1.15 mph
- 1 m/s = 2.24 mph
- 1 km/h = .621 mph

There is a lot of power in the wind. When wind strikes an object, it exerts a force on the object. When the object moves, work has been performed. Energy in wind is a function of its speed and mass. Higher wind speeds mean more energy. Lower speeds mean less energy. Wind power is dependant on air density, the area of interception of the wind, and wind speed. Air density decreases with an increase in temperature and increases with a decrease in temperature. Changes in elevation can result in big changes in air density. Power is directly related to the area of the blades intercepting the wind. Wind turbines with large blades intercept more wind and capture more wind power. Wind speed is the most important factor that affects the amount of wind power. This is why it is so important to site a wind turbine properly.

Wind speed and power vary with the height above the ground. There is "friction" close to the ground because of things like bushes, trees, buildings, and other obstructions. Wind speeds increase with height and are greater over hilly or mountainous areas and less over smoother terrains. Height is always included when reading and recording data on wind speeds. When reading wind speed data and no height is given, assume it is for 33 feet (10 m) above ground.

There is a lot of published wind data available. It is usually gathered near population centers. The main sources in the U.S. are the National Weather Service (NWS) and the Federal Aviation Administration (FAA). The wind speed data that these organizations have compiled are available to you at the National Climatic Data Center (NCDC) in Ashville, North Carolina (http://lwf.ncdc.noaa.gov/oa/ncdc.html). The National Renewable Energy Lab (NREL) website also has wind speed data available (http://www.nrel.gov/wind/).

When surveying the proposed site for a small wind turbine system, you should try to determine from the available wind speed data: the expected maximum instantaneous wind speed (wind gust), the average annual wind speed, and the distribution of wind speeds over a significant length of time, say a year. You could also see if there is already a wind turbine in that area. If so, talk to the owner and see what their results have been. You can also look around at trees and vegetation in the area. If they are low to the ground and somewhat distorted you can make up your mind that there is a significant amount of wind in the area on a regular basis. You could also find the nearest airport and see if they might have some wind data they will share with you.

The best way to survey a site to determine whether there will be enough wind is to use measuring instruments. Using an **anemometer**, mast and a recorder (Figure 22-60) will allow you to get an accurate measurement of wind speeds over a period of time. The anemometer and tower is easiest to install if the tower is hinged at the base and made of lightweight tubing. It is suggested that one year of wind speed data is accumulated before a decision is made about locating a small wind turbine on the site. Usually this amount of time is not practical so use four months minimum. Less time recording wind data will often result in a site that does not have the amount of wind to maximize a small wind turbine's electricity production.

Another good way to survey a site to determine the amount of wind activity is to go ahead and install a very small wind turbine. You can install a micro size wind turbine for about the same cost as an anemometer, recorder, and mast. Observe the micro-sized wind turbine's operation for a period of time and if it works properly, install a larger one and sell the smaller one.

FIGURE 22-60   An anemometer is a device used to measure wind speed.

© Cengage Learning 2012.

SMALL WIND TURBINE SYSTEM TOWERS

Towers are an integral part of the performance of a small wind turbine system. They are one of the few parts of a wind system where you have a choice. When installing a small wind turbine, choose a tower that is strong enough to withstand the strongest wind gusts that could be encountered.

The commercial success of medium and large size wind turbines in some areas of the country is mostly due to the use of higher towers. Taller towers make sense for smaller turbines as well. Manufacturers prefer taller towers because the greater wind speeds at the top of tall towers means their turbines will produce greater amounts of electricity. Using taller towers also mean more flexibility in siting. Taller towers allow placement closer to trees and buildings while still keeping the wind turbine high enough to not be affected buy the items around it.

As mentioned earlier in this chapter, for small wind turbines today's minimum recommended tower height is 80'. When trees are nearby, 100–120' is recommended. If the terrain is suitably open, shorter towers may work fine. Never install a wind turbine on a tower than is not at least 20–30 feet above any buildings or trees in the vicinity.

Towers must be strong enough to withstand the forces of a high wind. Towers are rated by the thrust load they can endure without buckling. Standards in the U.S. require withstanding a 120 mph wind with no damage. All towers flex and sway in the wind. Slender tubular towers are more flexible than truss towers. If the tower or turbine blades flex too much, the blades could touch the tower and shatter.

There are two common tower types: a freestanding tower and a guyed tower. Freestanding towers require a deep concrete foundation to properly support it. Guyed towers use several anchors and connecting cables to support the tower. Freestanding towers are more expensive but take up less space.

Freestanding towers can be broken down into two types: the lattice type (Figure 22-61) and tubular type (Figure 22-62). Lattice and tubular towers usually require a crane for installation. Some towers are hinged at the bottom and can be tipped up and into place (Figure 22-63).

Guyed towers are the most common for small wind turbine systems (Figure 22-64). They are supported by 3 or 4 **guy** wires installed at different levels. The guy wire radius is critical. It depends on the site, the loads imposed by the wind turbine, and the stiffness of the mast. Normally it is the manufactures who specify guy wire radius. A good rule of thumb is that the guy wire radius should not be less than ½ the height of the tower. It could be greater but not less.

CAUTION

CAUTION: Always follow the manufacturer's instructions when installing a tower that is supported with guy wires.

There are other mounting locations for small wind turbines but they are not recommended. Rooftop mounting can be noisy and rooftops create wind turbulence. The roof may also not be strong enough to support the weight of a small wind turbine. Do not mount a small wind turbine on a tree. This is an unsafe installation that could result in someone getting seriously hurt and tree mounting makes maintaining a wind turbine very hard to do. The turbine needs to be relatively stationary while operating and trees sway too much. Trees also have a habit of falling down in a high wind. Existing farm windmill towers could be used, but they are usually too short and not that strong at the base.

© Cengage Learning 2012.

FIGURE 22-61   A lattice-type freestanding tower.

© Cengage Learning 2012.

FIGURE 22-62   A tubular-type freestanding tower.

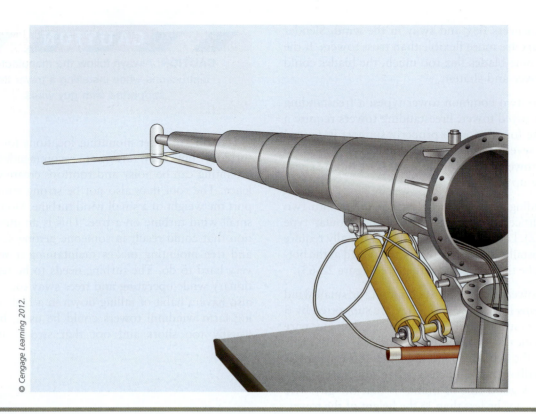

FIGURE 22-63   Some towers are hinged at the bottom and can be tipped up and into place.

FIGURE 22-64   A tower that uses guy wires to hold it securely in place.

Wood poles can be used if they are a Class 4 or higher. However, they are usually too short (40–50 feet) but if you can get a 70' pole or higher they do not provide a bad mounting means. Steel pipes are sometimes used because of their strength and cost. However, they are heavy and difficult to handle in long lengths.

When considering what type and make of tower, there are a few things to always think about. The amount of available space is a big consideration. Large open spaces allow the installation of guyed towers. Smaller spaces may require a freestanding tower. Maintenance is another consideration. When maintenance needs to be done on the turbine how will it be accessed? Hinged towers for small wind turbines work well. If a hinged tower is not possible consider looking for a rental company in your area with a lift that could reach the top of the tower you plan on installing. Always consider what type of tower will be the easiest to install. Always consider installing a tower that is as tall as possible. The last thing is to always install a top quality tower. Towers may cost as much or even more than the small wind turbine.

SMALL WIND TURBINE INSTALLATION

The actual mechanical installation of a small wind turbine system must be done according to the installation instructions supplied by the wind turbine and tower manufacturer. Very often it is a company that specializes in the mechanical construction of wind turbine systems who is hired to erect the tower and wind turbine. Then, an electrical contracting company who specializes in the wiring of small wind turbine systems is hired to install the electrical wiring. There are also many companies who do both the mechanical installation of a wind turbine system as well as the electrical wiring. This section will focus on the installation of the electrical wiring of a small wind turbine system.

All small wind turbine system wiring must be done according to the *NEC*® and any local rules. Article 694 in the 2011 *National Electric Code*® covers small wind turbine systems. The conductors from a wind system are like the power lines from a small generating station and this wiring can be installed overhead or underground. Most installations will use underground wiring from the base of the tower to a house.

Leads from the generator located in the nacelle must be connected to the wires running down the tower. The connections must be permanent and weatherproof. If the generator is stationary, the leads can be connected directly to the conductors coming down the tower. If the generator moves there must be a mechanism for transferring power from the moving platform of the wind machine to the stationary tower. Slip rings and brushes are usually used to accomplish this task.

On small wind turbines the power cable from the generator typically runs through the center of the tower. In a tubular tower the power cable hangs freely, but in a lattice type tower the power cable is secured to the tower at certain intervals. As the machine yaws (turns to face the wind), the cable will twist. The cable twists several times before it must unwind. The connections at the turbine itself are made in a number of ways including: split-bolts, wirenuts, compression connectors, or insulated terminal blocks.

The connection between the leads from the slip rings (if used) and the tower conductors should be made in a junction box that is sized according to Sections 314.16 or 314.28. If using a lattice type tower, the conductors are then often run down the inside in a conduit like EMT or PVC for added protection. If conductors are run on the outside of a tower, they need to be protected with conduit like RMC, EMT or PVC. You will need to use weatherproof couplings and connectors, and support the conduit according to the *NEC*®. If using a tubular tower, simply run the conductors down the inside.

There are very few installations where the wiring from the wind turbine to a house is run above the ground. For small wind turbines it is better to dig a trench, lay in PVC conduit, and pull in the conductors. The burial depths for the wiring method used to bring the conductors from the tower to the house are found in Table 300.5. The rules for installing the wiring both in the trench and as it emerges from underground is covered in Section 300.5. This information was covered earlier in this textbook.

Section 694.20 states that some means must be provided to disconnect all current-carrying conductors of a small wind electric power source from all other conductors in a building. The disconnecting means must comply with the rules in Section 694.22. The disconnecting means does not have to be suitable as service equipment but could be. It must consist of manually operated switches or circuit breakers and be compliant with all of the following requirements:

- Located where readily accessible
- Externally operable without exposing the operator to contact with live parts
- Able to plainly indicate whether in the open or closed position
- Have an interrupting rating sufficient for the circuit voltage and the current that is available at the line terminals of the equipment
- Suitable for the environment if located outside

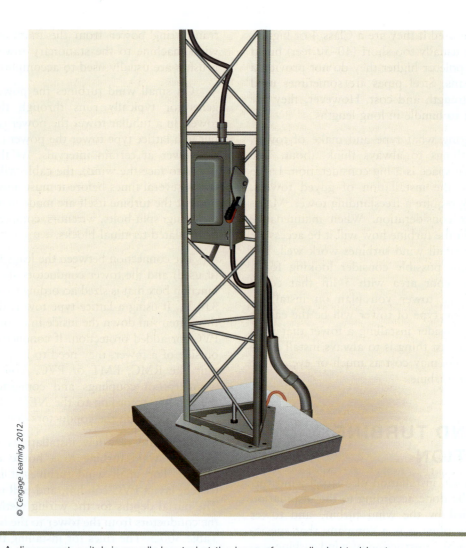

FIGURE 22-65 A disconnect switch is usually located at the base of a small wind turbine tower.

When all terminals of the disconnecting means may be energized in the open position, a warning sign must be mounted on or adjacent to the disconnecting means. The sign must be clearly legible and have the following words or equivalent:

WARNING

ELECTRIC SHOCK HAZARD. DO NOT TOUCH TERMINALS. TERMINALS ON BOTH THE LINE AND LOAD SIDES MAY BE ENERGIZED IN THE OPEN POSITION.

The small wind electric system disconnecting means must be installed at a readily accessible location either on or next to the turbine tower. It could also be installed on the outside of the house or inside the house nearest the point of entrance of the system conductors. A turbine disconnecting means is not required to be located at the nacelle or tower but most manufacturers recommend that a disconnect switch be located at the tower base (Figure 22-65). Each turbine system disconnecting means must be permanently marked to identify it as a small wind electric system disconnect. A plaque must be installed in accordance with Section 705.10.

Section 694.24 requires a means to be provided that will disconnect equipment such as inverters, batteries, and charge controllers from all ungrounded conductors from all sources. If the equipment is energized from more than one source, the disconnecting means must be grouped and identified. A single disconnecting means is permitted for the combined AC output of one or more inverters in an interactive grid-tie system.

The wind turbine's output wiring is connected to a control panel or an inverter. For grid-tie systems, wiring from the grid-tie inverter will be connected to a circuit breaker in the building's existing service panel. The conductor sizing is based on Section 694.12. The turbine output circuit current is the circuit current when the wind turbine is operating at maximum output power. The wind turbine output circuit is the circuit conductors between the internal components of a small wind turbine and other equipment. The maximum output power is defined as the maximum one minute average power output a wind turbine will produce during normal operation (instantaneous power output could be higher such as when there are high speed wind gusts). Section 694.12(B)(1) states that small wind electric system currents must be considered to be continuous. Therefore, the circuit conductors and overcurrent devices must be sized to carry not less than 125 percent of the maximum currents as calculated in Section 694.12(A). For example, if the rated maximum power output of a wind turbine was 5000 watts at a voltage of 240 volts, the current used to size the output circuit conductors would be 26 amps: 5000 watts / 240 volts = 20.83 amps; 20.83 amps × 1.25 = 26 amps. This would result in a wire size of at least 10 AWG assuming all terminations to be 75°C. The rating or setting of overcurrent devices is determined the usual way as outlined in Section 240.4. In this example, the overcurrent protection device used for the 10 AWG conductor size would be 30 amps.

The wiring method used is up to the electrician, but Section 694.30(A) states that it can be any of the raceway or cable wiring methods included in the *NEC*® or it could be some other wiring system and associated fittings specifically intended for use on wind turbines. However, when turbine output circuits operating at maximum voltages greater than 30 volts are installed in readily accessible locations, the circuit conductors must be installed in raceways. Section 694.30(B) states that if flexible cords and cables are used to connect the moving parts of turbines or used to permit ready removal for maintenance and repair, they must be of a type identified as a hard service cord or portable power cable; be suitable for extra-hard usage, listed for outdoor use, and water resistant. Cables exposed to sunlight must be marked sunlight resistant.

One other thing to consider from Section 694.30(C) is that if the output circuit from a turbine is DC and is installed inside a house, the wiring must be contained in metal raceways or metal enclosures from the point of penetration of the surface of the building or structure to the first readily accessible disconnecting means.

It is good wiring practice to use strain relief on the tower conductors at the point where they connect to the turbine and where they connect to a disconnect switch at the bottom of the tower. On guyed tubular towers, thread the power cables down the inside of the mast and use a strain relief at the attachment point at the top of the tower. Wind turbine manufacturers specify the cable size

and material type for several different turbine-to-load distances. There are many reasons why a wind turbine may underperform but voltage drop in the conductors is probably the number one problem. On gird-tie wind systems, the voltage drop between the wind turbine and the load is critical to proper operation. Acceptable voltage drops range from 1% for grid-tie systems. Other systems can operate with voltage drops of no more than 3%.

When installing conductors in a conduit wiring method as part of a small wind turbine system installation, the conduit fill rules as covered in Chapter 12 of this textbook must be followed. Remember that Informational Annex C is used when conductors are all the same size and have the same type of insulation. If you are using different combinations of conductor sizes or the conductor insulations are not the same, you will need to use the information in Tables 4 and 5 in Chapter 9 of the *NEC*®. Electricians usually prefer a conduit that is larger than the minimum size according to the *NEC*®. This makes pulling in the conductors, especially if the wind turbine is some distance away from a house, much easier. It also means that there is less chance of damaging the conductors during installation.

Grounding

Section 694.40(A) requires all exposed noncurrent-carrying metal parts of towers, turbine nacelles, other equipment, and conductor enclosures to be grounded. Attached metal parts like turbine blades and tail pieces that have no way to become energized are not required to be grounded.

Section 694.40(B) covers guy wire grounding. It says that guy wires used to support turbine towers do not have to be grounded. However, the Informational Note following this section points out that grounding of metallic guy wires may be required by lightning protection codes in your area.

Section 694.40(C) addresses tower grounding. It states that a wind turbine tower must be grounded with one or more auxiliary electrodes to limit voltages imposed by lightning. Auxiliary electrodes are permitted to be installed in accordance with Section 250.54. Ground rods are the auxiliary electrodes typically installed for this purpose. However, electrodes that are part of the tower concrete foundation and that meet the requirements for concrete encased electrodes are acceptable. The equipment grounding conductor and grounding electrode conductors must be connected to a metal tower by exothermic welding, listed lugs, listed pressure connectors, listed clamps, or other listed means. Devices such as connectors and lugs must be suitable for the material of the conductor and the structure to which they connect. All grounding connections must be accessible. Auxiliary electrodes and grounding electrode conductors are permitted to act as lightning protection system components if they meet applicable requirements. If separate, the tower lightning protection system grounding electrodes must be bonded to the tower auxiliary grounding electrode system. Guy wire lightning protection system ground electrodes are not required to be bonded to the tower auxiliary grounding electrode system.

Wind turbine systems are susceptible to damaging voltage spikes and are grounded to limit voltage surges from nearby lightening strikes. Grounding also ensures the fast operation of fuses and circuit breakers from electrical faults. Many manufacturers recommend the following grounding procedure. To ground the tower, drive copper clad 8' rods into the ground. On freestanding towers use two or more ground rods. On guyed towers drive a ground rod near the mast and also near each concrete anchor. Connect the ground rods to the tower to cable guys with heavy gauge copper wire and brass or bronze clamps. In areas of high lightning incidence or where the soil is dry and sandy a ground net is recommended. On guyed towers the mast and each anchor are wired to their own ground rod. Then, all ground rods are tied together with a buried ground wire. On freestanding towers there should be a ground rod for each leg of the tower.

Some additional items to remember when installing wind turbine wiring are:

- Properly mount all equipment so it is secure.

- Be sure to seal all holes or knock-outs that are open.

- Make sure there is enough clearance (working space) in front of the equipment.

- All terminals, connectors and conductors must be compatible and if using aluminum wire make sure it is prepared properly.

SMALL WIND SYSTEM INSTALLATION SAFETY

Safety must always be your primary concern during the assembly, installation, and operation of a small wind turbine system. As an installer you must always be aware of the risks involved with both the mechanical and electrical parts of the installation. Ideally, only a qualified and experienced small wind system installer should supervise the installation. And of course, always wear the proper personal protective equipment when installing a small wind turbine system.

Mechanical Safety Hazards

There are several mechanical hazards to consider during the installation. The main rotor is the most obvious and serious mechanical safety risk. When the turbine is operating the blades will be very difficult to see due to the speed of rotation. Never approach the turbine when it is operating. Always shut down the turbine by waiting until the turbine is stationary on a windless day. Follow the manufacturer's recommendation on how to stop the blades from turning for their specific small wind turbine. Working with tools of any kind can be dangerous. The assembly and installation of small wind turbines requires mechanical assembly with hand and power tools.

If a grid-tie inverter or batteries are part of the system, follow the same safety procedures as those covered in the PV system part of this chapter. Try to install a small wind turbine system during a calm day with wind speeds close to zero. Once the system is installed make sure to always ensure that the turbine is stationary when performing routine inspection or maintenance.

Tower Safety

The turbine itself must be installed on a tower. This may mean working at some height above ground. Small wind turbine towers pose no greater climbing risk than other similar poles and towers or even trees. Many wind turbine towers have a smooth surface, like a light pole, that is nearly impossible to climb. Those towers that are climbable can be equipped with devices that prevent falls. Always ensure that all personnel in the immediate vicinity are aware of any lifting or hoisting operations that will be occurring. Make sure that there are no loose parts or tools likely to fall and cause injury during the lifting operation. Where possible, all assembly work should be completed at ground level.

The following safety rules should be observed when installing or working on or around a tower.

- Only people who are directly involved with the tower installation should be allowed in the work area.

- All persons on the tower or in the vicinity must wear an OSHA approved hard hat.

- All tower work should be done under the supervision of a trained professional.

- Towers should never be erected near utility power lines.

- Workers should always use a full body harness and be tied off when climbing or working on a tower.

- Tool belts should always be used to properly hold the tools that you will be using.

- Never carry tools or parts in your hands when climbing a tower. Use a bucket and a hoist line to raise and lower tools and parts.

- Never stand or work directly below someone else who is working on the tower.

- Never work on a tower alone. Always work with at least one other person.

- Never climb a tower unless the turbine is furled and the alternator is locked out by short circuiting the output conductors in the disconnect switch located at the base of the tower.

- Never work on a tower during thunderstorms, high winds, tower icing, or severe weather of any kind.

> **CAUTION**
>
> **CAUTION:** Always install a wind turbine system tower according to the manufacturer's instructions. Towers should only be installed under the direct supervision of a trained and qualified professional.

Electrical Safety Hazards

Small wind turbines used in residential applications often produce electricity at a relatively high 240 volts. Some actually produce a three-phase alternating current voltage that is then transformed to a 240-volt single-phase voltage that can be readily connected to a house electrical system. At these voltage amounts there are inherent safety risks. Caution should always be used when connecting a small wind turbine to the electrical system. Be sure to install the proper size wiring that will bring the electrical power from the turbine to the house electrical system. If conductors of insufficient cross-sectional area are used, heat will build up in the wiring causing a potential fire hazard. Also, be sure to properly size and install overcurrent protection devices.

As in PV systems, batteries used in small wind turbine systems can deliver a large amount of current. A short circuit in the battery circuit can lead to hundreds of amps flowing through the battery cables. This will cause a heat build up and possibly an electrical fire. Batteries are also susceptible to explode when shorted out. Always use insulated electrical tools when working on the battery's electrical connections. A properly sized fuse or circuit breaker should be used in the cables connected to the battery. This will stop the risk of short circuit currents. Batteries are very heavy. Do not attempt to move heavy batteries by yourself. Get some help. Also, remember that liquid type lead-acid batteries can spill their electrolytes so always keep these batteries in an upright position. Do not allow the electrolyte to come into contact with your skin or face. Always follow the manufacturer's safety instructions when handling lead-acid batteries.

SUMMARY

Photovoltaic (PV) systems are solar energy systems that produce electricity directly from sunlight. Wind turbine systems produce electricity when wind hits the blades of a turbine and spins a generator that produces electricity. PV and wind turbine power systems are simply energy converters. They convert one form of energy (sunlight or wind) into another form of energy (electricity).

The use of fossil fuel sources to produce electricity has created many problems. Some of the problems include: global warming, acid rain, smog, water pollution, destruction of natural habitat from fuel spills, and loss of natural resources. The use of alternative energy systems like photovoltaic and small wind will help reduce the negative effects of using fossil fuels to produce electricity.

The photovoltaic solar cell is the basic building block of PV technology. A typical PV solar cell is about 4 inches across and produces about 1 watt of power in full sunlight at about .5 volts DC. A module, often called a panel, is a configuration of PV cells laminated between a clear outer superstrate (glazing) and an encapsulating inner substrate. Solar cells are wired together to produce a PV module. Modules (panels) are connected together to form a PV array. An array is wired to deliver a specific voltage. A charge controller regulates the battery voltage and makes sure that the PV system batteries (if used) are charged properly. Batteries are used to chemically store direct current (DC) electrical energy in a PV system. PV modules are connected to an inverter that "converts" the DC electricity produced by most solar arrays to the AC electricity commonly used in a house.

PV systems can either be stand-alone or grid-tied. A stand-alone system has no connection to the local electric utility's grid system and is used when a home owner wants to be totally independent from the electric utility. A Utility Grid Interconnected System, often called a Utility-Connected, Grid-Tie, Intertie, or Line-Tie system, is connected to the utility grid. They do not need battery storage because the utility grid is the power reserve.

A wind turbine system includes the wind machine, tower, and all associated equipment. Small wind turbine electric systems consist of one or more wind electric generators with individual systems up to and including 100 kW. These systems can include generators, alternators, inverters, and controllers. They can be interactive with other electrical power production sources or may be stand-alone systems. Wind turbine systems can have AC or DC output and can be used with or without batteries. The *NEC*® limits the voltage of a small wind turbine system to 600 volts in one- and two-family homes.

There are three main components to a wind turbine: the rotor, the alternator, and the tower. The wind turbine blades are attached to the rotor. The wind that strikes the blades will produce a turning force (torque) that turns the rotor. An alternator (generator) is connected to the rotor shaft, either directly or indirectly through some type of gearbox, and is made to rotate along with the rotor. The rotating alternator produces electricity. The nacelle is the enclosure housing the alternator and other parts of a wind turbine. The tower is a pole or other structure that supports a wind turbine and places the wind turbine high enough in the air to make sure that an adequate amount of wind will strike the blades. It is suggested that the tower height used in small wind turbine installation be no less than 80 feet. Higher towers can be used but many manufacturers suggest that there is not really much (if any) advantage to having a small wind turbine placed higher than 120 feet. Shorter towers can be used and often are. However, towers shorter than 80 feet will most likely result in a wind turbine system that is not producing anything close to its rated output.

This chapter has presented an introduction to solar photovoltaic (PV) systems and small wind turbine systems. Today's residential electrician works on the installation of these two systems on a regular basis. In the future, as more and more home electrical systems incorporate an alternative energy system, electricians will need to have the skills and knowledge needed to safely install such systems. It is a good idea for you to build on the knowledge you gained from studying this chapter by attending some specific training in photovoltaic system or small wind turbine system installation. PV and small wind turbine systems are not something that will be coming along in a few years. They are here now and should become an integral part of every house electrical system.

REVIEW QUESTIONS

Directions: Answer the following items with clear and complete responses.

1. Name the two types of alternative energy systems commonly installed in house wiring.

2. List five advantages of photovoltaic systems.

3. List three disadvantages of photovoltaic systems.

4. A typical PV solar cell is about 4 inches across and produces about _____ volts of DC electricity.

5. A _____ is a configuration of PV cells laminated between a clear outer superstrate (glazing) and an encapsulating inner substrate.

6. PV modules are connected together to form a PV _____ .

7. A _____ _____ regulates the battery voltage and makes sure that the PV system batteries (if used) are charged properly.

8. PV modules are connected to an _____ that "converts" the DC electricity produced by a solar array to the AC electricity commonly used in a house.

9. Describe a hybrid photovoltaic system.

10. Describe a stand-alone photovoltaic system.

11. Describe a grid-tie photovoltaic system.

12. Describe what net metering means when used in a photovoltaic or small wind turbine system.

13. The voltage production in a PV cell is called _____ _____ .

14. Name the three types of PV modules based on the composition of the silicon crystalline structure.

15. Name the two types of PV system trackers that are available.

16. Sealed lead-acid batteries have a valve to allow excess pressure from overcharging to escape and are called a _____ _____ _____ _____ battery.

17. Battery interconnection wires must be sized large enough to handle the high currents that batteries can produce. A recommendation is to use _____ AWG copper wire if you have a 500-watt or smaller inverter, _____ AWG copper wire for an 800-watt inverter, and _____ or larger for 1000-watt and larger inverters.

18. List at least five items to check to make sure you have installed the wiring for a PV system properly.

19. When working with PV modules and other system components you should be familiar with several safety precautions. List at least five.

20. There are many job site safety resources that an installer should have when installing a PV system. List at least five safety resources.

21. A _____ _____ _____ includes the wind machine, tower, and all associated equipment.

22. Wind turbines sized _____ or less are considered to be small wind turbines.

23. There are three main components to a wind turbine. Name them.

24. The _____ is the enclosure housing the alternator and other parts of a wind turbine.

25. An _____ is a device used to measure wind speed.

26. There are two common tower types. Name them.

27. Name the type of tower that is the most common type installed for a small wind turbine.

28. Article _____ in the 2011 *National Electrical Code*® covers small wind turbine systems.

29. Name at least five safety rules that should be observed when installing or working on or around a tower.

30. Article _____ in the 2011 *National Electrical Code*® covers solar photovoltaic (PV) systems.

Glossary

accessible (as applied to wiring methods) capable of being removed or exposed without damaging the building structure or finish, or not permanently closed in by the structure or finish of the building

accessible, readily (readily accessible) capable of being reached quickly for operation, renewal, or inspections without requiring a person to climb over or remove obstacles or to use portable ladders

ambient temperature the temperature of the air that surrounds an object on all sides

American Wire Gauge (AWG) a scale of specified diameters and cross sections for wire sizing that is the standard wire-sizing scale in the United States

ammeter (clamp-on) a measuring instrument that has a movable jaw that is opened and then clamped around a current-carrying conductor to measure current flow

ammeter (in-line) a measuring instrument that is connected in series with a load and measures the amount of current flow in the circuit

amorphous having no definite form or distinct shape

ampacity the current, in amperes, that a conductor can carry continuously under the conditions of use without exceeding its temperature rating

ampere the unit of measure for electrical current flow

analog meter a meter that uses a moving pointer (needle) to indicate a value on a scale

anemometer a device used to measure wind speed

antioxidant a special compound that is applied to exposed aluminum conductors; its purpose is to inhibit oxidation

approved when a piece of electrical equipment is approved, it means that it is acceptable to the authority having jurisdiction (AHJ)

arc the flow of a high amount of current across an insulating medium, like air

arc blast a violent electrical condition that causes molten metal to be thrown through the air

arc fault circuit interrupter (AFCI) a device intended to provide protection from the effects of arc

faults by recognizing characteristics unique to arcing and by functioning to de-energize the circuit when an arc is detected

arc fault circuit interrupter (AFCI) circuit breaker a circuit breaker intended to provide protection from the effects of arc faults by recognizing characteristics unique to arcing and by functioning to de-energize the circuit when an arc fault is detected

arc-flash a dangerous condition associated with the possible release of energy caused by an electric arc

architect's scale a device used to determine dimensions on a set of building plans; it is usually a three-sided device that has each side marked with specific calibrated scales

architectural firm a company that creates and designs drawings for a residential construction project

array a mechanically integrated assembly of modules or panels with a support structure and foundation, tracker, and other components, as required, to form a direct-current power-producing unit

attenuation the decrease in the power of a signal as it passes through a cable; it is measured in decibels;

often increases with frequency, cable length, and the number of connections in a circuit

auger a drill bit type with a spiral cutting edge used to bore holes in wood

auto-ranging a meter feature that automatically selects the range with the best resolution and accuracy

azimuth the sun's apparent location in the sky east or west of true south

back-to-back bend a type of conduit bend that is formed by two 90-degree bends with a straight length of conduit between the two bends

backboard the surface on which a service panel or subpanel is mounted; it is usually made of plywood and is painted a flat black color

backfeeding a wiring technique that allows electrical power from an existing electrical panel to be fed to a new electrical panel by a short length of cable; this technique is commonly used when an electrician is upgrading an existing service entrance

ballast a component in a fluorescent lighting fixture that controls the voltage and current flow to the lamp

balloon frame a type of frame in which studs are continuous from the foundation sill to the roof; this type of framing is found mostly in older homes

band joist the framing member used to stiffen the ends of floor joists where they rest on the sill

bandwidth identifies the amount of data that can be sent on a given cable; it is measured in hertz (Hz) or megahertz (MHz); a higher frequency means higher data-sending capacity

bathroom branch circuit a branch circuit that supplies electrical power to receptacle outlets in a bathroom; lighting outlets may also be served by the circuit as long as other receptacle or lighting outlets outside the bathroom are not connected to the circuit; it is rated at 20 amperes

bender a tool used to make various bends in electrical conduit raceway

bimetallic strip a part of a circuit breaker that is made from two different metals with unequal thermal expansion rates; as the strip heats up, it will tend to bend

blueprint architectural drawings used to represent a residential building; it is a copy of the original drawings of the building

bonding connected to establish electrical continuity and conductivity; the purpose of bonding is to establish an effective path for fault current that facilitates the operation of the overcurrent protective device

bonding jumper a conductor used to ensure electrical conductivity between metal parts that are required to be electrically connected

bottom plate the lowest horizontal part of a wall frame, which rests on the subfloor

box fill the total space taken up in an electrical box by devices, conductors, and fittings; box fill is measured in cubic inches

box offset bend a type of conduit bend that uses two equal bends to cause a slight change of direction for a conduit at the point where it is attached to an electrical box

branch circuit the circuit conductors between the final overcurrent

device (fuse or circuit breaker) and the outlets

break lines lines used to show that part of the actual object is longer than what the drawing is depicting

bridging diagonal braces or solid wood blocks installed between floor joists, used to distribute the weight put on the floor

bundled cables or conductors that are physically tied, wrapped, taped, or otherwise periodically bound together

buried box an electrical box that has been covered over with sheetrock or some other building material

bushing (insulated) a fiber or plastic fitting designed to screw onto the ends of conduit or a cable connector to provide protection to the conductors

cabinet an enclosure for a panelboard that is designed for either flush or surface mounting; a swinging door is provided

cable a factory assembly of two or more insulated conductors that have an outer sheathing that holds everything together; the outside sheathing can be metallic or nonmetallic

cable hook also called an "eyebolt"; it is the part used to attach the service drop cable to the side of a house in an overhead service entrance installation

carbon monoxide detector a device that detects the presence of carbon monoxide (CO) gas in order to prevent carbon monoxide poisoning; CO is a colorless and odorless compound produced by incomplete combustion

cartridge fuse a fuse enclosed in an insulating tube that confines the arc when the fuse blows; this fuse may be either a ferrule or a blade type

category ratings on the bandwidth performance of UTP cable; categories include 3, 4, 5, 5e, and 6; Category 5e is rated to 100 MHz and is often installed in houses

ceiling-suspended paddle fan a type of fan installed in a ceiling, usually having four or five blades, that is used for air circulation in residential applications; it can be installed close to a ceiling or hung from a ceiling with a metal rod; the fan often has a lighting fixture attached to the bottom of the unit; a special outlet box that is listed as being suitable for paddle-fan support is required

ceiling joists the horizontal framing members that rest on top of the wall framework and form the ceiling structure; in a two-story house, the first-floor ceiling joists are the second floor's floor joists

centerline a series of short and long dashes used to designate the center of items, such as windows and doors

charge controller equipment that controls dc voltage or dc current, or both, used to charge a battery

checkout the process of determining if all parts of a recently installed electrical system are functioning properly

chuck a part of a power drill used for holding a drill bit in a rigid position

chuck key a small wrench, usually in an L or T shape, used to open or close a chuck on a power drill

circuit breaker a device designed to open and close a circuit manually and to open the circuit automatically on a predetermined overcurrent without damage to itself when properly applied within its rating

circuit (electrical) an arrangement consisting of a power source, conductors, and a load

circuit tracer a type of tester that is used to determine which circuit breaker in an electrical panel protects a specific circuit; used when a circuit directory does not properly indicate what is protected by a specific breaker

circular mils the diameter of a conductor in mils (thousandths of inches) times itself; the number of circular mils is the cross-sectional area of a conductor

Class P ballast a ballast with a thermal protection unit built in by the manufacturer; this unit opens the lighting electrical circuit if the ballast temperature exceeds a specified level

coaxial cable a cable in which the center signal-carrying conductor is centered within an outer shield and separated from the conductor by a dielectric; used to deliver video signals in residential structured cabling installations

color rendition a measure of a lamp's ability to show colors accurately; the color rendering index (CRI) is a scale that compares color rendering for different light sources; the scale ranges from 1 (low-pressure sodium) to 100 (the sun); a CRI of 85 is considered to be very good

color temperature a measure of the color appearance of a light source that helps describe the apparent "warmth" (yellowish) or "coolness" (bluish) of that light source; light sources below 3200 K are considered "warm" while those above 4000 K are considered "cool" light sources; the letter, K, stands for Kelvin

combination switch a device with more than one switch type on the same strap or yoke

concentric knockout a series of removable metal rings that allow the knockout size to vary according to how many of the metal rings are removed; the center of the knockout hole stays the same as more rings are removed; some standard residential wiring sizes are ½, ¾, 1, 1¼, 1½, 2, and 2½ inches

conductor a material that allows electrical current to flow through it; examples are copper, aluminum, and silver

conduit a raceway with a circular cross section, such as electrical metallic tubing, rigid metal conduit, or intermediate metal conduit

conduit body a separate portion of a conduit system that provides access through a removable cover to the inside of the conduit body; a conduit body is used to connect two or more sections of a conduit system together; sometimes referred to as a "condulet"

conduit "LB" a piece of electrical equipment that is connected in-line with electrical conduit to provide for a 90-degree change of direction

connector a fitting that is designed to secure a cable or length of conduit to an electrical box

continuity tester a device used to indicate whether there is a continuous path for current flow through an electrical circuit or circuit component

copper-clad aluminum an aluminum conductor with an outer coating of copper that is bonded to the aluminum core

cord-and-plug connection an installation technique in which electrical appliances are connected to a branch circuit with a flexible cord with an attachment plug; the attachment plug end is plugged into a receptacle of the proper size and type

counter-mounted cooktop a cooking appliance that is installed in the top of a kitchen countertop; it contains surface cooking elements, usually two large elements and two small elements

crimp a process used to squeeze a solderless connector with a tool so that it will stay on a conductor

critical loads the electrical loads that are determined to require electrical power from a standby power generator when electrical power from the local electric utility company is interrupted

current the intensity of electron flow in a conductor

cutter a hardened steel device used to cut holes in metal electrical boxes

detail drawing a part of the building plan that shows an enlarged view of a specific area

deteriorating agents a gas, fume, vapor, liquid, or any other item that can cause damage to electrical equipment

device a piece of electrical equipment that is intended to carry but not use electrical energy; examples include switches, lamp holders, and receptacles

device box an electrical device that is designed to hold devices such as switches and receptacles

die the component of a knockout punch that works in conjunction with the cutter and is placed on the opposite side of the metal box or enclosure

digital meter a meter where the indication of the measured value is given as an actual number on a liquid crystal display (LCD)

dimension a measurement of length, width, or height shown on a building plan

dimension line a line on a building plan with a measurement that indicates the dimension of a particular object

dimmer switch a switch type that raises or lowers the lamp brightness of a lighting fixture

diode a semiconductor device that allows current to pass through in only one direction

dioxins a family of chlorinated chemicals; toxic under certain exposure conditions; emitted when combustion of carbon compounds is inefficient; associated with cancer of the stomach, sinus lining, liver, and lymph system

disconnecting means a switch that is able to de-energize an electrical circuit or piece of electrical equipment; sometimes referred to as the "disconnect"

DMM a series of letters that stands for "digital multimeter"

double-pole switch a switch type used to control two separate 120-volt circuits or one 240-volt circuit from one location

double insulated an electrical power tool type constructed so the case is isolated from electrical energy and is made of a nonconductive material

draft-stops also called "fire-stops"; the material used to reduce the size of framing cavities in order to slow the spread of fire; in wood frame construction, it consists of full-width dimension lumber placed between studs or joists

drip loop an intentional loop put in service entrance conductors at the point where they extend from a weatherhead; the drip loop conducts rainwater to a lower point than the weatherhead, helping to ensure that no water will drip down the service entrance conductors and into the meter enclosure

dry-niche luminaire a lighting fixture intended for installation in the wall of a pool that goes in a niche; it has a fixed lens that seals against water entering the niche and surrounding the lighting fixture

dual-element time-delay fuses a fuse type that has a time-delay feature built into it; this fuse type is used most often as an overcurrent protection device for motor circuits

duplex receptacle the most common receptacle type used in residential wiring; it has two receptacles on one strap; each receptacle is capable of providing power to a cord-and-plug-connected electrical load

dwelling unit one or more rooms arranged for complete, independent housekeeping purposes, with space for eating, living, and sleeping, facilities for cooking, and provisions for sanitation

eccentric knockout a series of removable metal rings that allow a knockout size to vary according to how many of the metal rings are removed; the center of the knockout hole changes as more metal rings are removed; common sizes are the same as for concentric knockouts

Edison-base plug fuse a fuse type that uses the same standard screw base as an ordinary lightbulb; different fuse sizes are interchangeable with each other; this fuse type may only be used as a replacement for an existing Edison-base plug fuse

EIA/TIA an acronym for the Electronic Industry Association and Telecommunications Industry Association; these organizations create and publish compatibility standards for the products made by member companies

EIA/TIA 570-B the newest standard document for structured cabling installations in residential applications

electrical drawings a part of the building plan that shows the electrical supply and distribution for the building electrical system

electrical shock the sudden stimulation of nerves and muscle caused by electricity flowing through the body

electric range a stand-alone electric cooking appliance that typically has four cooking elements on top and an oven in the bottom of the appliance

elevation drawing a drawing that shows the side of the house that faces in a particular direction; for example, the north elevation drawing shows the side of the house that is facing north

enclosure the case or housing of apparatus or the fence or walls surrounding an installation to prevent personnel from accidentally contacting energized parts or to protect the equipment from physical damage

Energy Star an international standard for energy efficient consumer products

equipment a general term including material, fittings, devices, appliances, luminaires (lighting fixtures), apparatus, machinery, and other parts used in connection with an electrical installation

equipment-grounding conductor the conductor used to connect the noncurrent-carrying metal parts of equipment, raceways, and other enclosures to the system-grounded conductor, the grounding electrode conductor, or both at the service equipment

exothermic welding a process for making bonding connections on the

bonding grid for a permanently installed swimming pool using specially designed connectors, a form, a metal disk, and explosive powder; this process is sometimes called "Cad-Weld"

extension lines lines used to extend but not actually touch object lines and that have the dimension lines drawn between them

F-type connector a 75-ohm coaxial cable connector that can fit RG-6 and RG-59 cables and is used for terminating video system cables in residential wiring applications; the most common styles used are crimp-on and compression

feeder the circuit conductors between the service equipment and the final branch-circuit overcurrent protection device

field bend any bend or offset made by installers, using proper tools and equipment, during the installation of conduit systems

fish the process of installing cables in an existing wall or ceiling

fitting an electrical accessory, like a locknut, that is used to perform a mechanical rather than an electrical function

flashing sheets of metal or other material used to weatherproof locations on exterior surfaces, like a roof, where penetrations have been made

floor joists horizontal framing members that attach to the sill plate and form the structural support for the floor and walls

floor plan a part of the building plan that shows a bird's-eye view of the layout of each room

fluorescent lamp a gaseous discharge light source; light is produced when the phosphor coating on the

inside of a sealed glass tube is struck by energized mercury vapor

footing the concrete base on which a dwelling foundation is constructed; it is located below grade

forming shell the support structure designed and used with a wet-niche lighting fixture; it is installed in the wall of a pool

foundation the base of the structure, usually poured concrete or concrete block, on which the framework of the house is built; it sits on the footing

four-way switch a switch type that, when used in conjunction with two three-way switches, will allow control of a 120-volt lighting load from more than two locations

foyer an entranceway or transitional space from the exterior to the interior of a house

framing the building "skeleton" that provides the structural framework of the house

fuel cell an electrochemical system that consumes a fuel like natural gas or LP gas to produce electricity; the consumption of the fuel gas is through an electrochemical process rather than a combustion process

furring strip long, thin strips of wood (or metal) used to make backing surfaces to support the finished surfaces in a room; furring refers to the backing surface, the process of installing it, or may also refer to the strips themselves; furring strips typically measure 1" x 2" or 1" x 3" and are typically laid out perpendicular to studs or joists and nailed to them, or set vertically against an existing wall surface

fuse an overcurrent protection device that opens a circuit when the fusible link is melted away by the extreme heat caused by an overcurrent

ganging joining two or more device boxes together for the purpose of holding more than one device

general lighting circuit a branch-circuit type used in residential wiring that has both lighting and receptacle loads connected to it; a good example of this circuit type is a bedroom branch circuit that has both receptacles and lighting outlets connected to it

general purpose branch circuit a branch circuit that supplies two or more receptacles or lighting outlets, or a combination of both; household appliances such as vacuum cleaners, small room air conditioners, televisions, or stereo equipment can also receive electrical power from this type of branch circuit

generator a rotating machine used to convert mechanical energy into electrical energy

girders heavy beams that support the inner ends of floor joists

green the process of designing and building a home that minimizes its impact on the environment both during construction and over its useful life

ground the earth

grounded connected to earth or to some conducting body that serves in place of the earth

grounded conductor a system or circuit conductor that is intentionally grounded

ground fault an unintended low-resistance path in an electrical circuit through which some current flows to ground using a pathway other than the intended pathway; it results when an ungrounded "hot" conductor unintentionally touches a grounded surface or grounded conductor

ground fault circuit interrupter (GFCI) a device that protects people from dangerous levels of electrical current by measuring the current difference between two conductors of an electrical circuit and tripping to an open position if the measured value exceeds approximately 5 milliamperes

ground fault circuit interrupter (GFCI) receptacle a receptacle device that protects people from dangerous amounts of current; it provides protection at the receptacle outlet where it is installed and can also provide GFCI protection to other "regular" receptacle devices on the same branch circuit; it is designed to trip when fault current levels are in the range of 4 to 6 milliamps or more

grounding an electrical connection to an object that conducts electrical current to the earth

grounding conductor a conductor used to connect equipment or the grounded conductor of a wiring system to a grounding electrode or electrodes

grounding electrode a conducting object through which a direct connection to earth is established

grounding electrode conductor the conductor used to connect the system grounded conductor to a grounding electrode or to a point on the grounding electrode system

guy a cable that mechanically supports a wind turbine tower

handy box a type of metal, surface-mounted device box used to hold only one device

hardwired an installation technique in which the circuit conductors are brought directly to an electrical appliance and terminated at the appliance

harmonics a frequency that is a multiple of the 60 Hz fundamental; harmonics cause distortion of the voltage and current AC waveforms

hazard a potential source of danger

heat pump a reversible air-conditioning system that will heat a house in cool weather and cool a house in warm weather

hidden line a line on a building plan that shows an object hidden by another object on the plan; hidden lines are drawn using a dashed line

high-intensity discharge (HID) lamp another type of gaseous discharge lamp, except the light is produced without the use of a phosphor coating

home run the part of the branch-circuit wiring that originates in the loadcenter and provides electrical power to the first electrical box in the circuit

horizontal cabling the connection from the distribution center to the work area outlets

hybrid system a system comprised of multiple power sources; the power sources may include photovoltaic, wind, micro-hydro generators, or engine-driven generators

hydraulic a term used to describe tools that use a pressurized fluid, like oil, to accomplish work

hydromassage bathtub a permanently installed bathtub with re-circulating piping, pump, and associated equipment; it is designed to accept, circulate, and discharge water each time it is used

hydronic system a term used when referring to a hot water heating system

icon a symbol used on some electrical meters to indicate where a selector switch must be set so the meter can measure the correct type of electrical quantity

IDC an acronym for "insulation displacement connection"; a type of termination where the wire is "punched down" into a metal holder with a punch-down tool; no prior stripping of the wire is required

impulse a type of transient voltage that originates outside the home and is usually caused by utility company switching or lightning strikes

incandescent lamp the original electric lamp; light is produced when an electric current is passed through a filament; the filament is usually made of tungsten

individual branch circuit a circuit that supplies only one piece of electrical equipment; examples are one range, one space heater, or one motor

inductive heating the heating of a conducting material in an expanding and collapsing magnetic field; inductive heating will occur when current-carrying conductors of a circuit are brought through separate holes in a metal electrical box or enclosure

insolation the term used for the measure of solar radiation striking the earth's surface at a particular time and place

insulated a conductor that is covered by a material that is recognized by the National Electrical Code® as electrical insulation

insulator a material that does not allow electrical current to flow through it; examples are rubber, plastic, and glass

interconnecting the process of connecting together smoke detectors so that if one is activated, they all will be activated

interrupting rating the highest current at rated voltage that a device is intended to interrupt under standard test conditions

intersystem bonding termination a device that provides a means for connecting bonding conductors for communications systems to the grounding electrode system

inverse time circuit breaker a type of circuit breaker that has a trip time that gets faster as the fault current flowing through it gets larger; this is the circuit breaker type used in house wiring

inverter equipment that is used to change voltage level or waveform, or both, of electrical energy; a device that changes dc input to an ac output; may also function as battery chargers that use alternating current from another source and convert it into direct current for charging batteries

irradiance a measure of how much solar power is striking a specific location

jack the receptacle for an RJ-45 plug

jacketed cable a voice/data cable that has a nonmetallic polymeric protective covering placed over the conductors

jet pump a type of water pump used in home water systems; the pump and electric motor are separate items that are located away from the well in a basement, garage, crawl space, or other similar area

jug handles a term used to describe the type of bend that must be made with certain cable types; bending cable too tightly will result in damage to the cable and conductor insulation; bending them in the shape similar to a "handle" on a "jug" will help satisfy NEC® bending requirements

junction box a box whose purpose is to provide a protected place for splicing electrical conductors

kilowatt-hour meter an instrument that measures the amount of electrical energy supplied by the electric utility company to a dwelling unit

knockout (KO) a part of an electrical box that is designed to be removed, or "knocked out," so that a cable or raceway can be connected to the box

knockout plug a piece of electrical equipment used to fill unused openings in boxes, cabinets, or other electrical equipment

knockout punch a tool used to cut holes in electrical boxes for the attachment of cables and conduits

lamp efficacy a measure used to compare light output to energy consumption; it is measured in lumens per watt; a 100-watt light source producing 1750 lumens of light has an efficacy (efficiency) of 17.5 lumens per watt (L/W)

laundry branch circuit a type of branch circuit found in residential wiring that supplies electrical power to laundry areas; no lighting outlets or other receptacles may be connected to this circuit

leader a solid line that may or may not be drawn at an angle and has an arrow on the end of it; it is used to connect a note or dimension to a part of the building

LED lamp a light-emitting-diode lamp is a solid-state lamp that uses light-emitting diodes (LEDs) as the source of light; because the light output of individual light-emitting diodes is small compared to incandescent and compact fluorescent lamps, multiple diodes are used together

legend a part of a building plan that describes the various symbols and abbreviations used on the plan

level perfectly horizontal; completely flat; a tool used to determine if an object is level

lighting outlet an outlet intended for the direct connection of a lamp holder, a luminaire (lighting fixture), or a pendant cord terminating in a lamp holder

line side the location in electrical equipment where the incoming electrical power is connected; an example is the line-side lugs in a meter socket where the incoming electrical power conductors are connected

load (electrical) a part of an electrical circuit that uses electrical current to perform some function; an example would be a lightbulb (produces light) or electric motor (produces mechanical energy)

loadcenter a type of panelboard normally located at the service entrance in a residential installation and usually containing the main service disconnect switch

load side the location in electrical equipment where the outgoing electrical power is connected; an example is the load-side lugs in a meter socket where the outgoing electrical power conductors to the service equipment are connected

lug a device commonly used in electrical equipment used for terminating a conductor

lumen the unit of light energy emitted from a light source

luminaire a complete lighting unit consisting of a lamp or lamps together with the parts designed to distribute the light, to position and protect the lamps and ballast (where applicable), and to connect the lamps to the power supply

Madison hold-its thin metal straps that are used to hold "old-work" electrical boxes securely to an existing wall or ceiling

magnetic declination the deviation of magnetic south from true south

main bonding jumper a jumper used to provide the connection between the grounded service conductor and the equipment-grounding conductor at the service

manual ranging a meter feature that requires the user to manually select the proper range

mast kit a package of additional equipment that is required for the installation of a mast-type service entrance; it can be purchased from an electrical distributor

Material Safety Data Sheet (MSDS) a form that lists and explains each of the hazardous materials that electricians may work with so they can safely use the material and respond to an emergency situation

maximum power point (MPP) indicates the maximum output of the module and is the result of the maximum voltage (Vmp) multiplied by the maximum current (Imp)

megabits per second (Mbps) refers to the rate that digital bits (1s and 0s) are sent between two pieces of equipment

megahertz (MHz) refers to the upper frequency band on the ratings of a cabling system

megohmmeter a measuring instrument that measures large amounts of resistance and is used to test electrical conductor insulation

meter enclosure the weatherproof electrical enclosure that houses the kilowatt-hour meter; also called the "meter socket" or "meter trim"

micro-hydro a small hydroelectric alternative energy system that uses water flow to turn a generator that produces electricity; the water flow can come from a stream or a small reservoir

micro-hydro electric system an alternative energy system that uses water pressure to move a turbine, which then drives a generator to produce electricity

mil 1 mil is equal to .001 inch; this is the unit of measure for the diameter of a conductor

module a complete, environmentally protected unit consisting of solar cells, optics, and other components, exclusive of tracker, designed to generate dc power when exposed to sunlight

motion sensor a device that upon detecting movement in a specific area will switch on a lighting load; once all movement stops and a short amount of time goes by the sensor device switches off the electricity to the lighting load; often used for outdoor lighting control in residential applications

multimeter a measuring instrument that is capable of measuring many different electrical values, such as voltage, current, resistance, and frequency, all in one meter

multiwire branch circuit a branch circuit that consists of two or more ungrounded conductors that have a voltage between them and a grounded conductor that has an equal voltage between it and each ungrounded conductor

multiwire circuit a circuit in residential wiring that consists of two ungrounded conductors that have 240 volts between them and a grounded conductor that has 120 volts between it and each ungrounded conductor

nacelle an enclosure housing the alternator and other parts of a wind turbine

nameplate the label located on an appliance that contains information such as amperage and voltage ratings, wattage ratings, frequency, and other information needed for the correct installation of the appliance

National Electrical Code® (NEC®) a document that establishes minimum safety rules for an electrician to follow when performing electrical installations; it is published by the National Fire Protection Association (NFPA)

National Electrical Manufacturers Association (NEMA) an organization that establishes certain construction standards for the manufacturers of electrical equipment; for example, a NEMA Type 1 box purchased from Company X will meet the same construction standards as a NEMA Type 1 box from Company Y

near end crosstalk (NEXT) electrical noise coupled from one pair of wires to another within a multi-pair voice or data cable

neutral conductor the conductor connected to the neutral point of a system that is intended to carry current under normal conditions; the neutral point is the midpoint on a single-phase, three-wire system; the voltage from either ungrounded "hot" wire in this system to the neutral point is 120 volts

new work box an electrical box without mounting ears; this style of electrical box is used to install electrical wiring in a new installation

nipple an electrical conduit of 2 feet or less in length used to connect two electrical enclosures

no-niche luminaire is a lighting fixture intended for installation above or below the water level; it does not

have a forming shell that it fits into but rather sits on the surface of the pool wall

noncontact voltage tester a tester that indicates if a voltage is present by lighting up, making a noise, or vibrating; the tester is not actually connected into the electrical circuit but is simply brought into close proximity of the energized conductors or other system parts

nonlinear loads a load where the load impedance is not constant, resulting in harmonics being present on the electrical circuit

object line a solid dark line that is used to show the main outline of the building

occupancy sensor a device that detects when a person has entered a room and switches on the room lighting; when everybody leaves a room and a short amount of time goes by the sensor device switches the room lighting off; often used for automatic control of room lighting in residential applications

Occupational Safety and Health Administration (OSHA) since 1971, OSHA's job has been to establish and enforce workplace safety rules

offset bend a type of conduit bend that is made with two equal-degree bends in such a way that the conduit changes elevation and avoids an obstruction

ohm the unit of measure for electrical resistance

Ohm's law the mathematical relationship between current, voltage, and resistance in an electrical circuit

ohmmeter an instrument that measures values of resistance

old work box an electrical box with mounting ears; this style of electrical box is used to install electrical wiring in existing installations

open circuit a circuit that is energized but does not allow useful current to flow on the circuit because of a break in the current path

open circuit voltage (Voc) the maximum voltage when no current is being drawn from the module

outlet a point on the wiring system at which current is taken to supply electrical equipment; an example is a lighting outlet or a receptacle outlet

outlet box a box that is designed for the mounting of a receptacle or a lighting fixture

overcurrent any current in excess of the rated current of equipment or the ampacity of a conductor; it may result from an overload, a short circuit, or a ground fault

overcurrent protection device (OCPD) a fuse or circuit breaker used to protect an electrical circuit from an overload, a short circuit, or a ground fault

overload a larger-than-normal current amount flowing in the normal current path

pad-mounted transformer a transformer designed to be mounted directly on a pad foundation at ground level; single-phase pad-mounted transformers are designed for underground residential distribution systems where safety, reliability, and aesthetics are important

panel a collection of modules mechanically fastened together, wired, and designed to provide a field-installable unit

panelboard a panel designed to accept fuses or circuit breakers used for the protection and control of lighting, heating, and power circuits; it is designed to be placed in a cabinet and placed in or on a wall; it is accessible only from the front

patch cord a short length of cable with an RJ-45 plug on either end; used to connect hardware to the work area outlet or to connect cables in a distribution panel

permanently installed swimming pool a swimming pool constructed totally or partially in the ground with a water depth capacity of greater than 42 inches (1 m); all pools, regardless of depth, installed in a building are considered permanent

personal protective equipment (PPE) equipment for the eyes, face, head, and extremities, protective clothing, respiratory devices, and protective shields and barriers

phantom load a load that consumes electricity even when the load is not performing its intended function; examples are battery chargers, internet routers, microwave ovens, and televisions

photocell a light-sensing device used to control lighting fixtures in response to detected light levels; no sunlight detected causes it to switch on a lighting fixture; sunlight detected causes it to switch off a lighting fixture

photovoltaic the name used to describe the effect of sunlight striking a specially prepared surface and producing electricity

pigtail a short length of wire used in an electrical box to make connections to device terminals

platform frame a method of wood frame construction in which the walls are erected on a previously constructed floor deck or platform

plot plan a part of a building plan that shows information such as the location of the house, walkway, or driveway on the building lot; the plan is drawn with a view as if you were looking down on the building lot from a considerable height

plug the device that is inserted into a receptacle to establish a connection between the conductors of the attached flexible cord and the conductors connected to the receptacle

plumb perfectly vertical; the surface of the item you are leveling is at a right angle to the floor or platform you are working from

polarity the positive or negative direction of DC voltage or current

polarized plug a two-prong plug that distinguishes between the grounded conductor and the "hot" conductor by having the grounded conductor prong wider than the "hot" conductor prong; this plug will fit into a receptacle only one way

pool cover, electrically operated a motor-driven piece of equipment designed to cover and uncover the water surface of a pool

porcelain standoff a fitting that is attached to a service entrance mast, which provides a location for the attachment of the service drop conductors in an overhead service entrance

power source a part of an electrical circuit that produces the voltage required by the circuit

pryout (PO) small parts of electrical boxes that can be "pried" open with a screwdriver and twisted off so that a cable can be secured to the box

pulling in the process of installing cables through the framework of a house

punch-down block the connecting block that terminates cables directly; 110 blocks are most popular for residential situations

raceway an enclosed channel of metal or nonmetallic materials designed expressly for holding wires or cables; raceways used in residential wiring include rigid metal conduit, rigid nonmetallic conduit, intermediate metal conduit, liquid tight flexible conduit, flexible metal conduit, electrical nonmetallic tubing, and electrical metallic tubing

rafters part of the roof structure that is supported by the top plate of the wall sections; the roof sheathing is secured to the rafters and then covered with shingles or other roofing material to form the roof

rain-tight constructed or protected so that exposure to a beating rain will not result in the entrance of water under specified test conditions

receptacle a device installed in an electrical box for the connection of an attachment plug

receptacle polarity when the white-colored grounded conductor is attached to the silver screw of the receptacle and the "hot" ungrounded circuit conductor is attached to the brass screw on the receptacle; this results in having the long slot always being the grounded slot and the short slot always being the ungrounded slot on 125-volt, 15- or 20-amp-rated receptacles used in residential wiring

reciprocating to move back and forth

redhead sometimes called a "red devil"; an insulating fitting required to be installed in the ends of a

Type AC cable to protect the wires from abrasion

reel a drum having flanges on each end; reels are used for wire or cable storage; also called a "spool"

resistance the opposition to current flow

RG-6 (Series 6) a type of coaxial cable that is "quad shielded" and is used in residential structured cabling systems to carry video signals such as cable and satellite television

RG-59 a type of coaxial cable typically used for residential video applications; EIA/TIA 570A recommends that RG-6 coaxial cable be used instead of RG-59 because of the better performance characteristics of the RG-6 cable

ribbon a narrow board placed flush in wooden studs of a balloon frame to support floor joists

ring one of the two wires needed to set up a telephone connection; it is connected to the negative side of a battery at the telephone company; it is the telephone industry's equivalent to an ungrounded "hot" conductor in a normal electrical circuit

ring wave a type of transient voltage that originates inside the home and is usually caused by home office photocopiers, computer printers, the cycling on and off of heating, ventilating, and air-conditioning equipment, and spark igniters on gas appliances like furnaces and ranges

riser a length of raceway that extends up a utility pole and encloses the service entrance conductors in an underground service entrance

RJ-11 a popular name for a six-position UTP connector

RJ-45 the popular name for the modular eight-pin connector used to terminate Category 5e and 6 UTP cable

Romex™ a trade name for nonmetallic sheathed cable (NMSC); this is the term most electricians use to refer to NMSC

roof flashing/weather collar two parts of a mast-type service entrance that, when used together, will not allow water to drip down and into a house through the hole in the roof that the service mast extends through

rough-in the stage in an electrical installation when the raceways, cable, boxes, and other electrical equipment are installed; this electrical work must be completed before any construction work can be done that covers wall and ceiling surfaces

running boards pieces of board lumber nailed or screwed to the joists in an attic or basement; the purpose of using running boards is to have a place to secure cables during the rough-in stage of installing a residential electrical system

saddle bend a type of conduit bend that results in a conduit run going over an object that is blocking the path of the run; there are two styles: a three-point saddle and a four-point saddle

safety switch a term used sometimes to refer to a disconnect switch; a safety switch may use fuses or a circuit breaker to provide overcurrent protection

scaffolding also referred to as staging; a piece of equipment that provides a platform for working in high places; the parts are put together at the job site and then taken apart and reconstructed when needed at another location

scale the ratio of the size of a drawn object and the object's actual size

schedule a table used on building plans to provide information about specific equipment or materials used in the construction of the house

sconce a wall-mounted lighting fixture

sectional drawing a part of the building plan that shows a cross-sectional view of a specific part of the dwelling

section line a broad line consisting of long dashes followed by two short dashes; at each end of the line are arrows that show the direction in which the cross section is being viewed

secured (as applied to electrical cables) fastened in place so the cable cannot move; common securing methods include staples, tie wraps, and straps

self-contained spa or hot tub a factory-fabricated unit consisting of a spa or hot tub vessel having integrated water-circulating, heating, and control equipment

service the conductors and equipment for delivering electric energy from the serving utility to the wiring system of the premises served

service center the hub of a structured wiring system with telecommunications, video, and data communications installed; it is usually located in the basement next to the electrical service panel or in a garage; sometimes called a "distribution center"

service disconnect a piece of electrical equipment installed as part of the service entrance that is used to disconnect the house electrical system from the electric utility's system

service drop the overhead conductors between the utility electric supply system and the service point

service entrance the part of the wiring system where electrical power is supplied to the residential wiring system from the electric utility; it includes the main panelboard, the electric meter, overcurrent protection devices, and service conductors

service entrance cable service entrance conductors made up in the form of a cable

service entrance conductors the conductors from the service point to the service disconnecting means

service entrance conductors, overhead the overhead service conductors between the service point and the first point of connection to the service conductors at the building or other structure

service entrance conductors, underground the underground service conductors between the service point and the first point of connection to the service entrance conductors inside or outside the building wall

service equipment the necessary equipment connected to the load end of the service entrance conductors supplying a building and intended to be the main control and cutoff of the supply

service head the fitting that is placed on the service drop end of service entrance cable or service entrance raceway and is designed to minimize the amount of moisture that can enter the cable or raceway; the service head is commonly referred to as a "weatherhead"

service lateral the underground conductors between the utility electric supply system and the service point

service mast a piece of rigid metal conduit or intermediate metal conduit, usually 2 or 2½ inches in diameter, that provides service conductor protection and the proper height requirements for service drops

service point the point of connection between the wiring of the electric utility and the premises wiring

service raceway the rigid metal conduit, intermediate metal conduit, electrical metallic tubing, rigid nonmetallic conduit, or any other approved raceway that encloses the service entrance conductors

setout the distance that the face of an electrical box protrudes out from the face of a building framing member; this distance is dependent on the thickness of the wall or ceiling finish material

shall a term used in the *National Electrical Code*® that means that the rule *must* be followed

sheath the outer covering of a cable that is used to provide protection and to hold everything together as a single unit

sheathing boards sheet material like plywood that is fastened to studs and rafters; the wall or roofing finish material will be attached to the sheathing

sheetrock a popular building material used to finish off walls and ceilings in residential and commercial construction; it is available in standard sizes, such as 4 by 8 feet, and is constructed of gypsum sandwiched between a paper front and back; often referred to as wallboard

short circuit an unintended low-resistance path through which current flows around rather than along a circuit's intended current path; it results when two circuit conductors come in contact with each other unintentionally

short circuit current (Isc) the maximum current output of a module under conditions of a circuit with no resistance (short circuit)

sill a length of wood that sets on top of the foundation and provides a place to attach the floor joists

sill plate a piece of equipment that, when installed correctly, will help keep water from entering the hole in the side of a house that the service entrance cable from the meter socket to the service panel goes through

single-pole switch a switch type used to control a 120-volt lighting load from one location

single receptacle a single contact device with no other contact device on the same strap (yoke)

small-appliance branch circuit a type of branch circuit found in residential wiring that supplies electrical power to receptacles located in kitchens and dining rooms; no lighting outlets are allowed to be connected to this circuit type

smart meter a meter that measures electrical consumption in more detail than a conventional kWH meter; it also has the ability to communicate that information back to the local electric utility for monitoring and billing purposes

smoke detector a safety device that detects airborne smoke and issues an audible alarm, thereby alerting nearby people to the danger of fire; most smoke detectors work either by optical detection or by ionization, but some of them use both detection methods to increase sensitivity to smoke

solar cell the basic photovoltaic device that generates electricity when exposed to light

sones an internationally recognized unit of loudness, measured in decibels (dB); doubling the sone value is equivalent to doubling the loudness; one sone is roughly equivalent to the sound of a quiet refrigerator

spa or hot tub a hydromassage pool or tub designed for immersion of users; usually has a filter, heater, and motor-driven pump; it can be installed indoors or outdoors, on the ground, or in a supporting structure; a spa or hot tub is not designed to be drained after each use

specifications a part of the building plan that provides more specific details about the construction of the building

spliced connecting two or more conductors with a piece of approved equipment like a wirenut; splices must be done in approved electrical boxes

split-wired receptacle a duplex receptacle wired so that the top outlet is "hot" all the time and the bottom outlet is switch-controlled

stab a term used to identify the location on a loadcenter's ungrounded bus bar where a circuit breaker is snapped on

standby power system a backup electrical power system that consists of a generator, transfer switch, and associated electrical equipment; its purpose is to provide electrical power to critical branch circuits when the electrical power from the utility company is not available

storable swimming pool a swimming pool constructed on or above the ground with a maximum water-depth capacity of 42 inches (1 m) or a pool with nonmetallic, molded polymeric walls (or inflatable fabric walls) regardless of size or water-depth capacity

STP an acronym for "shielded twisted pair" cable; it resembles UTP but has a foil shield over all four pairs of copper conductors and is used for better high-frequency performance and less electromagnetic interference

strap (yoke) the metal frame that a receptacle or switch is built around; it is also used to mount a switch or receptacle to a device box

strip to damage the threads of the head of a bolt or screw

structured cabling an architecture for communications cabling specified by the EIA/TIA TR41.8 committee and used as a voluntary standard by manufacturers to ensure compatibility

stub-up a type of conduit bend that results in a 90-degree change of direction

subfloor the first layer of floor material that covers the floor joists; usually 4-by-8-plywood or particleboard

submersible pump a type of water pump used in home water systems; the pump and electric motor are enclosed in the same housing and are lowered down a well casing to a level that is below the water line

supplemental grounding electrode a grounding electrode that is used to "back up" a metal water pipe grounding electrode

supported (as applied to electrical cables) held in place so the cable is not easily moved; common supporting methods include running cables horizontally through holes or notches in framing members, or using staples, tie wraps, or straps after a cable has been properly secured close to a box according to the *NEC®*

switch box a name used to refer to a box that just contains switches

switch loop a switching arrangement where the feed is brought to the lighting outlet first and a two-wire loop run from the lighting outlet to the switch

symbol a standardized drawing on the building plan that shows the location and type of a particular material or component

take-up the amount that must be subtracted from a desired stub-up height so the bend will come out right; the take-up is different for each conduit size

tempered treated with heat to maximize a metal's hardness

thermals rising currents of warm air that go up and over land during sunny daylight hours

thermostat a device used with a heating or cooling system to establish a set temperature for the system to achieve; they are available as line voltage models or low-voltage models; some thermostats can also be programmed to keep a home at a specific temperature during the day and another temperature during the night

thinwall a trade name often used for electrical metallic tubing

threaded hub the piece of equipment that must be attached to the top of a meter socket so that a raceway or a cable connector can be attached to the meter socket

three-way switch a switch type used to control a 120-volt lighting load from two locations

timer a device that controls the flow of electricity on a circuit for a certain amount of time; used in residential applications to switch off a bathroom fan or ceiling-suspended paddle fan after a specific length of time has gone by

tip the first wire in a pair; a conductor in a telephone cable pair that is usually connected to the positive side of a battery at the telephone company's central office; it is the telephone industry's equivalent to a grounded conductor in a normal electrical circuit

top plate the top horizontal part of a wall framework

torque the turning or twisting force applied to an object when using a torque tool; it is measured in inch-pounds or foot-pounds

tower a pole or other structure that supports a wind turbine

track lighting a type of lighting fixture that consists of a surface-mounted or suspended track with several lighting heads attached to it; the lighting heads can be adjusted easily to point their light output at specific items or areas

transfer switch a switching device for transferring one or more load conductor connections from one power source to another

transformer a piece of electrical equipment used by the electric utility to step down the high voltage of the utility system to the 120/240 volts required for a residential electrical system

transformer the electrical device that steps down the 120-volt house electrical system voltage to the 16 volts a chime system needs to operate correctly

transient voltage surge suppressor (TVSS) (receptacle or strip) an electrical device designed to protect sensitive electronic circuit boards from voltage surges

triplex cable a cable type used as the service drop for a residential service entrance; it consists of a bare messenger wire that also serves as the service grounded conductor and two black insulated ungrounded conductors wrapped around the bare wire

troffer a term commonly used by electricians to refer to a fluorescent lighting fixture installed in the grid of a suspended-ceiling

troubleshooting the process of determining the cause of a malfunctioning part of a residential electrical system

true RMS meter a type of meter that allows accurate measurement of AC values in harmonic environments

twistlock receptacle a type of receptacle that requires the attachment plug to be inserted and then turned slightly in a clockwise direction to lock the plug in place; the attachment plug must be turned slightly counterclockwise to release the plug so it can be removed from the receptacle

Type IC a light fixture designation that allows the fixture to be completely covered by thermal insulation

Type Non-IC a light fixture that is required to be kept at least 3 inches from thermal insulation

Type S plug fuse a fuse type that uses different fuse bases for each fuse size; an adapter that matches each Type S fuse size is required; this fuse type is not interchangeable with other Type S fuse sizes

utility box a name used to refer to a metal single-gang, surface-mounted device box; also called a handy box

utility pole a wooden circular column used to support electrical, video, and telecommunications utility wiring; it may also support the transformer used to transform the high utility company voltage down to the lower voltage used in a residential electrical system

UTP an acronym for "unshielded twisted pair" cable; it is composed of four pairs of copper conductors and

graded for bandwidth as "categories" by EIA/TIA 568; each pair of wires is twisted together

ventricular fibrillation very rapid irregular contractions of the heart that result in the heartbeat and pulse going out of rhythm with each other

volt the unit of measure for voltage

volt-ampere a unit of measure for alternating current electrical power; for branch-circuit, feeder, and service calculation purposes, a watt and a volt-ampere are considered the same

voltage the force that causes electrons to move from atom to atom in a conductor

voltage drop (VD) the amount of voltage that is needed to "push" the house electrical load current through the wires from the utility transformer to the service panel and back to the transformer; the amount of voltage drop depends on the resistance of the wires, the length of the wires, and the actual current carried on the wires

voltage tester a device designed to indicate approximate values of voltage or to simply indicate if a voltage is present

voltmeter an instrument that measures a precise amount of voltage

VOM a name sometimes used in reference to a "multimeter"; the letters stand for "volt-ohm milliammeter"

wall-mounted oven a cooking appliance that is installed in a cabinet or wall and is separated from the counter-mounted cooktop

wallboard a thin board formed from gypsum and layers of paper that is used often as the interior wall sheathing in residential applications; commonly called "sheetrock"

wall studs the parts that form the vertical framework of a wall section

wet-niche luminaire a type of lighting fixture intended for installation in a wall of a pool; it is accessible by removing the lens from the forming shell; this luminaire type is designed so that water completely surrounds the fixture inside the forming shell

wet location installations underground or in concrete slabs or masonry in direct contact with the earth in locations subject to saturation with water or other liquids, such as in unprotected areas exposed to the weather

Wiggy a trade name for a solenoid type of voltage tester

wind turbine a mechanical device that converts wind energy to electrical energy

wind turbine system a small wind electric generating system

wirenut a piece of electrical equipment used to mechanically connect two or more conductors together

wiring a term used by electricians to describe the process of installing a residential electrical system

work area outlet the jack on the wall that is connected to the desktop computer by a patch cord

Index

Note: Page numbers in **bold** indicate a figure or table.

A

AC (Alternating current), 243

Accessible, 326, 341
defined, 228

Accessible, readily, 242, **243**
defined, 228

Adjustable wrench, 90–91, **91**
use procedure, 113

AFCI. *See* Arc fault circuit interrupter (AFCI)

AHJ. *See* Authority having jurisdiction (AHJ)

Air barrier, 741–742

Air conditioning, 504–509
central air branch circuit installation, 505–506
heat pump, 477, 504
overcurrent protection device, 505, 508
room air branch circuit installation, 507–509
wiring requirements, basic, 537

Air-handling spaces, wiring and, 308, **308**

Air sealing penetrations, 742

Airtight boxes, 742, 742

Allen wrenches, 96, 97

Alternating current (AC), 243

Alternative energy systems, 750–808
installation of, 750–808
photovoltaic system, 752–794
wind turbine system, 794–805

Altitude, sun, 760, **764**

Aluminum conductors, 44
antioxidant and, 287, 289
copper-clad, 37, 44, 249–250
installation procedures, 73, 287, 289

Ambient temperature, 200, **201**
defined, 187

American National Standards Institute (ANSI)
Material Safety Data Sheet (MSDS), 26
symbols, 176, **178–180**, **181–182**
Symbols for Electrical Construction, 176

American Wire Gauge (AWG), 37, 44–45, 198

Ammeters, 139–141
clamp-on, 134, 139, 140–141, **140**, **141**, 156
in-line, 134, 139–140, **140**

Amorphous, 751

Ampacity, 45, 198–205, 201–204, 674–675
of a conductor, determining, 198–205, **201–204**
adjustment factors, 200, **201**
ambient temperature and, 198–200, **201**
defined, 37, 187, 198
derating, 200
examples of, 198–199, **203**
high temperature conductors, 198
insulated conductors, 198
minimum conductor size, **195**, **196**, **197**, 198, **199**
of nonmetallic sheathed cable, 48

Amperes
defined, 4
effects on human body, **7**

Analog meter, 134, 138, **138**

Analog ohmmeter, 142, 142

Anchors, 67–70, 67
installing, 75–78

Anemometer, 751, 798, 798

Antioxidant, 287, 289
defined, 37

Appliances, 736, **736**. *See also* Small appliance
circuit breakers and, 683
individual branch circuits for, 195–197, **195**, **196**, **197**

Approval, 39–40

Approved, 37

Arc, 4, 7, 604, 644

Arc blast, 4, 8

Arc burn, 7

Arc fault circuit interrupter (AFCI), 14, 63, **64**, 644, **674**, 679–680, **680**
defined, 14, 637
receptacles, 644, 648, 679
types of, 644, 679

Arc-flash, 4, 7

Architect's scale, 165, 174–176, **175**

Architectural drawings. *See* Building plans

Architectural firm, 165, 174

Architectural symbols, 176, **177**
abrasion protection, 372
bends in, 370, **371**
exposed runs, 367, **371**
installation requirements, 367–372, **371**
protection of, 369
redheads, 372, **372**
runs through horizontal framing members, 371, **372**

Architectural symbols (*continued*)
  securing and supporting, 371, **371**
  unsupported situations, 372
Array, photovoltaic, 751, 753,
  754, 765
  installation guidelines, 791
Audio system, 560. *See also* Video,
  voice, and data wiring
Auger, 84
Auger bit, 100, **101**, 121
Authority having jurisdiction
  (AHJ), 37, 39, 255, 327
Auto-ranging, 134, 138
AWG. *See* American Wire Gauge
Awl, 91, **91**
Azimuth, 751, 760

B

Back-to-back bend, 414, **414**,
  429–430
Backboard, **261**, 286
Backfeeding, **261**, 290–291
Ballast, 600–602, **601–602**. *See
  also* Lighting
  circuits, 600–602, **602**
  class P, 594, 600, 603, **602**
  defined, 594
Balloon frame, 165, 180, **183**
Band joist, 183, 183, 272
  defined, 165
Bandwidth, 559, 561
Baseboard heating installation,
  501, **502**, 504, **504**
Basement, receptacle outlet, 320
Bathroom
  branch circuit, 187, 191, 193, 194
  branch circuit installation, 495, **495**
  defined, 478
  exhaust fan in, 739–740, **739**
  lighting fixture installation, 608, **610**
  receptacle outlets, 319, **319**
Batteries, 753, **755**, 772–774,
  773–774. *See also* Inverter
  installation guidelines, 791
Battery-operated loads, 741
Bell boxes, **543**
Bender, 84, 94–95, **95**, 414, **415**
Bimetallic strip, 37, 63
Blueprints, 165, 167. *See also*
  Building plans

Body. *See* Human body
Bonding, 244–245, **248**. *See also*
  Grounding
  checklist of *NEC®* requirements, 706
  defined, 228
  of metal raceways/service equipment,
    251, 250, 251, **253**
  methods, 252–253
Bonding jumper, 253, **253**. *See
  also* Grounding
  main, 228, 244, **245**, 287
Bottom plate, 165, 180, **183**
Box fill, 326, 329, 331–332, 335
Box offset bend, 414, 414, 438–439
Branch circuits, 188–198,
  476–518. *See also* Ampacity;
    Overcurrent protection
  air conditioning installation, 504–509
  bathroom, 187, 191, **193**, 194
  bathroom installation, 495
  central air conditioning installation,
    505–506
  central heating, gas and oil installation,
    509–510
  circuit breaker installation for, 683–695
  cooking equipment, 195–197, **195**,
    196, 197
  counter-mounted cooktop installation,
    485–486
  countertop cook unit, 196–197, **197**
  defined, 187, 188, **189**
  dishwasher installation, 489–490
  electric dryer installation, 492–494
  electric heating installation, 500–504
  electric range, 195–196, **195, 196**
  electric range installation, 481–483
  garage installation, 523–527
  garbage disposal installation, 486–489
  general lighting, 187, 188–191, **189**
  general lighting installation, 478–479
  general purpose, 187, 188, **189**
  individual, 187, 194–198, **194**
  installation, 476–518
  laundry, 187, 191, **192**
  laundry installation, 490–491, **492**
  low-voltage chime circuit installation,
    514–517
  number and types of, 188–206
  outdoor wiring installation, 545
  overcurrent protection device, sizing,
    205–207
  ratings, 205–207
  room air conditioning installation,
    507–509
  small-appliance, 187, 191, **192**
  small appliance installation, 480–481
  smoke detector installation, 510–512
  swimming pool installation, 528–540
  types of, 188–206

wall-mounted oven installation,
    485–486
  water heater installation, 499, **500**
  water pump installation, 495–498
Break lines, 165, 174, **174**
Bridging, 165, 183, 183
Building plans, 130–149. *See also*
  Framing
  blueprints, 165
  detail drawings, 169, **171**
  electrical drawings, 165, 169–171,
    172, 327
  elevation drawings, 165, 166–169, **169**
  floor plan, 165, 167, **168**
  legend, 165, 176
  line types, 173–174, **174**
  overview, 167–176
  plot plan, 165, 167, **167**
  scale, 166, 174–176, **175**
  schedules, 166, 171, **172**
  sectional drawings, 166, 169, **170**
  specifications, 166, 171–173, **173**
  symbols, 176, **177–180**
  title block, 174, **175**
Bundled conductors/cables, 187,
  201
Bundling, **202**
Burial depths for underground
  service conductors, 238–239,
  240, 527, **542**
Bypass diodes, **772**
Buried box, 700, 708
Burns
  arc blast, 8
  electrical, 7
Bus bar, 213, 215, 681–682, **682**
Bushings, 234, 240, 252
  bonding, **251**, **253**, 275, **275**
  for ENT, 411
  insulated, **241**, **246**, **261**, 268, **268**,
    275, 411
  for RMC, 401, **403**
BX cable. *See* Armored-clad (Type
  AC) cable

C

Cabinet, 37, 65
Cable, 44, 47–52, 362–390. *See
  also* Service entrance cable;
    Structured cabling; *specific
    types of cable*
  armored-clad (Type AC), 49–50, **49**,
    367–372, **371**

categories, voice and data cables, 564, 573–575, **575**
color coding, 566, **566**
connectors, **51**
continuity of, 304
defined, 37, 44
fish, 363, **374**, 384–385
home run, 363
installation, 362–390
installation preparation, 374–376
installation requirements, 364–374
installing cable runs, 376–380, **378–379**, 387–390
installing in existing walls and ceilings, 384–385, **385**
installing through metal framing, 380, **380**
installing through studs and joists, 377–379, **378–379**
metal-clad (Type MC), 49–50, **49**, 372–373, **373**
nonmetallic sheathed (Romex™), 47–49, **49**, 364–367, **365–367**
optical fiber cable, 561
pulling in, 363, 376
reel, 363, 374
run, 363, 376
run installation, 376–380, **378–379**, 387–390
run securing and supporting, 381–384, **381–384**
securing and supporting, 363, 381–384
selecting, 364
service entrance, 50–51, **50**, **51**, 229, 232, 262–266, 373–374
service entrance supports, 240–241, **242**
spinner, 375, **375**
triplex, 229, 231, 262, **263**
types of, 47–52, 364
underground feeder (Type UF), 49, **49**, 367
underground (URD), 279
unrolling, 374–375, **375**, 377
Cable cutters, 96, **97**, 98, **98**, 118
Cable diagram, 176, **182**
Cable hook, **261**, 263, **263**, 272, 274
Cable strippers, 89–90, **89**, 98, **98**
Cable ties, 70, **71**
CAD. *See* Computer aided drafting (CAD)
Cadmium telluride, 766
Cad-weld. *See* Exothermic welding
Calculations, single-dwelling calculation examples, 208–221
Canadian Standards Association (CSA). *See* CSA International And Intertek Testing Services

Carbon dioxide (CO_2) fire extinguisher, 29
Cartridge fuses, 64, **65**, 672, 673, **674**
requirements, 677–678
Category (of structured cable), 559, 561, 563–566
Cautions. *See* Safety
Ceiling fan. *See* Ceiling-suspended paddle fan
Ceiling joists, 165, 180, 183
Ceiling, suspended. *See* Suspended ceiling
Ceiling-suspended paddle fan, 342–343, 342, 594
electrical box as support for, 315, **316**
installation, 620–621, **621**, 631–633
swimming pool, 528–540
switches, installation, 470–472, **471**
Centerline, 165, 174, **174**
Central air conditioning, 505–506
Central heating, branch circuit installation, 509–510
CFLs. *See* Compact fluorescent lamps (CFLs)
CFM. *See* Cubic feet per minute (CFM)
CFR. *See* Code of Federal Regulations (CFR)
Channel-lock pliers, 88, 88
Charge controller
installation guidelines, 751, 753, **755**, 792
Checkout, 701–710. *See also* Troubleshooting
defined, 700
determining if electrical system is working properly, 701–710, **707–710**
determining if *NEC*® requirements are met, 701
Chime circuit, installation of, 514–517
Chisel, 95, 96
Chuck, 84, 100
Chuck key, 84, 100
Circuit, 4, 5. *See also* Branch circuits
basic electrical, **5**
Circuitry checklist for *NEC*® requirements, **702–703**
electrical, human body and, **5**
multiwire, 37

Circuit breakers, 63, **63**, **64**, 264, **265–266**. *See also* Overcurrent protection
arc fault circuit interrupter (AFCI), 63, **64**, 644, 674, 679–680, **680**
defined, 37
disconnects, 783, **783**
entrance panel installation, 285–287, **286**
ground fault circuit interrupter (GFCI), 63, **64**, **674**, 678–679, **679**
HACR, 505
installation, 680–684
installation safety rules, 681
loadcenter sizing and, 213–215
requirements, 678
single-pole, two-pole, and twin/dual, 673, **673**
standard sizes, 205
Circuit tracer, 700, 712–714, 726–727
Circular mils, 37, 45
Circular saw, 126–127
CIS. *See* Copper indium selenium (CIS)
Clamp-on ammeter, 134, 139, 140–141, **140**, **141**, 146, **146**
use of, 156
Clearances
service drop conductors, 235–238, **236–239**
working spaces and, 283–285, **285**, 287
Clips, for service entrance cable, 263, **263**
Clothes dryer
branch circuit installation, 492–494, **494**
receptacle testing, 718–721
Clothes washer. *See* Laundry
Coaxial cable, 559, 561. *See also* Video, voice, and data wiring
Code. *See* National Electrical Code® (*NEC*®)
Code of Federal Regulations (CFR), 16, **16**. *See also* Occupational Safety and Health Administration (OSHA)
Color coding, for PV system wiring, 776–777, **776**
Combination switches, 62, **62**, 445, **446**
defined, 443
Common terminal, 60

Communication circuits/wiring. *See* Video, voice, and data wiring

Communications outlet, 321–322, **322**

Compact fluorescent lamps (CFLs), 603–604, **604**

Compass saw, 93, **94**

Computer/data wiring system, 568–569. *See also* Video, voice, and data wiring

Concentric knockout, 228, 253, **253**

Conductors, 5, 44–41. *See also* Service conductors; Wiring
aluminum, 44, 73
ampacity of, 45, 198–205, **201–204**, 311, 674–675
applications in residential wiring, **46**
circuit sizing, 777–780, **778**
color coding for, 46, 309–310
copper-clad aluminum, 37, 44, 249–250
defined, 4
derating, 198
direct burial, 308–309, 367
direct sunlight, 309
disconnects, 783–785
dual ratings, 311
equipment grounding, 228, 240, 244
grounding, 785–791
"hot", identification of, 310, **310**, 451
human body as, 5, **5**
identification/markings, 309–310, **310**
insulated, 300, 309
insulation types, **47**, 311
maximum number per electrical box, 311, 329
minimum length, 306–307, **306**
minimum size, **199**, 235, **235**, 281, **284**, 308
minimum size calculations, **195**, **196**, 198
neutral, 228, 235, 244
overcurrent protection, 780–783, **782**
photovoltaic system, 776–777
requirements for, 235–242, **235–242**, 301, 308–311, 309–310
size of, 45, **46**, **199**
size requirements, 674, **675**
sizing, 777–780
temperature limitation, 309
temperature ratings, **199**, 198–200
terminal identification markings, **45**
types of, 44–51, 239–240, 208–211
underground, 308–309

Conduit, 52–54, 394. *See also* Raceways; *specific types of conduit*
bending, 413–416, **414–415**, 421
bending a 90-degree stub-up, 427–428
bending a back-to-back bend, 429–430
bending a three-point saddle, 434–436
bending an offset bend, 431–433
bending box offsets, 437–439
bends, types of, 414, **414**
box offset bend, 394, 414, **414**, 437–439
clamps, 266, **267**
conduit "LB," **261**, 268, **268**
connector fittings, 267
cutting, 412–413, **412**, 422–426
cutting, threading, and reaming tools, 412, **412**
defined, 394
electrical, overhead service using, 266–268, **267**
threading, 413–416, 424–426

Conduit bender, 94–95, **95**, 413
defined, 71

Connectors, 37, 41. *See also* Fasteners
wire connectors, **48**

Construction basics. *See* Framing

Continuity testers, 135, **135**, **136**
defined, 134
troubleshooting with, 712, **713**, 724–725
use of, 151

Cooking appliances. *See* Counter-mounted cook top; Electric range; Wall-mounted oven

Copper indium selenium (CIS), 766

Copper-clad aluminum conductors, 37, 44, 249–250

Cord-and-plug connections, 489, 490, 538

Cordless drills, 102–103, **103**

Counter-mounted cook top
branch circuit, 196–197, **197**
branch circuit installation, 485–486

Countertop receptacles, 317–318, **318**

Crescent wrench, 90, 91

Crimp, 84, 88

Critical loads, 522, 548, 549

CSA International And Intertek Testing Services, 39
label, **40**

Cubic feet per minute (CFM), 739

Current
defined, 4
flow, 5–6
human body and, **5**, 6, **6**

Cutter, 84

D

D (penny), 67, **67**

Data wiring system, 568–569. *See also* Video, voice, and data wiring

Day use only systems, 753, 755

Dehumidistats, 739, 739

Derating, 198

Detail drawings, 165, 169, 171

Deteriorating agents, 298, 299

Device(s), 54–62, 636–670. *See also* Device boxes; Receptacles; Switches
combination, 62, **62**
defined, 37, 54
installation, 636–670
overcurrent protection, 63–65
receptacles, 55–58
switches, 58–62

Device boxes, 40–44, 327. *See also* Electrical boxes; Receptacles; Switches
ceiling fan, 215, **216**, 342, **342–343**
defined, 37
ganging, 40, **41**, 72, **328**
handy box, 37, 41, **42**
handy box installation, 341, 350–351
installation, 312, **312**, 337–338, **338**, 347–349
lighting outlet box, 341
lighting outlet box installation, 314, **314**, **315**, 341–342, **341–343**, 352–354
metal, 40–42, **41–42**, 327–328, **328**
metal installation, 311, **311**, 338–340, **339–340**, 347–349
new construction, installation in, 347–349
new work boxes, 42, 328, **329**
nonmetallic, 42–43, **42**, 327, **344**
nonmetallic installation, 312, **312**, 337–338, **338**, 347–349
old work boxes, 40, 328, **329**
old work boxes installation, 343–345, **344–345**, 355–359
outlet and junction boxes, 43–44, **43–44**
securing cable at, 381, **381–382**
selecting of, 328, **328–329**, **331–332**
special boxes for heavy loads, 44, 342
utility box, 38, 41, **42**

Diagonal cutting pliers, 88, **88**

Die, 84, 115

Digital meters, 134, 138, 139

Digital multimeter. *See* DMM (digital multimeter)

Digital voltage testers, 137, 137

Digital voltmeters, 138, 139
  use of, 154–155

Dikes, 88, **88**

Dimension, 165, 167

Dimension line, 165, 174, 174

Dimmer switches, 61, 61, 445, 445, 651
  defined, 443
  installation, 468–470, **469–470**

Diode, 751, 771

Dioxins, 734, 746

Direct current system, 753–754

Directories. *See* Green Book; Orange Book; White Book

Disconnect. *See* Disconnecting means

Disconnect switch, 66, **66**, **243**, 802, **802**

Disconnecting means, 242–243, **243**
  accessibility of, 24, **243**
  defined, 37
  main service, 264, **265–266**
  maximum number of, 243
  photovoltaic system disconnect, 783–785, **783–784**
  service disconnect, 187, 207
  switches, 60, 66, **66**

Dishwasher, branch circuit installation, 489–490, **489**

Distribution center. *See* Service center

DMM (digital multimeter), 144–145, 145
  defined, 134
  use of, 157–160

Door bell. *See* Low-voltage chime circuit

Door chime switches, 62, **62**

Double insulated
  defined, 4
  power tools, 17, 99

Double plate, 180

Double-pole switches, 60, **60**, 445, **444**. *See also* Switches
  defined, 443
  double-pole, double-throw switches, 455, **456**
  installation, 466–468, **468**

Draft-stops, 165, 180, **183**

Drawings. *See* Building plans

Drills, 100–103, 119–125
  cordless, 102–103, **103**
  hammer, 102, **102**, 124–125
  pistol grip, 100–101, **100–101**, 119–120
  right-angle, 101–102, **101**, 121–123

Drip loops, 228, 231, 242, **242**

Drive studs, 67, 68

Dropped ceiling. *See* Suspended ceiling

Dry chemical fire extinguishers, 29

"Dry locations," 300

Dry-niche luminaire, 522, 529, 531

Dryer, electric
  branch circuit installation, 492–494
  receptacle testing, 718–721

Dual-element time-delay fuses, 477, 497

Duplex receptacle, 55–56, **56**, 638, **638**. *See also* Receptacles
  defined, 637
  installation, 461–466, 647, 658–665
  parts of, **56**
  push-in terminals, 641, **641**
  switched duplex receptacle circuits, 461–466, **463–465**

Durability, 745

Dwelling unit, 187, 188
  lighting outlet requirements, 320–321, **321**
  receptacle outlet requirements, 311, 316–320
  single-dwelling calculation examples, 208–221
  wire sizes, 207

E

Eccentric knockout, 228, 253, **253**

Eddy currents, 147

Edison-base plug fuse, 64, 64, 672, 673, 674
  examples of, **677**
  requirements, 676–677

Efficacy, lamp, 594, 595, 596

Electric baseboard heating installation, 501, **502**, **503**

Electric clothes dryer
  branch circuit installation, 492–494
  circuit breakers and, 683

Electric furnace installation, 501

Electric heating, branch circuit installation, 415–418, 416–418

Electric power generation, on-site, 742–744, 743–744

Electric range, 394, 397
  branch circuit, 194, **195**, **196**, **197**, 198, **199**
  branch circuit installation, 481–483
  receptacle testing, 718–721

Electric water heater, 737–738, **737–738**. *See also* Water heater
  branch circuit installation, 499, **500**
  heat pump water heater, 738, **738**
  tankless, 738, **738**
  tank type, 738, **737**

Electrical Appliance and Utilization Equipment Directory (Orange Book), 39, **40**

Electrical boxes, 40–44, 311–321, 325–361. *See also* Device boxes; Ganging; Receptacles; Switches
  abrasion protection, 311
  accessibility or wiring, 315–316
  back-to-back, 307, **308**
  box extender, 647–648, **648**
  box fill, 311, 326, 329, **331–332**, **335**
  ceiling fan supported by, 315, **316**, 342–343, **342–343**
  components, 329
  existing walls and ceilings, installation in, 343–345, **344–345**
  ganging, 40, **41**, 72, **328**
  handy box masonry surface installation, 351
  handy box wood surface installation, 350
  installation in new construction, 347–349
  installation procedures, 325–361
  installation requirements, 311–321, **311–316**
  junction box, 37, 43–44, **44**, 327
  junction box installation, 341–343, **341–343**, 352–354
  lighting outlet box, 341, **341**
  lighting outlet box installation, 314, **314**, **315**, 341–343, **341–343**
  luminaires support by, 314, **315**
  maximum number of conductors, 311, 329–332, **333**
  metal, 40–42, **41–42**, 311, 327, **331–332**
  metal installation, 311, **311**, 338–340, **339–340**, 347–349
  mounting heights, 336

Electrical boxes (*continued*)
  mounting/set back from finished
    surface, 312–314, **313–314**
  new-work, 328, **329**
  nonmetallic, 42–43, **42**, 327
  nonmetallic installation, 312, **313**,
    337–338, **338**, 347–349
  old-work, 328, **329**
  old-work installation, 343–345,
    344–345, 355–359
  outlet boxes, 40, 43–44, **43–44**
  outlet boxes, adjustable bar hanger
    installation, 353–354
  outlet boxes, side-mounting bracket
    installation, 352
  requirements for, 307, **307**, **308**
  securing and supporting cable/raceway,
    311–312, **311**
  selection guide, **333**
  selection of, 328, **328–329**
  set back from finished surface,
    312–314, **313**, 337, 647–648
  setout, 326, 287
  sizes of, 327
  sizing of, 329–337, **330–337**
  support of, 312–314, **313**, **314**
  volume, total, 329, **331**
  walls and ceilings, installation in,
    343–345, **344–345**, 356–359
Electrical cabling diagram, 176,
  182, 701
Electrical circuit, 4, 5
  basic, **5**
  human body and, 5–6, **6**
Electrical conduit, overhead
  service using, 266–268, **267**
*Electrical Construction Equipment
  Directory* (Green Book), 39, 40
Electrical controls
  exhaust fan, 739
  installation of, 740–741
  lighting control equipment, 740
  phantom loads, 740–741, **741**
  programmable thermostat, 740, **740**
Electrical drawings, 165, 169–171,
  172, 327
Electrical equipment, installing,
  735–740
Electrical hazards, 5–8, 805. *See
  also* Hazards
Electrical metallic tubing (EMT).
  See EMT (Electrical metallic
  tubing)
Electrical nonmetallic tubing
  (ENT). *See* ENT (Electrical
  nonmetallic tubing)

Electrical panel, working space
  and, 11–12, **13**
Electrical shock, 4, 5–7, 32–33
  insulated tool requirements, 98
  from telephone line, 572
Electrical symbols, 176, **178–182**
Electrical system checkout,
  701–710, 707, 709
Electrical Testing Laboratories
  (ETL), 606. *See also* Intertek
  Testing Services (ITS)
Electrical wiring symbols, 176, 182
Electrician's bible. *See National
  Electrical Code®* (*NEC®*)
Electrician's hammer, 91, 91
Electrician's knife, 89–90, 90
  use in stripping insulation, 110–112
Electricity, 5–7
  safety rules for, 28
Elevation drawings, 165, 167–169,
  169
EMT (Electrical metallic tubing),
  52, 53, 395, 403–404, 404
  bends in/bending, 403, 414–416,
    414–415
  common use of, 395
  complete system installation, 404
  connectors and couplings, 417, **417**
  couplings and connectors, 417
  cutting/reaming, 412
  as equipment-grounding conductor,
    403, 404
  fittings and, **404**
  installation, 414–415, **414–415**
  installation requirements, 403
  number of conductors, 403
  raceways, 403
  securing and supporting, 404
  thinwall, 394
Enclosure
  meter, 228, 232
  service entrance, **261**, 264
Energized circuits, safety and, 7,
  98, 149
Energy efficiency, 735
Energy efficient appliances, 736
Energy guide label, **736**
Energy Recovery Ventilators
  (ERVs), 740
Energy Star, 734, 735, **735**
Energy Star rated models, 736, **736**
ENT (Electrical nonmetallic
  tubing), 53–54, **54**, 410–411

  abrasion protection, 411
  bends in, 410
  cutting/cut ends, 411
  fittings and, 411, **411**
  installation requirements, 410–411
  number of conductors, 410
  securing and supporting, 411, **411**
Equipment. *See also* Service
  equipment
  defined, 228
  grid-tie PV system installation, 786,
    787
  grounding, 785
  listing and labeling, 705
  personal protective, 7, 18–20, **18–19**,
    270
  stand-alone PV system installation, 786,
    786
Equipment-grounding conductors,
  240, 244. *See also* Grounding
  defined, 228
EMT as
  identification/marking of, 309–310
  RMC as, 404
ERVs. *See* Energy Recovery
  Ventilators (ERVs), 740
ETL (Electrical Testing Laborato-
  ries). *See* Intertek Testing
  Services (ITS)
Exam, licensing, 238
Excavating, OSHA requirements,
  17–18, 279
Exhaust fan, 739–740
Exothermic welding, 522, 532, 533
Expansion fittings, 409–410, **410**
Extension lines, 165, 174, **174**
Eyebolt. *See* Cable hook

F

F-type connector, 559, 562, 563,
  570
  installing on an RG-6 coaxial cable, 579,
    580
Fall protection, 19–20, **19**
Fan. *See* Ceiling-suspended paddle
  fan
Fasteners, 66–71, **66–71**
  anchors, 67–69, **68**
  installing lead (caulking) anchor, 77
  installing plastic anchors, 78
  installing tapcon masonry screws,
    79–81

installing toggle bolts, 75–76
nails, 67, **67**
nuts and washers, 69–70, **70**
screws, **69**, 69
tie wraps, 70–71, **71**
Feed-through, 712, **713**
Feeder, 207, 215, 288
defined, 187
detached garage, 527
sizing, 207, 215
Field bend, 394, 409, **409**
Files, metal, 95, 96
Fire, 28–30
classes of, 29, **29**
components of, 28, **28**
extinguishers, 28–30, **29**
fire-resistance rating, 307, **307**
wiring methods and, 307, **307**
Fire alarm systems. *See* Smoke
detectors
Fire extinguishers, 28–30, **29**
Fire-stops. *See* Draft-stops
Fish, 363, **374**, 384–385
Fish-eye trim ring, 611, 612,
613
Fish tape, 384–385, 419–420, **420**
metal vs. nonconductive, 421
and reel, 94, **94**, 420
Fittings
cable, 51, **51**
defined, 37
expansion, 409–410, **410**
RMC (Rigid metal conduit), 396, **396**
Flashing, 745
defined, 734
roof, 745
wall, 745
Flexible cords, 414–415
Flexible metal conduit. *See* FMC
(Flexible metal conduit)
Floor area, habitable, 188
Floor boxes, 315, **316**
Floor joists, 165, 183, **183**
Floor plan, 165, 167, **168**
Fluorescent lamps, 594, 600–604,
601–604, 604. *See also*
Lighting
FMC (Flexible metal conduit), 53,
53, 395, 404–406
common applications, **405**
cutting/cut ends, 405
"Greenfield," 46, 404
as grounding means, 406, **406**
installation requirements, 404
number of conductors, 405

raceways, 404
securing and supporting, 405
Foam fire extinguishers, 30
Folding rule, 91–92, **91**
Footing, 165, 183, **183**
Footwear, 19, 270
Forming shell, 522, 529, **532**,
537
Forstner bit, 101, **101**
Foundation, 165, 183
Four-way switches, 61, **61**, 459,
459, 651
defined, 443
installation, 457–461
switching circuits, 459–461, **460**
ungrounded circuit conductor and, 446
Foyer, 298, 320
Framing, 180, **183**. *See also*
Building plans
balloon frame, 165, 180, **183**
band joists, 165, 183, **183**, 272
bottom plate, 165, 180, **183**
bridging, 165, 183, **183**
ceiling joists, 165, 180, **183**
defined, 165
draft-stops, 165, 180, **183**
floor joists, 165, 183, **183**
footing, 165, 183, **183**
foundation, 165, 183
girders, 165, 183, **183**
metal framing members, cable
installation, 380, **380**
metal framing members, wiring and,
302, **302**
platform frame, 165, 180, **183**
rafters, 166, 180, **183**
residential basics, 180, **183**
sheathing boards, 166, 183, **183**
sill, 166, 183, **183**
subfloor, 166, 183, **183**
symbol, 176
top plate, 166, 180, **183**
wall studs, 166, 180, **183**
wood framing members, device box
installation and, 337–340, **338–340**
wood framing members, wiring and,
301–302, **301**
Freestanding towers, 799, **799**
Fuel cell, 751, 752
Fusible safety switch, 783, 783
Furnace. *See* Central heating;
Electric heating
Furring strip, 298, 303
Fuse puller, 97, 97
Fuses, 63–65, 64, 65, 673, 674.
See also Overcurrent protection

cartridge fuse, 64, **65**, 672, 673, **674**
cartridge fuse requirements, 677–678
defined, 37
Edison-base plug fuse, 64, **64**, 672,
673, **674**, **677**
Edison-base plug fuse requirements,
677
"maximum size fuse" on nameplate,
505
plug fuse requirements, 676–677
plug fuses, 64, **64**, 672, 673, **674**
standard sizes, 205
time-delay, 64, 551, 673
Type S plug fuse, 64, **64**, 672, 673,
674, **677**
Type S plug fuse requirements, 677

G

Ganging, 40. *See also* Device
boxes; Electrical boxes
defined, 37
metal device boxes, 40, **41**
one-, two-, and three-gang device
boxes, **328**
procedures, 72
Garage, 523–527
attached, 523–526
branch circuit installation, 523–527
detached, 526–527
electrical plan, **524**
lighting outlets, 320, 526, 527
receptacle outlets, 320, 525
Garbage disposal, branch circuit
installation, 525
*General Information for Electrical
Equipment Directory* (White
Book), 39, 40
General lighting circuits, 187,
188–190, 189. *See also*
Lighting
General purpose branch circuit,
187, 188, 189
Generators, 522, 546, 547. *See
also* Standby power system
GFCI. *See* Ground fault circuit
interrupter (GFCI)
Girders, 165, 183, 183
Gloves, 19, 270
Green, 734, 735
Green Book (*Electrical Construction
Equipment Directory*), 39, 40
Green product selection,
745–746

Green house wiring practices, 733–748
air barrier, 741–742
appliances, electric, 736
avoiding holes in air barrier, 742
durability, 745
energy efficiency, 735
equipment, electrical, 737–740
exhaust fan, 739–740, **739**
green product selection, 745–746
home owner education and reference manual, 747
installation instructions, manufacturer's, 739
installing electrical controls, 740–741
installing electrical equipment, 735–740
job site recycling program, 747
lighting fixtures installation, 742
lighting, 735–736
on-site electric power generation, 742–744, **743–744**
recycling and waste management, 746–747
ventilation fan, 739–740, **739**
waste management, 745

Greenfield, 53, 333. *See also* FMC (Flexible metal conduit)

Greenlee Textron, 68

Grid-tie inverters, 774, **775**

Grid-tie systems, 755–757, **759**
equipment/system grounding in, **787**
installation guidelines for, 791–792
NEC® requirements, 790–791
PV circuit types in, **778**

Ground, defined, 228. *See also* Grounding

Ground fault, 63, 299, 710. *See also* Grounding
defined, 37, 700
example of, **712**
troubleshooting, 710–714, **713**, 724–725

Ground fault circuit interrupter (GFCI), **13**, 63, **64**, **674**, 678–679, **679**. *See also* Grounding
defined, 4
operation of, **643**
receptacle, 57, **57**, 637, 642–644, **643**, 647
receptacle installation, 647, 648, 666
requirements for, 13, **703**
tamper-resistant receptacles, 642
testing of, 644, 678–679

Ground fault protection, 786–787

Ground mount systems, 768, **769**

Ground resistance meters, 44, **144**

Grounded conductor, 228, 244. *See also* Grounding
continuity of, 650–651

Grounded, defined, 228. *See also* Grounding

Grounding, 14–15, 243–253, **244**, 785–789, **785–789**
bonding, 244, **248**, 250, **251**, 251, **253**
checklist of *NEC*® requirements, 706
concrete-encased electrode, 245, **247**
connection points, **244**
definitions, 4, 228
electrodes permitted for, 243–253, 788
equipment, 785, **786**
equipment-grounding conductors, 228, 240, 244
ground fault protection, 786–787
ground ring as electrode, 245, 247–249
grounded conductor, 228, 244, 650–651
grounding electrode conductor, 228, 243
grounding electrode conductor connection methods, 251, **253**
grounding electrode conductor installation, 249–250, 287, **287**, **289**
grounding electrode conductor size, 249, 250, **252**
grounding electrode system, 244
grounding electrode system installation, 247–250
grounding electrodes permitted, 244–249
main bonding jumper, 228, 244, **245**
metal enclosures and raceways, 250, **251**, 251, **253**
metal frame use in, 245
module and rack, 785, **785**
neutral conductor, 228, 244
non-current carrying metal parts, 251–252, **253**
plaque requirements, 788–789
plate electrodes, 247, 248
requirements for residential service, 13, 243–253, **244**, 701
rod and pipe electrodes, 247, 248, **249**
"S" loops, 240
sizing equipment grounding conductor, 787–788
supplemental electrode, 2, 229, 233, **248**, **249**, **289**
surge suppression, 788, **789**
system, 786, **786**
three-prong electrical plug, 14, **14**
water pipe as electrode, 245, **246**, 248, 249

Grounding conductor, defined, 228. *See also* Grounding

Grounding electrode, 788. *See also* Grounding; Grounding electrode conductor
concrete-encased electrode, 245, **247**
defined, 228
ground ring, 245, 249
metal frame, 245
permitted electrodes, 245–249
plate electrodes, 247, 248
rod and pipe electrodes, 247, 248, **249**
sizing, 788
supplemental, 229, 232, **248**, 249, **249**, **289**
water pipe, 245, **242**, **248**, 249, 287, **287**, **289**

Grounding electrode conductor, 228, 243. *See also* Grounding
connection to grounding electrode, 250–251, **253**
installation, 249–250, 287, **287**, **289**
size, 249, 250, **251**

Grounding electrode system, 244, 247–250

Guy, 751

Guy wire, 277

Guyed towers, 799, **800**

H

Hacksaw, 93–94, **94**, 412
use procedure, 116

Hallway outlets, 320

Hammer, 91, **91**

Hammer drills, 102, **102**
use procedure, 124–125

Hand lines, 20, **22**

Hand tools, 85–92. *See also* Tools

Handy boxes, 41, 42, 331–332
defined, 37
installation, 341, 350–351

Hard hats, 18, 270

Harmonics, 134, 145

Hawkbill knife, 90, **90**

Hazard. *See also* Safety
defined, 4
electrical shock, 5–7, 787
in photovoltaic system, 792–794
physical, 792–793, **793**

"Heads", lighting, **622**

Heat pump, 477, 504

Heat Recovery Ventilators (HRVs), 740

Heating
  electric, branch circuit installation, 500–504
  gas and oil, branch circuit installation, 509–510
Hex key, 96, **97**
HID. *See* High-intensity discharge (HID) lamps
Hidden lines, 165, 174, 174
High-intensity discharge (HID) lamps, 594, 604–605, 604. *See also* Lighting
Hinge terminal, 60
Hole saw, 101, 101, 122–123
Home monitoring system, 560
Home run, 363
Horizontal cabling, 559, 563
Home, green,
  owner education and reference manual, 747
Horizon, **763**
"Hot" conductors, identification of, 310, **310**, 450
Hot tub. *See* Spas and hot tubs
HRVs. *See* Heat Recovery Ventilators (HRVs)
Human body. *See also* Personal protective equipment
  arc burns and, 7
  current flow/safety considerations, 5–6, **6**
  electrical burns and, 6–7
  electricity and, 5–6, **5**
Hybrid system, 751, **758**
  in parallel configuration, 758, **762**
  in series configuration, 758, **761**
Hydraulic, defined, 84
Hydromassage bathtubs, 552, 529, **531**, 540, **540**. *See also* Swimming pools

I

I-V curve, 766, **767**
IDC (Insulation displacement connection), 559, 561
IMC (Intermediate metallic conduit), 396, 402
  cutting, 412, **412**
  number of conductors, 402
  raceways, 402

Impulse, 637, 645. *See also* Receptacles
In-line ammeters, 134, 139–140, **139**
Incandescent lamps, 594, 596–598, **598**, **604**. *See also* Lighting
  installation, polarity and, 616
  safely removing a broken, 707
Individual branch circuit, 194–198, **194**
  clothes dryer, electric, 198
  cooking appliances, electric, 195–198, **195**, **196**, **197**
  defined, 187
  examples of, **195–197**
"Indoor use only," 300
Inductive heating, 298, 300
Injury. *See* Safety
Insolation, solar, 751, 760
Inspections, 17, 255
Instructions
  listing or labeling, 299
  manufacturer's installation, 606, 616
Insulated, defined, 37
Insulated bushings, **241**, **246**, **261**, 268, **268**, 275, 411
Insulated conductors, 234–235, 300, 309
Insulated tools, 98, **98**
Insulation displacement connection (IDC) *See* IDC (Insulation displacement connection)
Insulation, integrity of, 299
Insulators, 4, 5
Interconnecting (of smoke detectors), 477, 511
Intermediate metallic conduit (IMC). *See* IMC (Intermediate metallic conduit)
Interrupting rating, 672, 675–676
Intertek Testing Services (ITS), 35
Inverse time circuit breakers, 477, 497
Inverter, 751, 753, **756**, 774–776, **775**
  installation guidelines, 751, 753, **755**, 792
Irradiance, solar, 751, 760
Isc. *See* Short circuit current (Isc)

J

J-boxes. *See* Junction boxes
J-hooks, 574, **574**
Jack, 559, 563, 567, 569. *See also* Video, voice, and data wiring
Jet pump, 477, 495–498
Job site
  recycling program, 747
  safety rules, 794, **794**
Jug handles, 363, 365, **369**
Junction boxes, 37, 43–44, 44, 327. *See also* Device boxes; Electrical boxes
  defined, 37
  installation, 341–343, **341–343**, 352–354
  nonmetallic, 43, **43**
  swimming pool, installation, 537
  telephone, 568, **569**

K

Kcmil, 197
Keyhole saw, 93, **94**, 357
Kilowatt-hour meter, 146–147, **146**, **147**
  defined, 134
  reading, 161
Kitchen receptacle outlets, 316–318, **318**
Knife, electrician's, 89–90, 90, 110–112
Knockout (KO), 37, 41, 41
  concentric/eccentric, 228, 253, **253**
Knockout plug, 298, 300, **300**
Knockout punch, 84, 93, 93
  use procedure, 114–115
KO. *See* Knockout (KO)

L

Ladders, 20, **22–24**, 270
Lamps. *See* Lighting
Lateral (service lateral), 229, 233, **280**
Laundry
  branch circuit, 187, 191, **192**
  branch circuit installation, 490–491, **492**
  receptacle outlet, 320

Lead (caulking) anchors, 67, **68**
  installation, 77
Leader (line), 165, **174**
LED lamps, 594, 599, **599**
Legend, 165, 176
Levels, 84, 95, **95**
LFMC (Liquid-tight flexible metal conduit), 54, **54**, 395, 406–407
  common applications, **407**
  fittings and, **407**
  as grounding means, 407, **408**
  installation requirements, 406
  number of conductors, 406
  securing and supporting, 407
LFNC. *See* Liquidtight flexible nonmetallic conduit (LFNC)
Licensing exam, 238
Lifting, 20, **21**
Light switch. *See* Switches
Lighting, 592–635. *See also* Luminaires; Switches
  accent lighting, 606
  ballast, 594, 600–602, **601–602**
  basics of, 595–596, **596, 597**
  branch circuit, 187, 188–190, **189**
  branch circuit installation, 478–479
  calculations for circuits, 190
  ceiling-mounted, 342, 612, **614**
  ceiling-suspended paddle fan installation, 620–621, **621**, 631–633
  chandelier fixtures, 612, 614, **615**
  characteristics/comparison of lamps, **604**
  circuit, testing, 708
  color, 595, **596**
  color rendition, 594, 595
  color temperature, 594, 595, **597**
  compact fluorescent lamps (CFLs), 603–604, **604**
  control equipment, 740
  direct-mount, 612, **614**
  dwelling unit required outlets, 320–321, **321**
  electromagnetic spectrum, 595, **596**
  fixtures weighing more than 50 pounds, 342
  fluorescent lamps, 594, 600–604, **61–604, 604**
  general lighting, 606
  halogen lamps, 495
  "heads", **622**
  high-intensity discharge (HID) lamps, 594, 604–605, **604**
  incandescent lamps, 594, 596–598, **598, 604**
  installation, 614–622, **617–622**

installation, direct connection to ceiling, 617, **617–618**, 626
installation, direct connection to outlet box, 616–617, **616**, 625
installation, electrical equipment, 735–736
installation in bathtub and shower areas, 608, **610**
installation in clothes closet, 607–608, **608–610**
installation in dropped ceilings, 618–619, 629–630
installation instructions, manufacturer's, 606, 616
installation, outdoor luminaires, 551, 620, **620**
installation, safety considerations, 614–616
installation, strap to outlet box, 617, **618**, 627
installation, stud and strap connection to outlet box, 618, 628
label, lighting fixture, 606
lamp efficacy, 594, 596
LED lamps, 594, 599, **599**
lumen, 594, 595–596
mercury vapor lamps, 605
metal halide lamps, **604**, 605
minimum number of circuits, 190
minimum requirements for, 320
pendant fixtures, 614, **616**
performance of lighting system, 595
pole light installation, 545
recessed luminaires, 608–612, **611–613**
recessed luminaires installation, 618, **619**
sconces, 594, 612, **614**
security lighting, 606
selecting appropriate, 605–614, **608–614**
sodium vapor lamps, **604**, 605
surface-mounted luminaires, 612, **614–616**
task lighting, 606
track lighting installation, 621–622, **621–622**
troffer, 594, 619, 629–630
Type IC light fixture, 594
Type Non-IC light fixture, 594
types of lamps, 596–605, **598–604**
volt-amperes, 188
wall-mounted, 612, **614**
wall-switch controlled, 320
Lighting outlets, 341
  ceiling fan and, 315, **316**, 342–343, **342–343**
  defined, 326
  electrical box installation, 314, **315**, 341–343, **341–343**, 352–354
  garage, 320

requirements for dwelling units, 320–321, **321**
  sizing, 341
  suspended ceiling, installation in, 342–343, **342–343**
  symbols, **178**
  testing, 707–708
Line side, **261**, 273
Line terminals, 60
Line types in building plans, 173–174, **174**
Lineman pliers, 87, **87**, 108
Liquidtight flexible metal conduit (LFMC). *See* LFMC (Liquid-tight flexible metal conduit)
Liquidtight flexible nonmetallic conduit (LFNC), 54, **54**, 777
Listing and labeling, equipment, **705**
"Live" circuits
  identification of, 310, **310**
  safety and, 7, 98, 149
Load (electrical), defined, 4
Load side, **261**, 273
Load sizing, 190
Load terminals, 57, 60, **60**
Loadcenters, 65–66, **65–66**. *See also* Service panel
  defined, 37, 65
  feeders, 207, 215
  main breaker panel, 213, **214**
  main lug only (MLO), 213, **214**, 215
  panelboards, installation, 285–287, **286–287, 289**
  service panel trim-out, 671–696
  sizing, 213–215
  starting cable run from, 380, 387–390
  subpanel, 65–66, **66**, 215, 288–289, **290**
Lockout/tagout, 17, **17**, 32
Long-nose pliers, 87–88, **88**
Low-voltage chime circuit
  installation, 514–517
  switches, 63, **63**, 651
Low-voltage switches, 63, **63**, 651
Lugs, 187, 213
Lumen, 594, 595–596
Luminaires, 595. *See also* Lighting
  defined, 432, 594, 595
  dropped ceiling installation, 618–619, 629–630
  dry-niche, 522, 529, **531**, 537
  forming shell, 522, 529, **532**, 537

no-niche, 522, 529, 531
outdoor installation, 551, 620, **620**
recessed, 608–612, **611–613**
recessed installation, 618, **619**
sconces, 594, 612, **614**
support by electrical box, 314, **315**
surface-mounted, 612, **614–614**
swimming pool installation, 535, **536**, 537
type IC, 594, 611, **612**
type Non-IC, 594, 611, **612**
wet-niche, 522, 529, **531**, 537

M

Madison hold-its, 326, 344, **345**, 356, 357
Magnetic declination, 751, 760, **764**
Main bonding jumper, 228, 244, **245**, 287. *See also* Grounding
Main breaker panel, 213, **214**
Main lug only (MLO), 213, **214**, 215, 288
Manual ranging, 134, 138
Marking tape, 279
Masonry screw anchors, 68, **68**
Mast. *See* Service mast
Material handling, 20, **21–22**
Material Safety Data Sheet (MSDS), 4, 25–28
Mbps (Megabits per second), 559, 563
Maximum power point (MPP), 751, 766
Measuring tools, 91–92, **91**
Mechanical safety hazards, 804
Megabits per second (Mbps), 559, 563
Megahertz (MHz), 559, 563
Megger®, 143, **143**
Megohm (MΩ), 143
Megohmmeter, 143, **143**
    defined, 134
Metal-clad (Type MC) cable, 49–50, **49**, 299
    in attics, 373
    bends in, 373
    fittings, 373
    horizontal installation through framing members, 373
    installation requirements, 372–374
    protection of, 372–373
    securing and supporting, 373, **373**
    unsupported situations, 373, **374**
Metal device boxes, 40–42, **42–43**, 327. *See also* Receptacle; Switches
    depth gauge, 338, **339**
    ganging, 40, **41**, 72
    installation, 311, **311**, 338–340, **339–340**
    installation, step-by-step instructions, 347–349
    old work boxes, installation in sheetrock wall or ceiling, 357
    parts of, **41**
    selection guide, **333**
Metal files, 95, **96**
Metal framing members
    cable installation, 380, **380**
    wiring and, 302, **302**
Meter enclosure, 228, 232
Meter socket, **262**, 263, 270
    combination main service disconnect, 264, **266**
    connections for typical three-wire overhead service, 273, **273**, 276, **276**
    connections for typical three-wire underground service, **276**
    overhead service with service entrance cable, 270, **271**, **272**, **273**
    overhead service with service entrance mast, 276, **276**
    overhead service with service entrance raceway, 273, **276**
    with threaded hub, **262**, 263, 270, **271**
    underground service, 278, **278**
Meters and test instruments
    ammeters, 139–141, **140**, **142**, 156
    ammeters, clamp-on, 134, 139, 140–141, **140**, **141**, 156
    ammeters, in-line, 134, 139–140, **140**
    analog meter, 134, 138, **138**
    auto-ranging, 134, 138
    calibration, 148, **148**
    care and maintenance, 149
    clamp meter, 146, **146**
    continuity testers, 134, 135, **135**, 151
    digital, 134
    digital voltage testers, 137, **137**
    digital voltmeters, 138, **139**, 154–155
    DMM (digital multimeter), 134, 144–145, **145**, 157–160
    ground resistance meters, 44, **144**
    harmonics, 134, 145
    kilowatt-hour meter, 134, 146–147, **146**, **147**, 161
    manual ranging, 134, 138

Megger®, 143, **143**
megohmmeter, 134, 143, **143**
multimeters, 134, 144–146, **144**, **145**, 127–160
noncontact voltage tester, 134, 137, **138**, 153
ohmmeters, 134, 141–142, **142**
safety and, 148–149, **148**
testing receptacle outlets, 707, 716–717, 718–721
true RMS meter, 134, 145, **145**
voltage testers, 134, 136–137, **136–137**, 152–153
voltage testers, use in electrical system checkout, 707, 716–717, 718–721
voltmeters, 134, 137–139, **138**, **139**, 154–155
VOM (volt-ohm millimeter), 134
watt-hour meters, 146–148, **146**, **147**, 161
Wiggy, 134, 136, **136**
MHz (megahertz), 559, 563
Micro-hydro electric systems, 751, 752, 734, 744, **744**
Micro-invertors, **757**
Mil, 37
Module, photovoltaic, 751, 753, **754**
Molly bolts, 67, **68**
Momentary-contact switches, 62, **62**
MPP. *See* Maximum Power Point (MPP)
MSDS (Material Safety Data Sheet), 4, 25–28
Mud rings, 573
Multimeters, 144–146, **144**, **145**, **146**
    clamp meter, 146, **146**
    defined, 134
    DMM (digital multimeter), 134, 145, **145**
    true RMS meter, 134, 145, **145**
    use of, DMM, 157–160
    volt-ohm millimeter (VOM), 134, 144
Multiwire circuit, 46, 139
    defined, 37, 134

N

Nails 67, **67**
Nacelle, 751
Nameplate, 477, 486

National Electrical Code® (NEC®), 8–13. *See also NEC® Index*
applications, 13–15, **13**
defined, 4
determining if requirements are met, 702–703
finding information in, 31
history and development of, 8
layout of, **9**
licensing exam questions and, 238
organization of, 9–13, **9**
photovoltaic modules and, **767**
photovoltaic system wiring and, 776–791, **776–778, 782–789**
requirements, 789–791
scope of, 12
terms and definitions, 12
National Electrical Manufacturers Association (NEMA), 39
National Fire Alarm Code (NFPA 72), 510
National Fire Protection Association (NFPA), 7, 16
Nationally recognized testing laboratories (NRTL), 39–40, 705
NEC®. See National Electrical Code® (NEC®)
Needle-nose pliers, 87–88, **88**
NEMA. *See* National Electrical Manufacturers Association (NEMA)
Net metering system, 755, 757, **760**
Neutral conductor, 228, 235, 244
New work boxes, 35, 328, **329**
defined, 39
Nipple, 187, 202
NMSC. *See* Nonmetallic sheathed cable (NMSC)
No-niche luminaire, 522, 529, 531
Noncontact voltage tester, 134, 137, **138**
use of, 153
Nonlinear loads, 134, 145
Nonmetallic device boxes, 42–43, **42**, 327. *See also* Device boxes; Electrical boxes
installation, 312, **313**, 337–338, **338**
installation in sheetrock wall or ceiling, 358–359
installation, step-by-step instructions, 347–349
outlet and junction boxes, 43–44, **43–44**
special boxes for heavy loads, 44

Nonmetallic sheathed cable (NMSC), 47–49, **49**, 299, 301–305, **301–307**
ampacity of, 48
bends in, 365
color coding, 364
exposed, **365**
flat-type, 366, **370**
horizontal runs through framing members, 366, **371**
installation requirements, 364–365, **365**
jug handle, 365, **369**
non-supported situations, 366
protection of, 364–365, **367**
Romex™ trade name, 38, 47, 299
running boards and, 365, **366**
securing and supporting, 311, 366, **366**
types of, 47–49
in unfinished attic and basement, 365
NRTL. *See* Nationally recognized testing laboratories (NRTL)
Nut drivers, 93, **93**
Nuts and washers, 69–70, **70**

O

Object line, 165, 173, **174**
Occupational Safety and Health Administration (OSHA), 4, 16–18
Act of 1970, 16
electrical safety standards, **16**
initial inspections, tests, or determinations, 17
insulated tool requirement, 98
ladder regulations, 20
lockout/tagout, 17
personal protective equipment, 18–20
power tool regulations, 17
safety training, 16–17
scaffolding regulations, 21–24
trenching and excavating regulations, 17–18, 279
OCPD. *See* Overcurrent protection device (OCPD)
Offset bend, 414, **414**, 438–439
Ohm, defined, 4, 6
Ohmmeters, 141–142, **142**
defined, 134
Ohm's law, 4, 6
Old work boxes, 35, 328, **329**
defined, 38
installation, 343–345, **344–345**, 355–359

installation in sheetrock wall or ceiling (metal boxes), 357
installation in sheetrock wall or ceiling (nonmetallic boxes), 358–359
installation in wood lath and plaster wall or ceiling, 355–356
types of, 343–345, **344–345**
"Ondemand" water heaters, 738, **738**
Open circuit, 134, 139, 142, 710
defined, 700
example of, **711**
Open circuit voltage (Voc), 751, 766
Optical fiber cables (FO), 463
Orange Book (*Electrical Appliance and Utilization Equipment Directory*), 39, 40
OSHA. *See* Occupational Safety and Health Administration (OSHA)
Outdoor lighting installation, 540, 544–545
Outdoor receptacle installation, 540, 542–544
Outlet boxes, 40, 43–44, 327. *See also* Device boxes; Electrical boxes; Receptacles
adjustable bar hanger installation, 353–354
defined, 38
nonmetallic, **43**
side-mounting bracket installation, 352
Outlets, 55, **55**, 205. *See also* Receptacles; Switches
defined, 38, 55, 187
dwelling unit outlet requirements, 311, 316–320, **317–322**
symbols for receptacle outlets, **179**
testing of lighting, 707–708
testing of receptacles, 707, **707**, 716–717, 718–721
Overcurrent, defined, 38
Overcurrent protection, 63–65. *See also* Overcurrent protection device (OCPD)
AFCI circuit breakers, 63, **64**, 644, **674**, 679–680, **680**
ampacity of conductor and, 198–200, 674–675
bimetallic strip, 37, 63
branch circuit requirements, 205–207
cartridge fuse, 64, **65**, 672, 673, **674**
cartridge fuse requirements, 677–678
circuit breaker, 63, **63**, **64**, 678, **679–680**

circuit breaker installation, 285–287, **286**, 680–684

circuit breaker requirements, 678

Edison-base fuse, 64, **64**, 672, 673, **674**, **677**

Edison-base fuse requirements, 676–677

electric motor, 673

fuses, 37, 63–65, **64**, **65**

general requirements, 674–676, **675**

GFCI circuit breakers, 13, 63, **64**, **674**, 678–679, **679**

grid-tied systems, **782**

installing single-pole AFCI circuit breaker, 689–690

installing single-pole circuit breaker, 685–686

installing single-pole GFCI circuit breaker, 687–688

installing two-pole circuit breaker for a 120/240 volt branch circuit, 693–694

installing two-pole circuit breaker for a 240 volt branch circuit, 681, 691–692

interrupting rating, 672, 675–676

inverse time circuit breakers, 394, 411

in photovoltaic system, 780–783, **782**

placement, 780–781

plug fuse, 64, **64**, 672

plug fuse requirements, 676–677

requirements for, 674–680, **675**

sizing, 781–783

for stand-alone system, **782**

time-delay fuses, 64, 477, 497

Type S plug fuse, 64, **64**, 672, 673, **674**

Type S plug fuse requirements, 677

underground service conductor, 233

Overcurrent protection device (OCPD), 673–678, **673–677**. *See also* Overcurrent protection; *specific devices and circuits*

defined, 672

installation of, 680–684, **682–683**

interrupting rating, 672, 675–676

location of, 675

requirements for, 674–680, **675**

safety and, 676

sizing of, 205–207

types of, 673, **673**, 678–680, **679–680**

Overhead service, 230, **230**, **262**. *See also* Service entrance

conductors, 229, 232

equipment and materials, 262–270

general rules for, 241–242, **242**

installation, 270–278

raceways, 240 266, **267**

service locations, 270

temporary service entrance, 255, **256**

terms, **232**

using electrical conduit, 266–268, **267**

using entrance mast, 268–269, **269**, **270**, 276–278, **276**

using service entrance cable, 242, 262–266, **262**, 270–273

using service entrance raceway, 273–276, **274**, **275**

weatherheads, 242

Overhead service entrance. *See* Overhead service

Overload, 63, 139. *See also* Overcurrent protection

defined, 38

P

Panelboards, 65–66, 213. *See also* Loadcenters; Service panel

defined, 38

installation, 285–287, **286–287**, **289**

rating, 213

Panel, 751, 753. *See also* Loadcenters; Panelboards; Photovoltaic panels; Service panel

Parallax error, 138, 140

Patch cord, 559, 563, 569, 570, 586–588

Penny (d) nail sizes, 67, **67**

Personal protective equipment (PPE), 7, 18–20, **18–19**, 270

defined, 4

fall arrest system, **19**, 21

Personal safety rules, 793–794

Phantom loads, 734, 740–741, **741**

Phone wiring. *See* Video, voice, and data wiring

Photovoltaic (PV), defined, 734

Photovoltaic arrays, **765**

installation guidelines, 791

Photovoltaic modules, 762

diodes and, 771

I-V curve, 766

mounting photovoltaic, 766

in parallel configuration, 758, **762**

in series configuration, 758, **761**

types and characteristics, 765–766

photovoltaic panels, **743**

Photovoltaic shingles, 771

Photovoltaic systems, 742, 743, 752–757. *See also* Solar cells

advantages/disadvantages, 752–753

components of, 753, **754**, **755–759**, 761–776

electricity basics, 757–758

fundamentals of, 758–761

grid-tie systems, 755, 757, **759**

installation, 761

installation guidelines, 791–792

net metering, 755, 757, **760**

safety, 792–794

types, 753–755, **755–759**

wind turbine system, 794–805

wiring and *NEC*®, 776–791

wiring installation guidelines, 792

Physical hazards, 792–793, **793**

Pigtails, 326, 329, 647, **647**

Pipe, grounding with. *See* Water pipe

Pistol grip drills, 100–101, **100–101**

use procedure, 119–120

Plastic anchors, 67–68, **68**

installing, 78

Plate electrodes, 247, 248. *See also* Grounding

Platform frame, 165, 180, **183**

Pliers, 87–88, **87**, **88**

Plot plan, 165, 167, **167**

Plug fuses, 64, **64**, 673, **674**

Plugs, NEMA chart of, **59**. *See also* Receptacles

Plumb, 84, 95

Plumb bob, 95, **95**

PO. *See* Pryout (PO)

Point terminal, 60

Polarity, 136, 139

defined, 134

NEC® in PV module marks, **767**

receptacle, 700, 707

receptacle testing, 716–721

Polarized plug, **18**

defined, 4

power tools and, 17

Pole light installation, 545

Pole mount systems, 768, 766

Polyvinyl chloride conduit. *See* PVC (polyvinyl chloride) conduit

Pool. *See* Swimming pools

Pool cover. *See* Swimming pools

Porcelain standoff, 261, 269, 270, 277–278, 278

Pouch tools, 85

Power drills, 100–103, 119–125

Power saws, 103–105

Power source, 4, 5, 5

Power tools, 100–105
   circular saw, 103–104, **104**, 126–127
   cordless drills, 102–103, **103**
   double insulated, 17, 99
   drills, 100–103, 119–125
   hammer drills, 102, **102**, 124–125
   OSHA requirements, 17
   pistol grip drills, 100–101, **100–101**,
      119–120
   reciprocating saws, 104–105, **105**,
      128–129
   right-angle drills, 101–102, **101**, 121–123
   safety, 24–25, 99–100, 104
   saws, 103–105
PPE. *See* Personal protective
   equipment (PPE)
Pressurized water fire
   extinguisher, 29
Procedures
   adjustable wrench use, 113
   assembling a patch cord with RJ-45
      plugs and category 5e UTP cable,
      586–588
   bending a 90-degree stub-up, 427–428
   bending a back-to-back bend, 429–430
   bending a three-point saddle, 434–436
   bending an offset bend, 431–433
   box offsets, 437–439
   cable run, starting from a load center,
      387–390
   circuit breaker installation, 685–694
   circuit tracer use, 726–727
   circular saw use, 126–127
   clamp-on ammeter use, 156
   connecting a generator's electrical
      power to critical load branch circuits,
      458
   connecting wires together with a
      wirenut, 74
   continuity tester use, 151, 724–725
   cutting a hole in wooden framing with a
      hole saw and a right-angle drill,
      122–123
   cutting and threading, 424–426
   digital multimeter use, 157–160
   digital voltmeter use, 154–155
   drilling a hole in masonry with a ham-
      mer drill, 124–125
   drilling a hole in wooden framing with
      an auger bit and a right-angle drill,
      121
   electrician knife use, 110–112
   finding information in *NEC*®, 27
   ganging metal device boxes, 72
   hacksaw use, 116
   hammer drill use, 124–125
   handy box masonry surface installation,
      351
   handy box wood surface installation, 350
   installing aluminum conductors, 73

   installing a ceiling-suspended paddle
      fan with an extension rod and light
      kit, 631–633
   installing a light fixture directly to an
      outlet box, 625
   installing a light fixture directly to the
      ceiling, 626
   installing a light fixture in a dropped
      ceiling, 629–630
   installing a light fixture to an outlet box
      using a stud and strap, 628
   installing a light fixture to an outlet box
      with a strap, 627
   installing duplex receptacles in a metal
      electrical box, 661–662
   installing duplex receptacles in a
      nonmetallic box, 663–665
   installing F-Type connector on an RG-6
      coaxial cable, 579–580
   installing feed-through GFCI duplex
      receptacles, 666
   installing lead (caulking) anchor, 77
   installing light fixture to an outlet box
      with a strap, 627
   installing plastic anchors, 78
   installing RJ-45 jack on four-pair UTP
      category 5e cable, 581–585
   installing single-pole AFCI circuit
      breaker, 689–690
   installing single-pole circuit breaker,
      685–686
   installing single-pole GFCI circuit
      breaker, 687–688
   installing single-pole switches in multi-
      ganged nonmetallic box, 669
   installing single-pole switches in
      single-gang nonmetallic box,
      667–668
   installing tapcon masonry screws,
      79–81
   installing toggle bolts in a hollow wall
      or ceiling, 75–76
   installing two-pole circuit breaker for a
      120/240 volt branch circuit, 693–694
   installing two-pole circuit breaker for a
      240 volt branch circuit, 691–692
   kilowatt-hour meter, reading, 161
   knockout punch use, 114–115
   lockout/tagout, 32
   new construction device box installa-
      tion, 347–349
   noncontact voltage tester use, 153
   old-work metal electrical box, installa-
      tion in sheetrock wall or ceiling, 357
   old-work nonmetallic electrical box,
      installation in sheetrock wall or
      ceiling, 358–359
   old-work nonmetallic electrical box,
      installation in wood lath and plaster
      wall or ceiling, 355–356
   outlet box adjustable bar hanger
      installation, 353–354

   outlet box side-mounting bracket
      installation, 352
   pistol grip power drill use, 119–120
   reciprocating saw use, 128–129
   right-angle power drill use, 121–123
   screwdriver use, 106–107
   shock to a coworker, 32–33
   single-dwelling calculation examples,
      216–221
   stripping insulation with a knife,
      110–112
   terminal loops, using to connect
      conductors on a receptacle or
      switch, 656–657
   testing a standard three-way switch,
      722–723
   testing 120/240 volt range and dryer
      receptacles with voltage, 718–721
   testing 120-volt receptacles with a
      voltage tester, 716–717
   torque screwdriver use, 117
   torque wrench use, 117
   troubleshooting a receptacle circuit
      with a ground fault using a continuity
      tester, 724–725
   using lineman pliers to cut cable, 108
   using rotary armored cable cutter, 118
   voltage tester use, 152–153
   wire stripper use, 109
Programmable thermostat, 740,
   740
Pryout (PO), 38, 41, **41**
Pulling in, 363, 376
Pump pliers, 88, **88**
Punch-down block, 559,
   561, 563
Punch-down tool, 561, **561**
PVC (polyvinyl chloride)
   conduit, 52–53, **53**, 395,
      408–410, 746
   abrasion protection, 410
   bends in, **409**
   complete system installation, 409
   cutting/cut ends, 409
   cutting saws, **413**
   expansion fittings, 410, **410**
   field bends, 409, **409**
   heater box/heating blanket, 79, **79**
   installation requirements, 408
   schedule 40 PVC conduit, 408
   schedule 80 PVC conduit, 408
   securing and supporting, 381–383
   temperature changes and, 409

Q

"Quad wire," 563

R

Raceways, 51–54, 393–439. *See also* Conduit; *specific raceway types*
bending, 413–416, **414–416**
conductor installation, 227, 419–420
continuity of, 304
defined, 38, 395
electrical metallic tubing (EMT), 52, **53**, **395**, 403–404, **404**
electrical nonmetallic tubing (ENT), 53–54, **54**, 410–411
fastening/securement, 304, **305**
flexible metal conduit (FMC), 53, **53**, 395, 404–406
installation in a residential wiring system, 417–418, **418**
intermediate metal conduit (IMC), 52, **52**, 396, 402
liquidtight flexible metal conduit (LFMC), 54, **54**, 395, 406–407
metal, grounding and bonding of, **251**, 250, 251, **253**, 275
overhead service, 266, **267**, 274
overhead service installation, 273–276
polyvinyl chloride conduit (PVC), 52–53, **53**, 395, 408–410
rigid metal conduit (RMC), 52, **52**, 396–401
seals/sealants, 234, **234**
selecting raceway type and size, 395
service entrance, 229, 232, 234–235, 242, 266–267, **267**, 273–276, **274**, **275**
support/clamps, 266, **267**, 273
underground service, 279
Rafters, 166, 180, **183**
Rain-tight, 261
Readily accessible, 228, 242, 243
Reaming tools, 404, 412
Receptacles, 55–58, 55, 207, 638–642. *See also* Device(s); Device boxes; Electrical boxes; Switches
amperage ratings, 639–640, **639–640**, 642
arc fault circuit interrupter (AFCI), 644, 648
automatic timer, controlled by, 741
back of, **56**
back wiring, 641, **641**
basement, 320
bathroom, 319, **319**
configurations, NEMA chart of, **59**
countertop, 317–318, **318**, **319**
covers, 255, 650, **650**
defined, 55, **55**, 637

duplex, 55–56, **56**, 637, 638, **638**, 661–665
dwelling unit required outlets, 311, 316–322, **317–322**
fixed panels/room dividers and, 316–317, **317**
foyer, 298, 320
front and back of dwelling, 319, **320**
garage, 320
grade level, 319
ground fault circuit interrupter (GFCI), 57, **57**, 637, 642–644, **643**, 648, 666
grounded plug blade, long slot for, **640**, 641
hallway, 320
installation of, 316–322, **317–322**, 647–651, **648–651**
installing duplex receptacles, 461–466
installing duplex receptacles in a metal electrical box, 661–662
installing duplex receptacles in a nonmetallic box, 663–665
installing feed-through GFCI duplex receptacles in nonmetallic boxes, 666
kitchen, 316–318, **318**
laundry, 320
multiple (triplex), 638, **638**
outlet symbols, **179**
outlet versus, 55, **55**
pigtails, 647
placement of, 316, **317**
plug-in receptacle tester, 707, **707**
polarity, 700, 707
polarity testing, 716–721
selecting appropriate type, 638–642, **639–641**
single, 55, 637, 638, **638**
special types, 57–58, **58**
split-wired, 637, 641, **641**, 648, **648**
strap (yoke), 637, 639–640
switch installation, 652–654, **653–654**
tamper-resistant, 642
terminal loops, 641, 647
testing of, 707, **707**, 716–717, 718–721
transient voltage surge suppressor (TVSS), 645–647, **645–646**
troubleshooting with a continuity tester, 712, **713**, 724–725
voltage ratings, 639–640, **639–640**
wall space and, 316, **317**
Recessed lighting, 608–612, **611–613**. *See also* Lighting; Luminaires
example, 742
installation, 618, **619**
Reciprocating, defined, 84
Reciprocating saws, 104–105, **104**
use procedure, 128–129
Recycling and waste management, 746–747

Redheads, 372, **372**
Reel, 94, **94**, 363
Resistance, 4
human body and, 6, **6**
RG-6 (Series 6), 559
installing an F-type connector on, 580
RG-59, 559, 562
Ribbon, 166, 180
Right-angle drills, 101–102, **101**
use procedure, 121–123
Rigid metal conduit (RMC). *See* RMC (Rigid metal conduit)
Ring wave, 637, 645
Riser, 228, 233
RJ-11, 559, 562. *See also* Video, voice, and data wiring
RMC (Rigid metal conduit), 396–401
abrasion protection, 401
bending, 398
bushings, 401
cutting, 398
as equipment-grounding conductor, 396
fittings and, 396, **396**
installation requirements, 326–332
number of conductors, 396, **397**
Rod, as grounding electrode, 247, 248, **249**
Romex™, 38, 47–49, 299. *See also* Nonmetallic sheathed cable (NMSC)
Roof flashing, 261, 269–270, 270, 277, 745
Roof mount systems, 768–770, 770
Room air conditioning. *See* Air conditioning
Rotary BX cutter, 82, 82, 118
Rough-in, defined, 298
Rounding up, 185
Rule, folding, 91–92, 91
Running boards, 363, 364, 366, 367, 379

S

Saddle bend, 394, 414, 416, **416**
Safety, 1–35. *See also* safety and caution statements throughout text
circuit breaker installation, 681
electrical rules, 28
energized circuits, 98, 149

Safety (*continued*)
  fall protection, 19–20, **19**
  fires/fire extinguishers, 28–30
  general rules, 20–28
  hand tool use, 85
  insulated tools, 98
  ladders, 20, 270
  lifting, 20, **21**
  lighting fixture installation, 614
  lockout/tagout procedures, 17, 32
  material handling, 20
  Material Safety Data Sheet (MSDS), 25–28
  meters and test instruments, 148–149, **148**
  National Electrical Code® (*NEC*®), 8–13
  National Fire Protection Association (NFPA), 16
  Occupational Safety and Health Administration (OSHA), 16–18, 98
  one-handed working procedure, 7
  overcurrent protection devices, fault currents and, 676
  personal protective equipment, 18–20, 270
  photovoltaic system, 792–794, **793–794**
  power tools, 17
  removing a broken incandescent lamp, 707
  scaffolding, 21–24, 270
  shock, co-worker shock procedure, 32–33
  shock hazard, 5–7
  throwing objects, **18**
  tools, 24–25
  training, 16–17
Safety switches, 66, **66**
  defined, 38
Saws
  circular, 103–104, **104**, 126–127
  hacksaw, 93–94, **94**, 412
  keyhole, 93, **94**
  reciprocating, 104–105, **104**, 128–129
Sawzall (reciprocating saw), 104, **104**, 128–129
Scaffolding, 4, 21–24, **26**, 270
Scale, 166, 174–176, **175**
Schedules, 166, 171, **172**
Schneider Electric Xantrex XW inverter/charger, **755**, **775**
Sconces, 594, 612, **614**
Scratch awl, 91, **91**
Screwdrivers, 85–87, **85**, **86**, **87**
  use procedure, 106–107
Screws, 69, **69**
  screwhead shapes/types, **69**
  self-drilling ("Tek-Screws"), **69**, 69

SE. *See* Service entrance
Seal roofing, 745
Section line, 166, **174**, 174
Sectional drawings, 166, 169, **170**
Secured (electrical cable), 363, 366, **366**
Service, defined, 228, 231
Service call, 714
Service center, 563, **565**
Service conductors, 233, 235–242. *See also* Service drop conductors; Service entrance
  clearances, 235–238, **236–238**
  defined, 229
  disconnecting means, 242–243, **243**
  grounding of metal enclosures/raceways for, 251–252
  required dimensions, 235–236, **235**
  sizing, 207–213, 216–221
  trees and, **236**
Service connection, application for, 254
Service disconnect, 242–243, **243**, 187, 207, 283
  combination meter socket and, **266**, 267–268
  panel with main circuit breaker, 264, **265**
Service drop, 229, 231, 255
Service drop conductors, 235–238
  clearances, 235–238, **236–239**
  insulation, 235
  minimum clearances, 238
  minimum size, 235
  required dimensions, **235**
  service mast, 238, **239**
  vertical clearances, 236–238, **239**
Service entrance, 228–257, 260–292. *See also* Service conductors; Service entrance cable
  cable, 50–51, **50**, **51**, 229, 232
  cable supports, 241, **242**
  checklist of *NEC*® requirements, 705–706
  clearances, 235–238, **236–239**
  conductors, 229, 232, 235–242
  conductors, sizing of, 207–213, 216–221
  defined, 38
  diagram of, **232**, **246**
  disconnecting means, **232**, 242, **243**, 281
  drip loops, 228, 231, 242, **242**
  enclosure, **261**, 264
  grounding and bonding requirements, 13, 243–253, **244**
  grounding electrode, 228, 233, 244–249

grounding electrode, supplemental, 229, 233, **248**, 249, **249**
installation, 260–293
installation and grounding, typical, **246**
installation requirements, 233–243
loadcenter sizing, 213–215
local utility company, working with, 254–257
meter enclosure, 228, 232
minimum size of, 207
overhead mast, **232**, 232, **239**
overhead raceways, 241, 266, **267**
overhead service, 230, **230**, 241–242
overhead service conductors, 229, 232
overhead service equipment and materials, 262–270, **262–270**
overhead service installation, 270–278
overhead service locations, 241, **242**, 270
overload protection, 243
panelboard installation, 285–287, **286–287**, 289
raceways, 229, 232, 234–235, 241, 266, **267**, 273–276
readily accessible, 228, 242, **243**
riser, 228, 233
seals/sealants, 234, **234**
service drop, 229, 235–238, **235**, 262
service equipment, 229, 233, 264, **265–266**
service equipment installation, 281–287, **285–287**
service head (weatherhead), 229, 231, 242, **242**, 262, **263**, 266, **267**
service lateral, 229, 233
service mast, 229, 231, 238, **239**, 268–269, **269**, **270**, 276–278, **276**
service point, 229, 231
service raceway, 229, 232, 234–235
single-dwelling calculation examples, 207–213, 216–221
sizing of conductors, 207–213, 216–221
standoff, 233, **233**
subpanel, 215
subpanel installation, 288–289
temporary, 255, **256**, **257**
terms and definitions, 229, 231–233, **232**
transformer, 229, 233
trees and, **236**
underground service, 230–231, **231**
underground service conductors, 229, 232, 238–240, **240**, 241
underground service equipment, materials, and installation, 278–280, **278**, **280**
underground service overload protection, 243
underground service protection, 241
upgrading, 289–291
utility pole, 229, 233

wiring methods, 240
working spaces, 283–285, **285**, 287
Service entrance cable, 50–51, **50, 51**, 232, 262–266
cable hook, **261**, 263, **263**, 272
clips, 263, **263**
combination meter socket/main breaker disconnect, 264, **266**
defined, 229
feeder cable, 288
fittings and supports, 51, **51**, 240, **242**, 263, **263**
meter socket connections, 273, **273**
overhead service, 262–266
overhead service installation, 270–273, **271–273**
panel with main circuit breaker, 264, **265**
rain-tight, 264
"regular," 264
type SER, 50, **51**, 288
type SEU, 50, **50**, 232, 262, **262**, **271**, 271
type USE, 51, **51**, 232, 279, 279, **281**
URD (underground) cable, 279
as wiring method, 373
Service equipment, 229, 233, 264, **265–266**
installation, 281–287, **285–287**
Service head (weatherhead), 229, 231, 242, **242**
electrical conduit and, 266, **267**
service entrance cable and, 263, **263**
Service lateral, 229, 233, **280**
Service mast, 229, 231, 238, 239
entrance mast for overhead service, 268–269, **269**
mast kit, **270**
overhead service installation, 276–278, **276**, **277**, **278**
Service panel. *See also* Panelboards
circuit breaker installation, 680–684
installation, 285–287, **286–287**, 289
overcurrent protection devices and, 673–680
trim-out, 671–696
Service point, defined, 229
Service raceway, 229, 232, 234, 266–268, **267**, 273–276, 279
Setout, 326
Shall, 4, 11
Sheath, 38, 44
stripping tools, 98
Sheathing boards, 166, 183, **183**

Sheetrock (wallboard), 84, 91, 301
electrical box mounting and, 312
old-work metal electrical box, installation in, 357
old-work nonmetallic electrical box, installation in, 358–359
Shielded twisted pair. *See* STP (shielded twisted pair)
Shock, electrical, 4, 5–7
insulated tools and, 98
procedure when co-worker is being shocked, 32–33
shock hazard, 5–7
Shoes, 16, 270
Short circuit, 63, 142, 299, 710
defined, 38, 134, 700
example of, **711**
troubleshooting, 710–714
Short circuit current (Isc), 751, 766
Siding penetration, 745
Sill, 166, 183, **183**
Sill plate, **261**, 264, **264**
Single-pole switches, 58, **58**, 651. *See also* Switches
defined, 443
installation, 448–452, 667–669
switching circuits, 448–452, **449–452**
ungrounded circuit conductor and, 446
Single receptacle, 55, 637, 638, **638**. *See also* Receptacles
Six-in-one tool, 89, **90**
Skilsaw (circular saw), 103–104, **104**, 126–127
Sledgehammer, 96, **97**
branch circuit installation, 480–481
Smoke detectors, 477, 510–512
branch circuit installation, 510–512
interconnecting, 477, 511
recommended locations, **511**
requirements, 510
"Smurf tube". *See* ENT (Electrical nonmetallic tubing)
Snap switches. *See* Switches
Solar cell, 751, 752
Solar window, 761
Sones, 734, 739
Spacers, 650
Spas and hot tubs, 538–540. *See also* Swimming pools
defined, 538
emergency shutoff switch, 539
installation requirements, 539
Specifications, 166, 171–173, **173**

Spliced conductors, 40, 43, 48
Spliced, defined, 38
Split-wired receptacle, 637, 641, 641. *See also* Receptacles
installation, 648, **648**
Splitter, 572
Stabs, 672, 682, 682
Stairway lighting outlet, 320
Stand-alone inverter, 774, 775
Stand-alone system, 754, 759
equipment/system grounding in, **786**
installation guidelines for, 791–792
inverter input circuit sizing for, 780
NEC® requirements to, 789–790
PV circuit types in, **778**
Standby power system, 522, 546–550
Standoff, 233, **233**
Steel plates, 302–303, **305**
Step ladders, 20, **24**
Storage area lighting outlet, 320
Storage pool, 537–538. *See also* Swimming pools
Strap (yoke), 637, 639–640. *See also* Receptacles
Strip (cable sheathing/insulation), 98, 110–112
Strip (screws or bolts), 84, 87
Studs
metal, 245
wall, 166, 180, **183**
Subfloor, 166, 183, **183**
Subpanel, 65–66, 66. *See also* Loadcenters; Panelboards
differences from main service panel, 288, 291
installation, 288–289, **290**
location, 288
sizing, 215
Sun charts, 761, **766**
Sun rise/set, 760, **763**
Sunlight resistant cable, 309
Supplemental grounding electrode, 229, 233, **248**, 249, **249**, **289**
Supported (electrical cable), 363, 366, **366**
Surface transformer, 622
Surfaced mounted lighting. *See* Lighting; Luminaires

Surge suppressors. *See* Transient voltage surge suppressor (TVSS)

Suspended ceiling, 302, 342–343, **343**
  lighting fixture installation in, 620–621, 629–630

Swimming pools, 528–545
  branch circuit installation, 528–545
  ceiling-suspended paddle fan, 535, **536**, 539
  dry-niche luminaire, 522, 529, **531**
  exothermic welding and, 532
  forming shell, 522, 529, **532**, 537
  grounding and bonding requirements, 532, 533
  hydromassage bathtubs, 529, **531**, 540
  junction boxes, 537
  luminaire installation, 535, **536**, 537
  metal part bonding, 439–441, **443**
  no-niche luminaire, 522, 529, 531
  permanently installed, 529–537
  self-contained spa or hot tub, 538–540

Spas and hot tubs, 538–540
  storable/storage, 432, 438, **441**, 445–446, **446**
  wet-niche luminaire, 522, 529, **531**, 537

Switch box, 38, 40

Switch loop, 443, 450, **451**, 653

Switches, 59–62, 364–392, 651–654
  alternating current general-use snap, 447
  alternating current or direct current general-use snap, 368
  back-wired connections, 652, **653**
  ceiling fan installation, 470–472
  combination, 62, **62**
  current and voltage rating, 446
  dimmer, 61, **61**, 468–470, 651
  dimmer installation, 468–470
  direct current general-use snap, 447
  double-pole, 60, **60**, 466–468, **468**
  double-pole installation, 466–468, **468**
  duplex receptacle circuits, 461–466
  duplex receptacles installation, 461–466
  emergency shutoff, in spas and hot tubs, 447
  faceplates/switch covers, 654, **654**
  four-way, 61, **61**, 457–461, 651
  four-way installation, 457–461, 653
  four-way switching circuit, 459–461, **460**
  installation, 652–654, **653–654**, 667–669
  low-voltage, 62, **62**, 651
  momentary-contact, 62, **62**
  number of switching locations, 446

  push-in terminations, 652–653, **653**
  rocker (decorator), 652, **652**
  safety, 66, **66**
  selecting appropriate, 651–652, **652**
  single-pole, 58, **58**, 448–453, 651
  single-pole installation, 448–453, 667–669
  single-pole switching circuits, 448–452, **449–452**
  styles of, 652, **652**
  switch loop, 443, 450, **451**, 653
  symbols for, **180**
  terminal loops, 652, **652**
  testing for proper connections, 708–710, **709–710**, 722–723
  three-way, 60–61, **60**, 443, 453–457, 651
  three-way installation, 453–457, 653
  three-way switching circuit, 454–457, **455–456**
  three-way testing, 722–723
  toggle, 652, **652**

Symbols, 176, **177–183**
  architectural, 176, **177**
  defined, 166
  electrical, 176, **181**
  electrical switch, **180**
  electrical wiring, 176, **182**
  lighting outlet, **178**
  receptacle outlet, **179**

Symbols for Electrical Construction (ANSI), 176

T

T-stripper, 89, **89**

Take-up, 394, 416

Tank type electric water heaters, 738, 737

Tankless electric water heaters, 738, 738

Tap tool, 90, 90

TapCons, 68, 68
  installing, 79–80

Tape measure, 91–92, **91**

Technology. *See* Video, voice, and data wiring

Telephone wiring, 568, 568, 572, 573. *See also* Video, voice, and data wiring

Temperature ratings
  of conductors, **199**, 198–200, 309
  of terminals, 202–203, 213–215
  of wirenuts and wire connectors, **206**

Tempered, defined, 84

Temporary service entrance, 255, **256**, **257**

Terminal identification markings, **45**

Terminal loops, 641, 647, 652, **652**, 656–657

Terminals, 60–61
  temperature ratings, 202–203, 213–215

Test and measurement instruments. *See* Meters

Testers. *See* Meters

Thermals, 751, 797

Thermostat, 477, 500, 501
  control cable, **306**
  defined, 477

Thinwall, 52, 394, 403. *See also* EMT (Electrical metallic tubing)

Threaded hub, **261**, 263, **264**

Three-gang airtight nonmetallic electrical box, **742**

Three-point saddle bend, 434–436

Three-prong electrical plug, 14, **14**

Three-way switches, 60–61, **60**, 443, 453–457, 651
  defined, 443
  installation, 453–457
  switching circuit, 454–457, **455–456**
  ungrounded circuit conductor and, 446

Three-wire electrical system, 243, **244**

Tie wraps, 70–71, 71

Time-delay fuses, 64, 477, 497, 673

Tip (wire), 566

Toggle bolts, 67, 68
  installing, 75–76

Toggle switch. *See* Switches

Tongue-and-groove pliers, 88, **88**

Tool pouch, 92, **92**

Tools, 83–132. *See also* Power tools
  adjustable wrench, 90–91, **91**, 113
  awl, 91, **91**
  cable cutters, 96, **97**, 98, **98**, 118
  cable strippers/rippers, 89, **89**, 98, **98**
  care and safe use, 85
  channel-lock pliers, 88, **88**
  chisel, 95, **96**
  compass (keyhole) saw, 93, **94**
  conduit bender, 94–95, **95**
  crescent wrench, 90–91, **91**, 113

diagonal cutting pliers, 88, **88**
dikes, 88, **88**
electrician's hammer, 91, **91**
electrician's knife, 89–90, **90**, 110–112
files, metal, 95, **96**
fish tape and reel, 94, **94**
folding rule, 91–92, **91**
fuse puller, 97, **97**
hacksaw, 93–94, **94**, 116
hand tools, 85–92
hawkbill knife, 90, **90**
hex key (Allen wrench), 96, **97**
insulated, 98, **98**
keyhole saw, 93, **94**
knockout punch, 93, **93**, 114–115
levels, 95, **95**
lineman pliers, 87, **87**, 108
long-nose pliers, 87–88, **88**
multipurpose, 89, **90**
needle-nose pliers, 87–88, **88**
nut drivers, 93, **93**
OSHA insulation requirements, 98
pliers, **27**, 87–88, **87**, **88**
plumb bob, 95, **95**
power, 24–25, 99–105
power drills, 99–103, 119–125
power saws, 103–105
pump pliers, 88, **88**
PVC heater box/heating blanket, 95, **95**
rotary BX cutter, 118
safe use of, 24–25, **27**, 85, 99–100, 104
saws, 93–94, **94**
scratch awl, 91, **91**
screwdrivers, 85–87, **85**, **86**, **87**, 97–98, 98
six-in-one tool, 89, **90**
sledgehammer, 96, **97**
specialty, 92–98
tap tool/triple tap tool, 90, **90**
tape measure, 91–92, **91**
tongue-and-groove pliers, 88, **88**
torpedo level, 95, **95**
torque screwdriver, 96–97, **97**, 117
utility knife, 90, **90**
wire strippers, 89–90, **89**, 109
Top plate, 166, 180, **183**
Torpedo level, 95, 95
Torque, 38, 70, 84
Torque screwdriver, 96–97, 97, 289
use procedure, 117
Torque wrench, 97–98, **98**, 288–289
use procedure, 117
Tower safety, 804–805
Track lighting installation, 621–622, **621–622**
Track mount systems, 770, **771**

Training
power tool use, 99–100, 104
safety, 16–17
Transformer, 233
defined, 229
pole-mounted and pad-mounted, 279–280, **280**
Transient voltage surge suppressor (TVSS), 637, 645–647, **645–646**. *See also* Receptacles
Traveler terminals, 60, 453, 459
Trenching and excavating, OSHA requirements, 17–18, 279
Triple tap tool, 90, **90**
Triplex cable, 229, 231, 263, **263**
Troffer, 594, 619, 629–630
Troubleshooting, 710–714, **711**
defined, 700
True RMS meter, 134, 145, **145**
TVSS (Transient voltage surge suppressor), 637, 645–647, 645–646
Two-wire electrical system, 243, 244
Type IC light fixture, 594. *See also* Lighting; Luminaires
installation, 611, **612**
Type Non-IC light fixture, 594. *See also* Lighting; Luminaires
installation, 611, **612**
Type S fuses, 64, **64**, 672, 673, **674**, **677**
requirements, 677
Type UF cable, 309, 312

U

UL. *See* Underwriters Laboratories (UL)
Underground feeder cable, 49, **49**
installation requirements, 367
Underground service, 230–231, **231**, 278–280, **278**, **280**, **281**, **282**. *See also* Service entrance; Underground system service
entrance conductors
equipment, materials, and installation, 278–280, **278**, **280**, **281**, **282**
maximum allowable length, **284**
meter connections, **278**

minimum burial depths, 238–239, **240**
raceway, **282**
temporary service entrance, 255, **256**
terms, **232**
transformers and power from local electric utility, 279, **280**
type UF cable, 309
type USE cable, **281**
URD (underground) cable, 279
wiring requirements checklist, 706
Underground service entrance. *See* Underground service
Underground system service entrance conductors, 229, 232, 238–240
minimum burial depths, 238–239, **240**
minimum cover requirements, 238–240, **240**
protection of service conductors, **241**
Underwriters Laboratories (UL), 39, **39**
directories, 39, **39**
label, 606, 705
Unit load, 188–190
Utility box, 38, 41, **42**
Utility company, local
electric power from, underground service, **280**
working with, 254–257
Utility Grid Interconnected System, 754
Utility knife, 90, **90**
Utility pole, 229, 233

V

VD. *See* Voltage drop (VD)
Ventilation fan, 739–740, 739
Ventricular fibrillation, 4, 7
Video, voice, and data wiring, 558–588
bandwidth, 462, 463
binding posts, 475–476, **475**
coaxial cable, 559, 561
data wiring installation, 568–569
F-type connector, 559, 562, 563, 570
horizontal cabling, 559, 563
IDC (insulation displacement connection), 559, 561
installing F-Type connector on an RG-6 coaxial cable, 579–580
installing RJ-45 jack on four-pair UTP category 5e cable, 581–585
J-hooks, 574, **574**
jack, 559, 563, 567, 569

Video, voice, and data wiring (*continued*)
  megabits per second (Mbps), 559, 563
  megahertz (MHz), 559, 563
  mud rings, 573
  patch cord, 559, 563, 569, 570, 586–588
  punch-down block, 559, **561**, 563
  punch-down tool, 559, **561**
  RG-6 (Series 6), 559
  RG-59, 559, 562
  RJ-11, 559, 562
  splitter, coaxial cable, 572
  tip (wire), 566
Voc. *See* Open circuit voltage (Voc)
Volt, 4, 5
Volt-amperes, 187, 188
Volt-ohm millimeter (VOM), 134, 144
Voltage, 5
  defined, 4
  electrocution and, 6, **6**
  human body and, **6**
  spikes, 645
Voltage drop (VD), **261**, 280–281, **284**
  calculation, 780
Voltage testers, 136–137, **136–137**
  defined, 134
  noncontact, 134, 137, **138**, 153
  testing 120/240 volt range and dryer receptacles, 718–721
  testing 120 receptacles, 716–717
  use of, 152–153, 616
Voltmeters, 137–139, **138**, **139**
  defined, 134
  use of, 154–155
VOM. *See* Volt-ohm millimeter (VOM)

W

Wall flashing, 745
Wall space, 316, **317**
Wall studs, 166, 180, **183**
Wallboard, 84, 91
Warning tape/ribbon, 279
Washers, 69–70, **70**

Waste management, recycling and, 746–747
Water heater, 737–738, **737–738**
  types of, 737–738
Water management, 745
Water pipe, as grounding electrode conductor, 245, 246, 248, 249, 287, 287, 289
Watt-hour meters, 146–148, 146, 147
  reading, 161
Weather collar, **261**, 269–270, **270**, 277
Weatherheads, 229, 231, 242, **242**
  electrical conduit and, 266, **267**
  entrance mast and, 269, 277, **277**
  service entrance cable and, 263, **263**, 271
  service entrance raceway and, 273
Wet locations, 299
Wet-niche luminaire, 522, 529, **531**, 537
Whirlpool bath. *See* Hydromassage bathtubs
White Book (*General Information for Electrical Equipment Directory*), 39, **40**
Wiggy, 134, 136, **136**
Wind turbine system, 743–744, **744**, 751, 752, 794–805, **794–802**
  applications of, 796–797
  components of, 794–795, **795**
  installation, 801–804
  introduction, 794–796
  measurement of wind, 797–798, **798**
  safety installation of, 804–805
  towers, 798–801
Wire cart, 419, **420**
Wire connectors, 47, 48, 206
Wire nut, 47, 48
  connecting wires with (installing), 74
  defined, 38
  temperature ratings, **206**
Wire strippers, 89–90, **89**
  use procedure, 109
Wireless computer networks, 472

Wiring, 295–324, 431–450. *See also* Conductors; Video, voice, and data wiring
  air-handling spaces and, 308, **305**
  bundling, **202**
  cable/raceway fastening/support, 304–306, **305**, **306**
  checklist of *NEC®* requirements, 703–705, 706
  damage of parts, 300
  defined, 298
  deteriorating agents, 299
  electrical boxes and, 307, **307**, **308**
  electrical continuity, 303–304
  electrical installation requirements, 299–300
  fire/fire-stop methods and, 307, **308**
  framing members, cable parallel to, 303, **303**
  framing members, metal, installation through, 302, **302**
  framing members, wood, installation through, 301–302, **301**
  furring strip, 298, 303
  general requirements, 299–308
  green house wiring practices, 733–748
  grooves, cable/raceways in, 303, **302**
  inductive heating, 298, 300
  installation instructions, 299
  insulation integrity, 299
  methods, 300–308
  "neat and workmanlike" installation, 300
  photovoltaic system, 776–791
  suspended ceiling, 302
  symbols, 176, **182**
  underground, 706
  unused openings, 300, **300**
Wood framing members
  electrical box installation and, 337–340, **338–340**, 347–349
  wiring and, 301–302, **301**
Working spaces and clearances, 283–285, **285**, 287

Y

Yoke, 637, 639–640. *See also* Receptacles; Switches

NEC Index

Note: Page numbers in **bold** reference Figures
*2011 NEC change

S

Section 110.11, 299
Section 110.11 Informational
　Note 2, 299
Section 110.12, 300
Section 110.12(A), 300, **300**,
　704
Section 110.12(B), 300
Section 110.14, 705
Section 110.14(A), 704
Section 110.14(B), 704
Section 110.14(C), 202, **502**
Section 110.26, 283, 286, 288,
　506, 706
Section 110.26(A)(1), 283
Section 110.26(A)(2), 283
Section 110.26(A)(3), 285
Section 110.26(A)(3)
　Exception 1*, 285
Section 110.26(A)(3)
　Exception 2*, 285
Section 110.3, 11
Section 110.3(B), 11, 39, 70, 97,
　299, **508**
Section 110.30 , 283
Section 110.7, 299
Section 110.9, **246**, 675
Section 200.6*, 309
Section 200.6(B)*, 309, **310**
Section 200.7, 310
Section 200.7(C), 451, 653

Section 200.7(C)(1)*, 310
Section 210.11(C)(1), 191, **192**
Section 210.11(C)(2), 191,
　192, 490
Section 210.11(C)(3), **193**, 194
Section 210.12*, 481, 680, 703
Section 210.19(A)(1), 198
Section 210.19(A)(3), **195**,
　197, 198
Section 210.21(B)(1), 640
Section 210.3, 205, 675
Section 210.4, 683
Section 210.4(B), 683
Section 210.4(D)*, 683
Section 210.4(D) Exception, 684
Section 210.50(C), 490
Section 210.52, 316, 504,
　504, 702
Section 210.52 Informational
　Note, 504
Section 210.52(A)*, 316
Section 210.52(B), 191
Section 210.52(B)(3), 702
Section 210.52(C)*, 317, **319**
Section 210.52(D)*, 319
Section 210.52(E), 319
Section 210.52(F), 319
Section 210.52(G)*, 320,
　523, **524**
Section 210.52(H), 320
Section 210.52(I)*, 320
Section 210.70, 320, 444

Section 210.70(A)(1), 320
Section 210.70(A)(2), 320
Section 210.70(A)(2)(a), 523,
　524
Section 210.70(A)(3), 320
Section 210.8*, 642, 703
Section 210.8(A), 13, 57
Section 210.8(A)(1), 540
Section 210.8(A)(2), 523
Section 220.12, 188, 190, **190**,
　208, 216
Section 220.50, 209, 217
Section 220.51, 209, 216
Section 220.52(A), 191, 208, 216
Section 220.52(B), 191, 208, 216
Section 220.53, 209, 210, 216,
　218, 220
Section 220.54, 198, 209, 216,
　493
Section 220.55, 198
Section 220.60, 198
Section 220.61, 210, 218,
　220, 527
Section 220.82(B), 212, 219
Section 220.82(B)(1), 211, 219
Section 220.82(B)(2), 211, 219
Section 220.82(B)(3), 212, 219
Section 220.82(C), 212, 219
Section 225.26, 544
Section 230.10, 234
Section 230.22, 234
Section 230.22 Exception, 234

Section 230.23, 235
Section 230.24, 235, **246**
Section 230.24(A) Exception
    No. 2, 235, **237**
Section 230.24(A) Exception
    No. 3, 236, **238**
Section 230.24(A) Exception
    No. 4, 235, **237**
Section 230.24(A) Exception
    No. 5*, 236
Section 230.24(B), 236, **239**
Section 230.28, 238, **239, 246**
Section 230.42(B), 210, 212,
    217, 220
Section 230.43, 240
Section 230.50(B)(1)*, 241
Section 230.51, 241, **242**
Section 230.51(A), 263
Section 230.54*, 241, **246**, 705
Section 230.7, 234
Section 230.70, 242, 243, **246**
Section 230.70(A)(1), **243, 246**,
    286, 705
Section 230.70(A)(2), **246**, 705
Section 230.71(A), 243
Section 230.79, 243
Section 230.79(C), 210, 212, 217,
    220, 243, **246**
Section 230.8, 234
Section 230.9, 234
Section 230.9(B), 234
Section 230.90, 243
Section 240.20* Deleted, 780
Section 240.21, 786
Section 240.24, 675
Section 240.24(A), 675
Section 240.24(C), 675, 705
Section 240.24(D), 705
Section 240.24(E)*, 246, 705
Section 240.4, **196, 197**, 674
Section 240.4(B)*, 206, 674, 675,
    781, 788
Section 240.4(D), 49, 200, 206,
    675, 781
Section 240.50, 676
Section 240.50(B), 676
Section 240.50(C), 676
Section 240.51, 676

Section 240.53, 677
Section 240.54, 677
Section 240.6(A), 11, 205, 206,
    497, 499, 674, 675, 781,
    783, 788
Section 240.60, 677
Section 240.61, 677
Section 240.81, 678
Section 240.83, 678
Section 240.83(B), 678
Section 240.83(C), 678
Section 240.83(D), 678
Section 240.83(E), 678
Section 250.104(A)(1), 706
Section 250.110*, 785
Section 250.118, 395
Section 250.118(10)*, 784
Section 250.118(5)*, 406
Section 250.118(6)*, 407
Section 250.118(6)(b), 407
Section 250.118(6)(c), 407
Section 250.119, 309
Section 250.120(C)*, 788
Section 250.134, 704
Section 250.140, 374
Section 250.142(B), 289
Section 250.148, 704
Section 250.166, 788
Section 250.166(B), 788
Section 250.166(C), 788
Section 250.24(A), 243, 246
Section 250.24(B), 244
Section 250.24(C), 244
Section 250.24(C)(1)*, 211,
    218, 221
Section 250.28, 244, 246,
    247, 706
Section 250.32, 374
Section 250.32(A), 527
Section 250.32(A) Exception, 527
Section 250.32(B)(1)*, 527
Section 250.32(B)(1) Exception*,
    527
Section 250.4, 244, 785
Section 250.4(A), 303
Section 250.50, 245, 706
Section 250.50 Exception, 245
Section 250.52, 245, 706, 788

Section 250.52(A)(1) through
    (A)(7), 245
Section 250.53*, 247, 706
Section 250.54*, 803
Section 250.64, 246, 247, 249
Section 250.64(C) *, 246, 706
Section 250.66, 244, 246, 247,
    249, 250, 252, 528, 530,
    706, 788
Section 250.68*, 246, 250
Section 250.68 (A) Exception
    No. 1, 250
Section 250.68 (A) Exception
    No. 2, 250
Section 250.70, 246, 251
Section 250.80, 251
Section 250.92*, 251, 275
Section 250.92(B)*, 252
Section 250.94*, 280
Section 300.10, 303
Section 300.11, 576
Section 300.11(A), 303, 304, 305,
    342, 382
Section 300.11(B), 304, **305**, 306
Section 300.11(B)(2), **305**, 306
Section 300.11(C), **305**, 306
Section 300.12, 306
Section 300.13(B), 650–651
Section 300.14, 306, 306, 703
Section 300.15, 307, 307, 704
Section 300.18(A), 398
Section 300.21, 307, 307
Section 300.22*, 308
Section 300.22(C)*, 308, 308, 704
Section 300.23, 303
Section 300.3(A), 300
Section 300.3(B), 300, 301, 702
Section 300.4, 301, 365, 371–373
Section 300.4(A), 703
Section 300.4(A)(1), 301, 302, 411
Section 300.4(A)(2), 301, 302
Section 300.4(B), 302, 302
Section 300.4(B)(1), 302
Section 300.4(B)(2), 302
Section 300.4(C), 302
Section 300.4(D), 303, 304,
    365, 576
Section 300.4(F), 303

Section 300.4(G), 268, 275, 401, 410, 411, 705
Section 300.5*, 239, 367, 541, 706
Section 300.5(D), 547
Section 300.5(D)(3), 279
Section 300.5(D)(4), 279
Section 300.5(G), 234, 547
Section 300.5(H), 278, 279, 547
Section 300.5(J), 547
Section 300.7(A), 234, 234, 705
Section 300.9, 418
Section 300.22(C) Exception, 308, **308**, 704
Section 310.10(D)*, 309
Section 310.10(D)(2)*, 309
Section 310.10(F)*, 308
Section 310.110(A)*, 309
Section 310.110(B)*, 309
Section 310.110(C)*, 310
Section 310.120(A)*, 309
Section 310.15, 311
Section 310.15(A)(3), 309
Section 310.15(B)(3)(a)*, 200
Section 310.15(B)(3)(a)(2)*, 202
Section 310.15(B)(3)(c)*, 777
Section 310.15(B)(7)*, 675, 705
Section 310.106(A)*, 308
Section 314.16, 311, 329, 331, 331, 345, 704, 801
Section 314.16(A), 311
Section 314.16(B), 311
Section 314.17, 311
Section 314.17(A), 704
Section 314.17(B), 311, 311
Section 314.17(C), 312
Section 314.17(C) Exception, 312, **312**, 366, 703
Section 314.20, 312, 313, 337, 647, 648
Section 314.23, 304, 312, 703
Section 314.23(B), 313
Section 314.23(B)(1), 313, 314
Section 314.23(C), 314, 314
Section 314.23(D), 343
Section 314.23(E)*, 537, 542
Section 314.23(F), 542
Section 314.25, 704

Section 314.27*, 314, **315**, 342
Section 314.27(A)(1)*, 314
Section 314.27(B)*, 315
Section 314.27(C)*, 315, 316, 703
Section 314.28*, 801
Section 314.29, 315, 704
Section 314.4, 704
Section 320.15, 367
Section 320.17, 369
Section 320.23, 365, 370, 373
Section 320.24, 370
Section 320.30, 371
Section 320.30(C), 371
Section 320.30(D), 372
Section 320.40, 372
Section 330.17, 372
Section 330.23, 373
Section 330.24(B), 373
Section 330.30, 373
Section 330.30(B), 373
Section 330.30(C), 373
Section 330.30(D), 373
Section 330.40, 373
Section 334.10(1)*, 364
Section 334.15, 364
Section 334.17, 302, 365
Section 334.23, 365
Section 334.24, 365
Section 334.30, 366, **370**, 703
Section 334.30(A), 366
Section 334.30(B), 366
Section 338.10(B)*, 374
Section 338.10(B)(4)*, 374
Section 340.24, 367
Section 342.2, 402
Section 342.22, 402
Section 344.2, 396
Section 344.22, 396
Section 344.24, 398
Section 344.26, 398, **401**
Section 344.28, 398
Section 344.30(A), 400
Section 344.30(B), 400
Section 344.30(B)(4), 401
Section 344.46*, 401
Section 344.60, 401
Section 348.2, 404
Section 348.22, 405

Section 348.24, 405
Section 348.26, 405
Section 348.28, 405
Section 348.30, 405
Section 348.60*, 406
Section 350.2, 406
Section 350.22, 406
Section 350.24, 407
Section 350.26, 407
Section 350.30, 407
Section 350.60*, 407
Section 352.2, 408
Section 352.22, 409
Section 352.24, 409
Section 352.26, 409
Section 352.28, 409
Section 352.30, 409
Section 352.44, 410
Section 352.46, 410
Section 352.60, 410
Section 358.2, 403
Section 358.22, 403
Section 358.24, 403
Section 358.26, 404
Section 358.28, 404
Section 358.30, 404
Section 358.30(A) Exception No. 2, 404
Section 358.30(B), 404
Section 358.42, 404
Section 358.60, 404
Section 362.2, 410
Section 362.22, 410
Section 362.24, 411
Section 362.26, 411
Section 362.28, 411
Section 362.30, 411
Section 362.46, 411
Section 362.60, 411
Section 400.7(A)(8), 499
Section 404.14(A), 447
Section 404.14(B)*, 447
Section 404.15(A), 446
Section 404.2(A), 446
Section 404.2(B), 446
Section 404.2(C)*, 450
Section 404.9(B)*, 448, 653, 654, 704

Section 406.12*, 642
Section 406.4*, 702
Section 406.4(A), 640
Section 406.4(B), 640
Section 406.9*, 642
Section 406.9(A), 702
Section 406.9(B)*, 535, 535, 538, 542, 544, 702
Section 408.30, 213
Section 408.36(D), 790
Section 408.4*, 702
Section 410.10(A), 547
Section 410.10(D), 608
Section 410.115(C), 611
Section 410.116, 611, **612**, 705
Section 410.130(E), 600, 602
Section 410.16*, 11, 705
Section 410.16(A)*, 607
Section 410.16(B), 607
Section 410.16(C), 607
Section 410.2, 11, 607, 705
Section 410.22, 704
Section 410.24(B)*, 617, 618
Section 410.36(B), 612, 629
Section 410.36(G), 544
Section 410.40, 547
Section 410.50, 616
Section 422.10(A), 499
Section 422.11(A), 501
Section 422.11(E), 499
Section 422.11(F), 502
Section 422.12, 501, **502**, 702
Section 422.13, 499
Section 422.16(A), 499
Section 422.16(B), 487
Section 422.16(B)(1), 487
Section 422.16(B)(2), 490
Section 422.30*, 499, 500, 501
Section 422.31*, 510
Section 422.31(B)*, 501
Section 422.62(B)(1), 501
Section 430.102, 510
Section 430.22, 496
Section 430.24*, 209, 210, 217, 218, 220
Section 430.32, 497
Section 430.52, 496
Section 430.6, 496

Section 440.14, 506, 508
Section 440.60, 508
Section 440.64, 508
Section 445.13, 550
Section 517.2, 12
Section 590.4(J), 544
Section 590.6(A)*, 13
Section 590.6(B)* , 13
Section 680.2*, 529
Section 680.21(A), 531, 533
Section 680.22(A), 534, 535
Section 680.22(B)*, 535
Section 680.22(C)*, 535
Section 680.22(B)(2)*, 535
Section 680.23(A)*, **533**, 535
Section 680.23(B)*, **533**, 537
Section 680.23(F)*, **533**, 537
Section 680.23(F)(2) Exception, **533**
Section 680.24(A)*, **533**, 537
Section 680.24(B), **533**
Section 680.25(A)*, **533**
Section 680.25(A) Exception, **533**
Section 680.25(B), **533**
Section 680.26, 532
Section 680.26(B)*, 537
Section 680.31, 538
Section 680.32*, 538
Section 680.33(A)*, 537
Section 680.33(B)*, 538
Section 680.41, 539
Section 680.42*, 538
Section 680.43*, 539
Section 680.71, 540
Section 680.73*, 540
Section 680.74*, 540
Section 690.10(A), 789
Section 690.10(C), 790
Section 690.10(E)*, 790
Section 690.11*, 790
Section 690.13*, 783
Section 690.14(C), 784
Section 690.14(C)(1), 784
Section 690.14(C)(1) Exception, 784
Section 690.14(C)(2), 783
Section 690.14(C)(4), 784
Section 690.14(C)(5), 784

Section 690.14(D), 791
Section 690.15, 783
Section 690.17, 783
Section 690.2*, 774
Section 690.31(B), 777
Section 690.31(E)*, 784
Section 690.31(E)(1)*, 784
Section 690.31(E)(2)*, 784
Section 690.31(E)(3)*, 784
Section 690.31(E)(4)*, 784
Section 690.31(F)*, 776
Section 690.4(D), 790
Section 690.41, 786
Section 690.42, 786
Section 690.42 Exception, 786
Section 690.43*, 785
Section 690.43(A)*, 785
Section 690.43(C)*, 785
Section 690.43(D)*, 785
Section 690.43(E)*, 785
Section 690.43(F)*, 785
Section 690.45(A), 787
Section 690.46, 788
Section 690.47(C)*, 788
Section 690.5, 786
Section 690.5 Exception No. 1, 797
Section 690.56(A), 788
Section 690.56(B), 788
Section 690.60, 790
Section 690.64*, 790
Section 690.74*, 779
Section 690.8*, 777, 781
Section 690.8(A), 778
Section 690.8(A)(1), 774, 778, 788
Section 690.8(A)(2), 778, 788
Section 690.8(A)(4), 780
Section 690.8(B)*, 778
Section 690.8(B)(1)*, 781
Section 690.8(B)(1)(a)*, 778
Section 690.8(B)(2)*, 788
Section 690.9(A), 781
Section 694.20*, 801
Section 694.22*, 801
Section 694.24*, 802
Section 694.30(A)*, 803
Section 694.30(B)*, 803

Section 694.30(C)*, 803
Section 694.40(A)*, 803
Section 694.40(B)*, 803
Section 694.40(C)*, 803
Section 694.12*, 803
Section 694.12(A)*, 803
Section 694.12(B)(1)*, 803
Section 705.10, 802
Section 705.12*, 790
Section 705.12(D), 790
Section 705.12(D)(1), 790
Section 705.12(D)(2)*, 790
Section 705.12(D)(4), 790
Section 705.12(D)(5), 790
Section 705.12(D)(5) Informational Note, 790
Section 705.12(D)(6), 790
Section 705.12(D)(7), 790
800.156*, 321, 567
Article 100, 11, 12, 51, 55, 231, 315
Article 110, 10, 11, 283, 299
Article 110, Part II, 10
Article 110, Part III, 10
Article 210, 10, 311
Article 220, 10, 207, 213, 235, 243
Article 220, Part III, 207
Article 220, Part IV, 207
Article 230, 10, 233
Article 240, 673, 674
Article 250, 10, 14, 243, 244, 246, 256, 262, 267, 282
Article 250, Part III, 10
Article 285, 645
Article 300, 10, 300
Article 310, 308
Article 312, 285
Article 314, 311, 329
Article 320, 50
Article 330, 10, 50, 372
Article 334, 47, 364, 367, 374
Article 338, 50, 374
Article 340, 49, 367, 541
Article 342, 52, 402
Article 344, 52, 241, 396, 401

Article 348, 53, 54, 404
Article 350, 54, 406
Article 352, 52, 408
Article 356, 54
Article 358, 52, 395, 403
Article 362, 53, 410
Article 404, 447, 472
Article 408, 285
Article 422, 10
Article 430, 496, 497
Article 440, 505, 508
Article 517, 10, 12
Article 550, 10
Article 680, 10, 529, 538, 540
Article 680, Part II, 529
Article 680, Part IV, 538
Article 680, Part VII, 540
Article 690, 776
Article 694, 801
Article 725, 502, 506, 517
Article 800, 10, 576
Article 820, 576
Article 90, 9, 10, 12, 39
Figure 410.2, 11
Informative Annex C, 396, 397, 398, 402, 403, 405, 407, 409, 410
Note 1 Table 220.55, 195, 196
Note 4 Table 220.55, 195–197, 484, 485, 486
Table 1, Chapter 9, 396, 396
Table 110.26(A)(1), 283
Table 210.21(B)(2), 640
Table 210.21(B)(3), 640
Table 210.24, 205, 207, 674, 675
Table 220.12, 188, 190, 208, 216
Table 220.42, 208, 216
Table 220.55, 195–198, **197**, 209, 216, **483–486**, 486
Table 220.55 Note 4, 195–197, **484–486**, 486
Table 250.122, 408, 527, 528, 532, 533, 537, 541, 787, 788
Table 250.66, 211, 213, 218, 221, 244, 249, 250, 252, 527, 528, 530, 706, 788

Table 250.66 Note 2, 788
Table 300.5, 49, 238, 239, 240, 280, 527, 541, 542, 542,
Table 300.5 (Note 5), **542**, 547, 706, 801
Table 310.104(A)*, 46, 198, 239, 279, 300, 311, 541, 776
Table 310.106(A)*, 308
Table 310.15(B)(16)*, 46, 48–50, 195–197, 198, 199, 200–207, **201**, **203–205**, **211**, 218, 221, 309, 311, 374, 486, 496, 499, 502, 541, 550, 674, 675, 779, 780
Table 310.15(B)(17)*, 777
Table 310.15(B)(2)(a)*, 200, 200, 201, 202, 309, 779
Table 310.15(B)(3)(a)*, 200, 201, 201, 202
Table 310.15(B)(3)(c)*, 777, 779
Table 310.15(B)(7)*, 207, 208, 210–213, 217, 218, 220, 221, 246, 705
Table 314.16(A), 329–332, 333, 334, 341
Table 314.16(B), 329–332, 335, 336
Table 344.30(B)(2), 400, 402
Table 348.22, 405–407, 406
Table 352.30, 409, 410, 410
Table 352.44, 410
Table 4, Chapter 9, 398, 401, 402, 403, 405, 407, 409, 410
Table 400.4, 487
Table 400.5(A)(1), 550
Table 430.248, 496, 496, 497, 507
Table 430.251(A), **507**
Table 430.52, 496, 497, 497
Table 5, Chapter 9, 398, **399**
Table 8, Chapter 9, 10, 281, 283, 780, 781
Table 800.179, 576
Table 820.179, 577
Table C8, Informative Annex C, 396, **397**